Organic Solar Cells

Organic Solar Cells

Materials Design, Technology and Commercialization

Edited by Liming Ding

WILEY-VCH

Editor

Prof. Liming Ding
National Center for Nanoscience
and Technology
No.11 ZhongGuanCun
BeiYiTiao
100190 Beijing
China

Cover Image: © Outflow_Designs/
Shutterstock

Library of Congress Card No.: applied for

British Library Cataloguing-in-Publication Data
A catalogue record for this book is available from the British Library.

Bibliographic information published by the Deutsche Nationalbibliothek
The Deutsche Nationalbibliothek lists this publication in the Deutsche Nationalbibliografie; detailed bibliographic data are available on the Internet at <http://dnb.d-nb.de>.

Print ISBN: 978-3-527-34933-3
ePDF ISBN: 978-3-527-83364-1
ePub ISBN: 978-3-527-83366-5
oBook ISBN: 978-3-527-83365-8

Typesetting Straive, Chennai, India

Printed and bound by CPI Group (UK) Ltd, Croydon, CR0 4YY

C114656_310122

Contents

Preface

This is a fabulous time for me to write two prefaces for my first two books in my life! Time goes really fast. I think a lot about the old days, the choices, and interesting experiences. When I finished three years in Chemistry Department of Anhui University, I needed to choose my major, and there were three majors: chemical analysis, organic synthesis, and polymer science. I chose polymer, because I thought polymer would be more fantastic and useful; that was 1987. The last semester I chose lecturer Wei Yang (now emeritus professor in SCUT) as my thesis supervisor to do copolymerization aiming to obtain materials with good mechanical properties. The time with supervisor Wei Yang built my preliminary experience on chemicals, reaction, and polymerization. During 1990–1998, I did my MS, PhD, postdoc work on solid polymer electrolytes and electron transfer dynamics, in Changchun Institute of Applied Chemistry (CIAC), which was the biggest institute in Chinese Academy of Sciences. For my PhD period, I was a joint student in CIAC. My supervisors, Professors Zinan Zhou, Yunqing Lin (CIAC), Dezhu Ma, Chengzong Yang (USTC), Shaojun Dong, and Erkang Wang (CIAC), stimulated my spirit and kindly helped me, and these will stay in my memory forever. During my postdoc time, I really got interested in optoelectronic devices based on conjugated polymers. In October of 1997, Professor Olle Inganäs interviewed me in a hotel in Wangfujing, Beijing; we talked around 25 min, then Olle kindly invited me to a small restaurant nearby for lunch. Before Christmas I received the postdoc offer from Olle. Without any experimental background on devices, I did learn a lot from those old group members in Linköping University on how to cut ITO glass, etching, spin coating, vacuum evaporation, measurements, etc. All new to me. I quickly learned how to fabricate polymer solar cells and polymer light-emitting diodes. That time, I made solar cells with a trilayer structure (Synthetic Metals 2000, 110, 133–140). Sweden, a small kindom, stays really special, and a very beautiful land, people like Olle doing cool stuff there everyday, serious, quiet, and outstanding! Later, I moved to the United States and worked with Robert Haddon on CNTs, Frank Karasz on PLEDs and organic solar cells (OSCs), Mike Durstock and Liming Dai on OSCs, and Tom Russell on photoactive polymers, block copolymers and phase separation. I worked in Konarka Lowell as a senior scientist for one year on flexible polymer solar cells, doing blade coating, materials screening, and device optimization. It was cool to learn the structure, running and commercial operation of a facing-future high-tech company. In 2010, I came back to China and joined National Center for Nanoscience and Technology as a full professor. My research project was on OSCs. We have invented many building blocks and high-performance materials such as copolymer donors, non-fullerene small molecular acceptors, and fullerene acceptors. We have pushed the PCE record for OSCs from 14% to 18%. This jumping just took two years, really awesome! We hold the dream that one day single-junction OSCs will deliver PCEs over 20%. I think it is the right time to deliver a book on OSCs, a way to sum up the past and to face the future, to challenge the PCE limit and those commercialization issues. All the contributing authors have done excellent jobs in their respective direction. The topic goes from polymer donors, p-type materials, fullerene acceptors, non-fullerene small molecular acceptors, ladder-type heteroacenes, chlorination, all-polymer solar cells, single component, tandem structure, large-area fabrication, indoor application, interfacial design, morphology control, stability, to voltage loss, covering almost all research hotspots. This book would help those people who have interests in OSCs or who are working in the same field, no matter students or colleagues, academic or industry. Special thanks to Dr. Shaoyu Qian, without her trust and invitation, this won't happen! I would thank Kat Wong and Farhath Fathima for their nice assistance and patience. I appreciate the help from my team members: Ling Liu, Ke Jin, Xiaoyan Du, Zuo Xiao, Chuantian Zuo, Jiamin Cao and Shan Chen.

At last, I thank my mom for her eternal love.

Beijing, 28 September 2021 *Liming Ding*

1

Conjugated Polymer Donors for Organic Solar Cells

Xiaopeng Xu[1], Xiyue Yuan[2], Qunping Fan[3], Chunhui Duan[2], Maojie Zhang[3], and Qiang Peng[1]

[1] *Sichuan University, School of Chemical Engineering, State Key Laboratory of Polymer Materials Engineering, No. 24 South Section 1, Yihuan Road, Chengdu 610065, P.R. China*
[2] *South China University of Technology, Institute of Polymer Optoelectronic Materials and Devices, State Key Laboratory of Luminescent Materials and Devices, 381 Wushan Road, Tianhe District, Guangzhou 510640, P.R. China*
[3] *Soochow University, Laboratory of Advanced Optoelectronic Materials, College of Chemistry, Chemical Engineering and Materials Science, 199 Ren'ai Road, Suzhou 215123, P.R. China*

1.1 Introduction

Organic solar cells (OSCs) are the devices that can directly convert solar energy into electricity with the advantages of low cost, light weight, and the potential for roll-to-roll fabrication on flexible substrates, which make them promising candidates for inexpensive renewable power sources [1–3]. At present, the highest power conversion efficiency (PCE) of OSCs in laboratory has exceeded 18% [4]. Important progresses have been also made in scalability and device stability, which demonstrated the great application prospect of this technology [5, 6]. These progresses were greatly benefitted from the design and synthesis of donor polymers with wide spectral absorption coverage to the solar spectrum and matched energy levels. Based on the optical bandgaps (E_g^{opt}s), donor polymers employed in OSCs can be divided into three major types: low bandgap (LBG) polymers ($E_g < 1.6$ eV), medium bandgap (MBG) polymers (1.6 eV $< E_g <$ 1.8 eV), and wide bandgap (WBG) polymers ($E_g > 1.8$ eV) (Figure 1.1a) [7]. In the early stage of OSCs field, fullerene derivatives were the dominant acceptors, which are poor light absorbers. As a result, the donor polymers are responsible for light harvesting in the resulting OSCs. Correspondingly, the LBG polymers played a more important role than the MBG and WBG polymers at that stage [8]. Nevertheless, LBG polymers with reduced E_g^{opt} generally had a high-lying highest occupied molecular orbital (HOMO) energy level, which resulted in a limited open-circuit voltage (V_{oc}) in the corresponding OSCs [9]. MBG polymers could have more potential to deepen the HOMO level without significantly increasing the bandgap, which afforded higher V_{oc} without

Xiaopeng Xu, Xiyue Yuan, and Qunping Fan have contributed equally to this work.

Organic Solar Cells: Materials Design, Technology and Commercialization, First Edition. Edited by Liming Ding.

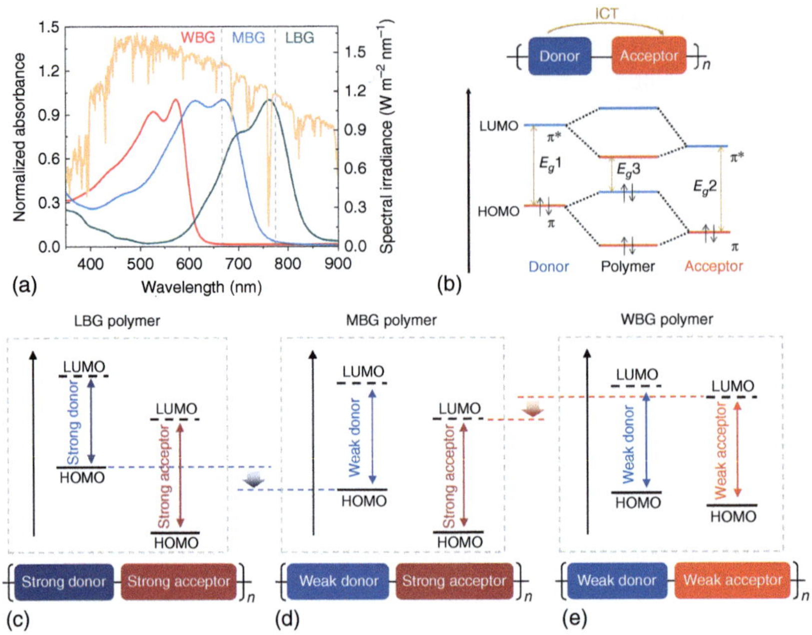

Figure 1.1 (a) UV–vis–near-infrared (NIR) absorption of different bandgap polymers. (b) Principle of bandgap lowering by donor–acceptor interaction. (c) LBG polymers constructed by using "strong donor-strong acceptor" configuration. (d) MBG polymers constructed by using "weak donor-strong acceptor" configuration. (e) WBG polymers constructed by using "weak donor-weak acceptor" configuration. Source: Xiaopeng Xu.

sacrificing short-current density (J_{sc}) in the resulted OSCs [9, 10]. Although the J_{sc}s of OSCs based on WBG polymers were limited by their narrow absorptions, they were also key components to complement the absorption spectra with LBG counterparts and achieve high V_{oc} in the multi-junction, ternary blend, or the currently prevailing non-fullerene OSCs [8–11].

Constructing donor–acceptor (D–A)-type copolymers has been demonstrated to be the most successful strategy to develop high-performance donor polymers (Figure 1.1b). With rational choice of electron-donating and electron-withdrawing units, the absorption ranges and frontier energy levels can be readily tuned. The "push-pull" effect between D and A units has distinct effect on the bandgaps of the resulted polymers: (i) the combination of "strong donor" and "strong acceptor" could be a feasible way to achieve a LBG polymer (Figure 1.1c); (ii) the pair of "weak donor" and "strong acceptor" could be employed to construct a MBG polymer (Figure 1.1d); and (iii) the utilization of "weak donor" and "weak acceptor" could be effective to build a WBG polymer (Figure 1.1e). Compared to the LBG polymer, the weak electron-donating ability of donor motif delivers a low-lying HOMO energy level of the MBG and WBG polymers, which is beneficial to realize high V_{oc} in the resulting polymer solar cells (PSCs) [10, 11]. Compared to the WBG polymer, the strong electron-withdrawing ability of acceptor block lowers the lowest unoccupied molecular orbital (LUMO) energy level of the MBG polymer, which enables a reduced bandgap *via* intramolecular charge transfer (ICT) [12]. In the following part, we will introduce the design strategies of LBG, MBG, and WBG polymers and analyze their photovoltaic properties.

1.2 LBG Polymers

LBG-conjugated polymers are usually obtained *via* the design strategies as follows [13, 14]. First, stabilization of the quinoidal resonance structure in conjugated polymers. Generally, two kinds of resonance structures, i.e. aromatic and quinoidal form coexist in conjugated polymers. As compared to the aromatic form, the quinoidal form has a smaller bandgap but is energetically less stable due to the destruction of the original aromaticity. Second, copolymerization of D unit and A unit alternatively to afford the so-called D–A copolymers [15]. Reduced optical gaps (E_g3 in Figure 1.1b) in an alternating copolymer can be achieved due to the molecular orbital hybrid of the electron-rich (D) and electron-deficient (A) moiety. Besides, the HOMO level of the polymers is mainly influenced by the donor units, while the LUMO level is determined by the acceptor units. Donor units employed in high-performance LBG polymers are mostly thiophene ring-containing building blocks with strong electron-donating capability. Compared to donor units, acceptor units play a more important role in constructing LBG polymers because the absorption spectra, energy levels, and aggregation properties can be more easily tuned by selecting appropriate acceptor units [16]. The common acceptor units with stronger electron-accepting ability include 2,1,3-benzothiadiazole (BT), isoindigo (IID), and diketopyrrolopyrrole (DPP). In this section, we will introduce the design strategies of LBG polymers at first. After that, several representative types of LBG polymers and their photovoltaic properties will be analyzed.

1.2.1 LBG Polymers Based on Benzothiadiazole (BT)

2,1,3-Benzothiadiazole (BT) is a widely used acceptor unit for constructing LBG polymers in a D–A-type framework (Figure 1.2). One of the representative LBG polymers is P1 [17, 18], a copolymer-based on cyclopentadithiophene (CPDT) and BT with an E_g of 1.47 eV. By replacing the carbon atom with silicon atom on CPDT, the polymer P2 [19, 20] with a higher crystallinity and improved charge transport properties was generated, which afforded a certified PCE of 5.2% when blended with a fullerene derivative in OSCs. When introducing P2 as the third component into the PTB7:$PC_{71}BM$ blend host, a higher fill factor (FF) of 77% was obtained due to the reduction of recombination losses. Correspondingly, a PCE improvement from 7.52% to 8.60% was achieved [21]. Replacing S with O in heteroaromatic rings tends to offer a more positive oxidation potential, which is beneficial for obtaining a higher V_{oc} in OSCs [18]. This was reflected by P3, which exhibited much deeper-lying HOMO level and higher V_{oc} than P1 in OSCs. However, this polymer could not generate photocurrent, which was possibly caused by the reduced driving force for exciton dissociation.

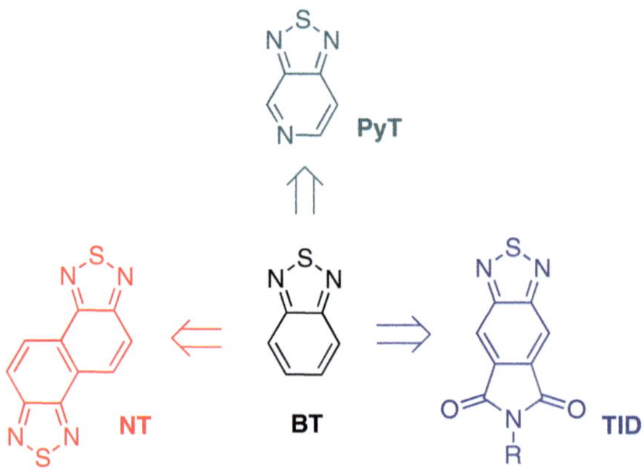

Figure 1.2 The chemical structures of BT and its derived building blocks. Source: Xiyue Yuan.

Fluorination is effective to improve the photovoltaic properties of conjugated polymers *via* tuning the frontier orbital energy levels and improving molecular ordering. However, fluorination can also bring negative effects on resulting polymers, such as poor solubility and excessive aggregation in organic solvents. Hence, the degree of fluorination shall be optimized to achieve a high-efficiency OSCs. Neher and coworkers [22] showed that the polymer with mono-fluorination on BT (P4) exhibited better photovoltaic performance than the polymers with non- and di-fluorination on BT (P1, P5) due to the stronger tendency to take face-on orientation when blended with $PC_{71}BM$. Similar fluorination effects were also observed by Jo et al. on the polymers P6–P8 [23]. Yang et al. [24] inserted a strong electron-donating oxygen atom into the CPDT unit to form the dithienopyran (DTP) unit for lowering E_g. The resulting polymer P9 exhibited a narrow E_g of 1.38 eV, and a deep HOMO level of −5.26 eV. Single-junction OSCs based on P9:$PC_{71}BM$ blend showed over 60% external quantum efficiency (EQEs) and effective optical response up to 900 nm. Moreover, a certified PCE of 10.6% in tandem OSCs based on this polymer was achieved. Guo and coworkers [25] reported a BT-based polymer P10 with an E_g of 1.46 eV by using a head-to-head substituted bithiophenes as the donor units, which afforded a promising PCE of 9.8% in fullerene-based OSCs. Replacing the phenyl ring on BT unit with pyridine ring leads to a new building block, thiadiazolo[3,4-*c*]pyridine (PyT), which possesses higher electron affinity and affords narrower bandgap in conjugated polymers. You and coworkers [26] reported a LBG polymer P11 based on PyT, which exhibited an E_g of 1.51 eV and a PCE of 6.3% in OSCs. When utilizing 4,4-bis(2-ethylhexyl)-4*H*-germolo[3,2-*b*:4,5-*b*′]dithiophene (DTG) as the electron-rich unit, the polymer P12 based on PyT exhibited a very low E_g of 1.32 eV and a decent PCE of 6.6% in inverted OSCs [27]. The chemical structures of P1–P12 are showed in Figure 1.3, and the corresponding photovoltaic parameters of OSCs are summarized in Table 1.1.

Fusing two BT units leads to the formation of naphtho[1,2-*c*:5,6-*c*′]bis[1,2,5]thiadiazole (NT), which possesses a larger planar structure and higher electron affinity than BT unit (Figure 1.4). The pioneering work based on NT was done by Cao and coworkers along with the creation of P13, which exhibited an E_g of

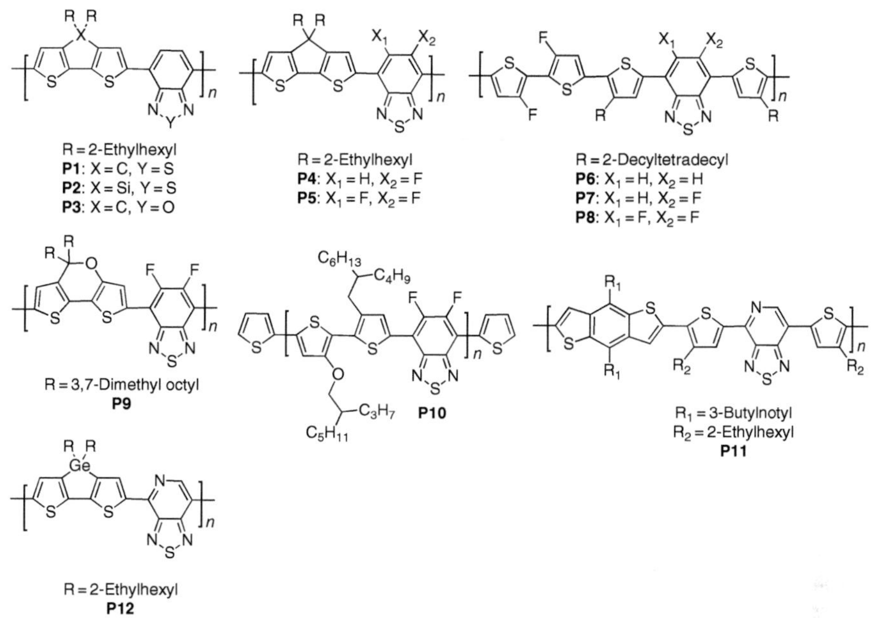

Figure 1.3 Chemical structures of the BT-derived conjugated polymers P1–P12. Source: Xiyue Yuan.

Table 1.1 Optoelectronic properties and device performances of P1–P12.

Polymer	E_g^{opt} (eV)	HOMO/ LUMO (eV)	Acceptor	V_{oc} (V)	J_{sc} (mA cm^{-2})	FF	PCE (%)	References
P1	1.47	−5.30/−3.55	$PC_{61}BM$	0.65	11.0	0.47	3.2	[17]
P2	1.45	−5.05/−3.27	$PC_{61}BM$	0.58	14.9	0.61	5.2	[20]
P3[a)]	1.47	−5.39/−3.71	$PC_{61}BM$	0.78	5.2	0.60	2.5	[18]
P4	1.45	−5.35/−3.90	$PC_{71}BM$	0.74	14.1	0.58	5.9	[22]
P5	1.51	−5.34/−3.52	$PC_{71}BM$	0.85	12.6	0.52	5.6	[24]
P6	1.57	−5.33/−3.76	$PC_{71}BM$	0.80	14.1	0.63	7.1	[23]
P7	1.57	−5.37/−3.80	$PC_{71}BM$	0.82	15.7	0.71	9.1	[23]
P8	1.59	−5.46/−3.87	$PC_{71}BM$	0.82	13.3	0.59	6.4	[23]
P9	1.38	−5.26/−3.61	$PC_{71}BM$	0.68	17.8	0.65	7.9	[24]
P10	1.46	−5.20/−3.74	$PC_{71}BM$	0.66	20.7	0.71	9.8	[25]
P11	1.51	−5.47/−3.64	$PC_{61}BM$	0.85	12.8	0.58	6.3	[26]
P12	1.32	−4.90/−3.60	$PC_{71}BM$	0.59	19.6	0.57	6.6	[27]

a) For P3, energy levels are determined from cyclic voltammetry ($E^{(Fc/Fc^+)} = -5.20$ eV).
Source: Based on Hendriks et al. [28].

Table 1.2 Optoelectronic properties and device performances of P13–P20.

Polymer	E_g^{opt} (eV)	HOMO/ LUMO (eV)	Acceptor	V_{oc} (V)	J_{sc} (mA cm^{-2})	FF (%)	PCE (%)	References
P13	1.58	−5.19/−3.26	$PC_{71}BM$	0.80	11.7	0.61	6.0	[29]
P14	1.54	−5.16/−3.77	$PC_{71}BM$	0.71	19.4	0.73	10.1	[30]
P15	1.52	−5.48/−3.65	$PC_{71}BM$	0.96	14.5	0.64	8.9	[34]
P16	1.56	−5.14/−3.46	$PC_{61}BM$	0.71	19.4	0.73	10.1	[33]
P17	1.60	−5.38/−3.53	$PC_{71}BM$	0.82	19.3	0.67	10.5	[33]
P18	1.42	−5.36/−3.48	$PC_{71}BM$	0.77	20.2	0.72	11.3	[32]
P19	1.52	−5.42/−3.48	$PC_{71}BM$	0.84	16.5	0.72	10.0	[32]
P20	1.40	−5.29/−3.40	$PC_{71}BM$	0.72	19.1	0.73	10.3	[31]

1.58 eV and a PCE of 6.0% when blended with $PC_{71}BM$ [29]. After that, more narrow bandgap polymers based on NT were developed for application in OSCs. For example, the polymer P14 bearing quaterthiophene donor units and NT acceptor units exhibited a high PCE of 10.1% in OSCs with an inverted structure at the active layer thickness of 290 nm due to the high crystallinity and hole mobility of the polymer [30]. More remarkably, high-performance thick-film OSCs were achieved by more NT-based polymers (P16–P20), which suggested the prospect of NT-based polymers in large-scale manufacturing of OSCs [31–33]. To reduce the energy loss (E_{loss}) of NT polymer-based OSCs, Takimiya and coworkers created a new building block naphtho[1,2-*c*:5,6-*c*′]bis[1,2,5]oxadiazole (NOz) by replacing the sulfur atom on NT unit with oxygen atom. The resulting polymer P15 exhibited similar E_g as P14 but significantly enhanced V_{oc} and greatly reduced E_{loss} in OSCs when blended with the same electron acceptor (≈0.85 versus ≈0.55 eV) [34]. The chemical structures of P13–P20 are showed in Figure 1.4, and the corresponding photovoltaic parameters of OSCs are summarized in Table 1.2.

[1,2,5]Thiadiazolo[3,4-*f*]isoindole-5,7-dione (TID) is another strong acceptor unit derived from BT (Figure 1.5). McCulloch and coworkers [35] reported a LBG polymer P21 based on this acceptor unit, which exhibited ideal frontier orbital energy levels for OSCs. This polymer can be solution-processed in a blend with $PC_{71}BM$ from a single solvent to reproducibly afford OSCs with the best PCE reaching 8.3%. Wang et al. [36] reported P22 with a low E_g of 1.49 eV, which was copolymerized from TID and the electron-rich unit indacenodithieno[3,2-*b*]thiophene (IDTT). The OSCs based on P22:$PC_{71}BM$ attained a good PCE of 6.7% with a high V_{oc} of 1.0 V, leading to the E_{loss} as low as 0.49 eV. In addition to TID, Facchetti and coworkers [37] reported a new electron-accepting building block benzo[1,2-*d*:4,5-*d*′]bis([1,2,3]thiadiazole) (iso-BBT). The resulting polymer P23 exhibited an E_g of 1.40 eV and afforded a PCE of 10.3% when blended with $PC_{61}BM$. Li and coworkers [38] introduced quinoidal structure design concept into the BT-derived unit and obtained the polymer P24 with an E_g of 1.45 eV. When employed in fullerene-based OSCs, P24 showed a PCE of

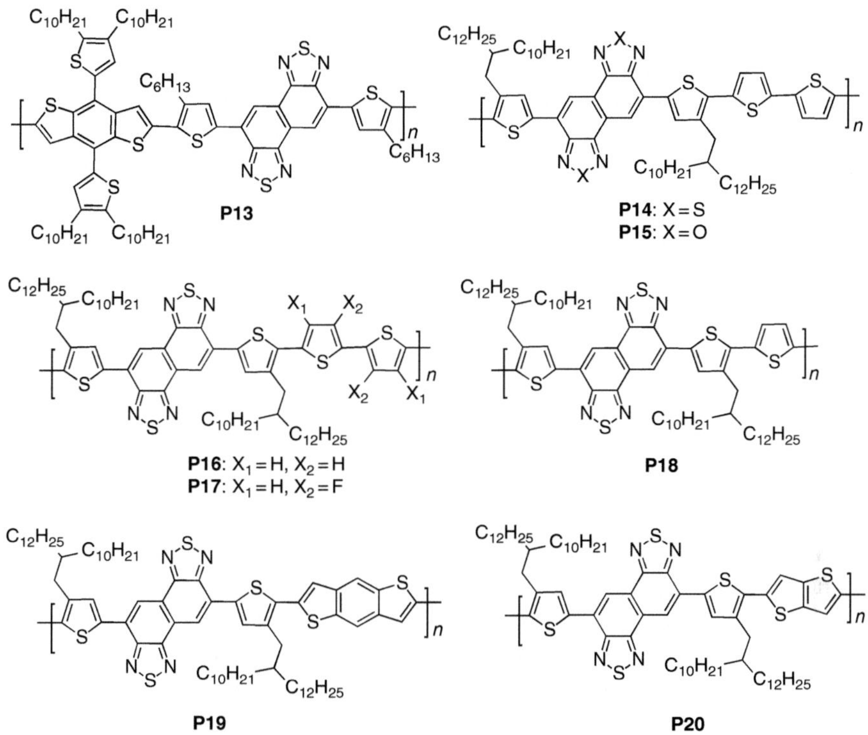

Figure 1.4 Chemical structures of the NT-derived conjugated polymers P13–P20. Source: Xiyue Yuan.

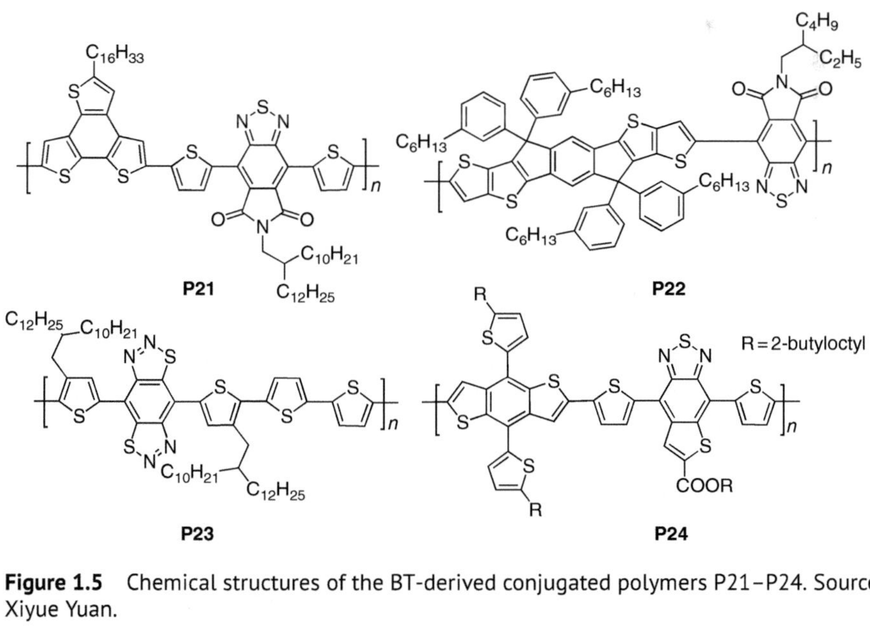

Figure 1.5 Chemical structures of the BT-derived conjugated polymers P21–P24. Source: Xiyue Yuan.

Table 1.3 Optoelectronic properties and device performances of P21–P24.

Polymer	E_g^{opt} (eV)	HOMO/ LUMO (eV)	Acceptor	V_{oc} (V)	J_{sc} (mA cm^{-2})	FF	PCE (%)	References
P21[a)]	1.50	−5.20/−3.70	$PC_{71}BM$	0.80	16.5	0.63	8.3	[35]
P22	1.49	−5.89/−3.84	$PC_{71}BM$	1.00	12.6	0.53	6.7	[36]
P23	1.40	−5.43/−4.03	$PC_{61}BM$	0.81	20.8	0.61	10.3	[37]
P24	1.45	−5.35/−3.39	$PC_{71}BM$	0.75	13.2	0.63	6.1	[38]

a) For P21, E_{LUMO} is determined by E_{HOMO} and E_g^{opt}.

6.1%. The chemical structures of P21–P24 are showed in Figure 1.5, and the corresponding photovoltaic parameters of OSCs are summarized in Table 1.3.

1.2.2 LBG Polymers Based on Isoindigo (IID)

Isoindigo (Figure 1.6a) is an isomer of the natural dye indigo, which was first known as a therapeutic agent for leukemia [39, 40]. In 2010, Reynolds and coworkers [41] introduced isoindigo-based materials into OSCs firstly. Although the highest PCE was only 1.76% achieved in that study, the unique electronic

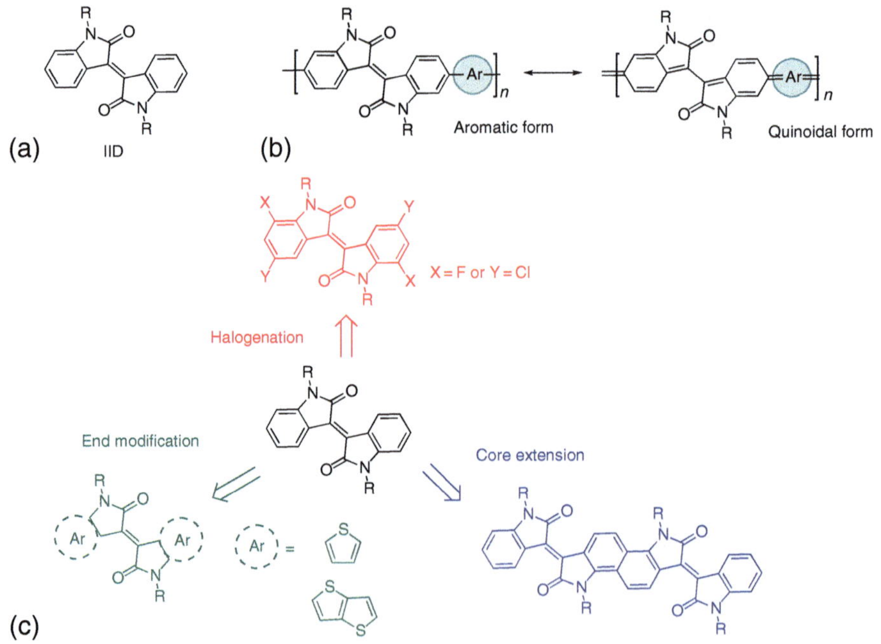

Figure 1.6 (a) The chemical structure of isoindigo. (b) The aromatic (left) and quinoidal form (right) of isoindigo-based polymers. (c) Structural modification based on isoindigo. Source: Xiyue Yuan.

Table 1.4 Optoelectronic properties and device performances of P25–P31.

Polymer	E_g^{opt} (eV)	HOMO/ LUMO (eV)	Acceptor	V_{oc} (V)	J_{sc} (mA cm^{-2})	FF	PCE (%)	References
P25	1.60	--	$PC_{61}BM$	0.91	9.1	0.54	4.5	[46]
P26	1.55	−5.43/−3.88	$PC_{71}BM$	0.84	3.9	0.53	1.7	[43]
P27	1.50	−5.69/−3.97	IT-M	0.72	14.6	0.66	6.9	[47]
P28	1.49	−5.34/−3.79	$PC_{71}BM$	0.77	9.3	0.65	4.7	[48]
P29	1.52	−5.43/−3.81	$PC_{71}BM$	0.72	7.6	0.60	3.2	[48]
P30	1.54	−5.20/−3.66	$PC_{71}BM$	0.71	7.9	0.34	1.9	[43]
P31	1.56	−5.71/−3.96	$PC_{61}BM$	0.94	8.6	0.50	4.0	[49]

properties and excellent light absorption capabilities made isoindigo an ideal building block for D–A conjugated polymers. The solubility of unsubstituted isoindigo is very poor in common organic solvents, but it can be improved by introducing various alkyl chains onto the amide nitrogen atoms. Due to its two lactam rings, isoindigo has strong electron-withdrawing properties [42]. Besides, the optoelectronic properties of isoindigo-based conjugated polymers can be tuned by structural modification including heteroatom substitution, end modification, and core extension (Figure 1.6c). Subsequently, conjugated polymers based on isoindigo and its derivatives were widely used in OSCs [43–45]. The comprehensive study on the energy levels of isoindigo-based polymers showed that LUMO levels were located on the isoindigo unit and the change of donor units could effectively manipulate the HOMO levels of isoindigo-based D–A polymers, leading to different V_{oc} values in OSCs (Figure 1.7, Table 1.4) [43, 46–52]. Meanwhile, the absorption spectra and the optical bandgaps of the resulting conjugated polymers could be readily tuned [43, 46–48].

Heteroatomic substitution (Figure 1.8) is another important method to tune the optoelectronic properties of isoindigo-based polymers, including light absorption, energy level, and solubility, etc. For example, the introduction of fluorine and chlorine atoms is effective to reduce the HOMO levels and improve charge mobility of the polymers [53]. Pei and coworkers [54] synthesized a set of fluorinated and chlorinated isoindigo-based polymers P32, P33 for using as the donor materials in OSCs. When blended with $PC_{71}BM$, the chlorinated polymer P33 performed much better than the fluorinated polymer P32 in J_{sc}. This was because that the smaller radius of F atoms endows the polymer P32 with better planarity and crystallinity, while the increased polymer crystallinity reduced the mixing between the polymer and $PC_{71}BM$. Yan and coworkers [55] studied the influence of fluorine substitution on aggregation characteristics and optical properties of the polymers in solution. The polymer P34 with fluorination on the donor unit exhibited the strongest extent of temperature-dependent aggregation, which led to a higher hole mobility for the polymer as compared to the fluorine-free polymer. Geng and coworkers [56] reported LBG polymers (P37–P39) based on dithienocarbazole and isoindigo. The HOMO

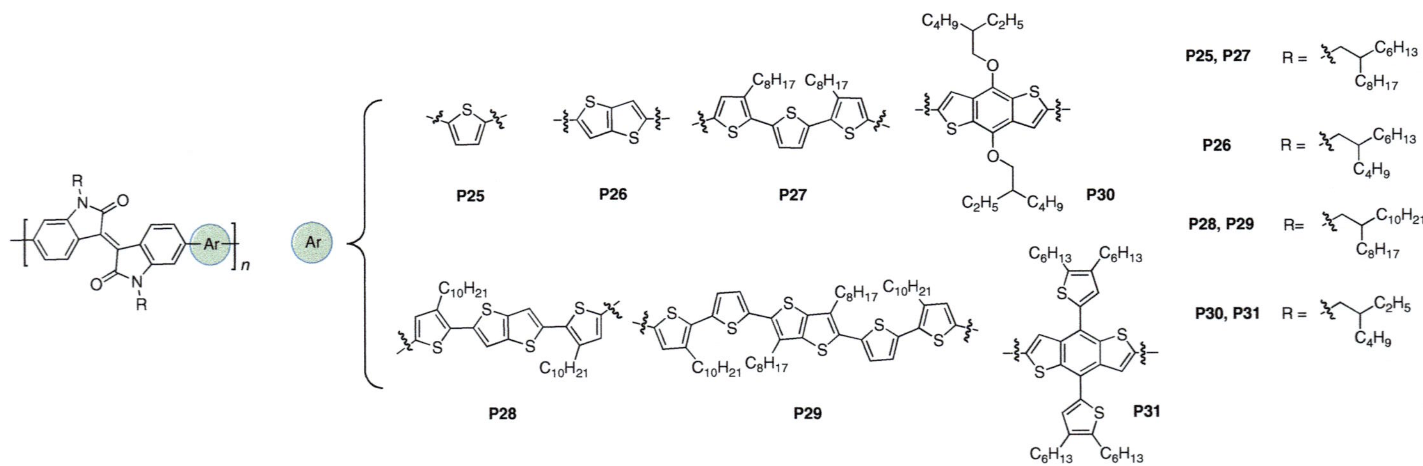

Figure 1.7 Chemical structures of the isoindigo-based conjugated polymers P25–P31. Source: Xiyue Yuan.

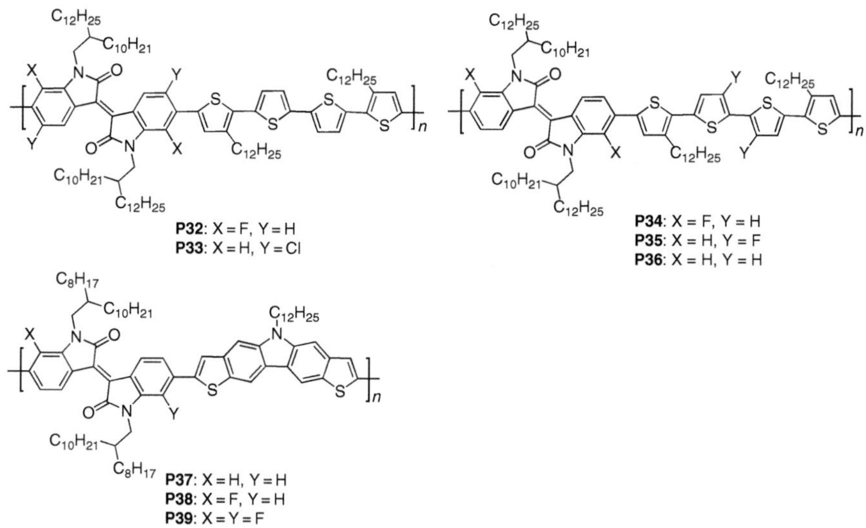

Figure 1.8 Chemical structures of the isoindigo-based conjugated polymers P32–P39. Source: Xiyue Yuan.

Table 1.5 Optoelectronic properties and device performances of P32–P39.

Polymer	E_g^{opt} (eV)	HOMO/ LUMO (eV)	Acceptor	V_{oc} (V)	J_{sc} (mA cm^{-2})	FF	PCE (%)	References
P32	1.45	−5.51/−3.92	$PC_{71}BM$	0.63	4.3	0.44	1.2	[54]
P33	1.50	−5.53/−3.85	$PC_{71}BM$	0.75	10.0	0.61	4.6	[54]
P34	1.51	−5.18/−3.67	$PC_{71}BM$	0.73	10.3	0.47	3.5	[55]
P35	1.52	−5.30/−3.78	$PC_{71}BM$	0.83	13.3	0.61	6.7	[55]
P36	1.51	−5.28/−3.77	$PC_{71}BM$	0.73	11.4	0.55	4.6	[55]
P37	1.61	−5.22/−3.54	$PC_{71}BM$	0.79	14.6	0.62	7.2	[56]
P38	1.59	−5.30/−3.67	$PC_{71}BM$	0.81	12.6	0.69	7.1	[56]
P39	1.59	−5.36/−3.74	$PC_{71}BM$	0.86	12.3	0.68	7.2	[56]

energy levels of the polymers were pushed down with the increase of F-substitution degree in isoindigo unit, which were helpful to obtain high V_{oc} in the resulting OSCs. However, the champion OSCs were achieved by the non-fluorinated polymer P37, which afforded the best PCE of 7.2% in conventional devices and 8.2% in inverted devices. The reason for the poorer device performance of fluorinated polymers was the poor phase separation. The chemical structures of P32–P39 are showed in Figure 1.8, and the corresponding photovoltaic parameters of OSCs are summarized in Table 1.5.

The other two types of isoindigo-based conjugated polymers involve skeleton changes of isoindigo (Figure 1.9): (i) changing the isoindigo terminal groups from

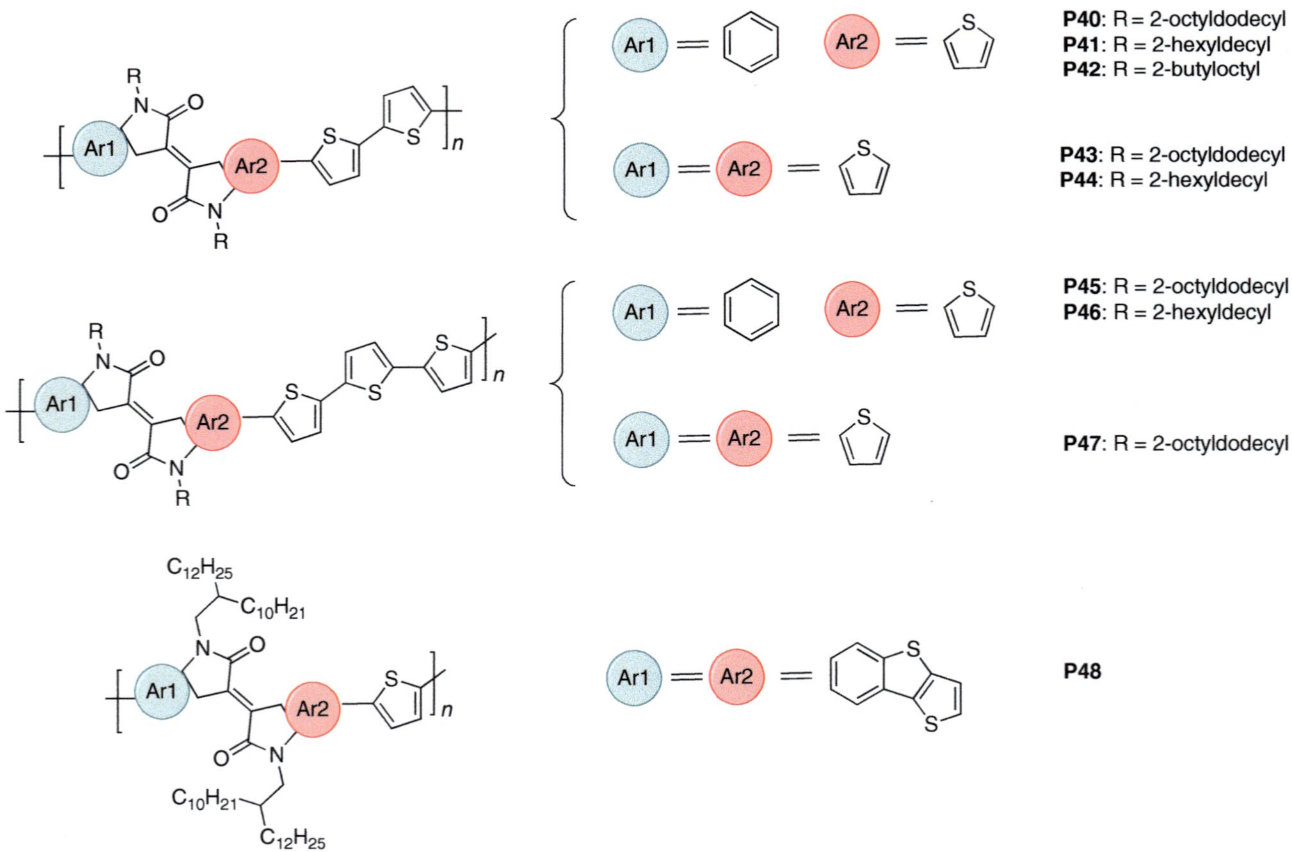

Figure 1.9 Chemical structures of the isoindigo-derived conjugated polymers P40–P48. Source: Xiyue Yuan.

phenyl rings to heterocyclic rings or fusing additional heteroaromatic rings onto the terminal phenyl rings; (ii) inserting additional conjugated units into the exocyclic double bond between the lactam rings in isoindigo core. Fréchet and coworkers [57] reported isoindigo-derived polymers with increased coplanarity by replacing isoindigo terminal phenyl rings with thiophene rings. Upon thin-film analysis by atomic force microscopy (AFM) and grazing incidence X-ray diffraction (GIXD), the thiophene-terminated isoindigo polymers favor a face-on orientation in thin films, while the phenyl-terminated isoindigo polymers favor an edge-on orientation. As a result, P45, the polymer with modest planarity, showed the highest PCE in OSCs due to the face-on-dominated orientation. McCulloch and coworkers [58] reported an eight-ring fused isoindigo-based acceptor unit thieno[3,2-*b*][1]benzothiophene isoindigo (TBTI) by fusing isoindigo core with thieno[3,2-*b*]thiophene and obtained a polymer P48 (Figure 1.9), which had a low E_g of 1.60 eV. Benefitted from its high hole mobility (0.31 $cm^2\ V^{-1}\ s^{-1}$) and suitable LUMO level (−3.50 eV), P48 afforded a high PCE of 9.1% in fullerene-based OSCs. The chemical structures of P40–P48 are showed in Figure 1.9, and the corresponding photovoltaic parameters of OSCs are summarized in Table 1.6.

Kelly and coworkers [59] used diisoindigo and ring-fused bisisoindigo dimer to study the effect of acceptor unit number and size on the optoelectronic properties of donor–acceptor–acceptor (D–A–A)-type polymers (Figure 1.10). It was found that the diisoindigo-containing polymers had lower LUMO levels as well as lower hole mobilities and higher electron mobilities than their isoindigo analogs. The polymers P51 and P52 containing bisisoindigo had the lowest LUMO levels and much smaller E_g than the isoindigo- and diisoindigo-containing polymers, consistent with their increased conjugation length. When incorporated into OSCs, the isoindigo-based polymers performed better than the diisoindigo analogs, and the bisisoindigo-based polymers performed poorer than the other polymers. The lower LUMO level of the diisoindigo- and bisisoindigo-based polymers reduced the energetic driving force for exciton dissociation at the polymer:$PC_{71}BM$ interface, leading to low J_{sc} and PCE

Table 1.6 Optoelectronic properties and device performances of P40–P48.

Polymer	E_g^{opt} (eV)	HOMO/ LUMO (eV)	Acceptor	V_{oc} (V)	J_{sc} ($mA\ cm^{-2}$)	FF	PCE (%)	References
P40	1.37	−5.54/−3.84	$PC_{71}BM$	0.54	10.4	0.57	3.2	[57]
P41	1.39	−5.43/−3.87	$PC_{71}BM$	0.52	13.6	0.57	4.0	[57]
P42	1.37	−5.33/−3.86	$PC_{71}BM$	0.45	7.6	0.41	1.4	[57]
P43	1.12	−5.22/−3.78	$PC_{71}BM$	0.19	4.0	0.53	0.4	[57]
P44	1.09	−5.04/−3.82	$PC_{71}BM$	--	--	--	--	[57]
P45	1.39	−5.30/−3.72	$PC_{71}BM$	0.49	14.7	0.66	4.7	[57]
P46	1.39	−5.21/−3.72	$PC_{71}BM$	0.42	10.8	0.51	2.3	[57]
P47	1.23	−5.06/−3.77	$PC_{71}BM$	0.18	5.4	0.45	0.4	[57]
P48	1.60	−5.10/−3.50	$PC_{71}BM$	0.72	17.7	0.71	9.1	[58]

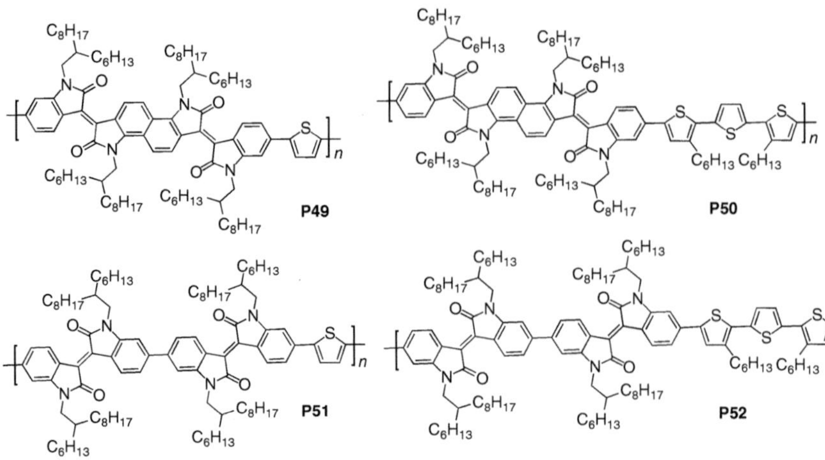

Figure 1.10 Chemical structures of the isoindigo-derived polymers P49–P52. Source: Xiyue Yuan.

Table 1.7 Optoelectronic properties and device performances of P49–P52.

Polymer	E_g^{opt} (eV)	HOMO/ LUMO (eV)	Acceptor	V_{oc} (V)	J_{sc} (mA cm^{-2})	FF	PCE (%)	References
P49	1.30	−5.83/−3.97	$PC_{71}BM$	0.80	0.8	0.54	0.3	[59]
P50	1.60	−5.90/−3.91	$PC_{71}BM$	0.86	4.2	0.50	1.8	[59]
P51	1.31	−5.67/−3.97	$PC_{71}BM$	0.66	2.4	0.58	0.9	[59]
P52	1.54	−5.68/−3.85	$PC_{71}BM$	0.77	5.9	0.61	2.8	[59]

values. The chemical structures of P49–P50 are showed in Figure 1.10, and the corresponding photovoltaic parameters of OSCs are summarized in Table 1.7.

1.2.3 LBG Polymers Based on Diketopyrrolopyrrole (DPP)

DPP (Figure 1.11) is a widely used electron-deficient building block for constructing LBG polymers with strong light absorption and high carrier mobility [60–64], which has two lactam rings with eight π-electrons. Generally, DPP polymers have E_gs lower than 1.50 eV with light absorption extended to 1000 nm. Figure 1.12 shows the chemical structures of a few DPP polymers (P53–P64) [28, 65–74], and the basic optoelectronic and photovoltaic properties of these polymers are summarized in Table 1.8. By changing electron-rich comonomers with different electron-donating ability, the E_g^{opt}s of the polymers can be finely tuned. Besides, the donor units have a profound influence on the morphology of donor:acceptor blends, thus affecting exciton dissociation and charge carrier transport [65]. In a set of DPP-based polymers with different donor units (P53–P55, P57, and P60), the polymers with poorer solubilities resulted in the formation of fibers with smaller widths (Figure 1.13), which

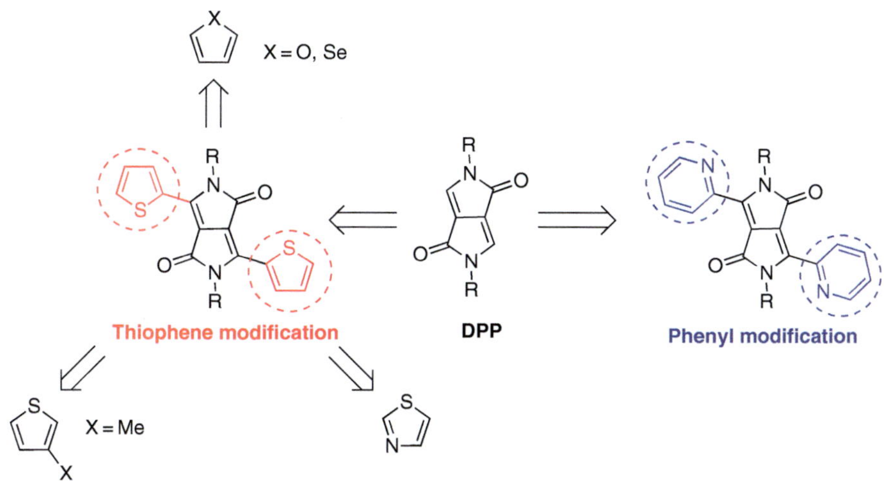

Figure 1.11 The chemical structure of DPP and its derived building blocks. Source: Xiyue Yuan.

thereby contributed to higher EQEs and PCEs in OSCs. The length of branched alkyl chains also had a critical influence on the device performance. For example, the PCE could be increased from 3.2% to 7.4% in blends of P65–P67 upon decreasing the length of the side chains [75]. This was caused by a progressive increase in the photocurrent due to the decreased phase separated domain size in the polymer:$PC_{71}BM$ blends (Figure 1.14).

Changing the flanking units is a general strategy to tune the properties of DPP polymers. A few representative DPP polymers with different flanking units are shown in Figure 1.15. Janssen and coworkers [65] reported a DPP polymer (P69) with two additional methyl substituents on the 3-position of the flanking thiophene rings to raise the LUMO energy level of the parent DPP polymer (P68) [18]. This small change led to improved EQEs (60% for P69 versus 50% for P68). To further improve the V_{oc} of DPP polymers, Janssen and coworkers [76] replaced the electron-rich thiophene (T) rings with electron-deficient thiazoles (Tz) and obtained the polymer P71. The polymer gave a very high V_{oc} of 0.96 V but a very low PCE of 1.1% when blended with $PC_{71}BM$. This was caused by the imbalanced electron–hole mobilities and reduced energetic driving force for exciton dissociation. Yang and coworkers [77] showed that the substitution of thiophene flanking units by furan and selenophene could lead to reduced E_g^{opt}s and enhanced hole mobilities (P72–P74). The polymer with the lowest E_g^{opt} (P74, E_g^{opt} = 1.38 eV) resulted in high-performance tandem and semi-transparent OSCs with PCEs of 9.5% and 4.5%, respectively. McCulloch and coworkers [78] reported a set of DPP-based polymers (P75–P77) with different chalcogenophene comonomers (thiophene, selenophene, and tellurophene) to study the effect of the heteroatom substitution on the optoelectronic and photovoltaic properties. Upon increasing the size of the chalcogen atom (S → Se → Te), the E_gs of the polymers were reduced due to the upshifting of the HOMO energy levels. When 2-pyridine was used as the flanking unit [79],

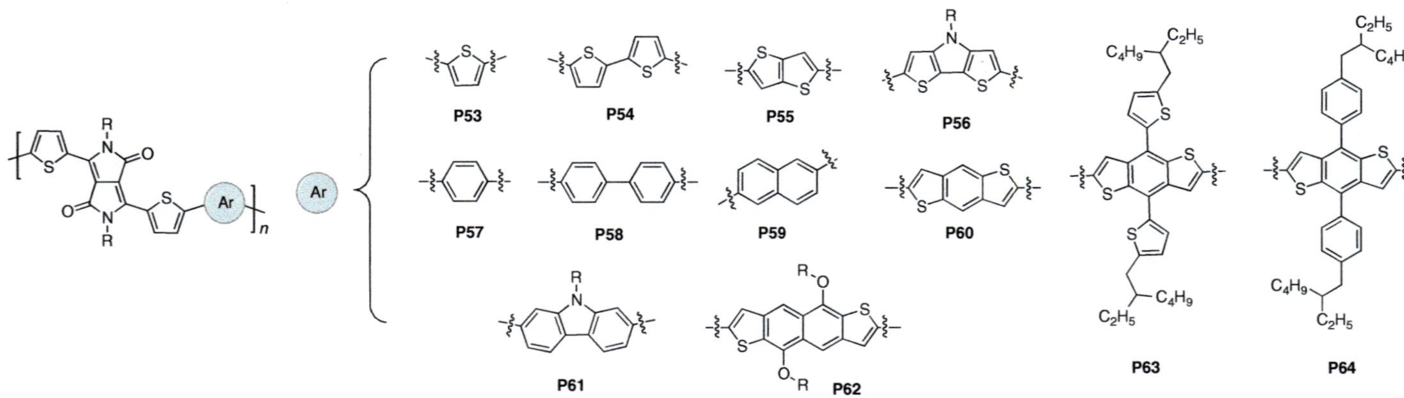

Figure 1.12 Chemical structures of the DPP polymers P53–P64. Source: Xiyue Yuan.

Table 1.8 Optoelectronic properties and device performances of P53–P67.

Polymer	E_g^{opt} (eV)	HOMO/ LUMO (eV)	Acceptor	V_{oc} (V)	J_{sc} (mA cm^{-2})	FF	PCE (%)	References
P53	1.33	−5.17/−3.61	$PC_{71}BM$	0.67	15.4	0.69	7.1	[28]
P54[a)]	1.43	−5.09/−3.64	$PC_{71}BM$	0.64	16.0	0.69	7.1	[65]
P55	1.35	−5.10/−3.68	$PC_{71}BM$	0.67	20.1	0.70	9.4	[66]
P56	1.23	−5.26/−3.68	$PC_{61}BM$	0.43	16.6	0.54	3.9	[67]
P57	1.53	−5.48/−3.66	$PC_{71}BM$	0.80	14.0	0.67	7.4	[28]
P58	1.63	−5.39/−3.59	$PC_{71}BM$	0.80	11.5	0.63	5.7	[68]
P59	1.55	−5.29/−3.79	$PC_{71}BM$	0.76	11.8	0.52	4.7	[69]
P60[a)]	1.51	−5.36/−3.81	$PC_{71}BM$	0.77	13.2	0.68	6.9	[65]
P61	1.57	−5.40/−3.90	$PC_{71}BM$	0.77	9.0	0.52	3.6	[71]
P62	1.36	−5.33/−3.74	$PC_{71}BM$	0.76	13.3	0.68	6.9	[72]
P63	1.44	−5.30/−3.63	$PC_{71}BM$	0.74	13.5	0.65	6.5	[73]
P64	1.46	−5.35/−3.56	$PC_{71}BM$	0.76	13.6	0.60	6.2	[74]
P65	1.52	--/--	$PC_{71}BM$	0.80	14.0	0.67	7.4	[75]
P66	1.52	--/--	$PC_{71}BM$	0.79	11.9	0.60	5.7	[75]
P67	1.52	--/--	$PC_{71}BM$	0.81	6.6	0.59	3.2	[75]

a) For P54 and P60, energy levels are determined from cyclic voltammetry $\left(E^{(Fc/Fc^+)} = -5.23\,\text{eV}\right)$.
Source: Based on Hendriks et al. [28].

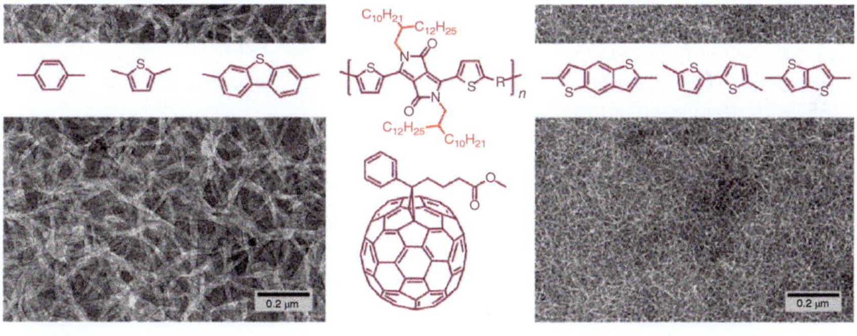

Figure 1.13 Bright field transmission electron microscopy (TEM) images ($1.2 \times 1.2\,\mu m^2$) of the polymer:$PC_{71}BM$ blend films based on several DPP polymers. Source: Li et al. [65] with permission of American Chemical Society.

good coplanarity in the resulting polymer (P78) could be guaranteed. The sp^2 hybridization nitrogen of pyridine also downshifted the HOMO energy level of the polymer and thereby led to high V_{oc} in resulting OSCs. The chemical structures of P68–P78 are showed in Figure 1.15, and the corresponding photovoltaic parameters of OSCs are summarized in Table 1.9.

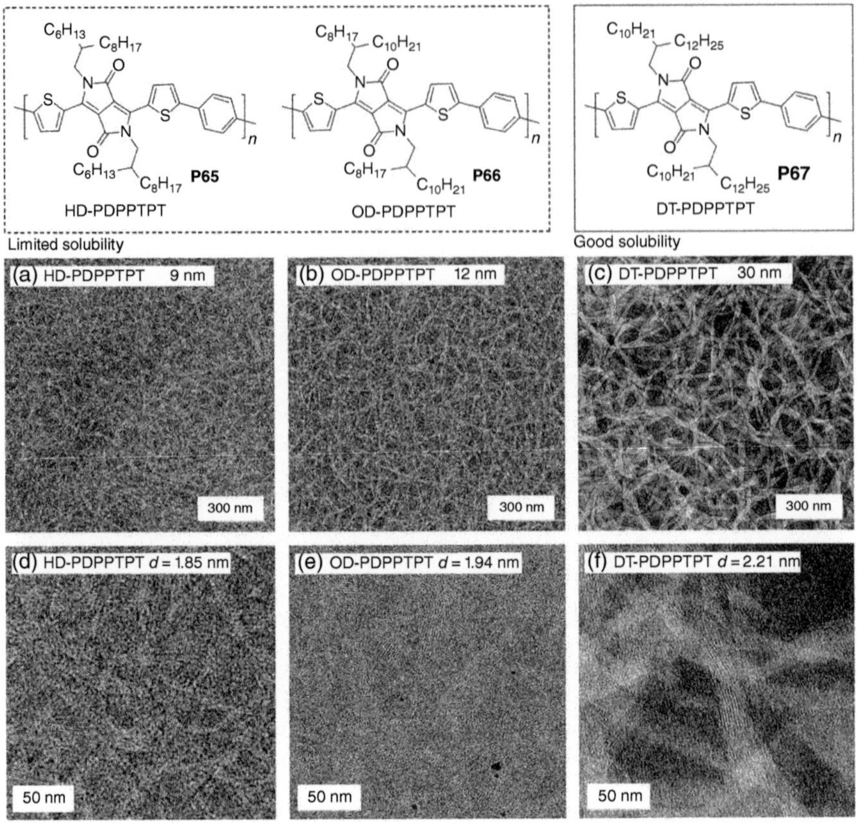

Figure 1.14 Bright field TEM images of the P65-P67:$PC_{71}BM$ blends for different length of the side chains. (a,b,c) Image size: $1.2 \times 1.2\ \mu m^2$. (d,e,f) Image size: $225 \times 225\ nm^2$. Source: Li et al. [65] with permission of John Wiley & Sons, Inc.

Besides, Janssen and coworkers reported a set of regular alternating D1-A-D2-A terpolymers and demonstrated its superior performance in fullerene-based OSCs compared to the parent D–A copolymers (Figure 1.16). Compared to the polymer with random alternation of D1 and D2 with A, the D1-A-D2-A regular polymer allows exact chemical composition [28, 82] and controlled local variations in HOMO/LUMO energy levels. Among these polymers, the polymer containing a terthiophene and a thiophene–phenylene–thiophene (TPT) donor segment (P79) showed optimal energy levels and optical bandgap as compared to the parent copolymers. As a result, this polymer afforded a high PCE of 8.0% when blended with $PC_{71}BM$. The chemical structures of P79–P81 are showed in Figure 1.16, and the corresponding photovoltaic parameters of OSCs are summarized in Table 1.10.

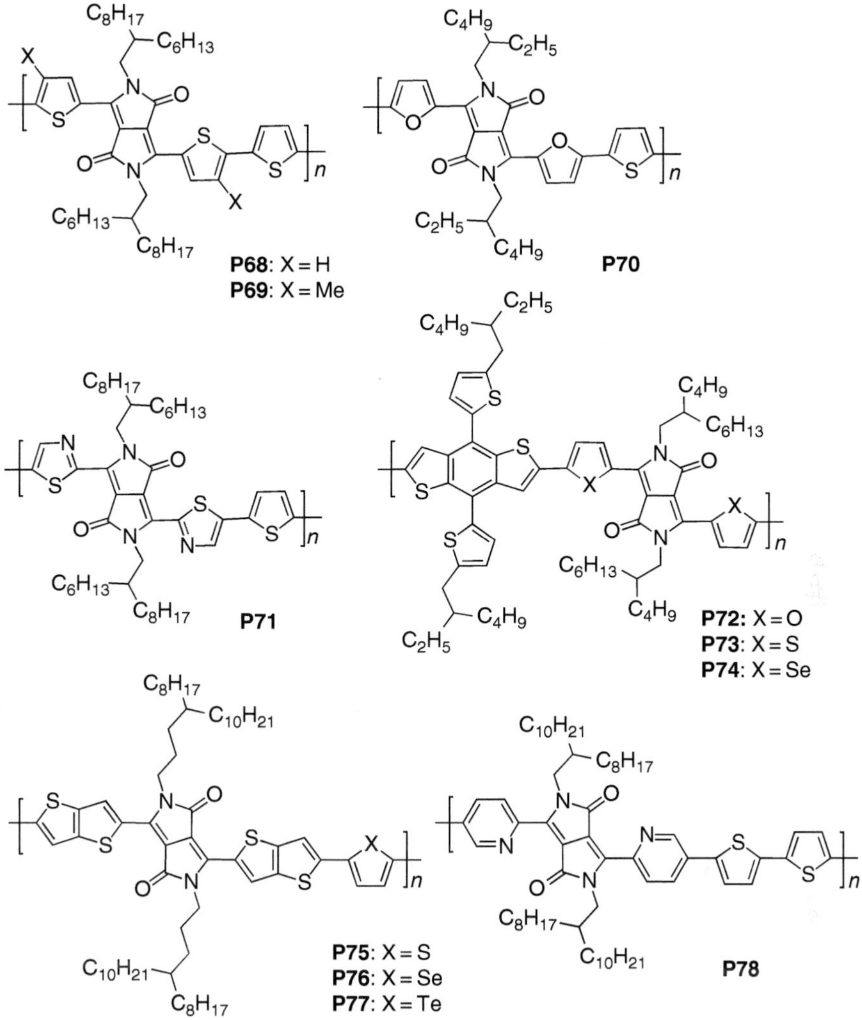

Figure 1.15 Chemical structures of the DPP polymers P68–P78. Source: Xiyue Yuan.

1.3 MBG Polymers

As discussed in the previous section, in the early stage of PSCs using fullerene derivatives as the electron acceptors, extensively researches on the development of conjugated donor polymers have been focused on design and synthesis of LBG polymers to extend the light absorption range over 800 nm and thereby improve the J_{sc} of resulted PSCs [10]. Nevertheless, LBG polymers with reduced E_g^{opt} generally had a

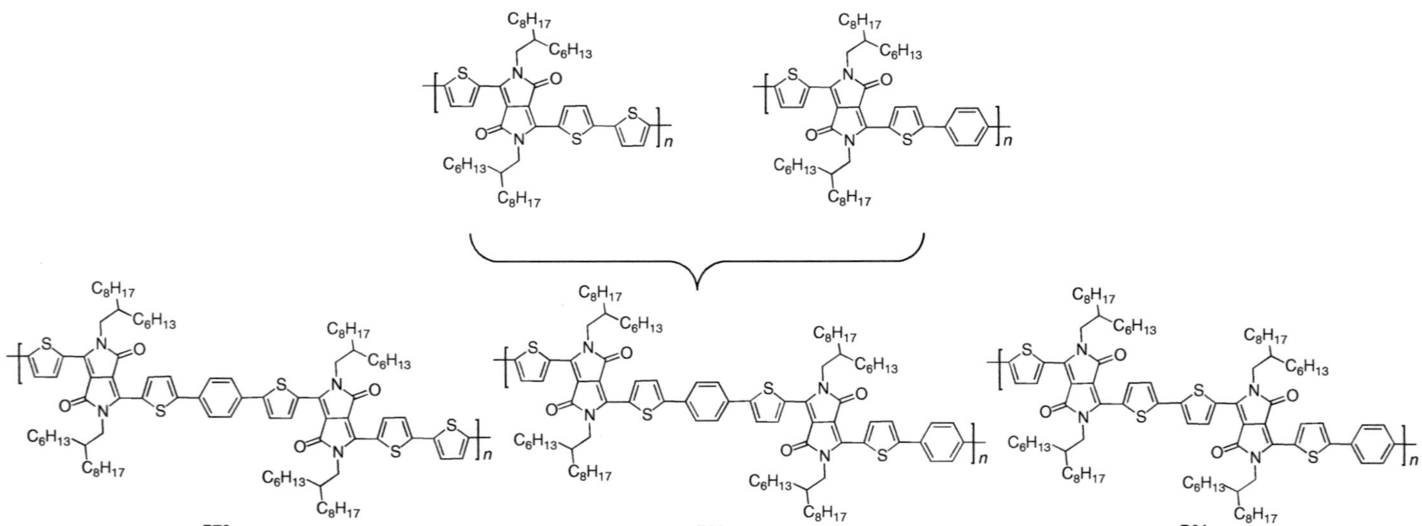

P79 P80 P81

Figure 1.16 Chemical structures of the DPP polymers P79–P81. Source: Xiyue Yuan.

Table 1.9 Optoelectronic properties and device performances of P68–P78.

Polymer	E_g^{opt} (eV)	HOMO/LUMO (eV)	Acceptor	V_{oc} (V)	J_{sc} (mA cm^{-2})	FF	PCE (%)	References
P68	1.33	−5.17/−3.61	$PC_{71}BM$	0.67	15.4	0.69	7.1	[80]
P69[a)]	1.30	−5.43/−3.68	$PC_{71}BM$	0.60	17.8	0.66	7.0	[65]
P70	1.40	−5.20/−−	$PC_{71}BM$	0.65	14.8	0.64	6.5	[81]
P71[a)]	1.44	−5.97/−4.07	$PC_{71}BM$	0.96	2.0	0.58	1.1	[76]
P72	1.51	−5.26/−3.64	$PC_{71}BM$	0.77	10.9	0.56	4.7	[77]
P73	1.46	−5.30/−3.63	$PC_{71}BM$	0.73	13.7	0.65	6.5	[77]
P74	1.38	−5.25/−3.70	$PC_{71}BM$	0.69	16.8	0.62	7.2	[77]
P75	1.39	−5.08/−3.69	$PC_{71}BM$	0.57	13.5	0.66	8.8	[78]
P76	1.37	−5.07/−3.70	$PC_{71}BM$	0.56	21.5	0.63	7.6	[78]
P77	1.32	−5.05/−3.73	$PC_{71}BM$	0.52	21.7	0.63	7.1	[78]
P78	1.68	−5.77/−3.86	$PC_{71}BM$	0.92	8.0	0.66	4.9	[79]

a) For P69 and P71, energy levels determined from cyclic voltammetry $\left(E^{(Fc/Fc^+)} = -5.23\text{ eV}\right)$.
Source: Based on Hendriks et al. [28].

Table 1.10 Optoelectronic properties and device performances of P79–P81.

Polymer	E_g^{opt} (eV)	HOMO/LUMO (eV)	Acceptor	V_{oc} (V)	J_{sc} (mA cm^{-2})	FF	PCE (%)	References
P79	1.47	−−/3.73	$PC_{71}BM$	0.75	15.9	0.67	8.0	[28]
P80	1.57	−5.79/−3.79	$PC_{71}BM$	0.80	13.8	0.67	7.5	[82]
P81	1.42	−5.69/−3.86	$PC_{71}BM$	0.74	6.4	0.67	3.2	[82]

high-lying HOMO energy level, which resulted in a limited V_{oc} and a large E_{loss} [9]. LBG polymers also intrinsically displayed narrow absorption range, which made them could hardly realize an ideal broad photon absorption to make full use of the solar spectrum [10]. In these regards, making a trade-off wisely between the J_{sc} and V_{oc} is quite essential to improve the performance of PSCs. Despite the limited light absorption (< 780 nm) of MBG polymers, the strategy of molecular design and morphology control could enable to improve exciton generation and charge transport for achieving high J_{sc}s that were comparable to LBG polymers. Importantly, MBG polymers could have more potential to deepen the HOMO level without increasing the bandgap, which afforded higher V_{oc} without sacrificing J_{sc} in the resulted PSCs [9, 10]. In addition, the MBG polymers have also played an important role to harvest high-energy photons in ternary and tandem devices. The pair of "weak donor" and "strong acceptor" is an effective way to construct an MBG polymer. Several typical strong acceptor blocks including BT, quinoxaline (Qx), thienopyrrolodione (TPD), and functionalized thieno[3,4-*b*]thiophene (TT) derivatives have been widely used to build MBG polymers. In the following part, we will introduce the design strategies of MBG polymers based on the above acceptor blocks and analyze their photovoltaic properties.

1.3.1 MBG Polymers Based on Benzothiadiazole (BT)

As aforementioned, BT is a widely used "strong acceptor" unit not only in LBG polymers but also in MBG polymers. On the other hand, benzo[1,2-*b*:4,5-*b*′]dithiohpene (BDT) is an ideal "weak donor" block for building MBG polymers. Both BT and BDT have high chemical stability, rigid planar structure, and ease of preparation. BDT–BT-based polymers (Figure 1.17) have shown promising PCEs over 10%, implying the combination of BDT and BT is one of the ideal pairs for high-performance PSCs [83, 84]. What's more, the optical, electrical, and morphological properties of the resulted polymers can be feasibly modulated by introducing various side chains of BDT and functional groups on BDT as well as π-bridges between BDT and BT, which enable the optimization of the final photovoltaic performances (Figure 1.17) [85]. In 2008, Hou et al. reported P82 (E_g^{opt} = 1.70 eV) based on the BT acceptor block and alkoxy-substituted BDT donor block, which produced a poor PCE of 0.90% [86]. By replacing the one-dimensional (1D) alkoxy side chains with two-dimensional (2D) alkylthienyl groups on the BDT block, the resulting polymer P83 exhibited lowered HOMO level (−5.45 versus − 5.10 eV) and extended light absorption than P82, which improved the V_{oc} from 0.68 V of P82 to 0.92 V of P83 and PCE up to 9.4% [87]. Yang and coworkers introduced thiophene units as the bridges between BDT and BT, which extended the conjugation length and developed P84 [88]. Peng et al. introduced alkoxy and fluorine atom, respectively, onto the BT block and developed P85 and P86 [89]. The fluorination effectively lowered the HOMO level of P86 (−5.41 eV), improving the V_{oc} (0.86 V) and PCE (6.21%).

Note that quite a few researches focused on the side chain modification of BDT units, including unsubstituted (P87) [90], alkyl (P88–P90) [91], alkylthienyl (P83–P86, P91) [84, 87–89], fluorinated alkylthienyl (P92, P93), chlorinated alkylthienyl (P94), alkylthiothienyl (P92) side chains, etc. Peng and coworkers combined fluorination and sulfuration on the alkylthienyl side chains and developed P96, which improved the absorption and lowered the HOMO level compared to the fluorinated polymer of P92 and sulfurated polymer of P95 [84]. The resulted devices achieved a much higher PCE of 10.69% in fullerene PSCs and 11.66%

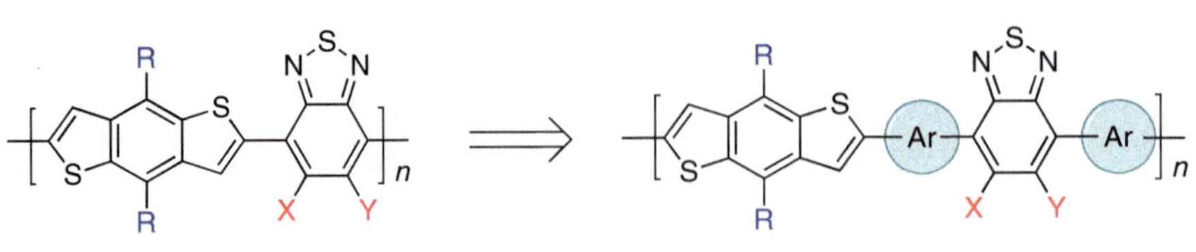

Figure 1.17 The chemical modification strategy of BDT-BT-based polymers. Source: Xiaopeng Xu.

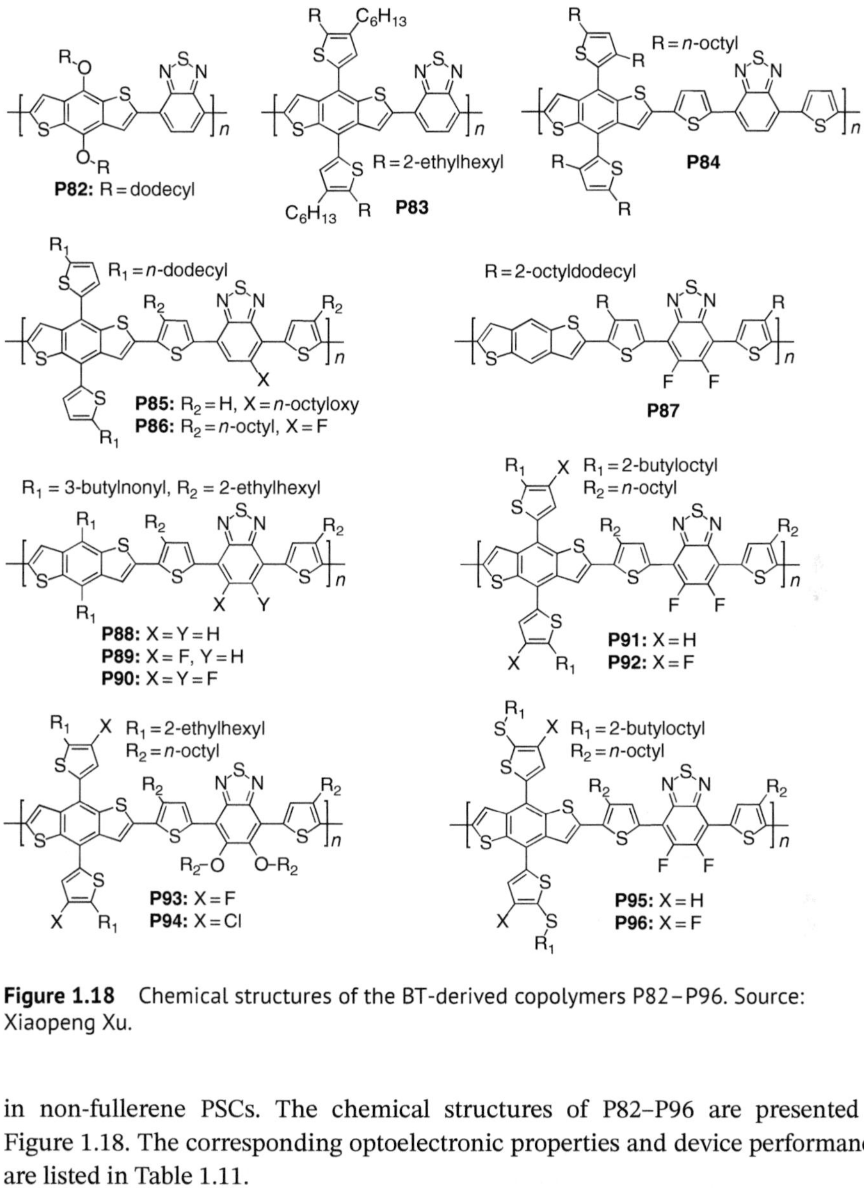

Figure 1.18 Chemical structures of the BT-derived copolymers P82–P96. Source: Xiaopeng Xu.

in non-fullerene PSCs. The chemical structures of P82–P96 are presented in Figure 1.18. The corresponding optoelectronic properties and device performances are listed in Table 1.11.

Compared to the thiophene unit, the benzene ring is featured with symmetry and lower electron-donating ability [93]. Using benzene ring as the side group, the overall properties of the resulted polymers would be modulated. Hou and coworkers [94] attached alkylphenyl groups onto BDT block and developed P97, which exhibited an E_g^{opt} of 1.70 eV and deep HOMO of −5.35 eV. The resulting devices produced a high V_{oc} of 0.88 V and a PCE of 8.07%. Sulfur and oxygen atoms could be inserted into the alkylphenyl group to modulate the energy levels and intermolecular interactions (P98–P101) [95, 96]. Furthermore, the alkoxyphenyl side chains could be further modified by introducing fluorine atom at the meta-position [96–98], ortho-position

Table 1.11 Optoelectronic properties and device performances of P82–P96.

Polymer	E_g^{opt} (eV)	HOMO/ LUMO (eV)	Acceptor	V_{oc} (V)	J_{sc} (mA cm^{-2})	FF	PCE (%)	References
P82	1.70	−5.10/−3.19	$PC_{61}BM$	0.68	2.97	0.44	0.90	[86]
P83	1.65	−5.45/−3.65	$PC_{71}BM$	0.92	15.4	0.66	9.4	[87]
P84	1.75	−5.31/−3.44	$PC_{71}BM$	0.92	10.70	0.575	5.66	[88]
P85	1.67	−5.32/−3.58	$PC_{71}BM$	0.82	12.53	0.549	5.64	[89]
P86	1.63	−5.41/−3.72	$PC_{71}BM$	0.86	12.05	0.599	6.21	[89]
P87	1.73	−5.40/−3.67	$PC_{71}BM$	0.85	14.56	0.751	9.29	[90]
P88	1.65	−5.42/−−	$PC_{61}BM$	0.78	11.7	0.476	4.33	[91]
P89	1.67	−5.48/−−	$PC_{61}BM$	0.85	11.4	0.506	5.28	[91]
P90	1.73	−5.53/−−	$PC_{61}BM$	0.90	12.2	0.621	7.16	[91]
P91	1.69	−5.14/−3.35	$PC_{71}BM$	0.79	15.41	0.682	8.31	[84]
P91	1.69	−5.14/−3.35	ITIC	0.82	13.73	0.638	7.15	[84]
P92	1.71	−5.34/−3.53	$PC_{71}BM$	0.86	16.47	0.687	9.74	[84]
P92	1.71	−5.34/−3.53	ITIC	0.90	15.38	0.663	9.17	[84]
P93	1.77	−5.39/−3.65	IT-4F	0.778	21.73	0.545	9.14	[92]
P94	1.78	−5.51/−3.60	IT-4F	0.782	21.03	0.70	11.60	[92]
P95	1.70	−5.20/−3.40	$PC_{71}BM$	0.84	16.28	0.683	9.33	[84]
P95	1.70	−5.20/−3.40	ITIC	0.89	14.63	0.659	8.58	[84]
P96	1.71	−5.41/−3.59	$PC_{71}BM$	0.92	16.60	0.70	10.69	[84]
P96	1.71	−5.41/−3.59	ITIC	1.03	17.09	0.663	11.66	[84]

[98] or more than one fluorine atom in each alkoxyphenyl side chain (P102–P104) [96]. Shin et al. [96, 98] found that introducing one fluorine at the ortho-position or two fluorine atoms onto each alkoxyphenyl side chain would enlarged the E_g^{opt} to more than 1.8 eV due to the large steric hindrance at ortho-position and high electron-withdrawing effect of fluorine atom. In contrast, introducing fluorine atom onto the ortho-position of alkoxyphenyl side chain did not induce any steric hindrance but lowered the HOMO level of the resulting polymers (P103, P104) [96, 98]. The π-conjugated side chains efficiently increased the electron density and interchain aggregation among conjugated polymers, thereby enhancing light absorption and charge transport properties of the related PSCs. Cho and coworkers [99] employed oligothienyl side chains to further extend the 2D conjugation lengths of conjugated copolymer systems and developed a series MBG polymers (P105–P107). With the thiophene numbers increasing, the E_g^{opt} was reduced, and the HOMO level was lowered. P106 with the optimum size of the chromophoric side chains exhibited an extended absorption spectrum and high carrier transport, leading to the highest PCE of 6.5%. Yang and coworkers [100] replaced the alkylthienyl side chains on BDT with the extended alkylbenzothienyl side chains and developed P108. Fusing benzene on the flanking thiophene ring of BDT side chains

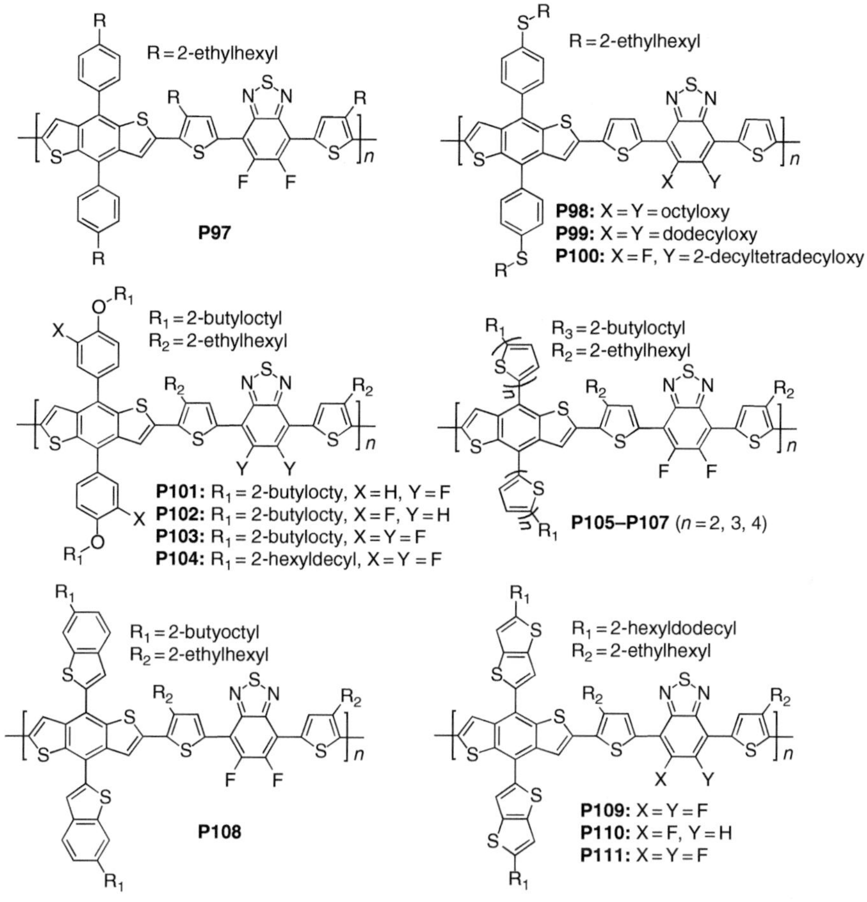

Figure 1.19 Chemical structures of the BT-derived copolymers P97–P111. Source: Xiaopeng Xu.

slightly red-shifted the absorption bands, obviously enhanced the intermolecular interactions, and pronouncedly downshifted the HOMO and LUMO energy levels of P110, which improved the PCE to 7.30%. Yang and coworkers [101] used alkyl-substituted thieno[3,2-*b*]thiophene as the extended side chains and developed P109–P111. Devices based on the three polymers maintained ~90% of the initial efficiency after annealing for 30 days at 100 °C, showing their long-term stability. The chemical structures of P97–P111 are presented in Figure 1.19. The corresponding optoelectronic properties and device performances are listed in Table 1.12.

Most of the BDT-based polymers are based on the symmetric 1D BDT and 2D BDT units. To combine the two kinds of BDTs, Yang and coworkers proposed a symmetry-breaking strategy to combine the advantages of both 1D and 2D symmetric BDT building blocks and constructed a series of high-performance donor polymers (P112–P119) [102, 103]. The asymmetric BDT blocks efficiently triggered a modulation of the preferential ordered microstructure, formed the

Table 1.12 Optoelectronic properties and device performances of P97–P111.

Polymer	E_g^{opt} (eV)	HOMO/ LUMO (eV)	Acceptor	V_{oc} (V)	J_{sc} (mA cm^{-2})	FF	PCE (%)	References
P97	1.70	−5.35/−3.34	$PC_{71}BM$	0.88	12.94	0.709	8.07	[94]
P98	1.79	−5.42/−3.63	$PC_{71}BM$	0.74	11.60	0.63	5.4	[95]
P99	1.78	−5.41/−3.63	$PC_{71}BM$	0.76	11.26	0.72	6.1	[95]
P100	1.76	−5.60/−3.84	$PC_{71}BM$	0.82	12.28	0.73	7.4	[95]
P101	1.70	−5.12/−3.42	$PC_{71}BM$	0.83	11.33	0.663	6.23	[93]
P102	1.73	−5.42/−3.69	$PC_{71}BM$	0.83	12.23	0.618	6.28	[96]
P102	1.73	−5.39/−3.66	$PC_{71}BM$	0.83	13.11	0.653	7.02	[97]
P103	1.77	−5.56/−3.79	$PC_{71}BM$	0.91	11.25	0.646	6.62	[98]
P104	1.76	−5.58/−3.77	$PC_{71}BM$	0.90	13.18	0.644	7.64	[96]
P105	1.73	−5.37/−3.61	$PC_{71}BM$	0.90	10.6	0.547	5.22	[99]
P106	1.70	−5.31/−3.62	$PC_{71}BM$	0.86	12.3	0.614	6.48	[99]
P107	1.68	−5.30/−3.63	$PC_{71}BM$	0.84	6.35	0.405	2.17	[99]
P108	1.76	−5.36/−3.60	$PC_{61}BM$	0.90	12.93	0.627	7.30	[100]
P109	1.71	−5.29/−−	$PC_{71}BM$	0.74	11.76	0.655	5.70	[101]
P110	1.69	−5.39/−−	$PC_{71}BM$	0.80	11.95	0.662	6.29	[101]
P111	1.75	−5.48/−−	$PC_{71}BM$	0.86	11.71	0.653	6.54	[101]

Table 1.13 Optoelectronic properties and device performances of P112–P119.

Polymer	E_g^{opt} (eV)	HOMO/ LUMO (eV)	Acceptor	V_{oc} (V)	J_{sc} (mA cm^{-2})	FF	PCE (%)	References
P112	1.73	−5.03/−3.30	$PC_{71}BM$	0.83	13.92	0.665	7.68	[102]
P113	1.72	−5.01/−3.29	$PC_{71}BM$	0.84	13.64	0.717	8.22	[102]
P114	1.76	−5.06/−3.29	$PC_{71}BM$	0.88	14.92	0.719	9.44	[102]
P115	1.75	−5.03/−3.28	$PC_{71}BM$	0.88	13.98	0.739	9.09	[102]
P116	1.76	−5.35/−3.59	$PC_{71}BM$	0.892	15.17	0.755	10.22	[103]
P117	1.75	−5.33/−3.58	$PC_{71}BM$	0.884	14.86	0.750	9.84	[103]
P118	1.75	−5.32/−3.57	$PC_{71}BM$	0.876	14.98	0.709	9.30	[103]
P119	1.74	−5.32/−3.58	$PC_{71}BM$	0.868	14.05	0.703	8.57	[103]

intense molecular stacking, and shortened the π–π stacking distance to enhance π–π overlap from the extended conjugated aryl units, which was beneficial for charge transport. Based on such strategy, a high PCE reaching 10.22% was realized when using $PC_{71}BM$ as the acceptor in PSCs. The chemical structures of P112–P119 are presented in Figure 1.20. The corresponding optoelectronic properties and device performances are listed in Table 1.13.

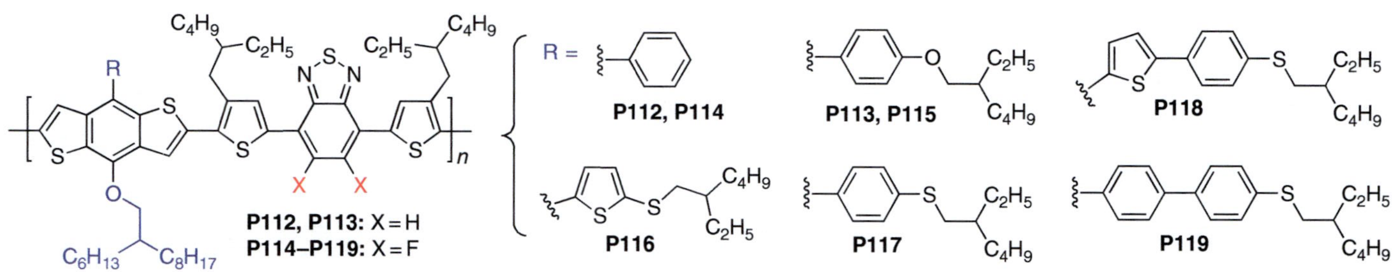

Figure 1.20 Chemical structures of the BT-derived copolymers P112–P119. Source: Xiaopeng Xu.

Modulation of the π-spacers between BDT and BT have also shown great potential in developing efficient MBG polymers. Introducing a bulky bridge between the BDT and BT blocks, such as a hexylthiophene-substituted thiophene bridge (P120), could introduce large steric hindrance into the polymer backbone, which had less effect on the E_g^{opt}, but effectively lowered the HOMO, thereby increasing the V_{oc} of the resulted PSCs [104]. Replacing thiophene with furan as the π-spacers between BDT and BT blocks could enlarge the E_g^{opt} and lower the HOMO of the resulting polymers (P121, P122) [105], while replacing thiophene with selenophene as the π-spacers could reduce the E_g^{opt} and increase the HOMO of the resulting polymers (P123, P124) [106]. Moreover, the BDT block could also be modulated by replacing the sulfur atoms with oxygen atoms. The thieno[2,3-*f*]benzofuran and benzo[1,2-*b*:4,5-*b*′]difuran-derived polymers exhibited reduced E_g^{opt}s and increased HOMOs (P125–P128). The chemical structures of P120–P128 are presented in Figure 1.21. The corresponding optoelectronic properties and device performances are listed in Table 1.14.

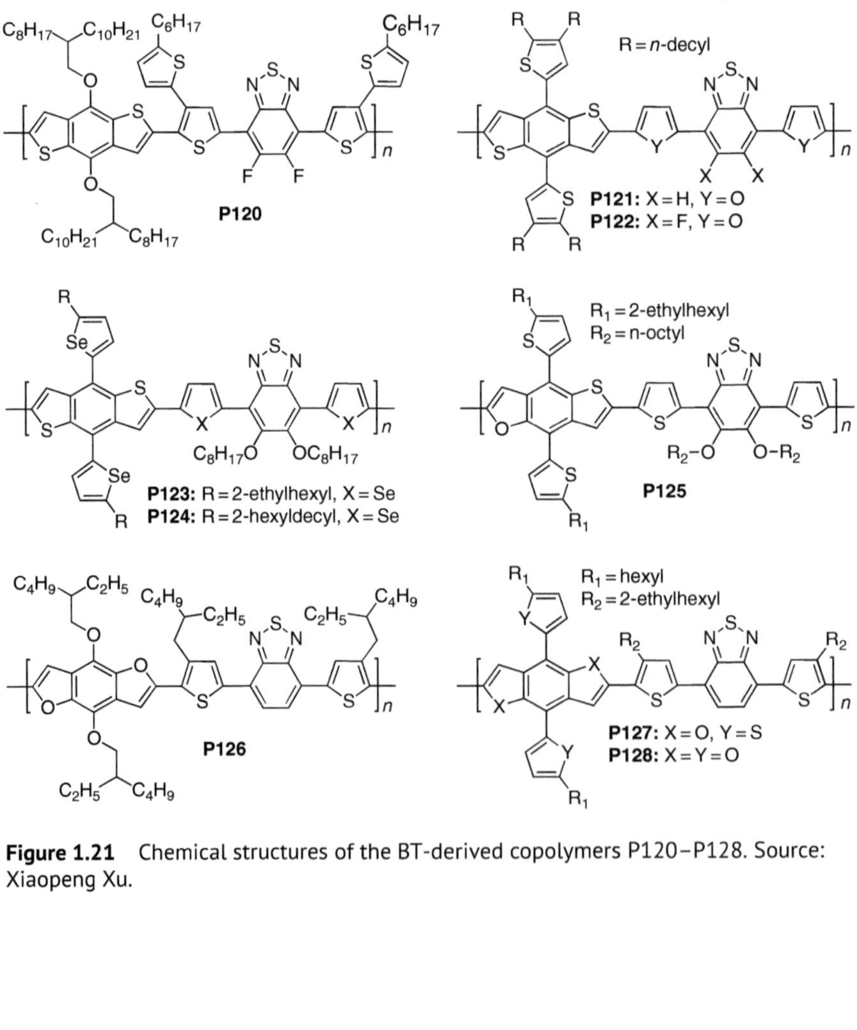

Figure 1.21 Chemical structures of the BT-derived copolymers P120–P128. Source: Xiaopeng Xu.

Table 1.14 Optoelectronic properties and device performances of P120–P128.

Polymer	E_g^{opt} (eV)	HOMO/ LUMO (eV)	Acceptor	V_{oc} (V)	J_{sc} (mA cm^{-2})	FF	PCE (%)	References
P120	1.70	−5.47/−3.74	$PC_{71}BM$	0.80	12.56	0.604	6.19	[104]
P121	1.76	−5.36/−3.59	$PC_{71}BM$	0.86	7.9	0.55	3.7	[105]
P122	1.78	−5.28/−3.60	$PC_{71}BM$	0.86	8.7	0.59	4.4	[105]
P123	1.69	−5.26/−3.57	$PC_{71}BM$	0.70	12.28	0.589	5.07	[106]
P124	1.63	−5.40/−3.77	$PC_{71}BM$	0.73	12.22	0.594	5.34	[106]
P125	1.63	−5.20/−3.57	$PC_{71}BM$	0.78	13.51	0.61	6.42	[107]
P126	1.60	−5.10/−3.24	$PC_{71}BM$	0.78	11.77	0.546	5.01	[108]
P127	1.68	−5.08/−3.39	$PC_{71}BM$	0.73	9.94	0.609	4.42	[109]
P128	1.61	−5.11/−3.60	$PC_{71}BM$	0.80	5.84	0.556	2.60	[109]

In addition, there also had many other BT-based polymers using various donor blocks. For example, Ashraf et al. employed indacenodithiophene (IDT) and it derivatives silaindacenodithiophene (SiIDT) and germaindacenodithiophene (GeIDT) as the donor blocks and developed P129–P131 [110, 111]. These polymers exhibited an MBG around 1.7 ~ 1.8 eV, a low HOMO around −5.2 ~ −5.3 eV. Jen and coworkers reported the IDTT-based ladder-type polymer of P132, which exhibited a medium E_g^{opt} of 1.78 eV and a high hole mobility of 2×10^{-2} cm^2 V^{-1} s^{-1} [112]. The resulting PSCs realized a high PCE of 7.03%. Yan and coworkers developed the terthiophene-based D–A polymer of P133 with an asymmetric arrangement of alkyl chains [113]. The asymmetric arrangement of alkyl chains enabled highly efficient thick-film PSCs with PCE up to 10.7% without the need of processing additives. Woo and coworkers introduced dialkoxylphenylene as the donor block and developed a series of semi-crystalline photovoltaic polymers (P134–P136) [114]. The noncovalent intramolecular interactions (N···S, N···H, O···S, O···H, F···S, F···H) served as conformational lock to enhance chain planarity, intermolecular ordering, and thermal stability without losing solution processability. The polymers formed well-distributed interpenetrating nano-fibrillar-networked morphologies with $PC_{71}BM$, showing well-balanced hole and electron mobilities. P136 exhibited a high PCE up to 9.39% in a 290-nm thick conventional single-cell device structure without any additional interfacial layer. Using P136 in the all-PSCs, a high PCE of 5.10% was realized [115]. Ding and coworkers introduced different lactam-containing tricyclic blocks and developed a series of BT-based polymers (P137–P139), which had medium E_g^{opt}s around 1.65 ~ 1.72 eV [116, 117]. Among them, the P138 displayed less bimolecular recombination and balanced charge transport in the active layer, P138:$PC_{71}BM$ solar cells gave a PCE of 10.16% with an active layer thickness of 267 nm. He and coworkers developed P141, in which a chlorine atom was introduced at the 4-position in the middle thiophene unit [118].

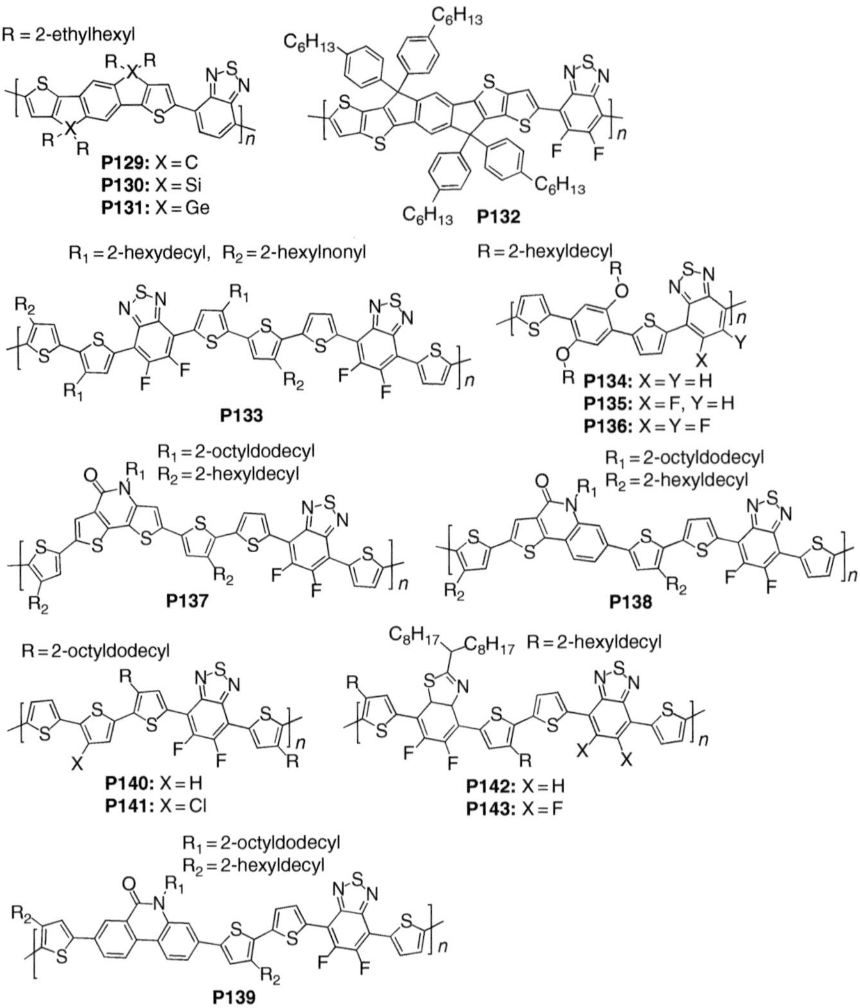

Figure 1.22 Chemical structures of the BT-derived copolymers P129–P143. Source: Xiaopeng Xu.

Compared to the non-chlorinated P140, the chlorination lowered the HOMO level of P60 (−5.33 versus −5.19 eV) and enhanced the intermolecular π–π interactions in the active layer film. The PSCs based on P141:$PC_{71}BM$ exhibited enhancements both in V_{oc} and FF, resulting in further improvement of PCE from 10.17% of P140:$PC_{71}BM$ to 11.18%. Feng et al. introduced the weak electron-withdrawing 5,6-difluorobenzo[*d*]thiazole (BTA-2f) and developed two new BT-based MBG polymers (P142, P143) [119]. P143 with difluorinated BT exhibited lowered HOMO level, enhanced ICT, and improved crystallinity. The PCE of resulting fullerene-free PSCs using Y6 as the acceptor was significantly improved from 2.61% (P142) to 16.08% (P143). The chemical structures of P129–P143 are presented in Figure 1.22. The corresponding optoelectronic properties and device performances are listed in Table 1.15.

Table 1.15 Optoelectronic properties and device performances of P129–P143.

Polymer	E_g^{opt} (eV)	HOMO/ LUMO (eV)	Acceptor	V_{oc} (V)	J_{sc} (mA cm^{-2})	FF	PCE (%)	References
P129	1.7	−5.3/−3.6	$PC_{71}BM$	0.80	14.48	0.56	6.5	[110]
P130	1.8	−5.3/−3.5	$PC_{71}BM$	0.91	12.18	0.52	5.8	[110]
P131	1.7	−5.2/−3.4	$PC_{71}BM$	0.85	13.95	0.55	6.5	[110]
P131	1.74	−−/−−	$PC_{71}BM$	0.86	10.1	0.58	5.02	[111]
P132	1.78	−5.30/−3.50	$PC_{71}BM$	0.95	12.21	0.61	7.03	[112]
P133	1.63	−5.31/−3.68	$PC_{71}BM$	0.82	18.7	0.683	10.5	[113]
P134	1.72	−5.29/−3.57	$PC_{71}BM$	0.81	10.4	0.61	5.08	[114]
P135	1.72	−5.35/−3.63	$PC_{71}BM$	0.81	10.2	0.62	5.11	[114]
P136	1.76	−5.45/−3.69	$PC_{71}BM$	0.86	11.4	0.74	7.26	[114]
P136	1.76	−5.45/−3.69	P(NDI2OD-T2)	0.85	11.9	0.51	5.03	[115]
P137	1.62	−5.26/−3.21	$PC_{71}BM$	0.82	15.35	0.696	8.75	[116]
P137	1.62	−5.26/−3.21	ITIC	0.90	14.18	0.593	7.58	[116]
P137	1.65	−5.44/−3.05	$PC_{71}BM$	0.84	14.64	0.703	8.61	[117]
P138	1.67	−5.41/−3.01	$PC_{71}BM$	0.82	17.27	0.721	10.16	[117]
P139	1.72	−5.48/−2.99	$PC_{71}BM$	0.89	13.60	0.699	8.47	[117]
P140	1.65	−5.19/−3.53	$PC_{71}BM$	0.75	19.23	0.705	10.17	[118]
P141	1.67	−5.33/−3.64	$PC_{71}BM$	0.80	18.71	0.746	11.18	[118]
P142	1.69	−5.45/−3.76	Y6	0.90	10.07	0.289	2.61	[119]
P143	1.73	−5.55/−3.82	Y6	0.81	27.25	0.727	16.08	[119]

1.3.2 MBG Polymers Based on Quinoxaline (Qx)

The quinoxaline (Qx) skeleton is a simple but effective system in organic photovoltaics and has been extensively studied as a strong electron deficient unit for building high-performance MBG copolymers [120, 121]. The Qx can be modified by introducing side chains on the 1,2- or 5,6-positions, introducing halogen atoms on the 5,6-positions and fusing aromatic rings to extend Qx core (Figure 1.23). BDT-Qx-based polymers are quite representative in this filed. The side chains on both BDT and Qx can be easily modified by introducing alkoxy, alkylthienyl, or alkoxyphenyl groups (P144–P151) [122–124]. In addition, the side chains of Qx block can be further modified by introducing fluorine atoms (P152–P154) [125–127]. For example, Zou and coworkers introduced the monofluorinated and difluorinated alkoxyphenyl groups, respectively, on the 1,2-positions of Qx as the side chains and developed P153 and P154 [126, 127]. P154 with difluorinated alkoxyphenyl groups exhibited blue-shifted absorption (E_g^{opt}, 1.76 versus 1.73 eV), lower HOMO level (−5.45 versus −5.36 eV) than those of P153 with monofluorinated alkoxyphenyl groups. Using ITIC as the acceptor in fullerene-free solar

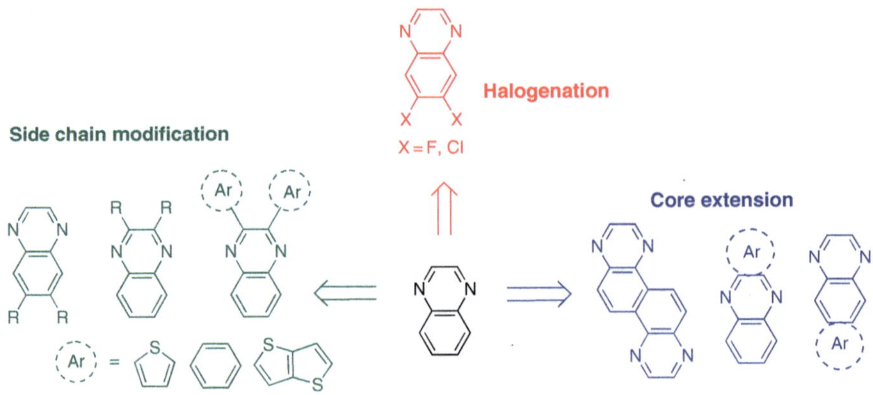

Figure 1.23 The chemical structure of quinoxaline and the structural modification based on quinoxaline. Source: Xiaopeng Xu.

Table 1.16 Optoelectronic properties and device performances of P144–P154.

Polymer	E_g^{opt} (eV)	HOMO/ LUMO (eV)	Acceptor	V_{oc} (V)	J_{sc} (mA cm^{-2})	FF	PCE (%)	References
P144	1.68	−4.79/−2.27	$PC_{71}BM$	0.78	13.2	0.708	7.29	[122]
P145	1.72	−4.83/−2.22	$PC_{71}BM$	0.88	11.4	0.759	7.61	[122]
P146	1.66	−4.76/−2.28	$PC_{71}BM$	0.72	13.4	0.708	6.82	[122]
P147	1.72	−4.77/−2.25	$PC_{71}BM$	0.85	11.3	0.757	7.25	[122]
P148	1.78	−5.25/−3.2	$PC_{71}BM$	0.82	12.34	0.601	6.08	[123]
P149	1.75	−5.29/−3.58	$PC_{71}BM$	0.85	12.13	0.635	6.54	[123]
P150	1.72	−5.20/−3.50	$PC_{71}BM$	0.86	12.77	0.699	7.68	[123]
P151	1.78	−5.51/−3.57	$PC_{71}BM$	0.94	11.28	0.647	6.9	[124]
P152	1.66	−5.52/––	$PC_{71}BM$	0.87	11.4	0.73	7.2	[125]
P153	1.73	−5.36/−3.56	ITIC	0.87	16.33	0.596	8.47	[126]
P154	1.76	−5.45/−3.50	ITIC	0.90	14.89	0.66	9.07	[127]

cells, the higher V_{oc} and FF values contributed to the improved PCE from 8.47% to 9.07%. The chemical structures of P144–P154 are presented in Figure 1.24. The corresponding optoelectronic properties and device performances are listed in Table 1.16.

Introducing halogen atoms on the Qx skeleton shows more impact on the energy levels of Qx-based polymers. Wang and coworkers developed P155–P157 by introducing different numbers of fluorine atom on the Qx block [128]. The E_g^{opt} was gradually enlarged, while the HOMO level was stepwise downshifted from the non-fluorinated P155 to the monofluorinated P156 and difluorinated P157. With number of fluorine atom increasing, the molecular packing orientation changed from edge-on direction to face-on direction, facilitating the vertical charge transport

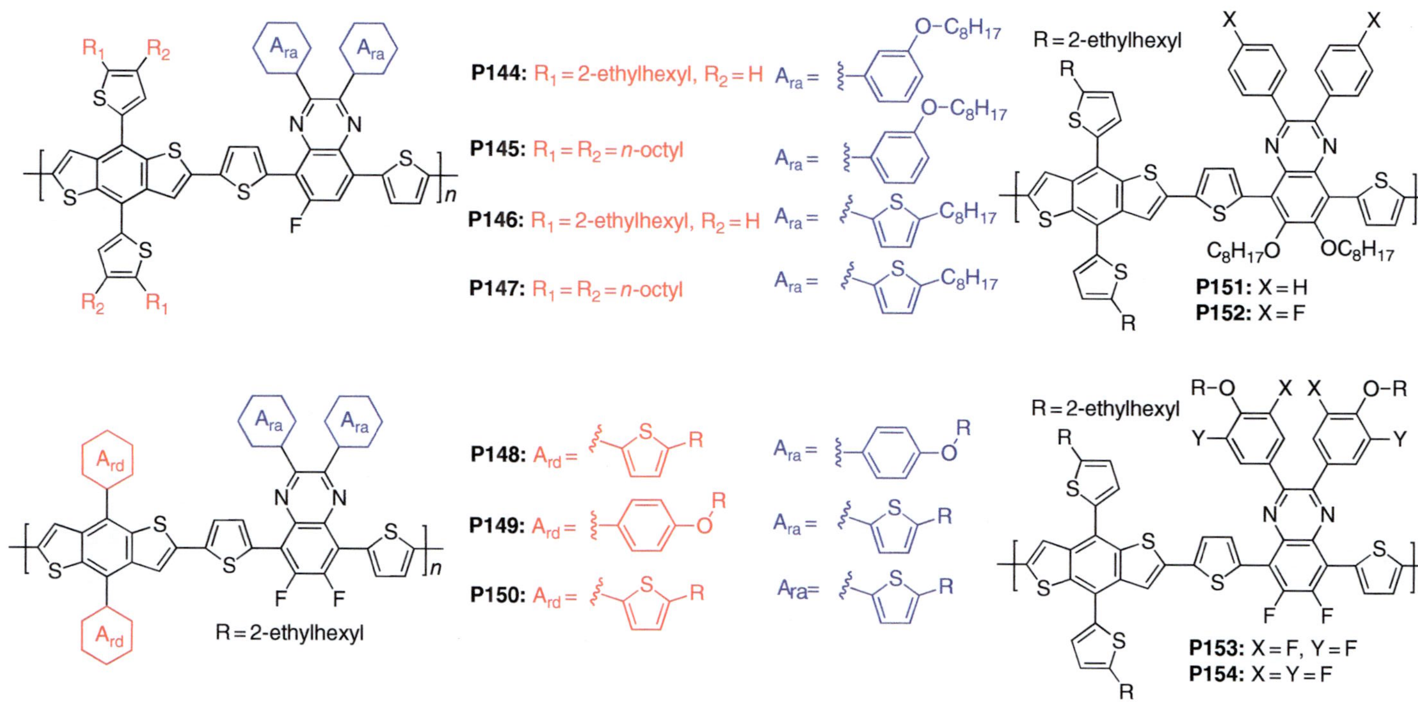

Figure 1.24 Chemical structures of the Qx-derived copolymers P144–P154. Source: Xiaopeng Xu.

and enhancement in hole mobilities. The significantly improved V_{oc} (from 0.60 to 0.77 V) and J_{sc} (from 9.10 to 12.62 mA cm^{-2}) boosted the PCE from 2.82% to 5.19%. Chang and coworkers introduced fluorine atoms on the side chains and/or the skeleton of Qx block and developed a series of Qx-based polymers (P158–P161) to elucidate the positional and populational effects of fluorine atoms on the properties of Qx-based polymers [129]. The gradual but noticeable improvements in the PCEs of these polymers in inverted-type PSCs were observed by increasing the number or changing the position of fluorine substituents on the Qx unit. Hou and coworkers reported P162–P164 by introducing fluorine atoms on the Qx block or on both the side chains of BDT block and the skeleton of Qx block [130]. The fluorination made the HOMO level offset and LUMO level offset between the P164 and ITIC acceptor to 0.24 and 0.04 eV, respectively. This energy level modulation effectively improved the V_{oc} from 0.69 to 0.95 V in the fullerene-free PSCs, without sacrificing the charge extraction. Moreover, the polymer with more fluorine tended to form stronger interchain π–π interaction when it was blended with ITIC, which was helpful for efficient charge transport. As a result, P164:ITIC devices yielded a high PCE of 11.34%. Zhang et al. replaced fluorine atom on P165 with chlorine atom and developed P166 [131]. The chlorination reduced the E_g^{opt} and lowered the HOMO level of P166. The simultaneously increased V_{oc}, J_{sc}, and FF resulted in the improved PCE from 6.12% to 8.16%. The chemical structures of P155–P166 are presented in Figure 1.25. The corresponding optoelectronic properties and device performances are listed in Table 1.17.

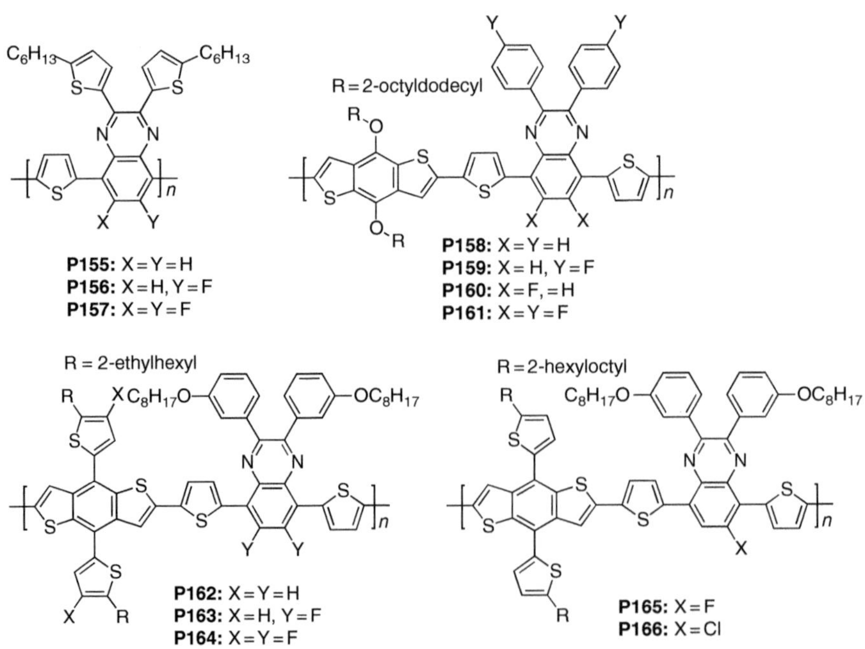

Figure 1.25 Chemical structures of the Qx-derived copolymers P155–P166. Source: Xiaopeng Xu.

Table 1.17 Optoelectronic properties and device performances of P155–P166.

Polymer	E_g^{opt} (eV)	HOMO/ LUMO (eV)	Acceptor	V_{oc} (V)	J_{sc} (mA cm^{-2})	FF	PCE (%)	References
P155	1.61	−5.10/−3.53	$PC_{71}BM$	0.60	9.10	0.518	2.82	[128]
P156	1.68	−5.18/−3.54	$PC_{71}BM$	0.68	11.05	0.549	4.14	[128]
P157	1.77	−5.33/−3.54	$PC_{71}BM$	0.77	12.62	0.531	5.19	[128]
P158	1.67	−5.29/−3.62	$PC_{71}BM$	0.61	8.08	0.637	3.14	[129]
P159	1.71	−5.35/−3.64	$PC_{71}BM$	0.78	8.76	0.661	4.52	[129]
P160	1.73	−5.36/−3.63	$PC_{71}BM$	0.82	9.62	0.705	5.56	[129]
P161	1.73	−5.47/−3.74	$PC_{71}BM$	0.91	10.15	0.715	6.60	[129]
P162	1.70	−5.18/−3.48	ITIC	0.69	16.16	0.599	6.68	[130]
P163	1.72	−5.34/−3.62	ITIC	0.83	17.16	0.645	8.90	[130]
P164	1.79	−5.49/−3.70	ITIC	0.95	17.87	0.668	11.34	[130]
P165	1.68	−5.31/−3.50	$PC_{71}BM$	0.93	12.40	0.53	6.12	[131]
P166	1.66	−5.33/−3.56	$PC_{71}BM$	0.95	14.08	0.61	8.16	[131]

The Qx block can also be further modified by fusing more aromatic rings to extend the π-conjugation. Bo and coworkers employed the fluorinated dibenzo[*a,c*]phenazine and acenaphtho[1,2-*b*]quinoxaline as the acceptor units, respectively, and developed P167 and P168 [132, 133]. These polymers exhibited MBGs and low HOMO levels. P167 exhibited a high hole mobility of 3.56×10^{-3} cm^2 V^{-1} s^{-1}. The resulting PSCs based on P167 realized a good PCE of 6.0%. Zhang et al. fused two pyrazine rings onto a benzene core and developed P169 [134]. P169 combined with $PC_{71}BM$ demonstrated decent J_{sc} and remarkable V_{oc}. Huang and coworkers fused two Qx blocks and developed quinoxalino[6,5-*f*]quinoxaline (NQx) with an angular shape [135]. By changing the alkoxy side chains on NQx block, a series of MBG polymers of P170–P172 were developed. With carefully device engineering, Peng and coworkers improved the PCE of P172 from 1.7% to 7.02% [136]. The chemical structures of P167–P172 are presented in Figure 1.26. The corresponding optoelectronic properties and device performances are listed in Table 1.18.

1.3.3 MBG Polymers Based on Thienopyrrolodione (TPD)

Thieno[3,4-*c*]pyrrole-4,6-dione (TPD) is a simple, compact, symmetric, and planar structure, which can be beneficial for electron delocalization when it is incorporated into various conjugated polymers [137]. TPD has strong electron-withdrawing ability and is able to form quinoidal resonance structure in the polymers. Using the N-bridged dithieno[3,2-*b*:2′,3′-*d*]pyrrole (DTP), which has strong electron-donating ability, as the donor block, the resulted polymers exhibited relatively LBGs (P173) [138, 139]. The bandgap of the resulting polymers can be enlarged and HOMO level can be downshifted by replacing the N-bridged

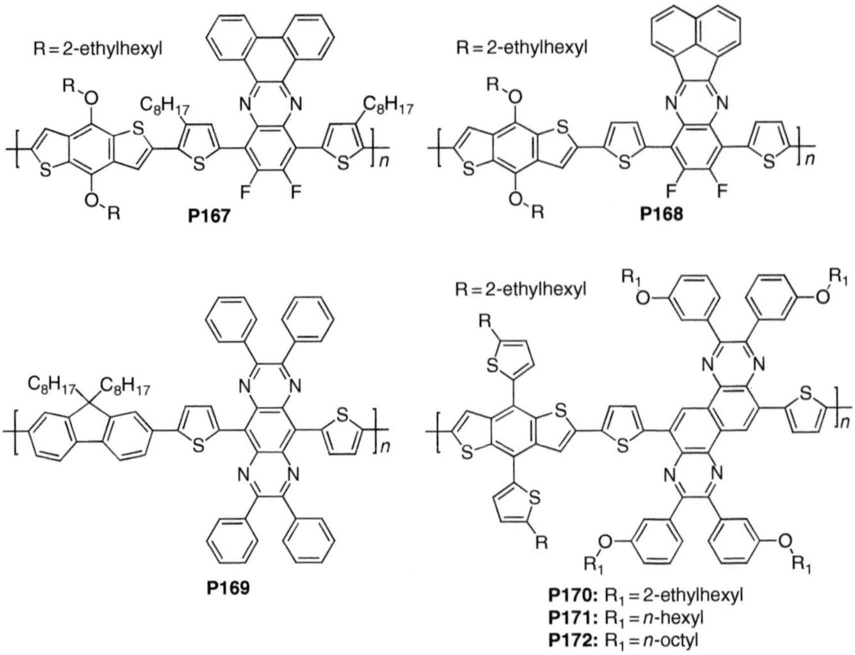

Figure 1.26 Chemical structures of the Qx-derived copolymers P167–P172. Source: Xiaopeng Xu.

Table 1.18 Optoelectronic properties and device performances of P167–P172.

Polymer	E_g^{opt} (eV)	HOMO/ LUMO (eV)	Acceptor	V_{oc} (V)	J_{sc} (mA cm^{-2})	FF	PCE (%)	References
P167	1.66	−5.30/−3.64	$PC_{71}BM$	0.74	12.50	0.65	6.0	[132]
P168	1.76	−5.54/−3.78	$PC_{71}BM$	0.77	10.75	0.57	4.72	[133]
P169	1.75	−5.70/−3.90	$PC_{71}BM$	0.83	4.9	0.41	1.70	[134]
P170	1.77	−5.16/−3.39	$PC_{71}BM$	0.88	9.1	0.48	3.8	[135]
P171	1.70	−5.16/−3.46	$PC_{71}BM$	0.79	8.0	0.42	2.7	[135]
P172	1.65	−5.16/−3.51	$PC_{71}BM$	0.80	4.9	0.43	1.7	[135]
P172	1.79	−5.36/−3.38	$PC_{71}BM$	0.82	14.24	0.601	7.02	[136]

DTP with C-bridged cyclopenta[2,1-*b*:3,4-*b*′]dithiophene (CPDT) and Si-bridged dithieno[3,2-*b*:2′,3′-*d*]silole (DTS) donor blocks (P174–P177) [139–143]. Compared to DTS, the dithieno[3,2-*b*:2′,3′-*d*]germole (DTG) with the larger Ge bridge could result in enhancement of ordering and electron delocalization [144]. The relative polymers exhibited lowered bandgaps and upshifted HOMO levels (P178) [144]. This variation trend could be further enhanced by introducing more germanium atoms, the diseleno[3,2-*b*:2′,3′-*d*]germole (DSG)-based P179 displayed a E_g^{opt}

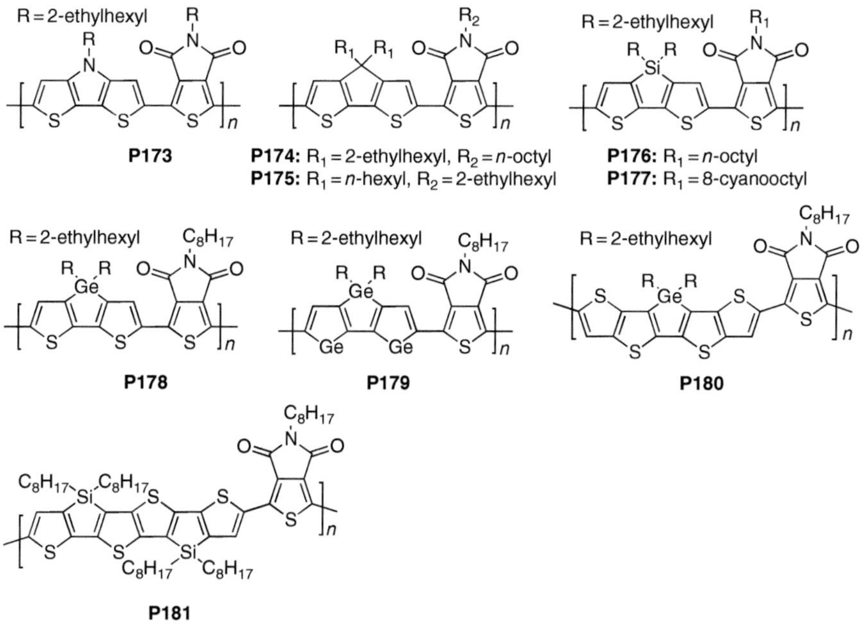

Figure 1.27 Chemical structures of the TPD-derived copolymers P173–P181. Source: Xiaopeng Xu.

of 1.60 eV and HOMO of −5.29 eV [145], which were smaller than the above DTS-based polymers. Extending the DTG by fusing two more thiophene afforded the dithienogermolodithiophene (DTTG). The resulted polymer P180 exhibited a medium E_g^{opt} of 1.75 eV and a deep HOMO level of −5.68 eV [146]. Fusing two DTS blocks produced the thieno[3,2-*b*]thienobis (silolothiophene) (Si4T) block, the resulted polymer P181 displayed slightly upshifted energy levels with minimal changed E_g^{opt} [147]. The chemical structures of P173–P181 are presented in Figure 1.27. The corresponding optoelectronic properties and device performances are listed in Table 1.19.

TPD-based polymers using thiophene derivatives as the donor block and spacers have been extensively studied. Their optoelectronic properties and energy levels can be finely tuned by introducing other heteroatoms (P182–P184) [148, 149], or fluorination on the thiophene blocks (P185) [150]. Moreover, Li and coworkers found that using bi-TPD block as the acceptor unit would be able to downshift the HOMO levels and redshift the absorptions (P186, P187) [151]. Along with suitable side chains introducing, the resulted polymer P187 based PSCs exhibited a high V_{oc} of 0.89 V and PCE of 9.08% [151]. They further lowered the energy levels of bi-TPD-based polymers by introducing the electron-deficient BT and fluorinated BT into the polymers, the resulted P188-based devices achieved a high V_{oc} of 1.02 V with a PCE of 6.46% [152]. While P189 with fluorinated BT could not match well with $PC_{71}BM$ to afford enough driving force for charge separation, resulting in poor device performance [152]. The chemical structures of P182–P189 are presented in Figure 1.28. The corresponding optoelectronic properties and device performances are listed in Table 1.20.

Table 1.19 Optoelectronic properties and device performances of P173–P181.

Polymer	E_g^{opt} (eV)	HOMO/ LUMO (eV)	Acceptor	V_{oc} (V)	J_{sc} (mA cm^{-2})	FF	PCE (%)	References
P173	1.62	−5.09/−3.42	$PC_{71}BM$	0.70	6.97	0.39	1.90	[138]
P174	1.67	−5.43/−3.25	$PC_{71}BM$	0.80	10.04	0.47	3.74	[139]
P175	1.61	−5.24/−3.55	$PC_{71}BM$	0.84	12.9	0.535	5.80	[140]
P176	1.70	−5.44/−3.17	$PC_{71}BM$	0.85	6.58	0.37	2.13	[139]
P176	1.73	−5.57/−3.88	$PC_{71}BM$	0.88	12.2	0.68	7.3	[141]
P176	1.73	−5.57/−3.88	$PC_{71}BM$	0.91	12.13	0.70	7.7	[142]
P177	1.73	−5.66/−3.68	$PC_{71}BM$	0.89	14.0	0.676	8.4	[143]
P178	1.69	−5.60/−3.50	$PC_{71}BM$	0.85	12.6	0.68	7.3	[144]
P179	1.60	−5.29/−3.40	$PC_{71}BM$	0.74	12.2	0.58	5.2	[145]
P180	1.75	−5.68/−3.88	$PC_{71}BM$	0.81	13.85	0.64	7.2	[146]
P181	1.70	−5.3/−3.6	$PC_{71}BM$	0.80	9.42	0.46	3.46	[147]

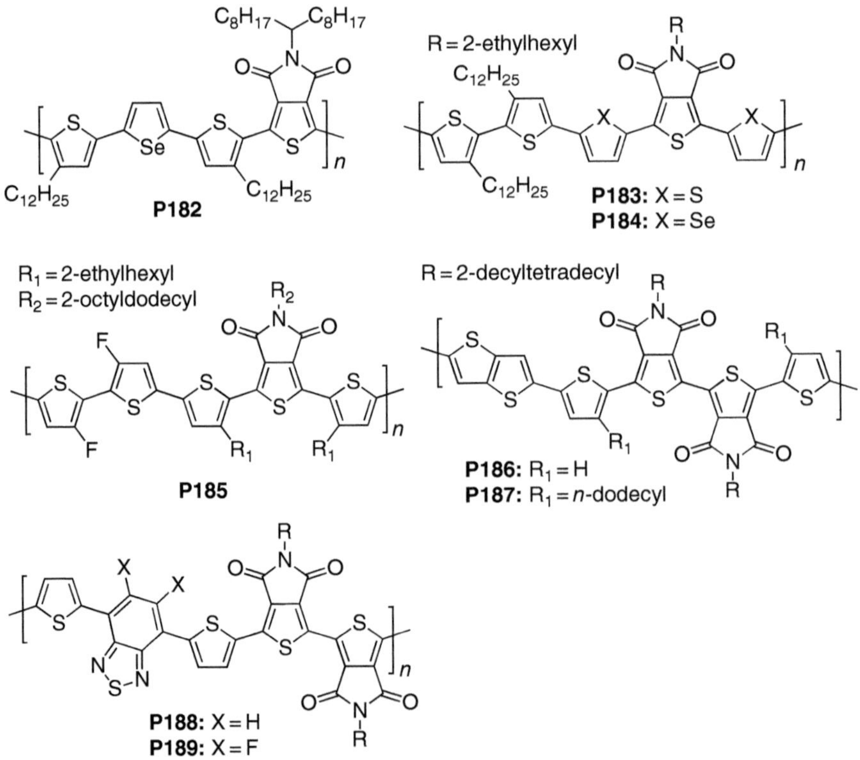

Figure 1.28 Chemical structures of the TPD-derived copolymers P182–P189. Source: Xiaopeng Xu.

Table 1.20 Optoelectronic properties and device performances of P182–P189.

Polymer	E_g^{opt} (eV)	HOMO/ LUMO (eV)	Acceptor	V_{oc} (V)	J_{sc} (mA cm^{-2})	FF	PCE (%)	References
P182	1.72	−5.49/−3.82	$PC_{71}BM$	0.88	10.74	0.62	5.80	[148]
P183	1.64	−5.50/−2.95	$PC_{71}BM$	0.92	4.73	0.461	2.04	[149]
P184	1.65	−5.52/−3.68	$PC_{71}BM$	0.80	4.53	0.420	1.52	[149]
P185	1.75	−5.55/−3.55	$PC_{71}BM$	0.96	10.50	0.514	5.52	[150]
P186	1.80	−5.61/−3.81	$PC_{71}BM$	0.91	13.45	0.672	8.22	[151]
P187	1.80	−5.59/−3.79	$PC_{71}BM$	0.89	14.15	0.721	9.08	[151]
P188	1.71	−5.72/−4.01	$PC_{71}BM$	1.02	9.48	0.668	6.46	[152]
P189	1.75	−5.85/−4.10	$PC_{71}BM$	0.96	2.55	0.636	1.56	[152]

BDT donor block has also been commonly used to pair with TPD acceptor block and develop high-performance conjugated polymers. The side chain effects on the overall properties of BDT–TPD-based polymers have been extensively studied (P190–P197) [153–155]. Beaujuge and coworkers found that linear side chains in BDT–TPD-based polymers could greatly impact the polymer self-assembling properties and device performances [154]. The N-alkyl substitutions on the BDT motifs triggered a critical change of preferential polymer orientation in thin films, resulting in a dramatic drop in bulk heterojunction (BHJ) device PCE [154]. A fine modulation of the number of aliphatic carbons in the linear N-alkyl-substituted TPD motifs did not significantly affect the preferential backbone orientation, yet this approach exhibited to be a main key to improve the device performance [154]. Among them, P195 with 2-ethylhexyloxy on the BDT block and heptyl on the TPD block achieved the optimized PCE of 8.5%. Zhu and coworkers reported a series of BDT–TPD-based polymers by introducing TT blocks into the polymer chains to extend the conjugation length and regulating their steric hindrance with side alkyl chains on BDT moiety (P198–P201) [156]. The additional alkyl chains on BDT moiety could significantly increase V_{oc} because of the decreased HOMO energy level. P200 with optimal length of alkyl side chain exhibited optimized thin-film morphology, contributing to higher PCE of 9.1% in the resulting PSCs. Recently, Huang and coworkers employed the bi-TPD block to mediate the crystallinity and developed P202 [157]. P202:Y6 BHJ blend film exhibited more favorable face-on backbone orientation and stronger crystallinity. As a result, charge transportation was enhanced, whereas the charge recombination was suppressed in the BHJ device, which contributed a high PCE of 14.2% in the fullerene-free devices. The chemical structures of P190–P202 are presented in Figure 1.29. The corresponding optoelectronic properties and device performances are listed in Table 1.21.

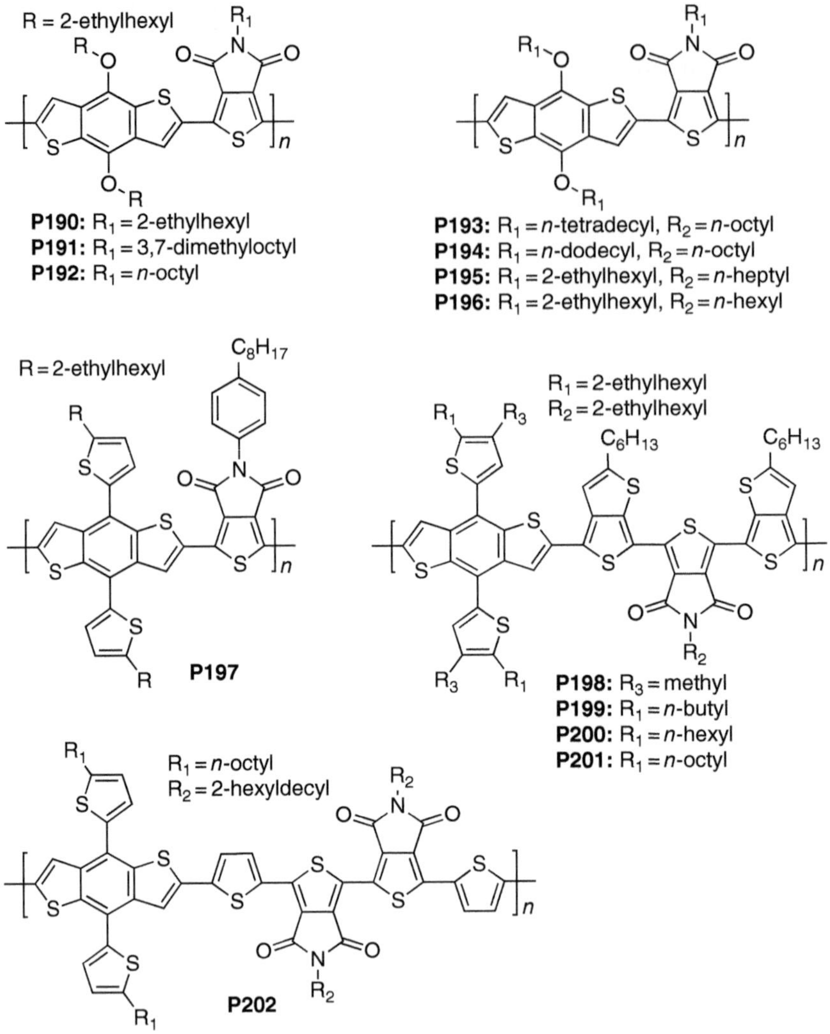

Figure 1.29 Chemical structures of the TPD-derived copolymers P190–P202. Source: Xiaopeng Xu.

1.3.4 MBG Polymers Based on Thieno[3,4-*b*]thiophene (TT)

BDT–TT-based polymers are quite representative high-performance polymers in the last decade (Figure 1.30). The TT monomer unit was designed because of its tendency to stabilizing the quinoidal structure that led to a medium energy bandgap around 1.6 ~ 1.8 eV [158]. This guaranteed an efficient light absorption in the solar spectrum with the highest photon flux around 700 nm. The quinoidal structure enhanced the planarity along the aromatic polymer backbone by providing a more rigid backbone as well. The extended π system in BDT unit enabled the polymer to form a better π–π stacking with a shorter distance between different polymer

Table 1.21 Optoelectronic properties and device performances of P190–P202.

Polymer	E_g^{opt} (eV)	HOMO/ LUMO (eV)	Acceptor	V_{oc} (V)	J_{sc} (mA cm^{-2})	FF	PCE (%)	References
P190	1.75	−5.48/−−	$PC_{71}BM$	0.87	8.1	0.56	4.0	[153]
P191	1.70	−5.57/−−	$PC_{71}BM$	0.81	9.7	0.67	5.7	[153]
P192	1.73	−5.40/−−	$PC_{71}BM$	0.85	11.5	0.68	6.8	[153]
P192	−−	−−/−−	$PC_{71}BM$	0.93	12.5	0.65	7.5	[154]
P193	−−	−−/−−	$PC_{71}BM$	0.93	8.3	0.53	4.1	[154]
P194	−−	−−/−−	$PC_{71}BM$	0.92	6.8	0.51	3.2	[154]
P195	−−	−−/−−	$PC_{71}BM$	0.97	12.6	0.70	8.5	[154]
P196	−−	−−/−−	$PC_{71}BM$	0.96	11.1	0.62	6.6	[154]
P197	1.78	−5.59/−3.62	$PC_{71}BM$	0.96	11.00	0.585	6.18	[155]
P198	1.61	−5.18/−3.26	$PC_{71}BM$	0.72	16.36	0.66	7.73	[156]
P199	1.63	−5.20/−3.27	$PC_{71}BM$	0.78	13.44	0.70	7.42	[156]
P200	1.64	−5.16/−3.26	$PC_{71}BM$	0.78	15.611	0.75	9.10	[156]
P201	1.61	−5.15/−3.27	$PC_{71}BM$	0.77	13.61	0.72	7.56	[156]
P202	1.78	−5.20/−3.45	Y6	0.83	25.6	0.667	14.2	[157]

Figure 1.30 The aromatic form and quinoidal form of BDT-TT based polymers. Source: Xiaopeng Xu.

backbones and finally led to a high hole mobility [158]. The electron-donating ability of BDT block can be modulated by introducing alkoxy (P203–P209) [159–162], alkylthienyl (P210, P211) [163, 164], or alkoxyphenyl (P212) [165] side chains, or fluorine atom at the backbone (P208, P209) [162]. The electron-withdrawing property of TT block can be finely tuned by introducing ester (P203, P204) [159, 160], carbonyl (P205, P206) [159], or sulfuryl groups (P207) [161] at the 2-position, and/or introducing halogen atom (P204, P206, P207, P209–P215) [159–167] at the 3-position. In addition, introducing 2D side chains on the BDT block (P210–P215) [163–165] and/or on the 2-position of TT block (P213–P215) [166, 167] also extended the π-conjugation length of the resulting polymers for improving charge transport properties. He and coworkers introduced 2D side chains on both BDT and TT blocks and developed P213, which resulted in a strong and broad absorption [166]. Moreover, the 2D side chains could induce a modulation of the preferential

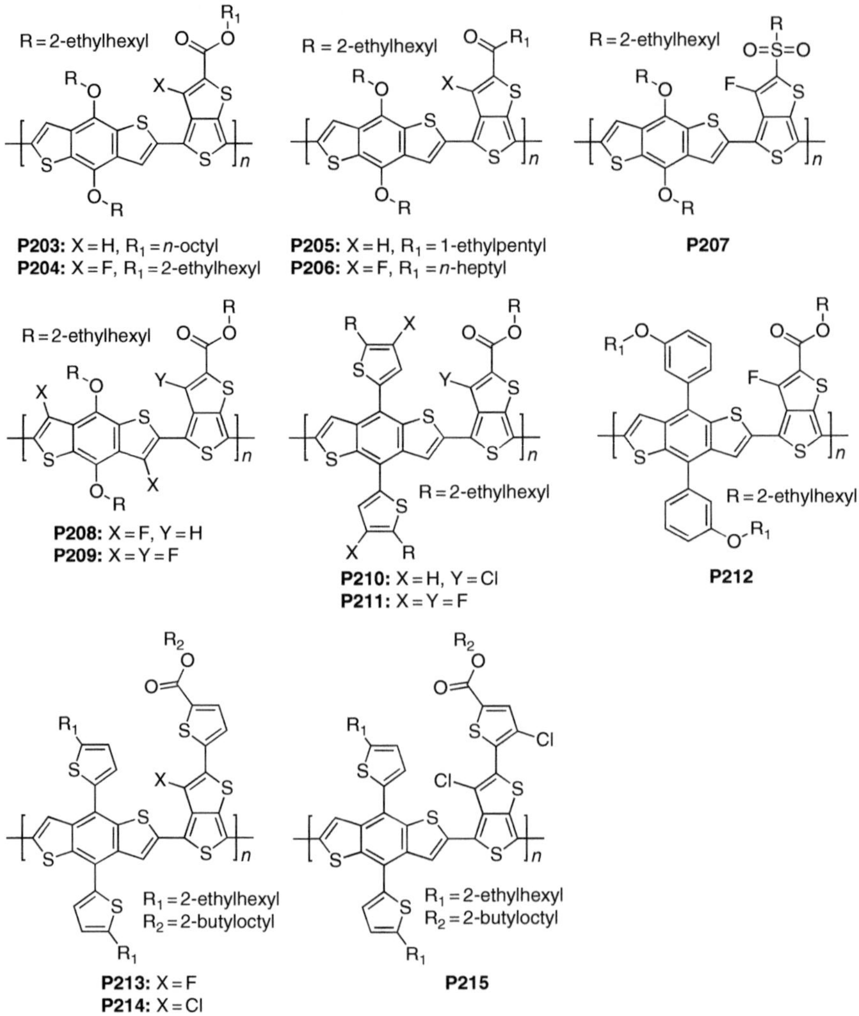

Figure 1.31 Chemical structures of the TT-derived copolymers P203–P215. Source: Xiaopeng Xu.

ordered microstructure that shortened the π–π distance to strengthen the π–π stacking between polymer backbones, which could be favorable toward higher J_{sc} and FF. As a result, P213:PC$_{71}$BM devices realized a high PCE of 9.13%. The chemical structures of P203–P215 are presented in Figure 1.31. The corresponding optoelectronic properties and device performances are listed in Table 1.22.

Introducing π bridges between BDT and TT blocks to extend the conjugation area is beneficial for face-to-face packing and significant increase of photovoltaic efficiency [168]. Hou and coworkers developed a series of thiophene-bridged polymers (P216–P219) by introducing fluorine on the BDT side chains and/or 3-position of TT block [169]. They found that fluorination on the donor and the acceptor moieties has little influence on the optical properties, but a synergistic effect on lowering

Table 1.22 Optoelectronic properties and device performances of P203–P215.

Polymer	E_g^{opt} (eV)	HOMO/ LUMO (eV)	Acceptor	V_{oc} (V)	J_{sc} (mA cm^{-2})	FF	PCE (%)	References
P203	1.61	−5.01/−3.24	$PC_{71}BM$	0.62	13.2	0.63	5.15	[159]
P204	1.63	−5.15/−3.31	$PC_{71}BM$	0.74	14.5	0.690	7.40	[160]
P205	1.61	−5.12/−3.35	$PC_{71}BM$	0.70	14.7	0.641	6.58	[159]
P206	1.61	−5.22/−3.45	$PC_{71}BM$	0.76	15.2	0.669	7.73	[159]
P207	1.65	−5.12/−3.49	$PC_{71}BM$	0.76	14.1	0.58	6.22	[161]
P208	1.75	−5.41/−3.60	$PC_{71}BM$	0.68	11.1	0.422	3.2	[162]
P209	1.73	−5.48/−3.59	$PC_{71}BM$	0.75	9.1	0.394	2.7	[162]
P210	1.71	−5.43/−3.55	ITIC	0.91	14.53	0.58	7.57	[163]
P211	1.64	−5.47/−3.56	ITIC	0.94	16.0	0.605	9.1	[164]
P212	1.62	−5.45/−3.34	$PC_{71}BM$	0.86	16.4	0.622	9.0	[165]
P213	1.61	−5.32/−3.60	$PC_{71}BM$	0.73	17.78	0.704	9.13	[166]
P214	1.72	−5.45/−3.60	ITIC	0.91	15.79	0.613	8.81	[167]
P215	1.66	−5.43/−3.63	ITIC	0.87	14.52	0.574	7.38	[167]

their molecular energy levels. The V_{oc} of the resulting devices increased from 0.56 to 0.78 V, and the PCE improved from 4.5% to 8.6%. On the other hand, replacing the alkylthienyl side chains with *meta*-alkoxyphenyl groups, the optical absorption, molecular packing, and charge transport properties of the polymer (P220) were little changed, while the HOMO level could be reduced distinctly, so V_{oc} of the PSC device could be significantly enhanced from 0.60 to 0.78 V [170]. Replacing the electron-rich thiophene bridges with electron-deficient thiazole spacers is also favorable in obtaining deep HOMO levels of the corresponding polymers [171]. Yang and coworkers developed the thiazole-bridged P221 and P222 by changing the orientations of the thiazole relative to the TT moiety [171]. P222 with 2-position of thiazole attached to TT block exhibited a lower HOMO energy level, and better planar molecular configuration, together with preferable phase domains and good intermixing with $PC_{71}BM$. Consequently, a remarkable PCE of 9.72% was realized. Further lowering the HOMO level by introducing sulfur atoms on the BDT side chains and fluorine atom on the TT block, the resulting polymers (P223, P224) realized higher V_{oc}s in devices [172]. P224:$PC_{71}BM$-based PSCs achieved an impressive V_{oc} of 0.97 V and PCE of 10.30%. The chemical structures of P216–P224 are presented in Figure 1.32. The corresponding optoelectronic properties and device performances are listed in Table 1.23.

Expanding the π-conjugation system in the rigid BDT unit is conducive to lower positive charge density and exciton binding energy [173]. Yu and coworkers employed dithieno[2,3-*d*:2′,3′-*d*′]benzo[1,2-*b*:4,5-*b*′]dithiophene (DBD) as the donor building block along with side chain engineering and developed

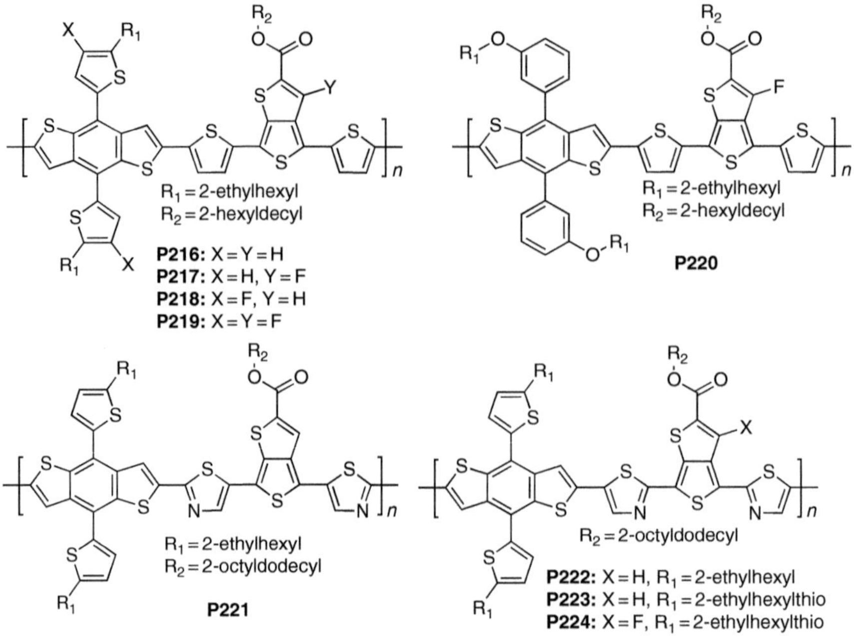

Figure 1.32 Chemical structures of the TT-derived copolymers P216–P224. Source: Xiaopeng Xu.

Table 1.23 Optoelectronic properties and device performances of P216–P224.

Polymer	E_g^{opt} (eV)	HOMO/ LUMO (eV)	Acceptor	V_{oc} (V)	J_{sc} (mA cm^{-2})	FF	PCE (%)	References
P216	1.61	−4.90/−3.10	$PC_{71}BM$	0.56	12.2	0.667	4.5	[169]
P217	1.65	−4.95/−3.12	$PC_{71}BM$	0.60	14.3	0.657	5.6	[169]
P218	1.64	−5.15/−3.27	$PC_{71}BM$	0.74	14.4	0.677	7.2	[169]
P219	1.64	−5.20/−3.30	$PC_{71}BM$	0.78	15.2	0.724	8.6	[169]
P240	1.70	−5.12/−3.49	$PC_{71}BM$	0.78	13.4	0.718	7.50	[170]
P241	1.73	−5.54/––	$PC_{71}BM$	0.83	16.84	0.695	9.72	[171]
P242	1.70	−5.43/––	$PC_{71}BM$	0.82	13.11	0.643	6.91	[171]
P243	1.71	−5.45/−3.52	$PC_{71}BM$	0.93	16.20	0.613	9.23	[172]
P244	1.74	−5.48/−3.51	$PC_{71}BM$	0.97	16.65	0.638	10.30	[172]

P225 and P226 [173]. Optimal structures of side alkyl chains in the polymers improved the compatibility of the polymer chains with $PC_{71}BM$, improving the morphology of the polymer:$PC_{71}BM$ blend film with concomitant enhancement of the crystallinity of the polymer chains in the blend film. Consequently, the device efficiency in P225 was enhanced up to 7.6%. Bathula et al. empolymed the linear shaped naphtho[2,3-*b*:6,7-*b*′]dithiophene (NDT) and "zigzag"

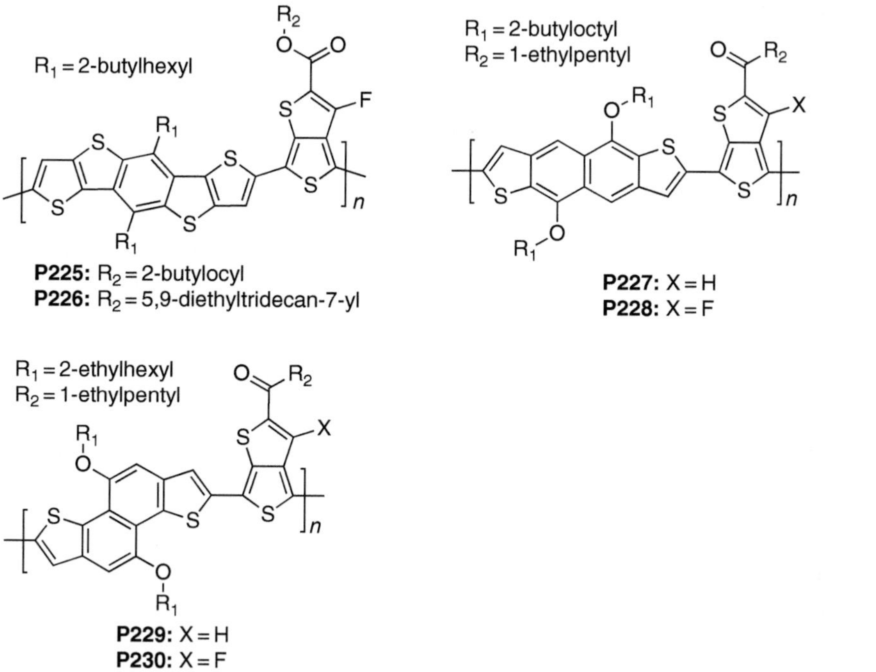

Figure 1.33 Chemical structures of the TT-derived copolymers P225–P230. Source: Xiaopeng Xu.

Table 1.24 Optoelectronic properties and device performances of P225–P230.

Polymer	E_g^{opt} (eV)	HOMO/ LUMO (eV)	Acceptor	V_{oc} (V)	J_{sc} (mA cm^{-2})	FF	PCE (%)	References
P225	1.68	−5.24/−−	$PC_{71}BM$	0.89	13.0	0.653	7.60	[173]
P226	1.67	−5.30/−−	$PC_{71}BM$	0.88	10.7	0.521	4.9	[173]
P227	1.65	−5.44/−3.79	$PC_{71}BM$	0.71	11.68	0.54	4.49	[174]
P228	1.64	−5.51/−3.87	$PC_{71}BM$	0.72	13.50	0.53	5.16	[174]
P229	1.60	−4.86/−3.26	$PC_{71}BM$	0.55	11.99	0.38	2.51	[175]
P230	1.69	−5.03/−3.34	$PC_{71}BM$	0.71	13.52	0.51	4.88	[175]

naphtho[1,2-*b*:5,6-*b*′]dithiophene (zNDT) as the donor blocks, respectively, and developed a series of MBG polymers (P227–P230) with extended π-conjugation [174, 175]. These polymers exhibited high field effect carrier mobilities up to ~10^{-3} cm^2 V^{-1} s^{-1}, showing the promising applications of this blocking in developing high-performance photovoltaics. The chemical structures of P225–P230 are presented in Figure 1.33. The corresponding optoelectronic properties and device performances are listed in Table 1.24.

1.4 WBG Polymers

As discussed in the introduction part, a very important molecular design strategy of WBG polymers is to introduce both the D unit with relatively weak electron-donating ability and the A unit with relatively weak electron-withdrawing ability due to a weak D unit will deep-shifts molecular HOMO level and a weak A unit will upshifts molecular LUMO level, respectively [176]. D units employed in high-performance WBG polymer donors are mostly thiophene or benzene fused building blocks with weak electron-donating ability, such as thiophene, BDT, IDT, and the related derivatives. Compared to D units, A units play a more important role in constructing WBG polymer donors due to their diversity [177, 178]. At present, the most common A units with weak electron-withdrawing ability for constructing high-performance WBG polymer donors include ester or cyano-substituted thiophene, thienopyrroledione, alkyloxy substituted benzothiadiazole, benzotriazole (BTA), benzodithiophene-4,8-dione (BDD), thiazole or pyrazine derivatives-based *N*-heterocycles (such as dithiazole, thiazolothiazole, benzodithiazole, thiadiazole, oxadiazole, pyrazine, quinoxaline, tetrazine, and et al.), and the related derivatives. In this contribution, the WBG polymer donors are divided into the following seven branching categories: (i) polythiophene (PT) derivatives; (ii) polymer donors with a backbone of benzodithiophene-*alt*-thiophene derivative; (iii) polymer donors based on benzothiadiazole derivatives; (iv) polymer donors based on BTA derivatives; (v) polymer donors based on thiazole, pyrazine, and their derivatives containing *N*-heterocycles; (vi) polymer donors based on BDD derivatives; and (vii) others. The following part will introduce the design strategies of these WBG polymer donors firstly, and then, some representative polymers and their photovoltaic properties will be analyzed.

1.4.1 WBG Polymers Based on Polythiophene (PT) Derivatives

Thiophene is one of the basic units of organic photovoltaic materials, and the corresponding PT and its derivatives have attracted great attention in the past decades due to their versatile and promising applications in many fields. Among them, as the most representative conjugated polymer donor, poly(3-hexylthiophene) (P231, $E_g > 1.9$ eV) with a simple chemical structure and low-cost synthetic pathway shows high hole mobility and crystallinity, which is beneficial to form a nanoscale interpenetrating fiber network in fullerene-based blend films and be considered as one of the most promising candidates for large-scale manufacturing of OSCs [179, 180]. However, P231 possesses a relatively high HOMO level of −4.90 eV due to its homopolymerized structure without electron-withdrawing unit, which makes it use to have a low V_{oc} in the $PC_{61}BM$- and $PC_{71}BM$-based OSCs [179]. Recently, the P231-based OSCs have achieved PCEs of $7 \sim 9\%$ by developing fullerene derivatives (such as ICBA [180]) and non-fullerene small molecule (SM) acceptors with high LUMO levels (such as TrBTIC [181] and ZY-4Cl [182]) to increase V_{oc}. On the other hand, to deep-shift HOMO level of PT derivatives, some molecular modification strategies have also been carried out, such as introducing 2D-conjugated side chains [183], ester or cyano groups [184], halogen atoms [185], and thiophene-fused electron-withdrawing units [150, 186, 187]. For example, a design concept of "side chain isolation" was proposed for developing a 2D-styryltriphenylamine-conjugated PT derivative P232 with red-shifted absorption spectrum and deep-shifted HOMO level. Matched with ICBA, the P232-based OSCs achieved a PCE of 3.6% with a high V_{oc} of 0.94 V [183]. Another strategy applied to decrease HOMO level of PT derivatives is to introduce electron-withdrawing groups into molecular backbone. In 2014, Zhang et al. synthesized a PT derivative P233 by attaching two electron-withdrawing carboxylate substituents on PT backbone [184], and the HOMO level of P233 (−5.26 eV) decreased by 0.36 eV in comparison with P231 (−4.90 eV). As a result, P233 achieved an improved PCE of 7.2% with a significantly increased V_{oc} of 0.91 V in the $PC_{71}BM$-based OSCs. Moreover, by using LBG SM acceptor ITIC instead of $PC_{71}BM$ in OSCs, P233 achieved a higher PCE of 10.16% with a J_{sc} of 16.5 mA cm^{-2}, which was the highest PCE in the OSCs based on PT derivatives [188]. Subsequently, a series of PT derivatives were designed and synthesized by introducing halogen atoms into the backbone of P233. The incorporation of chlorine or fluorine atoms into polymers both downshifted molecular HOMO levels, leading to higher V_{oc} in OSCs. Due to the suitable phase-separated morphology with favorable molecular packing and miscibility, chlorinated P236 achieved efficient exciton dissociation, improved charge collection, and weak charge recombination, resulting in the best PCE up to 12.38% [185]. In addition, constructing the D–A type PT derivatives by incorporating the thiophene-fused acceptor units has been proved to be an effective strategy to reduce molecular HOMO levels. In 2013, a PT derivative P238 with a lower HOMO level of −5.13 eV was synthesized by polymerizing thiophene-fused BDD acceptor unit with 2,2′-bithiophene (2T) donor unit, and an improved PCE of 6.88% with a higher V_{oc} of 0.83 V was achieved in its OSCs [186]. In addition to reduce molecular HOMO levels, the 3,3′-difluoro-2,2′-bithiophene (ff2T) unit has an improved planarity, more stable conformation, and additional noncovalent bonding interactions in comparison with the fluorine-free 2T unit, which leads to the improved crystallinity, extinction coefficient, and charge mobility, thus the related polymers can achieve higher V_{oc}, J_{sc}, and FF values simultaneously in OSCs [187]. In 2016, Fan et al. reported a ff2T-based PT derivative P239 and found that P239 shows a deeper HOMO level, improved extinction coefficient and crystallinity compared with that of non-fluorinated P238. Matching with $PC_{71}BM$, the OSCs yielded a PCE of 9.2% [187]. Subsequently, P239 was introduced as the hole-cascade material to optimize the blend morphology of PTB7-Th:N2200, and the ternary all-polymer OSCs achieved an improved PCE of 7.2% [189]. Wang et al. reported a TPD-based PT derivative P241 with low HOMO level of −5.55 eV by introducing the synergistic effects of fluorination and thiophene-fused acceptor unit (thienopyrroledione), and the corresponding $PC_{71}BM$-based OSCs achieved a PCE of 5.52% with a high V_{oc} of 0.96 V [150]. The chemical structures of representative PT derivatives are shown in Figure 1.34, and the corresponding photovoltaic parameters of OSCs are summarized in Table 1.25.

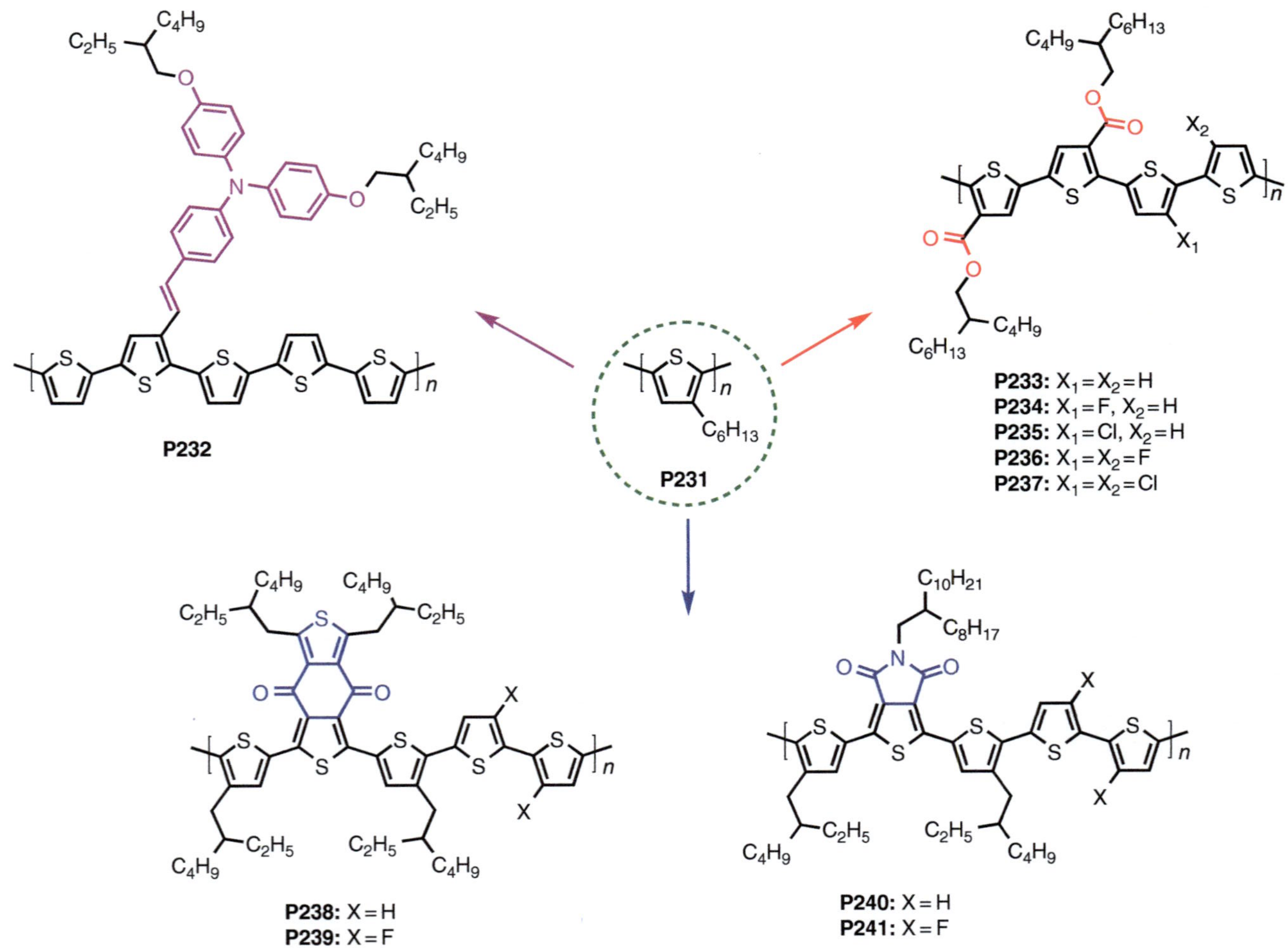

Figure 1.34 Chemical structures of the PT-derived copolymers P231–P241. Source: Qunping Fan.

1.4.2 WBG Polymers Based on Benzodithiophene-*alt*-Thiophene Derivatives

In 2008, BDT unit with a weak electron-donating, large and rigid coplanar fused-ring structure was firstly used in the synthesis of photovoltaic polymer materials [190]. Up to now, as one of the most popular and successful building blocks, BDT unit has been used in hundreds of high-performance photovoltaic materials [85]. Recently, the polymer donors with a backbone of benzodithiophene-*alt*-thiophene derivative have attracted great attentions due to their WBG of ~2.0 eV and low HOMO levels, which are complementary in absorption spectra and matching in energy levels with the booming LBG SM acceptors. Firdaus et al. developed a series of WBG polymers by polymerizing benzodithiophene and thiophene derivatives, such as non-fluorinated P242, difluorinated P243, and 2D-thienyl substituted P244 [191]. The synergistic effects of fluorination and 2D-conjugation decreased energy levels, improved absorption coefficient, and optimized molecular packing of polymers. As a result, the P244-based OSCs achieved the highest PCE of 9.8% with simultaneously improved V_{oc}, J_{sc}, and FF. In 2017, Park et al. reported a random WBG (2.02 eV) polymer donor P245 by introducing a methylcarboxyl substitution on thiophene unit [192]. Due to its good solubility, the P245-based OSCs fabricated by halogen-free solvents achieved a high PCE of 9.73%. Subsequently, Liu et al. synthesized a regioregular polymer P247 with a WBG of 1.99 eV [193]. Compared to its random analog P245, P247 has higher-order packing and improved absorption coefficient, which results in a higher PCE of 12.07% in its OSCs. Moreover, by introducing fluorine atom on the 2D-conjugated thienyl of polymer P245, Yao et al. developed polymer donor P246 with a similar WBG of 2.00 eV and a deeper HOMO level of −5.59 eV [194]. Matched with IT-4F, the P246-based OSCs achieved a high PCE of 14.7% due to the efficient charge transfer in device assisted by a large electrostatic potential difference between P246 and IT-4F. In the following work, Hou and coworkers reported a series of WBG polymer donors PDTB-EF-T (P248, P249, and P250) with deep HOMO level < -5.50 eV by introducing the synergistic electron-withdrawing effects of the fluorine atom and ester group [195]. Hence, a high V_{oc} of 0.90 V was achieved when matched with IT-4F, corresponding to a low E_{loss} of 0.62 eV. Moreover, the aggregation and molecular packing of PDTB-EF-T can be optimized well by side chain engineering of the ester, which results in the stronger interchain π–π interaction and ordering structure, and thus the improved charge transport and reduced recombination are achieved for the linear decyl-substituted PDTB-EF-T (P248)-based PSCs with a high PCE of 14.2%. Recently, Chen et al. reported a carboxylate-functionalized thienothiophene-based WBG polymer donor P251 [196]. Compared to its analog polymer 2TC-TT-BDTF T with two carboxylate groups on the outer thiophenes, P251 shows a higher planarity, stronger aggregation, red-shifted absorption, improved crystalline ordering, and larger charge carrier mobility. As a result, the P251-based OSCs demonstrated a greatly improved PCE of 11.15% in comparison with the 2TC-TT-BDTFT-based ones (9.65%). Very recently, Duan and coworkers introduced a cost-effective building block, 3,4-dicyanothiophene (DCT), for constructing a WBG polymer P253 with deep HOMO level for application as donor in OSCs [197]. Thanks to its stronger intermolecular interaction, more planar backbone, and larger dipole moment of the

Table 1.25 Optoelectronic properties and device performances of P231–P241.

Polymer	E_g^{opt} (eV)	HOMO/ LUMO (eV)	Acceptor	V_{oc} (V)	J_{sc} (mA cm^{-2})	FF	PCE (%)	References
P231	1.94	−4.90/−3.00	$PC_{61}BM$	0.61	9.4	0.53	3.0	[179]
P231	1.94	−4.90/−3.00	ICBA	0.87	11.35	0.75	7.4	[180]
P231	1.94	−4.90/−3.00	TrBTIC	0.88	13.04	0.719	8.25	[181]
P231	1.94	−4.90/−3.00	ZY-4Cl	0.88	16.49	0.65	9.46	[182]
P232	1.91	−5.10/−3.08	ICBA	0.94	6.55	0.584	3.6	[183]
P233	1.90	−5.26/−−	$PC_{71}BM$	0.91	11.0	0.72	7.2	[184]
P233	1.90	−5.26/−−	ITIC	0.94	16.50	0.6567	10.16	[188]
P234	1.93	−5.32/−2.98	ITIC-Th1	0.93	17.52	0.662	10.85	[185]
P235	1.92	−5.40/−3.03	ITIC-Th1	0.98	15.32	0.565	8.48	[185]
P236	1.91	−5.34/−3.01	ITIC-Th1	0.94	18.50	0.712	12.38	[185]
P237	1.95	−5.44/−3.00	ITIC-Th1	0.99	13.54	0.518	6.94	[185]
P238	1.77	−5.13/−3.23	$PC_{71}BM$	0.83	11.57	0.71	6.88	[186]
P239	1.74	−5.42/−3.53	$PC_{71}BM$	0.95	13.2	0.73	9.2	[187]
P240	1.84	−5.42/−3.31	$PC_{71}BM$	0.93	6.15	0.433	2.48	[150]
P241	1.75	−5.55/−3.55	$PC_{71}BM$	0.96	11.18	0.514	5.52	[150]

cyano group, P253 showed stronger aggregation in solution, higher π–π coherence length and higher relative dielectric constant in film compared to its cyano-free analog P252. Paired with SM acceptor IT-4F, the OSCs achieved PCEs of 2.3% and 11.2% for P252 and P253, respectively. Furthermore, by optimizing the alkyl chains of the DCT-containing polymers, an optimized polymer P254 was developed and achieved a higher PCE of 13.4%. The chemical structures of the polymers with a backbone of benzodithiophene-*alt*-thiophene derivatives are showed in Figure 1.35, and the corresponding photovoltaic parameters of OSCs are summarized in Table 1.26.

1.4.3 WBG Polymers Based on Benzothiadiazole (BT) Derivatives

BT with strong electron-withdrawing ability and stable quinoid structure is one of the most popular acceptor units in the synthesis of D–A-structured LBG and MBG photovoltaic materials [24, 198]. On the contrary, only few BT-based polymer donors with a WBG were successfully developed so far, by matching donor units with weak electron-donating ability [199], introducing alkyloxy substitution on BT units [199, 200], or designing thiophen-fused BT derivatives [4, 201] to decrease the ICT effects between D and A units. For example, Bo and coworkers reported a polymer donor P255 with a WBG of 1.97 eV by polymerizing fluorene with weak electron-donating ability and dodecyloxy-substituted BT derivative [199]. Matched with a WBG SM acceptor NI-T-NI, its OSCs obtained an impressively high V_{oc} of 1.30 V. On the other hand, attaching alkylphenyl with steric hindrance on the BDT unit and alkoxy chains on the BT unit could enlarge their E_g over 1.80 eV.

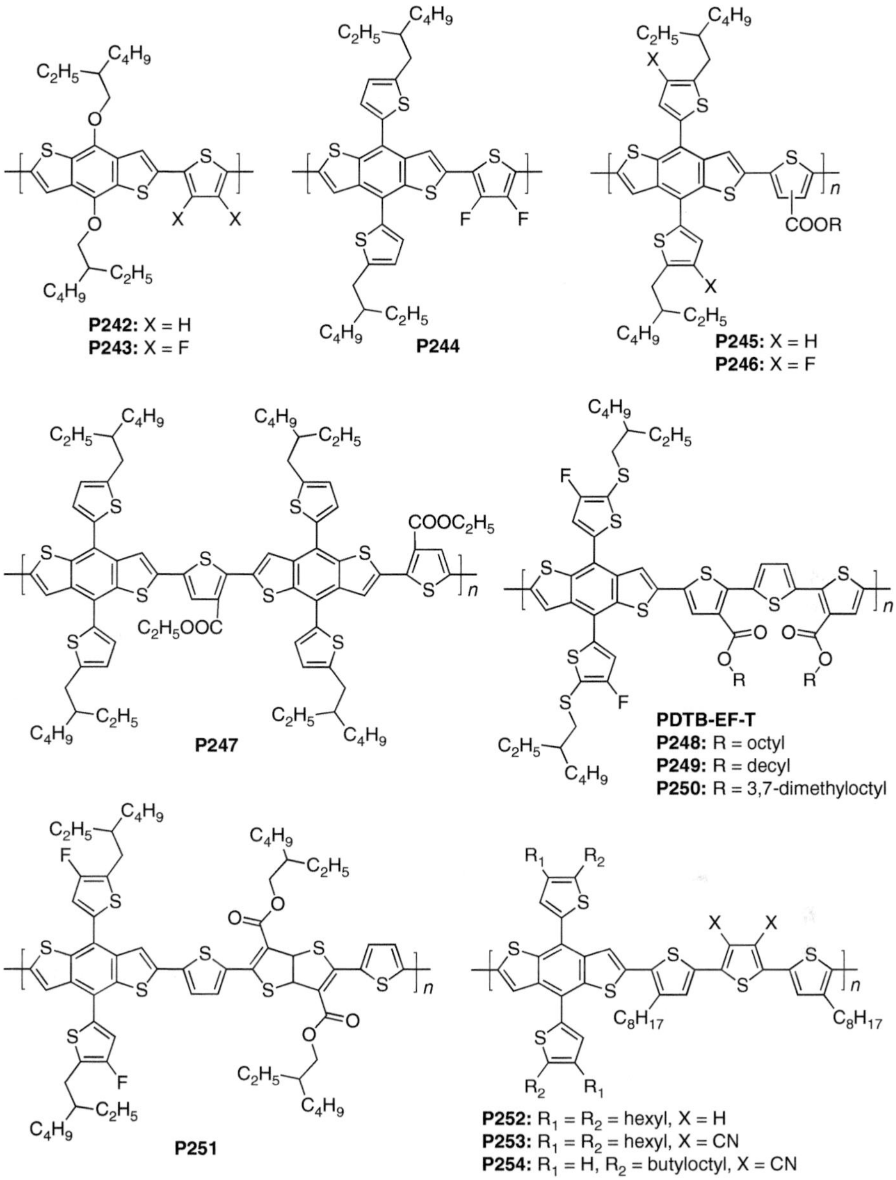

Figure 1.35 Chemical structures of the benzodithiophene-*alt*-thiophene-derived copolymers P242–P254. Source: Qunping Fan.

For example, Bo and coworkers synthesized a WBG (1.81 eV) polymer donor P256 with 4-alkyl-3,5-difluorophenyl substituted BDT as donor unit, alkyloxy-substituted BT as acceptor unit with two thiophenes as the spacers [200]. Matched with $PC_{71}BM$, its OSCs without any post-processing achieved a PCE of 8.24%. Using benzooxadiazole (BO) instead of BT, the corresponding polymer P257 showed a lower hole mobility, which resulted in a lower PCE of 5.67%. In 2017, Liu et al.

Table 1.26 Optoelectronic properties and device performances of P242–P254.

Polymer	E_g^{opt} (eV)	HOMO/ LUMO (eV)	Acceptor	V_{oc} (V)	J_{sc} (mA cm^{-2})	FF	PCE (%)	References
P242	2.1	−5.2/−−	ITIC	0.79	7.6	0.41	2.5	[191]
P243	2.1	−5.3/−−	ITIC	0.94	11.6	0.53	5.8	[191]
P244	2.0	−5.2/−−	ITIC	0.94	16.9	0.62	9.8	[191]
P245	2.02	−5.42/−3.40	ITIC	0.95	17.01	0.6008	9.73	[192]
P246	2.0	−5.59/−3.67	IT-4F	0.91	21.5	0.75	14.7	[194]
P247	2.07	−5.51/−3.46	FTIC	1.005	17.03	0.702	12.07	[193]
P248	1.93	−5.51/−3.59	IT-4F	0.896	20.05	0.64	11.5	[195]
P249	1.93	−5.50/−3.59	IT-4F	0.900	20.73	0.76	14.2	[195]
P250	1.94	−5.54/−3.60	IT-4F	0.904	20.31	0.61	11.2	[195]
P251	1.88	−5.41/−3.53	m-ITIC	0.97	16.57	0.6937	11.15	[196]
P252	2.00	−5.23/−2.89	IT-4F	0.78	7.1	0.41	2.3	[197]
P253	1.88	−5.62/−3.50	IT-4F	0.92	18.9	0.64	11.2	[197]
P254	1.88	−5.72/−3.52	IT-4F	0.91	21.2	0.69	13.4	[197]

reported a WBG polymer P258 based on a vertical-benzodithiophene (vBDT) building block [202]. Compared to traditional BDT-based polymers, the vBDT unit was connected *via* the phenyl group instead of the thiophene unit in P258, which led to larger torsion between the vBDT unit and the adjacent thiophenes, thus enlarged the bandgap of polymer and significantly changed the film morphology. Blended it with ITIC-Th, P258 achieved an improved PCE over 8%. At the same year, Chen and coworkers developed a highly crystalline polymer donor PBODT by polymerizing a BO-fused dithienobenzoxadiazole (DTBO) unit and tetra-thiophene (4T) unit [201]. Due to the weak electron-withdrawing ability of DTBO building block, P259 showed a WBG of 1.88 eV, which was significantly different from that of LBG polymer with a backbone of BO-*alt*-4T. The studies of variable temperature UV–vis, grazing-incident wide-angle X-ray scattering (GIWAXS), and organic field effect transistors indicated that P259 had strong self-aggregation effect, high crystallinity, and high hole mobility. Matched with SM acceptor ITIC and IDIC, the as-cast OSCs achieved the PCEs of 7.06% and 9.09% with high V_{oc} values of 1.00 and 0.93 V, respectively. Notably, when the OSCs processed with a single non-chlorinated solvent, a high PCE of 8.19% was still achieved. Very recently, Ding and coworkers synthesized a high-performance D–A copolymer donor P260 based on a BT-fused dithienobenzothiadiazole (DTBT) unit [4]. P260 shows a WBG of 1.98 eV, a deep HOMO level of −5.51 eV, and a high hole mobility of 1.59×10^{-3} cm^2 V^{-1} s^{-1}, which could be will matched with SM acceptor Y6. As a result, the P260:Y6-based OSCs gave a PCE of 18.22%. Then, Yan and coworkers replaced the fluorinated side chains with the chlorinated side chains and developed P261 [203]. They found the replacement of fluorine atoms with chlorine atoms did not change the conformation of the polymer backbone and the molecular packing so that the P261-based devices could

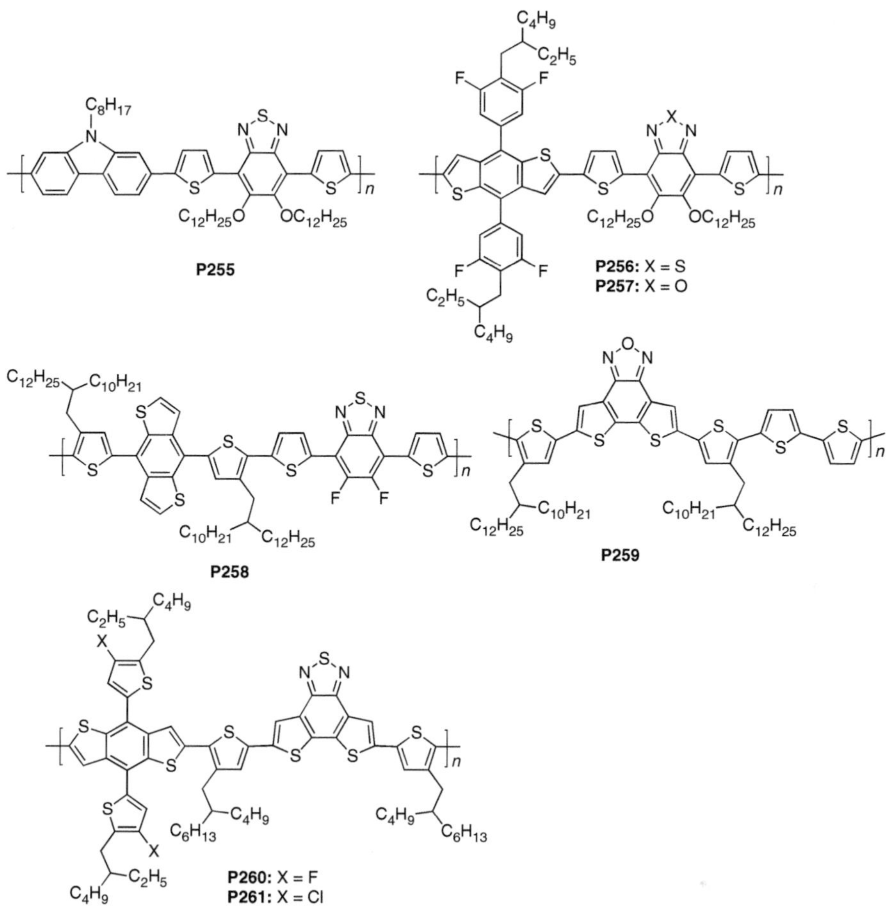

Figure 1.36 Chemical structures of the BT-derived polymers P255–P261. Source: Qunping Fan.

maintain a decent FF. P261:Y6:$PC_{71}BM$ devices achieved a higher PCE of 17.97% than P260:Y6:$PC_{71}BM$ devices (17.21%). The chemical structures of the polymer donors based on benzothiadiazole derivatives are showed in Figure 1.36, and the corresponding photovoltaic parameters of OSCs are summarized in Table 1.27.

1.4.4 WBG Polymers Based on Benzotriazole (BTA) Derivatives

BTA with a unique triazole structure has a relatively weak electron-withdrawing ability due to the lone-pair electrons on the 2-position of N contribute to the fused-triazole and its photovoltaic materials generally have strong self-aggregation and π–π stacking effects [204]. Therefore, the BTA-based polymer donors usually show a WBG, high crystallinity, high extinction coefficient, and good hole mobility. Consequently, the BTA-based polymer donors have developed rapidly and have been widely used for the efficient OSCs based on LBG SM acceptors in the past few years [205–208]. In the early studies, the high HOMO levels of BTA-based polymer donors limit their application in photovoltaic materials. To decrease the HOMO

Table 1.27 Optoelectronic properties and device performances of P255–P261.

Polymer	E_g^{opt} (eV)	HOMO/ LUMO (eV)	Acceptor	V_{oc} (V)	J_{sc} (mA cm^{-2})	FF	PCE (%)	References
P255	1.97	−5.23/−3.26	NI-T-NI	1.30	2.91	0.53	2.01	[199]
P256	1.81	−5.50/––	$PC_{71}BM$	0.89	12.67	0.73	8.24	[200]
P257	1.82	−5.62/––	$PC_{71}BM$	0.96	9.24	0.64	5.67	[200]
P258	1.94	−5.50/−3.56	ITIC-Th	0.93	15.2	0.56	8.1	[202]
P259	1.88	−5.44/−3.48	ITIC	1.00	10.7	0.664	7.06	[201]
P259	1.88	−5.44/−3.48	IDIC	0.93	14.1	0.693	9.09	[201]
P260	1.98	−5.51/−2.77	Y6	0.859	27.7	0.766	18.22	[4]
P261	2.00	−5.49/−3.49	Y6:$PC_{71}BM$	0.87	26.83	0.770	17.97	[203]

level of BTA-based polymers donors, You and coworkers synthesized a difluorinated benzotriazole (2fBTA) unit and developed a WBG polymer P263 [209]. Compared with its fluorine-free analog (P262), P263 obtained a dramatically improved PCE from ~4% to ~7%, owing to the higher V_{oc} and FF values. Recently, Li and coworkers designed and synthesized a series of thienyl BDT (BDT-T)-*alt*-2fBTA-structured high-performance polymer donors with WBG and deep HOMO level by modifying the 2D-conjugated thienyl of BDT unit [206–208, 210–212]. For example, polymer P265 was synthesized by introducing alkylthio substitutions into 2D-conjugated thienyls, targeting a deep-shifted HOMO level and improved crystallinity. Blending with ITIC, the P265-based OSCs obtained a PCE of 9.53% [206]. Subsequently, Fan et al. synthesized a polymer, P269 [207], *via* side chain engineering of P265. Compared with P265, P269 showed the increased steric hindrance on the BDT-T donor unit and the decreased steric hindrance on the 2fBTA acceptor unit, which resulted in stronger intermolecular π–π interactions, and more efficient charge separation and transport in its OSCs. Thus, the P269:ITIC-based OSCs achieved an improved PCE of 10.5%. Another 2fBTA-based polymer donor P266 was synthesized by attaching trialkylsilyl side chains into the 2D-conjugated thienyls [208]. Thanks to the σ* (Si)–π* (C) interaction deep-shifted the molecular HOMO level and increased absorption coefficient of P266, both higher V_{oc} and J_{sc} values were achieved in its OSCs. Fluorination of the 2D-conjugated thienyl on BDT unit has been convincingly established to be beneficial for deep-shifting molecular HOMO level and improving optical absorption, crystallinity, and charge transport of photovoltaic materials [169, 210, 211, 213]. Fan et al. reported a WBG polymer P268 based on difluorinated BDT-TF donor unit with mono-fluorinated thienyls and 2fBTA acceptor unit and found that the corresponding OSCs achieved improved photovoltaic parameters compared with those of the P270-based OSCs [210]. Subsequently, Xue et al. synthesized a tetra-fluorinated BDT-T2F donor unit with difluorinated thienyls and then developed another WBG polymer P267 with BTA unit [211]. In addition to deep-shift the molecular HOMO level and increasing absorption coefficient and hole mobility, fluorination on 2D-conjugated thienyls of the polymers could suppress triplet formation to reduce charge carrier recombination in OSCs. With these favorable advantages, the P267:*m*-ITIC-based OSCs achieved a high PCE of 11.63%. It is well known that the introduction of chlorine atom and silicon atom substitutions into molecular backbone are two important strategies to deep-shift molecular HOMO level, improve absorption coefficient, and increase charge mobility of photovoltaic materials [208, 214]. By combining the chlorine and alkylsilyl substitutions into the 2D-conjugated thienyls of BDT-T unit, a BTA-based WBG polymer P272 was synthesized by Fan and coworkers and achieved a high PCE of 12.8% in its OSCs [212]. Yang and coworkers synthesized two BTA-based WBG polymer donors 273 and P274 with alkyloxy- or alkylthio-substituted 2D-biphenyl BDT units [215]. Using alkylthio chains instead of alkoxy chains, P274 showed a deeper HOMO level by over 0.20 eV compared to its counterpart P273, resulting in a significantly improved V_{oc}. Blending with ITIC, P274 achieved a higher PCE of 12.09% in its OSCs compared to the counterpart polymer P273 (9.20%). In the same year, Yang and coworkers synthesized a 2D–2D asymmetric BDT monomer [216]. The studies showed that this asymmetric structure had unique advantages in comparison with the conventional 2D symmetric BDT unit. The 2D-conjugated thienylthiol side chain on the BDT unit could reduce molecular HOMO level, broaden absorption spectra, and improve solubility. The bare rigid aryl of naphthalene as another 2D-conjugated side chain could be regarded as lever arms to stir up well the elongated ITIC and weaken the entanglements among polymer chains during the spin-coating process, which enabled a favorable morphology without post treatment. As a result, the corresponding BTA-based WBG polymer donor P275 with a 2D–2D asymmetric structure yielded a high PCE of 11.56% in its as-cast OSCs [216]. Currently, most high-performance WBG polymer donors are based on BDT units. Recently, Sun and coworkers developed a benzodifuran (BDF)-based polymer P276 [217], in which chlorinated thienyl BDF and BTA building blocks were used as the electron-donating and electron-withdrawing units, respectively. Blending with TTPT-T-4F, P276 obtained a remarkably higher PCE of 14.0% in OSCs compared to its analog BDT-based polymer P271 (12.72%). Yan and coworkers reported a polymer P277 with a WBG of 2.05 eV and strong aggregation properties, which was based on a BDT donor unit and a BTA acceptor unit [218]. Different with conventional BDT-based polymers with a BDT unit linked *via* the α positions of the thiophene units, the BDT units in P277 were connected *via* the phenyl group of the BDT unit. In the P277:O-IDTBR blend film, both the donor and acceptor could maintain their crystallinity and form small domains. With these favorite properties, P277 achieved a PCE of 11.6% with a high V_{oc} of 1.08 V in its OSCs. By introducing a siloxane-terminated side chain into BTA building block and its BTA-fused derivative, Huang and coworkers developed two WBG polymer donors P278 and P279 [219]. The resulting two polymers were successfully used to fabricate efficient ternary all-polymer OSCs with a LBG polymer acceptor N2200 and achieved a PCE of 9.17% with an active layer thickness of 350 nm. Subsequently, the P279:N2200-based all-polymer OSCs obtained a significantly improved PCE of 11.76% by the meticulous optimization of blend morphology in Liu's group [220].

To match the deep LUMO levels of the LBG SM acceptors, Huang and coworkers further reduced the HOMO levels of the polymer donor PTzBI series consist of a BDT electron-donating unit and a pyrrolo[3,4-*f*]benzotriazole-5,7(6*H*)-dione (TzBI) electron-withdrawing building block by introducing two 2D-conjugated difluorophenyl units on the 3,6-positions of BDT unit. The resulting WBG polymer donor P280 with 3-branched alkyl side chains of the TzBI unit showed the optimized molecular orientation and crystallinity [221]. Matching with SM acceptor IT-4F, the P280-based OSCs achieved a certified high PCE of 12.25% with both improved J_{sc} and V_{oc} on a device area of 1 cm^2. Notably, Huang and coworkers selected another SM acceptor Y6 with a similar LUMO level but a lower bandgap to match P280 for more complementary absorption and well-matched energy levels, the corresponding OSCs obtained an impressive PCE over 16% [222]. Recently, two BTA units-fused naphtho[1,2-*c*:5,6-*c*]bis(2-octyl[1-3]triazole) (TZNT) was also used as electron-withdrawing building block to construct high-performance WBG polymer donors [223, 224]. Peng and coworkers synthesized two WBG polymers, non-fluorinated P281 and fluorinated P282 with a backbone of TZNT-*alt*-bithiophene (DT) [223]. Compared to P281, P282 had a higher crystallinity, stronger absorption coefficient, and deeper HOMO level, leading to the higher PCEs of 10.60% and 11.48% in binary and ternary blend OSCs, respectively. Soon afterward, Peng and coworkers developed the others two TZNT-based WBG polymer donors P283 and P284 by using 2-alkylthio-substituted thienyl BDT (BDT-TS) unit or 2-alkylthio-3-fluorine-substituted thienyl BDT (BDT-TSF) unit instead of DT unit [224]. The rigid planar backbone of BDT-T and TZNT units imparted high crystallinity and good molecular stacking properties to these polymers. Using IT-4F as SM acceptor, the P284-based OSCs obtained a high PCE of 13.25% with a high V_{oc} of 0.93 V and a low E_{loss} of 0.59 eV. Furthermore, the P284:IT-4F-based homo-tandem devices with improved light-harvesting ability achieved a further increased PCE of 14.52%, which was the best value for homo-tandem OSCs. The chemical structures of the polymer donors based on BTA derivatives are showed in Figure 1.37, and the corresponding photovoltaic parameters of OSCs are summarized in Table 1.28.

1.4.5 WBG Polymers Based on Thiazole, Pyrazine, and Their Derivatives Containing *N*-Heterocycles

N-heterocycles-based acceptor building blocks with a weak electron-withdrawing ability, strong intramolecular noncovalent bonds and intermolecular interaction, such as bisthiazole (BTz) [225–227], thiazolo[5,4-*d*]thiazole (TTz) [228–231], benzo[1,2-*d*:4,5-*d*']bis(thiazole) (BBTz) [232], 1,3,4-thiadiazole (TDz) [233], 1,3,4-oxadiazole (ODz) [234], pyrazine (PAz) [235], quinoxaline (Qx) [236, 237], dithieno[3,2-f:2′,3′-h]quinoxaline (DTQx) [238], 1,2,4,5-tetrazine (TAz) [239], and related derivatives, have been widely used for constructing high-performance WBG polymer donors with a good planarity, deep HOMO level, high absorption coefficient, and strong crystallinity. In 2011, Zhang et al. reported a polymer donor P285 with a WBG of 2.07 eV by combining BTz as electron-withdrawing unit and alkyl-substituted IDT as weak electron-donating unit [225]. Blending with $PC_{71}BM$, its OSCs achieved a low PCE of 2.77% due to the limited absorption spectrum.

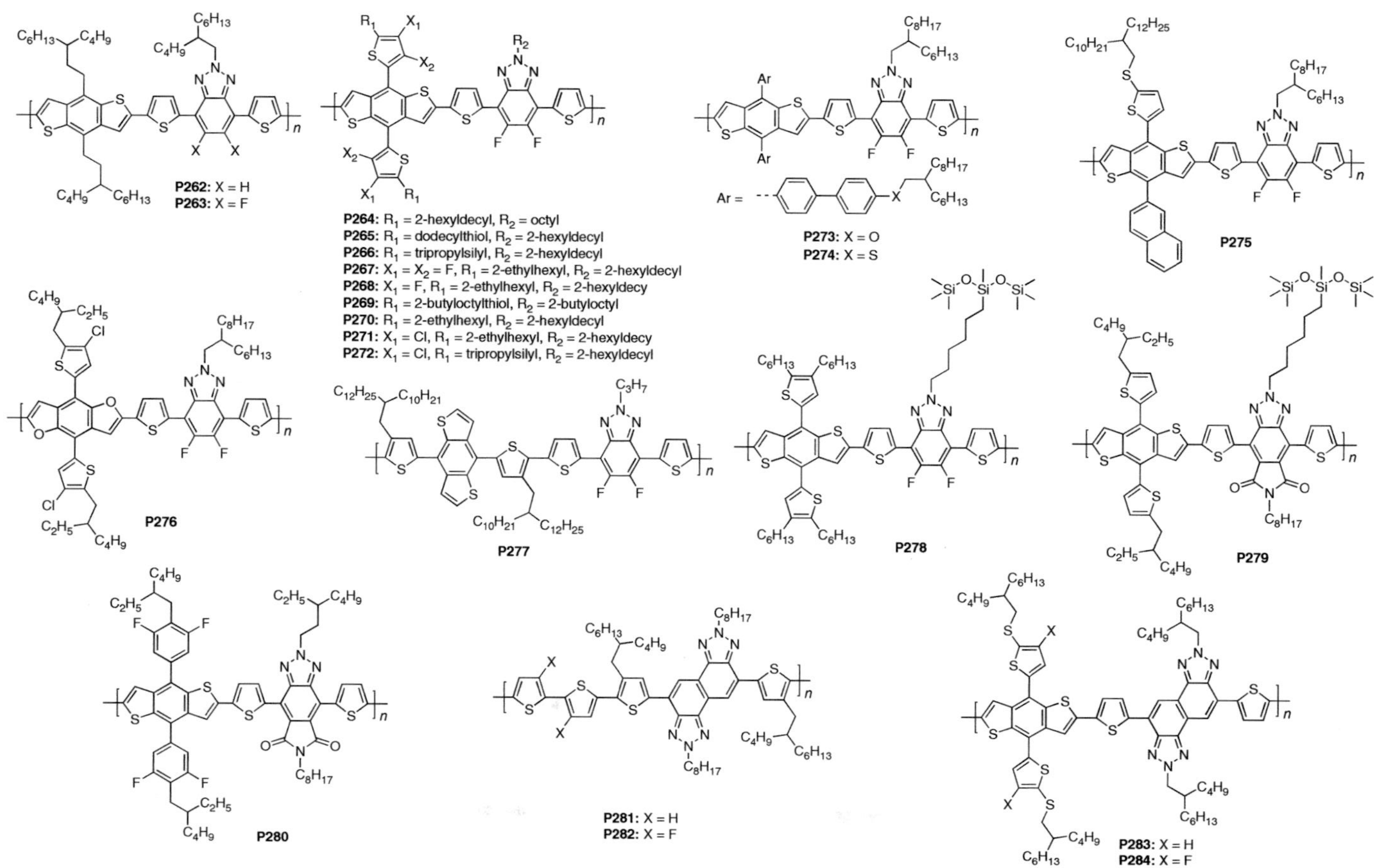

Figure 1.37 Chemical structures of the benzotriazole-derived polymers P262–P284. Source: Qunping Fan.

Table 1.28 Optoelectronic properties and device performances of P262–P284.

Polymer	E_g^{opt} (eV)	HOMO/ LUMO (eV)	Acceptor	V_{oc} (V)	J_{sc} (mA cm^{-2})	FF	PCE (%)	References
P262	1.98	−5.29/−2.87	$PC_{71}BM$	0.70	11.14	0.552	4.36	[209]
P263	2.00	−5.36/−3.05	$PC_{71}BM$	0.79	11.83	0.729	7.10	[209]
P264	1.94	−5.36/−3.42	ITIC	0.89	18.8	0.62	10.4	[210]
P265	1.93	−5.32/−3.08	ITIC	0.89	17.43	0.6148	9.53	[206]
P266	1.96	−5.40/−3.24	ITIC	0.94	17.32	0.6977	11.41	[208]
P267	2.00	−5.50/−3.02	m-ITIC	0.984	18.03	0.6554	11.63	[211]
P268	1.94	−5.36/−3.42	ITIC	0.89	18.8	0.62	10.4	[210]
P269	1.93	−5.37/−3.47	ITIC	0.89	19.0	0.62	10.5	[207]
P270	1.94	−5.18/−3.22	IT-4F	0.60	17.0	0.632	6.4	[212]
P271	1.94	−5.39/−3.41	IT-4F	0.80	17.7	0.683	9.7	[212]
P272	1.94	−5.56/−3.50	IT-4F	0.93	19.2	0.715	12.8	[212]
P273	1.93	−5.25/−3.49	ITIC	0.832	17.96	0.6156	9.20	[215]
P274	1.93	−5.46/−3.50	ITIC	0.914	18.52	0.7143	12.09	[215]
P275	1.91	−5.44/−3.49	ITIC	0.936	18.21	0.678	11.56	[216]
P276	1.86	−5.50/−3.39	TTPT-T-4F	0.86	22.17	0.736	14.0	[217]
P271	1.89	−5.48/−3.37	TTPT-T-4F	0.85	21.25	0.701	12.72	[217]
P277	2.05	−5.47/−3.42	O-IDTBR	1.08	16.26	0.636	11.6	[218]
P278	1.93	−5.23/−2.93	N2200	0.81	15.35	0.5638	7.00	[219]
P279	1.93	−5.31/−3.10	N2200	0.84	16.82	0.5187	7.34	[219]
P279	1.93	−5.31/−3.10	N2200	0.88	17.62	0.7578	11.76	[220]
P280	1.85	−5.38/−3.06	IT-4F	0.891	19.65	0.7409	12.96	[221]
P280	1.85	−5.38/−3.06	Y6	0.81	26.68	0.7411	16.02	[222]
P281	1.94	−5.16/−3.10	IT-M	0.75	10.15	0.581	4.42	[223]
P282	1.84	−5.24/−3.23	IT-M	0.80	17.33	0.725	10.05	[223]
P283	1.99	−5.39/−3.30	IT-4F	0.88	18.65	0.689	11.31	[224]
P284	1.97	−5.45/−3.41	IT-4F	0.93	19.23	0.741	13.25	[224]

Subsequently, using multi-alkyl thienyl-BDT unit instead of IDT unit, Zhang and coworkers reported another BTz-based WBG (1.89 eV) polymer donor P286 [226]. Compared to P285, P286 shows a red-shifted absorption onset from 600 to 650 nm, significantly improved planarity and crystalline structure, as well as increased hole mobility, leading to a higher PCE of 6.09% in the $PC_{71}BM$-based OSCs. Soon afterward, by replacing multi-alkyl thienyl-BDT unit with multi-alkoxyl phenyl-BDT unit in polymer backbone, Zhang and coworkers developed an efficient BTz-based polymer donor P287 [227]. Compared to P286, P287 exhibits a wider bandgap of 2.0 eV with strong absorption in the range of 300 ~ 620 nm and a lower-lying HOMO level of −5.36 eV. The OSCs based on P287:$PC_{71}BM$ showed a PCE of 8.1% with a high V_{oc} of 0.96 V, a J_{sc} of 10.9 mA cm^{-2} and a high FF of 76.7%, which was among the highest values for the OSCs combined fullerene derivatives and polymer donors with a bandgap near to 2.0 eV. Moreover, blending with a LBG SM acceptor ITIC, its OSCs achieved a higher PCE of 10.3% with both improved V_{oc} of 1.01 V and J_{sc} of 14.1 mA cm^{-2}.

Compared to BTz unit, TTz unit has a larger fused-ring rigid structure, resulting in the stronger intermolecular aggregation and interaction of the corresponding polymers [228–230]. In 2016, Zhang and coworkers synthesized a polymer donor named PTZ1 (P288) by polymerizing BDT-T electron-donating unit and TTz electron-withdrawing unit for photovoltaic applications [228]. P288 with high crystallinity and favorable backbone orientation shows a desirable WBG of 1.97 eV, a deep HOMO level −5.31 eV, and a relatively high hole mobility >10^{-3} cm^2 V^{-1} s^{-1}. Consequently, single-junction OSCs based on P288:$PC_{71}BM$ obtained a PCE of 7.7% with a high V_{oc} of 0.94 V. Moreover, its tandem PSCs based on P288 as a donor material in the front cell achieved an improved PCE of 10.3% with a high V_{oc} of 1.65 V. Blending P288 with IDIC, the fabricated OSCs achieved the highest PCE of 11.5%, while the OSCs still obtained a PCE of 9.6% with an active layer thickness of 210 nm and a relatively high PCE of 10.5% with a device area of up to 0.81 cm^2 [229]. By incorporating a sulfur atom into 2D-conjugated alkylthienyl groups, Su et al. reported another similar WBG (1.96 eV) polymer P289 [230]. The P289-based OSCs with SM acceptors ITIC or IDIC achieved PCEs more than 8%. Moreover, by inserting ITIC into the binary P289:IDIC system, the active layer shows smooth and gradient energy levels, improved crystallinity, and optimized morphologies, which results in efficient exciton separation, charge transport and collection. Consequently, the optimized ternary OSCs based on P289:ITIC:IDIC (1 : 0.1 : 0.9) obtained a higher PCE of 11.1% in comparison with those of binary OSC systems. Recently, Huang and coworkers reported a series of WBG polymer donor analogs (P290 and P291) composed of thienothiophene (TT) or TTz units [231] and studied their device performance with N2200 in all-polymer OSCs. Results show that all-polymer OSCs with the TTz-based P291 achieved a PCE as high as 8.4%, which largely outperform the analog P290-based ones with a PCE of only 0.7%. In 2019, Yang and coworkers developed a WBG (2.1 eV) polymer donor P292 with a backbone of BDT-T-*alt*-BBTz [232]. Compared to TTz unit, BBTz unit has a further enlarged fused-ring rigid structure by inserting a benzene ring into TTz unit, resulting in the weaker electron-withdrawing ability, deep HOMO level, and stronger intermolecular aggregation and interaction of the corresponding polymers. Blending P292 with a LBG SM acceptor ITIC-F, the OSCs obtained a high PCE of 13.3% with a high V_{oc} of 0.91 V, a high J_{sc} of 20.9 mA cm^{-2}, and an FF of 0.70. TDz unit with high polarizability and relatively weak electron-withdrawing ability, has been widely studied in optoelectronic field. By polymerizing BDT and TDz building blocks, Peng and coworkers synthesized a polymer donor P293 with a WBG over 2.07 eV and a low-lying HOMO level of −5.35 eV [233], which matches well with the typical LBG SM acceptor ITIC. By using a single green solvent of o-xylene, the as-cast OSCs achieved a large V_{oc} up to 1.10 V and an extremely low E_{loss} of 0.48 eV, as well as the desirable J_{sc} of 17.78 mA cm^{-2} and FF of 65.4%, leading to a high PCE of 12.80%. When adopting a homo-tandem device architecture, the PCE is further improved to 13.35% with an improved V_{oc} of 2.13 V. After that, by using ODz instead of TDz, Peng and coworkers reported another polymer donor P294 [234]. Compared to the TDz-based P293, P294 with a similar backbone has a wider bandgap of 2.12 eV and a deeper HOMO level of −5.68 eV. By adding 4% of copper(I) iodide as an additive to form coordination complexes with P294 for optimizing blend morphology, significantly improved device performances are achieved due to the improved absorption and crystallinity of photovoltaic materials. As a result, the PCE is elevated from 10.12% to 12.34%.

N-heterocyclic benzene derivatives, such as PAz [235], Qx [236, 237], DTQx [238], and TAz [239], also have been applied in the synthesis of WBG photovoltaic materials. Chen et al. firstly introduced F atom into PAz-based electron-withdrawing unit and successfully applied it in constructing a polymer donor P296 for OSCs [235]. Compared to its non-fluorinated counterpart P295, P296 had a better planarity, improved absorbance with red-shifted spectrum and higher absorption coefficient, reduced HOMO level, and optimized molecular packing and morphological properties. Blending with MeIC, the P296-based OSCs delivered a PCE of 12.3%, which was much higher than that of the P295-based control OSCs (10.2%). In addition to high efficiency and good stability, the application of OSCs requires the realization of low-cost devices. Sun et al. demonstrated a low-cost WBG polymer donor P297 with a simple backbone of thiophene-*alt*-(6,7-difluoro-2-(2-hexyldecyloxy)quinoxaline), which was synthesized by only two-step simple reactions with high overall yield of 87.4% from cheap raw materials [236]. More important, the P297-based OSCs achieved a high PCE of 12.70%, and the corresponding as-cast devices with thickness insensitivity also obtained a high PCE of 10.41%, which indicated that P297 was a promising polymer donor for commercial application of OSCs. By attaching an additional methyl substituent in the Qx unit of low-cost polymer donor P297, Sun et al. reported another similar polymer donor P298 [237]. Compared to P297, P298 showed an upshifted HOMO level of −5.52 eV, stronger molecular crystallization, and better hole transport capability. Impressively, the OSCs based on P298:TPT10 with near zero HOMO offset achieved a high PCE of 16.32% due to the efficient exciton dissociation and hole transfer. Recently, Hou and coworkers fused two thiophene units along with Qx to design DTQx, which is quite similar with DTBT [238]. The resulting polymer P299 achieved a high PCE of 18.0% in the ternary devices using BTA3 and BTP-eC9 as the acceptor materials. In 2019, Peng and coworkers synthesized a TAz-containing WBG polymer P300 [239]. The studies showed that P300 had a strong aggregation property, which resulted in serious phase separation and large domains when blending with Y6. To address the above defect, Peng and coworkers developed a new strategy of platinum(II) complexation to regulate molecular crystallinity and packing of TAz-based polymers, optimize blend morphology, and improve PCE of the resulting OSCs. Compared to binary polymer P300, the TAz-based ternary polymer P301 had bulky benzene ring on the platinum(II) complex to increase steric hindrance along the polymer main chain, which was beneficial to inhibit molecular aggregation strength and regulate blend morphology, and thus improved PCE to 16.35% in OSCs [239]. The chemical structures of the polymer donors based on thiazole, pyrazine, and their derivatives containing *N*-heterocycles are showed in Figure 1.38, and the corresponding photovoltaic parameters of OSCs are summarized in Table 1.29.

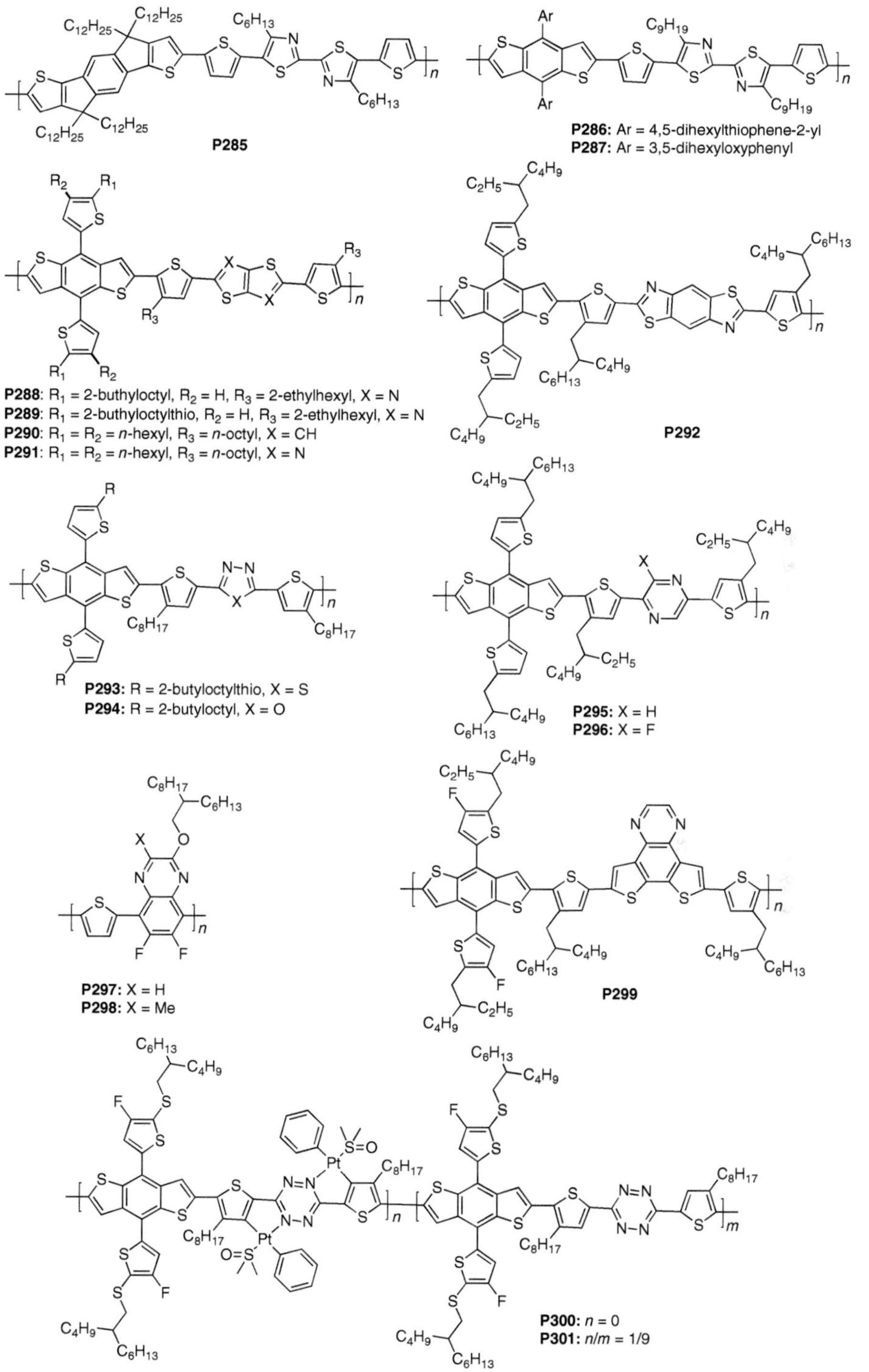

Figure 1.38 Chemical structures of the polymer donors based on thiazole, pyrazine, and their derivatives containing N-heterocycles P285–P301. Source: Qunping Fan.

Table 1.29 Optoelectronic properties and device performances of P285–P301.

Polymer	E_g^{opt} (eV)	HOMO/ LUMO (eV)	Acceptor	V_{oc} (V)	J_{sc} (mA cm^{-2})	FF	PCE (%)	References
P285	2.07	−5.26/−2.80	$PC_{71}BM$	0.90	7.43	0.414	2.77	[225]
P286	1.89	−5.21/−3.25	$PC_{71}BM$	0.92	10.53	0.629	6.09	[226]
P287	2.0	−5.36/−3.36	$PC_{71}BM$	0.96	10.9	0.767	8.1	[227]
P287	2.0	−5.36/−3.36	ITIC	1.01	14.1	0.723	10.3	[227]
P288	1.97	−5.31/−3.33	$PC_{71}BM$	0.95	11.3	0.72	7.7	[228]
P288	1.97	−5.31/−3.33	IDIC	0.92	16.4	0.762	11.5	[229]
P289	1.96	−5.44/−3.48	ITIC	1.01	15.4	0.523	8.13	[230]
P289	1.96	−5.44/−3.48	IDIC	0.928	14.7	0.591	8.06	[230]
P289	1.96	−5.44/−3.48	ITIC:IDIC	0.953	17.4	0.669	11.1	[230]
P290	2.01	−5.04/−3.03	N2200	0.75	2.0	0.46	0.7	[231]
P291	1.91	−5.20/−3.29	N2200	0.87	14.4	0.67	8.4	[231]
P292	2.10	−3.30/−5.40	ITIC-F	0.91	20.9	0.70	13.3	[232]
P293	2.07	−5.39/−2.80	ITIC	1.10	17.78	0.654	12.8	[233]
P294	2.12	−5.68/−2.89	ITIC-Th	1.06	17.1	0.681	12.34	[234]
P295	2.00	−5.44/−3.56	MeIC	0.95	15.54	0.688	10.2	[235]
P296	1.95	−5.48/−3.70	MeIC	0.96	17.29	0.738	12.3	[235]
P297	1.92	−5.54/−2.98	IDIC	0.969	17.81	0.736	12.7	[236]
P298	1.95	−5.52/−2.76	TPT10	0.88	17.25	0.748	16.32	[237]
P299	2.05	−5.39/−2.86	BTP-eC9	0.82	26.0	0.752	16.0	[238]
P299	2.05	−5.39/−2.86	BTA3:BTP-eC9	0.84	26.9	0.796	18.0	[238]
P300	1.93	−5.50/−3.50	Y6	0.80	25.1	0.649	13.03	[239]
P301	1.94	−5.53/−3.51	Y6	0.81	26.45	0.763	16.35	[239]

1.4.6 WBG Polymers Based on Benzodithiophene-4,8-dione (BDD) Derivatives

Benzodithiophene-4,8-dione (BDD) and its derivatives with a rigid large planarity and two fused electron-withdrawing C=O bonds have been widely used to construct high-performance polymer donors for photovoltaic applications [240–244]. Among these BDD-based polymers, the polymer donors (namely PBDB-T series) with a molecular backbone of BDT-*alt*-BDD display some unique properties, such as high absorption coefficient, WBG, temperature-dependent absorption spectra, deep-lying HOMO level, and high carrier mobility [240–251]. In 2012, Hou and coworkers developed a BDD-based WBG (1.80 eV) polymer donor with a 2D-conjugated thienyl BDT derivative, namely PBDB-T (P302) [241]. P302 displayed a strong aggregation effect in solution state, and the studies indicated that device efficiency of the $P302:PC_{61}BM$-based OSCs could be easily affected by the processing temperature used in solution preparation. By matching P302 with a LBG SM acceptor ITIC, Hou and coworkers reported a high PCE of 11.21% [252]. After that, the following-up reports using P302 congeners sprang up in fullerene-based and SM-acceptor-based OSCs. In 2015, Zhang et al. synthesized a PBDB-T derivative, namely PM6 (PBDB-TF, P303) [242], by attaching a fluorine atom on 2D-conjugated thienyl substituents of polymer. Compared to P302, fluorinated P303 showed obviously downshifted HOMO level, stronger crystallinity, and smaller π–π stacking distance. Matching with $PC_{71}BM$, its OSCs obtained a PCE of 9.2%. P303 also could match different types of LBG SM/polymer acceptors very well with respect to absorption, energy levels, and morphological compatibility. In this regard, blending P303 with typical SM acceptor ITIC, the OSCs achieved a PCE of 9.7% with a V_{oc} up to 1.04 V [253]; with a strongly crystalline SM acceptor IDIC, the as-cast OSCs achieved the PCEs exceeded 11% with different active layer thicknesses [254]; with a tetra-fluorinated SM acceptor IT-4F, the OSCs delivered a PCE of 13.5% [255]; and with polymerized small molecular acceptors (PSMAs) such as PZ1 [256], PF1-TS4 [257], PF2-DTSi [258], PF3-DTCO [259], PFBDT-IDTIC [260], PN1 [261], and PYT [262], the all-PSCs provided the PCEs of 9 ~ 13%. Notably, matching P303 with the star SM acceptors of Y6 series, the OSCs have achieved promising PCEs of up to 15 ~ 18% [263–265]. Compared to fluorine atom, chlorine atom is more easily introduced into photovoltaic materials by low-cost strategy [214]. Fan et al. developed another P302 derivative, namely PM7 (PBDB-TCl, P304) [243], by attaching a chlorine atom on 2D-conjugated thienyl substituents of polymer. Compared to the halogen-free P302, P304 showed lower HOMO level, higher absorption coefficient, enhanced crystallinity, and higher carrier mobility. Moreover, the P304:IT-4F-based OSCs achieved a high PCE of 13.1%, which was much higher than the P302:IT-4F-based ones (5.8%). By combining the advantages of sulfur atom and fluorine atom substituents on thienyl BDT, Hou and coworkers synthesized a WBG polymer donor P305 with high absorption coefficient and deep HOMO level, and the corresponding P305:IT-4F-based OSCs also achieved a PCE more than 13% [244]. Introducing halogen atoms, such as fluorine and chlorine, on the 4-position of thiophene π-bridges of P302, the synthesized polymers P307 [245] and P309 [246] exhibited deeper HOMO level and smaller bandgap of ~1.78 eV, leading to improved photovoltaic performance in OSCs. Aiming to enhance the carrier transport ability, Qin et al. enlarged the 2D-conjugated side groups of P302 from thiophene to benzo[*b*]thiophen and synthesized a new polymer P310 [247]. Due to P310 had good solubility and enhanced aggregation effect in tetrahydrofuran (THF), the THF-processed P310:IT-M-based OSCs achieved a PCE of 12.10%. Liu et al. synthesized a series of PBDT-T derivatives, P311–P313 and found that the subtle atom changes in 2D-benzene on BDT unit can profound effect the electronic structure and self-assembly of polymers [248]. Among these polymers, P311 achieved a much higher PCE of 10% with a record FF of 80.5% in the $PC_{71}BM$-based OSCs due to the formation of optimal interpenetrating network morphology. By further attaching two chlorine atoms on the thiophene π-bridges of P311, an optimized WBG polymer donor P314 with a deeper HOMO level and improved crystallinity was synthesized by Ye et al. and achieved a higher PCE of 12.7% [249]. Moreover, Li et al. found that introducing fluorination on 2D-benzene on BDT unit also could significantly reduce the HOMO level of polymers, and the resulting WBG polymer donor P315 with 2D-conjugated alkoxyl-fluorophenyl side chains obtained an increased PCE of 11.7% in comparison with non-fluorinated P312 (6.2%) [250]. Xu et al. designed a WBG polymer donor P316 based on 2D-conjugated *m*-alkoxyphenyl BDT and BDD units [251]. Matching with $PC_{71}BM$, the P316-based OSCs achieved a PCE of 8.4% with a high V_{oc} of 0.95 V. Consequently, the photovoltaic performance of P316 in devices was further improved by pairing it with ITIC, and the related OSCs achieved a PCE of 10.8% [266].

In addition to optimizing 2D-conjugated groups on BDT unit and introducing halogen atoms into thiophene π-bridges of P302, the modification of fused-ring building blocks of BDT and BDD units also have attracted great attentions. Chao et al. synthesized a new electron-withdrawing fused-ring building block of 2,3-bis(2-ethylhexyl)benzo[1,2-*b*:4,5-*c*′]dithiophene-4,8-dione (TTDO) *via* the design of cyclohexane-1,4-dione embedded into a thieno[3,4-*b*]thiophene [267]. Then, by polymerizing TTDO with fluorinated thienyl BDT, they developed a WBG (1.80 eV) polymer donor, P317. Blended with SM acceptor Y6, the P317-based OSCs obtained a PCE of 16.1% with a high FF of 77.1%. Compared to BDT unit, dithieno[2,3-*d*;2′,3′-*d*′]benzo[1,2-*b*;4,5-*b*′]dithiophene (DTBDT) with extended π-conjugation and larger molecular planarity could effectively improve the charge carrier transport of the related photovoltaic materials. Inspired these, Huo et al. reported a WBG (1.85 eV) polymer P318 by polymerizing an electron-rich DTBDT unit and an electron-deficient BDD unit [268]. P318 had a highly rigid backbone since both DTBDT and BDD units were rigid and planar, and its OSCs with a $PC_{71}BM$ acceptor achieved a high PCE of 9.7%, which was the highest value reported in the OSCs based on WBG polymer donors. By using NDT with extended π-conjugation and larger molecular planarity as electron-donating unit, Peng and coworkers developed WBG polymer donors P319 and P320 for efficient OSCs [269, 270]. Compared to its analogous polymers (P302 and P306), P319 and P320 showed higher crystallinity and stronger aggregation, which induced large phase separation with poorer morphology. To optimize the morphology, P319 and P306 were blended with SM acceptor Y6-T for forming the ternary blend devices. As expected, the optimized ternary OSCs achieved a PCE as high as 16.57%. Ternary copolymerization strategies have proven advantageous in boosting the photovoltaic performance of OSCs [271]. Cui et al. reported a WBG terpolymer P321 by incorporating an electron-withdrawing unit of ester-substituted thiophene into polymer P303 to downshift molecular HOMO level and broaden absorption [272]. In the P321:IT-4F-based OSCs, a high PCE of 15.1% was obtained. Soon afterward, Sun et al. synthesized a fluorine and ester-functionalized thiophene (FE-T) with electron-withdrawing ability and then developed another WBG terpolymer donor P322 by incorporating FE-T as the third component into polymer P303 to fine-tune the energy levels and crystallinity [273]. Blending P322 with an SM acceptor Y6, the OSCs achieved a high PCE of 16.4% with a small E_{loss} of 0.53 eV. Moreover, by incorporating 1 mol% concentration of iridium (Ir) complexes into the polymer-conjugated backbone of P303, Wang et al. developed a WBG terpolymer donor P323 [274]. The studies indicated that the introduction of small amount

Ir-complexes could rationally modify the molecular aggregations and carefully control the corresponding blend morphology and physical mechanisms. As a result, the P323:Y6-based OSCs delivered a high PCE of 17.32%, which was much better than that of the P303:Y6-based ones (15.39%). The chemical structures of the polymer donors based on BDD derivatives are showed in Figure 1.39, and the corresponding photovoltaic parameters of OSCs are summarized in Table 1.30.

1.4.7 Other WBG Polymers

In addition to the above-summarized WBG polymers with some typical acceptor units, there still has many other building blocks employed to construct high-performance WBG polymer donors (Figure 1.40, Table 1.31). Cao et al. synthesized a fused pentacyclic aromatic lactam building block (thieno[2′,3′:5,6] pyrido[3,4-g]thieno[3,2-*c*]isoquinoline-5,11(4*H*,10*H*)-dione [TPTI]) with weak electron-withdrawing ability and then developed a D-A conjugated polymer donor P324 with a backbone of TPTI-*alt*-thiophene [275]. P324 possessed a WBG (1.86 eV) and a deep HOMO level, and the corresponding P324:$PC_{71}BM$-based OSCs afforded an outstanding PCE up to 7.80% with a high V_{oc} of 0.87 V, which was one of the best PCEs in OSCs based on WBG polymer donor. Liao et al. reported a series of readily accessible and scalable polymer donors (P325–P328) with a backbone of BDT-*alt*-2,5-dithienyl TPD for application in OSCs blended with IT-4F [276]. All polymers readily dissolved in non-chlorinated solvents, and the corresponding active layers could be processed in ambient from xylene as solvent to fabricate OSCs with PCEs of 12 ~ 14%. In the P328:IT-4F-based OSC modules processed in ambient from a benign solvent, a certified PCE of 10.1% for a device area of 20.4 cm^2 was achieved. Moreover, the same module also delivered a power of ~40 $\mu W\,cm^{-2}$ (PCE ~22%) under indoor lighting. Liu et al. designed a fused building block of 5H-dithieno[3,2-*b*:2′,3′-*d*]pyran-5-one with weak electron-withdrawing ability and then synthesized a polymer donor P329 with a WBG of 1.96 eV and a deep HOMO level of −5.45 eV [277]. Blending P329 with Y6, the OSCs provided a PCE over 14%. To fully exploit the potential of P329 series polymers, Ding and coworkers developed another WBG polymer donor P330 with a thiolactone unit of 5H-dithieno[3,2-*b*:2′,3′-*d*]thiopyran-5-one [278]. From P229 to P330, the simple replacement of lactone with thiolactone enhanced molecular π–π stacking, gifting P330 a higher hole mobility. When matched with Y6, its OSCs gave a high PCE up to 16.72%. The mechanism that how noncovalent conformation locked to improve the photovoltaic performance of WBG polymer donors was also investigated. By developing a model system with three structure-related polymers P331–P333, a clear correlation was found between the locking strength in polymer backbone and the photovoltaic performance [279]. Among three polymers, P331 with the strongest S···O and F···H double-side locking showed improved planarity, packing, crystallinity, and hole mobility compared to these of P332 with a S···O single-side locking and P333 with a weakened S···O locking caused by Cl–H steric repulsion, which gave the highest PCE of 16.23% in the P331:Y6-based OSCs, while the OSCs based on P332:Y6 and P333:Y6 achieved PCEs of 11.73% and 0.92%, respectively. He

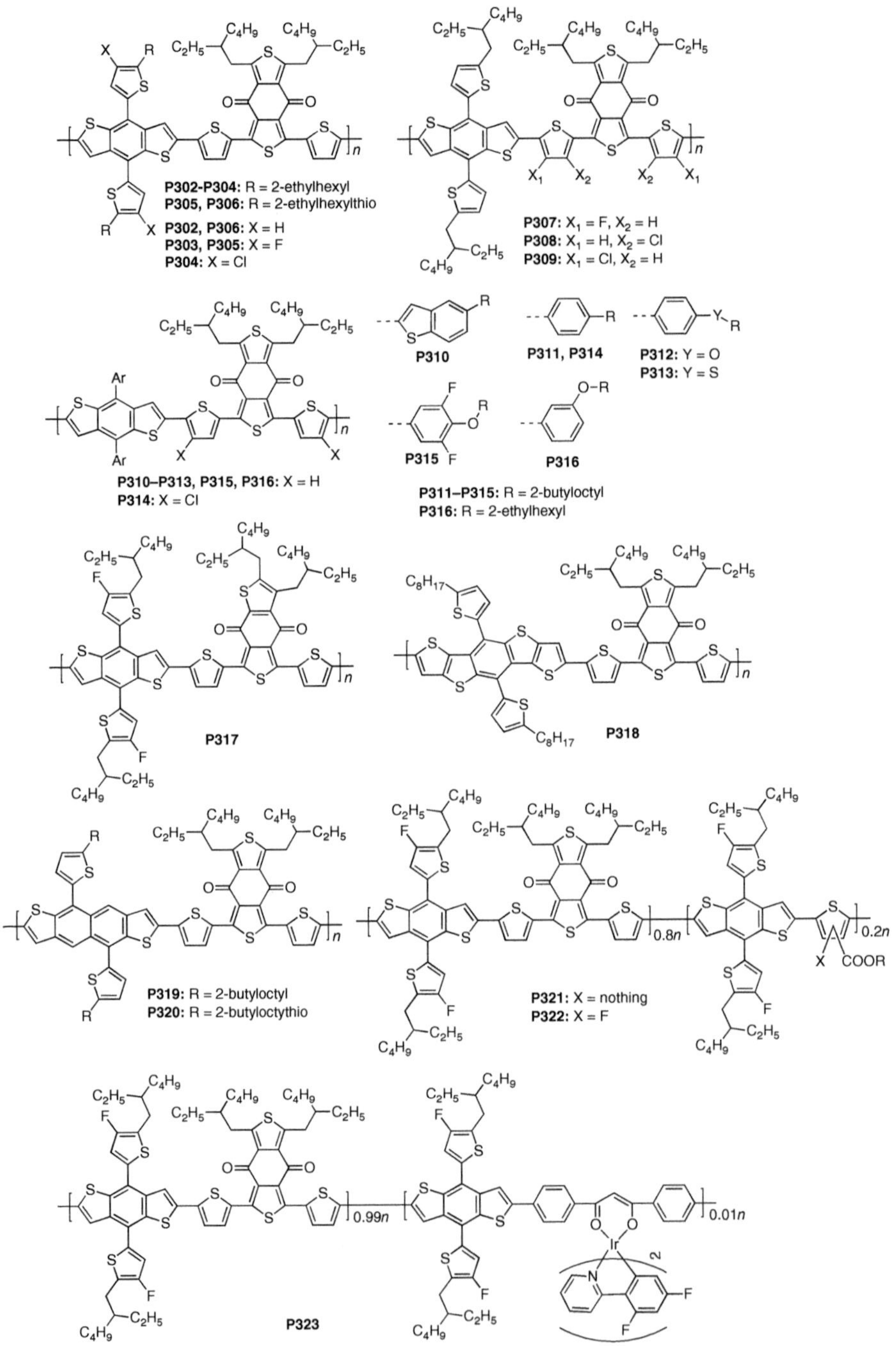

Figure 1.39 Chemical structures of the BDD derived polymers P302–P323. Source: Qunping Fan.

Table 1.30 Optoelectronic properties and device performances of P302–P323.

Polymer	E_g^{opt} (eV)	HOMO/ LUMO (eV)	Acceptor	V_{oc} (V)	J_{sc} (mA cm^{-2})	FF	PCE (%)	References
P302	1.80	−5.23/−3.18	$PC_{61}BM$	0.86	10.68	0.7227	6.67	[241]
P302	1.80	−5.23/−3.18	ITIC	0.899	16.81	0.742	11.21	[252]
P303	1.80	−5.50/−3.61	$PC_{71}BM$	0.98	12.7	0.74	9.2	[242]
P303	1.80	−5.50/−3.61	ITIC	1.04	16.0	0.58	9.7	[253]
P303	1.80	−5.50/−3.61	IDIC	0.97	17.8	0.69	11.9	[254]
P303	1.80	−5.50/−3.61	IT-4F	0.84	22.2	0.725	13.5	[255]
P303	1.80	−5.50/−3.61	PZ1	0.96	17.1	0.682	11.2	[256]
P303	1.80	−5.50/−3.61	PF2-DTSi	0.99	16.48	0.661	10.77	[258]
P303	1.80	−5.50/−3.61	PYT	0.93	21.78	0.6633	13.44	[262]
P303	1.80	−5.50/−3.61	Y6	0.83	25.3	0.748	15.7	[263]
P303	1.80	−5.49/−3.47	L8-BO	0.87	25.72	0.815	18.32	[265]
P304	1.80	−5.52/−3.57	IT-4F	0.88	20.9	0.711	13.1	[243]
P305	1.80	−5.40/−3.60	IT-4F	0.88	20.5	0.719	13.0	[244]
P307	1.78	−5.47/−3.46	C8-ITIC	0.94	19.6	0.72	13.2	[245]
P309	1.78	−5.48/−3.47	IT-4F	0.84	20.6	0.7109	12.33	[246]
P310	1.80	−5.36/––	IT-M	0.96	17.97	0.70	12.1	[247]
P311	1.84	−5.43/−3.36	$PC_{71}BM$	0.93	13.1	0.805	10.0	[248]
P312	1.82	−5.30/−3.30	$PC_{71}BM$	0.88	11.1	0.727	7.2	[248]
P313	1.83	−5.45/−3.37	$PC_{71}BM$	0.93	11.6	0.715	7.8	[248]
P314	1.85	−5.59/−3.53	IT-4F	0.85	19.74	0.76	12.7	[249]
P312	1.81	−5.38/−3.57	IT-4F	0.72	16.8	0.512	6.2	[250]
P315	1.86	−5.50/−3.64	IT-4F	0.89	20.4	0.645	11.7	[250]
P316	1.90	−5.42/−3.36	$PC_{71}BM$	0.95	12.4	0.71	8.4	[251]
P316	1.90	−5.42/−3.36	ITIC	1.01	18.1	0.59	10.8	[266]
P317	1.80	−5.48/−3.68	Y6	0.84	24.8	0.771	16.1	[267]
P318	1.93	−5.36/−3.43	$PC_{71}BM$	0.92	14.11	0.75	9.74	[268]
P306	1.83	−5.47/−3.57	Y6-T	0.909	24.04	0.758	16.57	[270][a)]
P320	1.81	−5.43/−3.54						
P321	1.83	−5.48/−3.63	IT-4F	0.899	21.5	0.78	15.1	[272]
P322	1.83	−5.52/−3.72	Y6	0.877	25.402	0.737	16.42	[273]
P323	1.83	−5.55/−3.60	Y6	0.845	26.15	0.784	17.32	[274]

a) Device parameters from the ternary devices based on P306:319:Y6-T.

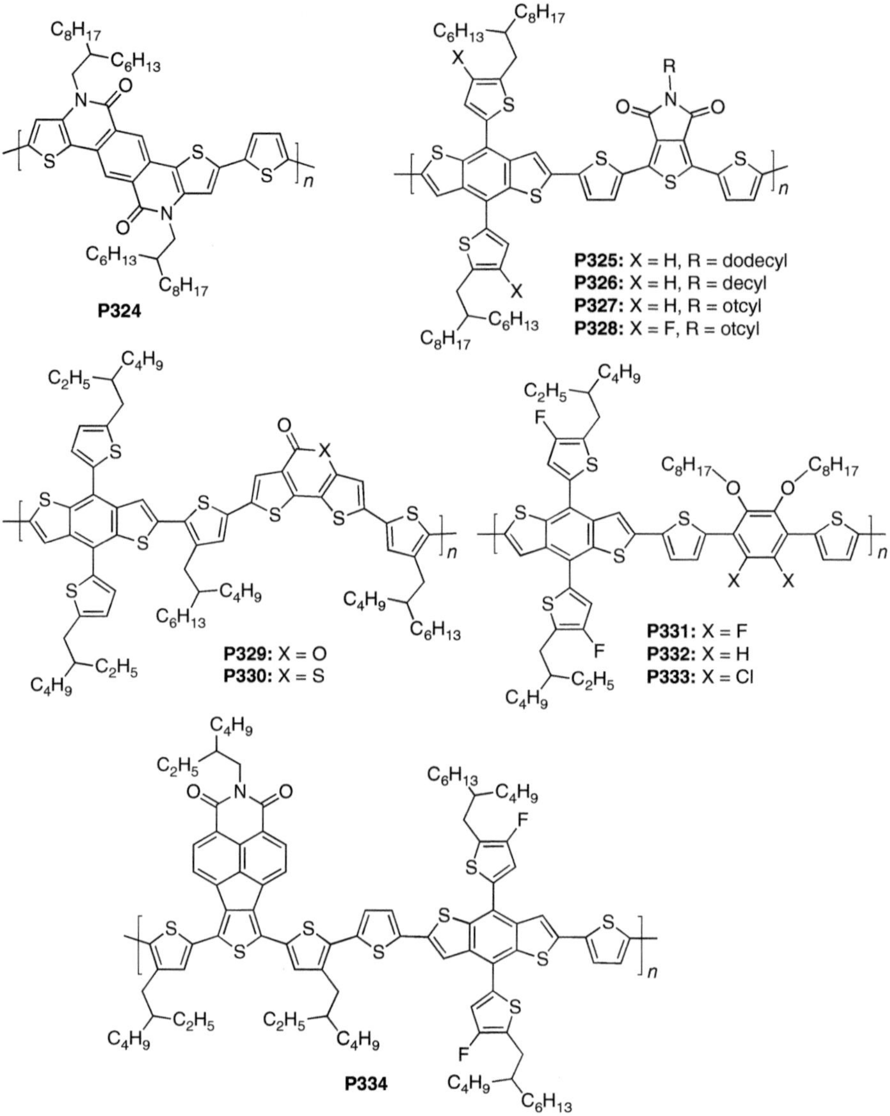

Figure 1.40 Chemical structures of the WBG polymer donors of P324–P334. Source: Qunping Fan.

and coworkers reported a new electron-deficient monomer naphthalenothiophene imide (NTI) and NTI-based copolymer P334 [280]. The strong electron-withdrawing ability of NTI could effectively lower the HOMO level of the polymer (−5.52 eV). The polymer chain modification resulted in the closed π–π stacking and ordered polymer packing of P334 and was proven to be an effective strategy to increase the device performance. OSC devices for P334:Y6 exhibited an efficiency of 16.72%, which was further enhanced to 17.35% by the third component $PC_{71}BM$.

Table 1.31 Optoelectronic properties and device performances of the WBG polymer donors of P324–P334.

Polymer	E_g^{opt} (eV)	HOMO/ LUMO (eV)	Acceptor	V_{oc} (V)	J_{sc} (mA cm^{-2})	FF	PCE (%)	References
P324	1.86	−5.42/−2.77	$PC_{71}BM$	0.87	13.69	0.656	7.8	[275]
P325	1.87	−5.48/−3.61	IT-4F	0.81	19.4	0.747	11.7	[276]
P326	1.88	−5.45/−3.57	IT-4F	0.81	19.6	0.742	11.8	[276]
P327	1.90	−5.49/−3.59	IT-4F	0.80	20.1	0.753	12.1	[276]
P328	1.90	−5.62/−3.73	IT-4F	0.92	21.6	0.723	14.4	[276]
P329	1.96	−5.45/−2.79	Y6	0.80	23.93	0.747	14.36	[277]
P330	1.96	−5.48/−2.83	Y6	0.85	25.41	0.749	16.22	[278]
P331	2.16	−5.44/−3.48	Y6	0.881	26.01	0.708	16.23	[279]
P332	2.18	−5.44/−3.48	Y6	0.846	22.44	0.618	11.73	[279]
P333	2.37	−5.04/−3.03	Y6	0.843	3.39	0.322	0.92	[279]
P334	1.79	−5.52/−3.43	Y6	0.872	26.31	0.73	16.72	[280]
P334	1.79	−5.52/−3.43	Y6:$PC_{71}BM$	0.875	26.45	0.75	17.35	[280]

1.5 Summary and Outlook

In this part, we have summarized several representative types of LBG, MBG, and WBG polymer donors and discussed the relationship between the chemical structures and the optical, electronic, and device performance. With the rapid development of non-fullerene acceptors, the number of publications of LBG and MBG polymers is far behind than WBG polymers in recent years, but it is undeniable that LBG and MBG polymers have made great contributions to the field of OSCs. Although WBG polymers have dominated the high performance in the current fullerene-free PSCs, LBG and MBG polymer donors will continue play an important role in OSCs, especially in semi-transparent, ternary, and tandem devices.

References

1 Yu, G., Gao, J., Hummelen, J.C. et al. (1995). *Science* 270: 1789.
2 Boudreault, P.-L.T., Najari, A., and Leclerc, M. (2011). *Chem. Mater.* 23: 456.
3 Thompson, B.C. and Frechet, J.M. (2008). *Angew. Chem. Int. Ed.* 47: 58.
4 Liu, Q.S., Jiang, Y.F., Jin, K. et al. (2020). *Sci. Bull.* 65: 272.
5 Lin, Y.B., Adilbekova, B., Firdaus, Y. et al. (2019). *Adv. Mater.* 31: 1902965.
6 Subbiah, J., Beaujuge, P.M., Choudhury, K.R. et al. (2009). *ACS Appl. Mater. Interfaces* 1: 1154.
7 Bin, H., Zhong, L., Zhang, Z.G. et al. (2016). *Sci. China Chem.* 59: 1317.
8 Cai, Y.H., Huo, L.J., and Sun, Y.M. (2017). *Adv. Mater.* 29: 1605437.

9 Jung, J.W., Liu, F., Russell, T.P., and Jo, W.H. (2015). *Adv. Mater.* 27: 7462.

10 Jung, J.W., Liu, F., Russell, T.P., and Jo, W.H. (2015). *Adv. Energy Mater.* 5: 1500065.

11 Zhou, H., Yang, L., Price, S.C. et al. (2010). *Angew. Chem. Int. Ed.* 49: 7992.

12 Zhou, H., Yang, L., Stoneking, S., and You, W. (2010). *ACS Appl. Mater. Interfaces* 2: 1377.

13 Dou, L., Liu, Y., Hong, Z. et al. (2015). *Chem. Rev.* 115: 12633.

14 Hildner, R., Köhler, A., Müller-Buschbaum, P. et al. (2017). *Adv. Energy Mater.* 7: 1700314.

15 Cheng, Y.-J., Yang, S.-H., and Hsu, C.-S. (2019). *Chem. Rev.* 109: 5868.

16 Brus, V.V., Lee, J., Luginbuhl, B.R. et al. (2019). *Adv. Mater.* 31: 1900904.

17 Mühlbacher, D., Scharber, M., Morana, M. et al. (2006). *Adv. Mater.* 18: 2884.

18 Bijleveld, J.C., Shahid, M., Gilot, J. et al. (2009). *Adv. Funct. Mater.* 19: 3262.

19 Chen, H.Y., Hou, J.H., Hayden, A.E. et al. (2010). *Adv. Mater.* 22: 371.

20 Scharber, M.C., Koppe, M., Gao, J. et al. (2010). *Adv. Mater.* 22: 367.

21 Gasparini, N., Jiao, X.C., Heumueller, T. et al. (2016). *Nat. Energy* 1: 16118.

22 Albrecht, S., Janietz, S., Schindler, W. et al. (2012). *J. Am. Chem. Soc.* 134: 14932.

23 Jo, J.W., Jung, J.W., Jung, E.H. et al. (2015). *Energy Environ. Sci.* 8: 2427.

24 You, J., Dou, L., Yoshimura, K. et al. (2013). *Nat. Commun.* 4: 1446.

25 Shi, S., Liao, Q., Tang, Y. et al. (2016). *Adv. Mater.* 28: 9969.

26 Zhou, H., Yang, L., Price, S.C. et al. (2010). *Angew. Chem. Int. Ed.* 49: 7992.

27 Yau, C.P., Fei, Z., Ashraf, R.S. et al. (2014). *Adv. Funct. Mater.* 24: 678.

28 Hendriks, K.H., Heintges, G.H.L., Gevaerts, V.S. et al. (2013). *Angew. Chem. Int. Ed.* 52: 8341.

29 Wang, M., Hu, X., Liu, P. et al. (2011). *J. Am. Chem. Soc.* 133: 9638.

30 Vohra, V., Kawashima, K., Kakara, T. et al. (2015). *Nat. Photonics* 9: 403.

31 Jin, Y., Chen, Z., Dong, S. et al. (2016). *Adv. Mater.* 28: 9811.

32 Jin, Y., Chen, Z., Xiao, M. et al. (2017). *Adv. Energy Mater.* 7: 1700944.

33 Kawashima, K., Fukuhara, T., Suda, Y. et al. (2016). *J. Am. Chem. Soc.* 138: 10265.

34 Kawashima, K., Tamai, Y., Ohkita, H. et al. (2015). *Nat. Commun.* 6: 10085.

35 Nielsen, C.B., Ashraf, R.S., Treat, N.D. et al. (2015). *Adv. Mater.* 27: 948.

36 Wang, C., Xu, X., Zhang, W. et al. (2016). *Adv. Energy Mater.* 6: 1600148.

37 Bianchi, L., Zhang, X., Chen, Z. et al. (2019). *Chem. Mater.* 31: 6519.

38 Zhou, P., Zhang, Z.G., Li, Y. et al. (2014). *Chem. Mater.* 26: 3495.

39 Wee, X.K., Yeo, W.K., Zhang, B. et al. (2009). *Bioorg. Med. Chem.* 17: 7562.

40 Yuen, J.D. and Wudl, F. (2013). *Energy Environ. Sci.* 6: 392.

41 Mei, J., Graham, K.R., Stalder, R., and Reynolds, J.R. (2010). *Org. Lett.* 12: 660.

42 Wang, E., Mammo, W., and Andersson, M.R. (2014). *Adv. Mater.* 26: 1801.

43 Zhang, G., Fu, Y., Xie, Z., and Zhang, Q. (2011). *Macromolecules* 44: 1414.

44 Wang, E., Ma, Z., Zhang, Z. et al. (2010). *Chem. Commun.* 47: 4908.

45 Stalder, R., Mei, J., Subbiah, J. et al. (2011). *Macromolecules* 44: 6303.

46 Ma, Z., Wang, E., Vandewal, K. et al. (2011). *Appl. Phys. Lett.* 99: 143302.

47 Ma, Z., Sun, W., Himmelberger, S. et al. (2014). *Energy Environ. Sci.* 7: 361.

48 Xu, X., Cai, P., Lu, Y. et al. (2013). *J. Polym. Sci., Part A: Polym. Chem.* 51: 424.
49 Cao, K., Wu, Z., Li, S. et al. (2013). *J. Polym. Sci., Part A: Polym. Chem.* 51: 94.
50 Fang, L., Zhou, Y., Yao, Y.-X. et al. (2013). *Chem. Mater.* 25: 4874.
51 Jung, E.H. and Jo, W.H. (2014). *Energy Environ. Sci.* 7: 650.
52 Xu, X., Li, L., Liu, B., and Zou, Y. (2011). *Appl. Phys. Lett.* 98: 063303.
53 Zhang, M., Gu, Y., Guo, X. et al. (2013). *Adv. Mater.* 25: 4944.
54 Zheng, Y.Q., Wang, Z., Dou, J.H. et al. (2015). *Macromolecules* 48: 5570.
55 Hu, H., Jiang, K., Kim, J.-H. et al. (2016). *J. Mater. Chem. A* 4: 5039.
56 Deng, Y., Liu, J., Wang, J. et al. (2014). *Adv. Mater.* 26: 471.
57 Chen, M.S., Niskala, J.R., Unruh, D.A. et al. (2013). *Chem. Mater.* 25: 4088.
58 Yue, W., Ashraf, R.S., Nielsen, C.B. et al. (2015). *Adv. Mater.* 27: 4702.
59 Randell, N.M., Radford, C.L., Yang, J. et al. (2018). *Chem. Mater.* 30: 4864.
60 Nielsen, C.B., Turbiez, M., and McCulloch, I. (2013). *Adv. Mater.* 25: 1859.
61 Mei, J., Graham, K.R., Stalder, R. et al. (2011). *Chem. Mater.* 23: 2285.
62 Kanimozhi, C., Yaacobi-Gross, N., Chou, K.W. et al. (2012). *J. Am. Chem. Soc.* 134: 16532.
63 Chen, M.S., Lee, O.P., Niskala, J.R. et al. (2013). *J. Am. Chem. Soc.* 135: 19229.
64 van Franeker, J.J., Heintges, G.H.L., Schaefer, C. et al. (2015). *J. Am. Chem. Soc.* 137: 11783.
65 Li, W., Hendriks, K.H., Furlan, A. et al. (2013). *J. Am. Chem. Soc.* 135: 18942.
66 Choi, H., Ko, S.-J., Kim, T. et al. (2015). *Adv. Mater.* 27: 3318.
67 Hendriks, K.H., Li, W., Wienk, M.M., and Janssen, R.A. (2014). *J. Am. Chem. Soc.* 136: 12130.
68 Li, W., Furlan, A., Roelofs, W.S. et al. (2014). *Chem. Commun.* 50: 679.
69 Sonar, P., Singh, S.P., Li, Y. et al. (2014). *Energy Environ. Sci.* 4: 2288.
70 Zoombelt, A.P., Mathijssen, S.G.J., Turbiez, M.G.R. et al. (2010). *J. Mater. Chem.* 20: 2240.
71 Jo, J., Gendron, D., Najari, A. et al. (2010). *Appl. Phys. Lett.* 97: 203303.
72 Peng, Q., Huang, Q., Hou, X.B. et al. (2012). *Chem. Commun.* 48: 11452.
73 Dou, L., You, J., Yang, J. et al. (2012). *Nat. Photonics* 6: 180.
74 Dou, L., Gao, J., Richard, E. et al. (2012). *J. Am. Chem. Soc.* 134: 10071.
75 Li, W., Hendriks, K.H., Furlan, A. et al. (2014). *Adv. Mater.* 26: 1565.
76 Li, W., Hendriks, K.H., Furlan, A. et al. (2015). *J. Am. Chem. Soc.* 137: 2231.
77 Dou, L., Chang, W.H., Gao, J. et al. (2013). *Adv. Mater.* 25: 825.
78 Ashraf, R.S., Meager, I., Nikolka, M. et al. (2015). *J. Am. Chem. Soc.* 137: 1314.
79 Jung, J.W., Liu, F., Russell, T.P., and Jo, W.H. (2013). *Chem. Commun.* 49: 8495.
80 Bijleveld, J.C., Zoombelt, A.P., Mathijssen, S.G.J. et al. (2009). *J. Am. Chem. Soc.* 131: 16616.
81 Yiu, A.T., Beaujuge, P.M., Lee, O.P. et al. (2012). *J. Am. Chem. Soc.* 134: 2180.
82 Hendriks, K.H., Li, W., Heintges, G.H.L. et al. (2014). *J. Am. Chem. Soc.* 136: 11128.
83 Feng, K., Yang, G., Xu, X. et al. (2018). *Adv. Energy Mater.* 8: 1602773.
84 Zhang, G., Xu, X., Bi, Z. et al. (2018). *Adv. Funct. Mater.* 28: 1706404.
85 Yao, H., Ye, L., Zhang, H. et al. (2016). *Chem. Rev.* 116: 7397.
86 Hou, J., Park, M.H., Zhang, S. et al. (2008). *Macromolecules* 41: 6012.

87 Subbiah, J., Purushothaman, B., Chen, M. et al. (2015). *Adv. Mater.* 27: 702.
88 Huo, L., Hou, J., Zhang, S. et al. (2010). *Angew. Chem.* 122: 1542.
89 Peng, Q., Liu, X., Su, D. et al. (2011). *Adv. Mater.* 23: 4554.
90 Chen, W., Xiao, M., Han, L. et al. (2016). *ACS Appl. Mater. Interfaces* 8: 19665.
91 Stuart, A.C., Tumbleston, J.R., Zhou, H. et al. (2013). *J. Am. Chem. Soc.* 135: 1806.
92 Dong, Y., Yang, H., Wu, Y. et al. (2019). *J. Mater. Chem. A* 7: 2261.
93 Shin, J., Kim, M., Lee, J. et al. (2015). *RSC Adv.* 5: 106044.
94 Zhang, M., Gu, Y., Guo, X. et al. (2013). *Adv. Mater.* 25: 4944.
95 Gong, X., Li, G., Li, C. et al. (2015). *J. Mater. Chem. A* 3: 20195.
96 Shin, J., Kim, M., Kang, B. et al. (2017). *J. Mater. Chem. A* 5: 16702.
97 Chen, W., Du, Z., Han, L. et al. (2015). *J. Mater. Chem. A* 3: 3130.
98 Shin, J., Kim, M., Lee, J. et al. (2017). *Chem. Commun.* 53: 1176.
99 Lee, J., Kim, J.-H., Moon, B. et al. (2015). *Macromolecules* 48: 1723.
100 Wang, J., Xiao, M., Chen, W. et al. (2014). *Macromolecules* 47: 7823.
101 Li, Z., Zhang, T., Xin, Y. et al. (2016). *J. Mater. Chem. A* 4: 18598.
102 Liu, D., Zhu, Q., Gu, C. et al. (2016). *Adv. Mater.* 28: 8490.
103 Liu, D., Wang, J., Gu, C. et al. (2018). *Adv. Mater.* 30: 1705870.
104 Ren, J., Bao, X., Han, L. et al. (2015). *Polym. Chem.* 6: 4415.
105 Duan, C., Furlan, A., van Franeker, J.J. et al. (2015). *Adv. Mater.* 27: 4461.
106 Kranthiraja, K., Gunasekar, K., Cho, W. et al. (2015). *J. Mater. Chem. C* 3: 796.
107 Fan, L., Cui, R., Guo, X. et al. (2014). *J. Mater. Chem. C* 2: 5651.
108 Huo, L., Huang, Y., Fan, B. et al. (2012). *Chem. Commun.* 48: 3318.
109 Zhang, Y., Gao, L., He, C. et al. (2013). *Polym. Chem.* 4: 1474.
110 Ashraf, R.S., Schroeder, B.C., Bronstein, H.A. et al. (2013). *Adv. Mater.* 25: 2029.
111 Fei, Z., Ashraf, R.S., Huang, Z. et al. (2012). *Chem. Commun.* 48: 2955.
112 Xu, Y.X., Chueh, C.C., Yip, H.L. et al. (2012). *Adv. Mater.* 24: 6356.
113 Hu, H., Jiang, K., Yang, G. et al. (2015). *J. Am. Chem. Soc.* 137: 14149.
114 Nguyen, T.L., Choi, H., Ko, S.J. et al. (2014). *Energy Environ. Sci.* 7: 3040.
115 Kang, H., Uddin, M.A., Lee, C. et al. (2015). *J. Am. Chem. Soc.* 137: 2359.
116 An, M., Xie, F., Geng, X. et al. (2017). *Adv. Energy Mater.* 7: 1602509.
117 Li, D., Xiao, Z., Wang, S. et al. (2018). *Adv. Energy Mater.* 8: 1800397.
118 Chen, H., Hu, Z., Wang, H. et al. (2018). *Joule* 2: 1623.
119 Feng, L.W., Chen, J., Mukherjee, S. et al. (2020). *ACS Energy Lett.* 5: 1780.
120 Xu, X., Li, K., Li, Z. et al. (2015). *Polym. Chem.* 6: 2337.
121 Xu, X.P., Wu, Y.L., Fang, J.F. et al. (2014). *Chem. Eur. J.* 20: 13259.
122 Wu, H., Zhao, B., Wang, W. et al. (2015). *J. Mater. Chem. A* 3: 18115.
123 Wang, M., Ma, D., Shi, K. et al. (2015). *J. Mater. Chem. A* 3: 2802.
124 Fan, Q., Xiao, M., Liu, Y. et al. (2015). *Polym. Chem.* 6: 4290.
125 Fan, Q., Jiang, H., Liu, Y. et al. (2016). *J. Mater. Chem. C* 4: 2606.
126 Yuan, J., Qiu, L., Zhang, Z.G. et al. (2016). *Nano Energy* 30: 312.
127 Xu, S., Feng, L., Yuan, J. et al. (2017). *ACS Appl. Mater. Interfaces* 9: 18816.
128 Qiao, Z., Wang, M., Zhao, M. et al. (2015). *Polym. Chem.* 6: 8203.
129 Putri, S.K., Kim, Y.H., Whang, D.R. et al. (2017). *Org. Electron.* 47: 14.
130 Zheng, Z., Awartani, O.M., Gautam, B. et al. (2017). *Adv. Mater.* 29: 1604241.

131 Zhang, Y., Ren, F., Li, Q. et al. (2018). *J. Mater. Chem. C* 6: 4658.
132 Li, G., Lu, Z., Li, C., and Bo, Z. (2015). *Polym. Chem.* 6: 1613.
133 Fang, T., Lu, Z., Lu, H. et al. (2015). *Polymer* 71: 43.
134 Zhang, F., Bijleveld, J., Perzon, E. et al. (2008). *J. Mater. Chem.* 18: 5468.
135 Liu, L., Zhang, G., He, B. et al. (2017). *Mater. Chem. Front.* 1: 499.
136 Yu, T., Xu, X., Li, Y. et al. (2017). *ACS Appl. Mater. Interfaces* 9: 18142.
137 Zou, Y., Najari, A., Berrouard, P. et al. (2010). *J. Am. Chem. Soc.* 132: 5330.
138 Hu, X., Shi, M., Zuo, L. et al. (2011). *Polymer* 52: 2559.
139 Zhang, Y., Zou, J., Yip, H.-L. et al. (2011). *J. Mater. Chem.* 21: 3895.
140 Li, Z., Tsang, S.W., Du, X. et al. (2011). *Adv. Funct. Mater.* 21: 3331.
141 Chu, T.Y., Lu, J., Beaupre, S. et al. (2011). *J. Am. Chem. Soc.* 133: 4250.
142 Chu, T.-Y., Lu, J., Beaupré, S. et al. (2012). *Adv. Funct. Mater.* 22: 2345.
143 Xu, B., Yi, X., Huang, T.-Y. et al. (2018). *Adv. Funct. Mater.* 28: 1803418.
144 Amb, C.M., Chen, S., Graham, K.R. et al. (2011). *J. Am. Chem. Soc.* 133: 10062.
145 Fei, Z., Ashraf, R.S., Han, Y. et al. (2015). *J. Mater. Chem. A* 3: 1986.
146 Zhong, H., Li, Z., Deledalle, F. et al. (2013). *J. Am. Chem. Soc.* 135: 2040.
147 Schroeder, B.C., Ashraf, R.S., Thomas, S. et al. (2011). *Chem. Commun.* 48: 7699.
148 Wang, D.H., Pron, A., Leclerc, M., and Heeger, A.J. (2013). *Adv. Funct. Mater.* 23: 1297.
149 Huang, J., Zhao, Y., He, W. et al. (2012). *Polym. Chem.* 3: 2832.
150 Wang, K., Xu, Z., Guo, B. et al. (2016). *RSC Adv.* 6: 63338.
151 Qiao, X., Chen, W., Wu, Q. et al. (2017). *Chem. Commun.* 53: 3543.
152 Qiao, X., Wu, Q., Wu, H. et al. (2017). *Adv. Funct. Mater.* 27: 1604286.
153 Piliego, C., Holcombe, T.W., Douglas, J.D. et al. (2010). *J. Am. Chem. Soc.* 132: 7595.
154 Cabanetos, C., Labban, A.E., Bartelt, J.A. et al. (2013). *J. Am. Chem. Soc.* 135: 4656.
155 Yuan, J., Dong, H., Li, M. et al. (2014). *Adv. Mater.* 26: 3624.
156 Zhang, C., Zhou, Z., Wu, H. et al. (2017). *Adv. Electron. Mater.* 3: 1700213.
157 Zhao, J., Li, Q., Liu, S. et al. (2020). *ACS Energy Lett.* 5: 367.
158 Lu, L. and Yu, L. (2014). *Adv. Mater.* 26: 4413.
159 Chen, H.Y., Hou, J.H., Zhang, S.Q. et al. (2009). *Nat. Photonics* 3: 649.
160 Liang, Y., Xu, Z., Xia, J. et al. (2010). *Adv. Mater.* 22: E135.
161 Huang, Y., Huo, L., Zhang, S. et al. (2011). *Chem. Commun.* 47: 8904.
162 Son, H.J., Wang, W., Xu, T. et al. (2011). *J. Am. Chem. Soc.* 133: 1885.
163 Wang, H., Chao, P., Chen, H. et al. (2017). *ACS Energy Lett.* 2: 1971.
164 Chang, C., Li, W., Guo, X. et al. (2018). *Org. Electron.* 58: 82.
165 Li, W., Guo, B., Chang, C. et al. (2016). *J. Mater. Chem. A* 4: 10135.
166 Chao, P., Wang, H., Qu, S. et al. (2017). *Macromolecules* 50: 9617.
167 Chao, P., Wang, H., Mo, D. et al. (2018). *J. Mater. Chem. A* 6: 2942.
168 Huang, Y., Guo, X., Liu, F. et al. (2012). *Adv. Mater.* 24: 3383.
169 Zhang, M., Guo, X., Zhang, S., and Hou, J. (2014). *Adv. Mater.* 26: 1118.
170 Zhang, M., Guo, X., Ma, W. et al. (2014). *Adv. Mater.* 26: 2089.
171 Liu, F., Zhou, Z., Zhang, C. et al. (2017). *Adv. Mater.* 29: 1606574.

172 Zhu, D., Bao, X., Ouyang, D. et al. (2017). *Nano Energy* 40: 495.
173 Son, H.J., Lu, L., Chen, W. et al. (2013). *Adv. Mater.* 25: 838.
174 Koti, R.S., Sanjaykumar, S.R., Hong, S.J. et al. (2013). *Sol. Energy Mater. Sol. Cells* 108: 213.
175 Bathula, C., Song, C.E., Badgujar, S. et al. (2013). *Polym. Chem.* 4: 2132.
176 Ma, Y., Kang, Z., and Zheng, Q. (2017). *J. Mater. Chem. A* 5: 1860.
177 Xu, X.P., Zhang, G.J., Li, Y., and Peng, Q. (2019). *Chin. Chem. Lett.* 30: 809.
178 Fan, Q., Méndez-Romero, U.A., Guo, X. et al. (2019). *Chem. Eur. J.* 14: 3085.
179 Kim, Y., Choulis, S.A., Nelson, J. et al. (2005). *Appl. Phys. Lett.* 86: 063502.
180 Guo, X., Cui, C.H., Zhang, M.J. et al. (2012). *Energy Environ. Sci.* 5: 7943.
181 Xu, X.P., Zhang, G.J., Yu, L.Y. et al. (2019). *Adv. Mater.* 31: 1906045.
182 Yang, C., Zhang, S., Ren, J. et al. (2020). *Energy Environ. Sci.* 13: 2864.
183 Zhang, Z.G., Zhang, S., Min, J. et al. (2012). *Macromolecules* 45: 113.
184 Zhang, M., Guo, X., Ma, W. et al. (2014). *Adv. Mater.* 26: 5880.
185 Wang, Q., Li, M., Zhang, X. et al. (2019). *Macromolecules* 52: 4464.
186 Qian, D., Ma, W., Li, Z. et al. (2013). *J. Am. Chem. Soc.* 135: 8464.
187 Fan, Q., Su, W., Guo, X. et al. (2016). *Adv. Energy Mater.* 6: 1600430.
188 Qin, Y., Uddin, M.A., Chen, Y. et al. (2016). *Adv. Mater.* 28: 9416.
189 Su, W., Fan, Q., Guo, X. et al. (2016). *J. Mater. Chem. A* 4: 14752.
190 Tumbleston, J.R., Collins, B.A., Yang, L. et al. (2008). *Nat. Photonics* 8: 385.
191 Firdaus, Y., Maffei, L.P., Cruciani, F. et al. (2017). *Adv. Energy Mater.* 7: 1700834.
192 Park, G.E., Choi, S., Park, S.Y. et al. (2017). *Adv. Energy Mater.* 7: 1700566.
193 Liu, Y., Lu, H., Li, M. et al. (2018). *Macromolecules* 51: 8646.
194 Yao, H.F., Cui, Y., Qian, D.P. et al. (2019). *J. Am. Chem. Soc.* 141: 7743.
195 Li, S., Ye, L., Zhao, W. et al. (2018). *J. Am. Chem. Soc.* 140: 7159.
196 Chen, J., Wang, L., Yang, J. et al. (2019). *Macromolecules* 52: 341.
197 Zhang, B., Yu, Y., Zhou, J. et al. (2020). *Adv. Energy Mater.* 10: 1904247.
198 Fan, Q., Su, W., Guo, X. et al. (2017). *J. Mater. Chem. A* 5: 5106.
199 Zhang, X., Zhang, J., Lu, H. et al. (2015). *J. Mater. Chem. C* 3: 6979.
200 Li, G., Gong, X., Zhang, J. et al. (2016). *ACS Appl. Mater. Interfaces* 8: 3686.
201 Jiang, H., Wang, Z., Zhang, L. et al. (2017). *ACS Appl. Mater. Interfaces* 9: 36061.
202 Liu, Y., Chen, S., Zhang, G. et al. (2017). *J. Mater. Chem. A* 5: 15017.
203 Zeng, A.P., Ma, X.L., Pan, M.A. et al. (2021). *Adv. Funct. Mater.* 31: 2102413.
204 Min, J., Zhang, Z.G., Zhang, S., and Li, Y. (2012). *Chem. Mater.* 24: 3247.
205 Gao, L., Zhang, Z.G., Xue, L.W. et al. (2016). *Adv. Mater.* 28: 1884.
206 Bin, H.J., Zhang, Z.G., Gao, L. et al. (2016). *J. Am. Chem. Soc.* 138: 4657.
207 Fan, Q., Su, W., Guo, X. et al. (2017). *J. Mater. Chem. A* 5: 9204.
208 Bin, H.J., Gao, L., Zhang, Z.G. et al. (2016). *Nat. Commun.* 7: 13651.
209 Price, S.C., Stuart, A.C., Yang, L. et al. (2011). *J. Am. Chem. Soc.* 133: 4625.
210 Fan, Q., Su, W., Meng, X. et al. (2017). *Solar RRL* 1: 1700020.
211 Xue, L., Yang, Y., Xu, J. et al. (2017). *Adv. Mater.* 29: 1703344.
212 Su, W., Li, G., Fan, Q. et al. (2019). *J. Mater. Chem. A* 7: 2351.

213 Li, G.D., Xu, Q.Q., Chang, C.M. et al. (2019). *Macromol. Rapid Commun.* 40: 1800660.
214 Zhao, Q., Qu, J., and He, F. (2020). *Adv. Sci.* 7: 2000509.
215 Chen, W., Huang, G., Li, X. et al. (2018). *ACS Appl. Mater. Interfaces* 10: 42747.
216 Liu, D., Zhang, K., Zhong, Y. et al. (2018). *J. Mater. Chem. A* 6: 18125.
217 Li, X., Weng, K., Ryu, H.S. et al. (2020). *Adv. Funct. Mater.* 30: 1906809.
218 Chen, S., Liu, Y., Zhang, L. et al. (2017). *J. Am. Chem. Soc.* 139: 6298.
219 Fan, B.B., Zhu, P., Xin, J.M. et al. (2018). *Adv. Energy Mater.* 8: 1703085.
220 Zhu, L., Zhong, W.K., Qiu, C.Q. et al. (2019). *Adv. Mater.* 31: 1902899.
221 Fan, B.B., Du, X.Y., Liu, F. et al. (2018). *Nat. Energy* 3: 1051.
222 Fan, B.B., Zhang, D.F., Li, M.J. et al. (2019). *Sci. China Chem.* 62: 746.
223 Tang, D.S., Wan, J.H., Xu, X.P. et al. (2018). *Nano Energy* 53: 258.
224 Feng, K., Yuan, J., Bi, Z. et al. (2019). *iScience* 12: 1.
225 Zhang, M., Guo, X., Wang, X. et al. (2011). *Chem. Mater.* 23: 4264.
226 Wang, K., Guo, X., Guo, B. et al. (2016). *Macromol. Rapid Commun.* 37: 1066.
227 Guo, B., Li, W., Guo, X. et al. (2017). *Nano Energy* 34: 556.
228 Guo, B., Guo, X., Li, W. et al. (2016). *J. Mater. Chem. A* 4: 13251.
229 Guo, B., Li, W., Guo, X. et al. (2017). *Adv. Mater.* 29: 1702291.
230 Su, W., Fan, Q., Guo, X. et al. (2017). *Nano Energy* 38: 510.
231 Cao, Z.X., Chen, J.L., Liu, S.J. et al. (2019). *Chem. Mater.* 31: 8533.
232 Wen, S., Li, Y., Rath, T. et al. (2019). *Chem. Mater.* 31: 919.
233 Xu, X.P., Yu, T., Bi, Z.Z. et al. (2018). *Adv. Mater.* 30: 1703973.
234 Xu, X.P., Li, Z.J., Bi, Z.Z. et al. (2018). *Adv. Mater.* 30: 1800737.
235 Chen, K., Ma, R.J., Liu, T. et al. (2020). *J. Mater. Chem. A* 8: 7083.
236 Sun, C.K., Pan, F., Bin, H.J. et al. (2018). *Nat. Commun.* 9: 743.
237 Sun, C., Qin, S., Wang, R. et al. (2020). *J. Am. Chem. Soc.* 142: 1465.
238 Xu, Y., Cui, Y., Yao, H.F. et al. (2021). *Adv. Mater.* 33: 2101090.
239 Xu, X.P., Feng, K., Bi, Z.Z. et al. (2019). *Adv. Mater.* 31: 1901872.
240 Zheng, Z., Yao, H.F., Ye, L. et al. (2020). *Mater. Today* 35: 115.
241 Qian, D.P., Ye, L., Zhang, M.J. et al. (2012). *Macromolecules* 45: 9611.
242 Zhang, M.J., Guo, X., Ma, W. et al. (2015). *Adv. Mater.* 27: 4655.
243 Fan, Q.P., Zhu, Q.L., Xu, Z. et al. (2018). *Nano Energy* 48: 413.
244 Zhao, W.C., Li, S.S., Yao, H.F. et al. (2017). *J. Am. Chem. Soc.* 139: 7148.
245 Fei, Z., Eisner, F.D., Jiao, X. et al. (2018). *Adv. Mater.* 30: 1705209.
246 Wu, Y., An, C., Shi, L. et al. (2018). *Angew. Chem. Int. Ed.* 57: 12911.
247 Qin, Y.P., Ye, L., Zhang, S.Q. et al. (2018). *J. Mater. Chem. A* 6: 4324.
248 Liu, T., Huo, L.J., Chandrabose, S. et al. (2018). *Adv. Mater.* 30: 1707353.
249 Ye, L., Xie, Y., Weng, K. et al. (2019). *Nano Energy* 58: 220.
250 Li, W., Li, G., Guo, H. et al. (2019). *J. Mater. Chem. A* 7: 1307.
251 Xu, Z., Fan, Q., Meng, X. et al. (2017). *Chem. Mater.* 29: 4811.
252 Zhao, W.C., Qian, D.P., Zhang, S.Q. et al. (2016). *Adv. Mater.* 28: 4734.
253 Wang, Y., Fan, Q., Guo, X. et al. (2017). *J. Mater. Chem. A* 5: 22180.
254 Fan, Q., Wang, Y., Zhang, M. et al. (2018). *Adv. Mater.* 30: 1704546.
255 Fan, Q.P., Su, W.Y., Wang, Y. et al. (2018). *Sci. China Chem.* 61: 531.
256 Meng, Y., Wu, J., Guo, X. et al. (2019). *Sci. China Chem.* 62: 845.

257 Fan, Q., Su, W., Chen, S. et al. (2020). *Angew. Chem. Int. Ed.* 59: 19835.
258 Fan, Q.P., Su, W.Y., Chen, S.S. et al. (2020). *Joule* 4: 658.
259 Fan, Q., Ma, R., Liu, T. et al. (2020). *Solar RRL* 4: 2000142.
260 Yao, H., Bai, F., Hu, H. et al. (2019). *ACS Energy Lett.* 4: 417.
261 Wu, J., Meng, Y., Guo, X. et al. (2019). *J. Mater. Chem. A* 7: 16190.
262 Wang, W., Wu, Q., Sun, R. et al. (2020). *Joule* 4: 1070.
263 Yuan, J., Zhang, Y.Q., Zhou, L.Y. et al. (2019). *Joule* 3: 1140.
264 Ma, R.J., Liu, T., Luo, Z.H. et al. (2020). *ACS Energy Lett.* 5: 2711.
265 Li, C., Zhou, J.D., Song, J.L. et al. (2021). *Nat. Energy* 6: 605.
266 Fan, Q., Xu, Z., Guo, X. et al. (2017). *Nano Energy* 40: 20.
267 Chao, P., Chen, H., Zhu, Y. et al. (2020). *Adv. Mater.* 32: 1907059.
268 Huo, L.J., Liu, T., Sun, X.B. et al. (2015). *Adv. Mater.* 27: 2938.
269 Xu, X.P., Feng, K., Lee, Y.W. et al. (2020). *Adv. Funct. Mater.* 30: 1907570.
270 Xu, X.P., Feng, K., Yu, L.Y. et al. (2020). *ACS Energy Lett.* 5: 2434.
271 Dang, D., Yu, D., and Wang, E. (2019). *Adv. Mater.* 31: 1807019.
272 Cui, Y., Yao, H., Hong, L. et al. (2019). *Adv. Mater.* 31: 1808356.
273 Sun, H.L., Liu, T., Yu, J.W. et al. (2019). *Energy Environ. Sci.* 12: 3328.
274 Wang, T., Sun, R., Shi, M.M. et al. (2020). *Adv. Energy Mater.* 10: 2000590.
275 Cao, J.M., Liao, Q.G., Du, X.Y. et al. (2013). *Energy Environ. Sci.* 6: 3224.
276 Liao, C.Y., Chen, Y., Lee, C.C. et al. (2020). *Joule* 4: 189.
277 Liu, J., Liu, L., Zuo, C. et al. (2019). *Sci. Bull.* 64: 1655.
278 Xiong, J., Jin, K., Jiang, Y.F. et al. (2019). *Sci. Bull.* 64: 1573.
279 Wang, T., Qin, J., Xiao, Z. et al. (2020). *Nano Energy* 77: 105161.
280 Zhang, G.Y., Ning, H.J., Chen, H. et al. (2021). *Joule* 5: 1.

2

p-Type Molecular Photovoltaic Materials

Qihui Yue[1,2] *and Xiaozhang Zhu*[1,2]

[1] *Chinese Academy of Sciences, Institute of Chemistry, Beijing National Laboratory for Molecular Sciences, CAS Key Laboratory of Organic Solids, Zhongguancun North First Street 2, Haidian District, Beijing 100190, P.R. China*
[2] *University of Chinese Academy of Sciences, School of Chemistry and Chemical Engineering, Yuquan Road No. 19(A), Shijingshan District, Beijing 100049, P.R. China*

2.1 Introduction

Organic solar cells (OSCs) have unique advantages such as light weight, flexibility, diverse material choices, and large-area processability and have received wide attention in recent years [1–3]. The research on active layer materials is the key and basis of the boost of OSCs. According to the type of active layer materials, OSCs can be divided into polymer type (PSCs) and small-molecule type (SMSCs). Compared with polymers, small molecules have distinct characteristics: (i) the well-defined molecular structure with definite molecular weight that avoids batch-to-batch problems, ensuring the device performance reproducibility, benefiting the understanding of structure–property relationship; (ii) the synthesis simplicity that makes the structure highly controllable; and (iii) the strong crystallinity that brings a high mobility, which is conducive to charge transport [4, 5]. The research on SMSCs can be traced back to 1959 [6]. Anthracene crystal was sandwiched between two electrodes, but due to the poor exciton dissociation, the device gave a low power conversion efficiency (PCE). Then, Tang reported a planar heterojunction (PHJ) device, which was the key breakthrough, in which copper phthalocyanine (CuPc) and perylenetetracarboxylic bis-benzimidazole (PTCBI) were vacuum-deposited layer by layer [7]. In this device, excitons were dissociated at the interface of the two layers, and PCE reached 1.0%, indicating that the combination of electron donors and electron acceptors could effectively increase the exciton dissociation and thus the PCEs. Since then, the development of *p*-type and *n*-type molecular photovoltaic materials, also called small-molecule donor materials (SMDs) and small-molecule acceptor materials (SMAs), has been a primary issue in SMSCs. In the early stage, researchers mainly developed materials, especially SMDs, for vacuum-deposited

Organic Solar Cells: Materials Design, Technology and Commercialization, First Edition. Edited by Liming Ding.

small-molecule solar cells (VD-SMSCs). These materials often exhibited good thermal stability and proper molecular weight. Then, since bulk heterojunction (BHJ) structure that could greatly increase the interface area between the donor and acceptor and benefit the exciton dissociation was proposed, solution-processed PSCs have developed rapidly [8]. However, for solution-processed small-molecule solar cells (SP-SMSCs), the poor film-forming property and low solubility as well as the low molar absorption coefficients of traditional SMDs limited the development, and much effort has been devoted to overcome these shortcomings and great progress was achieved. In the research of SMSCs based on fullerene acceptors, PCEs of over 10% have been realized in both VD- and SP-SMSCs [9–13]. Later, high-performance nonfullerene acceptors (NFAs) promoted the PCEs of SP-SMSCs to over 15% [14–18].

In this chapter, some representative SMDs used in fullerene-based SMSCs will be introduced, including their design, photoelectronic property, and photovoltaic performance. Current progress on SMSCs based on NFAs will also be stated. Finally, aiming at the future of SMSCs, challenges and possibility will be discussed.

2.2 *p*-Type Molecular Photovoltaic Materials Used in Vacuum-Deposited Solar Cells

Vacuum deposition refers to the process of vaporizing the material with thermal evaporation in vacuum, then depositing the material onto the substrate. This also reflects the requirements of materials used: suitable molecular weight and considerable thermal stability. Utilizing vacuum deposition has unique advantages: (i) VD-SMSCs can take advantage of the mature technology of organic light-emitting diodes (OLEDs) but the requirements for the mask need not to be so harsh; (ii) the film formation is simple, especially for multilayer devices. And the organic solvent use and liquid waste treatment can be avoided; (iii) the thickness of the active layers and interlayers can be precisely controlled, even with a thickness of only a few nanometers; (iv) the film morphology can be well controlled by regulating the deposition conditions; (v) the film is of good quality with high purity, compactness, and reproducibility, favoring high and steady performance; and (vi) good device stability can be expected [19].

In VD-SMSCs, fullerenes C_{60} and C_{70} as acceptors significantly improved the PCE because of their great thermal stability, high electron affinity, strong aggregation, as well as high electron mobility. The device structure optimization, such as BHJ and planar-mixed heterojunction (PM-HJ) structures, also did contribution. Nowadays, VD-SMSCs have realized PCEs of over 10% [9–11], and it deserves to be mentioned that Heliatek GmbH, which devotes much to the organic photovoltaic commercialization, announced a multijunction VD-SMSC with a high PCE of 13.2% in 2016 [20]. Among the various SMDs, oligothiophenes (OTs), D–A–A′ type SMDs (D and A represent electron-donating and -accepting moieties, representatively), borondipyrromethenes (BODIPYs) are the most common, which will be described in detail.

2.2.1 Oligothiophene-Based Molecular Donors

Thiophene is an electron-rich unit and OT derivatives are widely used in organic optoelectronic devices due to the great charge transport property. Quinoidization and D–A strategies can easily tune the photoelectronic properties of OT-based SMDs.

In A–D–A-type OTs, the push-pull effect between D and A can enhance the intramolecular charge transfer (ICT) effect and red-shift the absorption, favoring a high short-circuit current density (J_{sc}), and the highest occupied molecular orbital (HOMO) or the lowest unoccupied molecular orbital (LUMO) energy levels can be controlled separately by changing D or A, respectively, favoring for a high open-circuit voltage (V_{oc}), and the side chains can greatly influence the molecular stacking in the thin film. One representative material, DCV5T-$Bu_{2,2,4,4}$ with *n*-butyl-substituted quinquethiophene as D and dicyanovinyl (DCV) as A (the number in subscript indicates the substituent's position), was reported by Bäuerle and coworkers [21]. It exhibited an optical bandgap (E_g^{opt}) of 1.77 eV in the solid state and well-matched energy levels with C_{60}, yielding an outstanding PCE of 3.4% in the device, which was further promoted to 5.80% through buffer layer engineering [22]. To investigate the structure–property relationship more detailly, efforts were devoted to varying the conjugation length, and the side chain. Research on the terminal effect was also executed [23].

Pfeiffer and coworkers reported DCV*n*Ts-Bu ($n = 3_{2,2}, 4_{1,1,4,4}, 5_{2,2,4,4}, 6_{2,2,5,5}$) [22]. With the conjugation length increasing, LUMO energy level fluctuated tinnily but HOMO energy level was significantly upshifted, which might lead to a decrease in V_{oc} but more efficient charge generation. DCV3T-$Bu_{2,2}$-based and DCV4T-$Bu_{1,1,4,4}$-based devices gave poor PCEs, while the other two devices showed small differences in J_{sc} and fill factor (FF), but the high V_{oc} of DCV5T-$Bu_{2,2,4,4}$-based device led to its best PCE of 5.80%. Bäuerle designed DCV*n*Ts (n = 1–6) without any side chain [24]. With the n increasing, maximum absorptions red-shifted, accompanied with enhanced molar extinction coefficients. Since the energy levels of DCV*n*Ts (n = 4–6) were well matched with C_{60}, they were chosen to fabricate PHJ devices, and PCEs of 1.2%, 2.6%, 2.8% were realized, respectively. By using BHJ device structure, the PCE of DCV5T-based device was promoted to 5.2%. Bridging the thiophenes of DCV3T with nitrogen atoms offered DCV-SN5-1, -2, and -3, among which the difference was the substituent on the nitrogen (propyl/hexyl/tolyl) [25]. DCV-SN5 SMDs exhibited decreased E_g^{opt}s and increased molar extinction coefficients because of the planarization and rigidification of the central core, but DCV-SN5-3 exhibited a weaker and blue-shifted absorption compared with the other two, which was due to a distortion between the bulky substituent and the backbone. In blends, DCV-SN5-1-based one exhibited the most obvious interconnected networks, benefiting charge transport and leading to the best PCE of 6.5%. Linking two methyl and two ethyl groups on the terminal thiophenes of DCV4T, DCV4T-Me, and -Et were offered [26]. DCV4T-Me showed the most obvious red-shift between the absorptions of solution and film, indicating its better ordering in the film. Its single crystal showed the best coplanarity with strong π–π interactions while in the blend, DCV4T-Me could form a crystalline homo-phase with similar molecular packing, leading to a high PCE of 3.8%. Based on DCV5T-$Bu_{2,2,4,4}$, by replacing the *n*-butyl with ethyl, DCV5T-$Et_{2,2,4,4}$ had similar photoelectronic properties but lower hole mobility, leading to a lower PCE of 2.5% [27]. While replacing with methyl and regulating their position and number, DCV5T-Me($_{1,1,5,5}$, $_{2,2,4,4}$, and $_{3,3}$) were obtained [28]. Comparing the absorptions between the solutions and films, red-shifted maximum absorption peaks (50–72 nm) were observed, among which the red-shift of DCV5T-$Me_{3,3}$ was most obvious, indicating its highly ordered stacking. From the DCV5T-$Me_{1,1,5,5}$ to -$Me_{3,3}$-based blend films, crystallites of both donor and acceptor became bigger and phase separation became larger, favoring charge transport, and the PCE was promoted from 4.8% to 6.9% with significantly increased J_{sc}. And based on DCV5T-$Me_{3,3}$, a record PCE of 9.7% was realized in a triple-junction homo-tandem device [29]. DCV6T-$Et_{2,2,5,5}$ and -$Bu_{1,2,5,6}$ were both derived from DCV6T-$Bu_{2,2,5,5}$, giving PCEs of 3.8% and 4.9%, respectively [30, 31]. It is worth noting that heat substrate was needed during the fabrication of the devices above, aiming at the formation of better phase separation.

Another type is α-OT. α-6T is one of the most representative examples. α-6T exhibited high-lying energy levels, therefore, when it was matched with C_{60} with a low-lying LUMO energy level, due to the large energy offsets and limited absorption, the performance was not ideal, with a PCE of only 1.5% in the PHJ device [32].

OT-based SMDs (as shown in Figure 2.1 and Table 2.1) perform quite well in VD-SMSCs; however, the main weakness is that the film needs to be quite thin, otherwise FF will suffer. Therefore, the mobility should be increased to ensure that the active layer can harvest sufficient light while having little effect on FF.

2.2.2 D-A-A′-Type Molecular Donors

In D–A–A′-type SMDs, ICT effect can be adjusted to a desired degree by manipulating the three groups. These SMDs exhibit relatively large dipole moments along the backbone and tend to stack into an antiparallel dimeric configuration parallel to the substrate, facilitating charge hopping and vertical charge transport. In the blends, favorable phase separation can be formed due to the homo-aggregation of these SMDs.

In 2011, Wong and coworkers designed DTDCTB, in which the dihedral angle between the thiophene of diarylaminothienyl and benzothiadiazole (BT) units was only 5.5°, indicating the good coplanarity and enhancing the ICT [33]. The absorption of DTDCTB was extended to the near-infrared (NIR) region, greatly increasing the light harvest. And due to the strong electron-withdrawing ability of BT and DCV units, a low-lying HOMO energy level was observed, favoring a high V_{oc}. The relevant PM-HJ device achieved a PCE of 5.81% with a V_{oc} of 0.79 V, a J_{sc} of 14.68 mA cm^{-2}, and an FF of 0.50. Using a BHJ structure, an improved PCE of 8.0% was given [34]. Further applying it as the front cell in a triple-junction device that contains two DBP:C_{70}-based subcells, a PCE of 11.1% was achieved [9]. Modifying the diarylaminothienyl unit, DPDCTB, DTDCPB, and DPDCPB were obtained [35]. Compared with DTDCTB, DTDCPB with diarylaminophenyl as D unit exhibited a highly distorted conformation with a dihedral angle of 24.7°.

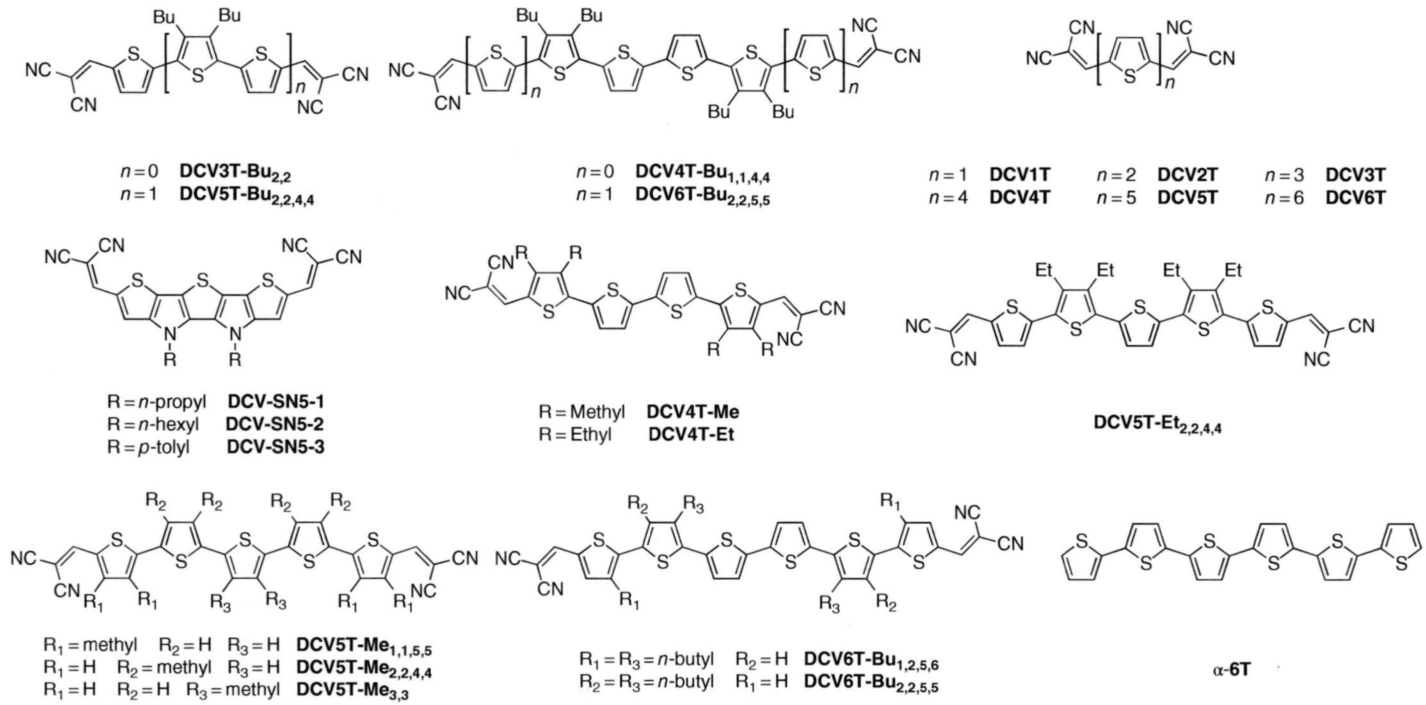

Figure 2.1 Molecular structures of the OT-based SMDs. Source: Qihui Yue.

Table 2.1 Performance of the OT-based SMDs.

Donor	E_g^{opt} (eV)	V_{oc} (V)	J_{sc} (mA cm^{-2})	FF	PCE (%)	References
DCV5T-$Bu_{2,2,4,4}$	1.77	0.98	10.6	0.49	3.4	[21]
	1.77	1.00	11.4	0.51	5.76	[22]
DCV4T-$Bu_{1,1,4,4}$	—	1.13	5.9	0.28	1.87	[22]
DCV6T-$Bu_{2,2,5,5}$	—	0.93	7.7	0.46	3.29	[22]
DCV4T	1.78	0.97	2.9	0.42	1.2	[24]
DCV5T	1.69	0.97	5.1	0.52	2.6	[24]
DCV6T	1.68	0.91	4.8	0.64	2.8	[24]
DCV-SN5-1	1.69	0.92	10.20	0.69	6.5	[25]
DCV-SN5-2	1.73	0.95	7.97	0.49	3.7	[25]
DCV-SN5-3	1.75	0.95	1.94	0.40	0.7	[25]
DCV4T-Me	1.77	0.98	6.5	0.59	3.8	[26]
DCV4T-Et	1.84	0.98	5.7	0.39	2.2	[26]
DCV5T-$Et_{2,2,4,4}$	—	1.00	8.4	0.40	2.5	[27]
DCV5T-$Me_{1,1,5,5}$	1.72	0.91	9.6	0.63	4.8	[28]
DCV5T-$Me_{2,2,4,4}$	1.69	0.95	9.4	0.62	4.8	[28]
DCV5T-$Me_{3,3}$	1.69	0.95	11.5	0.63	6.9	[28]
DCV6T-$Et_{2,2,5,5}$	—	0.90	6.5	0.64	3.8	[30]
DCV6T-$Bu_{1,2,5,6}$	1.70	0.89	10.9	0.61	4.9	[31]
α-6T	2.2	0.68	5.6	0.39	1.5	[32]

Due to the electron-rich nature and the better quinoidal propensity of thiophene, DPDCTB and DTDCTB showed red-shifted absorptions and higher lying HOMO energy levels compared with DPDCPB and DTDCPB, respectively. Also, due to the strong electron-donating ability of *p*-tolyl, DTDCPB and DTDCTB exhibited red-shifted absorptions and higher lying HOMO energy levels compared with DPDCPB and DPDCTB, respectively. Ultimately, DTDCPB-based device obtained a balance between V_{oc} and J_{sc}, delivering the highest PCE of 6.8%. By further optimizing the device structure, DTDCPB-based device achieved an outstanding PCE of 9.6% [34]. Combining it with a solution-processed cell, an excellent PCE of 15% was stated [36]. Wong and coworkers also reported series SMDs DTCPB, DTCTB, DTCPBO, and DPCTBO with cyano as terminal [37]. Due to the weaker electron-withdrawing ability of cyano, the LUMO energy levels were upshifted; meanwhile, the ICT effects were reduced, leading to blue-shifted absorptions and lower extinction coefficients. Compared with DTCPB and DTCTB, DTCPBO and DPCTBO exhibited red-shifted absorptions and lower-lying LUMO energy levels, attributed to the stronger electron deficient property of their A units. DTCPB-based blend showed the most suitable nanomorphology, reducing the recombination and demonstrating the highest FF, J_{sc}, and PCE (6.55%) in the device.

Coplanar dithienopyrrole (DTP) as D unit has also been introduced into this type of SMDs. Due to the lack of nitrogen contributions to the conjugated backbones, blue-shifted absorptions were observed, but larger absorption coefficients could compensate the light harvest loss. Linking *p*-tolyl substituted DTP with pyrimidine (PM)-DCV and BT-DCV, respectively, TDPM and TDBT were obtained, between which TDBT exhibited a smaller E_g^{opt} and a higher-lying HOMO energy level [38]. The relevant devices gave similar J_{sc}s and FFs, but the higher V_{oc} of TDPM-based one led to a higher PCE of 5.5%. Later, two isomers antiBTDC and synBTDC were offered by fusing a benzene ring with DTP and linking it with BT-DCV [39]. antiBTDC had a longer effective conjugation length, a smaller dihedral angle, and a smaller intermolecular interfacial distance and exhibited a cofacial π-stacking that made each D unit be close to an A unit in the dimer, leading to a more homogeneous electron density distribution. These facilitated intra- and intermolecular charge transfer, leading to a red-shifted absorption with an enhanced extinction coefficient and a higher-lying HOMO energy level. Also, the strong π–π interactions and the diminished molecular dipole of antiBTDC could facilitate the charge transport. The device based on antiBTDC delivered a PCE of 7.2%, with a V_{oc} of 0.91 V, a J_{sc} of 14.2 mA cm^{-2}, and an FF of 0.56. To further increase the V_{oc}, PYDC was designed by replacing the BT with PM, based on which the device gave a PCE of 5.1% with a high V_{oc} of 1.06 V. Starting from antiBTDC and varying the substituents offered *i*BuBTDC and *n*BuBTDC [40]. The two SMDs showed greater coplanarity and stacked more compactly, and such high stacking densities contributed to higher absorption coefficients. However, the orthogonal stack mode of *n*BuBTDC hindered the intermolecular charge transfer. *n*BuBTDC-based device exhibited comparable performance to antiBTDC-based one, while *i*BuBTDC-based one achieved a PCE of 9.3% with a higher FF of 0.60, attributed to the high ratio of face-on π–π stacking. Recently, bi-ternary devices with two BHJ layers: DTDCPB:C_{70} and *i*BuBTDC:C_{70} were constructed, achieving a PCE of 10% [11].

Molecular structures and performance parameters were shown in Figure 2.2 and Table 2.2. This type of SMDs often show higher J_{sc}S due to the smaller E_g^{opt}s, and the device performance has a great correlation with the molecular stacking, phase separation, as well as the device structure. The improvement of FFs is a focus and increasing hole mobility to realize more balanced carrier transport is favorable.

2.2.3 Borondipyrromethenes-Based Molecular Donors

BODIPYs exhibit a variety of modification positions, making the photoelectronic property highly tunable, and they show strong NIR absorptions and delocalized molecular orbitals, making them promising photovoltaic molecular donors.

Ph_2-benz-BODIPY is a typical example, and its absorption could reach 900 nm. And it gave PCEs of 3.8% and 2.3% in the opaque and semitransparent devices, respectively [41, 42]. To make better use of NIR light, Leo and coworkers extended the conjugation length, introduced trifluoromethyl, and synthesized BDP-H, BDP-Me, BDP-OMe [10]. Because of the stronger electron-donating ability from hydrogen to methyl and to methoxyl, solution absorptions were red-shifted, while

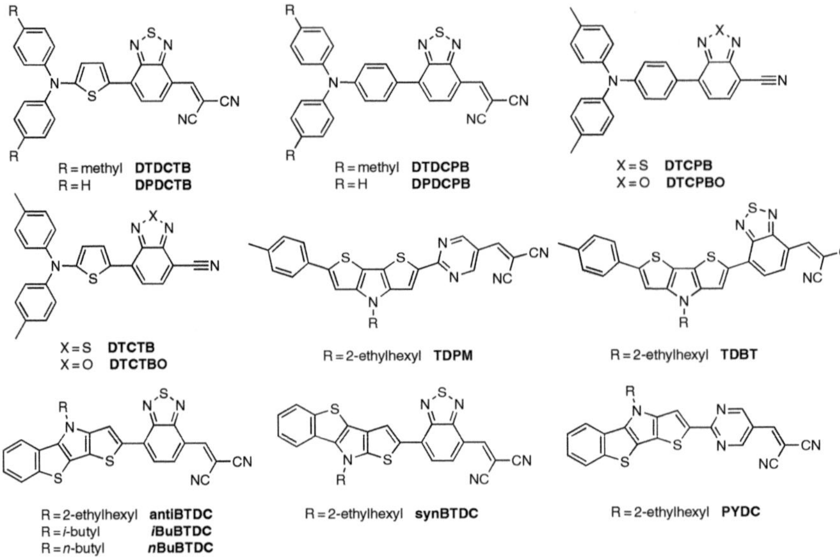

Figure 2.2 Molecular structures of the D–A–A′-type SMDs. Source: Qihui Yue.

Table 2.2 Performance of the D–A–A′-type SMDs.

Donor	E_g^{opt} (eV)	V_{oc} (V)	J_{sc} (mA cm^{-2})	FF	PCE (%)	References
DTDCTB	1.44	0.79	14.68	0.50	5.81	[33]
	1.44	0.80	16.7	0.60	8.0	[34]
DPDCTB	1.48	0.88	12.99	0.49	5.8	[35]
DTDCPB	1.66	0.93	13.48	0.53	6.8	[35]
	1.66	0.92	15.8	0.67	9.6	[34]
DPDCPB	1.73	1.00	11.45	0.49	5.6	[35]
DTCPB	2.04	0.81	10.11	0.53	4.40	[37]
DTCTB	1.86	0.90	11.15	0.65	6.55	[37]
DTCPBO	1.92	0.87	10.33	0.50	4.54	[37]
DPCTBO	1.83	0.95	10.52	0.60	5.96	[37]
TDPM	1.97	0.94	11.39	0.50	5.5	[38]
TDBT	1.66	0.76	11.63	0.46	4.1	[38]
synBTDC	1.66	1.01	11.6	0.52	6.1	[39]
antiBTDC	1.52	0.91	14.2	0.56	7.2	[39]
PYDC	2.05	1.06	10.1	0.48	5.1	[39]
*n*BuBTDC	1.52	0.92	15.7	0.52	7.5	[40]
*i*BuBTDC	1.52	0.94	16.5	0.60	9.3	[40]

in the solid state, the absorptions were obviously broadened and red-shifted, indicating the ordered stacking. BDP-H and BDP-OMe had more *J*-aggregation characters, but BDP-Me was more *H*-aggregated. BDP-OMe performed best in the device with a PCE of 6.1%, attributed to the enough interfaces and interpenetrating pathways in the blend, and when the device was matched with DCV5T-$Me_{3,3}$-based device in series, a PCE of 10% was achieved. Modifying BDP-OMe by replacing the trifluoromethyl with perfluorinated ethyl and *n*-propyl, BDP-2, and BDP-3 were obtained with *J*-aggregation characters [43]. They showed similar absorptions and energy levels, but because of the favorable morphology in the BDP-2-based blend, the device delivered a higher PCE of 6.4%. To investigate the effect of aggregation mode on the performance, BDP-H′ and BDP-J were designed [44]. They showed similar solution absorptions and energy levels, but in the solid films, BDP-H′ was more *H*-aggregated while BDP-J preferred *J*-aggregation. In the devices, BDP-J-based one gave a PCE of 4.2% with a V_{oc} of 0.90 V while BDP-H′-based one gave a PCE of 5.5% with a V_{oc} of 0.83 V. It was noting that BDP-J-based device exhibited a high external quantum efficiency (EQE) even at a small driving force (0.17 eV) and a lower nonradiative energy loss (0.32 eV).

Molecular structures of the BODIPYs-based SMDs mentioned earlier are shown in Figure 2.3. From Table 2.3, it can be found that these SMDs exhibit small E_g^{opt}s, making them promising candidates for the rear cell of a tandem device [10]. However, the J_{sc}s are not satisfactory, and the FFs also need to be improved, which need further optimization on molecular structure, morphology, and device engineering.

2.2.4 Other Molecular Donors

There are some other types of SMDs (as shown in Figure 2.4) used in VD-SMSCs, delivering good photovoltaic performance (as shown in Table 2.4).

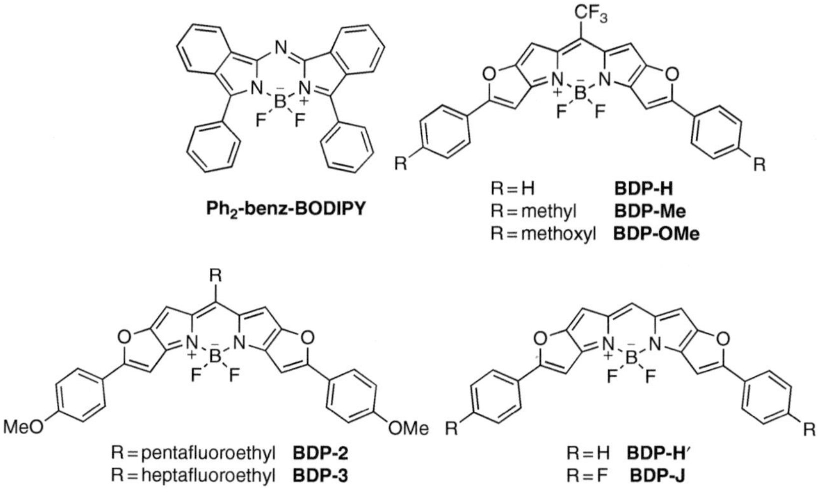

Figure 2.3 Molecular structures of the BODIPY-based SMDs. Source: Qihui Yue.

Table 2.3 Performance of the BODIPY-based SMDs.

Donor	E_g^{opt} (eV)	V_{oc} (V)	J_{sc} (mA cm^{-2})	FF	PCE (%)	References
Ph_2-benz-BODIPY	1.55	0.81	8.0	0.59	3.8	[41]
BDP-H	—	0.89	6.1	0.45	2.5	[10]
BDP-Me	1.37	0.85	9.9	0.54	4.6	[10]
BDP-OMe	1.32	0.73	13.3	0.63	6.1	[10]
BDP-2	1.41	0.76	13.6	0.62	6.4	[43]
BDP-3	1.42	0.75	13.5	0.60	6.1	[43]
BDP-H′	—	0.83	10.9	0.60	5.5	[44]
BDP-J	—	0.90	8.5	0.55	4.2	[44]

Merocyanines (MCs) are typical D–A-type dyes and the photoelectronic property as well as the molecular dipole moment strongly depend on the length of the polymethine chain and the kind of the D and A moieties. To compare the difference between VD- and SP-SMSCs based on MCs, Würthner and coworkers utilized HB194, of which the maximum absorption peak was at 616 nm, to match with C_{60} and $PC_{60}BM$ to construct BHJ OSCs, respectively [45]. Since C_{60} exhibited a higher electron affinity and slower growth conditions during the vacuum deposition led to more aggregation, the VD-SMSC gave a lower V_{oc} but a higher PCE of 4.5% (compared with the PCE of 3% in the SP-SMSC), attributed to the favorable blend morphology. Further optimizing the device structure and the ratio of donor and acceptor, the PCE was promoted to 6.1% with simultaneously increased parameters [46]. To investigate the effect of the buffer layers on the performance, the same group compared the performance of HB364/C_{60}-based PHJ devices with V_2O_5, MoO_3, and WO_3 buffer layers. V_2O_5 layer could induce a high degree of crystallization of HB364, leading to the highest V_{oc} and FF and the PCE (2.7%) [47]. Besides, they utilized ID583/C_{60} PHJ devices to explore the molecular orientation's effect on the performance by differing the substrate temperature [48]. ID583 preferred to stand along their long axis perpendicular to the substrate, however, when the substrate was heated at a temperature above the glass transition temperature, they tended to align their long molecular axes along 45° relative to the surface. The latter orientation improved the exciton dissociation, leading to a better performance.

Phthalocyanines (PCs) demonstrated excellent thermal stability and high absorptions, making them promising candidates for VD-SMSCs. For metal-containing PCs, copper, and zinc are most common centers, based on which the SMDs exhibit an 18 π-electrons coplanar structure. Forrest and coworkers constructed a PHJ device based on CuPc and C_{60}, which delivered a PCE of 3.6% [49]. Riede and coworkers found that in ZnPc:C_{60}-based BHJ devices, the hot substrate was favorable for the formation of phase separation, contributing to a high PCE of 3% (compared with the PCE of 1.4% obtained in the device based on a room-temperature substrate) [50]. To obtain a higher V_{oc}, the same group designed F_4ZnPc by introducing four fluorine atoms into ZnPc to downshift the HOMO energy level, based on which the device

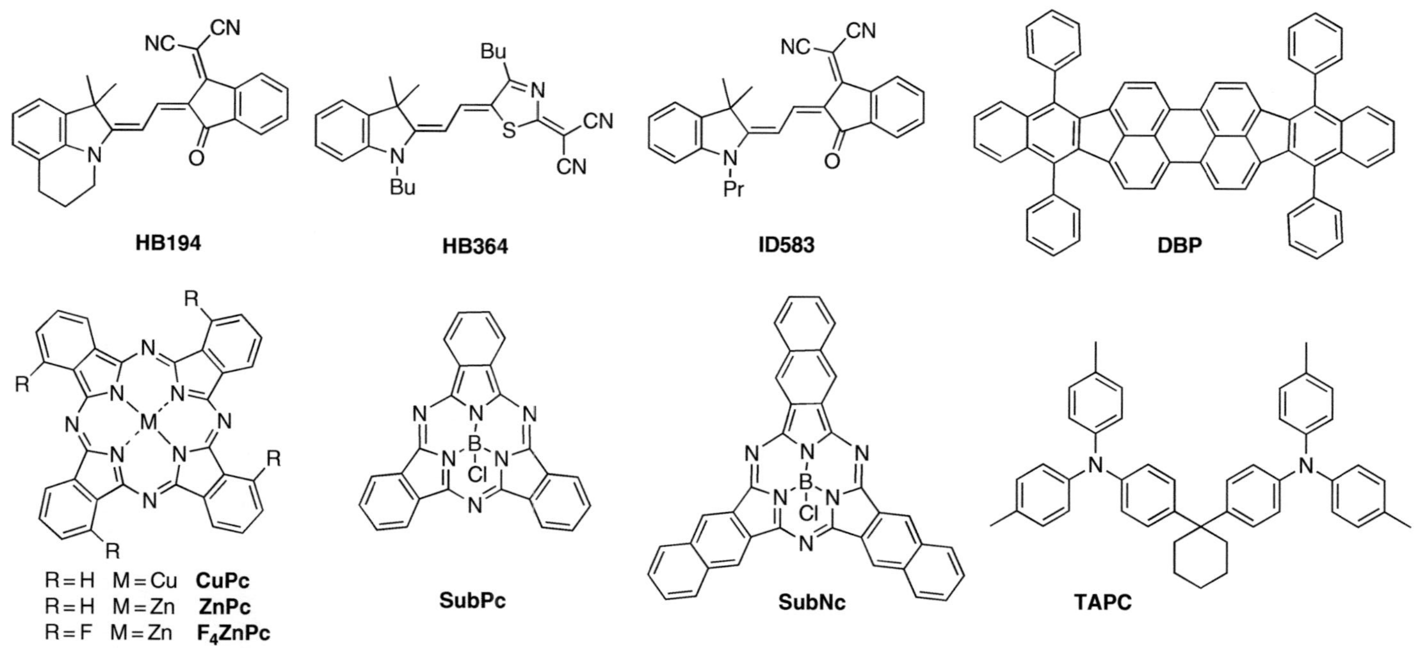

Figure 2.4 Molecular structures of other popular vacuum-deposited SMDs. Source: Qihui Yue.

Table 2.4 Performance of the SMDs mentioned in this part.

Donor	E_g^{opt} (eV)	V_{oc} (V)	J_{sc} (mA cm^{-2})	FF	PCE (%)	References
HB194	—	0.80	11.0	0.48	4.5	[45]
		0.96	12.6	0.47	6.1	[46]
HB364	—	0.74	5.3	0.68	2.7	[47]
ID583	2.05	1.06	5.2	0.70	3.9	[48]
CuPc	—	0.58	18.8	0.52	3.6	[49]
ZnPc	1.6	0.52	10.3	0.57	3.0	[50]
F_4ZnPc	—	0.69	12.1	0.55	4.6	[51]
SubPc	1.9	0.97	3.36	0.57	2.1	[52]
		1.08	8.50	0.49	4.5	[53]
SubNc	1.7	0.79	6.1	0.49	2.5	[54]
DBP	—	0.92	1.36	0.63	7.4	[34]
TAPC	—	0.91	11.43	0.50	5.23	[55]

gave comparable J_{sc} and FF but a higher V_{oc} of 0.69 V, leading to a better PCE of 4.6% [51]. However, further fluorination did not lead to an improved PCE due to the unfavorable morphology or the unmatched energy level [56]. On the other hand, SubPc, which is a 14 π-electrons aromatic macrocycle containing a boron core and a nonplanar cone-shaped structure with the tetrahedral boron center out of plane with the aromatic ligand, is one of the most representative metal-free PCs. Compared with CuPc and ZnPc, the lower-lying HOMO energy level of SubPc favored a higher V_{oc}. The SubPc/C_{60}-based PHJ cell gave a PCE of 2.1% with a V_{oc} of approximately 1 V and the PCE could be promoted to 4.5% after device structure optimization [52, 53]. However, the absorption of SubPc was in the range of 450–650 nm, therefore, to increase the light utilization, π-extended SubNc was designed, based on which a PCE of 2.5% with an enhanced J_{sc} was obtained [54].

There are other SMDs that also realized nice performance in the VD-SMSCs. For example, DBP-based BHJ device demonstrated a PCE of 7.4% [34], and TAPC-based device delivered a PCE of 5.23% even at a low donor concentration of 5% [55].

2.3 *p*-Type Molecular Photovoltaic Materials Used in Solution-Processed Solar Cells

After the proposal of BHJ structure, solution-processing became a new trend in the fabrication of OSCs [8]. The solution-processing method refers to forming a thin film on the substrate by spin-, blade-coating, or slot-die printing with material solutions. Compared with VD-SMSCs, SP-OSCs developed more rapidly because of the large material base (both small molecules and polymers) and the simple and low-cost fabrication process. However, due to the poor film-forming property of traditional small molecules, SP-SMSCs developed slowly. In recent years, through proper molecular design, morphology control, and device optimization, SP-SMSCs have achieved rapid progress. This section will focus on the solution-processed *p*-type molecular photovoltaic materials and introduce the molecular design and the device results.

In the early stage, the match of soluble PCs and hexabenzocoronene with perylenes was reported, but the performance was much worse than the polymer systems [57, 58]. Since then, there had been few related reports. Until 2006, star-shaped OTs appeared, of which the structure could increase the intermolecular interaction and obtain a high mobility, but the final performance was not ideal (PCE of 1%) due to the limited absorption and poor phase separation [59, 60]. Polycyclic arenes were also reported, but the similar problems limited the performance [61]. To improve the light utilization, D–A strategy was used. The early D–A molecules have been summarized in some excellent reviews [62, 63]. Some representative high-performance SMDs, including A-D-A-, D1-A-D2-A-D1- and porphyrin molecules, will be introduced in detail.

2.3.1 A–D–A-Type Molecular Donors

As mentioned above, D–A strategy is an effective way to reduce E_g^{opt}s to improve the sunlight utilization. Based on the same strategy, solution-processible donors are designed and synthesized. With regulation of the moieties including D/A moiety, π-linker and side chain, molecular solubility, photoelectronic properties, and stacking can be well adjusted. By optimizing the design of the A–D–A-type SMDs, coupled with device engineering, the related SMSCs have achieved PCEs of over 10% [12, 13].

2.3.1.1 Oligothiophene-Based A–D–A-Type Molecular Donors

Starting from DCV5T-$Bu_{2,2,4,4}$, Chen and coworkers introduced *n*-octyl side chains to offer DCN3T, DCN5T, and DCN7T [64, 65]. With increased conjugation length, the upshifted HOMO energy levels and the red-shifted absorptions were observed. DCN7T-based device gave a PCE of 3.70% with a poor FF of 0.34, which was due to the large phase separation. To further increase the solubility, cyanoacetate derivatives were used to offer DCAE7T, DCAO7T, and DCAEH7T, which all delivered PCEs of over 4% [66]. The significant improvements on FF were attributed to the improved morphology. Chen and coworkers also attempted many A moieties: when 3-ethylrhodanine (RD), a moiety with a stronger electron-accepting ability, was applied, DERHD7T showed a smaller E_g^{opt} of 1.69 eV compared with DCAO7T. And the device obtained a PCE of 6.10% with a high J_{sc} of 13.98 mA cm^{-2} [67]. Using dicyanomethylene group to replace thio group in RD, DRCN7T showed a red-shifted absorption and more ordered stacking in the film, and the PCE reached a record of 9.30%, with a V_{oc} of 0.91 V, a J_{sc} of 14.87 mA cm^{-2}, and a FF of 0.69 [68]. Further enhancing the electron-accepting ability of A unit, several SMDs with smaller E_g^{opt}s were obtained, but the performance varied from 0.6% to 5% [69]. In addition, examples of barbiturate as the A unit were reported, but the performance was not ideal [70]. These results indicated that photoelectronic properties can be effectively adjusted by regulating A moiety; however, the energy-level alignment and the blend morphology are also in great correlation to the final performance.

The adjustment of the central core is also of great importance. To investigate the influence of the conjugation length and spatial symmetry on the photoelectronic properties and the performance, based on DRCN7T, Chen and coworkers designed DRCN*n*T (n = 4–9), where n is the number of thiophene [12]. Compared with DRCN4T, DRCN5T–9T showed the same absorption peaks at 530 nm in solution, indicating that the quinquethiophene backbone had reached the effective conjugation length. However, in solid state, red-shifted absorptions with various patterns were observed, which was due to the different molecular stacking. DRCN5T showed a broader absorption with an obvious shoulder peak, indicating more effective molecular stacking. With the increase in n, these SMDs exhibited similar LUMO but upshifted HOMO energy levels. DRCN4T-based device only gave a low PCE of 0.24% due to the poor film-forming property, while DRCN5T–9T-based devices all achieved good results, among which the DRCN5T-based one delivered a champion PCE of 10.08%. And devices based on axisymmetric DRCN5/7/9T showed higher J_{sc}s due to the larger dipole moment differences between the ground state and the excited state ($\Delta\mu_{ge}$), which could lower the Coulombic binding energy of the exciton and facilitate charge dissociation. Besides, the introduction of substituents or new structural units is also an important aspect in the core adjustment. To investigate the effect of fluorination on the properties, Peng and coworkers designed TTH-D3TRh and TTF-D3TRh, among which TTF-D3TRh with fluorine atoms showed a deeper HOMO energy level, more favorable molecular stacking and a higher mobility, leading to a higher PCE [71]. Zhu and coworkers reported STDR-T*b*T by introducing a thieno[3,4-*b*]thiophene (T34*b*T)-based moiety into OT core [72]. Due to the quinoidization effect of T34*b*T, STDR-T*b*T showed a reduced E_g^{opt} of 1.60 eV compared with OT core-based STDR (E_g^{opt} of 1.70 eV). This increased light absorption, bringing a higher J_{sc} and PCE. Units like furan, selenophene and thieno[3,2-*b*]thiophene (T32*b*T) were also widely used to replace thiophene [73, 74].

Molecular structures and performance parameters of OT-based A–D–A-type SMDs mentioned above are showed in Figure 2.5 and Table 2.5. They exhibit wide E_g^{opt}s and when matched with fullerenes, the devices often give high V_{oc}s, but there is room for J_{sc} and FF improvement, which needs proper molecular design and device engineering.

2.3.1.2 Benzodithiophene-Based A–D–A-Type Molecular Donors

Benzodithiophene (BDT) with a large conjugated plane, good planar symmetry, and superior charge transport property is a popular unit used in SMDs. Design strategies on BDT-based A–D–A-type SMDs (Figure 2.6) are analyzed as follows.

Modification on BDT Modification on BDT unit includes the side chain engineering and the backbone adjustment, which will affect not only the solubility but also the photoelectronic properties and the molecular stacking.

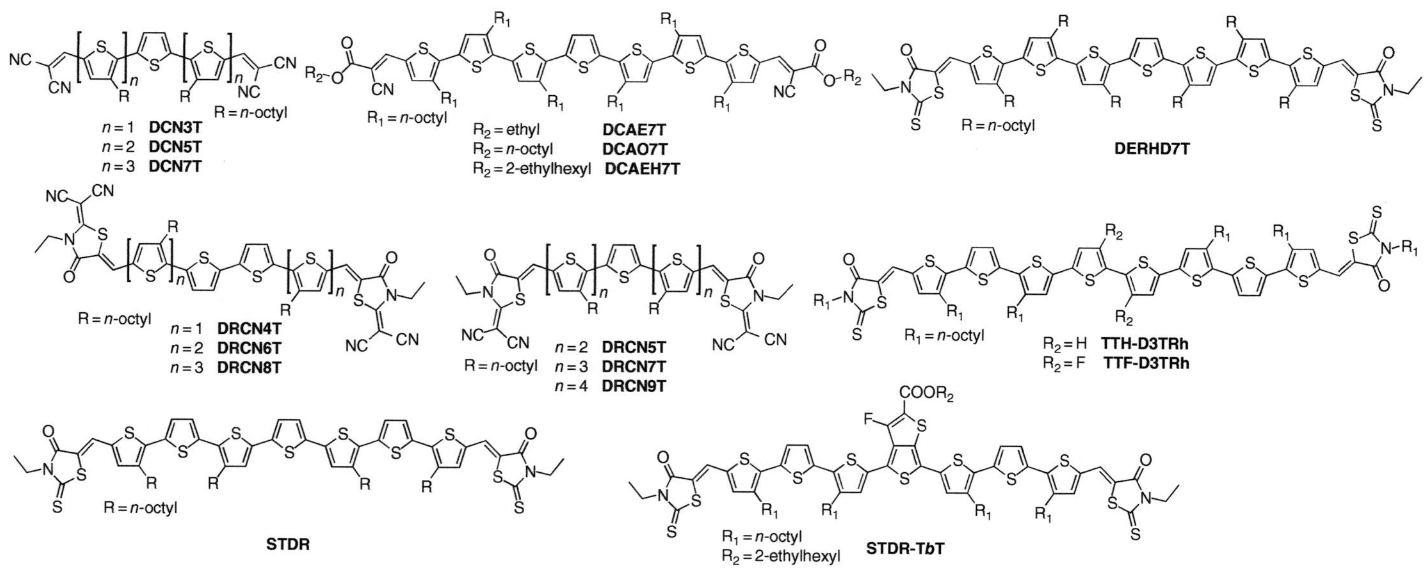

Figure 2.5 Molecular structures of the OT-based A–D–A-type SMDs. Source: Qihui Yue.

Table 2.5 Performance of the OT-based A–D–A-type SMDs.

Donor	E_g^{opt} (eV)	V_{oc} (V)	J_{sc} (mA cm^{-2})	FF	PCE (%)	References
DCN7T	1.68	0.88	12.4	0.34	3.70	[65]
DCAE7T	1.73	0.88	9.94	0.51	4.46	[66]
DCAO7T	1.74	0.86	10.74	0.55	5.08	[66]
DCAEH7T	1.75	0.93	9.91	0.49	4.52	[66]
DERHD7T	1.69	0.92	13.98	0.47	6.10	[67]
DRCN7T	1.62	0.91	14.87	0.69	9.30	[68]
DRCN4T	1.77	0.90	0.70	0.38	0.24	[12]
DRCN5T	1.60	0.92	15.88	0.69	10.08	[12]
DRCN6T	1.60	0.92	10.88	0.59	6.33	[12]
DRCN8T	1.61	0.87	10.98	0.68	6.50	[12]
DRCN9T	1.59	0.82	13.91	0.69	7.86	[12]
TTH-D3TRh	1.74	0.86	10.26	0.67	5.89	[71]
TTF-D3TRh	1.73	0.93	11.03	0.70	7.14	[71]
STDR	1.70	0.88	5.50	0.47	2.31	[72]
STDR-T*b*T	1.60	0.76	10.9	0.61	5.05	[72]

Chen and coworkers linked alkoxy, alkyl, and alkylthio to BDT to offer DR3TBDT, DR3TDOBDT, and DR3TSBDT [75–77]. They showed strong absorptions with obvious shoulder peaks in the range of 350–750 nm, indicating good molecular stacking. The device based on DR3TBDT gave a PCE of 7.38% [75]. Due to the weaker electron-donating ability of alkyl/alkylthio, HOMO energy levels of DR3TDOBDT and DR3TSBDT were deepened, leading to increased V_{oc}s. Through thermal annealing (TA) and solvent vapor annealing (SVA), interpenetrating networks were formed, benefiting exciton diffusion and charge transport, and the devices realized PCEs of 8.26% and 9.95%, respectively [76, 77]. It is worth noting that the DR3TSBDT-based one could achieve a PCE of 8% even at a thickness of 370 nm, attributed to its improved crystallinity. To form better π–π stacking and facilitate charge transport, it is effective to introduce conjugated side chains to make the π-electrons delocalized. Based on DR3TBDT, Chen and coworkers reported DR3TBDTT, DR3TBDTT-HD, and DR3TBDT2T with thiophene or bithiophene-based chains on BDT [78]. They showed similar absorptions and energy levels with DR3TBDT. And the device based on DR3TBDTT gave the best PCE of 8.12%, which could be further improved to 9.58% with SVA of CS_2 [79]. Moreover, when DR3TBDTT-E was added into this binary system, a PCE up to 10.02% was achieved [80]. The DR3TBDTT-HD-based device showed a high V_{oc} due to the weakened intermolecular interaction led by the bulky long chains, and DR3TBDT2T-based device delivered a nice FF, attributed to the enhanced π–π stacking. Using alkoxy-, and alkylthio-thienyl is also feasible, reaching PCEs of 6–9% [81, 82]. Compared with DR3TBDTT, DR3TBDTTF with

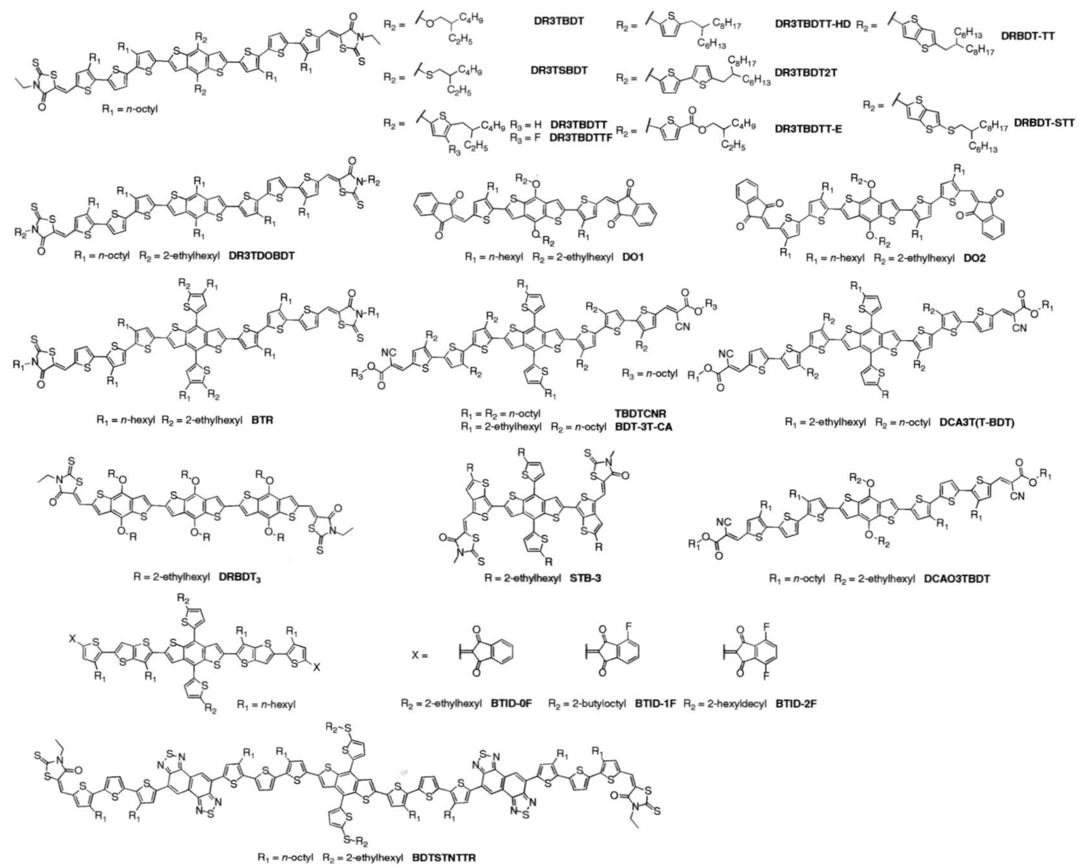

Figure 2.6 Molecular structures of the BDT-based A–D–A-type SMDs. Source: Qihui Yue.

fluorinated side chains exhibited a deeper HOMO energy level and enhanced crystallinity, favoring V_{oc} elevation and charge transport, and realized a PCE of 9.24%, which, after SVA, was promoted to 9.80% [83]. Chen and coworkers also introduced alkyl/alkylthio-substituted T32*b*T as side chains to get DRBDT-TT and DRBDT-STT, which both gave PCEs of over 8% [84]. A nematic liquid crystalline SMD, BTR, with *n*-hexyl-substituted alkylthienyl side chains was also reported, which achieved a PCE of 9.3% [85]. And even at a thickness of 400 nm, PCE remained 8%.

Dithienobenzodithiophene (DTBDT) and naphthodithiophene (NDT) derived from BDT are also popular cores. The extended conjugation backbone can enhance the planarity and π–π interactions, favoring charge delocalization and transport, leading to good photovoltaic performance [86, 87].

Design of the π-Linkers The design of π-linker also includes the substituent change and the structural variation, which in turn affects the solubility and crystallinity.

The most commonly used π-linker is thiophene-based unit. DO1 and DO2 with thiophene and bithiophene as π-linkers had similar E_g^{opt}s, but DO2 showed a stronger absorption and a higher mobility, leading to a higher PCE of 5.11% [88]. TBDTCNR, based on terthiophene π-linker with *n*-octyl on each thiophene, gave a PCE of 5.42% with high V_{oc} and FF but a moderate J_{sc} [89]. Removing alkyl chains on the π-linker of TBDTCNR offered BDT-3T-CA [90]. The improved planarity, strong intermolecular interactions, and highly ordered stacking led to its high FF of 0.75. By removing the *n*-octyl on the end thiophene of the π-linker, DCA3T(T-BDT) was reported and delivered a PCE of 8% [91]. Introducing T32*b*T into the π-linker also achieved good results [92].

Besides, with BDT as π-linker, $DRBDT_3$ exhibited a wide E_g^{opt} of 1.97 eV due to the unfavorable conjugation and stacking by the rigid structure of BDTs, giving a PCE of 4.09% [93]. STB-3 with T34*b*T as π-linker showed ordered molecular stacking, well-defined phase separation and high domain purity when matched with $PC_{71}BM$ and achieved a PCE of 9.26% [94]. Other π-linkers such as cyclopentadithiophene (CPDT) and phenoxazine (POZ) have also been reported [95, 96].

Terminal Moiety Variation Terminal unit has effects on energy levels, especially the LUMO energy level, and molecular stacking. Cyanoacetate, RD, 1H-indene-1, 3(2*H*)-dione (IN) are widely used.

Compared with DCAO3TBDT with cyanoacetate, DR3TBDT with RD showed a lower E_g^{opt} and an improved absorption coefficient. Attributed to the strong absorption and suitable phase separation, DR3TBDT-based device gave a higher PCE of 6.38% with simultaneously enhanced J_{sc} and FF than that of DCAO3TBDT-based device [75]. Wei and coworkers reported BTID-0F, -1F, and -2F with IN, fluorinated, and difluorinated IN [92]. With the increase of fluorine atoms, red-shifted absorption and deepened energy levels were observed. In the blends based on BTID-1F and -2F, hierarchical morphology existed, facilitating charge separation and transport; meanwhile, the favorable vertical phase distribution contributed to charge collection. Excellent PCEs of 10.4% and 11.3% were realized, respectively. Peng and coworkers designed a medium-sized SMD, BDTSTNTTR, with RD as

Table 2.6 Performance of the BDT-based A–D–A-type SMDs.

Donor	E_g^{opt} (eV)	V_{oc} (V)	J_{sc} (mA cm^{-2})	FF	PCE (%)	References
DR3TBDT	1.74	0.93	12.21	0.65	7.38	[75]
DR3TDOBDT	1.79	0.94	12.56	0.70	8.26	[76]
DR3TSBDT	1.74	0.92	14.61	0.74	9.95	[77]
DR3TBDTT	1.72	0.93	13.17	0.66	8.12	[78]
		0.89	14.21	0.76	9.58	[79]
DR3TBDTT-HD	1.76	0.96	11.92	0.59	6.79	[78]
DR3TBDT2T	1.76	0.92	12.09	0.72	8.02	[78]
DR3TBDTTF	1.73	0.93	14.10	0.75	9.80	[83]
DRBDT-TT	1.78	0.92	13.12	0.72	8.70	[84]
DRBDT-STT	1.80	0.91	12.40	0.71	8.01	[84]
BTR	1.82	0.90	13.90	0.74	9.3	[85]
DO1	1.59	0.91	9.47	0.48	4.15	[88]
DO2	1.60	0.92	8.58	0.65	5.11	[88]
TBDTCNR	1.75	0.90	9.08	0.66	5.42	[89]
BDT-3T-CA	1.87	0.88	6.30	0.75	4.16	[90]
DCA3T(T-BDT)	1.80	0.95	11.86	0.70	7.93	[91]
$DRBDT_3$	1.97	0.99	8.26	0.50	4.09	[93]
STB-3	1.69	0.93	14.24	0.70	9.26	[94]
DCAO3TBDT	1.84	0.95	8.00	0.60	4.56	[75]
BTID-0F	1.71	0.93	14.0	0.64	8.3	[92]
BTID-1F	1.70	0.94	15.3	0.72	10.4	[92]
BTID-2F	1.68	0.95	15.7	0.76	11.3	[92]
BDTSTNTTR	1.50	0.93	16.21	0.77	11.53	[13]

terminal, which showed a strong absorption in the range of 600–800 nm and a low-lying HOMO energy level. A record PCE of 11.53% was achieved with a V_{oc} of 0.93 V, a J_{sc} of 16.21 mA cm^{-2}, and an FF of 0.77 [13].

As shown in Table 2.6, BDT-based SMDs exhibit wide E_g^{opt}s, and the relevant devices can deliver high V_{oc}s and FFs, and they are also good choices for tandem devices [97]. Though the J_{sc}s have been promoted compared with that of OT-type SMDs-based devices, they are still far from satisfactory. Adjusting molecular structure, manipulating blend morphology, and optimizing device technology are urgently needed.

2.3.1.3 Dithienosilole-Based A–D–A-Type Molecular Donors

Dithienosilole (DTS) exhibits a strong electron-donating ability and a coplanar skeleton due to the reduced steric hindrance of the side chains by the long C—Si bond.

These make the related SMDs exhibit strong absorptions in the visible range and ordered stacking in solid state. Since the design strategies are similar with those in OT- and BDT-based SMDs, herein, we only give some representative examples as shown in Figure 2.7, and the parameters are listed in Table 2.7.

DCAO3TSi with octyl cyanoacetate terminal showed a strong absorption in the range of 450–750 nm, while DR3TDTS with RD terminal exhibited an obvious red-shifted absorption [98, 99]. And the PCEs were 5.84% and 8.02%, respectively, of which the better performance was attributed to the favorable absorption and morphology. C8-ester and C8-amide were designed with cyanoacetate and cyanoacetamide as terminals, respectively [100]. Due to the intermolecular hydrogen-bonding interactions, C8-amide showed denser stacking but was oriented more randomly, which led to good miscibility with fullerene. However, this resulted in the poor phase separation in blend, making the PCE unideal with a value of 3.22%. By using DCV terminal and bithiophene π-linker, SMD1 was obtained and gave a PCE of 3.81% with a high V_{oc} of 0.92 V [101]. RHO-tDTS containing three DTS units and two RD terminals exhibited a small E_g^{opt} of 1.67 eV, giving a PCE of 7.56% with a high J_{sc} of 12.69 mA cm^{-2} [102].

2.3.1.4 Dithienopyrrole-Based A–D–A-Type Molecular Donors

DTP has a strong electron-donating ability, leading to a high-lying HOMO energy level, and thus a low V_{oc}. By regulating the backbone, side chain and terminal, photoelectronic properties, and device performance can be well controlled.

Linking alkyl-substituted DTP with DCV and bithiophene, Bäuerle and coworkers reported various SMDs [103]. The results showed that SMDs with alkyl chains of the π-linker biased toward DTP often exhibited better solubility and lower-lying energy levels. And device based on SMD2 and SMD3 gave PCEs of 6.1%. Chen and coworkers utilized RD terminal and terthiophene π-linker to offer DR3TDTN [104]. It exhibited strong crystallinity, however, because the large domain was unfavorable for exciton dissociation, the device only gave a PCE of 3.03%. They then utilized carbazole (Cz) to replace the DTP and obtained DR3TCz [105]. Due to the weaker electron-donating ability of Cz, the absorption of DR3TCz was blue-shifted and HOMO energy level was down-shifted, and the device gave a PCE of 4.08% with a high V_{oc} of 0.94 V. Further fusing Cz with the inner thiophene of the terthiophene π-linker, DR2TDTCz was obtained, which showed a red-shifted absorption and an obvious shoulder peak, suggesting the effective stacking in the solid state. And the final PCE was 7.03% with a high FF of 0.75, attributed to the balanced charge transport. The molecular structures and the detailed parameters are shown in Figure 2.8 and Table 2.8, respectively.

2.3.2 D1-A-D2-A-D1-Type Molecular Donors

Inspired by the D–A copolymerization commonly used in polymers, various D1-A-D2-A-D1-type SMDs with a medium size have been designed. Taking the central D2 unit as the classification basis, they can be divided into three main categories.

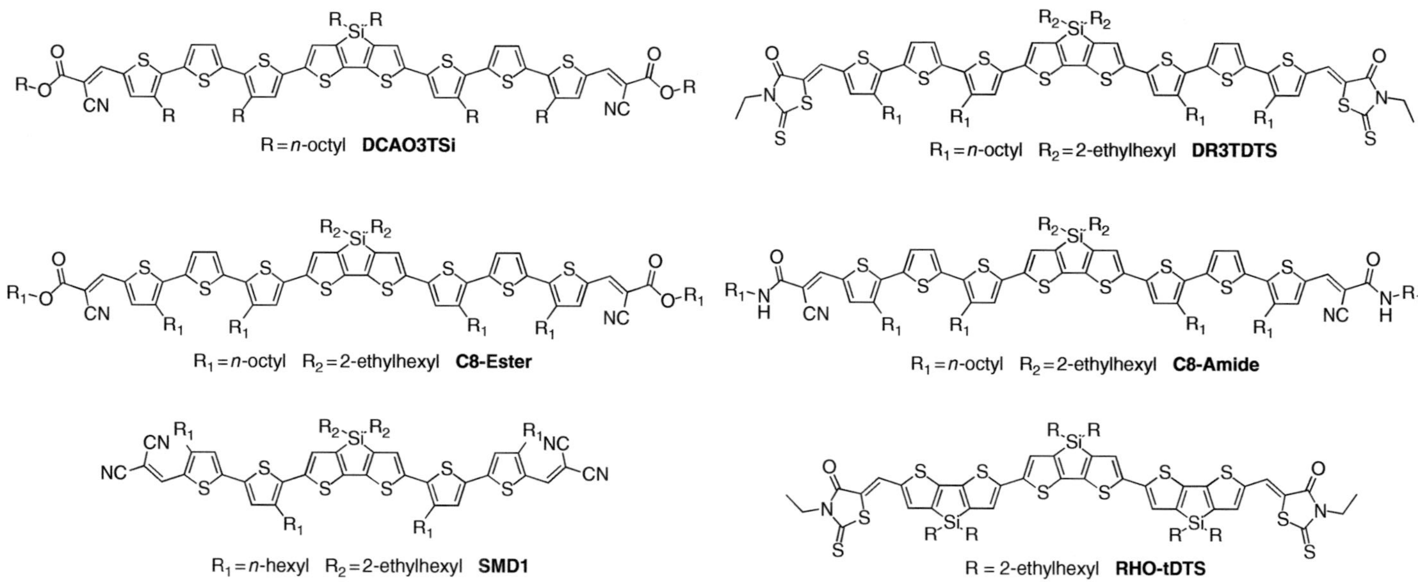

Figure 2.7 Molecular structures of the DTS-based A–D–A-type SMDs. Source: Qihui Yue.

Table 2.7 Performance of the DTS-based A–D–A-type SMDs.

Donor	E_g^{opt} (eV)	V_{oc} (V)	J_{sc} (mA cm^{-2})	FF	PCE (%)	References
DCAO3TSi	1.73	0.80	11.51	0.64	5.84	[98]
DR3TDTS	1.66	0.83	13.97	0.70	8.02	[99]
C8-Ester	1.75	0.82	9.79	0.54	4.31	[100]
C8-Amide	1.81	0.87	7.94	0.47	3.22	[100]
SMD1	1.72	0.92	8.73	0.48	3.81	[101]
RHO-tDTS	1.67	0.88	12.39	0.68	7.56	[102]

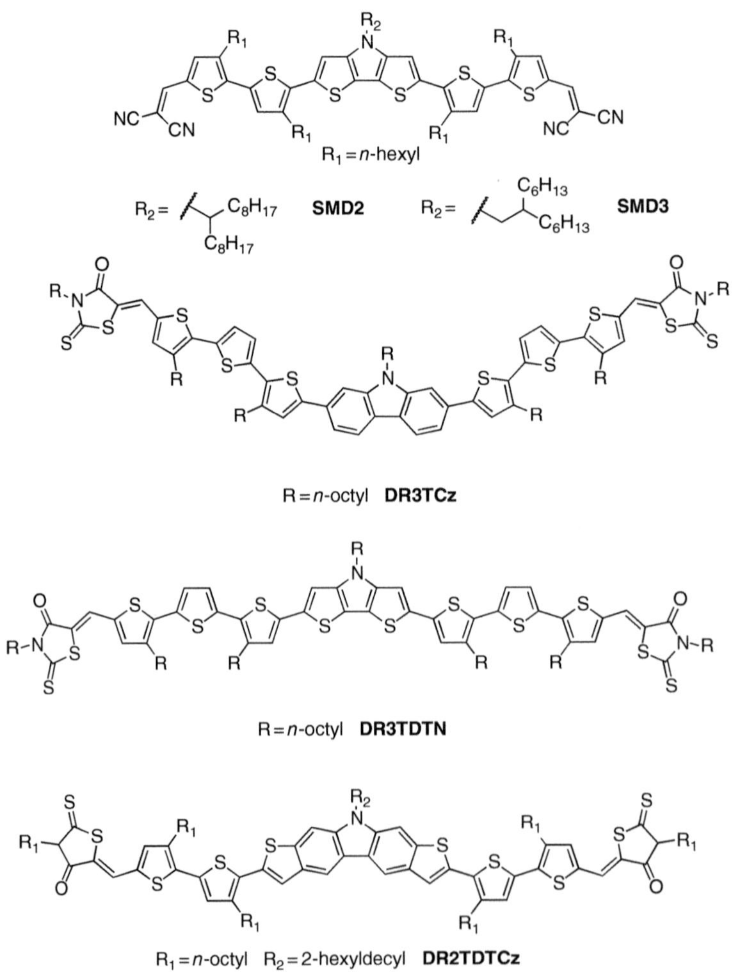

Figure 2.8 Molecular structures of the DTP-based A–D–A-type SMDs. Source: Qihui Yue.

Table 2.8 Performance of the DTP-based A–D–A-type SMDs.

Donor	E_g^{opt} (eV)	V_{oc} (V)	J_{sc} (mA cm^{-2})	FF	PCE (%)	References
SMD2	1.59	0.84	11.4	0.63	6.1	[103]
SMD3	1.58	0.84	10.1	0.72	6.1	[103]
DR3TDTN	1.49	0.67	8.22	0.55	3.03	[104]
DR3TCz	1.88	0.94	8.02	0.54	4.08	[105]
DR2TDTCz	1.82	0.90	10.34	0.75	7.03	[105]

2.3.2.1 Dithienosilole-Based D1-A-D2-A-D1-Type Molecular Donors

Bazan and coworkers took DTS as D2, pyridinethiadiazole (PT) or BT as A and bithiophene as D1 and obtained several SMDs through adjusting the position of the nitrogen of PT, which could obviously change the dipole moment and molecular stacking [106]. These SMDs showed broad and strong absorptions in the visible light region but due to the stronger electron-accepting ability of PT, the absorptions of *p*-DTS($PTTh_2$)$_2$, *d*-DTS($PTTh_2$)$_2$, and SMD4 were red shifted compared with BT-based SMD5, and the shoulder peaks indicated the ordered stacking. With MoO_3 as interlayer, SMD5-based device showed no obvious photovoltaic performance due to the poor crystallinity and phase separation, while the performance of PT-based SMDs was greatly related to the molecular symmetry. The PCEs of the devices based on symmetric *p*-DTS($PTTh_2$)$_2$ and *d*-DTS($PTTh_2$)$_2$ were 6.70% and 5.56%, respectively, while asymmetric SMD4-based device only gave a PCE of 3.16%. The reason was that the long ordering range of blends based on symmetric SMDs was favorable for charge transport. However, the lone pairs of electrons in PT pyridyl nitrogen were sensitive to acid, leading to a decreased PCE when using poly(3,4-ethylenedioxythiophene):poly(styrenesulfonicacid) (PEDOT:PSS) as interlayer. Therefore, Bazan and coworkers utilized fluorinated benzothiadiazole (FBT) to substitute PT to offer *p*-DTS($FBTTh_2$)$_2$, which achieved a PCE of 7.0% with PEDOT:PSS as interlayer [107], and further optimization made the PCE promoted to 9.02% [108]. Extending the conjugation offered a series of SMDs, among which AT1 showed a bad planarity due to the steric hindrance, and little crystalline in the blends, leading to a poor PCE of 1.30% [109], while X1, X2, and F1–F3 all showed relatively large $\Delta\mu_{ge}$, benefiting charge dissociation, and gave PCEs of approximately 6% [110]. By replacing the silicon of *p*-DTS($FBTTh_2$)$_2$ with germanium, *p*-DTG($FBTTh_2$)$_2$ was synthesized [111]. Good phase separation and well-connected domains were observed in the blend, and the PCE was comparable to that of *p*-DTS($FBTTh_2$)$_2$-based device. Moreover, conjugation-extended DTG(FBT_2Th_2)$_2$ realized a promoted PCE of 9.10% [112]. The structures of the SMDs mentioned above were shown in Figure 2.9.

From Table 2.9, DTS-based D1-A-D2-A-D1-type SMDs usually show small E_g^{opt}s, which is conducive to light harvest; however, when blended with fullerene, V_{oc}s were moderate with values of 0.70–0.80 V, showing room for improvement.

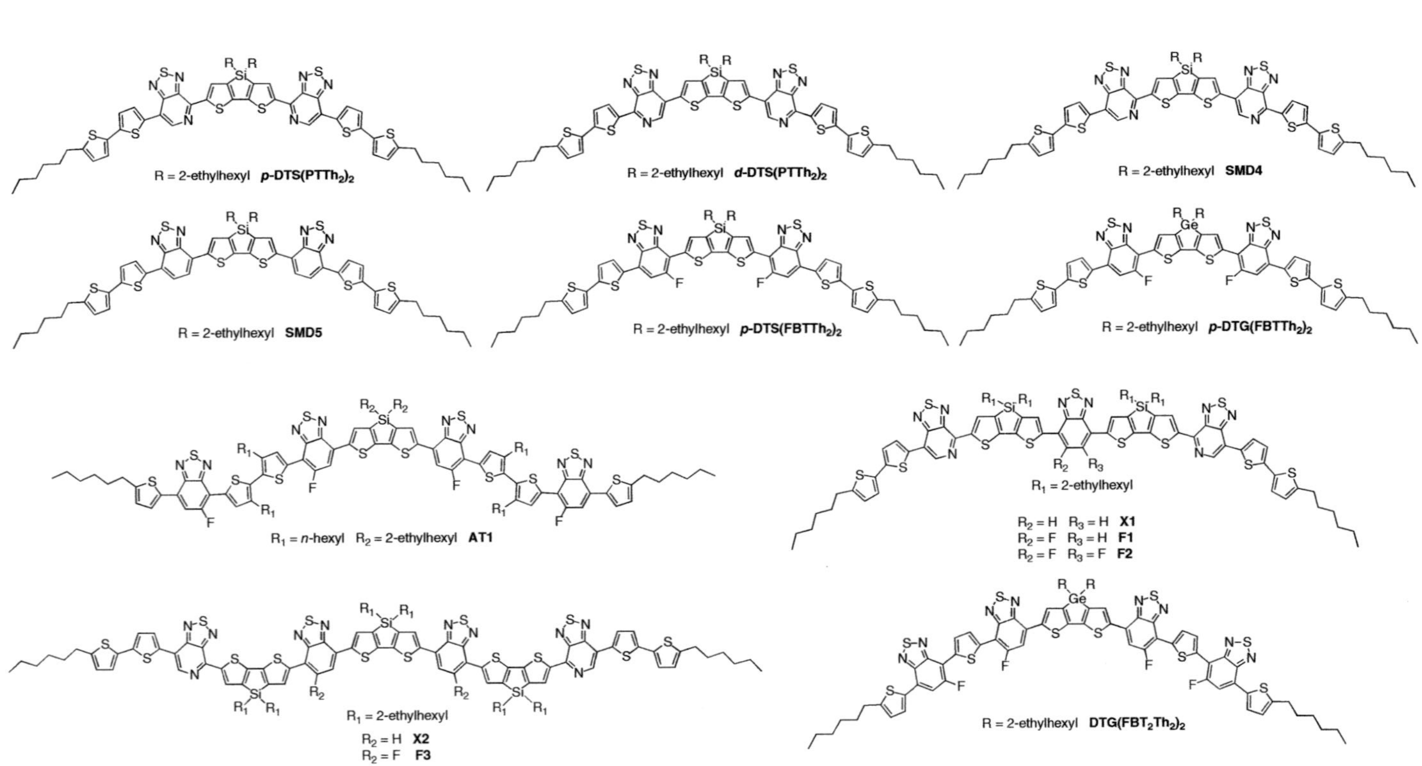

Figure 2.9 Molecular structures of the DTS-based D1-A-D2-A-D1-type SMDs. Source: Qihui Yue.

Table 2.9 Performance of the DTS-based D1-A-D2-A-D1-Type SMDs.

Donor	E_g^{opt} (eV)	V_{oc} (V)	J_{sc} (mA cm^{-2})	FF	PCE (%)	References
p-DTS(PTTh$_2$)$_2$	1.50	0.78	14.4	0.59	6.70	[106]
d-DTS(PTTh$_2$)$_2$	1.50	0.73	12.7	0.60	5.56	[106]
SMD4	1.52	0.72	9.8	0.45	3.16	[106]
SMD5	1.58	0.83	0.9	0.26	0.19	[106]
p-DTS(FBTTh$_2$)$_2$	1.55	0.81	12.8	0.68	7.0	[107]
		0.78	15.47	0.75	9.02	[108]
AT1	1.50	0.75	6.39	0.27	1.3	[109]
X1	1.44	0.71	13.6	0.60	5.8	[110]
X2	1.41	0.66	15.2	0.64	6.4	[110]
F1	1.47	0.75	12.9	0.65	6.3	[110]
F2	1.50	0.76	13.5	0.59	6.1	[110]
F3	1.45	0.71	14.2	0.65	6.5	[110]
p-DTG(FBTTh$_2$)$_2$	1.60	0.75	13.1	0.70	6.9	[111]
DTG(FBT$_2$Th$_2$)$_2$	1.62	0.79	15.9	0.73	9.1	[112]

2.3.2.2 Benzodithiophene-Based D1-A-D2-A-D1-Type Molecular Donors

With triisopropylsilylethynyl (TIPS)-substituted BDT as D2, FBT-TIPSBDT showed a planar structure, leading to effective π–π stacking and benefiting charge transport, and a PCE of 5.69% with a high V_{oc} of 0.94 V was delivered [113]. Taking bithiophene as D1, pyridopyrazine (PP) as A, BDT as D2, SMDs with different side chains were reported [114]. Among these SMDs, SM3, and SM4 showed obvious self-assembly capabilities, but π–π stacking of SM4 was denser. In blends, SM3-based one exhibited large phase separation, while SM4-based one showed bicontinuous interpenetrating networks, explaining the PCE difference (2.5% versus 6.5%). Müllen and coworkers designed several SMDs and studied the effects of the conjugation length of the D1 and the length of its alkyl chains on the final performance [115]. With D1 changing from thiophene to T32*b*T and to bithiophene, absorptions of M1, M2, and M3-C8 were red shifted with increased molar absorption coefficients, which was conducive to a high J_{sc}. And the devices gave PCEs of 1.1%, 3.7%, and 5.2%, respectively. With the alkyl chain on bithiophene changing from *n*-octyl to *n*-hexyl and to *n*-butyl, molecular stacking became more ordered, facilitating charge transport, and the PCEs were promoted from 5.2% to 5.5% and to 6.0%. Zhan and coworkers introduced triphenylamine and *n*-butyl-substituted benzene into D1 unit, and used diketopyrrolopyrrole (DPP) as A unit to offer DPA-diDPPBDT and *n*-Bu-diDPPBDT [116]. They showed similar E_g^{opt}s, but the stronger absorption of DPA-diDPPBDT and the favorable phase separation in the blend contributed to its higher PCE. Further replacing *n*-butyl of *n*-Bu-diDPPBDT with trifluoromethyl, BDTT-(DPP)$_2$-CF_3 was obtained, which delivered a PCE of over 6% with simultaneously increased J_{sc} and FF [117]. The structures of the molecules mentioned above are shown in Figure 2.10.

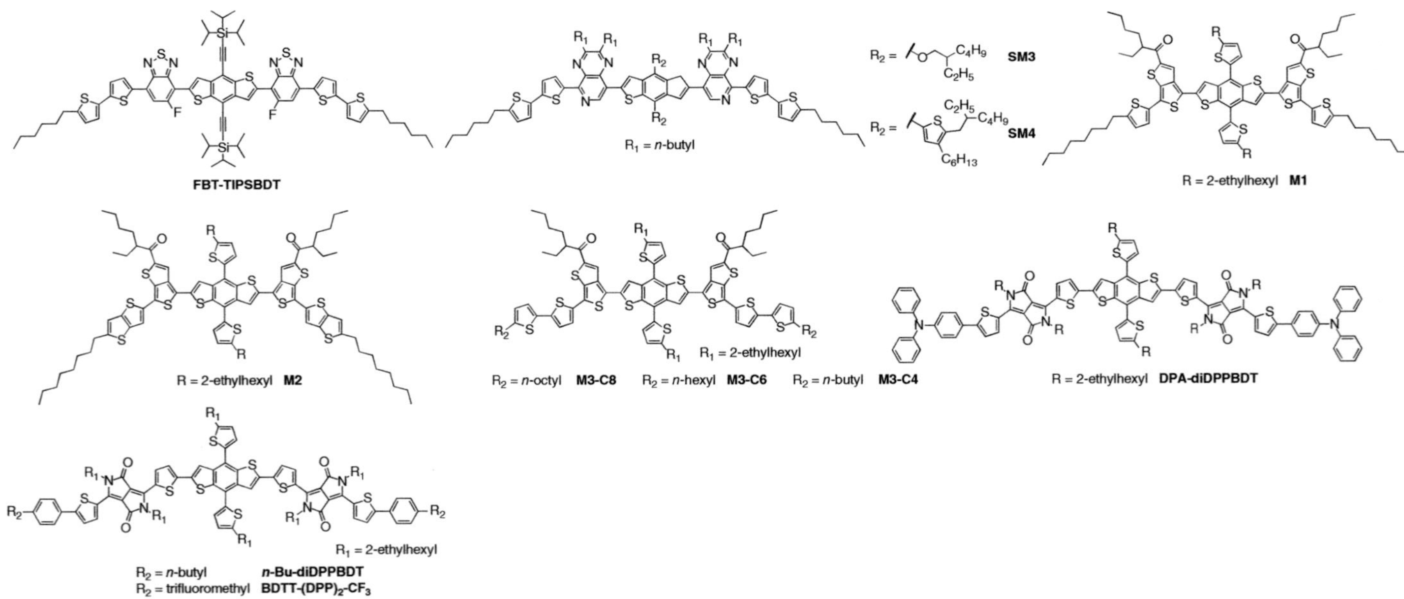

Figure 2.10 Molecular structures of the BDT-based D1-A-D2-A-D1-type SMDs. Source: Qihui Yue.

Table 2.10 Performance of the BDT-based D1-A-D2-A-D1-Type SMDs.

Donor	E_g^{opt} (eV)	V_{oc} (V)	J_{sc} (mA cm^{-2})	FF	PCE (%)	References
FBT-TIPSBDT	1.82	0.94	10.44	0.58	5.69	[113]
SM3	1.81	0.81	5.7	0.49	2.5	[114]
SM4	1.87	0.90	10.5	0.67	6.5	[114]
M1	1.7	0.87	2.66	0.47	1.1	[115]
M2	1.7	0.87	8.60	0.50	3.7	[115]
M3-C8	1.6	0.84	10.83	0.57	5.2	[115]
M3-C6	1.6	0.82	12.73	0.52	5.5	[115]
M3-C4	1.6	0.78	13.92	0.55	6.0	[115]
DPA-diDPPBDT	—	0.62	15.64	0.59	5.77	[116]
n-Bu-diDPPBDT	—	0.67	8.35	0.57	3.20	[116]
BDTT-$(DPP)_2$-CF_3	1.55	0.70	13.56	0.64	6.12	[117]

From the detailed photovoltaic performance parameters listed in Table 2.10, this type of SMDs often shows relatively low FFs, indicating that there may be some deficiency in the mobility or the blend morphology.

2.3.2.3 Indacenodithiophene-Based D1-A-D2-A-D1-Type Molecular Donors

Indacenodithiophene (IDT) owns a large planar conjugated skeleton, which can form ordered stacking in the solid state and facilitate charge transport, making it a widely used central electron-donating block in the design of SMDs (Figure 2.11).

Hou and coworkers designed IDT$(BTTh_2)_2$ with IDT as D2 [118]. It showed a strong absorption in 500–700 nm and the device gave a PCE of 4.25% with a high V_{oc} of 0.93 V, which was benefited from its deep HOMO energy level. Jen and coworkers took conjugation extended indacenodithienothiophene (IDTT) as D2 and reported four SMDs with increased dipole moment of D1-A group (IDTT-DFBT-BT, IDTT-FBT-TT, IDTT-FBT-BT, and IDTT-FBT-3T) [119]. A larger dipole moment could increase the self-assembly, enhance the order at the donor–acceptor interface, and increase the dielectric property that could lower the exciton binding energy and decrease charge recombination. Therefore, IDTT-FBT-3T achieved the best PCE of 6.54%.

BIT series reported by Wang and coworkers are representative examples. Compared with BIT, BIT-4F showed a lower-lying energy level due to the electron-withdrawing nature of difluorinated BT and a better planarity because of F-S or F-H interactions. By introducing a thiophene between D2 and A, BIT-4F-T showed a stronger absorption, higher π-electron delocalization, and more ordered stacking, also, $\Delta\mu_{ge}$ of BIT-4F-T was increased, and these all contributed to the high PCE of 8.1% [120]. Based on BIT-4F-T, BIT6F, BIT8F, and BIT10F were

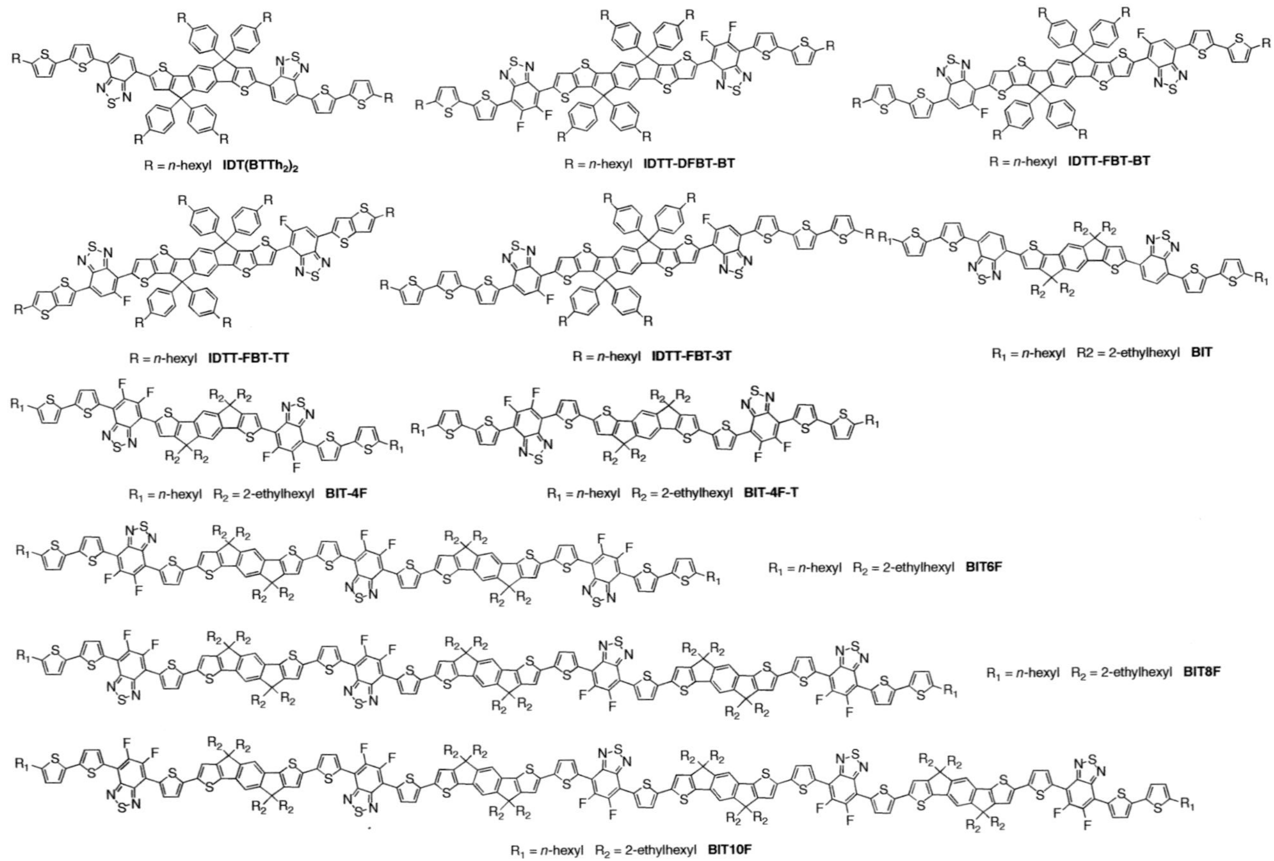

Figure 2.11 Molecular structures of the IDT-based D1-A-D2-A-D1-type SMDs. Source: Qihui Yue.

Table 2.11 Performance of the IDT-based D1-A-D2-A-D1-type SMDs.

Donor	E_g^{opt} (eV)	V_{oc} (V)	J_{sc} (mA cm^{-2})	FF	PCE (%)	References
IDT($BTTh_2$)$_2$	1.80	0.93	9.42	0.49	4.25	[118]
IDTT-DFBT-BT	1.78	0.93	10.0	0.46	4.37	[119]
IDTT-FBT-TT	1.82	0.99	11.7	0.50	5.75	[119]
IDTT-FBT-BT	1.77	0.96	12.0	0.53	6.12	[119]
IDTT-FBT-3T	1.69	0.95	12.4	0.55	6.54	[119]
BIT	1.77	0.90	7.2	0.38	2.5	[120]
BIT-4F	1.83	0.99	9.0	0.38	3.4	[120]
BIT-4F-T	1.81	0.90	11.9	0.76	8.1	[120]
BIT6F	1.79	0.89	13.44	0.76	9.09	[121]
BIT8F	1.78	0.89	10.90	0.69	6.78	[121]
BIT10F	1.77	0.91	11.72	0.66	7.10	[121]

reported by extending the conjugation length [121]. Conjugation expansion brought red-shifted absorption, increased molar absorption coefficient, up-shifted HOMO, and downshifted LUMO energy levels. These SMDs all showed face-on orientation, favoring charge transport. The devices based on BIT6F/BIT10F with centrosymmetric electron-rich IDT delivered higher PCEs than those of the devices based on BIT-4F-T/BIT8F with axisymmetric electron-deficient difluorinated BT. The highest PCE of 9.09% was achieved in BIT6F-based device.

From Table 2.11, this type of SMDs often show high V_{oc}s and FFs; however, their wide E_g^{opt}s limit the maximum obtainable J_{sc}. How to improve J_{sc} without sacrificing V_{oc} and FF is urgent, which requires optimization both on molecule and device.

2.3.3 Porphyrin-Based Molecular Donors

Porphyrins (Por), which are widely used in low-bandgap donors, own unique characteristics: (i) large conjugated planar structure with 26 π-electrons can facilitate π–π stacking, benefiting charge transport; (ii) wide absorption reaching NIR region and high extinction coefficient benefit light harvest and obtain a high J_{sc}; and (iii) multiple modification sites make the molecular properties highly tunable. Herein, some high-performance Por-based SMDs (Figure 2.12) will be introduced.

The most representative structure of Por-type SMDs is A-π-Por-π-A. Regulating the side chain on Por, Peng and coworkers took zinc Por (ZnP) as central unit, acetylene as π-linker, and DPP as terminal to offer DPPEZnP-based SMDs [122–124]. DPPEZnP-TEH, -TBO, and -THD all had broad absorptions from the ultraviolet (UV) to NIR region, attributed to the good conjugation planarity and the strong ICT effect. The devices achieved PCEs of 8.08%, 9.06%, and 9.41%, respectively. Among

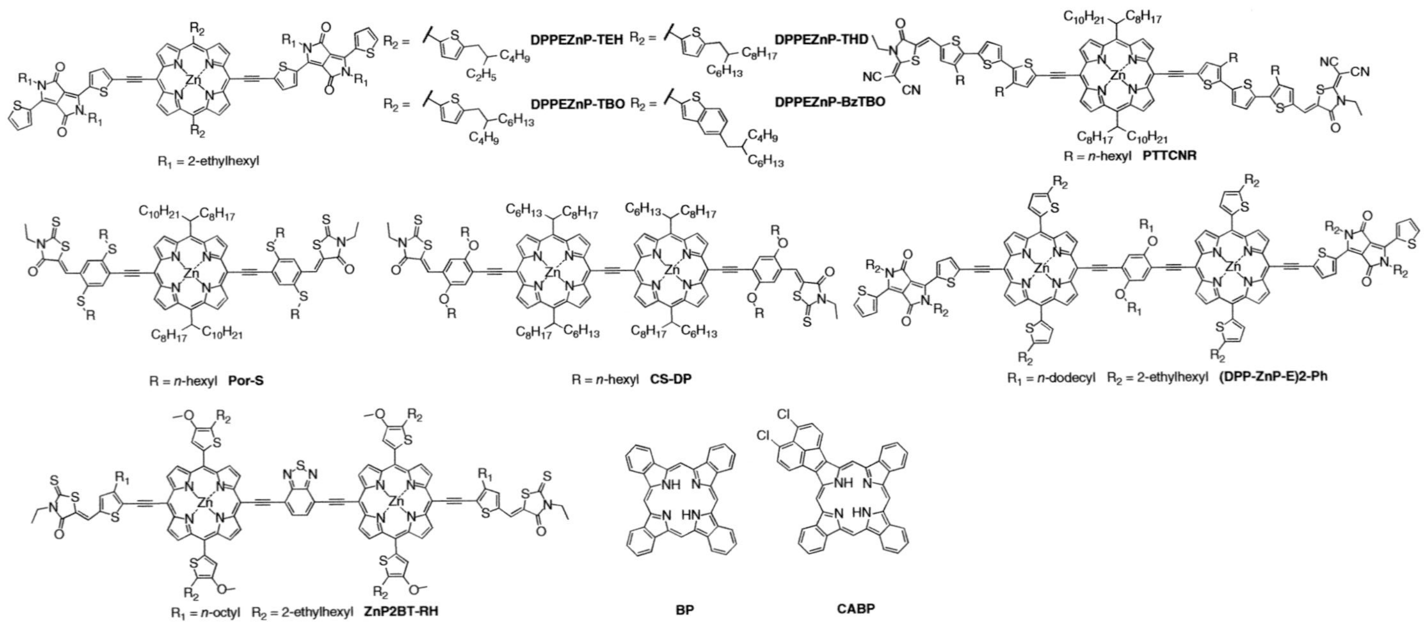

Figure 2.12 Molecular structures of the Por-based SMDs. Source: Qihui Yue.

them, DPPEZnP-TBO-based one achieved a respectable J_{sc} of 19.58 mA cm^{-2}, while DPPEZnP-THD-based one realized a high FF of 0.70. By replacing the thienyl of the side chain of DPPEZnP-TBO with benzothienyl with a weaker electron-donating ability, DPPEZnP-BzTBO was obtained with a lower lying HOMO energy level [125]. And the final device provided a similar PCE but a higher V_{oc} of 0.80 V. Inserting another conjugated unit between acetylene and terminal moiety is also an effective strategy. With RD derivatives as A, PTTCNR with terthiophene insertion showed extended conjugation and polymorphic molecular stacking and realized a PCE of 8.21% [126]. Por-S with alkylthio-substituted phenyl insertion showed a good self-assemble ability and preferred to J-aggregating in the blend, demonstrating a PCE of over 8% [127]. Utilizing CPDT and thienylenevinylenethiophene as insertion has also been reported with PCEs of approximately 8% [128, 129].

Medium-sized SMDs containing two Por units have also achieved excellent results. CS-DP with an A-π-Por-π-Por-π-A framework exhibited an ultra-narrow E_g^{opt} of 1.26 eV and delivered a PCE of 8.29% with a V_{oc} of 0.78 V. It is worth noting that the energy loss ($E_{loss} = E_g^{opt} - qV_{oc}$, where q represents the elementary charge) was only 0.48 eV, much smaller compared with most fullerene-based SMSCs [130]. In addition, A-π-Por-π-X-π-Por-π-A framework has been reported, where X can be electron-donating or -accepting unit. With alkoxy-substituted benzene as the X, (DPP-ZnP-E)2-Ph delivered a PCE of 6.42% [131], and ZnP2BT-RH with BT as the X gave a record PCE of 10.02% [132]. Both SMDs exhibited small E_g^{opt}s of approximately 1.40 eV and high molar extinction coefficients, indicating the strong ICT effect.

Besides, there are some metal-free Por-based SMDs such as BP and CABP. They owned similar structures, and when they were mixed at a ratio of 0.75 : 0.25 to form a solid solution, the PHJ device with fullerene as acceptor gave a PCE of 3.6%, higher than those of the devices using 100% BP or CABP. The proper energy levels of the solid solution could guarantee charge separation when matched with fullerene, and the partial disordering caused by the insertion of CABP into the BP crystal lattice could decrease the recombination, which both contributed to the performance enhancement [133].

As listed in Table 2.12, Por-based SMDs generally show narrow E_g^{opt}s, favoring high J_{sc}s, and are promising candidates for the rear cell of tandem devices [97]. However, the moderate V_{oc} of 0.7–0.8 V and FF of 0.6–0.7 reveal room for further improvement.

2.3.4 Other Molecular Donors

Compared with D–A copolymers, D–A oligomers show wider E_g^{opt}s, limiting the light harvest, but their high purity with defined structures and high crystallinity can facilitate charge transport (typical examples are shown in Figure 2.13). Wei and coworkers designed O-BDTdFBT by alternating BDT and difluorinated BT [134]. Compared with the polymer P-BDTdFBT, O-BDTdFBT exhibited a wider E_g^{opt} of

Table 2.12 Performance of the Por-based SMDs.

Donor	E_g^{opt} (eV)	V_{oc} (V)	J_{sc} (mA cm^{-2})	FF	PCE (%)	References
DPPEZnP-TEH	1.37	0.78	16.76	0.62	8.08	[122]
DPPEZnP-TBO	1.37	0.73	19.58	0.63	9.06	[123]
DPPEZnP-THD	1.37	0.74	18.2	0.70	9.41	[124]
DPPEZnP-BzTBO	1.40	0.80	16.82	0.68	9.08	[125]
PTTCNR	1.45	0.82	14.30	0.70	8.21	[126]
Por-S	1.74	0.91	13.19	0.67	8.04	[127]
CS-DP	1.26	0.78	15.14	0.70	8.23	[130]
(DPP-ZnP-E)2-Ph	1.41	0.68	16.34	0.58	6.42	[131]
ZnP2BT-RH	1.40	0.85	17.66	0.67	10.02	[132]
BP	1.73	0.77	5.9	0.68	3.1	[133]
CABP	1.64	0.65	2.5	0.52	0.9	[133]
BP:CABP	1.58	0.81	8.3	0.54	3.6	[133]

1.83 eV, but more obvious red-shifted absorption from the solution to the film indicated the more effective stacking. And the O-BDTdFBT-based blend showed higher crystallinity and proper phase separation, favoring exciton diffusion, charge separation, and transport, contributing to a higher PCE of 8.10%. Then, they reported O-1, O-2, M-1, and M-2 [135]. With the conjugation length increasing, red-shifted and stronger absorptions were observed. Compared with O-1 and O-2, M-1 and M-2 showed not only broadened absorptions and smaller E_g^{opt}s but also more ordered stacking due to the better planarity. The devices gave PCEs of 1.06%, 7.19%, 9.25%, and 8.91%, respectively, among which the poor performance of O-1-based one was due to the overmixed morphology.

In addition to the building blocks mentioned earlier, other units have showed good properties in SMDs (Figure 2.13). Nguyen and coworkers took DPP, thiophene, and benzofuran to construct DPP(TBFu)$_2$. It had matched both absorption and energy levels with PC$_{71}$BM. The optimized blend showed nanoscale phase separation and the device gave a PCE of 4.4% [136]. Sun and coworkers utilized POZ as the center, and linked BT, thiophene, and DCV on both sides of it in turn to offer POZ2, achieving a PCE of 7.44% [137]. Kido and coworkers regulated the side chain and the number of hydroxyl groups to offer squaraine (SQ) series SMDs, which well adjusted the photoelectronic properties and molecular stacking. Hydroxyl groups could downshift the HOMO energy level, while the linear side chain could improve the crystallinity. These SMDs showed different aggregation in the neat films, but they all J-aggregated when matched with PC$_{71}$BM, favoring a high J_{sc}. And the SQ2-based device demonstrated a champion PCE of 4.0% [138].

The relevant performance parameters are listed in Table 2.13.

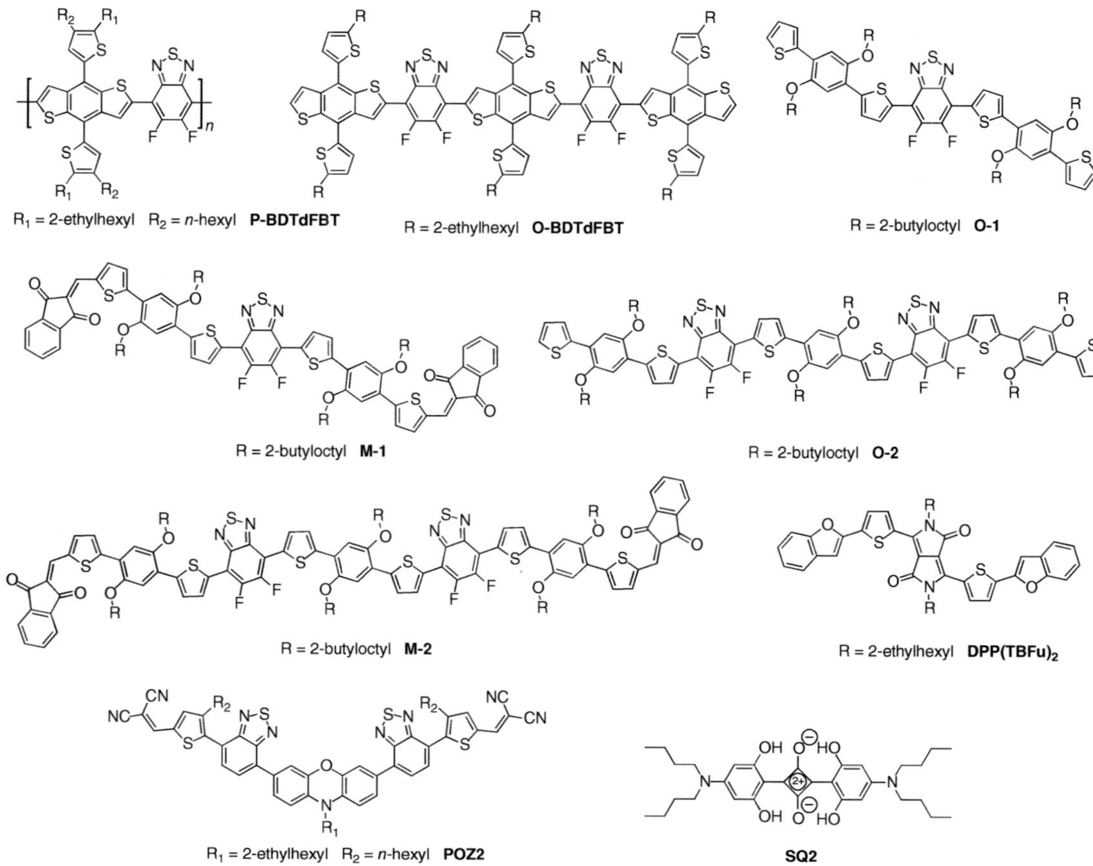

Figure 2.13 Molecular structures of the SMDs mentioned in this part. Source: Qihui Yue.

Table 2.13 Performance of the SMDs mentioned in this part.

Donor	E_g^{opt} (eV)	V_{oc} (V)	J_{sc} (mA cm^{-2})	FF	PCE (%)	References
O-BDTdFBT	1.83	0.97	11.48	0.70	8.10	[134]
O-1	1.94	1.04	3.92	0.25	1.06	[135]
O-2	1.85	0.90	11.2	0.70	7.19	[135]
M-1	1.75	0.94	13.5	0.71	9.25	[135]
M-2	1.76	0.92	13.3	0.70	8.91	[135]
DPP(TBFu)$_2$	1.7	0.92	10.0	0.48	4.4	[136]
POZ2	1.59	0.91	11.9	0.69	7.44	[137]
SQ2	1.66	0.86	9.4	0.49	4.0	[138]

2.4 Current Progress on Small-Molecule Solar Cells with Nonfullerene Acceptors

To get high-performance SMSCs, material combination and morphology control are of great importance. For material combination, materials with different E_g^{opt}s are needed, aiming to form complementary absorption to get a high J_{sc}. Also, the energy-level alignment should be proper to get a high V_{oc} meanwhile ensure the effective charge separation. Morphology control aims to form a balance between exciton dissociation and charge transport, and thus improve the performance. Fullerene-based SMSCs have achieved PCEs of over 10% as introduced [9–13]; however, fullerenes have some nonnegligible limitations such as synthetic difficulty, weak absorption, rigid energy levels, large E_{loss}, and morphology instability. In contrast, small-molecule nonfullerene acceptors (SM-NFAs) exhibit highly tunable E_g^{opt}s and energy levels, high absorption extinction coefficients, and high exciton diffusion coefficients; meanwhile, even a small driving force can drive efficient charge transfer in devices, and charges can transport through three-dimensional (3D) channels formed by strong π–π stacking and intermolecular interactions [139–141], bringing great breakthroughs in the performance of SMSCs [5].

However, due to the lack of proper materials, the development of NFA-based VD-SMSCs is slow, only several examples were reported. For example, when α-6T was matched with SubPc and SubNc, compared with C_{60}-based device, the improved energy-level alignment and broadened absorption led to significantly increased V_{oc}, J_{sc}, and thus PCEs (4.69% and 6.02%, respectively). Further depositing α-6T, SubNc, and SubPc sequentially, the PCE was promoted to 8.40%, attributed to the cascade-like energy levels that benefited exciton dissociation, and the energy transfer from SubPc to SubNc [142]. For SP-SMSCs, the choices of SM-NFAs are various. Through subtle molecular design, delicate morphology optimization, and device engineering, NFA-based SMSCs have achieved PCEs of over 15% [14–18].

Therefore, in this section, we mainly focus on the SP-SMSCs with NFAs, including binary and ternary devices.

2.4.1 Binary Solar Cells

Perylene diimide (PDI) derivatives have been introduced in SMSCs. However, due to the wide E_g^{opt}s that limit light harvest, and rigid planar structures that tend to form large aggregates, when matched with SMDs, PDI-based SMSCs only show moderate performance [143]. Another type of SM-NFAs, fused-ring electron acceptors (FREAs), which often exhibit planar A–D–A structures, and highly tunable photoelectronic properties, have also been applied. Nowadays, combining a wide-bandgap SMD and a small-bandgap FREA has been proved to be effective in constructing high-performance SMSCs. Some instructive examples will be introduced as follows.

IDIC, derived from ITIC [144], is one of the most well-known NFAs used in SMSCs [145]. It shows a strong absorption in 500–800 nm, which is complementary to the absorptions of most SMDs. Blending IDIC with DRTB-T delivered a high PCE of over 9%, attributed to the nanoscale phase separation and proper crystallinity [146]. Further pairing IDIC with various SMDs, such as H11, SM1, BDT(TVT-SR)$_2$, P2TBR, could achieve PCEs of 9–11% [147–150]. Fluorinating IDIC offered IDIC-4F, which is also popular in SMSCs. With OT-based D5T2F-P as donor, an impressive PCE of 9.36% was realized [151]. And with liquid-crystalline NDT-3T-R6 as donor, the PCE reached 10.7% with a high FF of over 0.70 [152]. Other ITIC derivatives, such as IT-4F and 6TIC, can also achieve excellent performance with PCEs of over 10% [153, 154], among which when 6TIC was matched with Por-based DPPEZnP-TBO, utilizing solvent additive and SVA could properly control crystallinity, improve phase separation and domain purity, contributing to an outstanding PCE of over 12% with a high J_{sc} of over 20 mA cm^{-2} and an FF of 0.74 [154].

Another popular NFA is Y6, which exhibits wide and strong absorption covering 600–950 nm, and realized a record PCE of 15.7% in a PSC [155]. Then, various BDT-based SMDs have been applied to match with it. By modifying BTR with chlorine, BTR-Cl was obtained, which maintained liquid crystalline character, but showed enhanced molecular order with an edge-on orientation [156]. When matched with Y6 that preferred a face-on orientation, 3D charge pathways formed, and significant phase separation was observed. These contributed to a high PCE of 13.6%. Using BTEC-2F, BSFTR, SM1-F, as donors could also obtain comparable or slightly better performance [157–159]. Matching ZR1 based on a DTBDT core with Y6, the PCE reached 14.34%, attributed to the hierarchical morphology that was conducive to charge separation and transport [160]. Blending SMDs with some Y6-derived SM-NFAs also delivered encouraging results. Combinations of B1:BO-4Cl [15] and BT-2F:N3 [18] both demonstrated PCEs of over 15%.

SMSCs based on other SM-NFAs, such as DR3TBDTT:O-IDTBR, DRCN5T:F-2Cl, and BSFTR:NBDTP-F$_{out}$, have also obtained inspiring PCEs [161–163].

2.4.2 Ternary Solar Cells

Compared with binary SMSCs, the introduction of the third component to construct ternary SMSCs (t-SMSCs) can individually or simultaneously (i) enhance light harvest, (ii) facilitate charge transfer, (iii) regulate morphology, and (iv) strengthen morphology stability [164]. According to the composition of NFA-based t-SMSCs, they can be divided into three main categories: two SMDs/one SM-NFA, one SMD/two SM-NFAs, and one SMD/one fullerene/one SM-NFA, and the state-of-the-art t-SMSCs have accomplished PCEs of over 15% and still show great potential [14, 16, 17].

The first category of t-SMSCs was seldom reported, which might be due to the difficulty in morphology control or the relatively low electron mobility of SM-NFAs. By adding DR3TBDTT-S-E into DCAO3TBDTT:IDIC system with a PCE of 9.49%, the PCE of the t-SMSC was improved to 10.04% with increased J_{sc} and FF. Such an enhancement was attributed to the fact that DR3TBDTT-S-E reduced the aggregations of the binary blend and established 3D charge transport pathways with mixed edge-on and face-on orientations [165]. For the second category, the two NFAs often exhibit similar structures with good compatibility. When adding 10% 4TIC into ZnP-TSEH:6TIC system, the crystallinity of the blend was effectively enhanced and proper multilength scale morphology formed, making the PCE promoted from 13.54% to 15.88% [16]. A small amount of Y7 can well modified the phase separation and molecular arrangement of B1:BO-4Cl-based system and thus promoted the PCE to a value of 16.28% [17].

From the perspectives of absorption complementarity, phase separation adjustment and electron transport ability, fullerene is a promising third component. In the BTR:$PC_{71}BM$ and BTR:NITI binary SMSCs, PCEs of 9.30% and 6.82% were realized, respectively; however, the t-SMSC yielded significantly promoted performance with an average PCE of 13.20% [166]. Besides the absorption contribution from NITI and the cascade-like energy level alignment, the more important reason was the formation of hierarchical morphology: $PC_{71}BM$ served as the framework, while BTR and NITI with good compatibility were filled in the vacancies of the framework, and SVA facilitated the phase separation between BTR and NITI, making NITI be extruded to the boundary of $PC_{71}BM$. After charge separation, holes and electrons transported mainly through BTR and $PC_{71}BM$ pathways. Such a morphology helped reduce the E_{loss}, suppress recombination, and balance charge transport. This work manifested the indispensability of fullerene and NFAs, and this strategy has been proved to be effective in achieving high-performance t-SMSCs [14, 167–169]. For example, adding a small amount of fullerene into BTR-Cl:Y6 blend, the PCE was promoted from 13.6% to 15.34% with an enhanced FF, attributed to the reduced bimolecular recombination and the better mixed donor/acceptor phase with reasonable domain sizes [14].

The molecular structures mentioned earlier are shown in Figures 2.14 and 2.15, and the performance parameters are listed in Table 2.14.

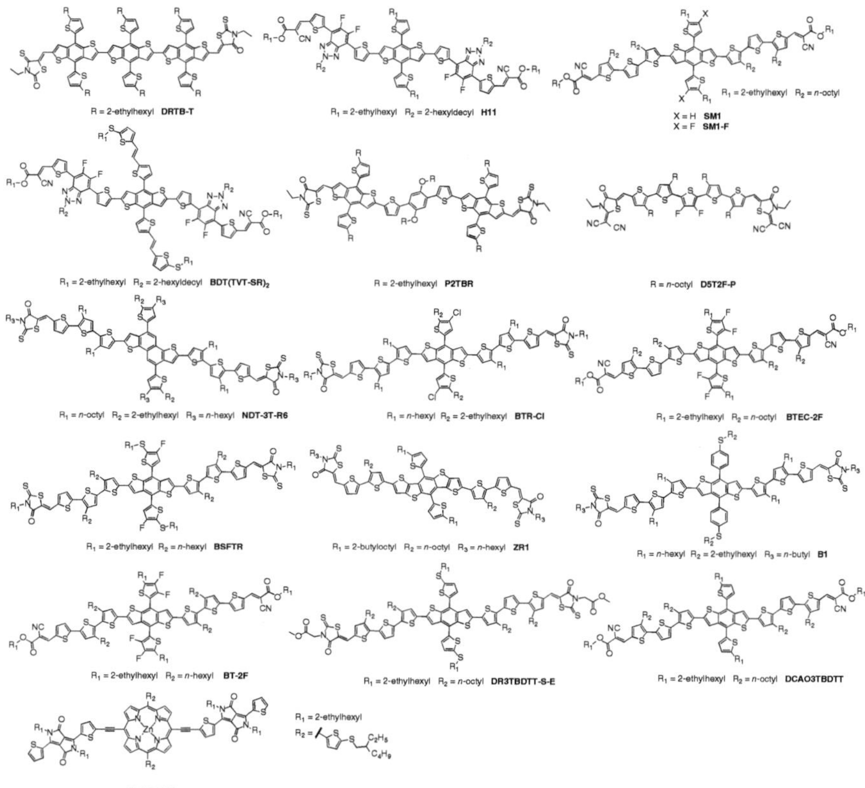

Figure 2.14 Molecular structures of the new SMDs mentioned in this part. Source: Qihui Yue.

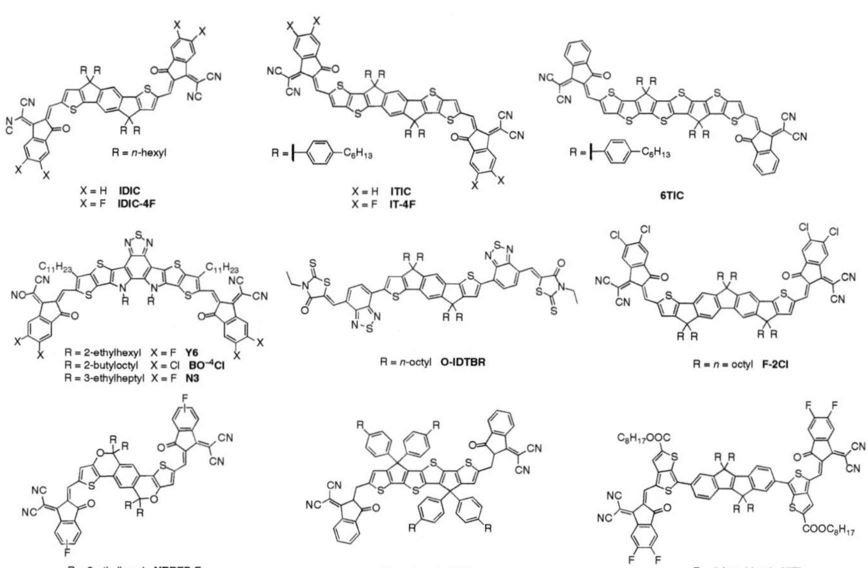

Figure 2.15 Molecular structures of the new SM-NFAs mentioned in this part. Source: Qihui Yue.

Table 2.14 Performance of the NFA-based SMSCs mentioned in this part.

Donor	V_{oc} (V)	J_{sc} (mA cm^{-2})	FF	PCE (%)	References
DRTB-T:IDIC	0.98	14.25	0.65	9.08	[146]
H11:IDIC	0.98	15.21	0.65	9.73	[147]
SM1:IDIC	0.91	15.18	0.74	10.11	[148]
BDT(TVT-SR)$_2$:IDIC	0.98	15.92	0.71	11.10	[149]
P2TBR:IDIC	0.94	17.5	0.70	11.5	[150]
D5T2F-P:IDIC-4F	0.86	16.85	0.63	9.36	[151]
NDT-3T-R6:IDIC-4F	0.78	18.89	0.73	10.7	[152]
DPPEZnP-TBO:6TIC	0.80	20.44	0.74	12.08	[154]
BTR-Cl:Y6	0.86	24.17	0.66	13.61	[156]
BTEC-2F:Y6	0.85	21.55	0.72	13.34	[157]
BSFTR:Y6	0.85	23.16	0.70	13.69	[158]
SM1-F:Y6	0.87	23.25	0.70	14.07	[159]
ZR1:Y6	0.86	24.34	0.68	14.34	[160]
B1:BO-4Cl	0.83	25.27	0.73	15.3	[15]
BT-2F:N3	0.85	24.28	0.75	15.39	[18]
DR3TBDTT:O-IDTBR	1.15	11.06	0.50	6.36	[161]
DRCN5T:F-2Cl	0.91	15.97	0.68	9.62	[162]
BSFTR:NBDTP-F$_{out}$	0.80	21.69	0.71	12.26	[163]
DCAO3TBDTT: DR3TBDTT-S-E:IDIC	0.91	16.37	0.67	10.04	[165]
ZnP-TSEH:6TIC:4TIC	0.81	25.95	0.76	15.88	[16]
B1:BO-4Cl:Y7	0.84	25.52	0.76	16.28	[17]
BTR:PC$_{71}$BM:NITI	0.94	19.5	0.74	13.63	[166]
BTR-Cl:Y6:PC$_{71}$BM	0.84	23.75	0.77	15.34	[14]

2.5 Summary and Outlook

p-Type molecular photovoltaic materials are promising candidates for highly efficient OSCs. The molecular innovation is one of the important factors that promote the development of SMSCs, leading to great breakthroughs. In addition, when they are added into some PSC systems, impressive performance results have been obtained [170–172]. In this chapter, we mainly focus on their applications in SMSCs, and representative examples have been reviewed systematically. Regardless of the SMDs applied in VD- or SP-SMSCs, the molecular design includes backbone engineering (electron-donating and electron-accepting moieties) and accessional group optimization (side chain and π-linker), which can affect not only photoelectronic properties but also morphology and thus performance. Nowadays, fullerene-based SMSCs have achieved PCEs of 10% in both VD- and SP-SMSCs, and NFA-based SMSCs have showed more excellent PCEs of over 15%. However, for the future of SMSCs, some challenges remain. Herein, we will discuss and provide some suggestions as follows.

For VD-SMSCs, how to achieve higher performance is the most challenging spot. Widely used C_{60} and C_{70} cannot take full advantages of the NIR light and often exhibit an large E_{loss}. Taking into account of the merits of solution-processable SM-NFAs, designing vacuum-processable SM-NFAs is promising to promote the performance of VD-SMSCs, which needs to simultaneously consider molecular weight and thermal stability. Also, with the new materials, morphology control methods may be changed, which needs further investigation. Another topic is how to reduce cost. The materials that can be obtained with economical ingredients, high yields, and simple synthesis and purification process are preferred. For SP-SMSCs, the pursuit of higher performance is still one of the development directions, which requires the cooperation of material innovation and morphology optimization. For cost reduction, the requirements are the same as above. Also, the inevitable thickness variation during large-area fabrication and the expectation of sufficient light harvest indicate that thick film is particularly needed. Therefore, high mobility is necessary. In addition, from the perspective of environmental friendliness, the use of halogenated solvents or even some halogen-free solvents such as toluene and xylenes should be prudent.

Overall, we believe that continued development of materials including SMDs and SM-NFAs, together with morphology optimization, device engineering, and deep investigation on device physics, will contribute to the advance of SMSCs.

References

1 Li, G., Zhu, R., and Yang, Y. (2012). *Nat. Photonics* 6: 153.
2 Lu, L., Zheng, T., Wu, Q. et al. (2015). *Chem. Rev.* 115: 12666.
3 Inganäs, O. (2018). *Adv. Mater.* 30: 1800388.
4 Lin, Y., Zhan, X., and Li, Y. (2012). *Chem. Soc. Rev.* 41: 4245.
5 Huo, Y., Zhang, H.-L., and Zhan, X. (2019). *ACS Energy Lett.* 4: 1241.
6 Kallmann, H. and Pope, M. (1959). *J. Chem. Phys.* 30: 585.
7 Tang, C.W. (1986). *Appl. Phys. Lett.* 48: 183.
8 Yu, G., Gao, J., Hummelen, J.C. et al. (1995). *Science* 270: 1789.
9 Che, X., Xiao, X., Zimmerman, J.D. et al. (2014). *Adv. Energy Mater.* 4: 1400568.
10 Li, T.-y., Meyer, T., Ma, Z. et al. (2017). *J. Am. Chem. Soc.* 139: 13636.
11 Li, Y., Sheriff, H.K.M. Jr., Liu, X. et al. (2019). *J. Am. Chem. Soc.* 141: 18204.
12 Kan, B., Li, M., Zhang, Q. et al. (2015). *J. Am. Chem. Soc.* 137: 3886.
13 Wan, J., Xu, X., Zhang, G. et al. (2017). *Energy Environ. Sci.* 10: 1739.
14 Hu, D., Yang, Q., Chen, H. et al. (2020). *Energy Environ. Sci.* 13: 2134.
15 Qin, J., An, C., Zhang, J. et al. (2020). *Sci. China Mater.* 63: 1142.
16 Li, N., Kan, Y., Gao, K. et al. (2020). *Joule* 4: 2223.
17 Jiang, M., Bai, H., Zhi, H. et al. (2021). *Energy Environ. Sci.* 14: 3945.

18 Ge, J., Hong, L., Song, W. et al. (2021). *Adv. Energy Mater.* 11: 2100800.
19 Burlingame, Q., Huang, X., Liu, X. et al. (2019). *Nature* 573: 394.
20 Heliatek GmbH. www.heliatek.com (accessed 21 September 2021).
21 Schulze, K., Uhrich, C., Schüppel, R. et al. (2006). *Adv. Mater.* 18: 2872.
22 Schüeppel, R., Schmidt, K., Uhrich, C. et al. (2008). *Phys. Rev. B* 77: 085311.
23 Mishra, A., Uhrich, C., Reinold, E. et al. (2011). *Adv. Energy Mater.* 1: 265.
24 Fitzner, R., Reinold, E., Mishra, A. et al. (2011). *Adv. Funct. Mater.* 21: 897.
25 Mishra, A., Popovic, D., Vogt, A. et al. (2014). *Adv. Mater.* 26: 7217.
26 Fitzner, R., Elschner, C., Weil, M. et al. (2012). *Adv. Mater.* 24: 675.
27 Schulze, K., Riede, M., Brier, E. et al. (2008). *J. Appl. Phys.* 104: 074511.
28 Fitzner, R., Mena-Osteritz, E., Mishra, A. et al. (2012). *J. Am. Chem. Soc.* 134: 11064.
29 Meerheim, R., Körner, C., and Leo, K. (2014). *Appl. Phys. Lett.* 105: 063306.
30 Wynands, D., Levichkova, M., Riede, M. et al. (2010). *J. Appl. Phys.* 107: 014517.
31 Wynands, D., Levichkova, M., Leo, K. et al. (2010). *Appl. Phys. Lett.* 97: 073503.
32 Sakai, J., Taima, T., and Saito, K. (2008). *Org. Electron.* 9: 582.
33 Lin, L.-Y., Chen, Y.-H., Huang, Z.-Y. et al. (2011). *J. Am. Chem. Soc.* 133: 15822.
34 Griffith, O.L., Liu, X., Amonoo, J.A. et al. (2015). *Phys. Rev. B* 92: 085404.
35 Chen, Y.-H., Lin, L.-Y., Lu, C.-W. et al. (2012). *J. Am. Chem. Soc.* 134: 13616.
36 Che, X., Li, Y., Qu, Y., and Forrest, S.R. (2018). *Nat. Energy* 3: 422.
37 Chen, C.-H., Ting, H.-C., Li, Y.-Z. et al. (2019). *ACS Appl. Mater. Interfaces* 11: 8337.
38 Lu, H.-I., Lu, C.-W., Lee, Y.-C. et al. (2014). *Chem. Mater.* 26: 4361.
39 Che, X., Chung, C.-L., Liu, X. et al. (2016). *Adv. Mater.* 28: 8248.
40 Che, X., Chung, C.-L., Hsu, C.-C. et al. (2018). *Adv. Energy Mater.* 8: 1703603.
41 Kraner, S., Widmer, J., Benduhn, J. et al. (2015). *Phys. Status Solidi A* 212: 2747.
42 Meiss, J., Holzmueller, F., Gresser, R. et al. (2011). *Appl. Phys. Lett.* 99: 193307.
43 Li, T.-y., Benduhn, J., Li, Y. et al. (2018). *J. Mater. Chem. A* 6: 18583.
44 Li, T.-y., Benduhn, J., Qiao, Z. et al. (2019). *J. Phys. Chem. Lett.* 10: 2684.
45 Kronenberg, N.M., Steinmann, V., Burckstummer, H. et al. (2010). *Adv. Mater.* 22: 4193.
46 Steinmann, V., Kronenberg, N.M., Lenze, M.R. et al. (2011). *Adv. Energy Mater.* 1: 888.
47 Ojala, A., Bürckstümmer, H., Hwang, J. et al. (2012). *J. Mater. Chem.* 22: 4473.
48 Ojala, A., Petersen, A., Fuchs, A. et al. (2012). *Adv. Funct. Mater.* 22: 86.
49 Peumans, P. and Forrest, S.R. (2001). *Appl. Phys. Lett.* 79: 126.
50 Pfuetzner, S., Mickel, C., Jankowski, J. et al. (2011). *Org. Electron.* 12: 435.
51 Meiss, J., Merten, A., Hein, M. et al. (2012). *Adv. Funct. Mater.* 22: 405.
52 Mutolo, K.L., Mayo, E.I., Rand, B.P. et al. (2006). *J. Am. Chem. Soc.* 128: 8108.
53 Pandey, R., Zou, Y., and Holmes, R.J. (2012). *Appl. Phys. Lett.* 101: 033308.
54 Verreet, B., Schols, S., Cheyns, D. et al. (2009). *J. Mater. Chem.* 19: 5295.
55 Zhang, M., Wang, H., Tian, H. et al. (2011). *Adv. Mater.* 23: 4960.
56 Brendel, M., Krause, S., Steindamm, A. et al. (2015). *Adv. Funct. Mater.* 25: 1565.

57 Petritsch, K., Dittmer, J.J., Marseglia, E.A. et al. (2000). *Sol. Energy Mater. Sol. Cells* 61: 63.
58 Schmidt-Mende, L., Fechtenkötter, A., Müllen, K. et al. (2001). *Science* 293: 1119.
59 Sun, X., Zhou, Y., Wu, W. et al. (2006). *J. Phys. Chem. B* 110: 7702.
60 Ma, C.-Q., Mena-Osteritz, E., Debaerdemaeker, T. et al. (2007). *Angew. Chem. Int. Ed.* 46: 1679.
61 Lloyd, M.T., Mayer, A.C., Subramanian, S. et al. (2007). *J. Am. Chem. Soc.* 129: 9144.
62 Mishra, A. and Bäuerle, P. (2012). *Angew. Chem. Int. Ed.* 51: 2020.
63 Walker, B., Kim, C., and Nguyen, T.-Q. (2011). *Chem. Mater.* 23: 470.
64 Liu, Y., Zhou, J., Wan, X., and Chen, Y. (2009). *Tetrahedron* 65: 5209.
65 Yin, B., Yang, L., Liu, Y. et al. (2010). *Appl. Phys. Lett.* 97: 023303.
66 Liu, Y., Wan, X., Wang, F. et al. (2011). *Adv. Energy Mater.* 1: 771.
67 Li, Z., He, G., Wan, X. et al. (2012). *Adv. Energy Mater.* 2: 74.
68 Zhang, Q., Kan, B., Liu, F. et al. (2014). *Nat. Photonics* 9: 35.
69 Long, G., Wan, X., Kan, B. et al. (2013). *Adv. Energy Mater.* 3: 639.
70 Demeter, D., Rousseau, T., Leriche, P. et al. (2011). *Adv. Funct. Mater.* 21: 4379.
71 Wang, Z., Li, Z., Liu, J. et al. (2016). *ACS Appl. Mater. Interfaces* 8: 11639.
72 Liu, F., Fan, H., Zhang, Z., and Zhu, X. (2016). *ACS Appl. Mater. Interfaces* 8: 3661.
73 Liu, Y., Yang, Y., Chen, C.-C. et al. (2013). *Adv. Mater.* 25: 4657.
74 Zhang, Q., Wang, Y., Kan, B. et al. (2015). *Chem. Commun.* 51: 15268.
75 Zhou, J., Wan, X., Liu, Y. et al. (2012). *J. Am. Chem. Soc.* 134: 16345.
76 Ni, W., Li, M., Wan, X. et al. (2014). *RSC Adv.* 4: 31977.
77 Kan, B., Zhang, Q., Li, M. et al. (2014). *J. Am. Chem. Soc.* 136: 15529.
78 Zhou, J., Zuo, Y., Wan, X. et al. (2013). *J. Am. Chem. Soc.* 135: 8484.
79 Li, M., Liu, F., Wan, X. et al. (2015). *Adv. Mater.* 27: 6296.
80 Wang, Z., Zhu, X., Zhang, J. et al. (2018). *J. Am. Chem. Soc.* 140: 1549.
81 Cui, C., Guo, X., Min, J. et al. (2015). *Adv. Mater.* 27: 7469.
82 Min, J., Cui, C., Heumueller, T. et al. (2016). *Adv. Energy Mater.* 6: 1600515.
83 Wang, Z., Xu, X., Li, Z. et al. (2016). *Adv. Electron. Mater.* 2: 1600061.
84 Kan, B., Zhang, Q., Liu, F. et al. (2015). *Chem. Mater.* 27: 8414.
85 Sun, K., Xiao, Z., Lu, S. et al. (2015). *Nat. Commun.* 6: 6013.
86 Song, H.G., Kim, Y.J., Lee, J.S. et al. (2016). *ACS Appl. Mater. Interfaces* 8: 34353.
87 Zhu, X., Xia, B., Lu, K. et al. (2016). *Chem. Mater.* 28: 943.
88 Shen, S., Jiang, P., He, C. et al. (2013). *Chem. Mater.* 25: 2274.
89 Patra, D., Huang, T.-Y., Chiang, C.-C. et al. (2013). *ACS Appl. Mater. Interfaces* 5: 9494.
90 Lin, Y., Ma, L., Li, Y. et al. (2014). *Adv. Energy Mater.* 4: 1300626.
91 Du, Z., Chen, W., Chen, Y. et al. (2014). *J. Mater. Chem. A* 2: 15904.
92 Deng, D., Zhang, Y., Zhang, J. et al. (2016). *Nat. Commun.* 7: 13740.
93 Ni, W., Li, M., Wan, X. et al. (2014). *Sci. China Chem.* 58: 339.
94 Xu, S., Zhou, Z., Fan, H. et al. (2016). *J. Mater. Chem. A* 4: 17354.

95 Kumar, C.V., Cabau, L., Koukaras, E.N. et al. (2015). *J. Mater. Chem. A* 3: 4892.
96 Cheng, M., Xu, B., Chen, C. et al. (2015). *Adv. Energy Mater.* 5: 1401720.
97 Li, M., Gao, K., Wan, X. et al. (2016). *Nat. Photonics* 11: 85.
98 Zhou, J., Wan, X., Liu, Y. et al. (2011). *Chem. Mater.* 23: 4666.
99 Ni, W., Li, M., Liu, F. et al. (2015). *Chem. Mater.* 27: 6077.
100 Kim, K.-H., Yu, H., Kang, H. et al. (2013). *J. Mater. Chem. A* 1: 14538.
101 Ye, D., Li, X., Yan, L. et al. (2013). *J. Mater. Chem. A* 1: 7622.
102 Li, W., Deng, W., Wu, K. et al. (2016). *J. Mater. Chem. C* 4: 1972.
103 Wessendorf, C.D., Schulz, G.L., Mishra, A. et al. (2014). *Adv. Energy Mater.* 4: 1400266.
104 Li, M., Ni, W., Feng, H. et al. (2015). *Chin. J. Chem.* 33: 852.
105 Feng, H., Li, M., Ni, W. et al. (2015). *J. Mater. Chem. A* 3: 16679.
106 Takacs, C.J., Sun, Y., Welch, G.C. et al. (2012). *J. Am. Chem. Soc.* 134: 16597.
107 van der Poll, T.S., Love, J.A., Nguyen, T.-Q., and Bazan, G.C. (2012). *Adv. Mater.* 24: 3646.
108 Gupta, V., Kyaw, A.K., Wang, D.H. et al. (2013). *Sci. Rep.* 3: 1965.
109 Lai, L.F., Love, J.A., Sharenko, A. et al. (2014). *J. Am. Chem. Soc.* 136: 5591.
110 Liu, X., Sun, Y., Hsu, B.B.Y. et al. (2014). *J. Am. Chem. Soc.* 136: 5697.
111 Sun, Y., Seifter, J., Huo, L. et al. (2015). *Adv. Energy Mater.* 5: 1400987.
112 Gupta, V., Lai, L.F., Datt, R. et al. (2016). *Chem. Commun.* 52: 8596.
113 Do, K., Cho, N., Siddiqui, S.A. et al. (2015). *Dyes Pigm.* 120: 126.
114 Wolf, J., Babics, M., Wang, K. et al. (2016). *Chem. Mater.* 28: 2058.
115 Duan, R., Cui, Y., Zhao, Y. et al. (2016). *ChemSusChem* 9: 973.
116 Tang, A., Zhan, C., and Yao, J. (2015). *Adv. Energy Mater.* 5: 1500059.
117 Kim, J.-H., Park, J.B., Yang, H. et al. (2015). *ACS Appl. Mater. Interfaces* 7: 23866.
118 Yong, W., Zhang, M., Xin, X. et al. (2013). *J. Mater. Chem. A* 1: 14214.
119 Intemann, J.J., Yao, K., Ding, F. et al. (2015). *Adv. Funct. Mater.* 25: 4889.
120 Wang, J.-L., Yin, Q.-R., Miao, J.-S. et al. (2015). *Adv. Funct. Mater.* 25: 3514.
121 Wang, J.-L., Liu, K.-K., Yan, J. et al. (2016). *J. Am. Chem. Soc.* 138: 7687.
122 Gao, K., Li, L., Lai, T. et al. (2015). *J. Am. Chem. Soc.* 137: 7282.
123 Gao, K., Miao, J., Xiao, L. et al. (2016). *Adv. Mater.* 28: 4727.
124 Gao, K., Deng, W., Xiao, L. et al. (2016). *Nano Energy* 30: 639.
125 Liang, T., Xiao, L., Gao, K. et al. (2017). *ACS Appl. Mater. Interfaces* 9: 7131.
126 Xiao, L., Chen, S., Gao, K. et al. (2016). *ACS Appl. Mater. Interfaces* 8: 30176.
127 Zhou, X., Tang, W., Bi, P. et al. (2018). *J. Mater. Chem. A* 6: 14675.
128 Arrechea, S., Aljarilla, A., de la Cruz, P. et al. (2017). *J. Mater. Chem. C* 5: 4742.
129 Arrechea, S., Aljarilla, A., de la Cruz, P. et al. (2016). *Nanoscale* 8: 17953.
130 Chen, S., Yan, L., Xiao, L. et al. (2017). *J. Mater. Chem. A* 5: 25460.
131 Lai, T., Chen, X., Xiao, L. et al. (2017). *Chem. Commun.* 53: 5113.
132 Xiao, L., Lai, T., Liu, X. et al. (2018). *J. Mater. Chem. A* 6: 18469.
133 Zhen, Y., Tanaka, H., Harano, K. et al. (2015). *J. Am. Chem. Soc.* 137: 2247.
134 Yuan, L., Zhao, Y., Zhang, J. et al. (2015). *Adv. Mater.* 27: 4229.
135 Yuan, L., Lu, K., Xia, B. et al. (2016). *Adv. Mater.* 28: 5980.
136 Walker, B., Tamayo, A.B., Dang, X.-D. et al. (2009). *Adv. Funct. Mater.* 19: 3063.

137 Cheng, M., Chen, C., Yang, X. et al. (2015). *Chem. Mater.* 27: 1808.
138 Chen, G., Sasabe, H., Sasaki, Y. et al. (2014). *Chem. Mater.* 26: 1356.
139 Zhang, J., Tan, H.S., Guo, X. et al. (2018). *Nat. Energy* 3: 720.
140 Wadsworth, A., Moser, M., Marks, A. et al. (2019). *Chem. Soc. Rev.* 48: 1596.
141 Yue, Q., Liu, W., and Zhu, X. (2020). *J. Am. Chem. Soc.* 142: 11613.
142 Cnops, K., Rand, B.P., Cheyns, D. et al. (2014). *Nat. Commun.* 5: 3406.
143 Liang, N., Meng, D., Ma, Z. et al. (2017). *Adv. Energy Mater.* 7: 1601664.
144 Lin, Y., Wang, J., Zhang, Z.-G. et al. (2015). *Adv. Mater.* 27: 1170.
145 Lin, Y., He, Q., Zhao, F. et al. (2016). *J. Am. Chem. Soc.* 138: 2973.
146 Yang, L., Zhang, S., He, C. et al. (2017). *J. Am. Chem. Soc.* 139: 1958.
147 Bin, H., Yang, Y., Zhang, Z.-G. et al. (2017). *J. Am. Chem. Soc.* 139: 5085.
148 Qiu, B., Xue, L., Yang, Y. et al. (2017). *Chem. Mater.* 29: 7543.
149 Guo, J., Bin, H., Wang, W. et al. (2018). *J. Mater. Chem. A* 6: 15675.
150 Li, X., Wang, Y., Zhu, Q. et al. (2019). *J. Mater. Chem. A* 7: 3682.
151 Duan, T., Gao, J., Babics, M. et al. (2020). *Solar RRL* 4: 1900472.
152 Li, H., Wu, Q., Zhou, R. et al. (2018). *Adv. Energy Mater.* 9: 1803175.
153 Yang, L., Zhang, S., He, C. et al. (2018). *Chem. Mater.* 30: 2129.
154 Gao, K., Jo, S.B., Shi, X. et al. (2019). *Adv. Mater.* 31: 1807842.
155 Yuan, J., Zhang, Y., Zhou, L. et al. (2019). *Joule* 3: 1140.
156 Chen, H., Hu, D., Yang, Q. et al. (2019). *Joule* 3: 3034.
157 Ge, J., Xie, L., Peng, R. et al. (2020). *Angew. Chem. Int. Ed.* 59: 2808.
158 Yue, Q., Wu, H., Zhou, Z. et al. (2019). *Adv. Mater.* 31: 1904283.
159 Qiu, B., Chen, Z., Qin, S. et al. (2020). *Adv. Mater.* 32: 1908373.
160 Zhou, R., Jiang, Z., Yang, C. et al. (2019). *Nat. Commun.* 10: 5393.
161 Yang, D., Wang, Y., Sano, T. et al. (2018). *J. Mater. Chem. A* 6: 13918.
162 Wang, Y., Wang, Y., Kan, B. et al. (2018). *Adv. Energy Mater.* 8: 1802021.
163 Wu, H., Yue, Q., Zhou, Z. et al. (2019). *J. Mater. Chem. A* 7: 15944.
164 Lu, L., Kelly, M.A., You, W., and Yu, L. (2019). *Nat. Photonics* 9: 491.
165 Chang, Y., Chang, Y., Zhu, X. et al. (2019). *Adv. Energy Mater.* 9: 1900190.
166 Zhou, Z., Xu, S., Song, J. et al. (2018). *Nat. Energy* 3: 952.
167 Liang, R.-Z., Zhang, Y., Savikhin, V. et al. (2019). *Adv. Energy Mater.* 9: 1802836.
168 Huo, Y., Gong, X.-T., Lau, T.-K. et al. (2018). *Chem. Mater.* 30: 8661.
169 Privado, M., Guijarro, F.G., Cruz, P.D.L. et al. (2021). *ACS Appl. Mater. Interfaces* 13: 6461.
170 Kumari, T., Lee, S.M., Kang, S.-H. et al. (2017). *Energy Environ. Sci.* 10: 258.
171 Yan, T., Ge, J., Lei, T. et al. (2019). *J. Mater. Chem. A* 7: 25894.
172 Miao, J., Meng, B., Ding, Z. et al. (2020). *J. Mater. Chem. A* 8: 10983.

3

Fullerene Acceptors

Zuo Xiao

National Center for Nanoscience and Technology, Center for Excellence in Nanoscience (CAS), Key Laboratory of Nanosystem and Hierarchical Fabrication (CAS), No.11 ZhongGuanCun BeiYiTiao, Beijing 100190, China

3.1 Introduction

In 1985, Kroto et al. observed the signal of C_{60} in mass spectrum when they tried to study the formation mechanism of long-chain carbon molecules in interstellar space and circumstellar shells [1]. This serendipity lead to the discovery of an important allotrope of carbon, the fullerenes. In 1990, Krätschmer et al. realized the macroscopic preparation of fullerenes [2]. From then on, fullerenes have attracted worldwide research interests, not only due to their fancy structures but also due to their unique physical and chemical properties. The past 35 years' research greatly expanded the knowledge on fullerenes. Fullerenes are cage-like molecules composed of 12 pentagons and a variable number of hexagons [3]. The I_h-symmetrical C_{60} and the D_{5h}-symmetrical C_{70} are the two most abundant fullerenes in the crude fullerene soot (Figure 3.1a) [2]. Fullerenes with fewer than 60 carbon atoms are called "smaller fullerenes" [9], and those with more than 70 carbon atoms are called "higher fullerenes" [10]. Although with low abundance, some smaller fullerenes and higher fullerenes were isolated and characterized in the experiment (Figure 3.1b) [4, 5, 11]. The cage of the fullerene molecules could be either empty or filled with atoms, molecules, or clusters. The latter are called endohedral fullerenes (Figure 3.1c) [6–8, 12]. The carbon atoms of fullerenes can also be replaced by heteroatoms (such as nitrogen and boron), affording so-called heterofullerenes [13]. Derivatives of azafullerenes (nitrogen-containing heterofullerenes) have been synthesized (Figure 3.1d) [14–16]. Fullerenes have abundant π-electrons and curved conjugated π-system, which render them special chemical reactivity and optoelectronic properties. Fullerenes can undergo characteristic [2+*n*] cycloadditions at the 6,6-double bonds (Figure 3.1e) [17]. These reactions change the sp^2-hybridized carbons into sp^3-hybridized ones and release the strain of fullerenes. Through these reactions, various functional groups can be introduced to fullerenes to make useful materials. Fullerene-based functional materials have been widely applied in optoelectronic devices and biological and medicinal fields [3, 18].

Organic Solar Cells: Materials Design, Technology and Commercialization, First Edition. Edited by Liming Ding.

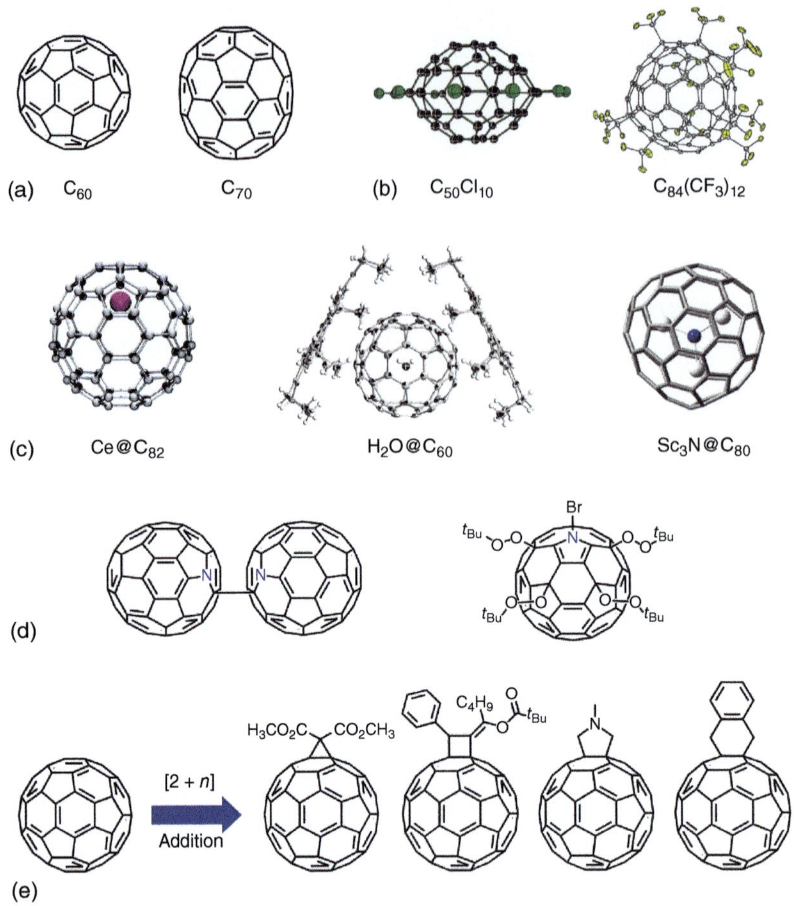

Figure 3.1 A brief survey of fullerenes. (a) C_{60} and C_{70}. Source: Zuo Xiao. (b) The smaller fullerene derivative $C_{50}Cl_{10}$ (Source: Chen [4]. Reproduced with permission of John Wiley & Sons.) and the higher fullerene derivative $C_{84}(CF_3)_{12}$ (Source: Kareev et al. [5]. Reproduced with permission of John Wiley & Sons.). (c) Endohedral fullerenes Ce@C_{82} (Source: Yamada et al. [6]. Reproduced with permission of (2010) American Chemical Society), H_2O@C_{60} (Source: Kurotobi and Murata [7]. Reproduced with permission of (2011) American Association for the Advancement of Science) and Sc_3N@C_{80} (Source: Zhang et al. [8]. Reproduced with permission of (2013) American Chemical Society). (d) Azafullerene derivatives $(C_{59}N)_2$ and $(C_{59}N)(OO^tBu)_4Br$. (e) Representative [2+*n*] addition fullerene derivatives. Source: Zuo Xiao.

Due to the strong electron-affinity, low-lying lowest unoccupied molecular orbital (LUMO) levels, and excellent electron transport capabilities, fullerenes and their derivatives have found important application as electron-accepting materials for organic solar cells (OSCs) [19, 20]. In 1992, Sariciftci et al. observed the ultrafast photoinduced electron transfer from a conjugated polymer MEH-PPV to the fullerene C_{60}, suggesting that fullerenes could be potential acceptors for organic photovoltaics [21]. In 1995, Heeger and coworkers invented bulk-heterojunction (BHJ) solar cells by blending the polymer donor MEH-PPV with a fullerene acceptor

(C_{60} or $PC_{61}BM$) [22]. In the active layer of a BHJ cell, the donor and acceptor undergo phase separation to form bicontinuous interpenetrating networks at the nanoscale. Such morphology generates sufficient interfaces between donor and acceptor, promotes exciton dissociation, and facilitates charge carrier transport. Under 430 nm illumination, the MEH-PPV:$PC_{61}BM$ solar cells showed ~90% internal quantum efficiency, which is much higher than that of pure conjugated polymer devices. The invention of BHJ solar cells opened a new avenue toward efficient and cost-effective solar cells, thus triggering extensive research [23]. One of the most popular directions is developing high-performance fullerene acceptor materials for OSCs. In the past 25 years, numerous fullerene acceptors have been designed and synthesized. In this section, representative fullerene acceptors for OSCs are summarized. These acceptors are categorized into five types: pristine fullerenes, fullerene monoadducts, fullerene bisadducts, fullerene multiadducts, and unconventional fullerenes. Molecular structural features, synthesis, material properties, and solar cells performance for different types of fullerene acceptors are introduced in detail.

3.2 Fullerene Acceptors for Organic Solar Cells

3.2.1 Pristine Fullerenes

Nowadays, bulk quantities of pristine fullerenes can be produced from either the vaporization of graphite or the controlled combustion of aromatic hydrocarbons. Krätschmer et al. first reported the macroscopic preparation of fullerenes from the soot of a vaporized graphite rod [2]. Smalley and coworkers improved this method by inventing a "contact arc" vaporization process and obtained C_{60} in gram quantities [24]. The content of fullerenes in the original soot reached 10%. Howard et al. reported an efficient combustion method for producing fullerenes [25]. By controlling the temperature, pressure, carbon/oxygen ratio, and residence time in the flame, they could obtain 3 g of fullerenes per kilogram of fuel carbon burned, with the C_{70}/C_{60} ratio varying from 0.26 to 5.7.

The above preparation methods make it possible to provide enough quantities of pristine fullerenes (predominantly C_{60} and C_{70}) for solar cells application. Due to the low solubility of pristine fullerenes, they were mostly applied in vacuum-deposited OSCs (Table 3.1). In 2004, Forrest and coworkers fabricated vacuum-deposited solar cells by using C_{60} as the acceptor and the copper phthalocyanine (CuPc) as the donor [26]. The cells delivered a power conversion efficiency (PCE) of 3.5% at 1 sun, indicating the good potential of pristine fullerenes for vacuum-deposited all-small-molecule OSCs. In 2005, Forrest and coworkers improved the PCE of CuPc:C_{60} cells to 5.0% by adopting a hybrid planar-mixed heterojunction device structure [27]. In 2009, Leo and coworkers reported vacuum-deposited BHJ cells based on a zinc phthalocyanine (ZnPc) donor and a C_{60} or C_{70} acceptor [28]. The ZnPc:C_{60} and ZnPc:C_{70} cells delivered PCEs of 2.27% and 2.87%, respectively. The better performance of the ZnPc:C_{70} cells is due to the higher short-circuit current density (J_{sc}), which is attributed to the stronger visible-light-absorption

Table 3.1 Performance of pristine C_{60}- and C_{70}-based solar cells.

Processing	Device type	Donor:Acceptor	V_{oc} (V)	J_{sc} (mA cm^{-2})	FF (%)	PCE (%)	References
Vacuum	Single junction	CuPc:C_{60}	0.50	15.4	46	3.5	[26]
Vacuum	Single junction	CuPc:C_{60}	0.54	15.0	61	5.0	[27]
Vacuum	Single junction	ZnPc:C_{60}	0.56	7.52	54.3	2.27	[28]
Vacuum	Single junction	ZnPc:C_{70}	0.56	9.88	52.2	2.87	[28]
Vacuum	Single junction	TPDCDTS:C_{60}	0.88	6.56	46	2.69	[29]
Vacuum	Single junction	TPDCDTS:C_{70}	0.83	9.53	48	3.82	[29]
Vacuum	Single junction	HB194:C_{60}	0.962	12.6	47	6.1	[30]
Vacuum	Single junction	DTDCTP:C_{70}	0.95	12.1	56	6.4	[31]
Vacuum	Single junction	DBP:C_{70}	0.86	10.26	59	5.19	[32]
Vacuum	Single junction	DBP:C_{70}	0.93	13.2	66	8.1	[33]
Vacuum	Single junction	iBuBTDC:C_{70}	0.94	16.5	60	9.3	[34]
Vacuum	Tandem	SubNc:C_{60} (Front) SubPc:C_{60} (Rear)	1.92	4.3	62	5.15	[35]
Vacuum	Tandem	F4-ZnPc:C_{60} (Front) DCV6T:C_{60} (Rear)	1.589	6.18	61.86	6.07	[36]
Vacuum + solution	Tandem	ZnPc:C_{70} (Front) DPSQ:C_{70} (Rear)	1.97	6.2	54	6.6	[37]
Vacuum	Tandem	DCV5T-Me:C_{60} (Front) DCV5T-Me:C_{60} (Middle) DCV5T-Me:C_{60} (Rear)	2.79	4.8	72.4	9.7	[38]
Vacuum + solution	Tandem	DTDCPB:C_{70} (Front) PCE-10:BT-CIC: TT-FIC (Rear)	1.56	13.8	71	15.4	[39]
Solution	Single junction	DTDCPB:C_{70}	0.95	13.4	46	5.9	[40]
Solution	Single junction	PTB7:C_{60}:C_{70}	0.67	12.8	62	5.3	[41]

of C_{70} than C_{60}. The superior light-harvesting capability of C_{70} is due to its lower molecular symmetry, which relaxes the dipole selection rules of low-energy transitions and enhances the oscillator strength [42]. This work demonstrated the superiority of C_{70} for vacuum-deposited OSCs. Beside phthalocyanine donors, researchers also developed other small-molecule donors to match C_{60} or C_{70}. In 2011, Wong and coworkers designed a D–π–A small-molecule donor TPDCDTS [29]. The vacuum-deposited TPDCDTS:C_{70} cells afforded a PCE of 3.82%, which is higher than that of the TPDCDTS:C_{60} cells (2.69%). Similar to Leo's results, the higher performance of the C_{70} cells came from the improved J_{sc}. In the same year, Würthner and coworkers reported a small-molecule donor based on a merocyanine dye HB194. The HB194:C_{60} cells gave the highest PCE of 6.1% [30]. In 2012, Wong and coworkers designed a D–A–A type small-molecule donor DTDCTP [31]. The vacuum-deposited planar-mixed heterojunction solar cells based on DTDCTP and C_{70} afforded a high PCE of 6.4%. In 2013, Yokoyama and coworkers reported a polycyclic aromatic molecule tetraphenyldibenzoperiflanthene (DBP) working as the donor for vacuum-deposited cells [32]. The DBP:C_{70} cells afforded a high PCE of 5.19%. By optimizing an electron filter buffer layer, Forrest and coworkers improved the PCE of DBP:C_{70} cells to 8.1% [33]. In 2018, Forrest and coworkers developed a series of D–A–A' type small-molecule donors based on an asymmetric heterotetracene donor unit [34]. The alkyl chain on the donor unit shows significant influence to the device efficiency. Among the D–A–A' donors, the iBuBTDC with an isobutyl chain exhibited the best performance. The vacuum-deposited iBuBTDC:C_{70} cells gave the highest PCE of 9.3%.

Besides being used in single-junction cells, pristine C_{60} and C_{70} were also applied in vacuum-deposited tandem cells and afforded higher efficiencies [43]. In 2010, Cheyns et al. reported C_{60}-based tandem solar cells with a high open-circuit voltage (V_{oc}) of 1.92 V and a PCE of 5.15% [35]. The tandem cells employed a chloroboron subnaphthalocyanine (SubNc) donor for the front cell and a chloroboron subphthalocyanine (SubPc) donor for the rear cell. In 2011, Riede et al. fabricated tandem cells with a fluorinated ZnPc donor F4-ZnPc and a C_{60} acceptor-based front cell and a sexithiophene-cored donor DCV6T and a C_{60} acceptor-based rear cell [36]. The tandem cells with a large effective area of 2 cm^2 gave a certified PCE of 6.07%. In 2012, Forrest and coworkers reported tandem cells based on a ZnPc:C_{70} front cell and a squaraine dye DPSQ and C_{70}-based rear cell [37]. A 6.6% PCE was achieved. In 2014, Leo and coworkers reported a triple-junction tandem cells based on a quinquethiophene-cored donor DCV5T-Me and a C_{60} acceptor [38]. The triple-junction cells afforded an extraordinary high V_{oc} of 2.79 V and a PCE of 9.7%. In 2018, Forrest and coworkers reported highly efficient tandem cells with a PCE of 15.4% [39]. The tandem cells adopted a vacuum-deposited DTDCPB:C_{70} front cell and a solution-processed PCE-10:BT-CIC:TT-FIC nonfullerene ternary rear cell.

Although less reported, C_{60} and C_{70} can also be applied in solution-processed BHJ solar cells. In 2014, Lin et al. reported solution-processed BHJ cells based on D–A–A type small-molecule donors and C_{70} [40]. By using 1,2-dichlorobenzene as solvent, DTDCPB as the donor, and a D/A ratio of 1 : 2.2, they could obtain homogeneous blend active layer *via* either spin-coating or bar-coating techniques.

The resulted DTDCPB:C_{70} cells delivered PCEs up to 5.9%, which is ~90% of that of vacuum-deposited reference. In 2017, Müller and coworkers realized a high glass transition temperature of 224 °C for a OSCs active layer by using a pristine fullerene acceptor alloy, C_{60}:C_{70} (1 : 1) [41]. The PTB7:C_{60}:C_{70} (2 : 1 : 1) cells not only afforded a decent PCE of 5.3% but also showed excellent thermal stability.

3.2.2 Fullerene Monoadducts

To make fullerene acceptors more adaptable for solution-processed OSCs, it is necessary to introduce solubilizing side chains and functional groups on to fullerenes. Pristine fullerenes can only undergo addition reactions because there is no place on fullerene for substitution. The [2+*n*] cycloadditions and the 1,4-addition are the typical reactions making fullerene derivatives [44]. In this section, fullerene monoadduct acceptors developed *via* these addition reactions are introduced sequentially (Figure 3.2).

3.2.2.1 [2+1] Addition Derivatives

The methanofullerene derivative $PC_{61}BM$ is the first and the most widely used fullerene acceptor in OSCs [22]. In 1995, Hummelen et al. reported the synthesis of $PC_{61}BM$ *via* the reaction between C_{60} and an *in-situ* generated diazo compound [45]. The reaction mechanism includes several steps: C_{60} undergoes the 1,3-dipolar addition with the diazo species to give an pyrazole intermediate; with thermal extrusion of nitrogen, the intermediate turned into a pair of isomers, the [6,6]methanofullerene ($PC_{61}BM$) and the [5,6]fulleroid; the [5,6]fulleroid is not stable and undergoes isomerization to give the more stable isomer $PC_{61}BM$. In 2003, Hummelen and coworkers synthesized C_{70} methanofullerene $PC_{71}BM$, which has stronger visible absorption than $PC_{61}BM$ [46]. Different from $PC_{61}BM$, $PC_{71}BM$ is a mixture of three regioisomers. Solar cells based on $PC_{71}BM$ and a conjugated polymer donor MDMO-PPV afforded a PCE of 3.0% (Table 3.2). $PC_{61}BM$ and $PC_{71}BM$ show high electron mobility (OFET mobility ~0.10 $cm^2\ V^{-1}\ s^{-1}$) [77] and are well soluble in many kinds of organic solvents. Due to the suitable miscibility with many kinds of donor materials (conjugated polymers and small molecules), $PC_{61}BM$ and $PC_{71}BM$ become the benchmark acceptor materials for organic photovoltaic research. The performance for some representative PCBM-based solar cells is listed in Table 3.2 [47–56]. In 2016, Yan and coworkers reported highly efficient solar cells based on a conjugated polymer donor PffBT4T-C_9C_{13} and $PC_{71}BM$ [53]. The cells afforded a PCE of 11.7%, which is the highest efficiency achieved by polymer:fullerene solar cells to date.

The success of $PC_{61}BM$ and $PC_{71}BM$ stimulated the development of PCBM-like fullerene acceptors. One direction is changing the aromatic part of the PCBM side chain. Troshin et al. developed $ThC_{61}BM$ by replacing the benzene moiety with a thienyl group [57]. Solar cells based on a donor polymer P3HT and $ThC_{61}BM$ gave a PCE of 3.7%. Jen and coworkers developed TPA-PCBM and MF-PCBM acceptors by replacing the benzene with the bulky aromatic groups triphenylamine (TPA) and 9,9-dimethylfluorene (MF), respectively [58]. TPA-PCBM and MF-PCBM

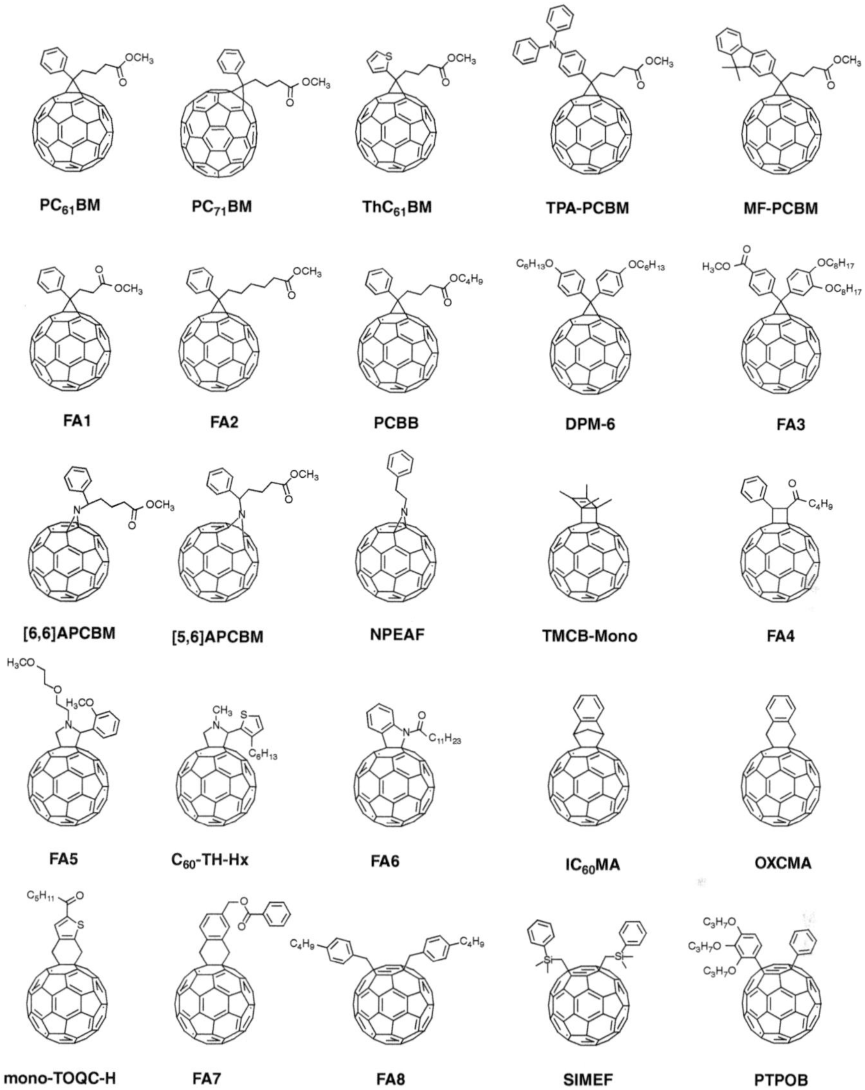

Figure 3.2 Representative fullerene monoadduct acceptors. Source: Zuo Xiao.

showed comparable performance to $PC_{61}BM$ in the P3HT-based solar cells but offered the cells much improved thermal stability. This is due to the much higher glass-transition temperatures of TPA-PCBM and MF-PCBM than $PC_{61}BM$. Another direction is modifying the aliphatic part of the PCBM side chain. Li and coworkers investigated the influence of the length of the middle alkyl chain to the performance of $PC_{61}BM$ [59]. They found that FA1 with the dimethylene chain and FA2 with the pentamethylene chain show slightly better performance than $PC_{61}BM$, delivering PCEs of 3.7% and 3.6% in P3HT-based solar cells, respectively. Cao and coworkers studied the effect of the ester alkyl chain to the performance of $PC_{61}BM$ [60]. By using a butyl to replace the methyl group, they developed the acceptor PCBB.

Table 3.2 Performance of fullerene-monoadduct-acceptors-based solar cells.

Donor:Acceptor	V_{oc} (V)	J_{sc} (mA cm^{-2})	FF (%)	PCE (%)	References
MDMO-PPV:$PC_{71}BM$	0.77	7.9	51	3.0	[46]
P3HT:$PC_{61}BM$	0.58	8.7	55	2.8	[47]
P3HT:$PC_{61}BM$	0.61	10.6	67.4	4.37	[48]
PTB7:$PC_{71}BM$	0.74	14.5	68.97	7.40	[49]
PTB7:$PC_{71}BM$	0.74	17.2	72	9.15	[50]
PTB7-Th:$PC_{71}BM$	0.80	15.73	74.3	9.35	[51]
PThTPTI:$PC_{71}BM$	0.87	13.69	65.6	7.80	[52]
PffBT4T-C_9C_{13}:$PC_{71}BM$	0.784	19.3	73	11.7	[53]
DERHD7T:$PC_{61}BM$	0.92	13.98	47.4	6.10	[54]
BTID-2F:$PC_{71}BM$	0.95	15.7	76.0	11.3	[55]
DPPEZnP-THE:$PC_{71}BM$	0.78	16.76	61.8	8.08	[56]
P3HT:$ThC_{61}BM$	0.60	10.6	58	3.7	[57]
P3HT:TPA-PCBM	0.65	9.9	62	4.0	[58]
P3HT:MF-PCBM	0.65	9.8	59	3.8	[58]
P3HT:FA1	0.564	10.8	60.3	3.7	[59]
P3HT:FA2	0.596	9.9	61.5	3.6	[59]
MEH-PPV:PCBB	0.85	6.08	43.0	2.84	[60]
P3HT:DPM-6	0.728	7.52	51	2.60	[61]
P3HT:FA3	0.61	5.83	57	2.06	[62]
P3HT:[6,6]APCBM	0.58	7.14	56	2.3	[63]
P3HT:[5,6]APCBM	0.58	8.10	60	2.8	[63]
P3HT:NPEAF	0.58	7.50	57	2.92	[64]
P3HT:TMCB-Mono	0.61	7.86	52	2.49	[65]
P3HT:FA4	0.53	8.60	65	3.00	[66]
PTB7:FA4	0.74	8.28	62	3.78	[66]
P3HT:FA5	0.66	7.85	66.2	3.44	[67]
Si-PCPDTBT:C_{60}-TH-Hx	0.675	10.6	56.1	4	[68]
P3HT:FA6	0.40	2.6	55	0.6	[69]
P3HT:$IC_{60}MA$	0.63	9.66	64	3.89	[70]
P3HT:OXCMA	0.63	9.63	59	3.60	[71]
P3HT:mono-TOQC-H	0.63	8.7	61	3.9	[72]
P3HT:FA7	0.65	11.3	57	4.2	[73]
P3HT:FA8	0.69	4.8	49	1.6	[74]
BP:SIMEF	0.75	10.5	65	5.2	[75]
P3HT:PTPOB	0.61	6.16	62	2.34	[76]

PCBB showed higher performance than $PC_{61}BM$ in the MEH-PPV-based solar cells. The diaryl-substituted methanofullerenes are another type of PCBM-like acceptors. Garcia-Belmonte et al. reported a methanofullerene DPM-6 with two 4-hexyloxybenzenyl groups [61]. Although DPM-6 shows identical LUMO level as $PC_{61}BM$, it delivered much higher V_{oc} than $PC_{61}BM$ in P3HT-based solar cells due to the reduced crystallinity of DPM-6. Shibata and coworkers developed a series of asymmetrical diaryl-substituted methanofullerene acceptors [62]. The FA3 delivered higher V_{oc} and comparable efficiency to $PC_{61}BM$ in P3HT-based solar cells.

The aziridine-flanked [2+1] monoadducts were also applied as acceptors for OSCs. In 2009, Wudl and coworkers synthesized the [6,6]APCBM and [5,6]APCBM by treating C_{60} with an azide compound [78]. These derivatives show good electron mobility up to 4.1×10^{-2} cm^2 V^{-1} s^{-1}. Solar cells based on P3HT:[6,6]APCBM and P3HT:[5,6]APCBM gave PCEs of 2.3% and 2.8%, respectively [63]. The [5,6]APCBM shows enhanced light absorption than [6,6]APCBM due to the lower molecular symmetry, thus contributing more photocurrent and offering high efficiency in solar cells. In 2012, Ding and coworkers developed a series of *N*-alkylaziridino[60]fullerene (NNAF) acceptors *via* the reaction of C_{60} and primary alkylamines with the aid of hypervalent iodine reagents [64]. All synthesis could be done under ambient conditions. These facile prepared NNAFs showed good performance in OSCs. Among them, the NPEAF afforded the highest PCE of 2.92% in P3HT-based solar cells.

3.2.2.2 [2+2] Addition Derivatives

In 2013, Swager and coworkers reported the synthesis of a series formal [2+2] derivatives *via* the Diels–Alder reactions between C_{60} and the *in situ* generated tetraalkylcyclobutadienes [65]. The monoadduct TMCB-Mono delivered a PCE of 2.49% in P3HT-based solar cells. Compared with $PC_{61}BM$, TMCB-Mono gave a higher V_{oc} due to its higher LUMO level. In 2016, Ueda et al. reported a regio- and stereo-selective intermolecular [2+2] cycloaddition of allenol esters with C_{60} [66]. With the aid of this new synthetic method, they obtained a cyclobutane-annulated fullerene acceptor FA4. Blending FA4 with the donor polymers P3HT and PTB7 delivered 3.00% and 3.78% PCEs in solar cells, respectively.

3.2.2.3 [2+3] Addition Derivatives

The 1,3-dipolar [2+3] addition of an azomethine ylide to fullerene is a classical functionalization method for fullerenes [79]. It is also called Prato reaction. The resulted fulleropyrrolidines can work as acceptors for OSCs. In 2010, Itoh and coworkers reported a series of fulleropyrrolidine acceptors [67]. The FA5 with a methoxyethoxyethyl and a 2-methoxyphenyl groups afforded the highest 3.44% PCE in P3HT-based solar cells, much better than that the reference $PC_{61}BM$ cells (2.53%). The V_{oc} for the fulleropyrrolidine-based cells are generally higher than that of $PC_{61}BM$ cells due to the higher LUMO levels of the fulleropyrrolidine acceptors. In the same year, Lee and coworkers reported another fulleropyrrolidine acceptor C_{60}-TH-Hx [68]. Solar cells based on C_{60}-TH-Hx and a low-bandgap polymer donor Si-PCPDTBT gave a PCE of 4%. In 2011, Chuang and coworkers reported a fulleroindoline derivative FA6 *via* the palladium-catalyzed [2+3] annulation of anilide with C_{60} [69]. FA6 delivered a PCE of 0.6% in P3HT-based cells.

3.2.2.4 [2+4] Addition Derivatives

The [2+4] Diels–Alder cycloaddition is an efficient method for preparing fullerene acceptors. Li and coworkers reported the Diels–Alder reaction between C_{60} and indene [70]. The monoadduct $IC_{60}MA$ delivered a PCE of 3.89% in P3HT-based solar cells, which is comparable to the reference $PC_{61}BM$ cells (3.88%). Due to the higher LUMO of $IC_{60}MA$ than $PC_{61}BM$, $IC_{60}MA$ cells showed a 0.05 V increase in V_{oc}. Kim and coworkers synthesized a monoadduct OXCMA through the [2+4] cycloaddition of an *in situ* generated *o*-xylylene to C_{60} [71]. OXCMA delivered a PCE of 3.60% in P3HT-based cells. Ding and coworkers reported a [2+4] monoadduct mono-TOQC-H through the reaction of C_{60} and a ketone functionalized thieno-*o*-quinodimethane [72]. The P3HT:mono-TOQC-H cells gave a PCE of 3.9%, which is better than that of the P3HT:$PC_{61}BM$ reference cells (3.6%). Fréchet and coworkers reported an ester group functionalized dihydronaphthyl fullerene FA7 [73]. Solar cells based on FA7 and P3HT gave the best PCE of 4.2%, which is slightly higher than that of P3HT:$PC_{61}BM$ cells.

3.2.2.5 1,4-Addition Derivatives

Besides [2+*n*] cycloaddition, fullerenes can also undergo 1,4-addition to give functionalized derivatives. Nakamura and coworkers prepared the 1,4-adduct FA8 through the quenching of C_{60} dianion with 1-(bromomethyl)-4-butylbenzene [74]. Solar cells based on FA8 and P3HT afforded a PCE of 1.6% with a relatively high V_{oc} of 0.69 V. Compared with the [2+*n*] derivatives, the 1,4-adducts possess higher LUMO levels, thus producing higher V_{oc} for OSCs. Nakamura and coworkers also developed a series of 1,4-adducts through an efficient two-step synthesis strategy [80]. First, the 1,2-addition of Grignard reagents to C_{60} can afford the (organo)-(hydro)[60]fullerenes, $C_{60}R_1H$, in high yields with the aid of dimethylformamide. Then, deprotonation of the $C_{60}R_1H$ followed by quenching the anion species $C_{60}R_1^-$ with alkyl halides (R_2X) give the 1,4-adducts $C_{60}R_1R_2$ in high yields. Among the synthesized $C_{60}R_1R_2$, the SIMEF acceptor shows the best photovoltaic performance. The solution processed p-i-n solar cells based on benzoporphyrin (BP) donor and SIMEF gave high PCEs up to 5.2% [75]. Wudl and coworkers also developed a series of 1,4-adducts with aryl functional groups [76]. The synthesis was accomplished by treating the 1,4-fullerenols with a variety of aromatic substrates in the presence of *p*-toluene sulfonic acid. Among these derivatives, the PTPOB afforded the best PCE of 2.34% when blending with P3HT.

3.2.3 Fullerene Bisadducts

Although a large amount of fullerene monoadduct acceptors have been developed after the first report of $PC_{61}BM$, they could seldom challenge PCBM acceptors in OSCs. In this context, more and more researches have focused on the development of fullerene bisadduct acceptors (Figure 3.3). In 2008, Blom and coworkers first

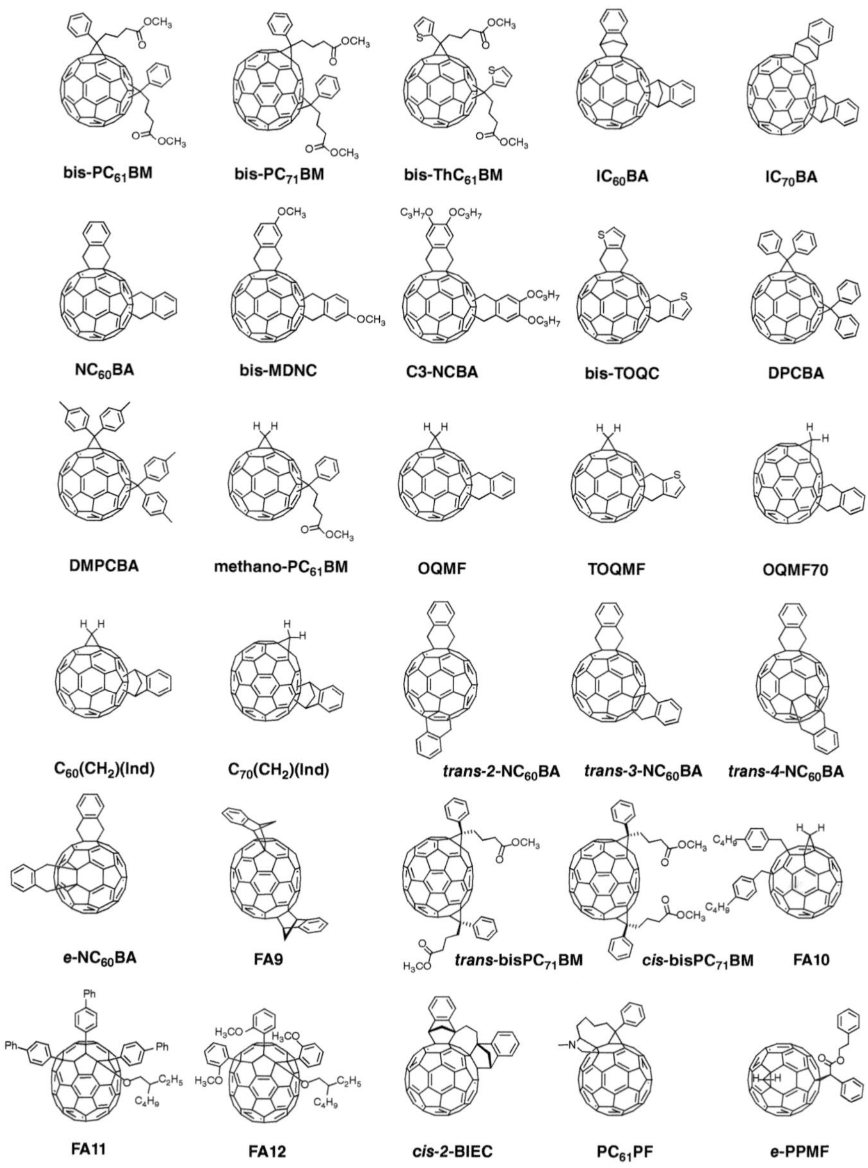

Figure 3.3 Representative fullerene bisadduct acceptors. Source: Zuo Xiao.

reported solar cells based on the bisadduct acceptor bis-$PC_{61}BM$ [81]. Compared with the monoadduct $PC_{61}BM$, bis-$PC_{61}BM$ shows a ~0.1 eV increase in the LUMO level, which can minimize the energy loss in the electron transfer from the donor to the acceptor. Consequently, the P3HT:bis-$PC_{61}BM$ cells gave a V_{oc} of 0.73 V, which is 0.15 V higher than the P3HT: $PC_{61}BM$ cells. Interestingly, the bis-$PC_{61}BM$ cells also showed similar J_{sc} and fill factor (FF) as the $PC_{61}BM$ cells, thus delivering a PCE of 4.5%, higher than that of the best $PC_{61}BM$ cells (3.8%) (Table 3.3). In 2009,

Table 3.3 Performance of fullerene-bisadduct-acceptors-based solar cells.

Donor:Acceptor	V_{oc} (V)	J_{sc} (mA cm^{-2})	FF (%)	PCE (%)	References
P3HT:bis-PC$_{61}$BM	0.724	9.14	68	4.5	[81]
P3HT:bis-PC$_{71}$BM	0.75	7.03	62	2.3	[82]
P3HT:bis-ThC$_{61}$BM	0.72	7.31	66	2.5	[82]
P3HT:IC$_{60}$BA	0.84	9.67	67	5.44	[70]
P3HT:IC$_{60}$BA	0.84	10.61	72.7	6.48	[83]
P3HT:IC$_{70}$BA	0.87	11.35	75	7.4	[84]
P3HT:NC$_{60}$BA	0.83	10.3	62	5.31	[71]
P3HT:NC$_{60}$BA	0.82	9.88	67	5.37	[85]
P3HT:NC$_{60}$BA	0.816	9.4	66	5.2	[86]
P3HT:bis-MDNC	0.83	8.67	63.6	4.58	[87]
P3HT:C3-NCBA	0.81	8.12	62	4.1	[88]
P3HT:bis-TOQC	0.86	7.7	66	5.1	[72]
P3HT:DPCBA	0.75	9.48	55.9	4.0	[89]
P3HT:DMPCBA	0.87	9.05	65.5	5.2	[89]
P3HT:methano-PC$_{61}$BM	0.69	8.03	69	3.81	[77]
P3HT:OQMF	0.81	11.19	63.4	5.74	[90]
P3HT:TOQMF	0.79	10.93	63.8	5.51	[90]
P3HT:OQMF70	0.81	12.42	68.41	6.88	[91]
P3HT:C$_{60}$(CH$_2$)(Ind)	0.78	10.3	73	5.9	[92]
P3HT:C$_{70}$(CH$_2$)(Ind)	0.79	11.1	73	6.4	[92]
P3HT:*trans*-2-NC$_{60}$BA	0.83	10.04	69	5.8	[93]
P3HT:*trans*-3-NC$_{60}$BA	0.88	10.21	71	6.3	[93]
P3HT:*trans*-4-NC$_{60}$BA	0.86	9.67	67	5.6	[93]
P3HT:*e*-NC$_{60}$BA	0.86	9.51	67	5.5	[93]
P3HT:FA9	0.86	10.3	67	5.9	[94]
P3HT:*trans*-bisPC$_{71}$BM	0.72	5.12	50	1.84	[95]
P3HT:*cis*-bisPC$_{71}$BM	0.69	4.88	51	1.72	[95]
P3HT:FA10	0.82	7.1	58	3.4	[74]
BP:FA11	0.80	6.17	68	3.37	[96]
BP:FA12	0.90	5.83	64	3.33	[96]
P3HT:*cis*-2-BIEC	0.80	6.60	53	2.8	[97]
P3HT:PC$_{61}$PF	0.70	8.1	58	3.3	[98]
P3HT:*e*-PPMF	0.72	9.88	63.3	4.51	[99]
PPDT2FBT:*e*-PPMF	0.86	14.60	64.5	8.11	[99]

Blom and coworkers further reported bisadducts bis-$PC_{71}BM$ and bis-Th$C_{61}BM$, which delivered PCEs of 2.3% and 2.5% in P3HT-based solar cells, respectively [82]. The development of indene C_{60} bisadducts ($IC_{60}BA$) is an important milestone in the development of fullerene acceptors. In 2010, Li and coworkers reported $IC_{60}BA$ *via* the cycloaddition of two indenes on to C_{60} [70]. Compared with $PC_{61}BM$, $IC_{60}BA$ shows a 0.2 eV increase in LUMO level. The P3HT:$IC_{60}BA$ cells delivered a high V_{oc} of 0.84 V, which is 45% higher than that of P3HT:$PC_{61}BM$ cells. With device optimization, the P3HT:$IC_{60}BA$ cells could afford the highest PCE of 6.48%, which is about twice that of P3HT:$PC_{61}BM$ cells [83]. Li and coworkers also reported $IC_{70}BA$ [100]. Owing to the stronger light absorption, the best P3HT:$IC_{70}BA$ cells could give an excellent PCE of 7.4% [84]. The success of ICBA acceptors triggered a lot of following research in the development of fullerene bisadducts acceptors. At about the same time, three different groups reported the development of dihydronaphthyl-based C_{60} bisadducts $NC_{60}BA$ [71, 85, 86]. PCEs up to 5.37% were achieved by the P3HT:$NC_{60}BA$ solar cells. In 2012, Xie and coworkers developed a methoxy-functionalized $NC_{60}BA$ derivative, bis-MDNC [87]. Solar cells based on P3HT and bis-MDNC showed a PCE of 4.58% with a V_{oc} of 0.83 V. In the same year, Wang and coworkers studied the alkyl chain length effect on the performance of dialkoxy-functionalized $NC_{60}BA$ acceptors [88]. They found that the C3-NCBA with a moderate-length propyl chain shows the best PCE of 4.1% in P3HT-based solar cells. Ding and coworkers reported a thiophene-functionalized fullerene bisadduct, bis-TOQC, by adding two thieno-*o*-quinodimethane addends to C_{60} [72]. P3HT:bis-TOQC solar cells afforded a decent PCE of 5.1%. Cheng et al. reported a series of diphenylmethano-C_{60} bisadduct acceptors [89]. The DPCBA and DMPCBA delivered 4.0% and 5.2% PCEs in P3HT-based solar cells, respectively. Meanwhile, due to the amorphous nature of DMPCBA, the P3HT:DMPCBA cells showed much improved thermal stability than the P3HT:$PC_{61}BM$ reference cells.

Although bisadduct acceptors can offer higher V_{oc} for solar cells, they generally deliver lower J_{sc} and FF as compared to the monoadduct acceptors. One reason is the increased steric hindrance from the addends, which inhibits the fullerene packing and lowers the electron mobility. In this regard, Li et al. used a very small 1,2-dihydromethano group (CH_2) working as the addend to develop a bisadduct acceptor, methano-$PC_{61}BM$ [77]. The methano-$PC_{61}BM$ shows a high electron mobility of 0.014 $cm^2\ V^{-1}\ s^{-1}$ and delivered a 3.81% PCE in P3HT-based solar cells. Ding and coworkers synthesized OQMF and TOQMF by adding the *in situ* generated *o*-quinodimethane and thieno-*o*-quinodimethane species to the methanofullerene $C_{60}CH_2$ [90]. Decent PCEs of 5.74% and 5.51% were achieved for P3HT:OQMF and P3HT:TOQMF solar cells, respectively. Ding and coworkers also reported the C_{70} bisadduct OQMF70, which delivered an impressive PCE of 6.88% in P3HT-based solar cells [91]. Matsuo et al. reported efficient acceptors $C_{60}(CH_2)$(Ind) and $C_{70}(CH_2)$(Ind) by adding indene on to $C_{60}CH_2$ and $C_{70}CH_2$, respectively [92]. $C_{60}(CH_2)$(Ind) and $C_{70}(CH_2)$(Ind) afforded decent PCEs of 5.9% and 6.4%, respectively, in P3HT-based solar cells.

Another factor leading to the low electron mobility of fullerene bisadducts is the isomeric mixture form of bisadducts, as the bis-functionalization on fullerenes is generally unselective. By using coarse-grained molecular dynamics simulation, Nelson and coworkers concluded that the "energetic disorder" brought from regioisomers of fullerene bisadducts is the major cause of reduction in electron mobility [101]. In this regard, considerable efforts have been paid to develop isomer-free bisadduct acceptors. An effective approach to obtain isomer-free bisadduct is employing high-performance liquid chromatography (HPLC) to separate the isomers. In 2014, Wang and coworkers reported the isolation of three isomers of $NC_{60}BA$: *trans*-2-$NC_{60}BA$, *trans*-3-$NC_{60}BA$, *trans*-4-$NC_{60}BA$, and *e*-$NC_{60}BA$ [93]. All isomerically pure isomers delivered higher efficiencies than the mixture $NC_{60}BA$. The *trans*-3-$NC_{60}BA$:P3HT cells gave the best PCE of 6.3%. In the same year, Wong and coworkers isolated a single isomer of $IC_{70}BA$ *via* multistep HPLC separation [94]. The structure of the isolated acceptor FA9 was confirmed by single-crystal X-ray diffraction (XRD) analysis. FA9 delivered a PCE of 5.9% in solar cells, which is higher than that of the mixture-$IC_{70}BA$-based cells (4.7%). Xie and coworkers reported the isolation of two bis$PC_{71}BM$ stereomeric isomers, the *trans*-bis$PC_{71}BM$ and the *cis*-bis$PC_{71}BM$ [95]. The structures of both isomers were identified by single-crystal XRD analysis. Solar cells based on *trans*-bis$PC_{71}BM$ and *cis*-bis$PC_{71}BM$ afforded 1.84% and 1.72% PCEs, respectively. This work indicates that stereomeric factors can also influence the photovoltaic performance of fullerene acceptors. Another effective approach is the development of regioselective synthesis of isomer-free fullerene bisadducts. In 2011, Nakamura and coworkers synthesized the isomerically pure acceptor FA10 *via* an oxidative cyclopropanation reaction of silylmethylfullerene [74]. Solar cells based on FA10 and P3HT gave a high V_{oc} of 0.82 V and a PCE of 3.4%. In 2012, Nakamura and coworkers reported the isomer isomerically pure 66π-electron C_{70} derivatives *via* regioselective synthesis [96]. The acceptors FA11 and FA12 delivered 3.37% and 3.33% PCEs in BP-based solar cells. "Tether-directed remote functionalization" is a useful strategy to obtain isomer-free fullerene multiadducts [102]. Using a molecular tether to carry two or more addition groups, multiaddition to fullerenes will present regioselectivity because the length, rigidity, and conformation of the tether can confine the reaction sites. In 2015, Imahori and coworkers reported an ethylene-tether-directed synthesis of an isomer-free indene-C_{60} bisadduct *cis*-2-BIEC [97]. Solar cells based on *cis*-2-BIEC and P3HT delivered a PCE of 2.8%. In the same year, Wong and coworkers reported a one-pot selective synthesis of a bisadduct $PC_{61}PF$ by directly using the side chain of $PC_{61}BM$ as the tether [98]. The isomer-free $PC_{61}PF$ afforded a 3.3% PCE in P3HT-based solar cells. In 2016, Ding and coworkers developed another method for synthesizing isomer-free fullerene bisadducts, the "prebisaddition-confined bisfunctionalization (PCB)" approach [99]. The synthesis was highly selective for bisadducts with an *e* configuration. An isomer-free bisadduct *e*-PPMF featuring a small CH_2 addend shows high electron mobility due to the eliminated energetic disorder from isomers and enhanced fullerene packing. *e*-PPMF showed high performance not only in P3HT cells but also in other D–A copolymer cells. PCEs of 4.51% and 8.11% were achieved by P3HT:*e*-PPMF and PPDT2FBT:*e*-PPMF solar cells, respectively. The 8.11% PCE is the highest efficiency achieved in fullerene-bisadduct-based solar cells to date.

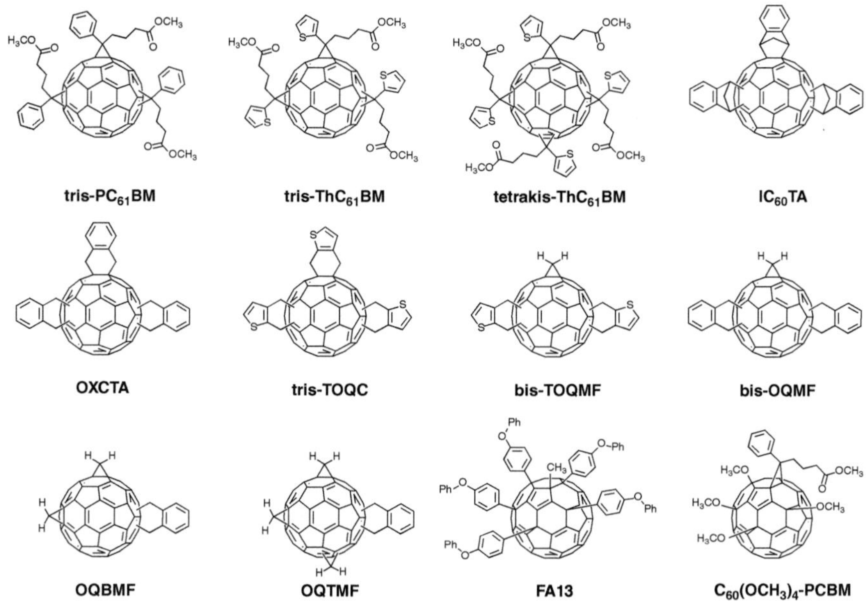

Figure 3.4 Representative fullerene multiadduct acceptors. Source: Zuo Xiao.

3.2.4 Fullerene Multiadducts

The success of fullerene bisadduct acceptors stimulated the exploration and development of fullerene multiadduct acceptors (Figure 3.4). Blom and coworkers first reported the photovoltaic performance of a fullerene trisadduct tris-PC_{61}BM [82]. However, tris-PC_{61}BM showed low electron mobility and only delivered a 0.21% PCE in P3HT-based cells (Table 3.4). In 2010, Fukuzumi and coworkers reported a trisadduct tris-ThC_{61}BM and a tetrakisadduct tetrakis-ThC_{61}BM, which afforded low PCEs of 0.34% and 0.09% in P3HT-based cells, respectively [103]. The indene C_{60} trisadduct (IC_{60}TA) reported by Kim and coworkers showed a better PCE of 2.25% as compared to the PCBM-type trisadducts [104]. Since that the indene moieties are much smaller than the PCBM side chains, they should have less influence to the fullerene packing, thus rendering the trisadducts better electron-transporting capability and higher photovoltaic performance. Kim et al. also reported an *o*-xylylene C_{60} trisadduct (OXCTA), which delivered a higher PCE of 2.63% in P3HT-based solar cells [71]. Ding and coworkers synthesized an acceptor tris-TOQC through the trisaddition of thieno-o-quinodimethane to C_{60} [105]. The P3HT:tris-TOQC cells afforded a PCE of 1.12%. Considering that the steric hindrance of the addends has significant influence to the performance of multiadducts, Ding and coworkers developed a series of efficient multiadduct acceptors by using compact methano addend CH_2. With one of the thieno-*o*-quinodimethane addends on tris-TOQC being replaced by a CH_2, the resulted bis-TOQMF showed much improved performance in P3HT-based cells. The P3HT:bis-TOQMF cells afforded a PCE of 4.56%, much higher than that of P3HT:tris-TOQC cells. Similarly, a trisadduct

Table 3.4 Performance of fullerene-multiadduct-acceptors-based solar cells.

Donor:Acceptor	V_{oc} (V)	J_{sc} (mA cm^{-2})	FF (%)	PCE (%)	References
P3HT:tris-PC$_{61}$BM	0.81	0.99	37	0.21	[82]
P3HT:tris-ThC$_{61}$BM	0.64	1.88	28	0.34	[103]
P3HT:tetrakis-ThC$_{61}$BM	0.57	0.48	32	0.09	[103]
P3HT:IC$_{60}$TA	0.88	6.27	40	2.25	[104]
P3HT:OXCTA	0.98	6.79	40	2.63	[71]
P3HT:tris-TOQC	0.91	3.60	34	1.12	[105]
P3HT:bis-TOQMF	0.94	8.09	58	4.56	[105]
P3HT:bis-OQMF	0.97	7.38	59.1	4.23	[106]
P3HT:OQBMF	0.95	9.67	70	6.43	[107]
P3HT:OQTMF	1.00	5.52	54	2.98	[107]
P3HT:FA13	0.76	2.8	51	1.08	[108]
P3HT:C$_{60}$(OCH$_3$)$_4$-PCBM	0.72	6.86	45.3	2.24	[109]

bis-OQMF delivered a PCE of 4.23% in P3HT-based cells [106]. Further replacing one o-quinodimethane addend with the CH_2, Ding and coworkers developed the trisadduct acceptor OQBMF [107]. Featuring two compact CH_2 groups, OQBMF exhibited outstanding performance in P3HT-based cells, offering a high V_{oc} of 0.95 V and a PCE of 6.43%. The 6.43% PCE is the highest efficiency achieved in fullerene-trisadduct-based solar cells to date. Ding and coworkers also synthesized a tetrakisadduct OQTMF featuring three CH_2 groups. The P3HT:OQTMF cells afforded a high V_{oc} of 1.00 V and a PCE of 2.98%. Isomer-free fullerene multiadducts were also developed for OSCs. Nakamura and coworkers discovered the highly selective and efficient pentaaddition of organocopper reagents to C_{60} [110]. The multiadduct FA13 synthesized *via* this reaction delivered a PCE of 1.08% in P3HT-based cells [108]. Xie and coworkers synthesized an isomer-free multiadduct $C_{60}(OCH_3)_4$-PCBM starting from $C_{60}Cl_6$ [109]. $C_{60}(OCH_3)_4$-PCBM gave a PCE of 2.24% in P3HT-based cells.

3.2.5 Unconventional Fullerenes

Unconventional fullerene acceptors were also developed and applied in OSCs (Figure 3.5). Martín and coworkers reported novel C_{60} and C_{70} homo- and heterodimers linked through 2-pyrazolino-pyrrolidino bridges [111]. Compared with the conventional PC$_{61}$BM acceptor, the fullerene dimers show larger absorption coefficients. The dimers FA14, FA15, and FA16 afforded 0.91%, 0.67%, and 0.91% PCEs in P3HT-based solar cells, respectively (Table 3.5). Open-cage fullerenes are novel fullerene derivatives generated *via* consecutive cleavage of σ- and π-carbon–carbon bonds on the fullerene cage [118]. The orifice on fullerene allows small atoms or molecules go inside the cavity to form interesting host–guest

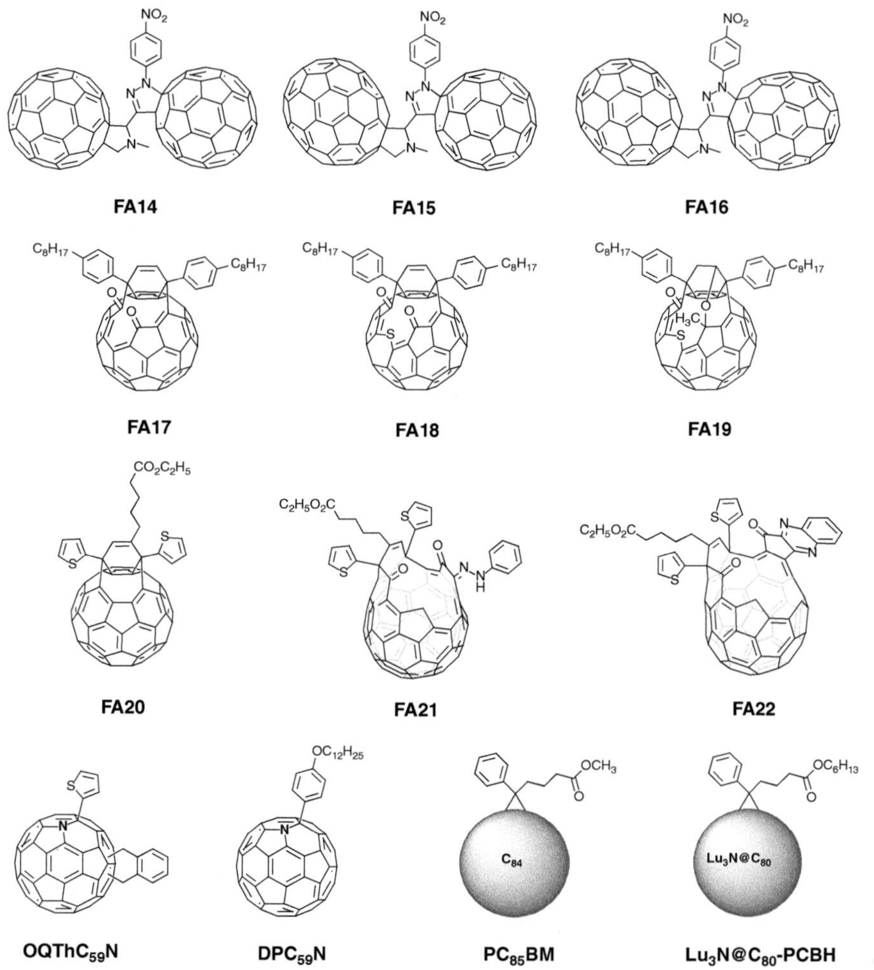

Figure 3.5 Representative unconventional fullerene acceptors.

complexes. Open-cage fullerenes were also applied as acceptors for OSCs. In 2011, Murata et al. synthesized a series of open-cage fullerenes FA17, FA18, and FA19, which delivered 1.35%, 2.49%, and 3.11% PCEs in P3HT-based solar cells, respectively [112]. In the same year, Chuang and coworkers reported a family of open-cage fullerene acceptors with various sizes of orifices [113]. The FA20, FA21, and FA22 afforded 2.9%, 1.7%, and 1.2% PCEs in P3HT-based solar cells, respectively. In the above two studies, open-cage fullerenes with larger orifices generally show lower electron mobility and poorer performance in solar cells. The cage-opened structures lead to inefficient packing of fullerenes, thus reducing the electron mobility. Researchers also developed novel azafullerene acceptors for OSCs. In 2014, Ding and coworkers first reported an azafullerene acceptor OQThC$_{59}$N [114]. Possessing a relatively high LUMO level, OQThC$_{59}$N gives a high V_{oc} of 0.78 V and a PCE of 4.09% in fullerene:P3HT solar cells. In 2015, Viterisi and coworkers developed another

Table 3.5 Performance of unconventional-fullerene-acceptors-based solar cells.

Donor:Acceptor	V_{oc} (V)	J_{sc} (mA cm^{-2})	FF (%)	PCE (%)	References
P3HT:FA14	0.44	4.27	48	0.91	[111]
P3HT:FA15	0.43	3.41	46	0.67	[111]
P3HT:FA16	0.43	5.08	42	0.91	[111]
P3HT:FA17	0.49	5.20	53	1.35	[112]
P3HT:FA18	0.62	6.49	62	2.49	[112]
P3HT:FA19	0.74	6.57	64	3.11	[112]
P3HT:FA20	0.62	7.1	64	2.9	[113]
P3HT:FA21	0.65	4.9	51	1.7	[113]
P3HT:FA22	0.57	3.5	56	1.2	[113]
P3HT:OQThC$_{59}$N	0.78	7.57	69.2	4.09	[114]
P3HT:DPC$_{59}$N	0.568	7.92	50	2.24	[115]
MDMO-PPV:PC$_{85}$BM	0.335	1.38	44.6	0.25	[116]
P3HT:Lu$_3$N@C$_{80}$-PCBH	0.81	8.64	61	4.2	[117]

azafullerene acceptor DPC$_{59}$N [115]. Compared with PC$_{61}$BM, DPC$_{59}$N shows stronger visible-light-absorbing capability and gave higher J_{sc} and external quantum efficiency in P3HT-based solar cells. A 2.24% PCE was achieved by P3HT:DPC$_{59}$N solar cells. Besides C_{60} and C_{70} derivatives, derivatives of higher fullerenes and endohedral fullerenes were also explored as acceptors for solar cells. In 2006, Hummelen and coworkers reported the PCBM-type acceptor PC$_{85}$BM based on C_{84} [116]. Compared with PC$_{61}$BM and PC$_{71}$BM, the LUMO level of PC$_{85}$BM is 0.35 eV lower, thus leading to a low V_{oc} of 0.335 V for the solar cells. The MDMO-PPV:PC$_{85}$BM only afforded a PCE of 0.25%. In 2009, Drees and coworkers reported a novel trimetallic nitride endohedral fullerene acceptor Lu$_3$N@C$_{80}$-PCBH [117]. The LUMO level of Lu$_3$N@C$_{80}$-PCBH is 0.26 V higher than PC$_{61}$BM. The P3HT:Lu$_3$N@C$_{80}$-PCBH cells delivered a high V_{oc} of 0.34 V and an impressive PCE of 4.2%.

3.3 Summary

In this section, representative fullerene acceptors, including pristine fullerenes, fullerene monoadducts, fullerene bisadducts, fullerene multiadducts, and unconventional fullerenes, and their performance in OSCs are summarized:

(1) For pristine fullerene acceptors, C_{60} and C_{70} have shown excellent performance in vacuum-processed single-junction and tandem solar cells. The highest PCEs of 9.3% and 15.4% were achieved by pristine-fullerene-based single-junction and tandem solar cells, respectively. C_{60} and C_{70} are not favored in solution-processed solar cells due to their low solubility. Nevertheless,

solution-processed cells based on pristine fullerenes afforded decent PCEs up to 5.9%.

(2) For fullerene monoadduct acceptors, $PC_{61}BM$ and $PC_{71}BM$ are still the most efficient and widely used acceptor in OSCs. Due to the good solubility and electron mobility, as well as suitable miscibility with many kinds of donor materials, $PC_{61}BM$ and $PC_{71}BM$ can generally guarantee decent performance of OSCs, thus being the benchmark acceptors for organic photovoltaic research. The highest PCE of 11.7% was achieved by $PC_{71}BM$-based cells.

(3) For fullerene bisadduct acceptors, the mixture-formed bisadducts represented by $IC_{60}BA$ and $IC_{70}BA$ have demonstrated their success in P3HT-based solar cells. High V_{oc} and PCEs were achieved for these cells. However, due to the steric hindrance from the addends and the energetic disorder from the regioisomers, mixture-formed bisadduct acceptors show lower electron mobility than monoadduct acceptors. This limits their photovoltaic performance. In this context, isomer-free bisadduct acceptors were developed and showed high performance in OSCs. The highest PCE of 8.11% was achieved by an isomer-free bisadduct *e*-PPMF, which was synthesized through a regioselective "PCB" method.

(4) For fullerene multiadduct acceptors, they generally show poor performance in solar cells due to the significantly decreased electron mobility caused by the bulky addends. By using the compact methano group CH_2 as the addend, Ding and coworkers developed a family of high-performance fullerene multiadduct acceptors. High V_{oc} up to 1.00 V and PCEs up to 6.43% were achieved by these CH_2-containing fullerene multiadducts.

(5) For unconventional fullerene acceptors, a bunch of fullerene dimers, open-cage fullerenes, azafullerenes, higher fullerenes, and endohedral fullerenes have been applied in OSCs. These novel materials delivered PCEs up to 4.2% in P3HT-based solar cells.

Although significant progress has been made in fullerene acceptors in the past 25 years, the poor light-harvesting capability is the critical defect of fullerene acceptors and it limits the potential of fullerenes in solar cells. Recent years, the rise of strong-light-absorbing nonfullerene acceptors has greatly pushed the efficiency of OSCs into the 18% level [119]. However, fullerene acceptors still show their advantages like high electron mobility as compared to the nonfullerene acceptors. In this regard, combining the advantages of fullerene and nonfullerene acceptors to develop highly efficient multi-component solar cells [120] and thick-film solar cells [121] could be a promising direction in the future.

References

1 Kroto, H.W., Heath, J.R., O'Brien, S.C. et al. (1985). *Nature* 318: 162.
2 Krätschmer, W., Lamb, L.D., Fostiropoulos, K., and Huffman, D.R. (1990). *Nature* 347: 354.

3 Langa, F. and Nierengarten, J.-F. (2007). *Fullerenes. Principles and Applications*. Cambridge: RSC.
4 Chen, Z. (2004). *Angew. Chem. Int. Ed.* 43: 4690.
5 Kareev, I.E., Kuvychko, I.V., Shustova, N.B. et al. (2008). *Angew. Chem. Int. Ed.* 47: 6204.
6 Yamada, M., Akasaka, T., and Nagase, S. (2010). *Acc. Chem. Res.* 43: 92.
7 Kurotobi, K. and Murata, Y. (2011). *Science* 333: 613.
8 Zhang, J., Stevenson, S., and Dorn, H.C. (2013). *Acc. Chem. Res.* 46: 1548.
9 Lu, X. and Chen, Z. (2005). *Chem. Rev.* 105: 3643.
10 Diederich, F. and Whetten, R.L. (1992). *Acc. Chem. Res.* 25: 119.
11 Xie, S.-Y., Gao, F., Lu, X. et al. (2004). *Science* 304: 699.
12 Popov, A.A., Yang, S., and Dunsch, L. (2013). *Chem. Rev.* 113: 5989.
13 Vostrowsky, O. and Hirsch, A. (2006). *Chem. Rev.* 106: 5191.
14 Hummelen, J.C., Knight, B., Pavlovich, J. et al. (1995). *Science* 269: 1554.
15 Brown, C.M., Cristofolini, L., Kordatos, K. et al. (1996). *Chem. Mater.* 8: 2548.
16 Zhang, G., Huang, S., Xiao, Z. et al. (2008). *J. Am. Chem. Soc.* 130: 12614.
17 Kadish, K.M. and Ruoff, R.S. (2000). *Fullerenes: Chemistry, Physics, and Technology*. New York: Wiley.
18 Cataldo, F. and Da Ros, T. (2008). *Medicinal Chemistry and Pharmacological Potential of Fullerenes and Carbon Nanotubes*. Berlin: Springer-Verlag.
19 Delgado, J.L., Bouit, P.-A., Filippone, S. et al. (2010). *Chem. Commun.* 46: 4853.
20 Li, C.-Z., Yip, H.-L., and Jen, A.K.-Y. (2012). *J. Mater. Chem.* 22: 4161.
21 Sariciftci, N.S., Smilowitz, L., Heeger, A.J., and Wudl, F. (1992). *Science* 258: 1474.
22 Yu, G., Gao, J., Hummelen, J.C. et al. (1995). *Science* 270: 1789.
23 Choy, W.C.H. (2013). *Organic Solar Cells: Materials and Device Physics*. Berlin: Springer-Verlag.
24 Haufler, R.E., Conceicao, J., Chibante, L.P.F. et al. (1990). *J. Phys. Chem.* 94: 8634.
25 Howard, J.B., McKinnon, J.T., Makarovsky, Y. et al. (1991). *Nature* 352: 139.
26 Uchida, S., Xue, J., Rand, B.P., and Forrest, S.R. (2004). *Appl. Phys. Lett.* 84: 4218.
27 Xue, J., Rand, B.P., Uchida, S., and Forrest, S.R. (2005). *Adv. Mater.* 17: 66.
28 Pfuetzner, S., Meiss, J., Petrich, A. et al. (2009). *Appl. Phys. Lett.* 94: 223307.
29 Lin, H.-W., Lin, L.-Y., Chen, Y.-H. et al. (2011). *Chem. Commun.* 47: 7872.
30 Steinmann, V., Kronenberg, N.M., Lenze, M.R. et al. (2011). *Adv. Energy Mater.* 1: 888.
31 Chiu, S.-W., Lin, L.-Y., Lin, H.-W. et al. (2012). *Chem. Commun.* 48: 1857.
32 Wang, Z., Yokoyama, D., Wang, X.-F. et al. (2013). *Energy Environ. Sci.* 6: 249.
33 Xiao, X., Bergemann, K.J., Zimmerman, J.D. et al. (2014). *Adv. Energy Mater.* 4: 1301557.
34 Che, X., Chung, C.-L., Hsu, C.-C. et al. (2018). *Adv. Energy Mater.* 8: 1703603.
35 Cheyns, D., Rand, B.P., and Heremans, P. (2010). *Appl. Phys. Lett.* 97: 033301.
36 Riede, M., Uhrich, C., Widmer, J. et al. (2011). *Adv. Funct. Mater.* 21: 3019.

37 Lassiter, B.E., Zimmerman, J.D., Panda, A. et al. (2012). *Appl. Phys. Lett.* 101: 063303.
38 Meerheim, R., Körner, C., and Leo, K. (2014). *Appl. Phys. Lett.* 105: 063306.
39 Li, Y., Lin, J.-D., Liu, X. et al. (2018). *Adv. Mater.* 30: 1804416.
40 Lin, H.-W., Chang, J.-H., Huang, W.-C. et al. (2014). *J. Mater. Chem. A* 2: 3709.
41 de Zerio Mendaza, A.D., Melianas, A., Nugroho, F.A.A. et al. (2017). *J. Mater. Chem. A* 5: 4156.
42 Arbogast, J.W. and Foote, C.S. (1991). *J. Am. Chem. Soc.* 113: 8886.
43 Ameri, T., Li, N., and Brabec, C.J. (2013). *Energy Environ. Sci.* 6: 2390.
44 Hirsch, A. (1999). *Top. Curr. Chem.* 199: 1.
45 Hummelen, J.C., Knight, B.W., LePeq, F. et al. (1995). *J. Org. Chem.* 60: 532.
46 Wienk, M.M., Kroon, J.M., Verhees, W.J.H. et al. (2003). *Angew. Chem. Int. Ed.* 42: 3371.
47 Schilinsky, P., Waldauf, C., and Brabec, C.J. (2002). *Appl. Phys. Lett.* 81: 3885.
48 Li, G., Shrotriya, V., Huang, J. et al. (2005). *Nat. Mater.* 4: 864.
49 Liang, Y., Xu, Z., Xia, J. et al. (2010). *Adv. Mater.* 22: E135.
50 He, Z., Zhong, C., Su, S. et al. (2012). *Nat. Photonics* 6: 591.
51 Liao, S.-H., Jhuo, H.-J., Cheng, Y.-S., and Chen, S.-A. (2013). *Adv. Mater.* 25: 4766.
52 Cao, J., Liao, Q., Du, X. et al. (2013). *Energy Environ. Sci.* 6: 3224.
53 Zhao, J., Li, Y., Yang, G. et al. (2016). *Nat. Energy* 1: 15027.
54 Li, Z., He, G., Wan, X. et al. (2012). *Adv. Energy Mater.* 2: 74.
55 Deng, D., Zhang, Y., Zhang, J. et al. (2016). *Nat. Commun.* 7: 13740.
56 Gao, K., Li, L., Lai, T. et al. (2015). *J. Am. Chem. Soc.* 137: 7282.
57 Troshin, P.A., Hoppe, H., Renz, J. et al. (2009). *Adv. Funct. Mater.* 19: 779.
58 Zhang, Y., Yip, H.-L., Acton, O. et al. (2009). *Chem. Mater.* 21: 2598.
59 Zhao, G., He, Y., Xu, Z. et al. (2010). *Adv. Funct. Mater.* 20: 1480.
60 Zheng, L., Zhou, Q., Deng, X. et al. (2004). *J. Phys. Chem. B* 108: 11921.
61 Garcia-Belmonte, G., Boix, P.P., Bisquert, J. et al. (2010). *J. Phys. Chem. Lett.* 1: 2566.
62 Sukeguchi, D., Singh, S.P., Reddy, M.R. et al. *Beilstein J. Org. Chem.*, 3762/bjoc.5.7.
63 Park, S.H., Yang, C., Cowan, S. et al. (2009). *J. Mater. Chem.* 19: 5624.
64 Liu, Y., Du, X., Cao, J. et al. (2012). *Synth. Met.* 162: 1271.
65 Han, G.D., Collins, W.R., Andrew, T.L. et al. (2013). *Adv. Funct. Mater.* 23: 3061.
66 Ueda, M., Sakaguchi, T., Hayama, M. et al. (2016). *Chem. Commun.* 52: 13175.
67 Matsumoto, K., Hashimoto, K., Kamo, M. et al. (2010). *J. Mater. Chem.* 20: 9226.
68 Kim, H., Seo, J.H., Park, E.Y. et al. (2010). *Appl. Phys. Lett.* 97: 193309.
69 Chen, C.-P., Luo, C., Ting, C., and Chuang, S.-C. (2011). *Chem. Commun.* 47: 1845.
70 He, Y., Chen, H.-Y., Hou, J., and Li, Y. (2010). *J. Am. Chem. Soc.* 132: 1377.
71 Kim, K.-H., Kang, H., Nam, S.Y. et al. (2011). *Chem. Mater.* 23: 5090.
72 Zhang, C., Chen, S., Xiao, Z. et al. (2012). *Org. Lett.* 14: 1508.

73 Backer, S.A., Sivula, K., Kavulak, D.F., and Fréchet, J.M.J. (2007). *Chem. Mater.* 19: 2927.
74 Zhang, Y., Matsuo, Y., Li, C.-Z. et al. (2011). *J. Am. Chem. Soc.* 133: 8086.
75 Matsuo, Y., Sato, Y., Niinomi, T. et al. (2009). *J. Am. Chem. Soc.* 131: 16048.
76 Varotto, A., Treat, N.D., Jo, J. et al. (2011). *Angew. Chem. Int. Ed.* 50: 5166.
77 Li, C.-Z., Chien, S.-C., Yip, H.-L. et al. (2011). *Chem. Commun.* 47: 10082.
78 Yang, C., Cho, S., Heeger, A.J., and Wudl, F. (2009). *Angew. Chem. Int. Ed.* 48: 1592.
79 Maggini, M., Scorrano, G., and Prato, M. (1993). *J. Am. Chem. Soc.* 115: 9798.
80 Matsuo, Y., Iwashita, A., Abe, Y. et al. (2008). *J. Am. Chem. Soc.* 130: 15429.
81 Lenes, M., Wetzelaer, G.-J.A.H., Kooistra, F.B. et al. (2008). *Adv. Mater.* 20: 2116.
82 Lenes, M., Shelton, S.W., Sieval, A.B. et al. (2009). *Adv. Funct. Mater.* 19: 3002.
83 Zhao, G., He, Y., and Li, Y. (2010). *Adv. Mater.* 22: 4355.
84 Guo, X., Cui, C., Zhang, M. et al. (2012). *Energy Environ. Sci.* 5: 7943.
85 Meng, X., Zhang, W., Tan, Z. et al. (2012). *Chem. Commun.* 48: 425.
86 Voroshazi, E., Vasseur, K., Aernouts, T. et al. (2011). *J. Mater. Chem.* 21: 17345.
87 Deng, L.-L., Feng, J., Sun, L.-C. et al. (2012). *Sol. Energy Mater. Sol. Cells* 104: 113.
88 Meng, X., Zhang, Z., Tan, Z. et al. (2012). *ACS Appl. Mater. Interfaces* 4: 5966.
89 Cheng, Y.-J., Liao, M.-H., Chang, C.-Y. et al. (2011). *Chem. Mater.* 23: 4056.
90 Ye, G., Chen, S., Xiao, Z. et al. (2012). *J. Mater. Chem.* 22: 22374.
91 He, D., Zuo, C., Chen, S. et al. (2014). *Phys. Chem. Chem. Phys.* 16: 7205.
92 Matsuo, Y., Kawai, J., Inada, H. et al. (2013). *Adv. Mater.* 25: 6266.
93 Meng, X., Zhao, G., Xu, Q. et al. (2014). *Adv. Funct. Mater.* 24: 158.
94 Wong, W.W.H., Subbiah, J., White, J.M. et al. (2014). *Chem. Mater.* 26: 1686.
95 Deng, L.-L., Li, X., Wang, S. et al. (2016). *Sci. Bull.* 61: 132.
96 Xiao, Z., Matsuo, Y., Soga, I., and Nakamura, E. (2012). *Chem. Mater.* 24: 2572.
97 Tao, R., Umeyama, T., Higashino, T. et al. (2015). *Chem. Commun.* 51: 8233.
98 Zhang, B., Subbiah, J., Lai, Y.-Y. et al. (2015). *Chem. Commun.* 51: 9837.
99 Xiao, Z., Geng, X., He, D. et al. (2016). *Energy Environ. Sci.* 9: 2114.
100 He, Y., Zhao, G., Peng, B., and Li, Y. (2010). *Adv. Funct. Mater.* 20: 3383.
101 Steiner, F., Foster, S., Losquin, A. et al. (2015). *Mater. Horiz.* 2: 113.
102 Diederich, F. and Kessinger, R. (1999). *Acc. Chem. Res.* 32: 537.
103 Choi, J.H., Son, K.-I., Kim, T. et al. (2010). *J. Mater. Chem.* 20: 475.
104 Kang, H., Cho, C.-H., Cho, H.-H. et al. (2012). *ACS Appl. Mater. Interfaces* 4: 110.
105 Chen, S., Ye, G., Xiao, Z., and Ding, L. (2013). *J. Mater. Chem. A* 1: 5562.
106 Chen, S., Xiao, Z., He, D. et al. (2014). *Asian J. Org. Chem.* 3: 936.
107 He, D., Du, X., Xiao, Z., and Ding, L. (2014). *Org. Lett.* 16: 612.
108 Niinomi, T., Matsuo, Y., Hashiguchi, M. et al. (2009). *J. Mater. Chem.* 19: 5804.
109 Deng, L., Xie, S., Yuan, C. et al. (2013). *Sol. Energy Mater. Sol. Cells* 111: 193.
110 Sawamura, M., Iikura, H., and Nakamura, E. (1996). *J. Am. Chem. Soc.* 118: 12850.
111 Delgado, J.L., Espíldora, E., Liedtke, M. et al. (2009). *Chem. Eur. J.* 15: 13474.

112 Murata, M., Morinaka, Y., Murata, Y. et al. (2011). *Chem. Commun.* 47: 7335.
113 Chen, C., Lin, Y., Horng, J., and Chuang, S. (2011). *Adv. Energy Mater.* 1: 776.
114 Xiao, Z., He, D., Zuo, C. et al. (2014). *RSC Adv.* 4: 24029.
115 Cambarau, W., Fritze, U.F., Viterisi, A. et al. (2015). *Chem. Commun.* 51: 1128.
116 Kooistra, F.B., Mihailetchi, V.D., Popescu, L.M. et al. (2006). *Chem. Mater.* 18: 3068.
117 Ross, R.B., Cardona, C.M., Guldi, D.M. et al. (2009). *Nat. Mater.* 8: 208.
118 Vougioukalakis, G.C., Roubelakis, M.M., and Orfanopoulos, M. (2010). *Chem. Soc. Rev.* 39: 817.
119 Liu, Q., Jiang, Y., Jin, K. et al. (2020). *Sci. Bull.* 65: 272.
120 Xiao, Z., Jia, X., and Ding, L. (2017). *Sci. Bull.* 62: 1562.
121 Qin, J., Zhang, L., Xiao, Z. et al. (2020). *Sci. Bull.* 65: 1979.

4

Non-fullerene Small-Molecule Acceptors for Organic Solar Cells

Wei Gao[1], Jun Yuan[2], Zhenghui Luo[1], Jinru Cao[3], Weihua Tang[3], Yingping Zou[2], and Chuluo Yang[1]

[1] *Shenzhen University, College of Materials Science and Engineering, Shenzhen Key Laboratory of Polymer Science and Technology, Shenzhen, 518060, P.R. China*
[2] *Central South University, College of Chemistry and Chemical Engineering, Changsha, 410083, P.R. China*
[3] *Xiamen University, Department of Materials Science and Engineering, College of Materials, Xiamen, 361005, P.R. China*

4.1 Molecular Design Principles

To overcome the shortcomings of fullerene derivatives, a potential non-fullerene small-molecule acceptor (SMA) should fulfill several fundamental requirements [1–3]:

(i) Wide absorption range with intense molar extinction coefficient in the visible and near infrared (NIR) region: SMA possessing wide and strong light absorption ability in the visible and NIR region can make them an even stronger light absorber than polymers and achieve significant photocurrent generation through hole transfer to the highest occupied molecular orbital (HOMO) of electron donor material after photoexcitation, which could significantly enhance the photocurrent as both channels (donor and acceptor) can contribute to free carrier generation;

(ii) Matched energy levels with donor: The lowest unoccupied molecular orbital (LUMO) and HOMO energy levels of SMA should be lower than those of electron donor in order to provide enough driving force for effective exciton separation. Typically, the energy offsets between HOMO of donor and HOMO of acceptor or LUMO of donor and LUMO of acceptor should be larger than 0.3 eV in theory.

(iii) High electron mobility: SMA should have high electron mobility that can match the high hole mobility of polymer to achieve balanced carriers transport, which is beneficial to reduce the charge recombination within devices.

(iv) Good solubility and proper crystallization: SMA should possess good solubility in commonly used solvent, which is convenient for solution processing and

Organic Solar Cells: Materials Design, Technology and Commercialization, First Edition. Edited by Liming Ding.

increases the miscibility with polymer to form uniform morphology. Moreover, the crystallization performance of SMA cannot be too strong or too weak. If the crystallinity of a SMA is too strong, it will induce a strong aggregation tendency that leads to excessively large domains with an inadequate interface between the donor and acceptor for efficient exciton dissociation. On the contrary, if the crystallinity of the SMAs is too weak, it is not conducive to ordered molecular packing and results in poor carrier transport performance.

In the past decades, the development of polymeric and small-molecule donor materials has constructed a large library of effective molecular structures and design strategies for organic photovoltaics (OPV) materials [4–6], which can be applied to the design of non-fullerene SMAs. In this chapter, we will mainly review perylene diimide (PDI)-based SMAs, acceptor–donor–acceptor (A–D–A)-type SMAs, and A–DA′D–A-type SMAs.

4.2 PDI-Based SMAs

Thanks to their exceptional optical and electronic properties, along with outstanding thermal/photo stability, low-cost synthesis and facile functionalization, PDIs-based organic semiconductor materials have been widely applied to organic field-effect transistors (OFETs), organic light-emitting diodes (OLEDs), perovskite solar cells (PVSCs), and organic solar cells (OSCs) [1–3, 7–15]. The PDI monomers have a strong aggregation tendency to form oversized domain sizes due to their large conjugated planar structures, which is detrimental for exciton dissociation [16–22]. Therefore, developing twisted PDI-based SMAs is essential for suppressing aggregation propensity [23–26]. Functionalization of PDI on different positions (α, β positions, or the N position) is a main strategy of designing high-performance PDI-based SMAs (Figure 4.1). According to the number of PDI units, we can divide the PDI-based SMAs into PDI monomers, PDI dimers, PDI trimers, and PDI tetramers.

4.2.1 PDI Monomers

Limited by low power conversion efficiencies (PCEs) of related OSCs, monomeric PDIs were rarely studied, although PDI-based SMAs were tremendously utilized in the field. Nevertheless, there are some noteworthy works. In 2013, Bazan and coworkers blended p-DTS(FBTTh$_2$)$_2$ with PDI (Figure 4.2) to fabricate all-small-molecule OSCs and yielded a PCE of 3.0% (Table 4.1) [27]. In the next year, Keivanidis's group achieved a PCE of 3.71% by combining PDI and PBDTTT-C-T, which is the highest values among PDI monomers functionalized only at N position [28].

To further reduce aggregation, substitutions at α or β positions on PDI monomer were introduced. Patrick E. Hartnett et al. developed a slip-stacked acceptor of phenyl-PDI by introducing a phenyl group at each α position, and the PBTI3T:phenyl-PDI device gave a PCE up to 3.67% [29]. Apart from introducing

Figure 4.1 Functionalization positions on a PDI. Source: Z. Luo and C. Yang.

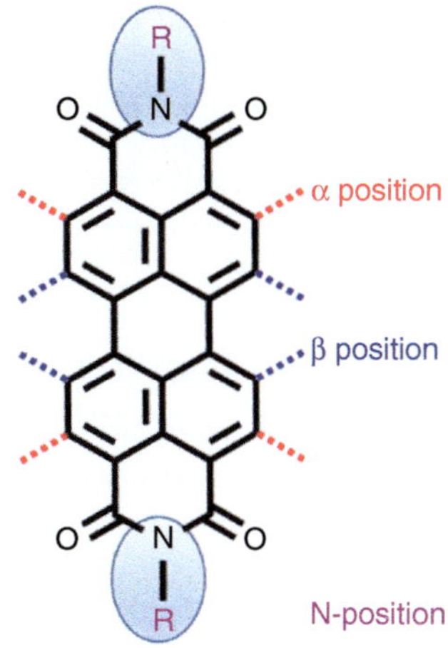

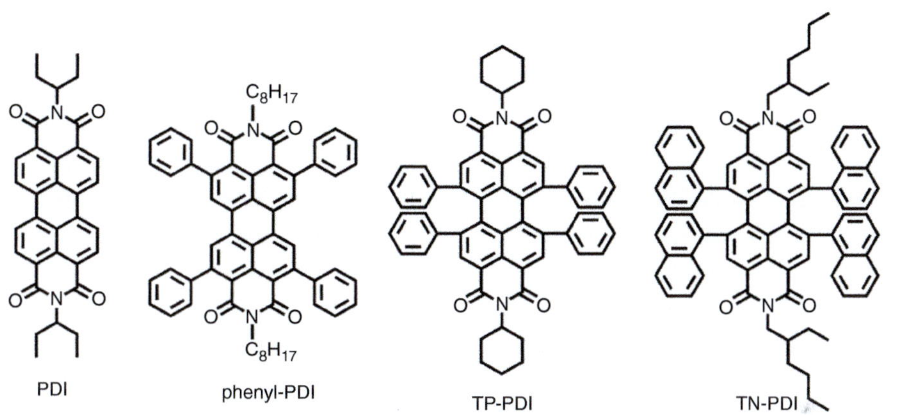

Figure 4.2 Chemical structures of monomeric PDIs. Source: Z. Luo and C. Yang.

phenyl groups at α position, adding benzene ring at β region was also reported. The SMA of TP-PDI realized a PCE of 4.1% when mixed with PTB7-Th [30]. In a later report, the same group reported a SMA of TN-PDI by introducing a naphthalene ring at each α position, and TN-PDI delivered a PCE of 3.0% when matched with a wide bandgap polymer donor PDBT-T1 [31].

Overall, the device performance of OSCs based on PDI monomers is not ideal, which is mainly related to the over-aggregation of PDI monomers. To alleviate the aggregation tendency, developing PDI dimers (or trimers or tetramers) is a good strategy.

4.2.2 PDI Dimers

In addition to modifying PDI monomers, a direct way to reduce the molecular planarity is to connect two PDIs together by a single bond or functional groups such as spirofluorene, benzene, and thiophene (Figure 4.3). The first PDI dimer (Per 1) was reported by Narayan's research team in 2012, who connected two

Table 4.1 Optical bandgap, energy levels, and photovoltaic parameters for PDI monomer- and dimer-based non-fullerene SMAs.

SMA	E_g^{opt} a) (eV)	HOMO b)/ LUMO c) (eV)	Donor	V_{oc} (V)	J_{sc} (mA cm^{-2})	FF	PCE (%)	References
PDI	—	—	*p*-DTS(FBTTh$_2$)$_2$	0.78	7.4	0.52	3.0	[27]
PDI	—	—	PBDTTT-C-T	0.80	8.10	0.519	3.71	[28]
phenyl-PDI	2.01	−5.92/−3.91	PBTI3T	1.024	6.56	0.5459	3.67	[29]
TP-PDI	1.87	−5.69/−3.82	PTB7-Th	0.87	10.1	0.464	4.1	[30]
TN-PDI	—	−5.60/−4.00	PDBT-T1	0.947	7.7	0.41	3.0	[31]
Per 1	1.80	−5.90/−4.10	PBDTTT-C-T	0.76	9.5	0.46	2.78	[32]
CP-V	2.18	−6.16/−3.98	PPDT2FBT	0.87	10.04	0.6016	5.28	[33]
T-PDI	2.18	−6.13/−3.95	PBDB-T	0.837	7.67	0.547	3.50	[34]
s-diPBI	2.08	−5.95/−3.87	PBDTTT-CT	0.73	10.58	0.4610	3.63	[35]
s-diPBI	—	−5.95/−3.87	PBDTBDD	0.87	8.26	0.611	4.39	[36]
s-diPBI	—	−6.13/−4.04	PBDTT-F-TT	0.80	11.98	0.59	5.90	[37]
SdiPBI-S	2.20	−6.05/−3.85	PDBT-T1	0.90	11.98	0.661	7.16	[38]
SdiPBI-Se	2.22	−6.09/−3.87	PDBT-T1	0.96	12.49	0.702	8.4	[39]
N-PDI	2.2	−6.00/−3.80	P3TEA	1.13	11.03	0.61	7.55	[40]
SdiPBI-BT	2.07	−5.86/−3.79	PDBT-T1	0.95	10.31	0.687	6.71	[41]
diPBI-BT	2.12	−5.79/−3.67	PDBT-T1	0.99	9.78	0.602	5.84	[41]
NDP	2.17	−6.08/−3.91	PTB7-Th	0.80	14.35	0.56	6.43	[42]
SDP	2.21	−6.11/−3.90	PTB7-Th	0.83	13.21	0.47	5.18	[42]
SF-PDI$_2$	2.05	−5.71/−3.71	P3HT	0.61	5.92	0.65	2.35	[43]
SF-PDI$_2$	—	—	P3TEA	1.11	13.27	0.643	9.5	[44]
PDI-T	1.95	−5.97/−4.02	PTB7-Th	0.88	10.20	0.41	3.68	[45]
FPDI-F	2.20	−6.00/−3.80	PTB7-Th	0.92	8.94	0.40	3.29	[45]
FPDI-S	2.22	−5.99/−3.77	PTB7-Th	0.94	12.48	0.58	6.72	[45]
FPDI-Se	2.22	−5.98/−3.76	PTB7-Th	0.92	11.40	0.55	5.77	[45]

a) Calculated from $E_g^{opt} = 1240/\lambda_{onset}$.
b) Calculated from the LUMO and E_g^{opt}, HOMO = LUMO − E_g^{opt}.
c) Calculated from CV curves.

nitrogen atoms at each PDI's imide position through a single bond [32]. Compared with PDI monomer, Per 1 has reduced planarity and aggregation behavior, blending with PBDTTT-C-T achieved a PCE of 2.77%. Other than single bond, a non-conjugated 1,1-diphenylcyclohexane linker was introduced into PDI dimer through functionalization of N position, thereby a V-type PDI dimer of CP-V was developed, and OSCs-based PTB7-Th:CP-V delivered a PCE of 5.28% [33]. Just recently, Yang's research group reported that two PDI units were connected by

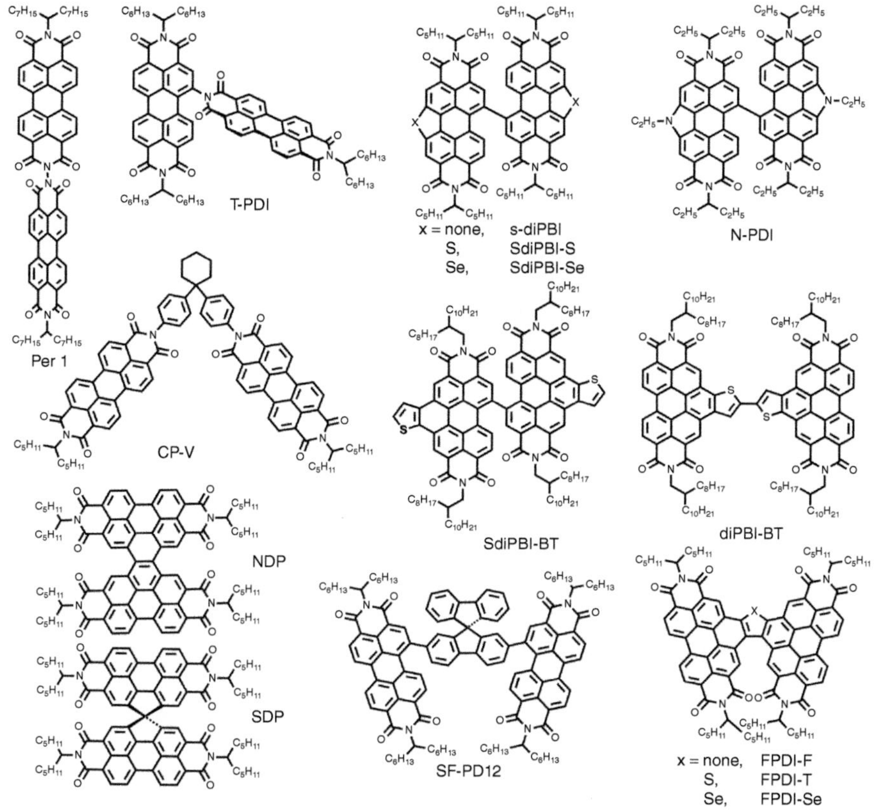

Figure 4.3 Chemical structures of selected PDI dimers. Source: Z. Luo and C. Yang.

β and amine positions to develop two PDI oligomers. Among them, the device efficiency of PDI dimer T-PDI and PBDB-T combination reached 3.5% [34].

As a matter of fact, the most researched are still β position linked PDI SMAs. In 2014, Wang and coworkers developed a PDI dimer of s-diPBI by connecting the bay positions of two PDIs and blended it with polymer donor PBDTTT-C-T, successfully achieving a PCE of 3.63% [35]. In next two reports, they improved the device efficiency by changing the polymer donor and the optimizing the device process. Consequently, the highest efficiency reached 5.90% [36, 37]. Later, they reported two derivatives of s-diPBI, the β positions at both ends of the dimer were capped with sulfur (SdiPBI-S) or selenium (SdiPBI-Se), which brought the PDI unit into a planar structure [38, 39]. Compared to s-diPBI, SdiPBI-S and SdiPBI-Se exhibit larger optical bandgaps and dihedral angles between the two PDI units. When combined with the medium-bandgap polymer PDBT-T1, SdiPBI-Se achieved a high PCE of 8.42%, while SdiPBI-S had a PCE of 7.16%, significantly higher than that of s-diPBI-based device. In addition to S and Se annulated PDIs, Hendsbee et al. synthesized a N-cyclized s-diPBI (N-PDI) at the gram level, requiring no column chromatography for purification [40]. The P3TEA:N-PDI-based device yielded a PCE of 7.55%. In the same year, Yang's group reported two thiophene ring-fused PDI dimers, namely SdiPBI-BT and diPBI-BT. The OSCs based on PDBT-T1:SdiPBI-BT exhibit a good PCE of 6.71%, which is higher than that of diPBI-BT-based devices (5.84%) [41]. In addition, Wang and coworkers employed mono-brominated PDI to synthesize two PDI dimers, NDP and SDP, using two different conditions in just one step (Figure 4.4) [42]. In comparison with the benzene-fused NDP, the spiro-fused SDP exhibits red-shifted absorption spectra due to the significant spiro-conjugation effect. Single-crystal X-ray diffraction analysis indicates that SDP shows a three-dimensional (3D) cruciform configuration. When blending the two SMAs with PTB7-Th, the device efficiency was 5.18% for NDP-based device, and 6.43% for SDP-based device, respectively.

In addition to connecting two PDIs through a single bond, Zhao and coworkers studied the effect of spacer units on the aggregation and phase separation between PDI-based SMAs and polymer donor to further tune the morphology of active layers [43]. They inserted a series of spacer units including thiophene, benzene, bithiophene, and spirobifluorene between two PDI units and found that the planar dimer with benzene ring as the spacer unit has the largest domain size and worst photovoltaic performance (PCE = 0.90%) due to the strong intermolecular interaction. In contrast, SF-PDI_2 with spirobifluorene as the spacer unit has the best photovoltaic performance (PCE = 2.35%), and the blend film with P3HT has a smaller domain size. In 2016, Yan and coworkers designed a new type of conjugated polymer P3TEA and combined it with SF-PDI_2 to fabricate OSC devices, eventually achieving an efficiency of 9.5% and an internal quantum efficiency of nearly 90%, along with high open-circuit voltage (V_{oc} of 1.11 V) and fill factor (FF of 0.64) [44]. The authors also calculated the energy loss (the difference between the bandgap of the donor material and the V_{oc} of the device), which was only 0.61 eV, indicating that non-fullerene OSCs have great potential for achieving low energy loss.

Ring-fusion strategy is an effective approach to enhance molecular packing, thereby improving charge carriers transport. In 2015, Jen's team developed a series of fused-ring PDI dimers bridged by furan, thiophene, and selenophene, namely FPDI-F, FPDI-T, and FPDI-Se [45]. Compared with the unfused dimer, the fused PDI dimers reduce the recombination energy and has an extended effective π-conjugation length, which is very conducive to exciton diffusion and charge transport. They found that the FPDI-T with a moderate torsion angle achieved a balance of various parameters, and thus delivering a highest PCE of 6.72%.

4.2.3 PDI Trimers

In order to further investigate the twisting strategies and design non-fullerene materials that can form 3D or quasi 3D molecular structures, three PDI units can be introduced into one molecule. At present, the efficiency of solar cells based on PDI trimers as acceptors can reach as high as 11.01% [46].

Hou and coworkers elongated Per1 to develop a PDI trimer of H-tri-PDI (Figure 4.5) [47]. When it is blended with an efficient narrow bandgap polymer donor PBDT-TS1, it achieves a PCE of 7.25% along with a high short-circuit current density (J_{sc}) of 16.52 mA cm^{-2} (Table 4.2). Employing a central core to exploit PDI

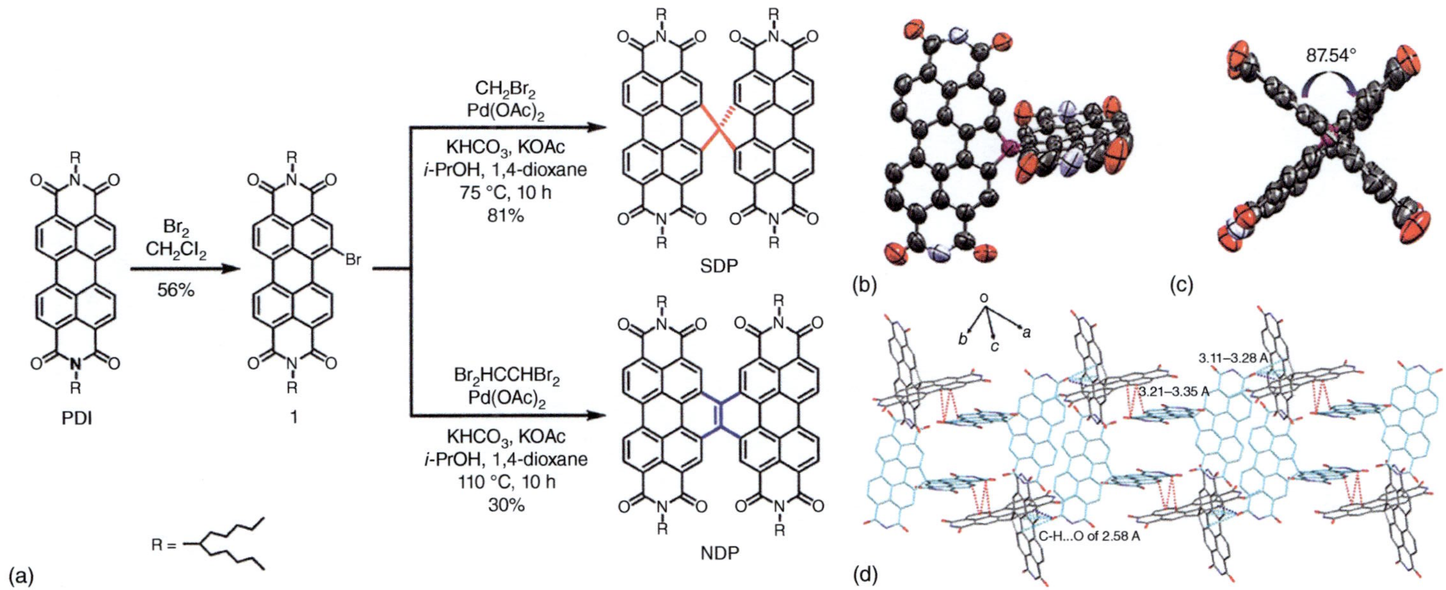

Figure 4.4 (a) The synthetic route of NDP and SDP; single-crystal structures of SDP with 50% thermal ellipsoids probability. The dark gray-, blue-, red-, and purple-colored atoms represent C, N, O, and spiro carbon, respectively. Hydrogen atoms and long alkyl chain groups are omitted for clarify. ORTEP structure form top (b) and side views (c) and packing arrangement (d). Source: Gao et al. [42]. Reproduce with permission of American Chemical Society

trimers is the most common method. In 2014, Zhan and coworkers synthesized a PDI trimer called S(TPA-PDI), in which three PDI units are connected by triphenylamine units [48]. Due to the presence of sp^3 hybridized nitrogen atoms in the core, the whole molecule is quasi 3D and non-planar structure, which inhibits intermolecular interaction and aggregation. When blended with PBDTTT-C-T, the PCE of the device is 3.32%, and its electron mobility is only 2.32×10^{-5} cm^2 V^{-1} s^{-1}, which explains the low FF (33.6%). In the same year, Chen's research team directly connected three PDIs to the benzene ring to synthesize a novel PDI trimer of B(PDI)3, and the OSCs based on PTB7-Th:B(PDI)3 achieved a PCE of 5.65% [49]. Soon after, Peng's group reported a PDI trimer Ta-PDI with triazine as the central core. Compared with analog with benzene as the core, they found that the triazine-based trimer has a smaller twisted structure and a stronger tendency to aggregate, thus a higher electron mobility [50]. Using PTB7-Th as the polymer donor, the Ta-PDI-based OSC showed a PCE of 9.18%, with a high J_{sc} of 17.1 mA cm^{-2} and a FF of 68.5%. In addition, they employed triphenylphosphine (TPP), triphenylphosphine monoxide (TPO), and triphenylphosphine sulfide as central core to develop three novel PDI trimmers, namely TPP-PDI, TPO-PDI, and TPS-PDI [46]. Compared with TPP-PDI and TPS-PDI, TPO-PDI shows the least twisted conformation due to the stronger electron-withdrawing ability of the oxygen atom, which improves the molecular packing of this SMA. Ultimately, OSCs based on PTTEA:TPO-PDI exhibit a PCE as high as 11.01%, significantly higher than those of device based on TPP-PDI (6.19%) and TPS-PDI (9.67%). It is worth mentioning that the PCE of 11.01% is the highest value for PDI-based OSCs. In the same year, the same group developed two novel PDI trimers, Alq_3-PDI and Alq_3-PDI2, are synthesized by utilizing a 3D tri(8-hydroxyquinoline)aluminum(III) (Alq_3) core, which can suppress the molecular aggregation properties [51]. In comparison with Alq_3-PDI, the extended π-conjugation of Alq_3-PDI2 leads to enhanced molecular packing, which broadens the absorption spectra and improves the domain size of blend film. Thus, the OSCs based on PTTEA:Alq_3-PDI2 yields higher J_{sc} and FF, thereby better device performance (PCE = 9.54%, J_{sc} = 15.74 mA cm^{-2}, FF = 0.713) relative to those of device based on Alq_3-PDI (PCE = 7.82%, J_{sc} = 13.56 mA cm^{-2}, FF = 0.663).

The fused-ring strategy can also be applied to develop PDI trimer-like PDI dimer. In 2015, Nuckolls and coworkers continued to extend NDP and developed a PDI trimer and a PDI tetramer, namely h-PDI3 and h-PDI4, and they achieved high PCEs in OSCs (7.9% for h-PDI3-based device and 8.3% for h-PDI4-based device), which were the highest values for non-fullerene bulk heterojunctions at that time [52]. In 2017, Chen and coworkers fused three PDI units into one benzotrithiophene, thus developing a fused-PDI trimer named Fused-TriPDI [53]. Compared with the non-fused precursor (TriPDI), Fused-TriPDI exhibits upshifted energy levels, better planarity, and enhanced electron mobility. As a result, the OSCs with PTB7-Th:Fused-TriPDI blend yielded higher V_{oc}, J_{sc}, and FF relative to those of device based on TriPDI, thus promoted PCE. In the same year, Wang's research group developed TPH-Se by fusing three selenium-hybridized PDI units to a benzene ring [54]. In comparison with TPH without Se, the integration of

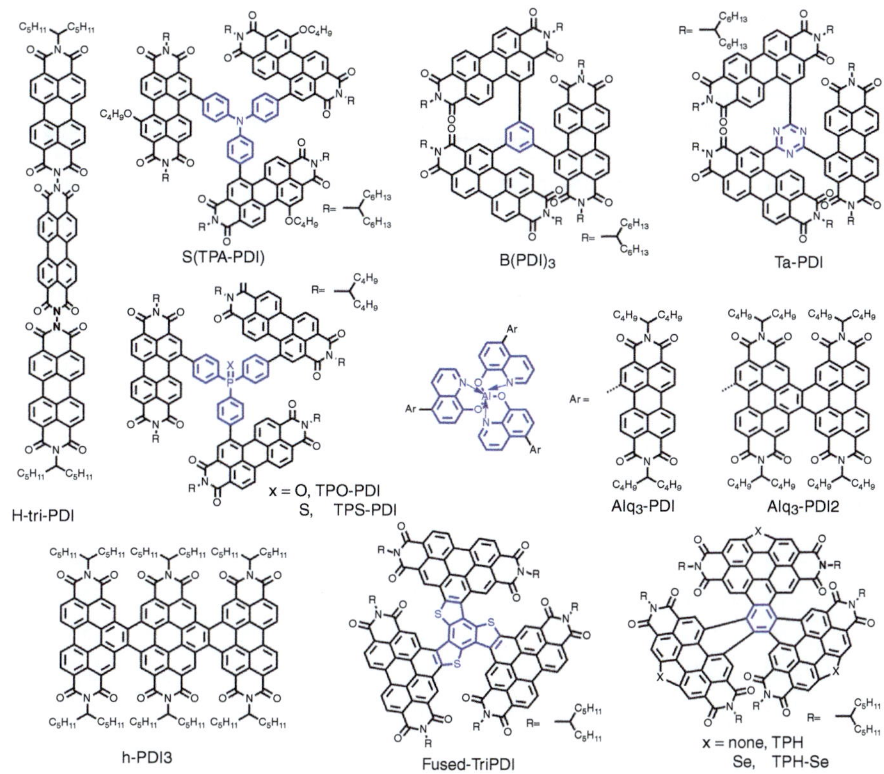

Figure 4.5 Chemical structures of selected PDI trimers. Source: Z. Luo and C. Yang.

Se to TPH endows TPH-Se higher LUMO energy level and enhanced absorption. Single-crystal X-ray diffraction results demonstrate that TPH-Se shows a more compact 3D network assembly and a more twisted propeller configuration relative to TPH, which can be ascribed to the Se· · ·O interactions (Figure 4.6). Consequently, TPH and TPH-Se acceptor-based OSCs gave high PCE of 8.28% and 9.28%, respectively. The better PCE for TPH-Se-based device is mainly attributed to the upshift LUMO value, enhanced mobility and absorption of TPH-Se, and more favorable morphology.

4.2.4 PDI Tetramers

In addition to the PDI trimers, we can also connect four PDI to the central cores to synthesize a PDI tetramers, which is beneficial to form 3D molecular structures, thereby facilitating to the charge transport.

In 2015, Yan and coworkers developed three PDI tetramers by connecting four PDI units with tetraphenylmethane, tetraphenylsilane, and tetraphenylgermane, namely TPC-PDI4, TPSi-PDI4, and TPGe-PDI4 (Figure 4.7), and they studied the effect of the central core on device performance and film morphology [55]. When blended them with the same polymer donor PffBT4T-2DT, it was found that devices

Table 4.2 Optical bandgap, energy levels, and photovoltaic parameters for PDI trimer-based non-fullerene SMAs.

SMA	E_g^{opt} [a] (eV)	HOMO[b]/ LUMO[c] (eV)	Donor	V_{oc} (V)	J_{sc} (mA cm^{-2})	FF	PCE (%)	References
H-tri-PDI	2.09	−6.02/−3.93	PBDT-TS1	0.732	16.52	0.6003	7.25	[47]
S(TPA-PDI)	1.76	−5.46/−3.70	PBDTTT-C-T	0.88	11.92	0.336	3.32	[48]
B(PDI)3	2.14	−6.00/−3.86	PTB7-Th	0.83	13.12	0.52	5.65	[49]
Ph-PDI	2.10	−5.88/−3.78	PTB7-Th	0.85	11.91	0.549	5.57	[50]
Ta-PDI	2.05	−5.86/−3.81	PTB7-Th	0.78	17.10	0.685	9.15	[50]
TPP-PDI	2.11	−5.75/−3.64	PTTEA	1.02	12.25	0.495	6.19	[46]
TPO-PDI	2.09	−5.79/−3.70	PTTEA	0.99	14.89	0.747	11.01	[46]
TPS-PDI	2.08	−5.82/−3.74	PTTEA	0.97	14.20	0.702	9.67	[46]
Alq_3-PDI	1.95	−5.71/−3.76	PTTEA	0.87	13.56	0.663	7.82	[51]
Alq_3-PDI	1.88	−5.76/−3.88	PTTEA	0.85	15.74	0.713	9.54	[51]
hPDI3	2.14	−5.96/−3.82	PTB7-Th	0.81	14.5	0.67	7.9	[52]
HPDI4	2.06	−5.90/−3.84	PTB7-Th	0.80	15.2	0.68	8.3	[52]
TriPDI	1.99	−5.91/−3.92	PTB7-Th	0.85	6.13	0.38	2.19	[53]
Fused-TriPDI	2.20	−6.15/−3.95	PTB7-Th	0.91	12.39	0.55	6.19	[53]
TPH	2.19	−6.02/−3.83	PDBT-T1	0.968	12.18	0.702	8.28	[54]
TPH-Se	2.17	−5.97/−3.80	PDBT-T1	1.00	12.99	0.715	9.28	[54]

a) Calculated from $E_g^{opt} = 1240/\lambda_{onset}$.
b) Calculated from the LUMO and E_g^{opt}, HOMO = LUMO − E_g^{opt}.
c) Calculated from CV curves.

based on TPC-PDI4 and TPSi-PDI4 exceeded 4% efficiency (Table 4.3), while the efficiency of the device based on TPGe-PDI4 is less than 2%. Other than the tetraphenylalkanes-like central core, spirobifluorene and spirocyclic dithiophene were used as central core to develop PDI tetramers, and the corresponding PDI tetramers are called SF-PDI4 and FSP [56, 57], respectively. The device efficiencies based on these two PDI tetramers as acceptors are 5.98% and 8.89%, respectively. Later, the PDI tetramer PBI-Por constructed with porphyrin as the core has a lower bandgap and weaker crystallinity than most PDI-based acceptors [58]. When mixing PBI-Por and the polymer PBDTBDD with complementary absorption, the device yields a PCE of 7.4%. In addition to liking PDI unit at β position, there is also a PDI tetramer of Me-PDI4 by connecting to the tetraphenyl carbon core unit through the imide region [59]. Combining it with the donor PBDTTT-C-T yielded a PCE of 2.35%. In 2017, Huang's team developed another PDI tetramer, P4N4, with 2,3,7,8-tetraphenylpyrazino[2,3-g]quinoxaline as the central core, using PDBT-T1 as the donor, the OSC with P4N4 as the acceptor achieved a PCE of 5.71% [60].

In 2015, Yan and coworkers developed a new PDI tetramer TPE-PDI4 (Figure 4.8) based on TPC-PDI4 by replacing carbon atoms with ethylene [61]. Because the

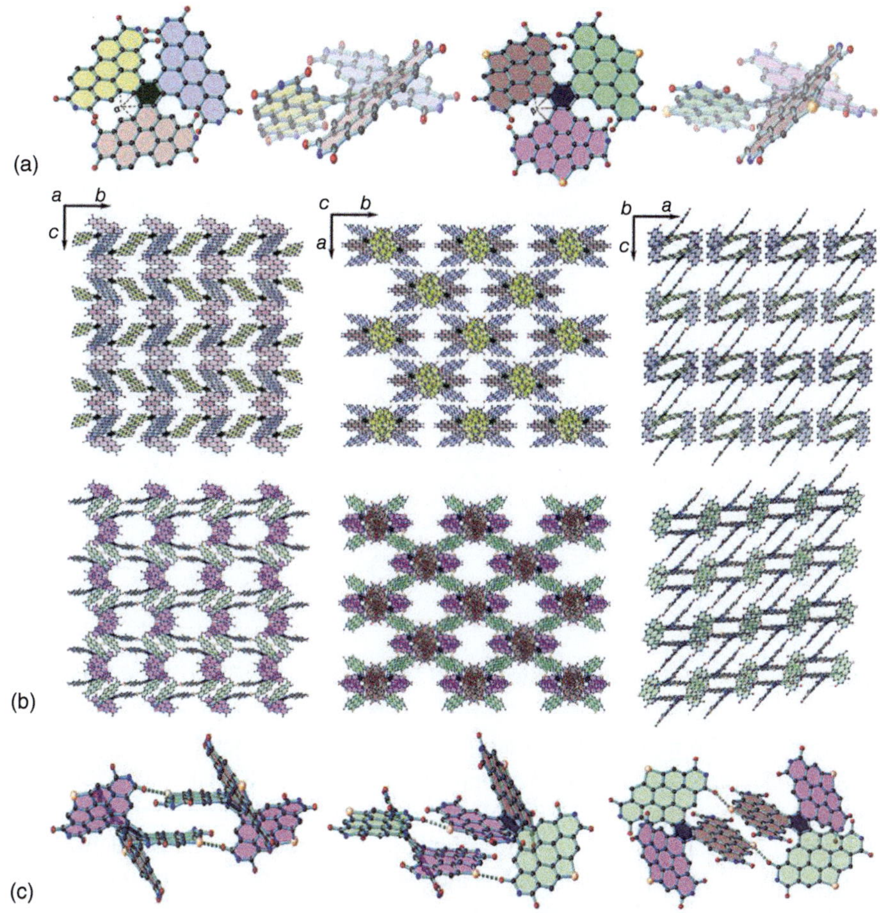

Figure 4.6 (a) X-ray molecular structures of TPH **4b** and TPH-Se **6a** (top view and side view). (The alkyl chains and hydrogen atoms are omitted for clarity). (b) The slipped 3D stacking mode of TPH **4b** (top) and TPH-Se **6a** (down). (c) Se· · ·O (3.0 Å) interactions between different PBI subunits of TPH-Se **6a**. Source: Meng et al. [54]. Reproduce with permission of American Chemical Society

tetrastyrene core has a four-leaf fan-like structure, its aggregation is weak. Blending with PffBT4T-2DT produces a suitable phase separation, and an efficiency of 5.53% was obtained. The following year, Yan and coworkers replaced the ethylene in TPE-PDI4 with pyrazine to synthesize the small-molecule TPPz-PDI4 [62]. The introduction of pyrazine reduced the torsion angle of the molecule, increased the electron mobility of the molecule, and further improved its efficiency (7.1%). From TPC-PDI4 to TPE-PDI4 to TPPz-PDI4, it is clear that reducing the degree of intramolecular distortion can improve electron mobility, fill factor, and device performance. Likewise, Yan and coworkers reported two PDI tetramers with fused and non-fused-ring structures based on tetrathienylbenzene as the core [63]. DFT calculations indicate that the fused FTTB-PDI4 shows a smaller twisted structure as

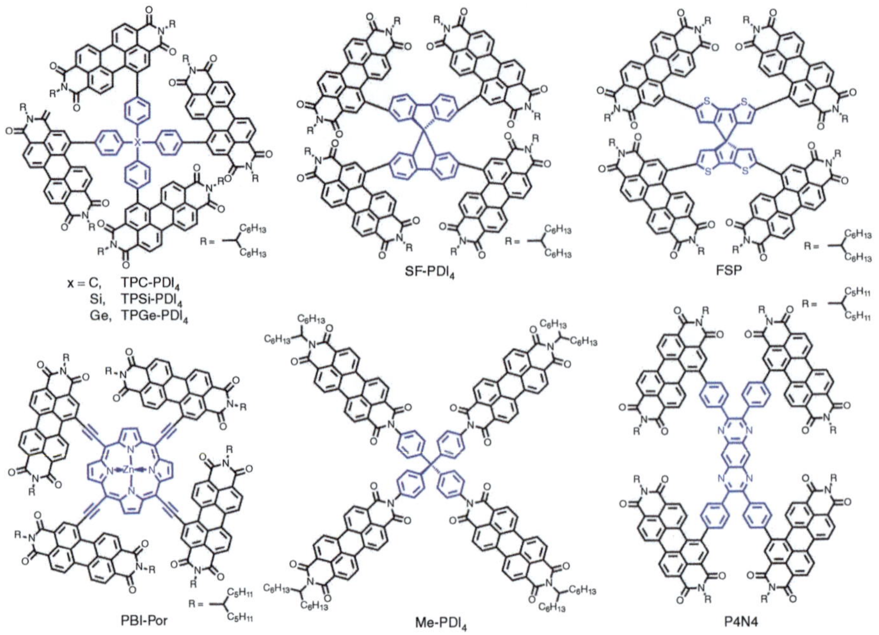

Figure 4.7 Chemical structures of selected PDI tetramers. Source: Z. Luo and C. Yang.

compared with the non-fused PDI tetramer (TTB-PDI4), which leads to enhanced molecular packing and improved domain purity in the P3TEA:FTTB-PDI4 blend (Figure 4.9). Additionally, FTTB-PDI4 displays enhanced absorption capability, upshifted LUMO energy levels and blue-shifted absorption. Consequently, the P3TEA:FTTB-PDI4 devices achieve the a PCE as high as 10.58% along with a small energy loss of 0.53 V, which is the highest value of solar cells based on PDI as the acceptor at that time.

In 2016, Yu and coworkers used BDT as the core to construct the α-substituted PDI tetramer of TPB [64]. Blending it with PTB7-Th achieved a PCE of 8.46%, along with a J_{sc} over 18 mA cm^{-2}. The next year, they employed the same central core to develop a β-substituted PDI tetramer of βTPB6 and developed βTPB6-C (the fused βTPB6) [65]. In comparison with the unfused βTPB6, βTPB6-C exhibits enlarged conjugated skeleton, enhanced absorption, more rigid molecular structure, and shallower energy level. As a result, the inverted OSCs based on βTPB6-C and PTB7-Th exhibited a better efficiency of 7.69% than that of βTPB6-based device.

Isomerization strategy plays a vital role in tuning absorption spectra, electronic structure, charge mobility, π–π stacking, eventually determining the device performance [69–71]. For instance, Yang and coworkers developed two isomeric PDI tetramers, namely BPT-Se and BPT-Se1 [66]. In comparison with BPT-Se, BPT-Se1 exhibits a shallower energy level, enhanced absorption profile, and improved molecular packing. As a result, the OSCs based on BTP-Se and polymer PDBT-T1 as donor gave a PCE of 9.54% with an outstanding FF of 0.732, which are higher than those of the device based on BTP-Se (PCE = 7.8%; FF = 0.675). In addition, Yan's

Table 4.3 Optical bandgap, energy levels, and photovoltaic parameters for PDI tetramer-based non-fullerene SMAs.

SMA	E_g^{opt} [a] (eV)	HOMO[b]/ LUMO[c] (eV)	Donor	V_{oc} (V)	J_{sc} (mA cm^{-2})	FF	PCE (%)	References
TPC-PDI_4	2.25	−6.00/−3.75	PffBT4T-2DT	0.96	9.2	0.49	4.3	[55]
TPSi-PDI_4	2.26	−6.01/−2.26	PffBT4T-2DT	0.94	8.5	0.53	4.2	[55]
TPGe-PDI_4	2.26	−5.94/−3.68	PffBT4T-2DT	0.92	5.0	0.37	1.6	[55]
SF-PDI4	2.05	−5.83/−3.78	P4T2FBT	0.93	11.03	0.511	5.27	[56]
SF-PDI4	2.05	−5.83/−3.78	PV4T2FBT	0.90	12.02	0.542	5.98	[56]
FSP	1.90	−5.60/−3.70	PTB7-Th	0.90	16.6	0.595	8.89	[57]
PBI-Por	1.48	−5.12/−3.68	PBDB-T	0.78	14.5	0.66	7.4	[58]
Me-PDI4	2.14	−5.96/−3.82	PBDTTT-C-T	0.77	7.83	0.45	2.73	[59]
P2O2	2.09	−5.87/−3.80	PDBT-T1	0.871	6.94	0.419	2.53	[60]
P2N2	2.08	−5.81/−3.73	PDBT-T1	0.950	7.95	0.511	3.86	[60]
P4N4	2.11	−5.80/−3.69	PDBT-T1	0.958	9.40	0.634	5.71	[60]
TPE-PDI_4	2.05	−5.77/−3.72	PBDTT-F-TT	0.91	11.7	0.52	5.53	[61]
BP-PDI_2	2.09	−5.81/−3.72	PBDTT-F-TT	0.93	8.6	0.39	2.42	[61]
TPPz-PDI_4	2.10	−5.86/−3.76	PffBT-T3(1,2)-2	0.987	12.6	0.57	7.1	[62]
TTB-PDI4	—	—/−3.68	P3TEA	1.05	12.06	0.529	7.11	[62]
FTTB-PDI4	—	—/−3.58	P3TEA	1.13	13.89	0.659	10.58	[63]
TPB	—	—/−3.89	PTB7-Th	0.79	18.2	0.59	8.47	[64]
βTPB6	—	—/−3.79	PTB7-Th	0.82	11.9	0.57	5.85	[65]
βTPB6-c	—	—/−3.75	PTB7-Th	0.92	14.9	0.56	7.69	[65]
BPT-Se	2.04	−5.89/−3.85	PDBT-T1	1.039	11.10	0.675	7.78	[66]
BPT-Se1	2.07	−5.88/−3.81	PDBT-T1	1.059	12.30	0.732	9.54	[66]
a-FTTN-PDI4	2.10	−5.74/−3.64	P3TEA	1.15	12.0	0.61	8.6	[67]
b-FTTN-PDI4	2.23	−5.83/−3.60	P3TEA	1.15	11.5	0.55	7.4	[67]
*o*CP-FPDI4	2.31	−5.94/−3.63	P3TEA	1.17	5.88	0.33	2.42	[68]
*p*CP-FPDI4	2.18	−5.81/−3.63	P3TEA	1.16	13.47	0.56	9.06	[68]

a) Calculated from $E_g^{opt} = 1240/\lambda_{onset}$.
b) Calculated from the LUMO and E_g^{opt}, HOMO = LUMO − E_g^{opt}.
c) Calculated from CV curves.

group synthesized two isomeric PDI tetramers, a-FTTN-PDI4 and b-FTTN-PDI4, with different functionalization sites (α and β positions) on a naphthalene ring, respectively [67]. Photoluminescence spectra and GIWAXS results revel that a-FTTN-PDI4 showed better π–π stacking than b-FTTN-PDI4, thereby leading to enhanced electron mobility. Consequently, the P3TEA:a-FTTN-PDI4-based devices achieved a higher PCE of 8.6% than b-FTTN-PDI4-based ones. The next year, the same group synthesized two regio-isomeric PDI tetramers (*o*CP-FPDI4

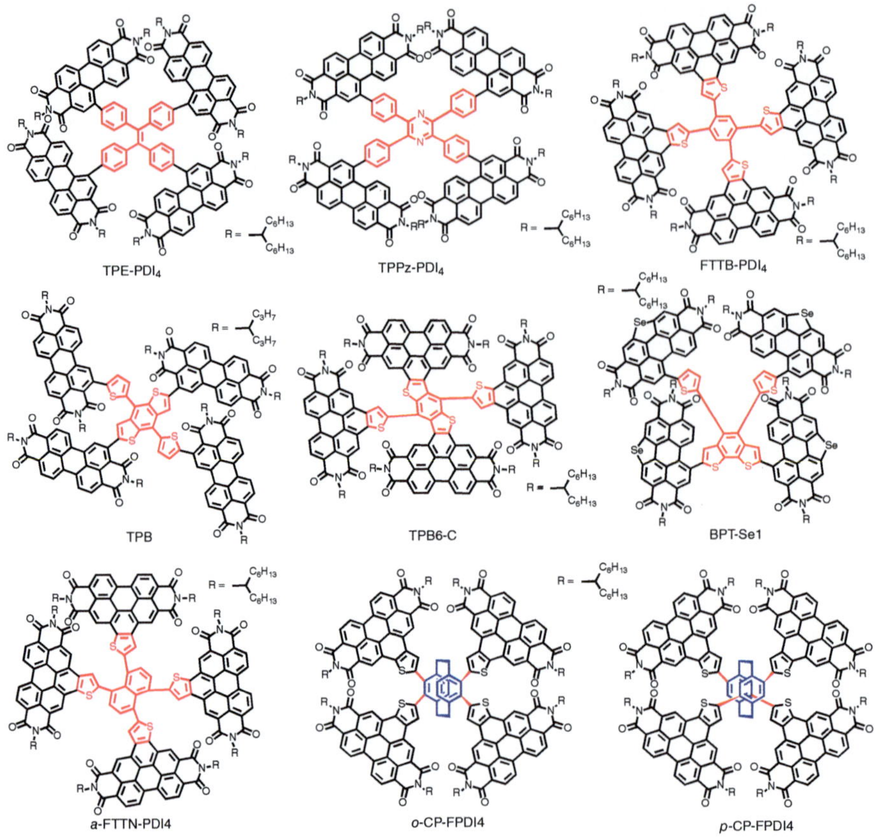

Figure 4.8 Chemical structures of selected PDI tetramers. Source: Z. Luo and C. Yang.

and *p*CPFPDI4) with a novel [2.2]paracyclophane central core [68]. Blending *p*CP-FPDI4 with polymer donor P3TEA realized a PCE of 9.06%, which is higher than that of *o*CP-FPDI4-based device (2.42%) due to the obviously reduced FF and J_{sc}. The enhanced FF and J_{sc} in *p*CP-FPDI4-based device are ascribed to the more suitable phase segregation, more effective charge dissociation, and balanced charge transport.

Although the device efficiency based on PDIs as acceptors is much lower than that of the A–D–A-type acceptors-based device, its intrinsic characteristics, including outstanding thermal/photo stability, low-cost synthesis, and facile functionalization, are still worthy of our attention. In addition, it can be used as an indoor photovoltaic or as a front cell for tandem devices. To further improve the device efficiency of PDI-based acceptors, some design guidelines are proposed: (i) the first and most important point is to broaden the absorption spectrum and enhance the light absorption capacity of PDIs; (ii) balancing the relationship between twisting angle and electron mobility plays an important role in producing suitable domain size/purity, thereby affecting the device performance; (iii) some commonly used molecular design strategies such as fusion strategy, isomeric strategy, and asymmetric strategy

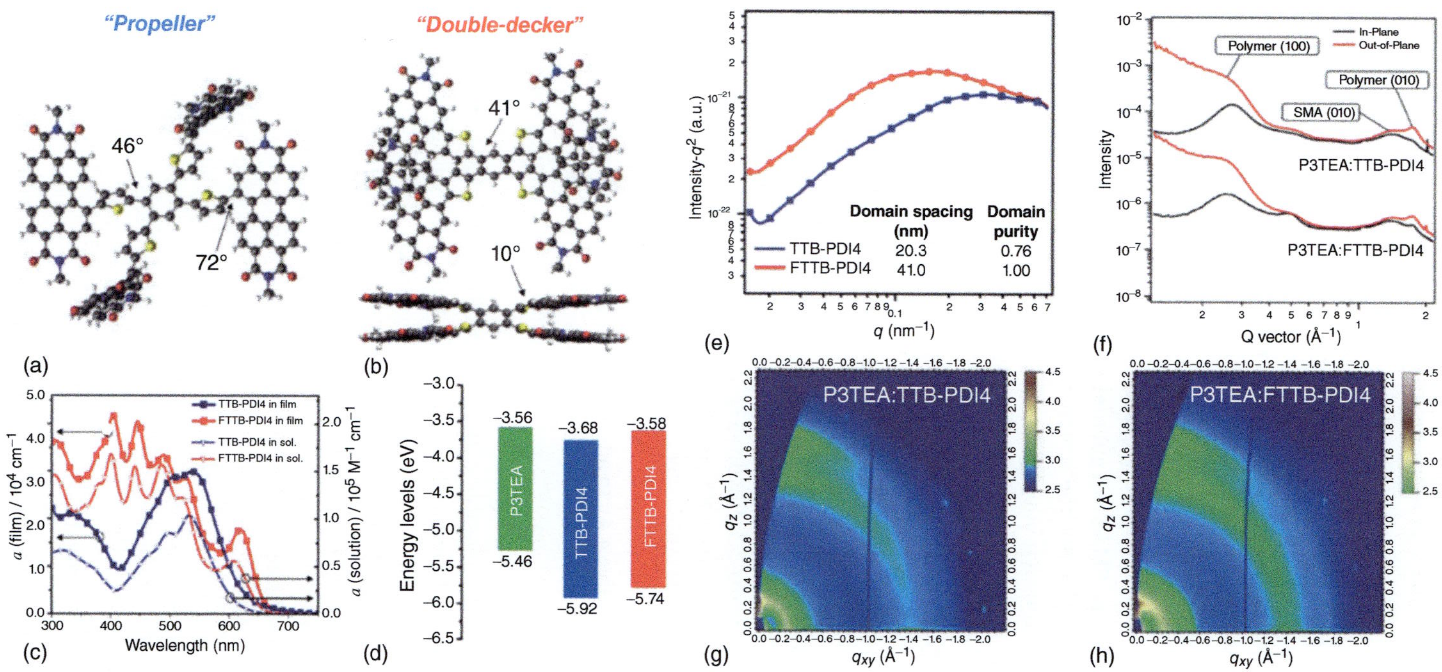

Figure 4.9 Optimized geometry of (a) TTB-PDI4 and (b) FTTB-PDI4; (c) absorption spectra of TTB-PDI4 and FTTB-PDI4 in solution and film; (d) energy levels of P3TEA, TTB-PDI4, and FTTBPDI4; (e) RSoXS profiles and (f) 1D GIWAXS profiles of the in-plane and out-of-plane directions of P3TEA:TTB-PDI4 and P3TEA:FTTBPDI4; (g) 2D GIWAXS pattern of P3TEA:TTB-PDI4 and (h) 2D GIWAXS pattern of P3TEA:FTTB-PDI4. Source: Zhang et al. [63]. Reproduced with permission of American Chemical Society

could be employed to optimize the molecular structure, thereby improving the physicochemical properties eventually increasing the device efficiency.

4.3 A–D–A-Type SMAs

In A–D–A-type SMA materials, A unit owns strong electron-withdrawing ability while D unit owns strong electron-donating ability. Such A–D–A configuration can produce strong intramolecular charge transfer (ICT) state between A and D units and endow SMAs with wide absorption range and low-lying HOMO and LUMO energy levels. Moreover, large π-conjugated backbone with planar and rigid conformation of A–D–A-type SMAs can guarantee good intermolecular interaction and high electron mobility. Usually, the structure of A–D–A-type SMAs can be divided into three parts: electron-rich donor core (D unit), electron-deficient end group (A unit), and side chains linked to D unit, which makes the modifiable sites of such type of SMAs become very rich. Among them, A and D units show very critical roles in regulating bandgap, energy levels, and crystalline properties of SMAs, and side chains flanking on D unit is mainly to improve the solubility and fine-tune the crystalline properties of SMAs and thereby optimizing micromorphology of active layer [1–3].

In 2015, Zhan and coworkers first reported a novel A–D–A-type SMA, namely ITIC, by combining a ladder-type donor (indacenodithieno[3,2-*b*]thiophene, IDTT) with two 2-(3-oxo-2,3-dihydro-1*H*-inden-1-ylidene)malononitrile (INCN) terminal groups (Figure 4.10) [72]. Due to the strong push-pull electronic effect between IDTT and INCN units, ITIC shows a narrow bandgap of 1.59 eV and meanwhile maintains deep-lying HOMO and LUMO energy levels of −5.48 and −3.83 eV, respectively (Table 4.4). In addition, the planar and rigid backbone of ITIC promotes the good intermolecular π–π stackings and thus obtains high electron mobility of up to 3.0×10^{-4} cm^2 V^{-1} s^{-1}. As a result, ITIC blending with PTB7-Th achieved a 6.8% PCE multiplying from a V_{oc} of 0.81 V, a J_{sc} of 14.21 mA cm^{-2}, and a FF of 0.591, which is comparable with that of PTB7-Th:$PC_{71}BM$-based OSCs and creates a new record for fullerene-free OSCs. Later, Hou and coworkers selected a wide bandgap polymer PBDB-T in their laboratory to blend with ITIC, achieving an outstanding PCE of up to 11.2% (a V_{oc} of 0.899 V, a J_{sc} of 16.81 mA cm^{-2} and a FF of 0.742) with excellent thermal stability. The significant PCE enhancement can be attributed to a more matched absorption spectrum and energy levels, and more compatible morphology in PBDB-T:ITIC blend than those in PTB7-Th:ITIC blend. Furthermore, an efficiency of 10.78% was certified based on PBDB-T:ITIC device with an area of 1.00 cm^2 [87]. The excellent photovoltaic performance of ITIC shows the great potential of A–D–A-type SMAs, opening a new door for the development of highly-efficient non-fullerene acceptors (NFAs).

4.3.1 Side Chains Optimization

The role of side chains is multiple for achieving high-performance SMAs. On one hand, they provide SMAs with good solubility to support its synthesis and

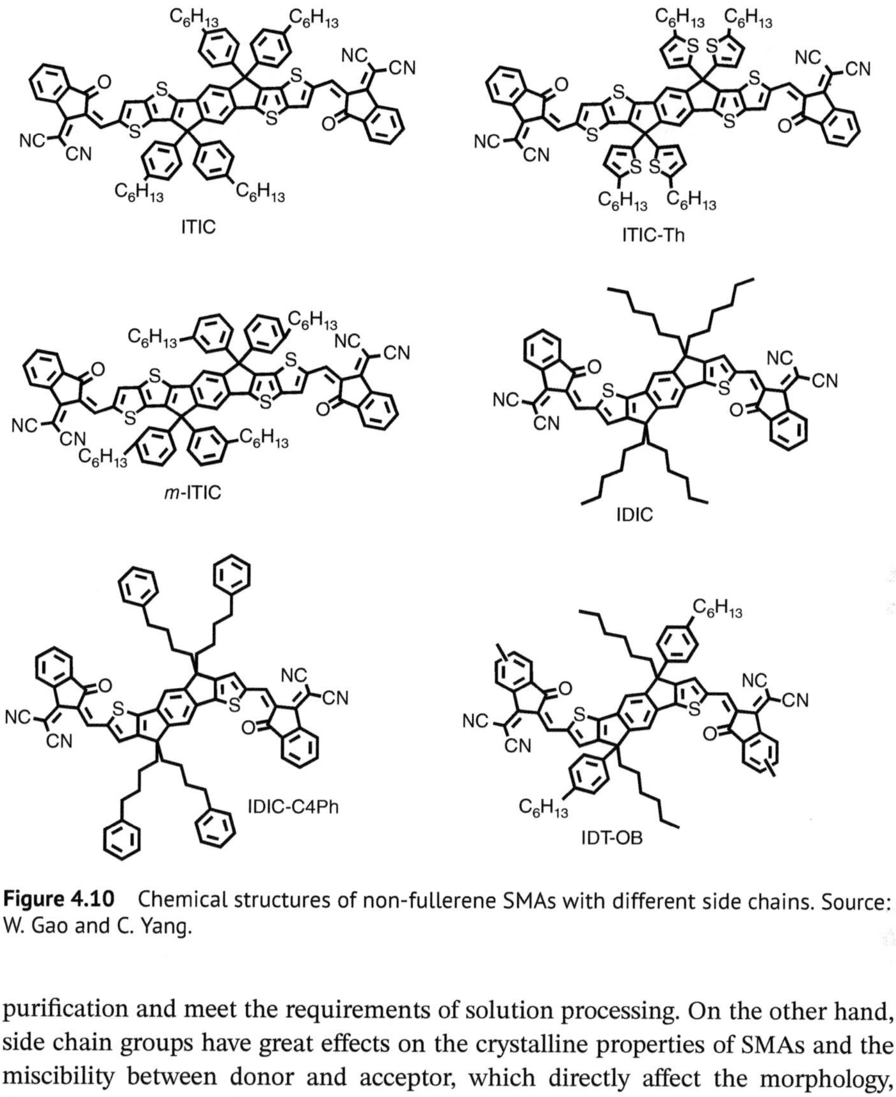

Figure 4.10 Chemical structures of non-fullerene SMAs with different side chains. Source: W. Gao and C. Yang.

purification and meet the requirements of solution processing. On the other hand, side chain groups have great effects on the crystalline properties of SMAs and the miscibility between donor and acceptor, which directly affect the morphology, charge transport, and recombination within active layer. However, the influence of side chains is also very complicated due to the diversity of side chain types and different effects; therefore, side chains optimization is necessary and important in the development of SMAs.

In ITIC, four 4-hexylphenyl are linked to IDTT core via sp^3 hybridized carbon atoms with twisted phenyl groups not conjugated to backbone of SMAs, which leads to little effect on absorption and energy levels of SMAs but can inhibit excessive crystallization of molecules and regulate the morphology of active layer.

ITIC-Th was modified from ITIC by replacing 4-hexylphenyl with 5-hexylthiophen [73]. Compared to ITIC, the LUMO energy level of ITIC-Th has been reduced from −3.83 to −3.93 eV due to the influence of δ-induced effect between ladder-type core (IDTT) and thiophene groups, and the HOMO energy level dropped from

Table 4.4 Optical bandgap, energy levels, and photovoltaic parameters for A–D–A-type SMAs with different side chains and end groups.

SMA	E_g^{opt} a) (eV)	HOMO/ LUMO b) (eV)	Donor	V_{oc} (V)	J_{sc} (mA cm^{-2})	FF	PCE (%)	References
ITIC	1.59	−5.48/−3.83	PTB7-Th	0.81	14.21	0.591	6.80	[72]
ITIC-Th	1.60	−5.66/−3.93	PDBT-T1	0.88	16.24	0.671	9.6	[73]
m-ITIC	1.58	−5.52/−3.82	J61	0.912	18.31	0.706	11.77	[69]
IDIC	1.62	−5.69/−3.91	PDBT-T1	0.89	15.05	0.65	8.71	[74]
IDIC-C4Ph	1.62	−5.70/−3.93	PBDB-TF	0.941	19.06	0.783	14.04	[75]
IDT-OB	1.66	−5.77/−3.87	PBDB-T	0.88	16.18	0.711	10.12	[76]
IT-M	1.60	−5.58/−3.98	PBDB-T	0.94	17.44	0.735	12.05	[77]
IT-OM-1	1.74	−5.50/−3.76	PBDB-T	1.01	12.31	0.51	6.3	[78]
IT-OM-2	1.63	−5.49/−3.86	PBDB-T	0.93	17.53	0.73	11.9	[78]
IT-OM-3	1.72	−5.52/−3.80	PBDB-T	0.97	16.38	0.68	10.8	[78]
IT-OM-4	1.68	−5.49/−3.81	PBDB-T	0.96	14.69	0.56	7.9	[78]
F-ITIC	1.56	−5.65/−4.09	PTPDBDT	0.94	14.1	0.66	8.8	[79]
Cl-ITIC	1.56	−5.70/−4.14	PTPDBDT	0.94	15.6	0.65	9.5	[79]
Br-ITIC	1.53	−5.73/−4.20	PTPDBDT	0.93	15.4	0.66	9.4	[79]
I-ITIC	1.55	−5.68/−4.14	PTPDBDT	0.95	14.5	0.65	8.9	[79]
IT-DM	1.63	−5.56/−3.93	PBDB-T	0.97	16.48	0.706	11.29	[77]
IT-4F	1.55	−5.66/−4.14	PBDB-T-SF	0.88	20.50	0.719	13.00	[80]
IT-4Cl	1.48	−5.75/−4.09	PBDB-T-2F	0.79	22.67	0.752	13.45	[81]
ITIC-6F	1.51	−5.78/−4.16	PBDB-TF	0.81	21.2	0.667	11.41	[82]
ITCPTC	1.58	−5.62/−3.92	PBT1-EH	0.95	16.5	0.751	11.8	[83]
ITCC	1.67	−5.47/−3.76	PBDB-T	1.01	15.9	0.71	11.4	[84]
MeIC	1.58	−5.57/−3.92	J71	0.918	18.41	0.742	12.54	[85]
ITC-2Cl	1.58	−5.58/−4.01	PBDB-TF	0.91	20.1	0.741	13.6	[86]

a) Calculated from $E_g^{opt} = 1240/\lambda_{onset}$.
b) Calculated from CV curves.

−5.48 to −5.66 eV. The deeper HOMO and LUMO energy levels match more suitable with those of polymer donors and enlarge the HOMO–HOMO and LUMO–LUMO energy gaps between donor and acceptor, which provides greater driving force for more efficient exciton dissociation. ITIC-Th shows a high electron mobility of 6.1×10^{-4} cm^2 V^{-1} s^{-1} due to better crystallization performance of ITIC-Th induced by sulfur–sulfur interaction between thiophene groups, which is found to be two times higher than that of ITIC pure film. As a result, OSCs based on PTB7-Th:ITIC-Th can achieve a 8.7% PCE, higher than that of PTB7-Th:ITIC-based OSCs.

Soon later, Li and coworkers changed the 4-hexylphenyl on ITIC into 3-hexylphenyl to synthesize *m*-ITIC [69]. ITIC and *m*-ITIC are isomers of each other with small difference lying in the relative positions of alkyl chains on phenyl. Compared with ITIC, *m*-ITIC exhibits nearly same absorption spectrum and HOMO and LUMO energy levels but significantly enhanced π–π stacking in neat film, and thus a higher electron mobility is achieved. As a result, OSCs based on J61:*m*-ITIC achieved a 11.77% PCE with a V_{oc} of 0.912 V, a J_{sc} of 18.31 mA cm^{-2}, and a FF of 0.706, which are all higher than those of J61:ITIC-based OSCs (a 10.57% PCE with a V_{oc} of 0.898 V, a J_{sc} of 17.97 mA cm^{-2}, and FF of 0.655). More importantly, *m*-ITIC-based OSCs show less thickness-dependent photovoltaic behavior than ITIC-based OSCs within film thickness range of 80–360 nm, which is beneficial to the fabrication of large-area devices.

In 2016, Zhan and coworkers designed and synthesized another famous SMA IDIC based on an indaceno[1,2-*b*:5,6-*b*']dithiophene (IDT) core with *n*-hexyl as side chains [74]. Due to the smaller steric hindrance effect of alkyl groups than aromatic phenyl groups, IDIC trends to become more planar, thus the intermolecular π–π stacking effects of IDIC is remarkably improved with a high electron mobility of up to 1.1×10^{-3} cm^2 V^{-1} s^{-1} in IDIC neat film. IDIC exhibits excellent photovoltaic performance that PDBT-T1:IDIC-based OSCs can achieve 8.71% PCE without any treatment and 10.37% PCE with 0.25% 1,8-diiodooctane (DIO) additive. When selecting fluorine-contained PM6 as polymer donor, OSCs based on PM6:IDIC can achieve 11.9% PCE without any treatment, which is a recorded value for as-cast OSCs at that time. Moreover, PM6:IDIC-based OSCs exhibit excellent thickness insensitivity property with over 11% PCE at film thickness of 250 nm [88].

By introducing a phenyl group into the tail of the side chain of IDIC, Yang and coworkers designed and synthesized an excellent SMA IDIC-C4Ph [75]. Moderately weakened crystallinity is observed in IDIC-C4Ph neat film compared to that of IDIC neat film, which gains a subtle balance between the crystallinity and miscibility. OSCs based on PBDB-T:IDIC-C4Ph obtain a 11.5% PCE, which is significantly higher than *n*-hexyl and 4-hexylphenyl substituted analogs-based OSCs. When changing polymer donor PBDB-T into PM6, as-cast OSCs based on PM6:IDIC-C4Ph achieved a 13.23% PCE with a high FF of 0.760; moreover, a 14% PCE with a FF of 0.783 can be recorded after annealing. This side chain modification effectively regulates the crystallization of SMAs to synergistically improve morphology and molecular orientations of active layer.

In addition to SMAs with symmetrical side chains, Bo group designed and synthesized a IDIC analog, called IDT-OB, with asymmetric side chains on IDT core, that is, one *n*-octyl and one 4-hexylphenyl are fixed on one substituted site [76]. This asymmetric alkyl substitution strategy achieves a fine tuning of crystalline properties of SMAs, leading to a better miscibility and phase separation between SMAs and polymer donors. OSCs based on PBDB-T:IDT-OB achieved a 10.12% PCE with a high and balanced carrier transport in active layer. In addition, IDT-OB also exhibits good film thickness insensitivity property and achieve a 9.17% PCE when the film thickness of OSCs based on PBDB-T:IDT-OB reaches 210 nm.

4.3.2 End Groups Engineering

In A–D–A-type SMAs, A unit is a kind of active methylene compounds with very strong electron-withdrawing ability, we usually call them end group. Such types of end groups are relatively limited and mainly can be divided into two categories: 2-(3-oxo-2,3-dihydro-1*H*-inden-1-ylidene)malononitrile (INCN) and its derivatives, and 2-(6-oxo-5,6-dihydro-4*H*-cyclopenta[*c*]thiophen-4-ylidene)malononitrile (CPTCN) and its derivatives. The absorption, energy levels, and crystallization properties of SMAs can be effectively regulated by adjusting the type, number, and relative position of substituents on end group.

Hou and coworkers modified INCN end group by adding one or two methyl groups onto 5 or/and 6 position of INCN and synthesized two SMAs IT-M and IT-DM, respectively [77]. Methyl as a weak electron-donating unit slightly reduce the electron-withdrawing ability of INCN. As the number of methyl group increases, the LUMO energy levels of SMAs are gradually elevated with absorption spectra gradually blue-shifted from ITIC to IT-M then to IT-DM. As a result, PBDB-T:IT-M- and PBDB-T:IT-DM-based OSCs achieved 12.05% and 11.29% PCEs with high V_{oc}s of 0.94 and 0.97 V, respectively. Moreover, methyl-substituted end group improves the crystallization performance of SAMs, and monomethyl-substituted INCN enables IT-M with a more suitable crystallization property than ITIC and IT-DM. Therefore, the morphology of PBDB-T:IT-M blend film was optimized and obtained a 74% high FF in devices (Figure 4.11).

After the success of methyl-substituted INCN, Hou and coworkers further modified the INCN with methoxy group due to the stronger electron-donating ability of oxygen atom than carbon atom and expected to obtain higher V_{oc} in OSCs. Interestingly, four isomers of INCN derivatives substituted by single methoxy at different position of INCN can be completely separated, and corresponding isomers of four SMAs (IT-OM-1, IT-OM-2, IT-OM-3, and IT-OM-4) were obtained [78]. IT-OM-1, IT-OM-2, IT-OM-3, and IT-OM-4 show high but different LUMO energy levels, which are −3.76, −3.86, −3.80, and −3.81 eV, and achieved 6.3%, 11.9%, 10.8%, and 7.9% PCEs with V_{oc}s of 1.01, 0.93, 0.96, and 0.97 V in devices, respectively, when blending with polymer donor PBDB-T. These results indicated that minor changes in the substitution position on INCN are demonstrated to not only influence the optoelectronic properties but also molecular packing of the isomers, leading to great difference in photovoltaic performance of four isomers-based OSCs.

Compared to adding electron-donating groups onto INCN, modifying INCN with electron-withdrawing units can enhance the electron-withdrawing ability of INCN. Li and coworkers substituted INCN with one fluorine (F), chlorine (Cl), bromine (Br), and iodine (I) atom onto 5 or 6 position of INCN and synthesized four SAMs F-ITIC, Cl-ITIC, Br-ITIC, and I-ITIC, respectively [79]. Compared to ITIC, halogen-substituted SMAs show red-shifted absorption spectra and dropped LUMO energy levels. As a result, OSCs based on wide bandgap polymer PTPDBDT and four SMAs achieved 8.8–9.5% PCE with significantly increased J_{sc}s but decreased V_{oc}s compared to PTPDBDT:ITIC-based OSCs, indicating single halogen substitution can effectively improve the photovoltaic performance of SMAs.

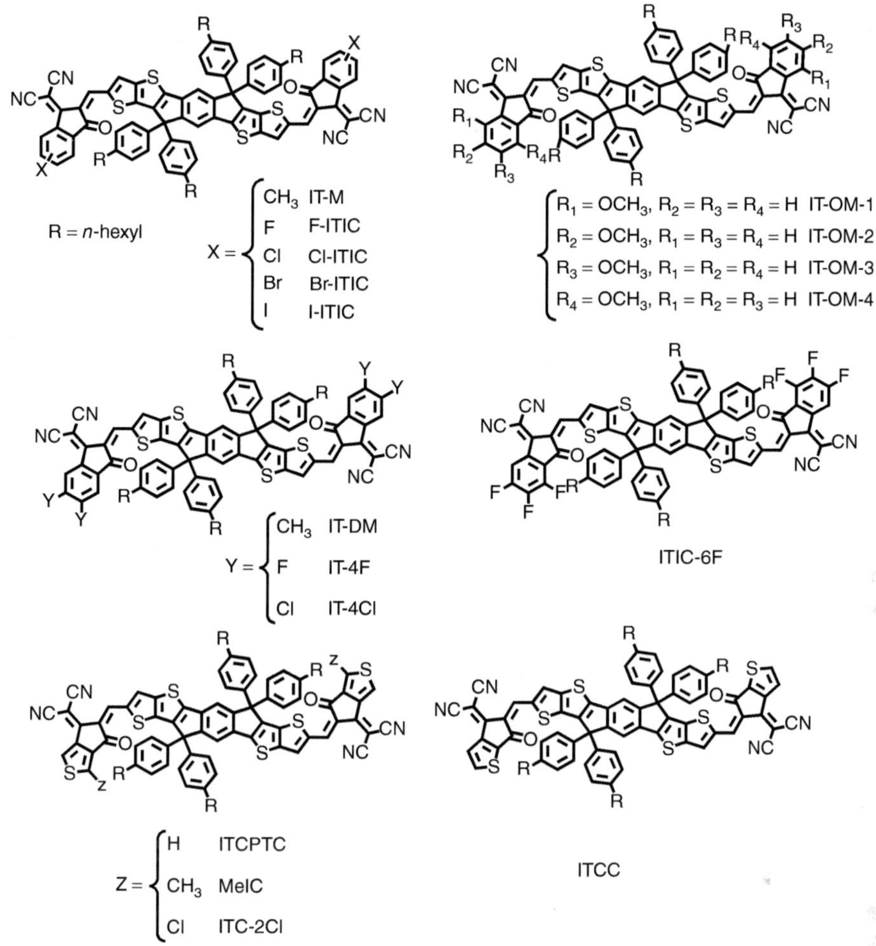

Figure 4.11 Non-fullerene small-molecule acceptors with different end groups. Source: W. Gao and C. Yang.

The single halogen-substituted INCN has two isomers that are difficult to separate, which will result in three isomers of SAMs in the final product. Hou and coworkers developed F and Cl double-substituted end group on 5 and 6 position of INCN and synthesized IT-4F [80] and IT-4Cl [81]. F atom with strong electronegativity and small radius can form weak interactions with electron-rich sulfur atom via non-covalent bonds, which enhances the intermolecular π–π stacking, thus promoting the carrier mobility in both neat and blend films. Though double fluorine-substituted INCN acting as end group significantly increase the J_{sc} of resultant SMAs-based OSCs, the V_{oc} of OSCs also drops sharply. To compensate the V_{oc} loss, PBDB-T was optimized by adding sulfanyl chain and fluorine atom onto the thiophene side chains to obtain a polymer donor PBDB-T-SF with lower HOMO energy level. As a result, PBDB-T-SF:IT-4F-based OSCs achieved a recorded PCE of over 13% for the first time. Theoretical calculations have revealed that the carbon-chlorine (C—Cl) bond has a larger dipole moment than the carbon-fluorine (C—F) bond, indicating that Cl-substituted INCN shows a stronger electron-withdrawing ability than F-substituted one and leading to IT-4Cl, a reduced optical bandgap compared to IT-4F. As a result, OSCs based on PBDB-TF:IT-4Cl also achieved an over 13% PCE with a further increased J_{sc} than that of PBDB-TF:IT-4F-based OSCs.

Three F atoms substituted INCN was first reported by Marks group, and corresponding six F atoms substituted SMA ITIC-6F was synthesized [82]. The bandgap of ITIC-6F can be further lowered to 1.51 eV compared to IT-4F with a bandgap of 1.55 eV. More importantly, the π–π stacking effect between SMAs increases as the number of F atoms on SMA increases. ITIC-6F obtains a more intense π–π stacking with a very short *d*-spacing of 3.28 Å than that of IT-4F, which promotes the electron mobility of ITIC-6F film and enabled OSCs based on PBDB-TF:ITIC-6F to achieve an over 12% PCE.

Thiophene as an important building block in the construction of photoactive materials show many unique advantages than benzene, such as stronger electron-donating ability, more intense intermolecular interaction, and smaller steric hindrance effect. End groups contained thiophene unit will exhibit different property. Yang group developed an efficient end group called CPTCN by fusing cyclopentane-1,3-dione onto 3,4-position of thiophene for the first time and synthesized a novel SMA ITCPTC [83]. It was found that the absorption spectrum of ITCPTC is slightly red-shifted compared to ITIC but with a slight lower LUMO energy level. In addition, introducing thiophene into the end group of SMA is able to form S–S non-covalent bond interaction, leading to a better π–π stacking and higher electron mobility of ITCPTC than those of ITIC in their neat films. As a result, PBT1-EH:ITCPTC-based OSCs achieved a PCE of 11.8% with a high FF of over 75%, which are significantly higher than those of PBT1-EH:ITIC-based OSCs under same fabrication conditions, indicating the great potential of CPTCN as end group.

At the same time, Hou and coworkers developed an isomer of CPTCN by changing the fusion site between cyclopentane-1,3-dione and thiophene from 3,4-position to 2,3-position [84]. Such structure change significantly weakens the electron-withdrawing ability of end group, which leads to bandgap of ITCC significantly increasing from 1.58 to 1.67 eV accompanied with LUMO energy level elevated from −3.96 to −3.76 eV in comparison with ITCPTC. As a result, OSCs based on PBDB-T:ITCC yielded an over 1 V V_{oc} and a 11.4% PCE.

Subsequently, Yang group functionalized the CPTCN by adding one methyl group or one Cl atom onto the thiophene unit of CPTCN and synthesized two SMAs MeIC [85] and ITC-2Cl [86]. These two CPTCN derivatives show that no isomer with the dicyanomethyl group orientates inversely with methyl group or chlorine atom, which is different from single methyl or Cl-substituted INCN. Compared to ITCPTC, methylated MeIC show almost no blue shift in absorption spectrum but greatly elevated LUMO energy level, while the absorption spectrum of ITC-2Cl exhibits a remarkably red shift of over 50 nm with LUMO energy level dropped. Methylated or chlorinated CPTCN both enhance the intermolecular π–π stacking between SMAs and promotes the carrier mobility of neat and blend films. As a result, OSCs based on J71:MeIC achieved higher PCE and V_{oc} than J71:ITCPTC-based OSCs, and PM6:ITC-2Cl-based OSCs achieved higher PCE and larger J_{sc} than PM6:ITCPTC-based OSCs.

4.3.3 Core Units Engineering

4.3.3.1 IDTT and Its Derivations

Compared with side chains and end groups, the electron-donating D cores are rich in type and variable in structure and play important roles in energy levels fine-tuning and backbone planarity manipulation. A library of novel D cores were thus developed via core structure innovations such as fused/non-fused cores, π-bridge, and asymmetric strategies.

In view of the successfully explored ITIC-based system above, the IDTT core was considered a potential building block to design efficient fused ring electron acceptors (FREAs). IDTT is composed of middle benzene and thieno[3,2-*b*]thiophene (TT) at both ends fused by cyclopentadiene. As such, vast efforts were devoted to modulating modules in IDTT, and a library of novel derived FREAs has been developed.

It is widely accepted that the match between donor and acceptor is critical to device performance. In order to escape the limitation of matching wide optical bandgap (E_g^{opt}) donor with narrow E_g^{opt} acceptor, Zhu and coworkers designed a FREA called NIDBT [89] by replacing TT with benzothiophene. The introduced benzene weakens the electron-donating ability for the expected E_g^{opt} of 1.84 eV. As planned, narrow E_g^{opt} donor PTB7-Th was blended with NIDBT, and the corresponding OSCs only exhibited a moderate PCE of 4.45% due to their almost overlapping absorption. On the contrary, flanking IDTT with another thiophene, the stronger electron-rich core 6,6,12,12-tetrakis(4-hexylphenyl)-indacenobis(dithieno[3,2-*b*;2′,3′-*d*]thiophene) (IBDT) was obtained. To investigate the number and position of substituted fluorine atoms on the end groups on device properties, four novel FREAs (INIC, INIC1, INIC2, and INIC3, Figure 4.12) were synthesized [70]. All these four FREAs have strong absorption at 600–750 nm with high extinction coefficients of $2.10–2.50 \times 10^5\ M^{-1}\ cm^{-1}$. Obviously, the fluorinated IC-based INIC1, INIC2, and INIC3 exhibited red-shifted absorption and deeper energy levels than these of INIC (without fluorine substitution). The resultant OSCs blending with FTAZ displayed the PCE of 7.70–11.50% with small E_{loss} (0.61–0.63 eV). In detailed, the device efficiency is gradually improved accompanied with the increased fluorine atoms in INIC series, and the highest PCE of 11.50% can be achieved in FTAZ:INIC3-based device due to fluorine-induced intermolecular interactions. Also, the FTAZ:INIC2-based OSCs presented a better property than FTAZ:INIC1, indicating the position of substituted fluorine significantly influences the device performance.

Coincidentally, through inserting pyrrole rings into IDTT, another electron-rich core named 5,5,12,12-tetrakis(4-hexylphenyl)-indacenobis(dithieno[3,2-*b*:2′,3′-*d*]pyrrol) (INP) [90] was developed by Tang and coworkers. The IC/IC-2F was end-capped for the synthesis of INPIC and INPIC-4F (structure shown in

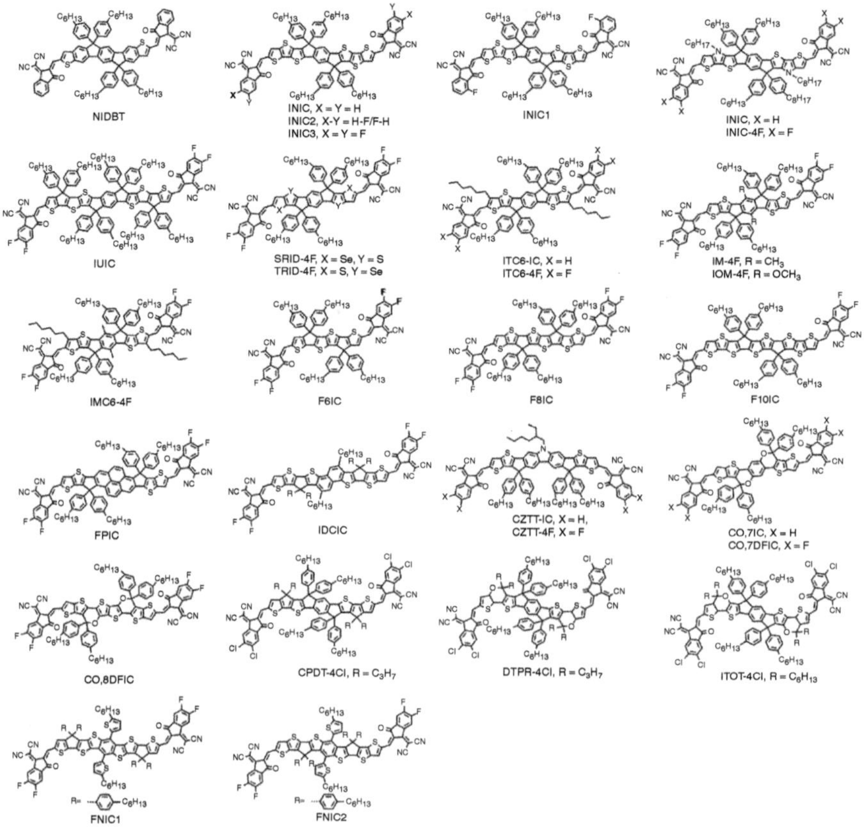

Figure 4.12 Chemical structures of IDTT-based FREAs and its derivatives. Source: J. Cao and W. Tang.

Figure 4.12), respectively. Compared with INIC3 mentioned above, the composite INPIC-4F behaves a red-shifted absorption onset (λ_{onset}) of 800 nm (c. 24 nm) with a narrower E_g^{opt} of 1.39 eV, powerfully suggesting stronger electron-donating INP block than IDTT and IBDT. Molecular simulation results show that both INPIC and INPIC-4F have coplanar π-conjugated frameworks. And the introduced fluorine endows INPIC-4F enhanced ICT effect, thereby facilitating charge transport for a high electron mobilites (μ_e) of 5.00×10^{-4} cm^2 V^{-1} s^{-1}. Moreover, the PBDB-T:INPIC-4F blend dominated a stronger face-on π–π stacking in out-of-plane (OOP) direction in comparison to PBDB-T:INPIC-based ones. The optimized PBDB-T:INPIC-4F devices afforded an outstanding PCE of 13.13% with a V_{oc} of 0.85 V, a J_{sc} of 21.61 mA cm^{-2}, and a FF of 0.715, which is the highest value reported in literatures at that time (Table 4.5). The E_{loss} was found to be as low as 0.56 eV. Further crystallization dynamics study reveals that INPIC-4F tends to form polycrystalline due to its fast crystallization rate, causing large phase separation in corresponding blend films as well as affected device efficiency. Molecular engineering and device process were then implemented to manipulate the crystallinity of

INPIC or prolong the crystallization time during the spin coating for efficient/stable OSCs [106–108].

Unlike the above two examples modifying IDTT core to enhance electron-donating ability, Zhan's group directly fused thiophene on both sides of IDTT via cyclopentadiene (namely IU). The coupling of the as-developed IU with electron-withdrawing 2F-IC generated IUIC [91]. Due to the large π-conjugation, IUIC exhibits strong absorption in NIR region (600–900 nm) with an E_g^{opt} of 1.41 eV. In comparison with the IDTT-cored ITIC4 (IT-4F, LUMO = −3.97 eV), IUIC shows the upshifted LUMO energy level of −3.87 eV, which is beneficial to the V_{oc} of the device. As a result, the as-cast OSCs blended with PTB7-Th achieved a PCE of 11.20%, much higher than that of ITIC4-based counterparts (8.18%). More interestingly, its NIR absorption can retain most of the visible light, providing IUIC with the potential for semitransparent organic solar cells (ST-OSCs). The resulting ST-OSC based on PTB7-Th:IUIC afforded a PCE of over 10.20% with an average visible transmittance (AVT) of 31% without any extra treatment.

Apart from extending core size, heteroatom substitution is another strategy to enhance electron-donating ability. By replacing sulfur with selenium, Lin et al. [92] reported two FREAs, namely, SRID-4F and TRID-4F. Besides, the structure–property relationship induced by region-specific selenium substitution was systematically investigated. Compared with sulfur atom, the more polarized selenium atom is surrounded by looser electron cloud delocalization, which makes the stronger orbital overlap between selenium and the π-conjugated system for narrower bandgap of selenium. Relative to IT-4F (1.51 eV), narrower E_g^{opt}s were thus observed in SRID-4F and TRID-4F (1.44 and 1.48 eV, respectively), implying more efficient ICT effect occurred in these two selenium substituted FREAs. Morphology studies showed that SRID-4F exhibits smaller (010) π–π stacking distance ($d_{\pi-\pi}$) of 3.4 Å than TRID-4F ($d_{\pi-\pi}$ = 3.9 Å) with preferential face-on orientation. And more ordered nanostructure induced by large dipole moment and O–Se non-covalent interaction between selenium atom and the carbonyl group of indanone are beneficial to charge transport. The corresponding PM6:SRID-4F-based devices achieved a higher PCE of 13.05% with a V_{oc} of 0.85 V, a J_{sc} of 20.21 mA cm^{-2} than its isomer TRID-4F-based ones (12.33%). The FFs of the devices are all over 0.750 (Table 4.5).

In A–D–A-type FREAs, the linkage between D cores and A end groups inevitably results in different steric isomers due to the rotatable single bond, whose uncertain impacts on device properties seriously hinder the development of novel FREAs. As depicted in Figure 4.13a, Tang's group [93] masterly attached hexyl chains into the both ends of D core, and a novel acceptor named ITC6-IC was explored (Figure 4.12). Molecular simulation calculation results (two rotamers TT-IC and TTC6-IC were selected as representatives of ITIC and ITC6-IC, respectively, Figure 4.13b) showed that the TT-IC has two dominant conformations at 0° and 180°, and their energy difference is only 4.35 kJ mol^{-1}, illustrating both conformations will possibly coexist. When the hexyl side chains were attached, the energy of TTC6-IC at 180° is increased to over 32 kJ mol^{-1}. The large energy effectively limits the appearance of other dominant conformations except for that at 0°, thus locking the conformation well. The

Table 4.5 Optical bandgap, energy levels, and photovoltaic parameters for A–D–A-type non-fullerene SMAs with different cores.

SMA	E_g^{opt} a) (eV)	HOMO/ LUMO b) (eV)	Donor	V_{oc} (V)	J_{sc} (mA cm^{-2})	FF	PCE (%)	References
ITIC4	1.52	−5.77/−3.97	PTB7-Th	0.69	16.66	0.715	8.18	[91]
NIDBT	1.84	−5.87/−3.91	PTB7-Th	0.83	11.45	0.467	4.45	[89]
INIC	1.57	−5.45/−3.88	FTAZ	0.96	13.51	0.579	7.70	[70]
INIC1	1.56	−5.54/−3.97	FTAZ	0.93	16.63	0.643	10.10	[70]
INIC2	1.52	−5.52/−3.98	FTAZ	0.90	17.56	0.668	10.80	[70]
INIC3	1.48	−5.52/−4.02	FTAZ	0.86	19.44	0.674	11.50	[70]
INPIC	1.46	−5.36/−3.82	PBDB-T	0.96	8.55	0.525	4.31	[90]
INPIC-4F	1.39	−5.42/−3.94	PBDB-T	0.85	21.61	0.715	13.13	[90]
IUIC	1.41	−5.45/−3.87	PTB7-Th	0.80	21.74	0.649	11.20	[91]
SRID-4F	1.44	−5.52/−3.90	PM6	0.85	20.21	0.752	13.05	[92]
TRID-4F	1.48	−5.52/−3.90	PM6	0.89	18.45	0.750	12.33	[92]
ITC6-IC	1.60	−5.73/−3.92	PBDB-T	0.97	16.41	0.730	11.61	[93]
ITC6-IC	1.60	−5.73/−3.92	PM7	1.01	13.60	0.598	8.21	[94]
ITC6-4F	1.52	−5.77/−4.03	PM7	0.90	21.64	0.743	14.47	[94]
IM-4F	1.46	−5.69/−4.19	PM6	0.88	22.12	0.728	14.17	[95]
IOM-4F	1.48	−5.72/−4.27	PM6	0.86	21.66	0.720	13.41	[95]
IMC6-4F	1.49	−5.71/−3.94	PM6	0.90	22.29	0.776	15.57	[96]
F6IC	1.36	−5.66/−4.02	PTB7-Th	0.61	18.07	0.640	7.10	[97]
F8IC	1.27	−5.43/−4.00	PTB7-Th	0.64	25.12	0.676	10.90	[97]
F8IC	1.27	−5.47/−4.06	PTB7-Th	0.65	23.30	0.634	9.55	[98]
F10IC	1.25	−5.26/−3.96	PTB7-Th	0.73	20.83	0.668	10.20	[97]
FPIC	1.63	−5.75/−3.97	PTB7-Th	0.76	15.30	0.731	8.45	[98]
IDCIC	1.45	−5.19/−3.72	FTAZ	0.87	21.98	0.710	13.58	[99]
CZTT-IC	1.56	−5.54/−3.97	PBDB-T	0.97	17.26	0.590	9.87	[100]
CZTT-IC	1.56	−5.54/−3.97	PM6	1.07	12.83	0.512	7.02	[100]
CZTT-4F	1.48	−5.61/−4.11	PBDB-T	0.88	19.06	0.585	9.81	[100]
CZTT-4F	1.48	−5.61/−4.11	PM6	0.94	19.73	0.651	12.07	[100]
CO_i7IC	1.60	−5.66/−3.92	PTB7-Th	0.82	13.12	0.589	6.36	[101]
CO_i7DFIC	1.55	−5.78/−4.06	PTB7-Th	0.67	17.36	0.719	8.32	[101]
CO_i8DFIC	1.26	−5.50/−3.88	PTB7-Th	0.68	26.12	0.682	12.16	[102]
IDTT-4Cl	1.49	−5.49/−3.91	PTB7-Th	0.61	18.34	0.643	7.70	[103]
CPDT-4Cl	1.35	−5.32/−3.78	PTB7-Th	0.74	23.24	0.691	12.15	[103]
DTPR-4Cl	1.30	−5.35/−3.85	PTB7-Th	0.68	24.97	0.618	10.75	[103]
ITOT-4Cl	1.28	−5.49/−3.90	PBDB-T	0.78	22.89	0.705	12.50	[104]
FNIC1	1.48	−5.61/−3.92	PTB7-Th	0.77	19.97	0.664	10.30	[105]
FNIC2	1.38	−5.56/−4.00	PTB7-Th	0.74	23.93	0.734	13.00	[105]

a) Calculated from $E_g^{opt} = 1240/\lambda_{onset}$.
b) Calculated from CV curves.

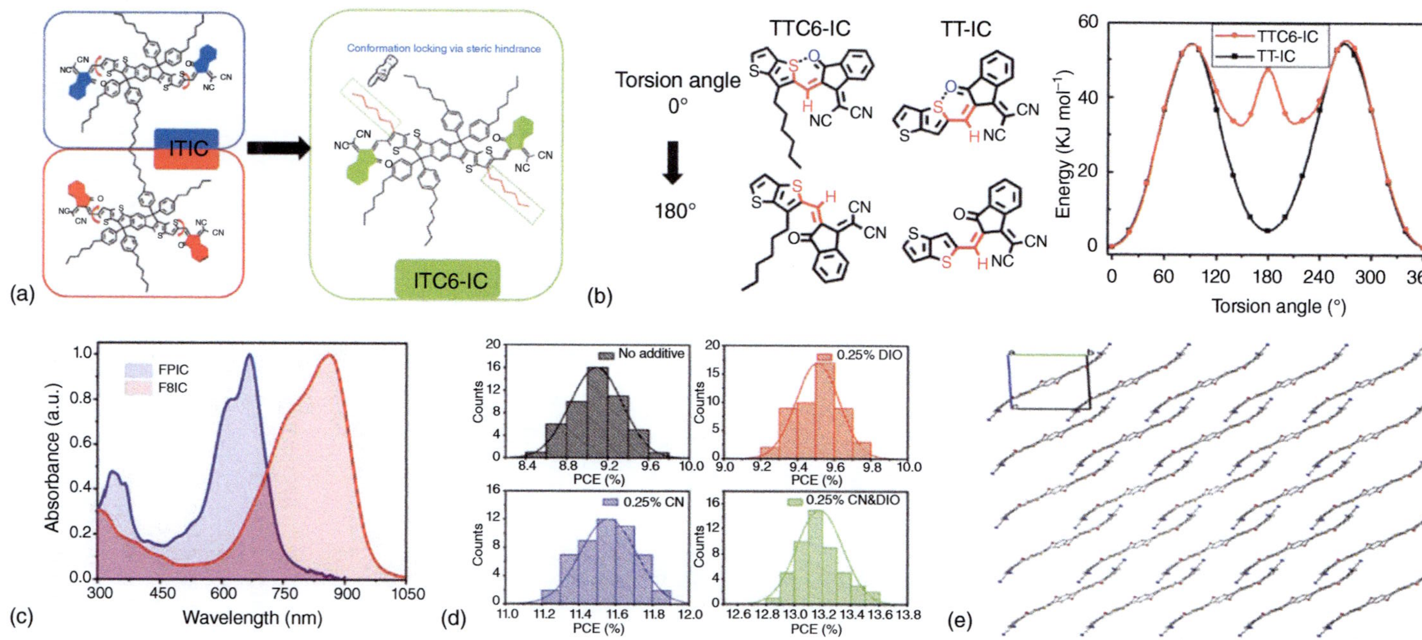

Figure 4.13 (a) Conformational locking of FREAs as demonstrated by ITC6-IC. (b) Possible rotamers and potential energy surface scan of TTC6-IC and TT-IC. Source: Zhang et al. [93]/Reproduced with permission of John Wiley & Sons. (c) Complementary absorption of FPIC and F81C. Source: Cai et al. [98]/Reproduced with permission of American Chemical Society. (d) Histograms of PCEs of FTAZ:IDCIC-based OSCs without or with additives. Source: He et al. [99]/Reproduced with permission of John Wiley & Sons. (e) The single-crystal structure of CO_i7IC: packing structure of CO_i7IC viewed along the *a*-axis (side chains are omitted for clarity). Source: Jin et al. [101]/Reproduced with permission of Royal Society of Chemistry.

as-developed ITC6-IC exhibited a slightly blue-shifted absorption but a higher maximum molar absorption coefficient of $2.64 \times 10^5\ M^{-1}\ cm^{-1}$ compared with that of ITIC ($1.72 \times 10^5\ M^{-1}\ cm^{-1}$). The steric hindrance of hexyl chains favors the formation of planar structure with uniform conformation and weakens the π–π stacking in ITC6-IC. The resulting blend films using PBDB-T and ITC6-IC as active layer feature fibrous nano-interpenetrating network for efficient charge transport. The derived PBDB-T:ITC6-IC-based device displayed a PCE of 11.61% with a V_{oc} of 0.97 V, a J_{sc} of 16.41 mA cm^{-2}, and a FF of 0.730, higher than ITIC-based counterparts (11.61% PCE with a V_{oc} of 0.91 V). The higher V_{oc} of 0.97 V can be attributed to upshifted LUMO energy level of ITC6-IC. Further fluorinated end group strategy leads to lower LUMO energy level and narrower E_g^{opt} for ITC6-4F [94]. Selecting PM7 with complementary absorption and deeper HOMO, the PCE of PM7:ITC6-4F control OSC was promoted to 14.47% (Table 4.5).

In spite of the modification on both ends of IDTT, the molecular engineering on the intermediate core is equally effective. Side chains are typically used to modify solubility, crystallinity as well as molecular aggregation behavior, occasionally accompanied with changes in absorption and energy level. Attaching substituted side chains on intermediate benzene, Tang and coworkers [95] reported two new FREAs, namely IM-4F (methyl-substituted) and IOM-4F (methoxy-substituted)(Figure 4.12). IM-4F exhibited upshifted HOMO/LUMO (−5.69 eV/−4.19 eV) than IDTT-cored IT-4F, and the lowest E_g^{opt} of 1.46 eV can be also found. Induced by p-π conjugative effect and large electronegativity of oxygen atom, IOM-4F only showed an E_g^{opt} of 1.48 eV. Featuring tempered nanoflake morphology with clear phase boundaries, the PM6:IM-4F-based device yielded a champion PCE of 14.17% with a V_{oc} of 0.88 V, a J_{sc} of 22.12 mA cm^{-2}, and a FF of 0.728. The following energy loss investigation suggested PM6:IM-4F cells behaved simultaneously reduced radiative/non-radiative recombination loss, thus a smallest E_{loss} of 0.67 eV, which provide a broader outlook with reduced E_{loss} via rational core modulation. Combining all three strategies of successfully developed acceptors ITC6-IC, ITC6-4F, and IM-4F above mentioned, a highly-efficient FREA called IMC6-4F [96] was designed and synthesized. As expected, the optimal PM6:IMC6-4F-based device contributed one of the record PCEs for binary OSCs of 15.57% (Table 4.5) with a V_{oc} of 0.90 V, a J_{sc} of 22.29 mA cm^{-2}, and a FF of 0.776, which is on a par with that of state-of-the-art acceptor Y6 [109]-based cells (15.70% PCE).

By replacing intermediate benzene with TT, Zhan and coworkers prepared a stronger electron-donating core, followed by the design of four FREAs (F6IC, F8IC, and F10IC, Figure 4.12) to systematically investigate structure–property relationship [97]. Increasing the core size from 6 to 10 fused rings upshifted the HOMO level from −5.66 to −5.26 eV and the LUMO level from −4.02 to −3.96 eV, leading to a decrease of E_g^{opt} from 1.36 to 1.25 eV. Three PTB7-Th blended films showed small root-mean-square (RMS) roughness (0.94–2.62 nm), and the smallest $d_{\pi-\pi}$ of 3.41 Å can be found in PTB7-Th:F8IC-based blend, contributing to its high fullerene-like mobilities ($\mu_e = 1.6 \times 10^{-4}\ cm^2\ V^{-1}\ s^{-1}$). The resulting devices based on PTB7-Th:F8IC displayed a PCE of 10.9% with a V_{oc} of 0.64 V, a J_{sc} of

25.12 mA cm^{-2}, and a FF of 0.676, higher than that of F6IC- and F10IC-based counterparts (7.1% and 10.2%, respectively). After adding 30% of the third component $PC_{71}BM$, double or triple peaks appeared in original position of lamellar and π–π peaks, indicating preferentially face-on oriented PTB7-Th and F8IC domains co-exist in PTB7-Th:F8IC:$PC_{71}BM$-based ternary OSCs, which improved the PCE to 12.30% with simultaneously improved V_{oc} of 0.67 V and J_{sc} of 25.61 mA cm^{-2}. However, similar absorption of F8IC and F10IC noted us extending core size has limited effect on performance promotion, and finding new alternative intermediate core is also an effective method. Then the acceptor named FPIC [98] was synthesized by the same group through replacing TT central core with pyrene, which features more planar structure and larger conjugation. Compared with the initial F8IC, weakened electron-donating core endows FPIC deeper HOMO of −5.75 eV and upshifted LUMO of −3.97 eV, corresponding to a wider E_g^{opt} of 1.63 eV. The blue-shifted absorption resulted in the poor J_{sc} of 15.3 mA cm^{-2}, thereby a moderate PCE of 8.45% when blended with common donor PTB7-Th. Considering complementary absorption of F8IC and FPIC (Figure 4.13c), FPIC was added into PTB7-Th:F8IC based binary OSCs as a third component. As expected, the J_{sc} thus got a substantial increase to 26.30 mA cm^{-2} with a champion PCE of 13.00%, 36% improvement relative to that of F8IC-based devices (9.55%) (Table 4.5). The above design strategy does not enhance electron-donating ability of central core but just enlarges conjugation length, which limits the extraction of charge for poor device parameters. In this case, Wang and coworkers [99] took the place of two benzene in pyrene with thiophene units, and more electron-donating naphtho[1,2-*b*:5,6-*b*′]dithiophene (NDT) core was applied for FREAs design. Due to the reinforced ICT effect, the as-developed IDCIC (structure shown in Figure 4.12) presents broad absorption at 600–830 nm with a narrow E_g^{opt} of 1.45 eV. Morphological study revealed that IDCIC tends to aggregate in DIO for poor solubility, while inhibited aggregation and excellent solubility with reduced domain size/purity can be found in chloronaphthalene (CN). Compromising the effect of the above two, binary additive of 0.25% CN&DIO was added and nanofiber structure with clear phase separation was observed in FTAZ:IDCIC-based blends for high J_{sc} and FF. As a result, the OSCs achieved a maximum PCE of 13.58% with a high J_{sc} of 21.98 mA cm^{-2} and a FF of 0.710 (Figure 4.13d). After that, another carbazole central core was employed by Tang's group, and two nonacyclic FREAs named CZTT-IC and CZTT-4F were synthesized [100]. CZTT-4F film exhibited broad absorption at 550–800 nm, c. 43 nm red-shifted than CZTT-IC. Owning to fluorinated end groups, CZTT-4F has deeper HOMO/LUMO energy levels (−5.61 eV/−4.11 eV) than CZTT-IC (−5.54 eV/−3.97 eV). In order to find a suitable donor for better device performance, two polymers (PBDB-T and PM6) were selected to match the acceptors. Contact angle result shows that CZTT-4F had a better miscibility with PM6 than that with PBDB-T, indicating the formation of excellent vertical phase separation for efficient charge transport. The derived OSCs delivered a PCE of 12.07% (Table 4.5), which is among the highest values for carbazole-based OSCs in the literature. On the contrary, the CZTT-IC blended well with PBDB-T and the corresponding devices obtained a 9.87% PCE with a high V_{oc} of 0.97 V. Surprisingly, PM6:CZTT-4F-based device also showed enhanced thermal and light-soaking stability with 90.1%/78.7% PCE retention after thermal annealing/light soaking for 100 hours.

In FREAs, cyclopentadiene is a very common structure used to condense the middle core and flanking units. However, this medium electron-donating C-bridged unit is adverse to design FREAs with stronger light-harvesting capability. As thus, Ding's group [101]attached electron-rich oxygen atoms into cyclopentadiene to construct carbon–oxygen-bridged (CO-bridged) ladder-type core, and two novel FREAs coupling with IC and IC-2F (CO_i7IC and CO_i7DFIC, respectively, Figure 4.12) were reported. Both CO_i7IC and CO_i7DFIC have wider E_g^{opt} (1.60 eV and 1.55 eV, respectively) than their carbon-bridged IDTT-based counterparts. Besides, single-crystal analysis revealed that the O–S non-covalent interaction exists between the carbonyl on the end group and the sulfur atom at both ends of the cores for conformation locking, and the introduced oxygen atoms caused the S-shaped backbone (Figure 4.13e). When blending with PTB7-Th, the CO_i7DFIC-based cells yielded a PCE of 8.32% with a decent FF of 0.719, higher than that of PTB7-Th:CO_i7IC-based ones (6.36% PCE). Then the CO_i8DFIC [102] was further developed via replacing the benzene with TT. The PCE was promoted to 12.16% with an impressive high J_{sc} of 26.12 mA cm^{-2}, which can be attributed to its strong light absorption at 600–1000 nm (E_g^{opt} = 1.26 eV). Different from Ding's strategy, Yan and coworkers [103] introduced oxygen atoms into flanked cyclopenta[2,1-*b*:3,4-*b*′]dithiophene (CPDT) blocks, and the new building block called dithieno[3,2-*b*:2′,3′-*d*]pyran (DTPR) was explored. For comparison, three IDTT-, CPDT-, and DTPR-based FREAs (IDTT-4Cl, CPDT-4Cl, and DTPR-4Cl, respectively) were synthesized by coupling with dichlorinated IC (IC-2Cl). The composite DTPR-4Cl behaved a λ_{onset} of 950 nm with a narrowest E_g^{opt} of 1.30 eV, powerfully indicating stronger electron-donating ability of DTPR block than IDTT and CPDT. Morphology study suggested that the PTB7-Th:CPDT-4Cl showed the compromised $d_{\pi-\pi}$ of 3.69 Å, crystal coherence length (CCL) of 13.4 Å and Flory–Huggins parameter (χ) of 0.95, indicating suitable phase separation and domain sizes to suppress charge recombination for high FF (Figure 4.14). Consequently, the PTB7-Th:CPDT-4Cl-based devices achieved a champion PCE of 12.15% with a FF of ~0.700, outperforming that of IDTT-4Cl- and DTPR-4Cl-based cells (7.70% and 10.75%, respectively). Furthermore, by swapping the position of oxygen atom and carbon atom (alkyl connected) in DTPR-4Cl as well as extended alkyl chains, the NIR FREA called ITOT-4Cl was designed [104]. Compared with DTPR-4Cl, ITOT-4Cl exhibited a narrower E_g^{opt} of 1.28 eV and deeper HOMO/LUMO energy levels (−5.49 eV/−3.90 eV). The OSCs choosing PBDB-T as donor displayed a PCE of 12.50% (Table 4.5). In order to further improve the performance of ITOT-4Cl-based device, the F–Br with strong crystalline was added into the binary active layer. Surprisingly, the additive of F–Br not significantly affected the CCL of PBDB-T:ITOT-4Cl-based blend (22.1 Å before and 22.4 Å after), indicating the preservation of order molecular packing. Consequently, the derived ternary OSCs with 40 wt% F–Br achieved enhanced V_{oc} of 0.80 V, J_{sc} of 24.40 mA cm^{-2}, and FF of 0.733, contributing to a record PCE of 14.29% among oxygen heterocycle containing NFAs-based devices.

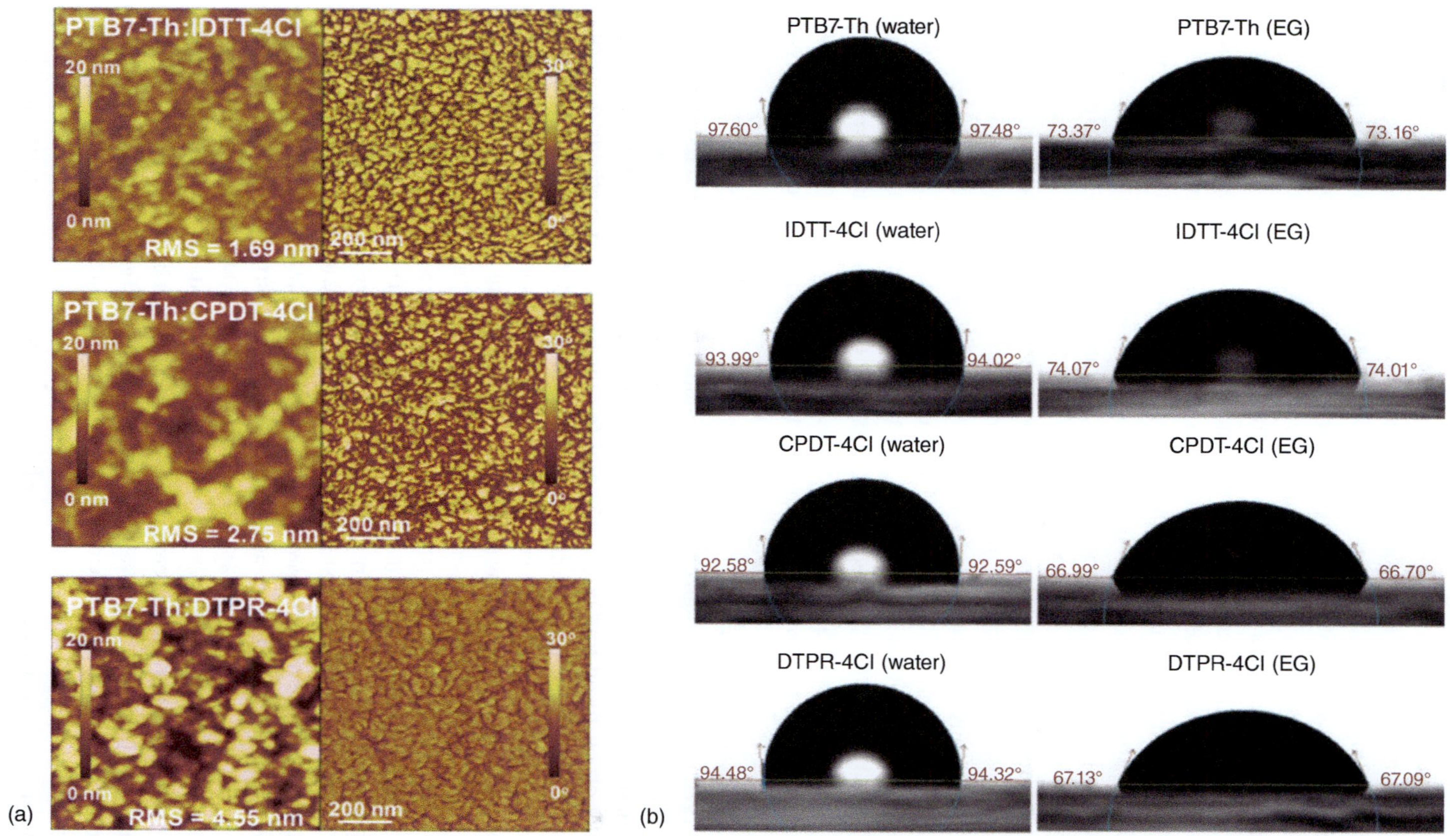

Figure 4.14 (a) AFM images (1 μm × 1 μm, left: height images and right: phase images) of PTB7-Th:IDTT-4Cl, PTB7-Th:CPDT-4Cl, and PTB7-Th:DTPR-4Cl blend films. (b) Contact angle measurements of PTB7-Th, IDTT-4Cl, CPDT-4Cl, and DTPR-4Cl neat films. Source: Zhang et al. [103]/Reproduced with permission of Royal Chemical Society.

Obviously, several strategies above such as extending the conjugation length of the D core and heteroatom substitution can effectively narrow E_g^{opt} for more photons harvest, thus a satisfactory device performance. Nevertheless, unlimited horizontal conjugation extension undoubtedly poses a huge challenge to molecule synthesis and its solubility. Similarly, obtaining a heteroatom-substituted monomer often requires complicated steps. In this case, vertical expansion of the conjugate structure has become an effective alternative, and two-dimensional (2D)-conjugated side chain strategy was applied into FREA design. For example, Zhan and coworkers [105] introduced benzo[1,2-*b*:4,5-*b*′]dithieno[3,2-*b*]thiophene (BDTT) building block as the central core, and two isomeric FREAs (FNIC1 and FNIC2, the structures shown in Figure 4.12) were developed to explore the effects of acceptors' isomerization on photoelectric properties. Both FREAs exhibited similar energy levels and broad absorption ranging from 600 to 900 nm. A better-defined bilayer interpenetrated structure can be observed in PTB7-Th:FNIC2 with excellent vertical phase segregation for reduced bimolecular recombination. As a result, the PTB7-Th:FNIC2-based blend showed a μ_e of 1.4×10^{-3} cm^2 V^{-1} s^{-1}, 60% higher than that of PTB7-Th:FNIC1-based one. The as-cast OSCs fabricated with PTB7-Th:FNIC2 yielded a higher PCE of 13.00% with a J_{sc} of 23.93 mA cm^{-2} and FF of 0.734 than FNIC1-based device (10.30%) (Table 4.5). The corresponding PTB7-Th:FNIC2-based ST-OSCs achieved PCEs varying from 9.51% to 11.60% (AVTs from 20.3% to 13.6%), outperforming PTB7-Th:FNIC1-based counterparts.

4.3.3.2 Spacer Unit Effects

FREAs generally feature high charge carrier mobility due to reduced reorganization energy induced by their rigid fused-ring-conjugated backbone. Besides, their energy levels and absorption can be fine-tuned via delicate molecular engineering. The modulation on fused-ring cores is one of the most common methods to develop diverse molecules, thus screening best-performing FREAs. However, modifying fused core including their core sizes and the increase or decrease of chemical groups also bring some defects such as complicated synthesis and difficult purification.

As thus, Zhan and coworkers [110] firstly designed and synthesized a novel SMA named DC-IDT2T (the structure shown in Figure 4.15) by inserting thiophene between IDT core and end group (π-bridge). The DC-IDT2T exhibits a strong absorption at 550–700 nm with a medium E_g^{opt} of 1.55 eV. When blending with PBDTTT-C-T, the large difference (1.26 eV) between DC-IDT2T's LUMO and PBDTTT-C-T's HOMO provides their corresponding device a high V_{oc} of 0.90 V. Disappointingly, the as-cast PBDTTT-C-T:DC-IDT2T-based OSCs only achieved a low PCE of 3.93% (Table 4.6). This terrible performance can be mainly attributed to low J_{sc} of 8.33 mA cm^{-2} caused by almost overlapping absorption of donor and acceptor. However, this work demonstrates π-bridge is a potential strategy to develop novel acceptors. After that, the same group replaced thiophene π-bridge with furan to generate DC-IDT2F [113]. The introduced furan weakened the electron-donating capability of core, leading to a μ_e of 4.1×10^{-4} cm^2 V^{-1} s^{-1} in PBDTTT-C-T:DC-IDT2F-based blend (1.5×10^{-3} cm^2 V^{-1} s^{-1} for PBDTTT-C-T:DC-IDT2T) for decreased J_{sc}, thus a lower

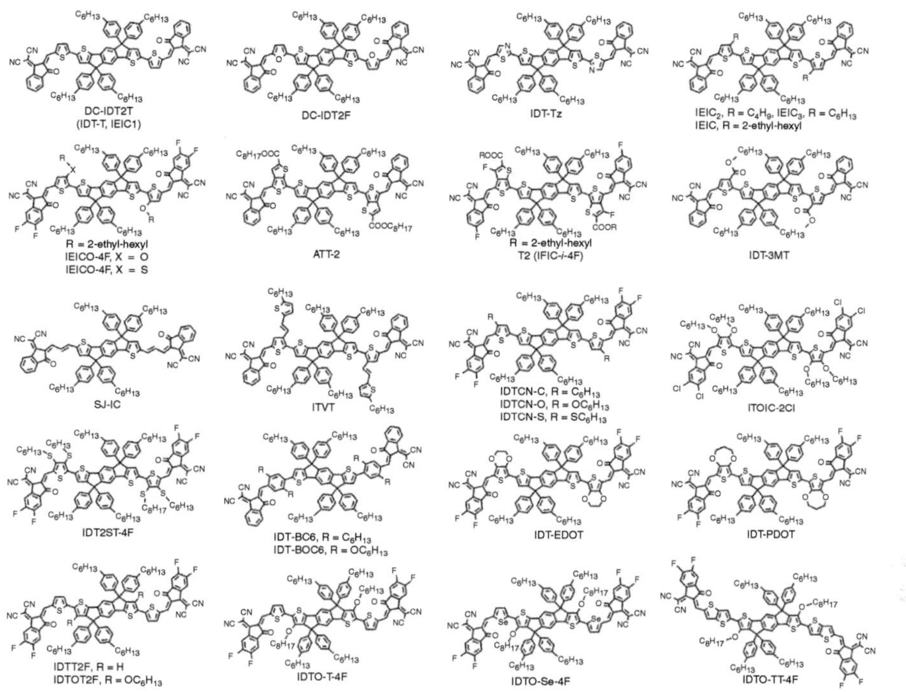

Figure 4.15 Chemical structures of some representative FREAs with space units. Source: J. Cao and W. Tang.

PCE of 3.08% for its corresponding OSCs. Furthermore, Huang and coworkers continue to optimize molecular structure by adding nitrogen atom into thiophene bridge for the design of IDT-Tz [111]. The composite IDT-Tz shows a λ_{onset} of 808 nm (c. 8 nm red-shifted than DC-IDT2T [also called IDT-T]), corresponding to a narrower E_g^{opt} of 1.53 eV. The selected donor PTB7-Th has a complementary absorption with IDT-Tz, promoting the light harvest for high J_{sc}. In addition, the intramolecular nitrogen-sulfur (N-S) non-covalent interaction between the nitrogen atom of thiazole and the sulfur atom of IDT core can lock conformation, which improves the rigidity and planarity of the backbone for efficient charge transport. As a result, the device based on PTB7-Th:IDT-Tz delivered a PCE of 8.52% with a V_{oc} of 0.88 V, an enhanced J_{sc} of 13.67 mA cm^{-2} and a FF of 0.708, much higher than that of PTB7-Th:IDT-T-based counterparts (4.08%).

Side chain is widely accepted can manipulate the solubility and intermolecular aggregation behavior, and rational side chain modification can induce the formation of better blend morphology for high PCE. As thus, side chain engineering was conducted into thiophene bridge of DC-IDT2T (also called IEIC1 there), and three SMAs were synthesized by introducing *n*-butyl (IEIC2), *n*-hexyl (IEIC3), and 2-ethyl-hexyl (IEIC), respectively (the structures shown in Figure 4.15) [112]. All three acceptors exhibited almost similar energy levels and absorption. However, due to the steric hindrance, different alkyl chain leads to diverse morphologies. The resonant soft

Table 4.6 Optical bandgap, energy levels, and photovoltaic parameters for A–D–A-type non-fullerene SMAs with different spacer units.

SMA	E_g^{opt} a) (eV)	HOMO/LUMO b) (eV)	Donor	V_{oc} (V)	J_{sc} (mA cm^{-2})	FF	PCE (%)	References
DC-IDT2T	1.55	−5.43/−3.85	PBDTTT-C-T	0.90	8.33	0.523	3.93	[110]
IDT-T	1.45	−5.41/−3.96	PTB7-Th	0.84	9.54	0.509	4.08	[111]
IEIC1	1.62	−5.44/−3.87	PTB7-Th	0.84	9.90	0.440	3.90	[112]
DC-IDT2F	1.54	−5.44/−3.88	PBDTTT-C-T	0.96	6.32	0.508	3.08	[113]
IDT-Tz	1.53	−5.62/−4.09	PTB7-Th	0.88	13.67	0.708	8.52	[111]
IEIC2	1.60	−5.45/−3.87	PTB7-Th	0.91	11.10	0.500	5.40	[112]
IEIC3	1.60	−5.46/−3.86	PTB7-Th	0.93	12.80	0.580	7.40	[112]
IEIC	1.61	−5.48/−3.85	PTB7-Th	0.95	12.30	0.490	6.00	[112]
IEICO-4F	1.24	−5.44/−4.19	PBDTTT-EFT	0.74	22.80	0.594	10.00	[114]
IEICS-4F	1.35	−5.43/−4.08	PTB7-Th	0.73	20.98	0.670	10.30	[115]
ATT-2	1.32	−5.50/−3.90	PTB7-Th	0.73	20.75	0.630	9.58	[116]
T2 (IFIC-i-4F)	1.30	−5.34/−3.96	PTB7-Th	0.65	24.85	0.670	10.87	[117]
IDT-3MT	1.52	−5.68/−4.16	PBDB-T	0.95	14.43	0.613	8.40	[118]
SJ-IC	1.49	−5.61/−3.94	J61	0.83	16.99	0.660	9.27	[119]
ITVT	1.48	−5.47/−3.59	PBZ	0.96	14.22	0.427	5.84	[120]
IDTCN-C	1.48	−5.59/−3.92	PBDB-T	0.84	20.33	0.696	11.92	[121]
IDTCN-O	1.53	−5.54/−3.80	PBDB-T	0.91	19.96	0.732	13.28	[121]
IDTCN-S	1.48	−5.57/−3.90	PBDB-T	0.85	19.04	0.657	10.60	[121]
ITOIC-2Cl	1.40	−5.62/−3.93	PBDB-T	0.88	17.10	0.616	9.37	[122]
IDT2ST-4F	1.43	−5.36/−3.71	PBDB-T	0.85	19.44	0.693	11.43	[123]
IDT-BC6	1.75	−5.55/−3.82	PBDB-T	0.92	5.63	0.440	2.30	[124]
IDT-BOC6	1.63	−5.51/−3.78	PBDB-T	1.01	17.52	0.540	9.60	[124]
IDT-EDOT	1.41	−5.43/−3.80	PBDB-T	0.86	21.34	0.620	11.32	[125]
IDT-PDOT	1.43	−5.39/−3.77	PBDB-T	0.85	5.26	0.489	2.18	[125]
IDTT2F	1.46	−5.57/−4.03	PBDB-T	0.81	18.51	0.590	8.85	[126]
IDTOT2F	1.44	−5.54/−3.94	PBDB-T	0.85	20.87	0.720	12.79	[126]
IDTO-T-4F	1.45	−5.49/−3.88	PBDB-T	0.86	20.12	0.727	12.62	[126]
IDTO-Se-4F	1.40	−5.48/−3.90	PBDB-T	0.83	18.55	0.692	10.67	[126]
IDTO-TT-4F	1.38	−5.39/−3.89	PBDB-T	0.86	17.21	0.694	10.21	[126]

a) Calculated from $E_g^{opt} = 1240/\lambda_{onset}$.
b) Calculated from CV curves.

X-ray scattering (R-SoXS) study revealed that the mode domain size of corresponding blends has a parabolic trend accompanied the carbon numbers of alkyl side chains from 0 to 8. And the PTB7-Th:IEIC3 exhibited the smallest domain size with 18 nm close to the exciton diffusion length (c. 10 nm) in organic materials, contributed to the high μ_e of $6.0 \times 10^{-3}\ cm^2\ V^{-1}\ s^{-1}$. The derived PTB7-Th:IEIC3-based OSCs afforded a champion PCE of 7.4% with a highest J_{sc} of $12.80\ mA\ cm^{-2}$, outperforming the other three ones (3.9–6.0%).

On the basis of IEIC, Hou and coworkers [114] inserted oxygen atoms in the 2-ethylhexyl branch, and the star π-bridged IEICO-4F was developed end capping with IC-2F (Figure 4.15). The O–S non-covalent interaction endows IEICO-4F good co-planarity and reduced reorganization energy, thus an ultra-narrow E_g^{opt} of 1.24 eV (the film absorption approaching 1000 nm) for enhanced ICT effect. The clear phase separation can be observed in PBDTTT-EFT:IEICO-4F-based blend film with a RMS of 1.22 nm. Consequently, an impressively high J_{sc} of $22.80\ mA\ cm^{-2}$, a V_{oc} of 0.74 V and a FF of 0.594 were obtained, contributed to a PCE of 10.00% for PBDTTT-EFT:IEICO-4F-based OSCs (Table 4.6). Moreover, benefiting from the donor evolvement and innovative device processes, the PCE of IEICO-4F-based binary OSCs has already been upgraded to 12.80% with a surprising J_{sc} of $27.30\ mA\ cm^{-2}$ [127]. The stunning J_{sc} also attracted us to adding IEICO-4F into binary system as a third component for highly-efficient ternary OSCs fabrication. Then utilizing sulfur to replace oxygen atoms, they further designed and synthesized IEICS-4F [115]. The IEICS-4F showed 0.11 eV wider E_g^{opt} than IEICO-4F and slightly upshifted HOMO/LUMO energy levels of −5.43 eV/−4.08 eV. To avoid spectral overlap, the PTB7-Th was selected as a matching donor. Morphological study revealed that sufficiently small domain size in PTB7-Th:IEICS-4F-based blend film induces the formation of highly crystalline domain, and this "near-optimum" morphology is beneficial to charge generation and minimized charge recombination. Besides, the preferential face-on orientation with high crystallinity of PTB7-Th:IEICS-4F boosts the charge transport. As a result, the PTB7-Th:IEICS-4F-based binary device displayed a PCE of 10.30% with enhanced FF of 0.670 (Table 4.6). The high AVT in visible range of IEICS-4F also contributed to the PCEs of 7.50% and 10.50% for binary (35% AVT in 370–740 nm) and tandem (AVT, 20%) ST-OSCs, respectively (Figure 4.16a).

It can be found that several SMAs mentioned above showed the NIR absorption and high AVT in visible range, which foresees its stupendous application potential in ST-OSCs and as rear cell in tandem OSCs. However, a dilemma also should be noted that the UV–vis–NIR was utilized to harvest more photons for high device performance, which simultaneously results in the reduced device transparency. Therefore, novel SMA was further explored to balance the properties and transparency, and the ATT-2 (Figure 4.15) with low E_g^{opt} of 1.32 eV was designed by introducing ester-substituted thieno[3,4-*b*]thiophene π-bridge [116]. The composite ATT-2 displays a strong absorption at 600–850 nm, which well preserves a large part of the visible area. Combining with PTB7-Th, the optimal 2% CN treated OSCs afforded a

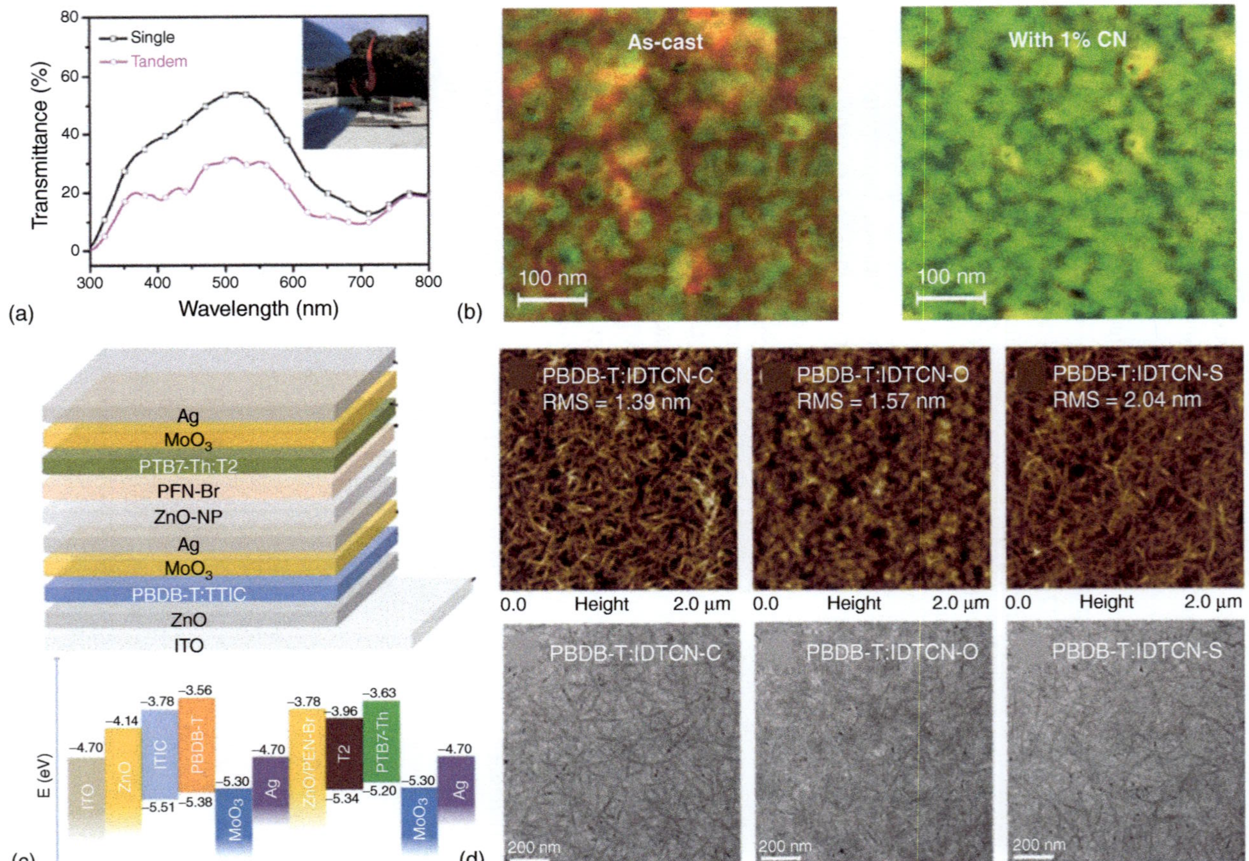

Figure 4.16 (a) Transmittance spectra of PTB7-Th:IEICS-4F-based single junction and tandem ST-OSCs. Inset: a photograph through a tandem ST-OSC. Source: Chen et al. [115]/Reproduced with permission of John Wiley & Sons. (b) PiFM images of the as-cast and 1% CN treated PTB7-Th:T2 blend films. (c) Schematic architecture of tandem OSCs and the energy levels of each layer. Source: Chen et al. [117]/Reproduced with permission of JOHN WILEY & SONS, INC. (d) TEM and AFM images of PBDB-T:IDTCN-C, PBDB-T:IDTCN-O, and PBDB-T:IDTCN-S-based blend films. Source: Liu et al. [121]/Reproduced with permission of JOHN WILEY & SONS, INC.

PCE of 9.58% and an outstanding J_{sc} of 20.75 mA cm^{-2} with an AVT of 37%. Considering the decreased AVT of Ag electrodes in NIR than in the visible region, the thickness of Ag electrode was reduced to increase the transmittance of visible light. As expected, the AVT exhibited an increasing trend at the expense of weakened J_{sc} and FF (Ag, 100–5 nm). At last, the PCE of 7.74% was obtained for PTB7-Th:ATT-2-based ST-OSCs (Ag, 20 nm), which is the highest value at that time with such high transmittance.

Also coveting the advantages of NIR SMAs, Li's group designed a SMA, namely, T2 (also called IFIC-i-4F) [117], through adding fluorine atom into ester-substituted thieno[3,4-*b*]thiophene π-bridge. End-capped with IC-F, T2 presented a 0.16 eV upshifted HOMO and 0.06 eV deeper LUMO than ATT-2 (−5.50 eV/−3.90 eV), corresponding to a narrower E_g^{opt} of 1.30 eV. Photo-induced force microscopy (PiFM) measurements revealed that the T2 can evenly mix with donor PTB7-Th, and proper phase separation can be observed with an average domain size (~20 nm) at top surface of their blend after processing with 1% (v/v) CN additive (Figure 4.16b). An optimal PCE of 10.87% was thus achieved with an outstanding J_{sc} of 24.85 mA cm^{-2}, a V_{oc} of 0.65 V and a FF of 0.670. As widely recognized that the tandem OSCs consist of two individual subcells, and its V_{oc} is the sum of the voltages of two subcells while the J_{sc} was determined by the lowest one. By chance, the low V_{oc} and high J_{sc} of PTB7-Th:T2-based device are just suitable for fabricating tandem OSCs. The resultant tandem device (PBDB-T:ITIC as front cell and PTB7-Th:T2 as rear cell, Figure 4.16c) afforded a surprisingly high PCE of 14.64% with a V_{oc} of 1.56 V, a J_{sc} of 13.22 mA cm^{-2}, and a FF of 0.710.

Further π-bridge engineering with methyl-3-thiophenecarboxylate generated a new SMA called IDT-3MT (Figure 4.15) [118]. Supramolecular O–S non-covalent interactions can lock the flat conformation of the IDT-3MT acceptor, which significantly extends the effective conjugation and broadens the absorption spectrum. Besides, the photoluminescence (PL)-quenching measurements showed that the PBDB-T:IDT-3MT-based blend displayed a higher fluorescence quenching efficiency of over 95% than IDT-T-based one (67%) when excited at 700 nm, suggesting efficient charge dissociation at the D/A interface. Consequently, the PBDB-T:IDT-3MT-based device achieved a highly improved PCE of 8.40% than IDT-T-based cells (6.36%) without any additives (Table 4.6).

Then the vinyl was inserted as the π-bridge into IDT-cored FREA IDT-IC, and the SJ-IC was designed and synthesized [119]. Compared with IDT-IC, the SJ-IC presents an obviously deeper LUMO of −3.94 eV and an upshifted HOMO of −5.61 eV, corresponding to a narrower E_g^{opt} of 1.49 eV for high J_{sc}. In addition, the increased structural complexity induced by the introduction of C=C bond weakens the crystallinity of SJ-IC, endowing the J61:SJ-IC-based blend suitable phased separation with pure acceptor domain after thermal annealing for efficient charge transport. A PCE of 9.27% was obtained by J61:SJ-IC-based OSCs with a V_{oc} of 0.83 V, a J_{sc} of 16.99 mA cm^{-2}, and a FF of 0.660, 25% higher than that of IDT-IC-based counterparts (6.95%). In order to absorb more photos at short wavelength, thienylenevinylene thiophene (TVT) was introduced between IDT core and IC end groups for the synthesis of ITVT [120]. The conjugated TVT side chains provide ITVT a strong absorption at 300–500 nm with a narrow E_g^{opt} of 1.48 eV. Featuring complementary absorption and matching energy levels with ITVT, the wide E_g^{opt} PBZ was selected as donor and their blend film exhibited a fibrillar structure with a μ_e of 1.85×10^{-4} cm^2 V^{-1} s^{-1}. The as-cast PBZ:ITVT-based OSCs (D/A ratio = 1 : 2) obtained a moderate PCE of 5.84% with a low FF of 0.427.

Different from the modification in inner position of thiophene above mentioned, Bo and coworkers [121] explored three novel SMAs by inserting diverse alkyl chain into outer position of thiophene bridge including hexyl (IDTCN-C), hexyloxy (IDTCN-O), and hexylthio (IDTCN-S) (Figure 4.15). All three SMAs show strong absorption at 550–800 nm in solution with molar absorption coefficient of 1.77–2.55 × 10^5 M^{-1} cm^{-1}, and IDTCN-O exhibited a wider E_g^{opt} of 1.53 eV than E_g^{opt}s of both IDTCN-C and IDTCN-S (1.48 eV). Theoretical simulation indicated that alkyl chains connected to the β position of thiophene restrict the free rotation of the terminal group for a stable dominant conformation. Moreover, compatible crystallization rate of donor and acceptor can form a good film morphology, and the IDTCN-O featuring strong crystallinity has a good miscibility with PBDB-T, which can be demonstrated by the decreased polymer fibrils in their blend (Figure 4.16d). As a result, the PBDB-T:IDTCN-O-based OSCs presented an outstanding PCE of 13.28% with a V_{oc} of 0.91 V, a J_{sc} of 19.96 mA cm^{-2}, and a FF of 0.732, much higher than that of IDTCN-C- and IDTCN-S-based devices (11.92% and 10.60%, respectively) (Table 4.6). The highest V_{oc} can be attributed to the highest LUMO energy levels of IDTCN-O (−3.80 eV), which can also be confirmed by the highest energy of charge transfer state (E_{ct}). The 13.28% PCE is the highest value reported in literature at that time.

From the above strategies, we can obviously summarize that the planarity is an important guarantee for efficient charge transfer. And the non-covalent interaction formed in inner position of π-bridge provides a planar backbone while that in outer position can lock the conformation between end groups and π-bridge. Combining the advantages of these two, Bo's group [122] reported a novel SMA named ITOIC-2Cl utilizing bis(alkoxy)-substituted thiophene as π-bridges (Figure 4.15). The better skeleton planarity can be obtained induced by double intramolecular sulfuroxygen (S–O) and oxygen–hydrogen (CH–O) interactions, which is not strong in solution to guarantee the solubility of SMAs. End-capped with strong electron-deficient di-chlorinated (IC-2Cl), the ITOIC-2Cl has a NIR film absorption approaching 900 nm, accordingly a narrow E_g^{opt} of 1.40 eV. The film blended with PBDB-T exhibited clear phase separation with uniformly distributed nano-sized structure, suggesting the formation of regular interpenetrating network for efficient exciton dissociation at D/A interface and charge transport. The resulting as-cast OSCs based on PBDB-T:ITOIC-2Cl delivered a PCE of 9.37% (Table 4.6).

Furthermore, replacing oxygen atom with sulfur atom and lengthening the alkyl chains, Zhang and coworkers [123] designed the IDT2ST-4F end-capped with IC-2F. In terms of the wider E_g^{opt} of 1.43 eV in relative to ITOIC-2Cl, the as-cast PBDB-T:IDT2ST-4F-based device only achieved a PCE of 9.27%. After annealing at 110 °C for five minutes, the enhanced crystallinity of IDT2ST-4F induced a more uniform fibrous morphology for improved J_{sc} of 19.44 mA cm^{-2} and FF of 0.693, thus a higher PCE of 11.43%. Coincidentally, Bo and coworkers [124] developed two novel SMAs using IDT as core and the bis(alkoxy)-substituted (IDT-BOC6) or dialkyl-substituted benzene (IDT-BC6) as conformational control units. Similar to ITOIC-2Cl and IDT2ST-4F mentioned above, the IDT-BOC6 presents a planar backbone due to dual non-covalent interaction, while the IDT-BC6 shows a distorted plane with large dihedral angles between π-bridge and IDT core/terminal groups. After treating with 1% DIO, the optimal PBDB-T:IDT-BOC6-based OSCs afforded a PCE of 9.60% with an impressive V_{oc} over 1.00 V. The irregular bulk phase separation morphology may be used to explain the poor PCE of PBDB-T:IDT-BOC6-based device (2.3%).

To further explore best-performing SMA, Bo and coworkers inserted 3,4-propylenedioxythiophene (PDOT) and 3,4-ethylenedioxythiophene (EDOT) bridging units for synthesizing two IDT-cored SMAs (IDT-PDOT and IDT-EDOT, respectively, Figure 4.15) [125]. The introduction of a tiny methylene group caused distinct structure–properties relationship. The IDT-EDOT has similar energy levels and absorption with IDT-PDOT, but about four times higher molar absorption coefficient of 1.95×10^5 M^{-1} cm^{-1} than IDT-PDOT (4.86×10^4 M^{-1} cm^{-1}). Morphological study revealed that the neat IDT-EDOT film showed a smaller π–π stacking distance of 3.45 Å than IDT-PDOT (3.64 Å), indicating more compact intermolecular packing for high electron mobility. In addition, the PBDB-T:IDT-EDOT-based blend presents clear nanofiber-like interpenetrating network structure, which increases the D/A contact interface for efficient exciton dissociation and charge transport. The optimal PBDB-T:IDT-EDOT-based OSCs (treated with 0.2% DIO) contributed to a salient PCE of 11.32% with a high J_{sc} of 21.34 mA cm^{-2}, a V_{oc} of 0.86 V and a FF of 0.620, surpassing the PBDB-T:IDT-PDOT-based ones (2.18%) and even the IEICO-4F-based devices (8.29%) (Table 4.6). The poor properties of IDT-PDOT-based cells can be attributed to inappropriate phase separation morphology. These results demonstrate that minor changes in molecule structures can seriously affect blend morphologies to cause unexpected intermolecular aggregation behavior, thus diverse device properties. And the best-performing SMAs can be explored via delicate molecule design strategies in π-bridge.

All the above molecular designs powerfully demonstrate the potential of π-bridge, boosting the PCE increasing from 2.30% to 13.28%. Among them, almost all the as-cast OSCs have poor or moderate properties, and many efficient systems need post-processing including thermal annealing and additives which reduces the reproducibility and stability. As thus, a new strategy was proposed by Bo and coworkers [126] to attach two alkoxyl substituents at the 3,8-positions of IDT core. Coupling with two thiophene π-bridge and IC-2F end groups generated the IDTOT2F (Figure 4.15). For comparison, the IDTT2F (without substituted alkyl) was also synthesized. IDTOT2F showed a better solubility than IDTT2F, and the middle-shrinking energy levels endow IDTOT2F a narrower E_g^{opt} for high J_{sc}. The derived as-cast PBDB-T:IDTOT2F OSCs presented a record PCE of 12.79%, much higher than that of IDTT2F- (8.85%) and IEICO-4F- (8.66%) based counterparts (Table 4.6).

Similarly using O–S non-covalent interactions, Liu and coworker [128] explored the effects of different π-bridges on device properties, and three novel SMAs were obtained by bridging thiophene (IDTO-T-4F), selenophene (IDTO-Se-4F) and thieno[3,2-*b*]thiophene (IDTO-TT-4F), respectively (Figure 4.15). All three SMAs exhibited similar LUMO energy levels around −3.89 eV while the IDTO-TT-4F has a highest HOMO of −5.39 eV. The strongest electron-donating ability provides IDTO-TT-4F enhanced ICT effect for a narrowest E_g^{opt} of 1.38 eV. GIWAXS results showed that the pristine SMAs films have diverse CCL (27.4 Å for IDTO-T-4F, 20.9 Å for IDTO-Se-4F and 19.0 Å for IDTO-TT-4F), the highest CCL of IDTO-T-4F is beneficial for the charge transport properties in corresponding film. In addition, the stronger (010) diffraction peak than the other two can be observed in PBDB-T:IDTO-T-4F-based blend, suggesting more compact π–π stacking for efficient charge transport. As a result, the PBDB-T:IDTO-T-4F-based device afforded a champion PCE of 12.62% with a V_{oc} of 0.86 V, a high J_{sc} of 20.12 mA cm^{-2} and a FF of 0.727, outperforming the other two counterparts (10.67% for PBDB-T:IDTO-Se-4F and 10.21% for PBDB-T:IDTO-TT-4F) (Table 4.6).

4.3.3.3 Asymmetric Cores

Unlike the widespread study of symmetric FREAs, asymmetric FREAs only shine from 2016. Featuring feasibility in chemical structure modulation and planar molecular backbones for efficient charge transport that symmetric FREAs owned, asymmetric FREAs additionally possess stronger intermolecular binding energy and larger dipole moment for high-performing OSCs. Generally, three main strategies including asymmetric core, asymmetric side chains, and asymmetric end groups were adopted into acceptors for structure optimization.

Thiophene and TT are the most common building blocks for molecule design, and the indacenodithiophene (IDT) and IDTT cores are successfully developed for efficient FREAs. Considering the advantages of asymmetric FREAs, Sun and coworkers [129] explored an asymmetric core named TPTT by flanking benzene with thiophene and TT unit. The π-conjugation length, fused-ring numbers, and electron-donating ability of the TPTT core are in between the IDT and IDTT cores. Coupling with IC generated the novel asymmetric acceptor TPTT-IC (Figure 4.17). Compared with IDT-cored IC-1IDT-IC [130] (−5.91 eV/−3.83 eV, E_g^{opt} = 1.70 eV) and IDTT-cored ITIC (−5.71 eV/−3.97 eV, E_g^{opt} = 1.60 eV), TPTT-IC has centered HOMO/LUMO of −5.78 eV/−3.95 eV and E_g^{opt} of 1.63 eV. Comfortingly, the PBT1-C:TPTT-IC blend films expectantly presented a higher μ_e of 3.23×10^{-4} cm^2 V^{-1} s^{-1} than PBT1-C:ITIC-based blend ($\mu_e = 2.85 \times 10^{-4}$ cm^2 V^{-1} s^{-1}), which may be attributed to favorable morphology induced by dipole–dipole intermolecular interaction of asymmetric TPTT-IC. The derived PBT1-C:TPTT-IC-based OSCs achieved a slightly higher PCE of 10.3% than ITIC-based ones (10.0%) (Table 4.7).

In order to enhance the light-harvesting ability of acceptor for higher J_{sc}, the asymmetric acceptor TPTTT-2F [131] (Figure 4.17) was synthesized through adding a thiophene unit and end-capping mono-fluorinated IC (IC-F). The extended core conjugation endows TPTTT-2F broader absorption, upshifted LUMO of −3.44 eV and more compact intermolecular π–π stacking (Figure 4.18a)

Table 4.7 Optical bandgap, energy levels, and photovoltaic parameters for A–D–A-type non-fullerene SMAs with asymmetric structures.

SMA	E_g^{opt} [a] (eV)	HOMO/LUMO [b] (eV)	Donor	V_{oc} (V)	J_{sc} (mA cm^{-2})	FF	PCE (%)	References
TPTT-IC	1.63	−5.78/−3.95	PBT1-C	0.96	15.50	0.694	10.50	[129]
TPTT-2F	1.58	−5.75/−4.04	PBT1-C	0.88	15.82	0.730	10.17	[131]
TPTTT-2F	1.56	−5.69/−4.01	PBT1-C	0.92	17.63	0.745	12.03	[131]
MeIC1	1.54	−5.59/−3.89	PBDB-T	0.93	18.32	0.741	12.58	[132]
SePTTT-2F	1.50	−5.55/−3.97	PBT1-C	0.90	18.02	0.759	12.24	[133]
IPT-2F	1.44	−5.51/−3.96	PBDB-T	0.86	22.40	0.724	14.00	[134]
IPTT-2F	1.42	−5.46/−4.04	PBDB-T	0.87	19.70	0.662	11.40	[134]
IPTTT-2F	1.43	−5.40/−4.07	PBDB-T	0.89	20.00	0.693	12.30	[134]
DTPPSe-IC	1.46	−5.38/−3.84	PBDB-T	0.90	17.32	0.634	9.88	[135]
DTPPSe-2F	1.40	−5.52/−4.05	PBDB-T	0.84	22.16	0.737	13.76	[135]
DTPPSe-4F	1.39	−5.53/−4.10	PBDB-T	0.78	21.18	0.728	12.03	[135]
IPT-4F	1.42	−5.57/−4.07	PM6	0.91	22.08	0.742	14.96	[136]
IPT-4F	1.41	−5.56/−4.05	PM6	0.88	22.15	0.750	14.62	[137]
IPT-4Cl	1.39	−5.58/−4.11	PM6	0.88	23.18	0.704	14.40	[136]
IPTBO-4F	1.41	−5.57/−4.07	PM6	0.92	22.08	0.725	14.67	[136]
IPTBO-4Cl	1.39	−5.64/−4.08	PM6	0.89	23.15	0.726	15.00	[136]
INO-4F	1.46	−5.64/−3.93	PM6	0.93	20.46	0.720	13.69	[137]
IN-4F	1.45	−5.59/−3.94	PM6	0.92	19.50	0.697	12.50	[137]
IPCl-4F	1.39	−5.60/−4.08	PM6	0.83	21.18	0.614	10.79	[137]
IPT2F-TCl	1.48	−5.59/−4.00	PBDB-T	0.86	20.59	0.775	13.74	[138]
IPT2F-Th	1.47	−5.58/−4.03	PBDB-T	0.86	19.65	0.712	12.52	[139]
IPT2F-Ph	1.46	−5.57/−4.00	PBDB-T	0.86	20.41	0.721	13.13	[139]
IPT2F-TT	1.45	−5.60/−4.00	PBDB-T	0.84	21.47	0.751	14.02	[139]
TTPTTT-IC	1.60	−5.64/−3.87	PBT1-C	1.00	12.47	0.637	7.91	[140]
TTPTTT-2F	1.54	−5.67/−4.04	PBT1-C	0.92	16.78	0.746	11.52	[140]
TTPTTT-4F	1.52	−5.69/−4.12	PBT1-C	0.86	19.36	0.721	12.05	[140]
a-BTTIC	1.43	−5.45/−3.83	PBDB-T	0.90	20.31	0.740	13.60	[141]
ITCNTC	1.68	−5.66/−3.92	J71	0.94	14.16	0.638	8.52	[142]
IDT6CN	1.63	−5.68/−3.97	PBDB-T	0.83	15.14	0.738	9.27	[143]
IDT6CN-Th	1.61	−5.71/−4.01	PBDB-T	0.81	16.75	0.767	10.41	[143]
IDT-OB	1.66	−5.77/−3.87	PBDB-T	0.88	16.18	0.711	10.12	[76]
IT-3F	1.55	−5.67/−4.09	PM6	0.90	20.02	0.733	13.83	[144]

a) Calculated from $E_g^{opt} = 1240/\lambda_{onset}$.
b) Calculated from CV curves.

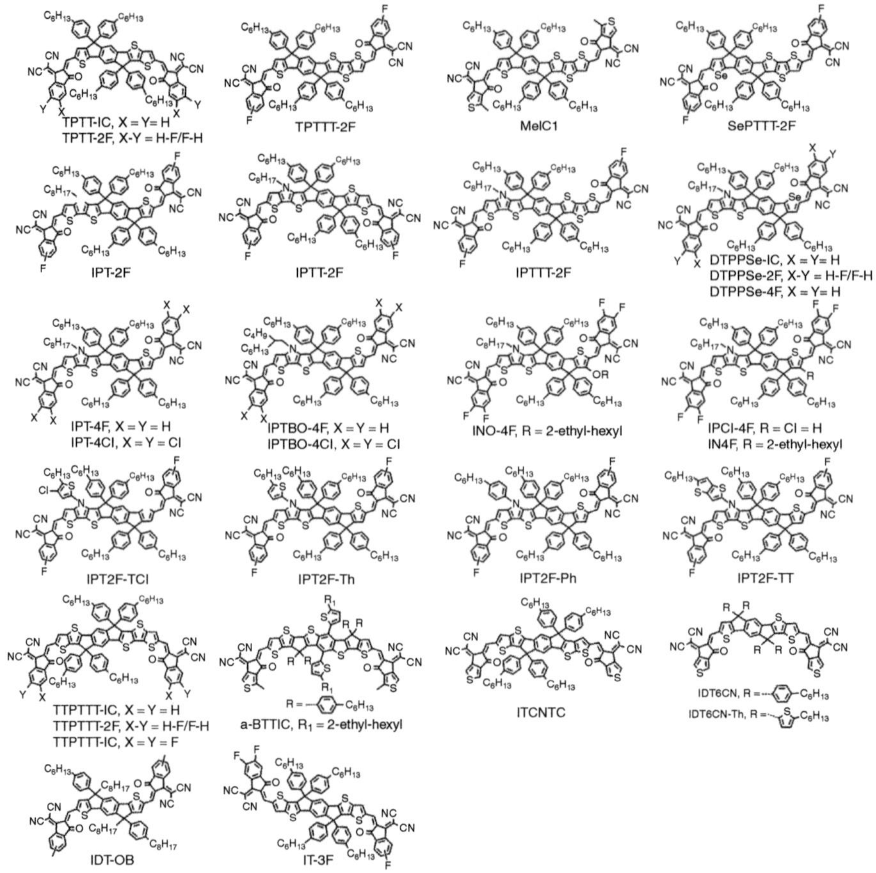

Figure 4.17 Chemical structures of asymmetric cored FREAs. Source: J. Cao and W. Tang.

when compared with TPTT-cored counterpart (TPTT-2F, with IC-F end group). Consequently, the PBT1-C:TPTTT-2F-based OSCs obtained a higher PCE of 12.03% than TPTT-2F-based ones (10.17%), along with a high V_{oc} of 0.92 V due to the energy level difference between donor and acceptor. After that, Yang and coworkers [132] replaced IC-F end group with weaker electron-withdrawing 2-(6-oxo-5,6-dihydro-4*H*-cyclopenta[*c*]thiophen-4-ylidene) malononitrile (CPTCN), and the classic FREA named MeIC1 was synthesized (Figure 4.17). Due to the upshifted LUMO energy levels of −3.89 eV and narrower E_g^{opt} of 1.54 eV in relative to TPTTT-2F, the resulting PBDB-T:MeIC1-based device achieved an improved PCE of 12.58% (Table 4.7). Given the absorption, energy levels and well-defined face-on orientation of MeIC1, many groups generally added MeIC1 as a third component to optimize corresponding binary systems for high-performance ternary OSCs. For further enhancing the electron-donating ability of core, the strategy of heteroatom substitution is applied to molecular design. The acceptor SePTTT-2F [133] was thus designed by replacing the previous thiophene with selenophenol. In comparison to thiophene, selenophenol has a stronger electron-donating ability and decreased aromaticity, which increased effective conjugation length for a narrower E_g^{opt} of 1.50 eV. In addition, the large size and great polarizability of selenium are easy to induce intermolecular Se-Se/selenium–aromatic interaction in corresponding selenophene-based FREAs. As a result, the SePTTT-2F and PBT1-C showed good miscibility with the formation of nanostructured interpenetrating networks in their blends, boosting the FF over 0.750. The OSCs fabricated with PBT1-C:SePTTT-2F active layer delivered a PCE of 12.24%, outperforming the TPTTT-2F-based counterparts (Table 4.7).

It is widely accepted that the crystallinity of FREAs seriously affects the morphology of blend films and packing behavior, thereby greatly influencing the device properties. In 2019, Tang's group successfully developed an INP-cored acceptor named INPIC-4F, and the derived device achieved the highest PCE of 13.13% at that time. However, strong crystallinity of INPIC-4F limits its photovoltaic performance and is also not conducive to solution-process large-area OSCs. In this case, the asymmetric strategy was conducted by replacing dithieno[3,2-*b*:2′,3′-*d*]pyrrole (DTP) block on one side of INP core with relatively weak electron-donating units including thiophene (T), thieno[3,2-*b*]thiophene (TT), or dithieno[3,2-*b*:2′,3′-*d*]thiophene (TTT). End-capped with IC-2F, three asymmetric FREAs (IPT-2F, IPTT-2F, and IPTTT-2F, the structures shown in Figure 4.17) were designed and synthesized, respectively [134]. Surprisingly, the weakest electron-donating IPT-cored IPT-2F presents the largest red-shifted absorption peaks (c. 36 nm) from solution to films, indicating a more well-aligned and compact intermolecular stacking in the solid state. Also, along with the expansion of π-conjugate system (from IPT, IPTT, to IPTTT), the HOMO energy levels gradually upshifted, while the LUMO energy levels exhibit a deepened trend. More interestingly, due to the O–S non-covalent interactions induced by the oxygen atoms on the end groups and the sulfur atom on donating core, three asymmetric FREAs show diverse preferred conformation (S-shaped conformations for IPT-2F and IPTTT-2F, C-shaped conformation for IPTT-2F, Figure 4.18b), which confirmed by theoretical calculations and electric potential simulations. Analysis of the active layer morphology found that S-shaped IPT-2F and IPTTT-2F assembled order textures with long nanofibrils for advantageous charge transport, while a heterogeneous morphology with irregular phase separation can be observed in C-shaped IPTT-2F-based blend. The OSCs based on PBDB-T:IPT-2F achieved a champion PCE of 14.00% with a V_{oc} of 0.86 V, a high J_{sc} of 22.40 mA cm^{-2}, and a FF of 0.724, higher than that of IPTT-2F- and IPTTT-2F-based counterparts (12.3% and 11.4%, respectively) (Table 4.7). To the best of our knowledge, 14% PCE is among the highest PCE values in literature for asymmetric FREAs-based binary OSCs at that time. The more dominant face-on orientation with a stronger (010) π–π stacking of IPT-2F-based blend can facilitate charge transfer in the vertical direction, which may be a reason of the higher FF and J_{sc} PBDB-T:IPT-2F obtained for better PCE in comparison to PBDB-T:IPTTT-2F. For further enhancing the electron-donating ability of core, the same group explored selenopheno[2″,3″:5′,6′]-*s*-indaceno[2′,1′:4,5]thieno[3,2-*b*]thieno[2,3-*d*]pyrrole (DTPPSe) core by replacing thiophene with selenophenol [135]. In order to systematically study the influence of terminal groups on molecular characteristics and

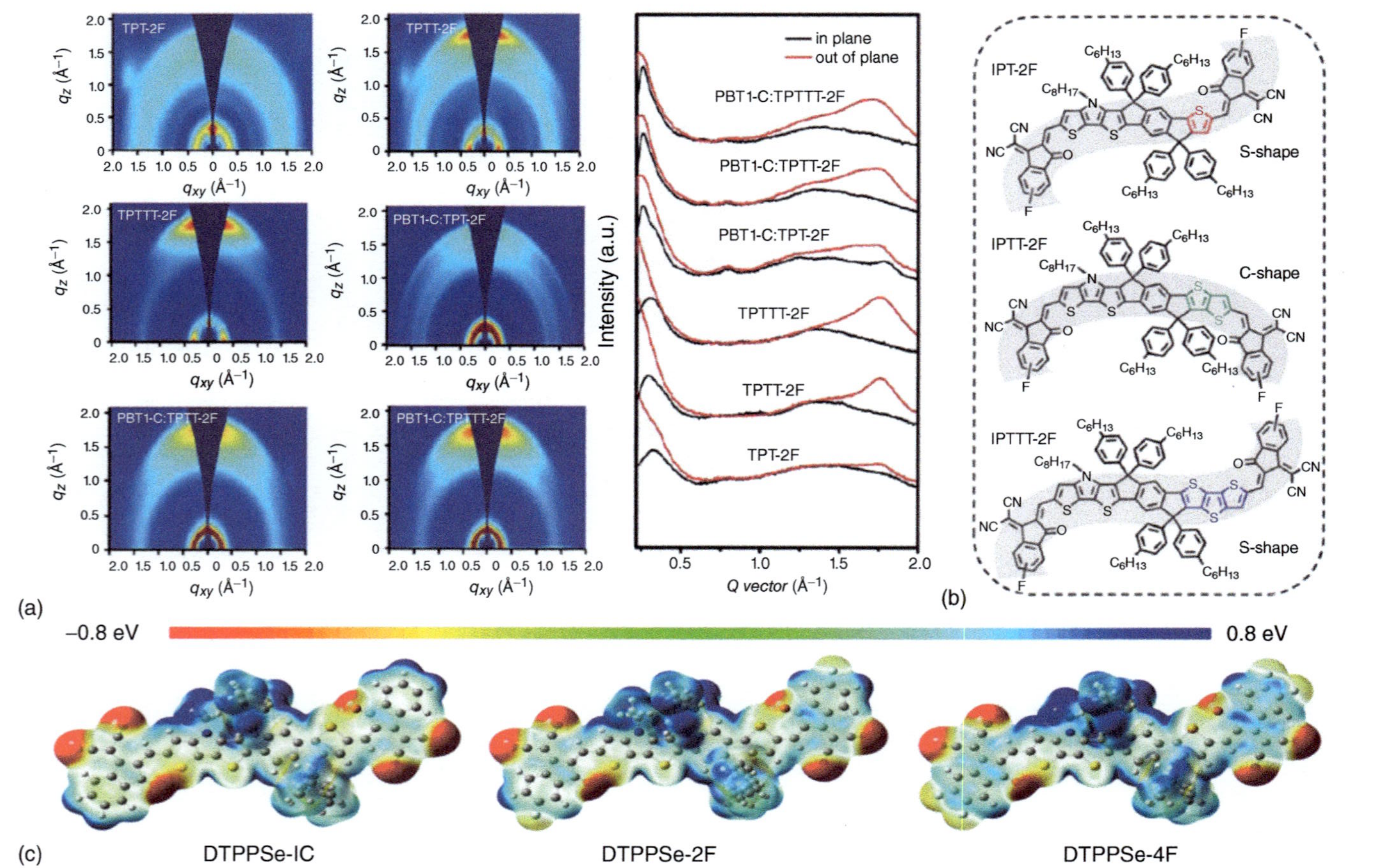

Figure 4.18 (a) GIWAXD patterns of TPT-2F, TPTT-2F, and TPTTT-2F and their corresponding blends. Source: Song et al. [131]/Reproduced with permission of Royal Society of Chemistry. (b) Diverse conformations of IPT-2F, IPTT-2F, and IPTTT-2F. Source: Yang et al. [134]/Reproduced with permission of Royal Society of Chemistry. (c) The molecular electrostatic potential (ESP) surface mapped at the 0.001 electron density of DTPPSe-based FREAs derived from DFT calculations. Source: Cao et al. [135]/Reproduced with permission of Royal Chemistry Society

device performance, three asymmetric FREAs were synthesized with enhanced electron-withdrawing end groups including IC (DTPPSe-IC), IC-F (DTPPSe-2F), and IC-2F (DTPPSe-4F) (Figure 4.17). The DTPPSe-2F showed the ~9 nm red-shifted λ_{onset} than IPT-2F, corresponding to a narrower E_g^{opt} of 1.40 eV. Besides, the gradually deepened energy levels and narrowed E_g^{opt} can be observed, along with the increasing electron-withdrawing abilities of end groups (IC < IC-F < IC-2F). Electrostatic potential (ESP) analysis shows that most parts of the surface have positive ESP, while the negative regions are only located at both ends of the acceptors (Figure 4.18c). Moreover, the increased fluorine atoms can expand the negative regions at both ends, which is more conducive to strengthen the intermolecular interactions with the donor to promote charge transfer. Surprisingly, the PBDB-T:DTPPSe-2F-based devices achieved the champion PCE of 13.76% among the three devices (Table 4.7), due to the better miscibility between donor and acceptor accompanied with uniformly fibrous phase separation morphology. The 13.76% PCE is the highest values for OSCs in literature using selenophene-containing FREAs.

Apart from heteroatom substitution, the strategy of halogenated end groups is another common strategy to narrow E_g^{opt} for enhanced intramolecular ICT. However, it can also easily induce excessive self-aggregation tendency. Alkyl- or aryl-based side chains engineering has proven to be quite effective in improving solubility and spontaneous molecular excessive aggregation, but too large or too small alkyl chains will also cause diverse intermolecular π–π stacking distance, thereby affecting charge transfer. In order to solve this dilemma, Tang and coworkers [136] further explored the relationship between the fluorinated/chlorinated end groups and octyl (C8)/2-butyl-1-octyl (BO) alkyl chains to modulate energy levels and balance aggregation and crystallinity. Four asymmetric FREAs were thus synthesized, namely, IPT-4F (fluorinated, C8), IPTBO-4F (fluorinated, BO), IPT-4Cl (chlorinated, C8) and IPTBO-4Cl (chlorinated, BO) (Figure 4.17). As presented in Figure 4.19a, the C8 chain located on DTP extends outward along the plane of backbone, and the rotation of C8 exerts negligible impact against the molecular aggregation. On the contrary, due to the steric hindrance with the phenylhexyl group on one side, one arm of the branched BO side chain exhibits an out-of-plane dominant conformation perpendicular to the backbone, which would effectively suppress molecular aggregation. On this basis, the maximum bare length of backbone (L_b) is assumed to quantify the combined effect of halogenated end groups and side chains on aggregation and stacking behavior (Figure 4.19b). Consequently, the L_b of IPT-4F, IPTBO-4F, IPT-4Cl, and IPTBO-4Cl is calculated to be 28.01, 24.31, 29.56, and 25.85 Å, respectively, suggesting that IPT-4F forms a more effective stacking than IPTBO-4F, while the IPTBO-4Cl get a better solubility and weakened crystallinity. In line with our prediction, by blending with PM6, the IPT-4F and IPTBO-4Cl-based devices outperformed IPTBO-4F- and IPT-4Cl-based ones, and the PM6:IPTBO-4Cl-based OSCs afforded the champion PCE of 15.0% with a V_{oc} of 0.89 V, a J_{sc} of 23.15 mA cm^{-2} and a FF of 0.726 (Table 4.7). The PCE defines a new record of asymmetric FREAs-based OSCs and verifies that dual molecular engineering for balanced crystallinity and solubility is an effective means to obtain best-performing OSCs.

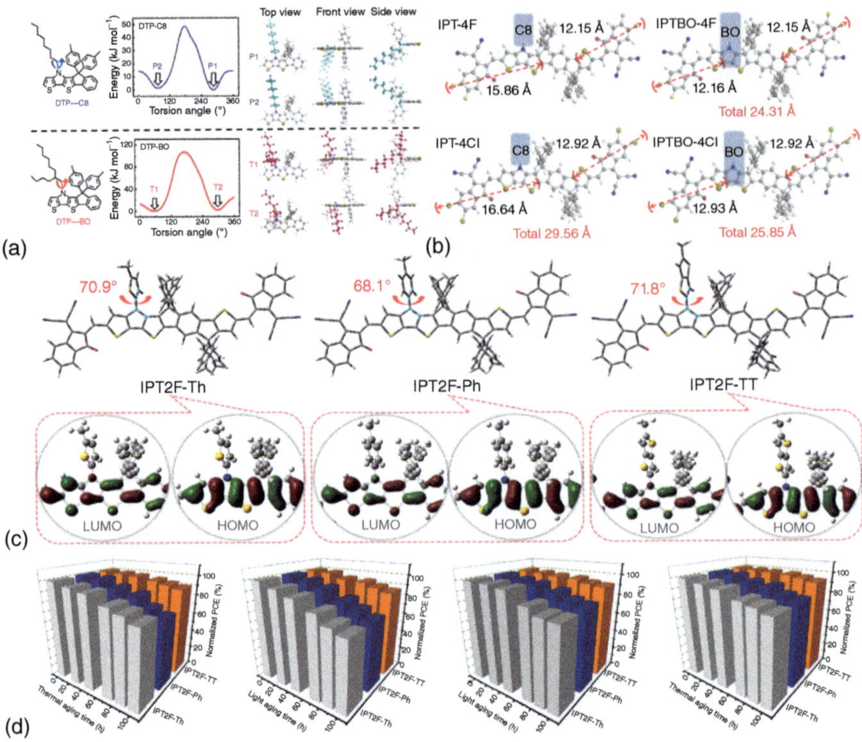

Figure 4.19 (a) Theoretical simulation for optimal *N*-alkyl chain conformations. (b) Diagrams of total calculated backbone bare lengths of IPT-4F, IPTBO-4F, IPT-4Cl, and IPTBO-4Cl. Source: Yang et al. [136]/Reproduced with permission of Royal Society of Chemistry. (c) Dihedral angles between 2D side chain and skeleton plane, and corresponding LUMO/HOMO distributions in 2D-conjugated areas for three FREAs. (d) Normalized PCE and FF of three FREAs-based devices as a function of thermal aging time at 85 °C or light aging time at room temperature. Source: Cao et al. [139]/Reproduced with permission of John Wiley & Sons.

Obviously, the thieno[2″,3″:5′,6′]-*s*-indaceno[2′,1′:4,5]dithieno[3,2-*b*:2′,3′-d] pyrrole (IPT) core is much potential to further develop efficient asymmetric acceptors. However, the IPT-4F and IPT-4Cl mentioned above remind us that enhancing the electron-withdrawing ability through halogenated end groups has only a negligible effect on the energy levels and E_g^{opt}s of the IPT-based acceptors, so that only two strategies can be conducted in IPT core: (i) modulation on flanked thiophene and (ii) *N*-functionalized side chains. Utilizing the first strategy, Tang and coworkers [137] tried to modulate the charge density of IPT core by introducing 2-ethylhexyloxy (INO-4F), 2-ethylhexyl (IN-4F), and chlorine (IPCl-4F), respectively, onto the β position of thiophene (Figure 4.17). With the enhanced electron-withdrawing ability of the substituents, the LUMO energy level of the acceptors gradually deepened while the HOMO energy level did not show a linear trend, corresponding to gradually narrowing E_g^{opt}s. All the films blended with PM6 displayed a face-on orientation with diversity intensities of (010) diffraction peaks, suggesting side chain engineering onto core seriously affects the intermolecular π–π stacking. In addition, the INO-4F and IPCl-4F exhibited poor miscibility along with their blends' enhanced π–π stacking trends for strong crystallinity, which may influence exciton dissociation and charge transport process. In contrast, PM6:IN-4F blend film exhibited CCL of 7.53 nm in in-plane direction, much smaller than that of PM6:IPT-4F-based blend (24.3 nm), indicating decreased crystallinity and restricted charge transport for poor J_{sc} and FF. As a result, 13.69% PCE was obtained with a V_{oc} of 0.93 V, a J_{sc} of 20.46 mA cm^{-2}, and a FF of 0.720 (Table 4.7). The moderate PCE in relative to PM6:IPT-4F-based OSCs (14.62%) reminds us the energy level matching, molecular stacking, and miscibility should also be taken into consideration while manipulating the charge density.

Inspired by the second strategy and the successful application of conjugated side chains on polymers, Tang and coworkers [138] reported a novel asymmetric FREA called IPT2F-TCl by replacing *n*-octyl with 2D chlorinated hexylthiophene side chain. In relative to IPT-2F, the IPT2F-TCl has deeper energy levels and wider E_g^{opt} resulting from the deflected 2D chlorinated side chain. When fabricated as a device, IPT2F-TCl and the polymer donor PBDB-T exhibit better miscibility with homogeneous and nanofibrous phase separation morphology than IPT-2F-based counterpart. Consequently, the PBDB-T:IPT2F-TCl-based OSCs delivered a PCE of 13.74% with an impressively high FF of 0.775, and the FF is among the highest values reported in the literature for asymmetric FREAs-based OSCs. The results demonstrate that 2D-chlorinated hexylthiophene side chain has great potential in designing FREAs for morphology manipulation to obtain a satisfactory FF.

Firstly applying *N*-functionalized 2D side chain into asymmetric FREAs design, the same group further developed three novel FREAs attaching different 2D side chains onto DTP block in IPT core, including thiophene (IPT2F-Th), benzene (IPT2F-Ph), and thieno[3,2-*b*]thiophene (IPT2F-TT) (the structures shown in Figure 4.17) [139]. All three 2D FREAs have strong absorption at 600–870 nm, corresponding to E_g^{opt} of 1.45 eV for IPT2F-TT, 1.46 eV for IPT2F-Ph, and 1.47 eV for IPT2F-Th, respectively. In addition, the three two-dimensional materials have almost the same energy levels. The twisting of *N*-conjugated side chains from backbone endows three FREAs with similar energy levels and light absorptions (~850 nm edge) (Figure 4.19c). When PBDB-T was selected as the donor, the PBDB-T:IPT2F-TT-based blend displayed higher charge mobility, more orderly phase separation, and more suitable intermolecular π–π stacking distance than other two counterparts. Consequently, the device based on PBDB-T:IPT2F-TT achieved a PCE of 14.02% and a FF of 0.751, which is better than PBDB-T:IPT2F-Ph (PCE of 13.13% and FF of 0.721), PBDB-T: IPT2F-Th (PCE of 12.52% and FF of 0.712), and PBDB-T: IPT-2F (PCE of 14.00% and FF of 0.724). The 14.02% PCE and 0.751 FF are the highest values of 2D-FREAs-based OSCs reported in the literature so far. More interestingly, the PBDB-T:IPT2F-TT-based devices surprisingly have high PCE retention rates of 90.22% and 86.67%, respectively, after 100 hours of continuous thermal aging or light aging, showing good thermal and light soaking stability (Figure 4.19d). This work proved that the *N*-functionalized 2D side chain modification on the IPT core can effectively improve device performance, and

TT is better than benzene and thiophene. Besides, diverse 2D side chains lead to differences in stability, which provides ideas for subsequent use of simple molecular engineering to improve device stability.

For further extending the length of fused-ring core for stronger absorption, a thiophene unit was added into the IDTT core to synthesize a new eight-membered asymmetric core. The end groups properties relationship was explored, and three asymmetric acceptors named TTPTTT-IC, TTPTTT-2F, and TTPTTT-4F were thus designed and synthesized (Figure 4.17) [140]. The increase of fluorine atoms on the end groups deepens the energy levels and narrows the E_g^{opt}s. Morphology studies showed that all three films blended with PBT1-C presented a face-on π–π stacking. And compared with TTPTTT-IC based one, the fluorinated TTPTTT-2F and TTPTTT-4F-based blends exhibited more clear nanofiber phase separation with small RMS of 1.46 and 1.26 nm, respectively. The resulting PBT1-C:TTPTTT-4F-based OSCs delivered a PCE of 12.05% with a V_{oc} of 0.86 V, a J_{sc} of 19.36 mA cm^{-2}, and a FF of 0.721, outperforming PBT1-C:TTPTTT-IC (7.91%) and PBT1-C:TTPTTT-2F (11.52%)-based counterparts (Table 4.7). The salient properties of PBT1-C:TTPTTT-4F can be attributed to the high J_{sc} induced by red-shifted absorption.

Different from extending the conjugation in the horizontal direction, Yang and coworkers [141] inserted a 2D alkyl-thiophene side chain into asymmetric core to generate the a-BTTIC (Figure 4.17). The introduction of 2D side chain not only extends asymmetrical conjugation backbone with a narrow E_g^{opt} of 1.43 eV but also offers a large dipole moment of 7.88 debye for stronger intermolecular dipole–dipole interaction and the promoted antiparallel molecule arrangement. Featuring clear phase separation and a well-developed nanostructured interpenetrating network, the PBDB-T:a-BTTIC-based OSCs achieved a high PCE of 13.60% with a V_{oc} of 0.90 V, a J_{sc} of 20.31 mA cm^{-2}, and a FF of 0.740, which is the highest PCE for commercialized PBDB-T-based OSCs (Table 4.7). The E_{loss} was found to be as low as 0.53 eV. More novelly, Yang and coworkers [142] convert IDTT symmetric core into asymmetric core by repalcing thieno [3,2-*b*]thiophene with thieno [2,3-*b*]thiophene. End capping with CPTCN, a novel asymmetric acceptor ITCNTC was synthesized. Compared with IDTT-cored ITCPTC (2.05×10^5 mol^{-1} cm^{-1}, E_g^{opt} = 1.58 eV), ITCNTC has a ca. 47 nm blue-shifted λ_{onset} with molar absorption coefficient of 1.80×10^5 mol^{-1} cm^{-1}, corresponding to a wider E_g^{opt} of 1.68 eV. In addition, the ITCNTC shows upshifted LUMO and down-shifted HOMO, that is, wider electrochemical bandgap. The ITCNTC featuring strong crystalline resulted in the oversized phase separation in J71:ITCNTC blends, which is not conductive to the charge transport and collection. The derived J71:ITCNTC OSCs only obtained a moderate PCE of 8.52% (Table 4.7). The poor PCE in relative to J71:ITCPTC (11.63%) can be ascribed to low J_{sc} and FF induced by weak absorption and poor morphology.

In terms of molecular design, side chains are critical as a manipulator in morphology and intermolecular stacking. For example, Yang's group performed side chain engineering on the asymmetric TPTT core mentioned above, and two FREAs were synthesized by embedding different side chains on cyclopentadiene, namely, IDT6CN (phenylhexyl substitution) and IDT6CN-Th (hexylthiophene substitution) (Figure 4.17) [143]. Obviously, c. 12 nm red-shifted λ_{onset} can be observed in neat IDT6CN-Th film with a slightly molar extinction coefficients of 2.10×10^5 mol^{-1} cm^{-1} in relative to IDT6CN film. Besides, IDT6CN-Th showed deeper HOMO/LUMO energy levels of −5.71 eV/−4.01 eV than IDT6CN (−5.68 eV/−3.97 eV). After adding 0.5% additives and thermal annealing at 110 °C for three minutes, the optimal PBDB-T:IDT6CN-Th-based device achieved a PCE of 10.41% with a V_{oc} of 0.81 V, a J_{sc} of 16.75 mA cm^{-2}, and an impressive FF of 0.767, higher than that of IDT6CN-based counterpart (9.27%) (Table 4.7). The excellent morphology accompanied with nanofibrous interpenetrating network accounts for the prominent FF.

Moreover, asymmetric side chains strategy was employed into molecule design to synthesize a novel FREAs named IDT-OB [76]. Compared to symmetric phenylhexyl substituted IDT-2B and *n*-octyl substituted IDT-2O, IDT-OB has similar energy levels and strong absorption in between. Due to the stronger dipole–dipole interaction of asymmetric acceptors, a more ordered intermolecular arrangement will be formed to promote charge transport, which provides the PBDB-T:IDT-OB blend a higher μ_e of 2.71×10^{-4} cm^2 V^{-1} s^{-1} and more balanced μ_e/μ_h of 1.30 for high J_{sc} and PCE. Morphology studies indicated that PBDB-T:IDT-2O and PBDB-T:IDT-OB blends behave prioritized face-on (010) π–π stacking in the OOP direction, while the face-on and edge-on crystallites co-exist in PBDB-T:IDT-2B-based blend film. In addition, the strong crystallinity of IDT-2O causes its corresponding blend strong aggregation for poor charge transport. In comparison, the fiber structure with a diameter of 10–20 nm can be observed in PBDB-T:IDT-OB blend films, suggesting the formation of excellent charge transport network. As a result, the as-cast device based on PBDB-T:IDT-OB exhibited a champion PCE of 10.12% with a highest J_{sc} of 16.18 mA cm^{-2}. More interestingly, the PBDB-T:IDT-OB-based OSC is insensitive to film thickness, and a PCE of 9.17% can be maintained when the thickness is over 210 nm.

Asymmetric side chains have been proved to be effective means to improve morphology, intermolecular packing and aggregation behavior, and photoelectric properties. However, the side chain modification has little effect on the energy levels, which limits us to screen suitable donors to match. In view of the electron-withdrawing A unit's powerful manipulation of energy levels as well as absorption and the advantages of asymmetric molecules, Hou and coworkers [144] used a one-pot method to prepare IDTT-cored asymmetric acceptor IT-3F (Figure 4.17) with asymmetric end groups in a 30% yield. For comparison, the other two symmetric FREAs were synthesized end-capping IC-F (IT-2F) and IC-2F (IT-4F). All three FREAs show strong absorption at 550–800 nm with similar molar absorption coefficient of $\sim 1.2 \times 10^5$ mol^{-1} cm^{-1}. The energy levels are deepened accompanied with the increased fluorine atoms. The derived binary PM6:IT-3F OSCs achieved a higher PCE of 13.83% than IT-2F- and IT-4F-based cells (12.65% and 13.62%, respectively) (Table 4.7). When IT-2F and IT-4F are selected to fabricate ternary OSCs, an enhanced PCE of 14.04% was obtained. The transmission electron microscope (TEM) results obviously showed that the ternary blend films have little difference from three binary ones, suggesting good compatibility of three above FREAs. In this case, diversity IDTT-cored FREAs (IT-M, ITIC, IT-2F, IT-4F, and IT-4Cl) are used to explore the preparation of high-performing multi-component OSCs. Consequently, the optimal six-component OSCs (PM6:IT-M:ITIC:IT-2F:IT-4F:IT-4Cl = 1 : 0.05 : 0.05 : 0.05 : 0.8 : 0.05) delivered a PCE of 13.26% with improved J_{sc} of 21.02 mA cm^{-2}.

4.3.3.4 Non-fused Cores

The development of FREAs typically adopted A–D–A type has promoted continuous breakthroughs in efficiency. Featuring fused ladder structures (usually over five rings) for potentially reducing the reorganization energy of molecules, the derived FREAs have expanded from 5 rings to 11 rings to utilize solar photos from visible to NIR. As thus, some efficient FREAs with medium/narrow E_g^{opt}s (ITIC [72], INPIC [90], IMC6-4F [94], etc.) have been developed and the PCE has exceed 18% [145]. However, the commercial application of organic photovoltaics needs to take some other factors into account such as synthesis complexity and cost consumption. Apart from considerable PCE, the FREAs' complex synthesis and high cost will undoubtedly limit its further development. For example, the synthesis of the simplest IDT-cored FREA already has over four steps, which may also face low yield. To solve this challenge, acceptors with non-fused cores have caught our attention and began to shine. Derived from fused IDT, a novel non-fused core was developed by flanking two thiophene units on the benzene. End capping with IC end groups, a non-fused-ring acceptor (NFRA) named Ph-IC was synthesized (Figure 4.20) [146]. In Ph-IC, long alkoxy chain ensures the solubility of acceptor, while O–H and O–S non-covalent interactions limit the rotation of chemical bond for good backbone co-planarity. In terms of optical properties, Ph-IC has a significantly deep HOMO of −5.72 eV and shallow LUMO of −3.79 eV, which is not conducive to hole transport and high $V_{oc.}$ The resultant OSCs blending with PBFTZ only presented a moderate PCE of 5.68% (Table 4.8).

Setting aside prejudices about low efficiency of Ph-IC, Chen and coworkers [147] adopted its simple structure for further molecular optimization. End-capped with IC-2F, several simple NFRAs were thus synthesized by introducing alkyl side chain at β position of thiophene including hydrogen (PTICH), hexyl (PTIC), and hexyloxy (PTICO) (Figure 4.20). As displayed in Figure 4.21a, single-bond-linked NFRAs will cause different conformations due to the rotation of the bond. For PT rotamer (connecting fragments of benzene and thiophene), O–H geometry has a small rotation energy barrier of 4.4 kJ mol^{-1} than O–S geometry, suggesting O–H geometry is the energetically preferential conformation. Nevertheless, this O–H dominant conformation is thermodynamically convertible, and the hydrogen bond can be broken when the PT fragmentation barrier is overcome by heating up, which has been proven by temperature-dependent ^{1}H NMR (Figure 4.21b). For TIC rotamer (connecting fragments of thiophene and 2F-IC), rotamer based on substituted thiophene and 2F-IC exhibited a higher energy barrier of 45 kJ mol^{-1} than thiophene-based counterpart (6.25 kJ mol^{-1}), and the terminal single bond shows less rotation tendency. When choosing polymer PM6 as a donor, the PM6:PTIC achieved a

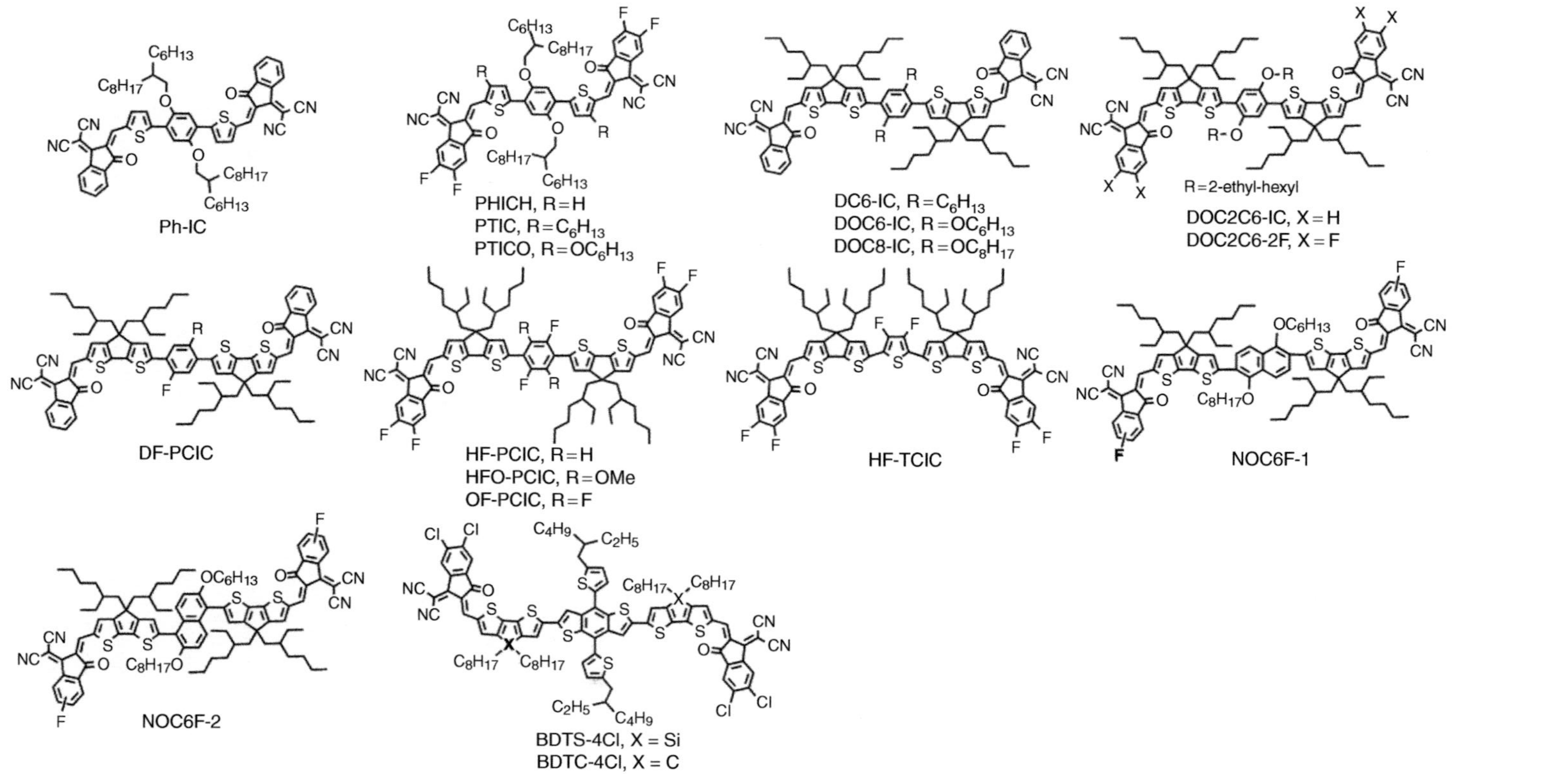

Figure 4.20 Chemical structures of non-fused acceptors. Source: J. Cao and W. Tang.

Table 4.8 Optical bandgap, energy levels, and photovoltaic parameters for A–D–A-type non-fullerene SMAs with non-fused cores.

SMA	E_g^{opt} [a)] (eV)	HOMO/ LUMO[b)] (eV)	Donor	V_{oc} (V)	J_{sc} (mA cm^{-2})	FF	PCE (%)	References
Ph-IC	1.62	−5.72/−3.79	PBFTZ	0.89	12.37	0.515	5.68	[146]
PTICH	1.60	−5.62/−3.80	PM6	0.92	8.22	0.540	4.08	[147]
PTIC	1.53	−5.59/−3.81	PM6	0.93	16.73	0.660	10.27	[147]
PTICO	1.56	−5.48/−3.66	PM6	1.01	12.60	0.520	6.62	[147]
DC6-IC	1.69	−5.53/−3.79	PBDB-T	0.99	11.19	0.622	6.87	[148]
DOC6-IC	1.43	−5.33/−3.72	PBDB-T	0.91	19.21	0.601	10.52	[148]
DOC8-IC	1.39	−5.36/−3.71	PBDB-T	0.92	17.74	0.577	9.41	[148]
DOC2C6-IC	1.44	−5.38/−3.73	PBDB-T	0.93	18.85	0.633	11.10	[148]
DOC2C6-2F	1.42	−5.49/−3.83	PBDB-T	0.85	21.35	0.732	13.24	[148]
DF-PCIC	1.59	−5.49/−3.77	PBDB-T	0.91	15.66	0.720	10.14	[149]
HF-PCIC	1.50	−5.53/−3.83	PM6	0.91	17.81	0.708	11.49	[150]
HFO-PCIC	1.48	−5.50/−3.81	PM6	0.93	12.62	0.710	8.36	[150]
OF-PCIC	1.59	−5.66/−3.86	PM6	0.91	13.76	0.734	9.09	[150]
HF-TCIC	1.44	−5.43/−3.83	PBDB-T	0.76	20.04	0.650	9.86	[151]
NOC6F-1	1.58	−5.55/−3.77	PBDB-T	0.95	17.08	0.658	10.62	[152]
NOC6F-2	1.68	−5.50/−3.72	PBDB-T	0.96	13.21	0.533	6.74	[152]
BDTS-4Cl	1.46	−5.45/−3.82	PBDB-T	0.83	9.80	0.459	3.73	[153]
BDTC-4Cl	1.42	−5.35/−3.75	PBDB-T	0.86	18.56	0.595	9.54	[153]

a) Calculated from $E_g^{opt} = 1240/\lambda_{onset}$.
b) Calculated from CV curves.

champion PCE of 10.27% with a high a V_{oc} of 0.93 V, a J_{sc} of 16.73 mA cm^{-2}, and an impressive FF of 0.660, outperforming PTICH- and PTICO-based counterparts (4.08% and 6.62%, respectively). The high J_{sc} can be attributed to the narrower E_g^{opt} of PTIC than the other two. However, the Stokes shift (SS) measurement showed the SS values of NFRAs are higher than that of FREA ID-4F in solution, suggesting NFRAs are prone to more non-radiation energy loss. And this loss is significantly reduced due to changes in molecular packing after film formation.

In order to adjust packing for low energy loss, selecting cyclopentadithiophene (CPDT) as flanked building blocks, Bo and coworkers [148] thus designed a series of NFRAs by attaching diversity alkyl chains into intermediate benzene including hexyl (DC6-IC), hexyloxy (DOC6-IC), octyloxy (DOC8-IC), and 2-ethyl-hexyl (DOC2C6-IC) (Figure 4.20). All three alkoxy-substituted NFRAs presented upshifted energy levels than DC6-IC. In addition, O–S intramolecular interaction locks planar conformation to promote π-electron delocalization for extended

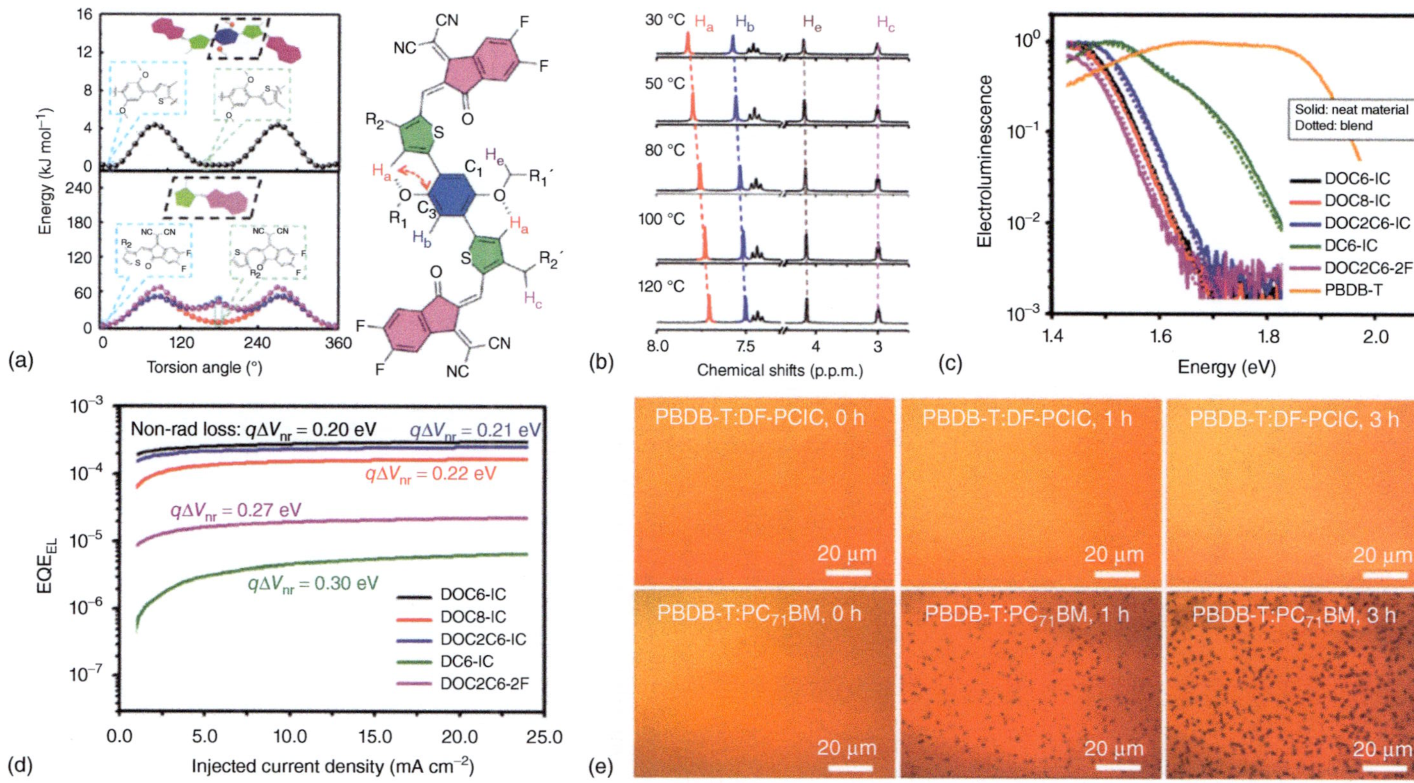

Figure 4.21 (a) Possible rotamers and their energy–torsion angle ($E-\theta$) curves, and the preferential conformation of PTIC. (b) Temperature-dependent NMR of PTIC in d_4-$C_6D_4Cl_2$. Source: Yu et al. [147]. Licensed Under CC BY- 4.0. (c) Electroluminescence from solar cells based on different acceptors, recorded using an injection current of 1 mA. (d) EQE_{EL} of the solar cells based on different acceptors, and the non-radiative recombination losses ($q\Delta V_{nr}$) calculated from EQE_{EL}. Source: Huang et al. [148]. Licensed Under CC BY- 4.0. (e) Optical microscopy images of PBDB-T:DF-PCIC and PBDB-T:PC_{71}BM-blended films after thermal treatment at 180 °C for various times. Source: Li et al. [149]/Reproduced with permission of JOHN WILEY & SONS, INC.

conjugation, endowing alkoxy-substituted NFRAs narrow E_g^{opt} (1.39–1.44 eV). Morphology analysis found that alkoxy-substituted NFRAs showed shorter π–π stacking distance of 1.7 Å^{-1} than DC6-IC (3.7 Å), and preferred face-on orientation can be observed only in DOC6-IC and DOC2C6-IC neat films. Besides, more efficient exciton dissociation can occur in PBDB-T:DOC6-IC and PBDB-T:DOC2C6-IC due to their smaller domain size than the other two devices. As a result, the PBDB-T:DOC2C6-IC-based OSCs delivered a highest PCE of 11.10% with a high V_{oc} of 0.93 V and a J_{sc} of 18.85 mA cm^{-2}, exceeding the PCEs (6.87–10.52%, V_{oc} over 0.90 V) of several others (Table 4.8). It can be noted that these high V_{oc}s at a solar cell with around 900 nm NIR absorption is rare, and the energy loss analysis was thus carried out. As shown (Figure 4.21c,d), emission of the neat film and their blends are approximative, and no apparent subgap absorption feature was observed, indicating the occurrence of small energy loss. The non-radiative recombination losses of NFRAs-based OSCs were finally calculated to be around 0.2 eV (0.3 eV for DC6-IC-based device), which is comparable to inorganic solar cells. Then the difluorinated end group strategy was employed to strengthen ICT effect and absorption, and an outstanding PCE of 13.24% was obtained by PBDB-T:DOC2C6-2F-based device with an enhanced J_{sc} of 21.35 mA cm^{-2} (Table 4.8).

Coincidentally, Chen's group [149] designed a novel NFRA named DF-PCIC (Figure 4.20) by inserting fluorine atoms onto the intermediate benzene. The DF-PCIC exhibited HOMO/LUMO energy levels of −5.48 eV/−3.77 eV and a narrow E_g^{opt} of 1.59 eV. Besides, DF-PCIC features a low crystallinity, which can be strengthened after the thermal annealing treatment for higher J_{sc} and FF induced by the formation of bicontinuous crystalline networks in corresponding blend films. The optimal PCE of 10.14% can be achieved under the condition of 0.5% CN additive and 130 °C annealing temperature. More interestingly, after thermal annealing at 180 °C for three hours, uniform film can be preserved in PBDB-T:DF-PCIC-based blend (Figure 4.21e), suggesting good thermal stability. In contrast, PBDB-T:$PC_{71}BM$ blend appeared granular agglomerations which is unbearable to stand the high temperature for a long time. For further extending photos harvesting, 2F-ICs are end-capped into the above core DF-PCIC used, and the HF-PCIC was developed [150]. In order to explore the effect of intermediate core unit on device properties, the other two NFRAs are synthesized as comparison by substituted two fluorine atoms (OF-PCIC) and two methoxy groups (HFO-PCIC) (Figure 4.20). In relative to HF-PCIC (−5.53 eV/−3.83 eV for HOMO/LUMO), OF-PCIC has deeper energy levels (−5.66 eV/−3.86 eV) due to the introduced two strongly electronegative fluorine atoms on intermediate benzene, while the methoxy slightly upshifted the energy levels of HFO-PCIC (−5.50 eV/−3.81 eV). As film formation, all three NFRAs have a strong absorption at 500–850 nm with molar absorption coefficient of ~2.00 × 10^5 M^{-1} cm^{-1}, and the HFO-PCIC shows the maximum λ_{onset} of 838 nm (E_g^{opt} of 1.48 eV). Single-crystal data indicated that the F–H non-covalent effect is the first choice, but it also includes the coexistence of other non-covalent effects, which affects the dihedral angles between benzene and CPDT for diversity aggregation behavior. When mixed with PM6, the highly crystalline HFO-PCIC caused a severe aggregation of the blend film, which destroyed the interpenetrating network constructed by the donor, resulting in the poor PCE of 8.36% in corresponding device. On the contrary, more smooth surfaces can be observed in HF-PCIC- and OF-PCIC-based blends with small RMS, and the PM6:HF-PCIC-based OSCs delivered a champion PCE of 10.97% with a V_{oc} of 0.92 V, a J_{sc} of 16.01 mA cm^{-2}, and a FF of 0.724. Considering the weak crystallinity of HF-PCIC and the low boiling point of chloroform (CF), CF was used instead of chlorobenzene (CB) for solution spin coating to improve crystallinity. The GIWAXS showed that the PM6:HF-PCIC blend film spin coating with CF presented stronger lamellar peaks, suggesting enhanced crystallinity for optimized J_{sc} of 17.81 mA cm^{-2}. The PCE was thus increased to 11.49% (Table 4.8). After that, a novel NFRA named HF-TCIC was developed by Chen and coworkers with 3,4-difluorothiophene as central core [151]. The electron cloud distribution displayed that HOMO mainly located at the intermediate cores and LUMO is determined by the end groups at both ends. Therefore, in relative to HF-PCIC, the weakened electron-donating core endows HF-TCIC deeper HOMO energy level (~0.1 eV), corresponding to a wider E_g^{opt} of 1.60 eV. With an optimal donor: acceptor weight ratio (1 : 1.2), the device fabricated with PBDB-T:HF-TCIC active layer achieved a moderate PCE of 9.86% after adding 0.5% 1-chloronaphthalene (CN) and appropriate thermal annealing at 120 °C.

It is noted from above that the modification of core is one of the most direct and effective methods for manipulating device properties. In addition, different from the rigid skeleton of FREAs, the modules making up the non-fused-ring skeleton are prone to a certain deflection due to steric hindrance. Therefore, side chain engineering on the core can also adjust the dihedral angle between the modules to change molecular packing for efficient NFRAs. Apart from above two, isomerization is also a strategy worth trying. As thus, linking CPDT at the 2,6- and 1,5-positions of naphthalene, two mutually isomer acceptors, NOC6F-1 and NOC6F-2 (Figure 4.20), were designed and synthesized by Bo and coworkers [152]. The steric hindrance induced by diverse connection modes provides these two isomers with distinguishing intermolecular π–π stacking, leading to different absorption in the solid state. The as-developed NOC6F-1 thus has a ca. 48 nm red-shifted absorption than NOC6F-2 (736 nm), corresponding to a narrower E_g^{opt} of 1.58 eV. Morphological study revealed that both films blended with PBDB-T exhibited nanostructured interpenetrating networks, and more compact π–π stacking can be observed in NOC6F-1-based blend. As a result, more efficient charge carrier generation and transportation were achieved for the PBDB-T:NOC6F-1, leading to a higher J_{sc} of 17.08 mA cm^{-2}, a FF of 0.658 and thus a higher PCE of 10.62% (Table 4.8). NOC6F-2-based OCSs obtained only 6.74% PCE. These results demonstrated that isomerization is also an effective way to explore high-performing NFRAs.

It is well known that the elements of the same main group have similar chemical properties, and the larger atomic number has a larger atomic radius, which is more conducive to electron delocalization. Therefore, 4,4′-dioctyl-dithieno[3,2-*b*:2′,3′-*d*]silole (DTS) was employed as flanked blocks. To prevent over-intermolecular aggregation, Chen and coworkers utilized 4,8-bis(5-ethylhexyl-2-thienyl)-benzo[1,2-*b*:4,5-*b*′]dithiophene (BDTT) as the central core, and two acceptors BDTS-4Cl and BDTC-4Cl (CPDT flanked) were synthesized (Figure 4.20) [153]. Both

NFRAs exhibited a strong absorption to NIR, and the BDTC-4Cl even has a λ_{onset} of 873 nm with a narrower E_g^{opt} of 1.42 eV. Consequently, more balanced $\mu_h/\mu_e = 1.61$ was obtained (3.13 for BDTS-4Cl-based blend), contributing to the PBDB-T:BDTC-4Cl-based OSCs a PCE of 9.54% with a V_{oc} of 0.86 V and a J_{sc} of 18.56 mA cm^{-2}. The poor efficiency of PBDB-T:BDTS-4Cl (3.73%) can be attributed to the low V_{oc} and J_{sc} induced by the deeper LUMO energy level and wider E_g^{opt} of BDTS-4Cl. After adding 33.3% $PC_{71}BM$ as a third component, the morphology with nanofibrous interpenetrating network is clearly visible, and the enhanced J_{sc} and FF increase the PCE to 12.19%.

4.4 A–DA′D–A–Type SMAs

As mentioned above, A–D–A-structured SMAs are extremely promising since they exhibit strong intramolecular charge transfer effect, superior energy level tunability, and good molecular crystallinity. Based on this successful design strategy, in 2017, Zou et al. developed a series of curve-shaped NFAs with an A–DA′D–A structure by inserting electron-withdrawing unit, such as benzo[*c*][1,2,5]thiadiazole (BT) and benzo[*d*][1,2,3]triazoles (BTA), into the donor core as its fused-ring central unit.

In order to distinguish the A–D–A-type molecules, Zou and colleagues summarize several important features of A–DA′D–A-type SMAs: (i) the single crystal of the A–DA′D–A molecules confirms that the molecules configuration is an excellent combination of a ladder-type electron-deficient-core-based central fused-ring (DA′D), electron-accepting end group (A), and two sp^2-hybridized nitrogen atoms in the pyrrole motif of the fused ring in a C2 symmetric manner. Thus, the alkyl side chains onto the *N*-atoms of the pyrrole motif could greatly influence the stacking of adjacent molecules; (ii) the slight twist DA′D core has to be sterically hindered to prevent over aggregation and, at the same time, maintain an efficient intramolecular interaction due to the multiple D–A interactions. Compared with the A–D–A molecules (like ITIC-series), it do not need to synthesize spiro-like structures; instead, the sterically hindered side chains adopt an orthogonal conformation to the main plane; (iii) with the specific molecular geometry of the A–DA′D–A family, the dimerization could be formed in a unit with a large quadrupole moment. Moreover, the relatively narrow density of state (DOS) both in the electron affinity and ionization energy could lead to a trap-free ambipolar transport. Such a high quantum efficiency in A–DA′D–A-type SMAs, coupled with high electroluminescence quantum efficiency, makes it possible to deliver high J_{sc}, high FF, and also low voltage losses at the same time under one sun conditions.

Now we know that one of the most important features in design strategy of A–DA′D–A-type SMAs is to introduce an electron-deficient segment in the middle of the central core to form a fused DA′D structure. Therefore, in this contribution, we will systematically review A–DA′D–A-type SMAs based on different electron-deficient segments (A′), such as BTA, BT, benzo[*c*][1,2,5]selenadiazole (BSe), and quinoxaline (Qx).

4.4.1 BTA-Based A–DA′D–A SMAs

Since the first A–DA′D–A pentacyclic SMA BZIC [154] by inserting an electron-withdrawing unit (benzo[*d*][1,2,3]triazoles, BTA) in the core of its fused-ring central unit reported in 2017, the electron-deficient-core-based fused ring opens a new door in the molecular design of high-performance acceptors. Compared to the classical A–D–A-type SMA ITIC [72], BZIC showed a red-shift absorption (c. 80 nm) and higher absorption coefficient. The wide absorption range makes the active layers absorb more photons, giving rise to a high J_{sc}.

After that, Yang and Zou extended the fused central ring by using a thieno[3,2-*b*] thiophene instead of a thiophene segment, and synthesized a new SMA Y1 (Figure 4.22) in collaboration with Prof. Gao's group from Linköping University [155]. By introducing the central fused ring from pentacyclic to heptacycilc, the HOMO energy level and absorption of Y1 match well with those of the commercial donor polymer PBDB-T. As a result, the PBDB-T:Y1-based device afforded a high PCE of 13.40% with an impressive J_{sc} of 22.44 mA cm^{-2}, a V_{oc} of 0.88 V, and an FF of 69.1% (Table 4.9). The high performance was attributed to the relatively high electroluminescence quantum efficiency (EQE$_{EL}$) of approximately 0.5×10^{-4}, which indicated low nonradiative recombination loss of the blend. Later, in order to further narrow the bandgap of Y1 to obtain more photons, Y2 and Y3 (Figure 4.22) were designed and synthesized by using the dicyanomethylene derivatives (CPTCN and INCN-2F) [156]. By modifying the end group of Y1, the target SMAs Y2 and Y3 exhibited red-shifted absorption onset of 25 and 43 nm, respectively. The Y2- and Y3-based devices showed the similar PCE of 13.40% and 13.24% with the J_{sc} of 23.56 and 24.10 mA cm^{-2} when matched with polymer donor PBDB-T and PM6, respectively. Although the red-shift of the SMA absorption can reduce the bandgap for high J_{sc}, the V_{oc} of the devices will also decrease concomitantly. Hence, balancing this trade-off between J_{sc} and V_{oc} is a challenge.

To compensate for the trade-off between J_{sc} and V_{oc}, side chain engineering is an effective strategy for manipulating energy levels and meanwhile maintain the similar absorption. Wu and Zou added two "wings" (undecyl side chains) onto the terminal of the DA′D core of Y3 to obtain a new SMA Y11 [157]. The optical absorption onset of Y11-based active layer is sharp, with an Urbach energy of 25–27 meV, which is closer to the properties of classic inorganic semiconductors, such as crystalline silicon or organometallic lead-halide perovskite. The devices exhibited a high EQE$_{EL}$ of 0.14%, corresponding to a low nonradiative recombination loss of 0.17 eV. As a direct consequence, an improved PCE of 16.54% and a higher V_{oc} of 0.85 V with a significantly decreased voltage loss (V_{oc} loss) of 0.51 eV were achieved in the Y11-based device. Moreover, Y11 has high electron mobility due to the stable molecular conformation and rigid backbone. The mobility in OSC devices influences charge trans-

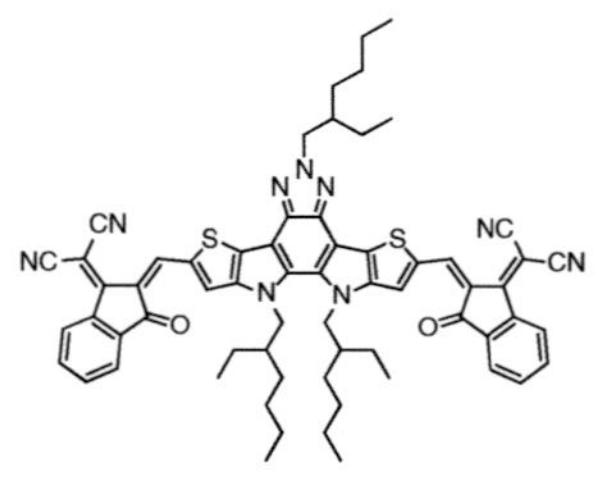

BZIC

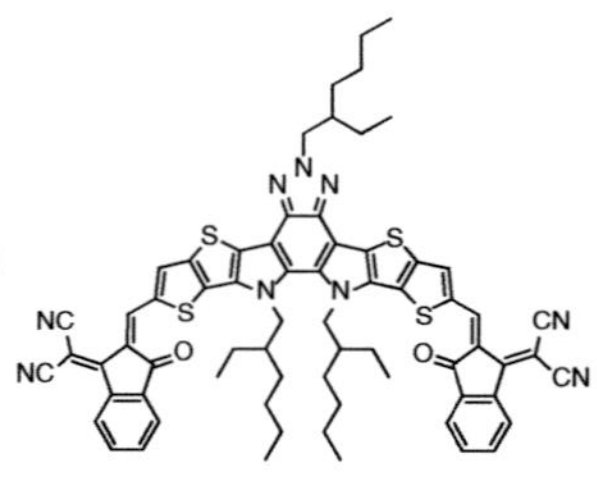

Y1

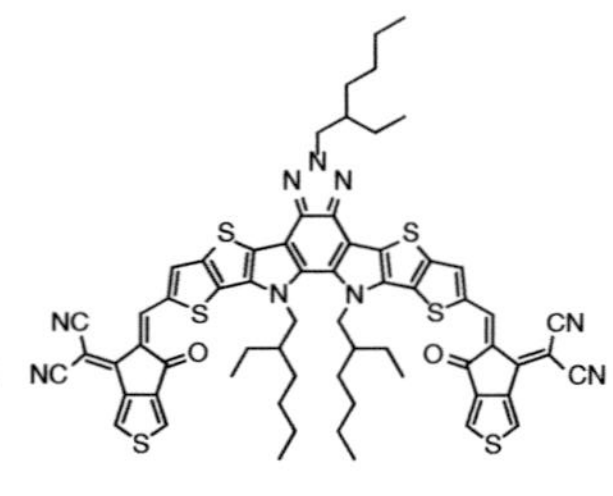

Y2

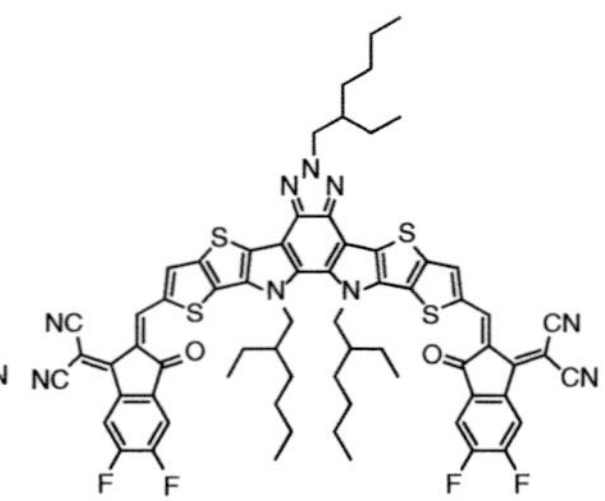

Y3

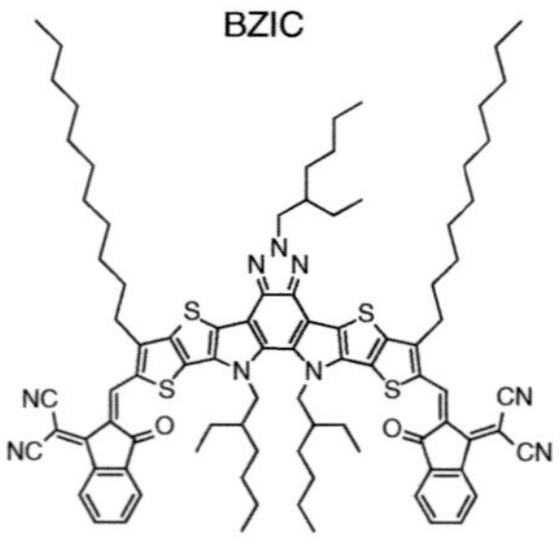

Y9

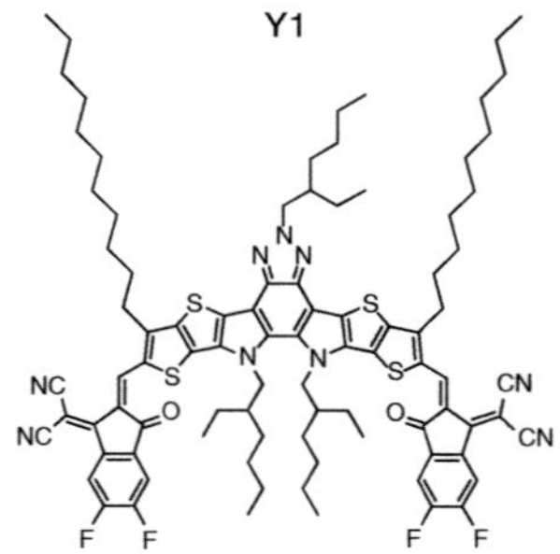

Y11

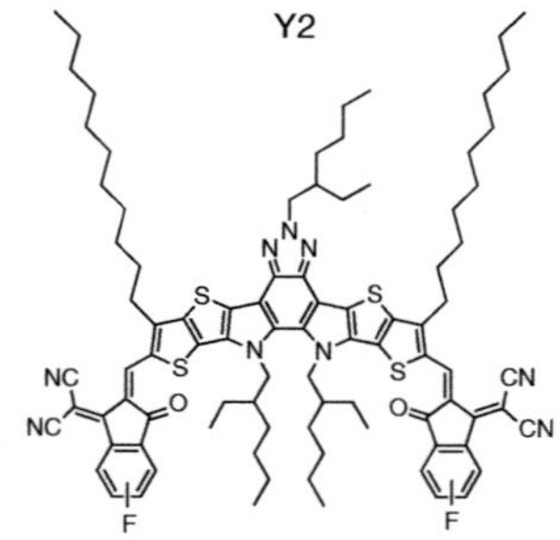

Y14

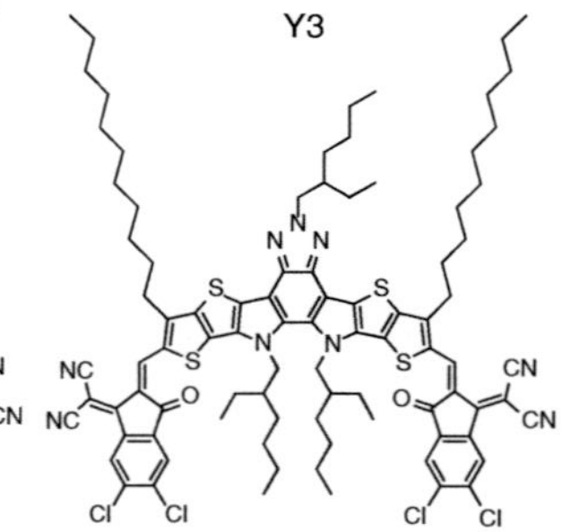

Y15

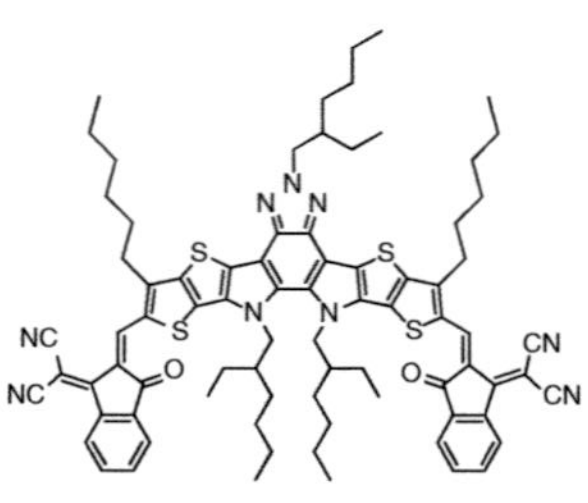

Y16

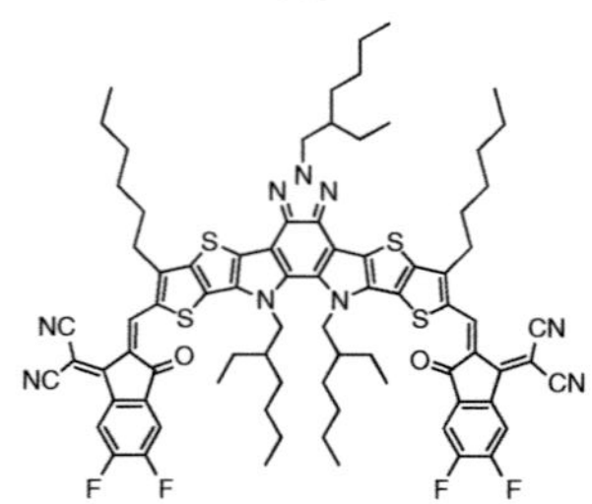

Y18

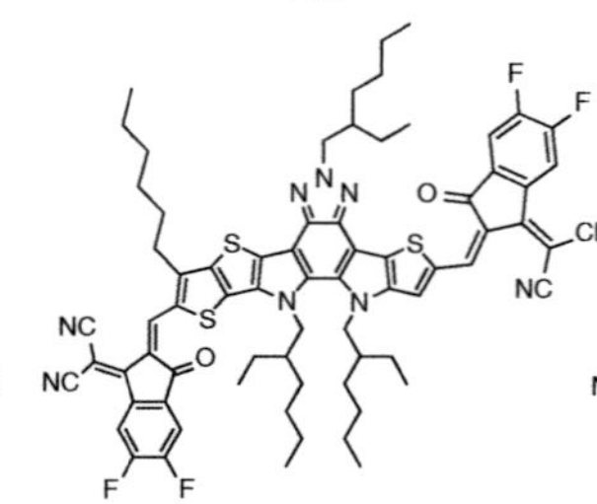

Y21

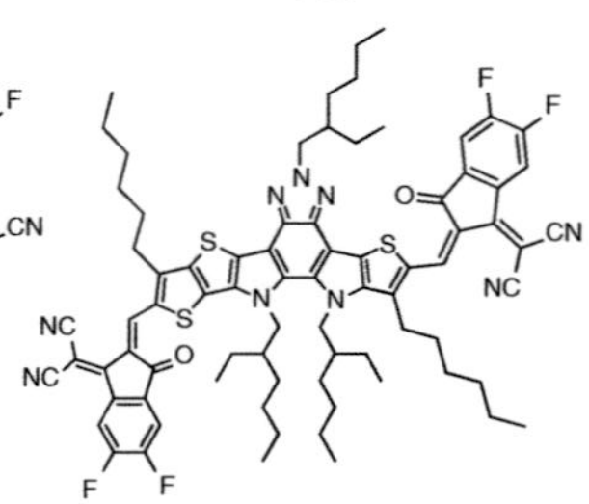

Y22

Figure 4.22 Chemical structures of BTA-based A–DA′D–A-type acceptors. Source: J. Yuan and Y. Zou.

Table 4.9 Optical bandgap, energy levels, and photovoltaic parameters for A–DA′D–A-type non-fullerene SMAs.

SMA	$E_g^{opt a)}$ (eV)	HOMO/LUMO[b] (eV)	Donor	V_{oc} (V)	J_{sc} (mA cm^{-2})	FF	PCE (%)	References
BZIC	1.45	−5.42/−3.88	HFQx-T	0.84	12.67	0.59	6.30	[154]
Y1	1.37	−5.45/−3.95	PBDB-T	0.87	22.44	0.69	13.42	[155]
Y2	1.34	−5.43/−4.04	PBDB-T	0.82	23.56	0.69	13.40	[155]
Y3	1.31	−5.56/−4.11	PM6	0.81	24.10	0.66	13.24	[156]
Y11	1.31	−5.69/−3.87	PM6	0.83	26.74	0.74	16.54	[157]
Y18	1.32	−5.58/−3.98	PM6	0.84	24.91	0.76	16.01	[158]
Y9	1.36	−5.59/−3.78	PBDB-T	0.90	23.28	0. 63	13.26	[159]
Y14	1.30	−5.56/−4.01	PBDB-T	0.79	26.15	0.71	14.92	[160]
Y15	1.30	−5.56/−3.93	PM6	0.88	23.79	0.68	14.13	[161]
Y16	1.30	−5.25/−3.66	PBDB-T	0.91	21.33	0.66	13.00	[162]
Y21	1.35	−5.65/−3.90	PM6	0.83	24.90	0.74	15.4	[163]
Y22	1.36	−5.69/−3.94	PM6	0.85	24.37	0.74	15.4	[164]
Y6	1.33	−5.65/−4.01	PM6	0.83	25.3	0.75	15.7	[109]
Y5	1.38	−5.55/−3.87	PBDB-T	0.88	22.80	0.70	14.10	[165]
BTP-4Cl	1.40	−5.68/−4.12	PM6	0.87	25.40	0.75	16.50	[166]
BTIC-4Br	—	−5.57/−4.11	PM6	0.85	20.67	0.69	14.03	[167]
BTP-2Br-m	—	−5.56/−4.07	PM6	0.88	25.03	0.73	16.11	[167]
TPT10	1.36	−5.52/−3.99	PTQ11	0.88	24.79	0.75	16.32	[168]
BTP-M	1.40	−5.61/−3.96	PM6	0.976	8.27	0.51	4.15	[169]
Y10	1.35	−5.56/−3.76	J11	0.89	21.21	0.71	13.46	[170]
BTTT-2Cl	1.36	−5.61/−3.98	PM6	0.90	23.80	0.65	13.83	[171]
BTTPC	1.39	−5.47-3.78	PBDB-T	0.89	22.25	0.73	14.51	[172]
BTTPC-Br	1.37	−5.54/−3.88	PBDB-T	0.86	24.71	0.71	15.22	[172]
BTPV-4F	1.21	−5.39/−4.08	PTB7-Th	0.65	28.30	0.66	12.10	[173]
BP-4F	—	−5.68/−3.88	PBDB-T-2Cl	0.88	27.13	0.76	14.62	[174]
BP5T-4F	1.34	−5.63/−3.88	PM6	0.89	24.60	0.76	16.70	[175]
ABP4T-4F	1.37	−5.65/−3.85	PM6	0.92	22.00	0.75	15.20	[175]
Y6-1O	1.43	−5.71/−3.84	PM6	0.89	23.2	0.78	16.10	[176]
Y6-2O	1.45	−5.73/−3.76	PM6	0.92	13.3	0.53	6.60	[176]
CH1007	1.30	−5.59/−3.97	PM6	0.82	27.03	0.72	15.96	[177]
N3	—	—/—	PM6	0.84	25.81	0.74	15.98	[178]
N4	—	—/—	PM6	0.82	25.01	0.70	14.31	[178]
N-C11	—	—/—	PM6	0.85	21.47	0.71	12.91	[178]
BTP-4F-12	—	−5.68/−4.06	PM6	0.85	25.30	0.76	16.40	[179]
DTY6	—	−5.67/−4.04	PM6	0.86	25.25	0.75	16.10	[180]
AQx-1	1.35	−5.59/−3.85	PM6	0.89	22.18	0.67	13.31	[181]
AQx-2	1.35	−5.63/−3.88	PM6	0.86	25.38	0.76	16.64	[181]
Y6Se	1.32	−5.70/−4.15	D18	0.84	27.98	0.75	11.70	[182]

a) Calculated from $E_g^{opt} = 1240/\lambda_{onset}$.
b) Calculated from CV curves.

port and recombination, consequently affecting the device performance, especially the FF. Although Y11 has made great success, the FF of Y11-based devices is only 73.1%, which still has much room for improvement. A high FF requires balanced mobilities of electrons and holes. To optimize the electron mobilities of Y11, Zou and coworkers performed molecular structure tailoring by reducing the alkyl chains from *n*-undecyl to *n*-hexyl on β position of the DA′D core that affords Y18 [158]. The reduced alkyl chains length of Y18-based BHJ blend showed more balanced hole and electron mobilities than Y11-based blend films, and therefore, resulting in the higher FF value of 76.4% in the OSCs. In order to further study the relation between the structure and properties, a series of alkylated BTA-based SMAs with various end groups (Y9, Y14, Y15, and Y16) were designed and synthesized [159–162].

The stability and cost issues are also important for commercial application of the OSCs. In the reduction of synthesis cost of the A–DA′D–A type SMAs, one way is the design and synthesis of molecules with more simple structure. Using the simple thiophene unit to replace the alkylated thieno[3,2-*b*]thiophene unit is the most intuitive approach to decrease the material cost of A–DA′D–A type molecules. Zou and coworkers reported six-membered-fused A–DA′D–A SMA Y21 by unidirectional replacement strategy to modify the DA′D core [163]. The six-membered-fused molecule Y21 can also determinate a broad absorption ranges with an optical gap of 1.35 eV. The OSCs based on Y21 blended with polymer donor PM6 exhibited a high PCE of 15.4% with a J_{sc} of 24.90 mA cm^{-2}, a V_{oc} of 0.83 V, and a FF of 0.744. Another six-membered-fused asymmetric SMA Y22 can achieve the same result [164].

4.4.2 BT-Based A–DA′D–A SMAs

The breakthrough in the efficiency of SMA-based OSCs emerged in early 2019. Zou and coworkers introduced the benzo[*c*][1,2,5]thiadiazole (BT) unit in the middle of the central fused core, to form a new BT-core-based fused unit dithienothiophen[3,2-*b*]-pyrrolobenzothiadiazole (TPBT), then, a star SMA molecule Y6 based on the TPBT central unit was designed (Figure 4.24) [109]. The Y6 molecules stimulate the progress of OSCs with remarkable PCE of 15–18%, pushing the OSCs filed to new phase [145, 183]. The single crystal of the Y6 molecule confirm that there are two types of π–π stacking between adjacent molecules, including the end group interactions and the central fused core interactions, as seen from the main view and side view of Figure 4.23 [184]. In addition, the specific molecular geometry (curve shape) of A–DA′D–A-type SMA Y6 could from a twisted 1D transport channel and slipped packing motif. Due to the dimerization in a crystal unit cell, the Y6 could form multi-transport channel, which is quite different from the ITIC-like molecules, resulting in a multi-dimension interpenetrating network and transport channels.

In recent years, significant efforts have been modified the molecular structure of Y6 to explore their photovoltaic applications and demonstrated their excellent morphological features, efficient charge generation, low voltage loss, and decent transport dynamic. The main optimizations of Y6 were focused on its end groups, central donor units, and side chains.

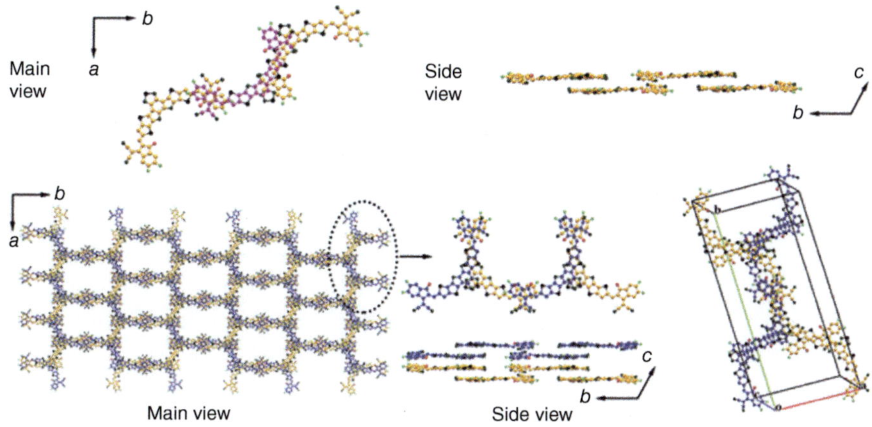

Figure 4.23 The single-crystal structure of Y6. Source: Zhu et al. [184]/Reproduced with permission of John Wiley & Sons.

The end groups play a vital role in determining their absorption, energy levels, and molecular packings, thus affecting the photovoltaic performance. The wide range of end groups developed for A–D–A-type SMAs provides useful guidelines for choosing the proper ending groups in the A–DA′D–A type SMAs. The current high-performance end groups applied in the A–DA′D–A type SMAs are featured with INCN and their similar structures. Since Zou and coworkers reported a SMA Y5 (composed of a DA′D core identical to Y6 and two INCN end groups, Figure 4.24) to achieve an impressive efficiency when blended with a common donor PBDB-T, the A–DA′D–A-type SMAs with different end groups (INCN derivatives) have sprung up for application in OSCs [165]. For instance, Hou and coworkers synthesized BTP-4Cl (Figure 4.24) containing the central core of Y6 and chlorinated INCN terminal that had a red-shifted absorption (c. 20 nm) and downshifted LUMO energy level (c. 100 meV) in comparison with Y6 [166]. The device based on PM6:BTP-4Cl showed a higher V_{oc} than Y6-based device, which was attributed to the lower non-radiative recombination loss. The simultaneously improved V_{oc} and J_{sc} led to an improved PCE of 16.5%. He and coworkers introduced bromination to further promote the halogen-substituted in Y-series molecules (Figure 4.24). However, BTIC-4Br exhibited a poor solubility in chloroform solution. Thus, BTIC-2Br-*m* was designed and synthesized by modulating the numbers of bromine substituents, and the device-based PM6:BTIC-2Br-*m* showed a high PCE of 16.11% [167]. In parallel, the similar SMA with mono-brominated INCN as the end groups, namely TPT10 (Figure 4.24), was reported by Li and cowokers. In order to match the energy levels of TPT10, they developed a new polymer donor PTQ11 [168]. The devices-based PTQ11:TPT10 films showed a high PCE of 16.32% in spite of a zero energy offset between HOMO energy levels of donor and acceptor, which was important in understanding the charge transfer process. Beside the electron-withdrawing halogen group, Chen and coworkers introduced the electron-donating methyl group on the INCN terminal and synthesized BTP-M (Figure 4.24) with higher energy levels and blue-shifted absorption [169]. By blending with the common donor

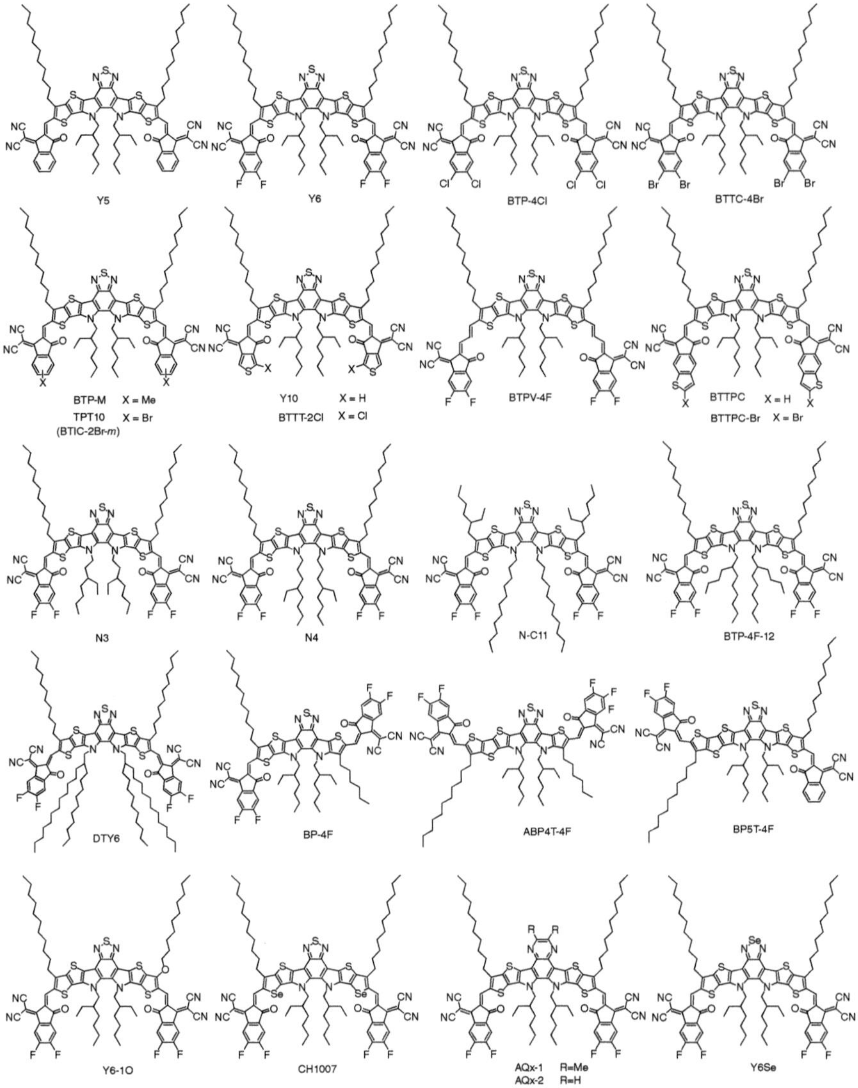

Figure 4.24 Chemical structures of BT/Qx/Bse-based A–DA′D–A-type acceptors–. Source: J. Yuan and Y. Zou.

PBDB-T, the devices showed a low J_{sc} and FF, and then a low PCE of 4.26%, though the V_{oc} is as high as 0.975 V. One possible reason leading to the low PCE is the photo-induced hole transfer from BTP-M to PM6 is insufficient, which is proved by the photoluminescence quenching experiments. However, BTP-M could form an alloy-like composite with Y6, which tunes the energy-level alignment and optimizes the blend morphology, and finally boosting the PCE to 17.03% for ternary OSCs based on PM6:Y6:BTP-M. In addition, Zou and coworkers employed a thienyl-fused indanone as end groups on the central core of Y6 and synthesized a new A-DA′D-A

SMA Y10 (Figure 4.24) [170]. Compared with Y6, Y10 presented an obviously elevated LUMO energy levels because of the introduction of the more easily polarized thiophene unit in terminal group, showing great potential in enhancement of V_{oc}. Meanwhile, Min and coworkers reported a SMA, called BTTT-2Cl by introducing the chlorine atom onto the thienyl-fused indanone terminal group (Figure 4.24) [171]. When combined with PM6, a decent PCE of 13.8% with a high V_{oc} of 0.89 V was demonstrated. Beside these, conjugated extension is another strategy for modifying the terminal group. Chen and coworkers reported two novel terminals by extending the conjugation of INCN with one more thiophene or brominated thiophene, and two SMAs BTTPC and BTTPC-Br were synthesized (Figure 4.24) [172]. BTTPC and BTTPC-Br exhibited small HOMO energy offsets in combination with PBDB-T and thus showed a faster hole transfer feature than Y5. As a result, high PCEs of 14.51% and 15.22% were achieved for BTTPC and BTTPC-Br-based devices, greater than that of Y5-based device. Tandem OSCs can boost efficiency by using a wider range of the solar spectrum. However, one main limitation for the development of tandem OSCs comes from the lack of efficient NIR absorption materials for rear cells. To settle this issue, Li and coworkers designed a new low bandgap A–DA′D–A-type SMA BTPV-4F by inserting another ethylene double bond between the central core of Y6 (Figure 4.24) [173]. BTPV-4F showed a significantly red-shifted absorption to 1050 nm (the optical bandgap near 1.21 eV). As a result, the OSCs based on PTB7-Th:BTPV-4F demonstrated a PCE of 13.4% with a superior J_{sc} of 28.9 mA cm^{-2}.

Generally, the molecular packing and interaction of SMAs are important factors in determining the morphology of the blend film, which are closely related to their molecular structures. Hence, controlling the morphology of active layer is important to obtain high-efficiency OSCs. Understanding the molecular conjugated length and isomerization of fused central core are critical in controlling the electronic structure and aggregation behavior of SMAs. Regarding the backbone modification, developing the SMAs with different conjugated length or isomerization of fused central core has shown great potential in achieving different molecular packing, energy levels, dipole moment, and hence the device performance. For example, Zhang and coworkers reported a new SMA BP-4F by incorporating one thiophene into dithiophen[3,2-*b*]-pyrrolobenzothiadiazole or removing one thiophene from dithienothiophen[3,2-*b*]-pyrrolobenzothiadiazole (Figure 4.24). The result showed that the six-membered fused electron-deficient-core-based acceptor BP-4F exhibited relatively blue-shifted spectra in comparison to Y6 [174]. As a result, a PCE of 14.62% was obtained by using BP-4F in combination with PBDB-T-2Cl as the donor polymer with a V_{oc} of 0.88 V, a J_{sc} of 21.73 mA cm^{-2}, and a FF of 76.5%. Gao et al. systematically investigated the conformation effects of Y6 and its derivatives, and two SMAs (BP5T-4F and ABP4T-4F) with their end groups orientating toward different directions were designed and synthesized (Figure 4.24) [175]. Among them, ABP4T-4F is an asymmetric isomer from Y6 derivatives (with optimized side chains) by replacing two thieno[3,2-*b*]thiophene with one thiophene and one dithieno[3,2-*b*:2′,3′-*d*]thiophene. BP5T-4F extends the conjugated length of Y6 derivatives by fusing one more thiophene onto one side of the central core.

The result showed that extending conjugated length of core does not significantly affect the absorption; however, the conformation changes have big impact on aggregation-induced red-shift of absorption spectrum. In addition, asymmetric ABP4T-4F and BP5T-4F can significantly suppress non-radiative recombination. Therefore, the device based on PM6:ABP4T-4F exhibited a PCE of 15.2% with V_{oc} of 0.92 V, and the PM6:BP5T-4F-based device showed a PCE of 16.7% with V_{oc} of 0.88 V.

In previous reports, alkoxy chains have a significant influence on the electronic and morphological properties of organic-conjugated molecules. It is due to the electron-donating effect of alkoxy groups, meanwhile the intramolecular conformational locking effect in some molecular structures. Along these lines, Yan and coworkers introduced alkoxy side chains on the β position of the central core and synthesized two new Y6 derivatives (Y6-1O and Y6-2O, Figure 4.24) [176]. However, Y6-2O exhibited a poor solubility (1.5 mg ml^{-1} in chloroform) and excessive aggregation property (due to the conformational locking effect of alkoxy group), resulting in low efficiency (PCE = 6.6%). Interestingly, from the absorption spectra of Y6-1O and Y6-2O, all of them showed the relatively blue-shifted spectra in comparison to that of Y6. Compared to PM6:Y6-based devices, although PM6:Y6-1O exhibit a slightly lower J_{sc} of 23.20 mA cm^{-2}, they can achieve an enhancement in V_{oc} of 0.89 V and FF of 78.3%, resulting in a high PCE of 16.1%. Substitution of chalcogen species from sulfur to selenium in the π-conjugation backbones can advantageously adjust both intrinsic optoelectronic properties and intermolecular interactions of the molecules. This is because the more polarizable selenium atom could produce stronger orbital overlap between selenium and the π-conjugation system, leading to a higher quinoid resonance character. Therefore, a new selenium-substituted Y6 derivative (CH1007, Figure 4.24) was reported by Lin et al. [177]. From the crystallographic study, due to the S–N interaction between BT moieties, both of two molecules possess very unique central core interaction, which facilitates further development of molecular packing to construct 3D networks. However, compared with Y6, CH1007 exhibited a red-shifted absorption (c. 60 nm), resulting in an extraordinary of J_{sc} of 27.48 mA cm^{-2} and a high PCE of 17.08% in PM6:CH1007:PC$_{71}$BM ternary OSCs.

Side chains, as a solubilizing group, are the most widely used in OSCs. Carefully modulation of the side chains is the key to balance the processability, charge transport, and the aggregation behavior of blend films. In order to understand the influence of alkyl chains on the electronic and photonic properties of Y6, Yan and coworkers designed a series of Y6 derivatives, N3, N4, and N-C11 with alkyl chains of different size and branching position (Figure 4.24) [178]. N-C11, which is designed by swapping the position of *n*-undecyl and 2-ethylhexyl alkyl chains on the Y6 molecule, exhibited a reduced PCE of 12.91% due to the decreased solubility and large over aggregated domains in the blend film. On the contrary, the other two acceptors N3 and N4 with 3rd-and 4th-position-branched alkyl chains showed a good solubility owing to the branched alkyl chains produced a twisted molecular backbone. Nonetheless, N4 exhibited excessive solubility, which also introduced undesirable large domains, lower domain purity, and edge-on orientation. Hou and coworkers adopted a simple alkyl chain modification strategy that changed 2-ethylhexyl (2-EH) on Y6 to 2-butyloctyl (2-BO) [179]. The obtained SMA, BTP-4F-12 (Figure 4.24), exhibited an enhanced crystallinity and electron mobility due to a better lamellar stacking in in-plane direction. More important, the enhanced solubility of BTP-4F-12 rendered a high PCE of 16.1% and 14.4% (active area of 1.07 cm^2) for spin coating and blade coating, respectively, by using T1 as the donor and low-toxic tetrahydrofuran (THF) as the solvent. Huang and coworkers developed a SMA DTY6 with long-branched alkyl chains (2-decyltetradecyl, 2-DT) on the nitrogen atoms of the pyrrole motif of Y6 (Figure 4.24) [180]. The steric hindrance effect of long-branched alkyl chain could improve the solubility and, meanwhile, suppress the molecule-excessive aggregation. As expected, DTY6-based film exhibited a suitable domain size and, thus, ensured efficient charge transport, leading to an impressive efficiency of 16.3% when blended with PM6. Encouragingly, PM6:DTY6 blend showed an ideal candidate for a large-area module device owing to their good processability in non-halogen solvent such as *o*-xylene (*o*-XY), toluene (TL), and 1,2,4-trimethyl-benzene (TMB). For the devices based on PM6:DTY6, a PCE of 16.1% was achieved under ambient fabrication conditions, after which modules with an area of 18 cm^2 were fabricated. The best module gave an outstanding PCE of 14.4% (certified 13.98%) with a V_{oc} of 5.11 V, a J_{sc} of 3.89 mA cm^{-2}, and a FF of 0.725.

4.4.3 BSe- and Qx-Based OSCs

As mentioned above, introducing the electron-deficient segment in the fused D unit of A–D–A-type molecules plays a key role in improvement of the performance of the OSCs. It is owing to the electronegative fused-ring core in A–DA′D–A type SMAs prefers quinoid resonance structures, resulting in low bandgap while maintaining deep energy levels matched well with high-performance polymer donors. However, limited by the synthetic difficulty of Cadogan reductive cyclization reaction, the main optimization performed for current A′ unit in A–DA′D–A-type SMAs were focused on BT, BTA, quinoxaline (Qx), and 2,1,3-benzoselenadiazole (BSe). Although all of them can be employed as an electron-deficient A′ core in the DA′D fused rings, their synthetic methods as well as electronic properties are different. In order to eliminate trade-off between the achievable V_{oc} and charge generation under small energy offset, Zhu and coworkers developed two types of Qx-based A–DA′D–A SMAs AQx-1 and AQx-2 for OSCs (Figure 4.24) [181]. By replacing the BT unit of Y6 with Qx unit, the corresponding quinoid resonance effect into heteroarene core could decrease the reorganization energy deriving from reduced free σ-bonds in the Qx-based SMAs, facilitating electron transport and intermolecular packing. The device based on the PBDB-TF (PM6):AQx-2 blend exhibited a PCE of 16.64% with a V_{oc} of 0.86 eV, J_{sc} of 25.38 mA cm^{-2}, and FF of 0.763. Due to the high Urbach energies (c. 25–50 meV) in reported A–DA′D–A type molecules, Lin and coworkers introduced a facile method of selenium substitution to reduce the Urbach energy of A–DA′D–A-type molecules. The molecule structure of selenium substitution SMA (Y6Se) is shown in Figure 4.24 [182]. Compared to the sulfur-containing analog Y6, Y6Se showed a lower Urbach energy, broader absorption, and higher electron mobility. As a result, the device based on as-cast D18:Y6Se showed a high J_{sc} of 27.98 mA cm^{-2}, a V_{oc} of 0.84 eV, and FF of 0.753, resulting in a high PCE of 17.7% (average PCE for eight devices: 17.5%).

References

1 Hou, J., Inganäs, O., Friend, R.H., and Gao, F. (2018). *Nat. Mater.* 17: 119.
2 Yan, C., Barlow, S., Wang, Z. et al. (2018). *Nat. Rev. Mater.* 3: 1.
3 Zhang, G., Zhao, J., Chow, P.C.Y. et al. (2018). *Chem. Rev.* 118: 3447.
4 Li, C., Liu, M., Pschirer, N.G. et al. (2010). *Chem. Rev.* 110: 6817.
5 Yao, H., Ye, L., Zhang, H. et al. (2016). *Chem. Rev.* 116: 7397.
6 Wu, J.-S., Cheng, S.-W., Cheng, Y.-J., and Hsu, C.-S. (2015). *Chem. Soc. Rev.* 44: 1113.
7 Lin, Y. and Zhan, X. (2016). *Acc. Chem. Res.* 49: 175.
8 Cheng, P. and Zhan, X. (2016). *Chem. Soc. Rev.* 45: 2544.
9 Heeger, A.J. (2014). *Adv. Mater.* 26: 10.
10 Li, Y. (2012). *Acc. Chem. Res.* 45: 723.
11 Søndergaard, R., Hösel, M., Angmo, D. et al. (2012). *Mater. Today* 15: 36.
12 Céspedes-Guirao, F.J., García-Santamaría, S., Fernndez-Lzaro, F. et al. (2009). *J. Phys. D: Appl. Phys.* 42: 105106.
13 Li, L., Guan, M., Cao, G. et al. (2010). *Appl. Phys. A: Mater. Sci. Process.* 99: 251.
14 Kim, S.H., Yang, Y.S., Lee, J.H. et al. (2013). *Opt. Mater.* 21: 439.
15 Weitz, R.T., Amsharov, K., Zschieschang, U. et al. (2008). *J. Am. Chem. Soc.* 130: 4637.
16 Dittmer, J.J., Marseglia, E.A., and Friend, R.H. (2000). *Adv. Mater.* 12: 1270.
17 Schmidt-Mende, L. (2001). *Science* 293: 1119.
18 Shin, W.S., Jeong, H.-H., Kim, M.-K. et al. (2006). *J. Mater. Chem.* 16: 384.
19 Mikroyannidis, J.A., Stylianakis, M.M., Suresh, P., and Sharma, G.D. (2009). *Sol. Energy Mater. Sol. Cells* 93: 1792.
20 Sharma, G.D., Balraju, P., Mikroyannidis, J.A., and Stylianakis, M.M. (2009). *Sol. Energy Mater. Sol. Cells* 93: 2025.
21 Rajaram, S., Armstrong, P.B., Bumjoon, J.K., and Fréchet, J.M.J. (2009). *Chem. Mater.* 21: 1775.
22 Sharma, G.D., Suresh, P., Mikroyannidis, J.A., and Stylianakis, M.M. (2010). *J. Mater. Chem.* 20: 561.
23 Li, C., Fu, H., Xia, T., and Sun, Y. (2019). *Adv. Energy Mater.* 9: 1900999.
24 Lin, Y. and Zhan, X. (2014). *Mater. Horiz.* 1: 470.
25 Liu, F., Hou, T., Xu, X. et al. (2018). *Macromol. Rapid Commun.* 39: 1700555.
26 Suman, S. and Singh, S.P. (2019). *J. Mater. Chem. A* 7: 22701.
27 Sharenko, A., Proctor, C.M., Van Der Poll, T.S. et al. (2013). *Adv. Mater.* 25: 4403.
28 Singh, R., Aluicio-Sarduy, E., Kan, Z. et al. (2014). *J. Mater. Chem. A* 2: 14348.

29 Hartnett, P.E., Timalsina, A., Matte, H.S.S.R. et al. (2014). *J. Am. Chem. Soc.* 136: 16345.
30 Cai, Y., Huo, L., Sun, X. et al. (2015). *Adv. Energy Mater.* 5: 1500032.
31 Cai, Y., Guo, X., Sun, X. et al. (2016). *Sci. China Mater.* 59: 427.
32 Rajaram, S., Shivanna, R., Kandappa, S.K., and Narayan, K.S. (2012). *J. Phys. Chem. Lett.* 3: 2405.
33 Park, G.E., Kim, H.J., Choi, S. et al. (2016). *Chem. Commun.* 52: 8873.
34 Luo, Z., Wu, K., Zhao, Y. et al. (2019). *Dyes Pigm.* 163: 356.
35 Jiang, W., Ye, L., Li, X. et al. (2014). *Chem. Commun.* 50: 1024.
36 Ye, L., Jiang, W., Zhao, W. et al. (2014). *Small* 10: 4658.
37 Zang, Y., Li, C.Z., Chueh, C.C. et al. (2014). *Adv. Mater.* 26: 5708.
38 Sun, D., Meng, D., Cai, Y. et al. (2015). *J. Am. Chem. Soc.* 137: 11156.
39 Meng, D., Sun, D., Zhong, C. et al. (2016). *J. Am. Chem. Soc.* 138: 375.
40 Hendsbee, A.D., Sun, J.-P., Law, W.K. et al. (2018). *Chem. Mater.* 28: 7098.
41 Zhang, C., Liu, T., Zeng, W. et al. (2017). *Mater. Chem. Front.* 1: 749.
42 Gao, G., Liang, N., Geng, H. et al. (2017). *J. Am. Chem. Soc.* 139: 15914.
43 Yan, Q., Zhou, Y., Zheng, Y.-Q. et al. (2013). *Chem. Sci.* 4: 4389.
44 Liu, J., Chen, S., Qian, D. et al. (2016). *Nat. Energy* 1: 16089.
45 Zhong, H., Wu, C.H., Li, C.Z. et al. (2016). *Adv. Mater.* 28: 951.
46 Zhang, G., Feng, J., Xu, X. et al. (2019). *Adv. Funct. Mater.* 29: 1906587.
47 Liang, N., Sun, K., Zheng, Z. et al. (2016). *Adv. Energy Mater.* 6: 1600060.
48 Lin, Y., Wang, Y., Wang, J. et al. (2014). *Adv. Mater.* 26: 5137.
49 Li, S., Liu, W., Li, C.-Z. et al. (2016). *J. Mater. Chem. A* 4: 10659.
50 Duan, Y., Xu, X., Yan, H. et al. (2017). *Adv. Mater.* 29: 1605115.
51 Zhang, G., Xu, X., Lee, Y.W. et al. (2019). *Adv. Funct. Mater.* 29: 1902079.
52 Zhong, Y., Trinh, M.T., Chen, R. et al. (2015). *Nat. Commun.* 6: 8242.
53 Wang, B., Li, H., Liu, W. et al. (2017). *J. Mater. Chem. A* 5: 9396.
54 Meng, D., Fu, H., Xiao, C. et al. (2016). *J. Am. Chem. Soc.* 138: 10184.
55 Liu, Y., Lai, J.Y.L., Chen, S. et al. (2015). *J. Mater. Chem. A* 3: 13632.
56 Lee, J., Singh, R., Sin, D.H. et al. (2016). *Adv. Mater.* 28: 69.
57 Sun, H., Song, X., Xie, J. et al. (2017). *ACS Appl. Mater. Interfaces* 9: 29924.
58 Zhang, A., Li, C., Yang, F. et al. (2017). *Angew. Chem. Int. Ed.* 56: 2694.
59 Chen, W., Yang, X., Long, G. et al. (2015). *J. Mater. Chem. C* 3: 4698.
60 Liu, X., Liu, T., Duan, C. et al. (2017). *J. Mater. Chem. A* 5: 1713.
61 Liu, Y., Mu, C., Jiang, K. et al. (2015). *Adv. Mater.* 27: 1015.
62 Lin, H., Chen, S., Hu, H. et al. (2016). *Adv. Mater.* 28: 8546.
63 Zhang, J., Li, Y., Huang, J. et al. (2017). *J. Am. Chem. Soc.* 139: 16092.
64 Wu, Q., Zhao, D., Schneider, A.M. et al. (2016). *J. Am. Chem. Soc.* 138: 7248.
65 Wu, Q., Zhao, D., Yang, J. et al. (2017). *Chem. Mater.* 29: 1127.
66 Luo, Z., Liu, T., Chen, Z. et al. (2019). *Adv. Sci.* 6: 1802065.
67 Zhang, J., Bai, F., Li, Y. et al. (2019). *J. Mater. Chem. A* 7: 8136.
68 Yu, H., Arunagiri, L., Zhang, L. et al. (2020). *J. Mater. Chem. A* 8: 6501.
69 Yang, Y., Zhang, Z.G., Bin, H. et al. (2016). *J. Am. Chem. Soc.* 138: 15011.
70 Dai, S., Zhao, F., Zhang, Q. et al. (2017). *J. Am. Chem. Soc.* 139: 1336.

71 Swick, S.M., Alzola, J.M., Sangwan, V.K. et al. (2020). *Adv. Energy Mater.* 10: 2000635.
72 Lin, Y., Wang, J., Zhang, Z.-G. et al. (2015). *Adv. Mater.* 27: 1170.
73 Lin, Y., Zhao, F., He, Q. et al. (2016). *J. Am. Chem. Soc.* 138: 4955.
74 Lin, Y., He, Q., Zhao, F. et al. (2016). *J. Am. Chem. Soc.* 138: 2973.
75 Li, Y., Zheng, N., Yu, L. et al. (2019). *Adv. Mater.* 31: 1807832.
76 Feng, S., Zhang, C., Liu, Y. et al. (2017). *Adv. Mater.* 29: 1703527.
77 Li, S., Ye, L., Zhao, W. et al. (2016). *Adv. Mater.* 28: 9423.
78 Li, S., Ye, L., Zhao, W. et al. (2017). *Adv. Energy Mater.* 7: 1700183.
79 Yang, F., Li, C., Lai, W. et al. (2017). *Mater. Chem. Front.* 1: 1389.
80 Zhao, W., Li, S., Yao, H. et al. (2017). *J. Am. Chem. Soc.* 139: 7148.
81 Zhang, H., Yao, H., Hou, J. et al. (2018). *Adv. Mater.* 30: 1800613.
82 Aldrich, T.J., Matta, M., Zhu, W. et al. (2019). *J. Am. Chem. Soc.* 141: 3274.
83 Xie, D., Liu, T., Gao, W. et al. (2017). *Sol. RRL* 1: 1700044.
84 Yao, H., Ye, L., Hou, J. et al. (2017). *Adv. Mater.* 29: 1700254.
85 Luo, Z., Bin, H., Liu, T. et al. (2018). *Adv. Mater.* 30: 1706124.
86 Luo, Z., Liu, T., Wang, Y. et al. (2019). *Adv. Energy Mater.* 9: 1900041.
87 Zhao, W., Qian, D., Zhang, S. et al. (2016). *Adv. Mater.* 28: 4734.
88 Fan, Q., Wang, Y., Zhang, M. et al. (2018). *Adv. Mater.* 30: 1704546.
89 Wang, P., Fan, H., Zhang, C., and Zhu, X. (2018). *Mater. Chem. Front.* 2: 136.
90 Sun, J., Ma, X., Zhang, Z. et al. (2018). *Adv. Mater.* 30: 1707150.
91 Jia, B., Dai, S., Ke, Z. et al. (2017). *Chem. Mater.* 30: 239.
92 Lin, F., Zuo, L., Gao, K. et al. (2019). *Chem. Mater.* 31: 6770.
93 Zhang, Z., Yu, J., Yin, X. et al. (2018). *Adv. Funct. Mater.* 28: 1705095.
94 Zhang, Z., Liu, X., Yu, J. et al. (2019). *J. Mater. Chem. C* 7: 13279.
95 Zhang, Z., Wang, H., Yu, J. et al. (2020). *Chem. Mater.* 32: 1297.
96 Zhang, Z., Guang, S., Yu, J. et al. (2020). *Sci. Bull.* 65: 1533.
97 Dai, S., Li, T., Wang, W. et al. (2018). *Adv. Mater.* 30: 1706571.
98 Cai, G., Xue, P., Chen, Z. et al. (2018). *Chem. Mater.* 31: 6484.
99 He, D., Zhao, F., Xin, J. et al. (2018). *Adv. Energy Mater.* 8: 1802050.
100 Wang, H., Zhang, Z., Yu, J. et al. (2019). *J. Mater. Chem. A* 7: 21903.
101 Jin, K., Deng, C., Zhang, L. et al. (2018). *Mater. Chem. Front.* 2: 1716.
102 Xiao, Z., Jia, X., Li, D. et al. (2017). *Sci. Bull.* 62: 1494.
103 Zhang, J., Li, Y., Peng, Z. et al. (2020). *Mater. Chem. Front.* 4: 1729.
104 Ke, X., Meng, L., Wan, X. et al. (2020). *Nano Energy* 75: 104988.
105 Wang, J., Zhang, J., Xiao, Y. et al. (2018). *J. Am. Chem. Soc.* 140: 9140.
106 Chen, M., Zhang, Z., Li, W. et al. (2019). *Sci. China Chem.* 62: 1221.
107 Li, W., Chen, M., Zhang, Z. et al. (2019). *Adv. Funct. Mater.* 29: 1807662.
108 Feng, H., Song, X., Zhang, Z. et al. (2019). *Adv. Funct. Mater.* 29: 1903269.
109 Yuan, J., Zhang, Y., Zhou, L. et al. (2019). *Joule* 3: 1140.
110 Bai, H., Wang, Y., Cheng, P. et al. (2015). *J. Mater. Chem. A* 3: 1910.
111 Yu, S., Chen, Y., Yang, L. et al. (2017). *J. Mater. Chem. A* 5: 21674.
112 Yan, C., Wu, Y., Wang, J. et al. (2017). *Dyes Pigm.* 139: 627.
113 Wang, Y., Bai, H., and Zhan, X. (2015). *J. Energy Chem.* 24: 744.
114 Yao, H., Cui, Y., Yu, R. et al. (2017). *Angew. Chem. Int. Ed.* 56: 3045.

115 Chen, S., Yao, H., Hu, B. et al. (2018). *Adv. Energy Mater.* 8: 1800529.
116 Liu, F., Zhou, Z., Zhang, C. et al. (2017). *Adv. Mater.* 29: 1606574.
117 Chen, F.X., Xu, J.Q., Liu, Z.X. et al. (2018). *Adv. Mater.* 30: 1803769.
118 Park, S.H., Park, G.E., Choi, S. et al. (2018). *J. Mater. Chem. C* 6: 7549.
119 Li, X., Yan, T., Bin, H. et al. (2017). *J. Mater. Chem. A* 5: 22588.
120 Li, X., Xu, Z., Guo, X. et al. (2019). *Dyes Pigm.* 160: 227.
121 Liu, Y., Li, M., Yang, J. et al. (2019). *Adv. Energy Mater.* 9: 1901280.
122 Feng, S., Lu, H., Liu, Z. et al. (2019). *Acta Phys. Chim. Sin.* 35: 355.
123 Wang, Z., Huang, K., Chen, X. et al. (2019). *J. Mater. Chem. C* 7: 14499.
124 Liu, Y., Zhang, Z., Feng, S. et al. (2017). *J. Am. Chem. Soc.* 139: 3356.
125 Jiang, P., Ming, S., Jia, Q.Q. et al. (2018). *J. Mater. Chem. A* 6: 21335.
126 Liu, Y., Li, M., Zhou, X. et al. (2018). *ACS Energy Lett.* 3: 1832.
127 Song, X., Gasparini, N., Ye, L. et al. (2018). *ACS Energy Lett.* 3: 669.
128 Liu, D., Kan, B., Ke, X. et al. (2018). *Adv. Energy Mater.* 8: 1801618.
129 Li, C., Xie, Y., Fan, B. et al. (2018). *J. Mater. Chem. C* 6: 4873.
130 Lin, Y., Li, T., Zhao, F. et al. (2016). *Adv. Energy Mater.* 6: 1600854.
131 Song, J., Li, C., Ye, L. et al. (2018). *J. Mater. Chem. A* 6: 18847.
132 Gao, W., An, Q., Zhong, C. et al. (2018). *Chem. Sci.* 9: 8142.
133 Li, C., Xia, T., Song, J. et al. (2019). *J. Mater. Chem. A* 7: 1435.
134 Yang, L., Song, X., Yu, J. et al. (2019). *J. Mater. Chem. A* 7: 22279.
135 Cao, J., Qu, S., Yu, J. et al. (2020). *Mater. Chem. Front.* 4: 924.
136 Yang, L., Hu, Z., Zhang, Z. et al. (2020). *J. Mater. Chem. A* 8: 5458.
137 Zhang, Z., Yang, L., Hu, Z. et al. (2020). *Mater. Chem. Front.* 4: 1747.
138 Cao, J., Qu, S., Yang, L. et al. (2020). *Sci. Bull.* 65: 1876.
139 Cao, J., Wang, H., Qu, S. et al. (2020). *Adv. Funct. Mater.* 30: 2006141.
140 Li, C., Song, J., Ye, L. et al. (2019). *Sol. RRL* 3: 1800246.
141 Gao, W., Liu, T., Li, J. et al. (2019). *J. Mater. Chem. A* 7: 11053.
142 Luo, Z., Li, G., Wu, K. et al. (2018). *Org. Electron.* 62: 82.
143 Gao, W., Zhang, M., Liu, T. et al. (2018). *Adv. Mater.* 30: 1800052.
144 Gao, B., Yao, H., Hou, J. et al. (2018). *J. Mater. Chem. A* 6: 23644.
145 Liu, Q., Jiang, Y., Jin, K. et al. (2020). *Sci. Bull.* 65: 272.
146 Li, X., Xu, Z., Guo, X. et al. (2018). *Org. Electron.* 58: 133.
147 Yu, Z.P., Liu, Z.X., Chen, F.X. et al. (2019). *Nat. Commun.* 10: 2152.
148 Huang, H., Guo, Q., Feng, S. et al. (2019). *Nat. Commun.* 10: 3038.
149 Li, S., Zhan, L., Liu, F. et al. (2018). *Adv. Mater.* 30: 1705208.
150 Li, S., Zhan, L., Zhao, W. et al. (2018). *J. Mater. Chem. A* 6: 12132.
151 Qin, R., Yang, W., Li, S. et al. (2019). *Mater. Chem. Front.* 3: 513.
152 Zheng, R., Guo, Q., Hao, D. et al. (2019). *J. Mater. Chem. C* 7: 15141.
153 Yi, Y.Q.Q., Feng, H., Zheng, N. et al. (2019). *Chem. Mater.* 31: 904.
154 Feng, L., Yuan, J., Zhang, Z. et al. (2017). *ACS Appl. Mater. Interfaces* 9: 31985.
155 Yuan, J., Huang, T., Cheng, P. et al. (2019). *Nat. Commun.* 10: 1.
156 Wang, R., Yuan, J., Wang, R. et al. (2019). *Adv. Mater.* 31: 1904215.
157 Liu, S., Yuan, J., Deng, W. et al. (2020). *Nat. Photonics* 14: 300.
158 Yuan, J., Zhang, C., Chen, H. et al. (2020). *Sci. China Chem.* 63: 1159.
159 Luo, M., Zhou, L., Yuan, J. et al. (2020). *J. Energy Chem.* 42: 169.

160 Luo, M., Zhu, C., Yuan, J. et al. (2019). *Chin. Chem. Lett.* 30: 2343.
161 Luo, M., Zhao, C., Yuan, J. et al. (2019). *Mater. Chem. Front.* 3: 2483.
162 Ma, X., Luo, M., Gao, W. et al. (2019). *J. Mater. Chem. A* 7: 7843.
163 Cai, F., Zhu, C., Yuan, J. et al. (2020). *Chem. Commun.* 56: 4340.
164 Cai, F., Peng, H., Chen, H. et al. (2020). *J. Mater. Chem. A* 8: 15984.
165 Yuan, J., Zhang, Y., Zhou, L. et al. (2019). *Adv. Mater.* 31: 1807577.
166 Cui, Y., Yao, H., Zhang, J. et al. (2019). *Nat. Commun.* 10: 2515.
167 Wang, H., Liu, T., Zhou, J. et al. (2020). *Adv. Sci.* 7: 1903784.
168 Sun, C., Qin, S., Wang, R. et al. (2020). *J. Am. Chem. Soc.* 142: 1465.
169 Zhan, L., Li, S., Lau, T.-K. et al. (2020). *Energy Environ. Sci.* 13: 635.
170 Zhang, Y., Cai, F., Yuan, J. et al. (2019). *Phys. Chem. Chem. Phys.* 21: 26557.
171 Yang, W., Luo, Z., Sun, R. et al. (2020). *Nat. Commun.* 11: 1218.
172 Qin, R., Wang, D., Zhou, G. et al. (2019). *J. Mater. Chem. A* 7: 27632.
173 Jia, Z., Qin, S., Meng, L. et al. (2021). *Nat. Commun.* 12: 178.
174 Gao, J., Gao, W., Ma, X. et al. (2020). *Energy Environ. Sci.* 13: 958.
175 Gao, W., Fu, H., Li, Y. et al. (2020). *Adv. Energy Mater.* 11: 2003177.
176 Chen, Y., Bai, F., Peng, Z. et al. (2020). *Adv. Energy Mater.* 11: 2003141.
177 Lin, F., Jiang, K., Kaminsky, W. et al. (2020). *J. Am. Chem. Soc.* 142: 15246.
178 Jiang, K., Wei, Q., Lai, J.Y.L. et al. (2019). *Joule* 3: 3020.
179 Hong, L., Yao, H., Wu, Z. et al. (2019). *Adv. Mater.* 31: 1903441.
180 Dong, S., Jia, T., Zhang, K. et al. (2020). *Joule* 4: 2004.
181 Zhou, Z., Liu, W., Zhou, G. et al. (2020). *Adv. Mater.* 32: 1906324.
182 Zhang, Z., Li, Y., Cai, G. et al. (2020). *J. Am. Chem. Soc.* 142: 18741.
183 Cui, Y., Yao, H., Zhang, J. et al. (2020). *Adv. Mater.* 32: 1908205.
184 Zhu, L., Zhang, M., Zhou, G. et al. (2020). *Adv. Energy Mater.* 10: 1904234.

5

Electron-Donating Ladder-Type Heteroacenes for Photovoltaic Applications: From Polymer Donor Materials to Small-Molecule Acceptor Materials

Qisheng Tu, Yunlong Ma, and Qingdong Zheng**

Chinese Academy of Sciences, Fujian Institute of Research on the Structure of Matter, State Key Laboratory of Structural Chemistry, 155 Yangqiao West Road, Fuzhou, 350002, P.R. China

5.1 Introduction

Solution-processed polymer solar cells (PSCs) have become one of the promising low-cost solar energy harvesting technologies with their rapidly grown power conversion efficiencies (PCEs) from around 1% to over 17% in the past two decades [1–3]. Bulk heterojunction (BHJ) active layers with p-type semiconducting polymers as electron donors and n-type semiconductors (fullerene derivatives or nonfullerene molecules) as electron acceptors play a key role in determining the PCEs of PSCs. In the early years of research on PSCs, fullerene derivatives (such as $PC_{61}BM$ and $PC_{71}BM$) were the dominant acceptors and most semiconducting polymers were designed to match the energy levels as well as the absorption profiles of the fullerene derivatives [2, 4]. Because the absorption bands of most fullerene derivatives locate at the short-wavelength region (300–600 nm), many low bandgap (LBG) polymers with enhanced light absorption bands in the long wavelength region have been developed by which the best PCE of the fullerene-based PSCs has exceeded 11% [5]. Nonetheless, to further improve the PCEs of single-junction PSCs based on the LBG polymers and the fullerene derivatives to far beyond 11% becomes impossible considering the trade-off between their short-circuit current density (J_{sc}) and open-circuit voltage (V_{oc}) values. Therefore, nonfullerene acceptors (NFAs) are emerging to replace the traditional fullerene derivatives. Unless otherwise stated, NFAs in the review refer to the small-molecule acceptor materials. In contrast with the traditional fullerene acceptors with intense absorption bands in the short-wavelength region, the absorption bands of NFAs can be tuned from the visible region to near-infrared (NIR) region [6]. To achieve complementary absorption, donor materials with relatively wide bandgaps are needed to achieve high-performance nonfullerene PSCs [7]. In the past two decades, most breakthroughs are related to the inventions of novel donor or acceptor materials. Take the donor materials as examples, the inventions of copolymers with

* Yunglong Ma and Qindong Zheng were contributed equally.

Organic Solar Cells: Materials Design, Technology and Commercialization, First Edition. Edited by Liming Ding.

thieno[3,4-*b*]thiophene and benzodithiophene (BDT) alternating units (PTB7-Th, PTB7) led to PSCs with PCEs over 10% and the inventions of terthiophene-based LBG copolymers afforded one of the highest PCEs of 11.7% among all the fullerene-based PSCs [4, 5]. Similarly, the invention of the benchmark acceptor–donor–acceptor (A–D–A) type NFA (ITIC), which used indacenodithieno[3,2-*b*]thiophene (IDTT) as the electron-rich core, yielded an initial PCE of 6.8% for the nonfullerene PSCs [8]. Using these IDTT-like fused-ring cores with various side chains, many other A–D–A type NFAs have been developed with excellent PCEs over 12% [9]. More recently, Zou and coworkers reported a series of outstanding NFAs (such as Y6 and Y11) with initial PCEs over 15%, which incorporated electron-deficient benzothiadiazole (or benzotriazole)-fused core into the NFAs to form the A–D–A′–D–A type configuration [10, 11]. This material innovation scenario indicates the importance of both donor and acceptor materials in determining the photovoltaic properties of PSCs.

For both donor and acceptor materials, the design of novel electron-donating units is essential. Among many electron-donating units, ladder-type heteroacenes, which consist of several "linearly overlapping" aromatic or heterocyclic subunits linked by sharing endocyclic carbon–carbon (C=C) double bonds, stand out as ideal candidates due to their salient features of good thermal, mechanical, and chemical stabilities [12]. Ladder-type heteroacenes have extended conjugated frameworks, which can elongate the effective π-conjugation length and facilitate the charge carrier mobility. At the same time, various electron-donating aromatic building blocks such as thiophene, benzene, pyrrole, cyclopentadienyl moieties can be incorporated into the ladder-type frameworks to afford heteroacenes with tunable properties [4]. Furthermore, solubilizing alkyl chains can be introduced to the ladder-type heteroacenes to endow the target donor or acceptor materials with good solution processability. Therefore, the research of ladder-type heteroacenes has greatly boosted the development of PSCs.

There have been some reviews on the BDT-based copolymers (such as PTB7-Th, PBDB-T, and PM6), which are the most-researched donor materials for both fullerene and nonfullerene PSCs in the past decade [13]. In this chapter, we will mainly summarize the recent developments of heteroacene-based polymer donor materials developed in our research group in the past decade. However, some important breakthroughs in the heteroacene-based materials will also be covered to let readers have a whole story on the development of ladder-type heteroacene-based materials for PSCs. This chapter is organized as follows: in Section 5.2, we focus on discussing the structures and photovoltaic performances of new donor–acceptor (D–A) copolymers based on the various ladder-type heteroacenes. The ladder-type heteroacenes will be classified according to their ring numbers including pentacyclic, hexacyclic, heptacyclic, nonacyclic etc. In Section 5.3, we focus on discussing the structures and photovoltaic performances of new A–D–A type NFAs which were synthesized by using ladder-type heteroacenes as the central cores. Finally, we conclude with a summary of the challenges and future molecular design strategies for the heteroacene-based semiconducting materials in the PSC application.

5.2 D–A Copolymers Based on Ladder-Type Heteroacenes

To achieve an enhanced light-harvesting ability (LBG) as well as a large V_{oc} of a PSC device, one of the most potent approaches is to design alternating D–A copolymers, where the electron donor unit can be used to provide a deep-lying highest occupied molecular orbital (HOMO) level and the electron-deficient unit can be used to tune the electronic bandgap of the polymers [14]. For the D–A type copolymers based on heteroacenes, their bandgaps, energy levels, carrier mobilities etc. can be finely tuned by using heteroacenes with different molecular structures. At the same time, the selection of acceptor units as well as π-bridges can also affect their optical, electrochemical, charge transport, and photovoltaic properties. Here, we mainly focus on the effect of the structure of the heteroacene core on the photovoltaic performance of the resulting D–A copolymers.

5.2.1 Pentacyclic and Hexacyclic Heteroacenes-Based D–A Copolymers

In 2010, Zheng et al. synthesized a D–A copolymer (PIFTBT6, Figure 5.1) based on the pentacyclic indenofluorene unit which contains two "linearly overlapping" fluorene segments [15]. As summarized in Table 5.1, despite a PCE of 0.97% for the PSCs based on PIFTBT6, a rather high V_{oc} of 0.98 V was achieved due to the relatively deep-lying HOMO energy level of PIFTBT6. Subsequently, our group further used branched 2-ethylhexyl side chains to replace the hexyl side chains on the indenofluorene backbone thus affording an optimized D–A copolymer (PIFTBT8)

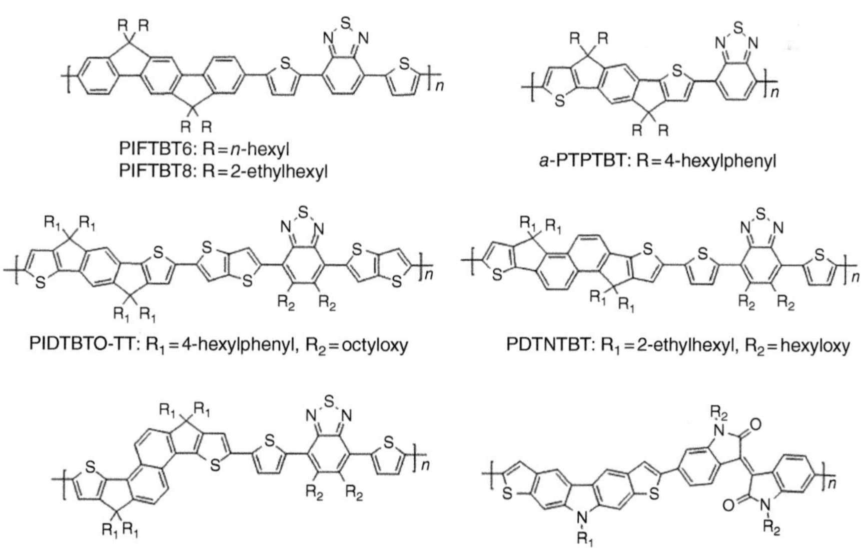

Figure 5.1 Chemical structures of typical copolymers based on pentacyclic and hexacyclic heteroacenes. Source: Qingdong Zheng.

Table 5.1 Summary of bandgaps, energy levels, and photovoltaic performances of copolymers shown in Figure 5.1.

Donor[a)]	E_g (eV)	HOMO/LUMO (eV)	V_{oc} (V)	J_{sc} (mA cm^{-2})	FF (%)	PCE (%)	References
PIFTBT6[b)]	1.97	−5.49/−3.46	0.98	3.25	31.0	0.97	[15]
PIFTBT8	1.97	−5.50/−3.50	1.04	9.74	50.1	5.05	[16]
a-PTPTBT	1.75	−5.36/−3.52	0.85	11.20	67.2	6.41	[17]
PIDTBTO-TT	1.87	−5.41/−3.51	0.91	14.77	60.9	8.15	[18]
PDTNTBT	1.78	−5.36/−3.42	0.86	10.29	53.8	4.78	[19]
PaDTNBTO	1.90	−5.40/−3.50	0.92	10.91	64.1	6.44	[20]
P(IID-DTC)	1.61	−5.22/−3.54	0.78	15.20	69.0	8.20	[21]

a) Blended with $PC_{71}BM$.
b) Blended with $PC_{61}BM$.

with an enhanced charge carrier mobility [16]. Therefore, the resulting PSCs based on PIFTBT8 showed a much higher PCE of 5.05% with a further increased V_{oc} of 1.04 V. These fluorene-derived units showed weak electron-donating nature thereby leading to the D–A copolymers with deep-lying HOMO energy levels. Compared to benzene, thiophene is more electron-rich; therefore, it can raise the HOMO energy level of the resulting material thereby leading to a reduced bandgap.

Replacing the phenyl groups at the ends of indenofluorene with thienyl groups affords another excellent example of ladder-type pentacyclic heteroacene, i.e. indacenodithiophene (IDT). The relatively strong electron-donating ability of IDT makes itself a good candidate for high-performance D–A copolymers with reduced bandgaps. Ting and coworkers synthesized an IDT-based D–A copolymer (a-PTPTBT) using benzothiadiazole as the acceptor unit. The copolymer (a-PTPTBT) showed an intriguing PCE of 6.41% with a V_{oc} of 0.85 V and a J_{sc} of 11.20 mA cm^{-2} [17]. Later, many IDT-based D–A copolymers were designed and synthesized by using different acceptor units for the copolymerization [4]. Nonetheless, none of these copolymers gave PCEs over 8.00%. In 2017, our group designed and synthesized an IDT-based copolymer (PIDTBTO-TT) using alkoxylated benzothiadiazole as the acceptor unit and thieno[3,2-*b*]thiophene as the π-bridge [18]. With its increased conjugated backbone planarity, PIDTBTO-TT showed a reduced π-π packing distance which is beneficial for more efficient charge transport. Thus, the single-junction device based on PIDTBTO-TT:$PC_{71}BM$ exhibited a high PCE of 8.15% with a V_{oc} of 0.91 V, a J_{sc} of 14.77 mA cm^{-2}, and a FF of 60.9%. When PIDTBTO-TT:$PC_{71}BM$ and PTB7-Th:$PC_{71}BM$ were used as active layer materials in the bottom and top sub-cells in a tandem device, respectively, a high PCE of 11.15% with a V_{oc} of 1.70 V, a J_{sc} of 9.68 mA cm^{-2}, and a FF of 67.8% was achieved. The efficiencies of 8.15% and 11.15% were the highest for the single-junction and tandem PSCs based on IDT-containing copolymers at that time, respectively. Substitution of the central phenyl group in IDT with naphthyl group leads to a hexacyclic dithienonaphthalene (DTN) building block, which contains two isomers, i.e. linear- and angular-shaped

DTNs. Using the linear- or angular-shaped DTN as a donor segment, and a benzothiadiazole derivative as an acceptor moiety, two isomeric DTN-based copolymers (PDTNTBT and PaDTNBTO) were synthesized in our group for the first time [19, 20]. The copolymer based on the angular-shape DTN (PaDTNBTO) showed a deeper-lying HOMO energy level and an improved charge transport compared to its linear-shaped analog. Thus, the PSCs based on PaDTNBTO:$PC_{71}BM$ exhibited a higher PCE of 6.44% with a larger V_{oc} of 0.92 V in comparison with those of the PDTNTBT-based devices (PCE = 4.78%, V_{oc} = 0.86 V). These results suggested that the structural isomerization on the backbone of heteroacene plays a vital role on the energy levels, bandgaps, charge mobilities, and photovoltaic performances of the resulting copolymers.

It should be noted that all the above-mentioned D–A copolymers are of an amorphous nature due to four bulky substituents flanked on their sp^3-hybridized bridging atoms, which will restrain a closely intermolecular packing. Therefore, it was expected that the removal of one sp^3-hybridized bridging atom in the conjugated backbone of the donor unit would increase the molecular packing order (or crystallinity) of the resulting D–A copolymer. In 2016, for the first time, Wang et al. reported an indenothiophene-based copolymer (PITFBT) with a semicrystalline feature [22]. Due to the weak electron-donating ability of indenothiophene, PIFTBT exhibited a relatively deep-lying HOMO energy level of −5.50 eV which is beneficial for achieving a high V_{oc} for the resulting PSCs. More importantly, with the semicrystalline feature of PITFBT, the PITFBT:$PC_{71}BM$ blend with a proper solvent additive exhibited a widely distributed nanofibrillar structure which is favorable for efficient charge transport. As a result, the PITFBT-based PSC showed an impressive PCE of 9.14% with a high V_{oc} of 0.90 V, a J_{sc} of 15.36 mA cm^{-2}, and a FF of 65.9%.

Dithienocarbazole (DTC), which is made by fusing two thiophenes on both sides of carbazole, was first developed by Geng and coworkers [21]. The copolymer of P(IID-DTC) was synthesized by the polymerization reaction between DTC and an isoindigo derivative [21]. P(IID-DTC) showed a low optical bandgap of 1.61 eV with an up-shifted HOMO energy level of −5.22 eV. The best-performance PSC based on P(IID-DTC):$PC_{71}BM$ blend exhibited an outstanding PCE of 8.20% with a V_{oc} of 0.78 V and a J_{sc} of 15.20 mA cm^{-2}.

5.2.2 Heptacene-Based D–A Copolymers

Ladder-type tetra-*p*-phenylene containing three "linearly overlapping" fluorene segments was used as a promising donor unit in high-performance D–A copolymers due to the extended π-conjugation length. Compared to fluorene or indenofluorene, the ladder-type tetra-*p*-phenylene derivative could broaden and strengthen the absorption band, thus resulting in an enhanced light-harvesting ability. Meanwhile, the relatively weak electron-donating feature of the ladder-type tetra-*p*-phenylene building block will lead to deep-lying HOMO energy levels of the resulting copolymers. In 2010, Zheng et al. first reported a D–A copolymer (P3FTBT6 in Figure 5.2) based on the ladder-type tetra-*p*-phenylene, which was also the first example of ladder-type heptacene-based D–A copolymers [15]. P3FTBT6 displayed

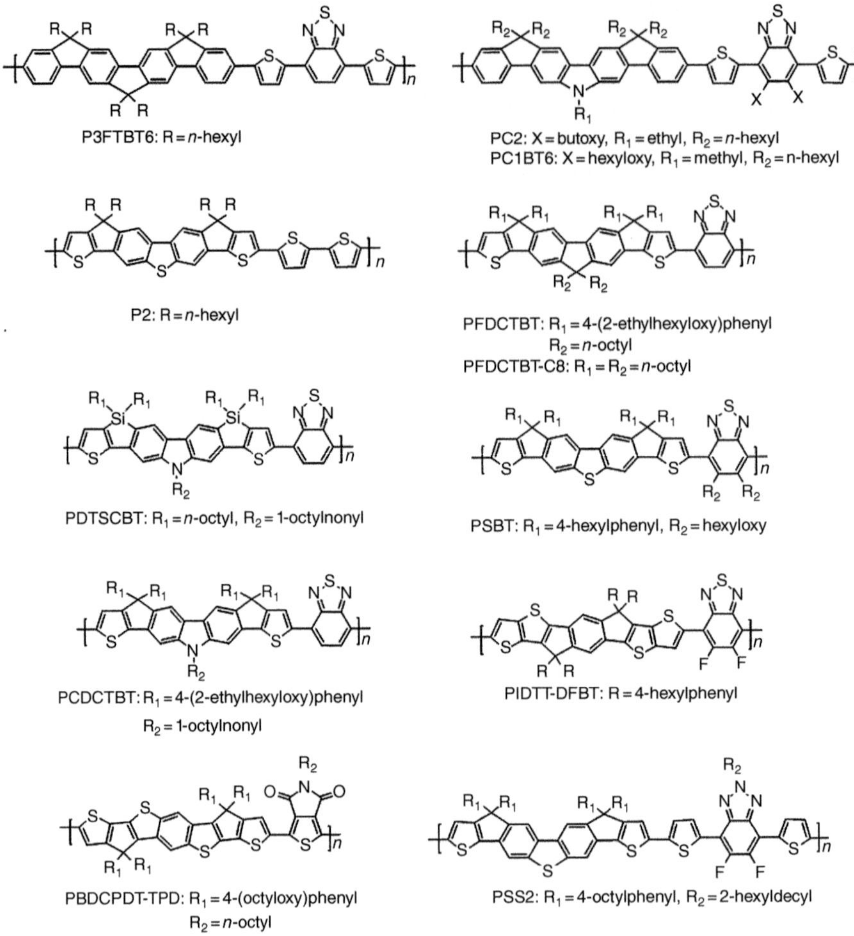

Figure 5.2 Chemical structures of some representative D–A copolymers based on heptacenes. Source: Qingdong Zheng.

a relatively deep-lying HOMO energy level of −5.45 eV (Table 5.2). Compared to the fluorene-based copolymer with the same acceptor segment, P3FTBT6 exhibited an enhanced and red-shifted absorption. When fabricated in ambient environment, the P3FTBT6:PC_{71}BM-based device showed a PCE of 4.50% with a large V_{oc} of 1.04 V. These exciting results reveal that the ladder-type fused arenes with extended conjugation are promising building blocks for constructing high-performance D–A copolymers. Nonetheless, the six out-of-plane side chains on the three sp^3-hybridized bridging atoms in the conjugated backbone limit their charge carrier mobilities when blended with the fullerene derivatives. Subsequently, the central sp^3-hybridized bridging atom on the ladder-type tetra-*p*-phenylene segment was substituted with nitrogen in our group to afford a new heteroheptacene, i.e. diindenocarbazole (DIC). Using the DIC and a dithienylbenzothiadiazole derivative as the donor and acceptor moieties, respectively, a wide-bandgap copolymer (PC2)

Table 5.2 Summary of bandgaps, energy levels, and photovoltaic performances of copolymers shown in Figure 5.2.

Donor[a)]	E_g (eV)	HOMO/LUMO (eV)	V_{oc} (V)	J_{sc} (mA cm^{-2})	FF (%)	PCE (%)	References
P3FTBT6	1.96	−5.45/−3.45	1.04	10.30	42.0	4.50	[15]
PC2	2.01	−5.43/−3.56	0.93	14.22	54.7	7.26	[23]
PC1BT6	2.01	−5.48/−3.59	0.95	14.00	55.2	7.34	[24]
PFDCTBT	1.87	−5.32/−3.55	0.77	9.50	38.0	2.80	[25]
PCDCTBT	1.88	−5.38/−3.61	0.80	10.70	43.0	3.70	[25]
PFDCTBT-C8	1.71	−5.30/−3.55	0.83	12.57	66.8	7.00	[26]
PDTSCBT	1.83	−5.35/−3.58	0.82	11.10	56.7	5.20	[27]
P2	2.09	−5.22/−3.13	0.79	9.20	57.4	4.17	[28]
PSBT	1.81	−5.40/−3.59	0.94	11.11	49.6	5.18	[29]
PSS2	2.01	−5.36/−3.54	1.03	15.70	58.3	9.46	[30]
PBDCPDT-TPD	1.88	−5.36/−3.20	0.87	12.21	62.2	6.60	[31]
PIDTT-DFBT	1.78	−5.30/−3.50	0.95	12.21	61.0	7.03	[32]

a) Except for PSS2 which was blended with EH-IDTBR, all the polymers were blended with $PC_{71}BM$.

with a deep-lying HOMO energy level of −5.43 eV was synthesized. PSC based on PC2 presented an intriguing PCE of 7.26% with a large V_{oc} of 0.93 V, a J_{sc} of 14.22 mA cm^{-2}, and a FF of 54.7% [23]. Further side-chain engineering on the DIC core led to copolymer PC1BT6, which showed a slightly enhanced PCE of 7.34% with a V_{oc} of 0.95 V and a J_{sc} of 14.00 mA cm^{-2} [24]. By substituting the two phenyl ending groups in DIC core with thienyl units, Zheng et al. also developed a more electron-rich building block of heteroheptacene (DTCC), which could be used to synthesize LBG copolymers [33]. These heteroheptacene-based copolymers delivered bathochromic absorption bands and moderate charge carrier mobilities, which are in favor of fabricating efficient PSCs. At the same time, Cheng and coworkers also reported a DTCC-based copolymer of PCDCTBT which showed a decent PCE of 3.70% [25]. Furthermore, they developed a new heteroheptacene by replacing the nitrogen atom with the sp^3-hybridized bridging atom in the central bridging atom to afford dithienocyclopentafluorene (DTCF) [25]. Through side-chain engineering in combination with device engineering, the DTCF-based copolymer (PFDCTBT-C8) showed an enhanced PCE of 7.00% [26]. When the other two sp^3-hybridized bridging carbon atoms of DTCC were replaced with silicon atoms in the PCDCTBT, the corresponding copolymer PDTSCBT exhibited an enhanced PCE of 5.20% in comparison with the PCE of 3.70% for PCDCTBT [27].

By replacing the central nitrogen atom in the DTCC with sulfur atom, a new ladder-type heteroheptacene building block, namely dithienocyclopentadibenzothiophene (DTCD), was reported by Zheng and coworkers [28]. The copolymer P2 based on DTCD and 2-2′-bithiophene afforded a decent PCE of 4.17% with a J_{sc} of

9.20 mA cm^{-2}, a V_{oc} of 0.79 V, and a FF of 57.4%. To reduce the optical bandgap, electron-deficient benzothiadiazole derivative was utilized as an acceptor moiety to afford the D–A copolymer of PSBT which has a deeper-lying HOMO energy level of −5.40 eV [29]. The optimal PSC based on PSBT:PC$_{71}$BM exhibited an increased PCE of 5.18% with a high V_{oc} of 0.94 V in comparison to that of P2-based device. More recently, using the DTCD and a difluorobenzotriazole derivative as the electron-rich and electron-deficient moieties, respectively, our group further developed a D–A copolymer (PSS2) which presented a deep-lying HOMO energy level of −5.36 eV and a wide optical bandgap of 2.01 eV [30]. When paired with a LBG NFA (EH-IDTBR), PSS2-based PSC afforded a PCE of 9.46% with a high V_{oc} of 1.03 V, a J_{sc} of 15.70 mA cm^{-2}, and a FF of 58.3%. It should be noted that the 9.46% PCE is the highest among all D–A copolymers based on these ladder-type heptacenes. In addition, after storage for 120 days, the optimal PSS2-based device exhibited good shelf-stability with a PCE of 9.03%, which is 95% of its initial value.

At the same time, Cheng and coworkers developed a heptacyclic benzodi (cyclopentadithiophene) (BDCPDT) building block, in which the thiophene is fastened on the two ends of BDT with sp^3-hybridized bridging atoms [31]. Using thieno[3,4-*c*]pyrrole-4,6-dione as the acceptor unit, the BDCPDT-based copolymer (PBDCPDT-TPD) afforded a decent PCE of 6.60%. Subsequently, Jen and coworkers extended the conjugation length of IDT unit with two additional thiophene rings on both ends of IDT to achieve a new heptacyclic heteroacene (IDTT), which later becomes a star building block for high-performance A–D–A NFAs [32]. A new D–A copolymer (PIDTT-DFBT) was synthesized by the copolymerization between the distannylated IDTT and a dibrominated benzothiadiazole derivative. Without any post-treatments, the PSC based on PIDTT-DFBT:PC$_{71}$BM showed an intriguing PCE of 7.03% with a large V_{oc} of 0.95 V.

5.2.3 D–A Copolymers Based on Heteroacenes with Nine or More Fused Rings

To further extend the conjugated length, Cheng et al. developed a ladder-type nonacyclic thienyl-phenylene-thienylene-phenylene-thienyl (TPTPT) building block which consists of alternating interfused benzene and thiophene segments [34]. The resulting D–A copolymer (PTPTPTBT in Figure 5.3) showed a relatively high hole mobility of 5.1 × 10^{-4} cm^2 V^{-1} s^{-1} with HOMO/LUMO energy levels of −5.24/−3.38 eV, and an optical bandgap of 1.73 eV (Table 5.3). The inverted PSC based on PTPTPTBT:PC$_{71}$BM delivered a PCE of 5.90% with a V_{oc} of 0.76 V, a J_{sc} of 12.1 mA cm^{-2}, and a FF of 64.0%. An undecyclic indacenodithiophene-biscyclopentadithiophene (IDTCPDT) with 11 aromatic rings in the conjugated backbone was developed by Jen and coworkers [35]. The IDTCPDT-based copolymer (PIDTCPDT-DFBT) showed a bathochromically shifted absorption band in comparison with the non-fused counterpart. It should be noted that PIDTCPDT-DFBT still exhibited suitable solubility despite its highly extended and rigid backbone. The best-performance field-effect transistor (FET) based on PIDTCPDT-DFBT showed an improved hole mobility of 2.4 × 10^{-2} cm^2 V^{-1} s^{-1} compared to that based on the

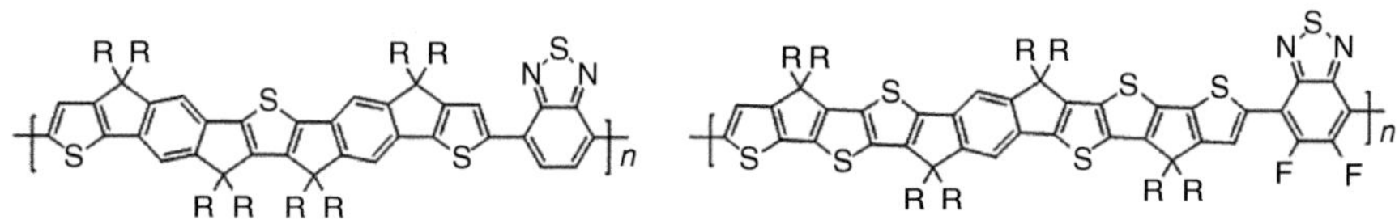

Figure 5.3 Chemical structures of PTPTPTBT and PIDTCPDT-DFBT. Source: Qingdong Zheng.

Table 5.3 Summary of bandgaps, energy levels, and photovoltaic performances of PTPTPTBT and PIDTCPDT-DTBT shown in Figure 5.3.

Donor[a)]	E_g (eV)	HOMO/LUMO (eV)	V_{oc} (V)	J_{sc} (mA cm^{-2})	FF (%)	PCE (%)	References
PTPTPTBT	1.73	−5.24/−3.38	0.76	12.10	64.0	5.90	[34]
PIDTCPDT-DFBT	1.59	−5.05/−3.28	0.75	14.59	59.0	6.46	[35]

a) Blended with $PC_{71}BM$.

non-fused counterpart (1.6×10^{-2} cm^2 V^{-1} s^{-1}). When blended with $PC_{71}BM$, the corresponding PSC showed a decent PCE of 6.46% which was lower than that of the counterpart with short conjugation length. The results suggest that the ladder-type donor unit should have an optimal conjugated length to achieve D–A copolymers with enhanced performance. Since too long conjugation length will decrease the solubility of the resulting copolymers, researchers have usually introduced more bulky side chains on the sp^3-hybridized bridging atoms which will inevitably lead to decreased charge mobilities.

5.3 A–D–A NFAs Based on Ladder-Type Heteroacenes

From 1995 to 2015, most efficient PSCs were fabricated by using a fullerene derivative as the electron acceptor material to blend with polymer donor or small-molecule donor materials. However, the highest PCE of these fullerene-based PSCs can barely exceed 12.00% [2]. This PCE limitation of fullerene-based PSCs is mainly attributed to the fact that the absorption bands of the fullerene derivatives are located in the short wavelength region and their energy levels can be tuned in a very small range. Therefore, NFA materials are emerging as promising alternatives to replace the traditional fullerene derivatives with their easy synthesis accessibility. More importantly, these NFAs can provide tunable bandgaps and energy levels in much wider ranges thereby leading to well-matched absorption profiles and energy levels between the NFAs and the donor materials. Although there have been a number of design strategies for NFAs, the acceptors with an A–D–A structural motif have been the most extensively explored in recent years since the first report of ITIC acceptor by Zhan and coworkers in 2015 [8]. In the A–D–A-structured acceptors, the electron-donating ladder-type heteroacenes have a great effect on the photovoltaic performance of the resulting NFAs. Thus, in the following, we will discuss the influence of the chemical structures of ladder-type heteroacenes on the performance of NFAs. It should be noted that both ending groups and π-bridges of NFAs can also affect their photovoltaic performance.

5.3.1 A–D–A NFAs Based on Heteropentacenes and Heterohexacenes

As a coplanar and symmetric configuration, the pentacyclic IDT moiety is a promising electron-rich unit for designing efficient A–D–A NFAs. The first example of IDT-based NFAs is DC-IDT2T (Figure 5.4), which was synthesized by Zhan and coworkers by using 2-(3-oxo-2,3-dihydroinden-1-ylidene)malononitrile (INCN)

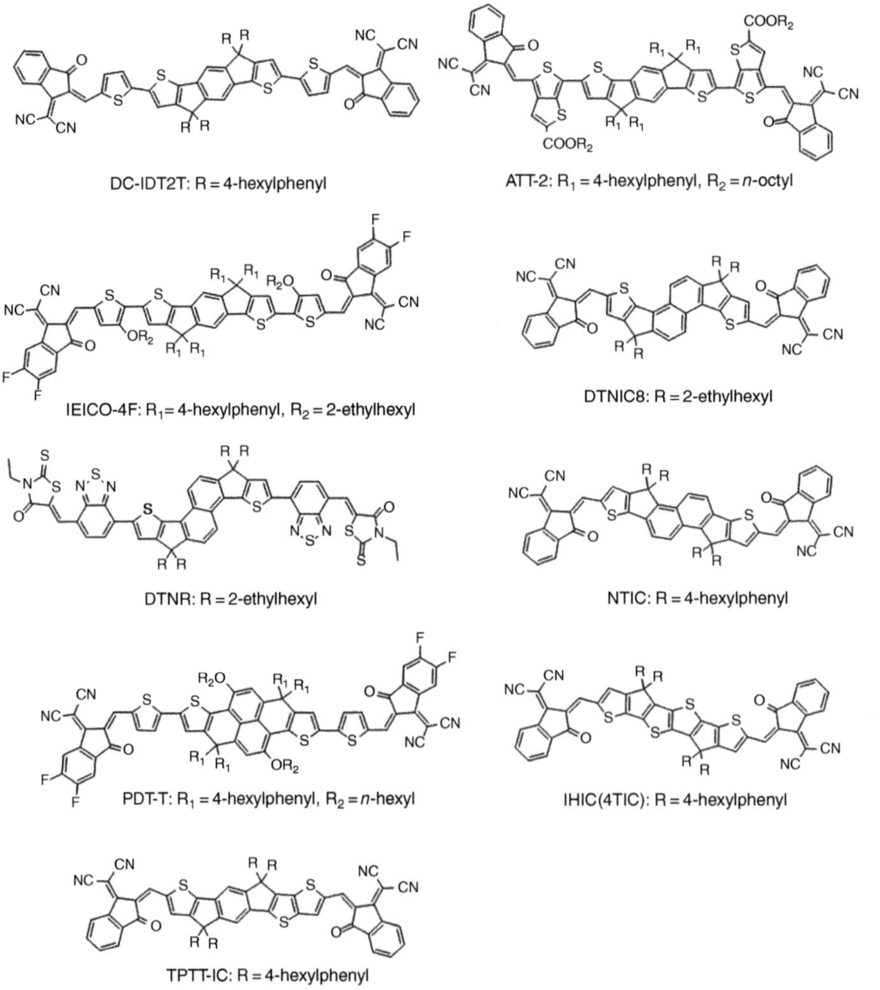

Figure 5.4 Chemical structures of some representative NFAs based on heteropentacenes or heterohexacenes. Source: Qingdong Zheng.

Table 5.4 Summary of bandgaps, energy levels, and photovoltaic performances of NFAs shown in Figure 5.4.

Acceptor	E_g (eV)	Donor	HOMO/LUMO (eV)	V_{oc} (V)	J_{sc} (mA cm^{-2})	FF (%)	PCE (%)	References
DC-IDT2T	1.55	PBDTTT-C-T	−5.43/−3.85	0.90	8.33	52.3	3.93	[36]
ATT-2	1.32	PTB7-Th	−5.50/−3.90	0.73	20.75	63.0	9.58	[37]
IEICO-4F	1.24	PTB7-Th	−5.44/−4.19	0.74	22.80	59.4	10.00	[38]
DTNIC8	1.73	PBDB-T	−5.91/−3.93	0.96	12.92	72.8	9.03	[39]
DTNR	1.72	PTB7-Th	−5.67/−3.75	1.08	15.72	56.0	9.51	[40]
NTIC	1.67	PBDB-T	−5.58/−3.76	0.94	13.55	68.1	8.63	[41]
PDT-T	1.44	PBDB-T	−5.43/−3.97	0.88	18.88	66.3	11.02	[42]
IHIC	1.38	PTB7-Th	−5.45/−3.93	0.75	19.01	68.1	9.77[a)]	[43]
TPTT-IC	1.63	PBT1-C	−5.78/−3.95	0.96	15.60	70.0	10.50	[44]

a) Measured from a semi-transparent device.

as the ending groups and thiophene as the π-bridges [36]. DC-IDT2T has a LBG of 1.55 eV with a relatively high-lying LUMO energy level. The PSC based on DC-IDT2T:PBDTTT-T-C showed a PCE of 3.93% (Table 5.4). Using a thieno[3,4-b]thiophene derivative as the π-bridge to substitute the thiophene bridge in DC-IDT2T, Zhu and coworkers designed and synthesized another IDT-based NFA (ATT-2) with the same ending group of INCN [37]. ATT-2 showed a reduced optical bandgap of 1.32 eV with a relatively deep-lying LUMO energy level of −3.90 eV. Because of the complementary absorption and matched energy levels between PTB7-Th and ATT-2, the corresponding PSC delivered an excellent PCE of 9.58% with a high J_{sc} of 20.75 mA cm^{-2} despite its relatively low V_{oc} of 0.73 V. To further reduce the bandgaps of the IDT-based NFAs and enhance the light absorption in the NIR region, alkoxy side chains and fluorinated ending groups were simultaneously employed in the conjugated backbone of DC-IDT2T [38]. Because of the conformational lock and fluorination effects, the resulting NFA (IEICO-4F) showed a down-shifted LUMO energy of −4.19 eV, and a red-shifted absorption band with an absorption onset of 1000 nm, corresponding to a rather low optical bandgap of 1.24 eV. A PCE of 10.00% with an impressive J_{sc} of 22.8 mA cm^{-2}, a V_{oc} of 0.74 V, and a FF of 59.4% was achieved for the PSC based on PTB7-Th:IEICO-4F. Through fine tuning the blend morphology, the best PCE of PTB7-Th:IEICO-4F-based PSCs was further improved to 12.80% with an enhanced J_{sc} of 27.3 mA cm^{-2} [45].

Using a heterohexacene-based building block developed for the D–A copolymers in our group, we further designed and synthesized a NFA (DTNIC8) with hexacyclic angular-shaped DTN as the electron-rich core and INCN as the ending groups [39]. DTNIC8 showed a deep-lying HOMO energy level of −5.91 eV with a medium bandgap of 1.73 eV. A well-defined phase separation morphology was observed in the blend film of PBDB-T:DTNIC8, yielding efficient exciton dissociation and transportation. Meanwhile, the hole and electron mobilities for the

PBDB-T:DTNIC8 blend were balanced with moderate values of 3.1×10^{-5} and 2.8×10^{-5} cm^2 V^{-1} s^{-1}, respectively. The best-performance PSC based on DTNIC8 delivered a PCE of 9.03% with a V_{oc} of 0.96 V, a J_{sc} of 12.92 mA cm^{-2}, and a FF of 72.8%. To further improve the photovoltaic performance, a new NFA (DTNR) was developed by using the ladder-type angular-shaped DTN flanked with the weak electron-deficient 3-ethylrhodanine moieties through benzothiadiazole-based bridging groups [40]. Compared to DTNIC8, DTNR showed an up-shifted LUMO energy level of −3.75 eV, which is beneficial to achieve a high V_{oc} and a decreased energy loss for the resulting PSC. When paired with the polymer donor of PTB7-Th, DTNR delivered an enhanced efficiency of 9.51% with an increased V_{oc} of 1.08 V as well as a rather low energy loss of 0.50 eV. Additionally, the DTNR-based device showed a higher PCE and better storage stability in comparison with the $PC_{71}BM$-based counterpart. At the same time, Chen and coworkers reported a NFA (NTIC) based on the linear-shaped DTN, which showed up-shifted HOMO/LUMO energy levels of −5.58/−3.76 eV and a reduced optical bandgap of 1.67 eV [41]. The best-performance PSC based on NTIC:PBDB-T showed a PCE of 8.63% which is lower than that of the device based on DTNIC8. Recently, Bo and coworkers also synthesized a naphthalene-based NFA (PDT-T), which has a LBG of 1.44 eV [42]. When PDT-T was blended with a wide bandgap polymer of PBDB-T, the photo-current responses were extended to 900 nm for the PDT-T-based PSC. Therefore, a high PCE of 11.02% with a J_{sc} of 18.88 mA cm^{-2} was obtained for the corresponding device. These results demonstrate that extending the conjugation length of ladder-type heteroacenes is helpful to improve the photovoltaic performance of NFAs. To strengthen the electron-donating ability of the central electron-rich core, thieno[3,2-*b*]thiophene was introduced to replace the naphthalene in DTN to afford a new NFA (IHIC) which has a reduced optical bandgap of 1.38 eV in comparison with DTNIC8 [43]. The PTB7-Th:IHIC blend showed a high electron mobility of 1.2×10^{-3} cm^2 V^{-1} s^{-1} which is favorable for obtaining a high PCE (9.77%) for the corresponding semi-transparent PSC. The same NFA was also reported in parallel and named as 4TIC by Jen and coworkers. The best-performance PSC based on PTB7-Th:4TIC delivered a PCE of 10.43% [46].

Using hexacyclic thiophene-phenylene-thieno[3,2-*b*]thiophene fused core, Sun and coworkers synthesized a new NFA (TPTT-IC) [44] which has a planar and asymmetric structure, and appropriate energy levels. When blended with a wide bandgap copolymer of PBT1-C, the resulting PSC based on TPTT-IC showed a high PCE of 10.50%, which is higher than that of the ITIC-based device under the same preparation conditions.

5.3.2 A–D–A NFAs Based on Heteroheptacenes

5.3.2.1 NFAs Based on Heteroheptacenes with sp^3-Hybridized Bridging Atoms

The heteroheptacenes are the most popular electron-rich core for constructing high-performance A–D–A NFAs considering their moderate conjugated lengths. Figure 5.5 shows the chemical structures of some representative NFAs based on heteroheptacenes. The optical, electrochemical, and photovoltaic properties of the

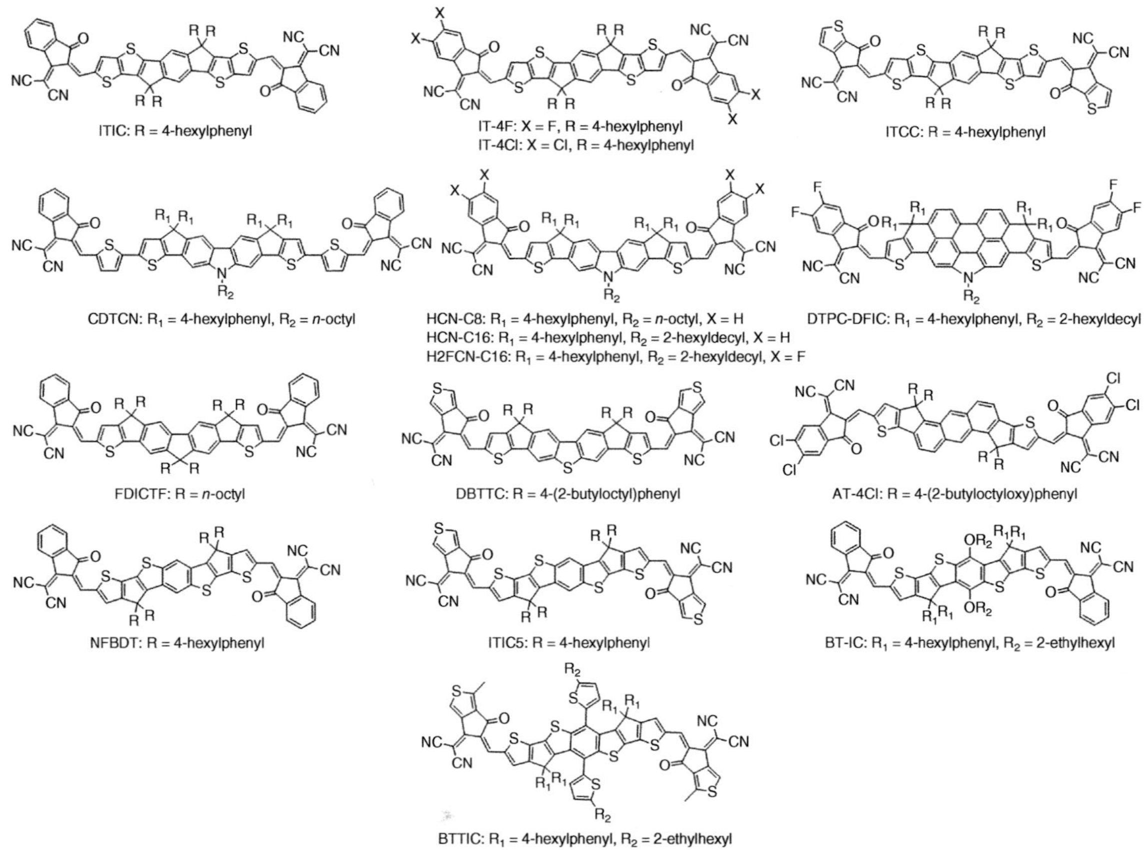

Figure 5.5 Chemical structures of NFAs based on heteroheptacenes with sp^3-bridged carbons. Source: Qingdong Zheng.

Table 5.5 Summary of bandgaps, energy levels, and photovoltaic performances of NFAs shown in Figure 5.5.

Acceptor	E_g (eV)	Donor	HOMO/LUMO (eV)	V_{oc} (V)	J_{sc} (mA cm^{-2})	FF (%)	PCE (%)	References
ITIC	1.59	PTB7-Th	−5.48/−3.83	0.81	14.21	59.1	6.80	[8]
IT-4F	1.51	PBDB-T-SF	−5.66/−4.14	0.88	20.88	71.3	13.10	[47]
IT-4F	1.51	T1	−5.66/−4.14	0.90	21.50	78.0	15.10	[48]
IT-4Cl	1.48	PM6	−5.75/−4.27	0.79	22.67	75.2	13.45	[49]
ITCC	1.67	PBDB-T	−5.47/−3.76	1.01	15.90	71.0	11.40	[50]
CDTCN	1.48	PBDB-T	−5.39/−3.80	0.96	11.26	57.6	6.23	[51]
HCN-C8	1.67	J71	−5.63/−3.54	1.01	6.24	37.8	2.38	[52]
HCN-C16	1.66	J71	−5.63/−3.57	1.03	12.30	43.5	5.51	[52]
H2FCN-C16	1.59	J71	−5.95/−3.99	0.90	18.62	66.7	11.18	[52]
DTPC-DFIC	1.21	PTB7-Th	−5.31/−4.10	0.76	21.92	61.3	10.21	[53]
FDICTF	1.63	PBDB-T	−5.43/−3.71	0.94	15.81	66.0	9.81	[54]
DBTTC	1.64	PBDB-T	−5.92/−4.03	0.97	17.25	67.0	11.25	[55]
AT-4Cl	1.60	PBDB-T	−5.71/−3.89	0.90	19.52	75.5	13.27	[56]
NFBDT	1.56	PBDB-T	−5.40/−3.83	0.87	17.85	67.2	10.42	[57]
ITIC5	1.53	J71	−5.48/−3.95	0.90	18.48	75.5	12.50	[58]
BT-IC	1.43	J71	−5.32/−3.85	0.90	17.75	65.7	10.46	[59]
BTTIC	1.47	PBDB-T	−5.59/−3.86	0.90	19.52	74.7	13.18	[60]

NFAs are summarized in Table 5.5. The benchmark NFA (ITIC) was first reported by Zhan and coworkers by using a ladder-type heptacyclic core (IDTT), which is formed by fusing two thiophene rings on the IDT unit [8]. The PSC based on PTB7-Th:ITIC only showed a modest PCE of 6.80%, which is mainly ascribed to the mismatched energy levels and overlapped absorption spectra between the LBG PTB7-Th donor and the LBG ITIC acceptor. When a wide bandgap polymer of PBDB-T was used to replace PTB7-Th, the resulting PSC showed a significantly enhanced PCE of 11.21% with a V_{oc} of 0.90 V, a J_{sc} of 16.81 mA cm^{-2}, and a FF of 74.2% due to the more complementary absorption and well-matched energy levels [61]. Fluorine atom has strong electronegativity, and it is widely used in the modification of ending groups in NFAs. When fluorine atoms were introduced to the ending groups of ITIC, the resulting NFA (IT-4F) [47] showed a bathochromically shifted absorption band and deeper-lying energy levels than ITIC. At the same time, IT-4F delivered a higher electron mobility of 5.05×10^{-4} cm^2 V^{-1} s^{-1} in comparison with ITIC, which is in favor of the enhanced charge transport in PSCs. When a wide bandgap polymer of PBDB-T-SF was used as the electron donor material, the resulting PSC showed a record-high PCE of 13.10% at that time with a high J_{sc} of 20.88 mA cm^{-2}. The PCE of IT-4F-based device was further improved to 15.10% when a ternary conjugated D–A copolymer with a finely tuned HOMO level was used [48]. Hou and coworkers further introduced chlorine atoms into the ending groups of ITIC to obtain a chlorinated NFA (IT-4Cl) with red-shifted absorption compared to IT-4F [49]. The LUMO energy level of IT-4Cl was down-shifted to −4.27 eV. Thus, a remarkable PCE of 13.45% with a high J_{sc} of 22.67 mA cm^{-2} and a relatively low V_{oc} of 0.79 V was achieved for the PSC based on PM6:IT-4Cl. When the phenyl in the ending group of ITIC was replaced with the thienyl unit, the resulting NFA (ITCC) exhibited an up-shifted LUMO energy level of −3.76 eV and an enhanced electron mobility of 9.26×10^{-4} cm^2 V^{-1} s^{-1} [50]. Thus, a PCE of 11.40% with an improved V_{oc} of 1.01 V was obtained for the PSC based on PBDB-T:ITCC.

Compared to IDTT, DTCC shows relatively weak electron-donating ability. Our group first reported a NFA (CDTCN) by using DTCC as the central core and two thiophene units as π-bridges [51]. With the strong intramolecular charge transfer (ICT) effect, CDTCN displayed a reduced bandgap of 1.48 eV with a large absorption coefficient of 2.9×10^5 M^{-1} cm^{-1}. PSCs based on CDTCN:PBDB-T showed a decent PCE of 6.23%. At the same time, Sun and coworkers also reported a DTCC-cored NFA with a similar PCE of 6.00% [62]. Subsequently, our group further designed and synthesized three DTCC-based NFAs (HCN-C8, HCN-C16, and H2FCN-C16) to investigate the influences of side-chain and end-group fluorination on the optical, electrochemical, and photovoltaic properties of the NFAs [52]. Although the side chain on the central nitrogen bridging atom barely affects the energy levels and bandgaps, the substitution of *n*-octyl with 2-hexyldecyl results in the molecular ordering change of the NAFs from crystalline (HCN-C8) to amorphous (HCN-C16), which can affect their miscibility with polymer donor materials. Furthermore, H2FCN-C16 with fluorinated ending groups (INCN2F) exhibited a reduced bandgap and deep-lying energy levels. The H2FCN-C16-based blend film exhibited a closer π–π packing with improved crystallinity thereby resulting in its enhanced charge transport. The best-performance PSC based on H2FCN-C16:J71 yielded an impressive PCE of 11.18% with a V_{oc} of 0.90 V, a J_{sc} of 18.62 mA cm^{-2}, and a FF of 66.7%. However, the best-performance PSCs based on HCN-C8 and HCN-C16 showed PCEs of 2.38% and 5.51%, respectively. These results suggest that the electron-rich DTCC is a promising building block for constructing high-performance NFAs. On the other hand, extending the π-conjugation of DTCC by substituting the central benzene units with naphthalene, a ladder-type dithienopicenocarbazole (DTPC) core was obtained. This two-dimensional conjugated DTPC with nitrogen bridging atom has a strong electron-donating ability, which could enhance the ICT effect and the intermolecular interaction of the resulting DTPC-derived NFAs. An ultra-narrow bandgap (1.21 eV) NFA (DTPC-DFIC) based on DTPC was designed and synthesized by Jen and coworkers [53]. When combined with a spectral-complementary electron donor material of PTB7-Th, the DTPC-DFIC-based PSC exhibited an efficiency of 10.21% with a high J_{sc} of 21.92 mA cm^{-2} as well as a quite low energy loss of 0.45 eV.

Compared with DTCC, the DTCF unit contains an additional sp^3-hybridized bridging atom where two bulky side chains are attached in the conjugated backbone, which would impede the close intermolecular π–π packing. Despite such deficiency, Chen and coworkers developed an excellent NFA (FDICTF) based on DTCF. PSCs based on FDICTF:PBDB-T displayed a PCE of 9.81% [54]. Using an electron-rich building block DTCD without the bulky side chain on the central bridging atom, our group designed and synthesized a new NFA (DBTTC) which is end-caped with 2-(6-oxo-5,6-dihydro-4*H*-cyclopenta[*c*]thiophen -4-ylidene)malononitrile [55]. DBTTC showed an optical bandgap of 1.64 eV with a LUMO energy level of −4.03 eV. The DBTTC:PBDB-T blend film exhibited clear phase separation with smooth surface, resulting in an improved charge transport. An impressive PCE of 11.25% with a high V_{oc} of 0.97 V and a J_{sc} of 17.25 mA cm^{-2} was obtained for the DBTTC-based PSC.

By replacing the naphthalene unit in linear-shaped DTN with the anthracene, Chen and coworkers designed and synthesized a new heteroheptacene (AT) with extended conjugation [56]. Using chlorinated INCN as the ending groups, a new NFA (AT-4Cl) was obtained which exhibited a decreased bandgap of 1.60 eV and an increased electron mobility 1.85×10^{-4} cm^2 V^{-1} s^{-1}. The PSC based on PM6:AT-4Cl afforded a high PCE of 13.27% with a V_{oc} of 0.90 V, a high J_{sc} of 19.52 mA cm^{-2}, and a FF of 75.5%.

With symmetrical configuration and high degree of coplanarity, the BDT unit is popular in constructing high-performance D–A copolymers. However, the BDT-based NFAs are rarely reported in comparison with a large number of BDT-based donor materials. Chen and coworkers synthesized a new NFA, namely NFBDT, based on a heptacyclic benzodi(cyclopentadithiophene) (FBDT) unit with a BDT-derived core [57]. The incorporation of BDT endows the target NFA (NFBDT) with a large conjugated backbone which would enhance the ICT effect. As a result, NFBDT showed a LBG of 1.56 eV with a strong absorption band in the range of 600–800 nm. Using the benchmark copolymer of PBDB-T as the donor material, the NFBDT-based PSC delivered an efficiency of 10.42% with a J_{sc} of 17.85 mA cm^{-2}, a V_{oc} of 0.87 V, and a relatively low FF of 67.2%. The result indicates that BDT is a promising candidate to design high-performance NFAs. To improve the photovoltaic property of FBDT-based NFAs, Zhan and coworkers replaced the phenyl of the INCN group in NFBDT with thienyl to obtain a new NFA (ITIC5) [58]. Compared to NFBDT, ITIC5 has a lower optical bandgap of 1.53 eV. After thermal annealing, ITIC5-based blend film exhibited fine and continuous network with enhanced crystallinity, which efficiently facilitated charge dissociation and transport. The corresponding PSC showed an improved PCE of 12.50% with an increased FF of 75.5%, a V_{oc} of 0.90 V, and a J_{sc} of 18.48 mA cm^{-2}. Besides the terminal group optimization, ladder-type NFAs could also be modified to extend their absorption bands to NIR region by introducing electron-donating side chain on the backbone to strengthen the ICT effect. Liao and coworkers developed new NFA (BT-IC) based on the FBDT building block [59]. An electron-donating side chain, 2-ethylhexyloxy, was employed on the BDT segment to decrease the bandgap via elevating the HOMO energy level without reducing the V_{oc}. BT-IC showed a low optical bandgap of 1.43 eV with an up-shifted HOMO energy level of −5.32 eV. When paired with J71, the BT-IC-based PSC exhibited a PCE of 10.46% with a high V_{oc} of 0.90 V. It should be noted that the HOMO driving force for this system is only around 0.1 eV which is much less than the empirical value of 0.3 eV to ensure the efficient charge dissociation. In combination with the modifications on ending group and side chain, Yang and coworkers reported a new FBDT-based NFA (BTTIC) with 2-(1-methyl-6-oxo-5,6-dihydro-4*H*-cyclopenta[*c*]thiophen-4-ylidene)malononitrile as terminal groups [60]. Meanwhile, 2-D conjugated side chains, 2-ethylhexylthiophene, were anchored on the BDT unit. BTTIC showed a wide absorption band with an optical bandgap of 1.47 eV. Benefiting from the favorable morphology of the PBDB-T:BTTIC blend, the corresponding device showed an impressive PCE of 13.18%.

5.3.2.2 NFAs Based on Heteroheptacenes Without sp^3-Hybridized Bridging Atoms

As previously reported, most ladder-type heteroheptacenes building blocks were usually fastened with sp^3-hybridized bridging atoms. However, the solubilizing side chains on the bridging atoms always stick out from the conjugated plane, which would increase the steric-hindrance effect and enlarge the intermolecular π–π packing distance, leading to a decreased carrier transport for the resulting NFAs. Hence, it is urgently needed to explore new A–D–A-type NFAs without the sp^3-hybridized bridging atoms in the ladder-type fused arenes core. Ma et al. developed two novel "M-series" LBG NFAs (M8 and M34 in Figure 5.6) based on ladder-type heteroheptacenes fastened with sp^2-hybridized nitrogen bridging atoms

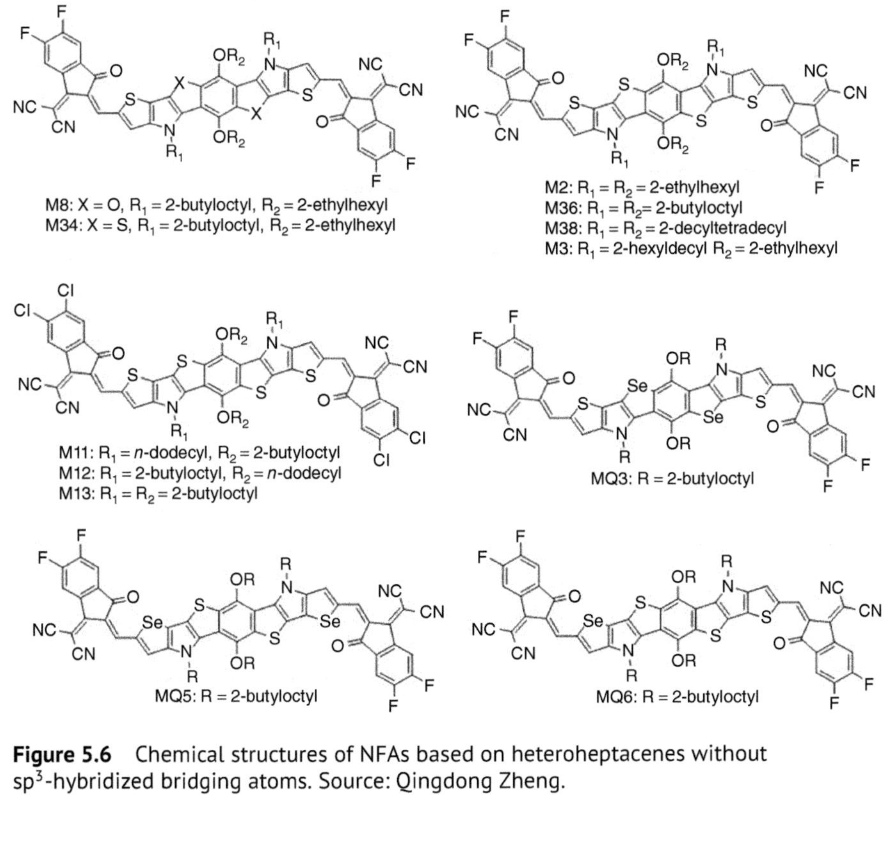

Figure 5.6 Chemical structures of NFAs based on heteroheptacenes without sp^3-hybridized bridging atoms. Source: Qingdong Zheng.

[63]. Both of them share the same side chains and ending groups, but they differ in the central electron-donating core. Replacing the benzodifuran with BDT in the heteroheptacene core resulted in a slightly blue-shifted absorption with a down-shifted HOMO energy level for M34. At the same time, a favorable morphology with a large portion of face-on orientation could be observed in the M34-based blend film. Therefore, PM6:M34 showed enhanced and more balanced charge transport with hole and electron mobilities of 6.25×10^{-4} and 2.65×10^{-4} cm^2 V^{-1} s^{-1}, respectively. The corresponding PSC based on M34:PM6 exhibited an outstanding PCE of 15.24% (Table 5.6) with a J_{sc} of 23.63 mA cm^{-2}, a V_{oc} of 0.91 V, and a FF of 70.7%, which was much higher than the 4.21% efficiency for the M8-based device. The π–π stacking distances and molecular orientations of the heteroheptacene-based NFAs can be effectively modulated by side-chain engineering to improve their photovoltaic performance. Therefore, the length of branched alkyl chain on the central benzo[1,2-*b*:4,5-*b*′]bis(4-*H*-dithieno[3,2-*b*:2′,3′-*d*] pyrrole (BDTP) building block of M34 was systematically optimized [64]. The blend based on M36 with 2-butyloctyl side chains showed a dominated face-on molecular orientation with a closely π–π stacking, leading to the best charge transport with hole/electron mobilities of $6.62 \times 10^{-4}/1.33 \times 10^{-4}$ cm^2 V^{-1} s^{-1}. However, enlarged π–π stacking distances were obtained for the blends based on M2 with 2-ethylhexyl side chains and M38 with 2-decyltetradecyl side chains. Consequently, the PSC based on PM6:M36 achieved an outstanding PCE of 16.00% with a J_{sc} of 24.63 mA cm^{-2}, a V_{oc} of 0.90 V, and a FF of 72.1%. Nevertheless, the PCEs were 11.16% and 8.89% for

Table 5.6 Summary of bandgaps, energy levels, and photovoltaic performances of NFAs shown in Figure 5.6.

Acceptor[a)]	E_g (eV)	HOMO/LUMO (eV)	V_{oc} (V)	J_{sc} (mA cm^{-2})	FF (%)	PCE (%)	References
M8	1.28	−5.49/−3.91	0.83	8.36	60.6	4.21	[63]
M34	1.39	−5.60/−3.91	0.91	23.63	70.7	15.24	[63]
M2	1.39	−5.60/−3.96	0.88	19.76	63.8	11.16	[64]
M36	1.39	−5.62/−3.95	0.90	24.63	72.1	16.00	[64]
M38	1.47	−5.65/−3.93	0.87	18.28	55.7	8.89	[64]
M11	1.39	−5.68/−3.98	0.73	6.47	47.6	2.26	[65]
M12	1.38	−5.68/−4.01	0.88	21.67	60.3	11.55	[65]
M13	1.38	−5.66/−4.00	0.88	22.34	67.0	13.14	[65]
M13[b)]	1.38	−5.66/−4.00	0.87	25.10	71.6	15.63	[65]
M3	1.41	−5.60/−3.85	0.91	24.03	76.2	16.66	[66]
MQ3	1.37	−5.63/−3.94	0.91	22.19	66.9	13.51	[67]
MQ5	1.34	−5.59/−3.95	0.86	24.48	74.3	15.64	[67]
MQ6	1.35	−5.61/−3.95	0.88	24.62	75.7	16.39	[67]

a) Blended with PM6.
b) Using Al as the electrode.

the PSCs based on M2 and M38, respectively. In addition to the length, the sizes and flanking positions of the alkyl chains were also systematically investigated, which later were shown to have a strong influence on the molecular orientation of "M-series" acceptors [65]. It was found that with linear *n*-dodecyl side chains on the nitrogen atoms, the resulting acceptor (M11) tends to adopt an "edge-on" orientation, thus leading to excellent horizontal charge transport with a FET mobility of 3.95×10^{-2} cm^2 V^{-1} s^{-1}. Interestingly, when the linear *n*-dodecyl chains on the nitrogen atoms were replaced with 2-butyloctyl chains, the dominant crystallite textures of the resulting acceptors (M12 and M13) were changed from an "edge-on" manner to a "face-on" manner. Consequently, PSCs based on M12 and M13 showed much higher vertical electron transport, and therefore higher PCEs, than those based on M11. Further optimization of the M13-based devices yielded an outstanding efficiency of 15.63% [65]. By synergistic optimization of the ending groups and side chains on the backbone of M13, an acceptor named M3 was produced and devices based on this material obtained a prominent PCE of 16.66% [66]. To the best of our knowledge, the efficiency of 16.66% is the highest value for single-junction PSCs based on all the A–D–A-type NFAs reported to date and it is also comparable to those based on the A–D–A′–D–A-type Y6-derivated acceptors [1, 3, 10].

Very recently, our group developed three novel NFAs (MQ3, MQ5, and MQ6) by using asymmetric and symmetric ladder-type heteroheptacene cores which have one and two selenophene heterocycles, respectively, to study the influence of noncovalent intramolecular interaction and molecular symmetry on the photovoltaic properties of the "M-series" acceptors [67]. Although MQ3 and MQ5 have the same number of selenophene heterocycles, the heteroheptacene core of MQ5 is end-capped with selenophene rings, while that of MQ3 is end-capped with thiophene rings. Due to the enhanced noncovalent interaction of O···Se compared to O···S, MQ5 shows a bathochromically shifted absorption band and greatly enhanced charge transport thereby leading to a higher PCE of 15.64% compared to 13.51% for MQ3. Moreover, MQ6 with an asymmetric heteroheptacene core shows a further reduced π–π stacking distance induced by the increased dipole moment thereby further improving the carrier transport of the resulting blend film. As a result, PSCs based on MQ6 exhibited the best performance among the three acceptors, with a maximum efficiency of 16.39%.

In addition to the above research, several other efforts, including end-group engineering and auxochromic group engineering, have also been devoted to the development of high-performance M-series NFAs and devices based on these acceptors have also achieved efficiencies over 15% [68–70]. All these results suggest that sp^3 carbon-free ladder-type skeletons with appropriate neighboring side chains are a promising class of building blocks to construct high-performance NFAs.

5.3.3 A–D–A NFAs Based on Heteroacenes with Eight or More Fused Rings

To extend the absorption band of NFAs, Ding and coworkers developed a carbon–oxygen-bridged NFAs (CO_i8DFIC, Figure 5.7) based on an octacyclic heteroacene

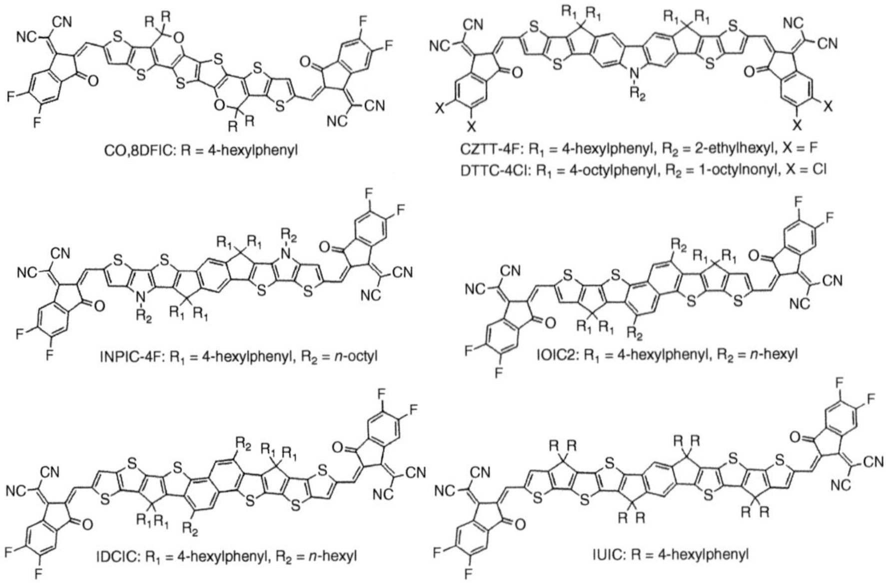

Figure 5.7 Chemical structures of NFAs based on ladder-type heteroacenes with eight or more fused rings. Source: Qingdong Zheng.

Table 5.7 Summary of bandgaps, energy levels, and photovoltaic performances of NFAs shown in Figure 5.7.

Acceptor	E_g (eV)	Donor	HOMO/LUMO (eV)	V_{oc} (V)	J_{sc} (mA cm^{-2})	FF (%)	PCE (%)	References
CO$_i$8DFIC	1.26	PTB7-Th	−5.50/−3.88	0.68	26.12	68.2	12.16	[71]
CZTT-4F	1.48	PM6	−5.61/−4.11	0.94	19.73	65.1	12.07	[72]
DTTC-4Cl	1.47	PM6	−5.72/−4.03	0.92	22.64	74.0	15.42	[73]
INPIC-4F	1.39	PBDB-T	−5.42/−3.94	0.85	21.61	71.5	13.13	[74]
IOIC2	1.55	FTAZ	−5.41/−3.78	0.90	19.70	69.3	12.30	[75]
IDCIC	1.45	FTAZ	−5.19/−3.72	0.87	21.98	71.0	13.58	[76]
IUIC	1.41	PTB7-Th	−5.45/−3.87	0.80	21.74	64.9	11.20	[77]

core [71]. CO$_i$8DFIC showed a rather low optical bandgap of 1.26 eV with a strong absorption band located at 600–1000 nm. When blended with PTB7-Th, CO$_i$8DFIC-based PSC showed a PCE of 12.16% (Table 5.7) with a high J_{sc} of 26.12 mA cm^{-2} and a relatively low V_{oc} of 0.68 V.

In 2019, Tang and coworkers synthesized a nonacyclic DTCC-based heteroacene (CZTT) with an extended, rigid, and coplanar structure by fusing a thiophene on both flanks of DTCC [72]. Using fluorinated INCN as ending groups, CZTT-based NFA (CZTT-4F) exhibited a LBG of 1.48 eV. Blended with PM6, CZTT-4F delivered a high PCE of 12.07% with a J_{sc} of 19.73 mA cm^{-2}. Compared to the fluorination in the ending groups, chlorination has also been proven to be effective to tune the photovoltaic performances of NFAs. Similar to CZTT, Hsu and coworkers developed a ladder-type nonacyclic heteroacene (DTTC) [73]. By using chlorinated ending groups, the NFA (DTTC-4Cl) exhibited red-shifted absorption [73]. The blend film based on DTTC-4Cl:PM6 showed a close π–π packing distance thereby leading to a remarkable PCE of 15.42% with a high J_{sc} of 22.64 mA cm^{-2} and a FF of 74.0% for the corresponding PSC. Tang and coworkers developed a nonacyclic core-based NFA (INPIC-4F) by inserting pyrrole in IDTT [74]. INPIC-4F showed a red-shifted absorption with a LBG of 1.39 eV. Benefiting from the preferred crystallinity and morphology, PSC based on INPIC-4F:PBDB-T exhibited an intriguing PCE of 13.13% with a high J_{sc} of 21.61 mA cm^{-2} and a relatively low V_{oc} of 0.85 V.

An octacyclic heteroacene based on naphtho[1,2-*b*:5,6-*b*′]dithiophene was developed by Zhan and coworkers, which was end-capped with INCN2F to get a NFA (IOIC2) [75]. IOIC2 exhibited a LBG of 1.55 eV with HOMO/LUMO energy levels of −5.41/−3.78 eV. The inverted PSCs based on FTAZ:IOIC2 blend gave a PCE of 12.30% with a V_{oc} of 0.90 V and a J_{sc} of 19.70 mA cm^{-2}. Extending the effective conjugation length of central core in NFAs is a useful strategy to strengthen the ICT effect and π–π interaction for reduced bandgaps and improved charge transport. Wang and coworkers extended the core of IOIC2 to decacyclclic heteroacene [76]. Thus, the corresponding NFA (IDCIC) showed a LBG of 1.45 eV with an increased absorption coefficient of 3.3×10^5 M^{-1} cm^{-1}. Due to the extended conjugation, FTAZ:IDCIC blend film showed an improved charge transport with more balanced hole and electron mobilities in comparison with the IOIC2-based counterpart. The best-performance IDCIC-based PSC showed an improved PCE of 13.58% with an increased J_{sc} of 21.98 mA cm^{-2}.

Zhan and coworkers further extended the core of IT-4F with undecyclic heteroacene to obtain a LBG NFA (IUIC) [77]. The strong electron-donating undecyclic heteroacene led to a broad absorption band up to 900 nm for IUIC. Because of the enhanced π-π interaction, the electron mobility of IUIC was improved to 1.1×10^{-3} cm^2 V^{-1} s^{-1}. And, the PSC based on PTB7-Th:IUIC blend delivered a decent PCE of 11.20%.

5.3.4 Other NFAs

Another distinctive A–D–A NFAs were derived from noncovalently locked building blocks where 2,5-difluorobenzene (DFB) and 2,5-dialkoxybenzene (DOB) were usually incorporated as central cores to create intramolecular noncovalent reactions such as S···O, S···F, and H···F. Chen and coworkers synthesized a new NFA (DF-PCIC in Figure 5.8) containing a DFB core and two cyclopentadithiophene (CPD) segments [78]. A nearly planar structure is achieved by the F···H noncovalent interaction. The optimized PSC based on PBDB-T:DF-PCIC yielded a decent PCE of 10.14% (Table 5.8) with a J_{sc} of 15.66 mA cm^{-2}. Later, Bo and coworkers developed a noncovalently fused-ring NFA (DOC2C6-2F) based on the DOB core and the CPD groups, which was formed by a ladder-like donor unit locked with O···S noncovalent interaction and flanked with two fluorinated ending

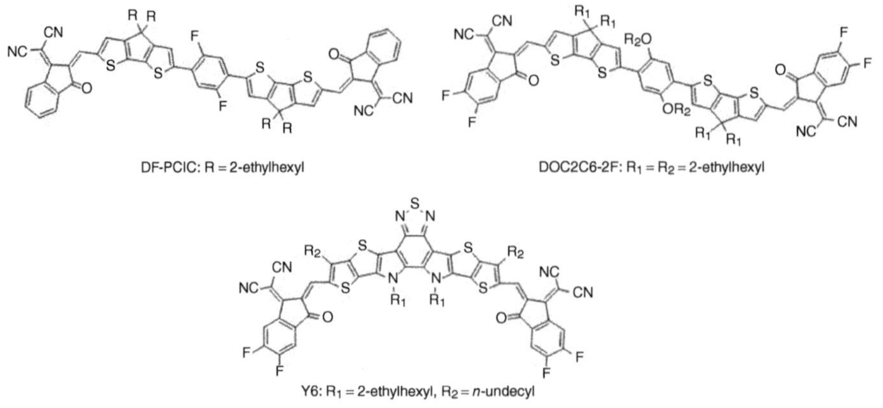

Figure 5.8 Chemical structures of some other NFAs. Source: Qingdong Zheng.

Table 5.8 Summary of bandgaps, energy levels, and photovoltaic performances of NFAs shown in Figure 5.8.

Acceptor	E_g (eV)	Donor	HOMO/LUMO (eV)	V_{oc} (V)	J_{sc} (mA cm^{-2})	FF (%)	PCE (%)	References
Y6	1.33	PM6	−5.65/−4.10	0.83	25.30	74.8	15.70	[10]
DF-PCIC	1.59	PBDB-T	−5.49/−3.77	0.91	15.66	72.0	10.14	[78]
DOC2C6-2F	1.42	PBDB-T	−5.49/−3.83	0.85	21.35	73.2	13.24	[79]

groups [79]. Compared with DF-PCIC, DOC2C6-2F showed markedly broadened absorption with a lower optical bandgap of 1.42 eV. The inverted PSC based on PBDB-T:DOC2C6-2F blend displayed an enhanced PCE of 13.24% with a high J_{sc} of 21.35 mA cm^{-2}.

Besides the A–D–A type NFAs, A–D–A′–D–A-type NFAs are emerging recently since Zou and coworkers reported a new family of nitrogen-bridged small-molecule acceptors (Y-series, such as Y6) using an electron-deficient unit as the central core [1, 10, 11]. Y6 exhibited a broad absorption band in the range of 600–930 nm with a low optical bandgap of 1.33 eV. When paired with a wide bandgap polymer of PM6, the best-performance PSC based on Y6 showed an outstanding PCE of 15.7%. The discovery of these Y-series NFAs greatly promoted the advances of organic solar cells, and since then many research groups used these Y-series NFAs (especially Y6) to successfully achieve PSCs with PCEs over 16.00% [1, 80].

5.4 Conclusions and Outlook

In the past decade, a large number of ladder-type heteroacenes have been designed and synthesized for both donor and acceptor materials. Extension of the conjugated backbones usually resulted in decreased solubilities of the resulting materials. A good balance between the solubility and crystallinity should be maintained. The electron-donating abilities as well as the conjugated lengths of the ladder-type heteroacenes can affect the optical, electrochemical, and photovoltaic properties of the resulting acceptor materials. Similar to the electron-donating groups, the electron-withdrawing groups and π bridges can also affect their optical, electrochemical, and photovoltaic properties. However, side chains mainly affect their crystallinity, molecular orientation as well as phase separation properties. So far, most ladder-type heteroacenes used bulky substituents flanked on their sp^3-hybridized bridging atoms to restrain the severe intermolecular aggregation induced by the extended planar π-conjugated core. For NFAs, the bulky side chains on the sp^3-hybridized bridging carbon atoms in the D unit will inevitably limit the efficient carrier transport of the resulting materials due to their decreased crystallinity. Therefore, it is in great need to explore novel A–D–A NFAs without the sp^3-hybridized bridging atoms in D unit. So far, A–D–A NFAs without any sp^3-hybridized bridging atoms in the conjugated backbone are much less developed compared to those with the sp^3-hybridized bridging atoms. Recently, our group developed a series of ladder-type heteroheptacene-based NFAs (M-series) which are all free of the sp^3-hybridized bridging atoms. In our work, two pairs of neighboring side chains on the heteroheptacene core were used to suppress the immoderate intermolecular aggregation thereby leading to suitable phase separation with optimal domain sizes. As a result, the benchmark NFA (M3) developed in our laboratory showed an outstanding PCE of 16.66%. To the best of our knowledge, the PCE of 16.66% is the highest for PSCs based on all A–D–A type NFAs reported to date and it is also comparable to those based on the Y-series (A–D–A′–D–A-type) NFAs.

However, to achieve PSCs with PCEs over 20%, there are still some remaining issues to be resolved: (i) Reducing the V_{oc} loss via synergistic molecular designs on both the donor and acceptor materials; (ii) Developing novel infrared NFAs with high absorptivity and wide absorption range; (iii) Achieving NFAs with face-on orientation and efficient electron transport; and (iv) Designing of new structures in the D, A, and the π-bridge which can be used to finely tune the properties of NFAs.

References

1 Li, S., Li, C.-Z., Shi, M., and Chen, H. (2020). *ACS Energy Lett.* 5: 1554.
2 Ma, Y., Kang, Z., and Zheng, Q. (2017). *J. Mater. Chem. A* 5: 1860.
3 Zhang, G., Zhao, J., Chow, P.C.Y. et al. (2018). *Chem. Rev.* 118: 3447.
4 Wu, J.-S., Cheng, S.-W., Cheng, Y.-J., and Hsu, C.-S. (2015). *Chem. Soc. Rev.* 44: 1113.
5 Zhao, J., Li, Y., Yang, G. et al. (2016). *Nat. Energy* 1: 15027.
6 Yan, C., Barlow, S., Wang, Z. et al. (2018). *Nat. Rev. Mater.* 3: 18003.
7 Zheng, Z., Yao, H., Ye, L. et al. (2020). *Mater. Today* 35: 115.
8 Lin, Y., Wang, J., Zhang, Z.G. et al. (2015). *Adv. Mater.* 27: 1170.
9 Dey, S. (2019). *Small* 15: 1900134.

10 Yuan, J., Zhang, Y., Zhou, L. et al. (2019). *Joule* 3: 1140.
11 Liu, S., Yuan, J., Deng, W. et al. (2020). *Nat. Photonics* 14: 300.
12 Chen, J., Yang, K., Zhou, X., and Guo, X. (2018). *Chem. Asian J.* 13: 2587.
13 Cui, C. and Li, Y. (2019). *Energy Environ. Sci.* 12: 3225.
14 Li, G., Zhu, R., and Yang, Y. (2012). *Nat. Photonics* 6: 153.
15 Zheng, Q., Jung, B.J., Sun, J., and Katz, H.E. (2010). *J. Am. Chem. Soc.* 132: 5394.
16 Yin, Z., Zheng, Q., Chen, S.-C., and Cai, D. (2013). *ACS Appl. Mater. Interfaces* 5: 9015.
17 Chen, Y.-C., Yu, C.-Y., Fan, Y.-L. et al. (2010). *Chem. Commun.* 46: 6503.
18 Ma, Y., Chen, S.-C., Wang, Z. et al. (2017). *Nano Energy* 33: 313.
19 Ma, Y., Zheng, Q., Yin, Z. et al. (2013). *Macromolecules* 46: 4813.
20 Ma, Y., Zheng, Q., Wang, L. et al. (2014). *J. Mater. Chem. A* 2: 13905.
21 Deng, Y., Liu, J., Wang, J. et al. (2014). *Adv. Mater.* 26: 471.
22 Wang, M., Cai, D., Yin, Z. et al. (2016). *Adv. Mater.* 28: 3359.
23 Wang, L., Cai, D., Yin, Z. et al. (2014). *Polym. Chem.* 5: 6847.
24 Tu, Q., Cai, D., Wang, L. et al. (2016). *J. Mater. Chem. C* 4: 6160.
25 Wu, J.-S., Cheng, Y.-J., Dubosc, M. et al. (2010). *Chem. Commun.* 46: 3259.
26 Chang, C.-Y., Cheng, Y.-J., Hung, S.-H. et al. (2012). *Adv. Mater.* 24: 549.
27 Wu, J.-S., Cheng, Y.-J., Lin, T.-Y. et al. (2012). *Adv. Funct. Mater.* 22: 1711.
28 Cai, D., Zheng, Q., Chen, S.-C. et al. (2012). *J. Mater. Chem.* 22: 16032.
29 Wang, L., Cai, D., Tang, C. et al. (2015). *J. Mater. Chem. C* 3: 5631.
30 Tu, Q., Tang, C., and Zheng, Q. (2019). *J. Mater. Chem. A* 7: 3307.
31 Chen, Y.-L., Chang, C.-Y., Cheng, Y.-J., and Hsu, C.-S. (2012). *Chem. Mater.* 24: 3964.
32 Xu, Y.-X., Chueh, C.-C., Yip, H.-L. et al. (2012). *Adv. Mater.* 24: 6356.
33 Zheng, Q., Chen, S., Zhang, B. et al. (2011). *Org. Lett.* 13: 324.
34 Cheng, Y.-J., Chen, C.-H., Lin, Y.-S. et al. (2011). *Chem. Mater.* 23: 5068.
35 Li, Y., Yao, K., Yip, H.-L. et al. (2014). *Adv. Funct. Mater.* 24: 3631.
36 Bai, H., Wang, Y., Cheng, P. et al. (2015). *J. Mater. Chem. A* 3: 1910.
37 Liu, F., Zhou, Z., Zhang, C. et al. (2017). *Adv. Mater.* 29: 1606574.
38 Yao, H., Cui, Y., Yu, R. et al. (2017). *Angew. Chem. Int. Ed.* 56: 3045.
39 Ma, Y., Zhang, M., Yan, Y. et al. (2017). *Chem. Mater.* 29: 7942.
40 Ma, Y., Zhang, M., Tang, Y. et al. (2017). *Chem. Mater.* 29: 9775.
41 Yi, Y.-Q.-Q., Feng, H., Chang, M. et al. (2017). *J. Mater. Chem. A* 5: 17204.
42 Jiang, P., Lu, H., Jia, Q.-Q. et al. (2019). *J. Mater. Chem. A* 7: 5943.
43 Wang, W., Yan, C., Lau, T.-K. et al. (2017). *Adv. Mater.* 29: 1701308.
44 Li, C., Xie, Y., Fan, B. et al. (2018). *J. Mater. Chem. C* 6: 4873.
45 Song, X., Gasparini, N., Ye, L. et al. (2018). *ACS Energy Lett.* 3: 669.
46 Shi, X., Zuo, L., Jo, S.B. et al. (2017). *Chem. Mater.* 29: 8369.
47 Zhao, W., Li, S., Yao, H. et al. (2017). *J. Am. Chem. Soc.* 139: 7148.
48 Cui, Y., Yao, H., Hong, L. et al. (2019). *Adv. Mater.* 31: 1808356.
49 Zhang, H., Yao, H., Hou, J. et al. (2018). *Adv. Mater.* 30: 1800613.
50 Yao, H., Ye, L., Hou, J. et al. (2017). *Adv. Mater.* 29: 1700254.
51 Wei, J., Tu, Q., and Zheng, Q. (2017). *Dyes Pigm.* 144: 133.
52 Tu, Q., Ma, Y., Zhou, X. et al. (2019). *Chem. Mater.* 31: 5953.

53 Yao, Z., Liao, X., Gao, K. et al. (2018). *J. Am. Chem. Soc.* 140: 2054.
54 Qiu, N., Zhang, H., Wan, X. et al. (2017). *Adv. Mater.* 29: 1604964.
55 Tang, C., Cai, D., and Zheng, Q. (2019). *J. Mater. Chem. A* 7: 9609.
56 Feng, H., Yi, Y.-Q.-Q., Ke, X. et al. (2019). *Adv. Energy Mater.* 9: 1803541.
57 Kan, B., Feng, H., Wan, X. et al. (2017). *J. Am. Chem. Soc.* 139: 4929.
58 Yan, C., Wang, W., Lau, T.-K. et al. (2018). *J. Mater. Chem. A* 6: 16638.
59 Li, Y., Zhong, L., Gautam, B. et al. (2017). *Energy Environ. Sci.* 10: 1610.
60 Gao, W., Liu, T., Ming, R. et al. (2018). *Adv. Funct. Mater.* 28: 1803128.
61 Zhao, W., Qian, D., Zhang, S. et al. (2016). *Adv. Mater.* 28: 4734.
62 Cao, Q., Xiong, W., Chen, H. et al. (2017). *J. Mater. Chem. A* 5: 7451.
63 Ma, Y., Cai, D., Wan, S. et al. (2020). *Angew. Chem. Int. Ed.* 59: 21627.
64 Ma, Y., Cai, D., Wan, S. et al. (2020). *Natl. Sci. Rev.* 7: 1886.
65 Wang, P., Ma, Y., Yin, P. et al. (2021). *Chem. Eng. J.* 418: 129497.
66 Ma, Y., Zhang, M., Wan, S. et al. (2021). *Joule* 5: 197.
67 Tang, C., Ma, X., Wang, J.-Y. et al. (2021). *Angew. Chem. Int. Ed.* 60: 19314.
68 Cai, D., Zhang, J., Wang, J.-Y. et al. (2020). *J. Mater. Chem. A* 8: 24543.
69 Wan, S., Ma, Y., Cai, D. et al. (2021). *Adv. Funct. Mater.* 31: 2010436.
70 Zheng, W., Ma, Y., Tu, Q. et al. (2022). *Chem. Eng. J.* 427: 131022.
71 Xiao, Z., Jia, X., Li, D. et al. (2017). *Sci. Bull.* 62: 1494.
72 Wang, H., Zhang, Z., Yu, J. et al. (2019). *J. Mater. Chem. A* 7: 21903.
73 Chen, T.-W., Peng, K.-L., Lin, Y.-W. et al. (2020). *J. Mater. Chem. A* 8: 1131.
74 Sun, J., Ma, X., Zhang, Z. et al. (2018). *Adv. Mater.* 30: 1707150.
75 Zhu, J., Ke, Z., Zhang, Q. et al. (2018). *Adv. Mater.* 30: 1704713.
76 He, D., Zhao, F., Xin, J. et al. (2018). *Adv. Energy Mater.* 8: 1802050.
77 Jia, B., Dai, S., Ke, Z. et al. (2018). *Chem. Mater.* 30: 239.
78 Li, S., Zhan, L., Liu, F. et al. (2018). *Adv. Mater.* 30: 1705208.
79 Huang, H., Guo, Q., Feng, S. et al. (2019). *Nat. Commun.* 10: 3038.
80 Liu, Q., Jiang, Y., Jin, K. et al. (2020). *Sci. Bull.* 65: 272.

6

Chlorinated Organic Solar Cells

Hui Chen, Mingrui Pu, and Feng He

Southern University of Science and Technology, Department of Chemistry, 1088 Xueyuan Blvd., Nanshan District, Shenzhen 518055, P.R. China

6.1 Introduction

Organic photovoltaic technology has attracted increasing interest over the past decade due to its synthetic versatility, low temperature processing, low material cost, light weight, and flexible application [1–6]. In recent years, based on material innovation and device technology progress, the power conversion efficiency (PCE) of organic solar cells (OSCs) has exceeded 18% [1, 3, 4, 6]. Among them, D–A (donor–acceptor) type polymer donors and new non-fullerene acceptors (NFAs) greatly promote the improvement of PCE. The efficient polymer donors with D and A structural units own strong absorption in visible region due to the enhanced intramolecular charge transfer (ICT) from the push–pull of electrons between unit D and unit A. Various strategies of molecular design for novel D–A type polymers have been developed [7–10].

In past decade, the polymer donors in OSCs have mainly been the D–A type copolymers. The band gap of these polymers is regulated by the backbone, side-chain and molecular interaction. The typical characteristics include the alternating D–A units and quinoidal resonance structures. Besides, the acceptor–donor–acceptor (A–D–A) type small-molecule acceptors greatly contributed photocurrent in OSCs resulting from their strong light-harvesting capability compared to the conventional fullerene derivatives. The novel NFAs such as ITIC [11] and Y6 [12] derivatives have the extensive absorption in visible and near infrared (NIR) region to generate high short-circuit current density (J_{sc}) and matched energy level to produce high open-circuit voltage (V_{oc}).

In design of materials, fluorination was extensively applied in donors and NFAs in recent years [13–15]. The fluorination lowers the highest occupied molecular orbital (HOMO) level of materials, improving the open-circuit voltage (V_{oc}) of their polymer solar cells (PSCs). Although the fluorination is a practical way to enhance

Organic Solar Cells: Materials Design, Technology and Commercialization, First Edition. Edited by Liming Ding.

performance of photovoltaic device, the low yields of products, fussy reaction steps, and use of hazardous regents are the deficiencies of fluorination [16, 17]. To conquer these issues, an alternative method is chlorination. Among halogens, fluorine is the most electrophilic and with the smallest atom size. Chlorine with a larger atom radius is the second most electrophilic. In analogous circumstance, the chlorination reveals the similar affects as the fluorination in polymers and solar cells; meanwhile, the synthesis of the chlorinated polymers is facile [18–22]. Besides, chlorine contains empty 3d orbitals that can accept lone pair electrons or π-electrons through Cl· · ·S or Cl· · ·π noncovalent interactions, while fluorine can not [21–26]. Moreover, the dipole moment of the carbon–chlorine (C—Cl) bond is larger than that of the carbon–fluorine (C—F) bond according to density function theory (DFT) calculation. The large dipole moment of C—Cl bond facilitates the charge transport due to the enhanced ICT effect. He and coworkers designed and prepared many chlorinated polymer donors and NFAs in PSCs [2, 23–37].

In this chapter, we first introduce the characteristics of the chlorination and fluorination. Then, the research progress of chlorinated polymer and small-molecule donors, and small-molecule NFAs are summarized.

6.2 Chlorination Versus Fluorination: A Comprehensive Study

The effects of chlorination are comprehensive on the polymer donors and small-molecule acceptors. The main characteristics of the chlorination versus fluorination are summarized in this section, which would be beneficial to deeply understand the chlorination in organic photovoltaics.

6.2.1 Synthesis

The main advantages of chlorination are high yield and low cost. The synthesis steps are few, the reaction conditions are mild, and the products are easy to purify [22, 38]. Introducing Cl into organic semiconductors is much easier than that of F. F are usually introduced through halogen-exchange reactions under strict conditions. And the yield is relatively low [39–42].

For example, the donor molecules PBDB-T-2Cl and PBDB-TF have very similar structures, and the difference is that PBDB-T-2F has fluorinated thiophene while PBDB-T-2Cl owns a side chain of chlorinated thiophene in a side chain. The synthetic routes of two polymers were systematically compared and analyzed in literature [22]. They were finally polymerized using the same method. The key precursors of PBDB-TF and PBDB-T-2Cl both rooted in BDT-T-based monomers in Figure 6.1. However, their syntheses are quite different. The fluorinated BDT-T-based monomer need four steps, resulting in high synthetic cost. Compared to the fluorinated BDT-T-based monomer, the chlorinated BDT-T-based monomer was facilely prepared from the low-cost 3-chlorothiophene raw material. The yield was high. From above, the synthesis cost of PBDT-T-2Cl is much lower than that of

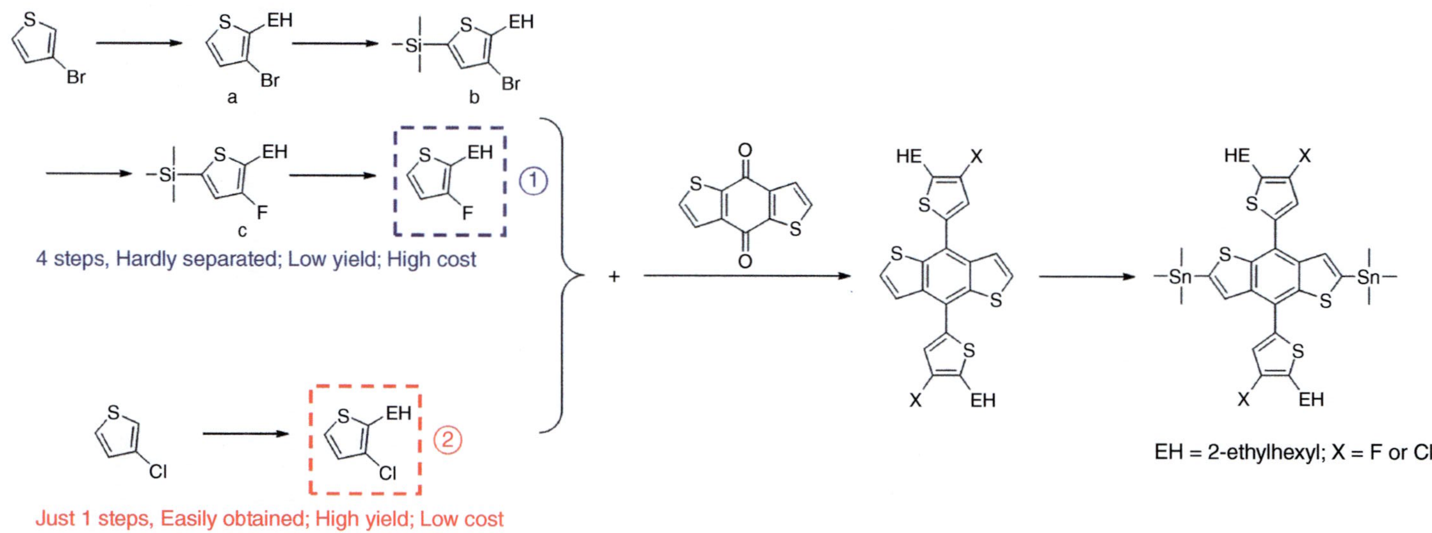

Figure 6.1 The key routes of PBDB-T-2Cl and PBDB-TF polymers. Source: Hui Chen, Mingrui Pu and Feng He.

PBDB-TF. Both as a promising photovoltaic material, the PBDB-T-2Cl has greater commercialization potential.

6.2.2 The Manipulation of Energy Level and Absorption

As halogen elements, the chlorine and fluorine had similar electronic structure and strong electronegativity. When they were introduced into the semiconductor molecule, the energy level and absorption spectra of material were largely changed. They could generate some similar effects on the HOMO and LUMO levels of molecule. And, the positions of chlorination and fluorination in molecule were diversified, optional sites could appear in donor or acceptor unit [15, 43, 44], even π bridge [45] in donor and acceptor materials. However, the final V_{oc} in device is up to the HOMO of donors and LUMO of acceptors [46, 47]. From our study, the influence of chlorine substitution is more obvious than that of fluorine substitution in terms of energy-level regulation. Although the electronegativity of fluorine was stronger than chlorine, the Cl with 3d empty orbitals could accommodate delocalized π-electrons [20]. In the benzo[*b*]thiophene material system, Bao and coworkers used molecular orbital diagrams to demonstrate that chlorination is more effective than fluorination in manipulation of HOMO and LUMO energy levels.

Changes in energy levels usually result in changes in the spectrum, and the effects of chlorination and fluorination on absorption are also complicated. In different material systems, similar positions of substitution produce very different results. For instance, the difluorothiophene in PBDT[2F]T and PGeTFDTBT [48, 49], the spectra shifts are disparate. Generally, the induced electron-withdrawing properties of fluorination bring about declined electron density in conjugated polymer, increasing the HOMO-LUMO bandgap [50–53]. Consequently, related absorption spectra present the blue-shift trend. But in some materials, as intramolecular F-H, F-F, and/or F· · ·S interactions induced the fluorination enhance conjugation length, leading to the red-shift of absorption spectra [54–60]. In some special cases, two interactions exist in molecule, the spectral shift is rooted in the result of the two competitive mechanisms. Taking account of the molecular aggregations, fluorination enhance the intermolecular interactions generating red-shift in solution, which could be demonstrated by temperature-dependent absorption spectra. Based on some results of reported fluorinated copolymer donors, the fluorination causes a spectral blue-shift in most cases [61–84]. As for chlorination, according to the reported work, the trend is like that of fluorination. Regarding chlorination, the trend is like that of fluorination due to the relative lack of reported literature. However, the causes and mechanisms of spectral changes are different. The increase in steric hindrance is a factor for the blue-shift of the chlorination [31–36]. In small-molecule acceptors, fluorination and chlorination on end groups are much simplified compared to polymers, they all cause a red-shift of the spectrum in most cases [85–88]. At present, there are many reasons for the redshift. The mainstream explanations include the enhanced IACT effect [36], the inductance effect on the empty 3d orbital of chlorine, the delocalization of electrons and the increase of dipole moment [89–95]. In terms of extinction coefficient, both chlorination and fluorination can improve the extinction coefficient of organic semiconductors.

6.2.3 The Steric Hindrance and Morphology

Because the van der Waals radius of fluorine is just 1.47 Å, the influence of steric hindrance could be ignored. But the van der Waals radius of chlorine is much larger, up to 1.76 Å. Generally, the strong electronegativity of F leads to the interaction between noncovalent molecules. With the different positions of chlorination, the situation of the chlorination reaction is more complicated and diverse, which determines the steric hindrance produced by the chlorination. In the case of large steric hindrance, the chlorination will distort the molecular configuration and reduce the crystallinity. Through some molecular designs, under the condition of reducing steric hindrance, the chlorination strengthens the interaction between molecules due to strong noncovalent interactions, such as Cl· · ·S and Cl· · ·π.

The dihedral angle between adjacent conjugated units is a key factor affecting the aggregation and morphology of conjugated polymers [54, 96, 97]. Introduction of chlorine atoms into the backbone of the conjugated polymer can increase the dihedral angle between adjacent conjugated units. For example, our previous research pointed out that in the PBDTClBT polymer, the dihedral angle of monochlorinated BT and Th that are close to each other is 53.13°, while the dihedral angle of monochlorinated BT and Th that are far apart is only 18.31°. Besides, the dihedral angle between Th near chlorine and adjacent BDT units is also just 20.05° [30]. A special case is that Pei and coworkers studied the strong preaggregation of F-IIDT in solution due to the interaction between the backbones of polymer, resulting in a large number of crystal fields without the preferred crystal orientation, and the chlorination changed the torsion angle to optimize crystallization of films and face-on orientation of molecular, generating a high PCE of Cl-IIDT-based devices [54].

Without considering other factors, the increase of the dihedral angle will destroy the coplanarity of the molecular framework and the IACT of the crystal. This is very important to D-A type organic semiconductors and is not conducive to the improvement of photoelectric performance. In order to reduce the influence of steric hindrance caused by chlorination, two methods containing chlorinated conjugated side chain and end group are usually adopted.

Hou and coworkers calculated the dihedral fluorination and chlorination of Th-based side group and BDT unit, the results pointed out that the two dihedral angles are the same, 53° in Figure 6.2 [22]. Chen and coworkers chlorinated the Th side chain of BDT in PBDT-Cl-3T2C polymer, the dihedral angle between backbone and side chain had a little increase [98]. Therefore, the effects on the dihedral angle between adjacent units were small.

Chlorination of end group is mainly used in non-fullerene small-molecule acceptor material systems. In this way, chlorinated materials can obtain a better

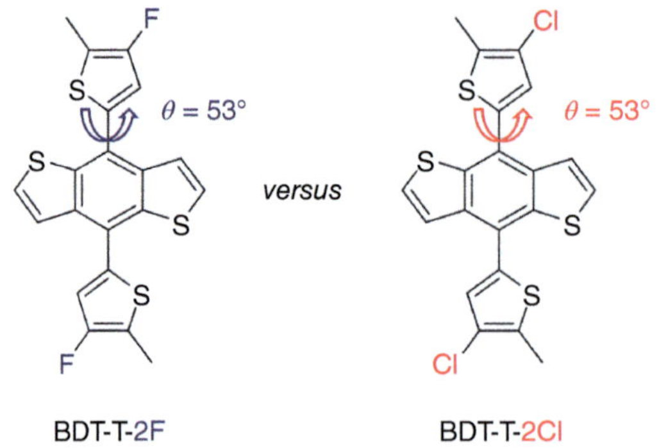

Figure 6.2 Theoretical calculation data, including twisting barriers of BDT-T-2F and BDT-T-2Cl and the four pairs of conjugated building blocks with F or Cl. Source: Hui Chen, Mingrui Pu, and Feng He.

coplanar structure [37, 99]. For example, Wei and coworkers used X-rays to analyze the single-crystal structure of *m*-ITIC-OR-4Cl, the chlorinated end group molecule showed an approximate coplanar geometric structure [99]. The study of Hou and coworkers showed that the chlorinated end groups did not cause much change to the optimized molecular configuration, indicating that the steric hindrance effect of the chlorine atoms on the end groups in some material systems is not obvious. He and coworkers prepared two new molecules chlorinated IDIC-4Cl and fluorinated IDIC-4F, respectively. They analyzed the single-crystal structure and systematically studied the influence of the addition of F and Cl on the molecular configuration and packing. The data of single-crystal analysis show that the dihedral angles between the IDT core of IDIC-4H, IDIC-4F, and IDIC-4Cl and the corresponding IC end groups are 14.9°, 16.3°, and 9.0°, respectively [35]. The addition of F slightly changed the distortion of the molecule, and the chlorination reduced the dihedral angle of the molecule. Although the atomic radius of Cl is larger than that of F, the chlorinated end groups will not cause distortion and accumulation of molecules. The π–π distance of IDIC-4Cl with an interlocked network configuration is 3.41 Å in Figure 6.3. Compared with the other two molecules, there is a stronger Cl···S interaction among the molecules. Moreover, the adjacent IDIC-4Cl molecules are almost completely parallel, while the dihedral angles between the IDT cores of IDIC-4H and IDIC-4F are 20.8° and 19.3°, respectively. The chlorinated DIC-4Cl molecule has good coplanarity, which is very conducive to the separation of excitons and the transfer of charges. There are also chlorine-substituted small molecules with smaller dihedral angles in other material systems [37].

6.2.4 The Synergistic Effect of Chlorination and Fluorination

Both chlorination and fluorination are effective strategies to improve the PCE of PSCs. The introduction of chlorine and fluorine atoms into certain material systems can combine their advantages. Some research groups have studied the synergistic effects of chlorination and fluorination.

The well-known D-A polymers (PTB) with TT and BDT units [100–103], such as PTB7 and PTB7-Th, were vastly studied. However, the V_{oc} of PTB7 and

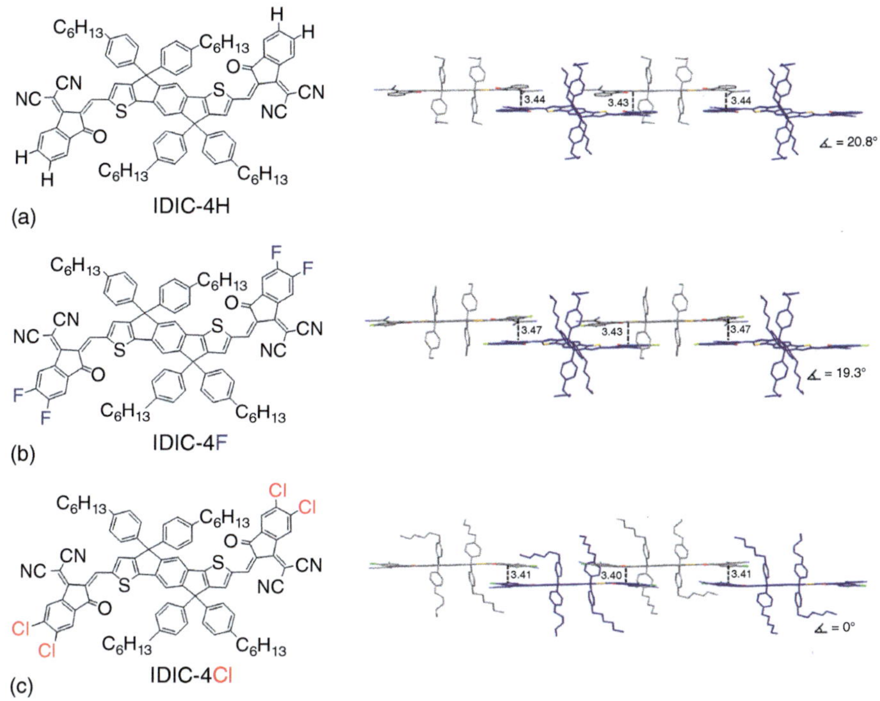

Figure 6.3 X-ray crystallographic structures of (a) IDIC-4H, (b) IDIC4F, and (c) IDIC-4Cl. (∡ represents the dihedral angles between IDT cores of adjacent molecules). Source: Hui Chen, Mingrui Pu and Feng He.

PTB7-Th-based PSCs are moderate, no more than 0.80 V in general. For this reason, some new series chlorinated polymers named $PBTCl_x$ were designed and synthesized in He's group [27], the corresponding molecular structure is shown in Figure 6.4. The chlorinated TT (TT-Cl) was firstly brought in PTB family conjugated polymers. The optoelectronic properties of polymers and corresponding performance parameters of PSCs are summarized in Table 6.1. Remarkably, the V_{oc} of the terpolymer $PBTCl_x$ solar cells could be finely tuned from 0.79 to 0.92 V via varying the content of TT-Cl units in terpolymer (Table 6.1). The effects of chlorination and fluorination on photovoltaic performance are different. F can lower the HOMO of the molecule without causing harmful steric effects. Cl can deepen the HOMO energy level more. However, it destroys the planarity of the polymer and affects the morphology, leading to poor J_{sc} and FF of devices. Considering the impact of chlorination and fluorination on the performance of the device, our group adopted a ternary polymerization strategy to synthesize a series of terpolymer $PBTCl_x$. With 25% Cl content in polymer, the PTBCl25-based device exhibited superior hole mobility and external quantum efficiency (EQE) response among all polymers. From the grazing incidence wide-angle X-ray scattering (GIWAX) analysis, the introduction of chlorine enlarges π–π stacking distance in the activelayer. According to the calculated full width at half-maximum (FWHM) of scattering peaks, the

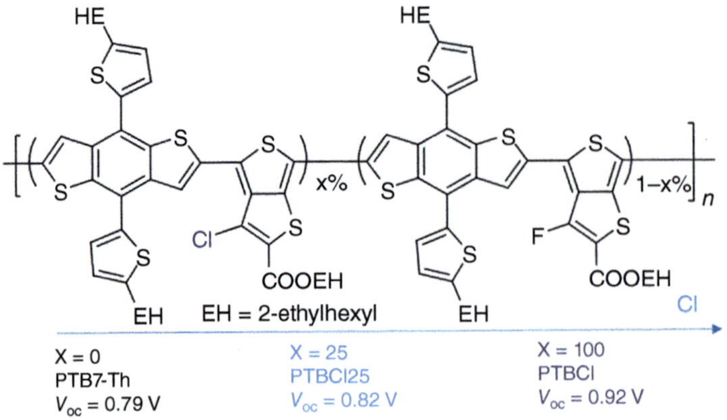

Figure 6.4 Schematic diagram of V_{oc} in PSCs based on PTBCl*x* and $PC_{71}BM$. Source: Hui Chen, Mingrui Pu and Feng He.

Table 6.1 Optical properties and HOMO of chlorinated polymers and device performance of corresponding PSCs.

Polymer	HOMO (eV)	E_g^{opt} (eV)	λ_{max}^{film} (nm)	V_{oc} (V)	J_{sc} (mA cm^{-2})	FF (%)	PCE (%)
PTB7-Th	−5.29	1.56	705	0.79	14.98	66.44	7.91
$PBTCl_{10}$	−5.35	1.61	702	0.80	14.07	69.28	7.79
$PBTCl_{20}$	−5.35	1.56	696	0.81	13.94	66.00	7.47
$PBTCl_{25}$	−5.36	1.57	691	0.82	15.31	66.19	8.31
$PBTCl_{50}$	−5.36	1.58	672	0.85	13.02	61.03	6.79
$PBTCl_{75}$	−5.50	1.63	644	0.89	11.55	57.86	5.90
PBTCl	−5.73	1.72	631	0.92	10.24	50.23	4.75

$PBTCl_{25}$-based film presents the highest crystallinity among all polymers. The optimal device based on $PBTCl_{25}$ had a decent efficiency of 8.31%, which was higher than the PTB7-Th control device (7.91%).

The synergistic effect of chlorination and fluorination has also been verified in other systems. Our group synthesized PCBT4T-2BO and PCBT4T-2OD by introducing chlorine into the position of the acceptor unit benthiadiazole on the backbone. PSCs based on these polymers and $PC_{71}BM$ were fabricated [29]. In terms of energy-level adjustment, chlorine can lower the energy level more effectively than fluorine, thereby further increasing V_{oc}. We also synthesized PCFBT4T-2OD, a polymer containing both fluorine and chlorine. The J_{sc} of PCFBT4T-2OD cells was further increased to 17.61 mA cm^{-2}, and the PCE was up to 8.84%. Compared with PCBT4T-2BO with monochlorine, the introduction of fluorine atoms into the polymer further enhances the π–π stacking, which is beneficial to the transport of charges in the active layer and improves the device performance of PSCs. These results all indicate that the synergistic effect of chlorination and fluorination is good for improving the efficiency of PSCs.

6.3 Recent Advances in Chlorinated Semiconductors

When chlorination just entered the field of organic optoelectronics, it was mainly used for the high and balanced bipolar transportation of organic field effect transistors (OFETs), because chlorination can lower the LUMO energy level, reduce the electrons of the charge injection barrier, and make electrons and holes could be transmitted at the same time. According to the research of Bao and coworkers, if the chlorination does not break the planarization of the conjugated core, the chlorinated molecule usually has a slightly smaller gap between the HOMO and LUMO energy levels and is even lower than the corresponding fluorine-containing LUMO, because Cl has an empty 3d orbital to fit the core of π electron conjugation [20], while the fluorine atom does not have these unique properties. Since 2015, the application of chlorination in OSCs has received increasing attention [54, 104]. In just five years, the PCE of chlorine-based OSCs increased from 1% to 18% [22, 105].

6.3.1 Chlorination on the Donor Units of Polymer Donors

Chlorination of polymer is an effective way to increase the V_{oc} of the device, because the chlorine atom has a strong electronegativity and the ability to accommodate delocalized electrons, which can effectively reduce the HOMO and LUMO energy levels. The current high-performance D–A copolymer donor materials usually consist of D units, A units and functional groups to improve the solubility of the polymer. In order to reduce the influence of steric hindrance, it is sometimes necessary to build π-bridge between the D and A units to improve the coplanarity of the molecules. In some cases, the coplanarity of backbone can be enhanced by modifying the conjugated side chain, thereby expanding the range of the absorption spectrum [100]. With the comprehensive study of chlorination reactions, chlorination is now introduced into D units, A units, π-bridges, and conjugated side chains, which will be discussed separately.

6.3.1.1 Chlorination of the Donor Units in Backbone of Polymer Donors

Benzo[1,2-*b*:4,5-*b*′]-dithiophene (BDT) is currently one of the most used D units in high-performance D–A polymers. In 2016, Lavarda and coworkers synthesized 24 kinds of PTB7 series derivatives [106], which were substituted by choosing different substituents and different positions on the PTB7 monomer in Chart 6.1. According to the characterization results of electronic and optical properties, chlorinated polymers have good application prospects in OSCs. Among the three chlorinated PTB7 derivatives, PTB7-Cl-12 and $PC_{61}BM$ have the best application prospects in OSCs because their parameters are close to the ideal values of E_{HOMO}, E_{LUMO} and ΔE_{HL}, which indicates that its V_{oc} is greater than PTB7.

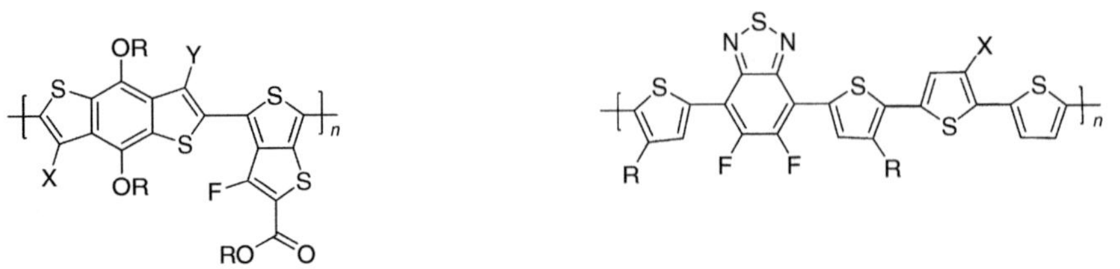

R = 2-ethylhexyl
Z = Cl, Br, F, OH, CN, NH_2, CH_3, CH=CH_2
PBT7: X = Y = H; PBT7-Z-2: X = H, Y = Z
PBT7-Z-X: X = Z, Y = H; PBT7-Z-XY: X = Y = Z

R = 2-octyldodecyl; PffBT4T-2OD: X = H
PBT4T-Cl: X = Cl

Chart 6.1 Polymer donor-based chlorinated D unit backbone. Source: Hui Chen, Mingrui Pu, and Feng He.

Thiophene (Th) is also general and typical D unit. He and coworkers chlorinated a thiophene in the middle position in the benzothiazolyl-based tetrathiophene polymer to synthesize a new polymer donor PBT4T-Cl (see Chart 6.1). OSCs based on PBT4T-Cl have a better performance than the control analog without Cl, the excellent PCE was up to 11.8% and the related stability was also outstanding [23]. This is mainly because the 4-position chlorination reaction of thiophene in the middle will reduce the HOMO level of the polymer, but the twist on the backbone is very small, which is conducive to maintaining a satisfactory morphology and main orientation. Through DFT calculation, it is found that the dipole moments of the ground state and the excited state of PBT4T-Cl are quite different, indicating that there is less Coulomb binding of excitons. Therefore, the hole and electron mobility of PBT4T-Cl-based devices have been improved from the space charge limited current (SCLC) test. Besides, the chlorination stabilizes the two-phase nanostructure in the active layer and ensure its stability.

6.3.1.2 Chlorination of the Donor Units in Side Chain of Polymer Donors

Compared with the chlorination on backbone of the D unit, the chlorination on the side chain has received more attention. Thienyl is a commonly used side chain of conjugated polymers. It is usually applied to BDT units (BDT-T) to improve the coplanarity of the backbone of polymer molecules [100]. The current chlorinated side chains are based on the thienyl group in BDT-T [22, 24, 43, 86, 87, 98, 107, 108]. As a donor component of polymer, BDT has the virtues of high degree of conjugation, high planarity, and low steric hindrance. Because the main feature of the BDT-T unit is high planarity, whether the chlorination with a large atomic radius will destroy the planarity is the first issue that needs attention. Hou and coworkers studied the torsion barriers in BDT-T-2F and BDT-T-2Cl from the DFT calculation [22]. According to the results of simulation calculations, BDT-T-2Cl and BDT-T-2F have similar optimal dihedral angles and torsional barriers.

In 2018, He and coworkers chlorinated BDT-T into the classical polymer donor PTB7-Th, replacing BDT-T unit, and resulted a new polymer donor PBClT (see Chart 6.2) [25]. Chlorinated BDT-T did not increase the steric hindrance of the conjugated backbone and disrupt the packing interaction. Because of the unique

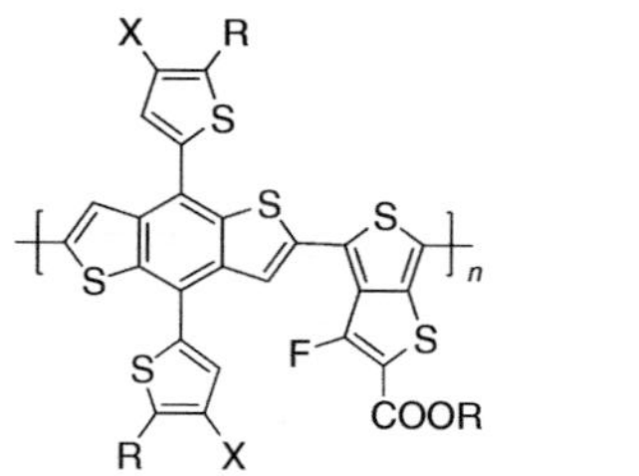

R = 2-ethylhexyl
PTB7-Th: X = H; PBClT: X = Cl

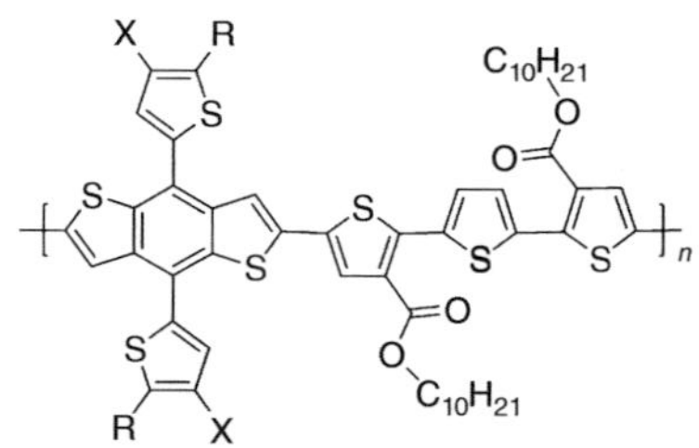

R = 2-ethylhexyl
PBDT-F-3T2C: X = F; PBDT-Cl-3T2C: X = Cl

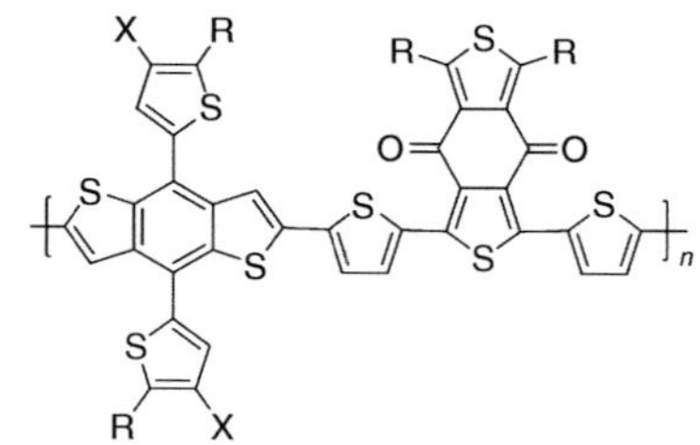

R = 2-ethylhexyl
PBDB-T: X = H; PBDB-T-2F or PM6: X = F
PBDB-T: 2Cl or PM7: X = Cl

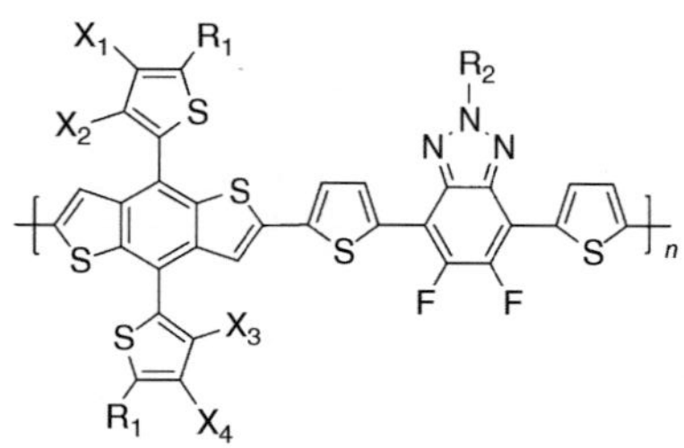

R_1 = 2-ethylhexyl, R_2 = 2-hexyldecyl
PBZ: $X_1 = X_2 = X_3 = X_4$ = H
PBT1Cl-Bz: X_1 = Cl, $X_2 = X_3 = X_4$ = H
PBT2Clas-Bz: $X_1 = X_2$ = Cl, $X_3 = X_4$ = H
PBT2Cls-Bz or PBZ-Cl: $X_1 = X_4$ = Cl, $X_2 = X_3$ = H

PBT4Cl-Bz: $X_1 = X_2 = X_3 = X_4$ = Cl
L68 or J52-Cl: $X_1 = X_4$ = Cl, $X_2 = X_3$ = H

R_1 = 2-ethylhexyl, R_2 = 2-ethylhexyl
L24: $X_1 = X_4$ = Cl, $X_2 = X_3$ = H

R_1 = 2-ethylhexyl, R_2 = 2-octyldodecyl
L810: $X_1 = X_4$ = Cl, $X_2 = X_3$ = H

R_1 = tripropylsilyl, R_2 = 2-hexyldecyl
PBZ-ClSi: $X_1 = X_4$ = Cl, $X_2 = X_3$ = H

Chart 6.2 Polymer donor-based chlorinated side chain of D unit. Source: Hui Chen, Mingrui Pu, and Feng He.

effect of chlorine atoms, it effectively reduces the level of HOMO and obtains a high V_{oc} over 1.0 V in PBClT-based OSCs. The results of this study show that the side chain chlorination reaction can well control the energy level of the polymer without destroying the π–π stacking interactions among the molecules.

BDTT-*alt*-DTBTz is a classic D–A molecular skeleton [109–111]. On this basis, side chain engineering and halogenation have been used to synthesize many materials with strong light absorption, ordered molecular packing, and high hole mobility. The chlorination of thiophene side chain in the molecular skeleton has been reported in some literatures [24, 87, 107, 108]. In these material systems, He and coworkers had carefully studied the relationship between the position of chlorination and the number of chlorine atoms and the performance of the device. The results showed that chlorination had a large effect on energy level, absorption coefficient, molecular dipole moment, molecular packing, and film morphology. These comprehensive factors ultimately determined PCE of the device. Four wide-band gap polymers were synthesized, named PBT1Cl-Bz, PBT2Cl^{as}-Bz, PBT2Cl^{s}-Bz (PBZ-Cl, L68 or J52-Cl), and PBT4Cl-Bz. They have a similar LUMO energy level. As the number of chlorine atoms increases, the energy level gets lower and lower. This is due to the electron- withdrawing effect of chlorine and the covalent interaction between molecules Cl···S and Cl···π causing the blue shift of the absorption. But the LUMO and HOMO energy levels of PBT2Cl^{as}-Bz and PBT2Cl^{s}-Bz are almost the same. When the number of chlorine atoms is the same, the position of the chlorine atom has little effect on the energy level. However, the temperature aggregation behavior caused by the difference in the position of chlorine atoms is obviously different. PBT4Cl-Bz exhibits the strongest shoulder peak in both solution and film conditions. Compared with the other three molecules, PBT4Cl-Bz has strong aggregation both in solution and film conditions. PBT2Cl^{as}-Bz and PBT4Cl-Bz showed obvious temperature-dependent aggregation characteristics, and their side chains were in the third and fourth positions of the chlorine-substituted side chain. The control of the morphology during the film preparation is very important. The PBT4Cl-Bz-based PSCs reached the PCE of 9.25%, high V_{oc} of 0.96 V, and maintained a good J_{sc} 16.04 mA cm^{-2}. Due to the deep HOMO energy level, the E_{loss} is only 0.54 eV. The single-crystal structure of BDT-T2Cl^{s} with symmetrical two chlorinated substituents and BDT-T4Cl with four symmetrical chlorinated substituents was analyzed by single-crystal X-ray diffraction (see Figure 6.5) and the effect of chlorination on geometry and molecular packing influence. The dihedral angle between Th dichloride and the BDT core in BDT-T4Cl is 67.58°, which is smaller than the dihedral angle of BDT-T2Cl^{s}. Moreover, the distance between adjacent surfaces of BDT-T4Cl on the BDT plane is shorter than that of BDT-T2Cl^{s}, indicating that although BDT-T4Cl has more Cl atoms, its molecular packing is closer than that of BDT-T2Cl^{s}.

When the chlorination strategy and side chain engineering work together, the energy level and morphology could be further optimized. Chen and coworkers synthesized three polymers containing BDT-2Cl and fluorobenzotriazole (FTAZ) unit with different alkyl chains [107]. Similar to the previous discussion, they

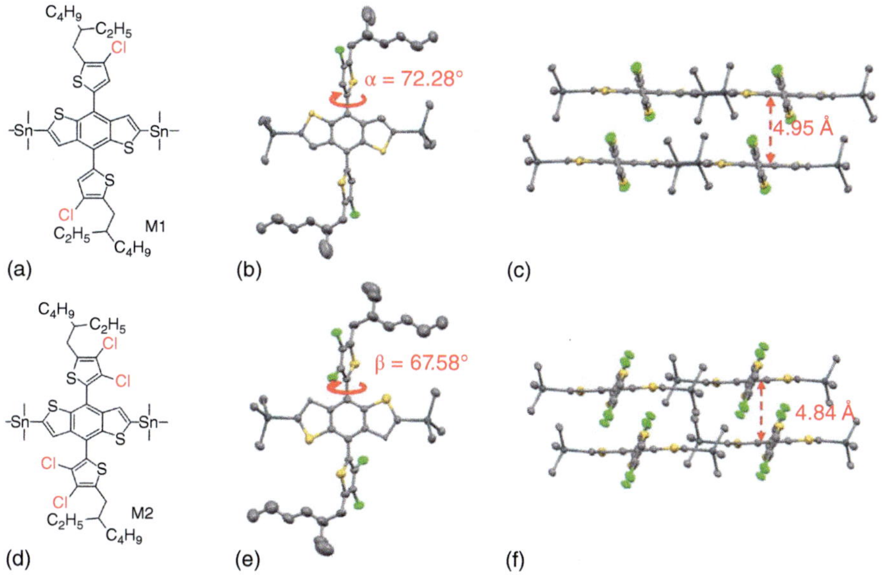

Figure 6.5 Structure of monomer M1 (BDT-T2Cl[s]) (a) and monomer M2 (BDT-T4Cl) (d), the crystal structure of monomer BDT-T2Cls front view (b) and the packing diagrams (c); monomer BDT-T4Cl front view (e) and the packing diagrams (f). Note: H atoms and 2-ethylhexyl chains were omitted for clarity. Source: Hui Chen, Mingrui Pu, and Feng He.

found that chlorination enhanced the planarity and energy level of the molecule and increased the polymerization activity of the monomer. By optimizing the side chain unit in the FTAZ group, they found that V_{oc}, J_{sc}, and FF vary greatly with the length of the side chain because the side chain has a significant effect on the energy level and molecular stacking orientation. Li and coworkers used alkylsilyl to replace the thienyl side chain on the BDT-T unit. The synthesized new polymer PBZ-ClSi (see Chart 6.2) has a deep HOMO energy level [87], which increases the absorption coefficient of the material and the charge mobility in related device. The PBZ-ClSi:IT-4F-based devices presented a high PCE of 12.8% with simultaneous enhanced V_{oc}, J_{sc}, FF. Amazingly, Zhou and coworkers obtained an outstanding V_{oc} of 1.24 V in the J52-Cl-based PSCs based on J52-Cl and a non-fullerene small-molecule BTA3 (see Figure 6.6) [108]. Interestingly, both polymers have the same benzotriazole unit, which they call the "same-A-strategy" (SAS).

BDT-T-*alt*-1,3-bis(4-(2-ethylhexyl)thiophen-2-yl)-5,7-bis(2-alkyl)benzo[1,2-*c*:4,5-*c'*]dithiophene-4,8-dione (BDD) is another typical backbone of conjugated polymers, and many efficiency polymer derivatives have been developed on this basis [22, 43, 86, 112, 113], such as PBDB-T-2Cl (PM7), which was synthesized by Hou and Zhang group independently. Compared to nonchlorinated PBDB-T, PBDB-T-2Cl had lower HOMO, enhanced absorption coefficient, improved crystallinity, and charge mobility. In addition, the PBDB-T-2Cl:IT-4F-based PSCs processed by nonhalogen solvent toluene [86], producing an excellent PCE of 13.1%, while the

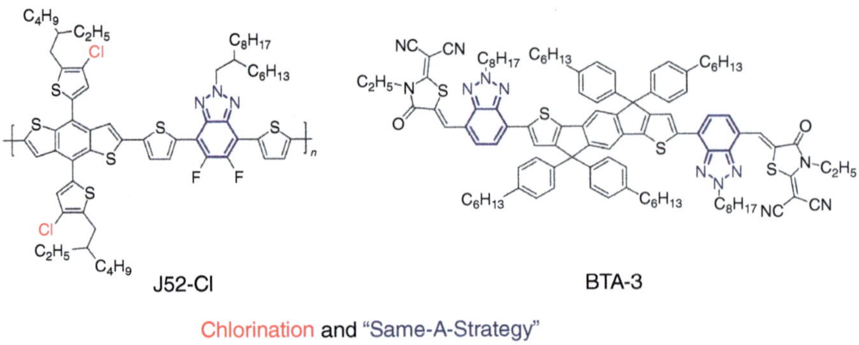

Figure 6.6 The scheme of chlorination and "Same-A-Strategy". Source: Hui Chen, Mingrui Pu, and Feng He.

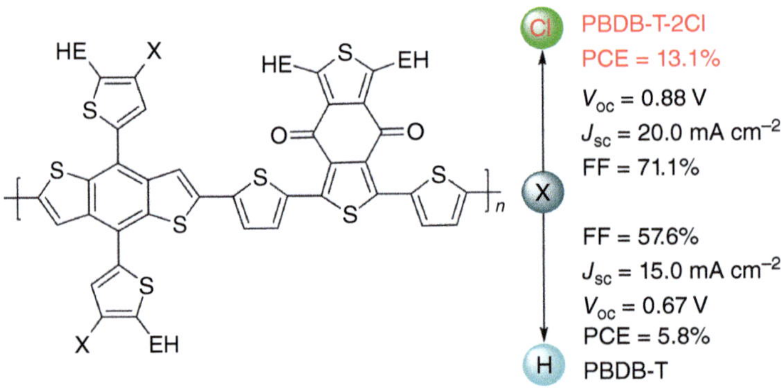

Figure 6.7 Molecular structure and photovoltaic performance of PBDB-T and PBDB-T-2Cl. Source: Hui Chen, Mingrui Pu and Feng He.

control device based on PBDB-T just owned a low PCE of 5.8% (see Figure 6.7), showing that chlorinated materials have the characteristics of eco-friendly solvent processing. Compared with the fluorinated PBDB-TF (PM6), the synthesis cost of PBDB-T-2Cl is low, and the calculated differences in molecular configuration, absorption spectrum, aggregation, morphology, and hole mobility are negligible, but its HOMO level is deep, resulting in high V_{oc} in corresponding device. The device processed by CB had a very high PCE 14.4%, which was the top PCE of chlorinated OSCs at that time. The research on PBDE-T shows that chlorination is more competitive than fluorination in the development of efficient donor polymers.

The chlorinated BDT-T donor and thiophene (T) units are copolymerized to synthesize D–A type copolymer PBDT-Cl-3T2C [98]. The Chen and coworkers also synthesized the control fluorinated polymer PBDT-F-3T2C, as shown in Chart 6.2. The effect of the chlorination on the energy level is similar to the other chlorinated functional units in the polymer. Compared with fluorine atom, chlorine atom has a stronger energy level deepening ability. The smaller size of fluorine atom has stronger self-assembly ability and smaller steric hindrance, which makes the crystallinity and aggregation of PBDT-F-3T2C in solution higher than that of PBDT-Cl-3T2C. Thus, the fluorinated polymer PBDT-F-3T2C has an improved absorption coefficient, favorable morphology, higher and more balanced charge mobility than PBDT-Cl-3T2C, and its PCE is also higher.

From above discussion, the chlorination and fluorination have similarities and unique features. The comparison results of the final photovoltaic performance of the fluorinated and chlorinated copolymer are different. Therefore, we can choose appropriate strategies according to different polymer systems and improve the performance of OSCs.

6.3.2 Chlorination on the Acceptor Units of Polymer Donors

Now the chlorinated A units are mainly on isoindigo (IIDT) [54], thiophene (TT) [26–28], and benzothiazole (BT) groups [29–31]. In this section, they will be summarized and discussed. The structure and electronic and photovoltaic properties of related molecules are shown in Chart 6.3 and Table 6.2, respectively.

In 2013, in organic field-effect transistors (OFET), the electron mobility of chlorinated IIDT polymers increased significantly, indicating that chlorination can control the electronic properties of conjugated polymers [114]. Later, Zheng et al. chlorinated isoindigo (Cl-IIDT) in related polymer and found that in addition to adjusting the energy level, chlorination also increased the distortion of the polymer backbone [54]. From the computed results of DFT, the dihedral angle of F-IIDT backbone was 11°, and the length between F and H atom was 2.27 Å, indicating that there is a strong difference between F and H The interaction, resulting in the large crystal domain mixed with $PC_{71}BM$ film and PCE is 1.19%. In contrast, chlorination distorted the Cl-IIDT framework, and the phenyl–phenyl torsion angle could reach 40°. Compared with fluorinated polymers, Cl-IIDT has suitable crystallinity and large planar conformation, which is beneficial to charge transfer. Cl-IIDT is fully mixed with $PC_{71}BM$ in the solution to form a good morphology in the film, and the PCE is 4.60%, obtaining a better photovoltaic performance at that time.

In order to reduce the torsion and steric hindrance of thiophene [3,*b*] caused by chlorination, He and coworkers adopted a 2D side chain design strategy to increase intermolecular interaction and aggregation [26]. 2D-PBTCl still has better blue shift absorption to match ITIC, and the HOMO level is deeper, and the V_{oc} of the device is higher. The PCE of ITIC and 2D-PBTC is 8.67%, which is higher than the PCE of PTB7-Th-based devices. These results indicated that the expansion of the conjugation 2D side-chain strategy and the chlorination had a synergistic effect in reducing steric effects and enhancing photovoltaic performance.

In 2017, He and coworkers chlorinated the BT unit of a special polymer, synthesized a series of asymmetric polymers, and studied the effect of chlorination on photovoltaic performance [30]. The chlorination on the BT unit would decline the HOMO levels, improving the V_{oc} of devices using $PC_{71}BM$ as acceptor and obtaining a high PCE of 9.11%. Although chlorination caused molecular distortion, due to the

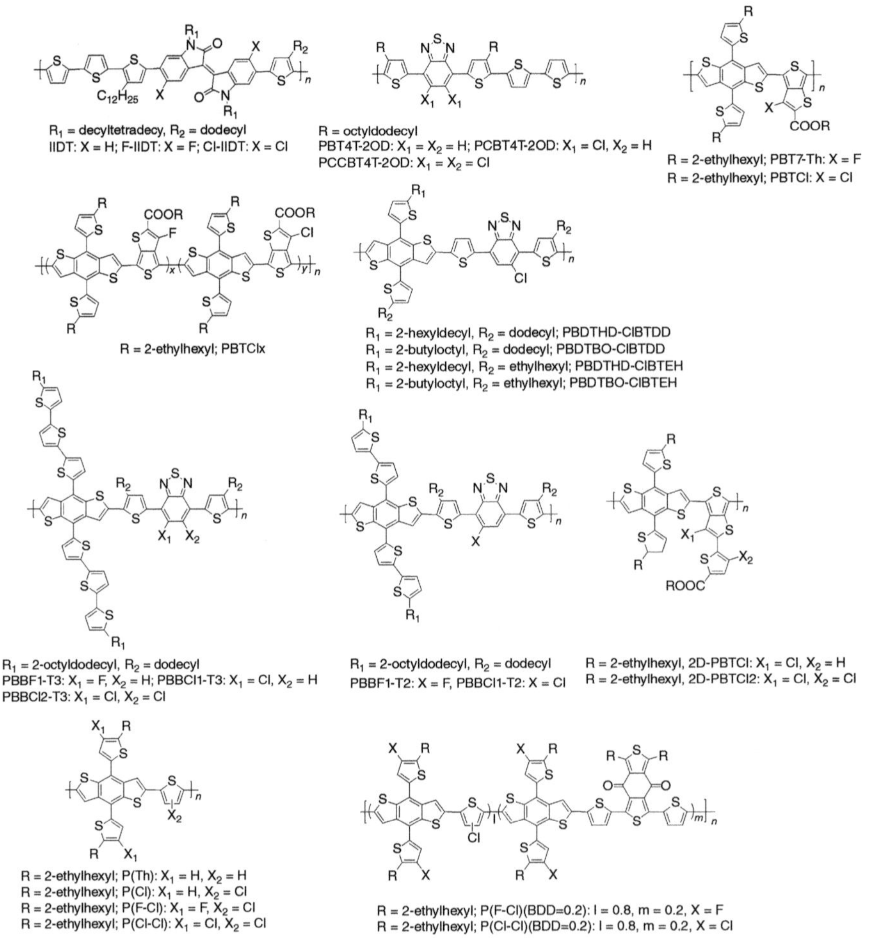

Chart 6.3 Polymer donor-based chlorinated backbone of A unit. Source: Hui Chen, Mingrui Pu, and Feng He.

large atomic size of the polymer, the polymer molecules could still maintain good orientation, thereby maintaining good photovoltaic performance.

Conjugated polymer donors with benzothiadiazole (BT) moiety were widely synthesized and developed in past several years. The Cl numbers on BT moiety gave some impressive influences in relative polymer photovoltaic performance. The monotonical V_{oc} of PSCs-based polymer with the number of chlorine atoms on the BT moieties were presented in Table 6.2. The initial polymer PBT4T-2OD-based PSCs showed a PCE of 4.89%, with low V_{oc} of 0.70 V and J_{sc} of 10.58 mA cm^{-2}. When added one Cl atom to BT in PCBT4T-2OD, the V_{oc}, J_{sc}, and FF were simultaneously enhanced, giving rise to a high PCE of 8.21%. As added two Cl atoms to BT in PCCBT4T-2OD [29], the expected high V_{oc} of 0.85 V was achieved, the disappointed J_{sc} and FF led to a moderated PCE of 6.12%. The chlorinated BT units could change the aggregation of the polymers from solution and film UV–vis absorption

Table 6.2 Electronic and photovoltaic performance of polymer donor-based chlorinated backbone of A unit.

Donor	HOMO (eV)	LUMO (eV)	Acceptor	V_{oc} (V)	J_{sc} (mA cm^{-2})	FF (%)	PCE (%)
F-IIDT	−5.51	−3.92	$PC_{71}BM$	0.63	4.26	44	1.19
Cl-IIDT	−5.53	−3.85	$PC_{71}BM$	0.75	10.00	61	4.60
PBT4T-2OD	−5.12	−3.63	$PC_{71}BM$	0.70	10.58	65.45	4.89
PCBT4T-2OD	−5.26	−3.59	$PC_{71}BM$	0.73	16.18	68.97	8.21
PCCBT4T-2OD	−5.32	−3.62	$PC_{71}BM$	0.85	11.93	60.14	6.12
PBT7-Th	−5.38	−3.70	ITIC	0.82	14.76	57.09	6.91
2D-PBTCl	−5.45	−3.60	ITIC	0.91	15.79	61.32	8.81
2D-$PBTCl_2$	−5.43	−3.63	ITIC	0.87	14.52	58.43	7.38
PBDTHD-ClBTDD	−5.53	−3.71	$PC_{71}BM$	0.76	16.79	71.69	9.11
PBDTBO-ClBTDD	−5.47	−3.70	$PC_{71}BM$	0.68	11.69	62.30	4.95
PBDTHD-ClBTEH	−5.53	−3.71	$PC_{71}BM$	0.79	13.33	63.12	6.88
PBDTBO-ClBTEH	−5.50	−3.73	$PC_{71}BM$	0.78	10.94	64.36	5.46
PBBF1-T2	−5.44	−3.20	$PC_{71}BM$	0.81	9.48	45.37	3.49
PBBCl1-T2	−5.49	−3.22	$PC_{71}BM$	0.87	8.44	49.49	3.64
PBBF1-T3	−5.45	−3.21	$PC_{71}BM$	0.73	12.74	66.94	6.21
PBBCl1-T3	−5.44	−3.38	$PC_{71}BM$	0.73	13.75	68.59	6.87
PBBCl2-T3	−5.50	−3.21	$PC_{71}BM$	0.84	9.90	63.94	5.33
P(Th)	−5.32	−3.32	ITIC-Th	0.779	8.8	37.3	2.6
P(Cl)	−5.47	−3.50	ITIC-Th	0.905	20.63	65.0	12.14
P(F-Cl)	−5.64	−3.66	IT-4F	0.879	20.3	66.0	11.8
P(Cl-Cl)	−5.70	−3.68	IT-4F	0.899	20.3	56.1	10.2
P(F-Cl)(BDD = 0.2)	−5.62	−3.74	IT-4F	0.838	21.3	71.0	12.7
P(Cl-Cl)(BDD = 0.2)	−5.67	−3.77	IT-4F	0.899	21.7	67.5	13.2

spectra. The improved steric hindrance from Cl atoms could produce blueshift in absorption. Astoundingly, the V_{oc} of PCBT4T-2OD is 0.73 V, which is a little improvement compared to PBT4T-2OD; the V_{oc} of PCCBT4T-2OD is 0.85 V, which is much boosted via addition of another Cl atom in PCBT4T-2OD. The HOMO levels of polymers were deepened with increasing numbers of chlorine atoms from the CV measurement, and the HOMO values of PBT4T-2OD, PCBT4T-2OD, and PCCBT4T-2OD were 5.12, 5.26, and 5.32 eV, respectively. The PCCBT4T-2OD with two Cl atoms reduces the π–π intermolecular interaction, the high V_{oc} could be obtained in PSCs device. The V_{oc} of benzodithiadiazole-based PSCs is able to finely tune through manipulation of Cl numbers on BT moieties.

In 2018, He and coworkers synthesized 2D chlorinated polymers using chlorinated BT units [31]. Compared with fluoropolymers, chlorinated polymers have a

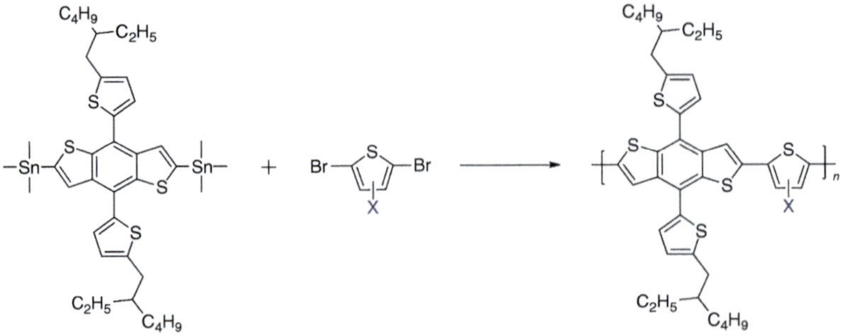

Figure 6.8 Synthesis routes of P(Th) and P(Cl). Source: Hui Chen, Mingrui Pu, and Feng He.

wider adsorption range. Among them, PBBCl1-T3:$PC_{71}BM$-based device had the highest PCE, which was 6.87%. Since the side chain reduced the twist of the polymer chain caused by the chlorination, the conjugated side chain becomes longer, and the advantage of the chlorination over the fluorination was hindered. Although the V_{oc} of dichlorinated PBBCl2-T3 is 0.87 V, its J_{sc} is not ideal. This may be due to the chlorination destroying the planarity of the polymer backbone and the interaction between molecules.

As a weak donor unit, thiophene exhibits the properties of an acceptor unit when it encounters a strong donor unit. The same is true for chlorinated thiophene (Cl-Th) [115–117]. Recently, with Cl-Th as acceptor unit, Moon and coworkers synthesized a D–A copolymer, P(Cl) [116]. The synthesis is simple, high yield, and low cost. The control molecule P(Th) without chlorination had a slight twist, and due to the larger atomic radius of Cl, P(Cl) had a larger twist from DFT computation. In addition, the molecular weight of the chlorination was higher, which enhances spectral absorption and red shifts, as presented in Figure 6.8. When mixed with ITIC-Th, the crystallinity of the film is higher and the stacking distance is smaller. The resulting device had an excellent PCE 12.14% and stability. Then, the Th side chain of the BDT unit was fluorinated and chlorinated to prepare P(F-Cl) and P(Cl-Cl), respectively (see Chart 6.3) [117]. The level of HOMO both had fallen, while the chlorinated polymer had fallen even more. Fluorination and chlorination will reduce solubility, thereby reducing molecular weight. In order to improve the solubility, they added a small proportion of BDD units, and the molecular weight and solubility were significantly increased. Using eco-friendly *o*-xylene as solvent to fabricated PSCs, P(F-Cl)(BDD = 0.2):IT-4F and P(Cl-Cl)(BDD = 0.2):IT-4F:PN (99.5:0.5, v/v%) obtained PCE of 12.70% and 13.97%, respectively.

6.3.3 Chlorination of the π-Bridge of the Polymer Donors

D–A type structure has been widely applied to design efficient polymer. The strong interaction between the D unit and the A unit can reduce the band gap of the material and increase the absorption of photons. With the deepening of research, π-bridge is derived as the key connection between D and a unit and can adjust the backbone of conjugated polymer (Chart 6.4). π-Bridge can effectively regulate the

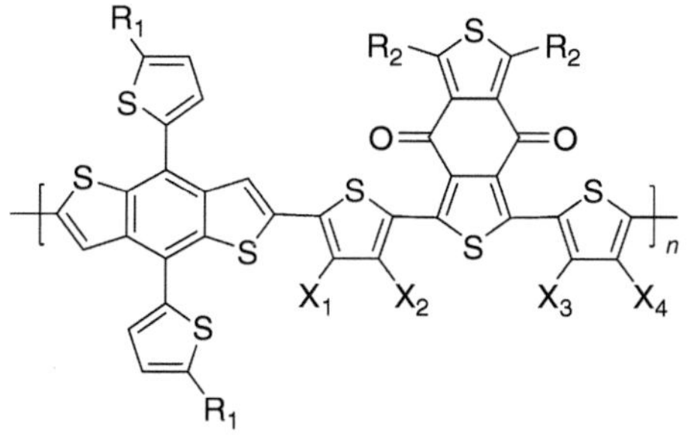

$R_1 = R_2$ = 2-ethylhexyl
PCl(4)BDB-T: $X_1 = X_4$ = Cl, $X_2 = X_3$ = H
PCl(3)BDB-T: $X_2 = X_3$ = Cl, $X_1 = X_4$ = H

R_1 = 2-butyloctyl, R_2 = 2-ethylhexyl
PBT1-C: $X_1 = X_2 = X_3 = X_4$ = H
PBT1-C-2Cl: $X_1 = X_4$ = Cl, $X_2 = X_3$ = H

Chart 6.4 Polymer donor-based chlorinated π-bridge. Source: Hui Chen, Mingrui Pu, and Feng He.

Table 6.3 Electronic and photovoltaic performance of polymer donor-based chlorinated π-bridge.

Donor	HOMO (eV)	LUMO (eV)	Acceptor	V_{oc} (V)	J_{sc} (mA cm^{-2})	FF (%)	PCE (%)
PCl(3)BDB-T	−5.54	−3.27	IT-4F	0.88	0.79	24.47	0.16
PCl(4)BDB-T	−5.48	−3.47	IT-4F	0.84	20.14	71.33	12.11
PBT1-C	−5.47	−3.46	IT-4F	0.79	19.70	70.5	10.9
PBT1-C-2Cl	−5.59	−3.53	IT-4F	0.85	19.74	76.0	12.7

photoelectric properties of D–π–A–π type conjugated materials. It is also a classic strategy to construct highly efficient conjugated polymers.

Sun and coworkers synthesized a polymer PBT1-C-2Cl by chlorinating the thiophene π-bridge of PBT1-C [45]. Through the study of DFT calculation, the dihedral angle between BDT and chlorinated Th unit increases due to the large size effect of Cl. The high electronegativity of chlorine and the molecular distortion play an important role in reducing the HOMO of PBT1-C-2Cl and help increase the V_{oc} of the related device. Using IT-4F as an acceptor, the PBT1-C-2Cl-based PSCs obtained a high PCE of 12.7% with increased V_{oc} and FF, the PCE of control device based on PBT1-C was just 10.9% in Table 6.3. According to the morphology analysis, the blend films have two different packing distances, owing to the interchain chalcogen–chalcogen interactions due to the chlorination. The mobility from the SCLC test also supported these results, the blend film with PBT1-C-2Cl had high and balanced carrier mobility, reducing the bimolecular recombination in device.

Hou and coworkers investigated different chlorinated thiophene orientation on polymer donors with D–π–A–π structure [118]. The two chlorinated polymer donors, PCl(3)BDB-T (P3) and PCl(4)BDB-T (P4), chlorinated Th group facing the BDD and BDT unit, respectively. Minor orientation changes result in huge differences in light absorption, molecular packing, carrier mobility, and device performance. In the chlorobenzene (CB) solution state, the absorption peaks of P3 and P4 are at 465 and 647 nm, respectively, indicating that the electron delocalization of P4 molecules is more effective than that of P3. According to the temperature-dependent light absorption experiment, P4 shows stronger aggregation than P3. Through DFT theoretical calculation and analysis, the P3 molecule has a highly twisted conjugated backbone, and the electrons on the P4 HOMO energy level are not delocalized on the BDD unit, which limits the IACT effect. In charge dynamics, the neat P3 and blend films had no obvious carrier transport characteristic from the SCLC measurements. However, neat film of P4 exhibited hole mobility of 2.57×10^{-5} cm^2 V^{-1} s^{-1}, and the blend film had balance hole and electron mobility of 5.74×10^{-6} cm^2 V^{-1} s^{-1} and 5.16×10^{-6} cm^2 V^{-1} s^{-1}, respectively. From morphology analysis of X-ray diffraction (XRD) and atom force microscope (AFM), the film with P4 owned well-organized and favorable phase-separation size. Thus, the PSCs based on P4:IT-4F generated a high PCE of 12.33%. In control conditions, the P3-based PSCs obtained a very low PCE of 0.18%.

6.3.4 Chlorinated Small Molecular Donors

Compared with PSCs, the current research on all-small-molecule OSCs is relatively small, and the device efficiency also lags PSCs, mainly due to the lack of high-performance small-molecule donors. However, polymers often suffer from batch-to-batch variation and low reproducibility, which hinders the commercialization of OSCs [119, 120]. Small molecules have broad application prospects in OSCs due to their high purity, definite molecular structure, and high reproducibility. At present, the efficiency of all-small-molecule OSCs also approaches 16% [119]. A small BTR molecule has the characteristics of liquid crystallinity, exhibit strong intermolecular interactions, which are conducive to maintaining the order of long-range directions, and have high carrier mobility. Liu and coworkers replaced the two large hexyl groups on the core of the BTR with two negatively charged chlorine (Cl) atoms to reduce its energy level in Figure 6.9. The chlorinated BTR is expected to match the small-molecule NFAs with low energy levels. At the same time, the chlorination can also reduce the dihedral angle of the thiophenes, increase the planarity of the backbone, and strengthen the interaction between molecules. Using Y6 as an acceptor, the fabricated OSCs reaching a record 13.6% PCE at that time. Chlorination enhanced the crystallinity and highlighted the importance of morphology control in all-small-molecule OSCs and demonstrated the great prospects of in OSCs for liquid crystal materials. Thus, chlorination provides a reliable way to produce efficient and reproducible OSCs and is useful for the commercial application of OSCs.

6.3.5 Chlorinated Small Molecular Acceptors

Small-molecule NFAs are receiving more and more attention due to their strong near-infrared absorption and high mobility. In the past few years, small-molecule NFAs based on OSCs have developed rapidly, and the efficiency of single junction

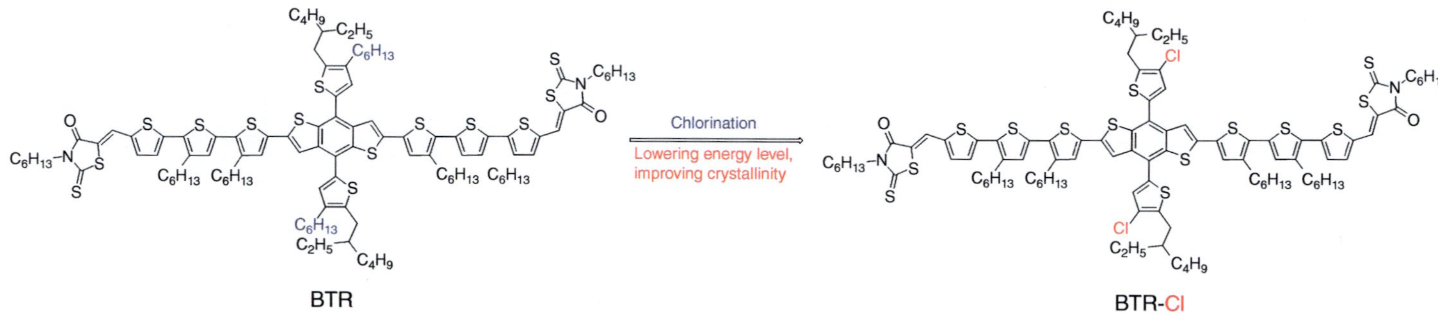

Figure 6.9 Molecular structures of BTR and BTR-Cl. Source: Hui Chen, Mingrui Pu, and Feng He.

PSCs has reached more than 18% after the structure of small-molecule NFAs has been optimized [1]. Halogenation is a strategy often used in organic semiconductors, such as chlorination and fluorination. When the small-molecule acceptor is chlorinated, some of its properties are consistent with the chlorinated polymer donor. For example, the HOMO levels have deepened to varying degrees. However, some properties are completely different, such as red and blue shift of light absorption. In this section, we will discuss the advances and photoelectric properties of chlorinated small-molecule acceptors.

6.3.5.1 Photovoltaic Performance of Chlorinated Small Molecular Acceptors

The IC (2-(3-oxo-2,3-dihydroindene-1-ylidene) dinitrile) is an electron-withdrawing unit and often used as an end group in low-band gap small-molecule acceptors with low-band gap, especially in the high-efficiency small-molecule acceptor-based PSCs in recent years. Chlorination treatment can effectively improve the electronic- withdrawal capability of IC. Small molecules with chlorinated IC as the terminal group extend the spectrum to the NIR.

In 2017, Li and coworker designed and synthesized a series of same conjugated core small molecules (see Chart 6.5) with or without halogens (F, Cl, I, and Br) and fabricated OFET and OSCs devices to study their photoelectric properties. They are all small molecules containing a single halogen atom, and the absorption spectrum ranges from 500 to 750 nm. The absorption peak shifts from H- to F-, Cl-, Br-, and I-ITIC in an orderly manner as the radius of the halogen atom increases, due to the induction effect and the delocalization of hetero electrons to the empty d orbitals of halogens. The HOMO and LUMO energy levels of F-, Cl-, and Br-ITIC have gradually dropped to a lower level, while I-ITIC shows a tiny high energy level compared with other halogenated ITIC. In addition, halogenated crystals have good crystallization properties. Using PTPDBDT as a donor, these halogenated ITIC derivative-based PSCs were fabricated. Except for fluorinated ITIC, the electron mobility of other halogenated molecules has improved. Due to the enhancement of J_{sc} and FF, the PCEs of all halogenated ITIC derivative-based devices are higher than that of H-ITIC based. Among these devices, Cl-ITIC-based device had the highest efficiency at 9.5% (see Table 6.4). In 2018, Chen and coworkers reported on 2,9-bis(2-methylene(3-(1,1-dicyanomethylene)indanone))-7,12-dihydro-4,4,7,7,12, 12-hexaoctyl-4*H*-cyclopenta [2″,1″:5,6;3″,4″:5′,6′]diindeno[1,2-*b*:1′,2′-b′]dithiophene (FDICTF, F-H) molecules also showed the effects of different halogenation, and the interaction of FF, F–Cl, and F–Br was presented by halogen atom substitution. After these end groups are halogenated, the NFA molecules exhibit spectral shifts and stronger absorption, increased crystallinity, and higher carrier mobility. However, the substitution of different halogen atoms has different effects on energy level and mobility.

So far, due to the uncertain qualitative nature of the Cl site, NFAs with monochlorinated IC end groups are usually a mixture of two isomers [88, 89]. In order to determine the role of Cl in the terminal position, He used recrystallization to separate IC-Cl- all kinds of monomers, and successfully synthesized ITIC-2Cl-δ and ITIC-2Cl-γ. The precise positioning of the chlorine atom helps to cultivate

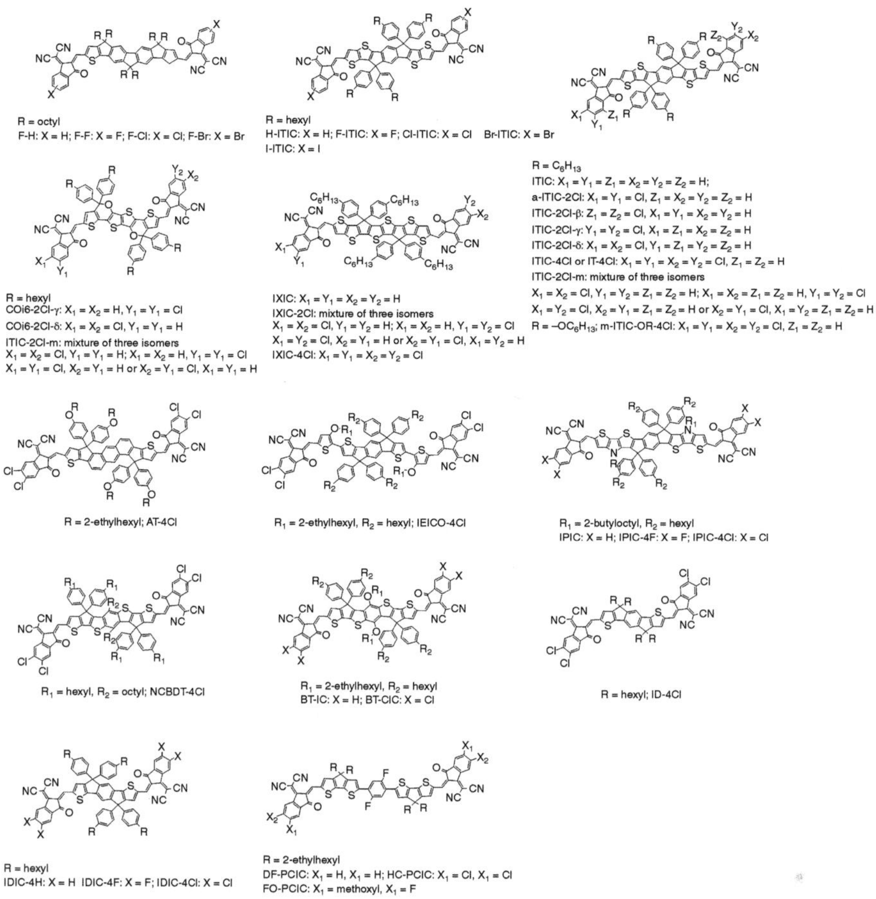

Chart 6.5 Small-molecule acceptor-based chlorinated IC end groups. Source: Hui Chen, Mingrui Pu, and Feng He.

single crystals, to study the properties of individual isomers and their influence on photovoltaic performance. Compared with ITIC-2Cl-δ, ITIC-2Cl-γ showed a high LUMO and a low HOMO, which resulted in a wider band gap and a slight blue-shift of the absorption spectrum. The LUMO and HOMO levels of ITIC-2Cl-m are between ITIC-2Cl-δ and ITIC-2Cl-γ. The higher LUMO level of ITIC-2Cl-γ, the higher V_{oc} was presented in corresponding devices. The precise position of Cl on the end group of ITIC-2Cl has a profound effect on the stacking of molecules. The ITIC-2Cl-γ had better coplanarity and shorter π–π packing distance from the single-crystal structure. In addition, ITIC-2Cl-γ formed a 3D rectangular interpenetrating network, which was conducive to multidirectional charge transport, while ITIC-2Cl-δ just had a linear packing structure [33]. The results showed that the PCE of ITIC-2Cl-γ:PBDB-TF-based PSCs was up to 13.03%, attributing to the overall increase in V_{oc}, FF, and J_{sc}. Moreover, the stability of ITIC-2Cl-γ- based device was also best among the other control acceptor-based device. This study

Table 6.4 Electronic and photovoltaic performance of small-molecule acceptor-based chlorinated IC end groups.

Acceptor	HOMO (eV)	LUMO (eV)	Donor	V_{oc} (V)	J_{sc} (mA cm^{-2})	FF (%)	PCE (%)
F-H	−5.42	−3.79	PBDB-T	0.94	15.02	67	9.59
F-F	−5.72	−4.13	PBDB-T	0.88	17.36	71	10.85
F-Cl	−5.62	−4.04	PBDB-T	0.87	17.61	75	11.47
F-Br	−5.62	−4.06	PBDB-T	0.87	18.22	76	12.05
H-ITIC	−5.61	−4.02	PTPDBDT	1.04	10.6	58	6.4
F-ITIC	−5.65	−4.09	PTPDBDT	0.94	14.1	66	8.8
Cl-ITIC	−5.70	−4.14	PTPDBDT	0.94	15.6	65	9.5
Br-ITIC	−5.73	−4.20	PTPDBDT	0.93	15.4	66	9.4
I-ITIC	−5.68	4.14	PTPDBDT	0.95	14.5	65	8.9
a-ITIC-2Cl	−5.29	−3.77	PBDB-T2F	0.88	18.91	73.50	12.23
ITIC-2Cl-β	−5.30	−3.71	PBDB-T2F	0.94	18.47	64.63	11.21
ITIC-2Cl-γ	−5.55	−3.92	PBDB-T2F	0.93	18.94	74.07	13.03
ITIC-2Cl-δ	−5.53	−3.94	PBDB-T2F	0.90	18.34	69.87	11.51
ITIC-2Cl-m	−5.54	3.93	PBDB-T2F	0.90	18.03	66.86	10.85
IT-2Cl	−5.68	−3.99	PBDB-T2F	0.922	19.08	74.8	13.16
IT-4Cl	−5.75	−4.09	PBDB-T2F	0.790	22.67	75.2	13.45
IT-4Cl	−5.75	−4.06	PBFTT	0.76	19.7	73.9	11.1
m-ITIC-OR-4Cl	−5.78	−4.05	BO2FC8	0.83	21.2	63.3	11.0
CO_i6-2Cl-γ	−5.46	−4.05	PTB7-Th	0.69	20.33	62.84	8.82
CO_i6-2Cl-δ	−546	−4.02	PTB7-Th	0.67	20.65	60.81	8.41
CO_i6-2Cl-m	−5.47	−4.02	PTB7-Th	0.69	20.78	64.32	9.22
IXIC	−5.13	−3.78	PBDB-T	0.82	20.9	65.4	11.3
IXIC-2Cl	−5.20	−3.90	PBDB-T	0.73	23.6	70.9	12.2
IXIC-4Cl	−5.20	−3.95	PBDB-T	0.69	22.9	71.2	11.2
AT-NC	−5.44	−3.80	PBDB-T	0.919	17.11	69.4	10.91
AT-4Cl	−5.71	−3.89	PBDB-TF	0.901	19.52	75.5	13.27
IEICO-4Cl	−5.56	−4.23	J52	0.700	23.8	60.7	10.1
IEICO-4Cl			PBDB-T	0.744	20.8	62.5	9.67
IEICO-4Cl			PTB7-Th	0.727	22.8	62.0	10.3
IPIC	−5.47	−3.82	PBDB-T	0.950	7.16	58.6	3.98
IPIC-4F	−5.54	−3.94	PBDB-T	0.835	19.8	67.1	11.1
IPIC-4Cl	−5.51	−3.95	PBDB-T	0.813	22.2	74.0	13.4
NCBDT-4Cl	−5.60	−4.02	PBDB-T-SF	0.851	22.35	74.3	14.1
BT-IC	−5.32	−3.85	PTB7-Th	0.81	17.5	59.6	8.3

Table 6.4 (Continued)

Acceptor	HOMO (eV)	LUMO (eV)	Donor	V_{oc} (V)	J_{sc} (mA cm^{-2})	FF (%)	PCE (%)
BT-CIC	−5.49	−4.08	PTB7-Th	0.70	22.5	71.0	11.2
ID-4Cl	−5.81	−3.97	PM6	0.767	17.87	74.76	10.25
IDIC-4H	−5.80	−3.87	PBDB-T2Cl	0.98	10.32	45.30	4.57
IDIC-4F	−5.95	−3.93	PBDB-T2Cl	0.88	13.00	61.91	7.10
IDIC-4Cl	−6.03	−4.02	PBDB-T2Cl	0.83	16.21	68.69	9.24
DF-PCIC	−5.49	−3.77	PTQ10	1.04	6.51	54.13	3.68
DF-PCIC	−5.49	−3.77	PBDB-T	0.89	15.28	61.87	8.43
FO-PCIC	−5.49	−3.75	PBDB-T	0.90	15.02	61.12	8.32
HC-PCIC	−5.54	−3.87	PTQ10	0.94	15.99	67.96	10.42
HC-PCIC	−5.54	−3.87	PBDB-T	0.73	17.53	69.08	9.03
HC-PCIC	−5.54	−3.87	PBDB-TF	0.89	18.13	72.06	11.75

implied that the 3D network structure may be a promising strategy to design better NFAs to improve PCE and the stability of PSCs. However, in different systems, the precise Cl position on the IC end group has different effects on device performance. For example, with CO_i6 as the electron-rich core, three new chlorinated acceptors [34], CO_i6-2Cl-γ, CO_i6-2Cl-δ, and CO_i6-2Cl-m, were designed and synthesized. The CO_i6-2Cl-γ:PTB7-Th-based device had a slightly higher V_{oc} than that of CO_i6-2Cl-δ-based device, and the CO_i6-2Cl-*m*-based device obtained a moderate V_{oc}. But the CO_i6-2Cl-*m*-based devices exhibited a highest PCE of 9.30% among these acceptor-based devices.

According to above discussion of monochlorination, the introduction of Cl has a profound effect on energy level, light absorption, crystallinity, and morphology, mainly because Cl has a strong electronegativity and empty 3d orbits of Cl to accommodate π-electrons. We will discuss that the number of introduced chlorine is increased and the effect of chlorination on the device is based on these acceptors.

Hou and coworkers designed and synthesized IT-4Cl and compared with IT-2Cl (Cl-ITIC) [21]. Compared with ITIC, the incorporation of Cl in IT-2Cl and IT-4Cl hardly affected the molecular packing and the distribution of electron cloud, but the HOMO and LUMO levels have brought significant changes, which are −5.63 and −3.53 eV of IT-2Cl and −5.71 and −3.62 eV of IT-4Cl, respectively, indicating that the number of Cl increases from one to two at each end, HOMO and LUMO are further deepen, and the gap between HOMO and LUMO is reduced, which implied the reason why its absorption spectrum of IT-4Cl is further red shifted than that of IT-2Cl. The increase of Cl did not bring about an obvious optimal conformational change, indicating that the final organization of Cl would not generate an additional steric effect, although atomic radius of Cl is much large.

Compared with the absorption spectra of solution, the films absorption of IT-2Cl and IT-4Cl is red-shift, especially IT-4Cl is about 40 nm red-shift related to IT-4F, attributing to the larger dipole moment of the chlorinated molecules and improved aggregation in film. Using PBDB-TF as donor, the PCE of corresponding are beyond 13%. Through the ternary strategy [90], a high PCE of 14.18% was obtained, the ratio of IT-2Cl and IT-4Cl was 2 : 8 in the active layer.

The introduction of end group with dichlorination into other acceptor cores can also obtain outstanding photovoltaic performance [35, 47, 90, 121, 122]. Chen and coworkers synthesized a small-molecule acceptor named NCBDT-4Cl, using PBB-T-SF as the donor, giving an efficient device with low E_{loss} of 0.55 eV [123]. The champion device achieved a high PCE of 14.1%. Another dichlorinated acceptor, IPIC-4Cl, has also been shown to improve device performance [121]. Compared with nonhalogenated and fluorinated, the blend of IPIC-4Cl and PBDB-T has higher crystallinity and better phase separation performance, which ensures the efficient dissociation of excitons and the collection of charges, and PCE with low E_{loss} of 0.51 eV reached 13.4%, the PCE was boosted up to 14.3% by ternary measurement of adding $PC_{71}BM$ as component.

Chlorination can also help OSCs work efficiently with a small driving force. Chen and coworkers also synthesized a dichlorinated small molecule, HC-PCIC. Compared with the related molecule without substituent, and with fluorination and methoxyl in IC group, chlorinated small molecule had a clear spectral red-shift and higher 0–0 peak, indicating that the chlorination has a stronger end-to-end arrangement. Because Cl has a larger atomic radius and dipole moment than H and F, it improves intermolecular interaction. By comparing the energy levels of three small-molecule acceptors and three polymer donors, a series of HOMO shifts are formed. The introduction of Cl enhanced the electron-withdrawing unit, enabling PSCs with small driving force to work efficiently. Therefore, the HC-PCIC-based PSCs obtained excellent PCE of 10.42% and 11.75%, and the HOMO offset was small, 0 and 0.06 eV, respectively.

6.3.5.2 The Investigation of the Crystal Structure of Chlorinated Small Molecular Acceptors

The crystal structure dispalys the specific configuration of the molecule, which has an important influence on the photoelectric properties. In order to further study the effect of chlorination on the stacking structure and photovoltaic performance, the single-crystal structure of some chlorinated NFAs was analyzed, and the relationship between structure and performance was deeply explored.

In 2018, He and coworkers analyzed the single-crystal structure of the small-molecule acceptor IDIC-4Cl and proved that the Cl⋯S interaction can achieve interlocking between molecules and form an effective linked molecular network [35]. The dihedral angle of IDIC-4Cl is very small, and the introduction of multiple chlorine atoms does not cause the distortion of the backbone, which makes the molecules more coplanar and more orderly. The parallel arrangement of IDIC-4Cl molecules forms a good J aggregation, which is conducive to charge transfer. The PSCs based on IDIC-4Cl:PBDB-T2Cl achieved a decent PCE of 9.24%.

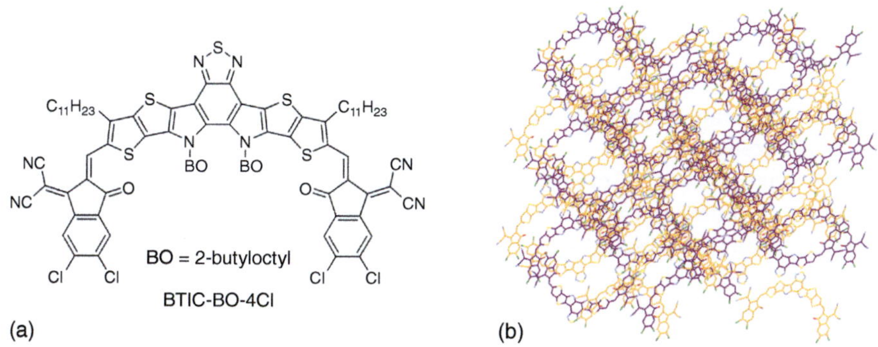

Figure 6.10 The molecule (a) and single-crystal, (b) structure of BTIC-BO-4Cl. Source: Hui Chen, Mingrui Pu, and Feng He.

For R10-4Cl molecule, single-crystal structure analysis shows that it has two different molecular orientations [36]. The molecules form a close parallel arrangement in one direction, and the distance between the end groups of adjacent molecules is only 3.32 Å. There is also an orientation effect that is inserted diagonally into these parallel molecules to interact with their end groups. The interpenetrating network formed from these two directions is conducive to the rapid transmission of carriers in multiple channels. Therefore, a high charge mobility is achieved in a device based on PBDB-T:R10-4Cl, and its PCE is 10.7%.

Recently, He and coworkers analyzed the single-crystal structure of the small-molecule acceptor BTIC-BO-4Cl in Figure 6.10 [124]. The S···O distance is 2.67 Å for a single BTIC-BO-4Cl molecule, which gives the acceptor a good planarity with a small dihedral angle of 7.6° between core and end group of chlorination. The two 2-butyloctyl chains are in completely opposite directions, which proves that the length of the alkyl chain has a great influence on molecular stacking. Although 2-butyl chain (BTIC-BO-4Cl) and 1-dodecyl chain (BTIC-C12-4Cl) have the same number of carbon atoms, the linear length of C12-4Cl of BTIC-C12-4Cl molecule is relatively long, the π–π distance is greater than that of BTIC-BO-4Cl molecules. Therefore, its efficiency is inferior to BTIC-BO-4Cl. The π–π distance between adjacent molecules of BTIC-BO-4Cl is relatively short, ranging from 3.29 to 3.53 Å, leading to J-aggregation of the entire molecule. Although there are two larger chlorine atoms on the terminal group, the distance between the end groups of π–π is 3.29 Å and almost completely overlaps, which is conducive to the efficient electronic hopping between these molecules. The short π–π distance of BTIC-BO-4Cl comes from the multiple interlocking effects of Cl···S and Cl···π interactions. This makes BTIC-BO-4Cl present a closed spiral loop microstructure composed of some layers and molecules in the vertical direction. Each ring interacts with two adjacent molecules through a strong interaction, and finally forms a 3D interpenetrating network, which can bless the electron transfer in multiple directions. Using PBDB-TF as donor, the fabricated PSCs obtained a high PCE of 16.43%. The single crystal and device results show that the synergistic effect of chlorination and alkyl chain can optimize the photovoltaic performance of the material.

6.3.5.3 The Semitransparent Organic Solar Cells Based on Chlorinated Small Molecular Acceptors

The chlorination of small-molecule acceptors leads to a red shift of the absorption spectrum, extending the spectrum to the NIR, so these acceptors have the potential to prepare semitransparent OSCs [77, 91, 125–127].

In 2016, Hou and coworkers designed and synthesized a small-molecule acceptor, IEICO-4Cl. The absorption spectrum of IEICO-4Cl is effectively red-shifted by the chlorination, and the absorption has shifted from 925 to 1010 nm compared with control molecule IEICO [127]. IEICO-4Cl mainly absorbs the region of 745–945 nm, which is very suitable for fabricating semitransparent OSCs. By optimizing the relationship between Au electrode thickness and average transmittance (AVT), devices with an average AVT 25.7% based on IEICO-4Cl and PTB7-Th have a PCE of 8.28%. When J52, PDBBT, and PTB7-Th are used as donors, respectively, the prepared blend films show different color changes, making IEICO-4Cl have greater application potential. This research showed that the chlorination would be a useful tool to fine-tune the energy level of the material to develop multifunction devices.

Next, Li and coworkers designed and synthesized an acceptor BT-ClC, with four Cl at a certain molecular site [126]. Compared to control molecule of BT-C, the red shift of the BT-ClC spectrum is about 60 nm, suggesting an increase in IECT. Under thin film conditions, the absorption of BT-CIC is mainly distributed between 650 and 1000 nm, which is very suitable for fabricating semitransparent PSCs. Cooperating BT-ClC with PTB7-Th, devices based on BT-ClC and PCE-10 showed a PCE of 11.2%. Taking 10 nm Ag as semitransparent cathode, the semitransparent PSCs with AVT of 43% presented a PCE of 7.1%. These results implied that chlorination would be a good strategy for developing high-performance semitransparent OSCs.

In addition to the IEICO-4Cl and BT-ClC discussed earlier, other chlorinated molecules such as ITIC-4Cl, ID-4Cl, and ITC-2Cl have also been reported to be used in the fabrication of efficient semitransparent OSCs. The chlorination is an effective method for synthesizing near-infrared materials. Compared with the control nonchlorinated molecules, the interaction between molecules is enhanced, the absorption coefficient is higher, and the absorption spectrum is red-shifted. Therefore, the chlorination provided a new way to fabricate high-performance semitransparent OSCs.

Recently, He and coworkers designed and synthesized low band-gap acceptor BTIC-2Cl-γCF_3 [128], the molecule structure is shown in Figure 6.11a. It has strong absorption in the near-infrared region for allowing the most of visible light to penetrate, which is a candidate to fabricate high-performance semitransparent OSCs. When balanced the PCE and the transparency of device, the BTIC-2Cl-γCF_3-based semitransparent PSCs with 24.45% AVT yields a high PCE of 13.06% from same halogen-free solvent the value of AVT is around 25% in Figure 6.11b. The determined color coordinates (x, y) of 10, 15, and 20 nm Ag ST devices were (0.29, 0.30), (0.26, 0.27), and (0.26, 0.26) in Figure 6.11c, respectively. Figure 6.11d exhibits

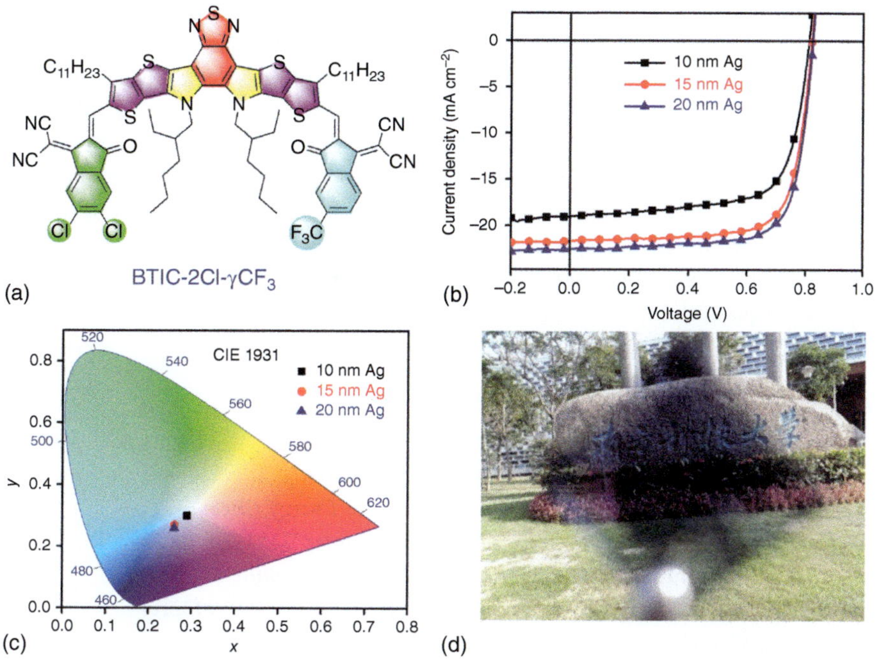

Figure 6.11 (a) Molecule structure of BTIC-2Cl-γCF_3, (b) *J–V* curves of toluene processed semitransparent tertiary devices with varied thickness of Ag cathode, (c) CIE coordinates of related devices, (d) outdoor image through the device. Source: Hui Chen, Mingrui Pu and Feng He.

an outdoor image through the related device, which shows good transparency, color sense, and transparency. These parameters indicated that BTIC-2Cl-γCF_3 is a very promising translucent photovoltaic material for semitransparent building integrated photovoltaic productions.

6.4 Conclusion and Outlook

In this chapter, the intrinsic properties of chlorination are introduced and compared with those of fluorination, the research progress of chlorinated polymers and small-molecule donors and small-molecule acceptors are summarized. Because chlorine atom has strong electronegativity (slightly lower than fluorine atoms) and unique 3d empty orbitals, which could accommodate lone pair electrons and promote delocalization of electrons, so chlorination can effectively deepen the energy level of material. In some specific material systems, it is more effective than fluorination in energy-level control. Although the chlorine atom has a larger atomic radius and steric hindrance, the influence of steric hindrance can be reduced through smart structural design. In addition, the C—Cl bond can form a larger dipole moment, expand the spectral absorption, and regulate the packing of molecules. The chlorination has a profound effect on the photoelectric properties of organic semiconductors. Generally, both the donor and acceptor molecules will simultaneously reduce their HOMO and LUMO levels. Compared with the effect on energy level, the effect of chlorination on light absorption is different for different materials. For the polymer donor, the chlorination reaction causes a blue shift of the spectrum, while for the small-molecule acceptor, the spectrum is red shifted. The effect of fluorination and chlorination on the photovoltaic performance of materials differs for different material systems. Therefore, it is necessary to choose a suitable strategy according to the specific material properties. However, considering the high yield and low cost in chemical synthesis, chlorination has greater commercial application potential.

References

1 Liu, Q., Jiang, Y., Jin, K. et al. (2020). *Sci. Bull.* 65: 272.
2 Chao, P., Johner, N., Zhong, X. et al. (2019). *J. Energy Chem.* 39: 208.
3 Hong, L., Yao, H., Wu, Z. et al. (2019). *Adv. Mater.* 31: 1903441.
4 Zhang, M., Zhu, L., Zhou, G. et al. (2021). *Nat. Commun.* 12: 309.
5 Yan, T., Song, W., Huang, J. et al. (2019). *Adv. Mater.* 31: 1902210.
6 Yu, R., Yao, H., Cui, Y. et al. (2019). *Adv. Mater.* 31: 1902302.
7 Yamamoto, T., Zhou, Z., Kanbara, T. et al. (1996). *J. Am. Chem. Soc.* 118: 10389.
8 Yu, W.L., Meng, H., Pei, J. et al. (1988). *Macromolecules* 31: 4838.
9 Hou, Q., Xu, Y., Yang, W. et al. (2002). *J. Mater. Chem.* 12: 2887.
10 Svensson, M., Zhang, F., Veenstra, S.C. et al. (2003). *Adv. Mater.* 15: 988.
11 Lin, Y., Wang, J., Zhang, Z. et al. (2015). *Adv. Mater.* 27: 1170.
12 Yuan, J., Zhang, Y., Zhou, L. et al. (2019). *Joule* 3: 1140–1151.
13 Zhang, Q., Kelly, M.A., Bauer, N., and You, W. (2017). *Acc. Chem. Res.* 50: 2401.
14 Zhang, M., Guo, X., Zhang, S., and Hou, J. (2014). *Adv. Mater.* 26: 1118.
15 Oh, J., Kranthiraja, K., Lee, C. et al. (2016). *Adv. Mater.* 28: 10016.
16 Berger, R., Resnati, G., Metrangolo, P. et al. (2011). *Chem. Soc. Rev.* 40: 3496.
17 Zhou, H., Yang, L., Stuart, A.C. et al. (2011). *Angew. Chem. Int. Ed.* 50: 2995.
18 Stephens, R. and Tatlow, J.C. (1962). *Rev. Chem. Soc.* 16: 44.
19 Bondi, A. (1964). *J. Phys. Chem.* 68: 441.
20 Tang, M.L., Oh, J.H., Reichardt, A.D., and Bao, Z. (2009). *J. Am. Chem. Soc.* 131: 3733.
21 Zhang, H., Yao, H., Hou, J. et al. (2018). *Adv. Mater.* 30: 1800613.
22 Zhang, S., Qin, Y., Zhu, J., and Hou, J. (2018). *Adv. Mater.* 30: 1800868.
23 Chen, H., Hu, Z., Wang, H. et al. (2018). *Joule* 2: 1623.
24 Chao, P., Liu, L., Zhou, J. et al. (2018). *ACS Appl. Energy Mater.* 1: 6549.
25 Chao, P., Mu, Z., Wang, H. et al. (2018). *ACS Appl. Energy Mater.* 1: 2365.
26 Chao, P., Wang, H., Mo, D. et al. (2018). *J. Mater. Chem. A* 6: 2942.
27 Qu, S., Wang, H., Mo, D. et al. (2017). *Macromolecules* 50: 4962.
28 Wang, H., Chao, P., Chen, H. et al. (2017). *ACS Energy Lett.* 2: 1971.
29 Hu, Z., Chen, H., Qu, J. et al. (2017). *ACS Energy Lett.* 2: 753.
30 Mo, D., Wang, H., Chen, H. et al. (2017). *Chem. Mater.* 29: 2819.

31 Yang, Z., Chen, H., Wang, H. et al. (2018). *Polym. Chem.* 9: 940.
32 Hu, Z., Chen, H., Zhong, X. et al. (2018). *Acta Polym. Sin.* 2: 273.
33 Lai, H., Chen, H., Zhou, J. et al. (2019). *J. Phys. Chem. Lett.* 10: 4737.
34 Liu, T., Zhao, Q., Wang, H. et al. (2019). *Mater. Chem. Front.* 3: 1859.
35 Qu, J., Chen, H., Zhou, J. et al. (2018). *ACS Appl. Mater. Interfaces* 10: 39992.
36 Qu, J., Zhao, Q., Zhou, J. et al. (2019). *Chem. Mater.* 31: 1664.
37 Lai, H., Chen, H., Zhou, J. et al. (2019). *iScience* 17: 302.
38 Zhang, Y., Yao, H., Zhang, S. et al. (2018). *Sci. China Chem.* 61: 1328.
39 Sakamoto, Y., Komatsu, S., and Suzuki, T. (2001). *J. Am. Chem. Soc.* 123: 4643.
40 Blaskovits, J.T., Bura, T., Beaupré, S. et al. (2016). *Macromolecules* 50: 162.
41 Kawashima, K., Fukuhara, T., Suda, Y. et al. (2016). *J. Am. Chem. Soc.* 138: 10265.
42 Liu, S., Kan, Z., Thomas, S. et al. (2016). *Angew. Chem. Int. Ed.* 55: 12996.
43 Fan, Q., Liu, T., Gao, W. et al. (2019). *J. Mater. Chem. A* 7: 15404.
44 Jung, J.W., Liu, F., Russell, T.P., and Jo, W.H. (2015). *Adv. Energy Mater.* 5: 1500065.
45 Ye, L., Xie, Y., Weng, K. et al. (2019). *Nano Energy* 58: 220.
46 Duan, C., Furlan, A., van Franeker, J.J. et al. (2015). *Adv. Mater.* 27: 4461.
47 Geng, R., Song, X., Feng, H. et al. (2019). *ACS Energy Lett.* 4: 763.
48 Firdaus, Y., Maffei, L.P., Cruciani, F. et al. (2017). *Adv. Energy Mater.* 7: 1700834.
49 Fei, Z., Shahid, M., Yaacobi-Gross, N. et al. (2012). *Chem. Commun.* 48: 11130.
50 Benjamin, I., Faraggi, E.Z., Avny, Y. et al. (1996). *Chem. Mater.* 8: 352.
51 Jung, J.W., Jo, J.W., Chueh, C.C. et al. (2015). *Adv. Mater.* 27: 3310.
52 Son, H.J., Wang, W., Xu, T. et al. (2011). *J. Am. Chem. Soc.* 133: 1885.
53 Guo, S., Ning, J., Körstgens, V. et al. (2015). *Adv. Energ. Mater.* 5: 1401315.
54 Zheng, Y.Q., Wang, Z., Dou, J.H. et al. (2015). *Macromolecules* 48: 5570.
55 Jiang, B., Du, C.C., Li, M.J. et al. (2016). *Polym. Chem.* 7: 3311.
56 Yum, S., An, T.K., Wang, X. et al. (2014). *Chem. Mater.* 26: 2147.
57 Nguyen, T.L., Choi, H., Ko, S.J. et al. (2014). *Energy Environ. Sci.* 7: 3040.
58 Tomassetti, M., Ouhib, F., Wislez, A. et al. (2015). *Polym. Chem.* 6: 6040.
59 Wang, Z., Zhao, J., Li, Y., and Peng, Q. (2014). *Polym. Chem. UK* 5: 4984.
60 Yang, Y., Wu, R., Wang, X. et al. (2014). *Chem. Commun.* 50: 439.
61 Jo, J.W., Jung, J.W., Wang, H.W. et al. (2014). *Chem. Mater.* 26: 4214.
62 Kim, I.B., Jang, S.Y., Kim, Y.A. et al. (2017). *ACS Appl. Mater. Interfaces* 9: 24011.
63 Shin, J., Kim, M., Kang, B. et al. (2017). *J. Mater. Chem. A* 5: 16702.
64 Shin, J., Kim, M., Lee, J. et al. (2017). *Chem. Commun.* 53: 1176.
65 Bronstein, H., Frost, J.M., Hadipour, A. et al. (2013). *Chem. Mater.* 25: 277.
66 Chochos, C.L., Leclerc, N., Gasparini, N. et al. (2017). *J. Mater. Chem. A* 5: 25064.
67 Gopal, A., Saeki, A., Ide, M., and Seki, S. (2014). *ACS Sustainable Chem. Eng.* 2: 2613.
68 Hwang, J., Park, J., Kim, Y.J. et al. (2017). *Chem. Mater.* 29: 2135.
69 Kim, J., Yun, M.H., Kim, G.H. et al. (2014). *ACS Appl. Mater. Interfaces* 6: 7523.

70 Kranthiraja, K., Kim, S., Lee, C. et al. (2017). *Adv. Funct. Mater.* 27: 1701256.
71 Zhan, L., Li, S., Xia, X. et al. (2021). *Adv. Mater.* 33: 2007231.
72 Li, Z., Lu, J., Tse, S.C. et al. (2011). *J. Mater. Chem.* 21: 3226.
73 Li, Z., Zhang, T., Xin, Y. et al. (2016). *J. Mater. Chem. A* 4: 18598.
74 Shi, S., Liao, Q., Tang, Y. et al. (2016). *Adv. Mater.* 28: 9969.
75 Wang, M., Ford, M., Phan, H. et al. (2016). *Chem. Commun.* 52: 3207.
76 Wang, X., Tang, A., Chen, Y. et al. (2016). *RSC Adv.* 6: 90051.
77 Cui, Y., Yang, C., Yao, H. et al. (2017). *Adv. Mater.* 29: 1703080.
78 He, X., Mukherjee, S., Watkins, S. et al. (2014). *J. Phys. Chem. C* 118: 9918.
79 Liu, P., Zhang, K., Liu, F. et al. (2014). *Chem. Mater.* 26: 3009.
80 Li, Y., Wang, J., Liu, Y. et al. (2016). *ACS Appl. Mater. Interfaces* 8: 26152.
81 Kim, J.H., Song, C.E., Kim, H.U. et al. (2013). *Chem. Mater.* 25: 2722.
82 Liu, D., Zhao, W., Zhang, S. et al. (2015). *Macromolecules* 48: 5172.
83 Verstappen, P., Kesters, J., Vanormelingen, W. et al. (2015). *J. Mater. Chem. A* 3: 2960.
84 Ko, S.J., Hoang, Q.V., Song, C.E. et al. (2017). *Energy Environ. Sci.* 10: 1443.
85 Lee, J.Y., Jo, J.W., and Jo, W.H. (2015). *Org. Electron.* 20: 125.
86 Fan, Q., Zhu, Q., Xu, Z. et al. (2018). *Nano Energy* 48: 413.
87 Su, W., Li, G., Fan, Q. et al. (2019). *J. Mater. Chem. A* 7: 2351.
88 Yang, F., Li, C., Lai, W. et al. (2017). *Mater. Chem. Front.* 1: 1389.
89 Wang, Y., Zhang, Y., Qiu, N. et al. (2018). *Adv. Energ. Mater.* 8: 1702870.
90 Chen, Y., Liu, T., Hu, H. et al. (2018). *Adv. Energ. Mater.* 8: 1801203.
91 Li, Y., Lin, J.D., Che, X. et al. (2017). *J. Am. Chem. Soc.* 139: 17114.
92 Dai, S., Zhao, F., Zhang, Q. et al. (2017). *J. Am. Chem. Soc.* 139: 1336.
93 Deng, D., Zhang, Y., Zhang, J. et al. (2016). *Nat. Commun.* 7: 13740.
94 Zhao, F., Dai, S., Wu, Y. et al. (2017). *Adv. Mater.* 29: 1700144.
95 Tang, M.L. and Bao, Z. (2011). *Chem. Mater.* 23: 446.
96 Zhang, S., Yang, B., Liu, D. et al. (2015). *Macromolecules* 49: 120.
97 Zhang, S., Qin, Y., Uddin, M.A. et al. (2016). *Macromolecules* 49: 2993.
98 Huang, S., Huang, X., Chen, L. et al. (2019). *Org. Electron.* 70: 86.
99 Lin, Y., Lu, Y., Tsao, C.S. et al. (2019). *J. Mater. Chem. A* 7: 3072.
100 Liao, S.H., Jhuo, H.J., Cheng, Y.S., and Chen, S.A. (2013). *Adv. Mater.* 25: 4766.
101 Liang, Y., Wu, Y., Feng, D. et al. (2009). *J. Am. Chem. Soc.* 131: 56.
102 Liang, Y., Feng, D., Wu, Y. et al. (2009). *J. Am. Chem. Soc.* 131: 7792.
103 Liang, Y., Xu, Z., Xia, J. et al. (2010). *Adv. Mater.* 22: 135.
104 Hiszpanski, A.M., Saathoff, J.D., Shaw, L. et al. (2015). *Chem. Mater.* 27: 1892.
105 Cui, Y., Yao, H., Zhang, J. et al. (2020). *Adv. Matter.* 32: 1908205.
106 Roldao, J.C., Oliveira, E.F., and Lavarda, F.C. (2016). *Org. Electron.* 33: 246.
107 Liao, Z., Xie, Y., Chen, L. et al. (2019). *Adv. Funct. Mater.* 29: 1808828.
108 Tang, A., Song, W., Xiao, B. et al. (2019). *Chem. Mater.* 31: 3941.
109 Bin, H., Gao, L., Zhang, Z.G. et al. (2016). *Nat. Commun.* 7: 13651.
110 Xue, L., Yang, Y., Xu, J. et al. (2017). *Adv. Mater.* 29: 1703344.
111 Fan, Q., Su, W., Guo, X. et al. (2017). *J. Mater. Chem. A* 5: 9204.
112 Zhao, W., Qian, D., Zhang, S. et al. (2016). *Adv. Mater.* 28: 4734.
113 Zhao, W., Li, S., Yao, H. et al. (2017). *J. Am. Chem. Soc.* 139: 7148.

114 Lei, T., Dou, J., Ma, Z. et al. (2013). *Chem. Sci.* 4: 2447.
115 Kini, G.P., Jeon, S.J., and Moon, D.K. (2020). *Adv. Mater.* 30: 1906175.
116 Jeon, S.J., Han, Y.W., and Moon, D.K. (2019). *Sol. RRL* 3: 1900094.
117 Jeon, S.J., Han, Y.W., and Moon, D.K. (2019). *Small* 15: 1902598.
118 Wu, Y., An, C., Shi, L. et al. (2018). *Angew. Chem. Int. Ed.* 57: 12911.
119 Nian, L., Kan, Y., Gao, K. et al. (2020). *Joule* 4: 1.
120 Chen, H., Hu, D., Yang, Q. et al. (2019). *Joule* 3: 3034.
121 Li, S., Zhan, L., Sun, C. et al. (2019). *J. Am. Chem. Soc.* 141: 3073.
122 Feng, H., Yi, Y.Q.Q., Ke, X. et al. (2019). *Adv. Energ. Mater.* 9: 1803541.
123 Kan, B., Feng, H., Yao, H. et al. (2018). *Sci. China Chem.* 61: 1307.
124 Mo, D., Chen, H., Zhou, J. et al. (2020). *J. Mater. Chem. A* 8: 8903–8912.
125 Li, X., Meng, H., Shen, F. et al. (2019). *Dyes Pigm.* 166: 196.
126 Su, W., Fan, Q., Guo, X. et al. (2019). *Phys. Chem. Chem. Phys.* 21: 10660.
127 Yao, H., Chen, Y., Qin, Y. et al. (2016). *Adv. Mater.* 28: 8283.
128 Chen, H., Lai, H., Chen, Z. et al. (2021). *Angew. Chem. Int. Ed.* 60: 3238.

7

Polymer–Polymer Solar Cells: Materials, Device, and Stability

Jianyu Yuan[1], Huiliang Sun[2,4], Yingjian Yu[3,5], Wanli Ma[1], Xugang Guo[2], and Jun Liu[3,5]

[1] *Soochow University, Institute of Functional Nano & Soft Materials (FUNSOM), Jiangsu Key Laboratory for Carbon-Based Functional Materials & Devices, 199 Renai Road, Suzhou, Jiangsu 215123, P.R. China*
[2] *Southern University of Science and Technology, Department of Materials Science and Engineering, 1088 Xueyuan Road, Shenzhen, Guangdong 518055, P.R. China*
[3] *State Key Laboratory of Polymer Physics and Chemistry, Changchun Institute of Applied Chemistry, Chinese Academy of Sciences, 5625 Ren Min Street, Changchun 130022, P.R. China*
[4] *Guangzhou University, School of Chemistry and Chemical Engineering, Center for Advanced Analytical Science, 230 Wai Huan Xi Road, Guangzhou, Guangdong 510006, P.R. China*
[5] *University of Science and Technology of China, 96 Jin Zhai Road, Hefei 230026, P.R. China*

7.1 Introduction

Organic photovoltaics (OPVs) based on a polymer–polymer (all-polymer) blend film are attractive due to not only their tunable optical, electrochemical, and structural properties, but also many superior features that are not readily available in conventional polymer-fullerene and polymer-nonfullerene molecule solar cells, including long-term stability, synthetic accessibility, and excellent film-forming properties suitable for large-scale manufacturing [1, 2]. During the past few years, we and a few other groups have developed efficient all-polymer solar cells (all-PSCs) through synergistic materials design and device engineering [2]. Quite recently, it has been shown that polymer acceptors can further extend the solar cell response spectrum, and power conversion efficiencies (PCEs) above 15% have been achieved in these all-polymer systems [3], demonstrating its great potential for future application.

Aromatic diimide-based conjugated polymers, especially containing naphthalene diimide (NDI) and perylene diimides (PDIs), have been proven to be the most efficient n-type polymers. For example, the benchmark n-type, P(NDI2OD-T2), Polyera ActivInk N2200 [4], a copolymer composed of NDI and bithiophene, shows an excellent electron mobility value of 10^{-3} cm^2 V^{-1} s^{-1} in SCLC and 0.85 cm^2 V^{-1} s^{-1} in the OTFT device, respectively. These are extremely desired for polymer acceptors in all-PSCs. The molecular structures of NDI polymers are largely modified from N2200, and the PCEs of the corresponding all-PSCs have been steadily improved. The PCEs of all-PSCs employing NDI polymers have achieved remarkable progress from below 3% in 2014 to approaching 12% now [5], which has surpassed state-of-the-art performance of fullerene OSCs. Except NDI and PDI, another

Organic Solar Cells: Materials Design, Technology and Commercialization, First Edition. Edited by Liming Ding.

member of aromatic diimide family, bithiophene diimide (BTI), was also adopted, which possesses relatively small five-membered thiophenes along with imide functionality bridged on bithiophene, significantly moderating steric hindrance. In addition, the open β-positions of thiophene ring in BTI offer excellent opportunities for structural modification and backbone expansion [6]. The recent invention of BTI fused ring oligomers (BTIn) yields a new series of polymer acceptors, which have a different electronic structure from NDI polymers and featured tunable LUMO levels, thanks to their delocalized topology along molecule backbones. BTI polymer acceptors have gifted greatly improved PCEs in all-PSCs from <1% in 2017 to >15% now [3, 7]. Some other aromatic diimides were also developed, resulting in promising performances. With the exploration of more novel aromatic diimide electron-deficient building blocks, even higher efficiency of all-PSCs can be expected.

Organic boron chemistry offers a new toolbox to tune optoelectronic properties of organic/polymer semiconductors. A successful example is the design of n-type organic/polymer semiconductors with boron–nitrogen coordination bond (B←N). In 2015, Liu and coworkers reported the first n-type polymer containing B←N unit for application as electron acceptor in all-PSCs [8]. Then they developed a new kind of electron-deficient building block based on B←N unit, double B←N bridged bipyridine (BNBP) [9]. A striking feature of BNBP-based polymer acceptors is their tunable LUMO energy levels, which leads to high open-circuit voltage of all-PSCs. Furthermore, these BNBP-based polymers exhibit strong absorption in the visible range and show high electron mobility. Their all-polymer photovoltaic devices exhibit PCE of exceeding 10% under solar light irradiation and PCE of exceeding 27% under indoor light irradiation [10, 11]. In 2019, Huang and coworkers reported another electron-deficient building block based on B←N unit, BNIDT, which affords all-PSCs with PCE of 8.78% [12]. Based on these research progresses, polymer acceptors containing B←N unit have emerged as a new kind of high-efficiency polymer acceptors for all-PSCs.

Meanwhile, the development of device engineering in all-PSCs becomes very flourishing since 2015. The PCEs of all-PSCs surpassed 5%, which was considered as a threshold value for OPVs [13]. Studies of the widely investigated "classical" n-type polymer N2200 have demonstrated how controlling the phase separation between donor and acceptor polymer translates to desirable exciton separation, charge transport, and overall photovoltaic performance in the devices [14, 15]. Lots of efforts have been dedicated into optimizing the morphology of N2200-based all-polymer bulk heterojunction (BHJ) blends, adopting strategies such as solvent engineering, thermal annealing, processing additive [16–19]. Afterward, modified device structures such as planar heterojunction, ternary and tandem structures have been incorporated to further enhance the light absorption and charge transport, leading to significantly improved PCE to over 10% [20–22]. Most importantly, multiple reports have shown that the presence of polymer-based acceptors in the photoactive layer represents a great assistance to device stability due to the intrinsic stability merits of conjugated polymers against thermal, photo, and mechanical stresses [2, 23]. Besides, these all-polymer blends exhibit reduced sensitivity to donor–acceptor (D–A) blend ratios, processing environment, and device area,

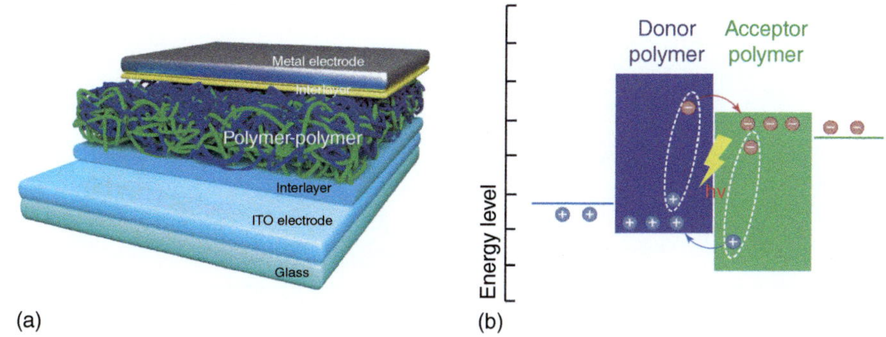

Figure 7.1 (a) Structure of single junction all-polymer BHJ solar cells, (b) energy levels of the donor and acceptor polymers where light absorption/excitation dissociation/charge collection take place in all-PSCs. Source: Jianyu Yuan.

compared with the polymer-fullerene and polymer-molecule systems. This progression demonstrates the unique potential of all-PSCs and its opportunities for further design to achieve efficient and stable solar cells for large-scale applications.

7.2 The Device Structure and Basic Principles of All-PSCs

The typical all-PSCs device structure is presented in Figure 7.1a. The device basically consists of an ITO glass substrate, transport electrode, two interfacial layers (one is electron transporting layer and the other is hole transporting layer), the active layer of all-polymer D–A BHJ blend, and the metal contact top electrode. There are two mainstream device structures applied in all-PSC fabrication: the first one is conventional structure and presents the architecture of metal electrodes/electron transporting layer/active layer/hole transport layer/ITO glass substrate; and the other has the inverted structure compared with the conventional one. For efficient operation, the all-polymer photoactive blend should develop an interpenetrating network of donor and acceptor polymer domains to provide effective percolation pathways connecting the acceptor regions to the cathode and donor regions to the anode.

As shown in Figure 7.1b, the basics principles of all-PSCs are similar to conventional polymer-fullerene solar cells, which is the photoexcited state of an electron–hole pair bound together by the Coulomb attraction. In order to achieve photocurrent, these excited electron–hole pairs need to be dissociated into free charge carriers. In organic semiconductors, due to their relatively low dielectric constant, the photoexcited Frenkel excitons have a much stronger binding energy. A large driving force from the D–A energy level offset is needed for the exciton dissociation [24]. Conventional fullerene acceptors are spherically symmetric materials, which can have close contact with the donor polymers in arbitrary directions to achieve efficient charge transfer. Meanwhile, fullerene derivatives tend to form highly crystalline domains with large charge delocalization, which will lower the Coulomb barrier and hence improve exciton dissociation [14]. Likewise, proper

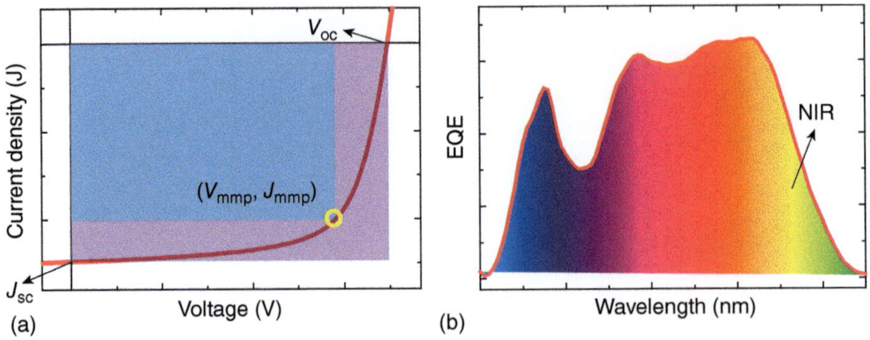

Figure 7.2 (a) *J*–*V* curve under illumination and (b) EQE versus wavelength for a typical all-PSC. Source: Jianyu Yuan.

packing orientation and degree of crystallinity in D–A polymers are helpful to efficiently dissociate electron–hole pairs into free charge carriers. Unfortunately, these all-polymer systems are anisotropic, to well control the orientation and crystallinity of both donor and acceptor crystals at the BHJ blend is a challenging task [25].

Figure 7.2a shows typical all-PSC current density (J) versus voltage (V) curve under illumination. Such plots are used to evaluate device performance parameters including: (i) the short-circuit current density (J_{sc}, the current flowing through an illuminated cell when there is no external resistance); (ii) the open-circuit voltage (V_{oc}, the cell voltage in sunlight when no current is flowing); (iii) the fill factor (FF; the ratio of maximum power output to theoretical power output); (iv) the PCE (the ratio of power output to power input). Figure 7.2b shows the external quantum efficiency (EQE) of a typical all-PSCs device. From the photocurrent spectrum, the ability of the solar cells to convert photons to electrons under irradiation at certain wavelengths and intensities can be obtained. In comparison with the conventional fullerene-based PSCs, the use of nonfullerene polymeric acceptor can further extend the photocurrent spectrum to near-infrared (NIR) region, potentially enhancing the J_{sc} of the PV device [2].

7.3 Materials Design Toward Efficient All-PSCs

7.3.1 Progress of N2200 and Its Derivative-Based All-PSCs

Aromatic diimide-based conjugated polymers are an important class of electron acceptor materials for all-PSCs [26]. For instant, the best-known NDI polymers have been regarded as the most efficient polymers. NDI groups have several advantages such as high electron affinity, electron-transporting capability, and excellent thermal and oxidative stability. Easy access to the functionalization of the NDI can be carried out at the carbon 2 and 6 positions for the preparation of regioselective dibrominated monomers and thus pure polymers [1]. In addition, side-chain engineering at the imide nitrogen position of NDI enables good solubility of polymers for solution processing and facile optimization of all-PSC

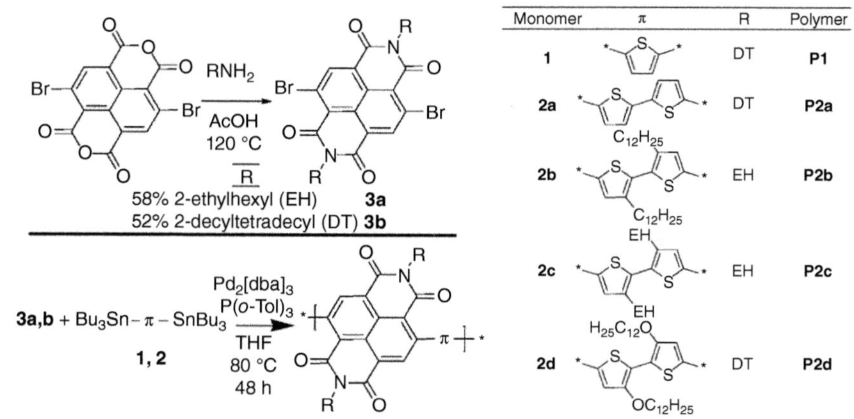

Figure 7.3 Synthesis of NDI polymers. Source: Guo and Watson [27] / Reproduced with permission of American Chemical Society.

blend morphology. For example, Guo and Watson pioneered the synthesis of NDI polymers by Stille coupling of regioisomerically pure dibromo-NDI monomer with various stannylated thiophene-based monomers. It is found that these polymers have low-lying LUMOs of c. 3.80 eV and narrow bandgap of c. 1.45 eV, which are indicative of their promising n-type characteristics (Figure 7.3) [27]. Among them, an NDI- and bithiophene-based copolymer (a.k.a. the benchmark n-type polymer, N2200), which opened a new era of n-type polymer research due to its highly electron mobility (Figure 7.4a) [6]. Since then, a large variety of NDI polymers have been designed and synthesized during the past decent. The performance of NDI polymers in all-PSCs has been steadily improved by optimizing polymer molecular weight, side-chain engineering, backbone modification, random copolymer strategy, etc. [28]. The PCEs of all-PSCs employing NDI polymers have achieved remarkable progress from below 3% in 2014 to approaching 12% in 2020, which has surpassed the state-of-the-art performance of fullerene OSCs. Recently, some new aromatic diimide-based polymer acceptors with distinct electron properties from NDI polymers have been developed and showed promising performances. In this section, we focus on some typically NDI and PDI polymers as well as emerging BTI polymer acceptors that have produced important advances in all-PSC performance and present guidelines for the design of polymer acceptor materials based on aromatic diimide toward high-performance all-PSCs.

7.3.1.1 Molecular Design Strategy for N2200 Derivatives

Our discussion begins from the N2200, which is still one of the best-performing polymer acceptors in all-PSCs. In most cases where the chemical structures of NDI polymer are derived from the N2200. In 2011, Loi and coworkers pioneered the fabrication of all-PSCs based on N2200 and P3HT, and an only PCE of <1% was obtained because of the poor polymer blend morphology and mobility [29]. Afterward, this development of all-PSCs was triggered by the invention of efficient benzodithiophene (BDT) polymer donors. For example, PTB7, PTB7-Th, PBDB-T,

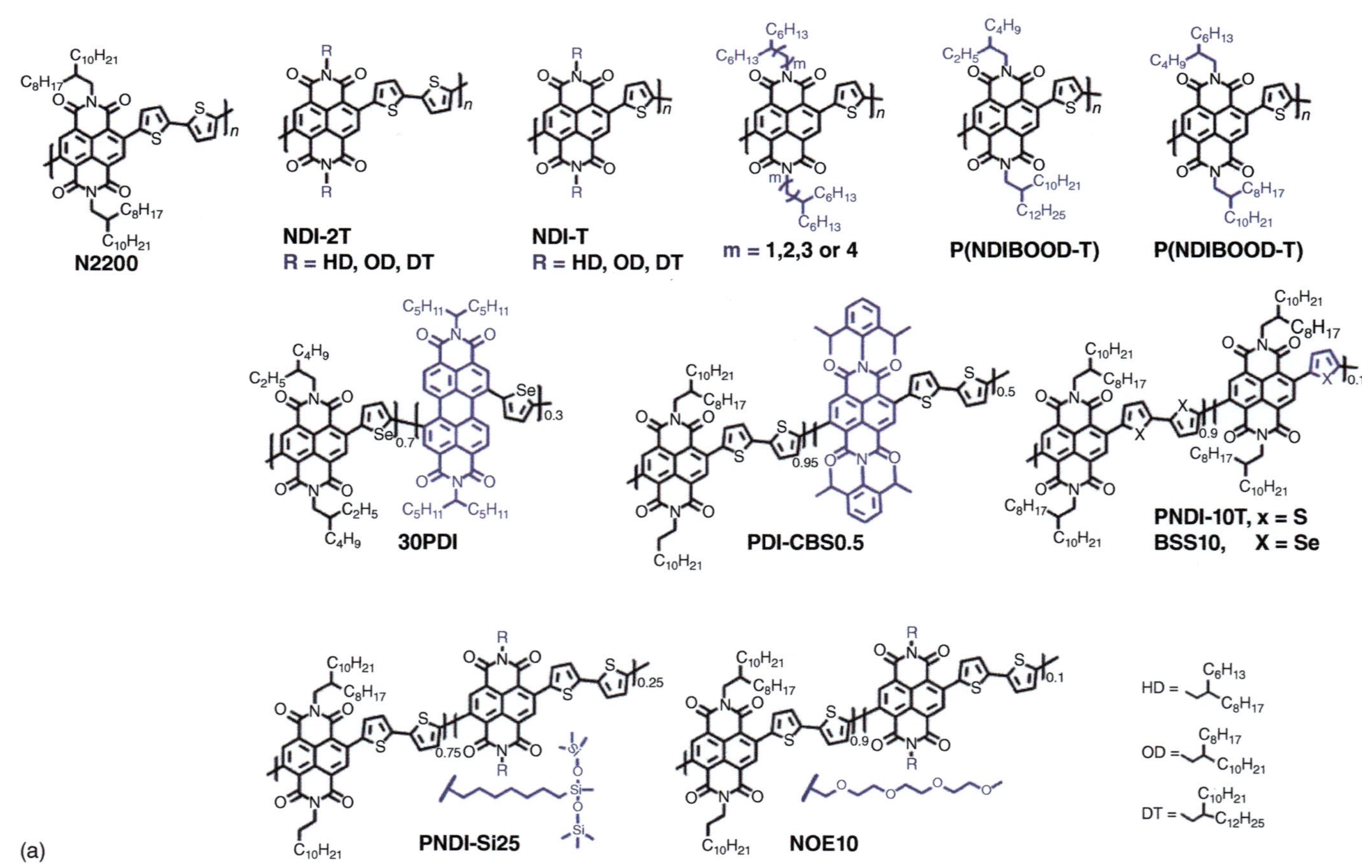

Figure 7.4 Chemical structure of representative aromatic diimide-based conjugated polymers based on NDI (a), PDI (b), and BTI (c). Source: Huiliang Sun.

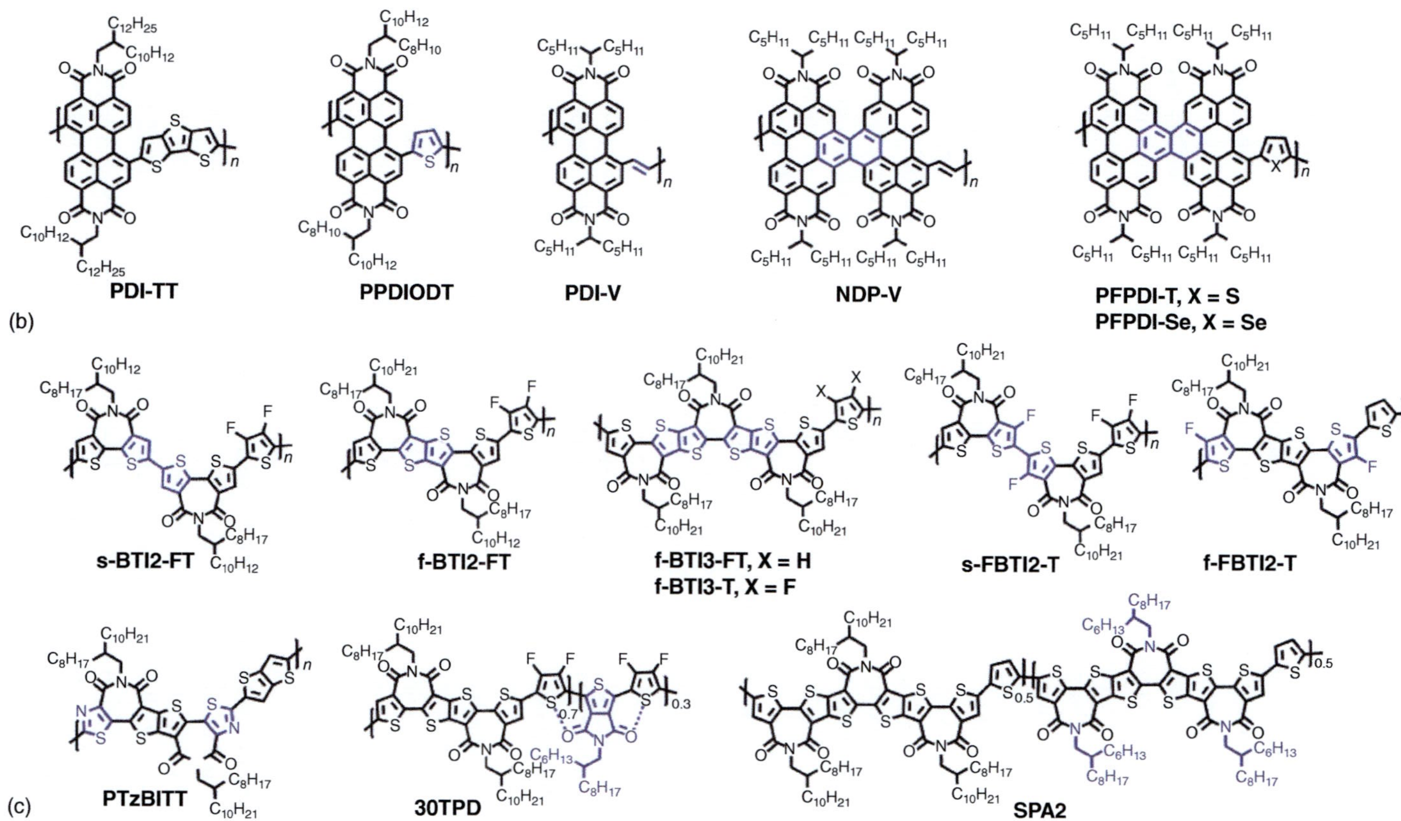

Figure 7.4 (*Continued*)

and J51. [30]. Recently, Huang and coworkers tailored the polymer donor with a siloxane-terminated side chain, which yields a PCE over 10% for the first time for all-PSCs [31]. Despite this, N2200 has high crystallinity and a strong tendency to aggregate, which always leads to a poor miscibility with donors and forms large domain. In this regard, side-chain engineering, heteroatom substitution, molecular-weight optimization, and random terpolymerization were the most successful strategies for the optimization of the crystallinity of NDI polymers and the blend morphology with various polymer donors.

Among them, side-chain engineering strategy is a simple but effective approach for tuning the crystallinity and blend morphology of all-PSCs without influencing on light absorption and energy levels. In 2016, Kim and coworkers investigated the effects of side-chain engineering of N2200. A series of NDI-bithiophene copolymers (NDI-2T) with different side chains of 2-hexyldecyl (2HD), 2-octyldodecyl (2OD), and 2-decyltetradecyl (2DT) is synthesized (Figure 7.4a) [32]. The results demonstrated that polymer equipped with the shortest side chain exhibited the highest electron mobility of 1.90 $cm^2\ V^{-1}\ s^{-1}$ and the PCE of 6.11% among the three polymers due to tighter molecule packing with a face-on orientation. A similar trend was observed in analogous series of NDI-monothiophene copolymers (NDI-T) (Figure 7.4a) [33]. In addition, the branching point of the side chains of polymer acceptors is another important factor effecting the performance of all-PSCs. For example, moving the branching point away from the NDI and its derivatives-based polymers was shown by Kim group and Zhou group to reduce steric hindrance around the polymer backbone and thus enhance intermolecular interactions, contributing to improved PCE (Figure 7.4a) [34]. The crystallinity of NDI polymers has also been tuned by using the asymmetry side chains. For example, asymmetry side chains with different lengths were incorporated into the NDI-T backbone to give the polymers, P(NDIEHDT-T) and P(NDIBOOD-T) (Figure 7.4a) [35]. Benefiting from the fine-tuning crystallinity of these copolymers and improved blend morphology, an all-PSC based on P-(NDIBOOD-T) yields a PCE of 6.89%.

Random terpolymerization is another effective approach to achieving a delicate balance between crystallinity and miscibility. The pioneering study was reported by Jenekhe and coworkers in 2015 [36]. A new series of random terpolymers composed of the NDI-biselenophene parent copolymer NDI-Se with bulky PDI as the third component were designed and synthesized to improve the blend compatibility enabled by the highly crystallinity of NDI polymers. As a result, the PCE of all-PSC based on 30PDI (30% PDI moiety, Figure 7.4a) was increased up to 6.3% from that (1.23%) of the parent copolymer. Recently, another bulky 2,6-diisopropyl-phenyl also incorporated into side chain of NDI terpolymer to tune aggregation and crystallinity by Bao and coworkers. An optimized blend morphology for the polymer PNDI0.5-based all-PSCs result in a significantly improved PCE from 5.1% to 8.5% (Figure 7.4a) [37]. A certain quantity of T2 in N2200 was replaced with T by Wang and coworkers, affording a series of random terpolymers. The increase of the amount of T in terpolymer led to disorder in the polymer backbone and greatly reduced crystallinity. Consequently, PNDI-T10 (10% NDI moiety, Figure 7.4a) yields the best all-PSC performance in those terpolymer acceptors due to the better miscibility with polymer donor, leading to a high PCE of 7.6% (Figure 7.4a) [38]. Recently, a selenophene analog of PNDI-T10, BBS10 (Figure 7.4a). The BSS10-based All-PSC shows a high internal quantum efficiency of 97% and a small energy loss (E_{loss}) of 0.59 eV, leading to a remarkable PCE of 10.1% [39].

The random terpolymerization has also been carried out to tune the crystallinity and packing structure of polymer acceptors by incorporating functional groups on the side chains. For example, Chen and coworkers designed and synthesized new the N2200-based terpolymer PNDI-Si25 (Figure 7.4a) by increasing the mol% of siloxane-terminated side chains, leading to a desirable face-on packing orientation of the polymer backbone [40]. An impressive PCE of 7.4% was attained in PNDI-Si25-based all-PSCs, which is higher than that of N2200. Recently, Huang and coworkers replaced a certain amount of the 2OD chain in N2200 with oligoethylene oxide (OE) chain to tune polymer packing structure and blend morphology. A resulting high PCE of 8.1% and a remarkable FF of 0.75 were attained by the polymer NOE10 (10% OE moiety, Figure 7.4a) [41].

7.3.1.2 Molecular Design Strategy for PDI Polymers

PDI polymers have received much attention as they possess many properties in common with NDI polymers including in high electron affinity and decent electron transport ability. In 2007, Zhan et al. synthesized the first PDI polymer acceptor (PDI-TT) (Figure 7.4b) [42], which triggers the development of PDI polymers in all-PSCs. Despite this, the high-performance all-PSCs based on PDI polymers have rarely been reported compared with the NDI polymers due to the more twisted backbone (50°–70°) enabled by severe steric hindrance in the PDI bay region. Therefore, molecular design strategy of PDI polymers for all-PSCs is highly consistent with that of NDI polymers. For example, Hou and coworkers designed and synthesized copolymer PPDIODT (Figure 7.4b) based on the PDI and the small T unit in the polymer backbone, yielding a high PCE of 6.58% in all-PSCs [43]. To further reduce the steric hindrance, Zhao and coworkers incorporated vinylene linkers into the PDI polymer and thus improved backbone planarity. Thus, polymer PDI-V (Figure 7.4b) showed a relatively high PCE of 7.57% in all-PSCs, which is much higher than those of the linker-free ones [43]. Where after, the same group synthesized fused PDI-V-based polymer NDP-V (Figure 7.4b) to further improve the efficiency of the PDI-V *via* decreasing conformational disorder due to the reduced number of rotatable C—C bonds in the backbone. Combing with higher absorption coefficient and more complementary absorption spectra with the polymer donor, the record PCE of 8.59% among the PDI polymers was achieved [44]. At the same time, Zhang and coworkers also developed a series of NDP-based polymers and achieved impressive PCEs [45]. To date, NDP-based polymer is the most successful PDI polymer.

7.3.1.3 Molecular Design Strategy for BTI Polymers

Despite the NDI and PDI polymers being the most successful in all-PSCs, they suffer from some disadvantages. For example, their absorption coefficient in the long-wavelength region is relatively low, and the absorption range is limited due to the high steric hindrance associated with NDI and PDI core and hence twisted polymer backbone. This limits the absorption of light absorption and thus photocurrent. In addition, low-lying LUMO energy level (3.90 eV) of NDI and PDI polymers typically results in increased E_{loss} in all-PSC. Therefore, developing polymer acceptors beyond NDI and PDI polymers to address the issues inherited in typically polymer acceptors may have a significantly impact in the advancement of all-PSCs [6]. Among these new polymer acceptors, BTI polymers have a different electronic structure from those of classical NDI and PDI polymers, showing tunable LUMO levels because of their delocalized FMO along polymer backbones. Such feature is highly desirable for improving V_{oc} in OSCs by minimizing the energy offsets between the polymer donor and acceptor, resulting in reducing E_{loss}. In addition, the open β-positions of thiophene rings in BTI provide opportunities for backbone expansion and structural modification (Figure 7.5) [46].

Thus, polymer acceptors based on BTI and its derivatives (BTIn) yielded significantly improved PCEs in all-PSCs from <1% in 2017 to ~14% in 2020 with high V_{oc}s. For example, Guo and coworkers reported two new BTI polymer acceptors, s-BTI2-FT and f-BTI2-FT, based on a single bond connected BTI unit and fused BTI unit, respectively. Benefiting from extended π-conjugation backbone and increased film crystallinity, f-BTI2-FT (Figure 7.4c) shows a much higher PCE of 6.85% compared with that (0.07%) of s-BTI2-FT (Figure 7.4c) in all-PSCs [47]. The same group also designed and synthesized a series of triimide-functionalized polymers, f-BTI3-T and f-BTI3-FT (Figure 7.4c). Compared with BTI2 polymers, the BTI3 polymers show a larger π-conjugation and narrower bandgaps as well as lower FMO levels. As a result, an increase PCE of 8.98% is achieved for the f-BTI3-T-based all-PSCs, which surpass the control device based on fullerene as electron acceptor (8.69%) [48].

Fluorination can be used to further optimize the FMO levels and improve the film morphology. Moreover, F atoms can trigger various non-covalent interactions, which can be beneficial for a self-planarized backbone and thus charge transport. However, incorporating F atoms into highly electron-deficient building block is greatly challenging due to the reduced chemical reactivity along with steric hindrance. Recently, Sun et al. devised an effective synthetic route to the dibrominated monomer s-FBTI2-Br by introducing F atoms first and then forming the imide functionality later, here, a novel thiophene derivative FE-T (Figure 7.5a) was designed and synthesized by overcoming the synthetic barrier, featuring a monothiophene unit functionalized with both an electron-withdrawing F atom and an ester group [49].

Based on FE-T, the first fluorinated BTI building block, s-FBTI2, was developed and the corresponding polymer acceptor s-FBTI2-FT (Figure 7.6b) delivers a dramatically increased PCE of 6.5% compared with s-BTI2-FT-based cell (0.07%). The PCE of all-PSCs based on f-FBTI2-T was further improved to 8.1% without using any additives *via* a synergistic strategy based on ring fusion and fluorination. Besides fluorination strategy (Figure 7.7) [50], Osaka and coworkers designed and synthesized a novel electron-deficient building block, TzBI, where the external thiophenes in f-BTI2 were replaced by thiazoles. The incorporation of TzBI into polymer enabled the resulting polymer PTzBITT (Figure 7.4c) with deep LUMO energy levels and

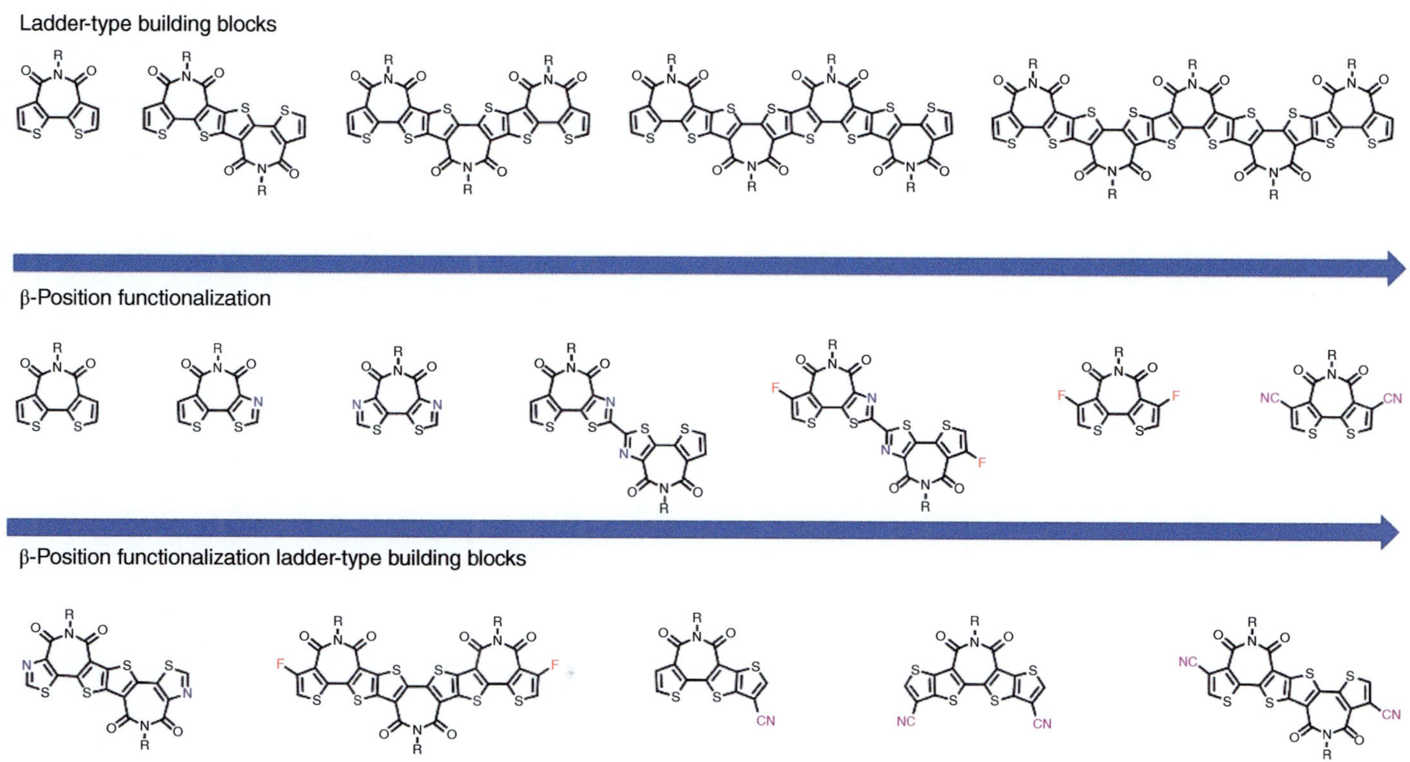

Figure 7.5 Three major strategies proposed to develop new BTI-based building blocks (BTIn) for polymer acceptors. Source: Huiliang Sun.

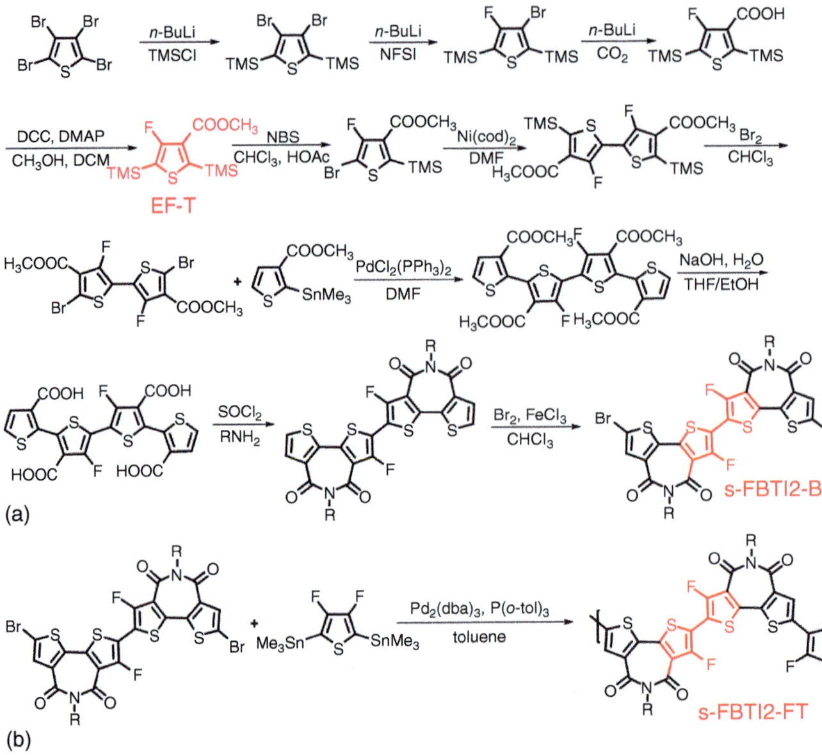

Figure 7.6 Synthetic route to (a) the fluorinated BTI monomers f-FBTI2-Br and (b) the corresponding polymer acceptor s-FBTI2-FT. Source: Huiliang Sun.

high backbone planarity. All-PSCs with polymer PTzBITT as the electron acceptor yielded a PCE of 3.3% [51].

The terpolymerization strategy has successfully optimized the chemical structures of NDI and PDI polymers. However, very few all-PSCs with terpolymer acceptors can achieve PCEs >7% because the key principle in the molecular design of NDI-based terpolymers is to introduce a third component to suppress their over aggregation and crystallinity, which can improve their compatibility with polymer donors. However, the irregular backbone of random terpolymers negatively influences the planarity and therefore inhibits charge transport. In order to minimize such effects and optimize all-polymer blend film morphology, posttreatments such as solvent vapor annealing and additives are usually required. These optimized approaches complicate device fabrication. Compared with the imide-functionalized NDI and PDI, the sulfur atom in the BTIn can promote intramolecular non-covalent interactions between (thienyl)S and (carbonyl)O or between S and other heteroatom on the neighboring unit, which not only promotes backbone coplanarity but also helps form more ideal blend morphology (Figure 7.8b). Recently, Guo and coworkers first incorporate the non-covalent interaction into polymer acceptor by the terpolymerization strategy (Figure 7.8b) [52], using thienopyrroledione (TPD) and f-BTI2 as the two acceptor units and 3,4-difluorothiophene (FT) as the donor co-unit. Thanks to

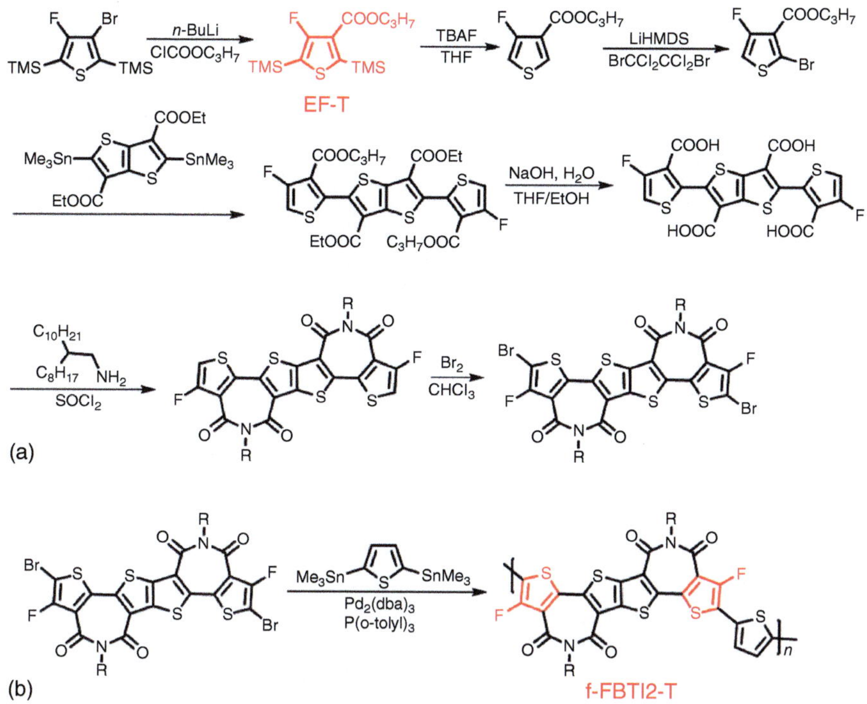

Figure 7.7 Synthetic route to (a) the fluorinated monomers f-FBTI2-Br *via* a synergistic strategy based on ring fusion and fluorination; (b) the corresponding polymer acceptor f-FBTI2-T. Source: Huiliang Sun.

the non-covalent S···O interactions triggered by the TPD, incorporating an optimal TPD content into the polymer backbone enables circumvent the disadvantages of molecular packing induced by the irregular sequence in the random polymer. More importantly, an optimal TPD context leads to increased face-on packing structure and improved miscibility with the donors. As a result, the best-performance terpolymer acceptor 30TPD shows a high PCE of 8.28% without any additive. 8.28% is the highest value among the additive-free all-PSCs based on a random polymer acceptor to date, which is much higher than those from the two-parent copolymer acceptor (4.4% for TPD polymer and BTI polymer, respectively).

7.3.1.4 BTI Polymers for High-Performance All-PSCs with Small E_{loss}

It was found that V_{oc} of All-PSC is typically lower than 0.9 V even for the most efficient ones based on NDI and PDI polymers. Large E_{loss} is the main reason for limiting V_{oc} and efficiency of all-PSCs. To this end, Sun et al. designed and synthesized two new BTI polymer acceptors SPA1 and SPA2 by combing the terpolymerization with side-chain engineering strategy. Benefiting from tunable LUMOs of BTI polymers, the E_{LUMO} offset between SPA1 and the polymer donor negligible to maximize the V_{oc}, resulting in All-PSCs with a remarkable V_{oc} of 1.15 V. However, the EQE was significantly limited because of a low degree of electron transfer and considerable charge recombination enabled by negligible E_{LUMO} offset. In contrast, SPA2 with

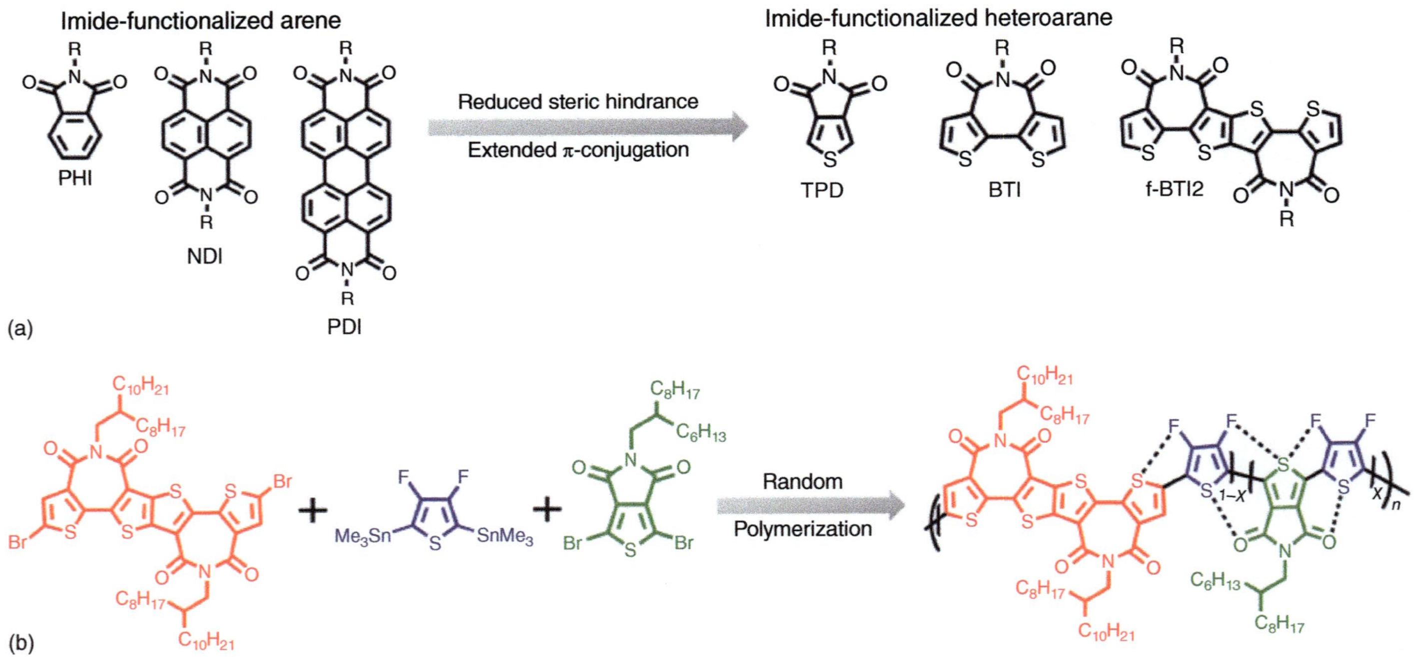

Figure 7.8 (a) Chemical structures of imide-functionalized arenes and heteroarenes; (b) BTIn-based terpolymers acceptor incorporating intramolecular non-covalent sulfur–oxygen interactions for additive-free all-PSCs. Source: Huiliang Sun.

relatively low-lying LUMO shows a large V_{oc} of 1.02 V and a high EQE over 70% in all-PSCs, resulting in a high PCE of 9.21% (Figure 7.9) [53]. More importantly, both all-PSCs based on BTI polymers have relatively high EQE of electroluminescence (10^{-3}), which are much higher than that (10^{-7}) of the NDI polymers, suggesting a greatly low nonradiative recombination loss ($E_{loss,nr}$) of <0.3 eV. This is comparable to that of the perovskite solar cells and has rarely been achieved in all-PSCs.

Taking advantages of BTI for reducing $E_{loss,nr}$, very recently, the incorporation of BTI into fused-ring electron acceptor (FREA)-based polymer leads to a new narrow-bandgap polymer acceptor L14 [7], featuring an acceptor–acceptor (A–A) type backbone. L14 not only shows a narrow bandgap and high absorption coefficient but also low-lying frontier molecular orbital levels, resulting in improved electron transfer characteristics, and unexpectedly, without sacrificing open-circuit voltage (V_{oc}), which is attributed to a small nonradiative recombination loss of 0.22 eV enabled by high EQE_{EL} of BTI polymer. As a result, a remarkable PCE of 14.3% is achieved, which is higher and its D–A analogue and also among the highest values for all-PSCs (Figure 7.10).

7.3.2 Progress of Polymer Acceptors Containing B←N Unit

In this part, we summarize the progress of polymer acceptors containing B←N unit. At first, we introduce the principle of B←N to downshift LUMO/HOMO energy levels (E_{LUMO}/E_{HOMO}) of conjugated polymers. Then we discuss the molecular design of electron-deficient building block based on B←N unit for polymer acceptors. Finally, we summarize how to tune the optoelectronic properties and optimize the phase separation morphology of polymer acceptors containing B←N unit toward high-performance all-PSCs.

7.3.2.1 Principle of B←N Unit

Low-lying E_{LUMO}/E_{HOMO} are key requirements for polymer acceptors of all-PSCs. The principle of B←N to downshift E_{LUMO}/E_{HOMO} is shown in Figure 7.11. Boron atom has the vacant p-orbital and the nitrogen atom with the lone pair electrons. The lone pair of electrons of the nitrogen atom occupies the vacant p-orbital of the boron atom, thus forming boron–nitrogen coordination bond (B←N). In B←N unit, the nitrogen atom shows cationic character (positive charge) and the boron atom shows anionic character (negative charge). When B←N unit is embedded in a π-conjugated system, the positive charge of the nitrogen atom can increase the static potential of the whole π-system, leading to enhanced electron-deficient nature of the π-system. Therefore, the π-system shows down-shifted E_{LUMO} and E_{HOMO}. B←N unit can be used as an effective strategy to design polymer acceptors with low-lying E_{LUMO} and E_{HOMO}.

7.3.2.2 Electron-Deficient Building Blocks Based on B←N Unit

Electron-deficient building blocks based on either one B←N unit or two B←N units have both been reported. As shown in Figure 7.12, the B←N bridged thienylthiazole (BNTT) is an electron-deficient building block containing one B←N unit. Since

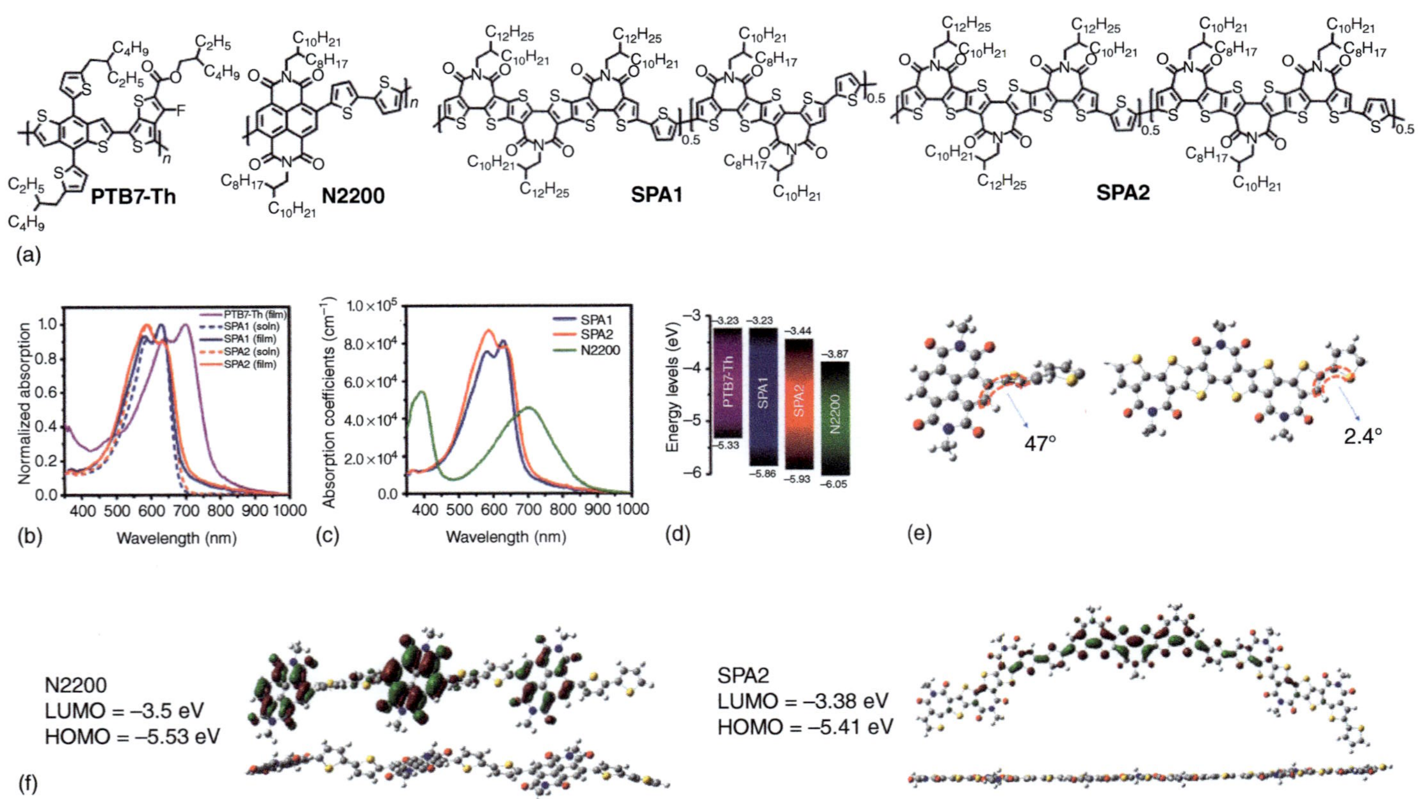

Figure 7.9 (a) Molecular structures of the polymer donor PTB7-Th, the benchmark polymer acceptor N2200, new polymer acceptors SPA1 and SAP2; (b) normalized UV–vis absorption spectra of the polymer donor and acceptors in solution and film state; (c) absorption coefficients of N2200, SPA1, and SAP2; (d) energy diagram of the photovoltaic materials used in this work; (e) chemical geometry of the molecular co-unit of 1 : 1; (f) Kohn–Shom LUMO profiles of the model compounds of polymers N2200 and SPA2. Source: Sun et al. [53] / Reproduced with permission of Springer Nature.

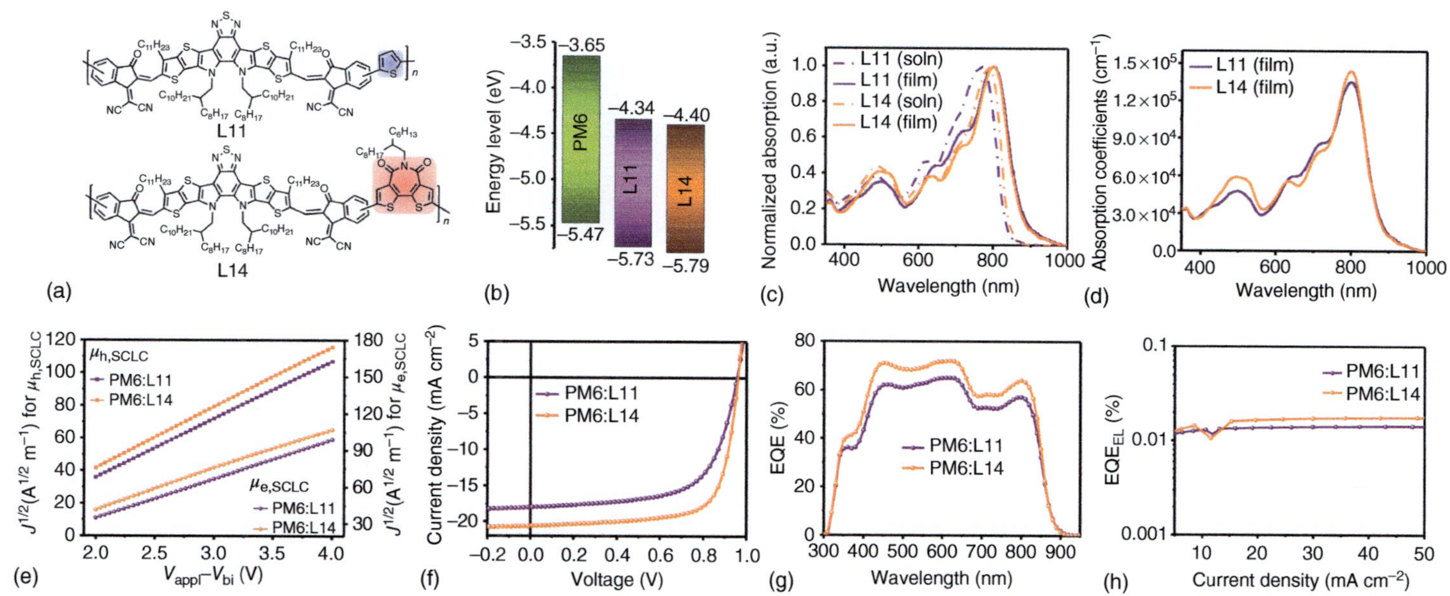

Figure 7.10 (a) Molecular structure of the A–A type polymer acceptor L14 with the donor–acceptor-type polymer acceptor L11 as control; (b) energy level diagram of the neat polymer films used in all-PSCs; (c) normalized UV–vis absorption spectra of L11 and LL14 in dilute chloroform solution (dash dot) and as thin film (solid); (d) UV–vis absorption spectra of L11 and L14 thin films; (e) the *J*–*V* curves of the electron-only and hole-only devices containing the optimal PM6:L11 and PM6:L14 active layer in dark; (f) *J*–*V* characteristics and (g) EQE spectra of the all-PSCs; (h) the EQE_{EL} of all-PSCs at various injection current densities [54]. Source: Huiliang Sun.

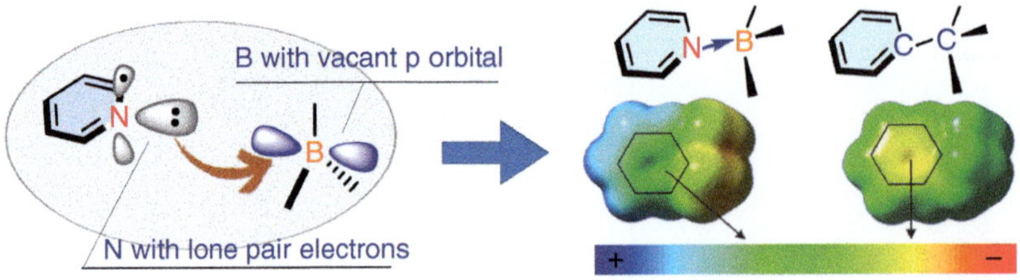

Figure 7.11 Principle of B←N unit. Source: Dou et al. [9] / Reproduced with permission of Springer Nature.

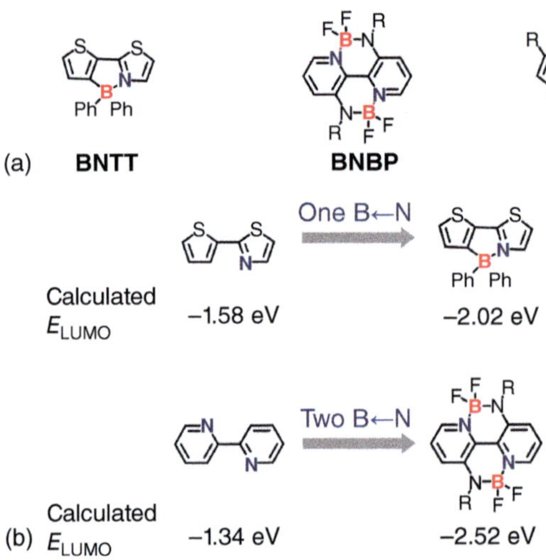

Figure 7.12 (a) Chemical structures of the model compounds of BNTT, BNBP, and BNIDT; (b) LUMO energy levels of the model compounds of BNTT, BNBP, and the corresponding compounds without B←N unit. Source: Yingjian Yu.

the electron-deficient capability of BNTT is not adequately strong, BNTT needs to be copolymerized with another electron-deficient building block (such as TPD, isoindigo (IID)) to give A–A-type polymer acceptors [55]. The double BNBP and the double B←N bridged 2,5-di(thiophen-2-yl)pyrazine (BNIDT) (see Figure 7.12) are electron-deficient building blocks containing two B←N units [9, 12]. The two B←N units lead to dramatic downshift of the E_{LUMO} of the building blocks themselves and endow them with superior strong electron-deficient capability. Therefore, BNBP and BNIDT can be copolymerized with typical electron-donating building blocks to afford donor–acceptor (D–A)-type polymer acceptors.

7.3.2.3 Optimizing E_{LUMO}

E_{LUMO} is a key parameter for polymer acceptors, which can greatly affect their applications in all-PSCs. For the classic polymer acceptors based on NDI unit, their E_{LUMO} are always low and could not be effectively tuned. This is because the LUMO is localized on the NDI units and that the E_{LUMO} is mainly decided by the NDI unit.

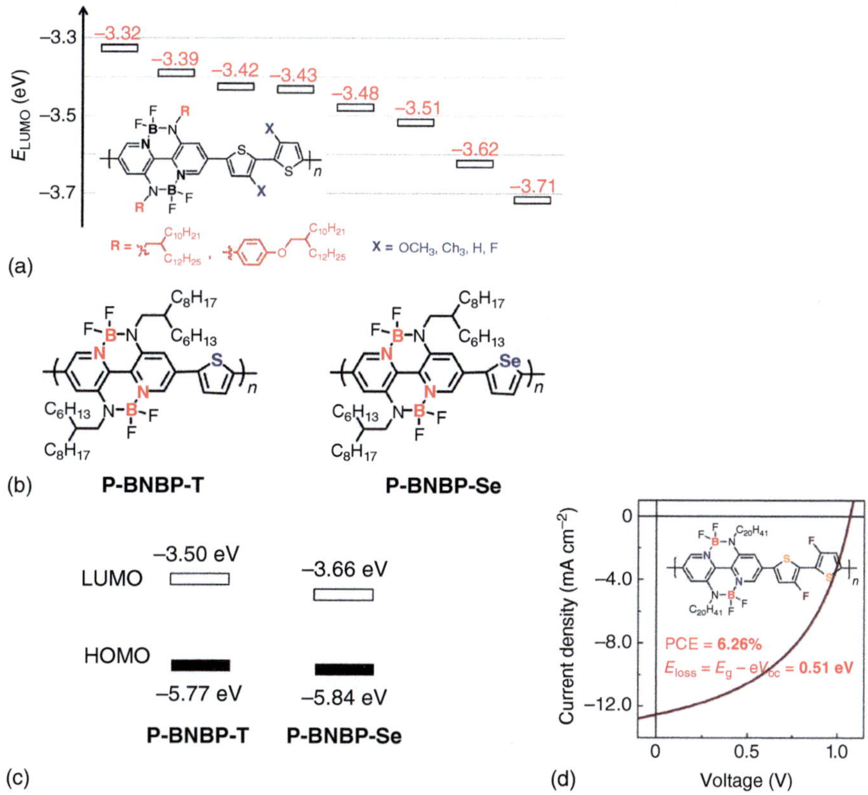

Figure 7.13 (a) E_{LUMO} alignments of eight polymers based on BNBP unit. Source: Long et al. [56] / Reproduced with permission of American Chemical Society.; (b) chemical structures of P-BNBP-T and P-BNBP-Se; (c) E_{LUMO}/E_{HOMO} alignment of P-BNBP-T and P-BNBP-Se. Source: Yingjian Yu; (d) chemical structure of P-BNBP-fBT and $J-V$ curve of the device based on the active layer of PTB7-Th:P-BNBP-fBT. Source: Long et al. [57] / Reproduced with permission of John Wiley & Sons.

In comparison, polymer acceptors based on BNBP unit show delocalized LUMOs over the polymer backbones. Therefore, their E_{LUMO} can be effectively tuned and optimized by changing the copolymerization units. Figure 7.13a shows eight D–A-type conjugated polymers with the same backbone but different substitutes on the A unit (BNBP unit) or the D unit (bithiophene unit). Their E_{LUMO} are finely tuned within a wide range from −3.3 eV to −3.7 eV by changing the substitutes on the D units and the A units. The tunable E_{LUMO} of polymer acceptors leads to open-circuit voltage as high as 1.0–1.3 V in all-PSCs [56]. For example, P-BNBP-Se (Figure 7.13b), a copolymer of the BNBP unit and selenophene unit, exhibits the E_{LUMO} of −3.66 eV, which is lower by 0.16 eV than that of the copolymer of the BNBP unit and thiophene unit (P-BNBP-T). As a result, the energy levels of P-BNBP-Se match well with the widely used polymer donor PTB7-Th. The device of P-BNBP-Se shows a V_{oc} of up to 1.03 V. While the all-PSC device based on the

PTB7-Th:P-BNBP-T blend shows a PCE of 2.27%, the corresponding device with P-BNBP-Se as the acceptor exhibited an enhanced PCE of 4.26% [58].

An outstanding polymer acceptor is the copolymer of BNBP unit and 3,3′-difluoro-2,2′-bithiophene (fBT) unit (P-BNBP-fBT, see Figure 7.13d). The electron-deficient fluoro substitutes of the fBT unit can lower the E_{LUMO} of the polymer. As a result, all-PSCs with P-BNBP-fBT as the acceptor and PTB7-Th as the donor show V_{oc} of as high as 1.06 V, PCE of 6.26% at the photon energy loss (E_{loss}) of only 0.51 eV [57].

7.3.2.4 Tuning Absorption Spectra

Polymer acceptors should have small optical bandgaps and wide absorption spectra, which improve sunlight harvesting of all-PSCs. The absorption spectra of polymer acceptors based on B←N unit could be effectively tuned and optimized by changing the copolymerization units. Their bandgaps could be tuned in the wide range of 1.5–1.9 eV. For example, P-BNBP-T (Figure 7.14a), a copolymer of BNBP unit and thiophene unit, shows narrow absorption spectra with the absorption peak at 622 nm and a medium bandgap of 1.92 eV. When the copolymerization unit is substituted by the thiophene-flanked diketopyrrolopyrrole (DPP) unit, the bandgap of P-BNBP-DPP is decreased to 1.56 eV. P-BNBP-DPP exhibits a broad absorption band from 500 to 800 nm. The maximum absorption wavelength is 760 nm, which is red shifted by c. 149 nm compared with that of P-BNBP-T [59].

Indoor photovoltaic (IPV) cells have recently received great attention because of their promising applications as power sources for devices of the Internet of Things (IoT) ecosystem. As shown in Figure 7.15b, the indoor artificial light sources, such

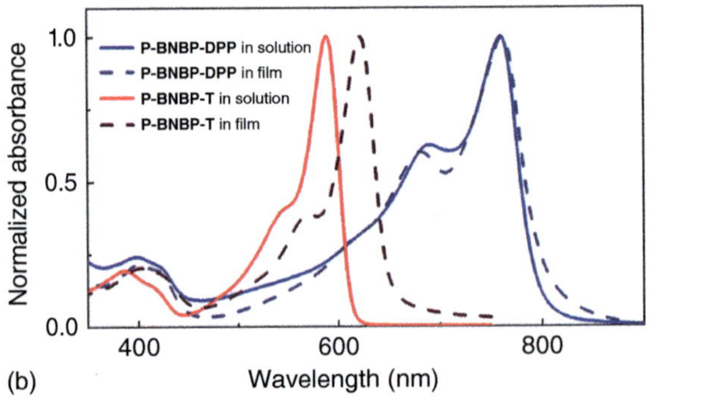

Figure 7.14 (a) Chemical structures of P-BNBP-T, P-BNBP-DPP; (b) absorption spectra of P-BNBP-T and P-BNBP-DPP. Source: Long et al. [59] / Reproduced with permission of Royal Society of Chemistry.

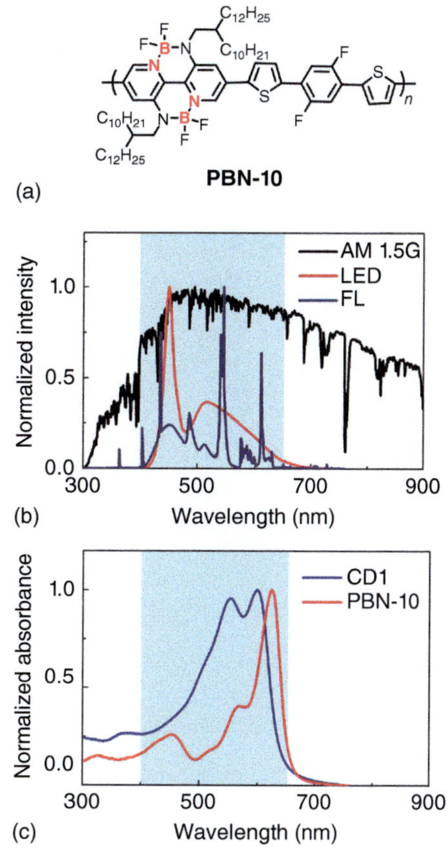

Figure 7.15 (a) Chemical structures of PBN-10; (b) the AM 1.5G solar spectrum and emission spectra of a white LED and FL; (c) absorption spectra of CD1 and PBN-10. Source: Ding et al. [11] / Reproduced with permission of Royal Society of Chemistry.

as white light emitting diodes (LEDs) and fluorescent lamps (FLs), exhibit emission spectra mainly in the narrow wavelength region from 400 to 650 nm. Medium bandgap is a basic requirement for polymer acceptors for IPV applications. According to the empirical prediction, the ideal bandgap of photoactive materials for IPVs is 1.9–2.0 eV. PBN-10 shows a medium optical bandgap of 1.95 eV, which is ideal for IPV applications. The all-polymer photovoltaic device based on PBN-10 shows a high PCE of 27.4% at a high V_{oc} of 1.16 V under FL illumination of 2000 lx [11]. This result indicates that polymer acceptors based on B←N units are very suitable for IPV applications.

7.3.2.5 Enhancing Electron Mobility

For polymer acceptors, high electron mobilities (μ_e) are necessary for efficient electron transport and consequently for the all-PSC device performance. The all-PSC device performance of P-BN-TPD (Figure 7.16a) is very poor because of its very low electron mobility. This is attributed to the steric hindrance of the pendant bulky phenyl groups on the boron atom, which prevent close π-stacking of conjugated polymer backbones. The steric hindrance effect of the pendant phenyl groups can be alleviated by extending the length of the repeating units. IID unit is much longer

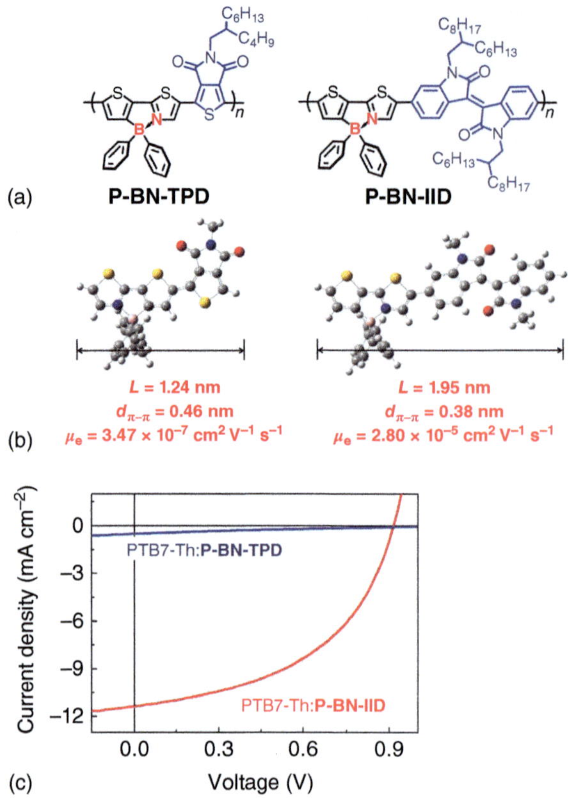

Figure 7.16 (a) Chemical structures of P-BN-TPD and P-BN-IID; (b) optimized molecular configurations of the repeating units of P-BN-TPD and P-BN-IID; (c) *J*–*V* curves of devices based on the active layers of PTB7-Th:P-BN-TPD and PTB7-Th:P-BN-IID. Source: Zhao et al. [55] / Reproduced with permission of John Wiley & Sons.

than the TPD unit. As shown in Figure 7.16b, when using IID unit to replace TPD as the copolymerization unit, the length of the repeating unit increases from 1.24 to 1.95 nm in DFT-optimized molecular configurations (B3LYP/6-31G*). While P-BN-TPD shows the π–π stacking distance ($d_{\pi-\pi}$) of the polymer backbones of 4.6 Å, P-BN-IID exhibits decreased $d_{\pi-\pi}$ of 3.8 Å, indicating improved compact stacking of the polymer backbones. As a result, the μ_e is enhanced by nearly two orders of magnitude from 3.40×10^{-7} cm^2 V^{-1} s^{-1} of P-BN-TPD to 2.80×10^{-5} cm^2 V^{-1} s^{-1} (estimated with the SCLC method). The resulting PCE of the all-PSC devices is enhanced from 0.12% of P-BN-TPD to 5.04% of P-BN-IID [55].

The "conjugated side chain" strategy, which is always used in molecular design of polymer donors, has also been used to enhance μ_e of polymer acceptors containing B←N units. Figure 7.17 shows the chemical structures of P-BNBPP-T with conjugated alkyloxyphenyl side chain and the reference polymer P-BNBP-T with typical alkyl side chain. The π–π stacking distance of P-BNBPP-T ($d_{\pi-\pi}$ = 3.63 Å) with conjugated side chains was smaller than that of P-BNBP-T ($d_{\pi-\pi}$ = 3.81 Å). Because of the smaller π–π stacking distance, the electron mobility of P-BNBPP-T ($\mu_e = 3.24 \times 10^{-4}$ cm^2 V^{-1} s^{-1}) is higher than that

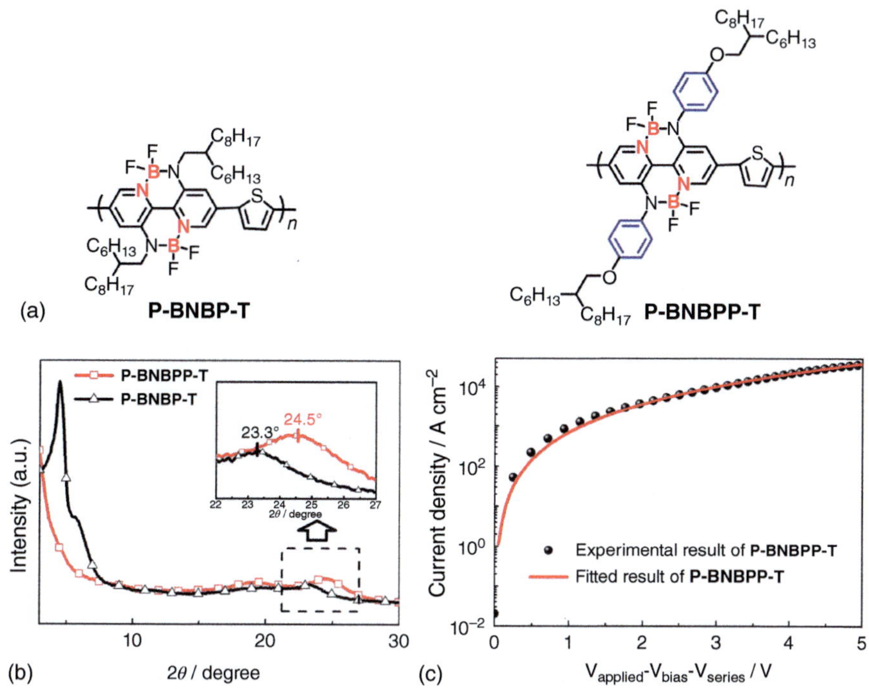

Figure 7.17 (a) Chemical structures of P-BNBP-T and P-BNBPP-T; (b) grazing-incidence X-ray diffraction patterns of P-BNBPP-T, P-BNBP-T; (c) SCLC fitting of the electron-only device based on the pristine film of P-BNBPP-T. Source: Zhao et al. [60] / Reproduced with permission of American Chemical Society.

of P-BNBP-T ($\mu_e = 7.16 \times 10^{-5}$ cm^2 V^{-1} s^{-1}) by about five times. The enhancement of μ_e contributes to the PCE increase of the resulting all-PSC [60].

7.3.2.6 Optimizing Active Layer Morphology

For efficient charge separation and charge transporting, ideal active layer morphology of all-PSCs should be bicontinuous interpenetrating network with high phase purity and phase domain size of c. 20 nm. Many factors influence active layer morphology of all-PSCs, such as crystallinity of polymers and aggregation tendency of polymer chains in solution. Figure 7.18 shows the chemical structures, two-dimensional (2D) grazing-incidence wide-angle X-ray scattering (2D-GIWAXS) patterns of P1 and PBN-12. PBN-12 shows the (100) and (200) reflection peaks in the out-of-plane direction and a strong (010) π–π stacking peak in both the in-plane and out-of-plane directions. However, P1 exhibits a (100) reflection peak in the out-of-plane direction and no obvious (010) π–π stacking peak in both the in-plane and out-of-plane directions, suggesting that the crystallinity of PBN-12 film is much stronger than that of P1 film. In the film-forming process,

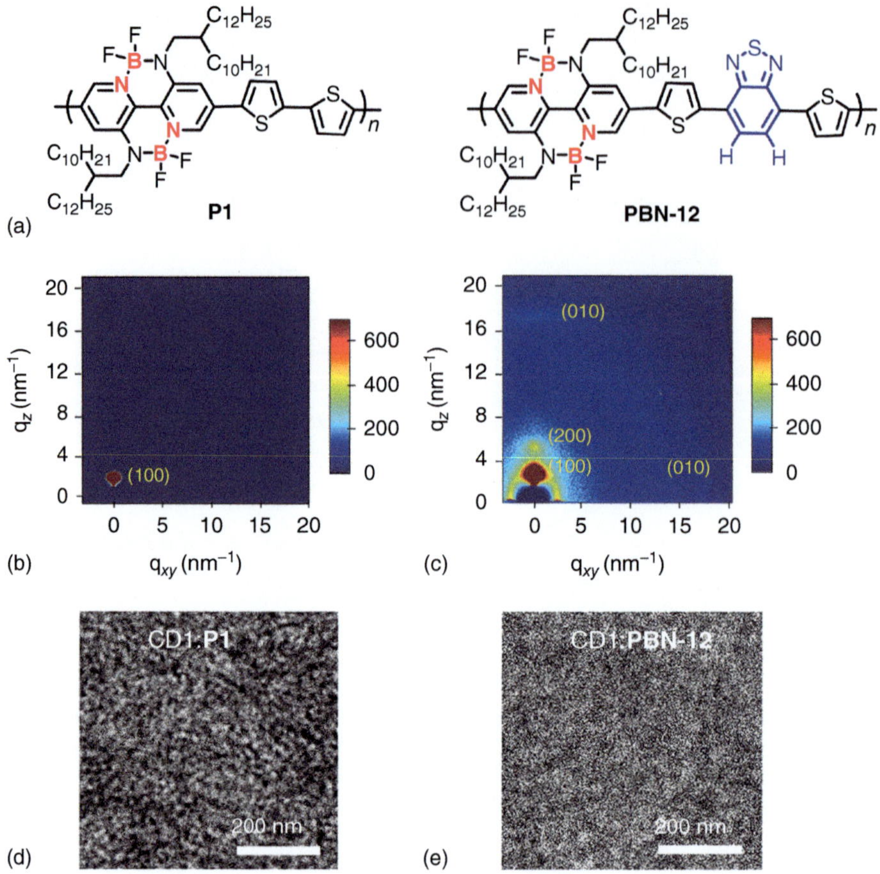

Figure 7.18 (a) Chemical structures of P1 and PBN-12; (b and c) 2D-GIWAXS patterns of P1 and PBN-12; (d and e) TEM images of CD1:P1 and CD1:PBN-12. Source: Zhao et al. [10] / Reproduced with permission of American Chemical Society.

the competition of crystallization of CD1 and PBN-12 leads to small-size phase separation and consequent excellent photovoltaic performance of the CD1:PBN-12 active layer [10].

It is difficult to manipulate active layer morphology of all-PSCs by conventional methods, such as thermal annealing, solvent annealing, and solvent additive. During all-PSC device fabrication, the regular method is to dissolve the polymer donor and the polymer acceptor together (DT method). Liu and coworkers reported an approach to improve active morphology of all-PSCs by dissolving polymer donor and polymer acceptor individually and then blending them immediately before spin-coating (DI method). The active layer with the DI method shows enhanced polymer ordering and higher domain purity than those with the DT method. This is due to the partially preserved polymer aggregates from solution. As the result, the PCE is enhanced from 5.36% to 7.09% [61]. Active layer morphology of all-PSCs based on CD1:PBN-12 blend can also be optimized by controlling the solution

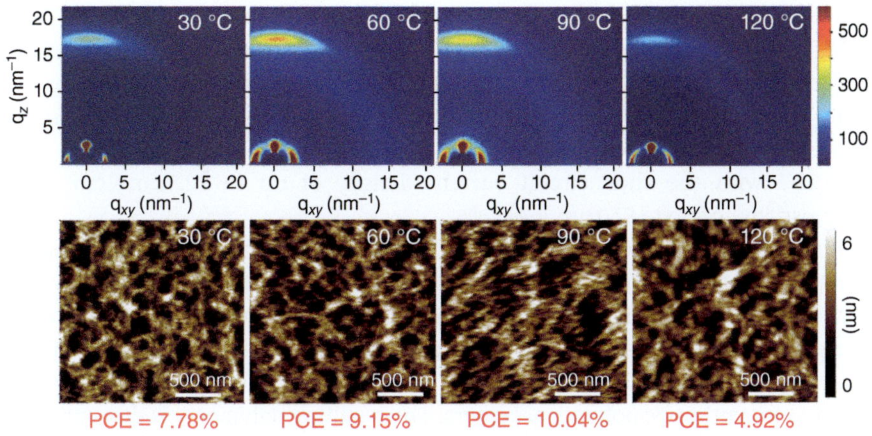

Figure 7.19 2D-GIWAXS patterns (top) and AFM height images (down) of CD1:PBN-12 blend films with different solution temperatures. Source: Wang et.al. [62] / Reproduced with permission of American Chemical Society.

temperature for spin coating. With the solution temperature increasing from 30 to 90 °C, the crystallinity of the polymer donor and the polymer acceptor in the active layer becomes stronger and the phase separation size becomes smaller. Consequently, the PCE of the all-PSC devices is enhanced from 7.78% to 10.04%. Further increase of solution temperature to 120 °C leads to decreased device efficiency. The effect of solution temperature on the phase-separation morphology of all-PSCs can be explained by the increased diffusion rate of the polymer chains in the film-forming process and the disaggregation of polymer chains in solution at high temperature [62] (Figure 7.19).

Most of the all-PSC devices based on polymer acceptors containing B←N unit suffer from low FF. Recently, Huang and coworkers reported a new polymer acceptor, BN-2fT. All-PSC device based on BN-2fT as the acceptor and PBDB-T as the donor show an outstanding FF of 70.4% possibly due to the excellent active layer morphology [12].

7.3.3 Progress of Polymer Acceptors Bearing Cyano Groups

Cyano group is a very strong electron-withdrawing substituent and has been used to design polymer acceptors with excellent all-PSC device performance. As shown in Figure 7.20, Guo and coworkers reported a narrow-bandgap polymer acceptor DCNBT-IDT based on 5,6-dicyano-2,1,3-benzothiadiazole (DCNBT) building block. DCNBT-IDT shows the low E_{LUMO} of −3.75 eV, a narrower bandgap of 1.43 eV, and a relatively high maximum absorption coefficient of 6.15×10^4 cm^{-1}. The all-PSC device with DCNBT-IDT as the polymer acceptor and PBDB-T as the polymer donor gave an outstanding PCE of 8.32% with a photoresponse of up to 870 nm [63]. In order to further extend the photoresponse wavelength, Guo and coworkers modified the chemical structure of DCNBT-IDT to design DCNBT-TPIC with ultra-narrow bandgap. The indacenodithiophene (IDT) group in DCNBT-IDT is replaced by the more electron-donating carbon–oxygen-bridged ladder-type unit in DCNBT-TPIC. DCNBT-TPIC shows an optical bandgap of 1.28 eV due to strong ICT character between the highly electron-deficient unit DCNBT and the highly electron-rich 5,5,11,11-tetraoctyl-5,11-dihydrothieno[2′,3′:5,6]pyrano[3,4-g]thieno-[3,2-*c*]isochromene (TPIC) unit. The all-PSC device based on DCNBT-TPIC exhibits a PCE of 10.22% with the short-circuit current density of up to 22.52 mA cm^{-2} and photoresponse of up to 950 nm [64].

Another strategy to design polymer acceptors is to use small molecular electron acceptors (SMAs) as building blocks to construct conjugated polymers. Many SMAs, in which an electron-donating core unit is endcapped with two cyano-based groups, have been reported. Organic solar cells based on these SMAs show PCE of higher than 15%–18% [65–68]. Li and coworkers firstly reported the approach to polymerize SMAs to synthesize polymer acceptors. The resulting polymer acceptors preserve the merits of the SMAs, such as proper energy levels, narrow bandgaps, high absorption coefficients, and high electron mobilities, and consequently exhibit excellent all-PSC device performance. As shown in Figure 7.21, Li and coworkers synthesized a polymer acceptor PZ1 with IDIC16 as the key building block and thiophene as the linking unit. PZ1 exhibits a broad absorption band with a narrow bandgap of 1.55 eV and high absorption coefficient (1.3×10^5 cm^{-1}). Compared with its corresponding SMA, IDIC-C16, the absorption spectrum of PZ1 in thin film is red shifted by c. 50 nm. The E_{HOMO} and the E_{LUMO} of PZ1 are slightly up-shifted than those of IDIC-C16 by 0.08 and 0.04 eV, respectively. The all-PSC device with PZ1 as acceptor and PBDB-T as donor achieved a high PCE of 9.19% [69]. By optimizing the polymer donors to match with PZ1, the PZ1-based all-PSC shows a PCE of as high as 11.2%, which is among the highest values for the all-PSCs in 2019 [70].

Following the molecular design strategy of PZ1, a series of polymer acceptors have been developed. Yan and coworkers reported a polymer acceptor PFBDT-IDTIC by replacing the thiophene linker in PZ1 by BDT unit. PFBDT-IDTIC exhibits a high absorption coefficient of 2.74×10^5 cm^{-1}. All-PSCs based on PFBDT-IDTIC as the polymer acceptor and PM6 as the polymer donor gave a PCE of 10.3% [71]. As shown in Figure 7.20, Wang and coworkers reported a series of polymer acceptors PF2-DTC, PF2-DTSi, and PF2-DTGe with the same IDIC16 building block as acceptor unit but different central bridging atoms in donor units. Among the three polymers, PF2-DTSi exhibits the highest PCE of 10.77% due to its improved exciton dissociation, charge transport, and optimized morphology [72]. Wang and coworkers also reported a non-conjugated polymer acceptor PF1-TS4 (Figure 7.20) with the IDIC16 building block as acceptor unit and a thioalkyl segment as the donor unit. PF1-TS4 shows a narrow bandgap of 1.58 eV and a high absorption coefficient of 1.25×10^5 cm^{-1}. The PF1-TS4-based all-PSC device shows a PCE of 8.63% with excellent morphological stability under thermal annealing [73].

The development of Y6-type SMAs, which was initially reported by Zou and coworkers, leads to further enhancement of OSCs [74]. Y6-type SMAs have the

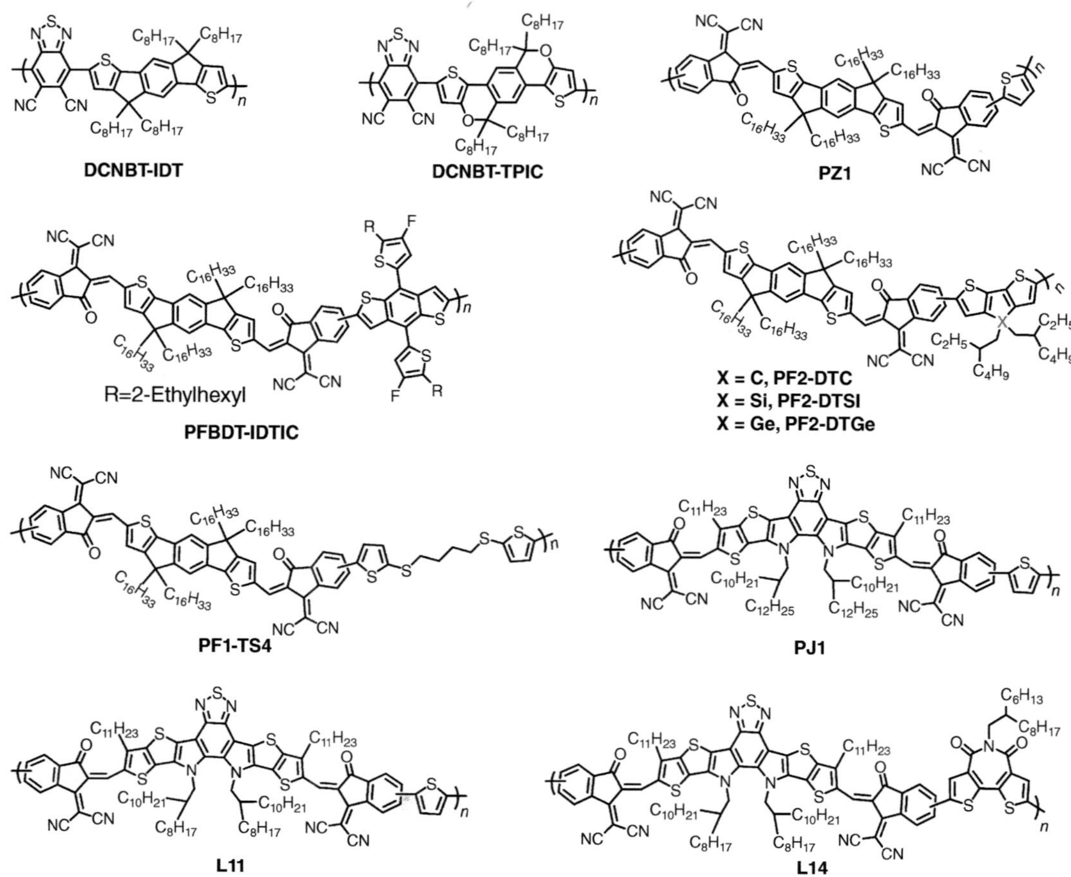

Figure 7.20 Chemical structures of polymer acceptors bearing cyano groups. Source: Yingjian Yu.

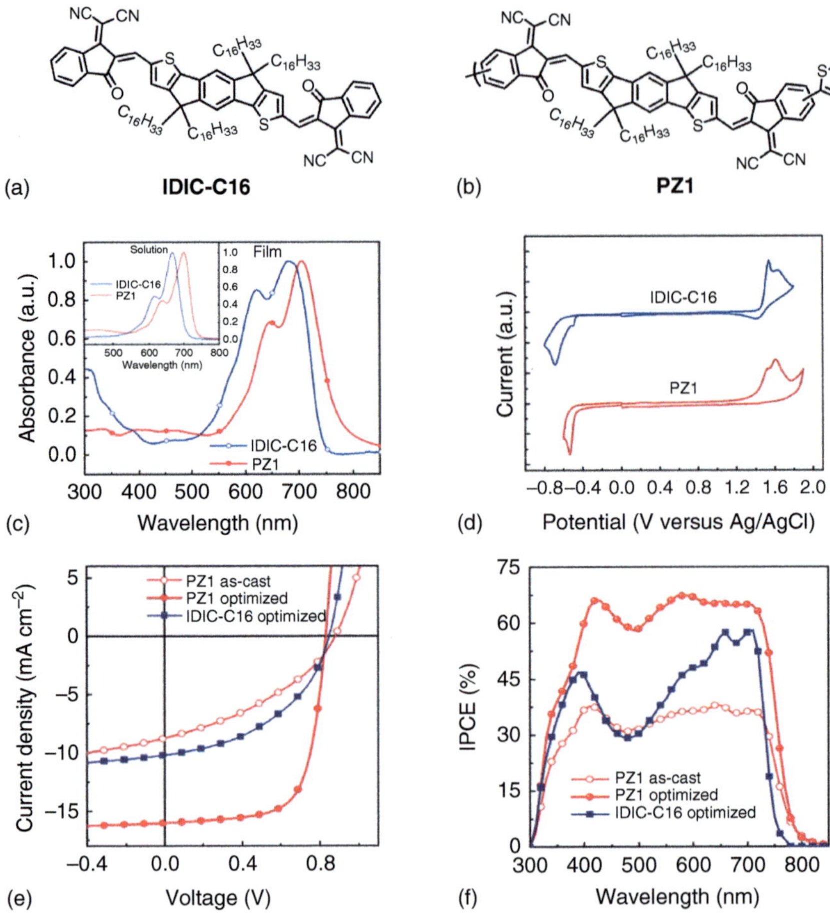

Figure 7.21 Chemical structures of (a) IDIC-C16 and (b) PZ1; (c) solution and film absorption spectra of PZ1 (red line) and IDIC-C16 (blue line). The inset shows their corresponding solution absorptions; (d) cyclic voltammograms of PZ1 (red line) and IDIC-C16 (blue line); (e) *J–V* curves of the PSCs based on PBDB-T:acceptor (1.5 : 1, w/w) under the illumination of AM 1.5G, 100 mW cm^{-2}; (f) IPCE spectra of the corresponding PSCs. Source: Zhang et al. [69] / Reproduced with permission of John Wiley & Sons.

great advantage of strong light absorption in the near-infrared region and small photon energy loss. To take advantage of Y6-type SMAs, several groups devoted to development of polymer acceptors based on Y6-type SMAs. As shown in Figure 7.22, Huang and coworkers reported a novel polymer acceptor PJ1 with a narrow bandgap of 1.4 eV and a high absorption coefficient (1.39×10^5 cm^{-1}) at high molecular weight. PJ1 is an alternating copolymer of TTPBT-IC unit and thiophene unit. When PJ1 is blended with a polymer donor PBDB-T, the all-PSC device shows a high PCE as high as 14.4%. The authors found that the molecular weights of PJ1 play an important role in the optoelectronic properties, active layer morphology, and all-PSC device performance [75].

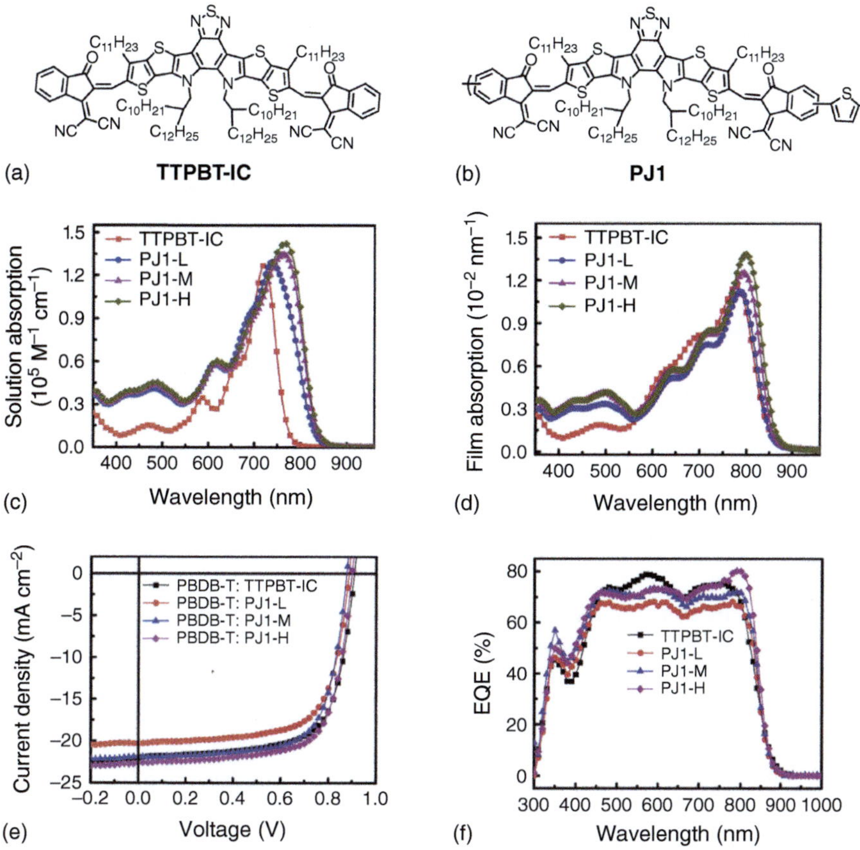

Figure 7.22 Chemical structures of (a) TTPBT-IC and (b) PJ1; UV–vis absorption spectra of acceptors in dilute chloroform solutions (c) and as thin films (d); (e) *J*–*V* curves of the PSCs based on PBDB-T:acceptor under the illumination of AM 1.5G, 100 mW cm^{-2}; (f) EQE curves of the corresponding PSCs. Source: Jia et al. [75] / Reproduced with permission of Elsevier.

Guo and coworkers designed a new A–A-type polymer acceptor L14 (Figure 7.20) by copolymerizing brominated TTPBT-IC unit with another electron-deficient distannylated BTI (BTI-Tin) unit. L14 exhibits a narrow bandgap of 1.39 eV, a high absorption coefficient of 1.44×10^5 cm^{-1}, and a very small nonradiative recombination loss ($E_{loss,nr}$) of 0.22 eV. When L14 is blended with polymer donor PM6, all-PSC device with a high PCE of 14.3% is achieved [7].

7.4 Device Performance and Stability of All-PSCs

7.4.1 Morphology Optimization and Device Engineering

The BHJ morphology improvement of all-polymer blends is one of the core reasons for the recent success of all-PSCs. In general, there are two major challenges in controlling the all-polymer blend morphologies: (i) demixing of crystalline D–A

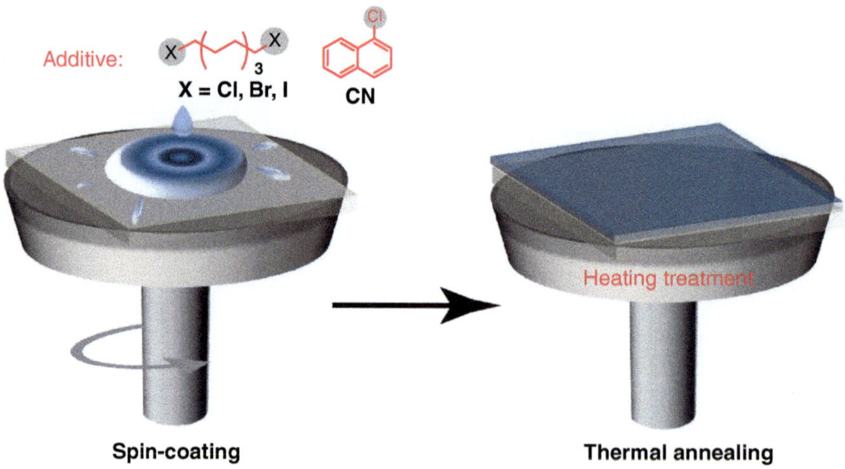

Figure 7.23 Spin-coating process of all-polymer BHJ blend and further thermal treatment. Source: Jianyu Yuan.

polymer domains; (ii) tuning polymer packing orientation at D–A interfaces. The development of efficient p-type and n-type conjugated polymers demonstrated how changing the chemical structures of donor and acceptor polymer translates to desirable morphologies. However, it is necessary to gain insight into the correlation between processing conditions and BHJ morphologies of all-PSCs, which in turn will not only aid the further molecular design, but also establish general guidelines for fabricating desired all-polymer blends.

For the polymer/fullerene system, optimizing the processing conditions of devices to achieve the desired active layer morphology is critical. A number of processing techniques, such as mixed solvents, additives, as well as thermal and solvent annealing, have been explored to control the bulk morphology to attain multiple structural requirements [76]. Among those reported processing approaches, thermal annealing and solvent additives are considered to be the most effective and versatile approaches as illustrated in Figure 7.23. In general, thermal annealing at temperatures beyond the glass-transition temperatures of donor or acceptor polymers will produce thermally induced morphology modifications and crystallizations in blend films as well as improvements of interfaces between active layers and electrodes [77]. Meanwhile, morphology control of the nanoscale structure of phase separation and polymer ordering in a blend film can be controlled through additive processing. Additives usually have a high boiling point and possess selective solubility to one of the active materials [16]. The addition of these additives can usually simultaneously enhance the light absorption and charge carrier mobility as well as suppress charge recombination, which contribute to higher J_{sc} and FF, ultimately enhancing the PCE. However, in all-PSCs, the morphology evolution after thermal annealing or use of additive is far less explored compared with that in polymer-fullerene solar cells. More importantly, the use of thermal annealing or additive process alone in general only allows for rudimentary morphology control.

The polymer acceptors with large planar structures such as NDI and PDI have stronger inter-chain interactions, which make their morphology more susceptible to posttreatment compared with the small molecules. In the system of N2200 based all-PSCs, we found that devices at the optimal annealing temperature of 160 °C exhibit over 30% enhanced PCE compared with those without any thermal treatment [78]. In addition, we also demonstrated widely adopted 1,8-di(R)octane (R = Cl, Br, and I) additives can be applied to multiple all-polymer devices with improved efficiency due to their general capability to increase the crystallinity and extend the effective time of film formation during deposition [17]. Interestingly, DBrO outperforms the widely used DIO in polymer-fullerene-based devices, obtaining a highest efficiency of 8.23% for the PTzBI/P(NDI2OD-T2)-based all-PSCs. In contrast, the addition of chloronaphthalene (CN) can alleviate the inter-chain interaction of polymers to prevent the formation of oversized domains, which make it especially efficient for systems using strongly aggregated polymers such as N2200. It is well known that BHJ morphology is superior to planar structure in organic solar cells. However, we first revealed that the P-i-N structure (planar heterojunction) can outperform the BHJ ones with a nearly 50% efficiency improvement in all-PSCs, reaching a PCE approaching 10% [22]. This planar structure can also provide an enhanced internal electric field and remarkably stable morphology under harsh thermal stress. We have further demonstrated generality of the P-i-N structure in several other all-polymer systems and revealed the improved charge generation in the P-i-N structured all-polymer blend [79]. Considering the adjustable polymer molecular weight and solubility, the P-i-N device structure can be more beneficial for all-polymer systems.

Researchers primarily focused on seeking an all-polymer pair that has desirable electrical properties and matched BHJ blend morphology. Engineering light absorption of all-polymer active layer (as shown in Figure 7.24) was relatively less explored in the early stages. Due to the intrinsically narrow absorption window of organic materials and the low thickness of all-polymer film, most binary all-PSCs utilize a relatively limited portion of sunlight. Therefore, ternary and tandem all-PSCs, which contain extra absorbing component, have been exploited as attractive approaches for broadening the absorption coverage of the active layers. The ternary approach, categorized into three classes: (i) two donors/one acceptor; (ii) one donor/two acceptors; and (iii) donor/acceptor/nonvolatile small-molecule additive, has similar simplicity in device fabrication, achieved by one-step solution processing [80, 81]. However, the introduction of the third component often has some effects on the carrier dynamics of all-PSCs, such as charge separation and transport, by altering the energy level alignment and morphology. Therefore, careful selection of the ternary components is critical for simultaneous optimization of all the photovoltaic parameters.

Tandem structure has been widely reported for enhancing the PCE in polymer-fullerene solar cells for better utilization of the solar spectrum. By using polymers with complementary absorption, tandem cell can extend the device absorption to the near-infrared region, covering significantly more solar spectrum. Meanwhile, the potential loss during photo-to-electron conversion process can be reduced by using tandem structure [82]. We reported an improved all-polymer

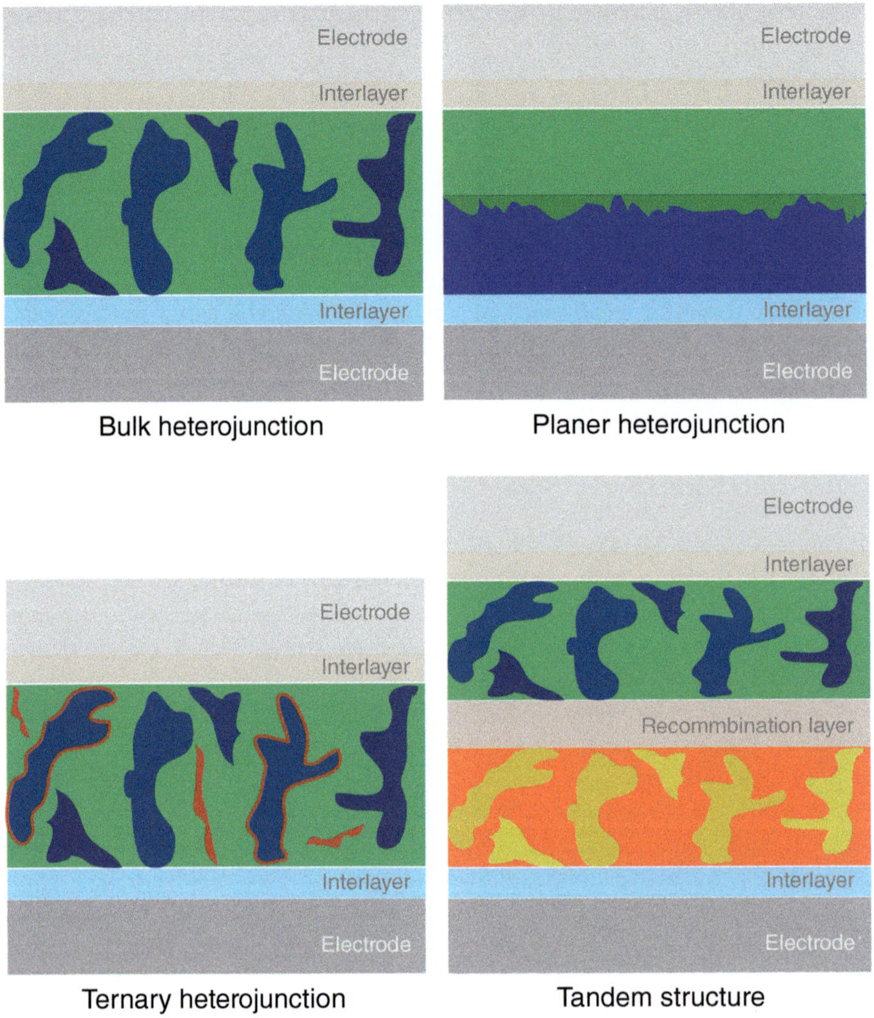

Figure 7.24 Device structures of all reported all-PSCs. Source: Jianyu Yuan.

tandem with a PCE of 8.3% [21], using complementary absorber layers for the first time in all-PSCs. A newly synthesized ternary conjugated polymer, PBFSF (bandgap of 1.55 eV), was used as the donor and N2200 as the acceptor in the back cell. For the front cell, a pair of wide-bandgap polymers PTP8 (bandgap of 1.80 eV) and P(NDI2HD-T) (bandgap of 1.85 eV) were used. Interestingly, homo tandem structures were also constructed to improve device absorption. Note that the optimal thickness of all-PSCs is usually around 100 nm since the carrier mobility of organic materials is relatively low, which thus limits the overall film absorption between 60% and 80%. Tandem structure with sub-cells using the same blend can equivalently double the film thickness. We reported a homo all-PSCs tandem structure using P2F-DO/N2200 blend. An improved PCE of 6.70% was achieved for the tandem device, which is over 43% higher than the 4.68% PCE of the optimized

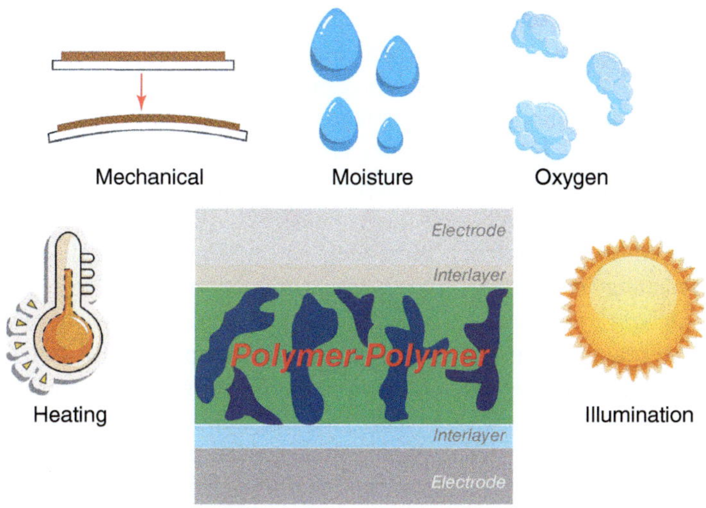

Figure 7.25 Stability enhancements in all-PSCs under different situations. Source: Jianyu Yuan.

single cell [82]. In 2018, another homo tandem adopting PTzBI-Si as donor and N2200 as acceptor was reported. The tandem cell demonstrated an improved overall absorption according to the increased J_{sc}, affording a PCE of 11% [31], while V_{oc} and FF were in agreement with the results of the single junction cell.

7.4.2 The Enhanced Stability of All-PSCs

The long-term stability of OPVs largely determines the prospect for future commercialization. In the past a few years, lots of efforts have been dedicated to improving stability of PSCs. And the stability of all-PSCs and polymer-PCBM systems has been compared. Through these investigations, as shown in Figure 7.25, it was found that the fullerene-based thin film exhibits thermal or photo-induced morphological instability due to the undesired endurance of environmental stress and fast molecular diffusion relative to the all-polymer blend [83]. In this section, we will discuss the long-term stability of all-PSCs under different situations such as thermal stress, ambient conditions, mechanical stress, and illumination (see Figure 7.4).

7.4.2.1 Thermal Stability

We first conducted a comparative study on two high-performance non-fullerene BHJ blends based on PBDBT:ITIC and PBDBT:N2200, with continuous heating at 80 °C in the N_2 glovebox without any encapsulation over 70 days [84]. It was found that the efficiency of PBDB-T:ITIC device deteriorated quickly to 76% of its initial value and further down to 54%. In contrast, the PBDB-T:N2200 all-PSCs showed significantly improved thermal stability with less than 10% PCE degradation after 1700 hours aging. More interestingly, we also reported that all-PSCs demonstrated high tolerance on D/A blend ratios in comparison with the small molecular-based nonfullerene solar cells. These results indicate that the variation of blend composition has less effect on all-PSCs performance, which is ideal for large-area devices. Cao and coworkers first reported all-PSCs based on PTzBI-Si:N2200 with high efficiency over 10%. Meanwhile, the PTzBI-Si:N2200 all-polymer blend exhibits excellent thermal stability [31], the device retained over 95% of its initial efficiency after continuous exposure at 80 °C. In addition, by introducing PNDIT-F3N as an interfacial cathode layer, they reported homo tandem all-PSCs with a record PCE over 11% and remarkable stability, remaining 93% of the initial efficiency after thermally annealing at 80 °C for 1000 hours. Quite recently, we reported efficient all-PSCs using a P-i-N planner structure, which still exhibited decent performance under continuous thermal stresses at a higher temperature of 160 °C for nearly 1000 hours, indicating exceptional stable morphology even under a harsh environment for these all-polymer blends [22]. The thermal stability originates from the innate properties of conjugated polymers, that is, low diffusion coefficients against aggregation. In contrast, heat promotes the fast diffusion and aggregation of small molecules, resulting in poor phase separation morphology.

7.4.2.2 Ambient Stability

Ambient stability of PSCs is desired for large-scale manufacturing of PV devices, considering the continuous exposure to ambient condition during printing process. In this regard, it was anticipated that the use of conjugated polymers would improve morphological stability under ambient condition due to their much more stable chemical structures. We reported superior ambient processability and stability of all-PSCs over fullerene-PSCs in 2019 [85], through a direct comparison of two N2200-based all-PSCs and fullerene-PSCs. In order to control the ambient humidity, we used an air glove box with controllable humidity. We chose two donors PBDB-T and PTzBI blended with PCBM or N2200. Different BHJ blends are tolerant of moisture to different extents. PCBM-based blends are effective under humidity from 0% to 20%, while all-polymer blends exhibit unprecedentedly stable performance (best PCE approaching 9%) under humidity from 0% up to 80%. We also conducted a long-term air stability characterization of unencapsulated all-PSCs, which exhibited excellent stability aged at ambient condition (~25 °C, relative humidity: 20%). These findings demonstrate that all-PSCs can achieve high device performance with ambient processing and show excellent long-term stability against oxygen and moisture, which situate them in an advantageous position for practical large-scale production of OSCs.

7.4.2.3 Mechanical Stability

To realize portable and wearable devices for real-life applications and endure various forms of mechanical stresses during the coating process such as roll-to-roll printing, mechanical integrity is an essential prerequisite of PSCs. In this regard, researchers have devoted significant efforts to examining the mechanical properties of photoactive materials for PSCs over the past decade. Unfortunately, the reported conventional fullerene-PSCs exhibited poor mechanical durability due to the stiffness of the

PCBM clusters. In 2015, Kim and coworkers first compared the mechanical properties between all-PSCs and fullerene-PSCs based on PBDTT-TPD:P(NDI2HD-T) and PBDTT-TPD:PCBM blends, respectively [86]. They investigated the tensile characteristics of the BHJ films through conducting a pseudo free-standing tensile test with a bar-shaped specimen, designed to exclude substrate effects and avoid complicated calculations. The elongation at break of the all-polymer blend (7.16%) was 60 times higher than that (0.12%) of the PCBM blend. The toughness of the all-polymer blend ($568.7\,J\,m^{-3}$) was 470 times higher than that ($1.2\,J\,m^{-3}$) of the PCBM blend. These outstanding results obtained from the PBDTT-TPD:P(NDI2HD-T) blends originated from intrinsic ductility of the polymers, the presence of tie chains and entanglements to yield strengthened interfaces within the active layer films. Although the superior mechanical properties of all-polymer blends over fullerene-polymer blends have been clearly demonstrated, the mechanical properties of the real all-PSCs should be examined in the future. Until now, stretchable device fabrication without using prepatterned substrates is very limited due to the difficulty in integrating all the ductile components of the electrode, interlayer, and substrate into the device. In addition, Intrinsic properties of conjugated polymers, such as chemical structure (i.e. rigid backbones and flexible side chains), molecular weight, dispersity, and thin-film morphology (i.e. phase separation and molecular packing structure), must be correlated with the mechanical properties of the corresponding all-PSCs [87].

7.4.2.4 Photostability

Stability under long-term solar irradiation should be guaranteed in all kinds of solar cells since exposure to light is inevitable during the operation of these devices. Lots of previous reports have demonstrated that the photo-degradation mechanism of fullerene-PSCs is due to the chemical instability of fullerene derivatives under irradiation stress [88]. This feature would lead to the formation of energetic traps and thus cause the reduction of charge generation and carrier mobility. The superior chemical stability of the NDI/PDI building blocks is attributed to their low-lying LUMO energy levels and can enable the fabrication of photo-stable all-PSCs. A systematical study on P(NDI2HD-T) and PCBM acceptors blending with the same polymer donor PBDTTPD was performed by Kim and coworkers [89], aiming to gain insight into their chemical stability under illumination in N_2 atmosphere. The two systems both exhibited similar initial PCE around 4% and pre-stored in high vacuum for more than 12 hours to maximally remove remaining solvents and additives to minimize their adverse effect on device stability. The encapsulated devices were exposed to 100 hours light irradiation under N_2, as it was observed, the PCE of PBDTTPD:PCBM polymer-fullerene system decreased rapidly to 50% of the initial value for the first 12 hours and then slightly reduced for the next 80 hours. In contrast, the PBDTTPD:P(NDI2HD-T) system showed significantly improved photo-stability after 100 hours illumination, keeping 80% of its initial device PCE. The TAS and SCLC measurement showed that the polaron yield deceased significantly and then created trapped states near the D/A interface in PCBM involved blend. While the P(NDI2HD-T) remained stable nearly for 24 hours illumination, mainly benefitting from its superior endurance against photo-illumination. While these findings show case the potential of photo-stable all-PSCs, the photo-stability of only a few all-PSCs has been examined. Therefore, more assessments on different all-polymer blend systems are required to develop an in-depth understanding of the photo-degradation mechanism in all-PSCs.

7.5 Conclusion and Outlook

In conclusion, great progress with the PCE improved from ~5% to over 15% has been achieved through the invention and application of NDI, B←N, and ITIC-based polymer acceptors for all-PSCs. To further enhance the PV performance to realize large-scale commercialization, continuous exploration of new classes of conjugated polymers is required [90]. In particular, developing polymeric acceptors with suitable light absorption, superior charge-carrier mobility, and minimal voltage loss remains very important. Meanwhile, further optimizing the polymeric donor to match up with these acceptors is also critical. Along with materials development, the enhancement of the PV performance in all-PSCs is closely related to desirable active layer morphologies and defined device architecture. Relative to fullerene and small molecule-based solar cells, the construction of efficient P-i-N, ternary and tandem structure in all-PSCs is less explored, which is important for achieving higher PCEs. In addition to the enhancement of efficiency in all-PSCs, further exploration of the stability is urgent. In the past several years, lots of pioneering efforts have demonstrated that all-PSCs possess superior thermal, ambient, mechanical, and photo stabilities over fullerene or small molecule-based solar cells. In this regard, all-PSCs have been considered as a promising long-term source to power the mechanically robust electronic devices. In future research, the interplay between molecular structure and stability in all-PSCs should be extensively investigated to open new avenues for designing conjugated polymers that exhibit commercially viable stability. Last but not least, lab-to-market transition of all-PSCs should be promoted in the near future. All-PSCs now become a promising young candidate in all kinds of OPV technologies and gain worldwide interests due to their exceptional stability. Nonetheless, several challenges still remain, and collaborations between academia and industry will be increasingly necessary to eventually realize the application of all-PSCs.

References

1 Facchetti, A. (2013). *Mater. Today* 16: 123.
2 Lee, C., Lee, S., Kim, G.U. et al. (2019). *Chem. Rev.* 119: 8028–8086.
3 Luo, Z., Liu, T., Ma, R. et al. (2020). *Adv. Mater.* 32: 2005942.
4 Yan, H., Chen, Z.H., Zheng, Y. et al. (2009). *Nature* 457: 679.
5 Zhu, L., Zhong, W., Qiu, C. et al. (2019). *Adv. Mater.* 31: 1902899.
6 Sun, H., Guo, X., and Facchetti, A. (2020). *Chem* 6: 1310.
7 Sun, H., Yu, H., Shi, Y. et al. (2020). *Adv. Mater.* 32: 2004183.

8 Dou, C., Ding, Z., Zhang, Z. et al. (2015). *Angew. Chem. Int. Ed.* 54: 3648.
9 Dou, C., Liu, J., and Wang, L. (2017). *Sci. China Chem.* 60: 450–490.
10 Zhao, R., Wang, N., Yu, Y., and Liu, J. (2020). *Chem. Mater.* 32: 1308.
11 Ding, Z., Zhao, R., Yu, Y., and Liu, J. (2019). *J. Mater. Chem. A* 7: 26533.
12 Li, Y., Meng, H., Liu, T. et al. (2019). *Adv. Mater.* 31: 1904585.
13 Ye, L., Jiao, X., Zhou, M. et al. (2015). *Adv. Mater.* 27: 6046.
14 Yuan, J., Guo, W., Xia, Y. et al. (2017). *Nano Energy* 35: 251.
15 Shi, S., Yuan, J., Ding, G. et al. (2016). *Adv. Funct. Mater.* 26: 5669.
16 Yuan, J. and Ma, W. (2015). *J. Mater. Chem. A* 3: 7077.
17 Yuan, J., Xu, Y., Shi, G. et al. (2018). *J. Mater. Chem. A* 6: 10421.
18 Shi, G., Yuan, J., Huang, X. et al. (2015). *J. Phys. Chem. C* 119: 25298.
19 Xu, Y., Yuan, J., Sun, J. et al. (2018). *ACS Appl. Mater. Interfaces* 10: 2776.
20 Ding, G., Yuan, J., Jin, F. et al. (2017). *Nano Energy* 36: 356.
21 Yuan, J., Ford, M.J., Xu, Y. et al. (2018). *Adv. Energy Mater.* 8: 1703291.
22 Xu, Y., Yuan, J., Liang, S. et al. (2019). *ACS Energy Lett.* 4: 2277.
23 Zhang, Z.G. and Li, Y. (2020). *Angew. Chem. Int. Ed.* 60: 4422.
24 Jin, F., Yuan, J., Guo, W. et al. (2018). *Adv. Funct. Mater.* 28: 1801611.
25 Schubert, M., Collins, B.A., Mangold, H. et al. (2014). *Adv. Funct. Mater.* 24: 4068.
26 Guo, X., Facchetti, A., and Marks, T.J. (2014). *Chem. Rev.* 114: 8943.
27 Guo, X. and Watson, M.D. (2008). *Org. Lett.* 10: 5333.
28 Kang, H., Lee, W., Oh, J. et al. (2016). *Acc. Chem. Res.* 49: 2424.
29 Fabiano, S., Chen, Z., Vahedi, S. et al. (2011). *J. Mater. Chem.* 21: 5891.
30 Zhu, P., Fan, B., Ying, L. et al. (2019). *Chem. Asian J.* 14: 3109.
31 Zhang, K., Xia, R., Fan, B. et al. (2018). *Adv. Mater.* 30: 1803166.
32 Lee, W., Lee, C., Yu, H. et al. (2016). *Adv. Funct. Mater.* 26: 1543.
33 Lee, C., Kang, H., Lee, W. et al. (2015). *Adv. Mater.* 27: 2466.
34 You, H., Kim, D., Cho, H.-H. et al. (2018). *Adv. Funct. Mater.* 28: 1803613.
35 Jia, T., Li, Z., Ying, L. et al. (2018). *Macromol. Rapid Commun.* 39: 1700765.
36 Hwang, Y.-J., Earmme, T., Courtright, B.A.E. et al. (2015). *J. Am. Chem. Soc.* 137: 4424.
37 Wu, Y., Schneider, S., Walter, C. et al. (2020). *J. Am. Chem. Soc.* 142: 392.
38 Li, Z., Xu, X., Zhang, W. et al. (2016). *J. Am. Chem. Soc.* 138: 10935.
39 Kolhe, N.B., Tran, D.K., Lee, H. et al. (2019). *ACS Energy Lett.* 4: 1162.
40 Feng, S., Liu, C., Xu, X. et al. (2017). *ACS Macro Lett.* 6: 1310.
41 Liu, X., Zhang, C., Duan, C. et al. (2018). *J. Am. Chem. Soc.* 140: 8934.
42 Zhan, X., Tan, Z.a., Domercq, B. et al. (2007). *J. Am. Chem. Soc.* 129: 7246.
43 Li, S., Zhang, H., Zhao, W. et al. (2016). *Adv. Energy Mater.* 6: 1501991.
44 Guo, Y., Li, Y., Awartani, O. et al. (2017). *Adv. Mater.* 29: 1700309.
45 Liu, M., Yang, J., Lang, C. et al. (2017). *Macromolecules* 50: 7559.
46 Shi, Y. and Guo, W.Y.X. (2019). *Acta Polym. Sin.* 50: 873.
47 Wang, Y., Yan, Z., Guo, H. et al. (2017). *Angew. Chem. Int. Ed.* 56: 15304.
48 Wang, Y., Yan, Z., Uddin, M.A. et al. (2019). *Sol. RRL* 3: 1900107.
49 Sun, H., Tang, Y., Guo, H. et al. (2019). *Sol. RRL* 3: 1800265.
50 Sun, H., Tang, Y., Koh, C.W. et al. (2019). *Adv. Mater.* 31: 1807220.

51 Teshima, Y., Saito, M., Fukuhara, T. et al. (2019). *ACS Appl. Mater. Interfaces* 11: 23410.
52 Sun, H., Liu, B., Koh, C.W. et al. (2019). *Adv. Funct. Mater.* 29: 1903970.
53 Sun, H., Liu, B., Yu, J. et al. (2020). *Sci. China Chem.* 63: 1785.
54 Sun, H., Liu, B., Ma, Y. et al. (2021). Regioregular narrow-bandgap n-type polymers with high electron mobility enabling highly efficient all-polymer solar cells. *Adv. Mater.* 33: 2102635.
55 Zhao, R., Dou, C., Xie, Z. et al. (2016). *Angew. Chem. Int. Ed.* 55: 5313.
56 Long, X., Dou, C., Liu, J., and Wang, L. (2017). *Macromolecules* 50: 8521.
57 Long, X., Ding, Z., Dou, C. et al. (2016). *Adv. Mater.* 28: 6504.
58 Ding, Z., Long, X., Dou, C. et al. (2016). *Chem. Sci.* 7: 6197.
59 Long, X., Wang, N., Ding, Z. et al. (2016). *J. Mater. Chem. C* 4: 9961–9967.
60 Zhao, R., Bi, Z., Dou, C. et al. (2017). *Macromolecules* 50: 3171.
61 Wang, N., Long, X., Ding, Z. et al. (2019). *Macromolecules* 52: 2402–2410.
62 Wang, N., Yu, Y., Zhao, R. et al. (2020). *Macromolecules* 53: 3325.
63 Shi, S., Chen, P., Chen, Y. et al. (2019). *Adv. Mater.* 31: 1905161.
64 Feng, K., Huang, J., Zhang, X. et al. (2020). *Adv. Mater.* 32: 2001476.
65 Tong, Y., Xiao, Z., Du, X. et al. (2020). *Sci. China Chem.* 63: 758.
66 Liu, Q., Jiang, Y., Jin, K. et al. (2020). *Sci. Bull.* 65: 272.
67 Zhang, Y., Liu, D., Lau, T.K. et al. (2020). *Adv. Funct. Mater.* 30: 1910466.
68 Yao, H., Wang, J., Xu, Y. et al. (2020). *Acc. Chem. Res.* 53: 822.
69 Zhang, Z.G., Yang, Y., Yao, J. et al. (2017). *Angew. Chem. Int. Ed.* 56: 13503.
70 Meng, Y., Wu, J., Guo, X. et al. (2019). *Sci. China Chem.* 62: 845.
71 Yao, H., Bai, F., Hu, H. et al. (2019). *ACS Energy Lett.* 4: 417.
72 Fan, Q., Su, W., Chen, S. et al. (2020). *Joule* 4: 658.
73 Fan, Q., Su, W., Chen, S. et al. (2020). *Angew. Chem. Int. Ed.* 59: 19835.
74 Yuan, J., Zhang, Y., Zhou, L. et al. (2019). *Joule* 3: 1140.
75 Jia, T., Zhang, J., Zhong, W. et al. (2020). *Nano Energy* 72: 104718.
76 Huang, Y., Kramer, E.J., Heeger, A.J., and Bazan, G.C. (2014). *Chem. Rev.* 114: 7006.
77 Ma, W., Yang, C., Gong, X. et al. (2005). *Adv. Funct. Mater.* 15: 1617.
78 Jin, F., Ding, G., Wang, Y. et al. (2017). *J. Phys. Chem. C* 121: 8804.
79 Liang, S., Lou, Z., Zhang, Q. et al. (2020). *J. Phys. Chem. C* 124: 25262.
80 Li, Z., Xu, X., Zhang, W. et al. (2017). *Energy Environ. Sci.* 10: 2212.
81 An, Q., Zhang, F., Zhang, J. et al. (2016). *Energy Environ. Sci.* 9: 281.
82 Yuan, J., Gu, J., Shi, G. et al. (2016). *Sci. Rep.* 6: 26459.
83 Savagatrup, S., Rodriquez, D., Printz, A.D. et al. (2015). *Chem. Mater.* 27: 3902.
84 Zhang, Y., Xu, Y., Ford, M.J. et al. (2018). *Adv. Energy Mater.* 8: 1800029.
85 Xu, Y., Yuan, J., Zhou, S. et al. (2019). *Adv. Funct. Mater.* 29: 1806747.
86 Kim, T., Kim, J.-H., Kang, T.E. et al. (2015). *Nat. Commun.* 6: 8547.
87 Chen, S., Jung, S., Cho, H.J. et al. (2018). *Angew. Chem. Int. Ed.* 57: 13277.
88 Rao, A.M., Zhou, P., Wang, K.-A. et al. (1993). *Science* 259: 955.
89 Kim, T., Younts, R., Lee, W. et al. (2017). *J. Mater. Chem. A* 5: 22170.
90 Li, S., Yuan, X., Zhang, Q. et al. (2021). *Adv. Mater.* 33: 2101295.

8

Organic Solar Cells with High Open-Circuit Voltage >1 V

Ailing Tang[1], Yuze Lin[2], and Erjun Zhou[1]

[1] *National Center for Nanoscience and Technology, CAS Center for Excellence in Nanoscience, CAS Key Laboratory of Nanosystem and Hierarchical Fabrication, No. 11 Zhongguancun BeiYiTiao, Beijing, 100190, P.R. China*
[2] *Institute of Chemistry, Chinese Academy of Sciences, Beijing National Laboratory for Molecular Sciences, CAS Key Laboratory of Organic Solids, No. 2 Zhongguancun North First Street, Beijing, 100190, P.R. China*

8.1 Introduction

Recently, the innovative development of photovoltaic materials has promoted rapid progress of the organic solar cells (OSCs), and now the power conversion efficiencies (PCEs) have reached over 18% [1–7]. As we know, the short-circuit current density (J_{sc}), the open-circuit voltage (V_{oc}), and the fill factor (FF) are the three main parameters that determine the PCEs of the photovoltaic devices. Till now, combined with the material design and device optimization, the internal quantum efficiencies approaching 100%, and the J_{sc} above 26 mA cm^{-2}, and the high FF near 0.80 have been achieved for the OSCs [8–14]. However, the V_{oc} of the state-of-the-art OSCs is still below 1.0 V, which restricts the PCE for a further breakthrough [15, 16]. The low V_{oc} usually is attributed to the large energy loss (E_{loss}). The E_{loss} is defined as $E_{loss} = E_g - qV_{oc}$, where q is elementary charge and E_g is the optical bandgap of the organic semiconductor having the smaller bandgap within active layer of solar cells. E_{loss} in OSCs is typically above 0.6 eV; in contrast, it is usually around 0.4 eV in Si-based solar cells or perovskite solar cells [17, 18]. There is a lot of room for improvement of the V_{oc} of OSCs, because the E_{loss} is much higher than the theoretical value predicted by Shockley–Queisser (SQ) theory [19]. Once the OSCs' energy loss is decreased, the V_{oc} will achieve a significant breakthrough and so will the device performance. Exploring the cause of the large E_{loss} and seeking solutions are hot topics of current research in the field of OSCs.

Usually, E_{loss} can also be expressed as the voltage loss ($V_{loss} = E_{loss}/q$) [18]. According to the detailed balance analysis, the voltage loss in solar cells can be divided into three factors: $qV_{loss} = \Delta E_1 + \Delta E_2 + \Delta E_3$ [18, 20]. ΔE_1 is the radiative recombination loss due to the absorption above the bandgap. ΔE_2 is the radiative recombination loss originating from the absorption below the bandgap.

Organic Solar Cells: Materials Design, Technology and Commercialization, First Edition. Edited by Liming Ding.

ΔE_3 is the non-radiative recombination loss, which can be quantified by the following formula: $\Delta E_3 = q\Delta V_{oc}^{non-rad} = -kT\ln(EQE_{EL})$, where k is the Boltzmann constant, T is absolute temperature, and EQE_{EL} is radiative quantum efficiency of the solar cell tested in the dark [21]. Among them, ΔE_1 is unavoidable for all of the solar cells, while the radiative recombination of ΔE_2 comes from the formation of the charge transfer (CT) state in the OSCs. To decrease the value of ΔE_2, it needs to reduce the energy offsets between the CT state and the singlet state of the donor or acceptor. ΔE_3 is a key loss channel in OSC devices, attributed to the low EQE_{EL} ($\sim 10^{-6}$–10^{-8}) of the OSC devices [22, 23]. On the one hand, Benduhn et al. have found that the $\Delta V_{oc}^{non-rad}$ drastically increased as the energy of the CT state (E_{CT}) decreases [24]. Recently, Eisner et al. proposed that $\Delta V_{oc}^{non-rad}$ could be reduced as the increase of the hybridization of the local excited (LE) and CT states based on their three-state model simulation [25]. When the energy of the LE and CT states is very close to each other, the oscillator strength from CT to ground state will increase due to the intensity borrowing mechanism, which will result in a considerable enhancement in EQE_{EL} [26], while the reduced energy offset between the LE and CT state may give rise to the slow and inefficient charge generation and further increase the overall radiative recombination rates, leading to the low FF and EQE. Thus, it is critical that fine-tuning the energy offsets at the D/A interface.

Accompanied by the appearance of non-fullerene acceptors, there is a large space to control the energy offsets at the D/A interface. Compared with the fullerene-based OSCs, the V_{oc} of the non-fullerene-based OSCs has been significantly improved, and the value of V_{oc} has increased to 1.3 V for the single-junction OSCs [27]. At the same time, the number of high-efficiency OSCs with V_{oc} above 1.0 V is increasing (as shown in Figure 8.1). Though the highest PCE is still below 14% [28], these high-voltage systems will provide good candidates for future device mechanism

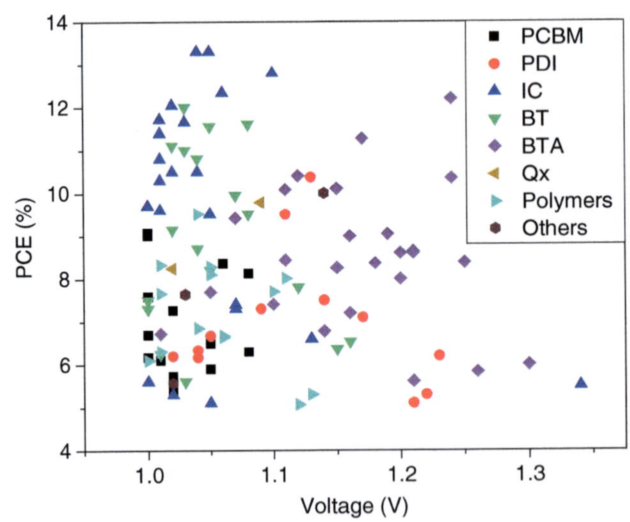

Figure 8.1 PCE against V_{oc} plots of the reported efficient organic solar cells with PCE above 5% and V_{oc} above 1.0 V. Source: Ailing Tang, Yuze Lin, and Erjun Zhou.

research, which will guide material design for further improvements in device performance. Moreover, the ideal tandem solar cells need each sub-cell to have a high V_{oc} and J_{sc} because the V_{oc} of the tandem device equals the sum of V_{oc} of each sub-cell while the lower J_{sc} value of the sub-cells determines the J_{sc} [29–31]. The development of the high-efficiency OSCs with high voltage is critical to improve the tandem cells' device performance. More importantly, the rapid development of the Internet of Things (IoT) ecosystem has promoted the research of indoor photovoltaics (iPV), which are expected to be a power supply source for the IoT systems [32–34]. The outdoor PV cell usually works under 1 sun illumination (AM 1.5G, 100 mW cm^{-2}). Simultaneously, the iPV can generate photoelectric conversion by harvesting artificial light, including fluorescent light (FL), white light-emitting diodes (LEDs), or the dim lighting environment [35]. Usually, the V_{oc} of the photovoltaic devices will decrease as the incident light density weakens [36]. iPVs need high V_{oc} values to offset the voltage drop under low light intensity. Besides, the active layer materials with medium bandgaps are also popular, which can match LEDs' emission and improve the light harvesting [37]. Thus, the studies of the high-efficiency OSCs with high voltage have important implications for future iPVs research.

Though significant progress has been achieved recently for the high-voltage OSCs, no review article focuses on them so far. Herein, we will summarize the recent development of the high-voltage OSCs, which mainly shows the PCEs above 8% with the V_{oc} higher than 1.0 V. We will address how the molecular design of small-molecule acceptors (SMAs) and polymer acceptors (PAs) affects the V_{oc} and performance of the device. These acceptors include the early reports of fullerene-based SMAs, non-fullerene SMAs containing perylene diimide (PDI) [18, 38], benzothiadiazole (BT) [39], benzotriazole (BTA) [40], 3,9-bis(2-methylene-(3-(1,1-dicyanomethylene)-indanone)) (IC) and its derivatives etc. [41], and PAs composed of PDI [42], bithiophene imide (BTI) [43], double B←N bridged bipyridyl [44], etc. Finally, we discuss the existing challenges and perspectives for the future development of high-voltage OSCs.

8.2 n-Type Small-Molecule Acceptors

8.2.1 Fullerene-Based SMAs

In 1995, soluble [6,6]-phenyl-C61-butyric acid methyl ester ($PC_{61}BM$, as shown in Figure 8.2) was used as an acceptor blending with polymer donors to construct polymer solar cells (PSCs) [45]. Since then, $PC_{61}BM$ and its corresponding C70 derivative [6,6]-phenyl-C71-butyric acid methyl ester ($PC_{71}BM$) have been widely applied to fabricate OSCs, and the PCEs of these OSCs have been increased to 11% now [46–51]. Owing to the low-lying lowest unoccupied molecular orbital (LUMO) levels of the fullerene acceptors (~−4.13 eV), the V_{oc} of the OSCs is below 1.0 V [51, 52]. Because of the difficulty of the structural modification for the fullerene acceptors, down-shifting the highest occupied molecular orbital (HOMO) levels of

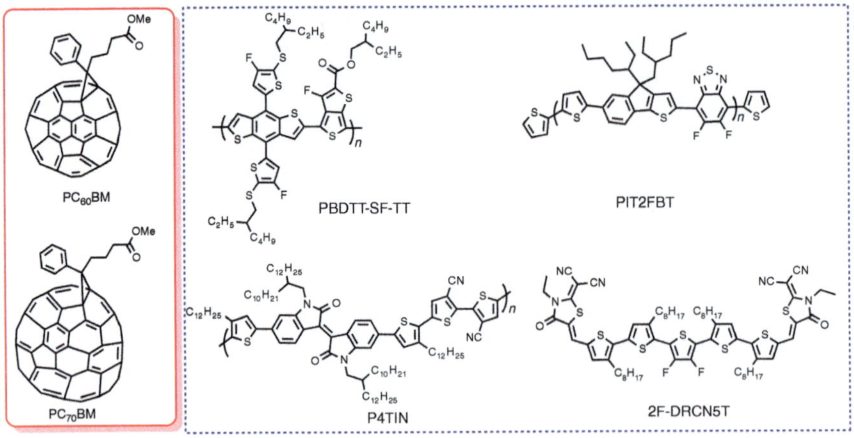

Figure 8.2 Molecular structures of fullerene acceptors and the reported polymer donors with the $V_{oc} > 1.0$ V and PCE > 8%. Source: Ailing Tang, Yuze Lin, and Erjun Zhou.

Table 8.1 Summary of the energy levels and device characteristics of fullerene-based OSCs with PCEs >8%.

Donor	HOMO (eV)	LUMO (eV)	Acceptor	V_{oc} (V)	J_{sc} (mA cm^{-2})	FF	PCE (%)	References
PBDTT-SF-TT	−5.54	−3.95	$PC_{71}BM$	1.00	14.79	0.61	9.07	[53]
PIT2FBT	−5.54	−3.66	$PC_{71}BM$	1.00	14.42	0.62	9.01	[54]
P4TIN	−5.77	−3.70	$PC_{61}BM$	1.06	15.17	0.52	8.36	[55]
2F-DRCN5T	−5.65	−3.94	$PC_{71}BM$	1.08	11.90	0.64	8.13	[56]

Source: Ailing Tang, Yuze Lin, and Erjun Zhou.

the donors is the main pathway toward increasing the V_{oc} of the fullerene-based OSCs. Based on this, several types of high-voltage fullerene-based OSCs have been reported (Figure 8.2). The energy levels and device performances are illustrated in Table 8.1.

In 2018, poly(4,8-bis(5′-((2″-ethylhexyl)thio)-4′-fluorothiophen-2′-yl)benzo[1,2-*b*:4,5-*b*′]dithiophene-2,6-diyl)-*alt*-2′-ethylhexyl-3-fluorothieno[3,4-*b*]thiophene-2-carboxylate (PBDTT-SF-TT) was designed by Yang and coworkers, which is composed of an alkylthiolated and fluorinated BDT motif and 2′-ethylhexyl-3-fluorothieno[3,4-*b*]thiophene-2-carboxylate (TT) unit. The $PC_{71}BM$-based fullerene PSCs for PBDTT-SF-TT showed a PCE of 9.07% with a high V_{oc} of 1.00 V and J_{sc} of 14.79 mA cm^{-2}. The high V_{oc} is attributed to the low energy loss of 0.59 eV. The results indicate the synergistic effect of fluorination and alkylthiolation on BDT moiety can achieve a good balance between high V_{oc} and large J_{sc} [53].

In 2016, Zheng et al. designed a novel wide-bandgap copolymer (PIT2FBT) by incorporating an asymmetric-indenothiophene donor and a difluorinated BT acceptor, yielding a deep-lying HOMO energy level of −5.54 eV. PIT2FBT:$PC_{71}BM$-based devices show a high PCE of 9.01% in combination with a large V_{oc} of 1.0 V [54]. In 2017, Chao and coworkers utilized dicyano-substituted bithiophene as an electron-donating unit and iso-indigo as an electron-accepting unit to synthesize a D–A copolymer, P4TIN. Benefiting from the deep HOMO level and the improved solubility, inverted solar cells fabricated with P4TIN:$PC_{61}BM$ have achieved a very high V_{oc} of 1.06 V and a high PCE of 8.36% [55].

Small molecules can serve as promising candidates for their polymer counterparts. In 2019, Beaujuge and coworkers reported new oligomer-like small molecules based on oligothiophenes by fluoridating the central thiophene, named 2F-DRCN5T. The fluorine substitution at the core unit could down-shift the HOMO levels, increase the energy of the CT state and reduce the non-radiative voltage loss. As a result, the solution-processed solar cells can yield a high V_{oc} of 1.08 V. After the thermal annealing and solvent-vapor annealing, the devices can achieve a high PCE of over 8% [56].

8.2.2 Non-fullerene SMAs

Over the past few years, compared with conventional fullerene acceptors, non-fullerene acceptors have received more and more attentions, benefitting from the enhanced photon harvesting, easily tunable molecular energy levels, and superior morphological stability. The careful energy level regulation of NFAs has promoted the fast development of high-voltage systems. According to the building blocks of the NFSMAs, we will divide them into six categories to review the recent progress of the high-voltage systems based on the NFSMAs.

8.2.2.1 PDI-Based SMAs

PDI is a promising class of electron-deficient build blocks due to their high absorption coefficients and electron mobility. At the same time, PDI is easily modified by introducing alkyl or aryl substituents at the *N*-positions and/or at the bay core positions. Some SMAs have been designed by linking with a wide variety of electron-donating units at the bay positions. The chemical structures of the PDI based SMAs are shown in Figure 8.3, and the energy levels and device properties are listed in Table 8.2.

In 2016, Yan and coworkers reported a PDI-based SMA named SF-PDI_2, composed of two PDI units flanked with the central spirofluorene unit. Though the P3TEA:SF-PDI_2-based OSC exhibits a negligible driving force, it could exhibit fast

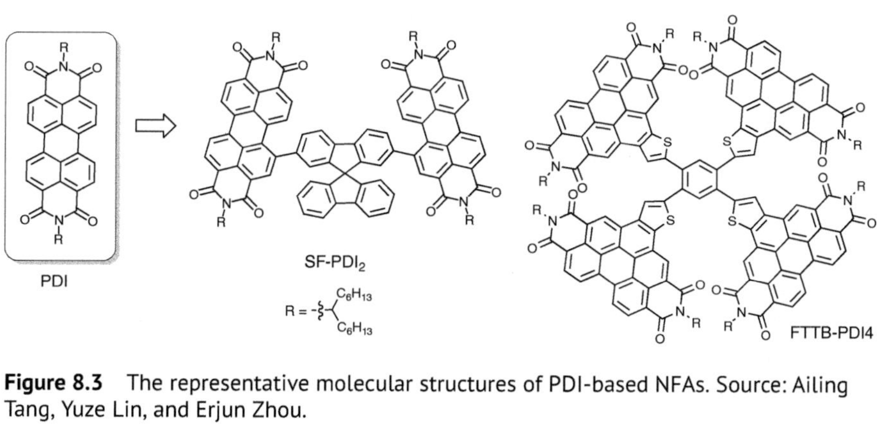

Figure 8.3 The representative molecular structures of PDI-based NFAs. Source: Ailing Tang, Yuze Lin, and Erjun Zhou.

Table 8.2 Summary of the energy levels and device characteristics of PDI and IC-based OSCs with PCEs >6%.

Acceptor	HOMO (eV)	LUMO (eV)	Donor	HOMO (eV)	LUMO (eV)	V_{oc} (V)	J_{sc} (mA cm^{-2})	FF	PCE (%)	References
SF-PDI_2	−5.99	−3.62	P3TEA	−5.37	−3.57	1.11	13.27	0.64	9.50	[18]
TTB-PDI4	−5.92	−3.68	P3TEA	−5.46	−3.56	1.05	12.06	0.53	6.67	[38]
FTTB-PDI4	−5.74	−3.58	P3TEA	−5.46	−3.56	1.13	13.89	0.66	10.37	[38]
ITCC	−5.47	−3.76	PBDB-T	—	—	1.01	15.9	0.71	11.4	[57]
ITIC	−5.50	−3.85	PBDB-TF	−5.48	−3.65	1.01	15.6	0.66	10.3	[58]
IT-M	−5.58	−3.93	PBDB-TF	−5.48	−3.65	1.02	15.78	0.65	10.5	[59]
C8-ITCC	−5.45	−3.85	PBDB-TF	−5.40	−3.64	1.04	16.1	0.63	10.5	[60]
IDT-BOC6	−5.51	−3.78	PBDB-T	−5.33	−2.92	1.01	17.52	0.54	9.60	[61]
ITIC	−5.60	−3.85	PBPD-Th	−5.42	−3.36	1.01	18.1	0.59	10.8	[62]
IDT6CN-M	−5.60	−3.87	PM6	−5.50	−3.61	1.04	16.6	0.77	13.3	[28]
IDT6CN-M	−5.60	−3.87	PM7	−5.52	−3.57	1.05	16.4	0.78	13.3	[28]
ITIC-Th	−5.78	—	PBDT-ODZ	−5.68	−2.89	1.06	17.10	0.68	12.34	[63]
ITIC	−5.48	−3.83	PBDT-TDZ	−5.35	−2.78	1.01	17.15	0.68	11.72	[64]
ITIC	−5.48	−3.83	PBDTS-TDZ	−5.39	−2.80	1.10	17.78	0.65	12.80	[64]
ITIC	−5.48	−3.83	PBDTSF-FBT	−5.41	−3.60	1.03	17.09	0.66	11.66	[65]
ITzN-C9	−5.62	−3.78	PBDB-TF	−5.58	−3.73	1.05	14.12	0.64	9.51	[66]
ITIC-Th	−5.66	−3.93	P1	−5.60	−3.65	1.01	17.89	0.63	11.39	[67]
IDIC	−5.70	−3.82	P3	−5.54	−2.98	1.00	15.99	0.61	9.70	[68]
IT-DM	−5.58	−3.82	J71	−5.40	−3.24	1.02	16.74	0.71	12.05	[69]

Source: Ailing Tang, Yuze Lin, and Erjun Zhou.

and efficient charge separation, leading to a high internal quantum efficiency (IQE) of near 90% and a promising PCE of 9.5%. In addition, the high EQE_{EL} afforded the solar cells very low non-radiative recombination loss. As a result, the V_{oc} of this blend is as high as 1.11 V [18]. They used the combination of P3TEA and SF-PDI_2 to construct a homo-tandem SMA OSC. The homo-tandem SMA OSCs exhibit a promising PCE of 10.8%, with a high V_{oc} over 2.1 V [70]. They synthesized two novel PDI tetramers with fused and nonfused structures based on a tetrathienylbenzene (TTB) core and PDI terminals. For nonfused PDI-tetramer, TTB-PDI4, thiophene is linked with TTB and four PDI units, while for the fused PDI-tetramer, FTTB-PDI4, thiophene rings are fused with PDI units. Compared with the single bonds of nonfused PDI-tetramer, the ring fusion can decrease the rotational freedom between the thiophene and PDI units, up-shift the LUMO level, and enhance light absorption. FTTB-PDI4 exhibits a "double-decker" geometry containing two pairs of mostly parallel PDI units. Consequently, the P3TEA:FTTB-PDI4 devices achieve the highest PCE of 10.58% with an improved V_{oc} of 1.13 V, J_{sc} of 13.89 mA cm^{-2}, and FF of

65.9%. In comparison, the highest PCE of the P3TEA:TTBPDI4 devices is only 7.11%, with a V_{oc} of 1.05 V, J_{sc} of 12.06 mA cm^{-2} and FF of 52.9% [38].

8.2.2.2 IC and Its Derivatives-Based A–D–A-Type SMAs

Besides the PDI-based SMAs mentioned above, A–D–A type linear small molecules with a fused ring ladder-type core have also achieved impressive device performance. In 2015, Zhan and coworkers firstly designed a novel acceptor ITIC based on a seven-ring fused core (indacenodithieno[3,2-*b*]thiophene, IT) and IC terminals [41]. Subsequently, the IDT-based SMA family rapidly grew, and the corresponding PCEs also increased from the initial 6.80% to 14% [71]. However, the strong in-chain CT between the donor and acceptor moieties in these types of SMAs usually affords them narrow bandgaps (~1.6–1.2 eV) and down-shifted LUMO levels, giving rise to a low $V_{oc} < 1$ V. With the development of polymer donors and careful design of SMAs, several high-voltage systems have emerged. Figure 8.4 depicts the chemical structures of the IC-based SMAs, and Table 8.2 summarizes the energy levels and device properties.

In 2017, Hou and coworkers reported a new A–D–A-structured SMA ITCC by simple end-group modification, which introduced thienyl-fused indanone as end groups on a previously reported SMA ITIC. They found ITCC showed a close π–π stacking distance, high electron mobility, and a high-lying LUMO level of −3.76 eV. The OSCs consisting of PBDB-T as the polymer donor obtain an impressive V_{oc} of greater than 1 V and a high PCE of 11.4% [57]. In 2018, they applied the fluorinated polymer (PBDB-TF) as the electron donor and ITIC as the electron acceptor to construct OSCs. Benefited from the deep HOMO level of PBDB-TF, the resulted devices showed a high V_{oc} of 1.01 V and a PCE of 10.3% [58]. Again, they used PBDB-TF as the donor to pair with the acceptor of IT-M, which contains a methyl modified IC as terminals. The corresponding devices also yield a high V_{oc} of 1.02 V and a similar PCE of 10.5% [59]. Yan and coworkers synthesized an A–D–A-type acceptor (C8-ITCC) with tetraoctyl-substituted IT unit as core and thienyl-fused indanone as end groups. With PBDB-TF as the donor, C8-ITCC-based devices can achieve a higher V_{oc} of 1.04 V and a PCE of 10.8% [60].

In 2017, Bo and coworkers designed a new SMA IDT-BOC6 by introducing conformational locking with IDT as a central core unit, bis(alkoxy)-substituted benzene ring as conformational control units and cyanoindanones as end-capping units. IDT-BOC6 exhibits a quite good planarity in the solid state, which endows a high fluorescence quantum yield and reduced energy loss. As a result, the devices showed a decreased non-radiative recombination loss, giving rise to a high V_{oc}. With PBDB-T as the electron donor, IDT-BOC6 based device achieved a PCE of 9.6% and a very high V_{oc} of 1.01 V [61]. Zhang and his coworkers synthesized a wide-bandgap polymer donor PBPD-Th composed of meta-alkoxyphenyl benzodithiophene-*alt*-thienyl benzodithiophene-4,8-dione and thienyl benzodithiophene-4,8-dione (BDD-Th). PBPD-Th shows strong absorption in the short wavelength of 300–650 nm with a bandgap of 1.91 eV and a deep HOMO level of −5.42 eV. As a result, the optimal PSCs based on PBPD-Th:ITIC showed a PCE of 10.8% with a V_{oc} of 1.01 V [62]. They also synthesized a D–A-type wide-bandgap polymer PSTZ by copolymerizing

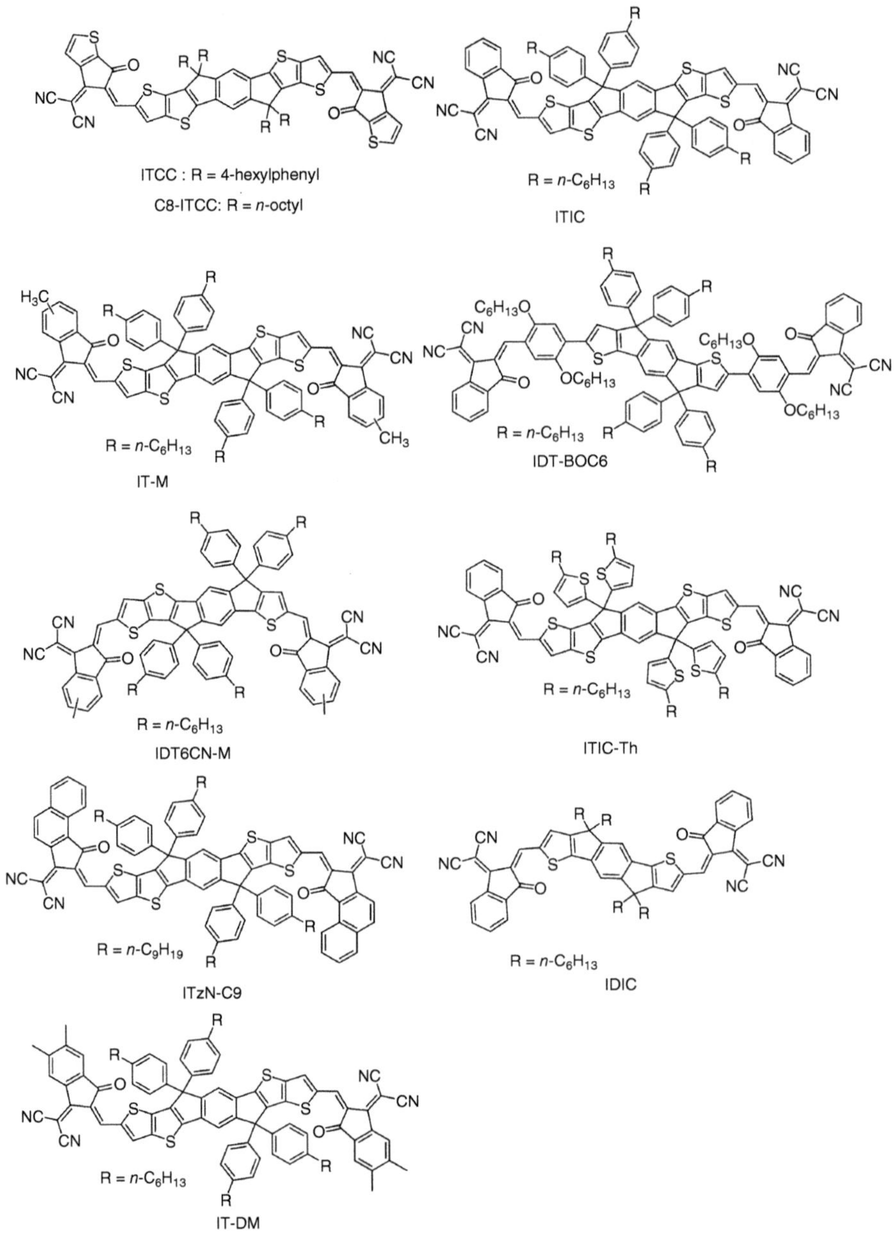

Figure 8.4 The representative molecular structures of SMAs containing IC units. Source: Ailing Tang, Yuze Lin, and Erjun Zhou.

2,5-bis(5-bromo-4-(2-ethylhexyl)thiophen-2-yl)-thiazolo[5,4-*d*]thiazole (TTz) and 4,8-bis(5-(2-butyloctylthio)thiophen-2-yl)-benzo[1,2-*b*:4,5-*b*′]dithiophene (BDTTS) to pair ITIC. Owing to the low HOMO level and high hole mobility of PSTZ, the PSTZ:ITIC-based PSCs achieved a PCE of 8.13% with a high V_{oc} of 1.01 V [72]. In 2019, they utilized an asymmetrical acceptor (IDT6CN-M) to design high-voltage PSCs. IDT6CN-M contains a ladder-type 4,4,9,9-tetrakis(4-hexylphenyl)-4,9-dihydro-*s*-indaceno[1,2-*b*]thiophene-*alt*-[5,6-*d*]thieno[3,2-*b*]thiophene (IDT6) core and methyl-substituted 2-(6-oxo-5,6-dihydro-4*H*-cyclopenta[*c*]thiophen-4-ylidene) malononitrile (CPTCN-M). The asymmetrical IDT6CN-M shows impressively large dipole moments that can enhance the intermolecular interaction and alter the molecular stacking form. Using the fluorinated polymer PM6 or chlorinated polymer PM7 as the donor, the small energy offsets (ΔE_{HOMO}) of 0.10 and 0.08 eV helped both the PM6:IDT6CN-M and PM7:IDT6CN-M-based PSCs achieve a high V_{oc} of 1.04 and 1.05 V, respectively [28].

Peng and coworkers developed a novel wide-bandgap copolymer of PBDT-ODZ based on BDT and 1,3,4-oxadiazole (ODZ) blocks, which possessed a wide bandgap of 2.12 eV and a low-lying HOMO level of −5.68 eV. Blending with 3,9-bis(2-methylene-(3-(1,1-dicyanomethylene)-indanone)-5,5,11,11-tetrakis(4-hexylthienyl)-dithieno[2,3-*d*:2′,3′-*d*′]-*s*-indaceno[1,2-*b*:5,6-*b*′]-dithiophene) (ITIC-Th), the PBDT-ODZ:ITIC-Th devices exhibit a large V_{oc} of 1.08 V and a PCE of 10.12%. After incorporating 4% of copper(I) iodide (CuI) as an additive, the PCE is increased to 12.34% [63]. By copolymerizing 1,3,4-thiadiazole (TDZ) and BDT building blocks, they attained the other two wide-bandgap copolymers, PBDT-TDZ and PBDTS-TDZ. PBDT-TDZ exhibits a low-lying HOMO level of −5.35 eV. Using ITIC as an acceptor, the resulting devices based on PBDT-TDZ exhibited a high V_{oc} of 1.01 V. PBDTS-TDZ with alkylthio side chains exhibits stronger optical absorption, lower-lying HOMO level, and higher crystallinity. PBDTS-TDZ:ITIC devices exhibited a higher V_{oc} of 1.10 V, higher J_{sc} of 17.78 mA cm^{-2}, giving rise to a higher PCE of 12.80% without any additive and posttreatment [64]. They attained a medium-bandgap polymer, PBDTSF-FBT, based on BDT unit modified with fluorinated alkylthienyl side chains as donor block and fluorinated BT as acceptor block. Owing to the fluorination and sulfuration strategies, PBDTSF-FBT exhibits a down-shifted HOMO energy level value and high crystallinity. PBDTSF-FBT:ITIC-based devices showed a high V_{oc} of 1.03 V and a high PCE of 10.69% [65]. In 2019, Sun and coworkers used a weak acceptor unit, methyl 2,5-dibromo-4-fluorothiophene-3-carboxylate (FE-T), to combine with BDT unit, affording a wide-bandgap polymer P1. P1 shows a much deep HOMO level of −5.60 eV. When blended with ITIC-Th, it can achieve a V_{oc} of 1.01 V and a high PCE of 11.39% [67]. They also designed two wide-bandgap polymers (P3 and P4) by incorporating thiophene-flanked phenylene as an electron-donating unit and quinoxaline as an electron-accepting co-unit. Paired with IDIC, P3 with two F atoms attains a large V_{oc} of 1.00 V and a large J_{sc} of 15.99 mA cm^{-2}, yielding a PCE of 9.7%. Further fluorination affords the polymer P4 the increased chain-twisting and down-lying HOMO level, leading to an increase of V_{oc} of 1.06 V. Because of the unsuitable HOMO level with IDIC, the corresponding device gave rise to the lowest PCE of 2.93% [68].

Tobin J. Marks designed two isomeric IDTT-based acceptors 3,9-bis(2-methylene-(3-(1,1-dicyanomethylene)-benz-(5,6)indanone))-5,5,11,11-tetrakis(4-nonylphenyl)-dithieno[2,3-*d*:2′,3′-*d*′]-*s*-indaceno[1,2-*b*:5,6-*b*′]di-thiophene (ITN-C9) and 3,9-bis (2-methylene-(3-(1,1-dicyanomethylene)-benz(6,7)indanone))-5,5,11,11-tetrakis (4-nonylphenyl)-dithieno[2,3-*d*:2′,3′-*d*′]-*s*-indaceno[1,2-*b*:5,6-*b*′]dithiophene (ITzN-C9). Compared with ITN-C9, ITzN-C9 has an elevated LUMO level. The PBDB-TF:ITzN-C9-based devices show a higher V_{oc} of 1.05 V than that of the devices based on PBDB-TF:ITN-C9-based devices (0.93 V). Single-crystal structures of ITN-C9 and ITzN-C9 reveal that the π-extended naphthyl end groups assemble in vertical columns with π-stacking distance as small as 3.31(1) Å, with electronic coupling integrals as large as 38 meV, and internal reorganization energies as small as 0.133 eV, which afford the devices a high PCE near 10% [66]. Yang and coworkers prepared an acceptor IT-DM with indacenodithieno[3,2-*b*]thiophene (IDTT) as the central core and 5,6-dimethyl-3-dicyanovinylindan-1-one as terminals. The J71:IT-DM-based device showed a V_{oc} of 1.02 V with a high PCE of 12.05% [69].

8.2.2.3 A_2-A_1-D-A_1-A_2-Type SMAs with BT as A_1 Units

Benzothiadiazole (BT) is a well-known building block used in p-type small molecules and polymers for its strong electron affinity and strong absorption in the visible spectrum when coupled with electron-rich units. Recently, BT as a building block has also been used for constructing the NFAs. Among them, A_2-A_1-D-A_1-A_2-type NFAs with BT as A_1 units have achieved a high V_{oc} above 1 V and a PCE of 11%. The relevant chemical structure of the BT-based SMAs is shown in Figure 8.5, and the energy levels and device properties are listed in Table 8.3.

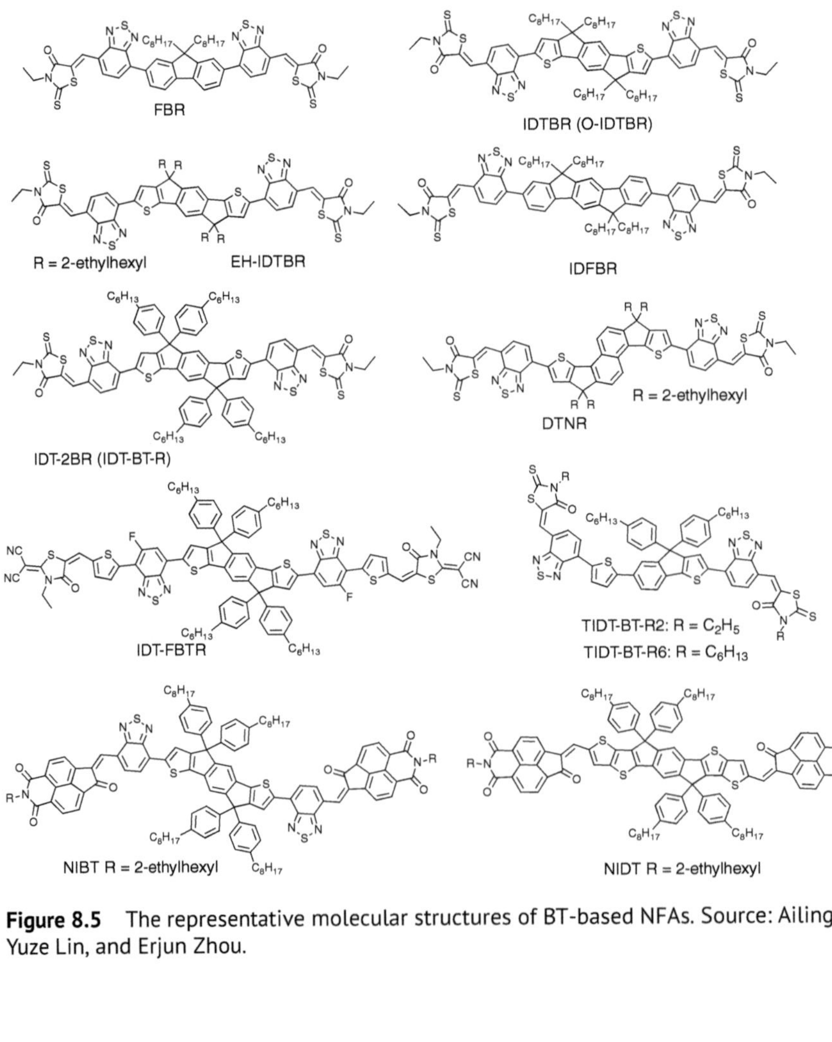

Figure 8.5 The representative molecular structures of BT-based NFAs. Source: Ailing Tang, Yuze Lin, and Erjun Zhou.

Table 8.3 Summary of the energy levels for BT based NFAs and the device characteristics of the OSCs with PCEs >7%.

Acceptor	HOMO (eV)	LUMO (eV)	Donor	HOMO (eV)	LUMO (eV)	V_{oc} (V)	J_{sc} (mA cm^{-2})	FF	PCE (%)	References
FBR	−5.83	−3.75	PffBT4T-2DT	−5.34	−3.70	1.12	11.5	0.61	7.8	[39]
IDTBR	−5.50	−3.90	PffBT4T-2DT	−5.34	−3.70	1.07	15.0	0.62	9.95	[39]
EH-IDTBR	−5.58	−3.90	PffBT4T-2DT	−5.34	−3.70	1.02	17.2	0.63	11.1	[73]
EH-IDTBR	−5.58	−3.90	PBDTTT-EFT	−5.24	−3.66	1.03	18.5	0.63	12.0	[74]
IDTBR:IDFBR	−5.75	−3.70	PCE10	−5.24	−3.66	1.03	17.2	0.6	11.0	[75]
O-IDTBR	−5.51	−3.88	PvBDTTAZ	−5.47	−3.42	1.08	16.26	0.64	11.6	[76]
EH-IDTBR: T2-ORH	−5.42	−3.60	PTB7-Th	−5.39	−3.41	1.05	17.51	0.63	11.55	[77]
IDT-2BR	−5.52	−3.69	PTB7-th	−5.20	−3.59	1.05	12.8	0.61	8.2	[78]
DTNR	−5.67	−3.75	PTB7-Th	−5.26	−3.62	1.08	15.72	0.56	9.51	[79]
IDT-FBTR	−5.47	−3.65	PTB7-Th	−5.26	−3.62	1.02	15.18	0.58	9.14	[80]
TIDT-BT-R2	−5.25	−3.65	PTB7-Th	−5.15	−3.44	1.04	13.1	0.64	8.7	[81]
NIBT	−5.38	−3.68	PBDB-T	−5.18	−3.37	1.04	17.34	0.60	10.8	[82]
NIDT	−5.34	−3.58	PBDB-T	−5.18	−3.37	1.14	13.28	0.66	10.0	[82]

Source: Ailing Tang, Yuze Lin, and Erjun Zhou.

In 2015, McCulloch and coworkers reported an A_2-A_1-D-A_1-A_2 type BT-based NFA, FBR, which contained 9,9-dioctylfluorene as the core, BT as A_1 unit, and 3-ethylrhodanine as A_2 unit [83]. Subsequently, they synthesized another NFA called IDTBR via replacement of the fluorene core in FBR with an n-octyl substituted indacenodithiophene unit [84]. In 2016, they used a low-bandgap polymer PffBT4T-2DT to pair with these two BT-based NFAs. The resulting PCEs for FBR and IDTBR are as high as 7.8% and 9.95%, respectively. The most remarkable aspects of the devices are the high V_{oc} of 1.12 V, with high J_{sc} (11.5 mA cm^{-2}) for PffBT4T-2DT:FBR and the high V_{oc} of 1.07 V with 75% EQE (J_{sc} = 15 mA cm^{-2}) for PffBT4T-2DT:IDTBR devices [39]. In addition, they applied PffBT4T-2DT to pair another acceptor, EH-IDTBR, which used a 2-ethylhexyl substituted indacenodithiophene as the core. Compared with chlorobenzene, using mesitylene as the solvent, an improved PCE of 11.1% with a high V_{oc} of 1.02 V was achieved without the need for pre- or post-treatments [73]. In 2018, they reported PBDTTT-EFT-based devices with EH-IDTBR as acceptor, which can achieve high efficiencies of 12% with a V_{oc} > 1 V. They found the charges can dissociate with negligible geminate recombination losses, resulting in a high EQE value around 90% and a high J_{sc} of 20 mA cm^{-2} [74]. They also synthesized a new indenofluorene analogue of IDTBR, named IDFBR. They used PCE10 as donor and IDTBR and IDFBR as the acceptors to fabricate the ternary OSCs. The higher-lying LUMO level of IDFBR than that of IDTBR will enhance the cascade effect in the ternary, and the devices can achieve a high V_{oc} of 1.03 V with a high PCE of 11.0% [75]. Zhan and coworkers used PTB7-Th as a donor to blend with IDT-2BR, resulting in a high V_{oc} of 1.05 V and a PCE of 8.5%. They further employed PDI-2DTT (perylene diimide moiety flanked with two dithienothiophene) as the third component in the active layer. PDI-2DTT showed lower LUMO and HOMO energy levels than those of IDT-2BR, which can be used as an energy driver to enhance the driving force for CT, and simultaneously achieved a minimal energy loss (0.55 eV). PTB7-Th/IDT-2BR blend showed a very high PCE (>10%) [78]. Yan and Ma designed a novel wide-bandgap (2.05 eV) polymer donor named PvBDTTAZ, composed of a BDT and a difluorinated triazole unit, to pair with the narrow-bandgap SMA (O-IDTBR). Differently from the conventional BDT-based polymers where the BDT units are linked into the polymer backbone via the α positions of the thiophene units, the BDT units in PvBDTTAZ are connected along the polymer backbone via the phenyl group of the BDT unit. The resulted OSCs showed a high V_{oc} of 1.08 V and a small voltage loss of 0.55 V. Benefited from the highly favorable film morphology with crystalline and small domains and effective CT, the devices showed a high EQE above 70%, giving rise to a high PCE of 11.6% [76]. Lim and coworkers utilized T2-ORH consisted of a bithiophene core and octylsubstituted rhodanine ends as the third component in ternary-blend solar cells with PTB7-Th and EH-IDTBR as host materials. The ternary device with a 1 : 1 : 1 blend ratio achieved a PCE of 11.55% with a V_{oc} of 1.05 V [77].

Zheng and coworkers utilized an angular-shaped dithienonaphthalene flanked by 3-ethylrhodanine units via BT bridges to attain a novel NFA, named as DTNR. It exhibited a high-lying LUMO level of −3.75 eV and strong absorption. Combined with a low-bandgap polymer, PTB7-Th, DTNR obtained a high PCE of 9.51% with a large V_{oc} of 1.08 V, owing to an extremely low energy loss of 0.50 eV [79]. Chen and Zhan designed a kind of "donor–acceptor (D–A) combined π-bridge" based on the monofluoro-substituted benzothiadiazole (FBT) to link a thiophene ring, which is used as bridge to link a IDT central core and 3-octyl-2-(1,1-dicyanomethylene)rhodanine terminal groups, and they constructed a new A-π-D-π-A-type small molecular acceptor (IDT-FBTR). This "offset effect" of the D–A combined π-bridge can regulate the HOMO-LUMO energy levels of the molecule, yielding a relatively high LUMO energy level of IDT-FBTR, which is helpful to increase the V_{oc}. As a result, the OSCs based on the PTB7-Th and IDT-FBTR showed a high V_{oc} of 1.02 V with a PCE of 9.14% [80]. Xu and coworkers designed two A_1-A_2-π-D-A_2-A_1 structural SMAs, TIDT-BT-R2 and TIDT-BT-R6, by using thiophene-indenothiophene (TIDT) as the core and benzothiadiazolerhodanine (BT-R) as the end groups. Compared with TIDT-BT-R6 with hexyl as alkyl chain, TIDT-BT-R2 with shorter alkyl chain (ethyl) showed enhanced photo-induced charge separation, better charge transport and collection properties. TIDT-BT-R2 exhibited a larger V_{oc} of 1.04 V and a higher PCE of 8.7% [81]. Jenekhe and coworkers designed an A=CH–D–CH=A type NFAs incorporating rylene-imide end-capping groups via methine bridges in D–A architectures. NIBT is composed of an IDT core and BT bridged bis(naphthalene-imide)arylenelidenes (BNIAs), which can attain a high V_{oc} of 1.04 V and high PCE of up to 10.8% with PBDB-T as the donor [82]. Compared with NIBT, NIDT without BT bridge shows a higher-lying

LUMO level of −3.58 eV. The resulting NIDT:PBDB-T-based devices show a higher V_{oc} (1.14 V) and high PCE of up to 10.0% [82].

8.2.2.4 A_2-A_1-D-A_1-A_2-Type SMAs with BTA or Qx as A_1 Units

Replacing the S atom in the BT unit with N atom affords another electron-withdraw building block, BTA. Compared with BT, the incorporation of N atom in BTA weakens its electron-withdrawing ability, which could up-shift its LUMO energy level. The design of the alkyl chain on the N atoms facilitates to fine-tune the solubility and crystallinity of the molecules. BTA has been employed to construct a series of medium-bandgap polymer donors. In 2017, Zhou and coworkers used BTA as A_1 unit to link the different middle D unit and terminal A_2 unit, affording the A_2-A_1-D-A_1-A_2 type SMAs. Benefiting from the weak electron-withdrawing ability of BTA, the resulted SMAs showed the up-lying LUMO levels and endowed the devices a high V_{oc}. To fine-tune the optoelectronic properties of SMAs, Zhou and coworkers have carefully designed the central core, terminals, and alkyl chains. Figure 8.6 lists the corresponding molecular structures, and the device parameters

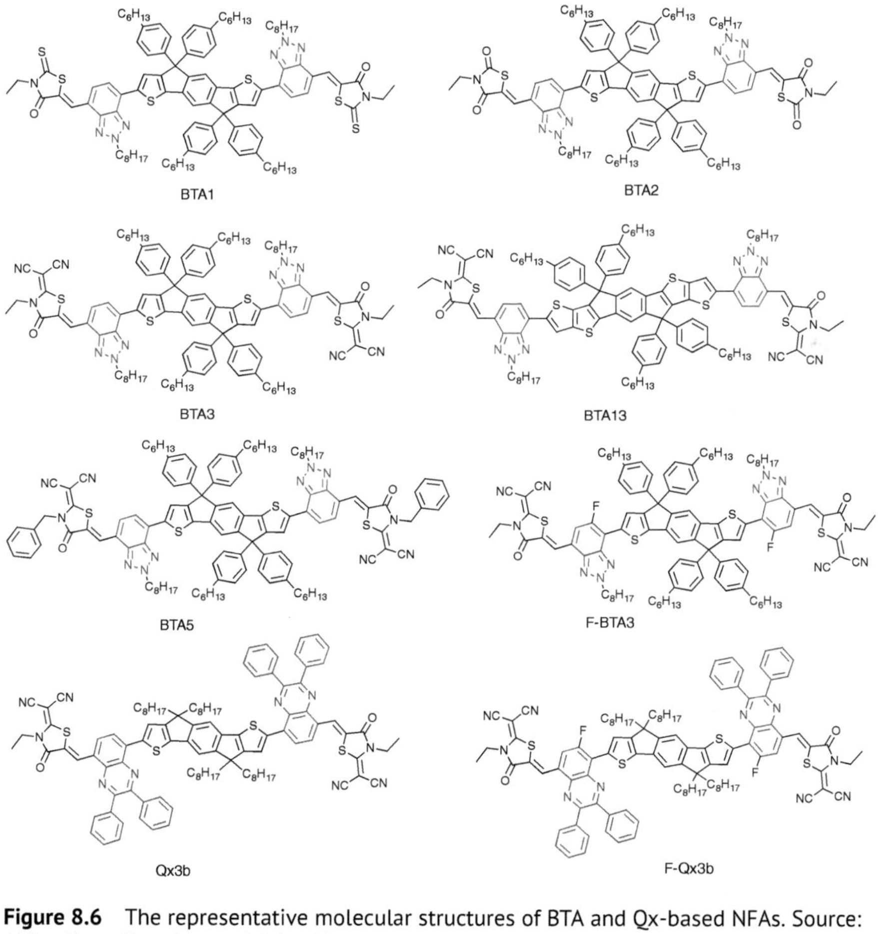

Figure 8.6 The representative molecular structures of BTA and Qx-based NFAs. Source: Ailing Tang, Yuze Lin, and Erjun Zhou.

Table 8.4 Summary of the energy levels for BTA based NFAs and the corresponding device characteristics of the OSCs.

Acceptor	HOMO (eV)	LUMO (eV)	Donor	HOMO (eV)	LUMO (eV)	V_{oc} (V)	J_{sc} (mA cm^{-2})	FF	PCE (%)	References
BTA1	−5.51	−3.55	P3HT	−4.9	−3.0	1.02	7.34	0.70	5.24	[85]
BTA2	−5.43	−3.38	P3HT	−4.9	−3.0	1.22	6.15	0.60	4.50	[86]
BTA3	−5.49	−3.73	J61	−5.32	−3.38	1.15	10.84	0.66	8.25	[87]
BTA3	−5.49	−3.73	PE1	−5.41	−3.53	1.11	11.95	0.64	8.43	[88]
BTA3	−5.49	−3.61	J71	—	—	1.20	10.39	0.69	8.60	[89]
BTA3	−5.49	−3.73	J52-Cl	−5.39	−3.45	1.24	13.16	0.67	10.35	[40]
BTA3	−5.49	−3.73	J52	−5.13	−3.22	1.07	14.62	0.60	9.41	[40]
BTA3	−5.49	−3.73	PfBTAZT-Cl	−5.44	−3.59	1.20	11.11	0.60	8.00	[90]
BTA3	−5.42	−3.61	PBQx-TCl	−5.39	−2.86	1.24	13.3	0.74	12.2	[91]
BTA13	−5.34	−3.62	J52-F	−5.29	−3.38	1.18	11.55	0.61	8.36	[92]
BTA3	−5.54	−3.69	J52-F	−5.36	−3.42	1.19	11.56	0.66	9.04	[93]
BTA5	−5.55	−3.62	J52-F	−5.36	−3.42	1.17	13.80	0.70	11.3	[93]
BTA5	−5.55	−3.71	PE31	−5.35	−3.54	1.11	13.68	0.66	10.1	[94]
BTA5	−5.55	−3.71	PE32	−5.22	−3.30	1.10	11.65	0.58	7.4	[94]
BTA5	−5.55	−3.71	PE33	−5.42	−3.55	1.16	12.68	0.61	9.0	[94]
BTA5	−5.55	−3.71	J52-Cl	−5.46	−3.58	1.21	11.36	0.63	8.6	[94]
BTA3	−5.59	−3.70	PBT1-C	−5.49	−3.42	1.21	10.89	0.57	8.6	[95]
F-BTA3	−5.59	−3.82	P2F-EHp	−5.41	−3.55	1.25	11.31	0.59	8.4	[96]
F-BTA3	−5.59	−3.82	J52-Cl	−5.39	−3.45	1.15	12.81	0.69	10.1	[97]
F-BTA3	−5.59	−3.82	F11	−5.23	−3.49	1.12	13.00	0.69	10.0	[97]
F-BTA3	−5.59	−3.82	PE52	−5.37	−3.49	1.14	10.53	0.56	6.8	[97]
Qx3b	−5.27	−3.63	PE61	−5.19	−3.44	1.02	12.52	0.65	8.2	[98]
F-Qx3b	−5.37	−3.69	PE62	−5.37	−3.48	1.09	13.11	0.70	9.8	[98]

Source: Ailing Tang, Yuze Lin, and Erjun Zhou.

are listed in Table 8.4. Firstly, by using P3HT as electron donor, BTA1 and BTA2 can realize high V_{oc} of 1.02 and 1.22 V with moderate PCEs of 5.24% [85] and 4.50% [86], respectively.

Subsequently, considering that the LUMO energy level mainly depends on the electron-withdrawing segment in the D–A-type copolymer, Zhou and coworkers chose the D–A copolymers containing BTA units to pair with their BTA-based SMAs, namely "Same-A-Strategy" (SAS), which may contribute to realize the close LUMO levels and decrease the voltage loss. As a result, they have achieved a series of high-voltage PSCs (>1.10 V). Firstly, they chose J61 containing alkylthio substituted BDT units and difluorine substituted BTA units as an electron donor to blend with BTA3. The resulted devices can yield a high V_{oc} of 1.15 V and a high PCE of 8.25% [87]. And then, they synthesized two new BTA-based polymers to pair BTA3. PE1 was derived from the polymer J61 by replacing the thiophene π-bridge with thieno[3,2-*b*]thiophene, and PE2 was synthesized by introducing fluorine atoms to the thiophene side chain in PE1. PE1 can realize a PCE of 8.43% with a V_{oc} of 1.11 V [88]. Furthermore, the low-lying HOMO level of J71 with alkylsilyl modified BDT unit afforded the PSCs based on J71:BTA3 a higher V_{oc} of 1.20 V with a PCE of 8.60% [89]. Then, chlorine atoms are introduced into the conjugated side chain of the BDT unit of the p-type polymer to fine-tune the optoelectronic properties. The chlorinated polymer J52-Cl, blended with BTA3, yields very small energy offsets and a decreased non-radiative recombination loss of 0.24 eV. J52-Cl:BTA3 device delivered a very high V_{oc} of 1.24 V, a J_{sc} of 13.16 mA cm^{-2}, and an FF of 66.62%, giving rise to a promising PCE of 10.5%, while the non-chlorinated counterpart (J52) gives rise to a V_{oc} of 1.07 V with a PCE of 9.41% [40]. In addition, a fused BTA-based copolymer (PfBTAZT-Cl) containing 6-fluoro-thieno[2′,3′:4,5]benzo[1,2-*d*]-[1,2,3]triazole (fBTAZT) and chlorine atoms substituted BDT unit, when blended with BTA3, can also attain a high V_{oc} above 1.20 V [90]. In 2021, Hou and coworkers reported a wide-bandgap polymer named PBQx-TCl to pair with BTA3. Benefited to the high EQE_{EL} approaching 2×10^{-3}, the PBQx-TCl:BTA3-based device yielded an impressively high V_{oc} of 1.24 V. The high carrier mobilities and suppressing bimolecular recombination afforded the PBQx-TCl:BTA3 a high PCE of 12.2%. They fabricated ternary cells with PBQx-TCl, BTA3 and BTP-eC9, which showed a PCE of 28.5% for a 1 cm^2 device and 26.0% for a 10 cm^2 device under 1000 lx LED illumination [91].

Compared with BTA3 with IDT core, BTA13 with IDTT as central unit shows up-shifted HOMO and LUMO levels. Blended with BTA13, the fluorinated polymer of J52-F shows a high PCE of 8.36% with an improved V_{oc} of 1.18 V and a J_{sc} of 11.55 mA cm^{-2} [92]. They also investigated the effect of the substituent groups in the A_2 unit on the molecular stacking model. BTA3 with ethyl substituent in RCN shows heterogeneous interaction between A_2 and A_1 unit while BTA5 with the benzyl substituted RCN exhibits homogeneous packing of A_2–A_2 model. The substitutes on the terminals have little influence on the energy levels. BTA3 and BTA5 blended with J52-F could achieve ultrahigh V_{oc} of 1.19 and 1.17 V, respectively. Benefited from the good molecular packing, BTA5 realized an improved PCE of 11.27%, obviously higher than that of BTA3 (PCE = 9.04%) [93]. They chose BTA5 as electron acceptor and four BTA-based polymer donors to construct OSCs with tetrahydrofuran (THF) as processing solvent. They investigated the influence of the substitutents (-H, -OCH_3, -F, -Cl) on the BTA units of the polymers. All of the four polymers can achieve a high V_{oc} above 1.10 V. PE31 with the unsubstituted BTA unit showed the efficient hole transfer and more balanced charge mobilities, thus leading to the highest PCE of 10.08% due to the efficient hole transfer and balanced carriers mobilities, while the methoxy-substituted polymer PE32 exhibited the lowest PCE of 7.40%. With the down-shifted HOMO levels of fluorinated PE33 and chlorinated J52-Cl, they showed the higher V_{oc} values of 1.16 and 1.21 V, respectively [94]. When using another kind of medium-bandgap D-π-A polymer (PBT1-C) as electron donor, PBT1-C:BTA3-based devices can also achieve high-energy CT state around 1.76 eV and low non-radiative recombination loss about 0.24 eV. As a result, the PBT1-C:BTA3-based devices can yield a high V_{oc} of 1.21 V [95].

To modify the molecular properties of BTA3, they further introduce fluorine atoms to BTA units and attain a new SMA, named F-BTA3. Compared with BTA3, F-BTA3 exhibits a higher molar extinction coefficient, larger CT driving force, and suppressed bimolecular recombination. Pyrrolo[3,4-*f*]benzotriazole-5,7-dione (TzBI)-based polymer is chosen as an electron donor, which possessed the low-lying HOMO levels. The resulted F-BTA3-based devices showed a very high V_{oc} of 1.25 V and a PCE of 8.38% [96]. In 2021, Zhou and coworkers investigated the effect of the electron-donating units (D) in polymer donors on the photovoltaic performance, with F-BTA3 as acceptor. They design three polymers (J52-Cl, F11, and PE52) with BDT unit, benzo[1,2-*b*:4,5-*b*′]difurans (BDF), and dithieno[2,3-d;2′,3′-*d*′]benzo[1,2-*b*;4,5-*b*′]dithiophenes (DTBDT) as the electron-donating unit, respectively. The tricyclic D units show better PCE than the pentacyclic D unit (10% versus 6.8%). F11 with BDF units exhibits a slightly lower V_{oc} of 1.12 V than those of the J52-Cl and PE52-based devices (1.14 V) [97].

In 2021, Zhou and coworkers chose quinoxaline (Qx) instead of BTA as the A unit to construct two new small molecular acceptors (Qx3b and F-Qx3b). They investigated the relative photovoltic performance by introducing the fluorine atom to the thiophene side chains of the p-type polymer and Qx units of the n-type small molecule. The photovoltaic device based on PE61:Qx3b achieved a PCE of 8.24% with a V_{oc} of 1.02 V, while the PE62:F-Qx3b-based device obtained a V_{oc} as high as 1.09 V with a slightly high PCE at 9.78%. That result indicated that the Qx unit is also a good choice to attain high-voltage solar cells [98].

8.3 n-Type Polymers

All-PSCs, fabricated with polymer donor and polymer acceptor, have attracted a lot of attention due to their tunable light harvesting, easy large-scale/large area manufacturing, and long-term stability. Recently, the PCE of the all-PSCs has reached up to 15.0% [99, 100]. Simultaneous adjustment of polymer donor and polymer acceptor energy levels can improve the light harvesting and increase the V_{oc}. As a result, an increasing number of all-PSCs have achieved a high V_{oc} above 1.0 V with the PCE above 5%. Figure 8.7 lists the relevant molecular structures of the polymer acceptors, and the device parameters are listed in Table 8.5.

Guo and coworkers designed a new electron-accepting building block, bithiophene imide, namely BTI. Thiophene ring fusion can extend the π-conjugation length, broaden the absorption, and improve the electron mobility. They designed a series of n-type terpolymers BTI2-xTPD by copolymerizing a fused BTI dimer, TPD unit with a difluorinated thiophene unit. The incorporation of a TPD unit facilitates balancing crystallinity and miscibility. The noncovalent S· · ·O interactions between TPD and thiophene unit increase the backbone planarity and improve the in-chain charge transport. BTI2-30TPD achieves an optimal PCE of 8.28% with a V_{oc} of 1.05 V [101]. They also developed a series of BTI2 and 3,4-difluorothiophene-based random copolymers PBTI2(xDT)-FT, by changing the ratios of 2-decyltetradecyl (DT) and

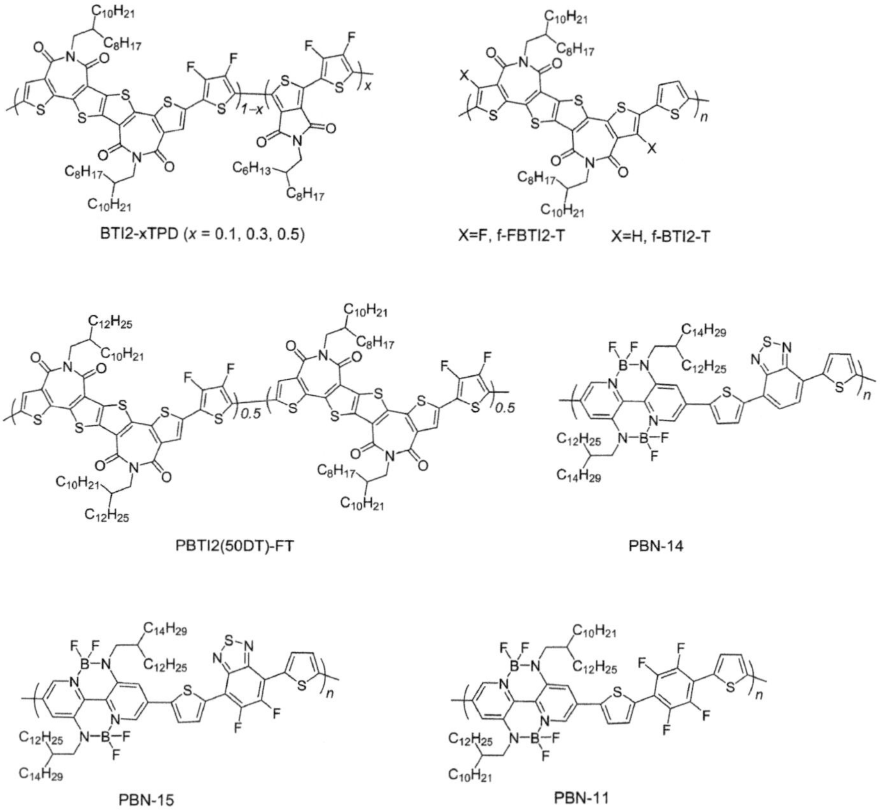

Figure 8.7 The molecular structures of the polymers acceptors. Source: Ailing Tang, Yuze Lin, and Erjun Zhou.

Table 8.5 Summary of the energy levels for the active layer in all-PSCs and the device characteristics of the all-PSCs with PCEs >5%.

Acceptor	HOMO (eV)	LUMO (eV)	Donor	HOMO (eV)	LUMO (eV)	V_{oc} (V)	J_{sc} (mA cm^{-2})	FF	PCE (%)	Ref-erences
BTI2-30TPD	−6.15	−3.41	PTB7-Th	−5.35	−3.32	1.05	13.56	0.58	8.28	[101]
PBTI2(50DT)-FT	−6.14	−3.43	PTB7-Th	−5.35	−3.32	1.01	13.96	0.59	8.32	[102]
f-FBTI2-T	−5.94	−3.46	PTB7-Th	−5.35	−3.32	1.05	13.60	0.57	8.1	[103]
PBN-14	−5.58	−3.53	BD3T	−5.11	−3.23	1.12	7.16	0.63	5.06	[44]
PBN-15	−5.78	−3.65	BD3T	−5.11	−3.23	1.04	13.82	0.66	9.51	[44]
PBN-11	—	—	DR3TBDTC	−5.18	−3.09	1.11	11.18	0.65	8.01	[104]

Source: Ailing Tang, Yuze Lin, and Erjun Zhou.

2-octyldodecyl (OD) side chains on BTI2. The side-chain engineering of PA could improve the blend film morphology and yield balanced charge carrier mobilities. The optimal polymer PBTI2(50DT)-FT blended with PTB7-Th achieved a promising PCE of 8.32% with a large V_{oc} of 1.01 V [102]. They further synthesized a fluorine atom functionalized *f*-BTI2 unit, namely di(fluorothienyl)thienothiophene diimide (*f*-FBTI2). By copolymerizing the *f*-FBTI2 and thiophene unit, they attained a high-performance n-type polymer, *f*-FBTI2-T, which showed a reduced bandgap and low-lying LUMO level. With PTB7-Th as polymer donor, the resulted all-PSCs based on *f*-FBTI2-T can yield a high PCE of 8.0% with a V_{oc} of 1.05 V [103].

Liu and coworkers have reported a new kind of amorphous polymer acceptors containing boron–nitrogen coordination bond (B←N unit), which can match various polymer donors to give efficient all-PSCs. They synthesized two amorphous polymer acceptors (PBN-14 and PBN-15), by copolymerizing an electron-deficient double B←N bridged bipyridine (BNBP) unit with an electron-deficient BT or difluorinated BT unit. They utilized a small molecular donor, BD3T, composed of carbazolyl substituted benzodithiophene core, 3,4′-dioctyl-2,2′:5′,2″-terthiophene (3T) bridge unit, and 3-ethylrhodanine terminals, to combine with PBN-14 and PBN-15, respectively. The BD3T:PBN-14-based devices showed a higher V_{oc} of 1.12 V and a medium PCE of 5.06%. Compared with PBN-14, PBN-15 possesses two extra fluorine substitutes on the BT unit, which afforded it a stronger aggregation tendency and lower LUMO level, suppressed bimolecular recombination. As a result, PBN-15-based devices showed a higher J_{sc} and FF, giving rise to a higher PCE of 9.5% with a decreased V_{oc} of 1.04 V [44]. Another polymer acceptor, PBN-11 containing BNBP and 2,2′-(perfluoro-1,4-phenylene)dithiophene (4fPT) units is chosen to blend with a small molecule donor DR3TBDTC. DR3TBDTC is composed of carbazolyl substituted BDT core, 3,3″-dioctyl-2,2′:5′,2″-terthiophene (3T) and 3-ethylrhodanine. The DR3TBDTC:PBN-11-based devices exhibit a high V_{oc} of 1.04 V. The active layer showed a small-size phase separation, contributed to a high PCE of 8.0% [104].

8.4 Conclusion and Outlook

One of the design rules for the high-voltage OSCs is to decrease the energy offset between the donor and acceptor molecular energy levels. With the development of the electron donors and electron acceptors, substantial progress has been achieved in the V_{oc} of the OSCs. The V_{oc} has been approaching 1.3 V with a decreased energy loss of about 0.5–0.6 eV, and the PCE of the high-voltage system has also exceeded 13%. There is still room for further improvements in V_{oc} and PCE. In this work, we summarized the recent progress of the electron acceptors in the high-voltage systems with V_{oc} above 1.0 V and PCE above 8%. Different molecular design strategies containing the core, end groups, and π-bridges, as well as their substituents, have been employed to tune the energy levels and improve the photovoltage. However, it is still unclear how these subtle structural changes influence voltage and device performance.

Compared with the increasingly enriched material systems, the relative lack of the investigation on the physical origins of different energy loss pathways in OSCs comes upon a growing problem. Among the energy loss, the CT loss and the non-radiative recombination loss are the main loss pathways. The non-radiative recombination loss can drastically decrease as the up-shift of the CT state while the up-shifted CT state will decrease the charge generation efficiency. However, some systems have achieved an excellent balance between the high V_{oc} and charge generation efficiency by designing highly luminescent material and suppressing the possible recombination pathways. Unfortunately, these design methods are materials-dependent in most cases. It is necessary to focus on the photovoltaic mechanisms in OSCs.

The current data have demonstrated a bright future due to the efforts devoted to new material design and device fabrication. Combining the in-depth analysis of the existing materials with the device physics will guide us to design the promising materials. In the future, it is expected that these high-voltage OSCs will open a new avenue for indoor light recycling applications.

References

1 Liu, Q., Jiang, Y., Jin, K. et al. (2020). *Sci. Bull.* 65: 272.
2 Cui, Y., Yao, H., Zhang, J. et al. (2020). *Adv. Mater.* 32: 1908205.
3 Li, C., Zhou, J., Song, J. et al. (2021). *Nat. Energy* 6: 605.
4 Qin, J., Zhang, L., Zuo, C. et al. (2021). *J. Semicond.* 42: 010501.
5 Jin, K., Xiao, Z., and Ding, L. (2021). *J. Semicond.* 42: 060502.
6 Lin, Y., Magomedov, A., Firdaus, Y. et al. (2021). *ChemSusChem* https://doi.org/10.1002/cssc.202100707.
7 Liu, F., Zhou, L., Liu, W. et al. (2021). *Adv. Mater.* 33: 2100830.
8 Hwang, Y.-J., Courtright, B.a.E., Ferreira, A.S. et al. (2015). *Adv. Mater.* 27: 4578.
9 Chen, J.D., Li, Y.Q., Zhu, J. et al. (2018). *Adv. Mater.* 30: 1706083.
10 Zheng, Z., Hu, Q., Zhang, S. et al. (2018). *Adv. Mater.* 30: 1801801.
11 Feng, L.-W., Chen, J., Mukherjee, S. et al. (2020). *ACS Energy Lett.* 5: 1780.
12 Xiao, Z., Jia, X., Li, D. et al. (2017). *Sci. Bull.* 62: 1494.
13 Park, S.H., Roy, A., Beaupré, S. et al. (2009). *Nat. Photonics* 3: 297.
14 Zeng, Y., Li, D., Xiao, Z. et al. (2021). *Adv. Energy Mater.*: 2101338.
15 Cui, Y., Yao, H., Zhang, J. et al. (2019). *Nat. Commun.* 10: 2515.
16 Zhu, C., Yuan, J., Cai, F. et al. (2020). *Energy Environ. Sci.* 13: 2459.
17 Yao, J., Kirchartz, T., Vezie, M.S. et al. (2015). *Phys. Rev. Appl.* 4: 014020.
18 Liu, J., Chen, S., Qian, D. et al. (2016). *Nat. Energy* 1: 16089.
19 Mark Gruber, J.W., Klein, K., Hörmann, U. et al. (2012). *Adv. Energy Mater.* 2: 1100.
20 Xie, Y., Wang, W., Huang, W. et al. (2019). *Energy Environ. Sci.* 12: 3556.
21 Azzouzi, M., Yan, J., Kirchartz, T. et al. (2018). *Phys. Rev. X* 8: 031055.
22 Saito, T., Natsuda, S.-I., Imakita, K. et al. (2020). *Solar RRL* 4: 2000255.
23 Ji, Y., Xu, L., Hao, X., and Gao, K. (2020). *Solar RRL* 4: 2000130.
24 Benduhn, J., Tvingstedt, K., Piersimoni, F. et al. (2017). *Nat. Energy* 2: 17053.
25 Eisner, F.D., Azzouzi, M., Fei, Z. et al. (2019). *J. Am. Chem. Soc.* 141: 6362.
26 Bixon, M., Jortner, J., and Verhoeven, J.W. (1994). *J. Am. Chem. Soc.* 116: 7349.
27 Zhang, Y., Guo, X., Guo, B. et al. (2017). *Adv. Funct. Mater.* 27: 1603892.
28 Fan, Q., Liu, T., Gao, W. et al. (2019). *J. Mater. Chem. A* 7: 15404.
29 Afshin Hadipour, B.D.B. and Blom, P.W.M. (2007). *J. Appl. Phys.* 102: 074506.

30 Meng, L., Zhang, Y., Wan, X. et al. (2018). *Science* 361: 1094.
31 Ameri, T., Li, N., and Brabec, C.J. (2013). *Energy Environ. Sci.* 6: 2390.
32 Freunek, M., Freunek, M., and Reindl, L.M. (2013). *IEEE J. Photovoltaics* 3: 59.
33 Dayneko, S.V., Pahlevani, M., and Welch, G.C. (2019). *ACS Appl. Mater. Interfaces* 11: 46017.
34 Cutting, C.L., Bag, M., and Venkataraman, D. (2016). *J. Mater. Chem. C* 4: 10367.
35 Mori, S., Gotanda, T., Nakano, Y. et al. (2015). *Jpn. J. Appl. Phys.* 54: 071602.
36 Reich, N.H., Van Sark, W.G.J.H.M., Alsema, E.A. et al. (2009). *Sol. Energy Mater. Sol. Cells* 93: 1471.
37 Cui, Y., Yao, H., Zhang, T. et al. (2019). *Adv. Mater.* 31: 1904512.
38 Zhang, J., Li, Y., Huang, J. et al. (2017). *J. Am. Chem. Soc.* 139: 16092.
39 Baran, D., Kirchartz, T., Wheeler, S. et al. (2016). *Energy Environ. Sci.* 9: 3783.
40 Tang, A., Song, W., Xiao, B. et al. (2019). *Chem. Mater.* 31: 3941.
41 Lin, Y., Wang, J., Zhang, Z.G. et al. (2015). *Adv. Mater.* 27: 1170.
42 Jiang, X., Xu, Y., Wang, X. et al. (2017). *Polym. Chem.* 8: 3300.
43 Wang, Y., Yan, Z., Guo, H. et al. (2017). *Angew. Chem. Int. Ed.* 56: 15304.
44 Miao, J., Meng, B., Ding, Z. et al. (2020). *J. Mater. Chem. A* 8: 10983.
45 Yu, J.G.G., Hummelen, J.C., Wudl, F., and Heeger, A.J. (1995). *Science* 270: 1789.
46 Deng, D., Zhang, Y., Zhang, J. et al. (2016). *Nat. Commun.* 7: 13740.
47 Zheng, Z., Zhang, S., Zhang, J. et al. (2016). *Adv. Mater.* 28: 5133.
48 Zhang, Y., Deng, D., Wang, Z. et al. (2017). *Adv. Energy Mater.* 7: 1701548.
49 Liu, Y., Zhao, J., Li, Z. et al. (2014). *Nat. Commun.* 5: 5293.
50 Chen, H., Hu, Z., Wang, H. et al. (2018). *Joule* 2: 1623.
51 Zhao, J., Li, Y., Yang, G. et al. (2016). *Nat. Energy* 1: 15027.
52 Vandewal, K., Tvingstedt, K., Gadisa, A. et al. (2009). *Nat. Mater.* 8: 904.
53 Du, Z., Bao, X., Li, Y. et al. (2018). *Adv. Energy Mater.* 8: 1701471.
54 Wang, M., Wang, Z., Ma, W. et al. (2016). *Adv. Electron. Mater.* 2: 1600340.
55 Liao, S.-F., Chen, C.-T., and Chao, C.-Y. (2017). *ACS Macro Lett.* 6: 969.
56 Babics, M., Duan, T., Balawi, A.H. et al. (2019). *ACS Appl. Energy Mater.* 2: 2717.
57 Yao, H., Ye, L., Hou, J. et al. (2017). *Adv. Mater.* 29: 1700254.
58 Yu, R., Yao, H., Hong, L. et al. (2018). *Adv. Energy Mater.* 8: 1802131.
59 Li, W., Ye, L., Li, S. et al. (2018). *Adv. Mater.* 30: 1707170.
60 Zhang, J., Li, Y., Hu, H. et al. (2019). *Chem. Mater.* 31: 6672.
61 Liu, Y., Zhang, Z., Feng, S. et al. (2017). *J. Am. Chem. Soc.* 139: 3356.
62 Fan, Q., Xu, Z., Guo, X. et al. (2017). *Nano Energy* 40: 20.
63 Xu, X., Li, Z., Bi, Z. et al. (2018). *Adv. Mater.* 30: 1800737.
64 Xu, X., Yu, T., Bi, Z. et al. (2018). *Adv. Mater.* 30: 1703973.
65 Zhang, G., Xu, X., Bi, Z. et al. (2018). *Adv. Funct. Mater.* 28: 1706404.
66 Swick, S.M., Zhu, W., Matta, M. et al. (2018). *Proc. Natl. Acad. Sci. U.S.A.* 115: E8341.
67 Tang, Y., Sun, H., Wu, Z. et al. (2019). *Adv. Sci.* 6: 1901773.

68 Yang, J., Uddin, M.A., Tang, Y. et al. (2018). *ACS Appl. Mater. Interfaces* 10: 23235.
69 Hao, M., Liu, T., Xiao, Y. et al. (2019). *Chem. Mater.* 31: 1752.
70 Chen, S., Zhang, G., Liu, J. et al. (2017). *Adv. Mater.* 29: 1604231.
71 Zhang, S., Qin, Y., Zhu, J., and Hou, J. (2018). *Adv. Mater.* 30: 1800868.
72 Su, W., Fan, Q., Guo, X. et al. (2017). *Nano Energy* 38: 510.
73 Wadsworth, A., Ashraf, R.S., Abdelsamie, M. et al. (2017). *ACS Energy Lett.* 2: 1494.
74 Baran, D., Gasparini, N., Wadsworth, A. et al. (2018). *Nat. Commun.* 9: 2059.
75 Baran, D., Ashraf, R.S., Hanifi, D.A. et al. (2017). *Nat. Mater.* 16: 363.
76 Chen, S., Liu, Y., Zhang, L. et al. (2017). *J. Am. Chem. Soc.* 139: 6298.
77 Oh, S., Song, C.E., Lee, T. et al. (2019). *J. Mater. Chem. A* 7: 22044.
78 Cheng, P., Zhang, M., Lau, T.K. et al. (2017). *Adv. Mater.* 29: 1605216.
79 Ma, Y., Zhang, M., Tang, Y. et al. (2017). *Chem. Mater.* 29: 9775.
80 Yang, Y., Wang, J., Xu, H. et al. (2018). *ACS Appl. Mater. Interfaces* 10: 18984.
81 Bai, W., Xu, X., Li, Q. et al. (2018). *Small Methods* 2: 1700373.
82 Kolhe, N.B., West, S.M., Tran, D.K. et al. (2019). *Chem. Mater.* 32: 195.
83 Holliday, S., Ashraf, R.S., Nielsen, C.B. et al. (2015). *J. Am. Chem. Soc.* 137: 898.
84 Holliday, S., Ashraf, R.S., Wadsworth, A. et al. (2016). *Nat. Commun.* 7: 11585.
85 Xiao, B., Tang, A., Zhang, J. et al. (2017). *Adv. Energy Mater.* 7: 1602229.
86 Xiao, B., Tang, A., Yang, J. et al. (2017). *ACS Macro Lett.* 6: 410.
87 Tang, A., Xiao, B., Wang, Y. et al. (2018). *Adv. Funct. Mater.* 28: 1704507.
88 Chen, Y., Zhang, Q., Du, M. et al. (2019). *ACS Appl. Polym. Mater.* 1: 906.
89 Xiao, B., Geng, Y., Tang, A. et al. (2019). *Solar RRL* 3: 1800332.
90 Chen, Y., Jiang, X., Chen, X. et al. (2019). *Macromolecules* 52: 8625.
91 Xu, Y., Cui, Y., Yao, H. et al. (2021). *Adv. Mater.* 33: 2101090.
92 Tang, A., Xiao, B., Chen, F. et al. (2018). *Adv. Energy Mater.* 8: 1801582.
93 Wang, X., Tang, A., Yang, J. et al. (2020). *Sci. China Chem.* 63: 1666.
94 Dai, T., Lei, P., Zhang, B. et al. (2021). *ACS Appl. Mater. Interfaces* 13: 21556.
95 An, N., Cai, Y., Wu, H. et al. (2020). *Adv. Mater.* 32: 2002122.
96 Nie, Q., Tang, A., Cong, P. et al. (2020). *J. Phys. Chem. C* 124: 19492.
97 Dai, T., Lei, P., Zhang, B. et al. (2021). *ACS Appl. Mater. Interfaces* 13: 30756.
98 Ji, H., Li, J., Du, M. et al. (2021). *J. Phys. Chem. C* 125: 10876.
99 Luo, Z., Liu, T., Ma, R. et al. (2020). *Adv. Mater.* 32: 2005942.
100 Peng, F., An, K., Zhong, W. et al. (2020). *ACS Energy Lett.* 5: 3702.
101 Sun, H., Liu, B., Koh, C.W. et al. (2019). *Adv. Funct. Mater.* 29: 1903970.
102 Sun, H., Liu, B., Wang, Z. et al. (2020). *J. Mater. Chem. C* 8: 4012.
103 Sun, H., Tang, Y., Koh, C.W. et al. (2019). *Adv. Mater.* 31: 1807220.
104 Zhang, Z., Miao, J., Ding, Z. et al. (2019). *Nat. Commun.* 10: 3271.

9

Single-Component Organic Solar Cells

Guitao Feng[1,3], Yiting Guo[2], and Weiwei Li[1]

[1]Beijing University of Chemical Technology, Beijing Advanced Innovation Center for Soft Matter Science and Engineering, State Key Laboratory of Organic-Inorganic Composites, ChaoYang District, BeiSanhuan East Road, 15, Beijing, 100029, P.R. China
[2]Henan University of Technology, College of Grain and Food Science, Lianhua Street, High-tech Zone, Zhengzhou, Henan, 450001, P.R. China
[3]Institute of Chemistry, Chinese Academy of Sciences, Zhongguancun North First Street 2, Beijing, 100190, P.R. China

9.1 Introduction

Organic solar cells (OSCs) as sustainable photovoltaic technology have shown promising commercial prospects, such as indoor energy for the Internet of Things, solar powered windows, and greenhouses [1, 2]. Besides, flexible and lightweight OSCs can be potentially applied to portable facilities, electronic skins, etc. [3–5]. All of these applications rely on three key parameters: power conversion efficiencies (PCEs), stability, and cost [6]. Currently, two-component photo-active layers containing physically blended electron donors (Ds) and acceptors (As), namely the "bulk-heterojunction" (BHJ) active layers (ALs), predominate the OSC field [7–12]. Due to the fast development of the fused-ring electron acceptors, the certified PCEs of bulk-heterojunction organic solar cells (BHJOSCs) are improved to an unprecedented and unexpected degree, and PCEs over 18% have been obtained in the state-of-the-art studies [13–15]. Nevertheless, the performance of BHJOSCs tends to be degraded under thermal and light stress, and the long-term stability in BHJOSCs is still difficult to realize in low-cost solution-processing manufacture. The degree of performance degradation in the OSCs depends on the stability of the conjugated materials in the photo-active layers, the interfacial layers, and the electrodes. It is acknowledged that, after a long-term operation, the donor/acceptor components tend to convert into thermodynamically-stable microstructures, typically consisting of large-scale self-aggregated domains, which are detrimental to charge separation and transport [16]. A straight approach to prevent self-aggregation in the BHJ systems is to covalently connect the donor with acceptor units, which is used to fabricate the so-called "single-component organic solar cells (SCOSCs)" [17]. Recent studies have demonstrated the significantly enhanced photo and thermal stability of the SCOSCs, compared with related

Organic Solar Cells: Materials Design, Technology and Commercialization, First Edition. Edited by Liming Ding.

BHJOSCs [18, 19]. Therefore, degradation of photovoltaic performance is alleviated in SCOSCs, suggesting a promising potential technology for the fabrication of large-area rugged devices.

In the early development stage of SCOSCs, pristine conjugated materials without the intramolecular heterojunction structures were introduced to manufacture the photovoltaic devices. However, the PCEs of the SCOSCs based on pristine conjugated materials were below 0.1% [20], illustrating that dissociation of the photo-induced exciton is inefficient. With the development of the BHJ concept, single component (SC)-conjugated materials with covalently linked electron-donating and electron-withdrawing groups started to attract attention and materials with the intramolecular heterojunction structures, such as molecular dyads, block copolymers (BCPs), and double-cable conjugated polymers (DCPs), were prepared for the high-performance SCOSCs. Although much progress has been achieved during the last two decades, the efforts spent on promoting development of the SCOSCs were less, and thus related cell performance is much lower than those of BHJOSCs. Herein, we will summarize the development of SCOSCs from the aspect of conjugated materials and microstructures, analyze the scientific challenges, and then discuss the opportunity of SCOSCs.

9.2 Single-Component Conjugated Materials for SCOSCs

Single-component conjugated materials with the intramolecular D–A structures, such as molecular dyads and conjugated polymers, have been widely used as the photo-active layers to fabricate SCOSCs, which exhibited the advantages of definite structures and excellent reproducibility [17]. Additionally, the development of donor units that appeared in SCOSCs was influenced by the development of conjugated molecules for BHJOSCs. As for the acceptor units, the fullerene and its derivatives as the efficient acceptor predominated the evolution of the performances of SCOSCs while the perylene bisimides (PBI) showed outstanding performance in some studies.

9.2.1 Molecular Dyads

9.2.1.1 Fullerene-Based "In-Chain" Molecular Dyads

Plenty of dyads combining fullerenes with molecular π-conjugated systems have been intensively studied for a long period. As shown in Figures 9.1 and 9.2, many fullerene-based "in-chain" molecular dyads have been reported. In the "in-chain" molecule dyads, the fullerene units were directly connected to the donor units.

Two fullerene-based dyads (**SM01** and **SM02**) [21, 22] reported by Nierengarte and coworkers were designed as SC materials for SCOSCs, in which two phenylenevinylene oligomers (n-PVs) were covalently linked with C_{60} through a pyrrolidine group. The best results were obtained with the PCE of approximately 0.3% under monochromatic irradiation. Scheme 9.1 illustrated the synthetic route of dyad **SM02**, and the formation of the donor segments with quantified aldehyde groups is the key process in the molecular synthesis.

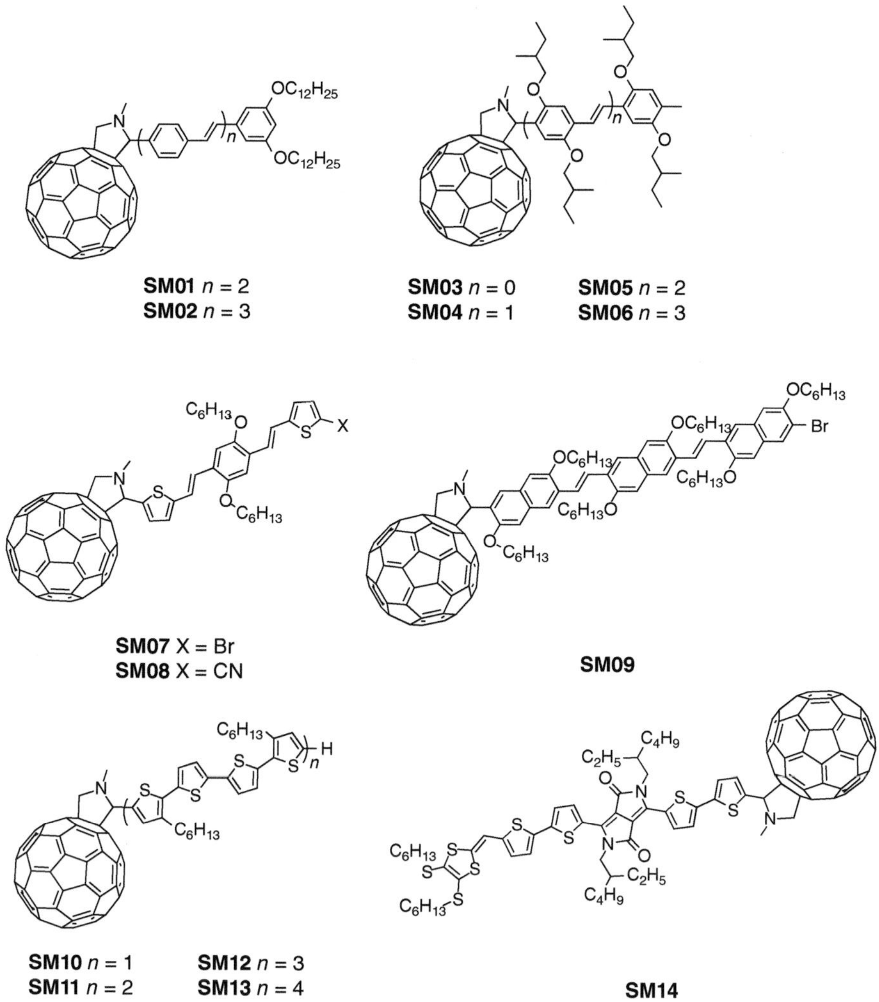

Figure 9.1 Chemical structures of fullerene-based "in-chain" molecular dyads.

Janssen and coworkers carried out a series of *n*-PV-fullerene dyads (**SM03–06**) for application in SCOSCs [23]. The best result was achieved in the **SM06**-based cells, providing a short-circuit current density (J_{sc}) of 0.235 mA cm^{-2} and the PCE of 0.04% under a white light with the power of 65 mW cm^{-2}. To improve the performance of SCOSCs, Sariciftci and coworkers developed three molecular dyads **SM07–09**, in which the thiophene or naphthalene groups were introduced to replace the benzene in the *n*-PV. The optimized devices showed the best PCE of 0.2% under the irradiation of white light (80 mW cm^{-2}) [24].

Although molecular dyads contained *n*-PV and the derivatives have been used for SCOSCs, the low performance indicates that PV was not a good backbone to construct these molecules. Compared with *n*-PVs, the oligothiophenes showed broader photoresponse, higher hole mobilities, and better crystallinity. Therefore, the

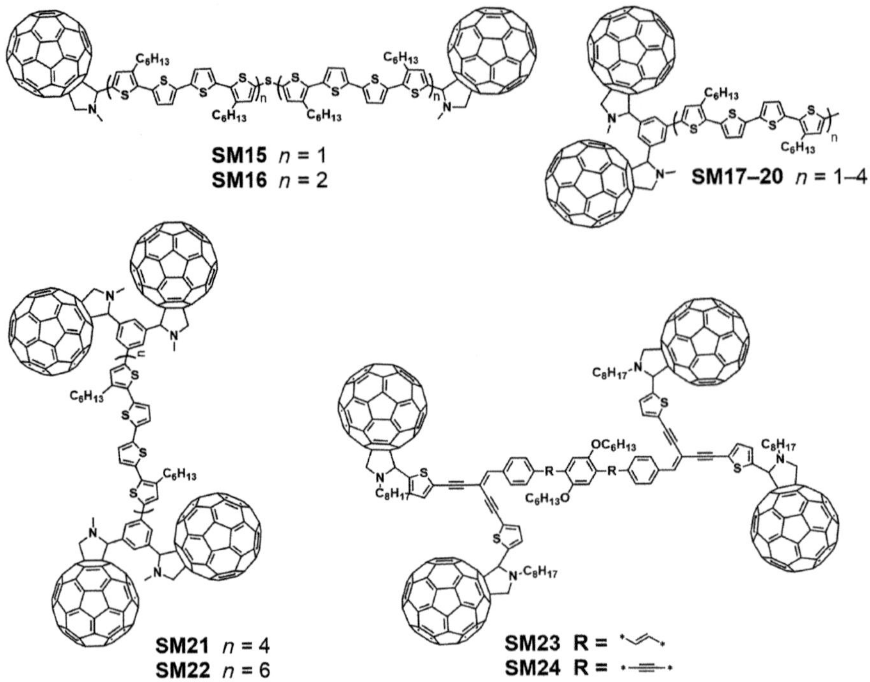

Figure 9.2 Chemical structures of multiple fullerene-based "in-chain" molecular dyads.

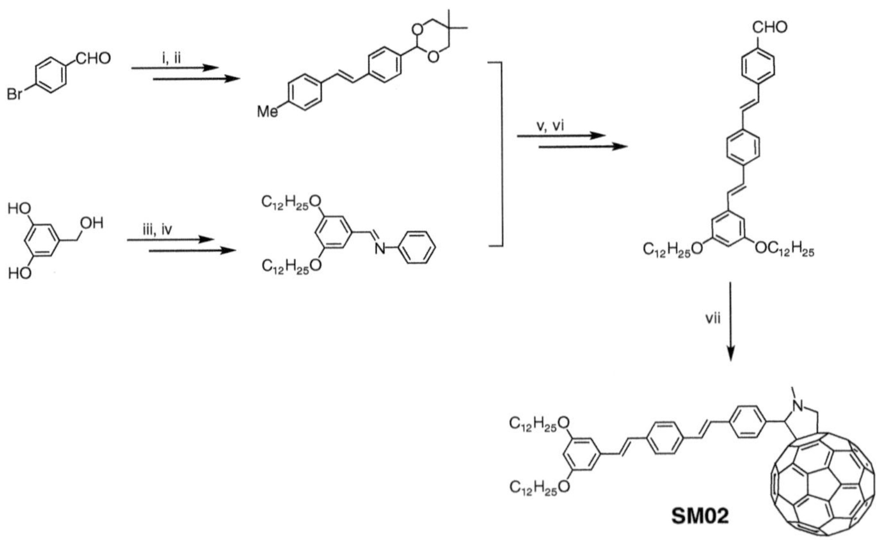

Scheme 9.1 The synthetic route of fullerene-based "in-chain" molecular dyad **SM02**. (i) *p*-methylstyrene, $Pd(OAc)_2$, tri-*o*-tolylphosphine, Et_3N, Xylene, 130 °C, 48 hours; (ii) 2,2-dimethylpropane-1,3-diol, *p*-TsOH, C_6H_6, reflux, Dean-Stark trap, 24 hours; (iii) MnO_2, CH_2Cl_2, rt, 1 hours; (iv) aniline, C_6H_6, reflux, Dean-Stark trap, 24 hours; (v) *t*-BuOK, DMF, 80 °C, 2 hours; (vi) CF_3CO_2H, CH_2Cl_2, H_2O, room temp., 4 hours; (vii) C_{60}, sarcosine, toluene, reflux, 16 hours.

molecular dyads based on oligothiophenes grafted with fullerene were investigated to improve the photovoltaic performance in SCOSCs. Yamashiro et al. reported the design and synthesis of molecular dyads that contained oligothiophenes as D unit and single C_{60} as A unit (**SM10–13**) [25, 26]. These molecules were used as photo-active layers to construct SCOSCs, in which the effect of the length of oligothiophenes on the photovoltaic performance was carefully studied. The research illustrated that a relatively high PCE of 0.4% was achieved with 16T-C_{60} (**SM13**) as the active layer in the SCOSC with a long conjugated backbone.

Singh and coworkers further introduced highly crystalline oligothiophene–diketopyrrolopyrrole (DPP) into the conjugated backbone to extend the absorption spectra to the near-infrared (NIR) region, and the corresponding molecular dyad **SM14** showed the broad absorption band in the 300–800 nm region [27]. The NIR photoresponse in the as-cast films of **SM14** was essentially dominated by the absorption of the bis-dithienyl-DPP with a maximum absorption peak at 650 nm. SCOSCs fabricated with the configuration of (indium tin oxide) ITO/MoO_x/**SM14**/Ca/Al were tested under monochromatic and white light irradiation. The photoresponse extends from 400 to 800 nm with a maximum external quantum efficiency (EQE) of 0.38 at 670 nm. Under simulated solar light, the cell gives a J_{sc} of 6.70 mA cm^{-2} and a PCE of 2.17%, which ranks among the highest values reported for molecular dyads.

Despite the molecular engineering of the donor segments, another rational method to improve the efficiency of the SCOSCs is to change the number of acceptors in SC materials. Under the guidance of this principle, two or more C_{60} were introduced into the molecular dyads, as shown in Figure 9.2. The synthesis of **SM15, SM16** was reported by T. Yamashiro and coworkers [28]. Also, N. Negishi et al. reported a series of oligothiophenes grafted with two C_{60} at the end (**SM17–20**) [29]. With the increasing content of C_{60}, the device containing **SM20** as the photo-active layer gave a PCE of 0.65%. Nonetheless, poor device performances were obtained when another two C_{60} were added (**SM21–22**) [30]. In another study, Janssen and coworkers reported the synthesis of molecular dyads based on the *p*-styrene (**SM23**) or *p*-phenylacetylene (**SM24**) oligomers as the donor core, with which four C_{60} were introduced [31]. A very low PCE of 0.015% was obtained in the SCOSCs with the structure of Glass/ITO/PEDOT:PSS/AL/LiF/Al. The photovoltaic properties have not been significantly improved as predicted with the increasing proportion of C_{60}, which may originate from unbalanced carrier mobilities and poor nanophase separation.

9.2.1.2 Fullerene-Based "Side-Chain" D–A Molecular Dyads

In the fullerene-based "in-chain" D–A molecules, conjugated donor parts were directly linked with C_{60} through the carbon–carbon single bond, which results in inappropriate phase separation and poor charge transport. Hence, researchers introduced flexible chains as linkers between the donor and acceptor segment to improve the photovoltaic performance in SCOSCs. Many studies emphasized that the nature and length of the flexible linkers are the determining factors for the nanophase separation of D and A parts in these materials.

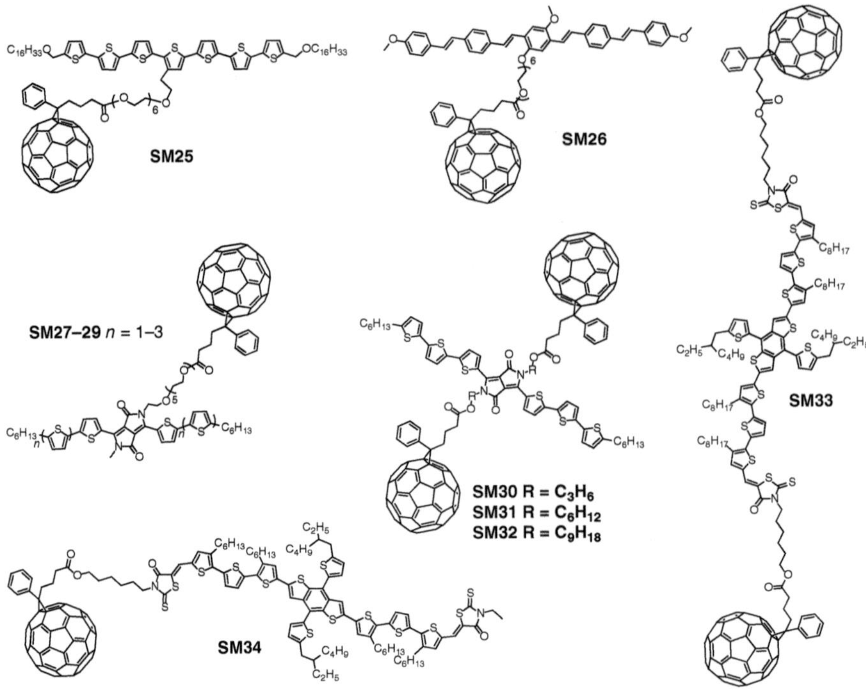

Figure 9.3 Chemical structures of fullerene-based "side-chain" D–A molecular dyads.

As shown in Figure 9.3, K. Hashimoto and coworkers described the synthesis of the molecular dyad **SM25**, combining the thiophene heptamer as donor with $PC_{61}BM$ attached at the median thiophene ring via an oligo(oxyethylene) chain as acceptor in 2007 [32, 33]. Interestingly, the absorption of **SM25** showed a blue shift in the as-cast film due to the formation of *H*-aggregate of oligothiophenes in the solid state. The device with the configuration of ITO/PEDOT:PSS/**SM25**/Al gave a J_{sc} of 0.93 mA cm^{-2} and a PCE of 0.15%. Furthermore, **SM26**, an *n*-PV containing five phenyl rings with $PC_{61}BM$ attached to the middle of the chain by a polyether linker, was developed by the same group. A PCE of 1.28% was achieved in the devices that contained **SM26** as the photo-active layer, and the sharply increased PCE compared with **SM25** confirms the importance of the molecular packing and crystallinity [34].

The influence of D/A ratio in SCOSCs was studied through a series of DPP-oligothiophene/fullerene dyads (**SM27–29**) developed by Tajima and coworkers [35]. Resulting from the oligothiophene elongated, the J_{sc} and FF (fill factor) of SCOSCs increased and the optimized device gave the best PCE of 1.11%. Besides, the length of the linkers between donor and acceptor units was investigated by B. Ma and coworkers. Several DPP-containing oligothiophene-fullerene triads (**SM30–32**) were designed and synthesized to verify the effect of varying lengths of linkers [36]. The charge-transport properties of **SM30–32** were determined in field-effect transistor (FET) devices. These three molecules exhibited the appropriate electron mobilities around 1.5×10^{-3} $cm^2\,V^{-1}\,s^{-1}$ while the hole mobilities were rarely obtained. The best PCE of 0.49% was achieved in the **SM32**-based devices. Despite the relatively poor performance, the results indicated that the performance of the device was associated with the alkyl spacers. And in the control experiment, the BHJOSCs with the P3HT as the donor and **SM30–32** as the acceptor were prepared and gave the best PCE of 2.41%. This result revealed that the excessive phenyl-C_{61}-butyric acid methyl ester (PCBM) in the SCOSC system adversely affected the aggregation of the donor units.

Nguyen et al. [37, 38] reported the synthesis of a different type of dumbbell molecule (**SM33**) consisting of a conjugated block with a central di(thiophen-yl) benzo[1,2-b:4,5-b′]dithiophene (DT-BDT) unit with two dioctyl terthienyl side blocks and two rhodanine terminal moieties on which PCBM was linked by 6-carbon alkyl spacer, and the synthetic route of **SM33** was illustrated in Scheme 9.2.

The **SM33** presents a broad absorption band in the 300–700 nm region with a maximum peak at 557 nm in solution. And SCOSCs with the structure of ITO/PEDOT:PSS/**SM33**/ZnO/Al were fabricated and evaluated under simulated solar light. The best devices provided a high $J_{sc} > 7.0\,\text{mA cm}^{-2}$ and a PCE of 2.31%. Stability tests performed at 80 °C under an inert atmosphere revealed that the **SM33**-based SCOSC presented excellent morphological stabilities, with no degradation of performance after annealing at 80 °C for 100 hours. Recently, Min and coworkers described the donor–acceptor type dyads (**SM34**) in which a single $PC_{71}BM$ unit as the acceptor was attached to the 3T–BDT–3T donor block [39]. With the decrease of fullerene content, the best devices with **SM34** gave a higher J_{sc} of 7.5 mA cm^{-2} and a PCE of 3.22% than that of cells based on **SM33**. The results demonstrate that the asymmetric molecular design is an efficient route to develop new molecular dyads toward high-performance SCOSCs.

9.2.1.3 PBI-Based Molecular Dyads

Besides the fullerene-based molecules, the perylene bisimide (PBI) and its derivatives were also used as the acceptor to design molecular dyads for SCOSCs. Figure 9.4 illustrates the chemical structures of dyads with PBIs for application in SCOSCs.

In 2009, Geng and coworkers reported the synthesis of three smectic liquid crystal molecules (**SM35–37**) with the structure of substituted fluorene-*alt*-bithiophene as the donor and single PBI as the acceptor [40] (Scheme 9.3).

Devices with the architecture of ITO/PEDOT:PSS/**SM37**/HBL/LiF/Al, where the hole-blocking layer (HBL) is *N*,*N*′-bis(1-ethylpropyl) substituted PBI, were fabricated. The PCE was significantly improved by thermal or solvent annealing processes, and the devices based on **SM37** with the longest conjugated backbone provided the best results. The PCE was 0.50% for the pristine thin film, increasing to 0.70% after thermal treatment, and to 1.50% with J_{sc} of 4.49 mA cm^{-2}, after solvent annealing and insertion of an HBL in the device. The same group also synthesized a series of similar conjugated materials (**SM38–40**) [41], in which the hexyl chain attached at the nitrogen atom of PBI was replaced by a branched neopentyl group. This modification improved the photovoltaic performance with an enhanced PCE of 1.75%. After that, Geng and coworkers extended their approach to the synthesis of related systems in which the terminal PBI was

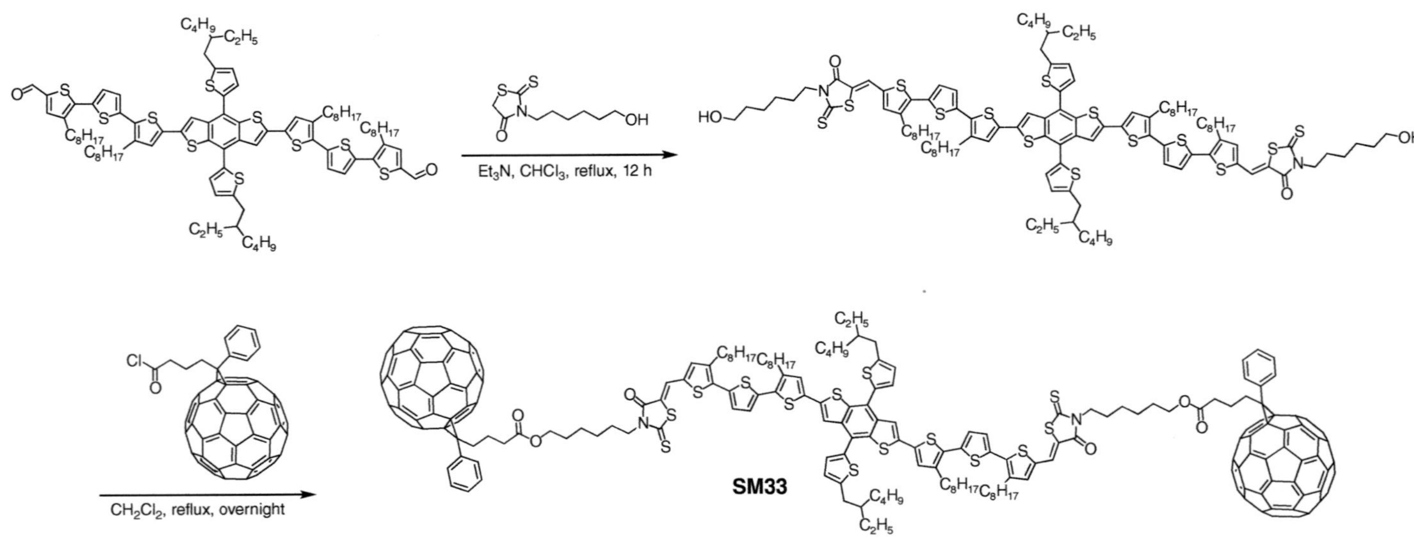

Scheme 9.2 The synthetic route of fullerene-based "in-chain" triad **SM33**.

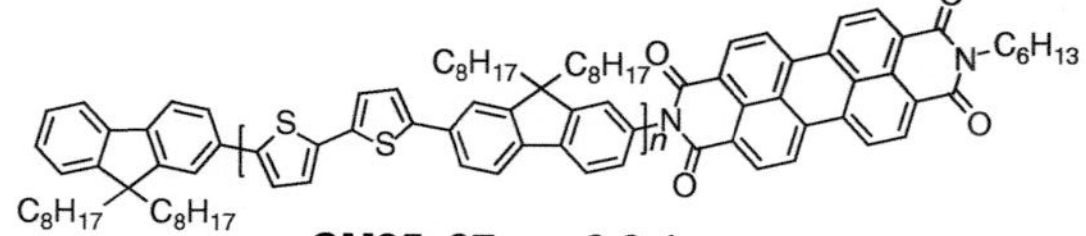

SM35–37 n = 2,3,4

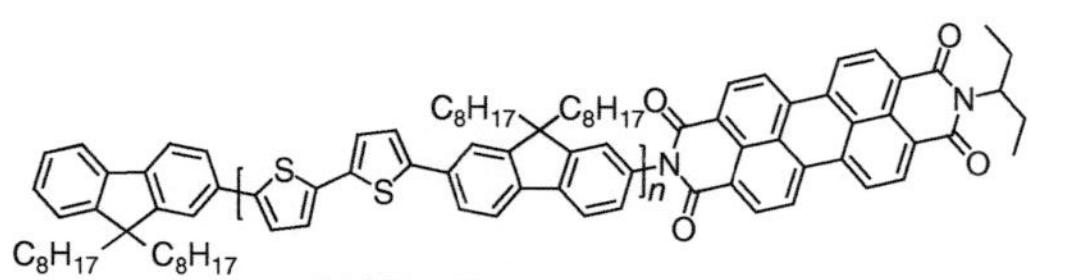

SM38–40 n = 2,3,4

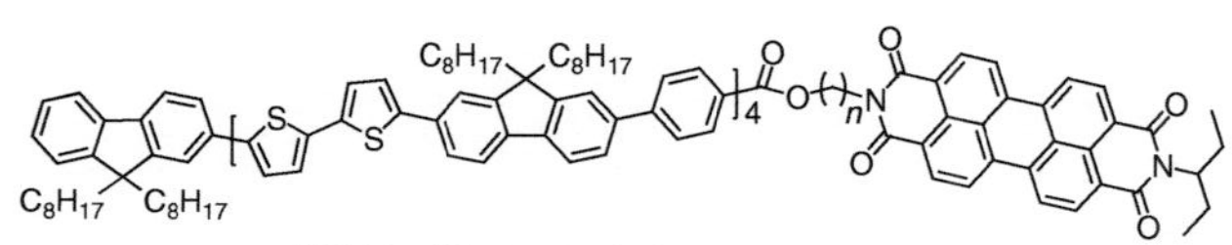

SM41–43 n = 2, 4, 6

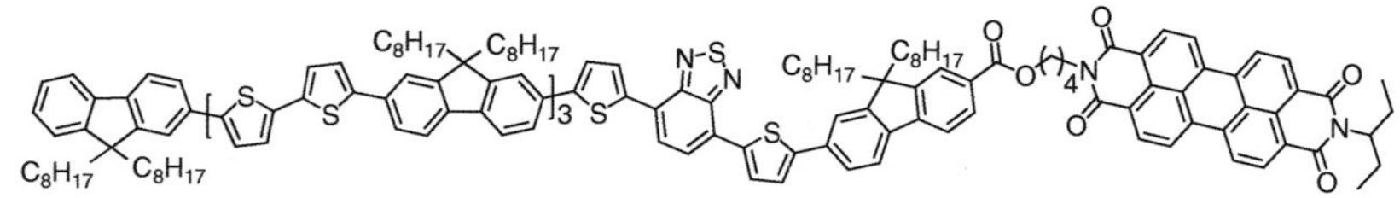

SM44

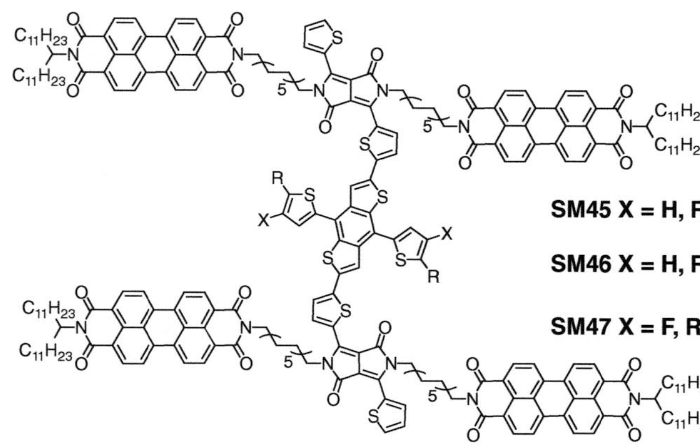

SM45 X = H, R =

SM46 X = H, R =

SM47 X = F, R =

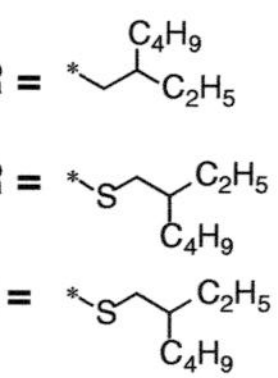

Figure 9.4 Chemical structures of PBI-based molecular dyads.

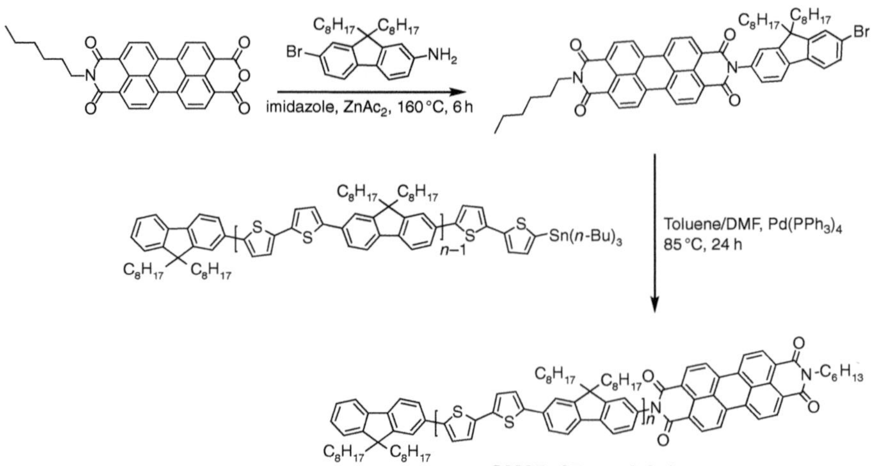

SM35–37 *n* = 2,3,4

Scheme 9.3 The synthetic route of PBI-based molecular system **SM35–37**.

attached to the hybrid conjugated backbones linked with alkyl spacers containing two, four, and six carbons (**SM41–43**) [42]. The absorption spectra of **SM43** in solution are identical to that of **SM40**, indicating that even without an alkyl spacer, the electronic interactions between the π-conjugated donor chain and the PBI acceptor are negligible in **SM40**. This is probably due to a large twist angle around the connecting bond due to steric interactions. The photovoltaic properties with **SM41–43** were characterized with the same type of device configuration as compounds **SM38**. Three compounds lead to rather comparable results with PCE values in the range of 2.0%, which is consistent with the EQE responses. To extend absorption of the conjugated donor, benzodithiazole (BT) was introduced (**SM44**), and this modification enhanced the absorption in the 500–600 nm region, resulting in a further improvement of performances with a J_{sc} of 5.32 mA cm^{-2} and a PCE of 2.70%. Recently, W. Li and coworkers have reported a new series of "double-cable type" molecules (**SM45–47**) with BDT–DPP–BDT as donor block and pendant PBIs as acceptor groups [43]. The pendant PBI groups were connected to the conjugated backbone by flexible dodecyl linkages while the effects of substitution of the lateral thienyl group of DT-BDT were investigated. These molecules were synthesized by *Stille* coupling using a dibromo-2T-DPP carrying two PBI side groups with three different distannyl-BDT compounds substituted by 2-(2-ethylhexyl)-thiophene, 2-((2-ethylhexyl)thio)-thiophene, and 2-((2-ethylhexyl)thio)-3-fluorothiophene. With systematic studies on the effect of the sulfur and fluorine atoms in the donor blocks, the results revealed that the additional sulfur atom in the side chain of 2T-DPP unit (**SM46**) enhanced the crystallinity of materials and elevated photovoltaic properties with a PCE of 2.52%.

9.2.2 Block Copolymers

BCPs have been widely studied due to their unique characteristics to obtain multifunction through self-assembly. As shown in Figures 9.5 and 9.6, BCPs with the intramolecular D–A structures were used as the light-sensitive layer in the SCOSCs. Besides, the morphologies have been meticulously investigated, and an encouraging SCOSC performance based on BCPs was achieved, exceeding 6% in the state-of-the-art study. Figure 9.5 presents the BCPs applied in the SCOSCs, which contain oligothiophenes or *n*-PVs as the donor blocks.

One of the first examples of D–A block polymers was reported in 2000 by Janssen and coworkers [44]. Two BCPs (**BCP1, BCP2**) with the *n*-PV blocks as the donor segment and PBI as the acceptor segment were prepared via the Pd-catalyzed *Suzuki* cross-coupling reaction. Further spectroscopic studies demonstrated that fast photo-induced charge transfer (CT) was obtained in the two systems both in solution and as-cast films while serious recombination also existed, especially in solution. The SCOSCs with the configuration of ITO/PEDOT:PSS/AL/LiF/Al gave a J_{sc} of 0.008 mA cm^{-2} (**BCP1**) and 0.012 mA cm^{-2} (**BCP2**), respectively. Another similar BCP (**BCP3**) was reported by Mikroyannidis et al. through the *Heck* coupling reaction [45]. Unfortunately, the photovoltaic properties of **BCP3** analyzed in SCOSCs only showed a PCE of 0.02%. Briefly, in the D–A copolymers present above, the alkyl linkages were used to connect the *n*-PVs and PBI macrosegments, leading to unappropriated microphase-separated nanostructures and poor performance.

A full-conjugated copolymer (**BCP4**) combining oligothiophenes as donor block and PDPP3T oligomers as acceptor block was developed by B.C. Thompson and coworkers [46]. For the first time, the donor and acceptor macrosegments of the block copolymer present two independent domains in the film state, indicating that this molecular design principle can be efficient to achieve high PCEs in BCP-based SCOSCs. This result also demonstrated that well-ordered molecule orientation and appropriate phase separation can be realized through rational structural designs.

Under the same guidelines, R. Verduzco and coworkers reported the synthesis of the all-conjugated block copolymer **BCP5** as the active material for SCOSCs [47]. **BCP5** was prepared through the GRIM (*Grignard* metathesis), in which a bromo end-functionalized P3HT macroreagent was synthesized and then *Suzuki–Miyaura* polycondensation was used to prepare the polymers. The morphology of **BCP5** films was further investigated by the X-ray scattering, in which the in-plane lamellar morphologies of alternating D and A domains with a diameter of 9 nm were obtained in the thin film. The photovoltaic performance was found to significantly depend on the thermal annealing (TA) treatment. The PCE increased from 1.5% after annealed for 20 minutes at 100 °C to 2.7% for 10 minutes at 165 °C, and the best device gave a J_{sc} of 5.20 mA cm^{-2} and a PCE of 3.10%, which is among the highest PCEs for SCOSCs based on diblock copolymers in that period. Interestingly, all these devices present high V_{oc} (open-circuit voltage) values in the range of 1.10–1.20 V.

A library of copolymers **BCP6** with various substituents on the fluorene and thiophene units were also developed by Verduzco and coworkers to analyze the effects on the solubility and packing of the polymers [48]. Despite a large number

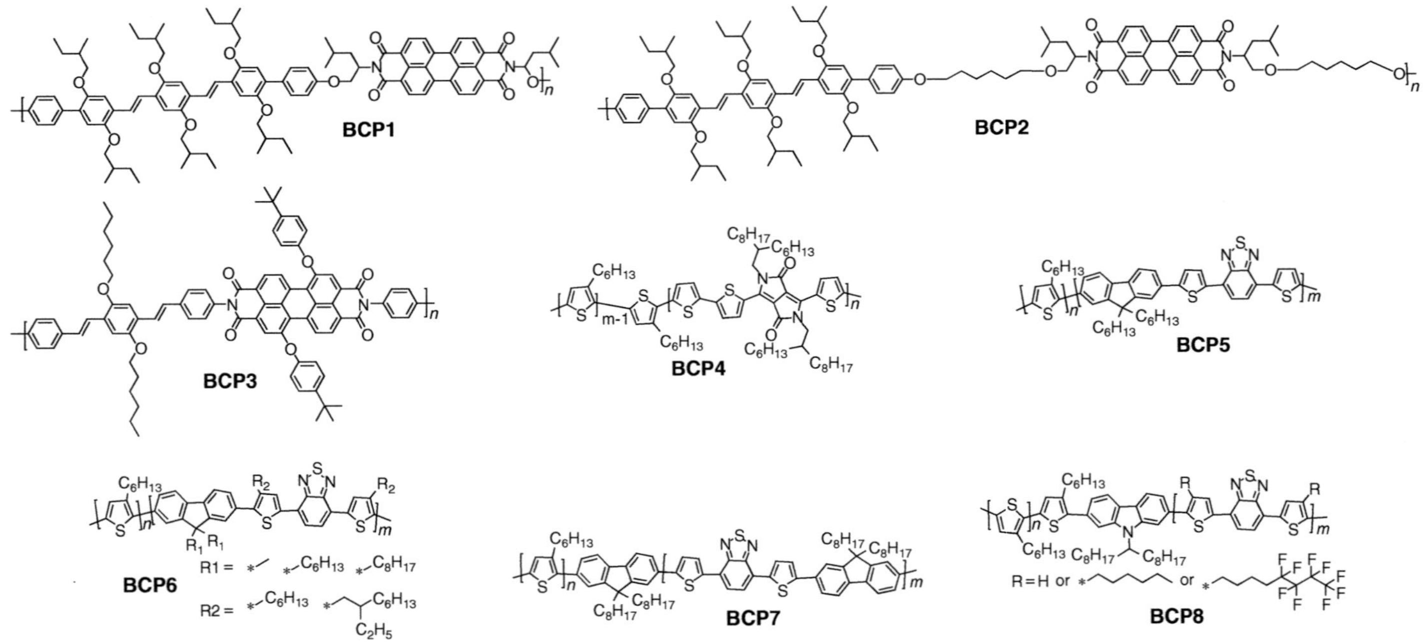

Figure 9.5 Chemical structures of block conjugated copolymers for SCOSCs (Part 1).

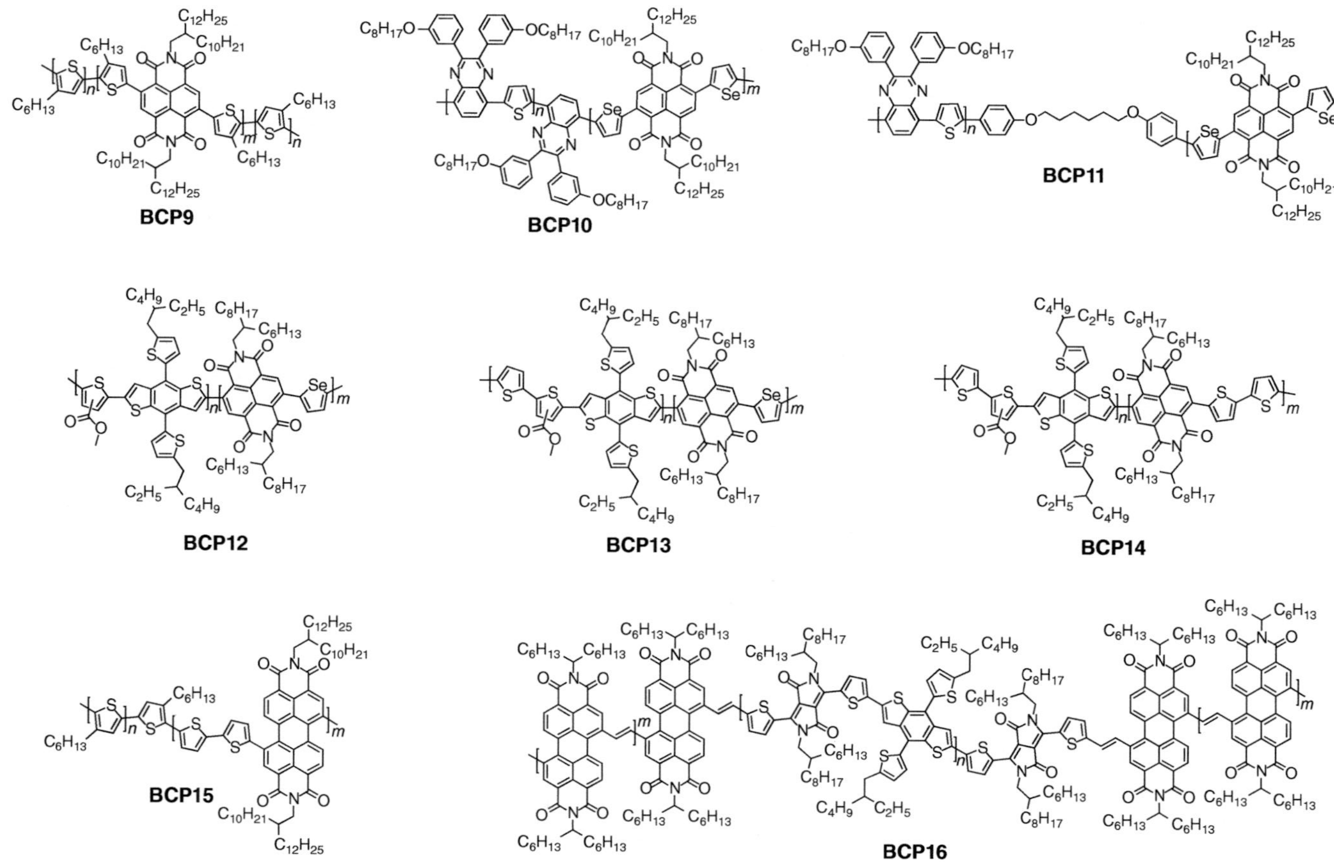

BCP9

BCP10

BCP11

BCP12

BCP13

BCP14

BCP15

BCP16

Figure 9.6 Chemical structures of block conjugated copolymers for SCOSCs (Part 2).

of copolymers, the added alkyl side units on the thiophene units corrupted the photovoltaic properties and the PCEs were inferior to 0.10%, indicating that the introduction of solubilizing alkyl chains on the thiophene rings has a detrimental effect on phase segregation.

Further studies on the effect of linkage between the D and A blocks were investigated by Verduzco's group and Gomez's group [49, 50]. A new copolymer (**BCP7**) in which the P3HT donor block and the dithienyl benzothiadiazole acceptor block were connected by a fluorene segment was compared with the same alkyl-substituted poly(3-hexylthiophene-2,5-diyl)-block-poly((9,9-dioctylfluorene)-2,7-diyl-*alt*-[4,7-bis(thiophen-5-yl)-2,1,3-benzothiadiazole]-2′,2″-diyl) (P3HT-FTBTF) (**BCP6**). Although **BCP6** exhibited lower molecular weight and higher P3HT content, both copolymers present rather similar absorption spectra. In the solar cell tests, the insertion of the fluorene bridge was reported to produce a large increase of PCEs from 0.05% for **BCP6** to 2.24% for **BCP7**, due to more efficient charge separation in the fluorene-bridged copolymer.

In another study, M. Sommer and coworkers described the synthesis of a new type of copolymer **BCP8** including the semifluorinated octyl-substituted dithienyl benzothiadiazole as the acceptor block and P3HT macroreagent as the donor block [51]. Compared with the alkyl-substituted copolymers, the introduction of partially fluorinated alkyl chains increased the *Flory–Huggins* interaction parameters between the donor and acceptor blocks, which led to almost full recrystallization of the P3HT block in the copolymers. SCOSCs with the configuration of ITO/PEDOT:PSS/active layers/Al were fabricated, and the **BCP8**-based device gave a higher PCE of 0.95%.

A recent study investigated by Gomez and coworkers presented the point that the electron transfer in the acceptor block can be the key parameter that limited the performance of SCOSCs. It was also emphasized that both donor and acceptor blocks should be carefully designed to achieve ambipolar charge transport and maximize the output power of photovoltaic cells. Under this principle, several D–A BCPs contained naphthalene diimide (NDI), or PBI groups were reported to enhance electron transport property as well as the nanophase separation.

In 2012, Nakabayashi et al. reported pioneering research in which "D–A–D" type BCPs (**BCP9**) with the end-functional P3HT as donor part and naphthalimide-dithiophene as acceptor part were prepared. A wide photoresponse ranging from 350 to 850 nm was obtained in the as-cast **BCP9** films, and the onset of absorption was enhanced to 890 nm after thermal annealing, suggesting an optimized self-aggregation in the post-processing condition. This work provided a fundamental example for the NDI-based copolymers and showed its potentiality for applications in SCOSCs.

Choi and coworkers demonstrated a new series of D–A BCPs without oligothiophenes as the donor blocks. Two types of copolymers composed of quinoxaline–thiophene and naphthalene diimide–selenophene blocks were synthesized, in which the donor and acceptor blocks are connected by flexible alkyl spacers (**BCP11**) or directly connected (**BCP10**) [52]. Besides, both the polymers contained a 1 : 1 ratio of donor and acceptor blocks measured by gas-phase chromatography. Photovoltaic cells with copolymers were fabricated in the inverted

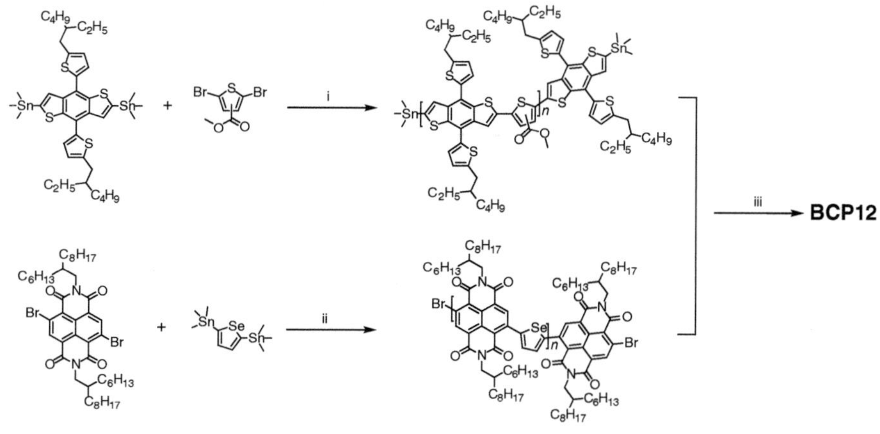

Scheme 9.4 The synthetic route of block copolymer **BCP12**. (i) $Pd_2(dba)_3$, P(*o*-tolyl)$_3$, toluene, and chlorobenzene under microwave irradiation at 180 °C; (ii) $Pd(PPh_3)_4$ in toluene at 100 °C, and (iii) $Pd_2(dba)_3$ and P(*o*-tolyl)$_3$ and toluene at 100 °C.

architecture ITO/ZnO/AL/MoO_3/Al. SCOSCs based on **BCP10** presented discouraging photovoltaic properties with $J_{sc} = 1.30\,mA\,cm^{-2}$, FF = 0.35, and PCE = 0.36%, while an enhanced performance with $J_{sc} = 4.04\,mA\,cm^{-2}$ and PCE = 1.54% in the **BCP11**-based devices. Further studies indicated that the insertion of a flexible linkage between D and A blocks in **BCP11** inhibited the charge recombination in the copolymer.

To obtain a complementary absorption and simplify the synthesis process, Choi and coworkers reported another example of conjugated BCP (**BCP12**) [53], in which a wide-bandgap donor block of DT-BDT–thiophenecarboxylate is connected to a narrow-bandgap acceptor block of naphthalene diimide–selenophene, and the detailed synthetic route was illustrated in Scheme 9.4. The copolymer was synthesized by *Stille* coupling of carefully prepared end-functional oligomeric D and A blocks. The M_n of D and A macroreagent and D–A copolymer clarified by the gel permeation chromatography (GPC) were 6.84, 10.7, and 18.1 kDa, respectively, with PDI (Polymer Dispersion Index) of 1.53, 2.16, and 2.31. Interestingly, the absorption of **BCP12** in the solution consisted of the features of D and A blocks. SCOSCs prepared with the configuration of ITO/ZnO/**BCP12**/MoO_3/Ag delivered a J_{sc} of $8.30\,mA\,cm^{-2}$, V_{oc} of 0.93 V, FF of 0.50, and the PCE close to 4.0%, while the cell based on the mixture of D and A oligomeric materials gave a suppressed performance with a lower PCE of 1.14% and decreased characteristics of J_{sc}, V_{oc}, and FF. The device stability based on **BCP12** in ambient conditions was also investigated, and the results indicated that SCOSCs prepared with **BCP12** still contained 80% of the original performance after 2000 hours of storage. The main-chain engineering was further studied by the same group.

With the aim to increase the crystallinity of the donor block and optimize the nanoscale morphology, an extra thiophene segment was introduced in the donor block of DT-BDT–thiophenecarboxylate, giving a new block copolymer **BCP13** [54]. The solar cells based on **BCP13** gave an enhanced J_{sc} of $10.77\,mA\,cm^{-2}$, FF of 0.63, and a PCE of 5.28% compared with the SCOSCs based on **BCP12**. Furthermore, selenophene units in the acceptor block were replaced by the bithiophene, and a new copolymer **BCP14** was obtained [55]. Due to the long π-conjugated backbone both in D and A blocks, SCOSCs based on the block copolymer **BCP14** provided a PCE of 6.43%, which is the best result among the reported SCOSCs based on BCPs so far.

Besides the outstanding works of NDI-based BCPs, Huang and coworkers synthesized block copolymers **BCP15** combining macrosegments of P3HT and PBI [56]. To remove the excess P3HT and reduce the polydispersity, preparative GPC was prepared to get several individual fractions of these copolymers. The research indicated that the optical and electrochemical properties strongly depend on the ratio of the P3HT and PBI blocks. The best results were obtained with a copolymer of the lowest PDI (1.23) containing 43% of P3HT, the PCE was increased from 0.06% to 1.00% after annealing at 150 °C. Recently, Sivula and coworkers developed a new type of fully conjugated "A–D–A" copolymer **BCP16** via the *Heck–Mizoroki* coupling [57]. Despite the relatively poor device performance with a PCE of 1.51%, the preparation of **BCP16** was carefully studied and the chemical structures of functional macrosegments poly{(*E*)-1,2-ethene-*alt*-5,12-dibromo-2,9-di(tridecan-7-yl)anthra[2,1,9-def:6,5,10-d'e'f']diisoquinoline-1,3,8,10(2H,9H)-tetraone} (pPDIV)–(H/V) and poly{2,6′-4,8-di(5-ethylhexylthienyl)benzo[1,2-b;3,4-b]dithiophene-alt-5-dibutyloctyl-3,6-bis(5-bromothiophen-2-yl)pyrrolo[3,4-c]pyrrole-1,4-dione} (pBDTTDPP)–(Br/Br) were verified through the matrix-assisted laser desorption/ionization (MALDI)-time-of-flight (TOF) mass spectra, nuclear magnetic resonance (NMR), and GPC measurements, which was contributed to the good air stability of pPDIV–(H/V) compared with the tributylstannyl-contained conjugated segments.

9.2.3 Double-Cable Conjugated Polymers

Double-cable conjugated polymer is one of the most important conjugated materials in SCOSCs. In 1996, Benincori et al. reported the first photovoltaic application of double-cable conjugated polymer **DCP1** via electrochemical polymerization (ECP) [58]. The C_{60} was directly grafted into the bridging carbon of cyclopentadithiophene to get the monomer **m1**, which was suitable for ECP. Ferraris et al. synthesized a new double-cable polymer **DCP2** via ECP, in which poly-bithiophene was used as the backbone and connected with pendant C_{60} as side unit via flexible alkyl chain spacer, as shown in Scheme 9.5 [59]. Both **DCP1** and **DCP2** exhibited photoelectrochemical and electrochemical properties of the conjugated backbones and the pendant fullerenes. However, poor solubility in common solvents makes them inappropriate materials in solution-processed OSCs. So, it is important to develop soluble double-cable conjugated polymers.

In 2001, the first solution-processed SCOSC based on double-cable polymer (**DCP3**) was reported by Janssen and coworkers through a *Sonogashira* coupling reaction (Scheme 9.6) [60]. Photovoltaic devices were fabricated, using spin-coated **DCP3** films as the active layer, ITO/PEDOT:PSS, and Al as electrodes. Under white-light illumination (100 mW cm^{-2}, air mass (AM) 1.5), a J_{sc} of 0.42 mA cm^{-2},

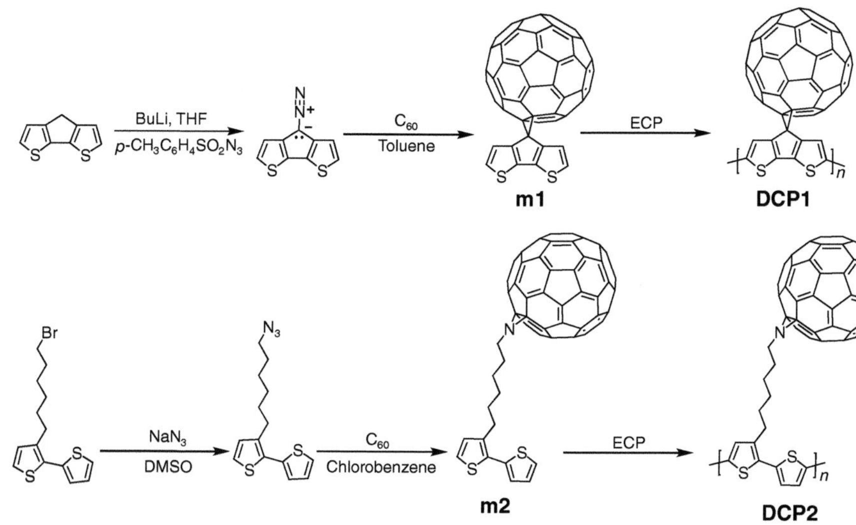

Scheme 9.5 Synthetic routes and chemical structures of **DCP1** and **DCP2**.

V_{oc} of 0.83 V, and FF of 0.29 were obtained. Although the PCE is only 0.1%, the result indicated a new type of material that can be used in SCOSCs. In the same year, Zhang et al. reported two double-cable conjugated polymers **DCP4** and **DCP5**, using poly-thiophene as the main chain and fullerene moiety as side groups, in which the content of fullerene is 7 and 14 mol% respectively, as shown in Scheme 9.6 [61]. Both **DCP4** and **DCP5** showed good solubility in chloroform, so that the polymer films could be prepared by spin coating from chloroform solution. The as-cast films showed an orange phase and then turned to dark blue after thermal annealing or solvent-vapor annealing (SVA) treatment. Absorption spectra showed that the onset of the two polymer films had red-shift (about 100 nm) to 700 nm after the annealing treatment, which is similar to the related polymers without C_{60} moiety, indicating that the small content of fullerene repeating unit has a negligible effect on the self-assembly of the poly-thiophene backbone. They prepared photodiodes by spin coating the film of **DCP4** or **DCP5** (70 nm) onto the top of ITO/PEDOT:PSS electrodes and Al electrodes (60 nm) were evaporated on the top of the polymer layer. The EQE spectra showed that **DCP5**-based devices had doubled response intensity. After treating the copolymer films with chloroform vapor, the EQE at long wavelengths was significantly enhanced due to the ordered state of the copolymers. The diode under the illumination of monochromatic light at 505 nm (0.1 mW cm^{-2}) showed a PCE of 0.6%, an FF of 0.25, and a V_{oc} of 0.75 V. The V_{oc} was decreased to 0.45 V after the treatment by chloroform vapor, which indicated a reflection of the diminished energy separation of the double-cable conjugated polymer.

In 2007, Li and coworkers designed and synthesized a new double-cable polymer (**DCP6**) with higher fullerene content and good solubility, as shown in Scheme 9.7 [62]. They selected polythiophene as the electron-donating backbone and fullerene as electron-accepting moiety, with an attached alkyl-phenylene group to ensure the

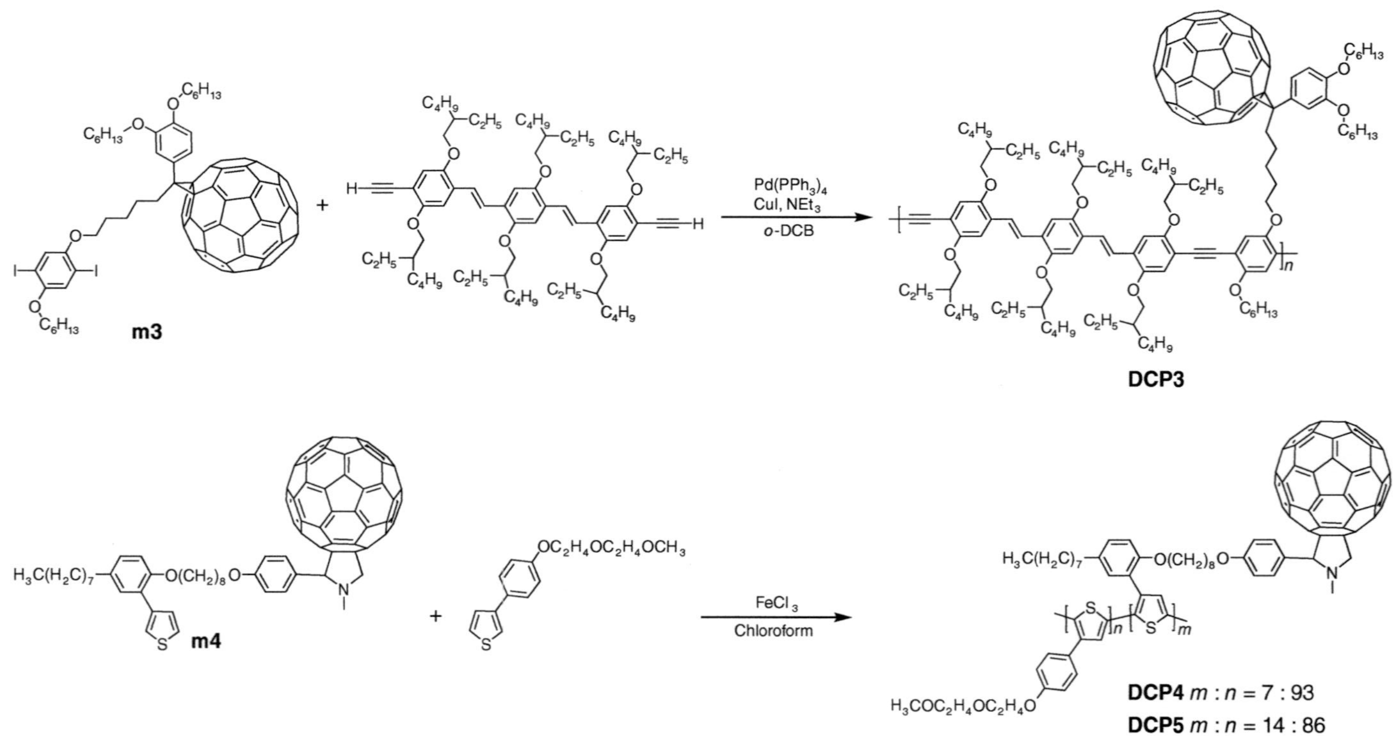

Scheme 9.6 Chemical structures and synthetic routes of **DCP3–DCP5**.

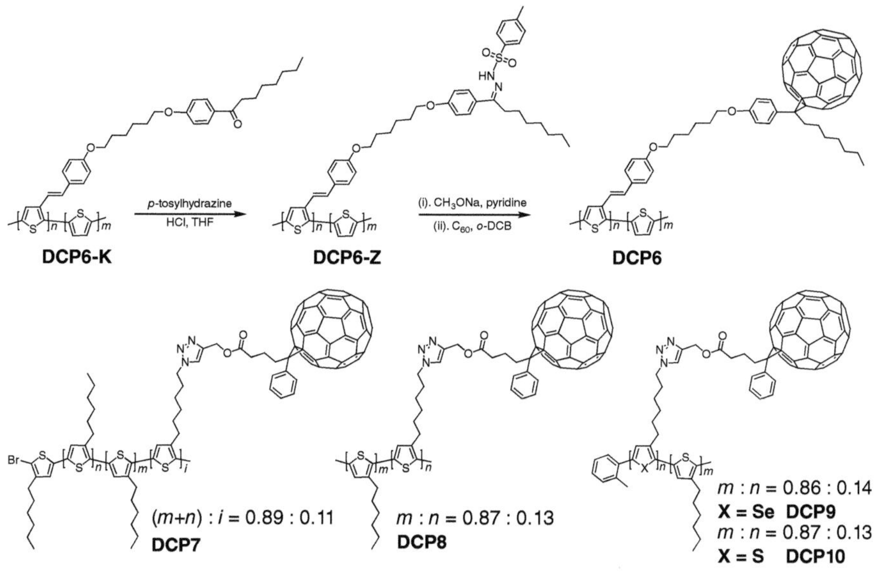

Scheme 9.7 Synthetic routes and chemical structures of **DCP6–10**.

solubility. They firstly synthesized the precursor polymer **DCP6-k** and then used two steps to graft the acceptor C_{60}. The content of the fullerene repeating unit was detected by elemental analysis of the content of carbon in the polymer, and the result showed that almost each repeated unit of backbone was substituted by one C_{60} molecule. With such a high content of C_{60}, **DCP6** was applied into SCOSCs with a configuration of ITO/PEDOT:PSS/**DCP6**/Ca/Al. Under the illumination of AM 1.5G, 100 mW cm^{-2}, the device showed a PCE of 0.52%, with a V_{oc} of 0.75 V and a J_{sc} of 2.41 mA cm^{-2}, which was among the best performance of double-cable polymer-based SCOSCs at that time and was more than five times higher than that of the related BHJ devices. Since then, this synthetic method of "grafting fullerene on electron-donating polymers" was widely used in the synthesis of double-cable conjugated polymers.

In 2010, Hashimoto and coworkers [63] synthesized two new double-cable polymers, including diblock polymer (**DCP7**) and random copolymer (**DCP8**), with poly(3-alkylthiophene) as backbone and PCBM as pendant unit, through a series of synthetic routes, including GRIM, Ni-catalyzed quasi-living polymerization, and the "click" cycloaddition reaction. The ratio of the fullerene attached thiophene unit was estimated by ^{1}H NMR spectra as 11% and 13% for the **DCP7** and **DCP8**, respectively, corresponding to the contents of fullerene in polymers of 38 and 41 wt%. SCOSCs were fabricated with the normal configuration (ITO/PEDOT:PSS/Polymer/Al). Under the illumination of simulated sunlight (AM 1.5G, 100 mW cm^{-2}), the device based on **DCP7** showed a PCE of 1.70% with V_{oc} of 0.54 V, J_{sc} of 6.15 mA cm^{-2}, and FF of 0.51, which was the best performance of SCOSCs at that time, much better than that based on **DCP8** (PCE of 0.48% with V_{oc} of 0.76 V, J_{sc} of 1.80 mA cm^{-2}, and FF of 0.35). The thermal stability of **DCP7** devices and the related BHJ devices of P3HT with PCBM were studied and showed a striking contrast under thermal annealing at 130 °C. The SCOSC based on **DCP7** showed only a slight decrease of PCE from 1.59% to 1.50% after 80 hours, while the BHJ device was decreased by almost 80% due to the PCBM aggregation. In 2012, the same group reported a higher PCE (2.5%, with FF of 0.63) of **DCP7**-based photovoltaic devices through controlling the synthesis method and optimizing the annealing condition of the devices. These results suggest that the application of single fullerene-attached diblock double-cable conjugated polymer could be an effective way to achieve efficient and stable SCOSCs. Five years later, Chen et al. introduced selenophene into the backbone of the P3HT and used similar methods to synthesize a new BCP with pendant PCBM (**DCP9**), and P3HT-based polymer (**DCP10**) was also synthesized for comparison [64]. Thermal annealing of the double-cable polymer films induced the microphase separation of approximately 30 nm. SCOSCs were fabricated using a reversed structure of ITO/ZnO/active layer/MoO_3/Ag. Under the irradiation of simulated solar light (AM 1.5G, 100 mW cm^{-2}), the devices based on the as-cast films showed negligible performance for the two polymers. After thermal annealing, both polymers showed the same trend of the decrease of J_{sc} and increase of FF, V_{oc}, and PCE without a large difference. The device fabricated from the film of **DCP9** annealed at 250 °C showed the best PCE of 1.52%, with V_{oc} of 0.51 V, J_{sc} of 5.02 mA cm^{-2}, and FF of 0.59.

As shown in Scheme 9.8, in 2017, Pierini et al. synthesized a new double-cable polymer poly[3-dodecylthiophene-*co*-3-(6-fullerenylhexyl)thiophene] (**DCP11**) based on polythiophene with pendant fullerene, using a strategy of "grafting after polymerization" [65]. According to the ^{1}H NMR spectrum, **DCP11** had a high content of the fullerene-grafted thiophene moiety (42% in molar ratio). More importantly, nanofibers of the double-cable polymer were produced by the technique of electrospinning and successfully integrated into SCOSCs. An optimized supramolecular and nanoscale structure of the active layer provided a high PCE of 5.58%. The dodecyl linker between the conjugated chain and fullerene reduced the charge recombination. The well-packed polymer chains with proper orientation generated the high performance of the devices. Two years later, the same group synthesized two new double-cable polymers (**DCP12, DCP13**), bearing different content of C_{60}-fullerene moieties and photo cross-linkable bromine groups at the end of a hexamethylenic side chain, from two homo-polymeric precursors with similar molecular weights but different regioregularity degrees, as shown in Scheme 9.8. When employed as active material in SCOSCs (ITO/PEDOT:PSS/active layer/Al), the polymer **DCP13** with higher regioregularity degree and lower content of fullerene showed better performance, giving a PCE of 5.21%, higher than 3.55% of the **DCP12**-based devices. Introduction of an Au nanoparticle thin layer between PEDOT:PSS and polymer layer further enhanced the performance of the devices and the PCE increased to 5.68%. Furthermore, the end group of bromines in these polymers can be cross-linked by the photochemical cleavage of C–Br bonds under ultraviolet (UV) light. Consequently, the UV-exposed devices showed better stability than the uncross-linked counterparts, after 80 hours of annealing at 150 °C. These results represent the highest performance of SCOSCs based on fullerene-grafted

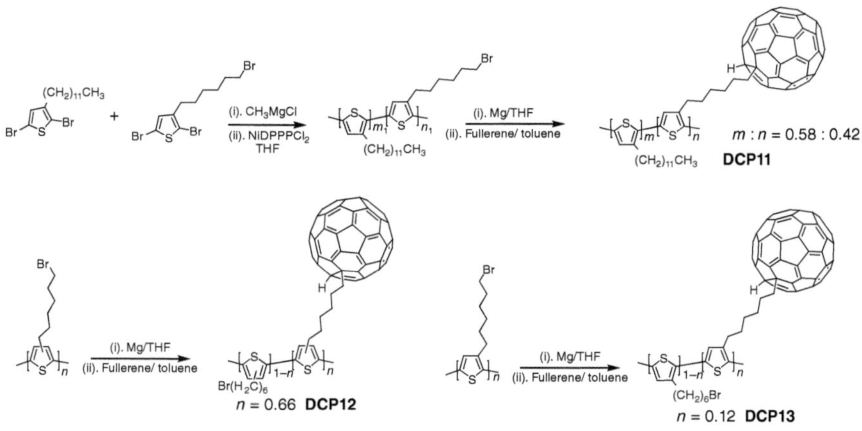

Scheme 9.8 The chemical structures and synthetic routes of **DCP11–13**.

double-cable polymers, which is comparable to most reported P3HT-based BHJ solar cells [66–69].

In addition to the above double-cable polymers grafted with fullerene derivatives, aromatic di-imide containing electron acceptors, such as NDI [70, 71], PBI [70, 72–81], and their derivatives [82], were also applied as pendant units in double-cable polymers for SCOSCs. Thelakkat and coworkers [69, 83] reported a DCP **DCP14**, as shown in Figure 9.7, with a poly(vinyltriphenylamine) segment as the electron donor for hole transport and a polyacrylate segment bearing PBI side groups as the electron acceptor for electron transport. The film of **DCP14** with a thickness of 250 nm was used as the active layer in a photovoltaic device with ITO/PEDOT:PSS and Al as electrodes. Under the irradiation of a white-light source (AM 1.5G, 77 mW cm^{-2}), the device gave a PCE of 0.07%, which was 10 times higher than that of the related BHJ devices. To improve the polymer's electrochemical stability, hole-carrier mobility, and to vary its highest occupied molecular orbital (HOMO) values, Thelakkat and coworkers [73] used bis(4-methoxyphenyl) phenylamine and *N,N′*-bis(4-methoxyphenyl)-*N,N′*-diphenyl-(1,1′-biphenyl)-4,4′-diamine as the hole-conducting moieties to take the place of bisphenyl-4-vinylphenylamine in **DCP14** and synthesized two new polymers **DCP15** and **DCP16**. The PCEs of the SCOSCs with **DCP15** and **DCP16** got fourfold to fivefold improvement, up to 0.26% and 0.32%, compared with **DCP14**. In the next year, the same group reported a new polymer **DCP17** with P3HT as the donor block and the same type of poly(perylene bisimide acrylate) as acceptor block to improve the crystalline of the donor part to give rise to rich phase behavior and improve the charge-carrier mobility. Russell and coworkers [84] used the same strategy to prepare a new polymer **DCP18**, which was used in SCOSCs and got a PCE of 0.49%. The unit of DPP with imide groups has a strong electron-withdrawing ability, which can reduce the bandgap of conjugated block. Koyuncu et al. [75] reported a new polymer **DCP19** by using DPP-introduced oligothiophene as low-bandgap donor moiety and the same poly(perylene bisimide acrylate) of **DCP18** as acceptor moiety. The polymer provided a strong and broad

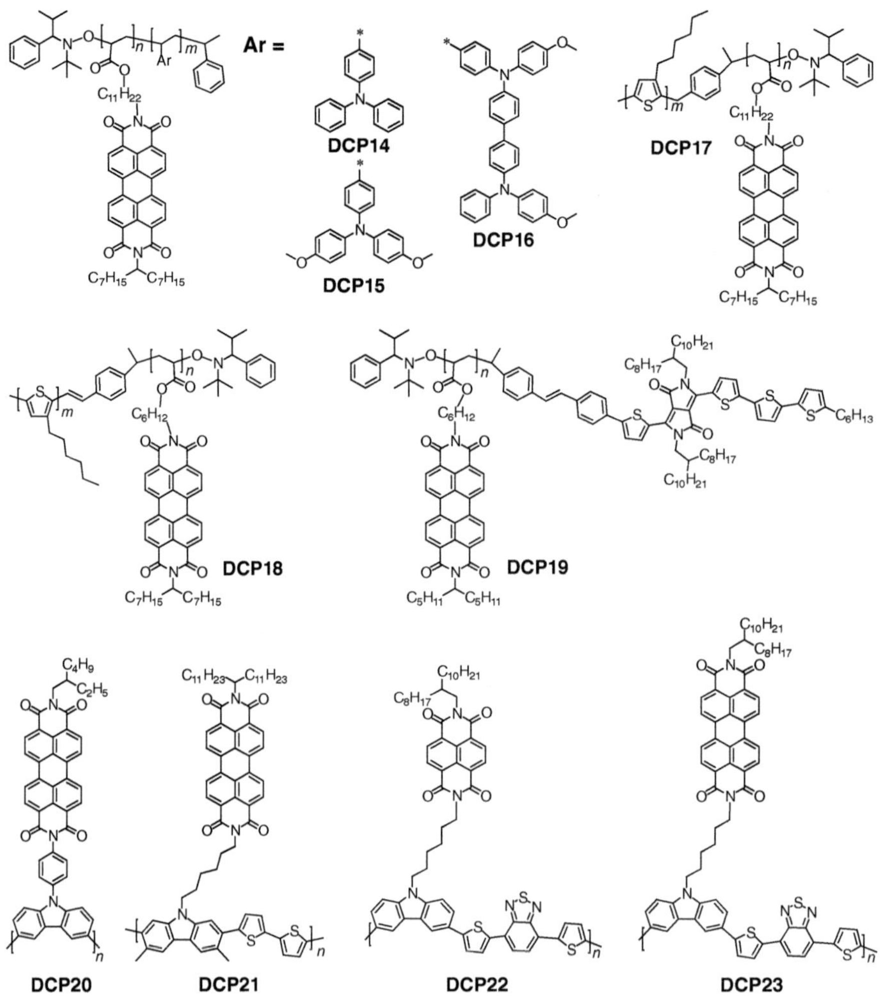

Figure 9.7 Chemical structures of the double-cable polymers **DCP14–DCP23**.

optical absorption, from 300 to 750 nm, with compatible HOMO–lowest unoccupied molecular orbital (LUMO) levels to **DCP18**. Through solvent annealing in dichloromethane, microphase separation can be found in the polymer film, resulting in an optimized PCE of 0.89% in SCOSC.

ECP is an efficient method to synthesize carbazole-contained polymers. Koyuncu et al. [85] synthesized a new double-cable polymer **DCP20** with a PBI-linked carbazole moiety by ECP. Remarkable fluorescence quenching can be detected in the films caused by the intramolecular charge transfer from the donor backbone of carbazole to the acceptor of pendant PBIs. However, the PCE of **DCP20**-based SCOSC was only 0.1%. In 2011, Mohamad et al. [86] reported the polymer **DCP21**, a new double-cable polymer with alternating 2,7-linked carbazole and dithienyl units as the backbone, PBI as pendant acceptor, and hexyl chain as the attached linker. They also found evidence about electron transfer from the conjugated backbone to the pendant PBI units, resulting in weak fluorescence and highly efficient quenching of excitons in both solutions and thin films. The study of steady-state photo-induced absorption spectroscopy confirmed the existence of long-lived polaronic states due to exciton dissociation on **DCP21**. However, these results did not provide an efficient SCOSC. The devices based on **DCP21** only showed a low EQE of around 0.43% at 410 nm.

In 2018, Ledwon and coworkers reported two new double-cable polymers **DCP22–23** containing NDI and PBI as acceptor side groups [70]. Both polymers used poly[3,6-carbazole-*alt*-5,5-(4′,7′-di-2-thienyl-2′,1′,3′-benzothiadiazole)] as the conjugated backbone. The introduction of benzothiadiazole to the backbone resulted in a broader absorption of polymers, covering the spectrum from 300 to 700 nm. Photovoltaic properties of the two polymers-based SCOSCs with the configuration of ITO/PEDOT:PSS/Polymer/Al showed a disappointing PCE of 0.03%.

Since 2017, Li and coworkers reported a series of double-cable polymers with PBI or its derivative, as shown in Figure 9.8 [76–82, 87–89]. They firstly synthesized three polymers **DCP24–26**, which used DPP and BDT as conjugated backbone and PBI as side unit with different length of alkyl chain as linkers. The DPP-contained backbone gave the polymers a narrow bandgap with wide absorption from 300 to 900 nm. The crystalline properties of the polymers were systematically studied. The results showed that longer alkyl (dodecyl) linkers can improve the crystallinity of the polymers compared with the shorter alkyl (hexyl) linkers, which could be further optimized by introducing sulfur atoms into the alkylthiophene side groups. As a result, the polymer **DCP26** was found to show better nanophase separation and get the best PCE of 2.74% in the SCOSCs, while **DCP24** and **DCP25**-based SCOSCs only provided low PCEs of 0.51% and 1.64% [76]. From this work, the authors were inspired to design new double-cable polymers with strong crystalline conjugated backbones to balance the contradictory crystallinity and phase separation requirements for charge generation and charge transport in the SCOSCs. Therefore, they brought out a new strategy of using linear conjugated backbone to adjust the phase separation of the double-cable polymers and reported other three new polymers **DCP27–29** [77]. They used linear conjugated backbone poly(benzodithiophene) (PBDTT) as the donor for hole transport, pendant PBI units as the acceptor for electron transport, and a dodecyl linker as the connected part. To further tune the energy levels and crystallinity, they also introduced sulfur and fluorine substituents to the donor parts of the polymers. From the grazing-incidence wide-angle/small-angle X-ray scattering (GIWAXS/GISAXS) results of the polymer films, the double-cable polymers showed a “face-on” orientation of both the BDT backbone and the pendant PBI units, which was favorable for charge transport in OSCs. Nanophase separation of the backbone and PBI side units was pointed out by the X-ray diffraction (XRD) and microscopy data, which is important to dissociate the excitons and transfer the resulting free charges to the electrodes. These advantages resulted in high quantum efficiency of EQEs up to 0.65 and internal quantum efficiencies (IQEs) up to 0.80 and high PCEs up to 4.18%, which is one of the highest performances in SCOSCs at that time. However, there is a significant drawback for the three polymers, in which the wide bandgap of the

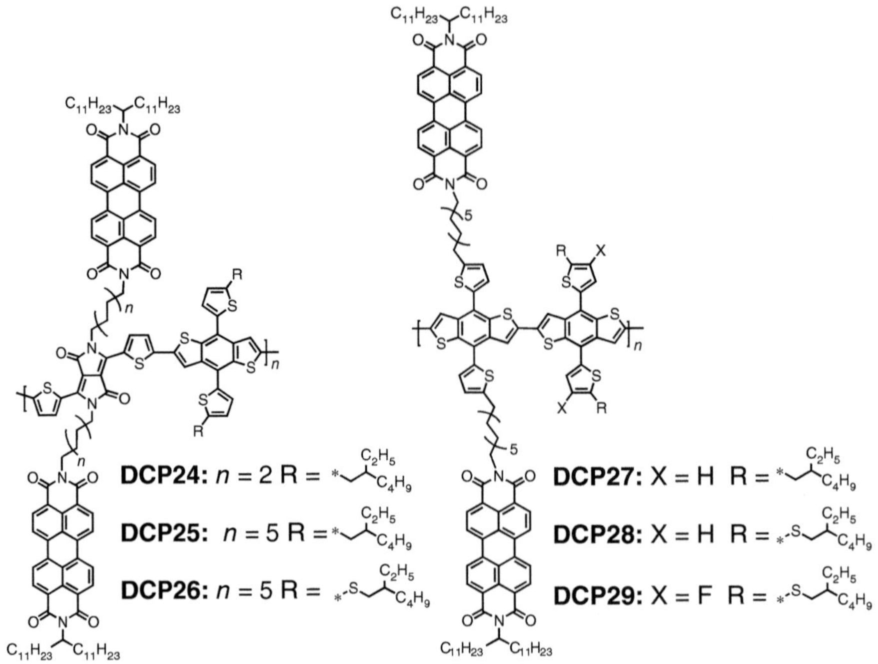

Figure 9.8 Chemical structures of the double-cable polymers **DCP24–DCP29**.

PBDTT backbone gives a small range of absorption of the polymer films, which limited the utilization range of sunlight.

Then, Li and coworkers selected other electron-deficient units based conjugated backbones to tune the bandgap of the PBI-based double-cable polymers, such as isoindigo (**DCP30**) [79], thieno[3,4-*c*]pyrrole-4,6-dione (TPD, **DCP31–33**) [87], and another DPP-based polymer (**DCP34**) [88] was simultaneously investigated. They successfully developed multiple double-cable polymers with narrow bandgap, and the onset of their absorption varied from 700 to 900 nm. However, they failed to get high quantum efficiency in the SCOSCs with these polymers as single active layer and most of the PCEs were below 3%. Besides, the influence of different content of the acceptor PBI in the double-cable polymers was also investigated; they synthesized two series of polymers using DPP, TPD, and BDT as the backbone and tuned the ratio of the pendant PBIs, as shown in Figure 9.9. Their studies suggested that the pendant PBI units can not only influence the energy level and the absorption spectrum but also improve the polymers' crystallinity and then alter the phase separation of the backbone and the side acceptors in the polymer films. Interestingly, these multifunctional double-cable polymers can act as electron donor or acceptor in BHJ solar cells or as the single active material in SCOSCs [78, 87].

In 2019, Li and coworkers found a new way to improve the PCE of SCOSCs through thermal-driven D/A lamellar phase separation. As shown in Figure 9.10, they designed a new double-cable polymer **DCP35** containing a strongly crystalline conjugated backbone (based on two-dimensional BDT alternating with 5,7-bis(2-ethylhexyl)benzo[1,2-*c*:4,5-*c*′]dithiophene-4,8-dione) as donor and PBI

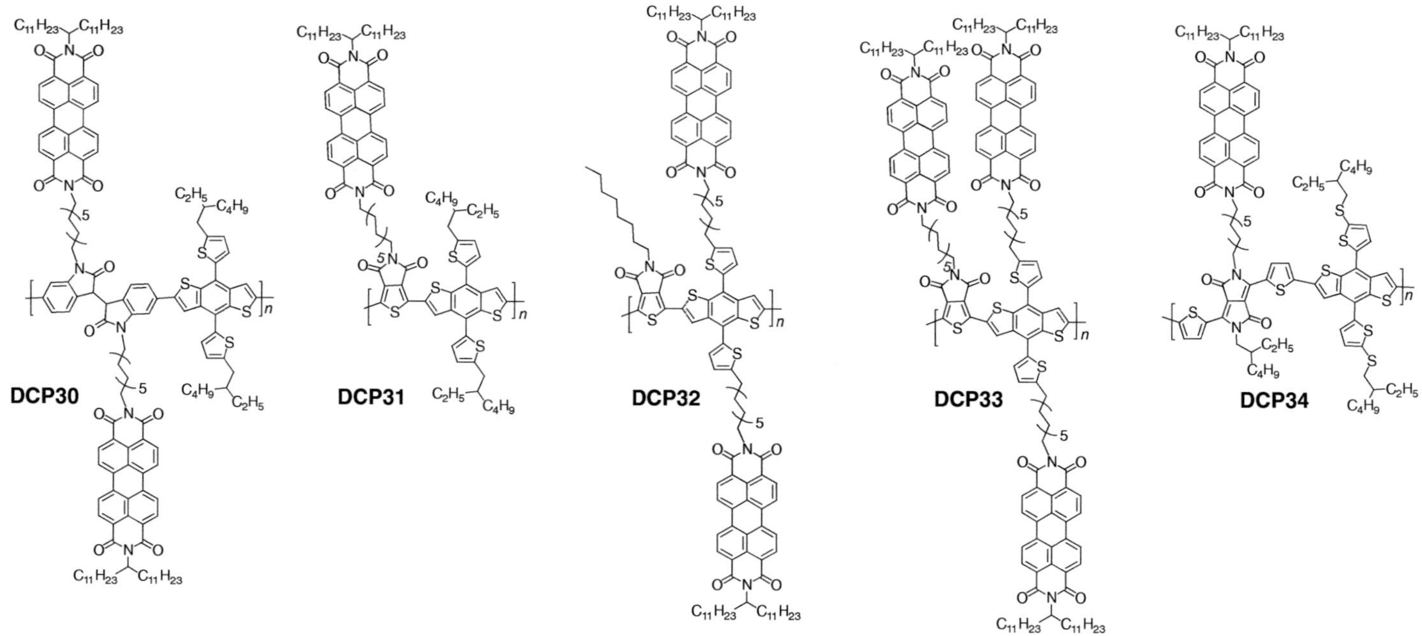

DCP30 DCP31 DCP32 DCP33 DCP34

Figure 9.9 Chemical structures of the double-cable polymers **DCP30–DCP34**.

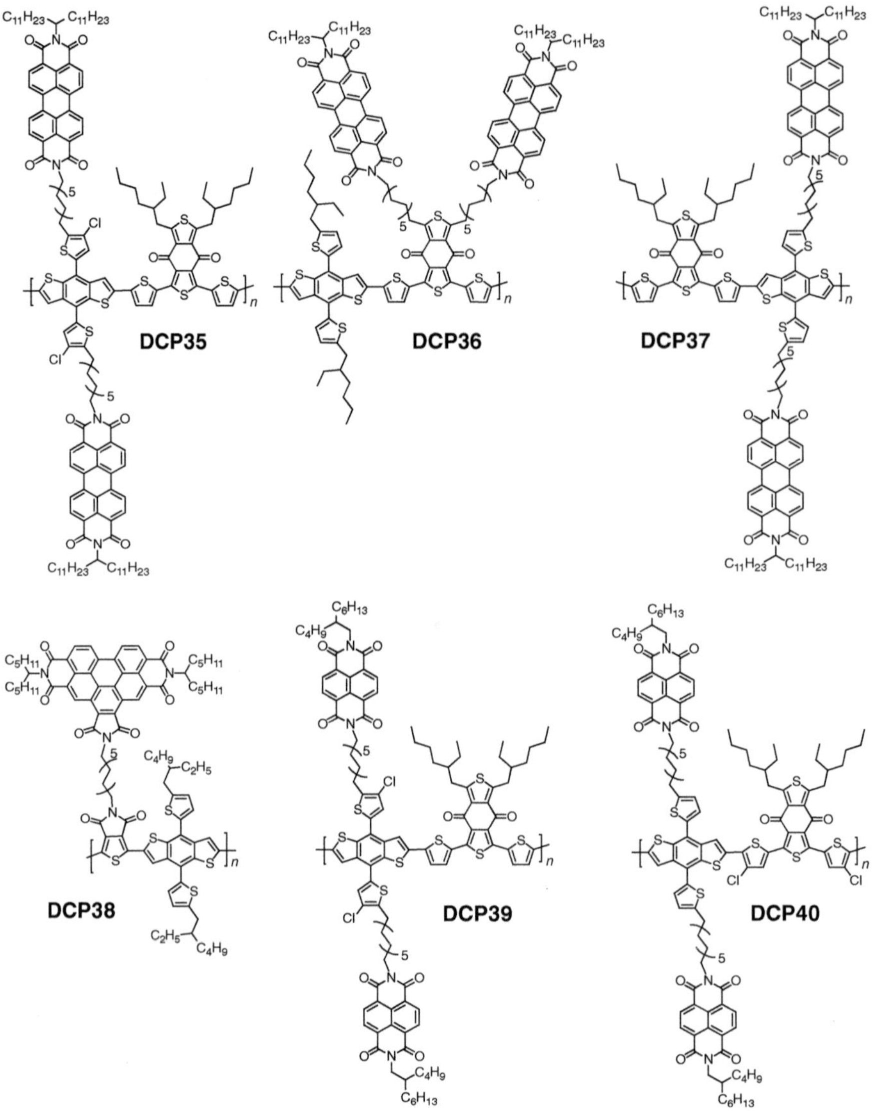

Figure 9.10 Chemical structures of the double-cable polymers **DCP35**–**DCP40**.

as the pendant acceptor. According to the results of thermal gravimetric analysis and differential scanning calorimeter, the polymer films were annealed at different temperatures for several minutes. After thermal stress, both the crystalline backbones and PBI side units tended to self-organize to form a well-ordered D/A lamellar nanostructure, especially at a high temperature of 230 °C, leading to highly efficient charge transport and low charge recombination. Therefore, a high PCE of 6.3% in SCOSCs can be obtained [80]. In the same year, they designed two new double-cable polymers with the same conjugated backbone and PBI side units, in which PBIs were positioned at one side (**DCP36**) and two sides symmetrically

(**DCP37**) along the conjugated backbones. After thermal annealing, both PBI side units and conjugated backbones in **DCP37** tended to self-organize to ordered nanostructures, resulting in good crystalline cooperativity, and the **DCP37**-based SCOSCs showed a higher PCE of 3.43%, while PBI units in **DCP36** hampered the crystallization of conjugated backbones resulting in poor crystalline cooperativity, and the **DCP36**-based SCOSCs only provided a PCE of 2.42% [81].

From **DCP24** to **DCP37**, they found that all the double-cable polymers have the same side pendant PBI units. In 2019, they designed a new polymer **DCP38** by choosing a new acceptor unit benzo[ghi]-perylene triimide (BPTI) to replace the PBI unit in **DCP29** to improve the electron transport efficiency in the polymer films. Consequently, they found the pendant units BPTI dominated the crystallinity of the conjugated backbone and enhanced the acceptors' contact region with each other. This could improve the electron transport of the new double-cable polymer **DCP38** in the thin film, as confirmed by space-charge limited current measurement. Therefore, the **DCP38**-based SCOSCs could provide a PCE of 4.34%, compared with 1.92% based on **DCP29** [82]. More recently, they brought out a new strategy to improve the exciton dissociation efficiency by controlling miscibility between conjugated backbones and aromatic side units in double-cable polymers. They designed two new polymers **DCP39, DCP40** with the same NDI side units and two almost identical conjugated backbones with extra Cl atoms on the different positions of the conjugated backbones. The polymer **DCP39** with Cl atoms positioned at the main chains formed the twist backbones, could get better miscibility between the backbones and NDI side units, which could contribute to the efficient exciton diffusion and charge separation. Consequently, **DCP39**-based SCOSCs gave a high PCE of 8.40%, with high EQEs up to 0.75 and FFs close to 0.70, which is also the highest PCE in SCOSCs until now [71].

According to the above paragraphs, Li and coworkers devoted themselves to developing SCOSCs by designing new double-cable polymers with different backbones, side units, and linkers since 2017. Several different strategies were under consideration to tune the energy levels, extend the absorption spectrum of the polymers, and adjust the nanophase separation of the polymer films. Finally, high quantum efficiencies were achieved in SCOSCs. These results will intrigue new conceptions about SCOSCs toward high PCEs and further application.

9.3 Morphological Studies of the Photo-Active Layers in the SCOSCs

The photovoltaic performances of OSCs are determined by several factors, such as microstructures of the active layer, charge carrier mobilities of the related materials, and the interface between the active layer and electrodes. Especially, the microstructures related to the surface morphologies and micro-phase separation of the photo-active layers predominate over the others. The relative studies of morphological control have been widely studied in the BHJOSCs that require biphasic films in which D and A domains should form two independent continuous networks as well as sufficient miscible regions. Besides, some feasible methods, including solvent additives, TA, and SVA, have been proven to be beneficial to form appropriate morphologies. Although the D–A heterojunction concept was also introduced to construct the photosensitive materials in the SCOSCs, which is the so-called "intramolecular heterojunction," the morphological formation in the single-component system has unique properties.

In the single-component system, chemical structures of photo-active materials have been restricted by the fully conjugated chemical structures or length-limited alkyl linkage between the D and A groups. Hence, the formation of suitable morphology was the synergistic crystallization of the donor, acceptor, and linkage. This unique characteristic needs to be carefully considered to obtain appropriate morphologies, the size of the D and A domains, which are closely related to the crystallinity and miscibility of the materials, should dimension commensurate with the exciton diffusion length while providing percolation pathways for the photogenerated charges. In this regard, the perfect morphology should well balance the crystallinity, the phase mixing, and the phase separation.

9.3.1 Morphological Studies of the Molecular Dyads in SCOSCs

Molecular dyads with fullerene or PBI as the electron acceptor have been widely used as the photovoltaic conversion medium in the SCOSCs, and the molecule aggregation in the photo-active films shows a significant correlation with the type of acceptor units, which means that fullerene-based dyads exhibit distinct characteristics in the nanoscale morphological separation compared with PBI-based systems. Therefore, a detailed discussion of the morphology control of the molecular dyads in SCOSCs was present depending on the classes of the acceptor.

In the fullerene-based "in-chain" SCOSC systems, the morphologies of the optimized **SM14** films were systemically studied through the 2D-GIXRD (grazing X-ray diffraction) (Figure 9.11d), atomic force microscope (AFM, Figure 9.11a–b), and transmission electron microscopy (TEM, Figure 9.11c). As shown in Figure 9.11d, the XRD diffractogram of **SM14** films shows peaks at $2q = 5.07°$ and $4.37°$, indicating an inter-plane spacing of 1.74 and 2.02 nm, which is similar to the aggregation in the blended DPP:PCBM films. And the disappearance of the diffraction peak at $2q = 24°$ revealed that the **SM14** films existed the edge-on orientation in the casting films. Furthermore, the crystalline size of **SM14** in the solid film is 24.4 and 41.5 nm for interplane and in-plane directions, respectively, and the large sheet structures were also observed in the AFM and TEM measurements. In the manuscript, the authors also emphasized that these long-range ordering microstructures are beneficial for high-performance OSCs.

As for the fullerene-based "side-chain" system, materials developed by Nguyen et al. (**SM33**) and Min and coworkers (**SM34**) were selected as the representatives. As shown in Figure 9.12, the as-cast **SM33** (also named BDTRh–PCBM) films showed a homogeneous and flat surface with a root-mean-square (RMS) roughness of 0.41 nm while the BDTRh:PCBM (BHJOSCs as control) blends exhibited severe phase separation with the relatively high RMS of 16.7 nm. The results indicated

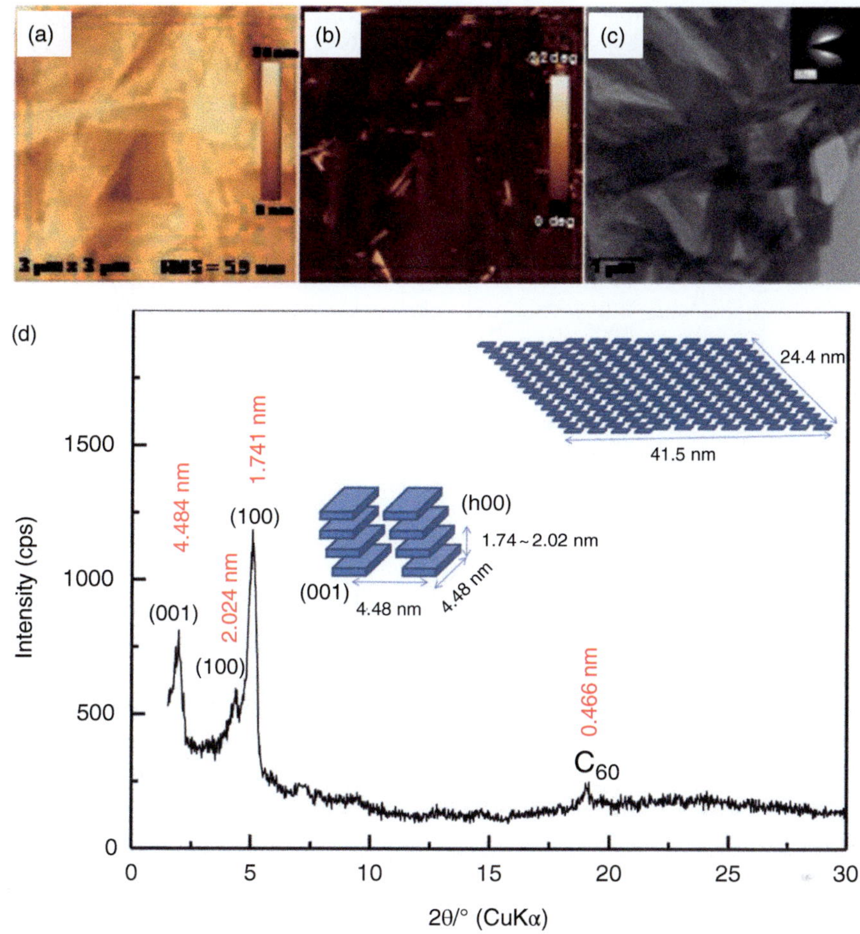

Figure 9.11 (a) AFM height and (b) phase images and (c) TEM images of the **SM14** based device, (d) 2D-GIXRD pattern of **SM14**. Lower inset: intermolecular and intramolecular distances, upper inset: crystalline size. Source: Narayanaswamy et al. [27]/Reproduced with permission of John Wiley & Sons.

that the self-aggregation tendency to form large molecular domains was suppressed when donor and acceptor parts were covalently connected.

Furthermore, the intermolecular packing and orientation of **SM33** in the SCOSC by comparison with BDTRh:PCBM blends in the BHJOSCs were also investigated through the GIWAXS measurements. Firstly, the pristine BDTRh films were prepared, and the results demonstrated a trend of aggregation with strong (100) out-of-plane interlamellar scattering peaks and (010) π–π stacking peaks in both in-plane and out-of-plane directions. When blended with PCBM, a hump-shape scattering peak appears due to aggregation of PCBM while the scattering features of neat BDTRh were also observed, which means the donor and acceptor still held their crystal domains associated with the large phase separation as shown in

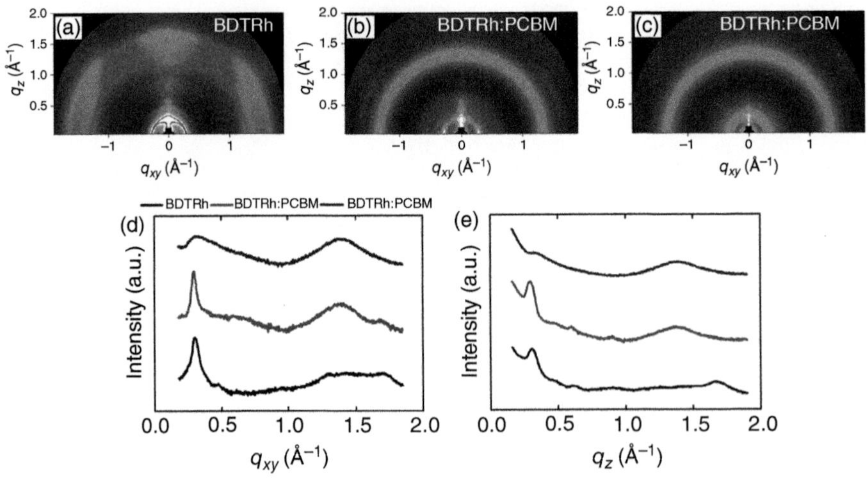

Figure 9.12 2D-GIWAXS images of (a) BDTRh, (b) BDTRh:PCBM, and (c) BDTRh-PCBM (**SM33**) films. Corresponding line-cuts in (d) in-plane and (e) out-of-plane directions. Source: Nguyen et al. [38]/Reproduced with permission of John Wiley & Sons.

the AFM measurements. In the **SM33** film, a weak and broad (100) interlamellar packing similar to the BDTRh:PCBM blends were obtained while the crystalline structure BDTRh was significantly hindered. The same phenomena were observed in the single-component molecular material **SM34** developed by Min et al.

However, another kind of scenery in microscopic phase separation in SCOSCs can be observed in the PBI-based dyads. Figure 9.13 shows the morphological studies investigated by Geng and coworkers.

To achieve appropriated micro-phase separation, the linear molecular dyads (**SM35–37**) with the liquid crystalline property oligo(fluorene-*alt*-bithiophene)s (OFbTs) as the donor and PBI units as the acceptor were prepared. The OFbTs were connected to the PBIs through a carbon–nitrogen bond and the strong π–π interactions qualify OFbTs and PBIs as the potential blocks for molecular self-assembly architectures. Through the TEM measurements, dark–bright stripes appeared in the films for all three co-oligomers after thermal annealing at 210 °C for two minutes while the ordered structure can be only observed in the **SM37** system. However, the molecular orientation in the annealed materials is still relatively weak.

To enhance the nanoscale self-aggregation, SVA is a promising method. Herein, the order degree of co-oligomers was increased with the CH_2Cl_2 vapor annealing. As shown in Figure 9.13d–f, the persistent lengths of the lamellae increase from 50 to 150 nm for thermally annealed films to 100–400 nm for solvent-vapor annealed films, and the crystalline sizes of the lamellae are 5.6, 7.4, and 8.9 nm for **SM35**, **SM36**, and **SM37**, respectively.

In a further study, flexible linkages with different lengths were introduced to connect the OFbTs and PBI segments, and the effect of the varied periods on the

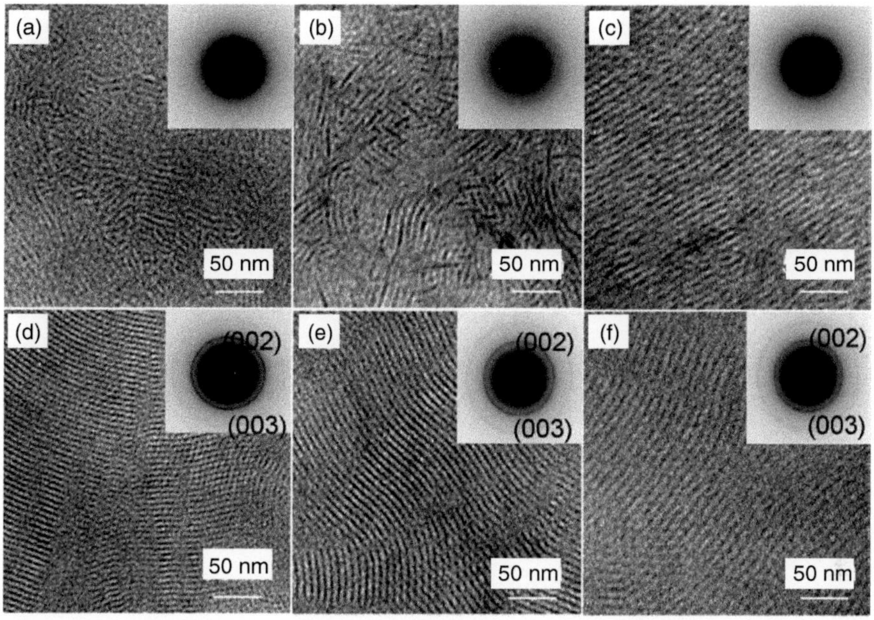

Figure 9.13 TEM images and SAED patterns (inset) of **SM35** (a, d), **SM36** (b, e), and **SM37** (c, f) thin films after thermal annealing (a–c) or solvent-vapor annealing (d–f). Source: Bu et al. [40]/with permission of American Chemical Society.

micro-phase formation was also investigated by the same group. As shown in Figure 9.14, the ordered structures within the powders of these co-oligomers were investigated by X-ray scattering and diffraction. A new diffraction peak appeared at $2\theta = 0.57°, 0.65°$, and $0.54°$ for **SM41**, **SM42**, and **SM43**, respectively, indicating the formation of lamellar nanostructures with a period of 15.5 nm for **SM41**, 13.6 nm for **SM42**, and 16.3 nm for **SM43**. Compared with the **SM37** (without alkyl linker between the donor and acceptor segments), the introduction of flexible spacers yielded a more regular microstructure. And the length of spacers also showed a significant effect on the molecular self-assembly. After thermal annealing at 160 °C for 10 minutes, the well-prepared films of the co-oligomers were studied by TEM. The dark-bright stripes are clearly present in the films with the periods of 14.2 ± 0.4, 15.0 ± 1.0, and 14.6 ± 0.7 nm for **SM41**, **SM42**, and **SM43**, respectively. Besides, the **SM42** showed multi-diffraction rings in the SAED patterns (selected area electron diffraction), and two of them with the *d*-spacings of 3.5 and 4.4 Å are attributed to the "π–π" stacking distances of the adjacent PDI units and donor segments, respectively. The tendency to form ordered structures was predominated by the PBI in the shorter linkage system **SM42**, and the aggregation properties of the donor segment were enhanced when the C_2-alkyl was replaced by C_6-alkyl. Combining all the above observations, appropriate nanostructures with the highest intermolecular packing order can be obtained in the **SM42**.

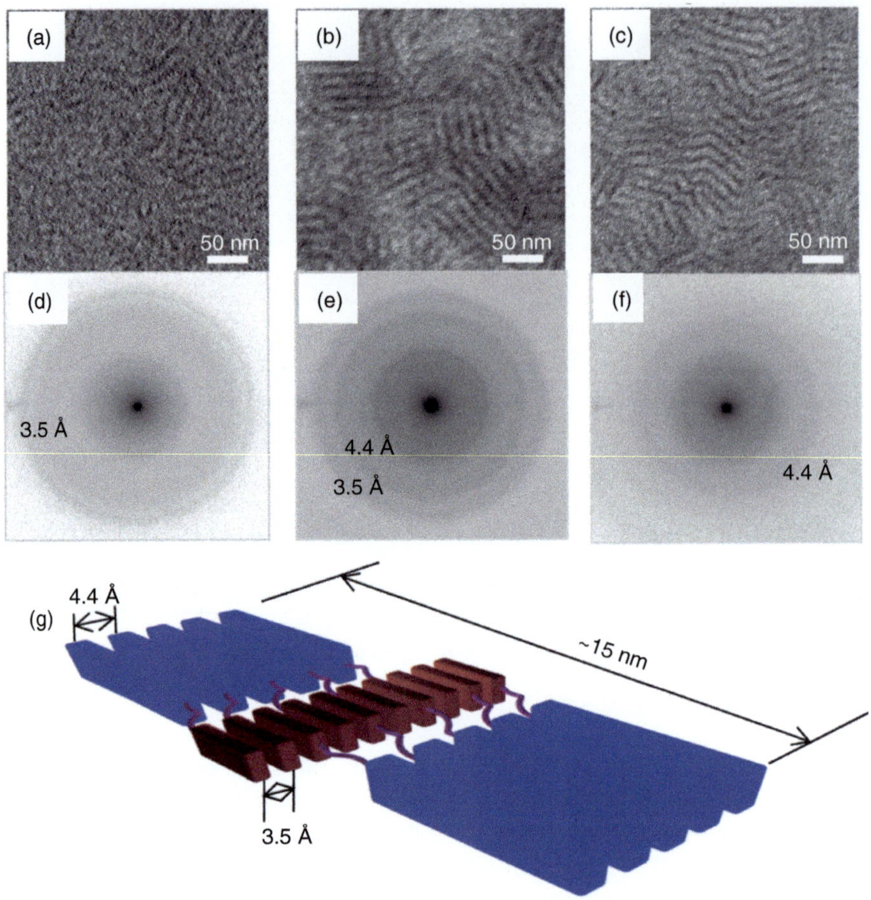

Figure 9.14 (a–c) TEM images and (d–f) SAED patterns of **SM41**, **SM42**, and **SM43** thin films with a thickness of ~70 nm (measured at room temperature). Schematic illustration of the lamellar nanostructure in the film (g). Source: Qu et al. [42]/Reproduced with permission of Royal Society of Chemistry.

9.3.2 Morphological Studies of the Block Copolymers in SCOSCs

It has been reported that BCPs could form well-ordered nanophase separation that is beneficial for charge separation via rationally controlling the crystallinity between donor and acceptor segments. The representative work was developed by Choi and coworkers, in which the formation of the appropriate morphologies in high-performance BCPs (**BCP12–14**) with the donor and acceptor blocks (DBs and ABs) as control was studied.

Figure 9.15 illustrated the study of morphologies and micro-separation within the BCPs developed by D.H. Choi and coworkers. The surface morphologies were measured by the AFM measurement, in which a smooth morphology that was rarely affected by the additive was obtained in the as-cast and optimized **BCP12** films, while larger surface domains can be observed in the DBs/ABs blended films. Furthermore, TEM was used to investigate the internal morphologies of the films in

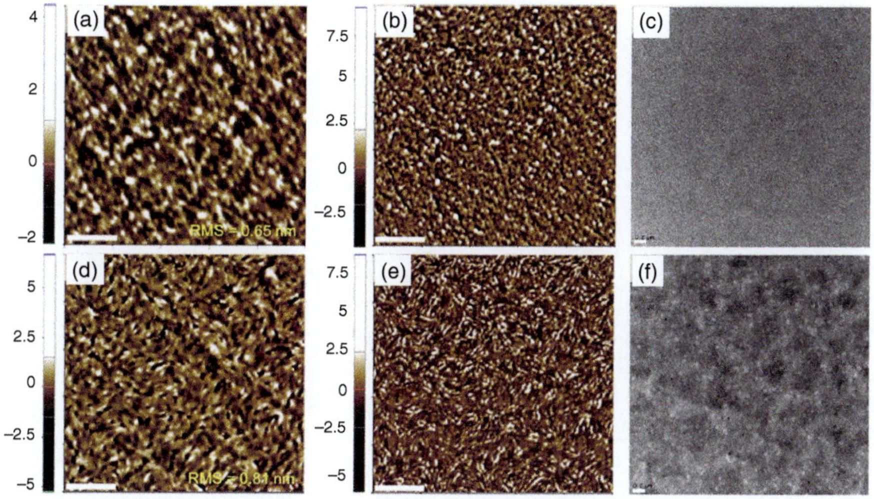

Figure 9.15 AFM images (scale bar = 1 μm) of the active layer in PSC devices fabricated with 1.0 vol% CN: topography of (a) **BCP12** and (d) DBs/ABs blend films and phase images of films fabricated with (b) **BCP12** and (e) DBs/ABs blend. TEM images (scale bar = 200 nm) of the active layer in PSC devices fabricated with 1.0 vol% CN: (c) **BCP12** and (f) DBs/ABs blend films. Source: Lee et al. [53]/with permission of American Chemical Society.

two different systems (**BCP12** film-based SCOSCs and DBs/ABs blend film-based BHJ OSCs), as shown in Figure 9.15c,f. The results indicated that the **BCP12** exhibited a uniform and homogeneous morphology, which is beneficial to efficient exciton diffusion and charge transport.

In the subsequent research, the additional thiophene unit was introduced to expand the conjugated system of the donor segment and the enlarged donor block can promote profound effects on the micro-phase separation. The crystallinity and orientation of the polymer chains were investigated through GIWAXS measurement, as shown in Figure 9.16. A high-intensity (010) diffraction peak appeared in the as-cast films and the corresponding *d*-spacing is 3.78 Å, indicating that the copolymer chains exhibited the face-on orientation. And the more uniform crystalline–crystalline phase of donor blocks and acceptor blocks can be formed after post-processing treatment of 10 minutes TA at 180 °C, revealed by the significantly enhanced (100) and (200) diffraction peaks in the q_z direction and the narrower full width at half maximum of (010) peak in q_{xy} direction. The same phenomena were also observed in the Kelvin probe force microscope (KPFM) test. Annealed polymer film also possessed highly continuous CPD (contact potential difference) bright/dark images with high contrast. The results demonstrated that the dispersed donor- and acceptor-rich domains were induced after annealing.

9.3.3 Morphological Studies of the Double-Cable Polymers in SCOSCs

Generally, the double-cable polymer is formed by three segments: the conjugated backbone, the pendant side units, and the covalent linkers. Compared with the block

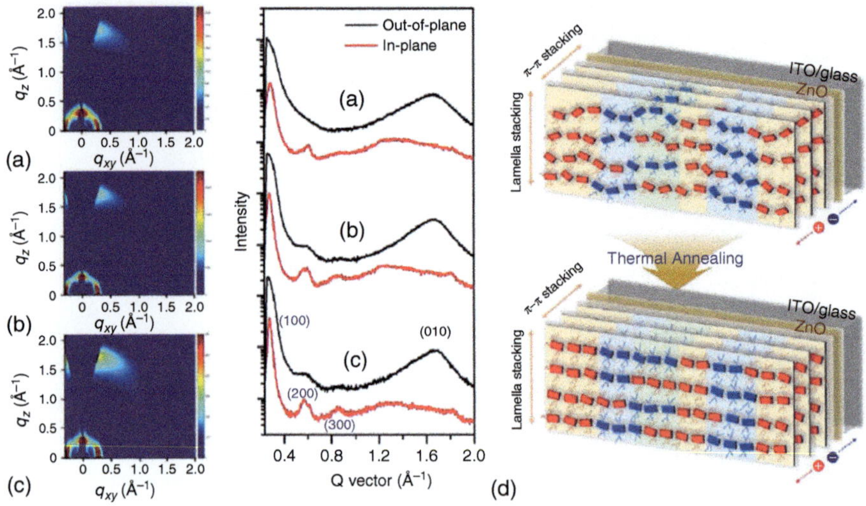

Figure 9.16 GIWAXS patterns of the as-cast and annealed **BCP13** films. Out-of-plane and in-plane diffraction profiles. (a) as-cast film, (b) annealed film at 120 °C, (c) annealed film at 180 °C. Annealing time = 10 minutes. (d) Schematic representation of the crystalline DBs and ABs in the as-cast and annealed films of **BCP14**. More uniform crystalline-crystalline phase of DBs and ABs can be formed with minor co-crystalline phase between DBs and ABs. Source: Park et al. [54]/Reproduced with permission of Royal Society of Chemistry.

copolymers, it is very difficult to control the morphology of the double-cable polymers, as both the main chain of conjugated backbones and the acceptor side units have strong crystallinity and aggregation tendency, which restrict each other, making it difficult to form a regular nanophase separation structure.

Li and coworkers have done a series of studies on tuning the nanophase separation of the double-cable polymers and made significant progress. For the polymer **DCP24–26**, as shown in Figures 9.8 and 9.17, the GIWAXS patterns revealed a clear enhanced crystallinity and face-on orientation from **DCP 24** to **DCP 26** by using longer alkyl (dodecyl) linkers and alkylthiothiophene side groups. Such changes are consistent with both enhanced planarity and rigidity of the π-conjugated backbone and stronger π-stacking interactions, which were beneficial for the charge transport efficiency and were reflected in the enhanced FF and PCE of the single-component photovoltaic devices of these polymers with an order of **DCP26** > **DCP25** > **DCP24** [76].

The study of **DCP27–29** revealed another strategy of using a linear conjugated backbone to improve the crystalline and adjust the phase separation of the double-cable polymers. Meanwhile, S and F atoms were also introduced into the alkylthiophene side groups. As shown in Figure 9.18, the AFM phase images showed small contrast with 10–20 nm surface morphology characteristics. The GIWAXS showed that the polymers had preferentially "face-on" orientation with the characteristic distance of about 5–6 nm from the peak (100) in the in-plane direction. And the GISAXS showed a *d*-spacing of about 5 nm, which was about twofold *d*-spacing of the distance between the backbone and acceptor units, as

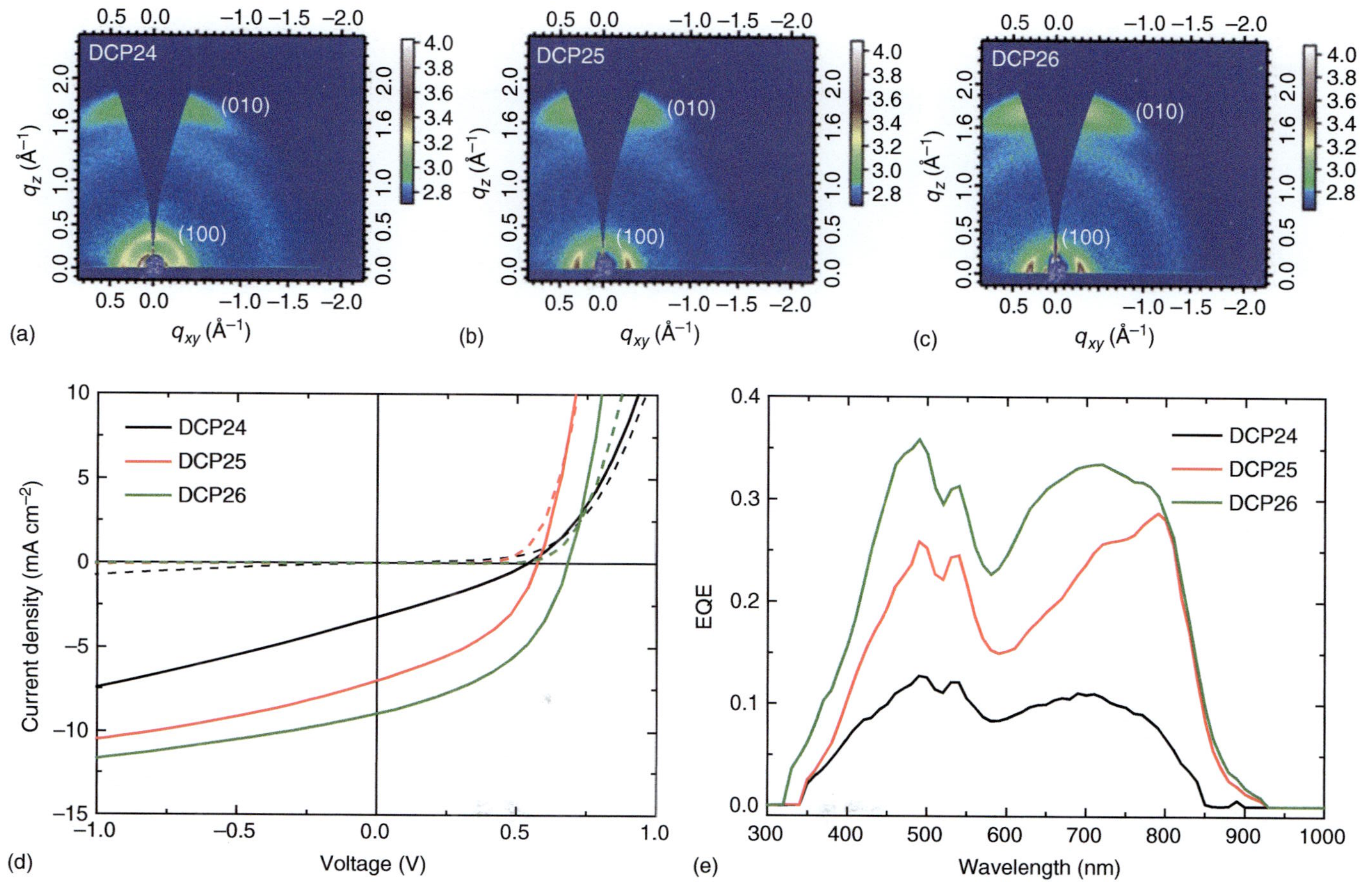

Figure 9.17 (a–c) GIWAXS patterns of **DCP24–26** in thin films. (d) *J–V* characteristics and (e) EQE spectra of the related polymer-based SCOSCs. Source: Lai et al. [76]/Reproduced with permission of American Chemical Society.

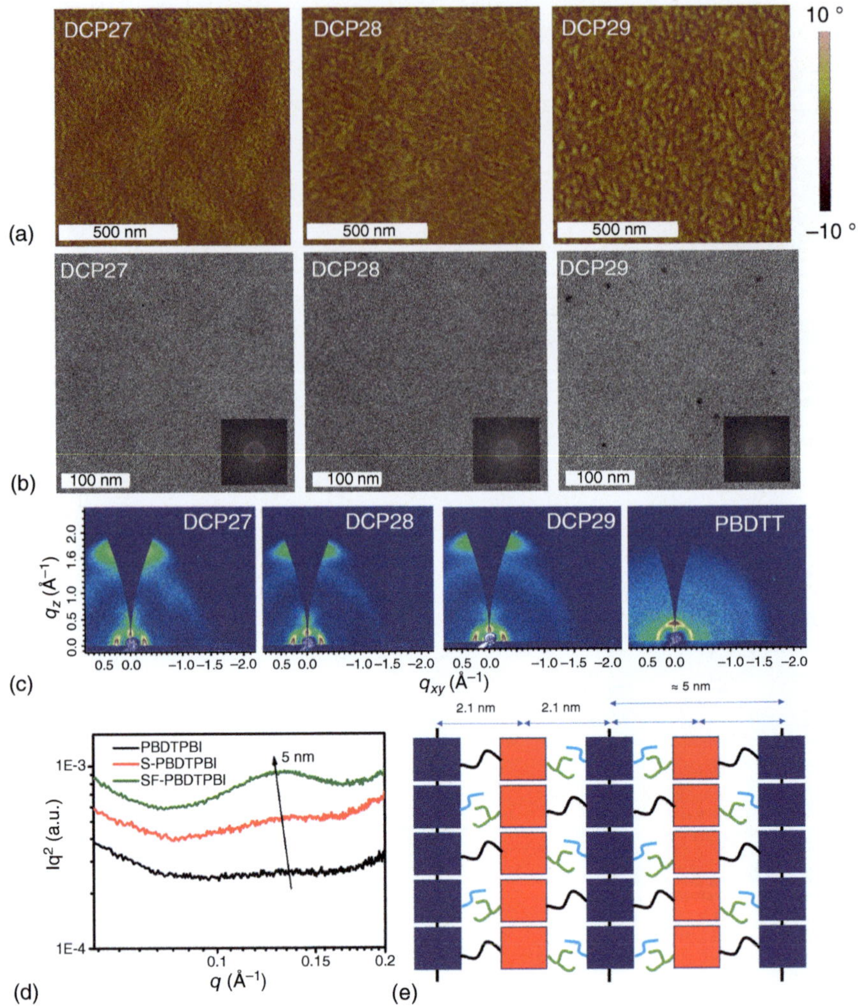

Figure 9.18 (a) AFM phase images, (b) TEM images, (c) GIWAXS patterns, (d) GISAXS profiles, and (e) the illustration of the polymers in nanostructures. Top view: BDT is blue color and PBI is red color. Source: Feng et al. [77]/Reproduced with permission of American Chemical Society.

confirmed by the TEM images. Then they illustrated the nanophase separation of the polymers, as schematically shown in Figure 9.18e, which could ensure efficient exciton dissociation into free charge carriers and provide percolating pathways for both electrons and holes.

To get better nanophase separation, they designed a new polymer **DCP35** with a strongly crystalline backbone. After annealing at a high temperature, both the backbone and the pendant acceptor units could self-aggregate into ordered nanostructures. As shown in Figure 9.19a,b, from the grazing-incidence medium-angle X-ray scattering (GIMAXS) profiles, four scattering peaks that can be observed in the thin film annealed at 280 °C, which is a typical scattering feature of lamellar

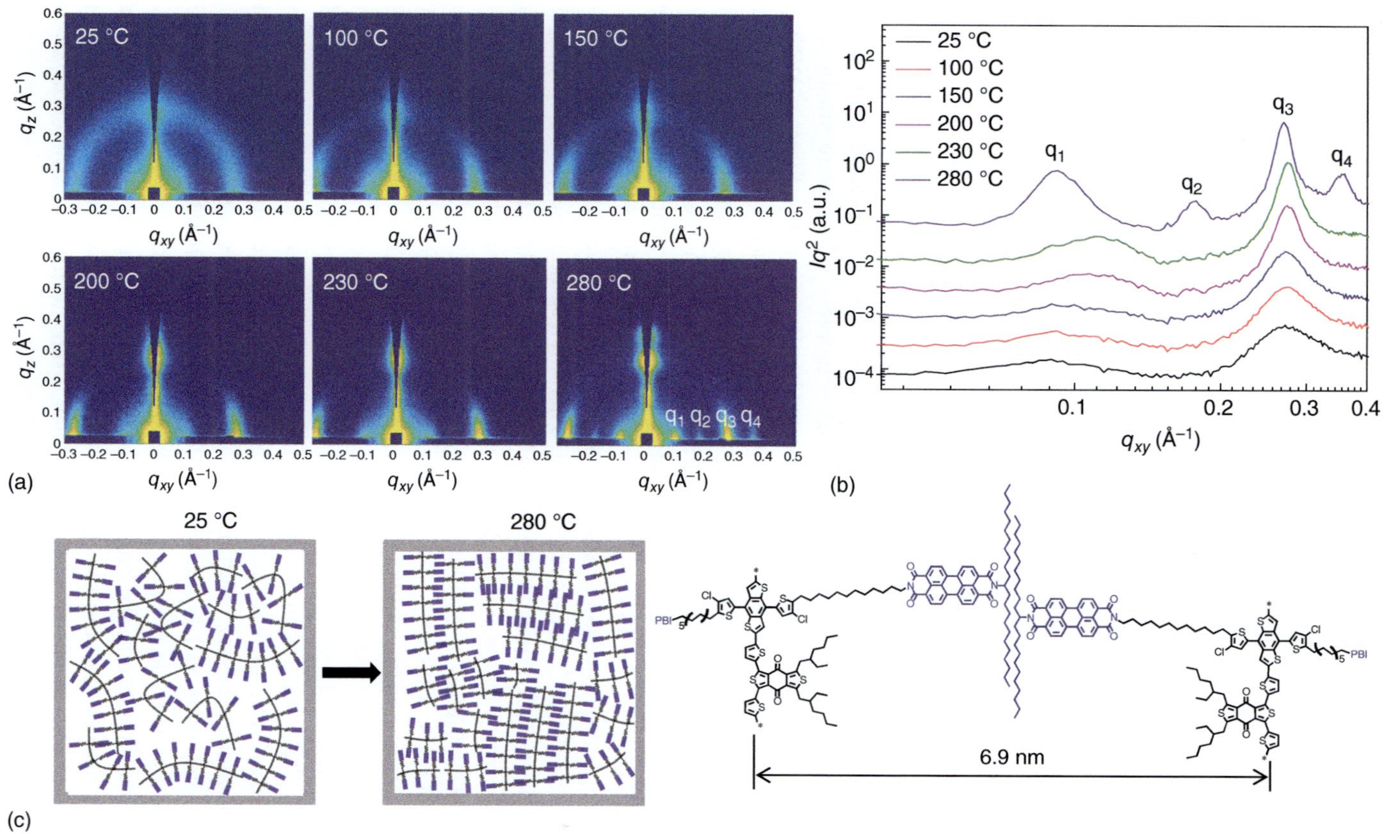

Figure 9.19 (a) GIMAXS profiles of thin films of **DCP35** with different annealing temperature, (b) the corresponding GIMAXS scattering curves (at Yoneda maximum). (c) Proposed illustration of the morphology evolution of the polymer films under thermal annealing from 25 °C to 280 °C. A simple scheme of *d*-spacing (6.9 nm) is also present. Source: Feng et al. [80]/Reproduced with permission of Elsevier.

structures with an alternating arrangement of donor and acceptor segments. And a 6.9 nm of *d*-spacing can be obtained from the peak *q*1, which should be the distance of the lamellar stacking between two backbones, as shown in Figure 9.19c. However, the peak *q*1 is obscure when the polymer thin films were annealed at low temperatures, especially those in the as-cast films indicated a highly amorphous and disordered nanostructure. After thermal annealing at high temperature, taking the GIWAXS results into consideration, they inferred that the degree of crystallization in the polymer film was enhanced by increasing the annealing temperature. Therefore, the nanophase separation was optimized, resulting in high charge transport efficiency, low charge recombination, and an excellent PCE of 6.3% in SCOSCs [80].

Li and coworkers also reported polymers **DCP36–37** and raised a new term of crystalline cooperativity in double-cable polymers. **DCP37** is a derivative of **DCP35**, from which Cl atoms on the alkyl-thiophene side groups were cleared away. As shown in Figure 9.20, the films of **DCP37** showed the same crystalline properties as **DCP35**, with good crystalline cooperativity between the conjugated backbones and the side units. However, **DCP36** presented completely different crystalline characteristics as its acceptor PBI units are on the same side of the backbones.

From the AFM height images, thin films of **DCP37** showed fibrous structures with a smooth surface and the RMS roughness values were small, only from 0.81 to 1.09 nm, even annealed at 200 °C. However, the films of **DCP36** presented notable granular crystals with a rough surface when annealed at high temperatures. For the GIWAXS, the coherence lengths (CLs) of (010) diffraction peaks in **DCP36** and **DCP37** increased with the rising annealing temperature, while CLs of (010) diffraction peak in their related donor polymer poly[(2,6-(4,8-bis(5-(2-ethylhexyl)thiophen-2-yl)-benzo[1,2-b:4,5-b′]dithiophene))-*alt*-(5,5-(1′,3′-di-2-thienyl-5′,7′-bis(2-ethylhexyl)benzo[1′,2′-c:4′,5′-c′]dithiophene-4,8-dione)] (PBDB-T) remained constant. This inferred that the enhanced CLs of (010) peaks were driven by the enhanced crystallinity of PBIs at high temperatures. Otherwise, when increasing annealing temperature, the *d*-spacings of (h00) peaks in **DCP36** are significantly enhanced, which may be caused by the large crystalline domain of **PBI**s, while they were almost unchanged in **DCP37**. In consideration of the absorption, mobility, and the results of AFM and GIWAXS, the PBIs in **DCP36** showed a strong tendency of crystallization and enhanced the crystallization of conjugated backbones, inducing poor crystalline cooperativity. This caused bad nanophase separation in **DCP36** thin films but large phase separation in **DCP37** thin films, so that they had completely different charge generation and recombination processes, resulting in different performances in their SCOSCs.

When they used NDI to replace the PBI side units in **DCP35** and synthesized two new double-cable polymers (**DCP39, DCP40**) with isomeric structures, they studied the influence of the miscibility of the donor and acceptor segments. The GIWAXS showed that the two polymers had similar crystallinity in the crystal region, as shown in Figure 9.21a–c. However, from the study of differential scanning calorimetry (DSC) and contact angle, they got the Flory–Huggins interaction parameter (χ), which could reflect the mixability between donor and acceptor segments in the amorphous region of the double-cable polymers. As a result,

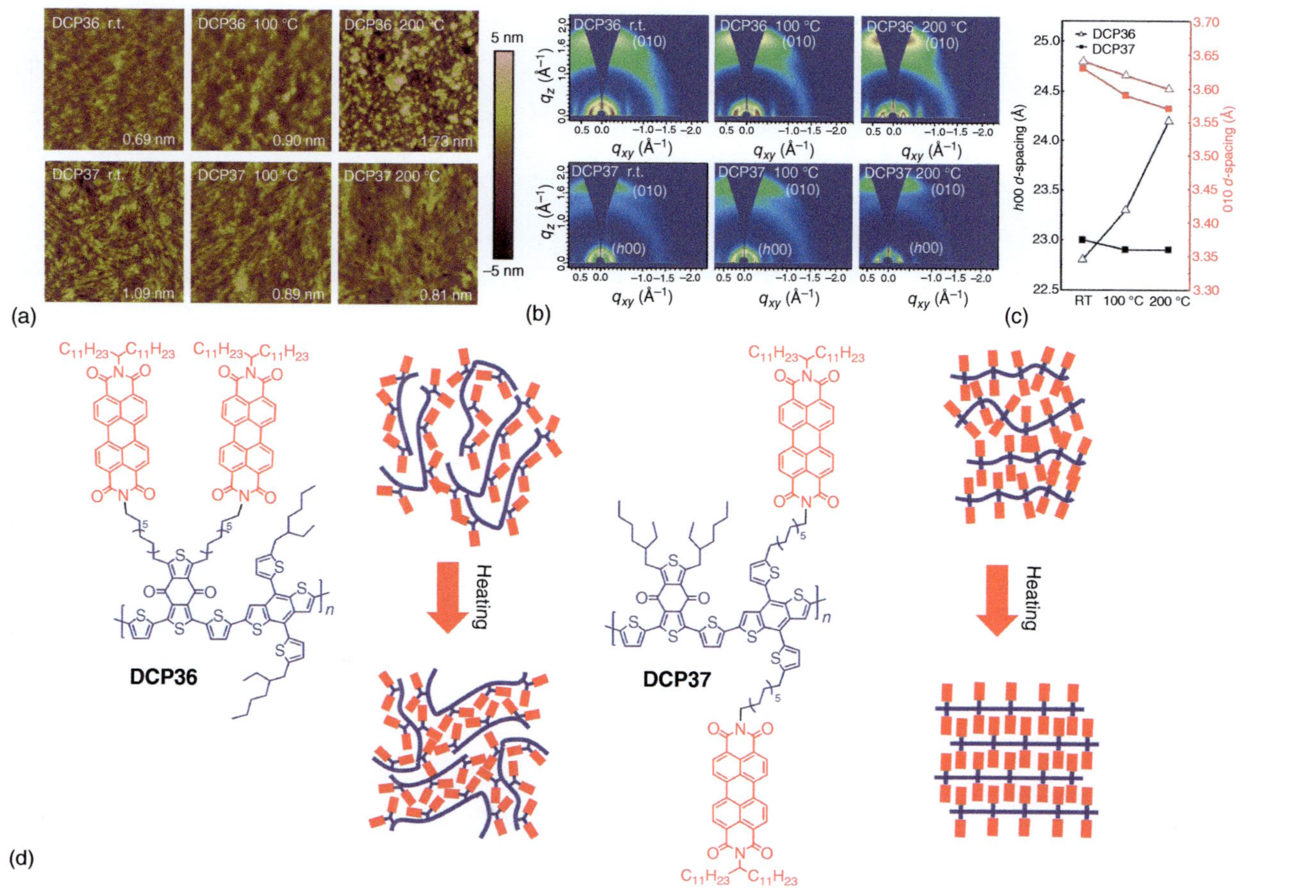

Figure 9.20 (a) AFM height images, (b) GIWAXS patterns, and (c) the related *d*-spacings of (h00) and (010) peaks of the polymer films with different annealing temperature. (d) The chemical structures and proposed phase separation models under thermal treatment. Source: Li et al. [81]/Reproduced with permission of John Wiley & Sons.

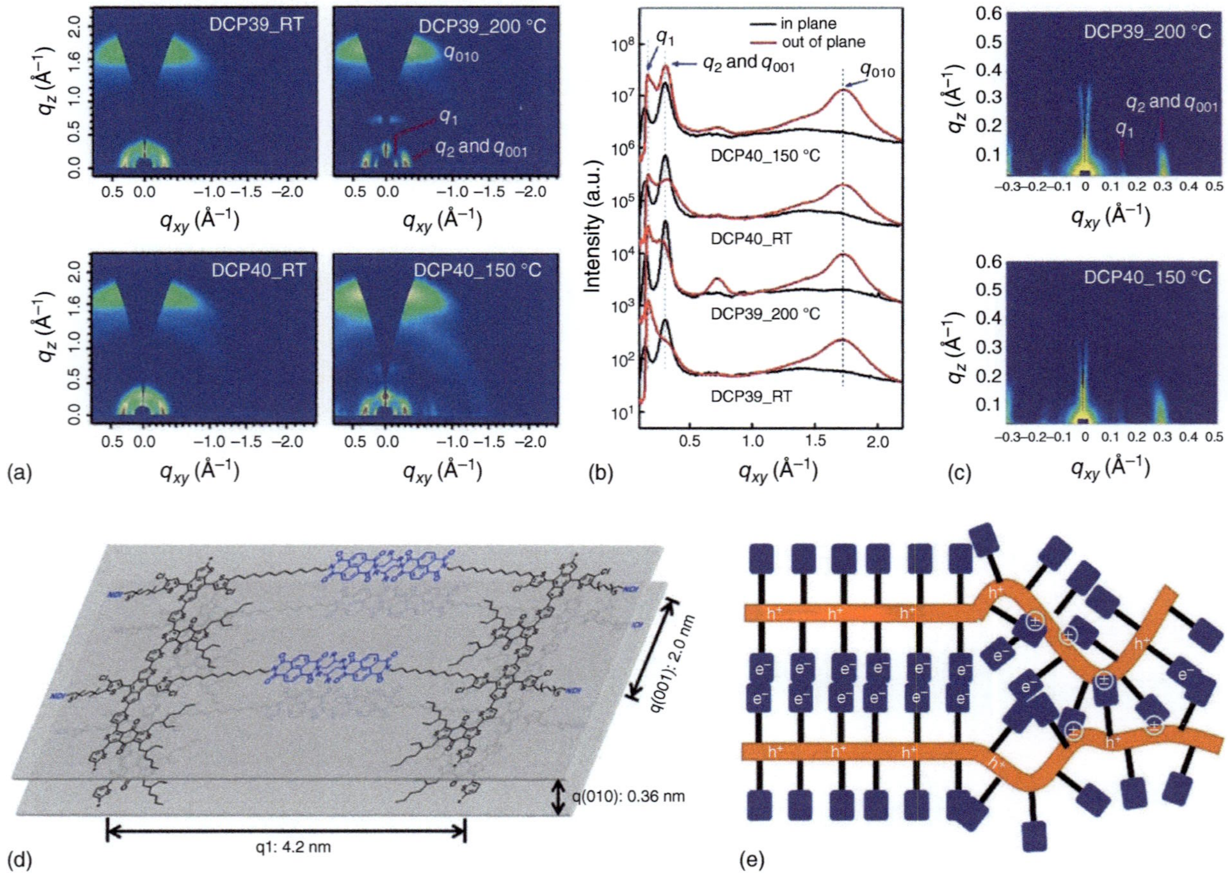

Figure 9.21 (a) GIWAXS patterns, (b) the OOP and IP cut-lines of the corresponding GIWAXS patterns, and (c) GIMAXS of the two polymers with different annealing temperature, (d) The idealized model to illustrate the crystalline region in **DCP39** film. (e) The crystalline and amorphous regions of double-cable polymers and the path for exciton separation and charge transport. Source: Jiang et al. [71]/Reproduced with permission of John Wiley & Sons.

DCP39 had a higher χ (1.02), which is almost twofold of the χ in **DCP40**, indicating that the conjugated backbones and NDIs in **DCP40** had better miscibility to create more interface to promote the exciton dissociation (Figure 9.21d). Consequently, **DCP40**-based SCOSCs could provide a record PCE of 8.4% with the highest EQE over 0.75, much better than the performance of **DCP39**.

Above all, these auspicious results provided several methods to tune the nanophase separation or miscibility in double-cable polymers and suggested promising guidelines to design high-performance SCOSCs.

9.4 Perspective and Challenges of SCOSCs

Although new materials with exhilarating photovoltaic performance have been already applied in the SCOSC field, there are still two essential challenges:

(1) Charge generation process in SCOSC system, including the exciton diffusion, charge separation, and transports. Only a few studies revealed that efficient charge separation could be obtained in some BCPs [53, 90, 91]. Despite the limited studies, some interesting results could also be obtained in double-cable polymers, in which the IQEs over 0.80 and FFs over 0.65 could be realized, suggesting highly efficient transport of charge carriers in the SC active layers [80]. However, the charge generation process in SCOSCs is far to be understood, such as how to balance the exciton separation/charge transport in nanoscale, how to define the interface of donor/acceptor (as face-on or edge-on) [92], the charge recombination mechanism, etc.
(2) V_{loss}, defined as the difference between V_{oc} and optical bandgap of the active layer, was found to be high, ~0.70 V, for the SCOSCs. For high-performance BHJOSCs, V_{loss} is often in the range of 0.4–0.5 V. In general, V_{loss} can be divided into three parts, which are the charge transfer (CT) voltage losses (ΔV_{CT}), i.e. losses associated with the driving force needed for efficient charge transfer from the exciton state to CT state; radiative recombination voltage losses (ΔV_{rad}), related to radiative recombination of charge carriers, and nonradiative voltage losses ($\Delta V_{non\text{-}rad}$). All of these losses in BHJOSCs are reduced thanks to the development of fused-ring non-fullerene electron acceptors. However, these acceptors are yet to be incorporated into SC polymers, and the intrinsic mechanism of V_{loss} in SCOSCs has been rarely studied.

Nevertheless, in this "dawn of SCOSCs" [17], more efforts should be provided to further enhance the performance of SCOSCs. The efforts included: (i) developing better designed SC conjugated materials, for instance, via introducing electron acceptors with the absorption spectra in the NIR region, (ii) understanding the evolution of microstructures in these materials by using new techniques, such as Raman and IR spectroscopy, Fourier transformed pump-probe spectroscopy in the fs regime, (iii) studying the charge generation process in SCOSCs, especially compared with that in BHJOSCs, (iv) stability and mechanical properties, and the study in large-area flexible SCOSCs should be paid special attention. Success in these areas will make SCOSCs more attractive and contribute more to the development of the field of OSCs.

List of Symbols and Abbreviations

J_{sc}	Short current density
V_{loss}	Voltage loss
V_{oc}	Open-circuit voltage
ΔV_{CT}	Charge transfer voltage losses
$\Delta V_{non\text{-}rad}$	Nonradiative voltage losses
ΔV_{rad}	Radiative recombination voltage losses
A	Acceptor
AFM	Atomic Force Microscope
Ag	Silver
AL	Active layer
Al	Aluminum
BCP	Block copolymer
BDT	Benzodithiophene
BHJ	Bulk-heterojunction
BHL	Hole blocking layer
BT	Benzodithiazole
CL	Coherence length
CT	Charge transfer
D	Donor
DCP	Double-cable conjugated polymers
DPP	Diketopyrrolopyrrole
EQE	External quantum efficiency
FET	Field-effect transistor
FF	Fill factor
GIMAXS	Grazing-incidence medium-angle X-ray scattering
GISAXS	Grazing-incidence small-angle X-ray scattering
GIWAXS	Grazing-incidence wide-angle X-ray scattering
IQE	Internal quantum efficiency
ITO	Indium tin oxide
KPFM	Kelvin probe force microscope
LiF	Lithium fluoride
MoO_x	Molybdenum oxide
NDI	Naphthalene diimide
NIR	Near-infrared
n-PV	Phenylenevinylene oligomers
OFbTs	Oligo(fluorene-*alt*-bithiophene)s
OSC	Organic solar cell
PBIs	Perylene bisimides
PCE	Power conversion efficiency

PDI	Polymer Dispersion Index
PEDOT:PSS	Poly(3,4-ethylenedioxythiophene)polystyrene sulfonate
RMS	Root-mean-square
SCOSCs	Single-component organic solar cells
SVA	Solvent vapor annealing
TA	Thermal annealing
TEM	Transmission electron microscopy
XRD	X-Ray Diffraction

References

1 Yan, N., Zhao, C., You, S. et al. (2020). *Chin. Chem. Lett.* 31: 643.
2 Needell, D.R., Phelan, M.E., Hartlove, J.T., and Atwater, H.A. (2021). *Energy* 219: 119567.
3 Wan, J., Xia, Y., Fang, J. et al. (2021). *Nanomicro. Lett.* 13: 44.
4 Pan, W., Han, Y., Wang, Z. et al. (2021). *J. Semicond.* 42: 050301.
5 Wang, Z.Y., Xu, M.C., Li, Z.L. et al. (2021). *Adv. Funct. Mater.* 31: 2103534.
6 Yang, W., Wang, W., Wang, Y. et al. (2021). *Joule* 5: 1209.
7 Zhu, C., Meng, L., Zhang, J. et al. (2021). *Adv. Mater.* 33: e2100474.
8 Liu, F., Zhou, L., Liu, W. et al. (2021). *Adv. Mater.* 33: e2100830.
9 Ma, J., Feng, G., Liu, F. et al. (2019). *Dyes Pigm.* 165: 335.
10 Yu, Y., Zhou, S., Wang, X. et al. (2017). *Org. Electron.* 47: 133.
11 Li, C., Zhang, A., Feng, G. et al. (2016). *Org. Electron.* 35: 112.
12 Feng, G., Xu, Y., Xiao, C. et al. (2016). *Polym. Chem.* 7: 164.
13 Cui, Y., Yao, H., Zhang, J. et al. (2020). *Adv. Mater.* 32: e1908205.
14 Karki, A., Gillett, A.J., Friend, R.H., and Nguyen, T.Q. (2020). *Adv. Energy Mater.* 11: 2003441.
15 Cai, Y., Li, Y., Wang, R. et al. (2021). *Adv. Mater.* 33: e2101733.
16 Wang, Y., Lee, J., Hou, X. et al. (2020). *Adv. Energy Mater.* 11: 2003002.
17 Roncali, J. and Grosu, I. (2019). *Adv. Sci.* 6: 1801026.
18 He, Y., Heumüller, T., Lai, W. et al. (2019). *Adv. Energy Mater.* 9: 1900409.
19 Liang, S., Jiang, X., Xiao, C. et al. (2021). *Acc. Chem. Res.* 54: 2227.
20 Chamberlain, G.A. (1983). *Sol. Cells* 8: 47.
21 Eckert, J.-F., Nicoud, J.-F., Nierengarten, J.-F. et al. (2000). *J. Am. Chem. Soc.* 122: 7467.
22 Nierengarten, J.-F., Eckert, J.-F., Nicoud, J.-F. et al. (1999). *Chem. Commun.*: 617.
23 Peeters, E., Van Hal, P.A., Knol, J. et al. (2000). *J. Phys. Chem. B* 104: 10174.
24 Guldi, D.M., Luo, C., Swartz, A. et al. (2002). *J. Org. Chem.* 67: 1141.
25 Yamashiro, T., Aso, Y., Otsubo, T. et al. (1999). *Chem. Lett.* 28: 443.
26 Negishi, N., Yamada, K., Takimiya, K. et al. (2003). *Chem. Lett.* 32: 404.
27 Narayanaswamy, K., Venkateswararao, A., Nagarjuna, P. et al. (2016). *Angew. Chem. Int. Ed.* 55: 12334.
28 Hirayama, D., Yamashiro, T., Takimiya, K. et al. (2000). *Chem. Lett.* 29: 570.
29 Negishi, N., Takimiya, K., Otsubo, T. et al. (2004). *Chem. Lett.* 33: 654.

30 Negishi, N., Takimiya, K., Otsubo, T. et al. (2005). *Synth. Met.* 152: 125.
31 Fernández, G., Sánchez, L., Veldman, D. et al. (2008). *J. Org. Chem.* 73: 3189.
32 Nishizawa, T., Tajima, K., and Hashimoto, K. (2007). *J. Mater. Chem.* 17: 2440.
33 Nishizawa, T., Tajima, K., and Hashimoto, K. (2008). *Nanotechnology* 19: 424017.
34 Nishizawa, T., Lim, H.K., Tajima, K. et al. (2009). *Chem. Commun.* 18: 2469.
35 Izawa, S., Hashimoto, K., and Tajima, K. (2011). *Chem. Commun.* 47: 6365.
36 Chen, T.L., Zhang, Y., Smith, P. et al. (2011). *ACS Appl. Mater. Interfaces* 3: 2275.
37 Liu, Y., Chen, C.-C., Hong, Z. et al. (2013). *Sci. Rep.* 3: 3356.
38 Nguyen, T.L., Lee, T.H., Gautam, B. et al. (2017). *Adv. Funct. Mater.* 27: 1702474.
39 Wang, W., Sun, R., Guo, J. et al. (2019). *Angew. Chem. Int. Ed.* 58: 14556.
40 Bu, L., Guo, X., Yu, B. et al. (2009). *J. Am. Chem. Soc.* 131: 13242.
41 Bu, L., Guo, X., Yu, B. et al. (2011). *Polymer* 52: 4253.
42 Qu, J., Gao, B., Tian, H. et al. (2014). *J. Mater. Chem. A* 2: 3632.
43 Xia, D., Yang, F., Li, J. et al. (2019). *Mater. Chem. Front.* 3: 1565.
44 Neuteboom, E.E., Meskers, S.C.J., Van Hal, P.A. et al. (2003). *J. Am. Chem. Soc.* 125: 8625.
45 Mikroyannidis, J.A., Stylianakis, M.M., Sharma, G.D. et al. (2009). *J. Phys. Chem. C* 113: 7904.
46 Khlyabich, P.P., Burkhart, B., Ng, C.F., and Thompson, B.C. (2011). *Macromolecules* 44: 5079.
47 Guo, C.H., Lin, Y.H., Witman, M.D. et al. (2013). *Nano Lett.* 13: 2957.
48 Smith, K.A., Lin, Y.-H., Mok, J.W. et al. (2015). *Macromolecules* 48: 8346.
49 Mok, J.W., Lin, Y.-H., Yager, K.G. et al. (2015). *Adv. Funct. Mater.* 25: 5578.
50 Le, T.P., Smith, B.H., Lee, Y. et al. (2020). *Macromolecules* 53: 1967.
51 Lombeck, F., Komber, H., Sepe, A. et al. (2015). *Macromolecules* 48: 7851.
52 Lee, D.H., Lee, J.H., Kim, H.J. et al. (2017). *J. Mater. Chem. A* 5: 9745.
53 Lee, J.H., Park, C.G., Kim, A. et al. (2018). *ACS Appl. Mater. Interfaces* 10: 18974.
54 Park, C.G., Park, S.H., Kim, Y. et al. (2019). *J. Mater. Chem. A* 7: 21280.
55 Park, S.H., Kim, Y., Kwon, N.Y. et al. (2020). *Adv. Sci.* 7: 1902470.
56 Wang, S., Yang, Q., Tao, Y. et al. (2016). *New J. Chem.* 40: 1825.
57 Rahmanudin, A., Yao, L., Sekar, A. et al. (2019). *ACS Macro Lett.* 8: 134.
58 Benincori, T., Brenna, E., Sannicolò, F. et al. (1996). *Angew. Chem. Int. Ed. Engl.* 35: 648.
59 Ferraris, J.P., Yassar, A., Loveday, D.C., and Hmyene, M. (1998). *Opt. Mater.* 9: 34.
60 Ramos, A.M., Rispens, M.T., van Duren, J.K. et al. (2001). *J. Am. Chem. Soc.* 123: 6714.
61 Zhang, F., Svensson, M., Andersson, M.R. et al. (2001). *Adv. Mater.* 13: 1871.
62 Tan, Z.A., Hou, J., He, Y. et al. (2007). *Macromolecules* 40: 1868.
63 Miyanishi, S., Zhang, Y., Tajima, K., and Hashimoto, K. (2010). *Chem. Commun.* 46: 6723.
64 Chen, P., Nakano, K., Suzuki, K. et al. (2017). *ACS Appl. Mater. Interfaces* 9: 4758.
65 Pierini, F., Lanzi, M., Nakielski, P. et al. (2017). *Macromolecules* 50: 4972.
66 Reyes-Reyes, M., Kim, K., and Carroll, D.L. (2005). *Appl. Phys. Lett.* 87: 083506.

67 Li, G., Shrotriya, V., Huang, J. et al. (2005). *Nat. Mater.* 4: 864.
68 Dang, M.T., Hirsch, L., and Wantz, G. (2011). *Adv. Mater.* 23: 3597.
69 Ma, W.L., Yang, C.Y., Gong, X. et al. (2005). *Adv. Funct. Mater.* 15: 1617.
70 Drewniak, A., Tomczyk, M.D., Hanusek, L. et al. (2018). *Polymers (Basel)* 10: 487.
71 Jiang, X., Yang, J., Karuthedath, S. et al. (2020). *Angew. Chem. Int. Ed.* 59: 21683.
72 Sommer, M., Lang, A.S., and Thelakkat, M. (2008). *Angew. Chem. Int. Ed.* 47: 7901.
73 Sommer, M., Lindner, S.M., and Thelakkat, M. (2007). *Adv. Funct. Mater.* 17: 1493.
74 Lindner, S.M., Hüttner, S., Chiche, A. et al. (2006). *Angew. Chem. Int. Ed.* 45: 3364.
75 Koyuncu, S., Wang, H.-W., Liu, F. et al. (2014). *J. Mater. Chem. A* 2: 2993.
76 Lai, W., Li, C., Zhang, J. et al. (2017). *Chem. Mater.* 29: 7073.
77 Feng, G., Li, J., Colberts, F.J.M. et al. (2017). *J. Am. Chem. Soc.* 139: 18647.
78 Li, C., Yu, C., Lai, W. et al. (2018). *Macromol. Rapid Commun.* 39: e1700611.
79 Yu, C., Xu, Y., Li, C. et al. (2018). *Chin. J. Chem.* 36: 515.
80 Feng, G., Li, J., He, Y. et al. (2019). *Joule* 3: 1765.
81 Li, C., Wu, X., Sui, X. et al. (2019). *Angew. Chem. Int. Ed.* 58: 15532.
82 Yang, F., Li, J., Li, C., and Li, W. (2019). *Macromolecules* 52: 3689.
83 Lindner, S.M. and Thelakkat, M. (2004). *Macromolecules* 37: 8832.
84 Zhang, Q., Cirpan, A., Russell, T.P., and Emrick, T. (2009). *Macromolecules* 42: 1079.
85 Koyuncu, S., Zafer, C., Koyuncu, F.B. et al. (2009). *J. Polym. Sci., Part A: Polym. Chem.* 47: 6280.
86 Mohamad, D.K., Fischereder, A., Yi, H. et al. (2011). *J. Mater. Chem.* 21: 851.
87 Yang, F., Wang, X., Feng, G. et al. (2018). *Sci. China Chem.* 61: 824.
88 Liang, S., Xu, Y., Jiang, X. et al. (2019). *Dyes Pigm.* 170: 107575.
89 Yu, P., Feng, G., Li, J. et al. (2020). *J. Mater. Chem. C* 8: 2790.
90 Johnson, K., Huang, Y.-S., Huettner, S. et al. (2013). *J. Am. Chem. Soc.* 135: 5074.
91 Yamamoto, S., Yasuda, H., Ohkita, H. et al. (2014). *J. Phys. Chem. C* 118: 10584.
92 Ran, N.A., Roland, S., Love, J.A. et al. (2017). *Nat. Commun.* 8: 79.

10

Tandem Organic Solar Cells: Recent Progress and Challenge

Lingxian Meng, Xiangjian Wan, and Yongsheng Chen

Nankai University, The Centre of Nanoscale Science and Technology and Key Laboratory of Functional Polymer Materials, College of Chemistry, 94 Weijin Road, Tianjin, 300071, P.R. China

10.1 Introduction

Organic solar cells (OSCs) have drawn great attentions with the advantages of low cost, flexibility, lightweight, and large-area printing [1–8]. In the past few years, benefiting from the new active layer materials design, mechanism understanding, and device optimization, OSCs have witnessed rapid development in power conversion efficiencies (PCEs), reaching over 17% for single-junction and tandem solar cells [9–11]. Although with great success, the PCEs of OSCs are still far behind than those of inorganic material-based solar cells such as silicon and recently developed perovskite solar cells [12–14]. For single-junction photovoltaic devices, the maximum theoretical PCE is defined by the Shockley–Queisser limit [15], which is independent on the particular type of materials. For OSCs, some intrinsic and special properties of organic semiconductor materials are responsible for the less efficiencies of OSCs. For example, the low charge mobilities and narrow absorption windows of most organic materials make OSC devices have only limited thickness of the active layers and low current in many cases [16, 17]. However, organic materials have their own intrinsic advantages. Unlike inorganic materials, organic materials have great varieties and their optical and physical properties could be tuned effectively from the molecule design [3, 18–20]. In principle, organic materials with any bandgaps could be designed and synthesized. Most importantly, organic materials could be solution processed using the existing industrial printing techniques such as screen printing and roll-to-roll printing in large scale. Therefore, how to take the advantages of organic materials and get high efficiencies have been extensively explored in OSC community. To this end, one of the effective strategies is to develop tandem cells with stacking two or more cells with complementary absorption spectra in series or parallel connection, which could harvest photons at the highest possible potential and simultaneously conquer thermalization losses for single-junction solar cells (Figure 10.1) [21–24]. The varieties and tunable bandgaps of organic materials make them ideal candidates for use in tandem cells. Since the physical connection of the

Organic Solar Cells: Materials Design, Technology and Commercialization, First Edition. Edited by Liming Ding.

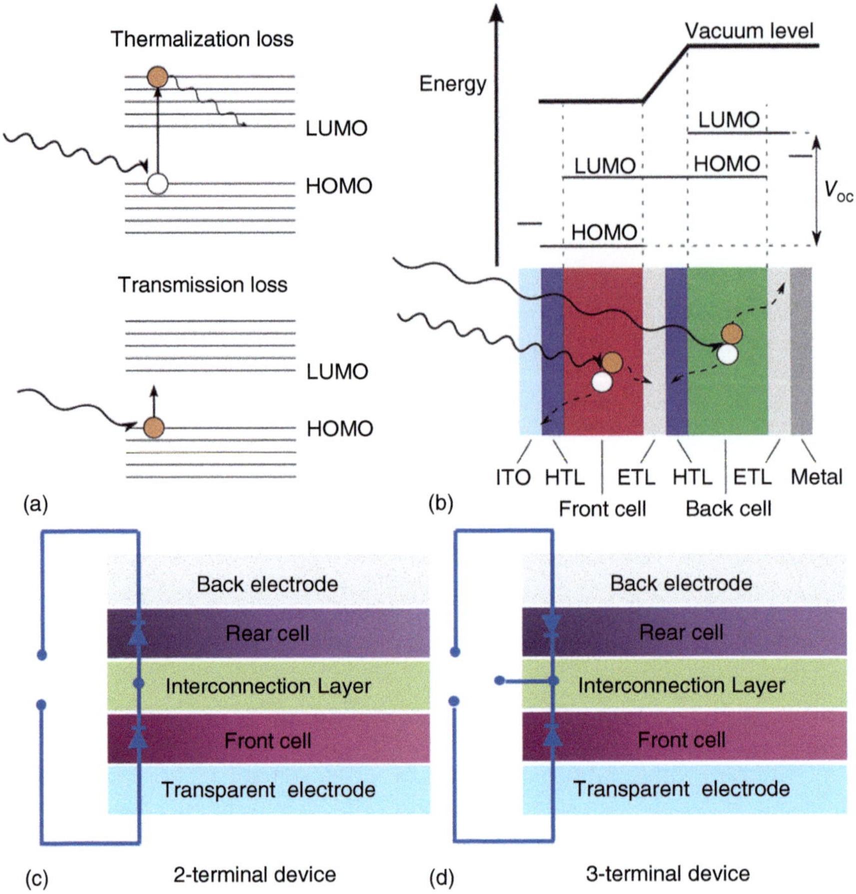

Figure 10.1 (a) Thermalization and transmission losses. (b) Arrangement of functional layers and energy levels in a tandem OSC. The ETL/HTL stack between the subcells forms the interconnection layer. Source: Di Carlo Rasi and Jonssen [21]. Reproduced with permission of John Wiley & Sons. (c) The device structure of tandem OSCs with series connection and (d) parallel connection. Source: Lingxian Meng, Xiangjian Wan, Yongsheng Chen.

subcells in parallel architectures is challenging and hard to compatible with future solution-processed production of OSCs, the vast majority of tandem OSCs are series connection, in which the active layers are stacked in series through interconnection layers (ICLs) (Figure 10.1c,d). Note that the tandem OSCs in this chapter all refer to the series-connection devices, i.e. 2-terminal tandem devices without special note.

To obtain high PCEs for the series-connection tandem devices, there are several factors that need to be considered simultaneously. Firstly, the subcells should have complementary and wide absorptions and thus well-balanced and high short-circuit current density (J_{sc}) of the tandem devices are expected. Secondly, the subcells should have minimal energy loss (E_{loss}) and high open-circuit voltage (V_{oc}) since the V_{oc} of the series-connection tandem device equals to the sum of the V_{oc} values

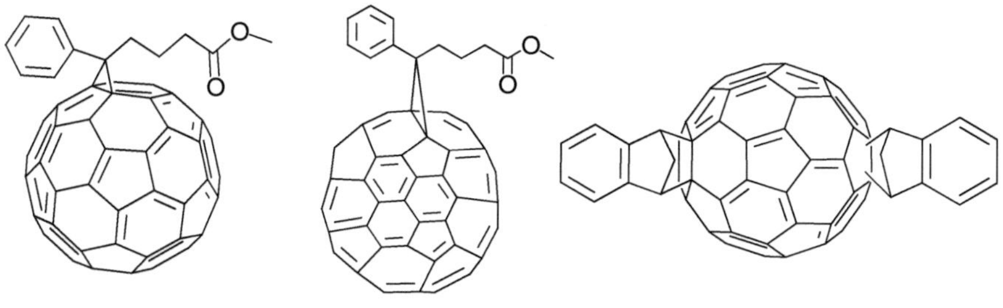

Figure 10.2 Fullerene acceptors used in tandem OSCs.

of each subcells if there is no potential loss in the ICLs. Thirdly, it is critical to have efficient and well-designed ICLs that can form ohmic contact with subcell active layers to ensure efficient charge extractions. In addition, the ICLs should be highly transparent and robust enough to protect underlying layers in the device fabrication process [22, 23].

In the past decade, great efforts have been devoted to tandem OSCs and the photovoltaic performance has reached 17% [11]. And some excellent and comprehensive reviews have been reported on this topic [21, 22, 25–30]. In this chapter, we will review the recent progress of tandem OSCs in the views of chemists. Firstly, we will review the recent progress of solution-processed tandem OSCs focusing on the subcell active layer materials. Secondly, the development and progress of ICLs will be discussed. Thirdly, the efficiency prediction for tandem OSCs will be discussed through a semi-empirical analysis model we reported earlier. Lastly, a summary and outlook with challenge analysis will be given.

10.2 Active Layer Materials in Tandem OSCs

Similar to the single-junction OSCs, active layer materials play the fundamental roles in the development of tandem OSCs. To exploit the full potential of the tandem OSCs, great efforts have been devoted to the subcell active layers, mainly focusing on the active materials design with different bandgaps to fulfill the requirement of complementary absorption range of tandem cells. With isotropic electron transfer capabilities and high electron mobilities, fullerene derivatives had been the standard acceptors since the invention of BHJ device in 1995 [31]. Thus, many wide- and low-bandgap donor materials have been designed and screened as front and rear active layers donors with fullerene derivative acceptors for fabricating tandem devices. And the PCEs increased from 6.5% in 2007 to 12.7% in 2017 for the tandem devices using fullerene derivatives as acceptors [2, 32]. However, some intrinsic properties of fullerene derivatives such as weak absorptions and hardly tunable energy levels significantly limit the further improvement performance of OSCs including the single-junction and tandem devices. To address the above issues, in recent years, various non-fullerene acceptors (NFAs) have been designed and made great success for single-junction and tandem devices [9, 19, 20, 30, 33–40].

To date, a high PCE of 17.3% has been achieved for the tandem device [11]. In this section, we will review the progress of tandem OSCs in the past decades, which is roughly divided into two parts based on the acceptors, i.e. fullerene acceptors and NFAs utilized in the tandem cells. The chemical structures, bandgap, and photovoltaic parameters of the representative materials used in tandem OSCs are listed in Table 10.1, 10.2, and Figure 10.2–10.5.

10.2.1 Tandem OSCs Based on Fullerene Acceptors

In the early years, the tandem OSCs were fabricated with the vacuum-deposited approach [22, 23, 28]. The first all solution-processed (except the top evaporated electrode) tandem OSC was reported by Kim et al. [2]. In the tandem device, PCPDTBT:PC_{61}BM and P3HT:PC_{71}BM were used as the front cell and rear cell, respectively. The complementary absorptions of PCPDTBT and P3HT made the whole cell covering the spectra range from 400 to nearly 900 nm and a PCE of 6.5% was achieved. This work demonstrated the great potential of the solution-processed tandem OSCs. However, in the following three years, little progress has been made on tandem OSCs, mainly limited by few active layer materials with suitable bandgaps, especially narrow bandgap materials. In 2010, Janssen and coworkers proposed a methodology by analyzing the optical absorption and electrical characteristics of individual single-junction subcells to identify the optimum device layout of the corresponding tandem device and could provide predicted optimal film thickness of the front and rear cells to obtain the better J_{sc} [41]. Guided by this methodology, a tandem device, comprising a wide-bandgap polymer PFTBT with a bandgap of 1.95 eV and a low-bandgap one (pBBTDPP2) with a bandgap of 1.40 eV mixed with PC_{61}BM as the electron acceptor, was fabricated, and eventually a PCE of 4.9% was achieved. In 2012, Yang and coworkers reported a low-bandgap polymer, PBDTT-DPP, with an optical bandgap of 1.44 eV and absorption onset of 858 nm [42]. The single-junction device of PBDTT-DPP:PC_{71}BM gave a PCE over 6%. P3HT and the indene-C60 bisadduct (IC_{60}BA) were selected as front cell materials owing to the high V_{oc} of the single-junction cell. Finally, a tandem OSC with inverted structure using P3HT:IC_{60}BA as the front cell materials and PBDTT-DPP:PC_{71}BM as the rear cell materials was fabricated and a certified PCE of 8.62% from NREL was achieved. Subsequently, they designed three polymers with low bandgap of 1.4–1.5 eV based on benzodithiophene (BDT) and diketopyrrolopyrrole (DPP) units [43]. The low bandgap and high performance of these polymer-based single-junction devices enabled them to be suitable as rear cells in tandem OSCs. With these polymers mixed with PC_{71}BM as the rear cells and P3HT:IC_{60}BA as the front cell, PCEs of tandem OSCs close to 9% were achieved. Subsequently, Janssen and coworkers reported a low-bandgap (1.46 eV) polymer of DPP derivative, PDPP5T, and used it as rear cell donor materials in tandem OSCs [44]. Despite with efficient PCE of 5.3%, PDPP5T:PC_{61}BM-based single-junction device only yielded a lower V_{oc} of 0.58 V. To ensure a high V_{oc} of the tandem device, they used PCDTBT:PC_{61}BM as the front cell with a high V_{oc} of 0.90 V. A PCE of 7.0% was achieved with a V_{oc} of 1.44 V, J_{sc} of 9.0 mA cm^{-2}, and fill factor (FF) of 0.54 for the tandem device. Later

Table 10.1 Photovoltaic performance of fullerene-based tandem organic solar cells covered by the review.

Front subcell			Back subcell				Tandem					
										PCE (%)		
Material	E_g (eV)	V_{oc} (V)	Material	E_g (eV)	V_{oc} (V)	Interconnection layer	V_{oc} (V)	J_{sc} (mA cm^{-2})	FF (%)	Average	Best	References
PCPDTBT:PCBM	1.40	0.66	P3HT:$PC_{71}BM$	1.90	0.63	TiO_x/PEDOT:PSS	1.24	7.8	67	/	6.5	[2]
PFTBT:$PC_{61}BM$	1.95	0.98	(pBBTDPP2):$PC_{61}BM$	1.40	0.61	ZnO/pH neutral PEDOT	1.58	6.0	52	/	4.9	[41]
P3HT:ICBA	1.9	0.85	PBDTT-DPP:$PC_{71}BM$	1.44	0.74	PEDOT:PSS/ZnO	1.56	8.62	66	/	8.62	[42]
P3HT:ICBA	1.9	0.85	PBDTP-DPP:$PC_{71}BM$	1.46	0.76	PEDOT:PSS/ZnO	1.58	8.41	64	/	8.6	[43]
P3HT:ICBA	1.9	0.85	PBDTT-FDPP:$PC_{71}BM$	1.55	0.77	PEDOT:PSS/ZnO	1.57	8.60	61	/	8.3	[43]
PCDTBT:$PC_{71}BM$	1.88	0.90	PDPP5T:$PC_{61}BM$	1.46	0.58	ZnO/PEDOT:PSS	1.44	9.0	54	/	7.0	[44]
P3HT:ICBA	1.85	0.83	PCPDTFBT:$PC_{71}BM$	1.37	0.74	Modified PEDOT:PSS/PEDOT:PSS (PH1000)/ZnO/C60-SAM	1.57	7.83	66.46	/	8.2	[45]
P3HT:ICBA	1.9	0.84	PDTP-DFBT:PCBM	1.38	0.68	PEDOT:PSS/ZnO	1.53	10.1	68.5	/	10.6	[46]
PDTP-DFBT:$PC_{71}BM$	1.38	0.68	PDTP-DFBT:$PC_{71}BM$	1.38	0.68	MoO_3/M-PEDOT:PSS/ZnO	1.36	11.5	65	/	10.2	[47]
PTB7-Th:$PC_{71}BM$	1.58	0.78	PTB7-Th:$PC_{71}BM$	1.58	0.78	ZnO/CPEPh-Na	1.54	11.11	66	/	11.3	[48]
PBDD4T-2F:$PC_{61}BM$	1.77	0.90	PDPP4T-2F:$PC_{71}BM$	1.37	0.79	MoO_3/Ag/PFN	1.68	11.29	61.30	11.32	11.62	[49]
PBDTFBZS:$PC_{71}BM$	1.81	0.88	PNDTDPP:$PC_{71}BM$	1.36	0.76	PEDOT:PSS/ZnO	1.59	9.10	65	/	9.40	[50]
PTIPSBDT-DFDTQX	1.85	0.81	PBPT-8:$PC_{71}BM$	1.56	0.71	*m*-PEDOT:PSS/PEIE	1.51	7.60	63	7.2	7.40	[51]
PDCBT:$PC_{71}BM$	1.90	0.80	PBDT-TS1:$PC_{61}BM$	1.51	0.80	MoO_3/Ag/ZnO NPs/PFN	1.60	11.55	54.47	10.02	10.16	[52]
PDTBTBz-$2F_{anti}$:$PC_{71}BM$	1.90	0.97	DT-PDPP2T-TT:$PC_{71}BM$	1.40	0.69	ZnO/Neutral PEDOT:PSS	1.58	9.89	66	/	10.3	[53]

(continued)

Table 10.1 (Continued)

Front subcell			Back subcell				Tandem					
Material	E_g (eV)	V_{oc} (V)	Material	E_g (eV)	V_{oc} (V)	Interconnection layer	V_{oc} (V)	J_{sc} (mA cm^{-2})	FF (%)	PCE (%) Average	PCE (%) Best	References
PTZ1:$PC_{71}BM$	1.97	0.94	PBDTTT-C-T:$PC_{71}BM$	1.58	0.74	MoO_3/Ag/PFN	1.65	9.6	65	10.1	10.3	[54]
PIDTBTO-TT:$PC_{71}BM$	1.87	0.91	PTB7-Th:$PC_{71}BM$	1.58	0.80	PDIN:PFN:/Al/MoO_3	1.70	9.68	67.69	10.63	11.15	[55]
P1(2F):$PC_{71}BM$	1.82	0.91	PTB7-Th:$PC_{71}BM$	1.58	0.80	ZnO/PEDOT:PSS	1.64	10.10	69	/	11.42	[56]
P2(4F):$PC_{71}BM$	1.82	1.00	PTB7-Th:$PC_{71}BM$	1.58	0.80	ZnO/PEDOT:PSS	1.72	9.28	63	/	10.05	[56]
DR3TSBDT:$PC_{71}BM$	1.77	0.910	PTB7-Th:$PC_{71}BM$	1.58	0.795	ZnO/*n*-PEDOT:PSS	1.689	10.51	64.6	11.21	11.47	[57]
DR3TBDTT:$PC_{71}BM$	1.72	0.881	PTB7-Th:$PC_{71}BM$	1.58	0.80	ZnO/*n*-PEDOT:PSS	1.643	10.30	61.5	10.52	10.73	[57]
DRBDT-TT:$PC_{71}BM$	1.78	0.910	PTB7-Th:$PC_{71}BM$	1.58	0.80	ZnO/*n*-PEDOT:PSS	1.671	10.24	59.6	10.28	10.43	[57]
SMPV1:$PC_{71}BM$	1.77	0.93	SMPV1:$PC_{71}BM$	1.77	0.93	CPE1/CPE2/M-PEDOT:PSS	1.82	7.7	72	/	10.1	[58]
PThBDTP:$PC_{71}BM$	1.86	0.95	DPPZnP-THE:PC61BM	1.37	0.78	PF3N-2TNDI/Ag/PEDOT:PSS 4083	1.67	9.85	69	/	11.35	[59]
DR3TSBDT:$PC_{71}BM$	1.77	0.914	DPPEZnP-TBO:$PC_{61}BM$	1.37	0.734	ZnO/*n*-PEDOT:PSS	1.623	12.17	63.3	12.28	12.50	[32]

Table 10.2 Photovoltaic performance of NFA-based tandem organic solar cells covered by the review.

Front subcell			Back subcell				Tandem					
Material	E_g (eV)	V_{oc} (V)	Material	E_g (eV)	V_{oc} (V)	Interconnection layer	V_{oc} (V)	J_{sc} (mA cm^{-2})	FF (%)	PCE (%) Average	PCE (%) Best	References
P3HT:SF(DPPB)$_4$	1.80	1.11	PTB7-Th:IEIC	1.55	0.95	MoO_3/Ag/PFN	1.97	8.28	52.0	8.21	8.48	[60]
P3TEA:SF-PDI$_2$	1.72	1.11	P3TEA:SF-PDI$_2$	1.72	1.11	PEDOT:PSS/ZnO	2.13	8.32	60.9	—	10.8	[61]
PBDTS-TDZ:ITIC	2.07	1.10	PBDTS-TDZ:ITIC	2.07	1.10	PEDOT:PSS/Ag/ZnO	2.13	9.77	64.1	12.92	13.35	[62]
PBDB-T:ITIC	1.59	0.92	PTB7-Th:4TIC	1.40	0.77	MoO_3/Ag/PFN	1.65	10.62	71.0	12.62	—	[63]
PTB7-Th:6TBA	1.52	0.98	PTB7-Th:4TIC	1.40	0.77	MoO_3/Ag/ZnO/PFN-Br	1.69	10.70	73	12.31	13.20	[64]
PBDB-T:IT-M	1.61	0.91	PTB7-Th:FOIC:F8IC	1.31	0.72	MoO_3/Mix-PEDOT:PSS/ZnO	1.62	10.6	73.9	12.7	13.3	[65]
PBDB-T:ITCC-M	1.70	1.00	PBDTTT-E-T:IEICO	1.34	0.92	ZnO/PCP-Na	1.80	12.0	63.9	13.1	13.8	[66]
J52-2F:IT-M	1.60	0.95	PTB7-Th:IEICO-4F	1.24	0.71	ZnO/PCP-Na	1.65	13.3	68.0	14.5	14.9	[67]
PBDB-T-2F:TfIF-4FIC	1.60	0.98	PTB7-Th:PCDTBT:IEICO-4F	1.24	0.69	PF3N-2TNDI/Ag/PEDOT:PSS	1.60	13.6	69.0	14.80	15.0	[68]
PM7:TfIF-4Cl	1.57	0.97	PTB7-Th:CO$_i$8DFIC:PC$_{71}$BM	1.20	0.69	ZnO:PEI/PEI/PEDOT:PSS	1.64	14.59	78	18.43	18.71	[69]
PBDB-T:IDTTA	1.75	0.98	PTB7-Th:IEICO-4F	1.24	0.71	MoO_3/Ag/ZnO	1.65	13.6	67	14.7	15.0	[70]
PBDB-T:F-M	1.65	0.98	PTB7-Th:NOBDT	1.39	0.77	ZnO/*n*-PEDOT:PSS	1.71	11.72	70.0	13.92	14.11	[71]
PBDB-T:F-M	1.65	0.99	PBDB-T:NNBDT	1.38	0.86	*m*-PEDOT:PSS/ZnO	1.82	10.68	74.7	14.03	14.52	[72]
DTDCPB:C$_{70}$	1.68	0.90	PTB7-Th:BT-ClC	1.38	0.69	Bphen:C60/Ag/PEDOT:PSS	1.59	13.3	71.0	—	15.0	[73]
DTDCPB:C$_{70}$	1.68	0.90	PTB7-Th:TT-FIC:BT-ClC	1.38	0.69	Bphen:C60/Ag/PEDOT:PSS	1.56	13.8	71	15.4		[74]
PBDB-T:F-M	1.65	0.94	PTB7-Th:O6T-4F:PC$_{71}$BM	1.20	0.69	*m*-PEDOT:PSS/ZnO	1.64	14.32	72.1	16.89	17.32	[11]
PM6:*m*-DTC-2F	1.61	0.99	PTB7-Th:BTPV-4F:PC$_{71}$BM	1.21	0.67	MoO_3/M-PEDOT:PSS/ZnO	1.65	14.5	68.5	15.9	16.4	[75]

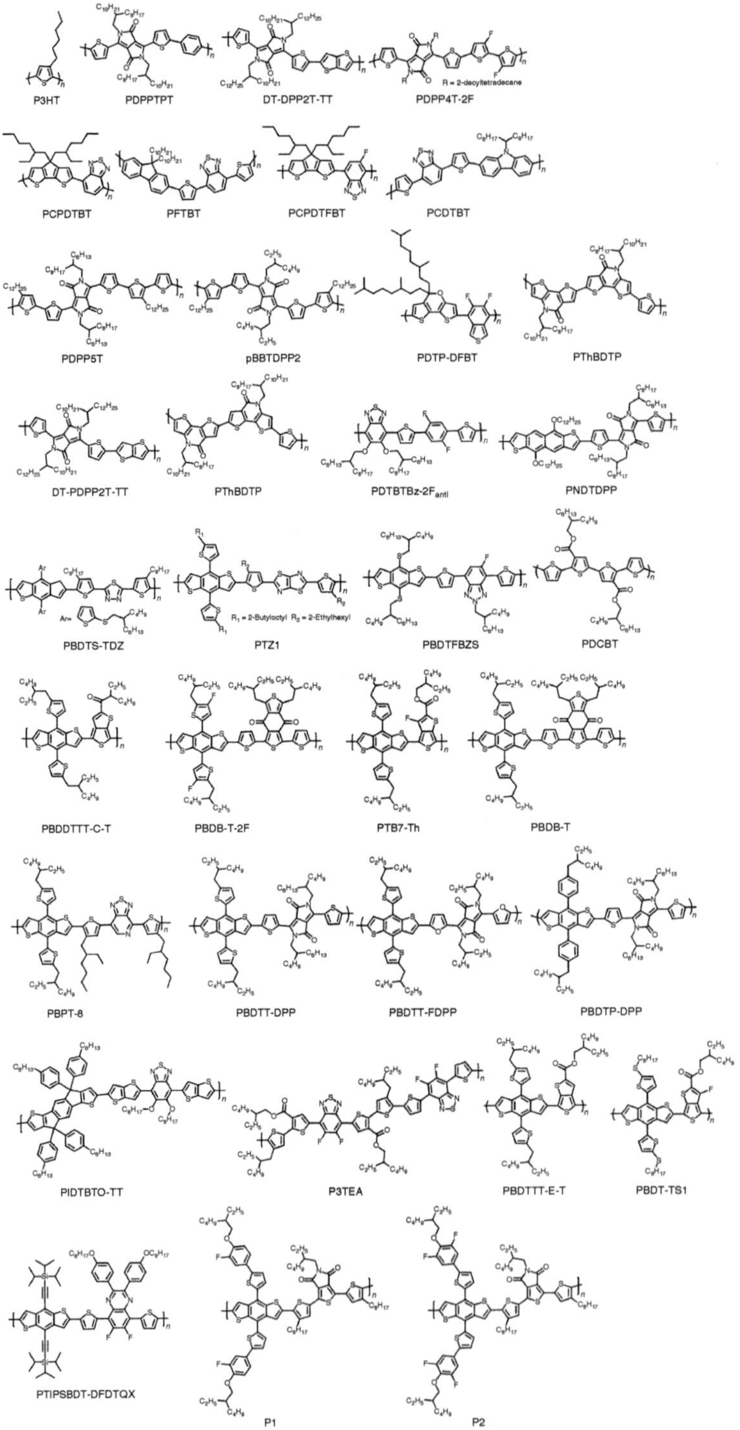

Figure 10.3 Polymer donor materials used in tandem OSCs.

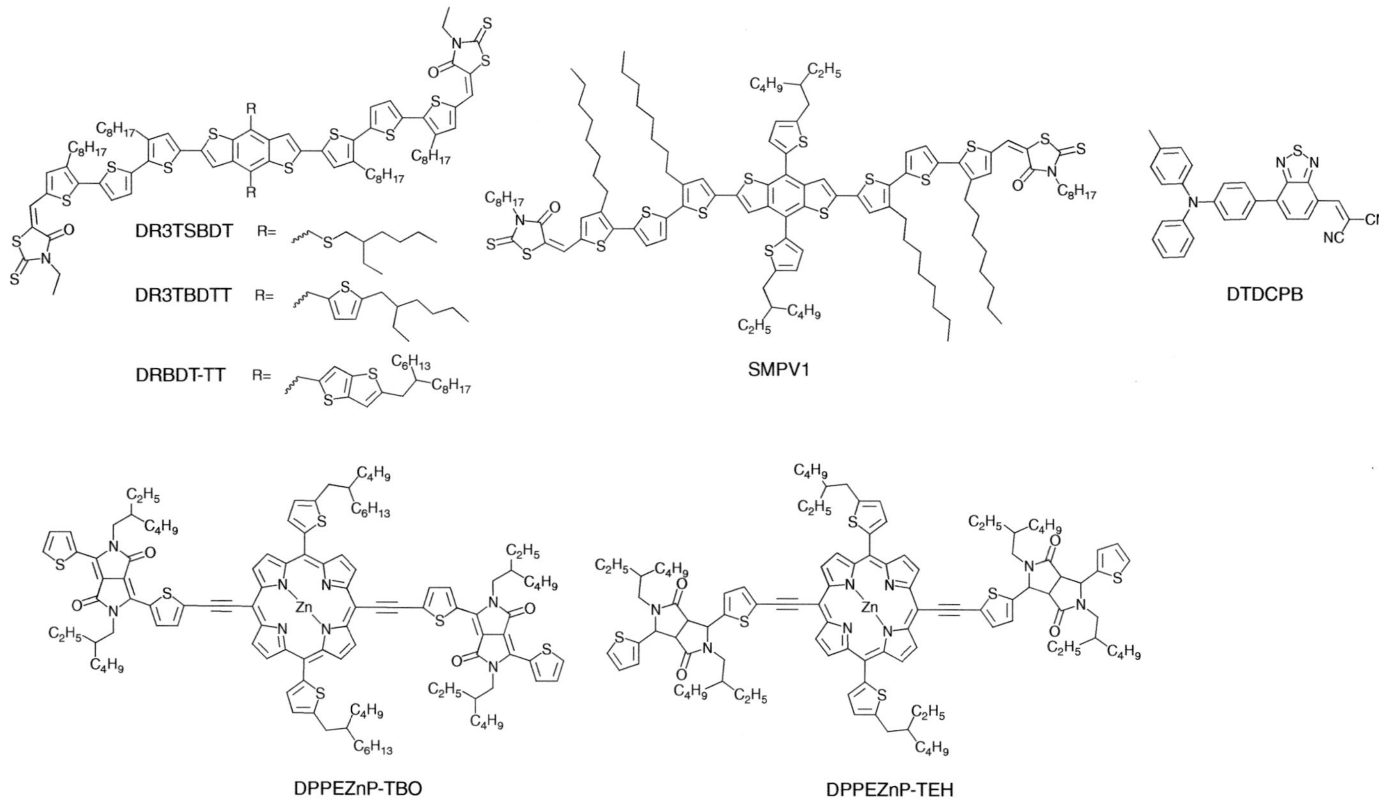

Figure 10.4 Small molecule donor materials used in tandem OSCs.

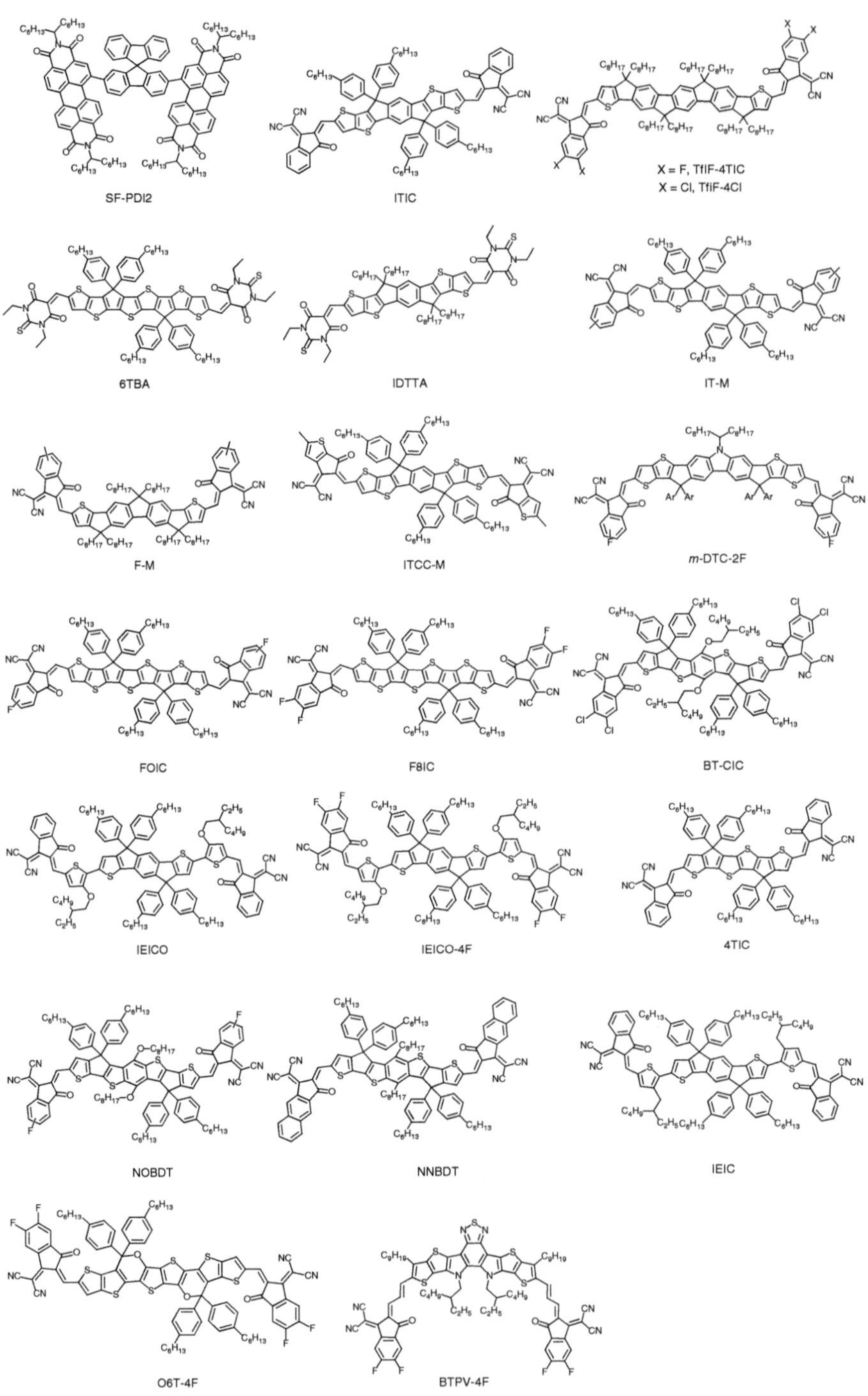

Figure 10.5 Small molecule acceptor materials used in tandem OSCs.

on, Jen et al. reported a fluoro-containing low-bandgap polymer PCPDTFBT and used it as rear cell donor material for constructing semitransparent solar cells and tandem solar cells with P3HT:IC_{60}BA as the front cell materials to give PCEs of 5.0% and 8.2%, respectively [45]. The above work demonstrated that the design of efficient rear cell donor materials with low bandgap is an effective strategy and also the prerequisite for constructing high-performance tandem OSCs. In 2013, Yang and coworkers designed a new polymer PDTP-DFBT with a low bandgap of 1.38 eV [46]. The single-junction device of PDTP-DFBT:PC_{71}BM showed a PCE of 7.9% with high external quantum efficiency (EQE) over 60% and spectral response that extended to 900 nm. The tandem solar cell with PDTP-DFBT:PC_{71}BM as the rear cell and P3HT:ICBA as the front cell gave a certified and high PCE of 10.6%. In their following work, a homo-tandem device using PDTP-DFBT:PC_{71}BM as the front and rear cells was constructed [47]. The homo-tandem device could harvest more efficient light compared with the single-junction device with maximum absorption increasing from 70% to 90% in visible region. A PCE of 10.2% was achieved for the homo-tandem device, indicating that the homo-tandem strategy is also an effective alternative strategy in enhancing the PCE for the corresponding single-junction device. In 2015, Heeger and coworkers reported a homo-tandem device based on PTB7-Th (also named PCE10):PC_{71}BM active layer using a novel ICL [48]. After careful optimization, a best PCE of 11.3% was achieved. In 2016, Hou and coworkers reported a new low-bandgap polymer named PDPP4T-2F, which showed an absorption onset at 900 nm, corresponding to an optical bandgap of 1.37 eV [49]. PDPP4T-2F:PC_{71}BM-based device showed a PCE of 7.58% and a high V_{oc} of 0.78 V owing to the deep highest occupied molecular orbital (HOMO) of the polymer. Then, using a wide-bandgap polymer PBDD4T-2F accompanied with PC_{61}BM as the front cell and PDPP4T:PC_{71}BM as the rear cell, a tandem device was fabricated and demonstrated an outstanding PCE of 11.62%.

Although P3HT:IC_{60}BA had showed decent single-junction device performance and achieved initial success as front cell in several high efficiency tandem devices as discussed above, its low V_{oc} and J_{sc} limit the further device improvement of the tandem OSCs using them as front cell. So, designing wide-bandgap and efficient donor materials is as important as that of low-bandgap polymers. Peng and coworkers designed a wide-bandgap (1.81 eV) polymer PBDTFBZS consisting of dialkylthiol-substituted benzo[1,2-*b*:4,5-*b*′]dithiophene donor block and monofluorinated benzotriazole acceptor block [50]. The PBDTFBZS:PC_{71}BM-based single-junction device showed good performance with a PCE of 6.18% and a high V_{oc} of 0.88 V. Tandem device using PBDTFBZS:PC_{71}BM as front cell and PNDTDPP:PC_{71}BM as rear cell afforded a PCE of 9.4%. Hwang and coworkers designed and synthesized two D–A copolymers, PTIPSBDTDTQX and PTIPSBDT-DFDTQX, based on the donor unit TIPS substituted benzodithiophene (TIPSBDT) and the acceptor quinoxaline-based units with or without fluorine substitution [51]. The introduction of F atoms on the quinoxaline unit is effective in lowering both the HOMO and lowest unoccupied molecular orbital (LUMO) energy levels of PTIPSBDTDFDTQX to get a high V_{oc} of 0.81 V and a PCE of 5.25% for the single-junction device. Finally, a PCE of 7.2% was achieved for the tandem device using PTIPSBDTDFDTQX:PC_{71}BM as the front cell and PBPT-8:PC_{71}BM as the rear cell. Later on, Hou and coworkers reported a polythiophene derivative named PDCBT [52]. The device based on PDCBT:$PC_{71}BM$ showed a V_{oc} of 0.80 V and high EQE values in the range 400–650 nm. A tandem device using PDCBT:$PC_{71}BM$ as front cell and PBDT-TS1:$PC_{61}BM$ as rear cell was fabricated. Although these two polymers PDCBT and PBDT-TS1 show some overlapping absorption in the range from 550 to 650 nm, symmetric current densities could still be realized by careful optimization of the thickness of the active layers. Finally, a PCE of 10.16% for the tandem device was obtained. Subsequently, other low bandgap polymers such as PDTBTBz-$2F_{anti}$ with a remarkably high V_{oc} of 0.97 V and PTZ1 with a high V_{oc} of 0.94 V were designed and used as front cell donor materials for constructing tandem devices and PCEs over 10% were achieved [53, 54]. In 2017, Zheng and coworkers reported a polymer, PIDTBTO-TT, with a bandgap of 1.87 eV and successfully applied it into the tandem OSCs [55]. By selecting PTB7-Th:$PC_{71}BM$ as the rear cell and PIDTBTO-TT mixed with $PC_{71}BM$ as the front cell, a high PCE of 11.15% for the tandem device was obtained. In the same year, Kim and coworkers reported two side chain fluorinated wide-bandgap P1 (2F) and P2 (4F) which had deep HOMO and thus high V_{oc} of 0.91V and 1.00 V, respectively [56]. Using the two polymers and $PC_{71}BM$ as front cell materials with PTB7-Th:$PC_{71}BM$ as rear cells, the optimized tandem devices gave PCEs of 11.42% and 10.05%, with high V_{oc} of 1.64 and 1.72 V, respectively. These results demonstrated the excellent examples for new wide band gap materials design approaches to advance the field of tandem OSCs.

It is clear that the progress of tandem OSCs is intimately connected with the active layer materials. The above results for solution-processed tandem OSCs are all based on polymer donors. As the counterpart of polymers, solution-processed small molecules have shown great potentials with advantages of no batch to batch, high mobility, and generally high V_{oc}, which enable them to be excellent alternative in tandem devices. In the past decade, various solution-processed small molecule donors have been designed and synthesized. Among them, molecules with the acceptor–donor–acceptor (A–D–A) architecture developed by our group have showed the best device performance and demonstrated completely comparable device performance in contrast to that of polymers [3, 20]. In recent years, we have reported series of solution-processed high efficiency small molecule donors and acceptors with the A–D–A type architecture. Some small molecule donors such as DRCN7T [76], DRCN5T [77], DR3TBDT [78], DR3TBDTT [79], and DR3TSBDT [80] designed and synthesized by our group have become the representative molecules and widely used in the OSC community. In view of their suitable absorption range, high V_{oc}, J_{sc} and good FF of the single junction devices, these A–D–A molecules would be ideal alternatives to substitute the widely-used P3HT as front cell materials to fabricate high-efficiency tandem OSCs. Thus, we selected three small molecules DR3TSBDT, DR3TBDTT, and DRBDT-TT as the front cell donor materials with $PC_{71}BM$ as acceptor to fabricate tandem devices. And PTB7-Th:$PC_{71}BM$ was utilized as the rear cell owing to the materials availability and high device repeat properties for the single-junction device [57]. Using an all solution-processed procedure except the back Al electrode, three tandem devices with the inverted structures were fabricated. Among them, the device based on DR3TSBDT showed a PCE of 11.47%, with a high V_{oc} of 1.689 V, J_{sc} of 10.51 mA cm $^{-2}$, and FF of 64.6%, which was among the highest values reported for tandem cells in 2016. The tandem OSCs based on DR3TBDTT and DRBDT-TT also showed high PCEs of 10.73% and 10.43%. Yang and coworkers also designed a small molecule SMPV1 and used it to fabricate homo-tandem device to afford a PCE of 10.1% with a high V_{oc} of 1.82 V and good FF of 72% [58]. In 2016, Huang and coworkers employed a porphyrin derivative molecule DPPZnP-THE as the rear cell donor material to fabricate tandem device. DPPZnP-THE covered a broad range from 300 to 900 nm with a low bandgap of 1.37 eV [59]. Selecting PThBDTP:$PC_{71}BM$ as the front cell, a PCE of 11.35% was achieved. The above results demonstrated the great potential of small molecule active layers for high performance solution-processed tandem OSCs.

In order to obtain higher efficiency, it is important for the subcells to have complementary absorptions and generate high and balanced (matched) currents. This requires a rather challenging effort to design and select suitable active materials for use in the subcells. In 2017, we and collaborators reported a high performance solution-processed tandem solar cell based on two small molecules DR3TSBDT and DPPEZnP-TBO, which offered efficient and complementary absorption when being used as electron donor materials in the front and rear subcells, respectively [32]. Through delicated film thickness adjustment, the J_{sc} of the tandem device reached 12.29 mA cm^{-2}. Eventually, an all small molecule tandem device with a PCE of 12.50% (verified 12.70%) was achieved, representing a new level of capability for solution-processed OSCs.

10.2.2 Tandem OSCs Based on Non-fullerene Acceptors

Although much progress has been made in fullerene derivative-based OSCs including single-junction and tandem OSCs, the intrinsic properties of fullerene derivatives such as weak absorptions and hardly tunable energy levels significantly limit the device performance. To address these issues, great efforts have been devoted to NFAs and substantial progress have been made in the OSC community [19, 30, 33]. Various types of NFAs including small molecules and polymers have been designed and evaluated. Among them, molecules with A–D–A architecture which have borrowed the design rationale from the A–D–A donor molecules have achieved big success. Detailed properties on the A–D–A architecture molecules can be seen in our recent review paper [20]. To date, the vast majority of single-junction devices using NFAs with high PCEs are fabricated using A–D–A type acceptors [19, 30, 33]. Compared with fullerene-based acceptors, NFAs, especially A–D–A type acceptors, exhibit excellent more merits and greater potential for constructing high-performance tandem OSCs, for example, (i) NFAs-based single-junction devices generally show low E_{loss}, which makes them obtain higher V_{oc} in the tandem devices, (ii) the easiness of bandgap tunability of NFAs presents more room to design subcells and thus high J_{sc} of the corresponding tandem OSCs are highly expected. With these advantages, many studies have been conducted on the NFA-based tandem OSCs and great progress has been made in recent years.

In 2016, Chen and coworkers demonstrated the first efficient non-fullerene tandem OSC by utilizing two NFA systems as the front and rear cells via solution processing [60]. Benefiting from the high V_{oc} of 1.11 V for the P3HT:SF(DPPB)$_4$-based front cell and 0.95 V for the PTB7-Th:IEIC-based rear cell, the tandem OSC showed a high V_{oc} of 1.97 V and a PCE of 8.48%. With the high V_{oc}, the tandem device was used for the solar-cell-driven water splitting tests, indicating the great potential of high-voltage tandem cells. However, the overlap absorption and inferior FF of the front and rear cell impeded the further improvement of photovoltaic performance. Later on, Yan and coworkers reported a homo-tandem device with an impressively high V_{oc} over 2.1 V [61]. They used P3TEA as the donor and SF-PDI$_2$ as the acceptor in the front and rear cell to construct the homo-tandem device. With device optimization, a PCE of 10.8% was achieved accompanied with the outstanding V_{oc} of 2.13 V. In 2017, Peng and coworkers designed and synthesized a wide-bandgap copolymer PBDTS-TDZ. When accompanied with ITIC as the acceptor, a PCE of 12.80% was achieved with a high V_{oc} of 1.1 V [62]. A homo-tandem device was also fabricated. Finally, an outstanding PCE of 13.35% was achieved through optimizing the thickness of the front and rear cells with a high V_{oc} of 2.13 V. The above results gave more confidence for the researchers in the OSC field to fabricate higher efficiency tandem devices. Subsequently, the study on tandem OSCs mainly focuses on the design and synthesis of NFAs, especially those A–D–A type small molecule acceptors with complementary absorptions. The wide bandgap materials can provide high V_{oc} and the low bandgap materials facilitate to increase J_{sc} of the tandem devices. And thus, high efficiency tandem devices can be expected.

In 2017, a low-bandgap acceptor 4TIC with absorption onset around 900 nm was designed and synthesized by Jen and coworkers [63]. The single-junction device with PTB7-Th as the donor material gave a PCE of 10.43% with J_{sc} of 18.75 mA cm^{-2}, FF of 0.72, and V_{oc} of 0.78 V. The high efficiency together with its small voltage loss made PTB7-Th:4TIC a suitable rear subcell for tandem solar cells. They fabricated a series- connection tandem OSCs incorporating this rear cell and PBDB-T:ITIC as front cell, and obtained a promising PCE of 12.62%. In their next work, they designed an acceptor, 6TBA. With PTB7-Th as donor material, 6TBA-based device exhibited a higher V_{oc} of 0.98 V and lower voltage loss of 0.54 V than that of PBDB-T:ITIC [64]. A tandem device was then fabricated using PTB7-Th:6TBA as the front cell and PTB7-Th:4TIC as the rear cell delivered a high PCE of 13.20%. In order to balance the tradeoff between the V_{oc} and J_{sc} in the tandem device, Yang and coworkers reported two NFAs named FOIC and F8IC based on the terthieno [3,2-*b*]thiophen (6T) core with one or two F atoms substituent on the end groups and used the ternary system PTB7-Th:FOIC:F8IC as rear cell materials for tandem cell [65]. The V_{oc} of the rear cells could be tuned in a wide range (0.67–0.75 V) by changing the ratio of the two acceptors. Finally, employing PBDB-T:IT-M as the front cell, a tandem device with a PCE of 13.3% was obtained.

Hou and coworkers reported a tandem device based on two A–D–A type acceptors and delivered an outstanding PCE of 13.8%, a V_{oc} of 1.79 V, a J_{sc} of 12.0 mA cm^{-2}, and an FF of 0.641 [66]. Firstly, an acceptor ITCC-M was designed based on the well-known acceptor ITIC through replacing the phenyl group in the end groups by α-methylthiophene groups. Compared with ITIC, the E_g of ITCC-M was enlarged by around 0.11 eV with the half-peak-height position in the long-wavelength direction shifting from 750 nm to 700 nm. Blending with PBDB-T as the donor material, ITCC-M-based device showed a PCE of 9% with a high V_{oc} over 1 V. Next, they selected a near-infrared absorption acceptor, IEICO and a polymer donor PBDTTT-E-T as the rear cell. Finally, the tandem device was fabricated using a ZnO/PCP-Na as the ICL and gave an impressive efficiency of 13.8%. The moderate EQE response and range of the rear cell together with the low FF limited the device performance. So, in their following work [67], to extend the absorption range and enable well complementary absorption of the subcells, IEICO-4F with absorption onset around 1000 nm was selected to replace IEICO as the rear cell acceptor. Meanwhile, an acceptor IT-M with a bandgap of 1.6 eV and a wide-bandgap polymer J52-F were selected as the front cell. The much more complementary absorption of the subcells contributed to the improved J_{sc} of 13.3 mA cm^{-2} and thus enhanced PCE of 14.9%. In 2019, Huang and coworkers reported a NFA TfiF-4FIC with E_g of 1.61 eV [68]. The single-junction device based on PM6:TfiF-4FIC delivered a PCE of 13.1% with a high V_{oc} of 0.98 V and a notable FF of 76%. The excellent photovoltaic parameters, together with the wide bandgap, enable PM6:TfIF-4FIC a good candidate as the front cell for fabricating tandem OSCs. A ternary combination PCE10:PCDTBT:IEICO-4F was optimized as the rear cell materials to ensure a high J_{sc} of 25.3 mA cm^{-2}. Finally, the tandem device afforded a PCE of 15.0%. In 2021, Huang and coworkers developed an efficient ICL of ZnO NPs:PEI/PEI/PEDOT:PSS [69]. They used PM7:TfiF-4Cl as the front cell and PCE10:CO$_i$8DFIC:PC$_{71}$BM as the rear cell to construct the tandem device. In order to suppress the charge recombination of the front cell with the increased thickness of the active layer to get higher FF, they increase the proportion of the TfiF-4Cl in the front cell to manipulate the optical field distribution and then increase the J_{sc}. With this stragety, the tandem device could simultaneously achieve higher J_{sc} and FF. Finally, a PCE of 18.71% was obtained. Recently, Anthopoulos and coworkers also synthesized a wide-bandgap acceptor, IDTTA. Using PBDB-T as the donor, the single-junction device of IDTTA exhibited a PCE of 10.8% with a V_{oc} of 0.98 V [70]. Employing PBDB-T:IDTTA as the front cell and PTB7-Th:IEICO-4F as the rear cell, the tandem device was fabricated to realize a PCE of 15% with a V_{oc} of 1.66 V and J_{sc} of 13.6 mA cm^{-2}.

Benzo[1,2-*b*:4,5-*b*′]dithiophene (BDT) is a widely used building block in the OSC community. Numerous donor materials including polymers and small molecules incorporating the BDT unit have been synthesized. In 2017, our group reported the first BDT-based acceptor NFBDT with a heptacyclic benzodi(cyclopentadithiophene) (FBDT) unit as the central unit, where the BDT core unit is covalently rigidified at its 3,7-positions with two external fused thiophene (T) units by the sp^3 -carbon bridges [81]. And PCE over 10% was achieved for the single-junction device based on PBDB-T:NFBDT. To meet the demands of rear cell in tandem device, we designed a new BDT core-based acceptor NOBDT through modifying the acceptor NFBDT by introducing F substituted end groups and BDT core incorporating oxygen alkyl chains [71]. NOBDT showed a near-infrared absorption with absorption onset around 900 nm. The single-junction device of

PTB7-Th:NOBDT demonstrated a PCE 10.55% with a V_{oc} of 0.77 V and energy loss 0.62 eV, indicating great potential in tandem OSCs. Meanwhile, we designed an acceptor F-M with absorption onset of 750 nm and the single-junction device based on PBDB-T:F-M exhibited a high V_{oc} around 1 V. By systematically optimizing the thickness of subcell active layers under the guidance of optical simulation, a high PCE of 14.11%, with a V_{oc} of 1.71 V, a J_{sc} of 11.72 mA cm^{-2}, and a satisfactory FF of 0.70 was achieved. In the following work, to balance the V_{oc} and J_{sc} of the tandem device, we reported a tandem device employing two subcells with the same donor PBDB-T and two acceptors F-M and NNBDT with complementary absorptions. An efficient tandem OSC with an optimized PCE of 14.52% was achieved with a high V_{oc} of 1.82 V, a notable FF of 74.7%, and a J_{sc} of 10.68 mA cm^{-2} [72]. This work demonstrates a promising strategy for fabricating high efficiency tandem OSCs through careful selection of the subcell active layer materials and tradeoff of the V_{oc} and J_{sc} of the tandem cells. The results also demonstrated that the BDT unit-based acceptors are indeed good candidates for fabricating high efficiency tandem devices, which is proved by the work reported by Forrest and coworkers [73]. Using a near infrared absorption acceptor BT-ClC blending with PCE-10 as the rear cell, and visible absorption vacuum deposited DTDCPB:C70 as the front cell, a tandem device was fabricated and delivered a PCE of 14.3% for 2 mm^2 device and a maximum PCE of 15% was afforded. In their following work, in order to improve the J_{sc} and efficiency of the tandem device, a low bandgap acceptor TT-FIC was added in the above rear cell PCE10:BT-ClC binary system [74]. Using the same front cell of DTDCPB:C_{70}, a tandem device was constructed and gave an improved PCE of 15.4%.

Despite that NFA-based tandem OSCs have made much progress, the overall efficiencies were still unsatisfactory. It is essential to synergistically design and screen the front and rear cell materials from the perspective of energy losses, absorption ranges, and the photovoltaic parameters such as FF and EQE of the subcell. In 2018, we have proposed a semi-empirical analysis for the possible but realistic PCE limit of two terminal monolithic tandem OSCs, which will be discussed in details in the following part. Under the guidance of the model analysis, one can not only get the possible PCEs for the defined subcells but also rationally design subcell active layers. Based on the model analyzed above, we constructed a tandem device using two A–D–A acceptors-based subcells with complementary absorptions covering a wide range from 300 to 1050 nm [11]. According to the semi-empirical model, firstly, a low bandgap A–D–A small molecule acceptor O6T-4F was selected as the rear cell acceptor material as it has an infrared absorption onset of ~1050 nm (optical bandgap E_g 1.20 eV) and the single-junction device based on it with PTB7-Th as donor and $PC_{71}BM$ as the secondary acceptor exhibited a high J_{sc} with EQE response value over 70% in the infrared absorption range, an FF of 69.7%, and a low E_{loss} of 0.51 eV. Next, PBDB-T:F-M was selected as front cell as the combination exhibits not only absorption onset close to the ideal absorption onset of 720 nm but also with optimal current ~16 mA cm^{-2} and high EQE response (average ~70% in the range of 300–750 nm), FF (69.8%). Using an all solution-processed procedure except the back electrode, a tandem device was constructed and achieved a record PCE of 17.3%. In 2021, in order to extend the absorption range of the rear cell, Li and coworkers designed and synthesized a low-bandgap acceptor BTPV-4F by inserting a double bond between the central core and end groups of the acceptor Y6 [75]. Benefiting from its extended conjugation length, the optical bandgap of the BTPV-4F is only 1.21 eV with an absorption onset of 1021 nm. When blending with PTB7-Th and $PC_{71}BM$, BTPV-4F based single-junction device demonstrated a higher J_{sc} of 28.9 mA cm^{-2} with a notable FF of 69.3%. Meanwhile, a wide-bandgap acceptor *m*-DTC-2F was synthesized with a bandgap of 1.61 eV as the front cell. PM6:*m*-DTC-2F-based device exhibits a PCE of 12.1% with a higher V_{oc} of 0.99 V. Finally, the tandem device with PM6:*m*-DTC-2F as the front cell and PTB7-Th:BTPV-4F:$PC_{71}BM$ as the rear cell achieved a promising PCE of 16.4% with a high J_{sc} of 14.5 mA cm^{-2}. The high efficiency and the semi-empirical analysis model indicate that OSC has much more potential than previously thought if using tandem cell. And PCE >25% could be achieved through careful active layer material design under the guidance of above semi-empirical analysis together with device optimization, including subcell active layer morphology control, thickness optimization and interconnecting layer engineering.

10.3 Interconnecting Layer Materials

In the tandem photovoltaic devices, the subcells are connected by the ICL materials, which extract charge carriers from the front and rear subcells and meanwhile provide a recombination region for the collected carriers. Thus, ICL materials are the critical components for realizing high performance for tandem OSCs [22, 23, 82]. For efficient ICLs, they should form ohmic contact with subcell active layers to ensure efficient charge extractions. And they should be highly transparent and robust enough to protect underlying layer during subsequent fabrication process. ICL is generally composed of a p-type hole transporting layer (HTL) and an n-type electron transporting layer (ETL). ICLs in tandem cells could be roughly divided into two major configurations: ETL–conducting layer–HTL (e-c-h-type) and ETL-HTL (e-h-type) in the conventional structure or h-c-e-type and h-e-type in the inverted structure [25, 83]. In fact, the ETL and HTL materials used in single-junction OSCs can be combined and used as ICL materials in tandem OSCs. For the e-c-h-type or h-c-e configuration, a conductive layer, generally very thin metal, is sandwiched between the ETL and HTL. The ETL and HTL are used to collect the electron and hole from the subcells, respectively, where the conductive layer is served as charge recombination center. At present, thin-film metals (i.e. Au, Ag, Al) are usually applied as conductive layer materials for most of e-c-h (or h-c-e) type ICLs. And the combination can satisfy the requirements for the ICLs, such as high transparency, minimized voltage losses, and functioned as an optical spacer to match the current between the subcells. In recent years, many e-c-h-type (or h-c-e-type) ICLs have been designed for tandem OSCs, such as TiO_2/Al/PEDOT:PSS [84], MoO_3/Al/ZnO [85], PEDOT:PSS/AgNW/ZnO [86], MoO_3/Ag/PFN [87, 88], ZnO/M-PEDOT:PSS/PMA or PMA:M-PEDOT:PSS/ZnO

[89], MoO_3/M-PEDOT:PSS/ZnO [65, 90], BPhen:C_{60}/Ag NP/PEDOT:PSS [74], MoO_3/Ag/ZnO/PFN-Br [64].

However, complicated evaporation steps and possible optical losses induced by the metal layer are often encountered for the h-c-e-type (or e-c-h-type) ICLs. Thus, many attentions have been drawn on the h-e-type (or e-h-type) ICLs, which can realize all solution process and demonstrate much advantages than the h-c-e-type (or e-c-h-type) ICLs. For the h-e-type (or e-h-type) ICLs, the hole and electron of the subcells are collected by the HTL and ETL, respectively, and the charge recombination occurs and completes at the ETL-HTL contact interface. Thus, a separate conductive layer is not needed. Currently, most of the HTL materials in the h-e-type (or e-h-type) ICLs are PEDOT:PSS or its modified products. PEDOT:PSS accompanied with ETL, especially n-type metal oxide, are the most popularly used h-e-type (or e-h) type ICL materials for tandem OSCs, such as PEDOT:PSS/ZnO [61, 91, 92], PEDOT:PSS/ZnO/CPE [93], PEDOT:PSS/SnO_2 [94], PEDOT:PSS/NPE [95], PEDOT:PSS/PEIE [96], M-PEDOT/CDQ PEI [97], H_xMoO_3/PEDOT:PSS/ZnO NPs [98], TiO_2/M-PDEOT [99], ZnO/N-PEDOT:PSS [71], PNDIT-EHP-F3N:PEI/PEDOT:PSS [100]. Other combinations without PEDOT:PSS were also reported, such as POM/ZnO [101], ZnO/CPE-K [48, 102], ZnO/PCP-Na [66], and MoO_x/SnO_x [103]. Device parameters of some representative tandem OSCs using various ICLs are also summarized in Tables 10.1 and 10.2.

10.4 The Semi-Empirical Analysis of Tandem OSCs

There are some theoretical calculations predicting the maximum efficiency of tandem OSCs. For example, as early as in 2008, Brabec and coworkers conducted a calculation that tandem devices based on PCBM derivatives as acceptors could gave efficiencies up to 15% [82]. In 2018, we proposed a semi-empirical model to direct the design and selection of subcell materials for tandem devices. With the guidance of this semi-empirical model, a record PCE of 17.3% has been achieved for tandem device, which is a new level for solution-processed OSCs [11]. The semi-empirical analysis was conducted on a 2-terminal monolithic tandem device with two subcells in series. Of course, it can be extended to be used in single-junction and multijunction series connected device. The details can be found in references [11, 104]. The following is the main content.

Based on some assumptions, the three photovoltaic parameters V_{oc}, $J_{sc,}$ and FF of a 2-terminal monolithic tandem device are obtained as follows:

(1) For a rear cell with absorption onset λ_1, the J_{sc} of the tandem cell could be obtained from Eq. (10.1).

$$J_{sc,tandem} = \frac{1}{2}\cdot\int_{300}^{\lambda_1}\frac{q\lambda}{hc}\cdot E(\lambda)\cdot \text{EQE}(\lambda)\cdot d\lambda = \int_{300}^{\lambda_2}\frac{q\lambda}{hc}\cdot E(\lambda)\cdot \text{EQE}(\lambda)\cdot d\lambda \quad (10.1)$$

where $E(\lambda)$ is the spectral irradiance in AM 1.5G, λ_1 and λ_2 are the absorption onset of rear and front cell, h is Planck's constant, c is the speed of light, and q is the elementary charge.

The absorption onset (λ_2) of a matched front cell could be determined from Eq. (10.2) under AM 1.5G light illumination.

$$J_{\text{sc,front}} = J_{\text{sc,rear}} = \frac{1}{2}\cdot\int_{300}^{\lambda_1} \frac{q\lambda}{hc}\cdot E(\lambda)\cdot \text{EQE}(\lambda)\cdot \text{d}\lambda = \int_{300}^{\lambda_2} \frac{q\lambda}{hc}\cdot E(\lambda)\cdot \text{EQE}(\lambda)\cdot \text{d}\lambda \tag{10.2}$$

(2) The V_{oc} of the subcells is determined by the following equations:

$$V_{\text{oc,front}} = \frac{1}{q}(E_{\text{g}} - E_{\text{loss}}) = \frac{1}{q}\left(\frac{1240}{\lambda_2} - E_{\text{loss}}\right) \tag{10.3}$$

$$V_{\text{oc,rear}} = \frac{1}{q}(E_{\text{g}} - E_{\text{loss}}) = \frac{1}{q}\left(\frac{1240}{\lambda_1} - E_{\text{loss}}\right) \tag{10.4}$$

Herein, the front and rear cell are considered to have the same E_{loss} for the calculation simplification. The E_{loss} is assumed to be 0.4–0.8 eV. The V_{oc} of the tandem cell is given by

$$V_{\text{oc,tandem}} = V_{\text{oc,front}} + V_{\text{oc,rear}} = \frac{1}{q}\left(\frac{1240}{\lambda_2} + \frac{1240}{\lambda_1} - 2E_{\text{loss}}\right) \tag{10.5}$$

So for the tandem solar cells, the PCE can be calculated from Eq. (10.6).

$$\text{PCE}(\%) = V_{\text{oc}}\cdot J_{\text{sc,tandem}}\cdot \text{FF}/P_{\text{in}} = \frac{1}{q}\left[\left(\frac{1240}{\lambda_1} - E_{\text{loss}}\right) + \left(\frac{1240}{\lambda_2} - E_{\text{loss}}\right)\right] \cdot\frac{1}{2}\cdot\int_{300}^{\lambda_1} \frac{q\lambda}{hc}\cdot E(\lambda)\cdot \text{EQE}(\lambda)\cdot \text{d}\lambda\cdot \text{FF}/P_{\text{in}} \tag{10.6}$$

According to the above assumptions, Figure 10.6a shows the achievable PCEs (14–28%) of the 2T tandem cells (double junctions) versus λ_{onset} (corresponding to E_{g}) of rear cell active layer, E_{loss}, EQE, at a fixed FF of 0.75. Furthermore, as shown in Figure 10.6b, if the $\lambda_{\text{onset,rear cell}}$ is up to 1100 nm with an average EQE of 75%, FF being 0.75 and a typical E_{loss} of 0.6 eV, a PCE of ~20% can be reached. Note that E_{loss} plays a great impact on the PCEs from the modeling results as shown in Figure 10.6c,d. For example, with the fixed FF of 0.75 and EQE of 75%, when E_{loss} equals to 0.5 eV, a PCE_{max} of ~22.6% was obtained with the $\lambda_{\text{onset,rear cell}}$ = 1060–1125 nm; when the E_{loss} is 0.6 eV, a 20.2% PCE_{max} was obtained with $\lambda_{\text{onset,rear cell}}$ = 1030–1100 nm; however, when the E_{loss} is as large as 0.7 eV, the PCE_{max} of 17.8% was obtained at a large range for $\lambda_{\text{onset,rear cell}}$ = 900–1100 nm. These modeling results also indicate that the optimum $\lambda_{\text{onset, rear cell}}$, i.e. the E_{g} of the rear sub cell active layer, is somewhat smaller than the optimum one in the SQ limit studies (of ~1127 nm) [105], probably due to the large E_{loss} of OSCs [106]. According to the model analysis, to get a high PCE of the tandem cell, each subcell should have high EQE response, small E_{loss} and high FF, and the absorption of them should be complementary and sufficiently broad.

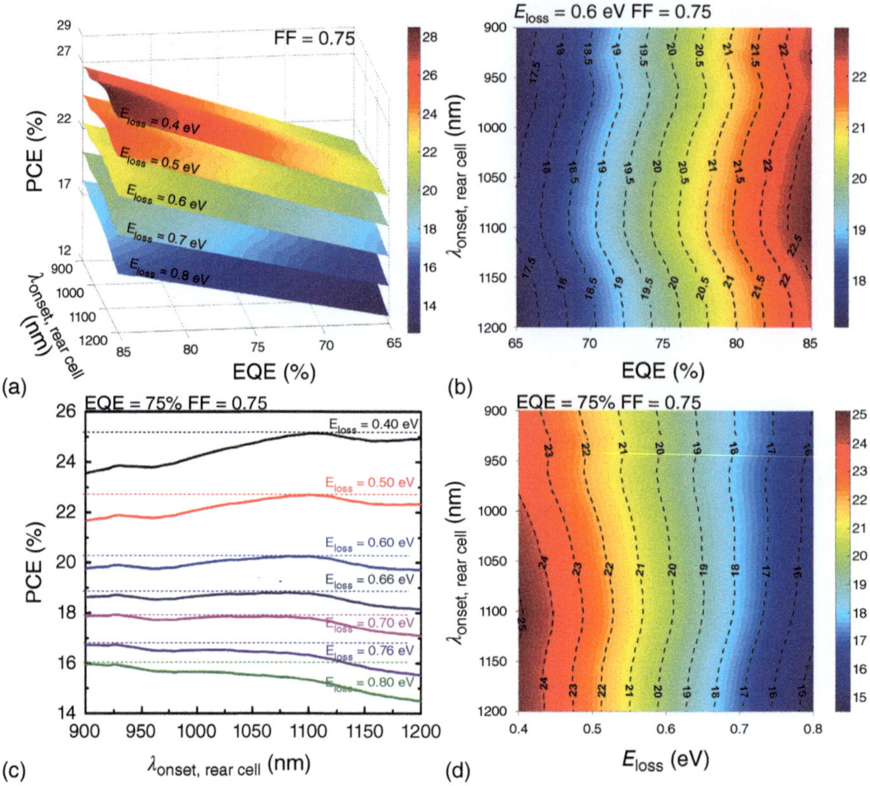

Figure 10.6 Predicted PCEs of 2T tandem solar cells based on semi-empirical analysis under AM 1.5G. (a) PCEs versus EQE and λ_{onset}, rear cell with assuming the E_{loss} of each subcell in the range of 0.4–0.8 eV and a fixed FF of 0.75. (b) PCEs versus EQE and λ_{onset}, rear cell with assumed E_{loss} of 0.6 eV and FF of 0.75. (c) PCEs versus λ_{onset}, rear cell with E_{loss} of 0.4, 0.5, 0.6, 0.66, 0.7, 0.76, and 0.8 eV, FF of 0.75 and EQE of 75%. (d) PCEs versus E_{loss} and λ_{onset}, rear cell with assumed EQE of 75% and FF of 0.75. Source: Meng et al. [11]. Reproduced with permission of The American Association for the Advancement of Science.

10.5 Conclusion and Outlook

In this chapter, we have summarized the recent progress of tandem OSCs mainly focusing on the active layer materials. Presently, the tandem OSCs have attained great progress with higher PCE over 17%. It should be noted that the PCEs of single-junction OSCs have also reached 17% and some even 18% in recent years, thanks to the active layer materials design and device optimization engineering. The results have demonstrated not only challenge but also opportunity for tandem OSCs. In the era of fullerene acceptor-based OSCs, tandem devices did not demonstrate higher efficiency than that of single-junction devices until suitable subcell materials with complementary absorption had been designed and screened. The history will repeat in the era of NFAs for tandem OSCs. It is strongly believed that higher efficiencies of tandem cells can be achieved through delicate active layer material design and screening together with the device optimization. Although triple and multiple junction cells can give higher efficiency in principle, it is more feasible to develop double junction tandem devices if considering the balance between device efficiency and device fabrication complexity. According to the semi-empirical model analysis discussed above, PCEs >25% are predicted for double junction tandem OSCs. Better subcell active layer materials are believed to be the key to realize the efficiency target. The present success of various donor and acceptor materials, especially new A–D–A molecules, will definitely give very valuable insights in designing better active layer materials.

Firstly, efficient rear cell with infrared absorption onset up to 1050–1150 nm is the prerequisite to get high performance tandem OSCs. However, currently, there are no rear cell active materials that can meet all the requirements such as absorption range, E_{loss} and EQE discussed in the above model analysis. Therefore, it is urgent to design rear cell materials with onset up to 1050–1150 nm and good photovoltaic performance with low E_{loss} and high EQE. Secondly, the front cell should have a complimentary absorption with an onset of around 720 nm in order to have maximized matched J_{sc} with the rear cell according to the above model analysis. Meanwhile, the two subcells should have minimal E_{loss} to have high V_{oc} and thus ensure high V_{oc} for the tandem devices. Besides, EQE and the FF of the subcells should be carefully considered for selecting the materials. Thirdly, it is also a tough task to develop efficient ICLs without incorporating PEDOT:PSS since the acidic and hydrophilic nature of PEDOT:PSS is not favorable for the stability of tandem devices.

According to the model analysis, assuming an E_{loss} of 0.4 eV, average EQE of 80%, and FF of 80%, the absorption onsets of 1100 and 720 nm for the rear and front cells, respectively, a PCE of 28.60% can be obtained for a 2T tandem device. It seems to be a greatly challenging PCE value. It is strongly believed that it can be realized since the single parameters such as E_{loss}, EQE, and FF have already been realized in the reported device. If they can be realized combinedly, the seemly impossible PCEs can come into reality for tandem OSCs.

Acknowledgments

The authors gratefully acknowledge the financial support from the National Natural Science Foundation of China (Grant No.: 21935007)

References

1 Blom, P.W.M., Mihailetchi, V.D., Koster, L.J.A., and Markov, D.E. (2007). *Adv. Mater.* 19: 1551.
2 Kim, J.Y., Lee, K., Coates, N.E. et al. (2007). *Science* 317: 222.
3 Chen, Y.S., Wan, X.J., and Long, G.K. (2013). *Acc. Chem. Res.* 46: 2645.
4 Huang, Y., Kramer, E.J., Heeger, A.J., and Bazan, G.C. (2014). *Chem. Rev.* 114: 7006.

5 Sun, Y.N., Chang, M.J., Meng, L.X. et al. (2019). *Nat. Electron.* 2: 513.
6 Park, S., Kim, T., Yoon, S. et al. (2020). *Adv. Mater.* 32: 2002217.
7 Wang, G.D., Zhang, J.Q., Yang, C. et al. (2020). *Adv. Mater.* 32: 2005153.
8 Jinno, H., Fukuda, K., Xu, X. et al. (2017). *Nat. Energy* 2: 780.
9 Liu, Q.S., Jiang, Y.F., Jin, K. et al. (2020). *Sci. Bull.* 65: 272.
10 Cui, Y., Yao, H.F., Zhang, J.Q. et al. (2020). *Adv. Mater.* 32: 1908205.
11 Meng, L.X., Zhang, Y.M., Wan, X.J. et al. (2018). *Science*: 1094.
12 Polman, A., Knight, M., Garnett, E.C. et al. (2016). *Science* 352: aad4424.
13 Yang, W.S., Noh, J.H., Jeon, N.J. et al. (2015). *Science* 348: 1234.
14 Yoshikawa, K., Kawasaki, H., Yoshida, W. et al. (2017). *Nat. Energy* 2: 17032.
15 Shockley, W. and Queisser, H.J. (1961). *J. Appl. Phys.* 32: 510.
16 Wurfel, U., Neher, D., Spies, A., and Albrecht, S. (2015). *Nat. Commun.* 6: 6951.
17 Burlingame, Q., Coburn, C., Che, X.Z. et al. (2018). *Nature* 554: 77.
18 Cheng, Y.J., Yang, S.H., and Hsu, C.S. (2009). *Chem. Rev.* 109: 5868.
19 Zhang, G., Zhao, J., Chow, P.C.Y. et al. (2018). *Chem. Rev.* 118: 3447.
20 Wan, X.J., Li, C.X., Zhang, M.T., and Chen, Y.S. (2020). *Chem. Soc. Rev.* 49: 2828.
21 Di Carlo Rasi, D. and Janssen, R.A.J. (2019). *Adv. Mater.* 31: 1806499.
22 Ameri, T., Dennler, G., Lungenschmied, C., and Brabec, C.J. (2009). *Energy Environ. Sci.* 2: 347.
23 Sista, S., Hong, Z.R., Chen, L.M., and Yang, Y. (2011). *Energy Environ. Sci.* 4: 1606.
24 Minnaert, B. and Veelaert, P. (2012). *Materials* 5: 1933.
25 Lu, S.M., Ouyang, D., and Choy, W.C.H. (2017). *Sci. China Chem.* 60: 460.
26 Li, G., Chang, W.-H., and Yang, Y. (2017). *Nat. Rev. Mater.* 2: 17043.
27 Lassiter, B.E., Renshaw, C.K., and Forrest, S.R. (2013). *J. Appl. Phys.* 113: 214505.
28 Ameri, T., Li, N., and Brabec, C.J. (2013). *Energy Environ. Sci.* 6: 2390.
29 Zhou, Y.H., Fuentes-Hernandez, C., Shim, J.W. et al. (2012). *Energy Environ. Sci.* 5: 9827.
30 Cheng, P., Li, G., Zhan, X.W., and Yang, Y. (2018). *Nat. Photonics* 12: 131.
31 Yu, G., Gao, J., Hummelen, J.C. et al. (1995). *Science* 270: 1789.
32 Li, M.M., Gao, K., Wan, X.J. et al. (2017). *Nat. Photonics* 11: 85.
33 Hou, J.H., Inganas, O., Friend, R.H., and Gao, F. (2018). *Nat. Mater.* 17: 119.
34 Li, S.X., Li, C.Z., Shi, M.M., and Chen, H.Z. (2020). *ACS Energy Lett.* 5: 1554.
35 Yao, J., Qiu, B.B., Zhang, Z.G. et al. (2020). *Nat. Commun.* 11: 2726.
36 Yuan, J., Zhang, Y.Q., Zhou, L.Y. et al. (2019). *Joule* 3: 1140.
37 Ma, Y.L., Zhang, M., Wan, S. et al. (2020). *Joule* 5: 1.
38 Zhan, L.L., Li, S.X., Lau, T.-K. et al. (2020). *Energy Environ. Sci.* 13: 635.
39 Zhu, L., Zhang, M., Zhou, G.Q. et al. (2020). *Adv. Energy Mater.* 10: 1904234.
40 Zhou, Z.C., Liu, W.R., Zhou, G.Q. et al. (2019). *Adv. Mater.* 32: 1906324.
41 Gilot, J., Wienk, M.M., and Janssen, R.A. (2010). *Adv. Mater.* 22: E67.
42 Dou, L.T., You, J.B., Yang, J. et al. (2012). *Nat. Photonics* 6: 180.
43 Dou, L.T., Gao, J., Richard, E. et al. (2012). *J. Am. Chem. Soc.* 134: 10071.
44 Gevaerts, V.S., Furlan, A., Wienk, M.M. et al. (2012). *Adv. Mater.* 24: 2130.

45 Chang, C.-Y., Zuo, L., Yip, H.-L. et al. (2013). *Adv. Funct. Mater.* 23: 5084.
46 You, J.B., Dou, L.T., Yoshimura, K. et al. (2013). *Nat. Commun.* 4: 1446.
47 You, J.B., Chen, C.C., Hong, Z.R. et al. (2013). *Adv. Mater.* 25: 3973.
48 Zhou, H.Q., Zhang, Y., Mai, C.K. et al. (2015). *Adv. Mater.* 27: 1767.
49 Zheng, Z., Zhang, S.Q., Zhang, J.Q. et al. (2016). *Adv. Mater.* 28: 5133.
50 Li, K., Li, Z.J., Feng, K. et al. (2013). *J. Am. Chem. Soc.* 135: 13549.
51 Kim, J.-H., Song, C.E., Kim, H.U. et al. (2013). *Chem. Mater.* 25: 2722.
52 Zheng, Z., Zhang, S.Q., Zhang, M.J. et al. (2015). *Adv. Mater.* 27: 1189.
53 Ko, S.-J., Hoang, Q.V., Song, C.E. et al. (2017). *Energy Environ. Sci.* 10: 1443.
54 Guo, B., Guo, X., Li, W.B. et al. (2016). *J. Mater. Chem. A* 4: 13251.
55 Ma, Y.L., Chen, S.C., Wang, Z.Y. et al. (2017). *Nano Energy* 33: 313.
56 Song, S., Kranthiraja, K., Heo, J. et al. (2017). *Adv. Energy Mater.* 7: 1700782.
57 Zhang, Q., Wan, X.J., Liu, F. et al. (2016). *Adv. Mater.* 28: 7008.
58 Liu, Y., Chen, C.C., Hong, Z.R. et al. (2013). *Sci. Rep.* 3: 3356.
59 Zhang, K., Gao, K., Xia, R.X. et al. (2016). *Adv. Mater.* 28: 4817.
60 Liu, W.Q., Li, S.X., Huang, J. et al. (2016). *Adv. Mater.* 28: 9729.
61 Chen, S.S., Zhang, G.Y., Liu, J. et al. (2017). *Adv. Mater.* 29: 1604231.
62 Xu, X.P., Yu, T., Bi, Z.Z. et al. (2017). *Adv. Mater.* 30: 1703973.
63 Shi, X.L., Zuo, L.J., Jo, S.B. et al. (2017). *Chem. Mater.* 29: 8369.
64 Zuo, L.J., Shi, X.L., Jo, S.B. et al. (2018). *Adv. Mater.* 30: 1706816.
65 Cheng, P., Liu, Y.Q., Chang, S.Y. et al. (2018). *Joule* 3: 432.
66 Cui, Y., Yao, H.F., Gao, B.W. et al. (2017). *J. Am. Chem. Soc.* 139: 7302.
67 Cui, Y., Yao, H.F., Yang, C.Y. et al. (2017). *Acta Polym. Sin.* 2: 135.
68 Liu, G.C., Jia, J.C., Zhang, K. et al. (2019). *Adv. Energy Mater.* 9: 1803657.
69 Liu, G.C., Xia, R.X., Huang, Q. et al. (2021). *Adv. Funct. Mater.* 31: 2103283.
70 Firdaus, Y., He, Q., Lin, Y. et al. (2020). *J. Mater. Chem. A* 8: 1164.
71 Zhang, Y.M., Kan, B., Sun, Y.N. et al. (2018). *Adv. Mater.* 30: 1707508.
72 Meng, L.X., Yi, Y.Q.Q., Wan, X.J. et al. (2019). *Adv. Mater.* 31: 1804723.
73 Che, X.Z., Li, Y.X., Qu, Y., and Forrest, S.R. (2018). *Nat. Energy* 3: 422.
74 Li, Y.X., Lin, J.D., Liu, X. et al. (2018). *Adv. Mater.* 30: 1804416.
75 Jia, Z.R., Qin, S.C., Meng, L. et al. (2021). *Nat. Commun.* 12: 178.
76 Zhang, Q., Kan, B., Liu, F. et al. (2014). *Nat. Photonics* 9: 35.
77 Kan, B., Li, M., Zhang, Q. et al. (2015). *J. Am. Chem. Soc.* 137: 3886.
78 Zhou, J.Y., Wan, X.J., Liu, Y.S. et al. (2012). *J. Am. Chem. Soc.* 134: 16345.
79 Zhou, J.Y., Zuo, Y., Wan, X.J. et al. (2013). *J. Am. Chem. Soc.* 135: 8484.
80 Kan, B., Zhang, Q., Li, M.M. et al. (2014). *J. Am. Chem. Soc.* 136: 15529.
81 Kan, B., Feng, H.R., Wan, X.J. et al. (2017). *J. Am. Chem. Soc.* 139: 4929.
82 Dennler, G., Scharber, M.C., Ameri, T. et al. (2008). *Adv. Mater.* 20: 579.
83 Yin, Z.G., Wei, J.J., and Zheng, Q.D. (2016). *Adv. Sci.* 3: 1500362.
84 Sista, S., Park, M.H., Hong, Z. et al. (2010). *Adv. Mater.* 22: 380.
85 Chou, C.H., Kwan, W.L., Hong, Z. et al. (2011). *Adv. Mater.* 23: 1282.
86 Li, N., Stubhan, T., Krantz, J. et al. (2014). *J. Mater. Chem. A* 2: 14896.
87 Martínez-Otero, A., Liu, Q., Mantilla-Perez, P. et al. (2015). *J. Mater. Chem. A* 3: 10681.
88 Zuo, L.J., Chang, C.Y., Chueh, C.C. et al. (2015). *Energy Environ. Sci.* 8: 1712.

89 Lu, S.M., Lin, H., Zhang, S.Q. et al. (2017). *Adv. Energy Mater.* 7: 1701164.

90 Wenchao, H., Sheng-Yung, C., Pei, C. et al. (2018). *Nano Lett.* 18: 7977.

91 Kouijzer, S., Esiner, S., Frijters, C.H. et al. (2012). *Adv. Energy Mater.* 2: 945.

92 Ho, C.H.Y., Kim, T., Xiong, Y. et al. (2020). *Adv. Energy Mater.* 10: 2000823.

93 Jo, J., Pouliot, J.R., Wynands, D. et al. (2013). *Adv. Mater.* 25: 4783.

94 Di Carlo Rasi, D., van Thiel, P.M.J.G., Bin, H. et al. (2019). *Solar RRL* 3: 1800366.

95 Lee, J., Kang, H., Kong, J., and Lee, K. (2014). *Adv. Energy Mater.* 4: 1301226.

96 Kang, H., Kee, S., Yu, K. et al. (2015). *Adv. Mater.* 27: 1408.

97 Rakwon, K., Sujung, P., Yun Kyung, J. et al. (2018). *Adv. Energy Mater.* 8: 1804416.

98 Zeng, W.W., Xie, C., Wang, W. et al. (2019). *Solar RRL* 4: 1900480.

99 Yang, J., Zhu, R., Hong, Z.R. et al. (2011). *Adv. Mater.* 23: 3465.

100 Zhang, K., Fan, B.B., Xia, R.X. et al. (2018). *Adv. Energy Mater.* 8: 1703180.

101 Vasilopoulou, M., Polydorou, E., Douvas, A.M. et al. (2015). *Energy Environ. Sci.* 8: 2448.

102 Prosa, M., Tessarolo, M., Bolognesi, M. et al. (2016). *Adv. Mater. Interfaces* 3: 1600770.

103 Becker, T., Trost, S., Behrendt, A. et al. (2018). *Adv. Energy Mater.* 8: 1702533.

104 Ke, X., Meng, L.X., Wan, X.J. et al. (2020). *J. Mater. Chem. A* 8: 9726.

105 Vos, A.D. (1980). *J. Phys. D: Appl. Phys.* 13: 839.

106 Benduhn, J., Tvingstedt, K., Piersimoni, F. et al. (2017). *Nat. Energy* 2: 17053.

11

Large-Area Flexible Organic Solar Cells

Shaorong Huang[1], Yufei Wang[2]*, Lintao Hou[2]**, and Lie Chen[1]***

[1]*Nanchang University, Institute of Polymers and Energy Chemistry (IPEC), Honggutan New District, Xuefu Avenue 999, College of Chemistry, Nanchang, 330031, Jiangxi, P.R. China*
[2]*Jinan University, Guangdong Provincial Engineering Technology Research Center of Vacuum Coating Technologies and New Energy Materials, Siyuan Laboratory, Department of Physics, Huangpu Avenue West 601, Guangzhou, 510632, Guangdong, P.R. China*

11.1 Introduction

Organic solar cells (OSCs), as a sustainable energy technology, have the characteristics of lightweight, low-cost, and flexibility and have unique advantages to manufacture large-area flexible devices via solution processing methods [1, 2]. Moreover, stability tests have shown that it is possible for OSCs to have a life span of up to 10 years, which shows their potential for practical application [3]. Therefore, OSCs have attracted much attention from both the academia and the industry recent years. However, compared with small-area OSCs, large-area devices exhibit much lower performance, especially the ones fabricated on large-area flexible substrates. Material requirements, fabrication methods, efficiency loss change drastically as the trend shifts from small-area to large-area OSCs.

OSCs generally are fabricated by spin coating in the laboratory [4], but such process is inappropriate for large-scale and flexible production. The experience gained by spin coating process has limited values for large-area preparation technology. Currently, the most potential mass production process for large-area OSCs has been proven to the roll-to-roll (R2R) technology. A various of film-forming methods, mainly including blade coating, slot-die coating, inkjet printing, spray coating, screen printing, relief printing and gravure printing, can be integrated into the R2R process to form a complete system. In addition, material requirements and efficiency loss of the large-area productions are quite different from those of the small-area ones. Therefore, a comprehensive consideration of these processes is highly required for large-area devices. This chapter focuses on large-area flexible organic solar cells (FOSCs), including material requirements, flexible electrode and

*Shaorong Huang and Yufei Wang contributed equally to this work.
**Lintao Hou and Lie Chen are corresponding authors.

Organic Solar Cells: Materials Design, Technology and Commercialization, First Edition. Edited by Liming Ding.

substrates, large-area flexible device fabrication, and efficiency loss in large-area devices and modules. Finally, we make a conclusion on the prospects and challenges of the development of large-area FOSCs.

11.2 Material Requirements for Large-Area Flexible Organic Solar Cells

The material requirements for large-area FOSCs are more complicated than those for small-area OSCs on glass substrate. Besides possessing excellent photovoltaic property, the materials need to be of good stretchable property and compatible to printing process. Due to the strong entanglement of the macromolecular chain, the films of polymer-based active layer often show good stretchability, which is a good property for flexible device. The power conversion efficiency (PCE) is observed to be very sensitive to the thickness of the active layer. For small-area OSCs, the thickness of the active layer prepared by spin coating is normally about 100 nm, an optimal result of comprehensive consideration of light absorption, morphology, charge transport and recombination, and other possible photophysical processes to achieve the best device performance. However, for the large-area devices, the thickness of active layers generally should be much greater than that of small-area devices due to the requirement of large-area coating and printing technology. Therefore, large-area manufacture process put forward new requirements for materials, that is, materials for large-area devices should work well in a thick film. The low charge mobility and high charge recombination in the bulk heterojunction (BHJ) films are found to be the key limitations to form a high-performance thick film.

In the past few years, many efforts have been made to develop high-quality films with good thickness tolerance. In this section, we mainly focus on the active layer materials for large-area FOSCs and will introduce four type systems for thick-film active layer: fullerene-based binary system, non-fullerene-based binary system, ternary system, and all-polymer system. At last, the design strategies of the materials for large-scale devices will be discussed.

11.2.1 Fullerene-Based Binary System

The fullerene-based binary system based on the fullerene and its derivatives ($PC_{61}BM$, $PC_{71}BM$ [5], indene-C60 bisadduct [ICBA], etc.) as acceptors with high efficiency and thick active layer has been reported in the past early years. Fullerene-based binary system can be divided into two types: polymer: fullerene system and small molecule: fullerene system. The structures and the device parameters are shown in Figure 11.1 and Table 11.1, respectively. The initial stage of the research on devices with thick-film polymer: fullerene binary started from the P3HT:PCBM system. P3HT has the characteristics of low price and simple synthesis, presenting the most potential for practical application. In addition, the regular chain structure of P3HT also can facilitate their self-organization into two-dimensional sheets by means of inter-chain stacking, promoting charge

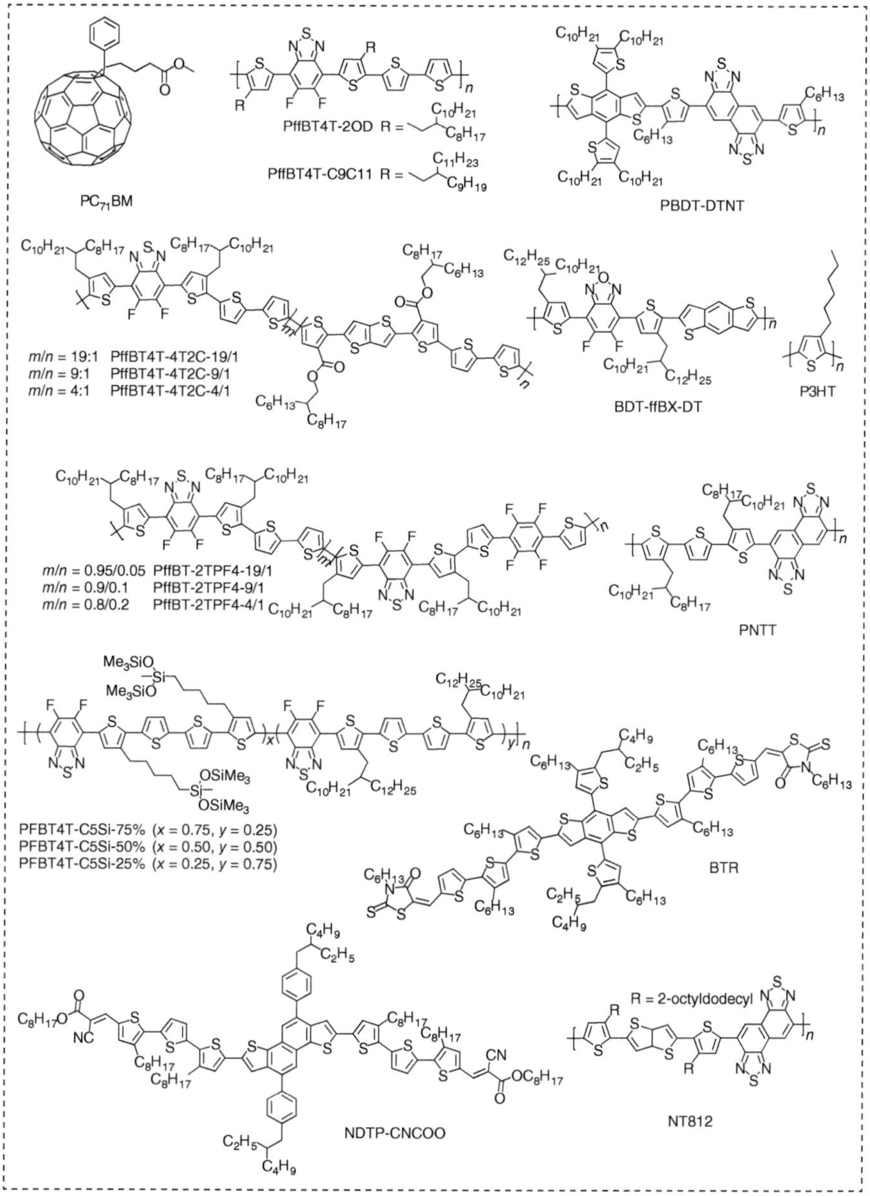

Figure 11.1 Polymer structures in thick-film polymer-fullerene binary devices.

transport. Combined with the ultrafast three-dimensional charge transfer channel of fullerene, P3HT:PCBM system possesses high charge mobility and favorable morphology of the blend film upon various optimal treatments, leading to a high thick-film performance. Yang and coworkers reported highly efficient polymer solar cells (PSCs) based on P3HT:$PC_{61}BM$ by controlling the active layer growth rate to increase hole mobility, balance charge transport, and improve absorption.

Table 11.1 The performance of thick-film polymer-fullerene binary devices.

Active layers	Thickness (nm)	V_{oc} (V)	J_{sc} (mA cm^{-2})	FF (%)	PCE (%)	References
P3HT:PC$_{61}$BM	~210–230	0.61	10.6	67.4	4.4	[6]
PffBT4T-2OD:PC$_{71}$BM	≈300	0.77	18.80	75.00	10.8	[7]
PffBT4T-C9C11:PC$_{71}$BM	≈300	0.79	20.20	74.00	11.70	[8]
PffBT-2TPF4-9/1:PC$_{61}$BM	168	0.769	17.18	70.8	9.4	[9]
PffBT4T-4T2C-9/1:PC$_{61}$BM	≈350	0.784	15.5	79.1	9.61	[10]
PFBT4T-C5Si-25%:PC$_{71}$BM	420	0.76	19.08	74.12	11.09	[11]
	600	0.75	18.32	71.74	10.15	
PBDT-DTNT:PC$_{71}$BM	280	0.74	17.62	66.10	8.62	[12]
	940	0.73	16.67	59.50	7.24	
NT812:PC$_{71}$BM	320	0.72	19.09	72.85	10.33	[13]
	1000	0.72	18.12	62.15	8.35	
BDT-ffBX-DT:PC$_{61}$BM	250	0.93	13.80	73.00	9.40	[14]
PNTT:PC$_{71}$BM	280	0.77	20.20	71.80	11.30	[15]
	1050	0.75	19.90	59.60	9.00	
NDTP-CNCOO:PC$_{71}$BM	302	0.940	10.77	71.10	7.20	[16]
BTR:PC$_{71}$BM	250	0.900	13.60	77.00	9.3	[17]
	400	0.890	12.92	≈70.00	≈8	

An outstanding PCE of 4.4% with a 210 ~ 230 nm thickness of active layer was achieved [6]. However, the poor light harvesting and high HOMO energy level of P3HT limit the further improvement in the device performance.

To expand the absorption range and down-shift the HOMO energy level, the design concept of narrow bandgap donor–acceptor (D–A) has been employed in developing efficient polymer donors for fullerene-based binary devices. Internal regulation of molecular structure and external morphological optimization of the film are often performed to improve the charge mobility and film morphology for thick film application. Internal regulation of molecular structure is aimed to improve the planarity of the backbone for efficient charge transport, such as incorporation of the planar building blocks, side chain engineering, π-bridge engineering, quinoidal units. External optimal treatments of the films are conducting extra pretreatment or posttreatment to induce a favorable growth of the active layer. These extra treatments mainly include thermal annealing, solvent annealing, additives, etc.

It has demonstrated that temperature-dependent aggregation effect is a key feature for donor polymers to develop favorable morphology and form thick active layers. A polymer with temperature-dependent aggregation property can dissolve in a certain solvent at high temperature and then gradually aggregate during the cooling process. Generally, temperature-dependent aggregation characteristic is the unique behavior of highly crystalline polymers. Regulating the transition between aggregation and disaggregation by solution temperature and concentration can readily fine-tune crystallinity and domain size, which is beneficial to form a high-quality thick film. In 2014, Yan and coworkers [7] reported three polymer donors (PffBT4T-2OD, PBTff4T-2OD, and PNT4T-2OD) based on polymultithiophene chain. These polymers have strong temperature-dependent aggregation properties, and the morphology can be well controlled by such temperature-dependent aggregation behavior. As a result, the PCE of over 10% with a fill factor (FF) of over 70% has been achieved by PffBT4T-2OD blended with $PC_{71}BM$. The devices also worked well with 300 nm-thick active layer. In 2017, the same group synthesized a new polymer PffBT4T-C9C11 [8] by changing the blanched alkyl chain in PffBT4T-2OD. Side chain engineering was demonstrated to further optimize the morphology of the active layer, and a breakthrough PCE of 11.7% with nearly 300 nm-thick active layer has been achieved. They also declared that the position of the alkyl chain and the fluorine atom in the polymer backbone was the main factor affecting the temperature-dependent aggregation of the materials.

The devices based on PffBT4T-2OD with different fullerene derivatives obtained high PCE, but the strong aggregation property of PffBT4T-2OD often leads to large and unsuitable domains during the room-temperature process. To ensure the solution spin-coating process, the active layer solution and indium tin oxide (ITO) substrate are both preheated to 70 ~ 110 °C before coating, which is detrimental for large-area device process. In 2017, Chen's group firstly reported that random copolymerization was a facile and effective method to regulate the aggregation of polymer PffBT4T-2OD. They synthesized new terpolymer donors (PffBT-2TPF4) [9, 18] by introducing tetrafluorinated benzene (PF4) as the third unit into polymer PffBT4T-2OD through random copolymerization. The aperiodic sequence distribution caused by random copolymerization effectively reduced the aggregation of the polymers, enabling a room-temperature process of the active layer. On the other hand, due to the self-assembly of the PF4 and careful control on the loading of PF4 into the backbone, the developed terpolymers still exhibited obviously temperature-dependent aggregation properties, beneficial for the formation of the well-defined film-state packing. Eventually, the room temperature processed terpolymers obtained a comparable performance to the high temperature processed PffBT4T-2OD. In 2019, they further introduced a self-assembly third unit of dimethyl 2,2′-(thieno[3,2-*b*]thiophene-2,5-diyl)bis(thiophene-3-carboxylate) (4T2C) into the PffBT4T-2OD matrix. Regulation of the loadings of the 4T2C unit, the morphology of active layer (PffBT4T-4T2C-9/1:$PC_{61}BM$) can be well-tuned to form a homogeneous fibril-like bicontinuous interpenetrating network with a preferred face-on orientation. As a result, the terpolymer-based device achieved a remarkable FF near 80%, which is the highest value of the fullerene-based solar cells without any treatment at that time. Moreover, an impressive PCE over 9.0% was obtained when the thickness of active layer over 300 nm [10]. Siloxane-terminated side chain of the polymers has been reported to have the ability to affect orientation of the molecules. In 2017, Chen and coworkers synthesized a terpolymer PFBT4T-C5Si-25% [11] by introducing siloxane-terminated side chains through random copolymerization. PFBT4T-C5Si-25% showed better molecular π–π stacking and enhanced vertical carrier transport than PffBT4T-2OD. PFBT4T-C5Si-25%:$PC_{71}BM$ displayed low film surface roughness, balanced charge transport, and small phase separation. High PCEs of 11.09% and 10.15% were obtained for the active layer with a thickness of 420 and 600 nm, respectively.

In addition to the PffBT4T-2OD family, fused ring moieties are also good building blocks to construct polymer donors for active layer with good thickness tolerance. In 2014, Huang and coworkers prepared the devices based on PBDT-DTNT:$PC_{71}BM$ with thick films. Incorporation of naphthalene unit was found to endow the PBDT-DTNT with a high mobility. The PBDT-DTNT:$PC_{71}BM$ obtained a PCE of 8.62% for a 280 nm-thick film and 7.24% for a nearly 1000 nm-thick film [12]. In 2016, they also introduced naphthalene unit into polymer NT812 [13]. A face-on orientation was detected in polymer NT812 film. When blended with $PC_{71}BM$, a PCE of 10.33% for a 320 nm-thick film and a PCE of 8.35% for a 1000 nm-thick film were achieved. In 2017, they synthesized a new polymer donor BDTffBX-DT with naked BDT unit. Non-halogenated solvent processed active layer based on BDTffBX-DT:$PC_{61}BM$ demonstrated a face-on orientation with a suitable phase separation. With a film thickness of 250 nm, BDTffBX-DT:$PC_{61}BM$ obtained a PCE of 9.4% [14]. The same group also successfully used naphthalene unit to improve the performance of PffBT4T-2OD family with thick film. They created a polymer donor PNTT with deep HOMO energy levels and high hole mobility [15]. The device based on the PNTT:$PC_{71}BM$ obtained the best efficiency of 11.3% at a film thickness of 280 nm and a decent PCE of 9% at a film thickness of 1050 nm.

Small-molecule semiconductors are also used as donor materials for thick-film fullerene-based OSCs. Wei and coworkers reported thick-film OSCs based on NDTP-CNCOO:$PC_{71}BM$ [16]. Due to the ordered stacking between molecules and favorable film morphology, the efficiency of 7.20% was obtained with a 300 nm-thick active layer. Wong and coworkers reported a small-molecule donor BTR with rigid and flat backbone and numerous flexible side chains. A PCE of 9.3% and 8.0% was achieved at a thickness of 250 and 400 nm, respectively [17]. Later, Kim and coworkers reported R2R slot-die coating device modules based on BTR:$PC_{71}BM$, obtaining a PCE of 7.46% and 4.83% for single cells and modules with an active area of 10 cm^2, respectively [19].

11.2.2 Non-fullerene-Based Binary System

Compared with the fullerene acceptor with narrow absorption and fixed energy levels, non-fullerene acceptors have been rapidly developed in recent years. Non-fullerene acceptors are normally referred to the conjugated small molecules with high electron affinity. They have the advantages of low cost, adjustable energy levels, and good light absorption making them promising alternatives to replace fullerene acceptors for OSCs. Zhan and coworkers have made important contributions to the design and synthesis of non-fullerene acceptors. In 2015, they designed the first non-fullerene acceptor ITIC with the absorption expanding to near-infrared light. Using polymer PTB7-Th with narrow bandgap as donor, a PCE of 6.8% was obtained [20]. In 2016, Hou and coworkers developed a wide-bandgap polymer PBDB-T and processed a non-fullerene-based device with PBDB-T:ITIC. A notable PCE of 11.21% was obtained [21], which is the first report that the efficiency of non-fullerene-based device exceeded that of fullerene-based one. The high performance of the non-fullerene-based device was ascribed to complementary absorption of donor and acceptor, appropriate energy level alignment, and favorable phase-separation morphology.

Similar to fullerene systems, for obtaining high-performance thick-film active layer, non-fullerene-based systems also should have high charge carrier mobility, wide light absorption, appropriate phase separation morphology, low bimolecular recombination, and so on. With the rapid development of the non-fullerene acceptors, the polymer donors with complimentary absorption have been also reported. The molecular structures and device parameters are shown in Figure 11.2 and Table 11.2. Huang and coworkers prepared thick-film devices based on a blend of wide-bandgap conjugated polymer PTzBI and ITIC. A PCE of 7.09% with a film thickness of 300 nm was achieved [22]. Similarly, Zhang and coworkers synthesized a wide-bandgap polymer PTZ1 based on thiazolothiazole unit. By optimizing the morphology of the active layer upon thermal annealing, the device with PTZ1:IDIC delivered the best performance of 9.5% at a film thickness of 210 nm [23]. It has been reported that the functional atoms, F and Cl atoms, not only have the ability of down-shifting HOMO energy level to boost the V_{oc}, but also can improve the crystallinity of the materials for efficient charge transport [33–35]. Hou and coworkers reported a polymer donor PBDB-T-SF with low energy levels, in which fluorine and sulfur atoms were introduced into the side chains of PBDB-T. When blended with non-fullerene acceptor IT-4F, the PBDB-T-SF:IT-4F-based devices exhibited a PCE of over 12% with a wide thickness range of 100–200 nm [24]. Zhang and coworkers synthesized two efficient polymer donors, PM6 and PM7. PM6 and PM7 have the same skeleton to the polymer PBDB-T, but only replace the H atom on thiophene side chain by F atom and Cl atom, respectively. Due to the improved inter-chain interaction and down-shifted HOMO levels favored by F and Cl atom, both PM6 and PM7 have a higher crystallinity, enhanced absorption, higher carrier mobility , and higher V_{oc} of the devices than PBDB-T. The device based on PM6:IDIC obtained a PCE of 11% at the active layer thickness of 255 nm [25]. The PM6:IT-4F blends achieved an excellent PCE of 12.2% with a 285 nm-thick film [26]. As for the PM7:IT-4F-based device, a PCE of 12.0% was obtained at a 200 nm-thick active layer [27].

Small molecular donors have also been employed for preparation of thick-film non-fullerene OSCs. In 2018, Hou and coworkers reported a highly efficient non-fullerene OSCs based on small molecular donors (DRTB-T-C2, DRTB-T-C4, DRTB-T-C6, DRTB-T-C8) and IT-4F acceptor. By prolonging the length of the end alkyl chain of the donors, the active layer preferred to form a face-on orientation. The devices based on DRTB-T-C4:IT-4F retained a high PCE of 10% with the active layer thickness up to 300 nm [28].

Although non-fullerene materials have many advantages, the lower charge mobility of non-fullerene materials than fullerene materials is a major problem when they are applied as thick film. Non-fullerene acceptor with a preferential face-on

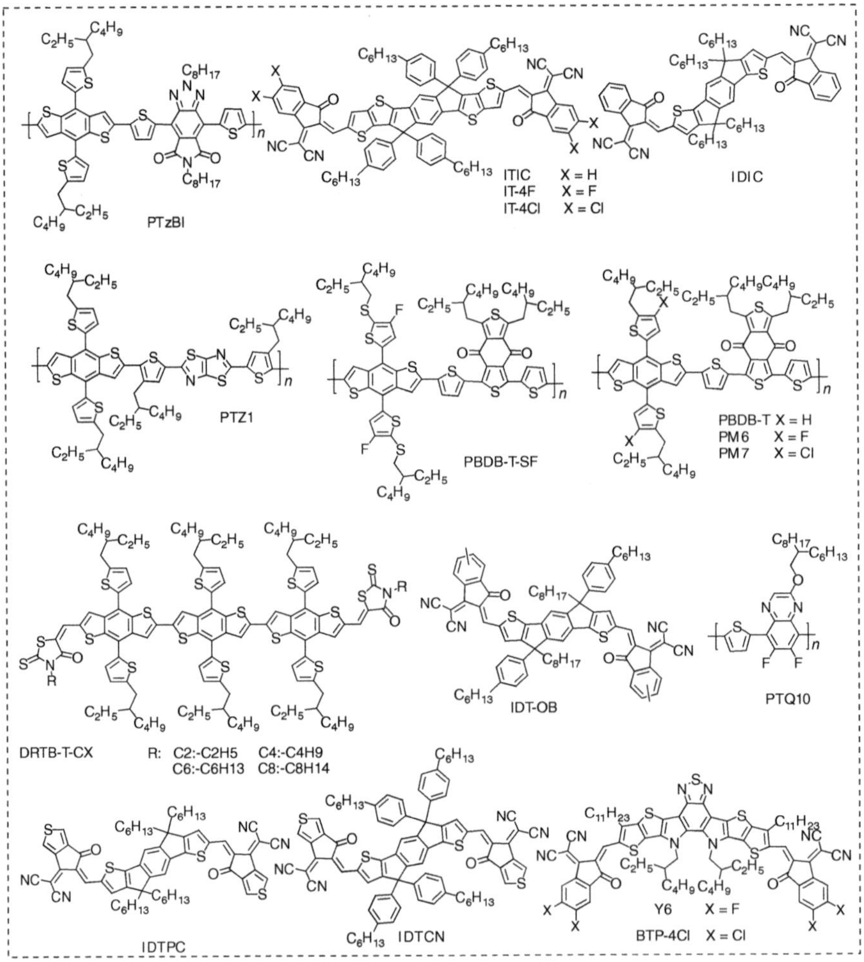

Figure 11.2 The chemical structures of polymers and non-fullerene small-molecule acceptors that perform well in thick-film polymer-non-fullerene binary solar cells.

orientation is very conductive to charge transport. Bo and coworkers synthesized a new non-fullerene electron acceptor IDT-OB with asymmetric side chains. The IDT-OB exhibited a face-on orientation, thereby leading to a high charge mobility. A notable PCE of 9.17% was obtained when the thickness of the active layer was 210 nm [29]. Luo et al. synthesized a new small molecule acceptor (IDTPC), which exhibited more ordered face-on π–π stacking and higher electron mobility when compared with its similar molecule IDTCN. The PTQ10:IDTPC-based device delivered a PCE of 10.0% at active layer thickness of 400 nm [30]. In addition, the design concept of halogenated polymers has also been regarded as a highly effective strategy for designing non-fullerene acceptors with improved charge mobility. A series of efficient non-fullerene acceptors with F and Cl endcap were synthesized, such as IT-4F, Y6, BTP-4Cl. In 2019, Zou and coworkers designed and synthesized a new small-molecule non-fullerene acceptor Y6, which has

Table 11.2 The performance of thick-film polymer-non-fullerene binary solar cells.

Active layers	Thickness (nm)	V_{oc} (V)	J_{sc} (mA cm^{-2})	FF (%)	PCE (%)	References
PTzBI:ITIC	300	0.81	17.54	49.91	7.09	[22]
PTZ1:IDIC	210	0.90	15.90	66.80	9.50	[23]
PBDB-T-SF:IT-4F	200	–	–	–	> 12.00	[24]
PM6:IDIC	225	0.97	17.20	68.00	11.30	[25]
PM6:IT-4F	285	0.83	22.60	64.80	12.20	[26]
PM7:IT-4F	200	0.87	19.70	63.90	11.00	[27]
DRTB-T-C4:IT-4F	300	0.893	18.68	61.00	10.18	[28]
PBDB-T:IDT-OB	210	0.88	15.64	66.90	9.17	[29]
PTQ10:IDTPC	400	0.913	17.9	61.3	10.0	[30]
PM6:Y6	200	0.83	25.8	66.9	14.3	[31]
PBDB-TF (PM6):BTP-4Cl	1000	0.830	26.3	46.3	10.1	[32]

multiple electronic push and pull effects, high carrier mobility, and broadened absorption. The devices based on PM6:Y6 achieved a remarkable PCE of over 15% with a thin active layer of ~100 nm and 14.3% with a thickness of 200 nm [31]. In 2020, Hou and coworkers processed 1000 nm thick-film devices employing PM6 as polymer donor and Cl-functionalized small-molecule BTP-4Cl as acceptor. The BTP-4Cl-based thick-film device was demonstrated to display low trap-assist recombination and appropriate phase separation morphology. PM6:BTP-4Cl-based device with 4 cm^2 area exhibited a PCE of 10.1% with an active layer thickness of 1000 nm [32]. Thereby, these materials show great potential for manufacturing large-scale FOSCs.

11.2.3 Ternary System

The ternary blending strategy of incorporating a third component into the active layer of binary OSCs is a facile and effective method to improve the device performance. The role of the third component in the ternary blending can be donor or acceptor to adjust the energy levels, broaden the absorption, and optimize the phase morphology. Working mechanism of ternary OSC includes charge transfer mechanism, energy transfer mechanism, parallel linkage mechanism, alloy model mechanism, etc. [36–39] Highly crystalline conjugated polymers or small molecules as the third component are in favor of improving the charge mobility and thickness tolerability. Molecules for the ternary system with thick films are shown in Figure 11.3, and Table 11.3 displays the corresponding performance for these systems.

Some ternary OSCs based on non-fullerene acceptors have performed with thick active layers. For example, Baran et al. reported P3HT:IDTBR:IDFBR-based devices, which exhibited a high PCE of 5.7% with a film thickness of 320 nm [40].

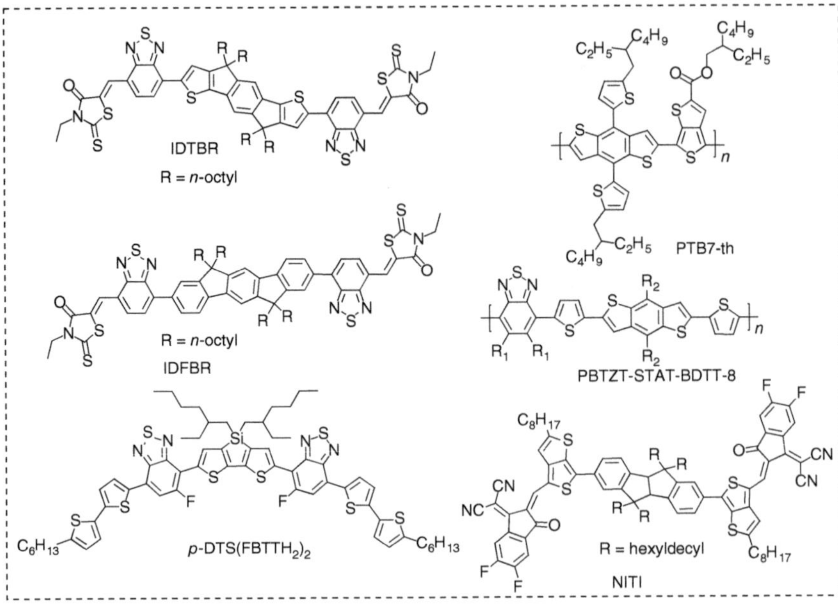

Figure 11.3 Chemical structures of the polymers capable of demonstrating high performance in thick-film ternary solar cells.

Table 11.3 The performance of thick-film ternary organic solar cells.

Active layers	Thickness (nm)	V_{oc} (V)	J_{sc} (mA cm^{-2})	FF (%)	PCE (%)	References
P3HT:IDTBR:IDFBR	320	0.780	11.30	64.00	5.70	[40]
PTB7-th:p-DTS $(FBTTH_2)_2$:$PC_{71}BM$	270	0.737	21.67	67.50	10.78	[41]
PTB7-th:PBTZT-STAT-BDTT-8:$PC_{71}BM$	≈300	0.770	19.35	74.05	11.03	[42]
PTB7-th:BTR:$PC_{71}BM$	250	0.751	21.40	70.00	11.40	[43]
BTR:NITI:$PC_{71}BM$	300	0.940	19.5	73.8	13.6	[44]
PM7:Y6:$PC_{61}BM$	300	0.802	26.8	66.7	14.3	[45]

In fact, taking advantage of the fullerene derivatives and non-fullerene molecules and using both as acceptors for constructing a thick-film ternary device is very meaningful. Wei and coworkers synthesized a molecule *p*-DTS$(FBTTH_2)_2$ for ternary devices with thick film. Addition of small molecule was found to improve the crystallinity and face-on orientation of the active layer. The PTB7-Th:*p*-DTS$(FBTTH_2)_2$:$PC_{71}BM$-based ternary device achieved a PCE of 10.78% with active layer thickness of 270 nm [41]. Ameri and coworkers processed ternary devices by adding a polymer PBTZT-STAT-BDTT-8 to the PTB7-th:$PC_{71}BM$ system. A PCE of 11.03% with active layer thickness of 300 nm was obtained [42]. In addition, nonradiative Förster resonance energy transfer from the PBTZT-STAT-BDTT-8 to PTB7-Th polymer bandgap polymer PTB7-Th was detected in the blend, contributing to the improvement of device performance. Zhang et al. added a liquid crystalline small-molecule donor BTR to the PTB7-Th:$PC_{71}BM$ system. The morphology of active layer was optimized, as revealed by the increased coherence length and domain purity with addition of BTR. As a result, a PCE 11.4% was obtained for the device with an active layer thickness of 250 nm [43]. Zhu and coworkers also prepared ternary devices with thick active layer by adopting BTR as donor, a small-molecule NITI and $PC_{71}BM$ as acceptors. For the BTR:NITI-based binary device, a PCE of 8.66% with a film thickness of 250 nm was obtained. After incorporation of $PC_{71}BM$ into the blend, the morphology of the active layer showed more appropriate phase separation. As a result, the BTR:NITI:$PC_{71}BM$-based ternary device yielded a PCE of 13.6% with a film thickness of 300 nm [44]. In 2020, Zhang and coworkers processed a ternary device based on PM7:Y6:$PC_{61}BM$. Addition of $PC_{61}BM$ into the active layer was observed to improve charge mobilities and reduce charge recombination, leading to an improved efficiency of 14.3% with a film thickness of 300 nm [45].

Controlling the morphology of large-area BHJ devices is still a tricky problem. To solve this issue, Chen and coworkers [46] adopted a pseudo-planar heterojunction (PPHJ) method to fabricate a large-area ternary with the fullerene derivative ICBA as the third component to enhance the PCE of the device with PM6:IT-4F. It can be found that ICBA promoted the vertical phase separation and the formation of PPHJ structure. Furthermore, the addition of ICBA also improved the crystallinity of PM6, thus accelerating the exciton dissociation and charge transfer. Consequently, 1.05 cm^2 PPHJ device based on PM6:ICBA:IT-4F with a ratio of 1 : 0.1 : 0.9 fabricated by blade-coating achieved a notable PCE of 14.25%, which was superior than both of the corresponding BHJ devices (13.73%) and the binary OSC (13.19%).

11.2.4 All-Polymer-Based System

All-PSCs have excellent thermal stability, flexibility, and mechanical stability features. Relative to small molecules, high-viscosity polymers are more conducive to large-scale manufacturing [47]. The typical molecular structures for all-PSCs with thick film are shown in Figure 11.4, and Table 11.4 displays their corresponding performance. N2200 acceptor is widely used in all-PSCs, which was first synthesized by Yan et al. [51]. Huang and coworkers reported a thick film all-PSC based on PTzBI:N2200 [48]. Manipulating the molecular weight of N2200 was found to effectively adjust the exciton generation, charge transfer, and recombination. An optimal PCE of 9.16% was obtained with a film thickness of 120 nm. When the film thickness increased to 230 nm, a PCE of 6.37% was obtained, together with a good thermal and mechanical stability. Later on, to further tailor the properties of the polymer PTzBI, they introduced a siloxane-terminated side chain into PTzBI and obtained a new donor PTzBI-Si. PTzBI-Si has high crystallinity with a face-on orientation, leading to high and balanced charge mobility. Consequently, by using the green solvent 2-methyltetrahydrofuran, a PCE of 10.05% was obtained for the active layer with a film thickness of 140 nm, while

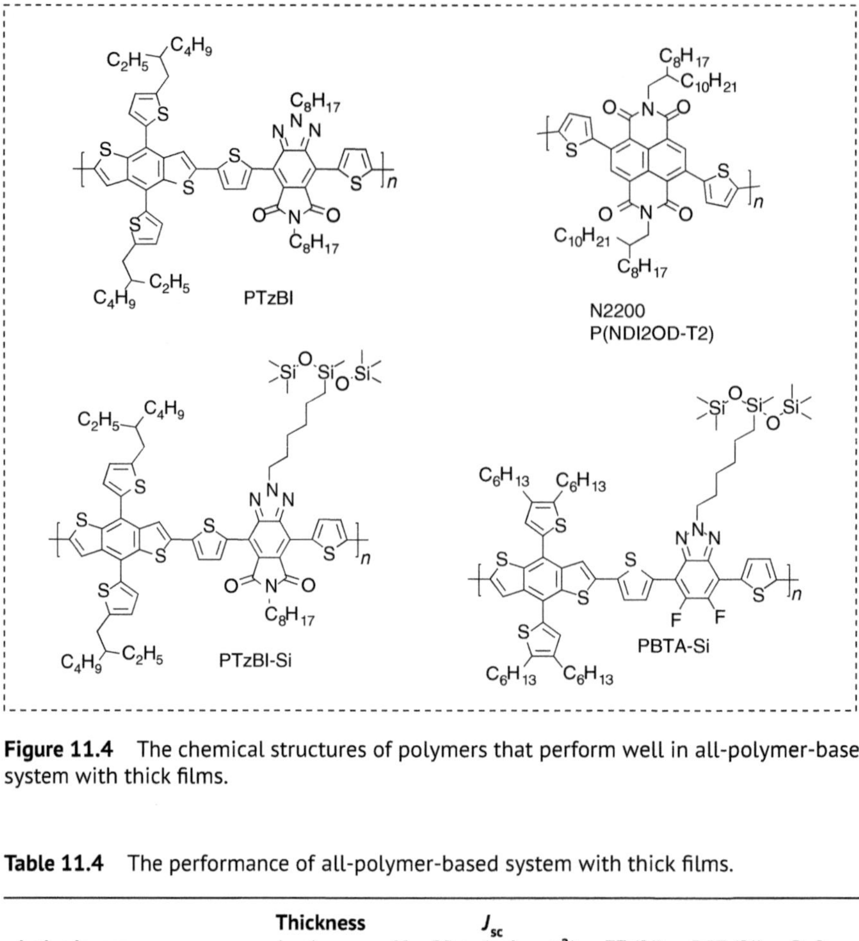

Figure 11.4 The chemical structures of polymers that perform well in all-polymer-based system with thick films.

Table 11.4 The performance of all-polymer-based system with thick films.

Active layers	Thickness (nm)	V_{oc} (V)	J_{sc} (mA cm^{-2})	FF (%)	PCE (%)	References
PTzBI:N2200	230	0.841	13.55	53.65	6.37	[48]
PTzBI-Si:N2200	180	0.857	14.95	68.73	8.81	[49]
PBTA-Si:PTzBI-Si:N2200	350	0.850	16.67	64.93	9.17	[50]
	420	0.830	16.45	60.77	8.34	

a PCE of 8.81% was obtained with the thickness of 180 nm [49]. In 2018, they further processed a ternary device based on PBTA-Si:PTzBI-Si:N2200 with PBTA-Si as the third component. Under a thickness of 350 and 420 nm, a high PCE of 9.17% and 8.34% was obtained, respectively [50]. Compared with the binary OSCs based on PTzBI-Si:N2200 blends, ternary all-polymer system has higher thickness tolerance.

11.2.5 Design Strategies of the Materials for Large-Area Devices

Considerable efforts and progress have been made on thick-film devices. It demonstrates that high charge carrier mobility, low bimolecular recombination, and favorable morphology with appropriate domain size and high phase purity are extraordinarily favorable to meet the requirements of large-scale device fabrication. Therefore, besides improving light harvesting, developing high-mobility materials is the most important to obtain high-performance thick films for large-scale application. Many design strategies, such enhancing molecular planarity of the backbone, improving crystallinity and inter-chain interaction by functional atoms effect (e.g. fluorine, chlorine, siloxane, and sulfur atoms) and weak non-covalent bonds, introducing the quinoidal units, are proven to be very effective for developing high-mobility materials. On the other hand, the optimization of BHJ morphology is also very important. Developing film morphology with ordered and preferential face-on orientation can guarantee high charge transport in a thick active layer. Moreover, external optimal treatments are also attractive methods to further optimize the morphology of the active layer.

As for the donor materials, small-molecule donors are more pure and lead to better batch-to-batch reproduction than polymer donors, while polymer donors are more flexible, easier to synthesis, and with better film-forming property, which makes them more attractive for large-area FOSCs. As for acceptor materials, fullerene and its derivatives have some fatal disadvantages, such as high cost, poor absorption, and limited ability to control the energy levels; thereby, they would not be ideal and extensively used acceptors for practical application. High-performance non-fullerene acceptors are more appealing than fullerene and its derivatives. On the other hand, fullerene has the characteristics of low reorganization energy and high reduction potential, which can accelerate the transfer of photo-induced charge and avoid the return of electrons simultaneously. Taking advantage of the fullerene and non-fullerene molecules to fabricate thick-film ternary devices is very promising. Furthermore, the state-of-the-art devices are based on the active layers that at least one component is a small molecule. In fact, compared with the device with small molecules, all-PSCs have superior thermal and mechanical stability; therefore they are more suitable for large-scale manufacturing, if their efficiency can be substantially improved.

11.3 Flexible Electrodes and Substrates

As one of the emerging solar cells, FOSCs have become a powerful rival to the traditional silicon solar cells owing to the advantages of light weight, flexible fabrication, and semitransparency. In order to achieve good mechanical property and high efficiency, it is essential to prepare highly transparent ultra-flexible electrodes and substrates to match high-performance photoactive layers.

11.3.1 Flexible Substrates

To maximize the absorption of the incidence sunlight in the active layers, the substrates used for the preparation of FOSCs should have good flexibility, transparency, and durability. Meanwhile, the ideal substrate must be compatible with the device structures, electrode types, and printing methods, which is favorable to achieve stable and long-life FOSCs. In addition, the substrate for FOSCs must possess a smooth surface in the nanoscale range to prevent spikes and irregularities of ultrathin films from penetrating through the device. Likewise, the influence of the substrate on the preparation process and stability of device must be considered, such as casting solvents intermiscibility, water/UV resistance, thermal stability, and shrinkage and dimensional stability. Generally, the commonly used substrates for FOSCs are mainly divided into plastics and metal foils.

The metal foil substrate is suitable for FOSCs with high-temperature process, and it can maintain good stability in the air (barrier to water and oxygen) [52–55]. In addition, the unique high conductivity of the metal foil can simultaneously act as the bottom electrode. Especially, stainless steel can withstand a high temperature up to 1000 °C, and its excellent high corrosion resistance and chemical stability are favored for FOSCs. However, metal foil substrates also have some drawbacks of non-transparency, poor flexibility, and large surface roughness, which can produce a great obstacle to achieving high-efficiency and ultra-flexible FOSCs. Actually, the surface roughness of the metal foil substrates can be solved by polishing or using an additional planarization layer [53]. But, a bad result could be produced when the conducting substrates as the back contact are completely isolated from the photovoltaic device through the thick additional planarization layer.

The transparent plastic substrate is mainly based on polymers and can be well applied in the process of R2R production due to its high flexibility. Common polymer substrates mainly include polyethylene terephthalate (PET), polyethylene naphthalate (PEN), polyethersulfone (PES), polyimide (PI) [56–62]. Except PI that can absorb part of visible light, other polymer substrates have high transmittance in the range of visible light, which is suitable for most FOSCs. Remarkably, the thermal resistance of PI to high temperatures up to 350 °C gives the ability of FOSCs bearing rigorous fabrication processing and working at high temperature.

There are some disadvantages existing in plastic devices, of course, such as the unstable dimensions during high-temperature processing and poor water and oxygen isolation. Therefore, an additional barrier coating is required for plastic devices to reduce deformation of plastic substrate and permeation of water and oxygen. Moreover, it improves the interlayer adhesion and the device resistance to chemicals. At present, the most mainstream plastic device substrates designed for large-area production of FOSCs are PET and PEN with PI substrates as supplement in a few cases. PET possesses the advantages of low cost and smooth surface, while PEN has good thermal stability [63]. PET was first introduced into FOSCs as substrate by Siegfried Bauer and coworkers, who designed high flexibility and lightweight

OSCs with thickness up to 1.4 μm (thinnest plastic substrate was 125 μm at that time) [64]. After Siegfried Bauer, the FOSCs with using PET substrates began to spread gradually.

11.3.2 Flexible Transparent Electrode Designs

The realization of high-performance FOSCs mainly depends on good transparency and high conductivity of flexible electrodes. In addition to these requirements, the flexible transparent electrodes (FTEs) should also have the properties of low cost, easy-to-use in large-area, excellent bending, and low-temperature preparation with energy-saving and time-saving. The traditional flexible transparent electrode for OSCs is ITO; however, ITO electrode for FOSCs is easy to be broken, and its conductivity will drop sharply when the number of bending increases, which leads to a poor device performance. Therefore, a series of candidates for the ITO electrode have emerged to enhance the mechanical properties and prolong device lifetime, which are mainly divided into conducting polymers, carbon nanotubes, graphene, metal nanowires, and hybrid films.

11.3.2.1 Conducting Polymers

Different from ITO film (forming on the plastic), which is brittle and expensive, conductive polymers have excellent flexibility and are solution-processable. Among them, the most representative conductive polymer of poly(3,4-ethylenedioxythiophene):poly(styrenesulfonate) (PEDOT:PSS) has been widely used as the electrode material and the electrode interface layer for low-cost large-area ultrathin organic electronics owing to the advantages of high transparency, thermal stability, and good solubility in water or polar solvents. The possibility of PEDOT:PSS as anode instead of ITO for FOSCs was evaluated by Zhang et al. in 2002, which indicated that highly transparent PEDOT:PSS has a great potential and prospect on fabricating plastic OSCs [65]. Since then, the investigation about transparent PEDOT:PSS electrode of the large-area FOSCs is not rare.

Na et al. [66] used PEDOT:PSS doped with dimethyl sulfoxide (DMSO) as an anode to prepare FOSCs based on P3HT:PCBM with plastic substrates (PET) and glass substrates considering the poor conductivity of PEDOT:PSS (Figure 11.5). Compared with ITO-based OSCs (glass 3.66% and flexible 2.99%), a comparable PCE (glass 3.27% and flexible 2.8%) was obtained in OSCs with PEDOT (doping with 5% DMSO) as the anode. Through further flexibility measurement, the PCE of FOSCs based on ITO electrode almost reaches 0 after 75 bending cycles, while the PCE of FOSCs with PEDOT anodes can still achieve 2.73%. This result showed that the FOSCs with PEDOT anode have excellent mechanical flexibility compared with ITO-based FOSCs. In addition, different from the fragile and expensive ITO, all ITO-free flexible devices show that low-cost PEDOT:PSS will be a promising anode or cathode material used for FOSCs, especially in the preparation of R2R flexible devices.

Similarly, the PEDOT:PSS anode with H_3PO_4 treatment was reported by Zhou and coworkers [67] who achieved a good result in all-plastic OSCs based on PES

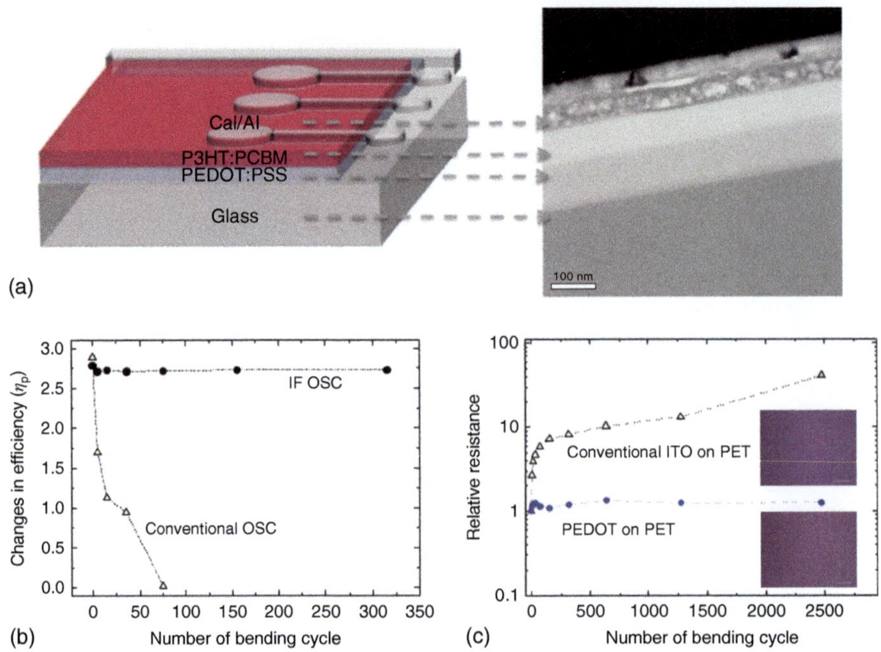

Figure 11.5 (a) Architecture of ITO-free based OSCs and its TEM cross-sectional image on the 100 nm scale bar. Curves of (b) efficiency and (c) relative resistance with bending numbers based on ITO and ITO-free FOSCs. Source: Na et al. [66]. Reproduced with permission of John Wiley & Sons.

substrate (Figure 11.6). After PEDOT was treated with H_3PO_4, the conductivity was significantly improved to 1460 S cm^{-1}. In addition, the PEDOT:PSS film with H_3PO_4 treatment could achieve a higher transmittance and a lower sheet resistance than the traditional H_2SO_4 treatment. More importantly, the toxicity of H_3PO_4 is far less than H_2SO_4, which was more suitable for large-area OSCs based on flexible substrates. The FOSCs with H_3PO_4 coated PES as a transparent bottom electrode were fabricated with excellent performance parameters in P3HT:ICBA including an open-circuit voltage of 0.84 V, a good FF of 60%, an optimal PCE of 3.3%. The result showed that H_3PO_4 treatment can become an effective way to improve the performance of FOSCs based on PEDOT anodes.

It is undeniable that the conductivity of PEDOT:PSS increases sharply after strong acid treatment (H_2SO_4), but the presence of residual strong acid may damage the device structure and lead to a poor stability. Yan and coworkers [68] developed a transfer-printing technique using mild acid to weaken the influence of strong acid on FOSCs (Figure 11.7). The conductivity, transmittance, and stability of transferred PEDOT:PSS film after two-step mild acid processing (CH_4SO_3 dipping and H_3PO_4 soaking) were greatly improved. Finally, the fabricated FOSCs based on PDMS/transferred PEDOT:PSS/PBDTT-S-TT:PC$_{71}$BM/Ca/Al exhibited a significant improvement on the performance and air stability. This study showed that the

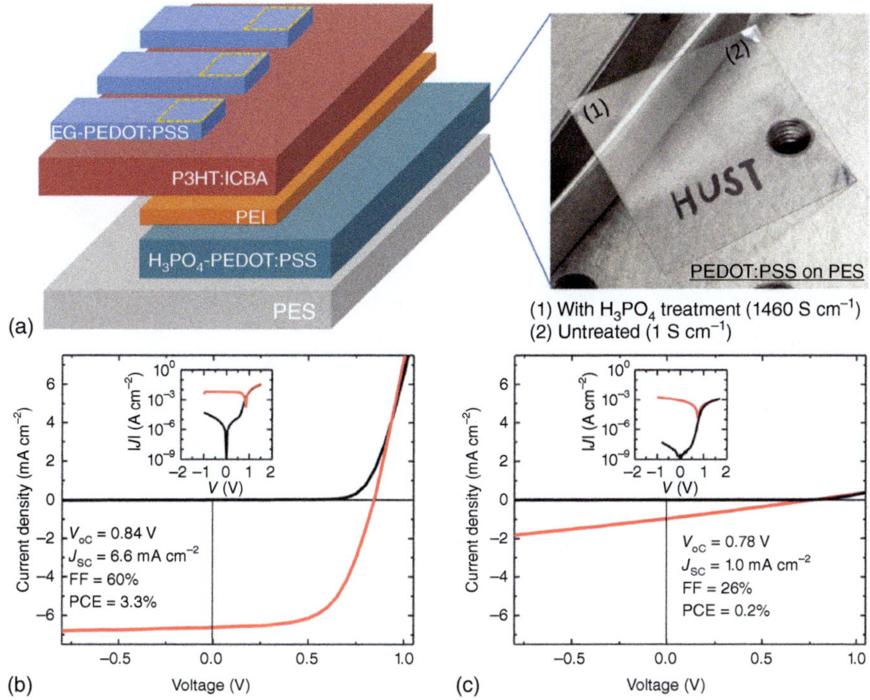

Figure 11.6 (a) Device structure of all-plastic OSCs, J – V characteristics of (b) representative solar cell and (c) reference cell in the dark and under 100 mW/cm^2 white light illumination in linear and semilog scale. Source: Meng et al. [67]. Reproduced with permission of American Chemical Society.

two-step weak acid treatment method could further improve the stability of FOSCs while maintaining high conductivity property of PEDOT:PSS.

Since the formation of metal oxide requires high temperature and high vacuum sputtering conditions, the use of transparent metal oxide electrodes in FOSCs undoubtedly increases the manufacturing cost. Ge and coworkers [69] reported a low-temperature all-solution-processed FOSC (Figure 11.8), which exhibited an excellent performance. After using the PEDOT:PSS film with CH_4SO_3 treatment as the transparent anode, the PCE of FOSCs based on PET/PEDOT:PSS(FTE)/PEDOT:PSS(buffer layer)/PBDB-T:IT-M/PDINO/Al structure reached 10.12% and the FF exceeded 70%. The performance improvement of FOSCs was attributed to the suitable work function formed on the transparent anode after PEDOT:PSS treatment with gentle acid under room temperature, which also endowed a better interface contact and hole transport ability. In particular, the fabricated FOSCs exhibited a good mechanical flexibility. The PCE of flexible devices still could maintain 94% of original PCE after 1000 bending measurement. The application of gentle acid on the PEDOT:PSS makes the fabrication of FOSCs possible at room temperature and can effectively facilitate the development of large-area fabrication methods.

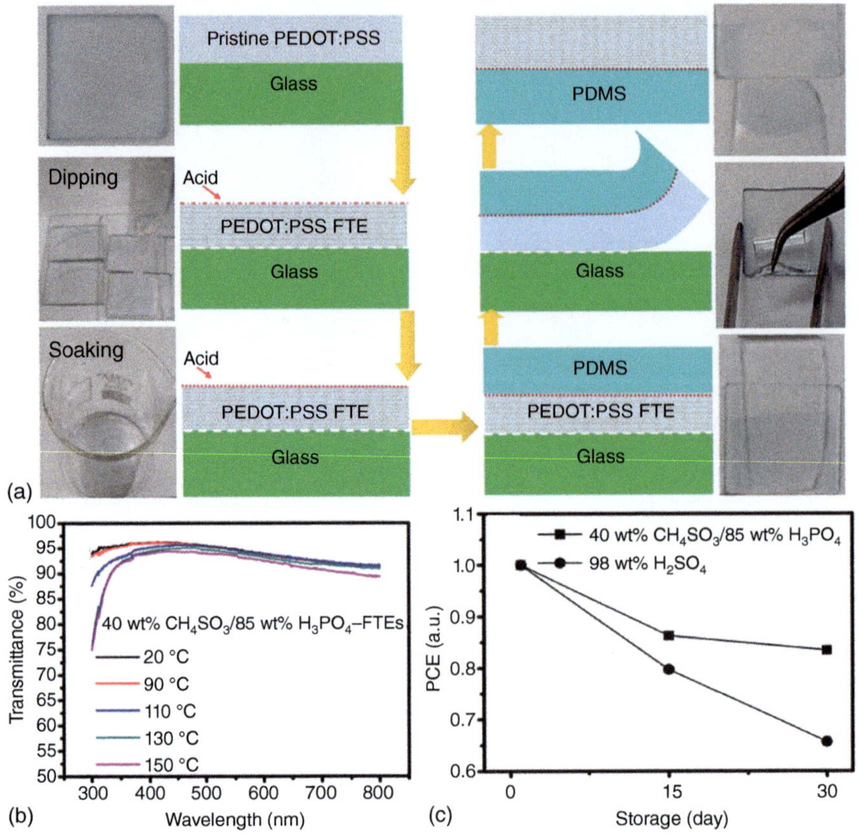

Figure 11.7 (a) Transfer-printing technique, (b) transmittance spectra of transferred PEDOT:PSS under different temperatures, (c) the stability measurement of FOSCs with different acids. Source: Fan et al. [68]. Reproduced with permission of American Chemical Society.

Although PEDOT:PSS as the transparent electrode has been successfully applied in FOSCs, the PEDOT:PSS has the probability of diffusion in device due to the acidic and hygroscopicity nature, which tends to destroy the structure and decrease the stability of devices. In addition, the conductivity of PEDOT:PSS film is relatively poor, which leads to a high interface contact resistance. Chen et al. reported a scalable and highly conductive PEDOT:PSS:CNTs transparent electrode which presents a high conductivity of 3264.27 S cm^{-1} with a high transmittance over 85%. The device (PBDTTT-C-T:PC_{71}BM) based on PEDOT:PSS:CNTs electrode achieves a high efficiency of 7.47%. Although the above problems can be solved with acid treatment (strong acid and mild acid), more or less acid residues still existed in PEDOT:PSS films after acid treatment, which may affect device performance. Moreover, the improvement space of the transmittance and conductivity for PEDOT:PSS is limited by itself. Thus, it is generally not the preferred transparent electrode material for high-efficiency FOSCs.

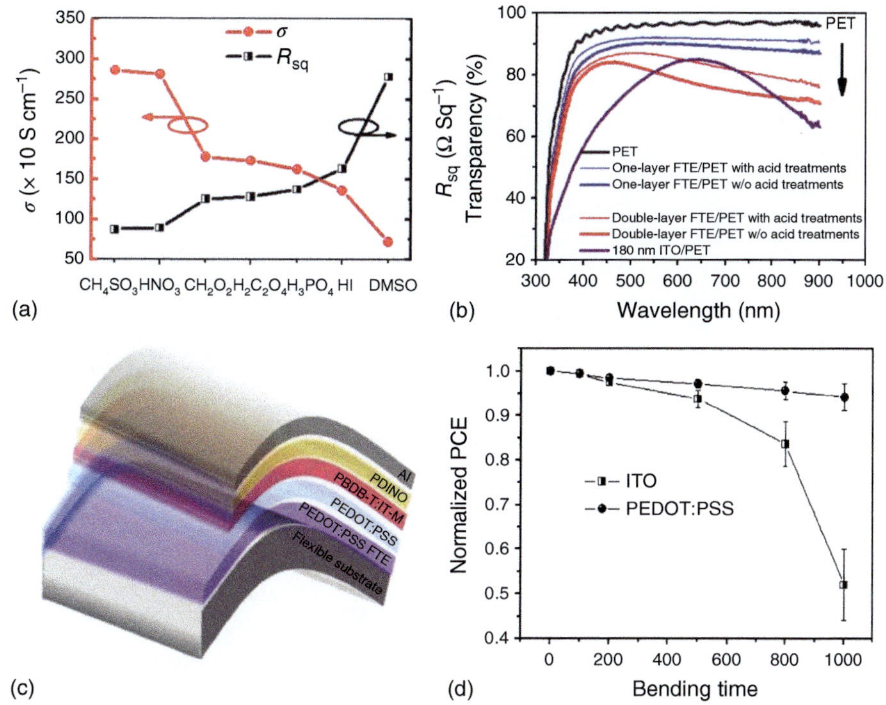

Figure 11.8 (a) Comparison of square resistance and conductivity of the PEDOT:PSS film with different acid treatments compared with DMSO doping at room temperature, (b) transmittance spectra of different substrates, (c) structure diagram of FOSCs, (d) the relationship between normalized PCE and bending numbers. Source: Song et al. [69]. Reproduced with permission of John Wiley & Sons.

11.3.2.2 Carbon Nanotubes

Carbon nanotubes have received extensive attention owing to the advantages of low resistance and super transmittance in the wavelength range of visible and infrared light, highly mechanical flexibility, and low-temperature deposition. In particular, single-walled carbon nanotubes (SWCNTs) possess a dominant position of large-area fabrication, easy synthesis, and suitable deposition on flexible substrates. Thus, SWCNTs can become an alternative to ITO bottom electrode which contains precious metal indium.

SWCNTs transparent electrode formed through printing method on the plastic PET substrate was reported by McGehee and coworkers (Figure 11.9) [70]. The fabricated SWCNTs film via the printing method exhibited a relatively smooth surface and uniform morphology with a high transmittance of 85% at the wavelength of 500 nm. Meanwhile, the flexible devices with PET/SWCNTs/PEDOT:PSS/P3HT:PC_{61}BM/Al structure were fabricated with a high PCE of 2.5% at that time. The photovoltaic performance of flexible device was close to that of ITO glass substrate (3%), which proved that SWCNTs have a potential application as the transparent electrode material in FOSCs. In addition, flexible devices based on SWCNTs exhibited a superior mechanical flexibility than the plastic substrate with ITO.

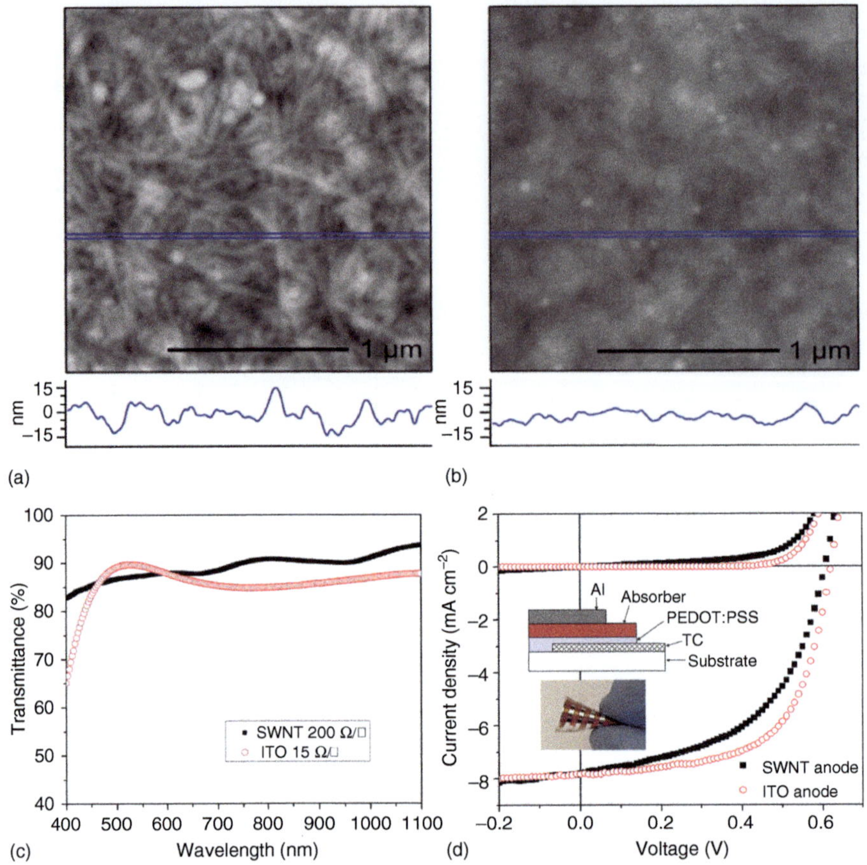

Figure 11.9 The AFM images of SWCNT film (a) before and (b) after PEDOT:PSS deposition, the corresponding the roughness values are 7 and 3.5 nm, respectively, (c) transmittance spectra of SWCNTs film and ITO, (d) current density–voltage curves of glass/ITO-based devices and PET/SWCNT-based devices. Source: Rowell et al. [70]. Reproduced with permission of AIP Publishing LLC.

Furthermore, Zarbin and coworkers [71] reported a highly conductive and transparent film based on carbon nanotube (Figure 11.10). By combining optimal concentration CNTs with high conductivity polyaniline (PANI) for secondary doping, the composite film (PANI/CNTs) has a superior transmittance of 89% at 550 nm and a sheet resistance of 295 Ω/□. In particular, the presence of CNT increased the chemical resistance of the film. Meanwhile, the composite electrode film could be easily fabricated on a flexible substrate and could maintain a good mechanical flexibility. Additionally, the conductivity and transmittance of PANI/CNTs film can be better maintained after bending test. Finally, the flexible device using the composite film electrode as the transparent bottom electrode achieved a better interface contact and a larger active area caused by regular rough surface of PANI/CNTs film. Thus, a better photovoltaic performance of device was obtained than the traditional ITO devices.

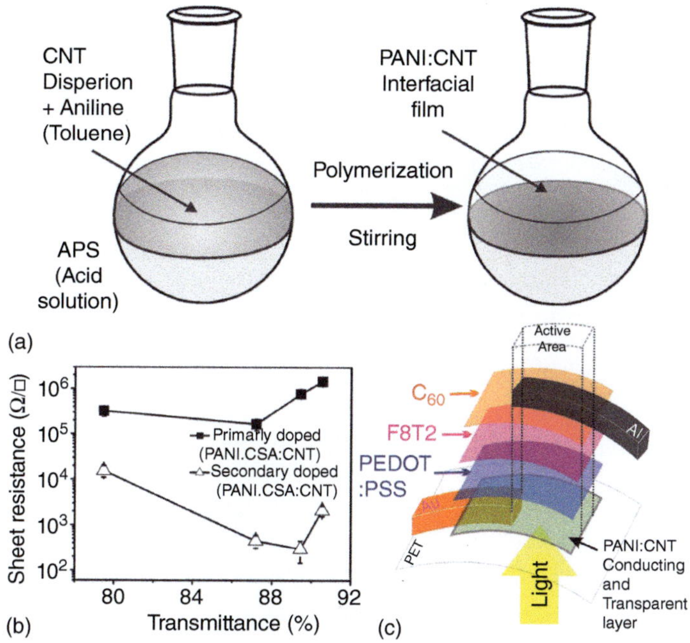

Figure 11.10 (a) Schematic diagram of polymerization in carbon nanotube dispersion, (b) the relationship between sheet resistance and transmittance at 550 nm for different PANI/CNTs films and (c) the configuration of flexible devices. Source: Salvatierra et al. [71]. Reproduced with permission of John Wiley & Sons.

Matsuo and coworkers fabricated SWCNT films with the direct drying deposition method, which exhibited better electrical property than CNT films prepared by other methods and regular ITO on flexible substrates (Figure 11.11) [72]. After the SWCNT film was doped with molybdenum trioxide by thermal annealing at 300 °C, the transmittance of films was superior to the pristine SWCNTs film, and the color of film was changed from green to deep-blue. The SWCNT film doped with MoO_3 as a transparent bottom electrode was firstly proposed in the narrow-band donor polymer PTB7 flexible OSC. Since the high short-circuit current density and the high shunt resistance were obtained compared with the devices with pristine SWCNTs films, the PCEs of 3.43% and 3.91% for flexible devices based on PI and PET were obtained, respectively. Especially, the PCE could still reach 3.82% for PET-based substrate devices after 10 bending tests as MoO_3-doped SWCNTs exhibited both high flexibility and high conductivity. This work confirmed that high-performance flexible devices based on carbon nanotubes could be achieved via a simple and stable method.

Although the CNT materials as transparent electrode for FOSCs have many advantages such as low absorption and high conductivity, there is a fundamental limitation existing in CNTs. Generally, a better conducting percolation network can be formed unless the density of CNTs satisfied certain threshold. Principally the high-density CNT films possess not only a low sheet resistance but a low transmittance. And most of the carbon nanotube films are limited by the synthesis methods

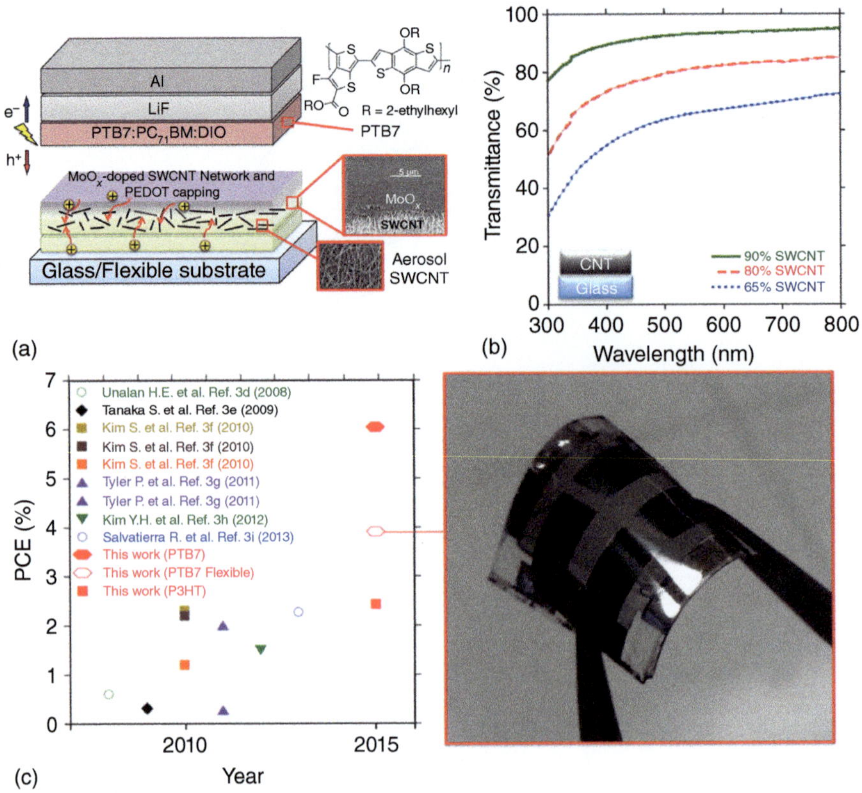

Figure 11.11 (a) Structures of substrate/MoO_x/SWCNT/MoO_x/PEDOT:PSS/PTB7:$PC_{71}BM$/LiF/Al, (b) transmittance spectra of pristine different-thickness SWCNTs films, (c) the reported PCE progress based on carbon nanotubes. Open symbol and closed symbol represent the glass-based device and flexible device, respectively. Source: Jeon et al. [72]. Reproduced with permission of American Chemical Society.

and have a large rough surface, which is incompatible with the large-area fabrication of highly efficient FOSCs.

11.3.2.3 Graphene

Two-dimension carbon nanomaterials, also known as graphene, possess super chemical and physical property like low contact resistance, high conductivity, good thermal stability and high flexibility, which offers a strong possibility of its application into FOSCs as the transparent electrode. The synthesis methods of graphene applied in FOSCs are majorly divided into chemical vapor deposition and reduction of graphite oxide.

Zhang and coworkers prepared a reduced graphene oxide film through chemical method and transferred it to the flexible PET substrates surface as the transparent bottom electrode. The performance of flexible devices was dependent on the efficiency of charge transport that determined by sheet resistance of reduced graphene oxide film when the optical transmittance of reduced graphene oxide

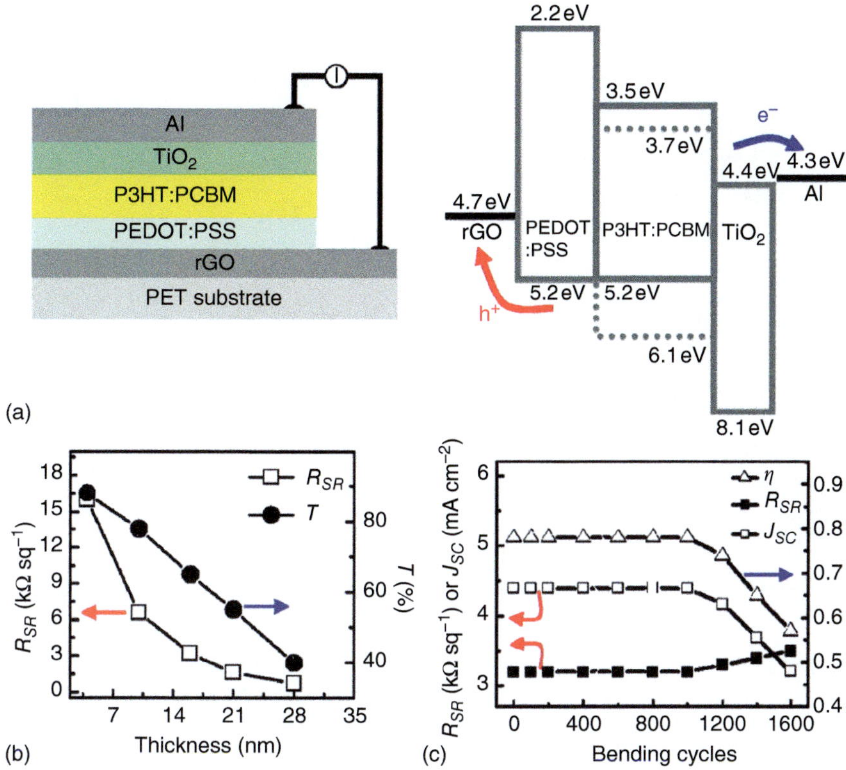

Figure 11.12 (a) Diagram of flexible device structure and related energy level, (b) the relationship between sheet resistance, transmission, and thickness of the reduced graphene oxide film, (c) short-circuit current density, power conversion efficiency, and sheet resistance as a function of bending numbers. Source: Yin et al. [73]. Reproduced with permission of American Chemical Society.

film is above 65%. However, the flexible device performance was greatly related to the transmission of reduced graphene oxide film when it is below 65%. Especially, the performance of flexible devices based on the reduced graphene oxide film still sustains 2.9% after 1000 times of bending test, which demonstrated the high possibility of constructing FOSCs based on chemically reduced graphene oxide thin films (Figure 11.12) [73].

Considering the formation of transparent electrode based on thin metals and ITO requires high vacuum and abundant time, the new type of visibly-transparent flexible (including plastic substrate and paper substrate) OSCs with all-graphene electrode was proposed by Kong and coworkers (Figure 11.13) [74]. The top electrode based on graphene was formed by using dry-transfer method at room temperature. The high optical transmission of 69% was reached at the wavelength of 550 nm by combining organic polymer with graphene electrode. The fabricated FOSCs with all-graphene electrode achieved the transmission of 54 ~ 61% in the visible light range with PCEs of 2.88 ~ 3.8%. Besides, the flexible devices with

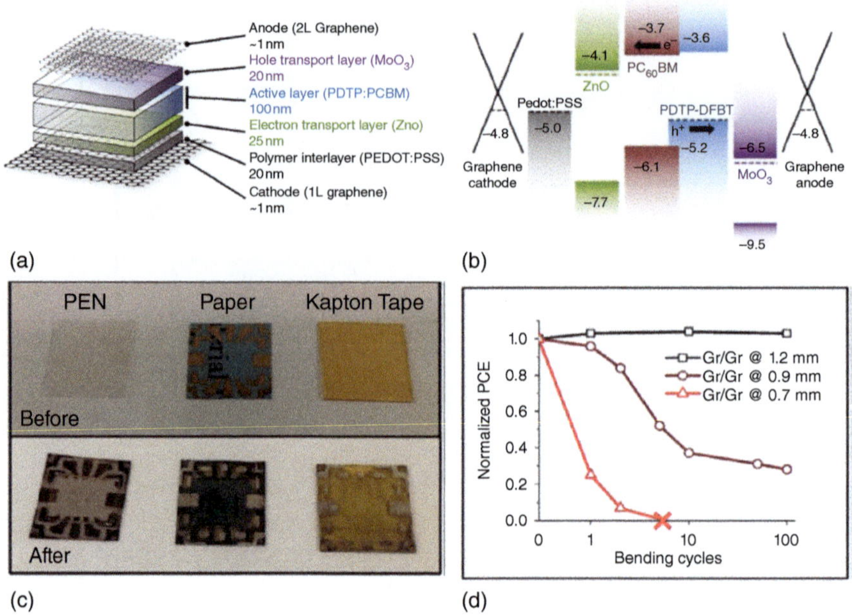

Figure 11.13 (a) The optimal device structure, (b) energy level diagram, (c) the photographs of substrates (PEN, paper, Kapton tape) before and after fabrication, (d) the relationship between normalized PCE and bending cycle number. Source: Song et al. [74]. Reproduced with permission of John Wiley & Sons.

superior all-graphene electrode are mechanically robust, and the performance of the devices is hardly changed after bending of 100 times.

R2R slot-die printing method with use of shear force and one-step reducing posttreatment was developed by Chen and coworkers to fabricate transparent PEDOT:PSS:reduced graphene oxide (RGO) composite electrode for flexible devices. The transmission and conductivity of PEDOT:PSS:RGO film after hydroiodic acid (HI) treatment were significantly improved to 82% and 1949 S cm^{-1}, as HI can simultaneously remove insulating PSS content and reduce GO. Furthermore, a super mechanical flexibility was found in the flexible composite electrode of PEDOT:PSS:RGO under various curvature radius conditions. In addition, the price of PEDOT:PSS:RGO electrode is much lower about 2.8 $ for per square meter than the traditional commercial ITO electrode. Ultimately, a higher PCE of 8.32% was achieved in large-area (1.01 cm^2 effective area) flexible devices with using PEDOT:PSS:RGO electrode and PTB7-Th:$PC_{71}BM$ active layer than the flexible device based on the ITO electrode (Figure 11.14) [75].

Despite graphene materials being successfully applied on the electrode flexible device, the result is modestly undesirable. The good stability and high conductivity of graphene material are urgently required to the application of high-efficiency flexible devices based on non-fullerene acceptors. Besides, most of graphene fabrication methods such as chemical vapor deposition are expensive, which is another key issue for the production of large-scale FOSCs.

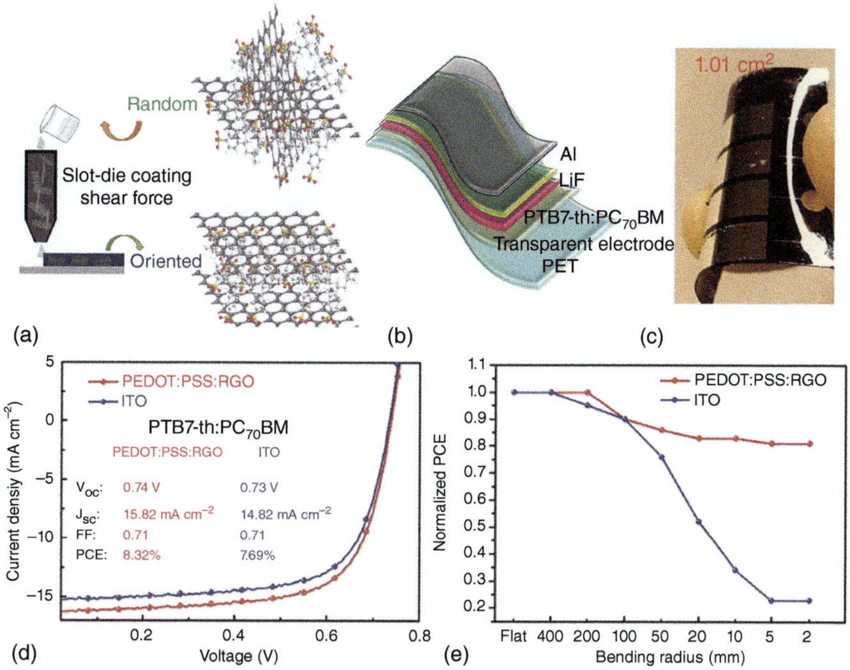

Figure 11.14 (a) Diagram of PEDOT:PSS:RGO film by slot-die coating, (b) structure and (c) photograph of flexible device with PEDOT:PSS:RGO electrode, (d) *J*–*V* curves of flexible devices with ITO electrode and composite electrode, (e) bending measurement of flexible devices. Source: Hu et al. [75]. Reproduced with permission of John Wiley & Sons.

11.3.2.4 Metallic Nanowires

The transparent bottom electrodes based on metallic materials have always occupied a large proportion in market owing to the super conductivity. The metallic materials as transparent electrodes are mainly divided into metal nanowire (NW), metallic mesh [76], metal grid [77], and ultrathin metal film [78]. Metallic nanowires that possess the structure of percolation network have exhibited promising electrical property and optical transparency, including copper, gold, cupronickel, and sliver [79–82]. After a random dispersed silver was firstly synthesized by P. Peumans and coworkers under low temperature condition [76], silver (Ag) nanowire has become the current mainstream of transparent electrodes used in the flexible OSCs owing to the advantages of high transmittance, low sheet resistance, and good mechanical flexibility.

Kang and coworkers proposed a high-efficiency flexible solar device with solution-processable Ag nanowire transparent bottom electrode (Figure 11.15) [83]. The Ag NWs/PET film exhibited a better performance such as the conductivity (sheet resistance≈13 $\Omega\,sq^{-1}$) and transparency (diffusive transmittance≈93.5% at 550 nm) and mechanical flexibility ($\Delta\Omega\ \Omega_0^{-1}\approx 1.1\pm 1\%$ under the bending radius≈200 μm) than the ITO/PET film (sheet resistance≈46 $\Omega\,sq^{-1}$, diffusive transmittance≈91.2%, $\Delta\Omega\,\Omega_0^{-1}\approx 64.6\pm 24\%$ under the bending radius of 200 μm). Furthermore, the flexible device with Ag NW electrodes based on the PTB7-F20:$PC_{71}BM$ active layer achieved

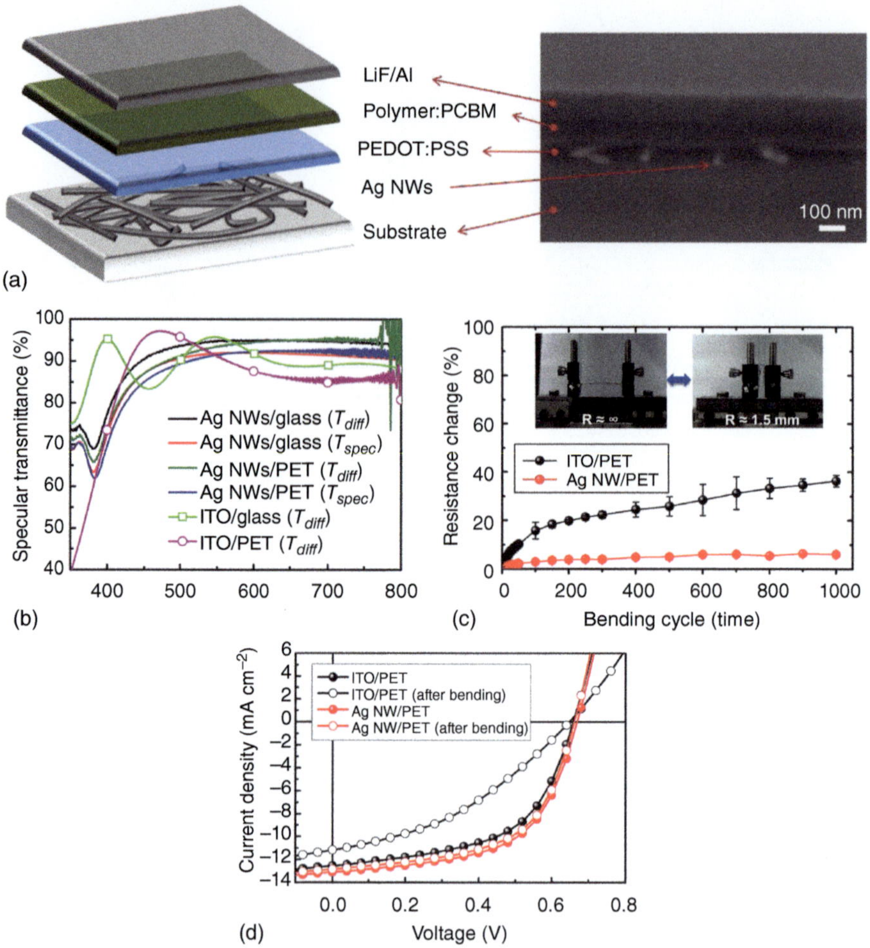

Figure 11.15 (a) Diagram of OSCs and cross-section SEM images with the structure of substrate/Ag NW/PEDOT:PSS/PBDTTT-C:PC_{61}BM/LiF/Al, (b) specular transmittances of various substrates with different electrodes, (c) the relationship between resistance change and bending number under bending radius of 1.5 mm, (d) J–V curves of flexible devices based on PBT7-F20:PC_{71}BM with different electrodes. Source: Song et al. [83]. Reproduced with permission of John Wiley & Sons.

a commensurate efficiency with the ITO/glass devices (5.02% versus 6.67%) and the Ag NWs/glass device (PCE ≈ 5.80%). Remarkably, the comparative value of PCE acquired from Ag NWs/PET flexible devices could still remain the initial value of 95% after undergoing the bending test for 1000 times, which was much higher than the flexible devices with ITO/PET structure (63% of initial efficiency after identical bending condition).

Since the challenge existed in the formation of large uniform interconnected network of Ag nanowire electrode, Seo and coworkers developed a cold isostatic-pressured (CIP) method to produce an interconnected network by forming the intimate contact between Ag nanowires (Figure 11.16) [84]. Thus, the

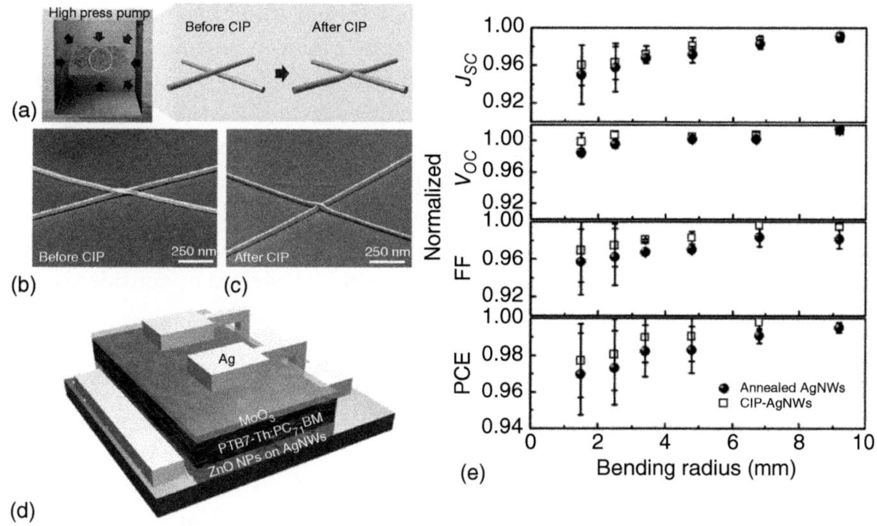

Figure 11.16 (a) Diagram of fabricated Ag NWs with CIP method, and the SEM images of Ag NWs (b) before and (c) after CIP treatment, (d) structure of flexible device, (e) the relationship between the performance parameters (V_{oc}, J_{sc}, FF, PCE) of flexible devices and bending radius. Source: Seo et al. [84]. Reproduced with permission of John Wiley & Sons.

performance of Ag nanowire fabricated through CIP method, such as conductivity, surface roughness, and flexibility, was extraordinarily improved. Likewise, a lower sheet resistance and a higher figure of merit were manifested for the Ag nanowire using CIP method than the regular Ag nanowire electrode. In addition, a larger FF of 63.5% was realized in FOSCs with CIP than regular FOSCs (53%) since the smooth surface roughness of Ag electrode by CIP generated a large short-circuit current density. Furthermore, the smoother surface of Ag nanowire electrode by CIP method realized a higher reproducibility of 100% than FOSCs with annealed Ag NW electrodes (75%). Besides, the PCE of prepared flexible devices with CIP could retain 95% of initial performance after bending cycle of 1000 times under extremely small radius of 4.8 mm.

Recently, a method using water to construct the Ag nanowire was reported by Chen and coworkers with the realization of a grid-like, smooth, flexible Ag nanowire electrode [85]. The new method that used water suspension of Ag NWs and polyelectrolyte (PSSNa) to fabricate flexible transparent electrode, is simple, and does not need any extra treatment. The fabricated Ag nanowire electrode possesses a superior performance such as transmission (92%), sheet resistance (10 Ω □$^{-1}$), surface roughness, and flexibility, which was extremely embodied in the flexible devices. Therefore, a superior performance was determined in the single-junction (13.1%) and tandem (16.5%) flexible (PET) devices, which was comparable to commercial ITO-based glass devices. An excellent durability was realized in the flexible devices with use of water-processed Ag nanowire. Especially, a low value of performance degradation was obtained under different bending radiuses and bending times. In addition, the Ag nanowire electrode that possesses high performance offered a great possibility of the application in organic light-emitting diodes and transistors. Meanwhile, it could also be used as a reference for study of other dispersions and flexible electrode materials.

The flexible transparent electrode based on Ag nanowire has been extensively applied on the flexible devices and exhibits a promising future comparing with other electrode materials (PEDOT:PSS, carbon nanotube, graphene) and becomes the current mainstream in flexible electrodes. The issues still existed in the length control, stability, and reproducibility of Ag nanowire, which should be considered and addressed.

11.3.2.5 Hybrid Films

Although transparent electrode materials such as conducting polymer (PEDOT:PSS), carbon nanotubes, graphene, metal nanowire (Ag nanowire) have their own unique fascination, all of them have modest shortcoming when used in high-efficiency flexible OSCs. To better utilize these transparent electrode materials in FOSCs, current researchers have combined some complementary electrode materials to achieve the aim as much as possible that the performance reaches or surpasses the level of traditional ITO glass device with identical active layer system.

Ge and coworkers combined the advantages of Ag nanowire electrode and graphene electrode to conceal the defects of Ag network and reported a graphene and Ag NWs composite electrode for FOSCs (Figure 11.17) [86]. The composite

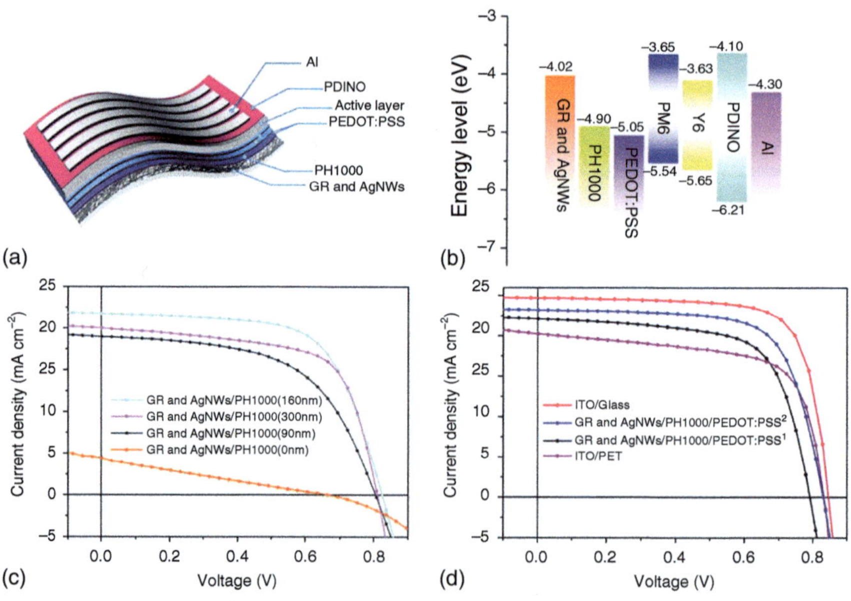

Figure 11.17 (a) Flexible devices framework, (b) energy level diagram, (c) *J–V* curves of flexible devices with different electrodes, (d) the bending measurement of flexible devices with different electrodes under various radiuses. PEDOT:PSS[1] and PEDOT:PSS[2] are single and double interface layer, respectively. Source: Zhang et al. [86]. Reproduced with permission of Royal Society of Chemistry.

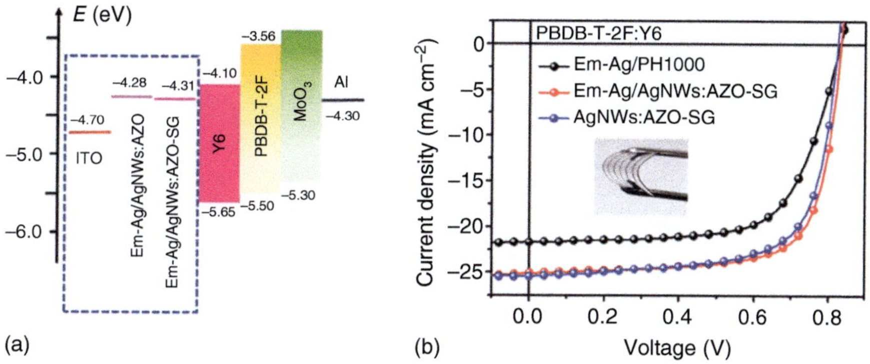

Figure 11.18 (a) Energy level diagram, (b) *J–V* curves of flexible OSCs with different composite electrodes. Source: Chen et al. [87]. Reproduced with permission of John Wiley & Sons.

electrode possessed a high transmission in the whole optical wavelength and a good conductivity. Meanwhile, the surface roughness and conductivity of graphene and Ag NWs composite electrode were further improved by spin-coating PH1000 film. In addition, the inappropriate work function between composite electrode and PEDOT:PSS was also ameliorated by PH1000 film. Thus, the fabricated flexible devices with graphene and Ag NWs/PH1000 electrode achieved a more excellent photovoltaic performance (13.44%) with the help of suitable energy level, uniform interface layer, and good ohmic contact. As the composite electrodes were resistant to bending, a super mechanical flexibility was realized in the prepared flexible device, and the photovoltaic performance could reach 84.6% of original efficiency after bending of over 1000 times (radius = 2 mm), which was favorable for larger-scale flexible devices used in commercial market.

Recently, Li and coworkers proposed a concept of “welding” to develop a flexible OSC with intimate binding of transparent bottom electrode and substrate (Figure 11.18) [87]. To solve the mismatch issue between transmittance/absorption spectra of bottom electrode and active layer based on non-fullerene system, a welding bottom electrode with the Em-Ag/Ag NWs:AZO-SG structure was designed for flexible devices. The composite electrode exhibited a super optical and electrical performance with smooth surface and good flexibility. A comparable result is achieved for the flexible devices compared with the ITO glass devices using identical active layer. In particular, two high PCE records were achieved in the FOSCs based on small-area (15.21%) and large-area (12.28%) single-junction system. Moreover, super mechanical robustness was found in the flexible devices with welding composite electrode structure, which provided an excellent reference for developing new-type transparent bottom electrodes and promoted the development of large-area flexible devices.

To achieve the optimal film performance in flexibility and electricity as well as optics, leaf-like biomimetic electrodes (PPZA) that possess efficient light capture and glossy surface were proposed by Tang and coworkers (Figure 11.19) [88]. The biomimetic electrodes were fabricated by mimicking the internal anatomy of the leaf using the contact film transfer technique, which effectively improved the

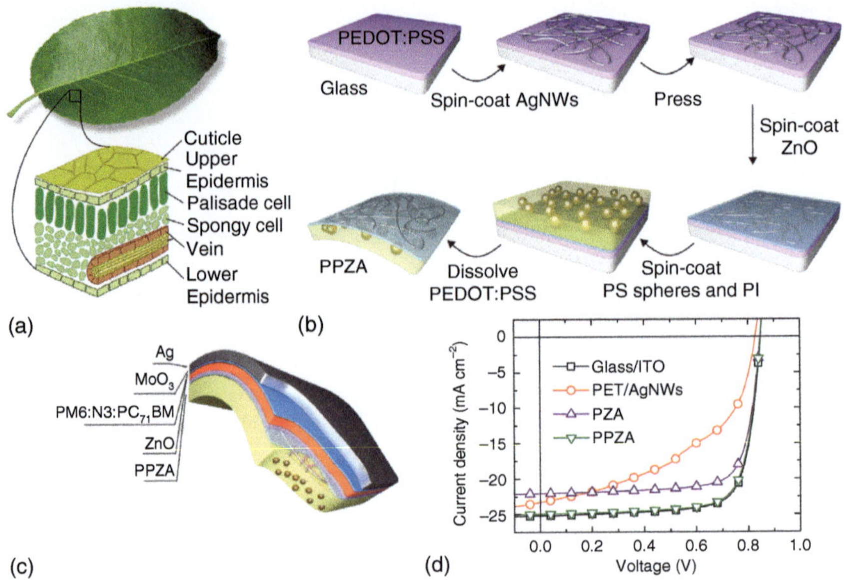

Figure 11.19 (a) Diagram of leaf internal structure, (b) fabrication route of the biomimetic electrode, (c) flexible device architecture, (d) *J–V* curves of flexible devices. Source: Qu et al. [88]. Reproduced with permission of John Wiley & Sons.

efficiency and mechanical robustness of flexible OSCs. Through the synergistic effect of these components (including conductive Ag NWs, transparent ZnO protecting layer, light-scattering PS spheres, and ultrathin flexible PI substrate), the prepared composite electrode PPZA exhibited a high optical transparency (88.2%), uniform surface (2.4 nm), good conductivity (23.4 $\Omega\ sq^{-1}$), and superior bendability. Furthermore, the flexible devices based on PM6:N3:$PC_{71}BM$ system with PPZA as the bottom electrode achieved an excellent photovoltaic performance (a record PCE value of 16.1%) and a good reproducibility and resistance to flexural strain. In addition, a super-high flexibility characteristic was embodied in these devices. The PCE of flexible OSCs could reserve 85% of pristine value after the 5000 times bending.

This kind of composite electrodes with complementary advantages have been successfully explored for flexible OSCs and have pushed PCE exceeding 16%. Currently, the simplification and exploitation of the preparation method to improve the reproducibility of composite electrodes are still the main goals. The final prospect is to achieve high-performance flexible devices, which can catch up with ITO glass-based devices using the same active layer, and then to realize the goal of replacing ITO with low-cost flexible electrodes.

11.4 Large-Area Flexible Device Fabrication

Commercial large-area flexible OSCs should be able to be manufactured through scale-up methods. Spin coating is the most commonly used method for active layer

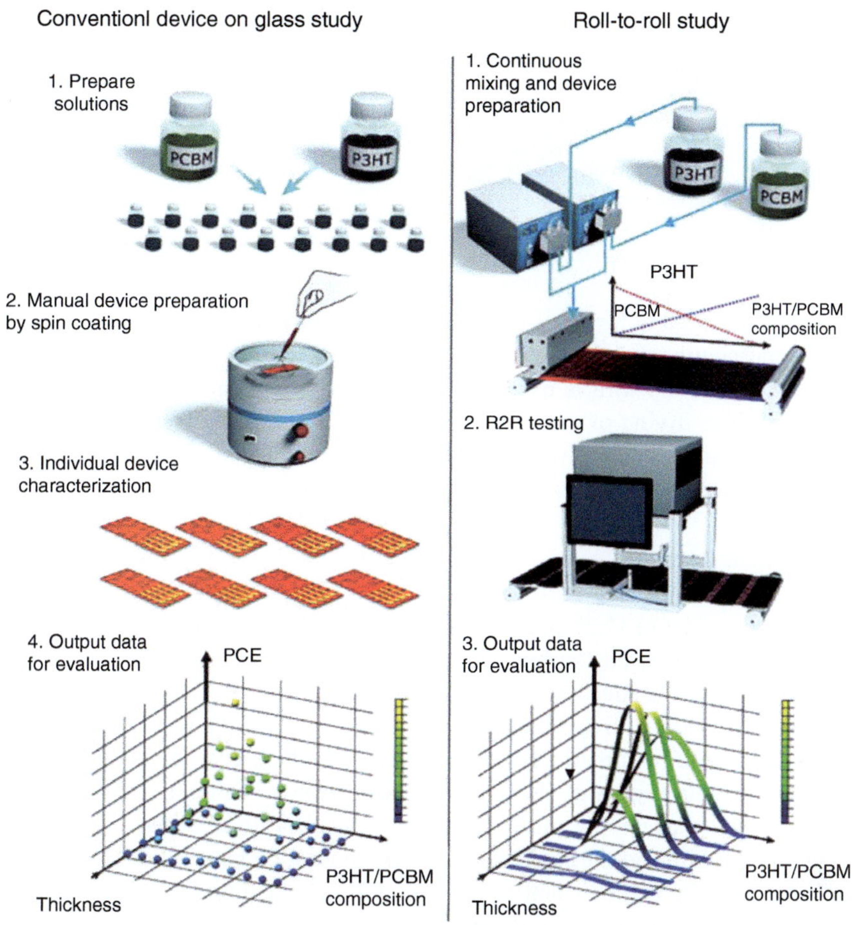

Figure 11.20 Comparison between conventional spin coating process on glass and large-area roll-to-roll process on flexible substrates of organic solar cells. Source: Alstrup et al. [89]. Reproduced with permission of American Chemical Society.

preparation of organic solar cells. In addition, the electrodes and most buffer layers of the state-of-the-art OSCs are fabricated under high vacuum condition in an inert atmosphere. The laboratory fabrication method is not suitable for low-cost and mass production, especially for the large-area devices on flexible substrates. R2R technology is the most prospective for manifesting high-performance large-area FOSCs. In R2R process, a flexible transparent plastic is used as a "web," which is usually unrolled to apply the coating, and then rolled again. Figure 11.20 shows the comparison of the conventional spin coating and R2R process [89].

11.4.1 Coating and Printing Methods

High-quality film of each layer in a device is a key factor for high-efficiency OSCs. For large-scale R2R process, either the printing methods or the coating methods

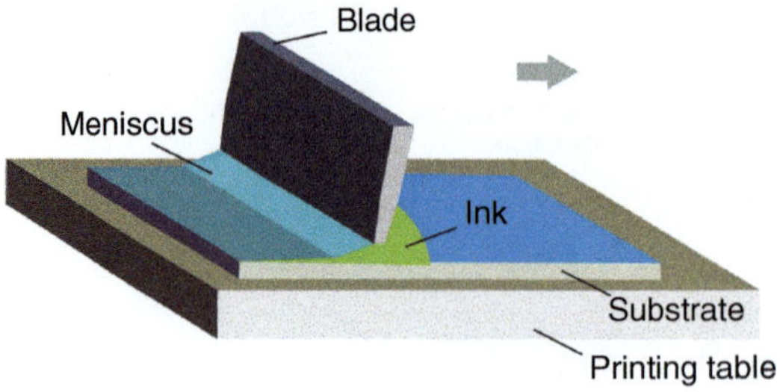

Figure 11.21 Schematic illustration of blade coating.

are usually employed to form a wet film. These methods are blade coating, slot-die coating, spray coating, inkjet printing, relief printing, gravure printing, screen printing, etc. However, dry film is used in OSCs, which is obtained by various other methods (such as drying or curing). The quality of the films processed by above-mentioned methods is strongly related to the processing speed, the viscosity of the ink, processing temperature, solvent evaporation rate, etc.

11.4.1.1 Blade Coating

Unlike spin coating, blade coating exhibits excellent compatibility to large-area R2R technology. Figure 11.21 shows schematic illustration of blade coating. When preparing the active layer, the ink is dropped in front of the blade, and then the blade moves parallel to substrate with a constant speed to form a thin film. The area can be easily scaled up, and the material usage is almost 100% with zero wastage in blade coating. The film thickness processed by blade coating is determined by the ink concentration and viscosity, the gap between blade and substrate, the blade speed, blade shape, the surface roughness of substrate, the meniscus formed between the wet film and the blade, and solvent evaporation rate. The film thickness of the final dry film can be calculated by the following equation, commonly in a wide range from nanometer to micrometer scale.

$$d = \frac{1}{2}\left(g\frac{c}{\rho}\right)$$

where "d" is film thickness, "g" is the distance between blade and substrate, "c" is ink concentration, "ρ" is film density. In addition, the film morphology developed by such method is also different from that prepared by spin coating, because spin coating depends on centrifugal force by different direction to form the thin films, but blade coating has the same direction centrifugal force during film process. The same direction centrifugal force can induce alignment of film along the blade moving direction. On the other hand, the slow solvent evaporation in the wet film by blade coating also can improve the crystallinity and optimize the morphology of active layer. In the past 10 years, the blade coating method has become a commonly used method for preparing large-area OSCs in the laboratory.

Initially, blade coating method was performed on fullerene-based active layer. In 2008, Mens et al. [90] fabricated a BHJ blend of MDMO-PPV:PC_{61}BM by the blade coating. They observed that PC_{60}BM showed higher crystallinity in blade-coated film than in spin-coated film. The slower solvent evaporation in blade coating than in spin coating resulted in the stronger aggregation of $PC_{61}BM$. Soon later, Meng and coworkers also used blade coating to prepare P3HT:$PC_{61}BM$ devices. They used the toluene with lower boiling point as processing solvent to replace high-boiling-point chlorobenzene (CB) or dichlorobenzene. Desired donor–acceptor self-organization was achieved without the slow drying process and high-boiling-point solvent. A PCE of 3.8% was obtained by blade coating, much higher than the 2.6% obtained by spin coating in toluene solution [91]. Meng and coworkers further processed large-area OSCs by accelerated blade coating. To obtain uniform film, the ink was blade-coated on a hot substrate with an acceleration of the blade motion to compensate the decrease in the wet film thickness because of the depletion of the solution. After the blade motion formed a wet film, hot air from the dryer was immediately directed from above to enhance drying and uniformity. In 2016, Yan and coworkers boosted the PCE of the fullerene-based OSCs up to 9.5% processed by blade coating [92].

With the emergence of non-fullerene acceptors, blade coating method has attracted great attention to be employed in large-area non-fullerene-based devices. Hou and coworkers reported blade-coated non-fullerene OSCs based on PBTA-TF:IT-M with environmentally friendly solvents, tetrahydrofuran/isopropyl alcohol (THF/IPA), a PCE of 10.6% with the device area of 1.0 cm^2 was obtained [93]. Zhao et al. also prepared PBTA-TF:IT-M-based device by blade coating method using environmentally friendly solvents that achieved a PCE of 11.7%. Ma and coworkers prepared PBDB-T:ITIC-based OSCs by blade coating method [94]. The device achieved a PCE of 10.03%, higher than the spin coating processed one. They also found that only 0.25% DIO additive was required for the blade-coated device, which is much lower than that for spin-coated one (1% DIO additive). The less imported additives in the film are beneficial for device stability. Ma's group also conducted blade coating to fabricate ternary device based on PBDB-T:PTB7-Th:FOIC. An excellent PCE of 12.02% was obtained, ascribing to the synergistic effect in optimizing morphology by the combination of blade coating and ternary strategy [95]. Recently, they reported PM6:IT-4F-based devices and found that substrate temperature affected IT-4F molecular stacking, the device obtained a PCE of 10.22% at film thickness of 400 nm [96].

Apart from producing BHJ films, blade coating also satisfies the needs of planar heterjunction (PHJ) OSCs using layer-by-layer method. In 2020, Min and coworkers reported PHJ OSCs with PM6 as polymer donor and Y6 as acceptor by layer-by-layer blade coating [97]. Due to suitable vertical phase separation for charge transport, the device obtained a PCE of 16.35%, higher than BHJ device (15.37%). They also adopted layer-by-layer method to process 11.52 cm^2 module, which obtained a record efficiency of 11.86% with a geometric fill factor (GFF) of over 90%. In 2020, Ma and coworkers reported double BHJ films PBDB-T:IT-M/PBDB-T:FOIC fabricated by sequential-blade coating. Because of the strong crystallinity of FOIC [98], the PBDB-T:FOIC film displayed a large phase separation, resulting in poor efficiency of the device. In sequential-blade coating system, the bottom film PBDB-T:IT-M with film thickness of 15 nm produces crystal nuclei for the top film PBDB-T:FOIC with film thickness 100 nm. Thereby, the balanced crystallize rate between PBDB-T and FOIC was obtained. Ultimately, PBDB-T:IT-M/PBDB-T:FOIC sequential-blade coating device achieved a PCE of 11.86%. In 2020, Huang and coworkers designed a small molecule acceptor DTY6 to apply in OSC module devices using non-halogen solvent o-xylene. By blade coating, the PM6:DTY6-based OSCs with an active layer area of 18 cm^2 achieved a PCE of 14.4% [99].

Blade coating shows great potential in the large-area fabrication of highly efficient OSCs. Film thickness and morphology control during blade-coating process is the most important issue that should be concerned to obtain high device performance. For example, during the blade coating process, the solvent evaporation is not as effective as the spin coating method; therefore, the improved crystallinity of film can facilitate the charge transport. However, for the highly crystalline materials, blade coating always leads to the excessive aggregation, unfavorable morphology, and unbalanced charge transport of the film, which also often happens in other large-scale coating and printing processes. In addition to regulating the key factors, which determine the thickness and morphology of films mentioned above, some assistant methods, e.g. preheating blade and substrate, fixing the blade and moving substrate, changing the ink supplementary pattern, have also been employed as effective approaches to improve the film quality.

11.4.1.2 Slot-Die Coating

Figure 11.22 shows the schematic illustration of slot-die coating. The coating motion of slot-die coating is similar to that of blade coating. Commonly used blade coating has a constant amount of solution beneath the blade gap, whereas slot-die coating can provide continuous supply of ink, which can be regarded as an upgrade blade coating in a certain sense. The slot-die head, the most important part in this process, has a gap formed by two blades, which is placed over the web. On the slot-die head, there is an ink container connecting to a piston pump or a pressure tank, which directly feeds the ink to the coating head. During the device process, a striped pattern can be printed by attaching a mask to the slot-die head. The thickness of dry film processed by slot-die coating can be evaluated by the following equation:

$$d = \frac{f}{Sw} \times \frac{c}{\rho}$$

Where "d" is thickness, "f" is the flow rate, "S" is web speed, "w" is the coated width, "c" is the ink concentration, and "ρ" is the film density. From the equation we can see

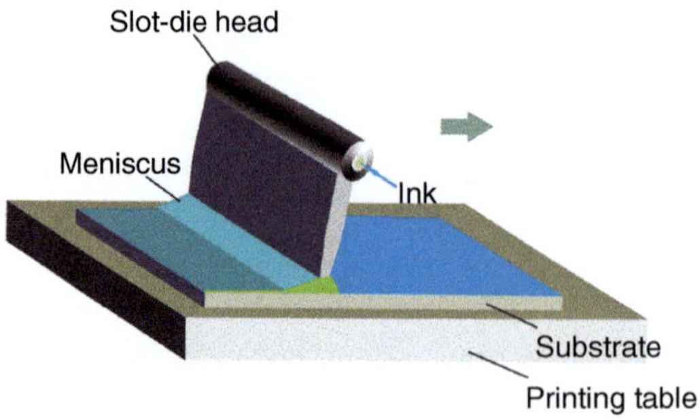

Figure 11.22 Schematic illustration of slot-die coating.

that the coating thickness strongly depends on the ink velocity and concentration, flow rate, the slot-die gap, web speed, etc. Additional steps are often required to control the meniscus and print a fine pattern. For example, the slot-die head can be kept at high temperature to control the viscosity and solubility of the ink. These coating parameters also exert great influence on the film morphology, similar to the case of blade coating.

Zimmermann et al. processed P3HT:PCBM-based device modules using the slot-die method and the active layers processed in ambient atmosphere. An FF of 64% was achieved by PET foil as substrate, which was close to the results on glass substrate [100]. Schrödner et al. also coated PEDOT:PSS and P3HT:PCBM layers by slot-die method on PET foils and compared device performance of different solvents processed active layers [101]. Brabec and coworkers reported slot-die coated active layers for solar cells processed from non-halogenated solvents, obtaining a PCE of 3% on flexible substrates [102]. Krebs and coworkers reported a slot-die coated PDTSTTz:PCBM solar cell with a PCE of 3.5%. The active layer thickness is optimized by controlling substrate speed [103]. They also established a completely ITO and vacuum-free R2R process, which allows the processing of all layers of the device without moving the substrate to different equipment.

For the non-fullerene-based system, Min and coworkers reported PBDB-T-SF:IT-4F-based devices by slot-die coating method, delivering a PCE of 12.9% [104]. Equipping the large-scale device module to a water splitting system yielded a solar-to-hydrogen efficiency of 6.15%. Moon and coworkers prepared a flexible device module based on a SMD2:ITIC. A PCE of 5.25% was obtained with an active area of 80 cm^2 [105]. Chen and coworkers applied the shear impulse to control the film morphology. The efficiency of flexible devices via slot-die printing achieves 9.10% for PTB7-Th:PC71BM and 9.77% for PBDB-T:ITIC based on 1.04 cm^2. Furthermore, 15 cm^2 flexible modules with effective efficiency up to 7.58% (PTB7-Th:PC71BM) and 8.90% (PBDB-T:ITIC) are demonstrated with satisfying mechanical flexibility and operating stability [106]. Recently, Ma and coworkers reported PM6:Y6-based device using halogenated (chlorobenzene [CB]) and hydrocarbon solvents 1,2,4-trimethylbenzene (TMB) and *ortho*-xylene (*o*-XY)) to fabricate photoactive devices. By controlling solution and substrate temperatures to obtain similar morphology of active layers, notable efficiency of 15.2%, 15.4%, and 15.6% was achieved for CB, TMB, and *o*-XY solvents, respectively [107]. Huang and coworkers reported a slot-die coated all-PSC based on PTzBI:N2200, a PCE of 9.10% was obtained [108].

Slot-die coating has been regarded as the greatest prospect toward the R2R production of solar panels. Nevertheless, due to the requirement of high-precision control over various coating parameters, slot-die coating is expensive at present. Additionally, a good pattern is normally required at a low speed of the web, preventing a rapid drying of multilayer structures in high-throughput continuous processing. Thereby, more efforts should further explore and understand operating parameters of slot-die coating, for minimization of efficiency loss during the transition from small-area devices to large-scale printed modules.

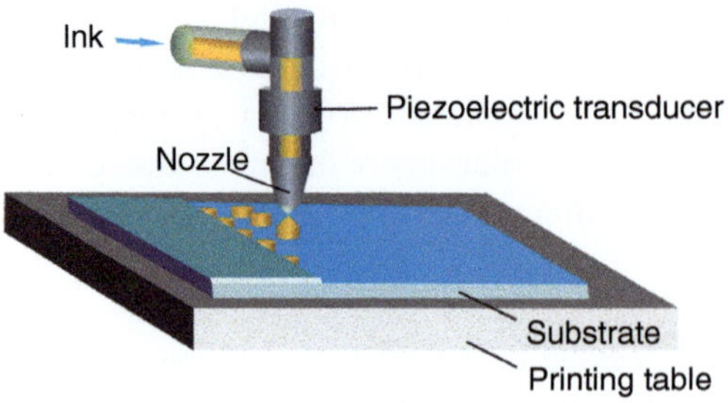

Figure 11.23 Schematic illustration of the inkjet printing process.

11.4.1.3 Inkjet Printing

Inkjet printing technology is widely used in daily life, and the use of printers is a good example. In recent years, inkjet printing has been applied in the field of OSCs. The most common inkjet printing methods are continuous inkjet printing and drop-on-demand inkjet printing. Drop-on-demand inkjet printing has higher material efficiency, which has proven to be more relevant to the electronics industry. As shown in Figure 11.23, it is a drop-on-demand inkjet printer. The film formation is composed of the droplet formation, ejection, positioning, spreading and coalescence on the substrate, and the final solidification. The printer head is mounted on the substrate, and either the substrate or the printer head is movable. The printer head is connected to a continuous solution supply pump. The pressure pulse controls the rate of droplet ejection from the printer head. The mechanical force of the piezoelectric transducer on the solution container generates the pulse. Generally, inkjet printing has a high demand on ink, and a low viscosity (4–30 cP) and high surface tension (typically >35 mN m^{-1}) are required for the operation. The film thickness is determined by four parameters: the droplet volume (V_d), the ink concentration (c), the number of droplets per unit area (N_d), and the density of the material (ρ).

$$d = N_d V_d \times \frac{c}{\rho}$$

Eom et al. studied the effect of additives on the efficiency of inkjet-printed devices, and the results showed that the surface morphology and charge collection were improved by adding various additives [109]. In 2014, an all inkjet-printed, all-air-processed OSC has been fabricated with a PCE of 5% [110]. Similarly, in 2015, Eggenhuisen et al. reported the use of environmentally friendly solvents to produce an all-inkjet-printed large-area (>1 cm^2) OSCs in air atmosphere. The device was fabricated using industrial-scale (512 nozzles) print heads, compatible with R2R technology and attained a PCE of 4.1% [111]. In 2019, Corzo et al. reported P3HT:O-IDTBR-based devices fabricated by the inkjet printing method. They studied the deposition temperature and drop spacing parameters. Finally, the PCEs of 6.47 and 6% were achieved for the 0.1 and 2 cm^2 OSCs. Furthermore, large-area marine-like devices were fabricated with a PCE of 4.76% [112].

The advantage of inkjet printing is a noncontact processing that effectively utilizes materials and has great flexibility in printing shape and thickness. However, it is difficult to retain these advantages in the case of high throughput. It is necessary to use a solution with appropriate dynamic viscosity, density, and surface tension

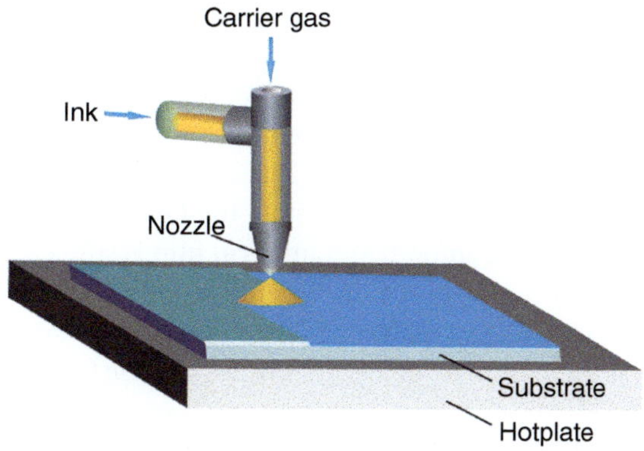

Figure 11.24 Schematic illustration of the spray coating process.

to print normally in the printer head. For the surface of the substrate, the solution needs to be adjusted to avoid surface interactions, for example, to dewetting or drying (coffee rings). Since it is difficult to form an ideal ink for inkjet printing, this printing technology still has great challenges in mass production in the future.

11.4.1.4 Spray Coating

Spray coating is a solution-based deposition technique that has been widely used in building painting and a number of industrial applications. It also has the advantages of low material waste, touch-free, and fast, compatible to the R2R process. Figure 11.24 shows an ultrasonic spray coater. The work mechanism of the spray coating is somewhat similar to that of inkjet coating, just the aerosol droplet instead of liquid droplet for film forming. The ink is forced through an airbrush nozzle to generate aerosol, then depositing on the substrate to form a film. Distance between the airbrush nozzle and substrate can be 10–30 cm, much longer than those of blade coating and slot-die coating. This method is universal, being capable of solution coating every layer of an OSC device, and a wide viscosity range for the ink is acceptable. The film thickness and morphology strongly depend on the ink concentration, ink viscosity, the choice of solvents, and the distance between the airbrush nozzle and substrate, etc.

Previous work on deposition of active layers by spray coating method mainly focused on P3HT:PCBM blends [113–118]. A device efficiency of lower than 3.0% was reported for P3HT:PCBM solar cells when the active layer was spray cast from a solution of two-solvent system. When the film was casted from a two-solvent system of *ortho*dichlorobenzene (*o*DCB) as the primary solvent and mesitylene as the secondary solvent, a booted PCE of 3.75% has been reported. This is because the secondary solvent with a lower boiling point and lower surface tension has been shown to promote the wettability and uniformity of the P3HT:PCBM film [114, 117]. Using a solvent mixture of DCB and CB, a PCE of 4.1% was achieved in P3HT:PCBM solar cells by airbrush spray coating [119]. In 2013, Wang et al. reported a spray-cast device based on a fullerene system, obtaining a PCE of 5.0% [120]. In 2015, Vak et al. deposited the active layer of P3HT:$IC_{60}BA$ by spraying method and achieved PCE of 4.2%. Besides, they used an LED–photodiode couple as an optical thickness monitor, and the spray deposition process operates in a manner analogous to a vacuum deposition process [121].

Spray coating not only can deposit the active layers, but also can deposit electrodes and buffer layers. There are some reports on successful deposition of the PEDOT:PSS transparent electrodes [122] and ZnO electron transporting layers [123, 124] by spray coating method. Even though the OSCs can indeed be processed by spray coating, obtaining a smooth and uniform film is still a big challenge. In the touch-free coating, the aerosol has to fly the long distance between nozzle and substrate before coating on the surface of the substrate, so the transition process from aerosol to film is difficult to be precisely control, leading to a great impact on the film morphology.

11.4.1.5 Screen Printing, Relief Printing, and Gravure Printing

When printing a defined area or pattern, the industry usually chooses screen printing, gravure printing, and relief printing technology. As shown in Figure 11.25a–c, engrave the image to be printed onto the image carrier, and then transfer it to the substrate. For screen printing technology, the image carrier is the mesh with printed pattern. In gravure printing, the printed image is engraved in the printing cylinder, and the ink to be printed is filled into the groove. The image of relief printing is also engraved on the cylinder, but it is a relief structure. Theoretically, screen printing, relief printing, and gravure printing can be used to fabricate any layer of a device if the materials can be processed by solution, and there are indeed some reported instances for printed active layers, buffer layers, and electrodes. At present, screen printing is mainly used to print large-scale electrodes of OSCs.

Figure 11.25a shows the screen printing method. Screen printing is conducted with a woven screen made of steel mesh or synthetic fiber. The pre-patterned screen is filled with the coating ink. A rigid squeegee brings the screen into a close

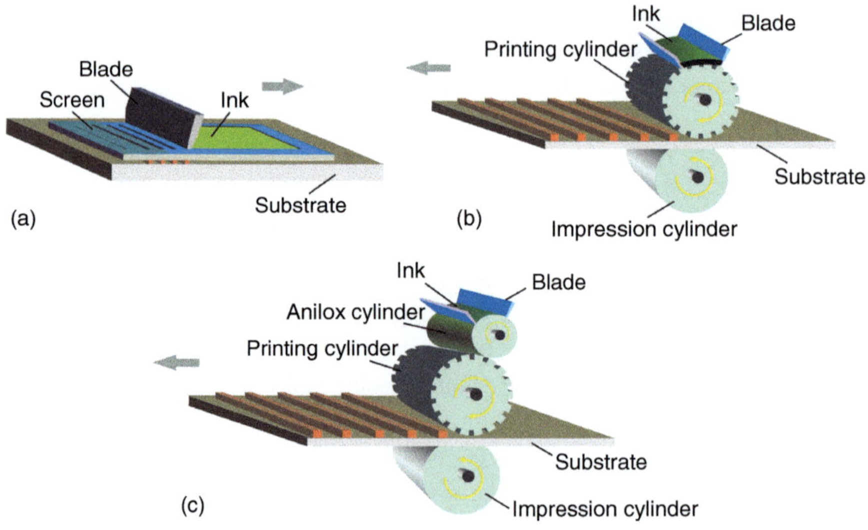

Figure 11.25 (a) Schematic illustration of screen printing, (b) gravure printing, (c) and relief printing.

contact with the substrate, then the squeegee moves across the screen to transfer the pattern onto the substrate. The film thickness is related to the ink concentration, viscosity, and theoretical paste volume of the screen. In 2011, Krebs et al. reported three different types of silver based screen-printable metal inks, including organic solvent-based, UV-curable, and water-based silver inks. Compared with the UV-curable ink and the water-based ink, the organic solvent-based ink had negative effect on the performance of the devices [125]. In addition, screen-printed silver grids have been embedded in the substrate to make electrodes for OPV devices [126–128]. Screen printing was also reported for coating active layers [129]. Therefore, screen printing has been proven to be one of the important printing technologies for large-scale device modules [130].

Figure 11.25c shows the relief printing method. The working process of relief printing is that the image is reproduced in relief, and then the printing cylinder transfers the ink to the substrate. In 2013, Hösel et al. used relief printing in anode printing silver bottom electrodes for device modules. The authors declared that relief printing was the best technique for process silver electrode, as it has a high production speed and consumes less ink [131]. In 2018, Alem et al. demonstrated a solar cell with the electron transport layer (ZnO) and the active layer (PCDTBT:$PC_{71}BM$) deposited by relief printing. The photovoltaic cells of $1\,cm^2$ with the active layer processed by relief printing showed an efficiency of up to 3.5%, which is comparable to those of other printing/coating techniques [132]. In 2019, Hart et al. successfully printed the anode (Ag) by relief printing [133].

Figure 11.25b shows the gravure printing method. The main difference between the gravure printing process and relief printing is that, in gravure printing, the transfer of ink is done through cavities, which is opposite to the relief. Using gravure printing to print transparents has also been reported. In 2015, Montanino et al. processed highly conductive and transparent films based on PEDOT:PSS doped with DMSO onto a flexible substrate PEN, by gravure printing method. The corresponding device based on PTB7:PCBM with the printed anode yielded a PCE value around 2% [134]. In 2018, Huang and Zhu reported that by tailoring the ink properties, printing conditions, and post-printing treatment, gravure printing enabled the printing of high-resolution, highly conductive Ag NW patterns in large area. The printed Ag NW patterns on the flexible substrate showed excellent flexibility under repeated bending [135]. In 2020, Wang and coworkers prepared uniformly printed Ag NW films. Owing to the good uniformity of the gravure-printed Ag NW electrode, the highest PCE of 13.61% was achieved for $1\,cm^2$ PSCs based on the gravure-printed flexible transparent electrodes [136]. Chen and coworkers reported a scalable and highly conductive PEDOT:PSS:CNTs transparent electrode by gravure printing. A large-area flexible electrode with a sheet resistance of $17\,\Omega\,sq^{-1}$ and a light transmittance of 80.7% was obtained by using the R2R technique [137].

11.4.2 R2R Methodology

R2R methodology is the best choice to be employed for high-performance large-area FOSCs. Figure 11.20 shows a small R2R process for OSCs, and Figure 11.26 shows

(a) (b)

Figure 11.26 R2R printing and coating machine for organic solar cells. (a) Appearance of the whole instrument, (b) an enlarged view of a part of the instrument.

R2R machine for large-scale devices. In R2R technique, a "web" is used. The web includes PET, PEN, polyethylene (PE), polypropylene (PP), and so on. PET is most commonly used in R2R manufacturing of OSCs. In the manufacturing process, the web is first unwound from the roll and then further processed by a printing/coating machine. After processing the pattern/ink to the web, it rolls up again. Apart from the printing/coating of the pattern/ink, other procedures (cleaning, speed control, surface treatments, heating, drying, UV curing, etc.) are also needed to obtain a complete OSC.

Coating steps can be divided into two categories. One is a discrete strategy: each step has a separate unwinder and rewinder. The other is the integrated strategy: all steps are integrated with each other in a single unwinding and rewinding process. In the case of high automation, the raw materials enter the machine at one end, and the completed solar cells can be output at the other end. Such an integration process will not only greatly reduce processing costs, but also help to obtain better product output. This is indeed an ideal process for the preparation of industrial large-scale OSCs. In actual production, the integrated system is very complicated, because each step requires different processing parameters. Therefore, it is best to use discrete methods to process such large-area OSCs, because each layer can be processed separately and after coating/printing, and the web can be moved to the next machine for processing.

The printing/coating process is carried out on a motor-driven roller or freely rotating due to the action of the web, guiding the web through the printing/coating unit. Methods such as blade coating, slot-die coating, inkjet printing, spray coating, and screen printing include at least one roller. Complex techniques such as relief printing and gravure printing involve at least two rollers, one roller containing the pattern and the web supported by another coating roller. Depending upon the system, more rolls can be utilized.

Krebs et al. established a complete manufacturing process of the devices and reported a detailed work on mass production from the lab to the fab [138]. Moreover, they also analyzed some neglected aspects during industrial production, such as labor, materials, and electricity costs. Therefore, a comprehensive consideration of material, modules, coating methods and cost is highly required for commercialization of the large-area devices method. Many efforts have been made in fabrication of the upscale devices, and some encouraged progresses have been achieved in the coating fabrication of active layers. Compared with that of spin-coated devices, the morphology of slot-die coating device is more difficult to modulate because of the complex processing parameters such as striping speed, inking speed, processing temperature, etc. Recently, Ma and coworkers realized an impressive PCE of 15.6% for PM6:Y6-based device by slot-die coating. Such high performance is almost equal to that fabricated by spin coating [107]. Slot-die coating method exhibits great compatibility with R2R setting and has been considered to be the most potential for industrial-scale manufacturing. However, high-performance large-area flexible devices and modules processed by all R2R technique are rarely reported, so there are still big challenges in a fundamental understanding of a complete manufacturing process.

11.5 Efficiency Loss in Large-Area Devices and Modules

Although the high-efficiency photovoltaic device has been explored in laboratory environment (refer to small-area devices), the route from large-area devices to module pattern is yet not easy. Herein, there is a broad cognition that PCE inevitably and significantly decreases with the increase of device area since the large-area electrode has large resistance (particular in modular solar cells). The importance of drying process conditions from small-area to large area R2R production was demonstrated by Krebs and coworkers, who also proposed the essentiality of the sacrificial solvent system and some basic problems such as uncontrolled electric energy in the later installation of OSCs [139]. As the connection part of each solar cell in module has no contribution to photovoltaic performance, a concept of GFF is proposed for evaluating the PCE loss from the geometric patterns, which is defined as the ratio between photoactive layer area and total area [140]. In addition, the regular processing solvents in high-efficiency OSCs are halogen solvents including chlorobenzene, *o*-dichlorobenzene, and chloroform, which harm the environment and deviate from the idea of green chemistry. Therefore, these processing solvents are often replaced by non-halogen solvents without sacrificing too much PCE.

Thereby, the PCE loss from small-area devices to module solar cells is mainly divided into electrical loss, geometric loss, optical loss, and additional loss. The electrical loss, geometric loss, and optical loss are mainly manifested in the reduction of J_{sc} and FF, while the additional loss resulted from the use of non-halogen solvents [19, 100, 141–144].

11.5.1 Electrical Loss

The electrical loss from small-area devices to modules can be reflected in the electrode resistance (R_s) and inter-contact resistance (R_{int}) between single cells. The cells between modules are commonly connected in series to acquire the vast voltage, but resulting in a higher internal resistance (R_i) in modules. Apart from above, the R_{int} that existed in module also possesses a large proportion in electrical loss, as displayed in Figure 11.27.

The sheet resistances (R_s) of top electrodes and bottom electrodes can be obtained through Eq. (11.1). Meanwhile, the R_i produced from module was acquired by Eq. (11.2). Herein, ρ refers to the resistivity, "d" is the thickness of active layer in single cell. "w" denotes the width of single OSCs, "L" and "I" stand for the length of active layer and inter-contact parts. Besides, an interior proportional relationship between R_s and R_i was existed in Eq. (11.3). Undoubtedly, the improvement of devices in area must correspond to the rise of electrode areas, including top and bottom electrodes. Specially, the value of R_s was increased drastically with the introduction of flexible substrates, which is much higher than the substrate based on ITO glass. Therefore, a higher internal resistance R_i in modules was achieved than devices based on small area. The R_i of OSCs also exhibits an intimate influence

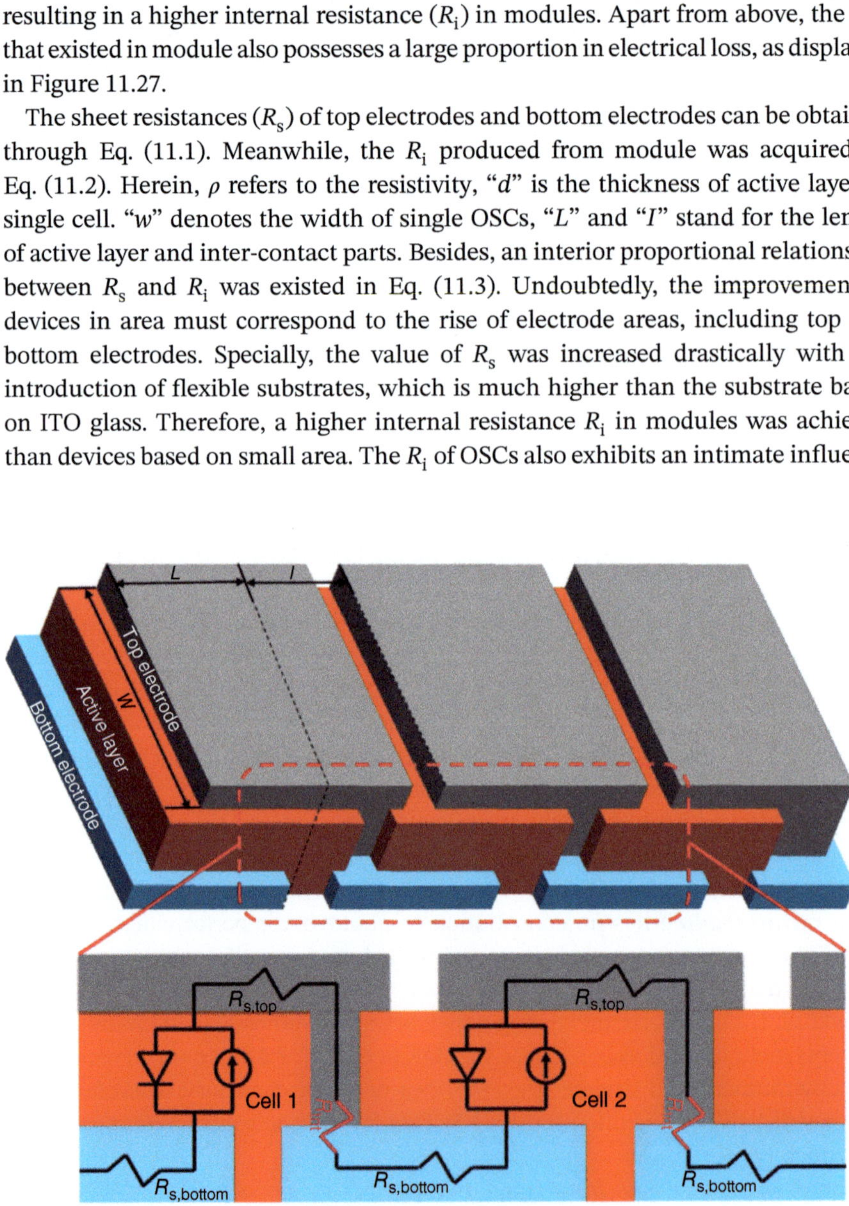

Figure 11.27 Cross section and the equivalent circuit diagram of OSC modules. "w" denotes the width of OSCs, "L" and "I" stand for the length of active layer and inter-contact parts. Source: Lucera et al. [144]. Reproduced with permission of John Wiley & Sons.

on the photovoltaic parameters (J_{sc} and FF), and the ultimate effects are presented on the values of PCE.

$$R_s = \frac{\rho}{d} \tag{11.1}$$

$$R_i = \frac{\rho}{d} \times \frac{L}{W} \tag{11.2}$$

$$R_i = R_s \times \frac{L}{W} \tag{11.3}$$

The R_{int} is introduced by the incomplete contact between adjacent cells in modules, which also brings a large "dead region" defined as the inactive part. Exactly, the R_{int} formed by ohmic contact between bottom and top electrodes (as shown in Figure 11.27) is determined by electrode materials, normal stress, and surface environment. Besides, the contact carrier at the interface is also affected by the work functions of bottom and top electrodes and the condition of contact surface. Generally, the smooth interface contact with homogeneous and high conductivity transparent electrodes is beneficial to obtaining the low values of R_{int}. A high PCE of module could be achieved by decreasing R_{int} [69].

11.5.2 Geometric Loss

The "dead area" between the neighboring cells (displayed in Figure 11.28) is originated from the inter-connect part among top and bottom electrodes and active layer, which does not have the ability to generate photocurrent. Thus, the geometric loss is proposed. Moreover, the GFF is defined as the ratio of effective active area to the total device area (refers modules):

$$\text{GFF} = \frac{\text{effective active area}}{\text{total area}} = \frac{n \times w \times L}{n \times w \times (L + l)} = \frac{L}{L + l}$$

Here, w represents the width of single OSCs, n is the number of OSCs, L and l refer to the length of effective active layer and dead area between inter-contact parts, respectively. The PCEs of module and cells satisfy the following relationship equation:

$$\text{PCE}_{\text{module}} = \text{PCE}_{\text{cell}} \times \text{GFF}$$

Therefore, the realization of low geometric loss, corresponding to a high GFF, always is critical for high-efficiency modules. Intrinsically, the small length of dead area possesses an essential role on the efficiency of module, because the small dead area denotes a high value of GFF [144].

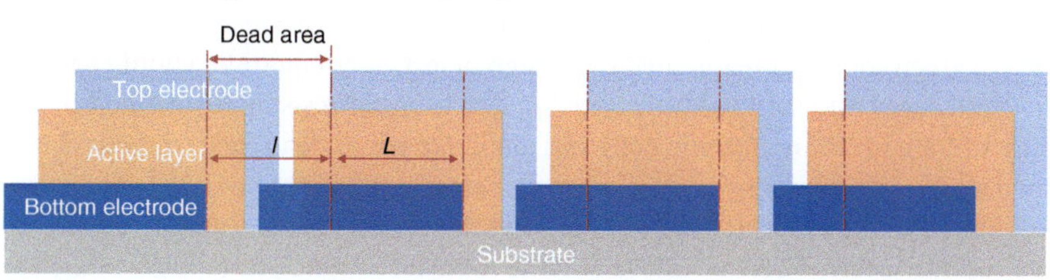

Figure 11.28 Cross-sectional diagram of module formed through single solar cells in series, where l and L are the length of dead area and effective active layer area. Source: Wang et al. [145]. Reproduced with permission of John Wiley & Sons.

11.5.3 Optical Loss

The optical loss is another key parameter influencing the module efficiency, because the transparency of large-area flexible electrode is lower than these rigid devices based on ITO glass. The optical loss is caused by the inadequate incident light (below 100%), which is the result of reflection and low transparency of flexible electrode in flexible module. Furthermore, compared with the small-area ITO devices, the adoption of large-area bottom electrodes also accompanies the large optical loss due to the strong absorption and reflection of incident light in thick functional-layer films of modules. The optical loss can be analyzed by the transfer matrix formalism, which can give the systematic and precise results [146–148].

11.5.4 Additional Loss

In addition to the abovementioned main loss in the module cells, there is the additional loss remaining that can be easily neglected in the transform process from small-area to large-area devices. The additional loss in large-area films comes from more defects than small-area devices, which can lead to a large current leakage in module consisting of multiple monolithic interconnecting cells in series and parallel. Hence, reducing defects of active layer by using the thick film is favorable for promoting the PCE loss. Meanwhile, it is a powerful way to diminish current leakage in modules. Apart from above loss, the introduction of free halogen is a key role in decreasing the toxicity of solvent (such as chlorobenzene, *o*-dichlorobenzene, etc.) in the large-area fabrication route [101]. The difference of 10% ~ 20% still existed when replacing the conventional toxic solvents to free-halogen solvents (such as xylene, toluene, and tetrahydronaphthalene, etc.). Thus, the decline of the performance from the alternative green solvent should be premeditated into the PCE loss.

11.5.5 Modular Designs

To minimize the PCE loss, the high-transparency and high-conductivity electrodes are used in module. And the active layer materials for modules should possess the good retention characteristics of photovoltaic performance when using the thick film and green solvent. Meanwhile, the design of large-area modules should be carefully considered and selected for achieving low internal resistance (affected by the width and length of monolithic cell) and small length of dead area (directly determined by the length of interconnected part), corresponding to the low electrical loss and high GFF. Indeed, the dead area in module has no contribution on photocurrent, and the interconnected part should be reduced as much as possible. Furthermore, the modular design owns a pretty decisive factor affecting the PCE loss. The high-efficiency solar module commonly has the intimate relation with optimal modular design.

11.6 Conclusion and Outlook

Different from small-area devices by spin coating, large-area flexible devices have a rigorous demand on the active layer materials, fabrication method, bottom electrode, and modular design. Roughly, the active layer materials for large-area need possess some certain tolerance to thickness; meanwhile, the mechanical and photovoltaic performances of FOSCs are also limited by flexibility and transparency of flexible electrodes.

Pursuing high-efficiency FOSCs is persistent and continuing. To better satisfy the diverse requirements of people, the distinctive advantages of ultrahigh flexibility and semitransparency of large-area FOSCs in the future should be positively developed. Moreover, the researchers who mainly focused on the photovoltaic performance are transforming their interests into the practical application of wearable device and smart windows. Through continuous innovation and development, the ultra-FOSCs will be widely used in the world.

References

1 Liu, T., Yang, T., Ma, R. et al. (2021). *Joule* 5: 914–930.
2 Sun, R., Wang, W., Yu, H. et al. (2021). *Joule* 5: 1548–1565.
3 Jorgensen, M., Norrman, K., Gevorgyan, S.A. et al. (2012). *Adv. Mater.* 24: 580.
4 Zhou, Y., Li, M., Lu, H. et al. (2021). *Adv. Funct. Mater.* 31: 2101742.
5 Lee, D., Hwang, H., Sin, D.H. et al. (2021). *ACS Energy Lett.* 6: 2610–2618.
6 Li, G., Shrotriya, V., Huang, J. et al. (2005). *Nat. Mater.* 4: 864–868.
7 Liu, Y., Zhao, J., Li, Z. et al. (2014). *Nat. Commun.* 5: 5293.
8 Hu, H., Chow, P.C.Y., Zhang, G. et al. (2017). *Acc. Chem. Res.* 50: 2519.
9 Liao, X., Zhang, L., Chen, L. et al. (2017). *Nano Energy* 37: 32–39.
10 Xie, Q., Liao, X., Chen, L. et al. (2019). *Nano Energy* 61: 228–235.
11 Liu, X., Nian, L., Gao, K. et al. (2017). *J. Mater. Chem. A* 5: 17619.
12 Hu, X., Yi, C., Wang, M. et al. (2014). *Adv. Energy Mater.* 4: 1400378.
13 Jin, Y., Chen, Z., Dong, S. et al. (2016). *Adv. Mater.* 28: 9811.
14 Wang, J., Wang, S., Duan, C. et al. (2017). *Adv. Energy Mater.* 7: 1702033.
15 Jin, Y., Chen, Z., Xiao, M. et al. (2017). *Adv. Energy Mater.* 7: 1700944.
16 Zhu, X., Xia, B., Lu, K. et al. (2016). *Chem. Mater.* 28: 943.
17 Sun, K., Xiao, Z., Lu, S. et al. (2015). *Nat. Commun.* 6: 6013.
18 Liao, X., Zhang, L., Hu, X. et al. (2017). *Nano Energy* 41: 27–34.
19 Heo, Y.J., Jung, Y.S., Hwang, K. et al. (2017). *ACS Appl. Mater. Interfaces* 9: 39519–39525.
20 Lin, Y., Wang, J., Zhang, Z.-G. et al. (2015). *Adv. Mater.* 27: 1170.
21 Zhao, W., Qian, D., Zhang, S. et al. (2016). *Adv. Mater.* 28: 4734.
22 Fan, B., Zhang, K., Jiang, X.-F. et al. (2017). *Adv. Mater.* 29: 1606396.
23 Guo, B., Li, W., Guo, X. et al. (2017). *Adv. Mater.* 29: 1702291.
24 Zhao, W., Li, S., Yao, H. et al. (2017). *J. Am. Chem. Soc.* 139: 7148.
25 Fan, Q., Wang, Y., Zhang, M. et al. (2018). *Adv. Mater.* 30: 1704546.

26 Fan, Q., Su, W., Wang, Y. et al. (2018). *Sci. China Chem.* 61: 531.
27 Fan, Q., Zhu, Q., Xu, Z. et al. (2018). *Nano Energy* 48: 413.
28 Yang, L., Zhang, S., He, C. et al. (2018). *Chem. Mater.* 30: 2129.
29 Feng, S., Zhang, C., Liu, Y. et al. (2017). *Adv. Mater.* 29: 1703527.
30 Luo, Z., Sun, C., Chen, S. et al. (2018). *Adv. Energy Mater.* 8: 1800856.
31 Yuan, J., Zhang, Y., Zhou, L. et al. (2019). *Joule* 3: 1140.
32 Ma, L., Zhang, S., Yao, H. et al. (2020). *ACS Appl. Mater. Interfaces* 12: 18777.
33 Huang, S., Wu, F., Liu, Z. et al. (2021). *J. Energy Chem.* 53: 63–68.
34 Zhang, H., Yao, H., Hou, J. et al. (2018). *Adv. Mater.* 30: 1800613.
35 Li, Y., Lin, J.-D., Che, X. et al. (2017). *J. Am. Chem. Soc.* 139: 17114.
36 Khlyabich, P.P., Rudenko, A.E., Thompson, B.C., and Loo, Y.-L. (2015). *Adv. Funct. Mater.* 25: 5557–5563.
37 Gokulnath, T., Choi, J., Park, H.-Y. et al. (2021). *Nano Energy.* 89: 106323.
38 Yang, L., Zhou, H., Price, S.C., and You, W. (2012). *J. Am. Chem. Soc.* 134: 5432–5435.
39 Ma, R., Liu, T., Luo, Z. et al. (2020). *ACS Energy Lett.* 5: 2711–2720.
40 Baran, D., Ashraf, R.S., Hanifi, D.A. et al. (2017). *Nat. Mater.* 16: 363.
41 Zhang, C., Zhao, K., Yin, Q. et al. (2017). *J. Am. Chem. Soc.* 139: 2387–2395.
42 Gasparini, N., Lucera, L., Salvador, M. et al. (2017). *Energy Environ. Sci.* 10: 885.
43 Zhang, J., Zhao, Y., Fang, J. et al. (2017). *Small* 13: 1700388.
44 Zhou, Z., Xu, S., Song, J. et al. (2018). *Nat. Energy* 3: 952.
45 Ma, L., Xu, Y., Zu, Y. et al. (2020). *Sci. China Chem.* 63: 21.
46 Liu, S., Chen, D., Hu, X. et al. (2020). *Adv. Funct. Mater.* 30: 2003223.
47 Fu, H., Li, Y., Yu, J. et al. (2021). *J. Am. Chem. Soc.* 143: 2665–2670.
48 Fan, B., Ying, L., Wang, Z. et al. (2017). *Energy Environ. Sci.* 10: 1243.
49 Fan, B., Ying, L., Zhu, P. et al. (2017). *Adv. Mater.* 29: 1703906.
50 Fan, B., Zhu, P., Xin, J. et al. (2018). *Adv. Energy Mater.* 8: 1703085.
51 Yan, H., Chen, Z., Zheng, Y. et al. (2009). *Nature* 457: 679.
52 Chuang, T.-K., Troccoli, M., Kuo, P.-C. et al. (2007). *Appl. Phys. Lett.* 90: 151114.
53 Hu, X., Song, Z., Liu, W. et al. (2012). *Appl. Surf. Sci.* 258: 5798–5802.
54 Jeong, J.K., Jin, D.U., Shin, H.S. et al. (2007). *IEEE Electron Device Lett.* 28: 389–391.
55 Xie, Z., Hung, L.-S., and Zhu, F. (2003). *Chem. Phys. Lett.* 381: 691–696.
56 MacDonald, W.A., Looney, M.K., MacKerron, D. et al. (2013). *Plast. Rubber Compos.* 37: 41–45.
57 MacDonald, W.A., Looney, M.K., MacKerron, D. et al. (2007). *J. Soc. Inf. Display* 15: 1075–1083.
58 Choi, M.-C., Kim, Y., and Ha, C.-S. (2008). *Prog. Polym. Sci.* 33: 581–630.
59 Vergnat, C., Landais, V., Combet, J. et al. (2012). *Org. Electron.* 13: 1594–1601.
60 Woo, K., Kim, D., Kim, J.S. et al. (2009). *Langmuir* 25: 429–433.
61 Lim, J.-W., Cho, D.-Y., Eun, K. et al. (2012). *Sol. Energy Mater. Sol. Cells* 105: 69–76.
62 Lu, Y., Hu, Z., Bian, J., and Wang, Y. (2011). *Adv. Mater. Res.* 239–242: 1211–1214.

63 Fahlteich, J., Fahland, M., Schönberger, W., and Schiller, N. (2009). *Thin Solid Films* 517: 3075–3080.

64 Kaltenbrunner, M., White, M.S., Glowacki, E.D. et al. (2012). *Nat. Commun.* 3: 770.

65 Zhang, F., Johansson, M., Andersson, M.R. et al. (2002). *Adv. Mater.* 14: 662–665.

66 Na, S.-I., Kim, S.-S., Jo, J., and Kim, D.-Y. (2008). *Adv. Mater.* 20: 4061–4067.

67 Meng, W., Ge, R., Li, Z. et al. (2015). *ACS Appl. Mater. Interfaces* 7: 14089–14094.

68 Fan, X., Xu, B., Liu, S. et al. (2016). *ACS Appl. Mater. Interfaces* 8: 14029–14036.

69 Song, W., Fan, X., Xu, B. et al. (2018). *Adv. Mater.* 30: 1800075.

70 Rowell, M.W., Topinka, M.A., McGehee, M.D. et al. (2006). *Appl. Phys. Lett.* 88: 233506.

71 Salvatierra, R.V., Cava, C.E., Roman, L.S., and Zarbin, A.J.G. (2013). *Adv. Funct. Mater.* 23: 1490–1499.

72 Jeon, I., Cui, K., Chiba, T. et al. (2015). *J. Am. Chem. Soc.* 137: 7982–7985.

73 Yin, Z., Sun, S., Salim, T. et al. (2010). *ACS Nano* 4: 5263–5268.

74 Song, Y., Chang, S., Gradecak, S., and Kong, J. (2016). *Adv. Energy Mater.* 6: 1600847.

75 Hu, X., Meng, X., Xiong, J. et al. (2017). *Adv. Mater. Technol.-US* 2: 1700138.

76 Lee, J.Y., Connor, S.T., Cui, Y., and Peumans, P. (2008). *Nano Lett.* 8: 689–692.

77 Tvingstedt, K. and Inganäs, O. (2007). *Adv. Mater.* 19: 2893–2897.

78 Zou, J., Li, C.Z., Chang, C.Y. et al. (2014). *Adv. Mater.* 26: 3618–3623.

79 Rathmell, A.R. and Wiley, B.J. (2011). *Adv. Mater.* 23: 4798–4803.

80 Sanchez-Iglesias, A., Rivas-Murias, B., Grzelczak, M. et al. (2012). *Nano Lett.* 12: 6066–6070.

81 Rathmell, A.R., Nguyen, M., Chi, M., and Wiley, B.J. (2012). *Nano Lett.* 12: 3193–3199.

82 Garnett, E.C., Cai, W., Cha, J.J. et al. (2012). *Nat. Mater.* 11: 241–249.

83 Song, M., You, D.S., Lim, K. et al. (2013). *Adv. Funct. Mater.* 23: 4177–4184.

84 Seo, J.H., Hwang, I., Um, H.D. et al. (2017). *Adv. Mater.* 29: 1701479.

85 Sun, Y., Chang, M., Meng, L. et al. (2019). *Nat. Electron.* 2: 513–520.

86 Zhang, W., Song, W., Huang, J. et al. (2019). *J. Mater. Chem. A* 7: 22021–22028.

87 Chen, X., Xu, G., Zeng, G. et al. (2020). *Adv. Mater.* 32: 1908478.

88 Qu, T.Y., Zuo, L.J., Chen, J.D. et al. (2020). *Adv. Opt. Mater.* 8: 2000669.

89 Alstrup, J., Jorgensen, M., Medford, A.J., and Krebs, F.C. (2010). *ACS Appl. Mater. Interfaces* 2: 2819.

90 Mens, R., Adriaensens, P., Lutsen, L. et al. (2008). *J. Polym. Sci., Part A: Polym. Chem.* 46: 138.

91 Chang, Y.-H., Tseng, S.-R., Chen, C.-Y. et al. (2009). *Org. Electron.* 10: 741–746.

92 Ro, H.W., Downing, J.M., Engmann, S. et al. (2016). *Energy Environ. Sci.* 9: 2835.

93 Zhao, W., Zhang, S., Zhang, Y. et al. (2018). *Adv. Mater.* 30: 1704837.

94 Zhang, L., Lin, B., Hu, B. et al. (2018). *Adv. Mater.* 30: 1800343.

95 Zhang, L., Xu, X., Lin, B. et al. (2018). *Adv. Mater.* 30: 1805041.

96 Zhang, L., Zhao, H., Lin, B. et al. (2019). *J. Mater. Chem. A* 7: 22265–22273.
97 Sun, R., Wu, Q., Guo, J. et al. (2020). *Joule* 4: 407.
98 Wang, Y., Wang, X., Lin, B. et al. (2020). *Adv. Energy Mater.* 10: 2000826.
99 Dong, S., Jia, T., Zhang, K. et al. (2020). *Joule* 4: 2004–2016.
100 Zimmermann, B., Schleiermacher, H.F., Niggemann, M., and Würfel, U. (2011). *Sol. Energy Mater. Sol. Cells* 95: 1587–1589.
101 Schrödner, M., Sensfuss, S., Schache, H. et al. (2012). *Sol. Energy Mater. Sol. Cells* 107: 283–291.
102 Machui, F., Lucera, L., Spyropoulos, G.D. et al. (2014). *Sol. Energy Mater. Sol. Cells* 128: 441–446.
103 Helgesen, M., Carlé, J.E., and Krebs, F.C. (2013). *Adv. Energy Mater.* 3: 1664–1669.
104 Wu, Q., Guo, J., Sun, R. et al. (2019). *Nano Energy* 61: 559–566.
105 Han, Y.W., Jeon, S.J., Lee, H.S. et al. (2019). *Adv. Energy Mater.* 9: 1970164.
106 Meng, X., Zhang, L., Xie, Y. et al. (2019). *Adv. Mater.* 31: 1903649.
107 Zhao, H., Naveed, H.B., Lin, B. et al. (2020). *Adv. Mater.* 39: 2002302.
108 Zhong, W., Hu, Q., Jiang, Y. et al. (2019). *Sol. RRL* 3: 1900032.
109 Eom, S.H., Senthilarasu, S., Uthirakumar, P. et al. (2009). *Org. Electron.* 10: 536.
110 Jung, S., Sou, A., Banger, K. et al. (2014). *Adv. Energy Mater.* 4: 1400432.
111 Eggenhuisen, T.M., Galagan, Y., Biezemans, A.F.K.V. et al. (2015). *J. Mater. Chem. A* 3: 7255.
112 Corzo, D., Almasabi, K., Bihar, E. et al. (2019). *Adv. Mater. Technol.* 4: 1900040.
113 Vak, D., Kim, S.-S., Jo, J. et al. (2007). *Appl. Phys. Lett.* 91: 081102.
114 Girotto, C., Rand, B.P., Genoe, J., and Heremans, P. (2009). *Sol. Energy Mater. Sol. Cells* 93: 454.
115 Hoth, C.N., Steim, R., Schilinsky, P. et al. (2009). *Org. Electron.* 10: 587.
116 Steirer, K.X., Reese, M.O., Rupert, B.L. et al. (2009). *Sol. Energy Mater. Sol. Cells* 93: 447.
117 Chen, L.-M., Hong, Z., Kwan, W.L. et al. (2010). *ACS Nano* 4: 4744.
118 Girotto, C., Moia, D., Rand, B.P., and Heremans, P. (2011). *Adv. Funct. Mater.* 21: 64.
119 Susanna, G., Salamandra, L., Brown, T.M. et al. (2011). *Sol. Energy Mater. Sol. Cells* 95: 1775.
120 Wang, T., Scarratt, N.W., Yi, H. et al. (2013). *Adv. Energy Mater.* 3: 505–512.
121 Vak, D., van Embden, J., Wong, W.W.H., and Watkins, S. (2015). *Appl. Phys. Lett.* 106: 033302.
122 Tait, J.G., Worfolk, B.J., Maloney, S.A. et al. (2013). *Sol. Energy Mater. Sol. Cells* 110: 98.
123 Noh, Y.-J., Na, S.-I., and Kim, S.-S. (2013). *Sol. Energy Mater. Sol. Cells* 117: 139.
124 La Notte, L., Bianco, G.V., Palma, A.L. et al. (2018). *Carbon* 129: 878–883.
125 Krebs, F.C., Sondergaard, R., and Jorgensen, M. (2011). *Sol. Energy Mater. Sol. Cells* 95: 1348.
126 Krebs, F.C., Alstrup, J., Spanggaard, H. et al. (2004). *Sol. Energy Mater. Sol. Cells* 83: 293.
127 Krebs, F.C., Spanggard, H., Kjaer, T. et al. (2007). *Mater. Sci. Eng., B* 138: 106.

128 Galagan, Y., Rubingh, J.-E.J.M., Andriessen, R. et al. (2011). *Sol. Energy Mater. Sol. Cells* 95: 1339.

129 Zhang, B., Chae, H., and Cho, S.M. (2009). *Jpn. J. Appl. Phys.* 48: 020208.

130 Krebs, F.C., Jorgensen, M., Norrman, K. et al. (2009). *Sol. Energy Mater. Sol. Cells* 93: 422.

131 Hösel, M., Søndergaard, R.R., Angmo, D., and Krebs, F.C. (2013). *Adv. Energy Mater.* 15: 995–1001.

132 Alem, S., Graddage, N., Lu, J. et al. (2018). *Org. Electron.* 52: 146–152.

133 Hart, A.S., Andersen, T.R., Griffith, M.J. et al. (2019). *RSC Adv.* 9: 42294–42305.

134 Montanino, M., De Girolamo Del Mauro, A., Tesoro, M. et al. (2015). *Polym. Compos.* 36: 1104–1109.

135 Huang, Q. and Zhu, Y. (2018). *Sci. Rep.* 8: 15167.

136 Wang, Z., Han, Y., Yan, L. et al. (2020). *Adv. Funct. Mater.* 31: 2007276.

137 Hu, X., Chen, L., Zhang, Y. et al. (2014). *Chem. Mater.* 26: 6293–6302.

138 Krebs, F., Tromholt, T., and Jørgensen, M. (2010). *Nanoscale* 2: 873.

139 Carlé, J.E., Helgesen, M., Hagemann, O. et al. (2017). *Joule* 1: 274–289.

140 Kubis, P., Li, N., Stubhan, T. et al. (2015). *Prog. Photovoltaics Res. Appl.* 23: 238–246.

141 Krebs, F.C., Gevorgyan, S.A., and Alstrup, J. (2009). *J. Mater. Chem.* 19: 5442.

142 Servaites, J.D., Yeganeh, S., Marks, T.J., and Ratner, M.A. (2010). *Adv. Funct. Mater.* 20: 97–104.

143 Kutsarov, D.I., New, E., Bausi, F. et al. (2017). *Sol. Energy Mater. Sol. Cells* 161: 388–396.

144 Lucera, L., Kubis, P., Fecher, F.W. et al. (2015). *Energy Technol.* 3: 373–384.

145 Wang, G., Adil, M.A., Zhang, J., and Wei, Z. (2019). *Adv. Mater.* 31: 1805089.

146 Pettersson, L.A.A., Roman, L.S., and Inganäs, O. (1999). *J. Appl.Phys.* 86: 487–496.

147 Lucera, L., Machui, F., Kubis, P. et al. (2016). *Energy Environ. Sci.* 9: 89–94.

148 Hösel, M., Søndergaard, R.R., Jørgensen, M., and Krebs, F.C. (2013). *Energy Technol.* 1: 102–107.

12

Organic Photovoltaics for Indoor Applications

Zhan'ao Tan[1], Yinglong Bai[2], and Shan Jiang[2]

[1] *Beijing University of Chemical Technology, Beijing Advanced Innovation Center for Soft Matter Science and Engineering, No. 15 North Third Ring East Road, Chaoyang District, Beijing, 100029, P.R. China*
[2] *North China Electric Power University, State Key Laboratory of Alternate Electrical Power System with Renewable Energy Sources, No. 2 Beinong Road, Huilongguan, Changping District, Beijing, 102206, P.R. China*

12.1 Introduction

Since the industrial revolution, more and more carbon dioxide has been released into the atmosphere, leading to a rise in global temperatures, much of which can be traced to energy production and the burning of fossil fuels [1–6]. Energy production and consumption have always been an important part of social development. The solution to the global energy crisis is to find clean energy to transform the supply of traditional energy [7]. At present, solar energy is an important clean energy with the advantages of low cost, abundant resources, renewable, and no environmental pollution [8]. There are many ways to utilize solar energy, among which, photovoltaic (PV) technology is considered as one of the most effective strategies to convert light energy into electric energy [9]. As a technology of light energy collection and utilization, PV cell is used worldwide as a single or matching power source in different energy consumption fields, which has taken center stage as a more desirable alternative [10–12]. At present, scientists have been trying to find efficient and low-cost PV technology. PV utilization technology can be divided into three generations, the first generation is based on chip solar cells, mainly crystalline silicon (c-Si) solar cells. The second generation is thin-film solar cells, such as those made of amorphous silicon (a-Si), CdTe and $CuInGaSe_2$ (CIGS), and other different inorganic film composed of III–V compound (GaAs) solar cells. The third generation of emerging solar cells includes dye-sensitized solar cells, perovskite solar cells, and organic solar cells [13, 14]. c-Si solar cells account for over 90% of the market, and thin-film solar cells account for the rest. Although currently emerging solar cells cannot compete with c-Si and thin-film solar cells in terms of efficiency, they are expected to achieve other applications that were not covered by the previous two generations of solar cells such as semi-transparent, ultra-flexible. This is also an important reason why the third-generation solar cell has always been a research hotspot.

Organic Solar Cells: Materials Design, Technology and Commercialization, First Edition. Edited by Liming Ding.

Due to the inferior stability of third-generation solar cells, it seems more feasible to use them in mild indoor environments than in standard outdoor sunlight, compared with commercialized c-Si panels. The development of PV technology for low light intensity and indoor environment applications has a broad application prospect in the field of household and building management systems. In recent years, flexible wearable devices, Internet of Things (IoTs), and other portable electronic devices have rapidly penetrated into our daily life [15, 16]. There is optimism that IoT technology will benefit our daily life in all sorts of smart ways. The success of IoT technologies will depend to a large extent on the capabilities and number of terminal nodes (sensors and actuators) that detects the state of the environment or equipment and communicate wirelessly with each other or connect to gateways. By 2020, the IoTs terminal node usage is expected to reach tens of billions. Figure 12.1a shows the comparison of the number of articles on IoTs and indoor photovoltaics (IPVs) on the Web of Science from 2005 to 2020 [20, 21]. These devices will be integrated with wireless devices and communication systems, although they are designed to operate periodically and consume little power (from microwatts to milliwatts, Figure 12.1b). Maintaining a stable power supply over the long term independent of the power grid is also a fairly challenging task. Therefore, the self-powered system of these electronic devices is a key factor in the success of IoT technology [22]. How to solve the power supply is undoubtedly an urgent problem to be solved. Due to the unique requirements of the IoTs, many people have accepted the scheme of using PV technology for energy collection under low light intensity or indoor conditions [23]. Efficient IPVs will be an excellent and friendly solution to provide sustainable energy to drive these devices by capturing ambient light energy from indoor lighting. The use of IPV can greatly reduce the maintenance cost of indoor installation for electronic equipments. Third-generation PV cells have great potential to meet the energy requirements of most indoor electronics, making them very promising power candidates. Figure 12.1c shows a typical wireless sensor network with organic photovoltaic (OPV) energy collection modules and the cross-section schematic diagram of OPV modules [17, 24]. As a result, IPV is increasingly becoming an important technology for the IoTs, the home, and a variety of electronic applications [25–27].

In the past few decades, OPV technology has made significant progresses. By synthesizing novel photoactive donor and acceptor materials (especially developing non-fullerene acceptors), engineering device structures, and optimizing blended layer phase-separation morphology, the power conversion efficiency (PCE) of single-junction solar cells has reached over 18% under air mass (AM) 1.5 G 1 sun irradiation [28]. At the same time, OPVs can be used as portable power sources to charge low-power electronic devices in indoor environments. Today, buildings with lights on 24 hours a day (hospitals, shops, airports, offices, and factories) are common, and artificial lighting is a common source of light [29, 30]. Indoor ambient lighting sources, such as white light-emitting diodes (LEDs), fluorescent lamps (FLs), and halogen lamps, can provide sufficient energy for most small and low-power indoor electronic devices. At present, primary/secondary batteries are the most widely available power supply option for these indoor electronic devices,

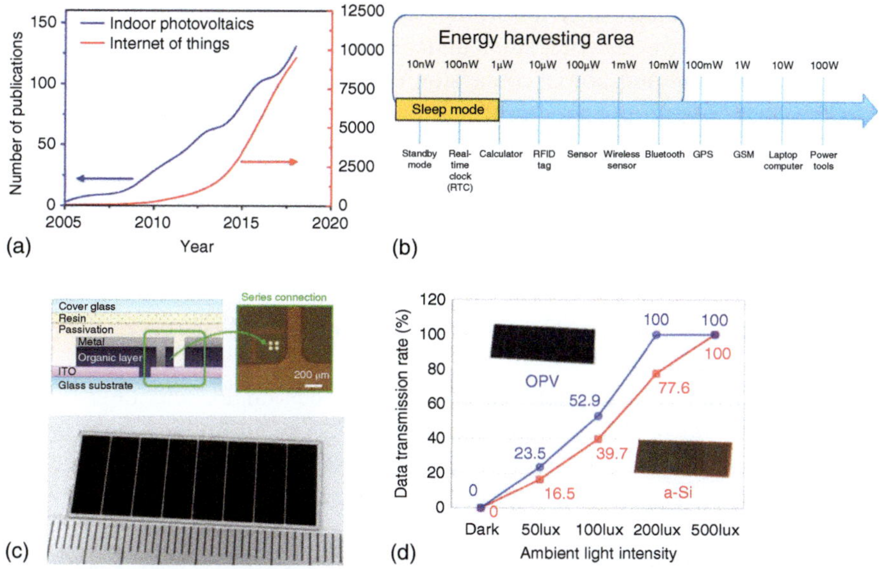

Figure 12.1 (a) The publication numbers of IoT and IPV from 2005–2020 in Web of Science. Source: Mainville and Leclerc [17]/Reproduced with permission of American Chemical Society. (b) The power levels for indoor applicated portable electronic devices and the energy-harvesting area for PV cells. Source: Cui et al. [18]/Reproduced with permission of American Chemical Society. (c) Organic photovoltaic modules used for wireless sensor network energy collection and their cross section schematic diagram. (d) The data transfer rate of wireless sensor nodes driven by OPV (blue line) or a-Si (red line) under environment illumination. Source: Aoki [19]/Reproduced with permission of Elsevier.

but battery power systems require regular replacement and maintenance, which adds to human and material costs [31]. Therefore, it is imperative to choose a semi-permanent indoor energy collection system, which can convert indoor light energy into electric energy and make low-power indoor electronic device work normally [32, 33]. In general, the PV performance of OPV devices is not as good as that of inorganic counterparts under sunlight. However, studies have shown that OPVs can convert indoor light into electricity more efficiently than other inorganic PVs (Figure 12.1d). For PV devices, efficient light collection is always the most important. Different from the wide spectrum of AM 1.5 G sunlight from ultraviolet–visible (UV–Vis) spectroscopy to infrared, indoor light is in the visible range, leading to the optimization of IPV without the pursuit of narrow bandgap photoactive materials. Freunek et al. have shown that when comparing different light sources, it is the spectral range of the light source, rather than the intensity of the light source, that determines the device PCE [34]. In addition, the maximum PCE available for PV devices depends on the structure of device and the bandgap of the photoactive layer. The active layer with a lower bandgap will produce a larger PCE under outdoor sun lighting, while the active layer with a higher bandgap will produce a larger PCE under indoor lighting. It can be predicted that the active layer bandgap for devices suitable for indoor lighting sources should be about 1.9–2.1 eV.

Organic materials have high absorption coefficient and adjustable absorption range, and the leakage current is small under dim lighting conditions, which makes the light absorption and power conversion of indoor lighting environment more effective. In addition, its lightweight and mechanical flexibility make it an ideal power source for indoor IoTs of wireless sensor networks [35, 36]. Therefore, OPVs have great potential in collecting environmental energy, which cannot be achieved with silicon-based inorganic PV. Based on this, OPVs is developing rapidly as an ideal indoor energy supply solution.

12.2 The Characteristics of Indoor Lighting Sources

At present, indoor lighting equipments mainly focus on FL and LED lighting sources. Figure 12.2a shows physical image of FL and LED. They have unique spectral characteristics and luminescence principles. Traditional FL, namely low pressure mercury lamp, is to use the mercury vapor at low pressure after the release of ultraviolet light, so that the phosphor gives off visible light luminescence, so it belongs to low pressure arc discharge light source. The electric energy consumed by FL in gas discharge is mainly converted into electromagnetic radiation in the ultraviolet range. About 3% of the energy is directly converted into visible light in the discharge, and the main wavelengths are 405 nm (blue–violet), 436 nm (blue), 456 nm (green), and 577 nm (yellow).

The color temperature range of FL for lighting is between 2700 and 6500 K, which can be divided into warm color light, warm white color light, and cool color light according to the different color temperature. The color temperature of warm color light is below 3300 K. Warm color light is similar to incandescent lamp, and red light has more components, which can give people a warm, healthy, and comfortable feeling, suitable for family, residence, dormitory, hotel and other places with low temperature. Warm white light also calls neutral color, with a color temperature between 3300 and 5300 K, because the light is downy, which makes people feel happy, comfortable and serene, suitable for shops, hospitals, offices, restaurants, waiting rooms, and other places. The color temperature of cool color light is above 5300 K, illuminant is close to natural light, having bright feeling, suitable for office, conference room, classroom, drawing room, design room, library reading room, exhibition window, and other places [18, 38].

The LEDs used for indoor lighting are white LEDs, which represent white light source with different color temperature. Generally speaking, indoor lighting sources have three color temperatures: warm white (2700–3000 K), natural white (4000–4500 K), and cold white (5000–6500 K). The higher the color temperature, the more blue light emitted in the white light. There are generally two ways to make white LEDs. For the most widely used phosphor LED, light is generated by the combined emission of GaN blue LED and yellow phosphor. In the blue region, there is a peak broadband spectrum. The color temperature of white LED can be adjusted by adjusting the thickness of phosphor coating. In addition, white light can be generated by red (R), green (G), and blue (B) LEDs, where three different red, green, and blue (RGB) colored LEDs are integrated into a module to generate white light by adding colors [18].

Interior lighting is a visual design, since the brightness of indoor light source is affected by human visual perception. When the human eyes observe a radiator from a certain direction, the intensity of human eyes vision depends not only on the radiation intensity of the radiator in that direction but also on the wavelength of the radiation. Human eyes can only see the electromagnetic radiation with a wavelength of 400–700 nm called visible light. Even within the range of visible light, human eyes have different visual sensitivity to different wavelengths of light. They are most sensitive to yellow and green light but less sensitive to red and purple light. They have no visual response to infrared and ultraviolet light other than visible light. In photometry, a function V(λ), called the "visual function" or "spectral optical efficiency," is defined to show the difference in the sensitivity of the human eyes to different wavelengths of radiation, as shown in Figure 12.2b. To the human eyes, the most sensitive wavelength located at 555 nm and the visibility function value is 1 ($V(\lambda@555) = 1$). Therefore, brightness is quantified using illuminance, which is measured by the spectral sensitivity of the human eyes [18]. For indoor light sources, the illuminance (light flux per unit area) is used as the unit of measurement after taking into account

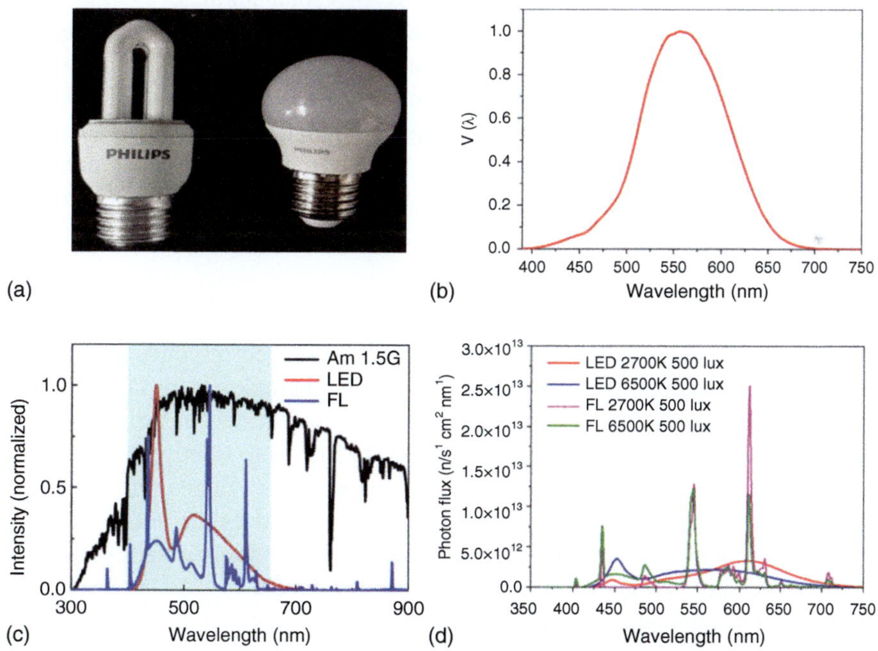

Figure 12.2 (a) The pictures of typical FL and LED light sources. Source: Yinglong Bai. (b) The standard sensitivity curve V(λ) of the human eye, known as the "visual function". Source: Cui et al. [18]/Reproduced with permission of American Chemical Society. (c) The emission spectra of AM 1.5 G sunlight and typical FL, LED light sources. Source: Ding et al. [37]/Reproduced with permission of Royal Society of Chemistry. (d) The radiative spectra of indoor light sources. Source: Cui et al. [18]/Reproduced with permission of American Chemical Society.

the different reactions of human eyes to light at different wavelengths, and the International System of Unit (SI) lux (lx) is adopted. The illumination of indoor lighting ranges from 200 to 1000 lx, which is 100–500 times lower than the standard AM 1.5 G illumination (Figure 12.2c,d) [19].

12.3 Testing System and Parameters for Indoor OPVs

For standard PV cells, the most direct evaluation of their performance is their PCE, which is measured by the short-circuit current density (J_{sc}), open-circuit voltage (V_{oc}), and filling factor (FF) under light conditions, concerning the accuracy of the AM 1.5G standard irradiance spectrum, the irradiance spectrum of the solar simulator, the spectral responsivity of the reference cell, and the spectral responsivity of the test devices. Obviously, solar simulators play an equivalent role with reference cells in the accurate measurements of PCEs. According to the IEC 60904-9 and ASTM E927-10 standards, spectral mismatch, irradiance spatial non-uniformity, and temporal instability are three important parameters to define a solar simulator. Therefore, the best way to accurately evaluate PCEs of OPVs is by choosing a combination of a spectral matched reference cell and a Class AAA solar simulator [39].

So far, there is no standard operating procedures for OPV test under indoor light conditions. In order to accurately evaluate PCE of a PV cells under a certain non-standard illumination condition, the incident light power must be carefully measured. The incident light power was usually evaluated by a lux meter, but this method is not suitable for PV measurement. The illuminance of a light source (E_V, in the unit of lux) can be calculated through the exact emission power spectrum as expressed by the following equation [23]:

$$E_V = K_m \int_0^\infty E_\lambda(\lambda) \cdot V(\lambda) d_\lambda \tag{12.1}$$

where the coefficient K_m is equal to 683 lm W^{-1}, and E_λ corresponds to emission spectra. $V(\lambda)$ is spectral luminosity factor for human photopic vision (Figure 12.2c). A low-precision lux meter does not have the ability to accurately measure the emission power spectrum and the illuminance from the lux meter is only a rough result. Furthermore, the commercially available lux meters are equipped with spherical probes, which are very different to the planar PV cells and thus not suitable for evaluating the incident light power. Therefore, it is particularly urgent to develop a standardized indoor light condition PV cell testing system.

IPV test and characterization system can simulate a variety of indoor light color, light intensity, and spectrum, along with large illumination area. Traceable to light intensity calibrated devices is key for accurately evaluating the indoor PV conversion efficiency [40]. Figure 12.3a,b demonstrates the schematic diagram with LED white light source and sectional view of the IPV cell testing system, respectively. Light intensity range can cover 200 ~ 10 000 lx. I–V characteristic curve measurement can be automatically conducted by source measure units, and the parameters of J_{sc}, V_{oc}, FF, and PCE can be derived according to the target light intensity. As shown

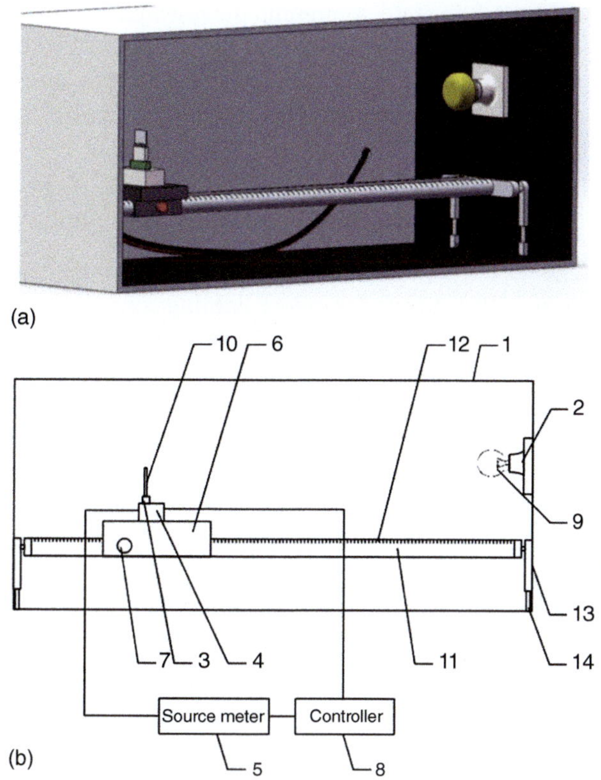

Figure 12.3 (a) Schematic diagram and (b) sectional view of the indoor photovoltaic cell testing system. Source: Zhan'ao Tan.

in Figure 12.3b, the numbers of 1–14 represent the imitated indoor test box, light source base, 3M test clip, 3M Base, digital source meter, sample stage, fasteners, controller, indoor light source, PV cell to be tested, slide rail, scale, bracket, and height adjustment parts, respectively.

12.4 Research Progresses

In order to widely applicate OPV devices in indoor environment, enhancement of the PCE is very important. Considering the intensity of indoor light source is lower and the spectrum is narrower than that of AM 1.5G standard sunlight, the optimization strategies for indoor OPV devices are different to the outdoor countparts. Firstly, the absorption band of the photoactive layer should match the spectrum of the indoor light source within 380–780 nm, and the external quantum efficiency (EQE) should be as high as possible to improve the efficiency of converting incident light into current. Secondly, since lower incident light intensity leads to lower carrier density, the trap assisted charge recombination must be effectively suppressed, which will increase the J_{sc} and FF of OPVs. Thirdly, considering the additional voltage loss (about 0.2 V) due to the significantly reduced light intensity, wide-bandgap materials should be selected as far as possible so that a higher V_{oc} can still be remained under indoor lighting [40]. At present, the research on OPVs for indoor applications mainly focuses on the following aspects. Figure 12.4a–c presents the molecular structures of the donor, fullerene acceptor, and non-fullerene acceptor mentioned in this chapter, and the PV performance of the devices is summarized in Table 12.1.

12.4.1 Fullerene-Based OPVs for Indoor Application

The decrease of V_{oc} under indoor lighting is the main limiting factor affecting the device efficiency under indoor lighting conditions. Increasing the V_{oc} of OPVs and matching the spectrum to improve J_{sc} are the key factors to maximize the performance of devices under indoor conditions. V_{oc} is directly related to the energy level difference between the highest occupied molecular orbital (HOMO) of the donor and the lowest unoccupied molecular orbital (LUMO) of the acceptor.

In 2016, Lee et al. compared the device performance with P3HT, poly[N-9′-heptadecanyl-2,7-carbazole-alt-5,5-(4′,7′di-2-thienyl-2′,1′,3′-benzothiadiazole)] (PCDTBT), and PTB7 as donor and fullerene derivative $PC_{71}BM$ as acceptor under indoor lighting conditions [41]. Among them, PCDTBT:$PC_{71}BM$-based device demonstrates the best performance, P_{max} of 13.9 μW cm^{-2}, PCE of 16.6%, and V_{oc} of 0.72 V under the FL illumination (300 lx). Although the PCE of PTB7:$PC_{71}BM$-based device is higher than that of PCDTBT:$PC_{71}BM$-based device under AM 1.5 G standard sunlight illumination, the PCE of the device under FL is lower due to the narrow bandgap of PTB7. Notably, the bandgap of PCDTBT is 1.9 eV, which matches well with the emission spectrum of FL [41]. In 2019, You et al. compared the IPV properties of PDTBTBz-2F_{anti}, P3HT, poly[(2,6-(4,8-bis(5-(2-ethylhexyl)thiophen-2-yl)-benzo[1,2-*b*:4,5-*b*′]dithiophene))-*alt*-(5,5-(1′,3′-di-2-thienyl-5′,7′-bis(2-ethylhexyl)benzo[1′,2′-*c*:4′,5′-*c*′]dithiophene-4,8-dione))] (PBDB-T), and PTB7, four donor polymers with different interband energies, emphasizing the importance of matching absorption spectrum of photoactive materials with the indoor light source spectrum. Figure 12.5a shows the device structure and thickness of each layer of a typical polymer:fullerene bulk heterojunction (BHJ) cell, and the energy level structures of four different donor polymers PDTBTBz-2Fanti, P3HT, PBDB-T, and PTB7 combined with fullerene acceptor $PC_{71}BM$ are displayed in Figure 12.5b [42]. Among these polymers, the absorption of PDTBTBz-2F_{anti} in the 2 eV wide-bandgap matchs well with the irradiation spectra of LEDs (Figure 12.5c). This spectral matching results in a high PCE of 23.1% for PDTBTBz-2F_{anti}:$PC_{71}BM$ based devices under LED (1000 lx) illumination (Figure 12.5d) [42]. Chen et al. reported a series of small molecular donors with finely regulated energy levels and bandgaps [43]. In the four molecules, the bandgap of 7-(4-(di-p-tolylamino)phenyl)benzo[c][1,2,5]thiadiazole-4-carbonitrile (DTCPB) is 2.04 eV, and the PCE is 6.55% under AM 1.5G 1 sun irradiation. Under FLs (1200 lx), DTCPB based devices can significantly increase the PCE to 16% due to efficiently enhanced harvesting light at visible wavelengths [43]. These results indicate that widegap materials (E_g = ~2 eV) are beneficial to achieve high PCE under indoor conditions.

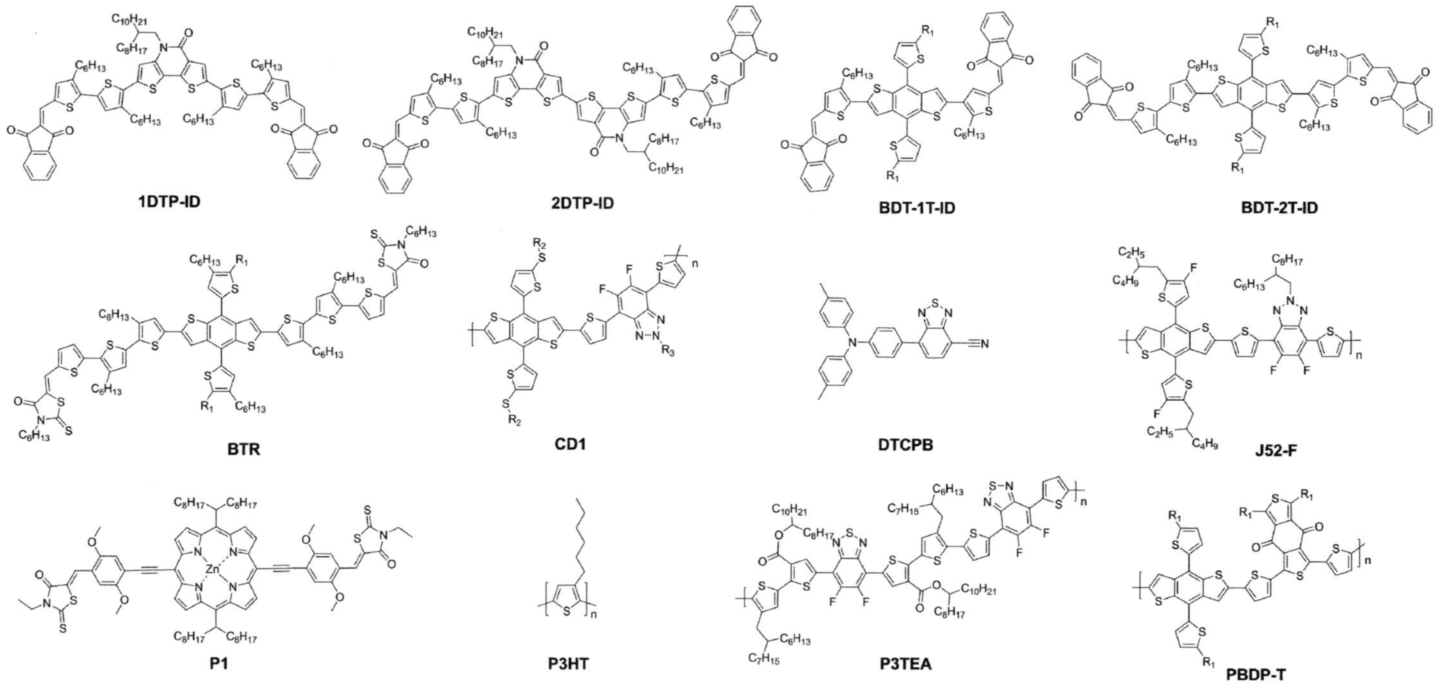

Figure 12.4 The molecular structures of (a) the donors, (b) fullerene acceptors, and (c) non-fullerene acceptors mentioned in this chapter.

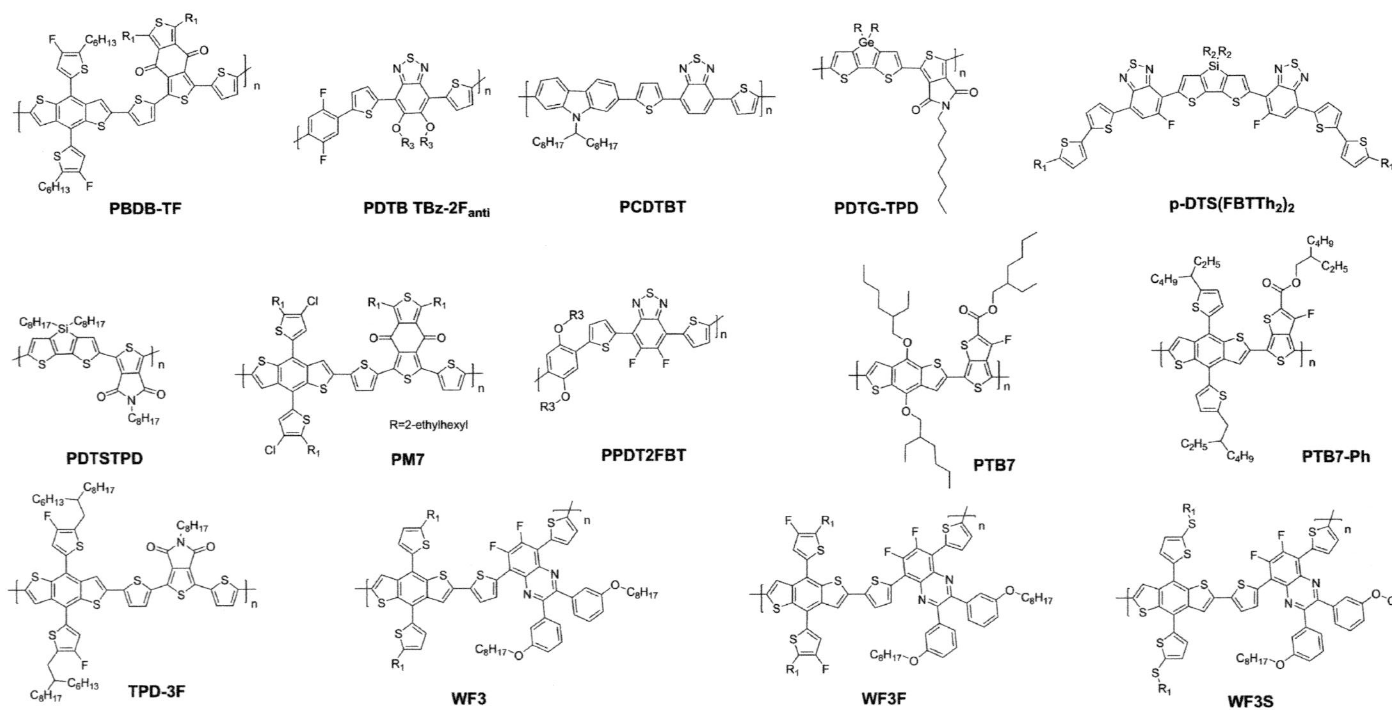

(a)

Figure 12.4 (*Continued*)

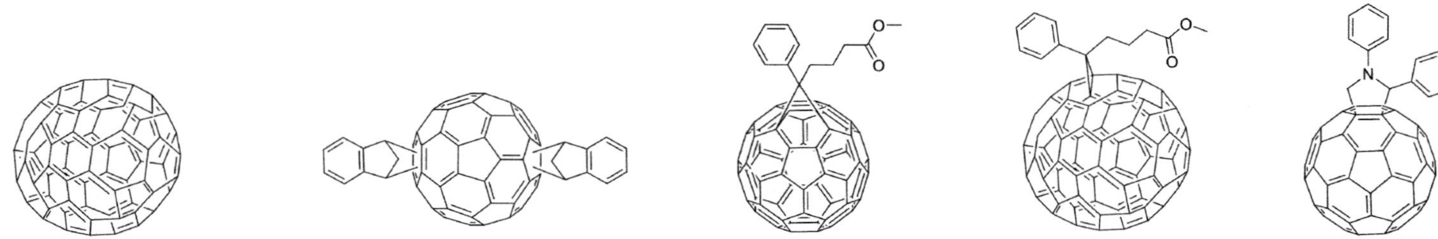

C70 ICBA $PC_{61}BM$ $PC_{71}BM$ PNP

(b)

Figure 12.4 (*Continued*)

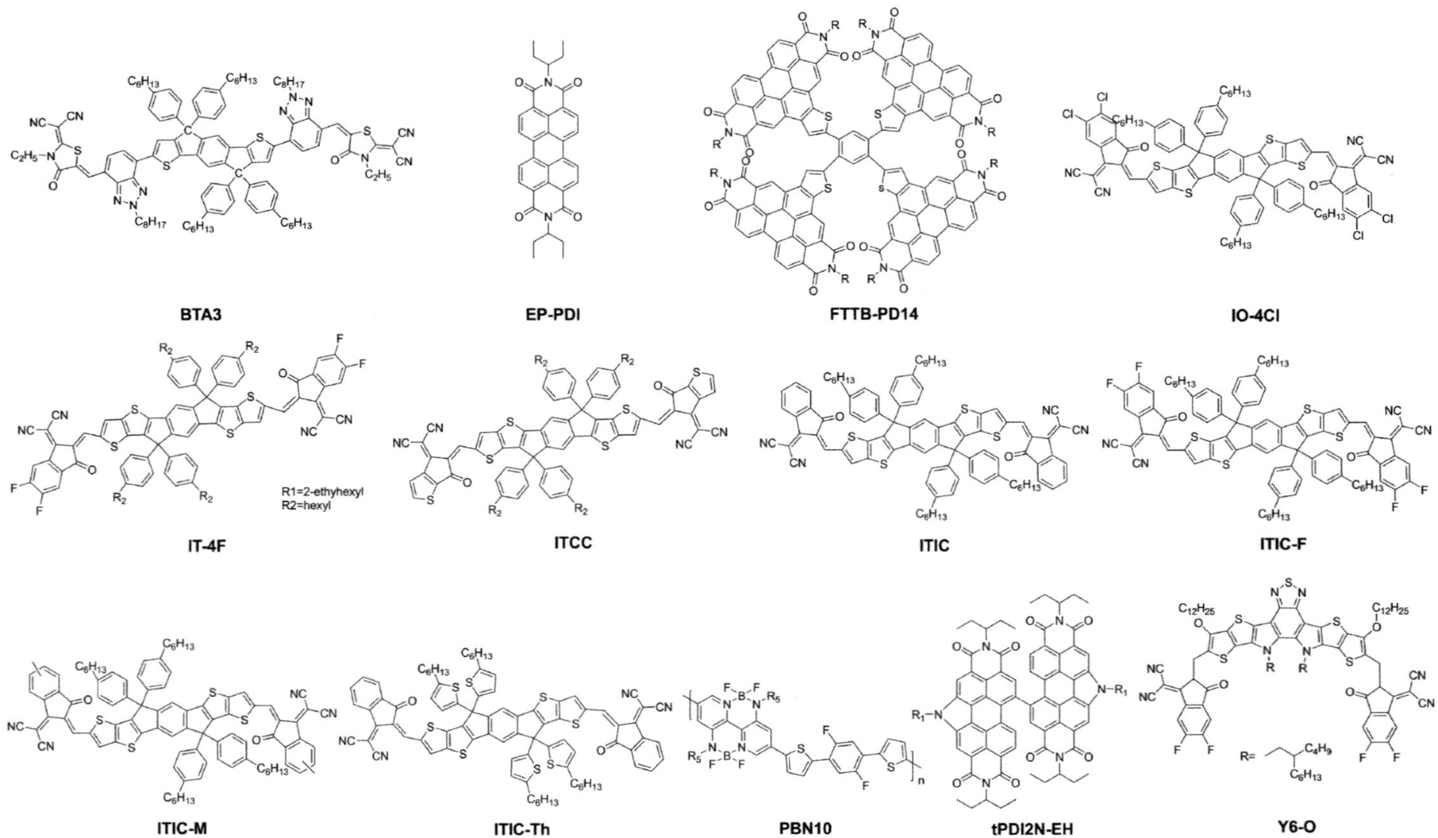

(c)

Figure 12.4 *(Continued)*

Table 12.1 The photovoltaic performance of the devices mentioned in this chapter.

Active layer	Light source and intensity (lx)	Device area (cm^2)	V_{oc} (V)	J_{sc} ($\mu A\,cm^{-2}$)	F (%)	PCE (%)	References
7:BTA3	LED (300)	0.04	1.08	22.1	57.0	15.12	[40]
J52-F:BTA3	LED (300)	0.04	0.96	25.1	64.2	17.18	[40]
J52-F:PM7:BTA3	LED (300)	0.04	1.00	26.5	68.1	20.04	[40]
P3HT:$PC_{71}BM$	FL (300)	0.15	0.41	20.6	56.6	5.80	[41]
PCDTBT:$PC_{71}BM$	FL (300)	0.15	0.72	27.7	69.3	16.60	[41]
PTB7:$PC_{71}BM$	FL (300)	0.15	0.61	28.6	69.5	14.60	[41]
PCDTBT:$PC_{71}BM$	FL (300)	100	4.87	314.0	61.2	11.20	[41]
PDTBTBz-2Fanti:$PC_{71}BM$	LED (1000)	0.1	0.82	112.4	70.4	23.10	[42]
P3HT:$PC_{71}BM$	LED (1000)	0.1	0.50	73.7	71.9	9.40	[42]
PBDB-T:$PC_{71}BM$	LED (1000)	0.1	0.67	90.2	71.3	15.30	[42]
PTB7:$PC_{71}BM$	LED (1000)	0.1	0.57	87.6	69.3	12.30	[42]
DTCPB:C_{70}	FL (200)	0.04	0.66	21.7	54.2	13.37	[43]
DTCPB:C_{70}	FL (400)	0.04	0.69	41.9	53.1	13.24	[43]
DTCPB:C_{70}	FL (600)	0.04	0.70	59.4	65.0	15.52	[43]
DTCPB:C_{70}	FL (800)	0.04	0.72	79.2	64.2	15.78	[43]
DTCPB:C_{70}	FL (1000)	0.04	0.73	99.2	62.8	15.67	[43]
DTCPB:C_{70}	FL (1200)	0.04	0.74	114.8	61.4	14.98	[43]
WF_3:$PC_{71}BM$ w/diphenyl ether (DPE)	LED (500)	0.15	0.57	58.3	64.2	12.83	[44]

Table 12.1 (Continued)

Active layer	Light source and intensity (lx)	Device area (cm^2)	V_{oc} (V)	J_{sc} ($\mu A\ cm^{-2}$)	F (%)	PCE (%)	References
$WF_3S:PC_{71}BM$ w/DPE	LED (500)	0.15	0.61	60.4	65.7	14.32	[44]
$WF_3F:PC_{71}BM$ w/DPE	LED (500)	0.15	0.69	63.6	67.4	17.34	[44]
$P1:PC_{71}BM$ (100 nm)	LED (300)	0.11	0.76	29.6	66.1	19.15	[45]
$P1:PC_{71}BM$ (200 nm)	LED (300)	0.11	0.76	28.2	67.2	18.43	[45]
$P3HT:PC_{61}BM$	LED (1000)	/	0.45	124.0	58.0	9.08	[46]
$P3HT:PC_{61}BM$	FL (1000)	/	0.45	124.0	58.0	9.94	[46]
P3HT:ICBA	LED (500)	/	0.73	50.0	63.0	13.05	[46]
P3HT:ICBA	FL (500)	/	0.73	50.0	62.0	13.76	[46]
PTB7:PNP	LED (200)	0.06	0.57	19.0	67.0	9.50	[14]
BDT-2T-ID:PNP	LED (200)	0.06	0.75	24.2	68.0	16.00	[14]
BDT-1T-ID:PNP	LED (200)	0.06	0.84	19.2	59.0	12.40	[14]
BDT-2T-ID:PNP	LED (200)	9.50	4.20	39.9	66.0	/	[14]
1 DTP-1D:PNP	LED (200)	/	0.69	26.4	68.0	17.70	[47]
2 DTP-1D:PNP	LED (200)	/	0.71	22.4	61.0	13.80	[47]
1 DTP-1D:PNP	LED (200)	9.60	4.05	35.1	/	/	[47]
$BTR:PC_{71}BM$	FL (1000)	0.15	0.79	133.1	75.2	28.10	[48]
PPDT2FBT:ITIC-M	LED (300)	0.09	0.53	20.8	57.0	6.90	[49]
PPDT2FBT:ITIC-F	LED (300)	0.09	0.29	34.8	31.3	3.50	[49]
PPDT2FBT:tPDI2N-EH	LED (300)	0.09	0.79	20.9	49.9	9.00	[49]
PPDT2FBT:ITIC-M	LED (1000)	0.09	0.62	68.5	54.6	7.50	[49]
PPDT2FBT:ITIC-F	LED (1000)	0.09	0.45	85.5	37.6	4.70	[49]
PPDT2FBT:tPDI2N-EH	LED (1000)	0.09	0.84	65.4	50.2	8.90	[49]

CD1:ITIC	FL (1000)	/	0.78	116.0	68.1	17.90	[37]
CD1:PBN-10	FL (1000)	/	1.14	120.0	66.2	26.20	[37]
CD1:ITIC	LED (1000)	/	0.77	107.0	67.5	15.40	[37]
CD1:PBN-10	LED (1000)	/	1.14	105.0	65.4	21.70	[37]
PBDB-TF:ITCC	LED (1000)	1.00	0.96	95.8	72.2	22.0	[50]
PBDB-TF:ITCC	LED (500)	1.00	0.95	47.8	70.6	21.2	[50]
PBDB-TF:ITCC	LED (200)	1.00	0.92	19.2	70.0	20.4	[50]
PBDB-TF:ITCC	FL (500)	1.00	0.94	39.5	73.1	22.0	[50]
PBDB-TF:$PC_{71}BM$	LED (1000)	1.00	0.782	94.1	74.1	18.1	[50]
PBDB-TF:$PC_{71}BM$	LED (500)	1.00	0.76	47.2	72.7	17.2	[50]
PBDB-TF:$PC_{71}BM$	LED (200)	1.00	0.71	18.9	71.3	15.9	[50]
PBDB-TF:IT-4F	LED (1000)	1.00	0.71	113.0	78.0	20.8	[50]
PBDB-TF:IT-4F	LED (500)	1.00	0.69	52.6	75.6	19.6	[50]
PBDB-TF:IT-4F	LED (200)	1.00	0.66	20.8	73.4	18.2	[50]
PTB7:$PC_{71}BM$	LED (500)	0.15	0.60	46.0	54.3	8.90	[51]
PTB7:$PC_{71}BM$:EP-PDI	LED (500)	0.15	0.65	57.0	68.5	15.40	[51]
PTB7-Th:PBDB-T:ITIC-Th:$PC_{71}BM$	LED (1000)	0.11	0.68	102.3	57.5	14.29	[52]
PTB7-Th:PBDB-T:ITIC-Th:$PC_{71}BM$	LED (500)	0.11	0.63	52.1	53.2	10.31	[52]
PTB7-Th:PBDB-T:ITIC-Th:$PC_{71}BM$	FL (1000)	0.11	0.68	110.5	55.5	9.87	[52]
PTB7-Th:PBDB-T:ITIC-Th:$PC_{71}BM$	FL (500)	0.11	0.64	52.8	58.2	13.24	[52]

Table 12.1 (Continued)

Active layer	Light source and intensity (lx)	Device area (cm^2)	V_{oc} (V)	J_{sc} ($\mu A\,cm^{-2}$)	F (%)	PCE (%)	References
PBDB-T:IT-M (100 nm)	LED (500)	1.00	0.88	54.2	75.4	22.8	[53]
PBDB-T:IT-M (300 nm)	LED (500)	1.00	0.86	48.8	74.9	20.0	[53]
PBDB-T:IT-M (500 nm)	LED (500)	1.00	0.85	44.8	71.8	17.2	[53]
PBDB-T:IT-M (1000 nm)	LED (500)	1.00	0.83	42.3	66.3	14.7	[53]
PBDB-T:IT-M (181 nm)	LED (500)	4.00	0.87	52.6	73.2	21.2	[53]
PBDB-TF:IO-4Cl	LED (200)	1.00	1.03	18.2	71.5	22.20	[54]
PBDB-TF:IO-4Cl	LED (500)	1.00	1.07	45.1	76.8	24.60	[54]
PBDB-TF:IO-4Cl	LED (1000)	1.00	1.10	90.6	79.1	26.10	[54]
PBDB-TF:IO-4Cl	LED (200)	4.00	1.02	18.3	74.3	23.00	[54]
PBDB-TF:IO-4Cl	LED (500)	4.00	1.05	45.0	74.8	23.40	[54]
PBDB-TF:IO-4Cl	LED (1000)	4.00	1.07	89.5	75.3	23.90	[54]
TPD-3F:IT-4F	FL (1000)	20.4	3.21	361	70.6	21.8	[55]

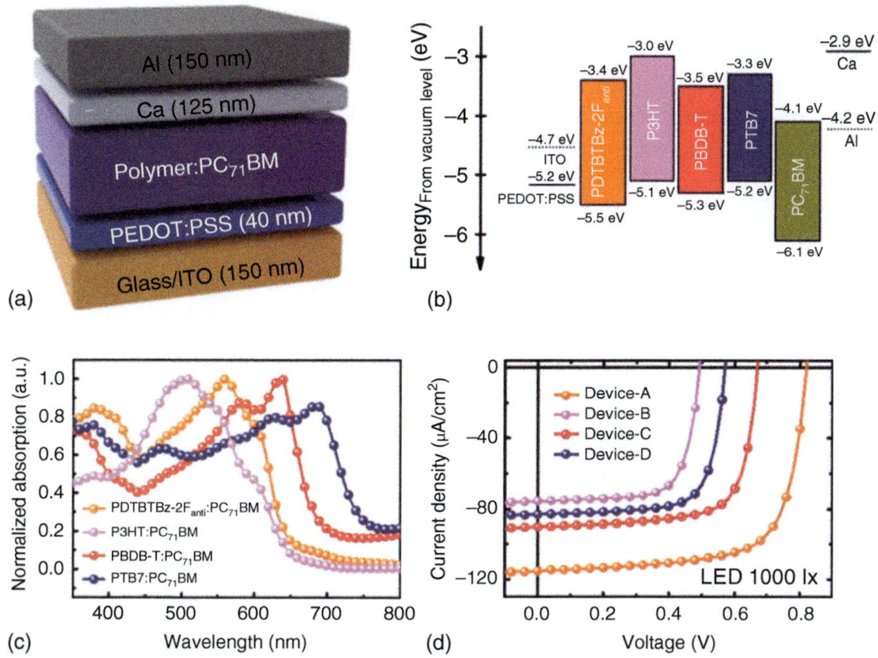

Figure 12.5 (a) The device structure of a typical organic photovoltaic cell for indoor applications. (b) Energy levels of four different donor polymers PDTBTBz-2F$_{anti}$, P3HT, PBDB-T, and PTB7 combined with fullerene acceptor $PC_{71}BM$. (c) Film absorption spectra of four donor polymers blended with fullerene acceptor $PC_{71}BM$. (d) *J–V* curves of corresponding devices under 1000 lx LED illumination. Source: You et al. [42]/Reproduced with permission of John Wiley & Sons.

In order to get a high V_{oc}, the HOMO of the donor must be deep enough. Singh et al. designed benzothiophene with alkyl (WF_3), alkyl sulfide (WF_3S), and fluorothiophene (WF_3F), demonstrating different HOMO levels [44]. The optical bandgaps of all three polymers range from 1.73 to 1.78 eV. Due to the electronegative substituent of fluorine, the HOMO of WF_3F is reduced to −5.62 eV. Thus, the PCE of the device under AM 1.5 G 1 sun and LED (500 lx) illumination reaches 9.44% and 17.3%, with V_{oc} of 0.88 and 0.69 V, respectively [44]. Yin et al. designed a porphyrin-based A–D–A donor material P1, which has a wide and strong absorption from visible light to near-infrared region, thus achieving a high V_{oc} of 0.9 V under 1 sun lighting condition. P1:$PC_{71}BM$ and PCDTBT:$PC_{71}BM$-based devices reach a PCE of 19.2% and 18.7% under LED lighting condition, with V_{oc} of 0.76 and 0.74 V, respectively [45].

Using acceptors with high LUMO levels is another way to increase V_{oc}. P3HT:indene-C60bisadduct (ICBA)-based device demonstrates higher V_{oc} of 0.89 V than that of P3HT:$PC_{61}BM$-based device with V_{oc} of 0.61 V under AM 1.5G 1 sun illumination, due to the higher LUMO level of ICBA (−3.74 eV) than that of $PC_{61}BM$ (−3.90 eV) [46]. Therefore, under TL5 FL (500 lx) illumination conditions, the PCE of P3HT:ICBA-based device reaches 13.8% with a V_{oc} of 0.73 V, significantly higher than P3HT:$PC_{61}BM$-based device with a PCE of 9.59% with a V_{oc} of 0.43 V. Although the J_{sc} of the P3HT:ICBA-based device (50 μA cm^{-2}) is lower than that of the P3HT:$PC_{61}BM$ device (62 μA cm^{-2}), a higher power output (P_{out}) of 22.57 μW cm^{-2} is achieved due to the higher V_{oc} [46]. Arai et al. reported two small molecular donors BDT-2T-ID and BDT-1T-ID, with a deep HOMO of −5.13 and −5.23 eV, respectively, and large E_g being 1.7–1.8 eV [14]. Compared with PTB7-based devices (PCE = 6.7%) under standard 1 sun illumination, BDT-2T-ID and BDT-1T-IDB-based devices show inferior PCE of 6.1% and 3.9%, respectively. However, their V_{oc} (around 1 V) is much higher than PTB7-based device (0.77 V) due to their deeper HOMO levels. Under LED illumination, the PCE of BDT-2T-ID and BDT-1T-ID-based devices significantly increases to 16.2% and 12.9%, respectively, three times higher than that of PTB7-based devices, because of their deep HOMO level results in a larger V_{oc} under LED illumination [14]. They also synthesized small-molecule light absorbers, 1DTP-ID and 2DTP-ID, and evaluated their OPV performances using a binary BHJ system with fullerene acceptor (*N*-phenyl-2-phenyl[60]fulleropyrrolidine [PNP]) under LED illumination with different illuminances (200–10 000 lx) and color temperatures (2900–7500 K) [47]. The 1DTP-ID-based OPVs, tailored for indoor applications demonstrate notably high power outputs of 11.3 μW cm^{-2} with PCE of 19.5% at 200 lx and 64.7 μW cm^{-2} with PCE of 22.4% at 1000 lx [47]. Lee et al. reported that BTR:$PC_{71}BM$-based device demonstrated a PCE of 28.1% under the FL (1000 lx) illumination and the P_{out} value reached 78.2 μW cm^{-2}, which is possible to charge a button-sized battery [48].

12.4.2 Non-fullerene-Based OPVs for Indoor Application

Compared to fullerene-based OPVs, non-fullerene-based OPVs have many advantages in further improving the PCE in indoor lighting condition. Fullerene derivatives have poor absorption in visible band, which can lead to loss of photocurrent. The tunability of their electronic structure is also very limited, which is hard to achieve high V_{oc} of the device. In contrast, non-fullerene acceptors can endow molecules with strong light absorption in the visible region through molecular design, and the energy levels can be easily adjusted to maximize V_{oc}, so many research groups have studied their indoor PV properties.

In 2019, Dayneko et al. fabricated PV devices with the donor polymer PPDT2FBT and four different acceptors, $PC_{61}BM$, 3,9-bis(2-methylene-(3-(1,1-dicyano methylene) – 5-methylindanone)– 5,5,11,11-tetrakis(4-hexylphenyl)-dithieno[2,3-d:2′,3′-d′]-s-indaceno[1,2-b:5,6-b′]-dithiophene (ITIC-M), 3,9-bis(2-methylene-((3-(1,1-dicyanomethylene)-6,7-difluoro)-indanone))-5,5,11,11-tetrakis(4-hexylphenyl)-dithieno[2,3-d:2′,3′-d′]-s-indaceno[1,2-b:5,6-b′]dithiophene (ITIC-F), and $tPDI_2N$-EH using a halogen-free ink formulation and spin-cast or slot-die coating methods. Under a standard illumination condition (AM 1.5G, 1 sun), devices with slot-die coated active layers produce PCE values from 6 to 9%, while the perylene diimide-based acceptor, $tPDI_2N$-EH, exhibits V_{oc} above 1 V. Under varied indoor LED illumination conditions (2800 and 5600 K, 10000–300 lx), a Si reference cell and devices using the PPDT2FBT:ITIC-F blend show large PCE drop with a decrease illumination intensity owing to poor spectral overlap.

PPDT2FBT:ITIC-M-based devices show less of a PCE drop owing to the ITIC-M acceptor having stronger absorption in the visible region. Both PPDT2FBT:$PC_{61}BM$ and PPDT2FBT:tPDI_2N-EH-based devices have minor PCE drop with decreasing the light intensity [49]. Ding et al. demonstrated all-polymer IPVs using a blend of polymer donor (CD1) and polymer acceptor (PBN-10) as the active layer. Owing to the medium bandgap (E_g = 1.93 eV) and proper energy level alignment of the two polymers, the all-polymer IPV cell shows a high PCE of 27.4% with a high V_{oc} of 1.16 V under 1000 lx FL illumination (Figure 12.6a) [37]. The large bandgaps of CD1 (1.93 eV) and PBN-10 (1.95 eV) provide absorption matching the emission spectrum of the indoor lighting source (Figure 12.6c). The HOMO and LUMO levels of CD1 and PBN-10 were −5.81 and −3.42 eV, respectively. A large energy shift between the LUMO of the acceptor and the HOMO of the donor is expected to increase the V_{oc} of the device (Figure 12.6b) [37].

Cui et al. used the wide-bandgap materials consisting of polymer donor poly[1-(5-(4,8-bis(5-(2-ethylhexyl)-4-fluorothiophen-2-yl)-6-methylbenzo[1,2-b:4,5-b′]dithiophen-2-yl) thiophen-2-yl)-5,7-bis(2-ethylhexyl)-3-(5-methylthiophen-2-yl)-4H,8H-benzo[1,2-c:4,5-c′]dithiophene-4,8-dione] (PBDB-TF) and non-fullerene acceptor 3,9-bis(4-(1,1-dicyanomethylene)-3-methylene-2-oxo-cyclopenta[b]thiophen)-5,5,11,11-tetrakis(4-hexylphenyl)-dithieno[2,3-d′:2,3-d′]-s-indaceno[1,2-b:5,6-b′]-dithiophene (ITCC) to absorb the indoor light. They also drilled the applications of fullerene derivative acceptor $PC_{71}BM$ and the low-bandgap non-fullerene acceptor IT-4F in the indoor applications, which aims to compare with the performance of non-fullerene-based OPV devices. As seen from Figure 12.6d, the PBDB-TF has absorbance range from 300 to 680 nm, which matches well with the LED radiative spectrum. Figure 12.6e shows the energy levels of donors and receptors containing PBDB-TF, $PC_{71}BM$, ITCC, IT-4F. The higher V_{oc} of indoor OPV devices may benefit from the low-lying ionization potential of ITCC, but a low V_{oc} and a large V_{oc} loss in the $PC_{71}BM$-based OPV device may be caused by the relatively large electron affinity of $PC_{71}BM$. The PBDB-TF delivers more prominent performance when combined with both fullerene and non-fullerene acceptors. To evaluate the IPV performance based on non-fullerene acceptor under ambient indoor light conditions, they constructed inverted OPV devices with structure of indium tin oxide (ITO) substrate/poly(3,4-ethylenedioxythiophene):poly(styrenesulfonate) (PEDOT:PSS)/active layer/poly[(9,9-bis(3′-((*N*,*N*-dimethyl)-*N*-ethylammonium)-propyl)-2,7-fluorene)-*alt*-2,7-(9,9-dioctylfluorene)] (PFN-Br)/Al, where the active layers are PBDB-TF:$PC_{71}BM$, PBDB-TF:ITCC, PBDB-TF:IT-4F, respectively. As a result, thanks to the higher V_{oc}, the PBDB-TF:ITCC-based OPV cells showed superior PV performance than the PBDB-TF:IT-4F-based devices at the same illumination conditions, and the PBDB-TF:ITCC-based device shows the maximum PCE of 22.0% under indoor light source of 1000 lx (Figure 12.6f). At the same time, the OPV device based on PBDB-TF:IT-4F also realizes a PCE of over 20%, which is also higher than the device based on fullerene acceptor of $PC_{71}BM$ with PCE of 18.1%. High performance of non-fullerene acceptor-based devices mainly benefits from the matching absorption spectrum, low V_{oc} loss, and low charge recombination [50].

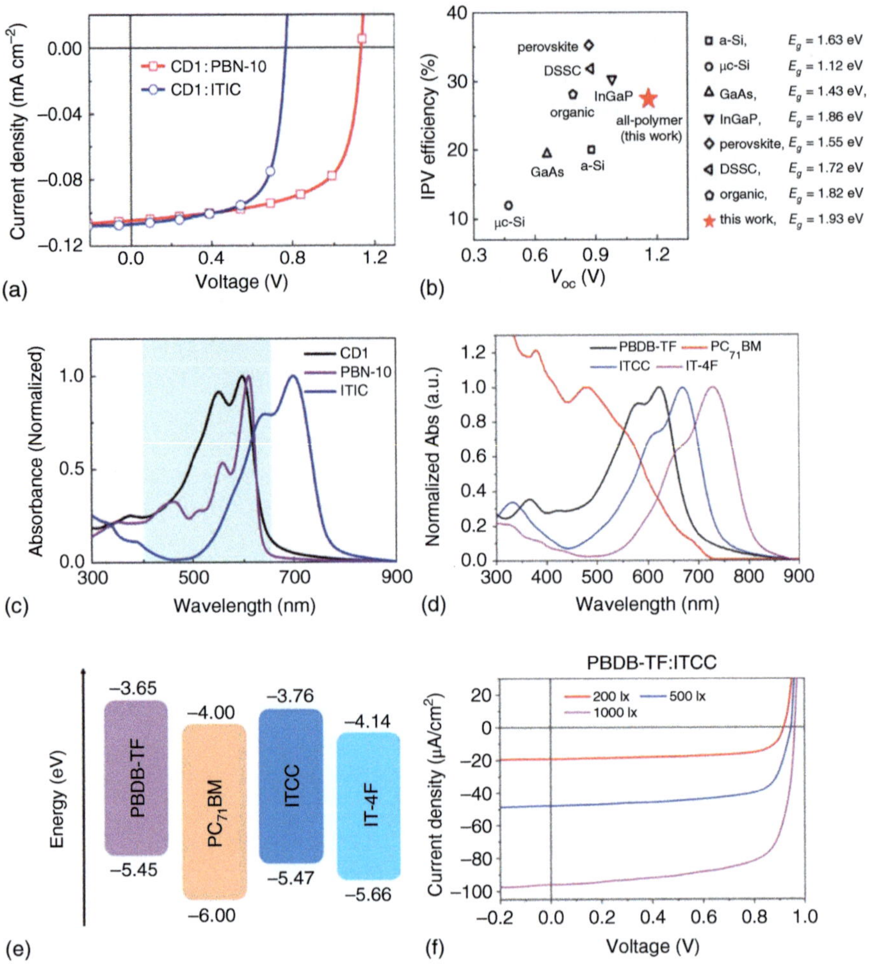

Figure 12.6 (a) *J*–*V* curves of all-polymer OPV under indoor lighting conditions. (b) The relation between PCE and V_{oc} for different types of PVs with different E_g. (c) Absorption spectra of CD1, PFN-10, and ITIC. Source: Ding et al. [37]/Reproduced with permission of Royal Society of Chemistry. (d) Absorption spectra of the donor and acceptors. (e) Energy levels of the donor and acceptors. (f) *J*–*V* curves of PBDB-TF:ITCC-based device under indoor illumination. Source: Cui et al. [50]/Reproduced with permission of John Wiley & Sons.

12.4.3 Multiple Blend OPVs for Indoor Application

In the binary OPVs, limited absorption mainly limits J_{sc}. In order to overcome this limitation, the ternary device structure has been applied to IPVs. The third component in the active layer can not only increase J_{sc} through complementary light absorption but also improve the FF value by adjusting the mixture morphology of BHJ. Singh et al. employed different non-fullerene acceptors which contain amorphous PDI4 and (4,4,9,9-tetrakis(4-hexylphenyl)-4,9-dihydro-s-indaceno

[1,b:5,6-b′]dithiophene-2,7-bis(methanylylidene))bis(1,3-diethyl-2-thioxodihydropyrimidine-4,6(1H, 5H)-dione) (IDT), and the 5,5,11,11-tetrakis(4-hexylphenyl)-5,11-dihydro-s-indaceno [1,2-b:5,6-b′]dithiophene-3, 9-bis(methanylylidene))bis (1,3-diethyl-2-thioxodihydropyrimidine-4,6(1H, 5H)-dione) (IDDT) as a crystalline modulators used in ternary-blend indoor OPV devices. Figure 12.7a shows the normalized absorption spectra of the materials used in photoactive layers containing poly [(5,6-difluoro-2,1,3-benzothiadiazol-4,7-diyl)-alt-(3,3‴-di(2-octyldodecyl)-2,2′,5′,2″,5″,2‴-quaterthiophen-5,5‴-diyl)] (OD), $PC_{71}BM$, PDI2, PDI4, IDT, and IDDT. Benefitting from the addition of non-fullerene acceptors, the OPV device realizes the efficient charge separation and transport, reduced energetic disorder, and the suppression of non-geminate recombination, which is originating from the formation of flatten film surface and smaller π–π stacking distance. To evaluate the IPV performance based on ternary blend films under ambient indoor light conditions, they constructed inverted OPV devices. As a result, the OPV devices with ternary blend films of OD:$PC_{71}BM$:PDI2, OD:$PC_{71}BM$:PDI4, OD:$PC_{71}BM$:IDT, and OD:$PC_{71}BM$:IDDT, showed PCEs of 20.5%, 18.27%, 21.13%, and 19.05%, respectively, higher than the device with binary-blend absorb layer showing PCE of 14.15%, testing under indoor light source (1000 lx, LED lamp) (Figure 12.7b) [51].

Recently, we used wide-bandgap photoactive materials, including the benzotriazole-based polymer donor (J52-F), chlorinated polymer donor (PM7), and A2-A1-D-A1-A2-structured acceptor (BTA3), to construct ternary OPVs for indoor light applications (Figure 12.7c) [40]. Benefitting from the introduction of PM7 as the third component in J52-F:BTA3-based blend, an enhanced PCE of 20.04% with a high V_{oc} of 1.00 V can be obtained under the test conditions with an illumination of 300 lx from a LED lighting source with a color temperature of 3000 K (Figure 12.7d). The excellent device performance is inseparable from the matched spectrum, enhanced light absorption, and the reduced E_{loss}. Meanwhile, the improved charge transport capability and suppressed carrier recombination also play an indelible role [40].

In order to further optimize the absorption of indoor light intensity, the quaternary bulk heterojunction (Q-BHJ) device is also proposed. This structure uses two donors and two acceptors to form the active layer. Nam et al. successfully fabricated Q-BHJ devices with wide absorption range and improved charge transfer performance [52]. Figure 12.7e shows the absorption spectra of PBDB-T, ITIC-Th, PTB7-Th, and $PC_{71}BM$ films. Under AM 1.5 G 1 sun illumination, the PCE of the PBDB-T:$PC_{71}BM$-based device is 7.38%, and the PCE of the PTB7-Th:ITIC-Th-based device is 7.50%. However, the PCE of the quaternary PTB7-Th:PBDB-T:ITIC-Th:$PC_{71}BM$ device is significantly improved to 9.46%, with V_{oc} of 0.81 V, J_{sc} of 19.4 mA cm^{-2}, and FF of 60.3%. The Q-BHJ device also shows better stability, and the encapsulated Q-BHJ device maintains 87.7% of the initial PCE after heating at 60 °C for 1020 minutes, while the PCE of PTB7-Th:ITIC-Th-based device decreases to 54.5% of the initial value. Q-BHJ devices obtained the best PCE of 14.29% at 1000 lx LED illumination (Figure 12.7f), showing great potential to

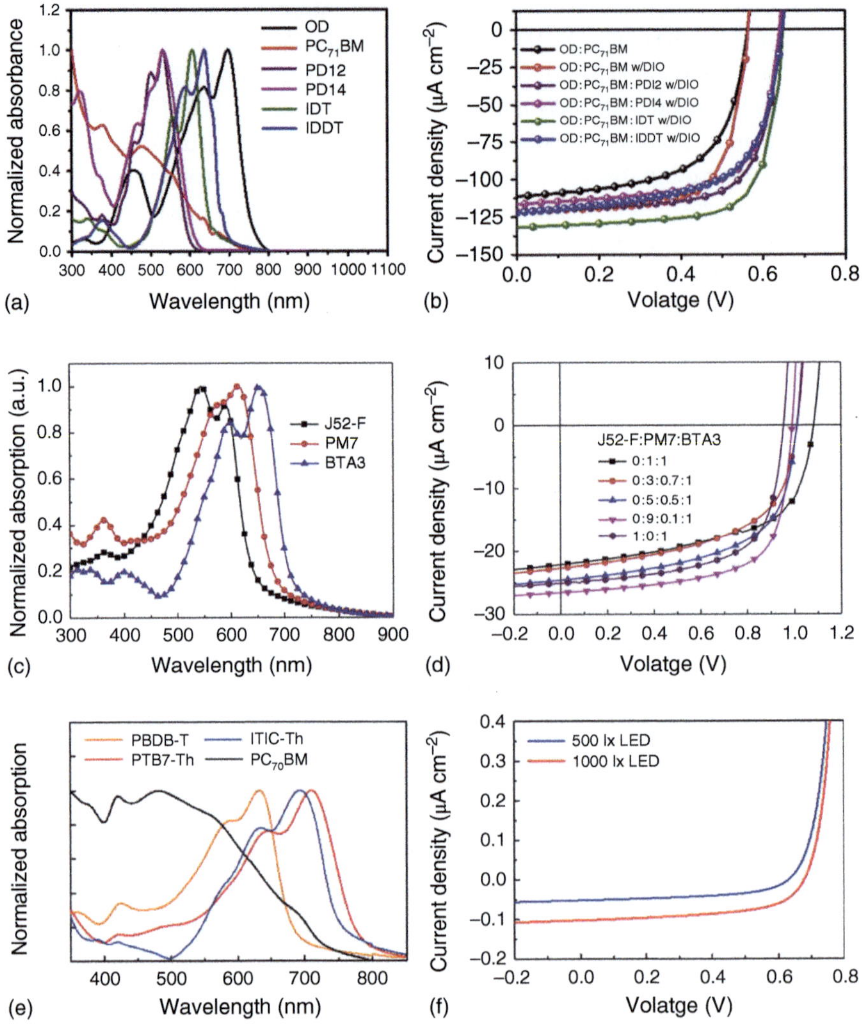

Figure 12.7 (a) Absorption spectra of the donor and acceptors. (b) The *J*–*V* curves of binary and ternary BHJ devices. Source: Singh et al. [51]/Reproduced with permission of Elsevier. (c) Absorption spectra of J52-F, PM7 and BTA3 films. (d) *J*–*V* curves of ternary BHJ device based on J52-F, PM7, and BTA3. Source: Yinglong Bai. (e) Absorption spectra of PBDB-T, ITIC-Th, PTB7-Th and $PC_{71}BM$ films. (f) The *J*–*V* curves of PBDB-T:ITIC-Th:PTB7-Th:$PC_{71}BM$-based quaternion OPV under indoor illumination. Source: Nam et al. [52]/Reproduced with permission of John Wiley & Sons.

achieve high PCE while improving device stability. In addition, they also demonstrated a semi-transparent device with high transparency and high PCE of 15.46% under LED lighting conditions (1000 lx) [52].

12.4.4 Interface Engineering of OPVs for Indoor Application

The interfaces between the electrodes and the photoactive layer significantly influence the efficiency and stability of OPVs. By choosing suitable interfacial materials, the energetic barrier height at the interface could be reduced to form an ohmic contact with less series resistance, inducing high charge collection efficiency of the corresponding electrodes for holes or electrons [56]. The conventional BHJ OPVs are composed of a photoactive layer sandwiched between an ITO/PEDOT:PSS positive electrode and a Ca/Al low-work-function negative electrode. However, both the PEDOT:PSS layer and the low-work-function metal electrode are known to degrade the device lifetime, since the low-work-function metal is susceptible to degradation by oxygen and water vapor, and the PEDOT:PSS layer is acidic and corrosive to the ITO electrode. To enhance the device stability, inverted device structure is proposed, where the ITO electrode is modified with low-work-function buffer layer and high-work-function stable interface layer is inserted between the stable metal electrode and photoactive layer [57]. The choice of electrode modification layer is mainly based on the electron energy level and charge transfer characteristics of the material. At the same time, the modified layer can adjust the roughness of the substrate, prevent the top electrode from diffusing to the active layer, and even block the invasion of oxygen and water. As the thickness of different active layers varies, the interface layer also has an important influence on the performance of the device, because it is directly related to R_s and R_{sh} in the device. Therefore, for IPV, the choice of appropriate interface modification layer plays a crucial role in improving the performance of devices.

Kyaw et al. studied the effect of calcium thickness on the performance of pDTS(FBTTh$_2$)$_2$:PC$_{71}$BM devices under different light intensities [58]. The FF is sensitive to the thickness of a calcium layer between the BHJ layer and the Al cathode, and the highest FF is 73% for 20 nm Ca thickness. Analysis of the *J–V* characteristics at various light intensities provides information on the different recombination mechanisms in the BHJ solar cells with different thicknesses of the Ca layer, finding that the *J–V* curves are dominated by first-order recombination from the short-circuit condition to the maximum power point and evolve to bimolecular recombination in the range of voltage from the maximum power point to the open-circuit condition in the optimized device with a Ca thickness of 20 nm [58].

Lechene et al. reported optimizing the performance of the PCDTBT:PC$_{71}$BM-based IPV device by varying the thickness of the electron transport ethoxylated polyethylene imide (polyethylenimine ethoxylated [PEIE]) layer. Figure 12.8a–c shows the *J–V*, FF, and PCE curves of OPVs with three different PEIE thicknesses. PEIE thickness has little effect on device performance under AM 1.5G 1 sun condition, while it has significant effect on device performance under indoor lighting condition. With thick PEIE, the PCE of device is 7.6% under indoor lighting condition with high R_{sh} of 500 KΩ, while the device with thin PEIE gives poor PCE of 0.7% due to significant leakage current [59]. Owing to the reduced current generation at low light intensity, possessing a sufficiently large shunt resistance associated with a low leakage current is of utmost importance for performance maximization. Goo and colleagues also reported optimizing P3HT:ICBA-based device performance by changing PEIE thickness [61]. The PV properties under different lighting condition were optimized by controlling the thickness of the PEIE layers, and the electrical, optical, and surface properties of the PEIE-modified ITO were investigated. Inverted OPVs based on P3HT:ICBA with an 8.5-nm-thick PEIE layer yielded a maximum PCE of 13.9% under an LED light with a luminance of 500 lx. Meanwhile, the same device showed poor performance with a PCE of 3.1% under 1 sun illumination owing to the insulating nature of PEIE [61]. Shin et al. conducted an in-depth study on electron-collecting layers comprising PEIE modification, ZnO nanoparticles (NPs), or a combination of the ZnO NPs and PEIE modification [62]. While poor PV behavior was observed for P3HT:ICBA-based OPVs with the ZnO NPs owing to the relatively high work function of the electron collection layer (ECL) under LED light in the absence of UV light, outstanding indoor performance with PCE of 14.1% under 1000-lx LED lamp was achieved by OPVs with PEIE modification as the PEIE contributed to work function reduction of the ECL [62].

Saeed et al. demonstrated the IPV performance of P3HT:ICBA-based flexible OPVs with PEDOT:PSS-based electrodes in comparison with reference OPVs with ITO electrodes [63]. The OPV with PEDOT:PSS (PH 1000) electrode produced an average PCE of 11.1%, whereas the reference device with the ITO electrode exhibited a PCE of 8.8% under a 500-lx LED illumination. The flexible OPVs also exhibited superior flexibility over reference OPVs by retaining their PCE up to 85% even after 500 bending cycles in air, whereas the reference OPVs declined to 80% of their PCE under the same bending conditions [63]. Lee et al. reported the indoor performance of flexible OPVs utilizing quasi-amorphous ZnO/Ag/ZnO as the transparent conducting electrode [64]. A ZnO/Ag/ZnO electrode with specific thickness values of 40/9/50 nm provides excellent transparent conducting electrode properties with transmittances up to 92% in the visible region, a sheet resistance of 4.8 Ω per square, and a root-mean squared surface roughness value of 2.1 nm. In addition, the micro-cavity effect and quasi-amorphous structural properties of the ZnO/Ag/ZnO electrode allow further enhanced light absorption and mechanical stability, respectively. P3HT:ICBA-based inverted OPVs with the ZnO/Ag/ZnO (40/9/50 nm) electrode yield an averaged PCE of 12.3% under a LED lamp with a luminance of 500 lx, which is 20% greater than the PCE value of the reference device with ITO electrode [64]. Ma et al. studied the effect of ECLs on OPV performance under the indoor low-light intensity [60]. Perylene-3,4:9,10-bis(dicarboximide) (PDI)-NO and PFN were selected as the electron transport layer (ETLs) and used in the two non-fullerene systems, P3TEA:FTTB-PDI4 and PM6:Y6-O. Under AM 1.5G 1-sun conditions, the device based on PM-6:Y6-O with perylene diimides with amino N-oxide as the terminal substituent (PDI-NO) and PFN exhibits very similar performances. However, under low-light intensity, however, the PFN-based devices suffer from much reduced performance, whereas PDI-NO can still maintain high efficiencies (Figure 12.8d). Devices based on PDI-NO and PM6:Y6-O can achieve high PCEs up to 31% under indoor lighting conditions. Further analysis shows that PDI-NO with a deep HOMO-based devices exhibits significantly lower leakage current and trap-assisted recombination under low-light intensity range (Figure 12.8e) [60].

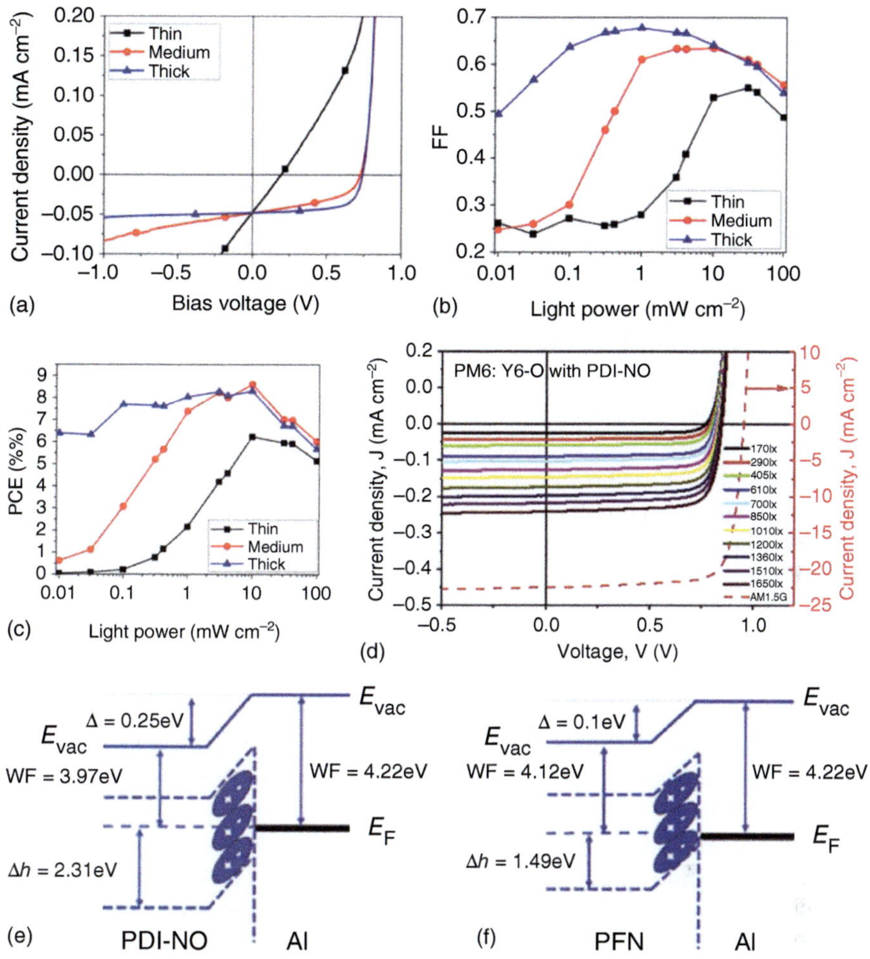

Figure 12.8 (a) $J-V$, (b) FF, and (c) PCE curves of OPVs with three different PEIE thicknesses. Source: Lechêne et al. [59]/Reproduced with permission of Elsevier. (d) Photovoltaic performance of PM6:Y6-O-based OPVs with PDI-NO ETL under different illumination conditions. (e) Energy level alignment at photoactive layer/Al interface with different ETLs. Source: Ma et al. [60]/Reproduced with permission of Elsevier.

12.4.5 Thick Film OPVs for Indoor Application

Because organic semiconductors have lower carrier mobility and shorter exciton diffusion length than inorganic semiconductors, the thicker active layer of 300–1000 nm often hinders the improvement of device PCE. Therefore, most of the previously reported high PCE results were obtained in active layers of 100–200 nm thickness. Although the absorption coefficient of organic semiconductors is much higher than that of inorganic semiconductors, and the active layer about 100 nm thick can absorb most of the incident light for OPVs, it is still challenge to fabricate such thin film with good uniformity by printing technology to ensure high manufacturing yields. Small et al. reported that space charge accumulation is one of the main reasons for the PCE reduction for polydithienogermole-thienopyrrolodione (PDTGTPD):$PC_{71}BM$-based devices with thickness >200 nm [65]. Vincent et al. proposed the OPV optimization design rule based on P3HT:ICBA, which realized the efficient absorption of indoor lighting sources [66]. P3HT:ICBA was chosen as the active layer because of its high overlap with the LED lighting source spectrum. In their experiments, they observed that the thickness of the optimal active layer varies with the light intensity. Under AM 1.5 G illumination, the device with a 130 nm thick active layer has a PCE of 4.2%, and under 1000 lx LED illumination, a device with a 230 nm thick active layer has a PCE of 5.4%. As the thickness of the active layer increases from 80 to 400 nm, the leakage current of the device decreases [66].

Shin and colleagues studied the PV characteristics of PPDT2FBT:$PC_{71}BM$ blended devices under indoor lights with varying photoactive layer thickness (120–870 nm) [67]. The well-matched absorption of photoactive layer and irradiance spectrum of indoor LED light substantially improved the indoor light harvesting (Figure 12.9a). As shown in Figure 12.9b, the PPDT2FBT:$PC_{71}BM$ device showed the best PCE of 16% and the maximum power of 44.8 $\mu W\,cm^{-2}$ under 1000 lx LED light, and the device with an 870-nm-thick photoactive layer still exhibited an excellent performance with a 12.5% PCE [67]. The indoor devices showed the noticeably enhanced thickness tolerance in the J_{sc} and FF values due to much lower series resistance/parallel resistance (R_s/R_p) ratio with a thick photoactive layer under the LED condition. These results indicate that the well-matched absorption with the irradiance spectrum and the large R_p associated with thick photoactive layer are prerequisite for achieving efficient indoor OPVs with high J_{sc} and FF values [67].

Liu et al. systematically studied the effect of R_s on the PBDB-TF:3,9-bis(2-methy lene-(3-(1,1-dicyanomethylene)– 5-methylindanone)– 5,5,11,11-tetrakis(4-hexylphenyl)-dithieno[2,3-d:2′,3′-d′]-s-indaceno[1,2-b:5,6-b′]-dithiophene (IT-M)-based device by introducing external R_s and tuning active layer thickness from 100 nm to 1 μm [53]. Compared with the AM 1.5G condition, the efficiencies were observably improved under low light, regardless of enlarging external R_s or increasing active layer thickness as shown in Figure 12.9c, indicating that the OPV cells under low light show dramatical restrained recombination processes, impressive exciton dissociation probability, and longer carrier lifetime differing with AM 1.5G condition. The optical modeling of the energy distribution for OPV cells with varied PBDB-TF:IT-M thicknesses was carried out, and the main signals are within 250 nm near the PEDOT:PSS side for 1 μm-thick OPV cell (Figure 12.9d), indicating that photons are completely absorbed within 250 nm [53].

12.4.6 Large-Area OPVs for Indoor Application

Upscaling of OPV cell has been a concern for outdoor applications due to the high R_s from large-area modules causing lower FF. However, for indoor applications, R_s is less critical. Instead R_p is more critical to the device performance. Via optimizing the active layers to have a good match with the indoor light source, 1 cm^2 OPV cells

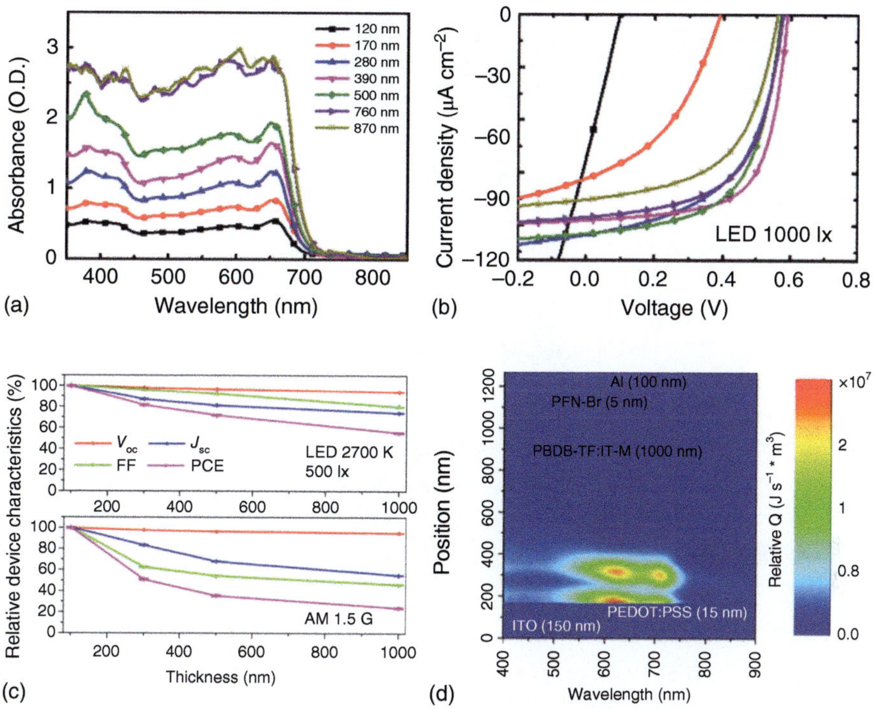

Figure 12.9 (a) Absorbance spectral changes of PPDT2FBT:$PC_{71}BM$ with varying photoactive layer thickness. (b) *J–V* characteristics of the PPDT2FBT:$PC_{71}BM$-based inverted OPVs with various photoactive layer thicknesses under 1000-lx LED. Source: Shin et al. [67]/Reproduced with permission of Elsevier. (c) The photovoltaic characteristics (V_{OC}, J_{SC}, FF, and PCE) with varied PBDB-TF:IT-M thicknesses under LED irradiation and AM 1.5G condition. (d) The simulated energy distribution in the OPV cells with 1 μm PBDB-TF:IT-M. Source: Liu et al. [53]/Reproduced with permission of Elsevier.

were fabricated and a top PCE of 22% under 1000 lx LED (2700 K) illumination was demonstrated by Cui et al. [54]. The best device showed very good stability under continuous illumination after 500 hours with over 95% of the original PCE maintained. They also fabricated PBDB-TF:IO-4Cl based OPV devices, and the PCE of the devices with area of 1 cm^2 (spin coating) and 4 cm^2 (squeegee coating) reached 26.1% and 23.9%, respectively, under 2800 K 1000 lx LED illumination [54]. The device also showed good stability under continuous indoor lighting condition, and maintained its PCE even after 1000 hours.

Toward real applications, OPV modules are also fabricated. As a demonstration, Lee et al. fabricated an 8-pixel module by spin coating PCDTBT:$PC_{71}BM$ blend solution on a substrate of 14 cm by 14 cm with an active area of 100 cm^2, as shown in Figure 12.10a [41]. The module generates J_{sc} of 314 μA, V_{oc} of 4.87 V, and FF of 61.2% under 300 lx. The R_p is even higher than that of the small-area devices which is probably due to an averagely thicker active layer prepared by spin coating a large film. The series resistance is increased compared with that of the small-area devices due

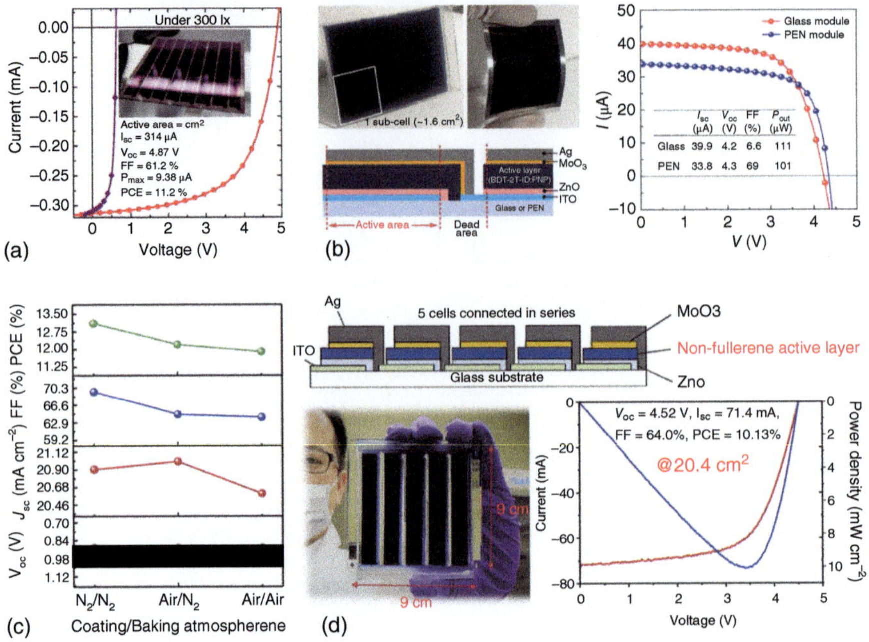

Figure 12.10 (a) The J–V curves of eight pixels connected in series (red circle) and a pixel (purple diamond), respectively, of the PCDTBT:$PC_{71}BM$ module. Source: Lee et al. [41]/with permission of AIP Publishing LLC. (b) Photographs of the BDT-2T-ID-based six series-connected modules fabricated on a rigid glass substrate (left) and on a flexible PEN substrate (right), device structure and corresponding devices performance. Source: Arai et al. [14]/with permission of American Chemical Society. (c) Photovoltaic parameters of TPD-3F-based devices tested under AM 1.5G solar irradiation at an intensity of 100 mW cm^{-2}. (d) Structure and J–V characteristics of the TPD-3F-based OPV module. Source: Liao et al. [55]/Reproduced with permission of ELSEVIER.

to longer ITO pathway throughout the 8-pixel. Although the active area is much bigger than the small-area device, the FF does not drop much [41]. Arai et al. fabricated OPV energy-harvesting modules utilizing the BDT-2T-ID:PNP photoactive layer, with a total active area of 9.5 cm^2 by connecting in series six subcells each with an area of ~1.6 cm^2, as shown in Figure 12.10b [14]. The OPV modules were prepared using both a rigid glass substrate and a flexible polyethylene naphthalate (PEN) substrate, by adopting the same structure used for the small-area devices. V_{oc} of both OPV modules scaled ideally with the number of the series-connected cells, achieving high V_{oc} of 4.2–4.3 V at 200 lx. The average V_{oc} for each subcell can be estimated as 0.70–0.72 V, indicative of a low degree of current leakage in these indoor OPV modules. The upscaling from 6 mm^2 for the small-area cells to 9.5 cm^2 for the practical modules caused only a negligible variation in FF from 67% to 66%–69%, resulting in considerably high maximum power outputs of up to 111 μW (PCE ≈ 15%) at 200 lx [14]. Liao et al. have synthesized and characterized a new series of benzo[1,2-b:4,5-b′]dithiophene(BDT)-2,5-dithienyl-thieno[3,4-c]pyrrole-4,6-dione (TPD) donor polymers and investigated alkyl chain length and fluorination effects on their BHJ-OPV performance with the small-molecule non-fullerene acceptor

IT-4F [55]. The TPD:IT-4F cell performance exhibited excellent tolerence to variations to polymer molecular weight, polydispersity, fabrication environment (ambient or N_2), and film deposition technique (spin coating or blade coating). Figure 12.10c shows the PV parameters of TPD-3F-based devices tested under AM 1.5G solar irradiation at an intensity of 100 mW cm^{-2}. They also demonstrated a large-area module (active area >20 cm^2, Figure 12.10d) with a PCE of 10.4% using a non-fullerene blend TPD-3F:IT-4F. This module also delivers a power of 40 μW cm^{-2} (PCE = 22%) under 1000 lx FLs indoor lighting, and the PCEs of 12%–14% were achieved for blends processed in ambient and/or without halogenated solvent. Compared with the PCEs achieved for the small cells (0.04 cm^2, 13% PCE), the present large module exhibits only a 20% cell-to-module loss.

12.5 Summary and Prospective

In conclusion, among various potential applications of OPVs, it seems more feasible to use them in mild indoor environments as power generation because of superior PCE and stability over outdoor AM 1.5 G 1 sun conditions. Commonly used indoor light sources (FLs and LEDs) have narrower emission spectra (400–700 nm) with lower intensity (200–1000 lx) as compared to the solar spectrum. Highly tunable bandgaps and energy levels, high absorption coefficients, and small leakage currents under dim light make OPVs promising candidates for indoor applications. The future large demand of sustainable energy source for low-power consumption applications due to the rapid growth of the IoTs will greatly stimulate the growth of IPV technologies. To commercialize the OPVs for indoor application, the key issues of efficiency, stability, and cost should be carefully considered.

For further improving the efficiency of organic IPVs, the molecular structures and the bandgaps of donor and acceptor materials should be finely tuned to be well matched with specific indoor light sources, especially for LEDs, which are more cost-effective and energy-saving than FL sources. Simultaneously, the energy level offsets between the donor and acceptor materials to ensure high V_{oc} and minimal energy loss as well as formation of appropriate interpenetrating network with suitable domain sizes in active layer should also be considered during material design. Fullerene-based acceptors are dominant in IPV applications due to their unique features of well-matched absorption and superior transport properties. However, the emergence of non-fullerenes acceptors may blaze a trail for promoting PCE of IPVs. Besides, ternary and multiple compositions are also helpful to finely optimize the morphology of blend film for further improving the performance. Furthermore, from viewpoint of indoor OPV devices, a high R_p is essential for high performance. Increasing the thickness of the photoactive layer and interface engineering with introducing the charge transport layer are effective strategies to enhance R_p with suppressed leakage currents [54]. In addition, because the R_s effect is not significant in indoor OPVs, it is easy to scale up for modules with high efficiency. By careful optimization, an over 40% PCE can be expected for organic IPVs, assuming EQE reaches 90% above the gap in the emission spectrum of the indoor light with energy loss of 0.50 eV [54].

Stability of the organic IPVs is another critical factor for commercialization, although the stability of organic IPVs should be better theoretically due to lower light intensity and mild working environment absence of UV light [55]. The target market of organic IPV is the replacement of small batteries powering IoT sensors to greatly reduce the maintenance of these wireless nodes. Therefore, more than 10 years lifetime with at least 80% of the initial efficiency is required for organic IPV devices to replace the battery market for the IoTs. Furthermore, the power output of the device should be around 30 $\mu W\,cm^{-2}$ under radiance of 500 lx with PCE above 20% for the targeted applications [17]. Many approaches, such as introducing photo-stable photoactive materials, constructing inverted device structure, using stable interfacial layers and metal electrodes, controlling phase-separated morphology, and involving physical encapsulation, have been proven to be useful for improving the stability of the OPVs [68].

Steady reductions in the production costs of individual cells or modules will speed up the commercialization process of organic IPVs. The overall cost of an OPV module mainly includes materials, capital, and labor, and the overall costs depend primarily on the materials cost, especially on the cost of photoactive materials [69, 70]. Therefore, the synthetic accessibility of photoactive materials should be as low as possible for cost cutting with less synthetic steps and scalable purification processes.

References

1 Collins, S. and Bell, G.J.N. (2004). *Nature* 431: 566.
2 Rogelj, J., Elzen, M.D., Höhne, N. et al. (2016). *Nature* 534: 631.
3 Scott, V., Haszeldine, R.S., Tett, S.F., and Oschlies, A. (2015). *Nat. Clim. Change* 5: 419.
4 Ashina, S., Fujino, J., Masui, T. et al. (2012). *Energy Policy* 41: 584.
5 Conibeer, G. (2017). *Mater. Today* 10: 42.
6 Jardine, C.N., Conibeer, G.J., and Lane, K. (2001). *EU PVSEC 2001* 5: 149.
7 Yan, N., Zhao, C., You, S. et al. (2020). *Chinese Chem. Lett.* 31: 643.
8 Pierro, M., Bucci, F., and Cornaro, C. (2015). *Prog. Photovoltaics* 23: 443.
9 Freitag, M., Teuscher, J., Saygili, Y. et al. (2017). *Nat. Photonics* 11: 372.
10 Radue, C., Van Dyk, E., and Macabebe, E. (2009). *Thin Solid Films* 517: 2383.
11 Hauch, J.A., Schilinsky, P., Choulis, S.A. et al. (2008). *Sol. Energy Mater. Sol. Cells* 92: 727.
12 Cueto, J.D., Rummel, S., Kroposki, B. et al. (2008). *2008 33rd IEEE Photovoltaic Specialists Conference.*
13 Fakharuddin, A., Jose, R., Brown, T.M. et al. (2014). *Science* 7: 3952.
14 Arai, R., Furukawa, S., Hidaka, Y. et al. (2019). *ACS Appl. Mater. Inter.* 11: 9259.
15 Gubbi, J., Buyya, R., Marusic, S., and Palaniswami, M. (2013). *Future Gener. Comput. Syst.* 29: 1645.
16 Atzori, L., Iera, A., and Morabito, G. (2010). *Comput. Networks* 54: 2787.
17 Mainville, M. and Leclerc, M. (2020). *ACS Energy Lett.* 5: 1186.
18 Cui, Y., Hong, L., and Hou, J. (2020). *ACS Appl. Mater. Inter.* 12: 38815.

19 Aoki, Y. (2017). *Org. Electron.* 48: 194.
20 Batista, D., Oliveira, L.B., Paulino, N. et al. (2019). *Sensors* 19: 1803.
21 Venkateswararao, A., Ho, J.K.W., So, S.K. et al. (2020). *Mater. Sci. Eng. R* 139: 100517.
22 Mathews, I., Kantareddy, S.N., Buonassisi, T., and Peters, I.M. (2019). *Joule* 3: 1415.
23 Minnaert, B. and Veelaert, P. (2014). *Energies* 7: 1500.
24 Sacco, A., Rolle, L., Scaltrito, L. et al. (2013). *Appl. Energy* 102: 1295.
25 Yue, X., Kiely, J., Gibson, D., and Drakakis, E.M. (2019). *IEEE Trans. Ind. Electron* 67: 2411.
26 C.M.F. Carvalho, N.F.S.V. Paulino. (2015). Springer
27 Wang, W.S., O'Donnell, T., Wang, N. et al. (2008). *ACM J. Emerg. Technol. Comput. Syst.* 6: 1.
28 Hou, X., Wang, Y., Lee, H.K.H. et al. (2020). *J. Mater. Chem. A* 8: 21503.
29 Cojocaru, L., Uchida, S., Tamaki, K. et al. (2017). *Sci. Rep.* 7: 1.
30 Yue, X., Kauer, M., Bellanger, M. et al. (2017). *IEEE Internet Things* 4: 2092.
31 G. Smart, J. Atkinson, J. Mitchell, M. Rodrigues and Y. Andreopoulos. (2016). International Conference on Telecommunications IEEE 1–6.
32 Cutting, C.L., Bag, M., and Venkataraman, D. (2016). *J. Mater. Chem. C* 4: 10367.
33 Bakytbekov, A., Nguyen, T.Q., Li, W. et al. (2020). *Energy Sci. Eng.* 8: 3883.
34 Freunek, M., Freunek, M., and Reindl, L.M. (2013). *IEEE J. Photovoltaics* 3: 59.
35 Çilingiroğlu, U., Tar, B., and Özmen, Ç. (2014). *IEEE Trans. Circuits Syst. Regul. Pap.* 61: 2491.
36 Mathews, I., Kantareddy, S.N.R., Sun, S. et al. (2019). *Adv. Funct. Mater.* 29: 1904072.
37 Ding, Z.C., Zhao, R.Y., Yu, Y.J., and Liu, J. (2019). *J. Mater. Chem. A* 7: 26533.
38 Biswas, S. and Kim, H.J.P. (2020). *POLYMERS-BASEL* 12: 1338.
39 Ye, L., Zhou, C., Meng, H. et al. (2014). *J. Mater. Chem. C* 3: 564.
40 Bai, Y., Yu, R., Bai, Y. et al. (2021). *Green Energy Environ.* 6: 920.
41 Lee, H.K.H., Li, Z., Durrant, J.R., and Tsoi, W.C. (2016). *Appl. Phys. Lett.* 108: 253301.
42 You, Y.J., Song, C.E., Hoang, Q.V. et al. (2019). *Adv. Funct. Mater.* 29: 1901171.
43 Chen, C.H., Ting, H.C., Li, Y.Z. et al. (2019). *ACS Appl. Mater. Inter.* 11: 8337.
44 Singh, R., Chochos, C.L., Gregoriou, V.G. et al. (2019). *ACS Appl. Mater. Inter.* 11: 36905.
45 Yin, H., Chen, S., Cheung, S.H. et al. (2018). *J. Mater. Chem. C* 6: 9111.
46 Yang, S.S., Hsieh, Z.C., Keshtov, M.L. et al. (2017). *Sol. RRL* 1: 1700174.
47 Arai, R., Furukawa, S., Sato, N., and Yasuda, T. (2019). *J. Mater. Chem. A* 7: 20187.
48 Lee, H.K.H., Wu, J., Barbé, J. et al. (2018). *J. Mater. Chem. A* 6: 5618.
49 Dayneko, S.V., Pahlevani, M., and Welch, G.C. (2019). *ACS Appl. Mater. Inter.* 11: 46017.
50 Cui, Y., Yao, H., Zhang, T. et al. (2019). *Adv. Mater.* 31: 1904512.
51 Singh, R., Shin, S.C., Lee, H. et al. (2019). *Chem. Eur. J.* 25: 6154.

52 Nam, M., Noh, H.Y., Cho, J. et al. (2019). *Adv. Funct. Mater.* 29: 1900154.
53 Liu, J., Cui, Y., Zu, Y. et al. (2020). *Org. Electron.* 85: 105798.
54 Cui, Y., Wang, Y., Bergqvist, J. et al. (2019). *Nat. Energy* 4: 768.
55 Liao, C.Y., Chen, Y., Lee, C.C. et al. (2020). *Joule* 4: 189.
56 Wang, F., Tan, Z.A., and Li, Y. (2015). *Energy Environ. Sci.* 8: 1059.
57 Tan, Z., Zhang, W., Zhang, Z. et al. (2012). *Adv. Mater.* 24: 1476.
58 Kyaw, A.K.K., Wang, D.H., Gupta, V. et al. (2013). *ACS Nano* 7: 4569.
59 B.P. Lechêne, M. Cowell, A. Pierre, J.W. Evans, P.K. Wright, A. Arias. (2016). 26, 631.
60 Ma, L.K., Chen, Y., Chow, P.C.Y. et al. (2020). *Joule* 4: 1486.
61 Goo, J.S., Shin, S.C., You, Y.J., and Shim, J.W. (2018). *Sol. Energy Mater Sol. Cells* 184: 31.
62 Shin, S.C., You, Y.J., Goo, J.S., and Shim, J.W. (2019). *Appl. Surf. Sci.* 495: 143556.
63 Saeed, M.A., Kim, S.H., Lee, S.Y., and Shim, J.W. (2020). *Thin Solid Films* 704: 138006.
64 Lee, B.R., Goo, J.S., Kim, Y.W. et al. (2019). *J. Power Sources* 417: 61.
65 Small, C.E., Tsang, S.W., Chen, S. et al. (2013). *Adv. Energy Mater.* 3: 909.
66 Vincent, P., Shin, S.C., Goo, J.S. et al. (2018). *Dyes Pigm.* 159: 306.
67 Shin, S.C., Koh, C.W., Vincent, P. et al. (2019). *Nano Energy* 58: 466.
68 Hsiao, P., Hung, W., Chen, Y. et al. (2020). *Renewable Energy* 152: 67.
69 Miranda, B.H.S., Corrêa, L.d.Q., Soares, G.A. et al. (2021). *Sol. Energy* 220: 343.
70 Lee, J.H., You, Y.J., Saeed, M.A. et al. (2021). *NPG Asia Mater.* 13: 43.

13

Interfacial Design for Efficient Organic Solar Cells

Yao Liu[1], *Menglan Lv*[2], *and Shengjian Liu*[3]

[1] *Beijing University of Chemical Technology, Beijing Advanced Innovation Center for Soft Matter Science and Engineering, State Key Laboratory of Chemical Resource Engineering, Beijing, 100029, P.R. China*
[2] *Guizhou Institute of Technology, School of Chemical Engineering, Guiyang, 550003, P.R. China*
[3] *South China Normal University, School of Chemistry, Guangzhou, 510006, P.R. China*

13.1 Introduction

Organic solar cells (OSCs) open another door for capturing solar energy as a sustainable and clean replacement of fossil fuel. This emerging photovoltaic technology has been the focus of academic and industrial community for decades, due to the great advantages in fabricating lightweight, flexible devices through inexpensive large area solution processing [1–5]. The last two decades have seen dramatic increases of power conversion efficiencies (PCEs) from 1% to >18% in single-junction solar cell devices due to tremendous efforts in device engineering and the synthetic tailoring on the chemical structure of organic semiconductors [6, 7]. Tang first proposed the fabrication of bilayer OSCs composed of vacuum-evaporated p-type and n-type organic semiconductors in 1986 [8]. Later, following the discovery of ultrafast electron transfer from poly[2-methoxy-5-(2-ethylhexyloxy)]-1,4-phenylenevinylene (MEH-PPV) to fullerene (C_{60}) in 1992, Heeger and coworkers pioneered the study of conjugated polymers as electron donors and fullerenes as electron acceptors in OSCs [9, 10]. Following this, Yu et al. and Halls et al. found that the use of bulk heterojunction (BHJ) in OSCs could improve exciton (tightly bound electron and hole pair) diffusion length in the active layer [11, 12]. Since then, the concept of BHJ OSCs has become a benchmark for efficient device design and optimization, ushering OSCs to rapidly advance the pace of third-generation photovoltaic research.

Generally, the fabrication of BHJ OSCs is realized by sandwiching a bicontinuous interpenetrating network of donor and acceptor between two charge-extracting electrodes [13, 14]. Charge extraction and transport are crucial for the contact between metal electrodes and organic semiconductors [15, 16]. The energy level mismatch at such contact usually impedes charge transport and collection due to additional energy barriers at the device interface [17–19]. For example, non-ohmic interfacial contact suppresses charge transport efficiency between metal electrodes

Organic Solar Cells: Materials Design, Technology and Commercialization, First Edition. Edited by Liming Ding.

and active layers, leading to additional charge carrier recombination at such interfaces and a resultant inferior device performance [20]. To overcome this undesirable effect, interface engineering, i.e. development of interlayer materials and their integration into devices, has emerged as an effective strategy to improve the efficiencies of solar cells [21]. Theoretically, during OSCs operation, when excitons dissociate into free holes and electrons, the anode will collect holes while electrons are extracted at the cathode, owing to the existence of built-in electrostatic field (V_{bi}) as induced by the asymmetric metal electrodes with different work-function (WF) [22]. In general, larger V_{bi} in solar cell devices is beneficial for more efficient extraction of photo-generated charges. The integration of interlayer materials in OSCs tunes the work-functions of metal electrodes, aiming to increase the V_{bi} of the devices and suppress charge recombination loss, to the benefit of increasing the open-circuit voltage (V_{oc}), short-circuit current density (J_{sc}), and fill factor (FF) of the solar cells [23, 24].

Solution-processed interlayer materials have recently come into the focus of fundamental research, and practical development efforts, since they allow for synthetic tailoring of their properties, are compatible with such device fabrication approaches as slot-die coating, ink-jet printing, spray-coating, etc., enabling all-organic device architectures [25, 26]. Thus, significant efforts have been devoted to exploring novel solution-processed interfacial materials to realize all-solution-processed multilayer solar cells. Such interlayer materials usually have orthogonal solubility relative to the photoactive layer materials and hold the potential as multifunctional interfacial layers to manipulate the mechanical and electrical contact in solar cell devices [27, 28].

In this chapter, we first introduce the mechanism and effect of interfacial design on device performance. Following this, we will review some representative cathode and anode interlayer materials, such as organic/polymer electrolytes, organic–inorganic hybrids, graphene oxides (GOs), transition metal oxides (TMOs), and self-assembled functional materials and their integration into efficient OSCs. We intend to describe our current understanding based on recent progress and outstanding issues of interlayer materials in OSCs and propose future directions and opportunities.

13.2 The Mechanism and Effect of Interfacial Design

13.2.1 The Role of Electrode Work-Function Difference

In a typical BHJ OSCs, the donor–acceptor active layer is loaded in-between the high work-function (Φ_{An}) anode and low work-function (Φ_{Cath}) cathode (Figure 13.1a). The anode–cathode work-function difference ($\delta\Phi_{AC}$) dictates the magnitude of built-in electrostatic field (V_{bi}), impacting the open-circuit voltage (V_{oc}), short-circuit current density (J_{sc}), and FF. At zero applied voltage bias, the internal electric field shows the largest magnitude. Under illumination, the drift of photo-generated electrons and holes toward the corresponding electrodes is fast, producing the maximum current density (Figure 13.1b). With the increasing

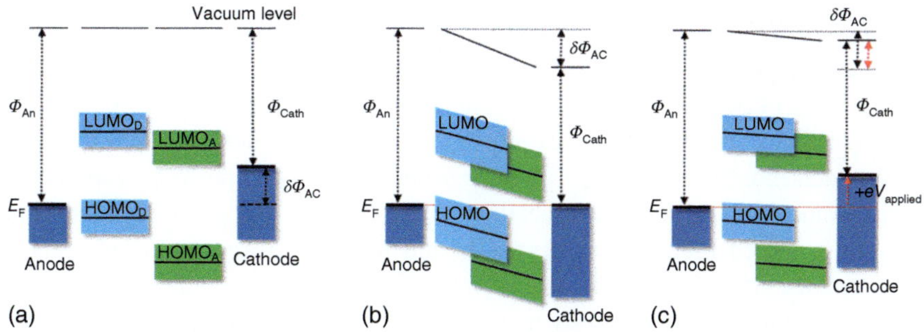

Figure 13.1 Schematic energy band diagrams: (a) anode, donor, acceptor, and cathode materials separately; OSCs in different regimes of device operation at (b) zero and (c) forward/positive ($+V_{applied}$) voltage biases. Source: Yao Liu.

of applied voltage in forward direction, the external electric field partially compensates the internal one, i.e. the net electric field decreases (Figure 13.1c). The charge extraction process gradually changes to a slower diffusion process due to the charge concentration gradient, changing from a large concentration at the donor–acceptor interface to a small concentration at the active layer/electrode interface for selective electrodes. Since it takes longer time for the charges to be removed, recombination may compete with extraction, and as a result, J_{sc}, FF, and V_{oc} will decrease. Monomolecular recombination dominates the charge losses in the short-circuit regime, while bimolecular recombination tends to play a critical role when approaching the open-circuit point [29]. The larger values of $\delta\Phi_{AC}$ afford faster charge extraction at short-circuit regime, which may reduce the recombination losses and increase J_{sc} when extraction and recombination time constants are comparable. When efficient charge extraction takes place, the range of external voltage biases is also extended toward the larger positive voltage biases, leading to higher FF values. Yet, the effect of $\delta\Phi_{AC}$ on V_{oc} is more intricate. For a relatively small $\delta\Phi_{AC}$, one can correlate the magnitudes of $\delta\Phi_{AC}$ with V_{oc} directly [30]. In another case, suppose the $\delta\Phi_{AC}$ is significantly larger than the difference between the energy levels of the highest occupied molecular orbital of donor ($\mathrm{HOMO_D}$) and the lowest unoccupied molecular orbital of acceptor ($\mathrm{LUMO_A}$), the V_{oc} of the devices can be evaluated by the following empirical relation [31].

$$V_{oc} = |\mathrm{HOMO_D}| - |\mathrm{LUMO_A}| - 0.3\,\mathrm{eV} \tag{13.1}$$

The origin of the factor 0.3 eV has been in the focus of many studies since its reduction offers a possibility for improving device performance [32]. This factor has been attributed to the generation of charge transfer (CT) state at the donor–acceptor interface. The difference between the quasi-Fermi levels in donor and acceptor components under a certain illumination level has been suggested to directly determine V_{oc}. Furthermore, the work-function offset between cathode and anode may directly affect the position of the quasi-Fermi level in the device active layer. Thus, the enhancement of V_{bi} in solar cell devices through interfacial design is crucial for achieving better device performance, especially for a higher V_{oc} [33].

13.2.2 The Interaction Between Metal Electrode and Interlayers

Energy level alignment at the contact between interlayers and metal electrodes has been the focus of tremendous research. The interfacial electronic structures at such contact are very complicated, and several semiempirical models were developed to describe such interfaces [34–38]. However, none of them can perfectly explain the experimental results of such interfacial contact. Thus, we will generally take some widely investigated materials as examples and highlight some important experimental results. Conjugated polyelectrolytes (CPEs) have been widely investigated as surface modification materials in organic electronics, which show thickness dependent effects on metal work-function modification. Specifically, when coating a thin layer of CPEs on metal surface (<10 nm), the interaction of charge moieties with metal electrode can generate interfacial dipoles, showing the power of tuning the work-function of metal electrodes (Figure 13.2a,b). However, since CPEs contain ionic side chains and mobile counter ions, one should consider the influence of mobile counter ions and ionic chain reorientation, when coating a relative thick CPE interlayer on metal surface (i.e. an interlayer thickness over the quantum mechanical tunneling limit [~10 nm]). According to the molecular reorientation model (Figure 13.2c), when applied an external electrical field on the devices, the counter ions of CPEs hold the potential to drift to the corresponding electrodes

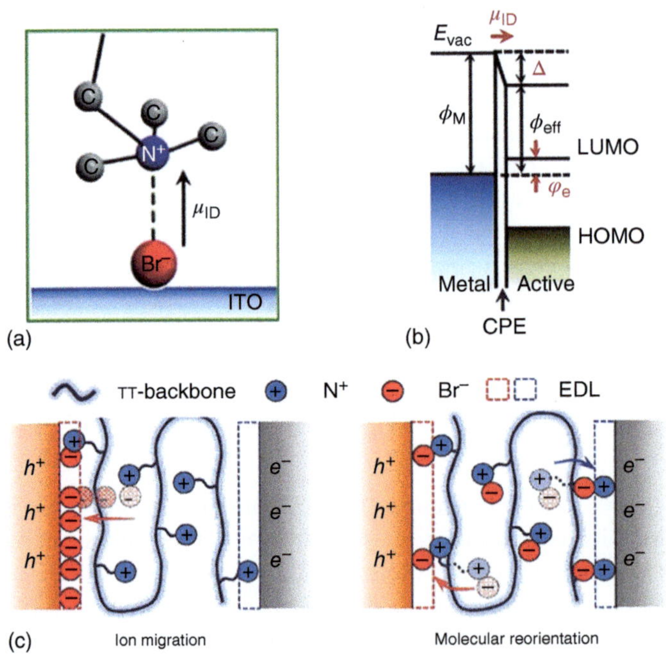

Figure 13.2 (a) Scheme of demonstrating interfacial dipole generated by ionic groups in CPEs; (b) Energy level scheme of the metal/CPE interlayer/active layers; (c) Schematics of the molecular reorientation model on surface modification. Source: Lee et al. [39]/Reproduced with permission of John Wiley & Sons.

while the ionic side chains of CPEs prefer reorganize to have more adjacent contact with metal electrodes. As a result, the mobile counter ions with CPEs are crucial for interfacial modification in organic electronics [39]. Additionally, the ionic species in CPEs play a critical role on interfacial dipole generation and orientation, affording various degrees and directions of vacuum energy level shifting. Thus, the polyelectrolyte interlayers generally tune metal work-function through the generation of interfacial dipole at the contact of interlayer/electrode [40]. As another important example demonstrated by Kippelen and coworkers, the modification of numerous metal electrodes can be realized by coating an ultrathin (<10 nm) layer of insulating ethoxylated polyethyleneimine (PEIE) or branched polyethyleneimine (PEI) [41]. The effect of this material system on interfacial modification can be mainly attributed to both (i) the orientation of the internal molecular dipole, induced by the lone pair of the nitrogen atom, and (ii) an interfacial dipole, due to partial charge transfer from the amine groups after physical adsorption onto the metal electrode surface. Interestingly, Emrick and coworkers also found that the combination of amine functionalities and electroactive molecules, such as fullerenes, affords the resultant interlayer materials with work-function pinning effect on metal surface, enabling work-function reduction of numerous metal electrodes to the same level. As a result, the amine functionalized interlayer materials have some universal capability to modulate the work-function of metal electrodes. This unique and interesting property most likely attributes to the nitrogen lone pair of the amine groups, which are capable to induce interfacial dipoles through partial charge transfer [42].

Zwitterions are charge neutral with dipole moments that influence the surroundings. This unique structure affords a good platform to study the impact of dipole orientation on interfacial modification [23] (Figure 13.3a). For example, Lee et al. calculated the permanent dipole moment of sulfobetaine (SB) zwitterions, which shows a value up to 15.2 Debye with the negative charge predominantly onto $-SO_3^-$ group and positive charge onto the $-N^+$ group [44]. Duzhko and coworkers proposed a classical electrostatic model of dipole–image dipole interaction to explain the impact of zwitterions on metal surface [43]. In this model, suppose an electrostatic dipole $\boldsymbol{p_o}$ near the metal surface at a distance $\boldsymbol{r}$. One can describe the surface polarization of such dipole by an image charge model [45] (Figure 13.3b). Ideally, the positive and negative sides of the dipole are mirrored across the plane of the metal surface by charges of the same value but opposite sign, inducing an image dipole, $\boldsymbol{p_i}$. Taking SB-functionalized polymers as example, the negative charges of the sulfonic group can be considered to locate further away from the polymer backbone when comparing with the positive charges on the nitrogen. As a result, the negative side of the dipole is more likely to reorient when applying electric field, while the positive side of the dipole can be regarded as relatively anchored (Figure 13.3a). In principle, suppose the dipole is anchored at a point of positive charge, the torque (τ) will drive the rotation of the original dipole. There are two possible quasi-stable orientations that are normal to the surface at $\theta = 0$ or $\theta = \pi$, i.e. the dipole is pointing away from or toward the surface (Figure 13.3c). Yet, the orientation away from the surface has a shorter distance, which is more energetically favorable than the opposite orientation

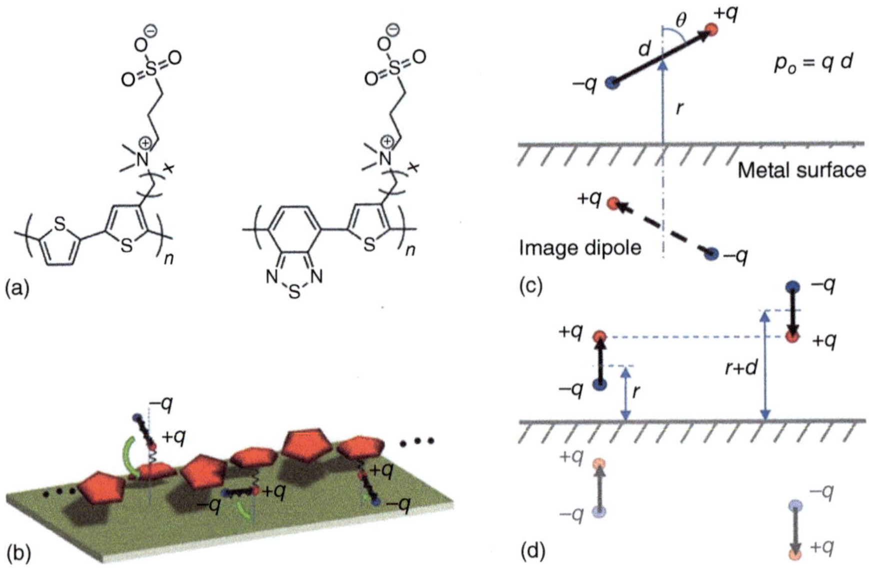

Figure 13.3 (a) Typical conjugated zwitterions (polythiophene-based zwitterions); (b) schematic depiction of the alignment of dipolar side chains of a zwitterionic polymer on a metal surface; (c) illustration of an "image" dipole where $\boldsymbol{p}_o = q\boldsymbol{d}$ (q is the elementary charge, d is a vector pointing from a negative charge, $-q$, to a positive charge, $+q$, which is equal to the separation distance between the charges) located a distance r from a metal surface; (d) two possible orientations of an electrostatic dipole fixed rigidly at the point of positive charge, where an upward orientation (left side) is a state with lower energy. Source: Liu et al. [43]/Reproduced with permission of John Wiley & Sons.

due to the potential interaction energy. Thus, in the interfacial dipole model, the self-induced torque of the dipole moment and energy minimization tend to drive the negative charges of sulfonic groups toward the metal surface. Such a net charge orientation generates an interfacial dipole (Δ) at the contact of metal electrode with zwitterions, which impacts the work-function of metal electrodes to the benefit of aligning interfacial energy levels in organic electronic devices [26] (Figure 13.3d).

13.2.3 Doping Effect on Energy Level Alignment

Interfacial doping is an efficient strategy to align the energy levels at the interface between electrodes and organic semiconductors (i.e. the hard–soft materials interface) to improve charge injection and extraction efficiency [46–55]. For example, contact doping at the organic/electrode interface has been developed to improve the electrical contact in solar cell devices as demonstrated by Gomez and coworkers [56]. A sulfonated poly(phenyl sulfone) (SPS) has been found to work as a strong polymeric acid, inducing interfacial doping of polymer donor in BHJ active layer. Such contact doping can efficiently overcome interfacial energy barrier to promote charge collection at the device interface. Although the SPS can perform as efficiently as poly(3,4-ethylenedioxythiophene)-poly(styrene sulfonate) (PEDOT:PSS) in solar cell devices, the intrinsically insulating property of SPS makes its processing less attractive when compared with PEDOT:PSS. Specifically, one has to optimize the degree of sulfonation on SPS and its thickness (<3 nm) to achieve optimal device performance. Yet, these results demonstrate the great potential of interfacial doping for making efficient solar cell devices. Kippelen and coworkers reported a solution-based method for electrical p-doping of conjugated organic semiconductors with a limited doping depth on their thin film surface. They achieved relatively controllable doping depth by simply immersing organic semiconductor thin films into a polyoxometalate solution (phosphomolybdic acid, PMA) in nitromethane [57]. This method can be used to numerous organic semiconductors, which increases their surface electrical conductivity and work-function and improves their photo-oxidative stability in air. This interfacial doping approach is applicable to various organic electronic devices, such as thin film photovoltaics and transistors, improving hole injection or collection at device interface. Ho and coworkers found a general strategy to realize solution-processed doped films over a wide range of work-functions (3.0–5.8 eV), by charge doping of CPEs and then internal ion exchange to generate self-compensated heavily doped polymers. This strategy can generate solution-processed ohmic contacts for high-performance organic light-emitting diodes (OLEDs), solar cells, photodiodes, and transistors, showing great power for aligning energy levels in organic electronic devices [47]. Self-doping interlayer materials were also designed and synthesized for solar cell applications [58]. Wu et al. developed n-type water/alcohol-soluble conjugated polymers (WSCPs) containing naphthalene diimides (NDIs) moieties. These NDI-based polymer interlayers show self-doping property, which are suitable for device fabrication and working as efficient electron transporting layers (ETLs) to generate doped interface between active layers and metal electrodes, affording high-performance solar cell devices with a broad interlayer thickness range (from 5 to 100 nm) [59]. Hu et al. designed and synthesized perylene diimides (PDIs)-based polyelectrolytes as interlayers in OSCs. These polymer interlayers show self-doping property, which is due to the doping effect of counter ions on PDI units. Thus, the superior electrical property of these interlayers affords the device interface with excellent electron-transporting properties in a wide interlayer thickness range. Interestingly, the counter ions play a critical role on the doping behavior, optoelectronic, and morphological properties of the resulting polyelectrolytes [60]. Emrick group covalently integrated planar electronically active PDI-based units into polyelectrolyte dopants. Impressively, the doping level of the target polymer can be tuned by varying the PDI-to-cation ratio, allowing chemical synthesis to optimize the doping efficiency of the resultant polymers. Thus, these PDI-based ionene polymer interlayers enable organic electronic devices with controllable interfacial doping and energy level alignment at the active layer/electrode interface through readily achieved solution processing method [46, 58]. A further study showed that some self-doping organic semiconductors (e.g. fullerene derivatives, CPEs) have the potential to improve the stability of lead halide perovskite solar cells [61–63]. Wang et al. reported that self-doped PDI derivatives show some kind of electron transfer process in solution. These self-doped PDI derivatives with stable n-type doping capability and facile solution processability can be used as electron transport layers in OSCs, affording high performance devices and showing great potential in organic electronics [64]. Sun et al. observed that the interface between the n-type interlayer and the donor polymers produces an extra contact surface for exciton dissociation. This interfacial contact also aligned energy levels between the n-type interlayer and the acceptor in the BHJ active layers, affording efficient electron extraction from the BHJ, which further improves the efficiencies of the solar cells [65].

13.2.4 Interface on BHJ Morphology and Device Stability

The morphology of BHJ active layer plays an important role on device performance [66, 67]. Several groups have demonstrated that the surface energy of the interfacial materials significantly impacts the morphology evolution and vertical phase separation of the BHJ active layers [68, 69]. As a result, the appropriate selection of interfacial materials holds the promise to improve their compatibility with the BHJ active layers and optimize the atop-BHJ active layer morphology, which is beneficial for the resultant charge dissociation and transport in solar cell devices. Fang and coworkers developed two hydrophilic small molecules and used them as cathode interlayers in inverted OSCs. They compared the effect of the conjugated units of underlying interlayer materials on the evolution of BHJ morphology. The more hydrophobic surface of the interlayer leads to the generation of better morphology upon the organic photoactive layer deposition, which is beneficial for the enhancement of V_{oc}, J_{sc}, and *FF* of inverted OSCs. Thus, the rational optimization of cathode interlayer could improve the bulk film morphology of the active layer to the benefit of achieving high-performance devices [70]. Wang and coworkers synthesized a novel polymer as underlying interlayer on PEDOT:PSS to optimize the morphology of the BHJ active layer in polymer solar cells. This novel polymer contains polar phosphonate moieties in the side chains, affording the PEDOT:PSS with more compatible surface energy with the active layer, which drives the donor polymer in the active layer to accumulate at the anode side and improve its crystallization. These interesting interfacial self-assembly results are beneficial for charge separation and transport, contributing to J_{sc} and PCE improvement of the solar cell devices [71]. To simply device fabrication, one blended poly(ethylene glycol) (PEG) derivatives into the active layer solutions to prepare *in situ* self-assembled PEG-based interlayers during the formation of BHJ active layer. It is due to the spontaneous migration of PEG-based derivatives to the surface of the active layers [72–76]. Additional functions of interlayer materials are their promise to improve device stability and durability. For example, some interlayers have the ability to work as additional protecting layers for the photoactive layers to prevent the penetration of oxygen and water. In addition, some interlayers hold the power to enhance the adhesion between the photoactive layers and the electrodes, affording a better device performance for flexible OSCs [77–79]. Thus, interface engineering is an efficient strategy to optimize the efficiency and stability of the solar cell devices.

13.2.5 Interfacial Morphology Characterizations

The morphology of solution processed interlayers is also an interesting topic in OSC research [80]. Numerous studies have reported that the surface feature of the interlayer is an important factor that impacts solar cell performance, especially when polyelectrolytes were used for surface modifications [14, 24, 43, 44, 81, 82]. Several experimental techniques, such as scanning electron microscopy (SEM) and atomic force microscopy (AFM), are the most common characterizations to provide surface morphology and roughness information at high resolution. In near-edge X-ray absorption fine structure (NEXAFS) measurements, polarized soft X-ray is used to investigate the orientation of covalent bond. NEXAFS has emerged as an important technique to reveal the chain orientation of organic molecules used for interfacial modification [14]. It is a technique for specific elements because the K-shell absorption edges of different elements have different energies. Specifically, the NEXAFS information can be revealed in Auger electron yield mode (~1 nm) or total electron yield (~10 nm), which provides structure information of different sample thicknesses. This technique is currently the most powerful measurement to show the molecular orientation at different sample depths near the surface. For example, the NEXAFS characterization was used to probe the aromatic backbone orientation of conjugated polymer zwitterions. In the NEXAFS spectra, the peak at 285.5 eV is characteristic of the thiophene carbon–carbon (double bond) π^* absorption. By varying the incidence angle of X-ray and combining with the change of signal intensity, the orientation of conjugated backbone can be analyzed [43].

Another example is the characterization of interfacial dipole. Generally, molecular structure and its arrangement at the interface dictate the interfacial electric dipole moment. The characterization of the interfacial dipole is important for understanding and exploring its influence on device interface engineering [80]. Ultraviolet photoelectron spectroscopy (UPS) is a highly surface-sensitive characterization because of the shallow probe depth of ultraviolet (UV) light. In UPS measurement, when coating a thin film of dipole layer on metal surface, one can extract the dipole moment from the shift of the secondary electron cutoff (E_{SEC}) in the high binding (low kinetic) energy region [83]. The dipole orientation plays a critical role on interfacial modification. For example, perpendicular alignment of sulfobetaine side chains to Au and indium tin oxide (ITO) surfaces was probed by the carbon K-edge NEXAFS measurements (Figure 13.4a,b). Note that the partial electron yield mode of NEXAFS that was used provides a surface-weighted signal, which is dominated by the outermost (2–4 nm) portion of the film. Hence, the degree of alignment was averaged in a 4 nm-thick film, and a top layer away from the underlying metal substrate was measured. Consistent with the UPS measurements that show no effect beyond the 3–4 nm, the degree of alignment was found to be significantly better in thinner film. The orientation of molecular dipole moments (toward or away from the metal surface) was also revealed by vibrationally resonant sum frequency generation

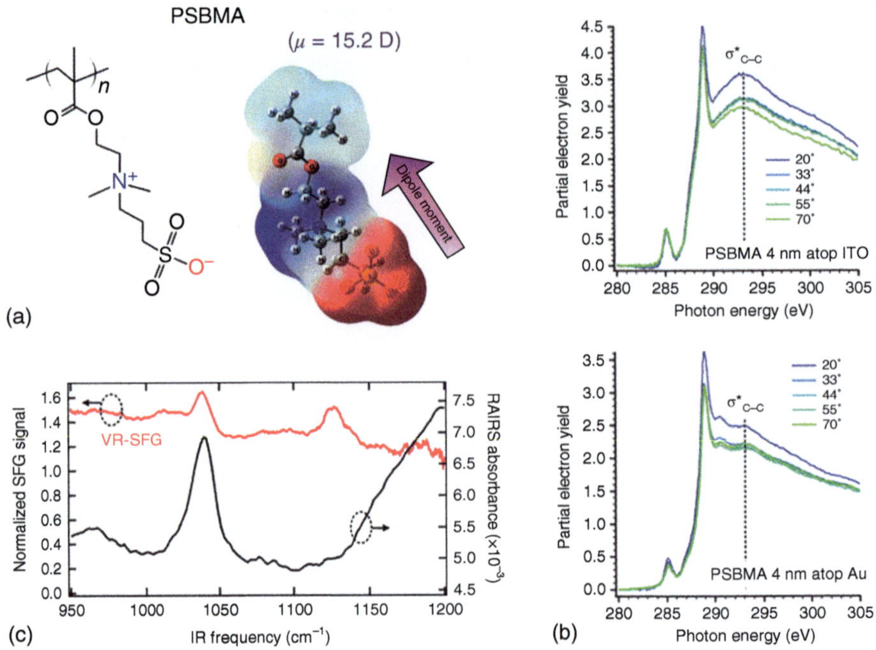

Figure 13.4 (a) Chemical structure and graphical charge distribution (DFT calculation) for one monomer unit of PSBMA; (b) carbon K-edge NEXAFS spectra of a 4 nm-thick PSBMA film on ITO or Au substrate; (c) VR-SFG spectra of the SO_3^- functional group of a 4 nm-thick PSBMA film on Au. Source: Lee et al. [44]/Reproduced with permission of American Chemical Society.

spectroscopy (VR-SFG) (Figure 13.4c). It has been shown that SO_3^- groups point toward the Au surface, i.e. the dipoles point away. Partial alignment of side chains that was concluded from both the NEXAFS and VR-SFG is consistent with the fact that both techniques probe the average orientation in the films. Similarly, for ITO surface, VR-SFG indicates that molecular dipole moments of fullerene derivatives are pointing away from the ITO surface, creating a net dipole moment that reduces the ITO work-function [44, 81].

13.3 Anode Interlayer Materials

13.3.1 PEDOT:PSS

PEDOT:PSS is the most widely used polymer hole transport layer (HTL) in both OSCs and OLEDs due to its unique solution processing property. Generally, PEDOT:PSS is made up of PEDOT and PSS, in which PEDOT acts as conductive polymer and PSS acts as a dispersing agent. PSS is attached to PEDOT by an oxidative polymerization of ethylenedioxythiophene (EDOT). In OSCs, the water solubility of PEDOT: PSS provides many advantages when it is employed as an HTL, such as: (i) sequential solution process of HTL and active layer could be realized due to the different solubility; (ii) water as a solvent for processing PEDOT:PSS is very environmentally friendly and beneficial for future applications; (iii) the conductivity of PEDOT:PSS could be easily adjusted by varying the ratio of PEDOT to PSS, which could satisfy various application conditions; (iv) the PEDOT:PSS possess high transparency via the near-infrared to visible light range, which could minimize the absorption loss in OSCs; (v) PEDOT:PSS has a work-function of 5.1 eV [20], which could not only match with HOMO levels of most donor materials, but also can form ohmic contact with ITO [84, 85].

Despite the extraordinary advantages as we mentioned above, some shortcomings of PEDOT:PSS still deserve attentions for further application in broader areas. Here are some issues that need to be addressed: (i) the acidity of PEDOT:PSS due to the styrene sulfonic acid constituents could corrode anodes such as ITO, which will lead to decreasing longtime stability of OSCs [86–88]; (ii) moisture and oxygen in the air could easily contaminate and diffuse into PEDOT:PSS since PEDOT:PSS is a water-soluble material, which could aggravate the degradation of OSCs [89, 90]; (iii) the relatively low conductivity still limits the application of PEDOT:PSS in large-scale fabrication process; (iv) the work-function of PEDOT:PSS should be deeper since more and more high-performance donor materials have been developed. Therefore, many ingenious works focused on the above issues have been conducted and achieved great progress.

As a polymer material, PEDOT:PSS (Scheme 13.1) is noted for its high conductivity. In OSCs, a highly conductive HTL guarantees minimum inner resistance and suppressed voltage drop in the interlayer, and an optimal internal field of the photoactive layer, therefore, results in enhanced exciton dissociation and charge extraction. However, the optimum thickness of PEDOT:PSS processed via spin coating in laboratory is about ~40 nm, and device performance fluctuations could happen when PEDOT:PSS thickness changes [91]. Also it is difficult to realize precise control of PEDOT:PSS thickness when using industrial fabrication process

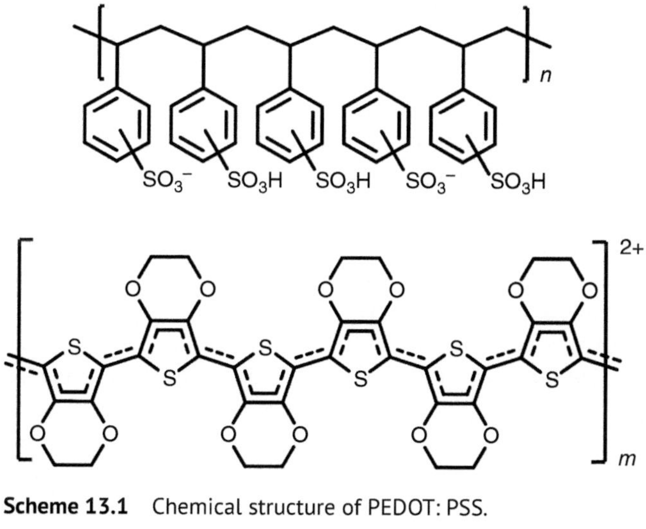

Scheme 13.1 Chemical structure of PEDOT: PSS.

such as roll-to-roll (R2R) printing, blade coating, inkjet printing, and so on. To ensure a better applicability of PEDOT:PSS in large-scale OSCs, improving the conductivity is the most straightforward way.

Generally, the film morphology, chemical and physical characteristics, which could be modified through a variety of posttreatments, determine the electrical properties of PEDOT:PSS. For example, it has been demonstrated that thermal treatment [92], which now is common knowledge in research community, is an effective way to enhance the conductivity of PEDOT:PSS. Besides, in some reports, effects of light treatment on PEDOT:PSS also have been investigated and found that the number of charge trapping defects could be reduced by UV or UV-ozone [93]. There are several other methods that are reported to enhance the conductivity of PEDOT:PSS, such as the addition of organic compounds such as ethylene glycol (EG), dimethyl sulfate, or ionic liquid into PEDOT:PSS solution and the surface treatment of PEDOT:PSS films with polar organic solution of salt, inorganic acid, or cosolvent. In 2002, Joo and coworkers [94] found that using dimethyl sulfoxide (DMSO), *N,N*-dimethyl formamide (DMF), and tetrahydrofuran (THF) separately as solvent additives could increase the PEDOT:PSS film conductivity. It is found that about two-order enhancement of conductivity was obtained when DMSO was added. From then on, adding polar organic solvents into PEDOT:PSS solution for improving the conductivity has gained great attention, and conductivity ranging from 0.61 to 677 $S\,cm^{-1}$ could be realized by employing various kinds of polar organic solvents as additives [95]. Ouyang et al. achieved more than two orders of magnitude enhancement of conductivity through adding solvents such as ethylene glycol (EG), *meso*-erythritol and 2-nitroethanol into the PEDOT:PSS aqueous solution [96].

Solvent mixture and surface posttreatment with polar solvent are the most common ways to enhance the conductivity of PEDOT:PSS. For example, in 2011, Kim et al. observed a high conductivity of 1418 $S\,cm^{-1}$ for PEDOT:PSS films by using solvent additive and solvent posttreatment. The posttreatment induced PEDOT:PSS showed lower PSS to PEDOT ratio, which contributes to the higher conductivity. Furthermore, ITO-free OSCs were realized using modified PEDOT:PSS as transparent electrode [97]. Other posttreatments with fluoro compound such as hexafluoroacetone (HFA) [98] were also reported to enhance the conductivity of PEDOT:PSS by simply dropping HFA aqueous solution onto PEDOT:PSS films; as a result, a conductivity of 1325 $S\,cm^{-1}$ was obtained. Eom et al. found that when mixing glycerol with PEDOT:PSS solution, glycerol could promote the aggregation of PEDOT phase, and an increase of conductivity from 0.782 to 152 $S\,cm^{-1}$ could be obtained [99]. When the other common solvent 2-propanol was used, the resultant PEDOT:PSS films could also get an enhanced conductivity [100].

Adding materials with high conductivity into PEDOT:PSS has also been proved to be effective to increase the conductivity of PEDOT:PSS. As an example, multiwalled carbon nanotubes (MWCNTs) are exploited as additives, and researchers found that a net-like structured conductive channels could be formed via self-assembly of MWCNTs. Wang et al. processed PEDOT:PSS/MWCNTs films by the polymerization of EDOT with MWCNTs and PSS solution [101]. The resultant

PEDOT:PSS/MWCNTs films exhibited special structure characteristics in which MWCNTs act as the core wrapped with PEDOT:PSS. The conductivity could be dramatically enhanced by the π–π conjugation conducted by the chains of MWCNTs and PEDOT:PSS. Yet, the material preparation of PEDOT:PSS/MWCNTs is more suitable for supercapacitors or thermoelectric electronic other than OSCs. Fortunately, with the development of two-dimensional (2D) materials such as WS_2 or MoS_2, exfoliation and functionalization of 2D nanosheets or nanoflakes are exploited and investigated in PEDOT:PSS-based composite materials for hole transporting in OSCs. In 2017, Huang and coworkers [102] proposed a facial and scalable protocol method for efficient exfoliation of bulk transition metal dichalcogenides (TMD) and graphite assisted with PEDOT:PSS solution, for preparing single- and few-layer nanosheets. It is worth noting that single crystalline characteristic of bulk materials could be retained by these TMD nanosheets, which is essential for application in OSCs. Consequently, the conductivity of PEDOT:PSS/WS_2 up to 1.40×10^{-3} S cm^{-1} was obtained, which is higher than that of pristine PEDOT:PSS. Furthermore, solar cell devices based on PEDOT:PSS/WS_2 showed more superior J_{sc} and PCE than those with bare PEDOT:PSS. Also, the PEDOT:PSS/WS_2 HTL fabrication is simply realized by processing hybrid PEDOT:PSS/WS_2 ink prepared by a simple centrifugation.

The corrosivity and hygroscopicity of PEDOT:PSS could accelerate device degradation. Therefore, it is of great importance to reduce or eliminate the negative impact of the corrosivity and hygroscopicity of PEDOT:PSS. Zhang and coworkers [103] demonstrated a protocol for fabrication of PEDOT:PSS/MoO_3 composite films by *in situ* formation of MoO_3 in aqueous PEDOT:PSS solutions. Through the combination of high conductivity of PEDOT:PSS and the ambient stability of MoO_3, high-performance HTL based on PEDOT:PSS/MoO_3 with smooth surface and considerably reduced hygroscopicity was obtained. Due to the low hygroscopicity of PEDOT:PSS/MoO_3 films, the longtime stability of the solar cell devices is greatly improved. In 2018, Zhou and coworkers [104] reported a facile and effective method for developing high-performance PEDOT:PSS/WO_x composite HTLs by simply mixing WO_x nanoparticles with PEDOT:PSS solution. The authors provided sufficient evidence to prove that besides effective charge extraction properties, finely tuned active layer morphology could also be obtained by the surface free energy difference between PEDOT:PSS and WO_x. The high FF of the optimal devices with PEDOT:PSS/WO_x as HTL is mainly attributed to the superior carrier extraction along with longer carrier lifetime. Choi and coworkers [105] incorporated photo-cross-linking agent into PEDOT:PSS by a fast photo-polymerization process. As a result, hydrophobic PEDOT:PSS film, a water-resistant HTL, was obtained. The resultant PEDOT:PSS films were employed in both perovskite and polymer solar cells. The PCEs and stability of the devices are simultaneously improved. Since the hydrophilic characteristics of PEDOT:PSS were suppressed and the modified one showed water-resistant property, the penetration of water into PEDOT:PSS was reduced. As a result, devices with hydrophobic PEDOT:PSS retained 60% of original PCEs after 570 hours of storage. Besides, the MeOH treatment could remove the PSS content and improve the surface conductivity as many early literatures have reported. Chen's group [106] introduced an evaporated MoO_3 layer to form a MoO_3/PEDOT:PSS bilayer for protecting the ITO anode. The ultrathin MoO_3 interlayer combined with PEDOT:PSS also plays an important role in enhancing the light intensity and charge generation in the photoactive layer. Greatly improved device stability was realized due to the reduced moisture and oxygen adsorption of the modified PEDOT:PSS.

Although various methods were used to reduce the negative impacts of the acid characteristics of PEDOT:PSS, new HTL materials are still in demand since the corrosion of ITO in OSCs is inevitable with the existence of PSS. Thus, the redundant sulfonic acid group of PEDOT:PSS is an alternative method. With the basic cognition that the high acidity of PEDOT:PSS is from the anionic dispersant PSS, in 2016, Li and Zeng [107] successfully developed a water-soluble sulfobutylated phenol formaldehyde resin (BSPF) with flexible alkyl sulfonic acid groups. Water-soluble PEDOT:BSPF was obtained by using BSPF as dopant. Notably, when using PEDOT:BSPF as HTL to modify ITO in OSCs, the device performance was improved significantly. The authors demonstrated that BSPF may be of promising potential as dopant of polymer semiconductors, especially for PEDOT-based organic electronics. In 2019, another dopant was developed and applied to dope PEDOT to obtain new conductive polymer by Li's group [108]. In this work, the authors continued their study on investigating the charge extraction property of PEDOT:PSS via the regulation of structure homogeneity. To enhance the structural homogeneity of PEDOT derivatives, the authors applied a doping process to PEDOT with a conductive dispersant poly(diphenylamine-4-sulfonic acid) (PDAS), a new HTL material PEDOT:PDAS was obtained. It is worth noting that in comparison to PEDOT:PSS, PEDOT:PDAS exhibits a much higher work-function (5.38 eV) and a lower acidity (pH $\sim$ 7) along with outstanding film formation and structural homogeneity. Moreover, the conductivity of PEDOT:PDAS was much higher than that of PEDOT:PSS-4083 although with a low PEDOT ratio in PEDOT:PDAS. For solar cell applications, PEDOT:PDAS exhibited comparable charge extraction ability with the PEDOT:PSS control films, resulting in comparable device performance. More interestingly, the synthesis method of the dispersant PDAS is very simple, and the raw materials are cheap and readily available, making PEDOT:PDAS a very promising HTL material.

Huang's group [109] focused on reducing the acid group of PEDOT:PSS by performing a facile doping of PEDOT:PSS with dopamine hydrochloride (DA·HCl) (Figure 13.5). PEDOT:PSS-DA was obtained after the reaction between DA·HCl and the sulfonic acid of PSS. Compared with PEDOT:PSS, the PEDOT:PSS-DA possesses higher work-function and superior conductivity. Consequently, when using PEDOT:PSS-DA as HTL, OSCs based on different photoactive layers observed an enhanced PCEs, indicating that doping strategy for PEDOT:PSS with amino derivatives is of great potential to develop novel HTLs. Very recently, the same group [110] proposed a secondary doping protocol by employing novel oxoammonium salts ($TEMPO^+X^-$, 2,2,6,6-tetramethylpiperidine-1-oxoammonium, X = (Br^-, $BF4^-$, OTf^-, $TFSI^-$) as efficient, facile p-type dopants. Increased carrier density along with enhanced conductivity was achieved when $TEMPO^+X^-$ salts were introduced into PEDOT:PSS. Moreover, the surface energy and work-function of the resultant HTL

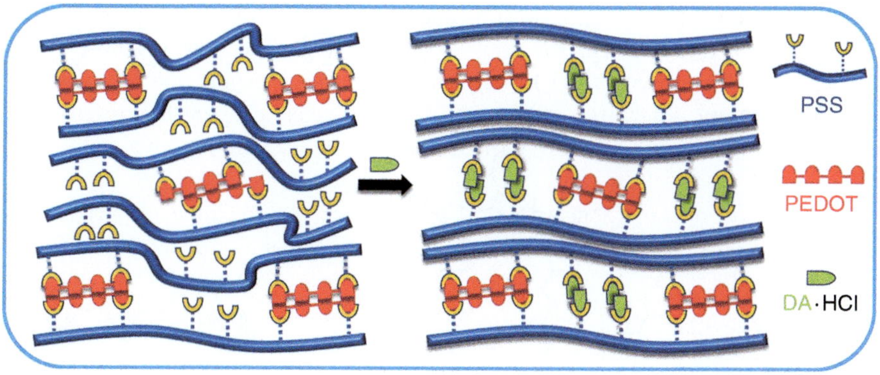

Figure 13.5 The schematic self-assembly process of PEDOT:PSS-DA film after the doping of dopamine hydrochloride in PEDOT:PSS. Source: Zeng et al. [109]/Reproduced with permission of John Wiley & Sons.

were finely tuned with the evolution of the counter ions of $TEMPO^{+}X^{-}$, resulting in better energy level alignment and improved charge collection. The authors also demonstrated the universality of this doping strategy via applying this method to other conjugated polymers. The improved conductivity and tunable work-function could be similarly realized.

13.3.2 Conjugated Polyelectrolytes

CPEs (Scheme 13.2) as another substitute for PEDOT:PSS have attracted widespread attention. Ke et al. prepared a new solution-processable self-doped sulfonic acid ring-substituted polyaniline (SPAN) as the HTL material in inverted OSCs. SPAN shows higher conductivity and better film quality when compared with PEDOT:PSS [111]. Lacher et al. developed a novel polymer HTL material (EDOT-SuNa) based on conductive EDOT with the property of tunable work-function and hydrophobicity. Therefore, the devices with the EDOT-based polymer HTL exhibited a higher V_{oc} when compared with the devices containing regular PEDOT:PSS [112]. Li et al. synthesized carboxyl-functionalized polythiophene (P3CPenT), which was used as HTL in OSCs. P3CPenT exhibited several ideal properties as an HTL, including suitable solubility in some polar solvents and better energy level alignment with the active layer [113]. Bazan and coworkers reported a series of pH neutral CPEs (CPE-K, CPE-Na, and CPE-TBA) as HTL materials in OSCs, and the CPE-K showed excellent film formation and solution processing property. The low contact resistance and improved *FF* of OSCs attribute to the self-doping effect between the side chains of CPE-K and conjugated backbone. Self-doping effect eliminates the inhomogeneity of the charge transport, and the device containing CPE-K exhibited an excellent performance [114–116]. Meanwhile, Cui et al. reported another pH-neutral CPE with wide bandgap (PCP-Na), which exhibited an electrical conductivity of 1.66×10^{-3} S cm^{-1}. The PCP-Na modified ITO exhibited a work-function of 5.22 eV, which is beneficial for hole collection in OSCs [117]. Cai et al. demonstrated the efficient hole transport ability of PEDOT-S as HTL in OSCs. Therefore, PEDOT-S showed high conductivity and optical transparency in OSCs. With a UV-Ozone treatment, the work-function

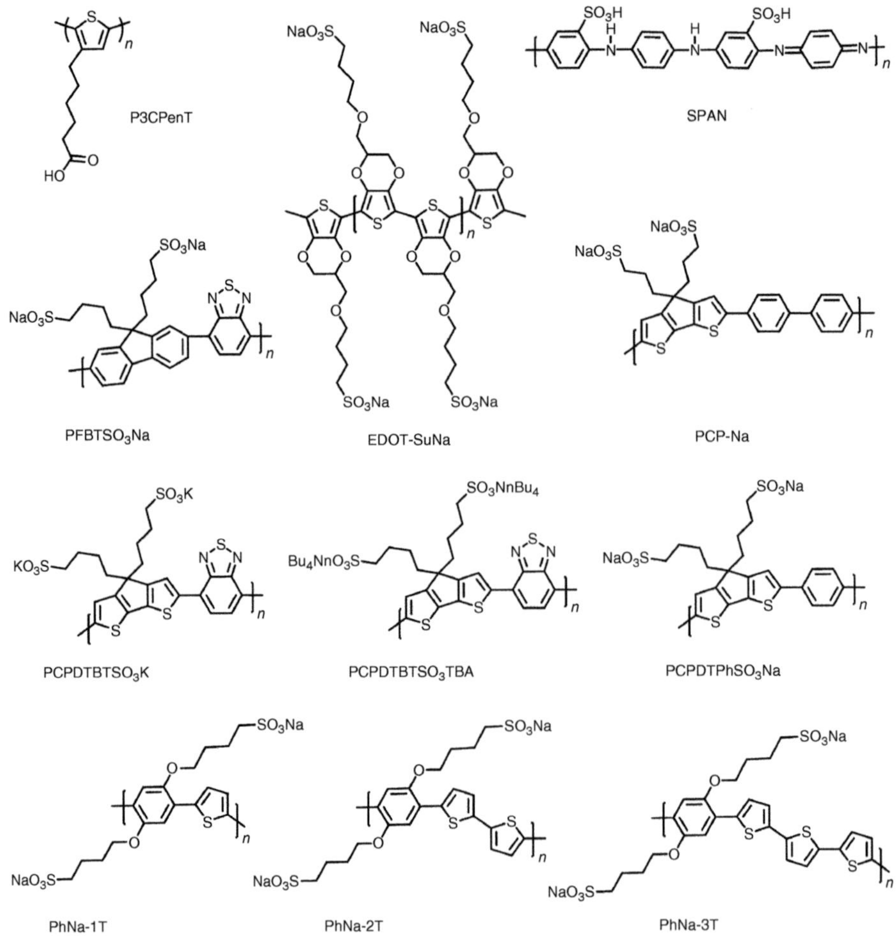

Scheme 13.2 The chemical structure of some typical CPEs.

of PEDOT-S can increase from 4.9 to 5.2 eV, which enables its efficient hole extraction ability for devices. The experimental results show that PEDOT-S as HTL can operate well in a wide range of thicknesses, which is beneficial for the printing production [118]. Jo et al. prepared a series of self-doped CPEs (PhNa-1T, PhNa-2T, and PhNa-3T). Those self-doped polymers show high conductivity and excellent hole transport property in the vertical direction of their films. Meanwhile, PhNa-1T has a low roughness surface, which shows a good contact with the active layer [119].

13.3.3 Cross-Linkable Polymers

Cross-linkable polymers represent another class of promising HTL materials (Scheme 13.3), which can be thermally or photochemically cross-linked to form HTLs. The cross-linked HTL materials are sufficiently solvent resistance, which is benefit for the fabrication of multilayer OSCs through solution processing technology. Hains and coworker reported a cross-linkable polymer material

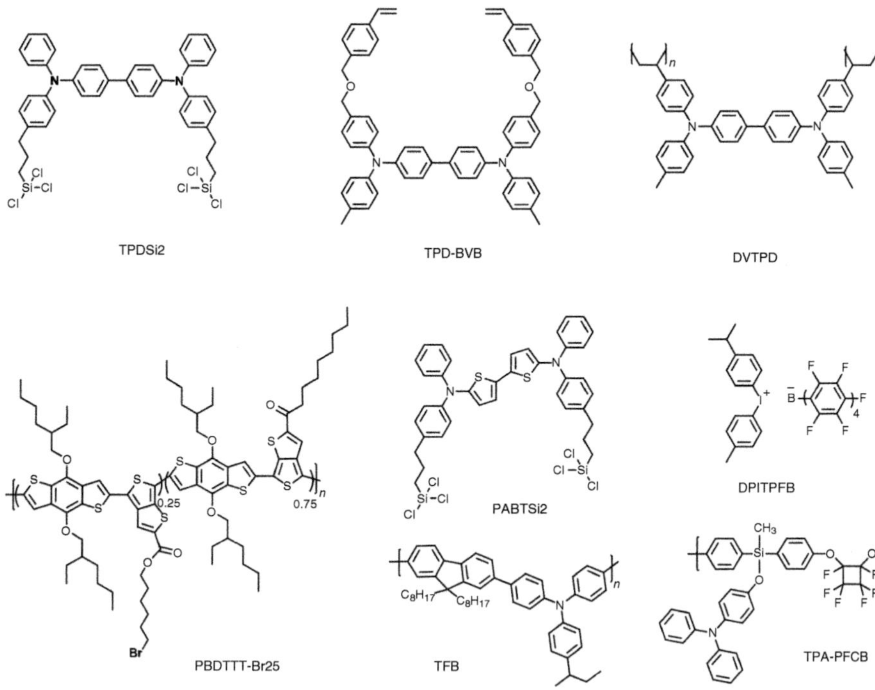

Scheme 13.3 The chemical structure of cross-linkable hole transporting materials.

(TPDSi2:TFB) as HTL. TPDSi2:TFB showed strong electron-blocking ability, good hole-transporting characteristics, and high transmittance. The solar cell devices with TPDSi2:TFB HTL exhibited high PCE and superior thermal stability compared with the devices using regular PEDOT:PSS as HTL [120]. Subsequently, the same group investigated the design, synthesis, characterization of a novel cross-linkable HTL material, PABTSi2:TFB. This cross-linkable material exhibited a higher HOMO level than TPDSi2:TFB [121]. Sun et al. also reported a solution-processed cross-linkable HTL material, TPD-BVB. The solar cell devices using TPD-BVB as HTL showed comparable efficiency and superior long-term air stability when compared with the devices fabricated with PEDOT:PSS [122]. Xu et al. reported a cross-linked D–A copolymer (PBDTTT-Br25) as an additional hole collection layer, which was inserted between a PEDOT:PSS layer and the photoactive layer. Inserting the PBDTTT-Br25 layer can provide an additional donor/acceptor interface, which facilitates hole transport, thereby reducing the possibility of charge carrier recombination [123]. Tsai et al. developed a styryl-functionalized triaryl-amine-based material (DVTPD), which can be thermally cross-linked at 180 °C for 30 minutes to form a solvent-resistant thin film. The film exhibited good coverage onto the substrate in the fabrication of multilayer OSCs. Meanwhile, the hole mobility of DVTPD can be significantly enhanced by incorporating an ionic dopant (DPITPFB) [124]. Sun et al. reported a hole transport and electron-blocking material containing triphenylamine, polyperfluorocyclobutane (TPA–PFCB), which was inserted between a PEDOT:PSS layer and the ITO anode. The solar cell devices containing TPA–PFCB showed a higher PCE compared with the control devices made from regular PEDOT:PSS [125]. Sun et al. studied the optical, electrical, and electrochemical properties of a new class of thermally cross-linkable hole transport materials. The conductivity and optical transparency of these HTL materials were improved by P-doping. The bilayer hybrid HTL structure with one neutral layer on the bottom of the doped layer has been used to provide both efficient hole-transporting and electron-blocking properties in the devices [126].

13.3.4 Graphene Oxides (GOs)

GO as another promising type of HTL material for OSCs can be simply processed in solution and has excellent light transparency. Also, its work-function is comparable to PEDOT:PSS. Li et al. reported that the PCEs of solar cell devices containing 2 nm GO HTLs are comparable to those with PEDOT:PSS as HTLs [127]. Kim et al. investigated the influence of the concentrations of GO dispersion on device performance. When using 0.15 wt% GO dispersion, they can get nearly complete coverage of GO on the substrates. Meanwhile, incorporating a more conductive material, single-walled carbon nanotubes (SWCNTs) in GO could improve the vertical conductivity of the HTL [128]. Smith and colleagues presented large-area solar cell devices using GO as the HTLs [129]. However, GO is insulating, which will increase the internal resistance of the devices with the increase of the GO film thickness, leading to a rapid PCE decline, which is unfavorable for the printing production. In GO, because most of carbon atoms are sp^3 hybridized, which were bonded with oxygen, it destroys the sp^2 conjugation. Hence, the substantial oxygen groups in GO make it an insulating material. The chemical reaction can partially reduce the GO, which will decrease the number of the oxygen groups bonded with carbon atoms and increase the sp^2 hybrid of GO. Therefore, this strategy has been an effective method to improve the conductivity of GO martial. Yang et al. demonstrated that GO treated with O_2 plasma showing a high work-function of 5.2 eV and the solar cell device performance was improved [130]. Li et al. reported that GO with precise oxidation (pr-GO) showed adjustable work-function, large interfacial dipole, and excellent film-forming properties. Meanwhile, the pr-GO exhibited excellent hole extraction capability, which can enhance device efficiency [131]. Kim et al. found a simple method to reduce GO and introduced fluorine group in GO. The functionalization of graphene oxide (FrGO) was performed by one-step reduction. The FrGO showed ideal properties as HTL in OSCs, such as high work-function and good conductivity ($3.5\,S\,m^{-1}$) [132]. Ding et al. demonstrated that a few-layered graphene quantum dots (F-GQDs) can be used as HTL for OSCs. The 4 nm F-GQDs with a high content of periphery COOH groups have a high work-function (5.26 eV) and excellent film-forming property. The solar cells using F-GQD as the HTLs exhibited a higher PCE than its counterparts based on the GO or PEDOT:PSS HTL. These results showed the great potential of F-GQD as HTL for OSCs [133]. Liu et al. designed and prepared sulfated GO-OSO_3H. Owing to the high work-function, GO-OSO_3H exhibited an excellent performance as a HTL in solar cell devices [134]. Zhao et al. prepared a 2D π-conjugated material

$G-SO_3H$ by introducing sulfonic acid to graphene scaffold ($G-SO_3H$) and obtained a work-function of 4.8 eV, which matches well with the HOMO of the donor polymer in active layer [135]. Stratakis et al. showed a photochemical method for partial reduction and doping of GO films. They can tailor the work-function of the GO-Cl from 4.9 to 5.23 eV by controlling the reaction time to match the HOMO of polymer donors [136]. Xia et al. reported a method to partially reduce GO by direct UV–ozone treatment. Since the ozone molecules can oxidize GO into CO_2 and generate highly conductive graphene particles, the UV–ozone treatment can improve the conductivity of GO HTLs. The J_{sc}, FF, and PCE of the solar cells were significantly improved by using the modified GO HTLs. However, chemical reduction method is inadequate for mass production because of the toxicity of reductant. The other method to produce GO is thermal method, which is a green method without using reductant [137]. By thermal annealing treatment of GO, Jeon et al. prepared a re-GO with a conductivity of 1.8 S cm^{-1}, which worked well as HTLs in solar cells [138].

13.3.5 Metal Oxides (MOs)

TMOs have tunability in terms of their electronic properties and inherent stability of oxidation, which are considered as promising HTL materials in OSCs. Kim et al. reported the use of a solution-processed molybdenum oxide nanoparticle layer (MoO_3) as an HTL in OSCs. The MoO_3 nanoparticles can reduce the charge recombination and resistance of OSCs [139]. Zhang et al. proposed a double-layer anode buffer layer composed of PTFE and MoO_3. The PTFE/MoO_3 double layer can form an interfacial dipole that increases the work-function of ITO anode and thus facilitates hole extraction [140]. Cheng et al. prepared solution-processed MoO_3 (s-MoO_3) as a HTL in solar cell devices. The s-MoO_3 ultrathin films treated with UV annealing showed better transmittance, surface morphology, and film quality [141]. Ji et al. sprayed the molybdenum oxide (MoO_3) microarray on the active layer by ultrasonic spraying at a very low temperature of 80 °C. Compared with devices without MoO_3, the solution-processed MoO_3 microarray improved the charge carrier transport and the PCE of the devices [142]. Kim et al. reported the use of low-temperature solution-processed vanadium oxide (s-VOx) and poly(4-styrene sulfonic acid) (PSS) composite material as HTLs. The s-VOx:PSS showed a work-function of 5.0 ~ 5.3 eV, affording an enhanced charge injection/extraction process at the interface due to the reduction of the energy barrier between the HTL and the organic active layer. Compared with the original s-VOx film, the s-VOx:PSS-based solar cells showed significantly improved performance [143]. Bao et al. reported a simple O_2 plasma treatment method for preparing vanadium oxide (V_2O_5) anode buffer layer on ITO. Compared with the traditional PEDOT:PSS HTL, solar cells using the V_2O_5 (O_2 plasma) anode buffer layer produced higher efficiencies [144]. Jiang et al. reported the application of $V_2O_5 \cdot nH_2O$ film as HTL in OSCs. The $V_2O_5 \cdot nH_2O$ layer was prepared directly from the vanadium oxide powder via the fusion quenching sol–gel method. When using $V_2O_5 \cdot nH_2O$ as HTL, the PCEs of the solar cells increased significantly [145]. Xu et al. proposed a method using a solution-treated vanadium oxide (s-VO_X) as HTLs in OSCs. The devices with s-VOx as HTLs were more efficient and showed better storage stability than the OSCs with PEDOT:PSS layer [146]. Irwin et al. did comprehensive studies on the electrical properties, microstructure, electronic structure, and optical properties of nickel oxide (NiO) thin films. The NiO can form a nearly transparent interlayer on the ITO substrate, showing a smooth, crystalline, and oriented morphology [147]. Xue et al. developed an indium-free transparent electrode with a NiO/Ag/NiO structure. This indium-free transparent electrode not only has excellent optical and electrical properties but also has suitable surface morphology, good thermal, humidity, and environmental stability [148]. Yang et al. prepared NiO_x films at room temperature from aqueous solution, which were used as the anode buffer layers in OSCs [149]. Yu et al. fabricated CuO_x anode buffer layer from copper acetylacetonate precursor aqueous solution at room temperature. Compared with OSCs with PEDOT:PSS anode buffer layer, the PCE of the devices with CuO_x anode buffer layers reached a much higher value [150].

13.4 Cathode Interlayer Materials

13.4.1 Organic Small Molecules

Various organic small molecules have been developed as cathode interlayers, including neutral or charged and conjugated or non-conjugated materials. To improve the efficiency and stability of OSCs, efficient small molecule-based cathode interlayer materials were designed by tailoring molecular structure and preparation method (Scheme 13.4). Wudl and coworkers demonstrated that $Na^+QHSO_3^-$ can work efficiently as cathode interlayers in OSCs, which mainly improved FF and PCE of the devices. Compared with the devices without $Na^+QHSO_3^-$ interlayer, the electron mobility of the optimal device was increased by two times [151]. Su and coworkers developed a unique set of star-like interlayer materials based on fluorene with neutral phosphonate pendent chains (3TPA-FEP) and triphenylamine (TPA) core. The molecule tends to aggregate due to the integration of polar phosphonate groups, resulting in a rough surface morphology. The J_{sc} of the devices was improved because the rough interlayer increased the contact area with metal cathode [152]. Later Zhang et al. reported small-molecule interlayers (PDIN and PDINO) based on n-type PDI derivative with amino terminal substituent. PDIN showed superior electrical conductivity of $8.0 \times 10^{-6}\ S\ cm^{-1}$, three orders of magnitude higher than PFN of $6.9 \times 10^{-9}\ S\ cm^{-1}$. Due to the π-stacks enhanced by self-organizing, the devices with PDIN showed significantly improved PCEs. Furthermore, the device performance was insensitive to the metal cathode, and high efficiencies were obtained in devices with Al, Ag, or Au cathodes [153]. Hou and coworkers reported NDIO based on NDI as cathode interlayer. The solar cell efficiency has been increased by 80% compared with the devices without interlayer [154]. Kim and coworkers demonstrated efficient cathode interlayer materials based on fluorene and phenothiazine trimer backbones with quaternary ammonium bromide pendant groups. These interlayers can efficiently improve the performance of the solar cell devices [155, 156]. Fang and coworkers designed two different conjugated small molecules with polar *N,N*-dimethyl amino end group, benzo[*c*][1,2,5]thiadiazole (FBF-N) and bisthiophenyl-benzo[*c*][1,2,5]thiadiazole (FTBTF-N), for use in inverted OSCs. Polar amine groups are beneficial to decrease ITO workfunction and increase V_{oc} of the solar cells [70]. Su and coworkers demonstrated a star-shaped small molecule based on a triazine core (TzPyBr) for use in OSCs as cathode interlayers [81]. It is proposed that the deep HOMO of TzPyBr effectively blocked hole transport toward the cathode, thus suppressing the interface recombination [157]. Ouyang et al. designed a small-molecule electrolyte (MSAPBS) with an LUMO level of −2.31 eV and a high electron mobility of 1.18×10^{-4} cm^2 V^{-1} s^{-1}, which allowed optimal photon capture and provided excellent photo-generated charge carrier collection in OSCs [158]. Yang and coworkers also developed a series of star-like

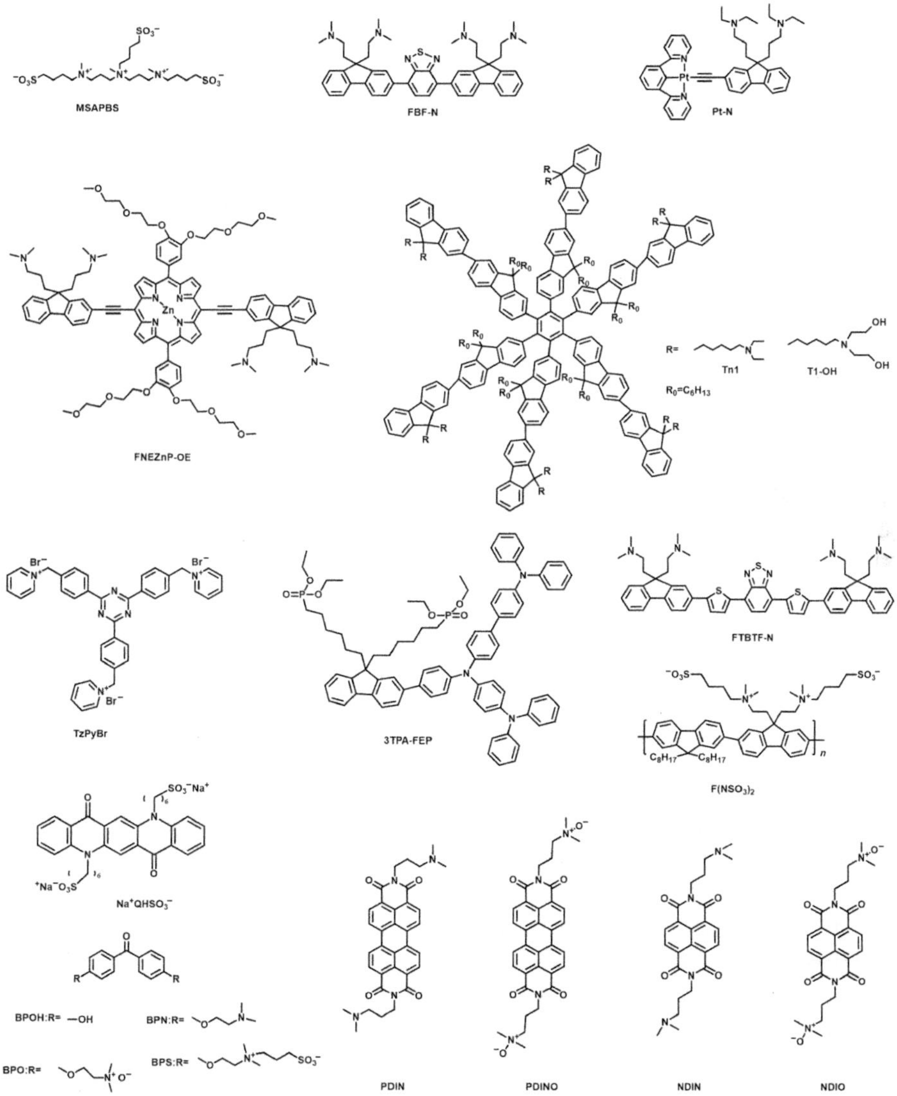

Scheme 13.4 Molecular structure of small-molecule interlayer materials.

small molecules based on triazine core (Tn1) or hexakis(fluoren-2-yl)benzene (HFB) (T1-OH). Taking T1-OH as an example, the diethanolamine groups in this molecule reduced the work-function of the cathode electrode to enhance electron collection. Also, these functional groups made T1-OH soluble in polar solvents (methanol or water), which was required for solution-processed OSCs [159, 160]. Huang and coworkers synthesized amino-functionalized organic platinum (II) complex (Pt-N) for solar cell interlayer applications. The wide bandgap of Pt-N also acts as an exciton blocker and inhibits the quenching of excitons at the cathode interface. In addition, metal–metal supramolecular interactions forming a highly conductive electron transport channel can increase the strong π–π stacking in organic platinum (II) complex (Pt-N) films [161]. Later, Huang and coauthors reported that two fluorene units were connected to the porphyrin nuclei by acetylene bond to form a planar-structured interlayer material (FNEZnP-OE). The fine coplanarity of the molecule significantly enhanced the electron mobility of the interlayer material, affording high-performance solar cell devices [162]. In addition to above methods of increasing molecular interactions, Li and coauthors applied aggregation-induced emission units into cathode interlayer materials. The J_{sc}, V_{oc}, and FF of the devices were dramatically improved, leading to a much higher PCE [163]. Ge and coworkers designed and synthesized a series of small-molecule interlayer materials with benzophenone (BP) unit as a core bonding different polar group, including hydroxyl (BPOH), neutral amino (BPN), amino N-oxide (BPO), and sulfobetaine (BPS) ions. OSCs containing BPN, BPO, BPS, and PFN interlayers showed enhanced PCEs [164]. Very recently, Zhang et al. reported a self n-doped hydrogen-bonding interlayer material, aliphatic amine-functionalized perylene-diimide (PDINN), which simultaneously down-shifts the work-function of the air-stable cathodes (silver and copper) and maintains good interfacial contact with the active layer through hydrogen bonding with CF bond. The OSCs based on PDINN engineered silver cathode demonstrate a high efficiency of 17.23% ($V_{oc} = 0.847$ V, $J_{sc} = 25.89$ mA cm^{-2}, FF = 78.59%) [165].

Fullerene derivatives as another kind of efficient interlayer materials are based on fullerene acceptor moiety, and they can be dissolved in a range of polar solvents. Fullerene derivatives have been the most widely used electron acceptor in OSCs for several decades, due to their superior electronic and processing properties. Fullerene derivatives have similar energy levels and charge transport capabilities with the acceptors of active layers in OSCs, which would be promising and efficient interlayer materials for OSCs (Scheme 13.5). Jen and coauthors reported the use of solution-processed self n-doped fulleropyrrolidinium iodide (FPI) and FPI-doped PCBM with high conductivity as interlayers in OSCs. It was proposed that when the iodides on the FPI and the buckyballs undergo electron transfer, effective n-doping and high conductivity are obtained. They further proposed that tetrabutylammonium salts (TBAX, $X^- = F^-, Br^-, I^-$) can also effectively dope fullerenes by electron transfer between the anions of TBAX and the buckyball. Thus, the anions of these alkyl ammonium-based surfactants also hold promise for doping the electron acceptor when in contact with the active layer to optimize interfacial contact and improve solar cell efficiency [166]. Chen and coauthors reported two kinds of PCBM derivatives (PCBDAN and PCBDANI) and used them

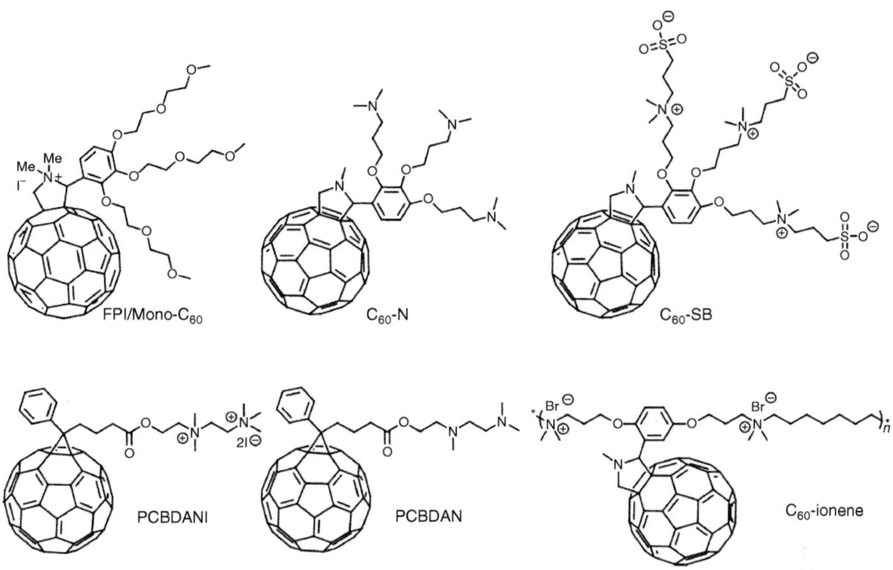

Scheme 13.5 Molecular structure of some typical fullerene interlayer materials.

as cathode buffer layer materials in OSCs. In comparison with the traditional low work-function active metal Ca cathode, both PCBDAN and PCBDANI cathode buffer layers demonstrated identical or slightly improved photovoltaic performance in PSCs. Meanwhile, they presented a strategy to fabricate polymer solar cells in inverted geometry, where PCBDAN was used as an additive in active layer and worked as cathode interlayers by self-organization. Furthermore, PCBDAN could be used both as acceptor and cathode interlayer in similar inverted OSCs [167–170]. In 2014, Emrick and coworkers used amine-functionalized fullerene (C_{60}-N) and sulfobetaine-substituted zwitterionic fullerene (C_{60}-SB) as cathode-independent buffer layer in conventional solar cell devices. Specifically, when coating a thin layer of C_{60}-N, the work-function of common metal electrodes, such as Ag, Cu, and Au, can be significantly reduced to 3.65 eV. The integration of C_{60}-N into solar cell devices can result in excellent ohmic contact for electron injection and a large built-in potential for efficient charge extraction. Later, the same group further reported that C_{60}-SB can modify the work-function of ITO from 4.5 to 4.0 eV and act as an electron transport layer in inverted OSCs to improve device performance [42, 171]. Chen and coworkers synthesized ZnO-C_{60} as a cathode interlayer material by the reaction of hydroxyl containing fullerene derivative, PCBE-OH, with zinc. The ZnO-C_{60} interlayers enhanced the function of electron collection in the devices, produced a fullerene-rich cathode surface, and promoted compatibility of the BHJ layer at the interface [172]. Siao and coworkers reported that an azide-cross-linked fullerene interlayer (C-PCBN3) can significantly improve the J_{sc}, V_{oc}, and FF of the devices compared with those with a TiO_2 interlayer or no interlayer [173]. Liu et al. reported fullerene polymer ionenes that involved dibromo-alkanes and C_{60}-diamine. The resulting cationic polymer interlayer C_{60}-ionene can support the utilization of a variety of high work-function metals such as Ag, Cu, and Au as the cathode for high performance OSCs [174].

13.4.2 Polymer Cathode Interlayer Materials

Over the past decades, water/alcohol-soluble conjugated polymers (WSCPs) have achieved great success as cathode interlayer materials for various organic electronic devices (Scheme 13.6). By taking advantage of the difference in solubility between WSCPs interlayer materials and traditional organic semiconductors, multilayered devices can be prepared by solution processing. These WSCPs can effectively reduce the work-function of the metal cathode and improve the electron transport in the devices by forming interfacial dipole at the interface between active layer and metal electrode.

In 2009, Cao and coworkers first demonstrated the use of WSCPs as cathode interlayer in OSCs. They found that the V_{oc} and PCE of the devices were improved remarkably after incorporating a thin layer of poly[(9,9-bis(3-(*N,N*-dimethylamino) propyl)-2,7-fluorene)-*alt*-2,7-(9,9-dioctylfluorene)] (PFN). The observed enhancement of V_{oc} was ascribed to the improved built-in field due to the interfacial dipole generated by the interlayer [175]. Later, they reported that the PFN interlayer could simultaneously enhance V_{oc}, J_{sc}, and *FF* of solar cells, leading to a high efficiency of both conventional and inverted device. The presence of interfacial dipoles enhances the intrinsic potential of the whole device, improves the electron transport characteristics, eliminates the accumulation of space charge, and reduces the energy loss due to the increase in the intrinsic field and carrier mobility [24, 176]. The work-function of cathode can also be regulated by non-CPEs with charged ionic groups in the structure. In 2012, Kippelen and coauthors successfully used non-CPEs PEI and PEIE as neutral surface modifiers and found that these materials are of great significance in single-junction and tandem photovoltaic devices. These materials successfully adjust the work-functions of the metal electrodes, increase build-in potential, and reduce the interfacial energy barrier [41, 177]. Except for the amino-containing groups, Cao and coworkers reported a novel conjugated polymer, PCP-EP, in which the hydrophilic phosphate ester side chain could not only effectively improve the solubility of polycarbazole materials but also effectively regulate the work-function of ITO electrode by forming an interface dipole. Thus, it can improve the electron transport and PCEs of photovoltaic devices [178]. In 2013, Liu et al. reported a novel cathode interlayer material, Hg-containing derivative of amino-functionalized polymer (PFEN-Hg). This interlayer material showed excellent performance in a wide range of thicknesses window, due to the existence of Hg–Hg interactions and the improved π–π stacking. Further investigation indicated that PFEN-Hg showed some attractive properties, such as good film formation, effective work-function adjustment, low optical absorption, and excellent electron collection [179]. Lv et al. reported a hyperbranched soluble polymer HBPFN by one-pot "A2+B3" Suzuki polymerization and used it as a cathode interlayer in OSCs. This work demonstrated that the structure ordering of the conjugated polymer main chains is not necessary for the interfacial materials to perform effectively [180]. Zhang and coauthors reported an n-type conjugated polymer containing dibenzothiophene-S,S-dioxide (FSO), bispyridinium, and fluorene scaffolds in the backbone (PFSOPyCl) and used this polymer as cathode

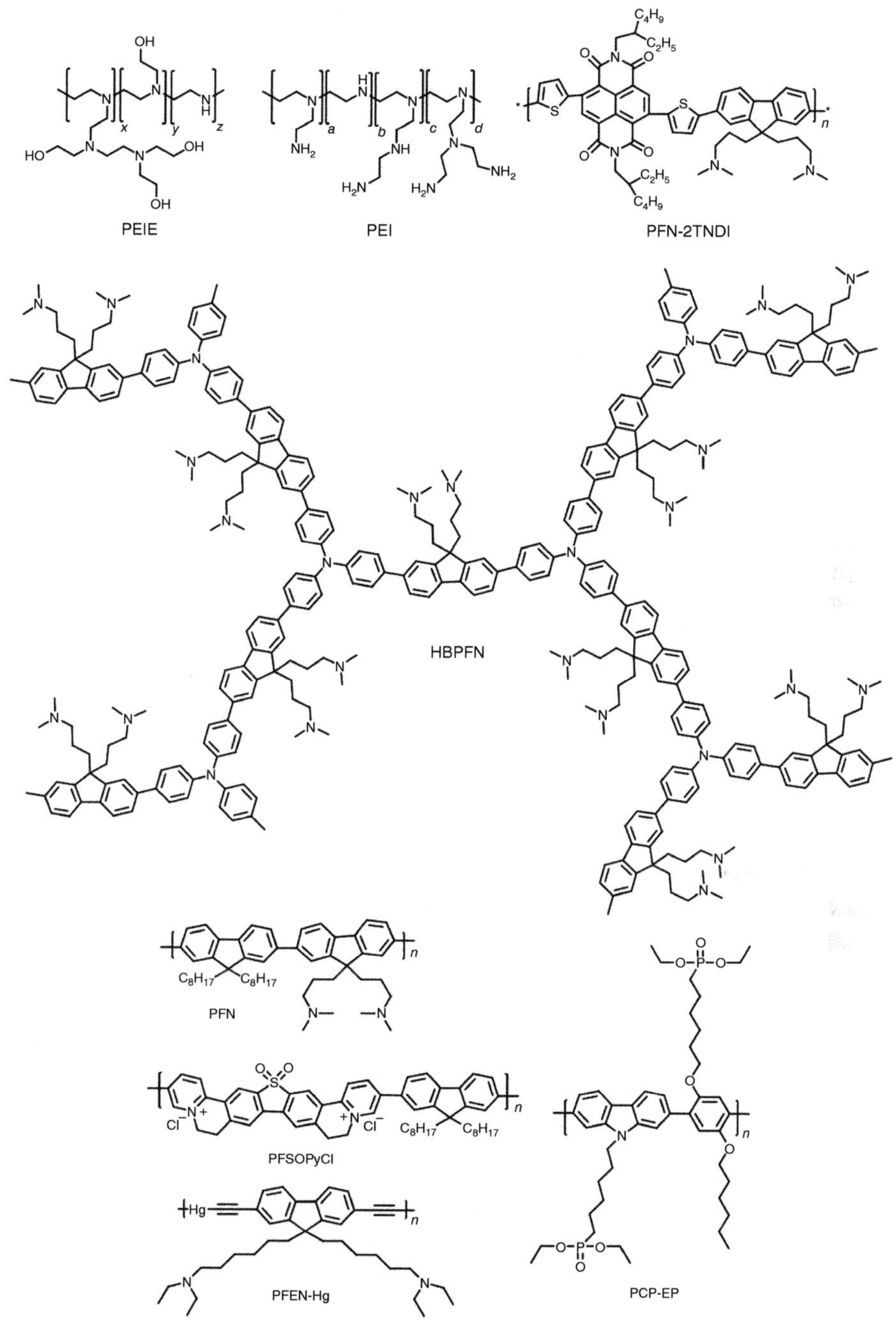

Scheme 13.6 Molecular structure of some typical polymer cathode interlayer materials.

interlayer in OSCs. The solar cell devices exhibited simultaneous enhancement in V_{oc}, J_{sc}, and *FF* relative to the device with bare Al electrodes [181]. Recently, Huang and coauthors developed self n-doped WSCPs. Two n-type WSCPs were reported, and both contain an NDI-based conjugated main chain, with highly polar amino- or ammonium-functionalized side chains, respectively. OSCs based on these self-doped WSCP ETLs exhibit significantly improved device performance; especially they can be processed with a broad thickness range of 5–100 nm [59]. Sun et al. reported that using *n*-type PFN-2TNDI as cathode interlayer in conventional fullerene-free solar cells resulted in much higher efficiency. The results show that the contact between the n-type cathode interlayers and the donor in the active layer provided an additional interface for charge dissociation, and the energy level matching between the interlayer and the acceptor in the active layer can be effectively optimized, enhancing electron extraction from the BHJ active layer at the interface. Compared with PFN, the PFN-2TNDI layer enhanced the photocurrent generation due to the appropriate energy level offset between the donor layer and the cathode interlayer, and it is suitable for exciton dissociation [65]. Many groups found n-doping behavior between non-conjugated polymer interlayer materials bearing polar moieties and a typical acceptor. This intermolecular n-doping effect may be a general phenomenon and could play an important role for those polymer cathode interlayer materials to work effectively in solar cell devices [182, 183].

13.4.3 Graphene Derivatives and Other Emerging Alternatives

Graphene and its derivatives have a high specific surface area, remarkable thermal/electrical properties, high charge carrier mobility, transparency, mechanical flexibility, and compatibility with solution processability, and hence they have been used in electronic devices. In addition, the semi-metallic band structure of graphene affords widely variable work-functions based on its continuously tunable Fermi energy. Dai and coworkers demonstrated an excellent electron-transporting cesium-neutralized GO(GO-Cs) layer, and the periphery –COOH groups in GO are replaced by –COOCs groups [184, 185]. Conventional and inverted OSCs with the GO-Cs cathode interlayer exhibited the same performance compared with those that used the most advanced cathode interlayer materials. Then Heeger and coworkers discovered a method of dealing with GO. Graphene was stamped and subsequently oxidized with HNO_3, and a transferable GO layer was obtained with a work-function of 4.3 eV [186]. Because this GO cathode interlayer material can make electron extraction/transport from the BHJ active layer to the Al cathode more efficient, the J_{sc} was improved and the series resistance was reduced. Yang and coauthors reported that a new graphene–fullerene composite (reduced graphene oxide (rGO)-pyrene-PCBM) was attached onto rGO via the noncovalent functionalization approach. The pyrene-PCBM moiety was synthesized via a facile esterification reaction, and pyrene was used as an anchoring bridge to link rGO and PCBM. rGO-pyrene-PCBM was successfully applied as electron extraction layer for solar cell devices [187]. Lv and coauthors developed a solution-processable n-doped graphene cathode interlayer material PDINO-G for OSCs by adding graphene to the classic alcohol/water-soluble interlayer material PDINO. The n-doped graphene-PDINO composite (PDINO-G) interlayer possessed increased conductivity, lower work-function, reduced charge recombination, and increased charge extraction rate. Previously, n-doped graphene was prepared only on the surface of chemical vapor-deposited (CVD) graphene films without solution processability. The effect of this cathode interlayer on photovoltaic performance of OSCs was studied by selecting the photovoltaic model system. The solar cell efficiency was significantly enhanced in comparison with that of the devices without graphene modification on the PDINO. The universality of the high-performance PDINO-G interlayer was verified in two other photovoltaic systems [188]. Later they reported that two three-dimensional (3D) interlayer materials, POSSFN and ADMAFN, are excellent surfactants for dispersing graphene in ethanol. The two 3D-surfactant-based graphene composites (POSSFN-G and ADMAFN-G) can work as high-performance cathode interlayers in OSCs [189].

13.5 Conclusion and Outlook

We summarized recent progress on interfacial design in OSCs, mainly focusing on the mechanism of interfacial modification, the strategies of performing device engineering, and the typical interlayer materials that have been developed for device optimization. Despite the great progress on the design of efficient interlayer materials and the understanding of the mechanism on interfacial design, much more study still lies ahead to get a comprehensive description of their multiple functionalities, resolve outstanding problems, and propose a view of future directions.

(1) Due to the diverse design on interlayer materials, it is challenging to draw a clear picture to understand the role of interlayers on device operation. Many models were developed to describe the energy level alignment at the contact between interlayer and metal electrode. However, since now, no one model can explain all the experiment results. Thus, one has to develop more powerful characterizations to probe the correlation of interfacial morphology and electronic structures at the device interface, including molecular packing, dipole orientation, and their influence on energy level alignment at the interface.
(2) The developed organic interlayer materials generally contain some unique functionalities (amine, zwitterions, electrolytes…). Some of them have the potential to degrade the organic semiconductors in the active layers. For example, Zhou and coworkers found that the recently developed non-fullerene acceptors can react with amine groups in PEIE, leading to device degradation [190]. Thus, the community has to give more emphasis on the chemical stability at the device interface.
(3) The morphology stability of the interlayer plays a critical role on the durability of device operation. Some small molecular interlayer materials hold the potential to diffuse into the active layer, degrading themselves as well as the active layer. On the other hand, most of the efficient interlayers to date are generally

hydrophilic, while the active layers are just the opposite. Thus, the long-term stability of the devices is quite challenging due to the potential de-wetting effect of the interlayers on the active layers.

References

1 Krebs, F.C., Espinosa, N., Hösel, M. et al. (2014). *Adv. Mater.* 26: 29.
2 Li, C., Zhou, J., Song, J. et al. (2021). *Nat. Energy* 6: 605.
3 Liu, G., Xia, R., Huang, Q. et al. (2021). *Adv. Funct. Mater.* 31: 2103283.
4 Yan, C., Barlow, S., Wang, Z. et al. (2018). *Nat. Rev. Mater.* 3: 18003.
5 Li, G., Zhu, R., and Yang, Y. (2012). *Nat. Photonics* 6: 153.
6 Liu, Q., Jiang, Y., Jin, K. et al. (2020). *Sci. Bull.* 65: 272.
7 Zhang, G., Zhao, J., Chow, P.C.Y. et al. (2018). *Chem. Rev.* 118: 3447.
8 Tang, C.W. (1986). *Appl. Phys. Lett.* 48: 183.
9 Sariciftci, N.S., Smilowitz, L., Heeger, A.J., and Wudl, F. (1992). *Science* 258: 1474.
10 Kraabel, B., Lee, C.H., McBranch, D. et al. (1993). *Chem. Phys. Lett.* 213: 389.
11 Yu, G., Gao, J., Hummelen, J.C. et al. (1995). *Science* 270: 1789.
12 Halls, J.J.M., Walsh, C.A., Greenham, N.C. et al. (1995). *Nature* 376: 498.
13 Liu, F., Gu, Y., Jung, J.W. et al. (2012). *J. Polym. Sci. Polym. Phys.* 50: 1018.
14 Liu, F., Gu, Y., Shen, X. et al. (2013). *Prog. Polym. Sci.* 38: 1990.
15 Peng, X., Hu, L., Qin, F. et al. (2018). *Adv. Mater. Interfaces* 5: 1701404.
16 Ma, H., Yip, H.L., Huang, F., and Jen, A.K.Y. (2010). *Adv. Funct. Mater.* 20: 1371.
17 Chueh, C.C., Li, C.Z., and Jen, A.K.Y. (2015). *Energy Environ. Sci.* 8: 1160.
18 Hu, Z., Zhang, K., Huang, F., and Cao, Y. (2015). *Chem. Commun.* 51: 5572.
19 Hu, Z., Zhong, Z., Chen, Y. et al. (2016). *Adv. Funct. Mater.* 26: 129.
20 Yip, H.L. and Jen, A.K.Y. (2012). *Energy Environ. Sci.* 5: 5994.
21 Duan, C., Zhang, K., Zhong, C. et al. (2013). *Chem. Soc. Rev.* 42: 9071.
22 He, Z., Wu, H., and Cao, Y. (2014). *Adv. Mater.* 26: 1006.
23 Liu, Y., Duzhko, V.V., Page, Z.A. et al. (2016). *Acc. Chem. Res.* 49: 2478.
24 He, Z., Zhong, C., Huang, X. et al. (2011). *Adv. Mater.* 23: 4636.
25 Gu, Y., Liu, Y., and Russell, T.P. (2020). *ChemPlusChem* 85: 751.
26 Wu, Y., Liu, Y., Emrick, T., and Russell, T.P. (2020). *Prog. Polym. Sci.* 103: 101222.
27 Lai, Y.Y., Cheng, Y.J., and Hsu, C.S. (2014). *Energy Environ. Sci.* 7: 1866.
28 Yin, Z., Wei, J., and Zheng, Q. (2016). *Adv. Sci.* 3: 1500362.
29 Cowan, S.R., Roy, A., and Heeger, A.J. (2010). *Phys. Rev. B* 82: 245207.
30 Mihailetchi, V.D., Blom, P.W.M., Hummelen, J.C., and Rispens, M.T. (2003). *J. Appl. Phys.* 94: 6849.
31 Brabec, C.J., Cravino, A., Meissner, D. et al. (2001). *Adv. Funct. Mater.* 11: 374.
32 Elumalai, N.K. and Uddin, A. (2016). *Energy Environ. Sci.* 9: 391.
33 Vandewal, K., Tvingstedt, K., Gadisa, A. et al. (2009). *Nat. Mater.* 8: 904.
34 Braun, S., Salaneck, W.R., and Fahlman, M. (2009). *Adv. Mater.* 21: 1450.

35 Bao, Q., Liu, X., Wang, E. et al. (2015). *Adv. Mater. Interfaces* 2: 1500204.
36 Li, C.Z., Chueh, C.C., Ding, F. et al. (2013). *Adv. Mater.* 25: 4425.
37 Zhang, K., Zhong, C., Liu, S. et al. (2014). *ACS Appl. Mater. Interfaces* 6: 10429.
38 Dong, S., Hu, Z., Zhang, K. et al. (2017). *Adv. Mater.* 29: 1701507.
39 Lee, B.H., Jung, I.H., Woo, H.Y. et al. (2014). *Adv. Funct. Mater.* 24: 1100.
40 Kang, R., Oh, S.H., and Kim, D.Y. (2014). *ACS Appl. Mater. Interfaces* 6: 6227.
41 Zhou, Y., Hernandez, C.F., Shim, J. et al. (2012). *Science* 336: 327.
42 Page, Z.A., Liu, Y., Duzhko, V.V. et al. (2014). *Science* 346: 441.
43 Liu, F., Page, Z.A., Duzhko, V.V. et al. (2013). *Adv. Mater.* 25: 6868.
44 Lee, H., Puodziukynaite, E., Zhang, Y. et al. (2015). *J. Am. Chem. Soc.* 137: 540.
45 van Reenen, S., Kouijzer, S., Janssen, R.A.J. et al. (2014). *Adv. Mater. Interfaces* 1: 1400189.
46 Liu, Y., Cole, M.D., Jiang, Y.F. et al. (2018). *Adv. Mater.* 30: 1705976.
47 Tang, C.G., Ang, M.C.Y., Choo, K.K. et al. (2016). *Nature* 539: 536.
48 Lüssem, B., Keum, C.M., Kasemann, D. et al. (2016). *Chem. Rev.* 116: 13714.
49 Salzmann, I., Heimel, G., Oehzelt, M. et al. (2016). *Acc. Chem. Res.* 49: 370.
50 Duzhko, V.V., Dunham, B., Rosa, S.J. et al. (2017). *ACS Energy Lett.* 2: 957.
51 Xu, Y., Sun, H., Liu, A. et al. (2018). *Adv. Mater.* 30: 1801830.
52 Reilly, T.H., Hains, A.W., Chen, H.Y., and Gregg, B.A. (2012). *Adv. Energy Mater.* 2: 455.
53 Lüssem, B., Riede, M., and Leo, K. (2013). *Phys. Status Solidi A* 210: 9.
54 Dai, A., Zhou, Y., Shu, A.L. et al. (2014). *Adv. Funct. Mater.* 24: 2197.
55 Li, C., Chang, C.Y., Zang, Y. et al. (2014). *Adv. Mater.* 26: 6262.
56 Mor, G.K., Jones, D., Le, T.P. et al. (2014). *Adv. Energy Mater.* 4: 1400439.
57 Kolesov, V.A., Hernandez, C.F., Chou, W.F. et al. (2016). *Nat. Mater.* 16: 474.
58 Cole, M.D., Sheri, M., Bielicki, C., and Emrick, T. (2017). *Macromolecules* 50: 7535.
59 Wu, Z., Sun, C., Dong, S. et al. (2016). *J. Am. Chem. Soc.* 138: 2004.
60 Hu, Z., Xu, R., Dong, S. et al. (2017). *Mater. Horiz.* 4: 88.
61 Liu, Y., Renna, L.A., Page, Z.A. et al. (2016). *Adv. Energy Mater.* 6: 1600664.
62 Liu, Y., Page, Z.A., Zhou, D. et al. (2018). *ACS Cent. Sci.* 4: 216.
63 Li, Y., Cole, M.D., Gao, Y. et al. (2019). *ACS Appl. Energy Mater.* 2: 1634.
64 Wang, Z., Zheng, N., Zhang, W. et al. (2017). *Adv. Energy Mater.* 7: 1700232.
65 Sun, C., Wu, Z., Hu, Z. et al. (2017). *Energy Environ. Sci.* 10: 1784.
66 Rivnay, J., Mannsfeld, S.C.B., Miller, C.E. et al. (2012). *Chem. Rev.* 112: 5488.
67 Bisquert, J. and Belmonte, G.G. (2011). *J. Phys. Chem. Lett.* 2: 1950.
68 Guo, X., Zhou, N., Lou, S.J. et al. (2013). *Nat. Photonics* 7: 825.
69 Xu, Z., Chen, L.M., Yang, G. et al. (2009). *Adv. Funct. Mater.* 19: 1227.
70 Zhang, W., Wu, Y., Bao, Q. et al. (2014). *Adv. Energy Mater.* 4: 1400359.
71 Meng, B., Fu, Y., Xie, Z. et al. (2014). *Macromolecules* 47: 6246.
72 Chen, F.C. and Chien, S.C. (2009). *J. Mater. Chem.* 19: 6865.
73 Chen, F.C., Chuang, M.K., Chien, S.C. et al. (2011). *J. Mater. Chem.* 21: 11378.
74 Jung, J.W., Jo, J.W., and Jo, W.H. (2011). *Adv. Mater.* 23: 1782.
75 Tai, Q., Li, J., Liu, Z. et al. (2011). *J. Mater. Chem.* 21: 6848.
76 Shi, Y., Tan, L., Chen, L., and Chen, Y. (2014). *J. Mater. Chem. C* 2: 8054.

77 Jørgensen, M., Norrman, K., and Krebs, F.C. (2008). *Sol. Energy Mater. Sol. Cells* 92: 686.
78 Tung, V.C., Kim, J., Cote, L.J., and Huang, J. (2011). *J. Am. Chem. Soc.* 133: 9262.
79 Lee, Y.Y., Tu, K.H., Yu, C.C. et al. (2011). *ACS Nano* 5: 6564.
80 Chen, Q., Wang, C., Li, Y., and Chen, L. (2020). *J. Am. Chem. Soc.* 142: 18281.
81 Lee, H., Stephenson, J.C., Richter, L.J. et al. (2016). *Adv. Mater. Interfaces* 3: 1500852.
82 Li, J., Rochester, C.W., Jacobs, I.E. et al. (2015). *ACS Appl. Mater. Interfaces* 7: 28420.
83 Page, Z.A., Duzhko, V.V., and Emrick, T. (2013). *Macromolecules* 46: 344.
84 Yao, H., Cui, Y., Yu, R. et al. (2017). *Angew. Chem. Int. Ed.* 56: 3045.
85 Wang, Z., Zhu, X., Zhang, J. et al. (2018). *J. Am. Chem. Soc.* 140: 1549.
86 Škraba, P., Bratina, G., Igarashi, S. et al. (2011). *Thin Solid Films* 519: 4216.
87 Wong, K.W., Yip, H.L., Luo, Y. et al. (2002). *Appl. Phys. Lett.* 80: 2788.
88 de Jong, M.P., van Ijzendoorn, L.J., and de Voigt, M.J.A. (2000). *Appl. Phys. Lett.* 77: 2255.
89 Norrman, K., Madsen, M.V., Gevorgyan, S.A., and Krebs, F.C. (2010). *J. Am. Chem. Soc.* 132: 16883.
90 Kawano, K., Pacios, R., Poplavskyy, D. et al. (2006). *Sol. Energy Mater. Sol. Cells* 90: 3520.
91 Xu, H., Yuan, F., Zhou, D. et al. (2020). *J. Mater. Chem. A* 8: 11478.
92 Huang, J., Miller, P.F., de Mello, J.C. et al. (2003). *Synth. Met.* 139: 569.
93 Benor, A., Takizawa, S., Chen, P. et al. (2009). *Appl. Phys. Lett.* 94: 193301.
94 Kim, J.Y., Jung, J.H., Lee, D.E., and Joo, J. (2002). *Synth. Met.* 126: 311.
95 Shi, H., Liu, C., Jiang, Q., and Xu, J. (2015). *Adv. Electron. Mater.* 1: 1500017.
96 Ouyang, J., Chu, C.W., Chen, F.C. et al. (2005). *Adv. Funct. Mater.* 15: 203.
97 Kim, Y.H., Sachse, C., Machala, M.L. et al. (2011). *Adv. Funct. Mater.* 21: 1076.
98 Xia, Y., Sun, K., and Ouyang, J. (2012). *Energy Environ. Sci.* 5: 5325.
99 Eom, S.H., Senthilarasu, S., Uthirakumar, P. et al. (2009). *Org. Electron.* 10: 536.
100 Xia, Y. and Ouyang, J. (2012). *ACS Appl. Mater. Interfaces* 4: 4131.
101 Wang, H., Hsu, J.H., Yi, S.I. et al. (2015). *Adv. Mater.* 27: 6855.
102 Xing, W., Chen, Y., Wu, X. et al. (2017). *Adv. Funct. Mater.* 27: 1701622.
103 Shao, S., Liu, J., Bergqvist, J. et al. (2013). *Adv. Energy Mater.* 3: 349.
104 Zheng, Z., Hu, Q., Zhang, S. et al. (2018). *Adv. Mater.* 30: 1801801.
105 Ha, S.R., Park, S., Oh, J.T. et al. (2018). *Nanoscale* 10: 13187.
106 Hu, X., Chen, L., and Chen, Y. (2014). *J. Phys. Chem. C* 118: 9930.
107 Li, Y. and Zeng, W. (2016). *Macromol. Mater. Eng.* 301: 133.
108 Guo, B., Yin, Q., Zhou, J. et al. (2019). *ACS Sustainable Chem. Eng.* 7: 8206.
109 Zeng, M., Wang, X., Ma, R. et al. (2020). *Adv. Energy Mater.* 10: 2000743.
110 Tang, H., Liu, Z., Hu, Z. et al. (2020). *Sci. China Chem.* 63: 802.
111 Ke, W.J., Lin, G.H., Hsu, C.P. et al. (2011). *J. Mater. Chem.* 21: 13483.
112 Lacher, S., Obata, N., Luo, S.C. et al. (2012). *ACS Appl. Mater. Interfaces* 4: 3396.
113 Li, W., Worfolk, B.J., Li, P. et al. (2012). *J. Mater. Chem.* 22: 11354.

114 Mai, C.K., Zhou, H., Zhang, Y. et al. (2013). *Angew. Chem. Int. Ed.* 52: 12874.
115 Zhou, H., Zhang, Y., Mai, C.-K. et al. (2014). *Adv. Mater.* 26: 780.
116 Zhou, H., Zhang, Y., Mai, C.-K. et al. (2015). *ACS Nano* 9: 371.
117 Cui, Y., Xu, B., Yang, B. et al. (2016). *Macromolecules* 49: 8126.
118 Cai, W., Musumeci, C., Ajjan, F.N. et al. (2016). *J. Mater. Chem. A* 4: 15670.
119 Jo, J.W., Jung, J.W., Bae, S. et al. (2016). *Adv. Mater. Interfaces* 3: 1500703.
120 Hains, A.W. and Marks, T.J. (2008). *Appl. Phys. Lett.* 92: 023504.
121 Hains, A.W., Ramanan, C., Irwin, M.D. et al. (2010). *ACS Appl. Mater. Interfaces* 2: 175.
122 Sun, Y., Gong, X., Hsu, B.B.Y. et al. (2010). *Appl. Phys. Lett.* 97: 193310.
123 Xu, Q., Wang, F., Qian, D. et al. (2013). *ACS Appl. Mater. Interfaces* 5: 6591.
124 Tsai, C.E., Liao, M.H., Chen, Y.L. et al. (2015). *J. Mater. Chem. C* 3: 6158.
125 Sun, Y., Wang, M., Gong, X. et al. (2011). *J. Mater. Chem.* 21: 1365.
126 Sun, Y., Chien, S.C., Yip, H.L. et al. (2011). *Chem. Mater.* 23: 5006.
127 Li, S.S., Tu, K.H., Lin, C.C. et al. (2010). *ACS Nano* 4: 3169.
128 Kim, J., Tung, V.C., and Huang, J. (2011). *Adv. Energy Mater.* 1: 1051.
129 Smith, C.T.G., Rhodes, R.W., Beliatis, M.J. et al. (2014). *Appl. Phys. Lett.* 105: 073304.
130 Yang, D., Zhou, L., Chen, L. et al. (2012). *Chem. Commun.* 48: 8078.
131 Li, C., Yang, X., Zhao, Y. et al. (2014). *Org. Electron.* 15: 2868.
132 Kim, S.H., Lee, C.H., Yun, J.M. et al. (2014). *Nanoscale* 6: 7183.
133 Ding, Z., Hao, Z., Meng, B. et al. (2015). *Nano Energy* 15: 186.
134 Liu, J., Xue, Y., and Dai, L. (2012). *J. Phys. Chem. Lett.* 3: 1928.
135 Zhao, F.G., Hu, C.M., Kong, Y.T. et al. (2018). *ACS Appl. Mater. Interfaces* 10: 24679.
136 Stratakis, E., Savva, K., Konios, D. et al. (2014). *Nanoscale* 6: 6925.
137 Xia, Y., Pan, Y., Zhang, H. et al. (2017). *ACS Appl. Mater. Interfaces* 9: 26252.
138 Jeon, Y.J., Yun, J.M., Kim, D.-Y. et al. (2012). *Sol. Energy Mater. Sol. Cells* 105: 96.
139 Kim, J.H., Park, E.K., Kim, J.H. et al. (2016). *Electron. Mater. Lett.* 12: 383.
140 Zhang, P., Xu, X., Dang, Y. et al. (2016). *ACS Sustainable Chem. Eng.* 4: 6473.
141 Cheng, F., Wu, Y., Shen, Y. et al. (2017). *RSC Adv.* 7: 37952.
142 Ji, R., Cheng, J., Yang, X. et al. (2017). *RSC Adv.* 7: 3059.
143 Kim, J., Kim, H., Kim, G. et al. (2014). *ACS Appl. Mater. Interfaces* 6: 951.
144 Bao, X., Zhu, Q., Wang, T. et al. (2015). *ACS Appl. Mater. Interfaces* 7: 7613.
145 Jiang, Y., Xiao, S., Xu, B. et al. (2016). *ACS Appl. Mater. Interfaces* 8: 11658.
146 Xu, W., Liu, Y., Huang, X. et al. (2016). *J. Mater. Chem. C* 4: 1953.
147 Irwin, M.D., Servaites, J.D., Buchholz, D.B. et al. (2011). *Chem. Mater.* 23: 2218.
148 Xue, Z., Liu, X., Zhang, N. et al. (2014). *ACS Appl. Mater. Interfaces* 6: 16403.
149 Yang, W., Yu, Z., Liu, W. et al. (2017). *J. Polym. Sci. Polym. Chem.* 55: 747.
150 Yu, Z., Liu, W., Fu, W. et al. (2016). *J. Mater. Chem. A* 4: 5130.
151 Pho, T.V., Kim, H., Seo, J.H. et al. (2011). *Adv. Funct. Mater.* 21: 4338.
152 Chen, D., Zhou, H., Liu, M. et al. (2013). *Macromol. Rapid Commun.* 34: 595.
153 Zhang, Z., Qi, B., Jin, Z. et al. (2014). *Energy Environ. Sci.* 7: 1966.
154 Zhao, K., Ye, L., Zhao, W. et al. (2015). *J. Mater. Chem. C* 3: 9565.

155 Do, T.T., Hong, H.S., Ha, Y.E. et al. (2015). *Macromol. Res.* 23: 367.
156 Do, T.T., Hong, H.S., Ha, Y.E. et al. (2014). *Synth. Met.* 198: 122.
157 Chen, D., Zhou, H., Cai, P. et al. (2014). *Appl. Phys. Lett.* 104: 053304.
158 Ouyang, X., Peng, R., Ai, L., and Zhang, X. (2015). *Nat. Photonics* 9: 520.
159 Zou, Y., He, Z., Zhao, B. et al. (2015). *Sci. Rep.* 5: 17329.
160 Liu, Y., Zhang, W., Zou, Y. et al. (2016). *Org. Electron.* 30: 182.
161 Liu, S., Zhang, G., Lu, J. et al. (2015). *J. Mater. Chem. C*: 3.
162 Zhang, L., Liu, C., Lai, T. et al. (2016). *J. Mater. Chem. A* 4: 15156.
163 Wang, C., Zhiyang, L., Li, M. et al. (2017). *Chem. Sci.* 8: 3750.
164 Liu, Z., Li, W., Peng, R. et al. (2017). *J. Mater. Chem. A* 5: 10154.
165 Yao, J., Qiu, B., Zhang, Z.G. et al. (2020). *Nat. Commun.* 11: 2726.
166 Li, C., Chueh, C.C., Yip, H.L. et al. (2013). *Adv. Mater.* 25: 2457.
167 Ma, D., Lv, M., Lei, M. et al. (2014). *ACS Nano* 8: 1601.
168 Jiao, W., Ma, D., Lv, M. et al. (2014). *J. Mater. Chem. A* 2: 14720.
169 Li, S., Lei, M., Lv, M. et al. (2013). *Adv. Energy Mater.* 3: 1569.
170 Lv, M., Lei, M., Zhu, J. et al. (2014). *ACS Appl. Mater. Interfaces* 6: 5844.
171 Liu, Y., Page, Z., Ferdous, S. et al. (2015). *Adv. Energy Mater.* 5: 1500405.
172 Liao, S.H., Jhuo, H.J., Cheng, Y.S., and Chen, S.A. (2013). *Adv. Mater.* 25: 4766.
173 Chang, C., Huang, W., Chang, Y. et al. (2015). *Chem. Mater.* 27: 1869.
174 Liu, Y., Sheri, M., Cole, M.D. et al. (2019). *Angew. Chem. Int. Ed.* 58: 5677.
175 Luo, J., Wu, H., He, C. et al. (2009). *Appl. Phys. Lett.* 95: 043301.
176 He, Z., Zhong, C., Su, S. et al. (2012). *Nat. Photonics* 6: 593.
177 Zhou, Y., Fuentes-Hernandez, C., Shim, J.W. et al. (2012). *Energy Environ. Sci.* 5: 9827.
178 Zhu, Y., Xu, X., Zhang, L. et al. (2012). *Sol. Energy Mater. Sol. Cells* 97: 83.
179 Liu, S., Zhang, K., Lu, J. et al. (2013). *J. Am. Chem. Soc.* 135: 15326.
180 Lv, M., Li, S., Jasieniak, J.J. et al. (2013). *Adv. Mater.* 25: 6889.
181 Chen, G., Liu, S., Xu, J. et al. (2017). *ACS Appl. Mater. Interfaces* 9: 4778.
182 Lv, M., Li, Y., Wei, X. et al. (2019). *ACS Appl. Energy Mater.* 2: 2238.
183 Kang, Q., Ye, L., Xu, B. et al. (2019). *Joule* 3: 227.
184 Liu, J., Xue, Y., Gao, Y. et al. (2012). *Adv. Mater.* 24: 2228.
185 Liu, J., Durstock, M., and Dai, L. (2014). *Energy Environ. Sci.* 7: 1297.
186 Wang, D.H., Kim, J.K., Seo, J.H. et al. (2013). *Angew. Chem. Int. Ed.* 52: 2874.
187 Qu, S., Li, M., Xie, L. et al. (2013). *ACS Nano* 7: 4070.
188 Pan, F., Sun, C., Li, Y. et al. (2019). *Energy Environ. Sci.* 12: 3400.
189 Pan, F., Bai, S., Wei, X. et al. (2020). *Sci. China Mater.* 64: 277.
190 Xiong, S., Hu, L., Hu, L. et al. (2019). *Adv. Mater.* 31: 1806616.

14

Morphological Characterization and Manipulation of Organic Solar Cells

Wei Li[1], *Long Ye*[2], *and Tao Wang*[3]

[1] *Swansea University, Sustainable Advanced Materials (Ser̂ SAM), Department of Physics, Singleton Park, Swansea, Wales SA2 8PP, UK*
[2] *Tianjin University, Tianjin Key Laboratory of Molecular Optoelectronic Sciences, Institute of Advanced polymer Materials, School of Materials Science and Engineering, 135 Yaguan Road, Tianjin 300350, P.R. China*
[3] *Wuhan University of Technology, School of Materials Science and Engineering, 122 Luoshi Road, Wuhan 430070, P.R. China*

14.1 Introduction

Modern organic solar cells (OSCs) are based on a sandwiched structure with a bulk-heterojunction (BHJ) active layer consisting of a mixture of electron donor and electron acceptor [1, 2]. At the heart of OSC, BHJ active layer plays a significant role in the power conversion process, which associates with light-harvesting, exciton dissociation, charge transport and charge collection. During this process, both electron donor and electron acceptor in the active layer could absorb photons and create excitons and the produced excitons can then dissociate into holes and electrons when they diffuse to the interfaces of electron-donor and electron-acceptor domains. However, due to the low dielectric permittivity of π-conjugated organic semiconductors, the excitons are tightly bounded by Coulombic force and only show a short diffusion length of around 5–10 nm [3]. Hence, the domain sizes of donor and acceptor should be confined in 10–20 nm to ensure excitons having rich donor–acceptor interfaces for exciton dissociation. On the other hand, after excitons dissociating into holes and electrons, a bi-continuous interpenetrating donor/acceptor network with high domain purity is required for efficient charge transport [4]. On top of that, exploring the nanoscale morphology of BHJ active layers while understanding the correlation between morphology and performance is a significant driving force for advancing OSCs.

The general way to fabricate the BHJ active layers of OSCs is solution processing; herein, conjugated electron donors and acceptors are first dissolved in a solution and then cast into a blend film. During the film-forming process, the donor and acceptor materials will partially demix to lower the free energy of the system and cause phase separation, finally forming a series of heterojunctions throughout the active layer. The donor and acceptor materials used here are p-type and n-type

Organic Solar Cells: Materials Design, Technology and Commercialization, First Edition. Edited by Liming Ding.

semiconductors, respectively, e.g. conjugated polymer, conjugated small molecule, and fullerene derivatives. With different chemical structures, these semiconducting materials can form various structural orders ranging from micron-scale to nanoscale in the BHJ active layer. It is also evident that this nanoscale morphology can be adjusted by film-processing methods, such as the choice of solvents, post-deposition treatments, and casting environments. Consequently, the morphological metrics within OSCs deliver a rich area to study, allowing researchers to explore and optimize the structural order, miscibility, and phase purity of blend components, as well as component distribution in the direction normal to the film. In this chapter, we first introduce a range of morphology characterization techniques including microscopic, scattering, and depth profiling methods (see Figure 14.1). We then discuss typical methods for morphology manipulation and address the correlation between domain size, miscibility, domain purity with J_{sc}, FF, and V_{oc} of an OSC. Finally, we compare the traditional fullerene and the novel non-fullerene

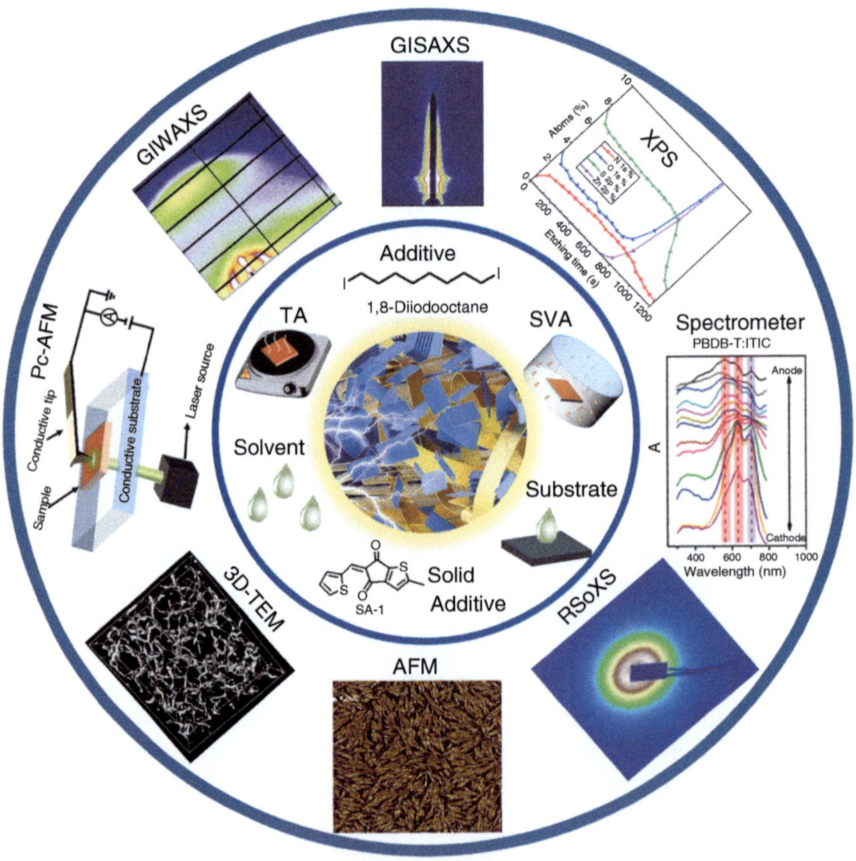

Figure 14.1 Overview of morphology characterization and manipulation methods used in OSCs. Source: Bavel et al. [5]/with permission of Royal Society of Chemistry; Liu et al. [6]/with permission of ELSEVIER; Bi et al. [7]/ Reproduced with permission from Elsevier.

acceptors on the basis of morphology and discuss the existing challenges and future prospects in morphology characterization and manipulation in OSCs toward high performance.

14.2 Morphological Characterization of Organic Solar Cells

14.2.1 Microscopic Methods

Microscopic method is the most intuitive method to examine the morphology of OSCs from nanometer to micrometer scales. There are three mainstream microscopic techniques: optical microscopy, scanning probe microscopy (SPM), and electron microscopy. As optical microscopes use lenses and visible light to magnify the sample, their resolution is limited to approximately 0.2 μm and not suitable for characterizing the nanoscale morphology of OSCs. In this section, we will focus on the SPM and electron microscopy techniques.

SPM was founded in as early as 1980s. It is a surface-sensitive technique that can investigate the surface information by scanning the sample with a physical tip. By applying different tips toward different interactions between tip and sample, the surface morphology, conductivity, potential, stiffness, elastic modulus, and thermal properties of the sample can be determined. Among SPM, atomic force microscopy (AFM) is the most commonly used technique in OSCs. As shown in Figure 14.2a, AFM consists of a sharp tip, a laser beam deflection system, and a position-sensitive detector. During operation, the tip can raster scan the sample through the van der Waals interaction between sample and tip, whilst the laser deflection can be used in a feedback loop to control force and tip position for imaging and measuring. As the crystalline and amorphous domains in organic films exhibit distinct contrast in topology and elastic modulus, topography is the most widely used information in OSCs. For example, Sun [8] group employed tapping mode AFM to study the surface morphology of a high crystalline polymer PBT1-S film (Figure 14.2b,c). Herein, the height and phase images in AFM are related to the difference of topology and material stiffness. It can be found that fibril-like crystals of PBT1-S are clearly resolved in both height and phase images across the entire area. For BHJ blend films, AFM topography provides information on the phase-separated domain size on the film surface, which sometimes cannot be simply applied to the bulk, especially when there is a surface wetting layer formed [10–13].

In addition, AFM can be extended to conducting atomic force microscopy (c-AFM), which is capable of probing the properties of the film beyond the top surface. As shown in Figure 14.2d, c-AFM works in conjunction with an external voltage source and probes the sample with a conductive tip. When c-AFM is scanning, the voltage source applies a potential difference between the tip and the conductive substrate, while the conductive tip can examine current and record the topography and current images simultaneously at the same location of the photoactive layer. As c-AFM represents the local electrical properties of the blend

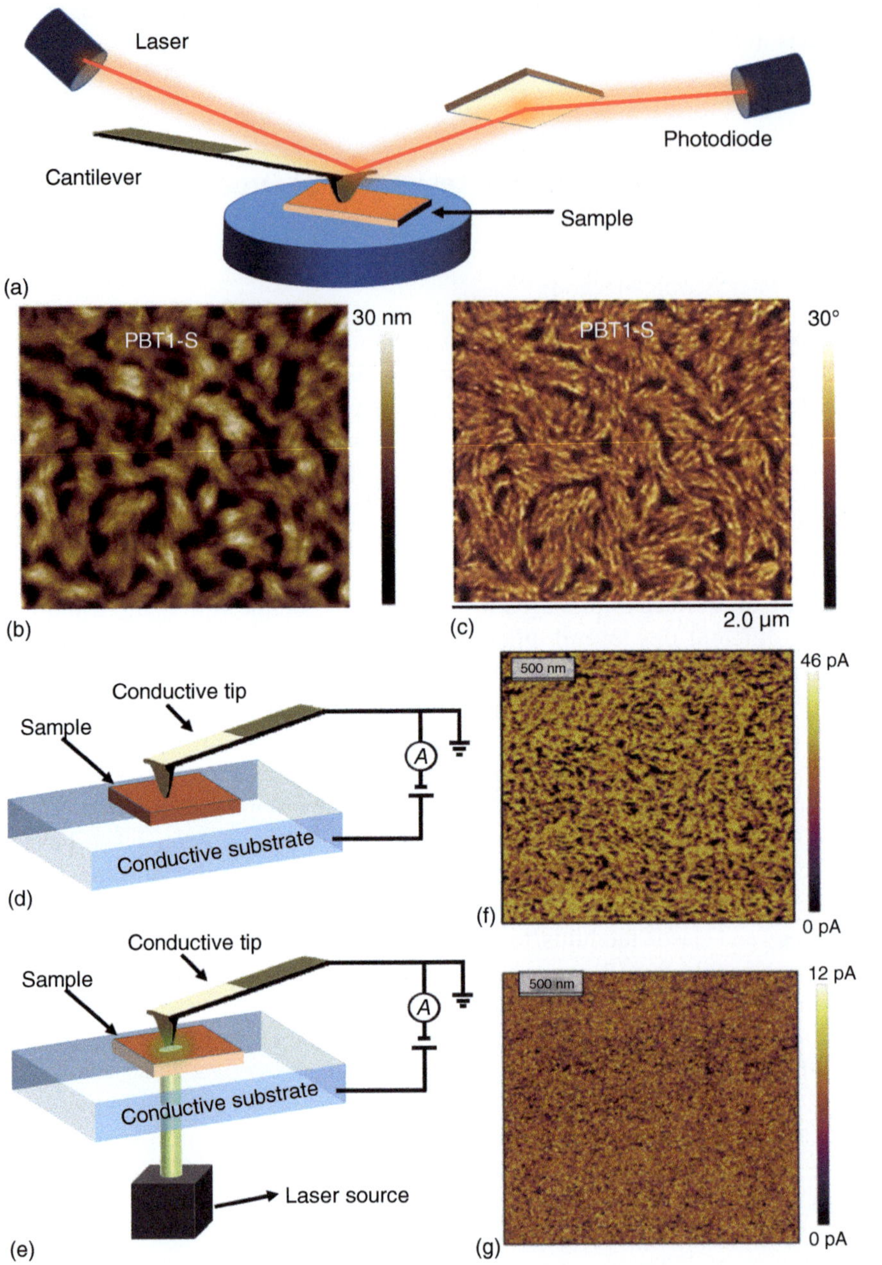

Figure 14.2 (a) Schematic illustration of AFM. AFM (b) height and (c) phase images of PBT1-S film. Source: Liu et al. [8]/with permission of JOHN WILEY & SONS, INC. Schematic illustration of (d) c-AFM and (e) pc-AFM. pc-AFM images of (f) 1% LMWF PM6:Y6 and (g) 52% LMWF PM6:Y6 films measured under white light illumination with no bias. Source: Karki et al. [9]/with permission of Royal Society of Chemistry.

film, the dark current images at positive and negative tip can reflect hole and electron transports [14]. Under forward bias (conductive substrate is positively biased), electrons are injected from the sharp tip to the sample and the bright region in c-AFM images is related to the electron injection which represents the electron-acceptor domain. Similarly, under reverse bias (conductive tip is positively biased), the c-AFM images thus show hole transport and the bright region is related to donor domain [15]. However, it should be noted that due to the different geometries of c-AFM and the planar device in mobility measurement, the traditional Mott–Gurney law is not suitable for c-AFM to quantify the carrier mobility. Herein, a semiempirical modified law (Eq. (14.1)) with the consideration of tip-sample geometry was proposed by Ginger group [16] to estimate the carrier mobilities in c-AFM:

$$J = \frac{9}{8}\varepsilon\varepsilon_0\mu_0 e^{0.89\gamma\left(\frac{V}{L}\right)^{\frac{1}{2}}}\frac{V^2}{L^3}\delta\left(\frac{L}{d}\right)^{1.6\pm0.1} \tag{14.1}$$

where J and V are the average current density and applied voltage of the AFM tip, respectively, ε and ε_0 are the relative dielectric constant of active layer and permittivity of free space, respectively, μ_0 is the charge carrier mobility, γ is the charge density dependence of the mobility, L is the thickness of the active layer, d is the diameter of the contact area between the tip and the active layer, and δ is the empirical dimensionless parameter of the offset between c-AFM calculations and the planar device data which is 7.8±1.

c-AFM can be further extended into photoconductive atomic force microscopy (pc-AFM) when it is performed in conjunction with an external light source. As shown in Figure 14.2e, a diffraction-limited laser beam illuminates the sample from the bottom, while a conductive tip can measure electrical properties simultaneously. Compared to AFM and c-AFM, pc-AFM can provide local photocurrents and illustrate how the individual nanostructures contribute to the device performance in OSCs. For example, Nguyen group applied a white light to illuminate the PM6:Y6 film while measuring the photocurrent. As shown in Figure 14.2f [9], the PM6:Y6 film with 1% low molecular weight fraction (LMWF) PM6 exhibits a continuous interpenetrating network of hole- (PM6, bright area) and electron (Y6, dark area)-rich domains with high photocurrent. However, the hole- and electron-rich domains are not continuously separated in the 52% LMWF PM6:Y6 film and only obtained a low photocurrent (Figure 14.2g), consistent with their device performance. Furthermore, although c-AFM and pc-AFM have been widely used and provided valuable information for the solar cell community, two concerns should be noted. First, performing c-AFM or pc-AFM with high bias could induce local oxidation or reduction in the blend film, which means they should be carried out in vacuum or inert atmospheres. Second, electrical currents only follow the path of the active layer with the lowest resistance, which can also straight down through the film, implying that the map of electrical conductance is not just simply correlated with the surface morphology and but should also be analyzed with caution.

In addition, combining AFM with a tunable infrared laser (IR-AFM) to produce a chemical contrast has also been studied to characterize morphology of the BHJ active layers. To conduct IR-AFM, the IR laser focuses on the interaction region between the AFM tip and sample. After the sample absorbed IR photons, an electric dipole can be generated from the vibrational motion of the IR-resonant bonds of the sample, enabling an attractive force between the AFM tip and sample molecules. For example, PII-2T-PS and PPDI-T possess different characteristic peaks at 1453 and 1706 cm^{-1} in their FTIR spectra, respectively. By activating the PII-2T-PS:PPDI-T blend film at 1453 and 1706 cm^{-1}, IR-AFM is able to spatially map the donor and acceptor domains with a resolution of about 10 nm [17].

Compared to the traditional optical microscopy, electron microscopy applies electrons for imaging, enabling superior spatial resolution to magnify the sample. Among electron microscopy techniques, transmission electron microscopy (TEM) is the most widely used method for organic semiconductor films. TEM is often operated in a bright-field (BF) imaging mode, where a cluster of electron beam goes through the thin film and run into the atoms in the sample, the collided electrons can then scattered into different angular and form bright and dark parts in the image with the mass, density, and thickness variations of the sample. In general, the contrast of a TEM image in OSCs is related to the crystallinity or the aggregation of components in the film. For example, as shown in Figure 14.3a, when $PC_{70}BM$ molecules aggregated strongly in the film, a dark ball-like domain with a domain size of 200 nm can be found due to the higher electron density of $PC_{70}BM$ aggregation [18]. Similarly, in Figure 14.3b, CO_i8DFIC molecules obtained higher crystallinity than PTB7-Th after solvent annealing, which allows us to observe CO_i8DFIC lamellar crystals with the size of about 250 nm in the film [19].

One extension to the traditional bright-field TEM is combining a series of 2D TEM to construct 3D TEM image. As shown in Figure 14.3c,d, 3D-TEM can be realized by taking a series of 2D-TEM projections at different angles by tilting the sample (specimen) with respect to the electron beam and subsequently reconstructing to the 3D image on the basis of computerized tomography. Figure 14.3e [5] shows a reconstructed 3D nanoscale polymer network of a PF10TB:$PC_{60}BM$ film. It can be found that the bright region belongs to PF10TB domain, which demonstrates an interconnected polymer-rich strand forming a truly three-dimensional network.

Since the commonly used donor and acceptor materials in OSCs are π-conjugated, organic semiconducting materials, it is not easy to obtain decent contrast for TEM if the donor and acceptor possess similar chemical structure. Hence, researchers have combined electron energy-loss spectroscopy with TEM (EF-TEM) to probe the organic films at different energy windows to enhance contrast of the components in the BHJ active layer [22]. For example, Rechberger et al. [20] have successfully applied EF-TEM to resolve the morphology of a ternary PIDTTQ:Si-PCPDTBT:$PC_{70}BM$ BHJ film. Herein, the zero-loss filtered image is the same to the regular bright-field TEM, and the contrast in the film is only attributed to their local density differences. While probing the film at 30 ± 4 eV could maximize the contrast between polymers (PIDTTQ:Si-PCPDTBT) and

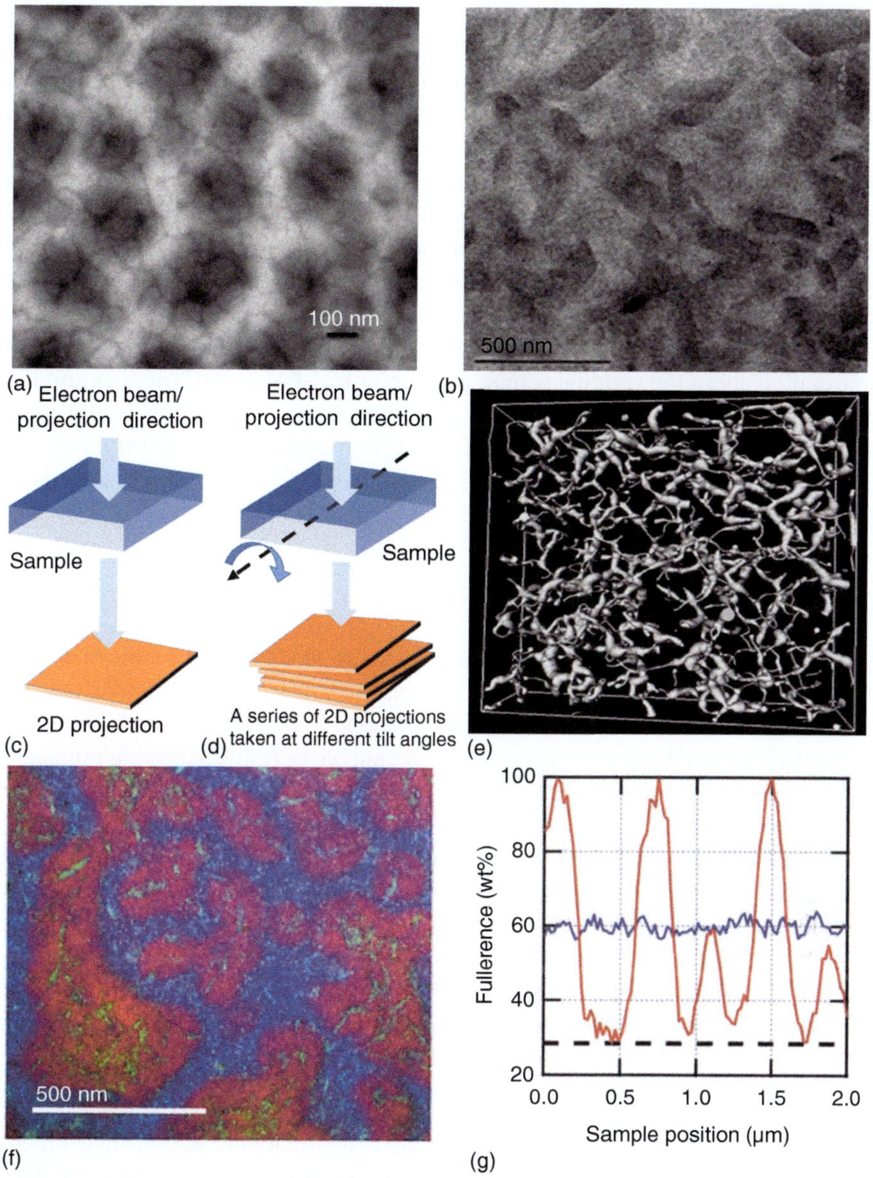

Figure 14.3 (a) TEM image of PTB7:$PC_{70}BM$ blend film cast from chlorobenzene (CB). Source: Liu et al. [18]/with permission of JOHN WILEY & SONS, INC. (b) TEM of PTB7-Th:CO_i8DFIC film treated with solvent annealing. Source: Li et al. [19]/with permission of ELSEVIER. Scheme of the principle of (c) traditional 2D-TEM image and (d) 3D-TEM techniques. (e) A snapshot of the reconstructed 3D TEM image of PF10TB:$PC_{60}BM$ film. Source: Bavel et al. [5]/with permission of Royal Society of Chemistry. (f) EF-TEM image of PIDTTQ:Si-PCPDTBT:$PC_{70}BM$ (0.5 : 0.5 : 2) film displaying in RGB color representing the three materials. (PIDTTQ: blue, Si-PCPDTBT: green, and $PC_{70}BM$: red). Source: Rechberger et al. [20]/with permission of JOHN WILEY & SONS, INC. (g) Histograms of composition variation profiles taken from STXM images of PTB7:$PC_{70}BM$ films cast from CB and CB:DIO solutions. The red and blue lines represent the $PC_{70}BM$ content in the PTB7:$PC_{70}BM$ films formed from CB solution or CB:DIO solution respectively. Source: Collins et al. [21]/Reproduced with permission of John Wiley & Sons.

fullerene ($PC_{70}BM$), probing the film at an energy loss of 267 eV with an energy window of 30 eV could maximize the structure-sensitive contrast between PIDTTQ and Si-PCPDTBT. Overall, a RGB color can be established on top of the above two views to visualize the nanomorphology of PIDTTQ, Si-PCPDTBT, and $PC_{70}BM$ in the film (Figure 14.3f).

Scanning transmission X-ray microscopy (STXM) is a powerful synchrotron-based technique, which can utilize a variable-energy focused X-ray to measure the sample and quantitatively map the chemical composition of organic films at a wavelength with high contrast. In addition to the regular morphological metrics such as shape and domain size, STXM can also provide information on miscibility and domain purities. In 2009, Watts et al. [23] successfully applied STXM to reveal quantitative information about the residual weight fraction of $PC_{60}BM$ in P3HT matrix and first discovered the limited purity in BHJ active layer. Later, as shown in Figure 14.3g, Collins et al. [21] further studied the composition profiles of PTB7:$PC_{70}BM$ films cast from CB or CB:DIO solution and demonstrated the effect of domain purity on device performance in solar cells.

14.2.2 Scattering Methods

Although traditional microscopic techniques can provide the information of domain sizes, shapes, and purities, due to the less-ordered structure and similar electron density of donor and acceptor in organic semiconductor films, it is still difficult to get sufficient spatial resolution and thus not capable of fully examining molecular packing and domain purity in a BHJ film. Therefore, scattering techniques based on X-ray or neutron are often used as a complementary technique to comprehensively characterize the nanoscale morphology of the BHJ active layer at multiple length scales. Scattering is a reciprocal space technique, which can probe the structure of the sample by utilizing scattering effect to generate patterns that can be correlated with nanoscale morphology. To enhance interactions between X-ray/neutrons and the thin BHJ active layers, scattering techniques are always carried out with synchrotron radiation to enable high flux and collimated X-ray or neutron photons to enhance intensity. Figure 14.4 shows a typical schematic view of scattering measurements processed in three different geometries:

(1) Grazing incidence scattering geometry: Grazing incidence scattering geometry is the most commonly used geometry for film samples. Here, a collimated X-ray or neuron impinges on the sample under a grazing angle k_i to minimize the background signal while enhancing the scattering intensity. The scattering beam can be then collected by a 2D charge-coupled device (CCD) and simply interpreted by the Bragg equation $n\lambda = 2d\sin\theta$. From Figure 14.4a, it is clear to see that the scattering angle k_f is inversely proportional to the distance (d) between the sample and the CCD. Therefore, depending on the distance between the sample and the CCD, the related experiments can be typically referred to wide-angle scattering (WAS) or small-angle scattering (SAS) which could reflect the morphology information associated with large or small k_f. It should be noted that WAS and

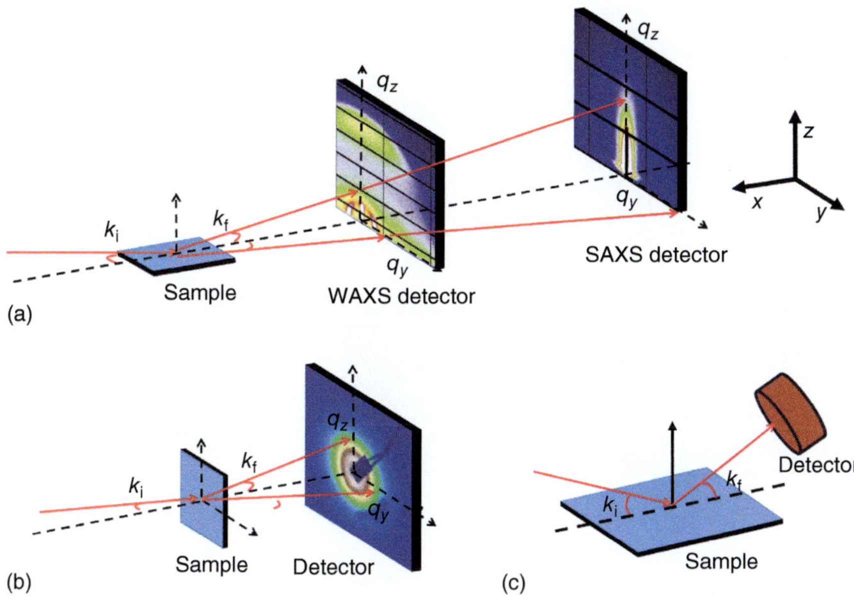

Figure 14.4 Schematic view of the typical (a) grazing incidence, (b) transmission, and (c) reflection geometry for X-ray and neutron scattering measurements.

SAS measurements are also suitable for the transmission and reflectance scattering geometries.

(2) Transmission scattering geometry: In transmission geometry, the X-ray or neutron impinges the sample and goes through directly under an incident angle of k_i, while the scattering signal can be collected by a CCD (Figure 14.4b). Compared to the grazing incidence scattering geometry, the thickness of the sample used here should be thin enough for the X-ray or neutron to penetrate and form decent scattering intensity on CCD.

(3) Reflectance geometry: In contrast to the grazing incidence geometry where the specular reflection is blocked and only the diffuse scattering being detected, reflectance geometry only measures the specular reflection at q_z direction and the scattering at q_y direction is null. Therefore, reflectance geometry scattering is often used for detecting the depth information of films. Here, by changing the perpendicular wavevector transfer of the incident angle (k_i), the resulting scattering intensity can reveal information at different depths (Figure 14.4c). For $k_i < k_c$ (critical angle), highly sensitive surface information can be achieved within several nanometers. For $k_i > k_c$, the scattering intensity reflects the averaged information of the film.

Grazing incidence wide-angle X-ray scattering (GIWAXS) can be understood as wide-angle X-ray scattering (with the scattering θ large than 2°) in a grazing incidence geometry. According to the Bragg equation, it can be found that the scattering signal obtained in GIWAXS is associated with the sub-nanometer structure (0.5–2 nm), which reflects the information of the molecular packing

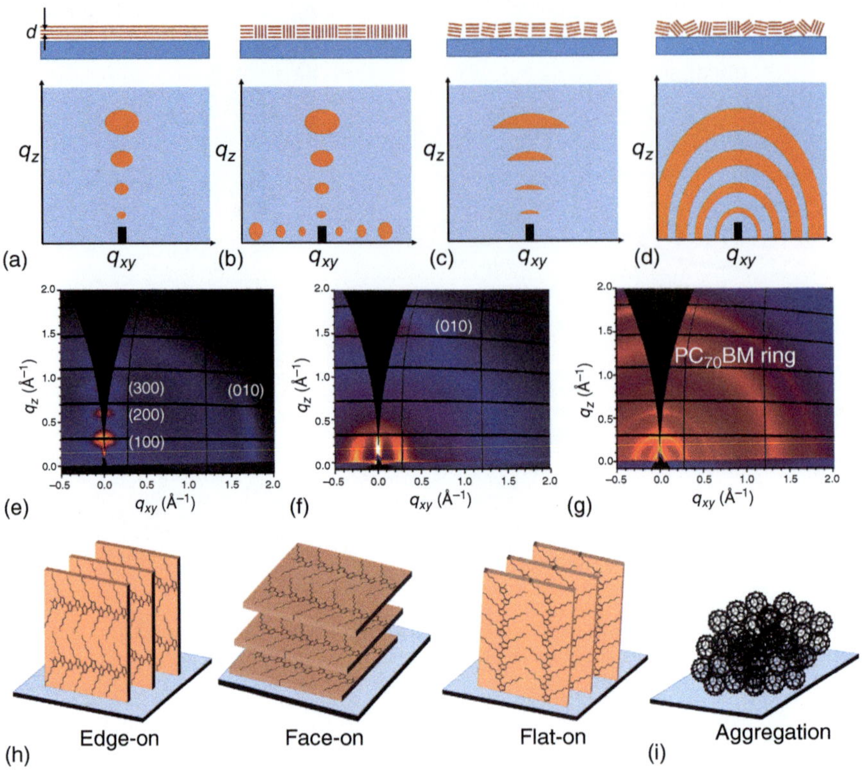

Figure 14.5 Schematic illustration of typical 2D GIWAXS images demonstrating film with different molecular packing. (a) Vertical lamellar stacking, (b) vertical and horizontal lamellar stacking, (c) vertically oriented domains, (d) isotropic domains without preferential orientation. 2D GIWAXS patterns of (e) PffBT4T-2OD, (f) PTB7-Th and (g) PffBT4T-2OD:$PC_{70}BM$. Source: Li et al. [11]/ Reproduced with permission from John Wiley and Sons. (h) The crystalline structure of a widely studied conjugated polymer (i.e. P3HT). (i) Aggregation of fullerenes in the film.

and crystallinity of the active layer [24]. Figure 14.5a–d summarizes four typical scenarios of GIWAXS patterns in organic semiconducting films. Here, the q_{xy} direction in 2D GIWAXS patterns only reflect the in-plane information (lateral morphology parallel to the film) of the film, while the q_z direction in 2D GIWAXS patterns relates to the out-of-plane information (vertical morphology normal to the film) of the film. As shown in Figure 14.5a, when there is a highly crystalline film (crystal spacing: d) with an orientation normal to the substrate, scattering peaks will appear along the vertical q_z direction (out-of-plane). In Figure 14.5b, scattering peaks could emerge along both horizontal q_{xy} (in-plane) and vertical q_z (out-of-plane) directions if there are two highly crystalline domains with an orientation parallel and perpendicular to the substrate. Besides, when the film is more textured and the domains inside are oriented with an angular distribution around the out-of-plane, scattering peaks in out-of-plane will be broadened (Figure 14.5c). In the case of a film with isotropic crystals, diffraction peaks will smear out into Debye–Scherrer-like rings (Figure 14.5d).

To ensure strong π–π ordering and good solubility, organic semiconducting materials always feature a larger planar conjugated structure and alkyl side chains in their chemical structures [25]. Herein, we will start with two commonly used high efficiency materials PffBT4T-2OD and PTB7-Th to examine the crystalline structure of polymer donors. Figure 14.5e shows the 2D GIWAXS pattern of a pure PffBT4T-2OD film cast from CB; it can be found that its GIWAXS pattern exhibits a strong crescent pattern at $q_z = 0.30\,\text{Å}^{-1}$, which is attributed to the (100) lamellar packing of PffBT4T-2OD. A lamellar spacing ($d = 2.1$ nm) of PffBT4T-2OD can also be calculated by $d = 2\pi/q$. Meanwhile, in the $q_z = 0.6$ and $0.9\,\text{Å}^{-1}$, two relatively weak crescent patterns can also be observed, which are attributed to the second- (200) and third- (300) diffraction peaks of the lamellar packing of PffBT4T-2OD. Moreover, in the in-plane direction, a peak can be observed at $q = 1.75\,\text{Å}^{-1}$, which is ascribed to the π-π stacking of PffBT4T-2OD with the spacing of 0.36 nm. Overall, we can conclude that the side chains and backbone of PffBT4T-2OD are perpendicular to the substrate orderly, indicating that PffBT4T-2OD molecules mainly exhibit the so-called "edge-on" crystalline structure in the film (Figure 14.5h, note that the chemical structure here is a scheme only). On the contrary, as shown in Figure 14.5f, the 2D GIWAXS pattern of PTB7-Th film exhibits (100) diffraction peak in in-plane, while the (010) π–π stacking diffraction peak in out-of-plane, indicating that the conjugated backbone of PTB7-Th is parallel to the substrate. As depicted in Figure 14.5h, this crystalline structure has been defined as "face-on" molecular packing. "Edge-on" and "face-on" are the most common molecular orientation in polymer donor materials. It has been widely accepted that "face-on" molecular order is more favorable for OSCs as holes and electrons can transfer more efficiently through π–π stacks along the out-of-plane, while "edge-on" molecular order is more favorable for transistor device because holes and electrons are transported in in-plane [26]. In addition, "flat-on" molecular order with ordered backbone in out-of-plane also exists in organic films (Figure 14.5h). However, due to the unstable structure of this crystalline structure, it is not commonly observed.

Fullerene electron acceptors possess a ball-like geometry structure, as shown in Figure 14.5i, which does not have complicated molecular orientations like conjugated polymers and always pack isotopically in the film. Figure 14.5g shows a 2D GIWAXS image of the PffBT4T-2OD:$PC_{70}BM$ blend film, in addition to the "edge-on" crystalline structure of PffBT4T-2OD, a Debye–Scherrer-like ring locating at $1.4\,\text{Å}^{-1}$ can also be found and relates to the aggregation of $PC_{70}BM$. This Debye–Scherrer-like ring shows the same intensity in either in-place or out-of-plane, suggesting that $PC_{70}BM$ exists in the film isotopically with domain spacing of 2 nm. Due to the isotopic feature of $PC_{70}BM$, the charge generation in polymer:fullerene system is relatively simple. As shown in Figure 14.6a,b, the charge generation efficiency mainly depends on the aggregation state of $PC_{70}BM$, attempting to suppress $PC_{70}BM$ aggregation to provide more donor:acceptor interfaces in BHJ active layer has been accepted as an efficient routine to improve charge generation. In contrast to fullerene acceptor, the geometry structure of the modern non-fullerene acceptors is more complicated, which consists of three parts, alkyl chains, an electron-donating backbone core, and two electron-accepting

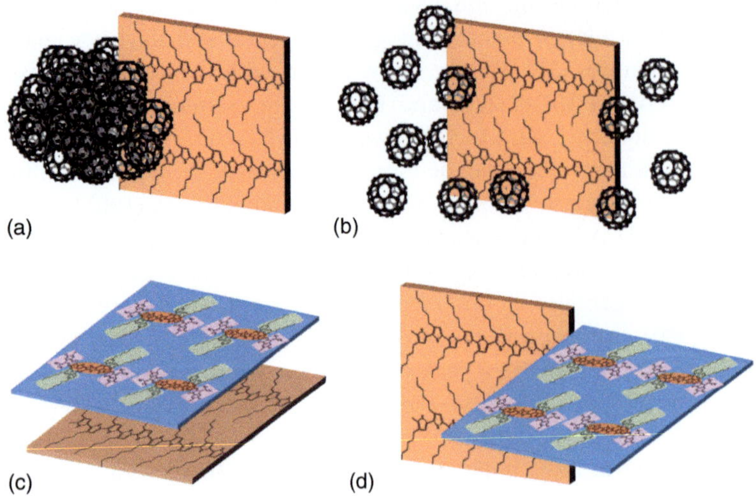

Figure 14.6 Schemes of the polymer: fullerene films with (a) strong and (b) weak fullerene aggregation. Schemes of the polymer:non-fullerene films with (c) both polymer and non-fullerene acceptor exhibiting face-on molecular orientation and (d) polymer showing edge-on but non-fullerene acceptor displaying face-on molecular orientation.

end-groups. Non-fullerene acceptors can exhibit "face-on," "edge-on," and "flat-on" molecular orientations in a film, sometimes simultaneously. This leads to much complicated morphological geometry for charge generation [27]. As shown in Figure 14.6c, when both polymer donor and non-fullerene acceptors exhibit face-on orientation, charges can be generated at the interfaces between the conjugated π–π backbones of polymer donor and non-fullerene acceptors, which has been proved to be an efficient process [19]. However, when polymer donor and non-fullerene acceptors exhibit edge-on and face-on molecular packing respectively, charge generation might occur between the alky chain of polymer and π–π backbone of non-fullerenes and cause different charge generation fundamentals [28] (Figure 14.6d). In addition to molecular order, GIWAXS can also be used to calculate the crystalline correlation lengths (crystalline size) of the crystals in the BHJ active layer by the Scherrer equation:

$$D_{hkl} = 2\pi K/\Delta q_{hkl} \tag{14.2}$$

where D_{hkl} could be roughly translated as the mean size of the crystalline domains which is smaller or equal to the grain size, K is the shape factor of the related crystals in the film, a typical value of about 0.9 is commonly used for organic materials, and Δq_{hkl} is half of the maximum intensity of the diffraction peak in 1D GIWAXS profile.

In contrast to GIWAXS that possesses a relatively short sample-detector distance (centimeters) to collect scattering signal at large angle region ($2\theta > 5°$), grazing incidence small-angle X-ray scattering (GISAXS) has a much longer sample-detector distance (several meters) and collects scattering signal at small scattering angle region ($2\theta < 5°$). Accordingly, all the pixels on the detector collect signal from a small angular range, and GISAXS can be easily regarded as zooming in the GIWAXS pattern in the small angle regime. Due to the small scattering angle in GISAXS, GISAXS is usually used to investigate the size scale from 10 to 500 nm. The scattering signal of GISAXS is from the variations of the mean electronic density and the mean electron scattering length of the components in the film. In OSCs, GISAXS could reflect the information of size, shape, phase separation, and uniformity of the distributions of donor and acceptor components. Since the GISAXS experiments are based on the intensity of the Fourier transform of the electron density, the received 2D GISAXS data must be interpreted to yield relevant morphological information. However, due to the complicated scattering process in the BHJ layers, there is no versatile formula to model all BHJ layers and it is difficult to assign a scattering pattern to a specific morphology. Therefore, multiple form factors and structure factors are often used to interpret the morphological information. Consequently, GISAXS is often used as a supplemental tool to fully characterize morphology in conjunction with electron microscopy.

Figure 14.7a shows a typical 2D GISAXS pattern of a BHJ film. The most distinct feature in GISAXS pattern is the horizontal scattering strip in the region about $q_z = 0.03\ \text{Å}^{-1}$, which is often used to model the lateral morphological structure. Wu et al. [29] applied GISAXS to reveal the domain size of $PC_{60}BM$. Figure 14.7b shows the 1D profiles of a P3HT:$PC_{60}BM$ film during the heating process to 150 °C; it can be found that the as-cast P3HT:$PC_{60}BM$ film does not show any Bragg diffraction peaks but exhibits a weak broad scattering peak in the range from 0.004 to $0.1\ \text{Å}^{-1}$. This broad peak becomes more pronounced with the increased annealing temperature, which is attributed to the aggregation of $PC_{60}BM$ particles during annealing. Hence, a conventional polydisperse spherical model with Schultz size distributions was applied to model the domain size of $PC_{60}BM$ particles. As a result, they found that the average size of $PC_{60}BM$ clusters increased from 10 to 17.5 nm when the temperature increased from 30 to 150 °C and become stable after prolonging the thermal annealing (TA) at 150 °C (Figure 14.7c).

In addition, Su and coworkers [30] also established a GISAXS model to study the domain size, shape, and aggregation in a PCPDTBT:$PC_{70}BM$ film. They found that the GISAXS intensity in its 1D profile can be divided into two regions which mainly come from different phase: (i) an upturn in the low-q region (<$0.01\ \text{Å}^{-1}$) following power-law behavior, which is associated with the polymer domain and can be fitted by the Debye–Anderson–Brumberger (DAB) model; (ii) a shoulder in the middle-q region (0.01–$0.03\ \text{Å}^{-1}$) which can be fitted by the fractal model and its signal mainly come from the aggregation of $PC_{70}BM$ phase (Figure 14.7d). Overall, the GISAXS data can be well fitted by the DAB-Fractal model:

$$\mathrm{I}(q) = \frac{A_1}{\left[1 + (q\xi)^2\right]^2} + A_2\langle P(q, R)\rangle S(q, R, \eta, \mathrm{D}) + B \tag{14.3}$$

$$S(q) = 1 + \frac{\sin\left[(D-1)\tan^{-1}(q\eta)\right] b \pm \sqrt{b^2 - 4ac}}{(qR)^D} \frac{D\Gamma\,(D-1)}{\left[1 + \frac{1}{(q\eta)^2}\right]^{(D-1)/2}} \tag{14.4}$$

where the first item represents the amorphous $PC_{70}BM$ dispersed PCPDTBT phase, A_1 is an independent fitting parameter, and ξ is the average correlation length of the

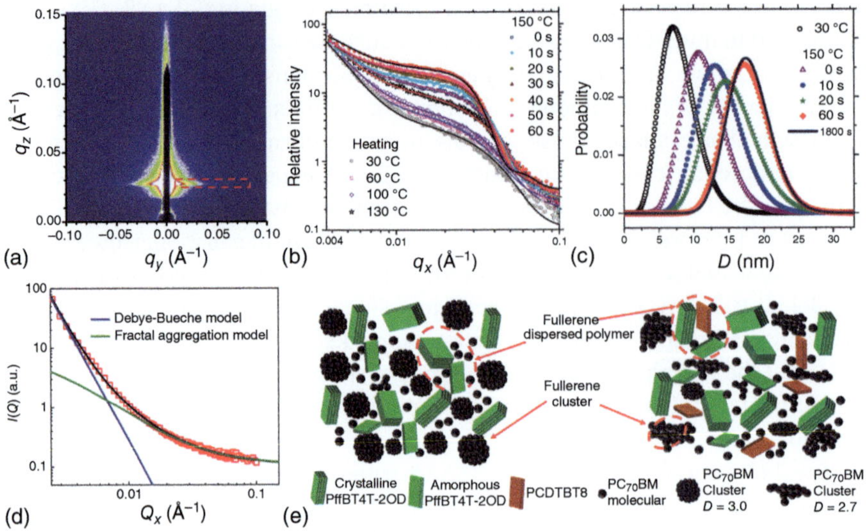

Figure 14.7 (a) A typical 2D GISAXS pattern of the BHJ active layer film. (b) 1D GISAXS profiles of P3HT:$PC_{60}BM$ film extracted from the corresponding 2D patterns. P3HT:$PC_{60}BM$ film was first thermal annealed at 30, 60, 100, 130, and 150 °C, while their related GISAXS data are recorded. Under the temperature of 150 °C, the GISAXS data were recorded at intervals of 10s. (c) The domain size distribution of P3HT:$PC_{60}BM$ obtained from modeling the GISAXS 1D profile with a polydisperse spherical model with Schultz size distributions. Source: Wu et al. [29]/Reproduced with permission of American Chemical Society. (d) GISAXS intensity calculated by model fitting constituted of the intensities determined by Debye–Bueche model (blue line) and Fractal aggregation model (green line), Source: Liao et al. [30]/Reproduced with permission of Royal Society of Chemistry. (e) The schematic illustration of PffBT4T-2OD:$PC_{70}BM$ film with different fullerene aggregation. Source: Li et al. [31]/Reproduced with permission of Elsevier.

polymer domain. The second item $S(q)$ represents the contribution from fractal-like aggregations of $PC_{70}BM$, where $P(q, R)$ and $S(q, R, \eta, D)$ are the form and structure factor of $PC_{70}BM$, R is the mean spherical radius of primary $PC_{70}BM$ molecules, and D is the dimensionality of $PC_{70}BM$ which describes the interaction between primary particles in this fractal-like aggregation system. The average domain size of the $PC_{70}BM$ domain can be then roughly characterized by the Guinier radius (R_g) of the network:

$$R_g = \sqrt{\left(\frac{D(D+1)}{2}\right)\eta} \tag{14.5}$$

Furthermore, Wang group [31] further applied this model to fit a highly crystalline PffBT4T-2OD:$PC_{70}BM$ system. As shown in Figure 14.7e, they found that when an amorphous polymer PCDTBT8 was introduced into the highly crystalline PffBT4T-2OD:$PC_{70}BM$ blend, the dimensionality of the $PC_{70}BM$ aggregation can decrease from 3 to 2.7, delivering a loose aggregation to $PC_{70}BM$ and providing more PffBT4T-2OD:$PC_{70}BM$ interfaces for charge generation. In addition to the above polydisperse spherical and DAB-Fractal model for GISAXS modeling, other methods such as distorted-wave Born approximation (DWBA) model [32], Guinier approximation [33], polydisperse ellipsoid with hard sphere [34], and Porod approximation [35] including different form and structure factors have also been employed to characterize the morphology of fullerene- and non-fullerene-based OSCs.

In addition, GISAXS can be transferred into grazing incidence small-angle neutron scattering (GISANS) when X-ray is replaced by a neutron source. Different to GISAXS probing matter through the electron, GISAXS interacts with the sample through nuclear interactions, which enable higher penetration toward most of the elements and results in a better signal-to-noise ratio in detection. Hence, compared to GISAXS, GISANS could probe nanostructure possessing low or non-contrast for X-rays, and the possibility to study vertical composition and even buried structure [36, 37]. For example, Muller-Buschbaum group [38] applied a time of flight (TOF) mode of GISANS to investigate the structure of the P3HT:PC_{60}BM film. Since the critical angle for GISANS depends on the wavelength of neutron, applying GISANS at different neutron wavelengths under a fixed incident allows probing the scattering information of different depths. At small wavelength, the scattering intensity is relating to information of the bulk film. At large wavelength, the incident angle is smaller than the critical angle and therefore is related to structure information at the near surface region. Figure 14.8a–d shows the TOF-GISANS image of P3HT:PC_{60}BM film measured at the neutron wavelength of 3.7, 4.5, 6.7, and 9.9 Å. Here, three main features of a specular reflection peak, a Yoneda peak, and transmitted signal hidden by the beam stop can be observed. Normally, the intensity in the Yoneda region along the q_y is used to interpret the inner nanostructures of the sample. With the increased neutron wavelength, it can be found that the intensity side maxima at the Yoneda peak position first expand to a high-q_y range and then disappeared at long neutron wavelengths (9.25 Å), suggesting that the average domain size on the film surface is bigger than that of the inner film. However, it should be noted that performing neutron scattering requires neutron–scattering facilities, which is not easily accessible for users, and neutron scattering always feature low flux, which requires longer measurement time compared to X-ray scattering.

As organic semiconductors always show energy dependence in their optical constants, applying the different energy dependence of the donor and acceptor in the BHJ active layer can also be used to enhance the scattering contrast in a scattering experiment. Resonant soft X-ray scattering (RSoXS) can measure the sample at specific resonant energies and enhances the scattering intensity by several orders of magnitude. Like GISAXS, RSoXS is also processed at small scattering angles. Differently, RSoXS is usually operated in a transmission geometry under vacuum, which is difficult to realize *in situ* measurements like GISAXS. More importantly, in addition to measuring the multiple length scales of a BHJ film, the best advantage of RSoXS is that it allows measuring the domain purity of components. Figure 14.8e [4] shows a typical 2D RSoXS pattern of a BHJ film. Here, a line cut is extracted from the 2D pattern and used for modeling, and the total scattering intensity (TSI) of the film can be described by the equation below:

$$\mathrm{TSI} = \int_0^\infty I(q)\, q^2 \mathrm{d}q = 2\pi^2 \Delta\rho_{12}^2 \phi_1 \phi_2 \mathrm{V} \tag{14.6}$$

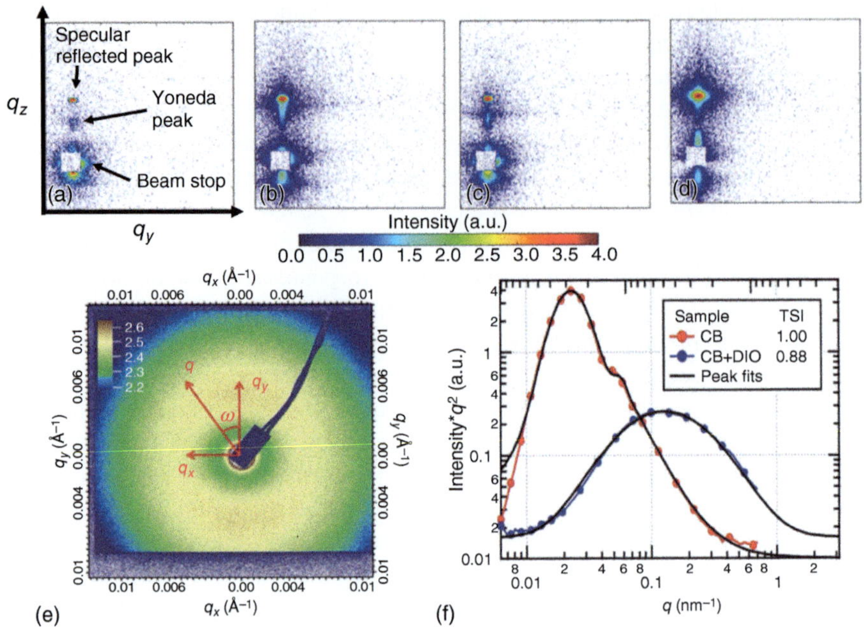

Figure 14.8 TOF-GISANS patterns of P3HT:PC_{60}BM film measured at different wavelength of (a) 3.7, (b) 4.5, (c) 6.7, (d) 9.9 Å. Source: Guo et al. [38]/Reproduced with permission of American Chemical Society. (e) Typical 2D RSoXS image of a BHJ film. Source: Jiao and Ade [4]/Reproduced with permission of John Wiley & Sons. (f) Integrated scattering profiles and the domain purity of PTB7:PC_{70}BM films cast from CB and CB:DIO solutions. Source: Collins et al. [21]/Reproduced with permission of John Wiley & Sons.

where the $\Delta\rho_{12}$ is the contrast of the electron density between acceptor and donor, V is the total volume of the film being illuminated during scattering, and ϕ_1 and ϕ_2 are the volume fraction of donor and acceptor domains. By considering the straightforward relation between ϕ_1 and ϕ_2, the normalized total integrated scattering intensity (TSI) values can be calculated to evaluate the average domain purity over all length scales for blends. It should note that the average domain purity received here is not the absolute domain purity. A specimen with the highest TSI is defined as domain purity of 1, while the domain purity of other specimens can be calculated by the difference of TSI. As shown in Figure 14.8f [21], the PTB7:PC_{70}BM films formed with or without 1-8-diiodooctane (DIO) additive obtained average domain purity of 1.0 and 0.8, indicating that more mixed region and diffuse interfaces existing in the PTB7:PC_{70}BM film processed with DIO.

14.2.3 Depth Profile

Since the architecture of OSCs features stacks consisting of multiple functional layers, surfaces and interfaces in OSCs play an important role in the device performance. Consequently, characterizing the vertical stratifications of electron donors and acceptors in a BHJ active layer is also significant for further optimizing morphology and obtaining higher PCE.

X-ray photoelectron spectroscopy (XPS) is a spectroscopic technique that can quantify the elements, identify the chemical state and the electronic structure of the film surface by the photoelectric effect, which has widely been used in surface characterization in OSCs. In polymer:fullerene solar cells, as polymer donors often contain sulfur and carbon elements whilst fullerene acceptor only contains carbon element, the sulfur signal S (2p) can be used to represent the content of polymer donors; therefore, the S (2p)/C (1s) signal can be applied to deduce the polymer/fullerene ratio in the film. For example, Franky group found that the S (2p)/C (1s) peak-area ratio in the bottom surface is much large than that in the top surface, implying that there is a PDTS-BTD-rich layer in the bottom surface but a $PC_{70}BM$-rich layer in the top surface. Besides, due to the limited inelastic-free path of electrons, the regular XPS can only measure the depth of the surface <10 nm. To fully characterize vertical morphology, Wang group [39] employed XPS with etching technique to probe the whole vertical morphology information of a PBDB-T:ITIC film. During their measurements, they first cast PBDB-T:ITIC blend film on the ZnO substrate and then etched the blend film while conducting XPS to measure the distribution of elements at different thickness. Figure 14.9a shows the N (1s), O (1s), S (2p), and Zn (2p) ratio of the PBDB-T:ITIC film at different etching steps. As all N atoms are from ITIC and most of the S atoms are from PBDB-T, it can be concluded that there is a PBDB-T rich layer at the first 10 nm of the film and the PBDB-T:ITIC ratio becomes stable and uniform in the rest of the thickness.

In addition to XPS, secondary-ion mass spectrometry (SIMS) can also probe the surface information of a film by sputtering the film with a focused primary ion beam (Ar_+) while collecting the ejected secondary ions for mass spectrometry analysis. As early as 2004, Janssen group [42] has applied TOF-SIMS to study the depth information of a MDMO-PPV:$PC_{60}BM$ device, indicating the possibility of achieving good resolution in layer-to-layer depth profiling. Later, since the $^2H^-$ secondary ion is not subject to interference from $^{32}S^-$ and $^{16}O_2{}^-$ that has similar mass, more researchers [43] used deuterated acceptors to replace the regular acceptors to enhance the contrast of SIMS and have revealed the vertical morphology of fullerene and non-fullerene systems more accurately. Besides XPS and SIMS, other surface-sensitive techniques such as Auger electron spectroscopy (AES), energy-dispersive X-ray spectroscopy (EDS), and ultraviolet photo-electron spectroscopy (UPS) can also quantify the distribution of elements in a film and have been proved to be efficient methods for characterizing the depth profiles of OSCs.

Near-edge X-ray absorption fine structure (NEXAFS) can measure the transition between the core states to unoccupied orbitals of molecules, which can provide quantification of the mass composition at the surfaces and the buried interface of thin films. In a typical NEXAFS measurement, an X-ray beam will bombard the surface of the film and can be then absorbed when X-ray radiation is polarized in the direction of the transition dipole moment (TDM) of the electron orbitals of a molecule. As materials possessing different compositions give different spectra,

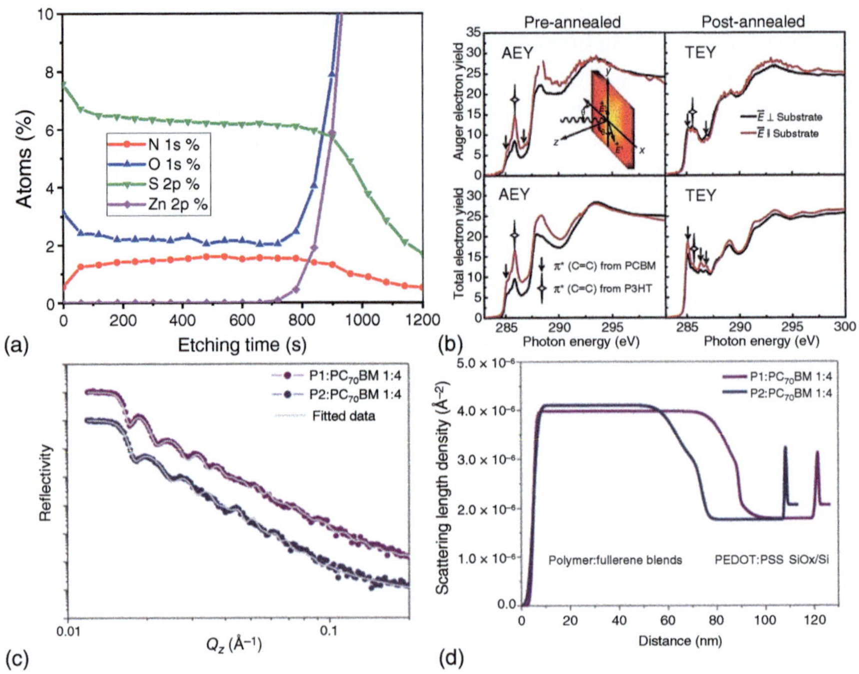

Figure 14.9 (a) Depth profile of PBDB-T:ITIC film measured by XPS with etching technique. (b) Carbon K-edge NEXAF spectra of P3HT:$PC_{60}BM$ films treated with pre-annealing and post-annealing. Source: Chen et al. [40]/Reproduced with permission of American Chemical Society. (c) Neutron scattering reflectivity and (d) scattering length density profiles of P1:$PC_{70}BM$ and P2:$PC_{70}BM$ films. Source: Wang et al. [41]/Reproduced with permission of John Wiley & Sons.

NEXAFS is a good method to study the chain orientation of polymer molecules and the composition near a film surface. Figure 14.9b [40] shows the NEXAFS spectra of P3HT:$PC_{60}BM$ pre-annealed and post-annealed films measured in Auger (AEY) and total (TEY) electron yields. Here, two peaks located at 285.0 and 286.46 eV are from the C=C π^* resonances of $PC_{60}BM$, and the peak at 285.85 eV is from the C=C π^* resonance of P3HT. In the pre-annealed film, it is clear that the C=C π^* resonances from P3HT exhibit stronger intensity than the one from $PC_{60}BM$ in the pre-annealed film and become reverse in the post-annealed film, suggesting that there is a P3HT-rich layer in the top surface (10 nm) of pre-annealed P3HT:$PC_{60}BM$ film but a $PC_{60}BM$ rich layer in the top surface of the post-annealed P3HT:$PC_{60}BM$ film. Furthermore, as AEY and TEY probe the film at the first 1 and 10 nm, respectively, it allows analyzing the depth information. By comparing the AEY and TEY spectra of the post-annealed film, we can also conclude that the average $PC_{60}BM$/P3HT ratio in the first 10 nm is higher than that in the first 1 nm.

Unlike XPS or DSIMS that can quantify the elements components of the film at different depths, X-ray reflectivity (XRR) and neutron reflectivity (NR) techniques can utilize the difference of electron densities of the components in the film to provide depth information, which allows us to get the depth profile with sub-nanoscale resolution. In a typical NR measurement, neutrons impinge on a sample and are collected by the CCD in a reflection geometry. During the measurement, the ratio of the elastically scattered to incident neutrons is measured as the function of the momentum transfer vector, while the specular reflection could be used for the investigation of the vertical morphology. Unlike XPS or DSIMS that can quantify the elements directly, XRR and NR need further modeling and analysis of the original data. Generally, in the NR modeling, the scattering length densities (SLDs) of the components in the film as well as the substrate should be obtained before fitting, and the film will be fitted into multiple layers with each layer being assigned a thickness, a roughness, and an SLD. The resulting SLD profile is essentially the chemical/density distribution profile as a function of depth perpendicular to the sample surface, and the volume fraction of one component at a particular depth can be easily converted from the fitted SLD value. Wang et al. [41] have studied the vertical morphology of P1:PC_{70}BM and P2:PC_{70}BM films by NR. Figure 14.9c,d shows the 1D NR profile, and it can be found that the NR profile of both P1:PC_{70}BM and P2:PC_{70}BM films is well fitted by a slab model that consists of multiple thin recursive layers. Here, the high SLD of SiO_x is evidenced by a distinct peak located between the PEDOT:PSS and Si layers, and the location of the Si/SiO_x, PEDOT:PSS, and PCDTBT:PC_{70}BM interfaces can also be easily identified. In addition, compared to the NR measurements, XRR possesses higher incident energy which only requires a short measurement time. However, XRR also features low contrast of the components; hence, deuterated acceptor or donor materials are always utilized in XRR measurement.

Ellipsometry is a non-destructive characterization technique that can provide the information of optical constants by measuring the changes in the polarization state of the light upon reflection from a thin film. As different components in the BHJ layer always have different optical constants, ellipsometry could be used to measure the vertical-component distribution in the film with further analysis. In 2008, Nelson group [44] used variable-angle spectroscopic ellipsometry (VASE) to study the concentration–depth profiles of as-cast and solvent-annealed P3HT:PC_{60}BM blend films. They first obtained the optical constants of the neat P3HT and PC_{60}BM films and then applied an effective medium approximation (EMA) to build the sub-layers whilst simulate the vertical-component distributions of P3HT and PC_{60}BM. It should also be noted that the accuracy of the VASE technique for depth profile is highly dependent on the used model. Hence, multi-sample and multi-angle measurements are recommended, together with other complementary characterizations to eliminate the deviation.

In addition, Lu group [7] recently introduced a plasma-etching method to probe the film-depth-dependent light absorption and has successfully resolved the distribution of the vertical components in a polymer:non-fullerene PBDB-T:ITIC film. They applied a low-pressure oxygen plasma to remove the surface layer of the film without damaging the underneath nanomorphology, while measuring the absorption of the blend with different etching degree. As shown in Figure 14.10, PBDB-T and ITIC possess different absorption range; the absorption spectra of each sublayer at different depths can be deduced from the changes of the absorption spectra during step etching.

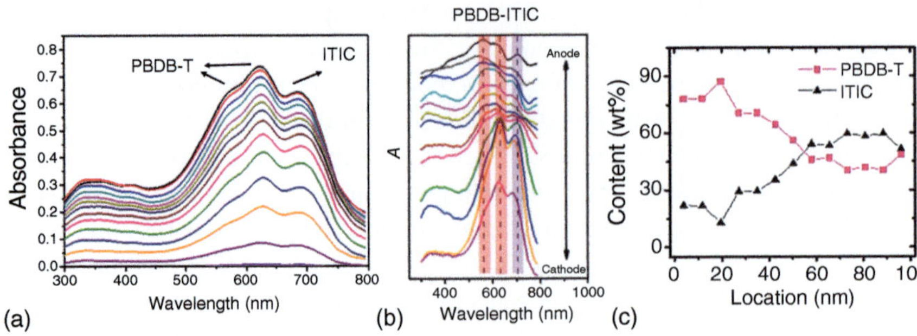

Figure 14.10 (a) The absorption spectra and (b) film-depth-dependent light absorption spectra of PBDB-T:ITIC film upon plasma etching. (c) PBDB-T and ITIC content profiles at different film-depth by fitting the evolution of spectroscopy as shown in (a). Source: Bi et al. [7]/Reproduced with permission of John Wiley & Sons.

14.3 Morphological Manipulation of Organic Solar Cells

14.3.1 Thermal Annealing

As thermal energy is a basic driving force for the molecule movement and morphological rearrangement in a film, heating the BHJ active layer is an effective method for morphology manipulation and has been widely studied. Figure 14.11a shows a typical process of heating the active layer-putting the as-cast active layer film on a hotplate under certain temperature for a certain time, which is also named TA. As early as 2003, Padinger et al. [47] have observed a PCE improvement from 1% to 3.5% after placing the P3HT:PC_{60}BM device on a hotplate with a temperature of 75 °C for 30 minutes. Later, Heeger group [48] further increased the TA temperature to 150 °C and obtained a PCE approaching 5% in P3HT:PC_{60}BM device with improved FF and J_{sc}. They also found that the improvement of the device performance can be attributed to the enhanced crystallinity of P3HT, which leads to prominent phase separation and therefore facilitating charge transport in OSCs. Afterward, TA has been utilized to improve the performance of various high crystalline polymer:fullerene (PffBT4T-2OD:PC_{70}BM) [12], small-molecule-donor:fullerene (BQR:PC_{70}BM) [49] systems. It is widely accepted that TA can further encourage molecular alignment and improve phase separation between donors and acceptors toward favorable charge transport and enhanced PCE. On the other hand, it is worth noting that TA could be detrimental to the device performance for some amorphous systems. For instance, Wang et al [50] found that the π–π stacking coherence length of PCDTBT is reduced upon TA even at low temperatures, which leads to a decreased hole mobility and device performance. Samuel group [51] also observed that TA PTB7-Th:PC_{70}BM film at 150 °C could not affect the crystallinity and orientation of PTB7-Th but could induce strong aggregation of PC_{70}BM in the blend film which results in excessive phase separation and reduced exciton dissociation in the device.

In non-fullerene systems, TA has also been demonstrated to be an effective method to optimize the BHJ morphology and improve the efficiencies in some

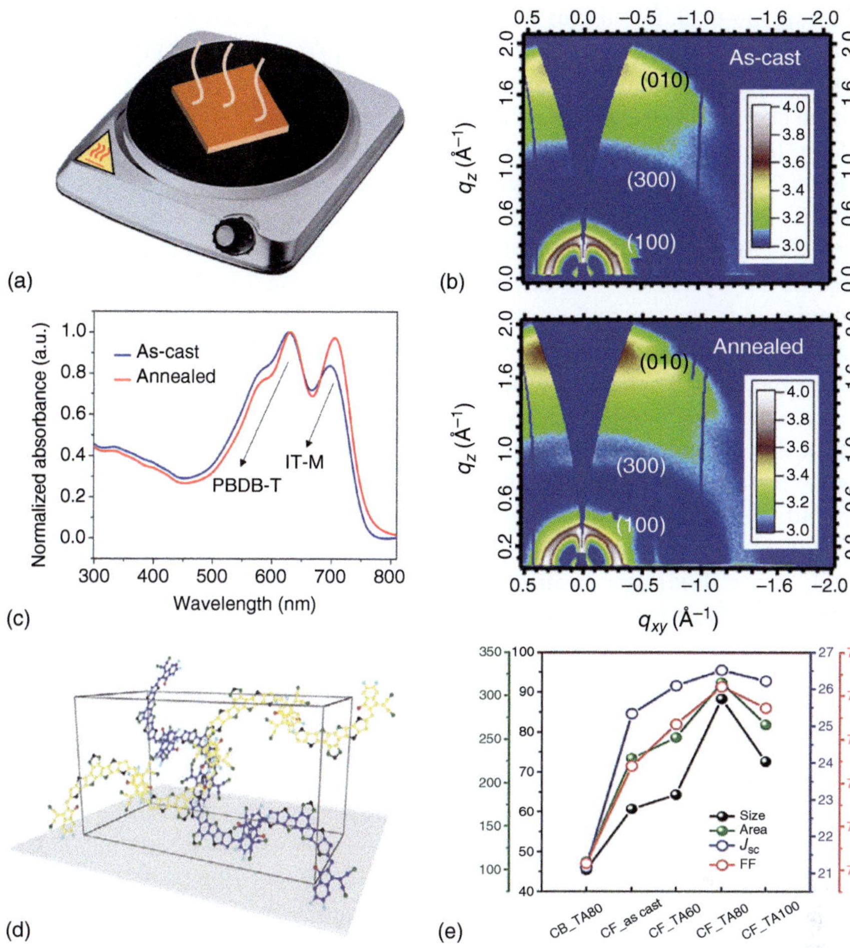

Figure 14.11 (a) Schematic illustration of thermal annealing. (b) 2D GIWAXS patterns of as-cast and thermal-annealed PBDB-T:IT-M films. (c) The absorbance of as-cast and thermal-annealed PBDB-T:IT-M films. Source: Liu et al. [45]/Reproduced with permission of Royal Society of Chemistry. (d) Molecular packing structure of Y6 in the blend film. (e) The (11-1) crystal size (black), (11-1) diffraction peak area (green), J_{sc}, and FF of PM6:Y6 devices cast with CB and CF upon thermal annealing at different temperatures. Source: Zhu et al. [46]/Reproduced with permission of John Wiley & Sons.

state-of-the-art non-fullerene OSCs including PBDB-T:ITIC, PBDB-T-2F:IT-4F, and PM6:Y6. Compared to the fullerene systems, TA not only could adjust the crystalline structure of the electron donor but can also rearrange the molecular packing of non-fullerene acceptors. For example [45] (Figure 14.11b,c), when PBDB-T:IT-M film is upon TA, not only the (*h00*) alkyl chain stacking of PBDB-T is improved but the π–π stacking of IT-M at 1.7 $Å^{-1}$ also exhibits a significant enhancement. With the enhanced π–π stacking of IT-M, absorbance is enhanced and contributes to achieving higher J_{sc}. Furthermore, the effects of TA on the recent high efficiency PM6:Y6 OSCs has been studied by Liu [46] group and Marks

[52] group. Liu et al. found that Y6 exhibits a unique 2D packing structure with a polymer-like conjugated backbone oriented normal to the substrate (Figure 14.11d), which is favorable for vertical charge transport. When the PM6:Y6 film is thermally annealed at temperatures of 60 and 80 °C, the crystal size and diffraction peak area of Y6 are increased, together with enhanced mobility and therefore improved J_{sc} and FF of OSCs. However, further increased TA temperature to 110 °C is detrimental to the crystalline structure of Y6 and results in reduced device performance (Figure 14.11e). In addition, in Marks's work, although they did not observe any change in the crystalline structure of PM6:Y6 films upon TA, they found that Y6 could migrate to the top surface of the BHJ film and facilitate a favorable vertical structure in the inverted PM6:Y6 devices. As a result, thermal-annealed OSCs obtained effective carrier mobility, leading to improved FF and PCE.

It should be emphasized that the TA temperature for different systems could be different due to the unique glass transition temperature (T_g) of each BHJ active layer. For instance, the optimized TA temperature for the traditional P3HT:PC_{60}BM is around 150 °C; however, it ranges from 80 to 110 °C for some non-fullerene OSCs. Furthermore, in addition to TA, other methods such as microwave annealing [53] and pulsed laser heating [54] can also deliver thermal energy to the BHJ layer and have been proved to promote the crystallization as well as device performance.

14.3.2 Solvent Vapor Annealing

In addition to TA, solvent vapor annealing (SVA) is another method to enhance crystallinity of the BHJ active layer. As its name implies, SVA relates to keeping a film in a solvent atmosphere, allowing the rearrangement of molecules toward an equilibrium state (Figure 14.12a). In 2005, Yang group first reported utilizing SVA to control the growth rate of the P3HT:PC_{60}BM active layer to enhance the device performance. They found that SVA could encourage P3HT to form a highly ordered π-conjugated structure in the film, which not only resulted in facilitated charge transport but also enhanced aggregation of P3HT and consequently improved absorption at long wavelength. The effect of SVA on the morphology of fullerene acceptors has been investigated by Tang et al. [57] They found that solvent annealing of the P3HT:PC_{60}BM film with tetrahydrofuran (THF) could increase the domain size of PC_{60}BM clusters to 30 nm; however, it can be reduced to 20 nm after SVA with carbon disulfide (CS_2). This can be explained by the better solubility of CS_2 to PC_{60}BM, which partially dissolves the PC_{60}BM clusters and leads to reduced aggregation. In addition to the polymer:fullerene system, Jones group [58] recently reported that SVA can tune the molecular orientation of a small-molecule donor and fullerene system (BTR:PC_{70}BM), delivering more face-on molecular orientation of BQR in the film and leading to enhanced charge mobility and suppressed recombination in OSCs.

For non-fullerene acceptors, Gurney et al. [55] systemically investigated the effect of solvents used for SVA on the morphology of PTB7-Th:ITIC films. In their work, a range of solvents including Acetone, CS_2, and CB were applied with the

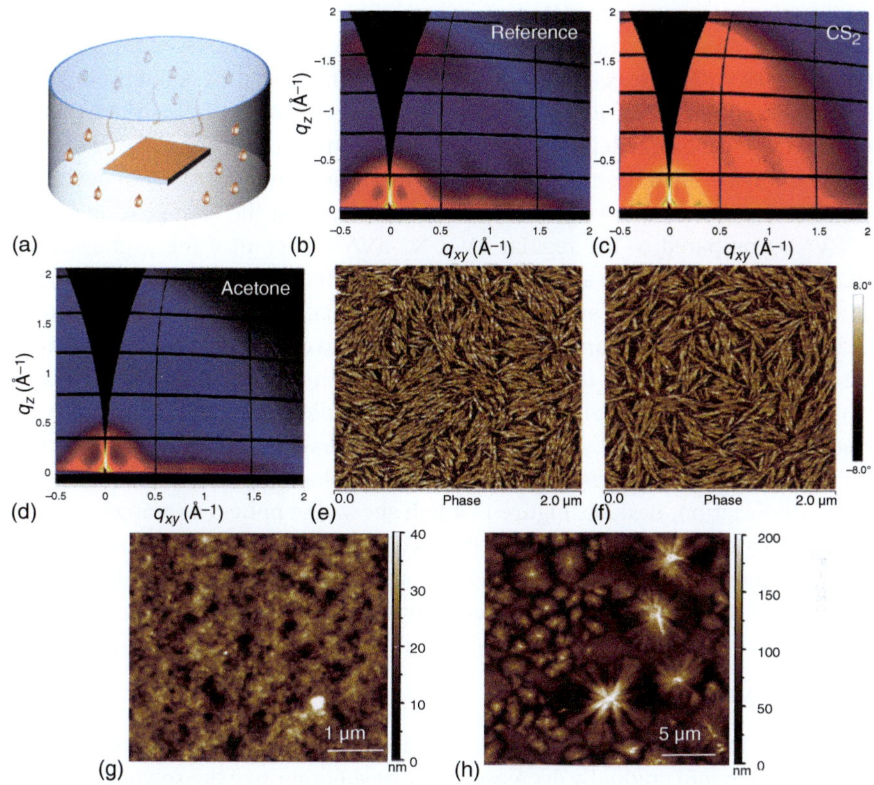

Figure 14.12 Schematic illustration of (a) typical solvent vapor annealing process. 2D GIWAXS scattering patterns of PTB7-Th:ITIC blend films (b) without any posttreatment, (c) with CS_2 solvent annealing, and (d) with acetone solvent annealing. Source: Gurney et al. [55]: reproduced with permission from Elsevier. AFM phase images of D18:Y6 films (e) without SVA and (f) with SVA. Source: Reprinted from Liu et al. [6]/with permission of ELSEVIER. Optical microscopy images of (g) as-cast and (h) solvent annealed films. Source: Li et al. [56]/with permission of JOHN WILEY & SONS, INC.

same SVA procedure. As shown in Figure 14.12b–d, they found that by exposing the PTB7-Th:ITIC film to CS_2 for 60 seconds after casting, the film obtained a big increase in the lamellar stacking of PTB7-Th in the in-plane as well as the π–π stacking in both PTB7-Th and ITIC in the out-of-plane direction. However, when acetone, a solvent with high vapor pressure but poor solubility was used for solvent annealing, the PTB7-Th:ITIC film shows a reduction in molecular ordering alongside with less change in the domain size and spacing. As a result, they concluded that the solvent for SVA should possess good solubility and high vapor pressure, to penetrate the active layer and promote molecular ordering. Very recently, Ding group [6] also applied SVA to a high-efficiency D18:Y6 system. As shown in Figure 14.12e,f upon SVA, the high crystalline D18 polymer donor exhibits improved crystallinity and increased crystal size, leading to an unprecedented PCE of 18.2%.

In addition to the regular SVA, Cao group [59] has combined TA and SVA for a BIT6F:$PC_{70}BM$ polymer:fullerene film and has successfully improved the efficiency from 7% to 9% with an excellent FF of 77%. They found that SVA could promote the nucleation and growth of BIT-6F crystals in the as-cast film, while the further TA treatment could be beneficial in driving better continuous morphology and maintain small phase separation in the film. Furthermore, Zhang et al. [60] developed an *in situ* SVA method to solvent-anneal the film during the spin-coating process (SC-SVA). Compared to the regular SVA, SC-SVA can control the aggregation of ITIC small molecules more effectively, avoid the formation of large-scale phase separation in the active layer, and make the device obtain higher exciton dissociation efficiency, resulting in enhanced charge transport and device performance (with the PCE from 6.25% to 7.21%). In addition, it should be noted that SVA could be detrimental for some high crystalline fullerene and non-fullerene acceptors. For example, Yang group [61] found that when the SVA time increases to more than one minute, P3HT will crystalize excessively and leads to large phase separation and inefficient exciton dissociation. Besides, Figure 14.12g,h shows the optical microscopy images of the as-cast and solvent-annealed PBDB-T:INPIC-4F films [56]. It can be found that SVA promotes the formation of INPIC-4F spherulites and causes large phase separation in the film, leading to reduced device performance.

14.3.3 Solvent

During the film-coating process of the BHJ active layers, the surface energy of the solution system will gradually decrease and the evaporation of the solvent will make the solid phase separates from the solution. Consequently, the final morphology of the film is highly dependent on the solvent evaporation rate, viscosity, surface tension, and miscibility of donor and acceptor in a solution system. In 2001, Hummelen et al [62] first studied the effect of solvents on the morphology and performance of MDMO-PPV:$PC_{60}BM$ OSCs. They found that the film cast by toluene exhibits a large phase separation and coarse surface, while the film cast by chlorobenzene shows a homogeneous film and smooth surface, which is attributed to the better solubility of $PC_{60}BM$ in chlorobenzene. Lee et al. [63] found that P3HT tends to form edge-on molecular orientation in the film when casting from the CF solvent but inclined to form face-on molecular orientation when casting from the DCB solvent (Figure 14.13a,b). This is due to the different boiling points of the solvent; DCB possesses a higher boiling point and leads to a slow film-drying process, which allows the alkyl chain of P3HT to orderly organize in the out-of-plane. However, CF has a lower boiling point and enables fast-drying process only to allow the backbone of P3HT to organize in the out-of-plane. Very recently, the effect of the solvent on the morphology of Y6 has also been studied by Liu group [46]. As shown in Figure 14.13c,d, Y6 film cast from CF exhibits a lamellar stacking peak at 0.27 $Å^{-1}$ in in-plane and a strong π–π stacking peak at 1.75 $Å^{-1}$ in out-of-plane, indicating that Y6 molecules take a face-on molecular orientation in the film. However, when CB was used to cast the film, the π–π stacking of Y6 shows a significantly reduced intensity, whilst a strong dispersive diffraction peak can be observed at 1.2 $Å^{-1}$

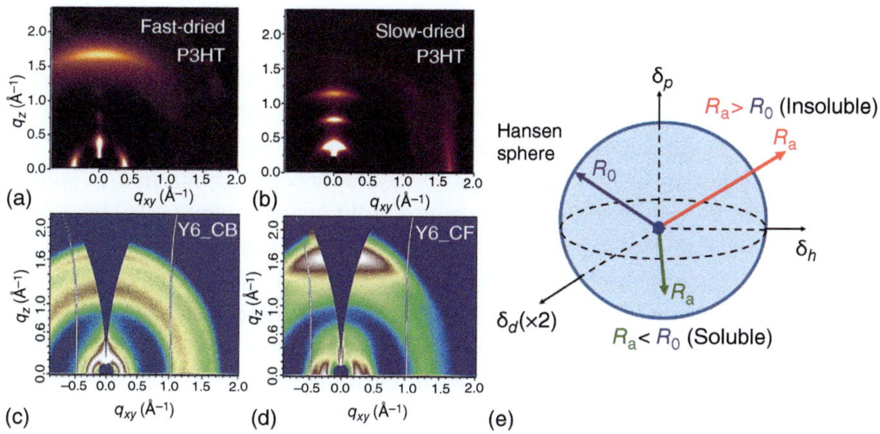

Figure 14.13 2D GIWAXS images of P3HT films cast from (a) CF and (b) DCB Source: Lee et al. [63]: reproduced with permission from Springer Nature. 2D GIWAXS images of Y6 neat films cast from (c) CB and (d) CF solvents. Source: Zhu et al. [46]/Reproduced with permission of John Wiley & Sons. (e) Schematic illustration of Hansen space which can be used to determine the solubility between a solvent and a solute.

which is related to the isotropic amorphous Y6. As a result, the PM6:Y6 device fabricated by CF obtained superior efficiency (16.5%) than that from CB.

In addition, employing mixed solvent to optimize the morphology and enhance the performance of OSCs has also been studied. For example, Ye et al. [64] designed an efficient ternary mixed solvent for PDPP3T:PC_{70}BM OSCs by introducing CF and DIO into the host DCB solvent. In their work, PDPP3T:PC_{70}BM film cast from neat DCB solvent possesses large phase separation and rougher interfaces between PDPP3T donor and PC_{70}BM acceptor, resulting in inferior exciton dissociation efficiency and low J_{sc} in OSCs. When CF is combined with DCB, the film obtained improved domain purity. Although it enabled enhanced FF, the J_{sc} is still low due to the relatively large domain size and sharp interfaces. Moreover, when DCB, CF, and DIO are used as mixed solvents simultaneously, the PDPP3T:PC_{70}BM film obtained relatively fuzzy donor–acceptor interfaces with small phase separation between PDPP3T and PC_{70}BM, consequently to both enhanced FF and $J_{sc.}$ Recently, Wang group [65] also applied a binary solvents approach to optimize the morphology of a polymer:non-fullerene PBDB-T:INPIC-4F system and obtained an improved efficiency from 8.1% to 13.1%. They found that casting film with different solvents possessing different boiling points enables the regulation of molecular ordering and phase-separated nanoscale morphology. When the film is cast using CB, the relatively high boiling point of CB could allow INPIC-4F molecules to have enough time to grow up and finally forms large spherulites, resulting in poor exciton dissociation and low J_{sc} of solar cells. In contrast, casting film from the low boiling point solvent CF leads to relatively low crystallinity and inferior charge transport, resulting in reduced FF. Encouragingly, when combining CB and CF as a mixed solvent (CB:CF [1.5 : 1, v/v]), small phase separation for charge dissociation and continuous network for charge transport can be realized.

Toxic halogenated solvents such as CB, DCB, and CF can dissolve most of the organic semiconducting materials and deliver favorable morphology in OSCs, which are the most commonly used solvents for BHJ deposition. However, when it comes to the commercialization of OSCs, "halogen-free" green solvents are more attractive as they are environmentally friendly and compatible with sustainable and mass industrial manufacturing. In an attempt to rationally select an alternative solvent to substitute halogenated solvents, Hansen solubility parameters (HSPs) have been utilized to predict and calculate the quality of a solvent. As shown in Figure 14.13e [66], HSP describes the total cohesion energy of the solvent, which is consisting of three parts: dispersion ($\delta_{d,}$), polarity (δ_p), and hydrogen bonding (δ_h). With the calculation of these three parameters, the miscibility, solubility, and tendency of phase separation between solute and solvent can be examined by:

$$R_a^2 = 4(\Delta\delta_d)^2 + (\Delta\delta_p)^2 + (\Delta\delta_h)^2 \tag{14.7}$$

where R_a is the Euclidean distance which can define a 3D Hansen space, and R_0 is the interaction radius of the dissolvability measured experimentally. $R_a > R_0$ means the solute could dissolve in the solvent, while $R_a < R_0$ suggests the solvent and solute are unlike to mix. Although HSP could not interpret factors such as crystal structures or π–π stacking, it can be used as an instruction to narrow down the selection range of processing solvents. For example, DCB is a good solvent for P3HT:$PC_{60}BM$ system, an environmentally friendly solvent system consisting of acetophenone (AP) and mesitylene (MS) can be used to mimic the solubility and similar volatility of DCB based on HSPs and achieve comparable device efficiency. In addition, Yan group [67] also developed a hydrocarbon solvent system 1,2,4-trimethylbenzene (TMB) and 1-phenylnaphthalene (PN) to substitute the traditional CB and DIO solvent system in a temperature-dependent aggregation polymer:fullerene PffBT4T-C_9C_{13}:$PC_{70}BM$ system and obtained a favorable morphology as well as enhanced device efficiency.

In non-fullerene OSCs, Hou group [68] also demonstrated two environmentally friendly solvent mixtures, *o*-xylene/PN and THF/IPA which can replace the CB:DIO solvent system and exhibit promising prospects for spin-coating and blade-coating-based PBTA-TF:IT-M OSCs.

14.3.4 Solvent Additive

Besides using mixed solvents for adjusting the morphology of BHJ active layer in OSCs, incorporating a few percentages of secondary solvent as the additive to the primary host solvent has also been found effective for morphology optimization. DIO is the most widely used solvent additive and has made significant contributions to the OSC field. As early as 2007, Peet et al. [69] reported that introducing 2% of DIO in the host CB solvent allows PCPDTBT to obtain optimal alky chain conformations in the PCPDTBT:$PC_{70}BM$ blend film during the film-drying process, leading to enhanced aggregation and absorbance of PCPDTBT and improved device efficiency from 2.8% to 5.5%. Later, Yu group [70] further achieved a record efficiency of 7.4% with the assistance of DIO in the PTB7:$PC_{70}BM$ system and found that the improved performance is due to the selective dissolvability of DIO to fullerene, which can reduce the fullerene aggregation in the BHJ active layer (Figure 14.14a,b). Since then, with the utilization of GIWAXS, GISAXS, and TEM, more and more researchers revealed that with a small amount of DIO in the host solvent (e.g. CB, DCB), DIO could increase the crystallinity of the polymer whilst decrease the domain size of fullerene aggregation, which could improve the exciton dissociation efficiency and charge transport ability simultaneously. After that, DIO has been a primary solvent additive for optimizing morphology in polymer:fullerene OSCs (e.g. PTB7-Th:$PC_{70}BM$, PffBT4T-2OD:$PC_{70}BM$, PBDB-T:$PC_{70}BM$). In polymer:non-fullerene systems, DIO has also been proved to be effective for morphology optimization in a variety of systems (e.g. PBDB-T:ITIC, PTB7-Th:ITIC, PM6:IT-4F, PTB7-Th:IEICO-4F). For example, Ye et al. [71] have demonstrated that DIO can enhance the π–π coherence length of IT-M small molecular acceptor whilst improving the domain purity in PBDB-T:IT-M OSCs, resulting in facilitated charge transport (Figure 14.14c,d). Wang group [73] reported that DIO can promote the J-aggregation of CO_i8DFIC in PTB7-Th:CO_i8DFIC blend film, leading to a significant red-shift of absorbance and improved J_{sc} and PCE.

Due to the success of DIO, researchers developed a variety of alkanes, mercaptans, and naphthalene solvent additives to assist the micro-nanomorphology optimization in OSCs (Figure 14.14e). In general, there are three guidelines for selecting the additive and host solvent: (i) Host solvent has high solubility to both donor and acceptor, but the solvent additive has selective solubility to one component only. (ii) The host solvent has a lower boiling point, but the solvent additive possesses a higher boiling point. (iii) The solvent additive only accounts for the 0.2–5 vol% in the solvent system. On top of these features, host solvent and additives could have different vaporizing speed, allowing donor and acceptor molecules to have different organization manner during the film-drying process. Besides, it is worth noting that the role of solvent additive can be varied depending on their interactions with donor or acceptor materials. For example, octanedithiol additive can increase the crystallinity in the P3HT:$PC_{60}BM$ system but could reduce the crystallinity in P3HT:$IC_{70}BA$ system [74, 75]. Baran group [76] observed that DIO has a contrasting effect on the microstructure and device performance of PTB7-Th:$PC_{70}BM$ and PTB7-Th:ITIC OSCs. Despite the positive impact of DIO (with 3 vol% in CB) in PTB7-Th:$PC_{70}BM$ to weak fullerene aggregation, DIO can increase the phase separation between PTB7-Th and ITIC resulting in unfavorable exciton dissociation. Furthermore, the content of the additive to the host solvent is crucial, with slight change could lead to contrasting morphology and device performance. For example, Bazan group noticed that incorporating 0.5% or 1% of DIO in CF host solvent could reduce the fullerene aggregation and encourage small phase separation in the PSFSiF:$PC_{70}BM$ film, but with the incorporation of 5% DIO could allow a slower crystallization of the active layer and result in large phase separation.

Furthermore, employing multiple additives to optimize the morphology and device performance has also been studied. Ma and coworkers [72] have combined ODT and DIO as a binary additive to work with CB host solvent. By applying *in situ* GIWAXS, TEM, and R-SoXS measurements, they found that ODT tends to work at the beginning of the film-forming process which could control the nucleation of

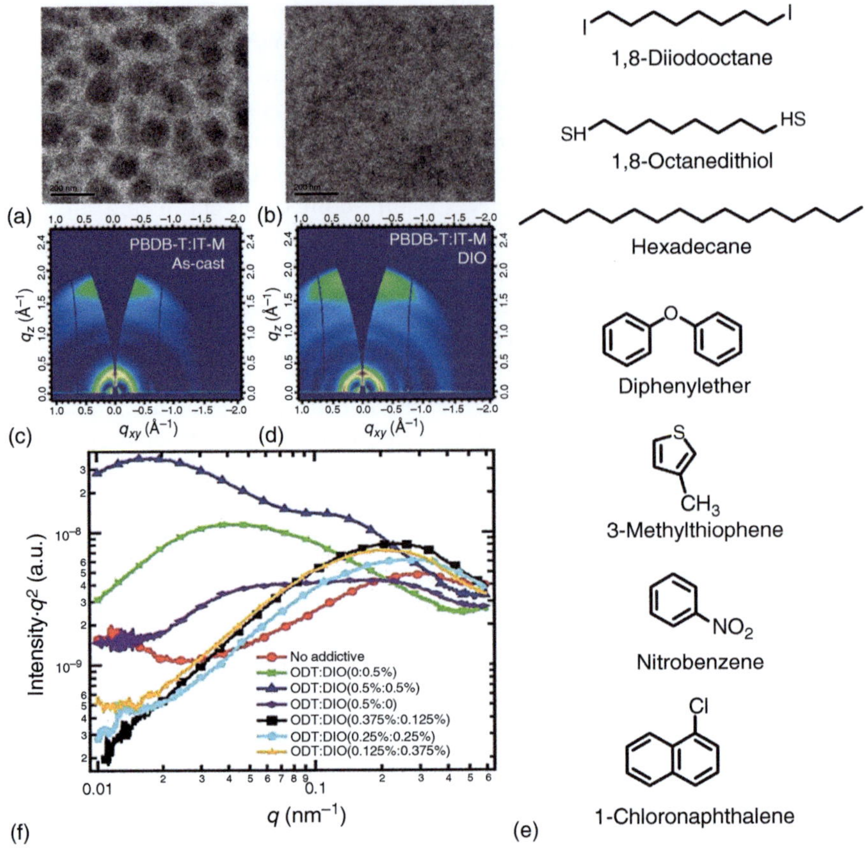

Figure 14.14 TEM images of PTB7:$PC_{70}BM$ films cast from CB (a) without or (b) with DIO. The scale bar is 200 nm Source: Liang et al. [70]/with permission of JOHN WILEY & SONS, INC. 2D GIWAXS images of PBDB-T:IT-M film cast from CB (c) without or (d) with DIO. Source: Ye et al. [71]/Reproduced with permission of John Wiley & Sons. (e) Chemical structures of typical solvent additives for BHJ morphology optimization. (f) Lorentz corrected RSoXS data of FTAZ:ITIC-Th film processed without and with various ratios of ODT and DIO. Source: Chen et al. [72]/John Wiley & Sons/CC BY 4.0.

the crystals, while DIO works throughout the whole film-forming process which can control the crystal growth. They received an active layer that could not only maintain the small domain size like as-cast film but also get enhanced crystallinity (Figure 14.14f). In addition to the morphology optimization, solvent additive has been reported to enhance the morphology stability in OSCs. For example, Zhan's group [77] has applied a solvent additive 4,4′ bisphenol (BPO) in a variety of fluoropolymer:fullerene systems. They found that BPO not only can improve the crystallinity of fluoropolymer whilst enhancing its light absorption ability and carrier transport ability but could also form hydrogen bonds between fluoropolymer polymers and fullerene. The formation of hydrogen bonds could lock the polymer donor and fullerene acceptor in the active layer and prevent them to aggregate, leading to improved morphology stability. As a result, the PTB7-Th:$PC_{70}BM$ device fabricated with BPO could maintain 80% PCE after heat treatment at 130 °C, exhibiting superior morphology stability than the device fabricated without BPO (50% drop in PCE).

14.3.5 Solid Additive

Although solvent additives are very effective for morphology optimization and have been widely used in the current high-performance solar cells, they could be trapped in the BHJ layer due to their high boiling points and have adverse effects on the device stability and reproducibility in large-scale production [78, 79]. Consequently, solid additive has been studied to adjust the morphology of OSCs. As early as 2014, Moon et al. [80] designed a small-molecule 2,2,3,3,4,4,4-heptafluoro-*N*-phenyl-butyr-amide (F-ADD) as solid additive to tune the vertical morphology of the P3HT:$PC_{60}BM$ film. They found that F-ADD is not immiscible with P3HT, but highly miscible with $PC_{60}BM$ due to the π–π stacking between the core of $PC_{60}BM$ and the benzene moiety of F-ADD. Due to this selective interaction with $PC_{60}BM$, F-ADD can tune the surface energy of F-ADD and $PC_{60}BM$ mixture and induce a gradient nanostructure alignment in the vertical direction. Later, Park and co-authors [81] further designed a solid additive, tetrabromothiophene (Br-ADD), which could selectively interact with PTB7 to increase the surface energy of PTB7 (from 25.04 to 30.16 mN m^{-1}), reaching values comparable to that of $PC_{70}BM$ (36.36 mN m^{-1}). As a result, Br-ADD not only delivered a smoother surface and well-mixed heterojunction but also changed the PTB7 and $PC_{70}BM$ distribution in the vertical direction in the film.

The current non-fullerene acceptors are featured with an A–D–A structure consisting of a planar electron-donating backbone, two electron-withdrawing terminal groups, and alkyl chains. Due to the strong steric hindrance effect from the side chains, it is difficult for the electron-donating backbone to form intermolecular stacking for efficient π-orbital overlaps; therefore, the electron transport mainly occurred between the terminal groups of two adjacent A–D–A molecules. On top of this principle, Hou group [82] designed a volatilizable solid additive SA-1 that can volatilize out of the active layer after TA. As shown in Figure 14.15, SA-1 possesses good miscibility and similar structure to the electron-withdrawing end groups of the A–D–A type IT-4F acceptor. Upon TA, the volatilization of SA-1 in the PBDB-TF:IT-4F blend film can induce the rearrangement of IT-4F to form strong intermolecular π–π interaction, leading to enhanced charge transport and PCE. More importantly, compared to the conventional DIO additive which exhibits bad morphology stability as slow evaporation of the residual DIO in the film could induce large aggregation of IT-4F, SA-1 can volatilize out of the film and lead to superior morphology stability. Later, Hou and Ma group [83] have further investigated the effect of the volatilization of those solid additives on morphology optimization. As shown in Figure 14.15, they found that although both SA-4 and SA-7 could facilitate ordered molecular arrangements, their different volatilization leads to different morphology features and device performance. SA-4 possesses

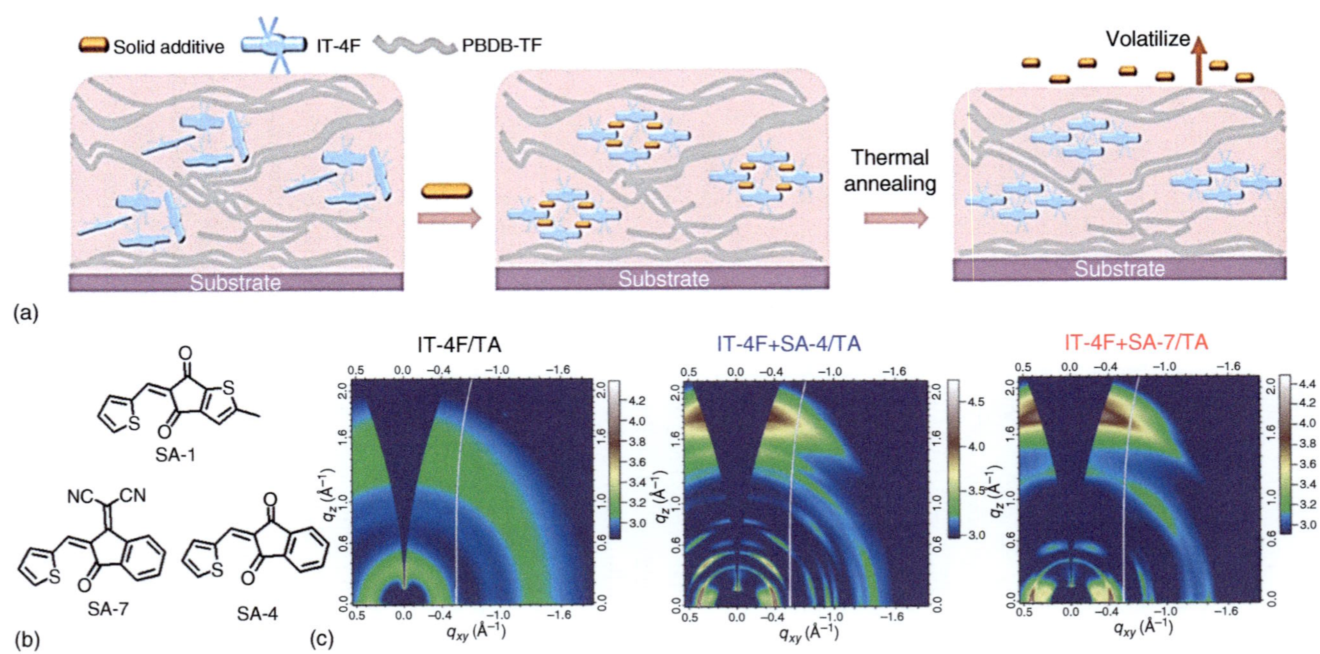

Figure 14.15 (a) Schematic images of the working mechanism of SA-1 solid additive. Source: Yu et al. [82]/Springer Nature/CC BY 4.0. (b) Chemical structures of solid additives SA-1, SA-4 and SA-7. (c) 2D GIWAXS patterns of as cast, cast with SA4, and cast with SA7 PBDB-TF:IT-4F films. Source: Yu et al. [83]/Reproduced with permission of John Wiley & Sons.

stronger volatility, which allows it to volatilize out from the active layer completely; however, SA-7 possesses inferior volatility; therefore, part of SA-7 could be trapped in the BHJ active layer and cast a negative effect on the morphology stability. Very recently, Cai et al. [84] designed three volatilizable additives with different degrees of fluorination. They found that although these additives can be evaporated upon moderate TA, more fluorine atoms on their chemical structures could lead to enhanced intermolecular π–π stacking, encouraging the PCE increased from 15.2% to 16.5% in PM6:Y6 OSCs. With further molecular dynamic simulations, they revealed the attractive interaction between these additives and Y6.

14.3.6 Substrate Effect

During the film-forming process of a BHJ film, the substrate of the casting solution can also affect the morphology. This is associated with the physical and chemical interactions between the substrate and the solution, which could modulate molecular arrangement in the film. As early as 2007, Campoy-quiles et al. [44] observed that casting the P3HT:PC_{60}BM film on PEDOT:PSS or a hydrophobic substrate could induce different vertical segregation of P3HT and PC_{60}BM components in the BHJ layer. The result can be well explained by the different surface energy of P3HT, PC_{60}BM as well as PEDOT:PSS and hydrophobic substrates. P3HT possesses lower surface energy than PC_{60}BM, P3HT will be driven to the interface with air to lower the energy of the free surface when casting on PEDOT:PSS substrate. In contrast, when the P3HT:PC_{60}BM film was cast on the hydrophobic substrate, P3HT is more compatible with the hydrophobic substrate having high surface energy and preferentially attracted to the bottom surface. In addition to surface energy, Bjorstrom et al. [85] reported that the nature of the substrate could also affect the vertical composition distributions in the APFO-3:PC_{60}BM film. In their work, although silicon substrate exhibits similar surface energy to gold substrate, a PC_{60}BM-rich phase can be found close to the bottom substrate when the blend film is cast on silicon, while the bottom interface is enriched with APFO-3 when the blend film is cast on gold. This is attributed to the chemical interactions between APFO-3 and gold, as the thiophene groups in the APFO-3 backbone could self-assembly on gold substrate and drive the polymer to the bottom surface.

In non-fullerene systems, the effect of substrates on the three-dimensional morphology and device performance has also been studied. As shown in Figure 14.16a,b, Wang group [86] has applied XPS with etching technique to characterize the PBDB-T:IT-M films cast on PEDOT:PSS and TiO_2 substrate with different TA conditions. They found that casting film on PEDOT:PSS or TiO_2 at room temperature or 80 °C has a similar effect to the vertical separation of PBDB-T and IT-M in the BHJ film: a PBDB-T-rich layer can be clearly seen in the first 10 nm of the top surface. However, when the active layers were thermally-annealed at high temperature (160 °C), this PBDB-T-rich surface layer remained on the ZnO substrate based film but disappeared on the PEDOT:PSS substrate-based film. This phenomenon can be explained by the interaction between PEDOT:PSS and PBDB-T. During the high TA

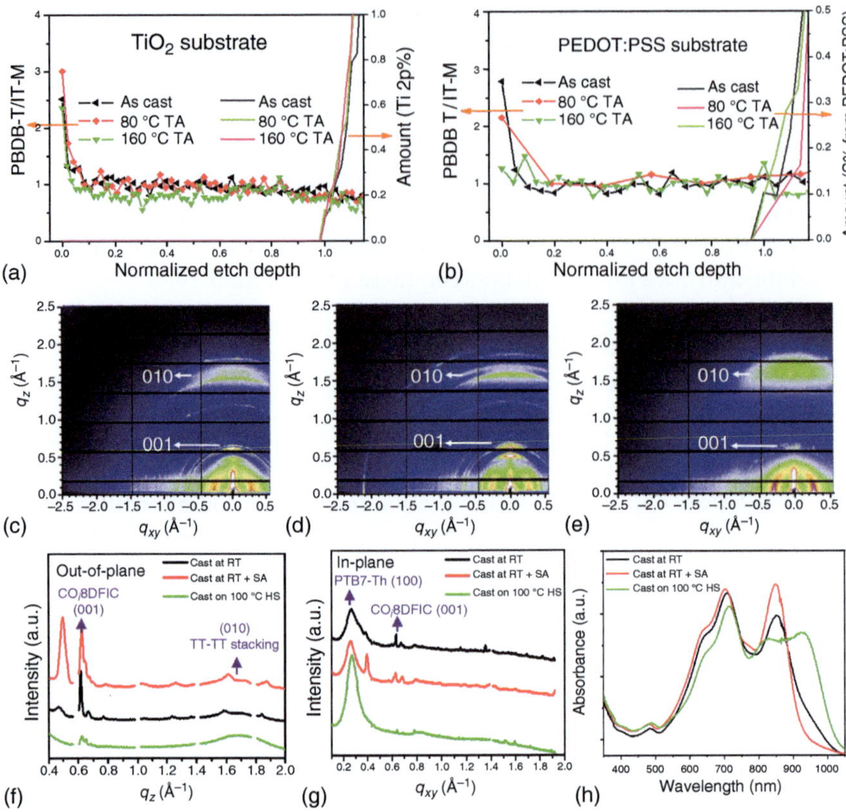

Figure 14.16 XPS depth profiles of PBDB-T:IT-M films cast on (a) TiO_2 and (b) PEDOT:PSS substrates following thermal annealing at different temperatures. Source: Li et al. [86]/Reproduced with permission of John Wiley & Sons. 2D GIWAXS patterns of PTB7-Th:CO_i8DFIC films cast with different methods: (c) as cast, (d) as-cast following with solvent annealing, and (e) cast on 100 °C hot substrate. (f) 1D out-of-plane and (g) in-plane GIWAXS profiles of PTB7-Th:CO_i8DFIC with different casting methods. (h) Absorbance of PTB7-Th:CO_i8DFIC films with different casting methods. Source: Li et al. [19]/Reproduced with permission of Elsevier.

process, PBDB-T can diffuse to the bottom surface and trapped by the existence of PEDOT:PSS.

In addition, the substrate effect on the film-forming process has also been demonstrated to affect the molecular order and aggregation of the donor and the acceptor in a BHJ active layer. For example, Chen and Yan group [87, 88] reported a high crystalline conjugated polymer electron-donor PffBT4T-2OD which exhibits a strong temperature-dependent aggregation behavior in solution. When the PffB4T-2OD:PC_{70}BM blend film is cast on a substrate at relatively low temperature (90 °C), the strong aggregation of PffBT4T-2OD will make it form a face-on molecular orientation and result in facilitated hole mobility as well as enhanced PCE. In contrast, when the PffB4T-2OD:PC_{70}BM blend film is cast on a substrate with high temperature (130 °C), the polymer chains of PffBT4T-2OD can be disaggregated during film formation, allowing the alky chains of PffBT4T-2D organize lamellar stacking in the out-of-plane and this converts the molecular order of PffBT4T-OD from face-on to edge-on. Very recently, Wang group [19, 56] also reported that casting film on a 100 °C hot substrate can inhibit the self-organization of several high crystalline non-fullerene acceptors during the film-forming process. As shown in Figure 14.16c–h, casting the PTB7-Th:CO_i8DFIC film at room temperature makes CO_i8DFIC has enough time to self-organize, allowing CO_i8DFIC molecules to form large fibrils with edge-on and flat-on molecular orientations (More pronounced when solvent annealing is assisted to increase the crystallinity). Inversely, casting PTB7-Th:CO_i8DFIC film on a 100 °C hot substrate could suppress the crystallization of CO_i8DFIC effectively whilst convert CO_i8DFIC lamellar crystallites into face-on-oriented *H*- and *J*- type aggregates, which not only improves the charge transport ability but also blue-shifts and red-shifts the CO_i8DFIC absorbance simultaneously and therefore enhances the light absorption ability.

14.4 Conclusion

In this chapter, we have reviewed a wide range of morphological characterization methods covering microscopic, scattering, and spectroscopic techniques that are capable of investigating the domain size, phase separation, miscibility, domain purity, and depth profiles of the BHJ active layer from microscale to nanoscale. We have also summarized a range of morphological manipulation methods for both fullerene- and non-fullerene-based BHJ active layers. The principle of those morphological manipulation methods as well as their correlation with the morphology development and device performance were discussed in detail. This chapter has shown the significance of understanding and further optimizing the BHJ morphology for the eventual success of OSC technology.

As we discussed above, compared to the traditional ball-like fullerene acceptors, the novel non-fullerene acceptors possess similar chemical structure to the electron donor. The use of nonfullerene acceptors not only reduces contrast for morphology characterization using such as TEM and GISAXS but also renders more complicated morphological metrics and photovoltaic processes in polymer:non-fullerene OSCs. This brings challenges as well as opportunities to manipulate the morphology and performance of non-fullerene OSCs.

With the rising of non-fullerene acceptors, OSCs have experienced rapid development in the last six years and have stepped into the door of commercialization. More studies on morphologies in OSCs should emphasis on the transition from laboratory based to industrially relevant processes. Designing environmentally friendly solution-processing techniques while achieving ideal morphology to realize efficient, thick-junction and large-area OSCs would emerge shortly as the top priority in the community.

References

1 Armin, A., Li, W., Sandberg, O.J. et al. (2021). *Adv. Energy Mater.* 44: 2003570.
2 Gurney, R.S., Lidzey, D.G., and Wang, T. (2019). *Rep. Prog. Phys.* 82: 036601.
3 Meredith, P., Li, W., and Armin, A. (2020). *Adv. Energy Mater.* 10: 2001788.
4 Jiao, X., Ye, L., and Ade, H. (2017). *Adv. Energy Mater.* 7 (18): 1700084.
5 Bavel, S.V., Sourty, E., With, G. et al. (2009). *J. Mater. Chem.* 19: 5388–5393.
6 Liu, Q., Jiang, Y., Jin, K. et al. (2020). *Sci. Bull.* 65 (4): 272–275.
7 Bi, P., Xiao, T., Yang, X. et al. (2018). *Nano Energy* 46: 81–90.
8 Liu, T., Huo, L., Chandrabose, S. et al. (2018). *Adv. Mater.* 30 (26): 1707353.
9 Karki, A., Vollbrecht, J., Gillett, A.J. et al. (2020). *Energy Environ. Sci.* 13: 3679–3692.
10 Li, G., Shrotriya, V., Huang, J.S. et al. (2005). *Nat. Mater.* 4: 864–868.
11 Li, W., Yan, Y., Gong, Y. et al. (2018). *Adv. Funct. Mater.* 28 (5): 1704212.
12 Yan, Y., Li, W., Cai, F. et al. (2018). *ACS Appl. Energy Mater.* 1 (7): 3505–3512.
13 Liu, Y., Xian, K., Peng, Z. et al. (2021). *J. Mater. Chem. A* 9 (35): 19874–19885.
14 Douheret, O., Lutsen, L., Swinnen, A. et al. (2006). *Appl. Phys. Lett.* 89 (7): 032107.
15 Bull, T.A., Pingree, L.S.C., Jenekhe, S.A. et al. (2009). *ACS Nano* 3: 627–636.
16 Reid, O.G., Munechika, K., and Ginger, D.S. (2008). *Nano Lett.* 8 (6): 1602–1609.
17 Gu, K.L., Zhou, Y., Morrison, W.A. et al. (2018). *ACS Nano* 12: 1473–1481.
18 Liu, F., Zhao, W., Tumbleston, J.R. et al. (2014). *Adv. Energy Mater.* 4: 1301377.
19 Li, W., Chen, M., Cai, J. et al. (2019). *Joule* 3: 819–833.
20 Rechberger, S., Gasparini, N., Singh, R., and Kim, M. (2020). *Solar RRL* 4: 2000114.
21 Collins, B.A., Li, Z., Tumbleston, J.R. et al. (2013). *Adv. Energy Mater.* 3: 65–74.
22 Pfaff, M., Klein, M.F.G., Mueller, E. et al. (2012). *Microsc. Microanal.* 18: 1380–1388.
23 Watts, B., Belcher, W.J., Thomsen, L. et al. (2009). *Macromolecules* 42: 8392–8397.
24 Mueller-Buschbaum, P. (2014). *Adv. Mater.* 26: 7692–7709.
25 Rand, B.P. and Richter, H. (2014). *Organic Solar Cells: Fundamentals, Devices, and Upscaling*. CRC Press.
26 Xiao, Y. and Lu, X. (2019). *Mater. Today Nano* 5: 100030.
27 Zhou, K., Wu, Y., Liu, Y. et al. (2019). *ACS Energy Lett.* 4 (5): 1057–1064.
28 Yang, L., Zhang, S., He, C. et al. (2018). *Chem. Mater.* 30 (6): 2129–2134.
29 Wu, W.R., Jeng, U.S., Su, C.J. et al. (2011). *ACS Nano* 5: 6233–6243.
30 Liao, H.C., Tsao, C.S., Shao, Y.T. et al. (2013). *Energy Environ. Sci.* 6: 1938–1948.
31 Li, W., Cai, J., Cai, F. et al. (2018). *Nano Energy* 44: 155–163.
32 Ruderer, M.A., Guo, S., Meier, R. et al. (2011). *Adv. Funct. Mater.* 21: 3382–3391.
33 Bi, P., Zheng, F., Yang, X. et al. (2017). *J. Mater. Chem. A* 5: 12120–12130.
34 Jhuo, H.J., Liao, S.H., Li, Y. et al. (2016). *Adv. Funct. Mater.* 26 (18): 3094–3104.
35 Liao, H.C., Tsao, C.S., Lin, T.H. et al. (2011). *J. Am. Chem. Soc.* 133: 13064–13073.
36 Ruderer, M.A., Meier, R., Porcar, L. et al. (2012). *J. Phys. Chem. Lett.* 3: 683–688.
37 Yin, W. and Dadmun, M. (2011). *ACS Nano* 5: 4756–4768.

38 Guo, S., Cao, B., Wang, W. et al. (2015). *ACS Appl. Mater. Interfaces* 7: 4641–4649.
39 Chen, M., Liu, D., Li, W. et al. (2019). *ACS Appl. Mater. Interfaces* 11: 26194–26203.
40 Chen, D., Nakahara, A., Wei, D. et al. (2011). *Nano Lett.* 11: 561–567.
41 Wang, T., Scarratt, N.W., Yi, H. et al. (2013). *Adv. Energy Mater.* 3: 505–512.
42 Bulle-Lieuwma, C.W.T., van Gennip, W.J.H., van Duren, J.K.J. et al. (2003). *Appl. Surf. Sci.* 203: 547–550.
43 de Villers, B.J.T., MacKenzie, R.C.I., Jasieniak, J.J. et al. (2014). *Adv. Energy Mater.* 4: 1301290.
44 Campoy-Quiles, M., Ferenczi, T., Agostinelli, T. et al. (2008). *Nat. Mater.* 7: 158–164.
45 Liu, X., Ye, L., Zhao, W. et al. (2017). *Mater. Chem. Front.* 1 (10): 2057–2064.
46 Zhu, L., Zhang, M., Zhou, G. et al. (2020). *Adv. Energy Mater.* 10 (18): 1904234.
47 Padinger, F., Rittberger, R.S., and Sariciftci, N.S. (2003). *Adv. Funct. Mater.* 13: 85–88.
48 Ma, W.L., Yang, C.Y., Gong, X. et al. (2005). *Adv. Funct. Mater.* 15: 1617–1622.
49 Bourque, A.J., Engmann, S., Fuster, A. et al. (2019). *J. Mater. Chem. A* 7 (27): 16458–16471.
50 Wang, T., Pearson, A.J., Dunbar, A.D.F. et al. (2012). *Adv. Funct. Mater.* 22: 1399–1408.
51 Jagadamma, L.K., Sajjad, M.T., Savikhin, V. et al. (2017). *J. Mater. Chem. A* 5 (28): 14646–14657.
52 Zhu, W., Spencer, A.P., Mukherjee, S. et al. (2020). *J. Am. Chem. Soc.* 142 (34): 14532–14547.
53 Ko, C.J., Lin, Y.K., and Chen, F.C. (2007). *Adv. Mater.* 19: 3520–3523.
54 Okraku, E.W., Gupta, M.C., and Wright, K.D. (2010). *Sol. Energy Mater. Sol. Cells* 94: 2013–2017.
55 Gurney, R.S., Li, W., Yan, Y. et al. (2019). *J. Energy Chem.* 37: 148–156.
56 Li, W., Chen, M., Zhang, Z. et al. (2019). *Adv. Funct. Mater.* 29 (5): 1807662.
57 Tang, H., Lu, G., Li, L. et al. (2010). *J. Mater. Chem.* 20: 683–688.
58 Schwarz, K.N., Geraghty, P.B., Mitchell, V.D. et al. (2020). *J. Am. Chem. Soc.* 142: 2562–2571.
59 Wang, J.L., Liu, K.K., Yan, J. et al. (2016). *J. Am. Chem. Soc.* 138: 7687–7697.
60 Zhang, L., Lin, B., Ke, Z. et al. (2017). *Nano Energy* 41: 609–617.
61 Li, G., Yao, Y., Yang, H. et al. (2007). *Adv. Funct. Mater.* 17 (10): 1636–1644.
62 Shaheen, S.E., Brabec, C.J., Sariciftci, N.S. et al. (2001). *Appl. Phys. Lett.* 78: 841–843.
63 Lee, H., Lee, D., Sin, D.H. et al. (2018). *NPG Asia Mater.* 10: 469–481.
64 Ye, L., Zhang, S., Ma, W. et al. (2012). *Adv. Mater.* 24 (47): 6335–6341.
65 Chen, M., Zhang, Z., Li, W. et al. (2019). *Sci. China Chem.* 62: 1221–1229.
66 Sanchez-Lengeling, B., Roch, L.M., Perea, J.D. et al. (2019). *Adv. Theor. Simul.* 2: 1800069.
67 Zhao, J., Li, Y., Yang, G. et al. (2016). *Nat. Energy* 1 (2): 15027.
68 Zhao, W., Zhang, S., Zhang, Y. et al. (2018). *Adv. Mater.* 30 (4): 1704837.

69 Peet, J., Kim, J.Y., Coates, N.E. et al. (2007). *Nat. Mater.* 6: 497–500.

70 Liang, Y., Xu, Z., Xia, J. et al. (2010). *Adv. Mater.* 22 (20): 135–138.

71 Ye, L., Zhao, W., Li, S. et al. (2017). *Adv. Energy Mater.* 7 (7): 1602000.

72 Chen, J., Bi, Z., Xu, X. et al. (2019). *Adv. Sci.* 6 (6): 1801560.

73 Zhang, X., Wang, H., Li, D. et al. (2020). *Macromolecules* 53 (10): 3747–3755.

74 Chen, H.Y., Yang, H., Yang, G. et al. (2009). *J. Phys. Chem. C* 113: 7946–7953.

75 Sun, Y., Cui, C., Wang, H., and Li, Y. (2011). *Adv. Energy Mater.* 1: 1058–1061.

76 Song, X., Gasparini, N., and Baran, D. (2018). *Adv. Electron. Mater.* 4 (10): 1700358.

77 Cheng, P., Yan, C., Lau, T.K. et al. (2016). *Adv. Mater.* 28 (28): 5822–5829.

78 Ye, L., Jing, Y., Guo, X. et al. *J. Phys. Chem. C* 117 (29): 14920–14928.

79 Villers, B.J.T., O'Hara, K.A., Ostrowski, D.P. et al. (2016). *Chem. Mater.* 28 (3): 876–884.

80 Moon, B.J., Lee, G.Y., Im, M.J. et al. (2014). *Nanoscale* 6: 2440–2446.

81 Kim, G.W., Lee, G.Y., Moon, B.J. et al. (2014). *Org. Electron.* 15: 3268–3273.

82 Yu, R., Yao, H., Hong, L. et al. (2018). *Nat. Commun.* 9: 4645.

83 Yu, R., Yao, H., Chen, Z. et al. (2019). *Adv. Mater.* 31 (18): 1900477.

84 Cai, J., Wang, H., Zhang, X. et al. (2020). *J. Mater. Chem. A* 8 (8): 4230–4238.

85 Bjorstrom, C.M., Nilsson, S., Bernasik, A. et al. (2007). *Appl. Surf. Sci.* 253: 3906–3912.

86 Li, W., Cai, J., Yan, Y. et al. (2018). *Solar RRL* 2 (9): 1800114.

87 Chen, Z., Cai, P., Chen, J. et al. (2014). *Adv. Mater.* 26 (16): 2586–2591.

88 Liu, Y., Zhao, J., Li, Z. et al. (2014). *Nat. Commun.* 5: 5293.

15

Operational Stability and Built-in Potential in Organic Solar Cells

Weixia Lan[1], Bo Wu[2], and Furong Zhu[3]

[1] Shanghai University, School of Mechatronic Engineering and Automation, 99 Shangda Road, Shanghai, 200444, P.R. China
[2] Nanjing Tech University, Institute of Advanced Materials, 30 South Puzhu Road, Nanjing, 211816, P.R. China
[3] Hong Kong Baptist University, Department of Physics, 224 Waterloo Road, Kowloon Tong, Hong Kong, P.R. China

15.1 Introduction

Organic solar cells (OSCs) are an attractive alternative to traditional inorganic solar cells. The development of solution-processable OSCs has attracted great interest due to its low-cost non-vacuum process technology. The organic semiconductors also exhibit a remarkable flexibility in turning the material properties such as molecular weight, light absorption, energy bandgap for new device concepts. The organic photovoltaic technology creates a pathway to design new semiconductor materials that can be integrated with organic–organic and nanostructure–organic composites for novel applications. A variety of organic optoelectronic systems are being explored for application in organic light-emitting diode (OLED) displays and lighting [1], organic field effect transistors [2], organic photodetectors [3, 4], OSCs [5, 6], organic photonics in communication [7], all-polymeric opto-couplers [8–10], chemical sensing [11, 12], biosensing [13], refractometers [14], memory [15], organic optical sensors [16], and optical integrated systems [17].

Organic electronic technology is highly promising as it can produce higher profitability in advanced OLED-based flat panel displays, OLED solid-state lighting, and OSCs. For example, OSCs are a potentially attractive alternative to inorganic counterparts and the development of solution-processable OSCs has attracted great interests due to its low-cost non-vacuum process photovoltaic technology. Power conversion efficiency (PCE) of >18% for both single junction and tandem OSCs [18, 19] has been reported due to the development of low bandgap photoactive organic semiconductors and improved device design [20, 21].

A broad range of distinct device technologies based on small organic molecules and polymeric photoactive materials are rapidly developing. Apart from the encouraging results in achieving high PCE for OSCs, realizing stabile performance of OSCs over a long operation lifetime also attracts a significant research

Organic Solar Cells: Materials Design, Technology and Commercialization, First Edition. Edited by Liming Ding.

effort. The commonly used bulk heterojunction (BHJ) OSCs, forming a regular configuration OSC, have a donor/acceptor blend sandwiched between a front transparent indium tin oxide (ITO) anode and a reflective rear metal cathode. In regular configuration OSCs, a hole-transporting buffer layer of poly(3,4-ethylene dioxythiophene):(polystyrene sulfuric acid) (PEDOT:PSS) is often deposited on an ITO surface to assist in hole transport and hole extraction at the ITO/organic interface. It has been shown that the use of acidic PEDOT:PSS hole-transporting layer is not the best choice for efficient operation of OSCs over a long period of time due to the deterioration in the contact property at the ITO/PEDOT:PSS interface [22]. The upper metallic cathode in regular configuration OSCs often undergoes a gradual oxidation resulting in a thermally unstable cathode/organic interface [23].

A comprehensive study on the degradation mechanisms of the regular configuration OSCs has been performed. The unbalanced charge mobility in the photoactive layer, due to oxygen-induced charge traps, is one of the degradation mechanisms [24]. In a previous work, two distinguishable degradation pathways in the regular configuration OSCs were identified [25]. One is associated with a localized failure related to the moisture encroachment in the charge generation area, inducing the charge recombination and thus eliminating the photocurrent. The other is dominated by an inevitable initial oxidation at the organic/cathode interface due to the presence of residual moisture and oxygen, leading to lowering PCE of OSCs. The degradation caused by the interfacial passivation can be avoided by the removal of the low work function cathode. This implies that the reverse configuration with a relatively environmental sensitive cathode/organic interface away from the top contact in the cell is preferred compared to the regular configuration OSCs.

Inverted or reverse configuration OSCs, having an organic functional photoactive layer sandwiched between a front transparent cathode and a rear anode, are preferable for efficient cell operation. In this device configuration, a PEDOT:PSS hole-transporting layer is nonexistent, avoiding the use of acidic PEDOT:PSS on ITO surface. In inverted OSCs, ITO surface is often modified with a suitable low work function electron extraction layer (EEL), for example, solution-processable ZnO [5], TiO_2 [26] and Cs_2CO_3 [27], low work function metals of Ca [28] and Al [29], and surface dipole-inducing materials [30].

In parallel to the development of new functional organic semiconductors, the detailed understanding of photo-generated exciton dynamics, charge transport, and recombination processes is another key to improving the performance of OSCs. The excited states in the photoactive organic films have been studied by the steady state photo-induced absorption using pump-probe technique [31]. For example, relaxation processes and dynamics of charge separation (from exciton) to charge transfer (CT) exciton at the donor (D)/acceptor (A) interface and the succeeding dissociation to free charges have been reported [32–34]. It has been shown that a significant amount of CT excitons can be recombined at the D/A interface before they are able to completely separate to form mobile charges [32, 34]. Various processes such as thermal annealing [33] or new syntheses route [35] can alter the morphology of the functional organic semiconductor layer and subsequently impact the relaxation dynamics and free charge yields. The recent progresses on electrode/organic interfacial engineering, interfacial exciton dissociation at the metal/organic interface, charge collection, and operational stability of solution-processed OSCs are discussed in the following sections.

15.2 Bimolecular Recombination in Organic Solar Cells

The emerging organic photovoltaic technology creates a pathway of manufacturing solar cells from conventional vacuum deposition methods to solution-processable techniques. Exciton dissociation at the donor/accepter interface governs the photocurrent generation and that at the organic/electrode interfaces are critical for charge collection in OSCs. Electronic properties at the metal/organic interface in OSCs have been investigated using *in situ* photoelectron spectroscopy. The rapid exciton dissociation at the metal/organic interface also has a profound effect on charge collection and a long-term stability of OSCs. However, it cannot be revealed using these approaches. It also is hardly observed in the measurement of current density–voltage (*J–V*) characteristics under the steady state conditions, for example, measured under AM 1.5G illumination of 100 mW cm^{-2}. The intensity of the solar simulator used for the measurement of OSCs is usually calibrated using a silicon reference cell (with a KG-5 visible color filter) to minimize the spectral mismatch.

In this section, performance of BHJ OSCs made with the blend system of poly[[4,8-bis[(2-ethylhexyl)oxy]benzo[1,2-*b*:4,5-b′]dithiophene-2,6-diyl][3-fluoro-2-[(2-ethylhexyl)carbonyl]thieno[3,4-*b*]thiophenediyl]] (PTB7): [6,6]-phenyl-C71-butyric acid methyl ester ($PC_{71}BM$) is discussed. The charge recombination processes, the distribution of charge density under different operation conditions, and charge collection at the organic/electrode interfaces in PTB7:$PC_{71}BM$-based OSCs are studied systematically using a combination of theoretical calculation, transient photocurrent (TPC) measurements, morphology analyses, and device optimization. The charge transport and recombination properties in the BHJ OSCs are investigated using the photo-induced charge extraction by linearly increasing voltage (photo-CELIV) method. Combined with light intensity-dependent *J–V* characteristic and external quantum efficiency (EQE), it is shown that the use of ZnO EEL has a profound effect on enhancing charge collection efficiency and thereby improving the overall performance of OSCs. The origin of the improvement in the cell performance is mainly associated with the improved charge collection efficiency. Light intensity-dependent *J–V* characteristics and the photo-CELIV results suggest that the charge recombination at the organic/cathode interface can be effectively suppressed by inserting an ultrathin ZnO EEL, leading to a significant improvement in the charge collection efficiency.

15.2.1 Effect of Metal Oxide Interlayer on Cell Performance

Despite the rapid growth during the last decade, the performance of OSCs, associated with low charge mobility and unstable organic/electrode interfacial properties, have placed a limitation on practical applications. The diffusion length of the excitons is only within a few nanometers. Exciton dissociation at the D/A interface determines the photocurrent generation [36]. To facilitate efficient separation of photo-induced excitons, the BHJ is introduced to minimize the domain size on nanometer scale, thereby to facilitate the efficient exciton dissociation [37, 38]. However, the high interfacial area in BHJ could also lead to increasing recombination of the photo-excited carriers, which is one of the main loss mechanisms that limits the photocurrent generation in the cells [39]. The processes of charge collection and recombination in OSCs are critical factors influencing the short-circuit current density (J_{sc}) and fill factor (FF), hence the PCE of OSCs [35, 40]. The interfacial properties at the organic/electrode interfaces have a profound impact on charge collection, efficiency, and operation stability of OSCs. To improve the performance of OSCs, it is important to understand charge recombination, transport, and collection processes in the cells, and the effects on the cell performance.

The loss of the photo-generated carriers due to recombination includes monomolecular and bimolecular recombination processes. Monomolecular recombination is considered as the first-order recombination process, where the charges in the excited states are still bounded due to the Coulomb interaction. It takes place between charges from the same excitons [39, 41, 42]. Monomolecular recombination process occurs in the OSCs that are operated typically in the voltage range between the short-circuit condition and maximum power point, due to the low carrier mobility and strong Coulomb interaction between the charges in the excited states in the organic BHJ blend.

The bimolecular recombination is the recombination process that occurs between the photo-generated electrons and holes before they are collected by the respective electrodes in the cells [40, 43]. Bimolecular recombination is one of the main losses affecting the charge transport in the BHJ and charge collection at the organic/electrode interface, resulting in inefficient charge extraction. Bimolecular recombination process has a strong impact on the FF. It is dominated in the OSCs operated at a relatively low effective potential across the active region, e.g. corresponding to the cells that are operated over the voltage range between the maximum power point and open-circuit condition.

Recombination of the charges prior to the charge collection at the organic/electrode interface has been identified as one of the most critical factors limiting the device performance [36, 44]. Different charge extraction layers, including EEL and hole extraction layer (HEL), are used to engineer the organic/electrode interfaces to minimize the recombination loss and therefore to enhance the charge collection efficiency. The use of the appropriate charge extraction layer also helps to modify the surface of the electrodes forming a desired ohmic contact at the BHJ/electrode interface, avoiding the formation of unfavorable interface dipoles and carrier traps for efficient device operation [45–49]. An ultrathin ZnO EEL is one of the most commonly applied cathode interlayers, which can be prepared by the different solution fabrication processes [20, 50–52]. It has been reported that the morphology and surface properties of ZnO EEL have a great impact on the cell performance. The performance of the cells also is influenced by the process conditions of ZnO EEL, e.g. ZnO formulation [53], process condition [54], and post-deposition treatment [55].

The effect of ZnO EEL on charge recombination and collection process and thereby the performance reproducibility of the solution-processed OSCs are analyzed systematically using a combination of photo-CELIV, TPC, and light intensity-dependent *J–V* characteristic measurements. In the study, a 30-nm-thick PEDOT:PSS (Baytron PVP AI4083) was first spin-coated onto the pre-cleaned ITO substrate and annealed at 140 °C in air for 10 minutes. A 90-nm-thick PTB7:PC_{71}BM layer was then formed on the surface of PEDOT:PSS by spin-coating the polymer blend solution at 2000 rpm for 120 seconds inside glove box. After drying at room temperature for ~1 hour, a 20-nm-think ZnO EEL was spin-coated on the surface of the active layer. Finally, samples were transferred to an adjacent vacuum evaporator for the deposition of a 100-nm-thick Al top contact. OSCs thus made have a device structure of ITO/PEDOT:PSS (30 nm)/PTB7:PC_{71}BM (90 nm)/ZnO (20 nm)/Al (100 nm). A structurally identical control cell without the ZnO EEL was made for comparison studies.

The *J–V* characteristics of OSCs with a ZnO EEL and control cell measured under AM 1.5G illumination (100 mW cm^{-2}) are plotted in Figure 15.1a. The cell parameters obtained for both types of cells are summarized in Table 15.1. It can be found in Figure 15.1 that an obvious improvement in FF is obtained when an ultrathin ZnO EEL is used, which is achieved without any compensation in J_{sc} and open-circuit voltage (V_{oc}) as compared to that of the optimized control cell. The cell with a 20-nm-thick ZnO EEL produces a high FF of 72.94% as compared to that of the control cell (67.23%). This leads to improving the PCE from 7.03% to 7.69%. As shown in Table 15.1, the OSC with a ZnO EEL also has a relatively lower series resistance (R_S) and higher shunt resistance (R_{SH}) compared to that of the control cell. R_S and R_{SH} seen in the OSC are associated with the charge transport and extraction at the organic/electrode interface, affecting FF of the cell. The results suggest that the use of a ZnO EEL improves the contact quality of the active layer/Al interface, leading to a significant reduction in the leakage current,

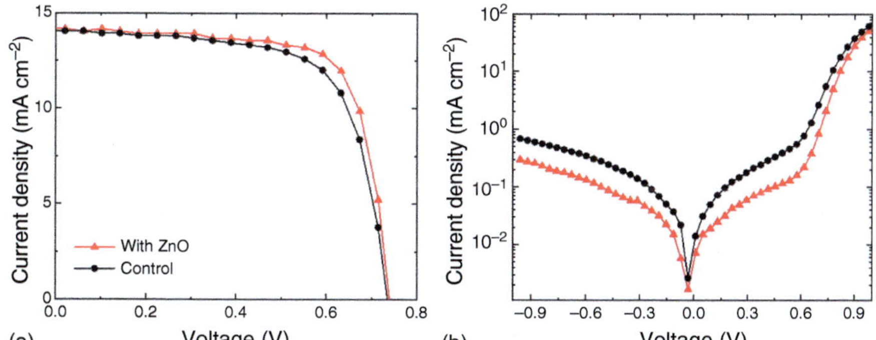

Figure 15.1 (a) *J–V* characteristics and (b) dark *J–V* characteristics measured for OSCs with a 20-nm-thick ZnO EEL and a structurally identical control cell without the ZnO EEL. Source: Weixia Lan.

Table 15.1 Summary of the parameters obtained for a set of structurally identical OSCs made with and without the ZnO EEL.

	PCE (%)	J_{sc} (mA cm^{-2})	FF (%)	V_{oc} (V)	R_S (Ω cm^2)	R_{SH} (Ω cm^2)
With ZnO	7.69	14.22	72.94	0.74	4.8	1513.8
Control	7.03	14.13	67.23	0.74	5.9	987.1

seen in the dark *J*–*V* characteristics in Figure 15.1b [24]. As a result, the use of the ZnO EEL contributes to the enhanced charge extraction and collection capability in the cells.

15.2.2 Charge Recombination Processes in Organic Solar Cells

The charge mobility and recombination process are important factors determining the performance of OSCs. The effect of the ZnO EEL on bimolecular recombination coefficient and concentration of the photo-excited charges in the OSCs are studied using photo-CELIV measurements. Photo-CELIV is a powerful technique for analyzing the extracted density of carriers by varying the delay time, providing an efficient way to investigate both the carrier mobility and recombination dynamics [24, 56–60]. In the photo-CELIV measurement, a laser pulse is introduced to generate the photo-carriers. The photocurrent extracted under the linearly increasing voltage pulse can then be recorded using an oscilloscope. A reverse bias linearly increasing voltage pulse with a voltage rise speed $A = \mathrm{d}U/\mathrm{d}t$ is used in the measurement. The photo-generated charges can be extracted by adjusting the delay time between the laser pulse and the voltage pulse. The time taken to reach the extraction current maximum (t_{max}) and the maximum current due to photoexcitation (Δj) can be estimated from the photo-CELIV transients. The charge mobility (μ) can be calculated using Eq. (15.1) as given below:

$$\mu = \frac{2d^2}{3At_{max}{}^2\left[1 + 0.36\frac{\Delta j}{j(0)}\right]} \tag{15.1}$$

Equation (15.1) is corrected for the electric field re-distribution during the charge extraction process. By selecting the proper voltage rise speed *A* and the forward offset voltage U_{offset} in the photo-CELIV measurements, the photo-CELIV transients are recorded for OSCs with a thin ZnO EEL and the control cell over a delay time range from 5 to 1000 μs. U_{offset} is set to be −0.6 V, which is very close to the built-in potential in the cell.

Figure 15.2 presents the typical photo-CELIV transients measured for the OSCs using different delay times of 5 and 10 μs. The carrier mobility results, calculated using Eq. (15.1), show that the effective carrier mobility μ in the device with a ZnO/Al cathode is higher than that of the control cell. For example, with a delay time of 5 μs used in the photo-CELIV measurement, the mobility for the carriers in the device with a ZnO/Al contact is estimated to be 2.0×10^{-3} cm^2 V^{-1} s^{-1}, which

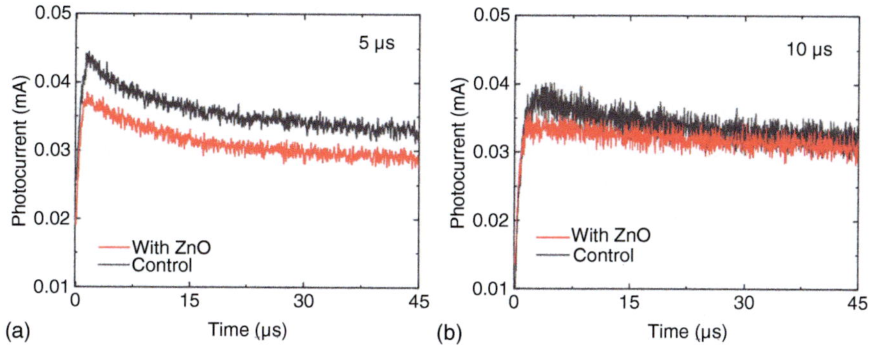

Figure 15.2 Typical photo-CELIV transients measured for an OSC with a 20-nm-thick ZnO EEL and the control cell using two different delay times of (a) 5 μs and (b) 10 μs. Source: Weixia Lan.

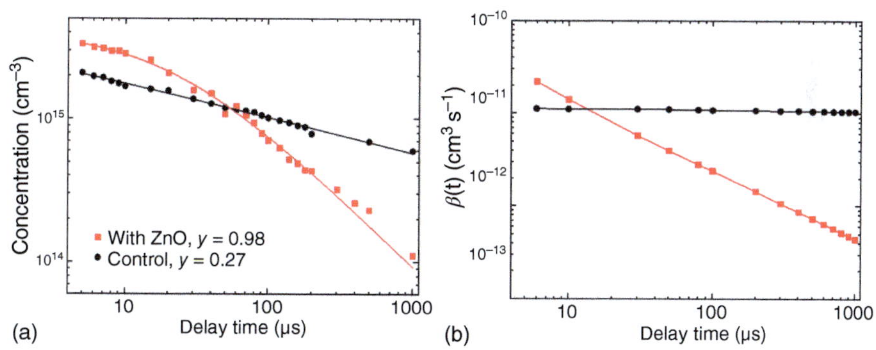

Figure 15.3 Double logarithmic plot of delay time-dependent (a) carrier concentration characteristics and (b) bimolecular recombination coefficients obtained for OSC with a 20-nm-thick ZnO interlayer and the control cell. Source: Weixia Lan.

is higher than that of the control cell (9.7×10^{-4} cm^2 V^{-1} s^{-1}). This reveals that the incorporation of a thin ZnO interlayer promotes the efficient carrier transport in the cells.

Figure 15.3a exhibits the double logarithmic plot of delay time-dependent density of the carriers in the blend layer, obtained by integrating the photo-CELIV transients over time (dots) as well as the corresponding curve fitting data (solid lines). The carrier density includes the total number of the remaining photo-generated charges after the recombination. The number of the extracted carriers varies with the delay time used in the photo-CELIV measurements. Two distinguishable different delay time-dependent $n(t)$ behaviors can be seen in Figure 15.3a. First, the density of total extracted charges decreases with the increase in the delay time for both devices. This behavior reflects that the longer the delay time is applied, the lesser the photo-generated carriers can be collected due to the recombination processes. Second, both types of devices exhibit different delay time-dependent carrier density characteristics. The fitting results suggest the power law dependence between the density of the carriers and the delay time for the control device. While for the OSC with a 20-nm-thick ZnO interlayer, $n(t)$ exhibits a much faster decay as the delay time increases. At a shorter delay time, the density of the carriers in OSC with a ZnO EEL is higher than that of the control cell, while it is much lower when delay time is in the order of several hundred μs, suggesting higher-order dependence between delay time and carrier density in OSC with a ZnO EEL [60].

The different delay time-dependent $n(t)$ behaviors in OSCs having dissimilar contacts were observed in the transient absorption spectrum (TAS) experiments reported for a number of different polymer/PCBM systems [61, 62]. A multiple trapping model, involving free and localized holes, is used to explain the phenomena. For the fast decay phase, it is assigned to the bimolecular recombination due to the presence of a large population of free holes in the highest occupied molecular orbital (HOMO), which is usually observed more clearly with high carrier densities when the density of photo-generated charges exceeds the density of localized states. Conversely, the slower phase is due to the limited bimolecular recombination of dissociated charges due to the trap states. Furthermore, $n(t)$ decay dynamics can also be influenced by the nonuniform charge distribution located at the organic/electrode interfaces. The modification of the contact properties is mainly responsible for the change in recombination processes observed for both types of the cells, as seen in Figure 15.3, since the fabrication process conditions of the cells are identical.

Based on the bimolecular recombination model, the delay time-dependent density of electrons $n(t)$, and density of holes $p(t)$, can be expressed by [56]:

$$n(t) = p(t) = \frac{n(0)}{1 + (t/\tau_B)^\gamma} \tag{15.2}$$

where $n(0)$ is the initial carrier density, t is the delay time, τ_B is the bimolecular recombination lifetime, and γ $(0 < \gamma \leq 1)$ is the dispersion parameter. Solid curves in Figure 15.3a are the theoretical fitting using Eq. (15.2), indicating that $n(t)$ derived by photo-CELIV measurements are in good agreement with the results obtained by the bimolecular recombination model. The fitted results of $n(0)$, τ_B, and γ for different cells are summarized in Table 15.2. The γ values calculated for the control device and cell with a 20-nm-thick ZnO EEL are 0.26 and 0.98, respectively. The small γ value derived from the photo-CELIV measurements implies that a highly dispersive recombination process dominates in the control cell, whereas a high γ of 0.98 suggests that an almost nondispersive recombination (for $\gamma = 1$) is anticipated in the OSCs with a thin ZnO EEL-modified cathode [56].

According to the delay time-dependent bimolecular recombination equation:

$$\frac{dn}{dt} = \frac{dp}{dt} = -\beta(t)np \tag{15.3}$$

The bimolecular recombination coefficient $\beta(t)$ as a function of carrier density $n(t)$ can be expressed as follows:

$$\beta(t) = -\frac{dn(t)/dt}{n^2(t)} \tag{15.4}$$

Table 15.2 Values of bimolecular recombination lifetime τ_B and dispersion parameter γ obtained for both types of the cells, acquired by the curve fitting using Eq. (15.2).

	$n(0)$ (cm^{-3})	γ	τ_B (μs)
ZnO/Al	4.12×10^{15}	0.98	21.41
Al	1.27×10^{15}	0.26	0.01

Combining Eqs. (15.2) and (15.3), $\beta(t)$ can also be expressed as:

$$\beta(t) = a\gamma n_0^{-1}(at)^{\gamma-1} \tag{15.5}$$

where $a = 1/\tau_B$ and τ_B is defined as the "effective" bimolecular recombination lifetime $\left(\tau_B = \left\{\frac{\gamma}{[n(0)\beta_0]}\right\}^{1/\gamma}\right)$. The "effective" bimolecular recombination lifetime τ_B in the OSC with a ZnO interlayer is 2.1×10^{-5} second, obtained from the curve fitting using Eq. (15.5). This is 4 orders of magnitude higher than that of the control cell (9.6×10^{-9} second). One crucial factor for a high bimolecular recombination coefficient of an OSC is that the bimolecular recombination lifetime of the carriers should be higher than their transit time in the intrinsic electric field. The improved bimolecular recombination lifetime is beneficial for efficient carrier transport and collection, and thereby a higher bimolecular recombination coefficient of a solar cell.

The relationship between $\beta(t)$ and t can then be achieved by substituting the previously fitted parameters $n(0)$, τ_B and γ into Eq. (15.5), which is presented in Figure 15.3b. The results reveal different power law dependences between $\beta(t)$ and delay time for two cells with different cathode structures. For OSC with a 20-nm-thick ZnO interlayer, the bimolecular recombination coefficient is almost independent on the delay time, with a constant β value of ~1.1×10^{-11} cm^3 s^{-1}. While for the control device, β shows a significant dependence on the delay time. A decrease by 2 orders of magnitude in β is observed for the control cell when the delay time in the photo-CELIV measurements increases from 5 to 100 μs. The similar delay time dependency on β was also observed in the P3HT:PCBM and other blend systems, measured by TAS measurements [61]. The behavior of delay time-dependent bimolecular recombination coefficient in an OSC is closely associated with the trap-limited recombination processes. Therefore, compared to the power law dependency between the delay time and $n(t)$ observed in the OSC with a ZnO interlayer, the slower decay in $n(t)$ seen in the control cell is predominantly due to the trap-limited recombination processes. While the faster decay in $n(t)$ observed in the ZnO EEL-modified cell suggests that the pure bimolecular recombination is the major loss mechanism. As a result, it is seen clearly that the effect of the ZnO interlayer helps to suppress the trap-limited recombination leading to the improvement of the charge collection efficiency.

15.2.3 Bias-Dependent Charge Collection

The presence of a metal oxide interlayer between the active layer and the metal contact plays an important role in determining the cell performance. The charge transport, extraction properties, and the interfacial exciton dissociation at the electrode/organic interface in solution-processed OSCs are investigated using light intensity-dependent J–V characteristic measurements.

The charge transport and recombination characteristics in OSCs can be analyzed using space charge limited current (SCLC) model. Figure 15.4a,b show the double logarithmic plot of J_{ph}($J_{ph} = J_l - J_d$, where J_l and J_d are the photocurrent density and dark current density) versus light intensity measured for two types of OSCs, showing the power law dependence of J_{ph} on light intensity, $J_{ph} \propto I^{\alpha}$ [63]. The power exponent, α, of both OSCs are about 1 at a relatively high effective voltage V_{eff}, $V_{eff} = V_0 - V_b$, where V_0 is the built-in potential measured at $J_{ph} = 0$, and V_b is the applied bias, suggesting that almost all of the charges are collected at the electrode at high V_{eff}. The power exponents of J_{ph}–I characteristics measured for all cells approach to one at $V_{eff} > 0.25$ V, which is around the maximum output power point. This suggests that most photo-generated charges can be efficiently collected without charge accumulation in the cell at high V_{eff}. As the charge mobility in organic materials is field-dependent, less efficient charge transport and collection in the OSCs would occur at the low V_{eff}, leading to the accumulation of space charges that gives rise to higher recombination, and thereby a lower power exponent in the J_{ph}–I characteristics. As shown in Figure 15.4, when V_{eff} decreases to a much lower value of 0.087 V, α drops to 0.67 for the control device, while higher exponent value of 0.85 can still be obtained for OSC with a ZnO interlayer. The different value of α at low V_{eff} suggests the different strengths of recombination in the two types of devices, indicating stronger recombination in the control device, leading to the poor charge transport and collection efficiency.

Figure 15.5 shows the double logarithmic plot of J_{ph} versus V_{eff}. The photocurrents become independent of the applied voltage and saturate at V_{eff} of 0.5 V. The

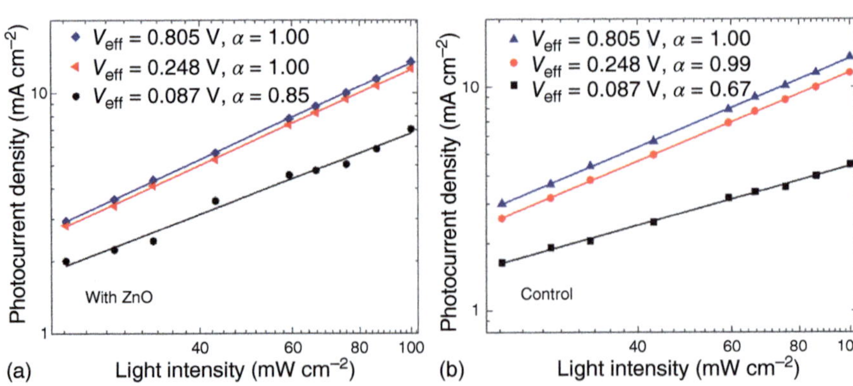

Figure 15.4 Double logarithmic plot of the photocurrent density as a function of light intensity for (a) an OSC with a 20-nm-thick ZnO interlayer, and (b) the control device under different effective biases. Source: Weixia Lan.

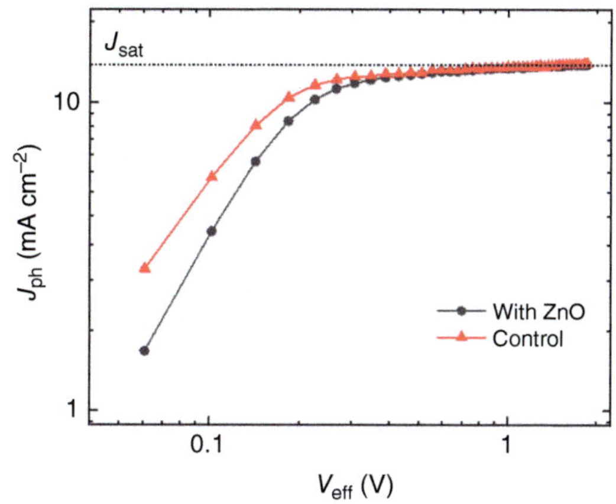

Figure 15.5 Double logarithmic plot of $J_{ph}-V_{eff}$ characteristics obtained for an OSC with a 20-nm-thick ZnO EEL and the control device. Source: Weixia Lan.

saturated photocurrent, J_{sat}, also is shown in Figure 15.5. As V_{eff} decreases, charge recombination would increase and not all the photo-generated carriers are collected. Thus, under specific V_{eff}, charge collection efficiency (η_{CC}) can be expressed as [64]:

$$\eta_{CC}(I, V_{eff}) = {}^{J_{ph}(I,V_{eff})}/_{J_{ph,sat}(I)} \tag{15.6}$$

At high V_{eff} (>0.5 V), η_{CC} approaches unity and gives rise to almost 100% collection of the photo-generated charges. In this regime, recombination is negligible. The recombination becomes increasingly important at low V_{eff} as P decreases with V_{eff}. As shown in Figure 15.5, charge collection efficiency of the control cell decreases much faster than that of the OSC with a 20-nm-thick ZnO interlayer at $V_{eff} < 0.5$ V. The loss in photocurrent due to recombination is obviously much lower in the OSCs with a ZnO interlayer than that in the control cell, revealing improvement in charge collection efficiency. As OSCs with a 20-nm-thick ZnO EEL and the control cell have the identical anode/organic contact and blend system, the charge accumulation and recombination behaviors at the anode contact and in the active layer could be considered the same. Hence, the variation in exponent of J_{ph}–I characteristics and the change in charge collection efficiency are attributed to the different charge accumulation behaviors at the organic/cathode interfaces in the devices. The difference in the J_{ph}–I characteristics suggests that the charge accumulation and recombination loss at the organic/cathode interface in the control cell is much stronger than that in the OSCs with a ZnO interlayer. The J_{ph}–V_{eff} results also indicate that the use of a 20-nm-thick ZnO EEL apparently favors an efficient charge collection process. It shows that insertion of a ZnO interlayer between the BHJ and metal contact helps to suppress the charge recombination.

15.3 Metal/Organic Interfacial Exciton Dissociation in Organic Solar Cells

Exciton dissociation at the D/A interface dominates the photocurrent generation in OSCs [65, 66]. The processes of exciton dissociation at the organic/electrode interface are critical for charge collection in OSCs. For regular OSCs with a typical configuration of ITO/PEDOT:PSS/organic photoactive layers (100–200 nm)/metal cathode, the exciton dissociation at the organic/cathode interface is relatively less investigated as compared to that at the D/A and anode/organic interfaces [67, 68]. Electronic properties at the organic/electrode in OSCs have been investigated using *in situ* X-ray photoelectron spectroscopy (XPS) [69]and ultraviolet photoelectron spectroscopy (UPS) [70]. The correlation between the charge transport/recombination and the cell performance parameters (lifetime, efficiency) have also been studied using SCLC [71] and photo-CELIV [24]. The exciton dissociation at the opaque metal/organic cathode interface in OSCs, however, cannot be elucidated fully using these approaches. TPC technique is used for measuring recombination kinetics of photo-generated carriers in OSCs [72, 73]. The photocurrent, generated due to interfacial exciton dissociation in the absence of internal built-in potential in the bulk is observed at ITO/organic interface. A clear understanding of the exciton dissociation at organic/cathode interface and its impact on the performance of OSCs are desired.

An Nd:YAG pulsed laser with pulse duration of 5 ns and the wavelength of 355 nm was used to generate the transient signals in the TPC measurements. The TPC, taking the difference between the current (I_L) excited by the laser and the dark current (I_D), for the devices at different biases can be expressed as:

$$I_{TPC} = I_L - I_D \tag{15.7}$$

where I_{TPC} in Eq. (15.7) includes the sum of photo-induced current contributed by (i) the drift current formed under the internal electric field (E_{int}), and (ii) the exciton dissociations at D/A and organic/electrode interfaces that are not dependent on E_{int}. Assuming a uniform electric field crossing the stack of the functional active layers in OSCs, E_{int} can be expressed as:

$$E_{int} = (V_0 - V_b)/d \tag{15.8}$$

where V_0 in Eq. (15.8) is the intrinsic built-in potential in the OSCs, V_b is the applied external bias, and d is the thickness of the stack of the functional layers. Usually, contributions to I_{TPC} due to exciton dissociations at D/A heterojunction and at organic/electrode cannot be easily distinguished or identified in the TPC measurements. In order to better understand the processes of exciton dissociation at the organic/electrode interface, the contribution in I_{TPC} due to the drift current arising from D/A interface under E_{int} can be offset by applying an opposite external bias on devices in the TPC measurements. This allows to exam the interfacial exciton dissociation in the absence of internal built-in potential. Therefore, the information of exciton dynamics and recombination kinetics of photo-generated carriers at organic/electrode interface can be obtained by measuring I_{TPC} for samples at different biases.

15.3.1 Charge Collection in Regular Configuration Organic Solar Cells

Regular OSCs with a configuration of ITO/PEDOT:PSS (30 nm)/PTB7:PC_{71}BM (90 nm)/Al and reverse configuration OSCs with a structure of ITO/Al

(~1.2 nm)/PTB7:$PC_{71}BM$ (90 nm)/MoO_3 (2.0 nm)/Ag were made. The pre-patterned ITO substrates, with a sheet resistance of ~15 Ω per square, were cleaned by ultrasonication sequentially with acetone, ethanol, deionized water, and isopropanol for 10 minutes, followed by an *ex situ* oxygen plasma prior to the cell fabrication. For reverse configuration OSCs, the surface of ITO/glass substrates was modified with ~1.2-nm-thick Al, deposited by thermal evaporation in vacuum with a base pressure of $<1.0 \times 10^{-4}$ Pa at an evaporation rate of 0.1 Å s^{-1}. The Al-modified ITO substrates were then transferred to an adjacent glove box with O_2 and H_2O levels below 0.1 ppm, which is connected to the evaporator. A pre-prepared polymer blend of PTB7 (1 Material) and $PC_{71}BM$ (NanoC) in a weight ratio of 1 : 1.5, dissolved in chlorobenzene solution, was spin-coated on substrates. Different OSCs have an identical active area of 3.0 mm × 3.0 mm. They are encapsulated with a glass lid in the glove box with both oxygen and moisture levels less than 0.1 ppm.

In order to analyze the exciton dissociation at the organic/cathode interface, samples consisting of a 400-nm-thick acceptor layer in a regular cell configuration of ITO/PEDOT:PSS (30 nm)/$PC_{71}BM$ (400 nm)/Al (30 nm) were prepared for the TPC measurements. The use of a thick photoactive layer, e.g. four times thicker than the actual active layer used in OSCs, is to suppress the influence of the drift current generated in the bulk. Samples with a ZnO interlayer at the cathode/organic interface, e.g. ITO/PEDOT:PSS (30 nm)/$PC_{71}BM$ (400 nm)/ZnO (20 nm)/Al (30 nm), were also made for comparison studies. A 20-nm-thick ZnO thin layer was coated on $PC_{71}BM$ surface using solution process, following the procedure described in a previous work [74]. In the experiments, I_{TPC} of the samples was measured using a small optical perturbation, provided by a pulsed Nd:YAG laser from the 30-nm thick semitransparent Al cathode side. Under this condition, very small amount of photons can be absorbed at the PEDOT:PSS/$PC_{71}BM$ anode interface after they penetrate through the semitransparent Al cathode and a 400-nm-thick $PC_{71}BM$ layer. Therefore, the contribution to I_{TPC} due to the exciton dissociation at the anode/organic interface can then be neglected, and the exciton dissociation at $PC_{71}BM$/Al interface in the devices can be seen clearly.

TPCs measured for the samples with a configuration of ITO/PEDOT:PSS/$PC_{71}BM$ (400 nm)/Al (30 nm) at different biases from 0.3 to 0.5 V are plotted in Figure 15.6a. The positive I_{TPC} measured for samples without the bias is always observed and is predominantly contributed by the drift current under optical perturbation, due to the difference in the work function between Al and ITO (V_0 of ~ 0.5 V). As the drift current dominates in the TPC measurement, the contribution to the I_{TPC} due to the exciton dissociation at the $PC_{71}BM$/Al interface is hardly observed. When a positive bias was applied between the ITO anode and Al cathode, e.g. from 0.3 to 0.5 V, the strength of the built-in potential inside the device is offset and decreases gradually with increase in the external bias. Therefore, the fraction in I_{TPC} arising from the drift current (E_{int}-dependent) decreases, while the contribution due to the exciton dissociation at the $PC_{71}BM$/Al interface (not E_{int}-dependent) increases.

Figure 15.6a illustrates clearly that the fast negative TPC is observed within a time frame of ~ns after the optical perturbation, a slower positive I_{TPC} is then appeared after a longer period of several tens of ns. The change in the polarity of the TPC

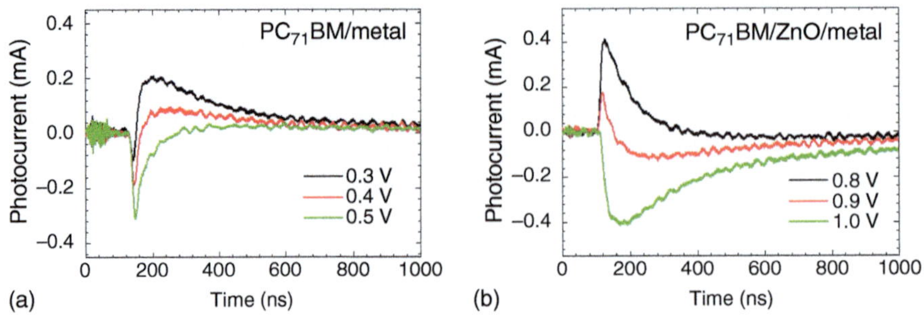

Figure 15.6 Transient photocurrents measured for samples consisting of a 400-nm-thick acceptor layer in a regular cell configuration of (a) ITO/PEDOT:PSS/$PC_{71}BM$ (400 nm)/Al (30 nm) and (b) ITO/PEDOT:PSS/$PC_{71}BM$ (400 nm)/ZnO (20 nm)/Al (30 nm) at different biases. Source: Bo Wu.

reflects the contribution of exciton dissociation at the $PC_{71}BM$/Al interface. The swift negative I_{TPC} signal shown in Figure 15.6a originates from exciton dissociation at the $PC_{71}BM$/Al interface, which is very sensitive to the photoexcitation (within 1 ps), and the slower positive I_{TPC} signal comes from the carriers that drift with E_{int} in the bulk and decrease with the increase in the positive bias. Figure 15.6a reveals clearly that the exciton dissociation at the $PC_{71}BM$/Al interface in regular OSCs produces an inherent adverse effect on electron collection. Photo-generated electrons, drifted with the internal electric field, must be compensated first by the electrons that flow in the opposite direction, due to unfavorable exciton dissociation at the $PC_{71}BM$/Al interface, before being collected by the cathode. The rapid negative TPC is apparently disadvantageous for electron collection in regular cells, but it is hardly observed in the measurement of photo-generated current under the steady state conditions, e.g. J–V characteristics of OSCs.

Device with a configuration of ITO/PEDOT:PSS/$PC_{71}BM$/ZnO (20 nm)/Al (30 nm), with a built-in potential V_0 of ~0.8 V, is used for the TPC measurements. TPCs measured for ITO/PEDOT:PSS (30 nm)/ $PC_{71}BM$ (400 nm)/ZnO (20 nm)/Al (30 nm) are plotted in Figure 15.6b. It shows clearly that the rapid negative TPC signal is not perceived in I_{TPC} measured for samples under different biases. Instead, a rapid positive TPC is observed at a bias $V_b = 0.9$ V within a time frame of ~ns after the optical perturbation. A slower negative I_{TPC} signal is observed, arising from the carriers drifted with E_{int}, e.g. $V_0 - 0.9 = -0.1$, in the bulk. This suggests that the direction of transient current, created by the exciton dissociation at the cathode contact of $PC_{71}BM$/ZnO/Al, is the same as that of the drift current in the bulk. TPC measurements reveal that the inevitable adverse interfacial exciton dissociation at organic/Al interface in regular OSCs, as shown in Figure 15.6a, can be eliminated by interposing a thin solution-processed ZnO interlayer, revealing its impact on preventing the detrimental interfacial exciton dissociation, thereby improving the electron collection efficiency.

In an actual BHJ OSC, a D/A blend layer is sandwiched between the anode and the cathode. The possible exciton dissociation at the contact between PTB7 and Al is also examined by the TPC measurements. The transient currents measured for samples consisting of a 400-nm-thick PTB7 donor layer in a regular configuration of ITO/PEDOT:PSS/PTB7 (400 nm)/Al (30 nm) and that recorded for ITO/PEDOT:PSS/PTB7 (400 nm)/ZnO (20 nm)/Al (30 nm) at different biases were also studied. It is found that there is no rapid TPC observed in I_{TPC} measured for both types of samples, regardless of whether a ZnO interlayer is present or not at the interface between PTB7 and Al contact. This suggests no exciton dissociation can be observed at PTB7/Al interface.

15.3.2 Charge Collection in Inverted Organic Solar Cells

TPCs measured for samples with a reverse configuration of ITO/Al (10 nm)/$PC_{71}BM$ (400 nm)/MoO_3 (2 nm)/Ag and ITO/Al (10 nm)/ZnO (20 nm)/$PC_{71}BM$ (400 nm)/MoO_3 (2 nm)/Ag under different biases are shown in Figure 15.7. The rapid positive TPCs, observed for the reverse configuration samples at different biases, originate from the exciton dissociation at Al/$PC_{71}BM$ (Figure 15.7a) and ZnO/$PC_{71}BM$ (Figure 15.7b) interfaces, while the negative slower photocurrents are the drift current generated in the bulk of the samples under different internal electric fields, E_{int}. It is clear that unfavorable exciton dissociation that impedes the electron collection at Al/$PC_{71}BM$ interface in regular OSCs, as shown in Figure 15.6a, does not occur in their reverse geometry cells. The inverse exciton dissociation at the Al/$PC_{71}BM$ interface in regular and inverted OSCs comes from the difference of the contact properties, depending on the sequence of forming an organic photoactive layer on cathode contact or depositing metal (e.g. Al) cathode on organic layer. When an organic/metal interface is formed by depositing the organic layer on the cathode in the reverse configuration cells, a relatively clear metal/organic interface is expected. Organic/Al interface is often formed by depositing Al contact on organic layer. The incoming energetic metal atoms can easily interrupt the relatively weak van der Waals interaction between molecules in the organic layer. The penetration of energetic metal atoms into the underlying functional organic layer is primary responsible for the complex interfacial properties, forming interfacial defects that could result in adverse exciton dissociation at the organic/metal interface in regular OSCs. However, for the inverted OSCs, the organic/metal interfaces are formed by

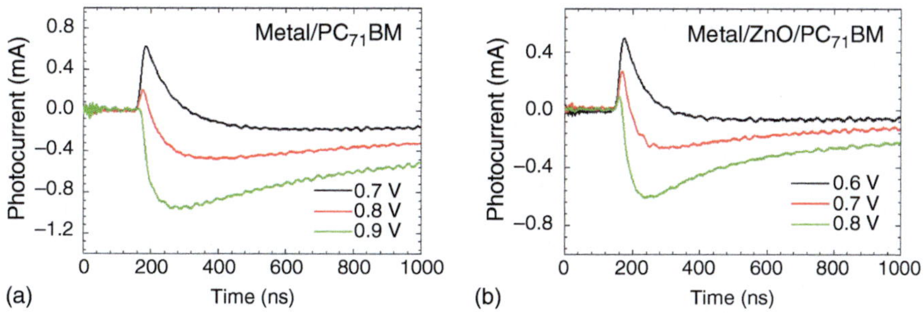

Figure 15.7 Transient photocurrents measured for samples with a reverse configuration of (a) ITO/Al (10 nm)/ $PC_{71}BM$ (400 nm)/MoO_3 (2 nm)/Ag and (b) ITO/Al (10 nm)/ZnO (20 nm)/ $PC_{71}BM$ (400 nm)/MoO_3 (2 nm)/Ag at different biases. Source: Bo Wu.

coating organic layers onto predeposited Al, resulting in a smooth contact without any risk of high-temperature damage or metal penetration.

Metal/organic material interfaces are crucial and play important roles in determining both the charge collection property and stability of OSCs. Much work has been reported to investigate the interfacial properties at Al/fullerene or Al/$PC_{71}BM$ interfaces. An electron transfer from Al to fullerene due to Fermi level-pinning at the Al/fullerene interface was reported [75, 76]. The sublevel electronic states in the bandgap of organic layer, formed due to the interface defects, film roughness, and inhomogeneous contact, are some of the factors that can influence or dominate carrier flow into and out of OSCs. Schematic diagram illustrating the processes of exciton dissociation at $PC_{71}BM$/Al and $PC_{71}BM$/ZnO/Al interfaces is shown in Figure 15.8a,b.

Upon photon excitation, adverse exciton dissociation at the $PC_{71}BM$/Al interface results in an accumulation of holes at these sub-band levels. They are likely collected by Al generating a negative TPC as shown in Figure 15.7a. Photo-generated electrons, drifted with the internal electric field, are compensated partially by the electrons created by unfavorable interfacial exciton dissociation, leading to the poor electron collection at the $PC_{71}BM$/Al interface. The valance band maximum (−7.2 eV) of ZnO is much lower than HOMO of $PC_{71}BM$ (−6.1 eV). When a ZnO interlayer is present at the interface between $PC_{71}BM$ and Al, the exciton dissociation takes place at the $PC_{71}BM$/ZnO interface, the possible flow of holes toward Al can be blocked effectively [77], as shown in Figure 15.8b. Therefore, the electrons that created by adverse exciton dissociation at $PC_{71}BM$/Al interface are suppressed at $PC_{71}BM$/ZnO/Al contact generating a positive TPC as shown in Figure 15.6b. The above analyses, based on the TPC measurements, provide an insight on understanding the origin of inverse exciton dissociation at the metal/organic interface in inverted and regular OSCs.

The relevance of above analyses made with TPC measurements can be seen by measuring the performance of regular or inverted OSCs having different cathode contacts. For regular structure, it is demonstrated that the FF and PCE can be improved by insertion of a 20-nm-thick ZnO EEL between BHJ and cathode, as shown in Figure 15.6a. While for reverse configuration OSCs, the performance of the cells made with or without the ZnO EEL is almost the same, suggesting that the unfavorable electron collection observed in the regular configuration OSCs is not present in the inverted cells.

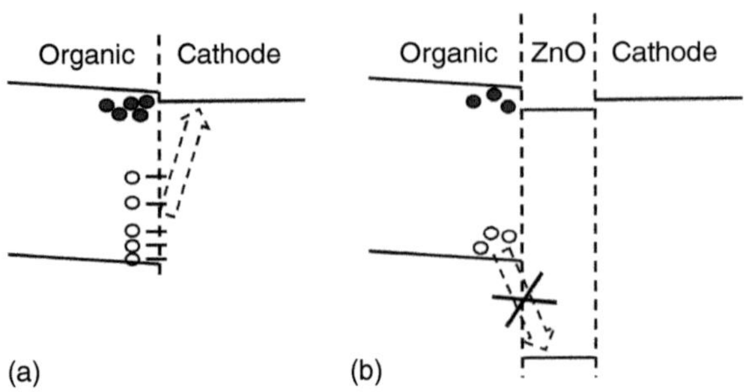

Figure 15.8 Schematic diagram illustrating the processes of exciton dissociation at (a) $PC_{71}BM$/Al and (b) $PC_{71}BM$/ZnO/Al interfaces. Source: Bo Wu.

15.4 Improvement of Charge Collection and Performance Reproducibility

In the previous section, it shows that inverted OSCs are more favorable than corresponding regular configuration cells. ZnO is one of the commonly used EEL due to its low cost and simple solution-based preparation process. The morphology of the EEL, e.g. solution-processed oxide interlayer, plays an important role in determining the contact quality at the cathode/active layer interface in the cells. The effect of the surface morphology of the oxide interlayer on the morphology evolution of the organic functional active layer in the inverted OSCs is analyzed. It demonstrates that the inverted OSCs are favorable for forming smaller domains in the active layer as compared to that in the active layer of the cells having a regular configuration. The improvement in the performance of inverted devices associates with the reduced domain size, which could then be closer to the exciton diffusion length in the polymer blend, favoring efficient operation of the OSCs [78].

In addition to the surface morphology, the electronic properties of the ultrathin ZnO EEL are essential for the development of high-performance solution-processed inverted OSCs [79]. A thin film ZnO EEL interlayer, prepared by the sol–gel process at a sintering temperature of 200 °C, has been used for application in inverted OSCs. Surface engineering of a sol–gel processed ZnO with a thin conjugated polyelectrolyte interfacial layer has a significant impact on enhancing the cell performance, achieved due to an improved charge transport via suppressed bimolecular recombination [80]. A cross-linkable water/alcohol-soluble conjugated polymer interlayer was also adopted for tuning the surface work function of ZnO interlayer in high-efficiency ZnO-based inverted OSCs [81]. It has been reported that the associated fluctuation in morphological and surface electronic properties of ZnO-modified ITO cathode, e.g. changes due to the experimental conditions of ZnO formulation [53], film fabrication [54], and post treatment [55], also place a practical challenge limiting the reproducibility of the performance of OSCs. However, the report on the effect of the sub-gap states, induced by the defects in the ZnO interlayer, on charge collection behavior and thereby the performance reproducibility of the solution-processed OSCs is rather rare. For commercialization of OSCs, there is a need to develop a robust and an effective ZnO modification approach that can be easily repeatable for application in making high-performance OSCs via a solution-processable route.

15.4.1 Effect of Metal Oxide Interlayer on Cell Performance

The inverted OSCs made with Al-modified ZnO interlayer have a device configuration of ITO/ZnO (20 nm)/Al (1.2 nm)/PTB7:$PC_{71}BM$ (90 nm)/MoO_3 (2.0 nm)/Ag. The structurally identical control OSCs with a single cathode interlayer of ZnO (20 nm) also were prepared for comparison. Pre-patterned ITO-coated glass substrates, with a sheet resistance of ~15 Ω per square, were cleaned by ultrasonication sequentially with acetone, ethanol, deionized water, and isopropanol for 10 minutes each. The ZnO interlayer is solution-prepared following the processes described in

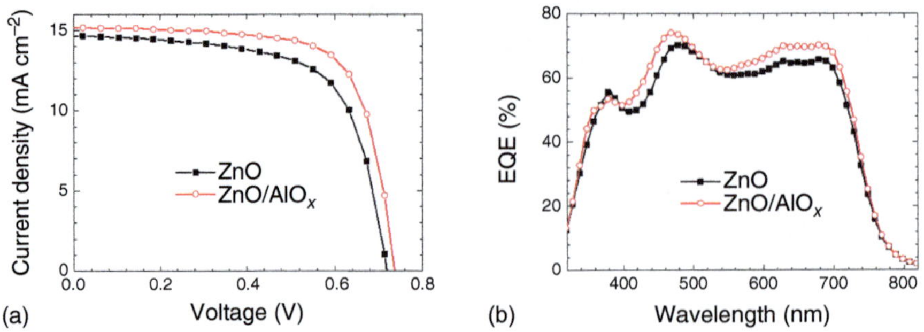

Figure 15.9 (a) *J–V* characteristics and (b) EQE spectra measured for the inverted OSCs of ITO/ZnO/Al (1.2 nm)/PTB7: $PC_{71}BM$ (100 nm)/MoO_3 (2 nm)/Ag (100 nm) (open circle, Al-modified ZnO), and ITO/ZnO/ PTB7: $PC_{71}BM$ (100 nm)/MoO_3 (2 nm)/Ag (100 nm) (solid square, ZnO). Source: Bo Wu.

a previous work [74]. For OSCs with Al-modified ZnO interlayer, ~1.2-nm-thick Al was deposited on ZnO/ITO surface by thermal evaporation in a vacuum chamber with a base pressure of $<1.0\times10^{-4}$ Pa at an evaporation rate of 0.1 Å s^{-1}. The Al-modified ZnO/ITO substrates were then transferred to an adjacent glove box connected to the evaporator with O_2 and H_2O levels below 0.1 ppm.

The *J–V* characteristics and EQE spectra of the OSCs with the Al-modified ZnO and pure ZnO interlayers are shown in Figure 15.9 [82]. Inverted OSCs with an Al-modified ZnO interlayer possess PCE of 8.0%, compared to the PCE of 6.9% for the control cell with a ZnO interlayer. Results in Figure 15.9 reveal that V_{oc}, J_{sc}, and FF are all enhanced. The improvement in FF is the most evident, increasing from 65% (ZnO) to 70% (Al-modified ZnO). The improvement in FF due to the use of Al-modified ZnO interlayer in the OSCs arises from improved charge collection efficiency at the organic/cathode interface.

The results show that the R_S of the cells made with an Al-modified ZnO interlayer and that of the control cell are similar. However, R_{SH} of the control cell is much larger for the OSCs made with an Al-modified ZnO interlayer. OSCs with an Al-modified ZnO interlayer also exhibited an improved EQE in almost the whole spectrum region, demonstrating an enhancement in the charge collection at the interface between functional organic layer and Al-modified ZnO as compared to that at organic/ZnO interface in the control cell. The device parameters obtained for two types of the inverted OSCs are summarized in Table 15.3.

It is known that the performance of the OSCs is also dependent on the surface morphology of the ZnO EEL [78]. In parallel to the analyses of the *J–V* characteristics of the OSCs, the possible change in the surface morphology of ZnO, caused by the modification of an ultrathin Al layer, was also examined by the scanning electron microscope (SEM) measurements. The SEM images measured for the surfaces of a pure ZnO and an ultrathin Al (1.2 nm)-modified ZnO EELs, deposited on ITO/glass using the identical condition for cell fabrication, are shown in Figure 15.10. It is clear that there is no observable change in the surface morphology, revealing the

Table 15.3 The performance of the inverted OSCs made with the Al-modified ZnO and pure ZnO interlayers. The values of J_{sc} and *PCE* calibrated using *EQE* measurements are also given in the parentheses for comparison.

OSCs	V_{oc} (V)	J_{sc} (mA cm^{-2})	FF (%)	PCE (%)	R_S (Ω cm^2)	R_{SH} (Ω cm^2)
Al-modified ZnO	0.74 ± 0.01	15.1 (15.3) ± 0.4	70 ± 2	7.9 (8.0) ± 0.1	4.8 ± 0.4	1524 ± 102
Pure ZnO	0.72 ± 0.01	14.5 (14.5) ± 0.3	65 ± 2	6.9 (6.9) ± 0.2	4.5 ± 0.4	1001 ± 97

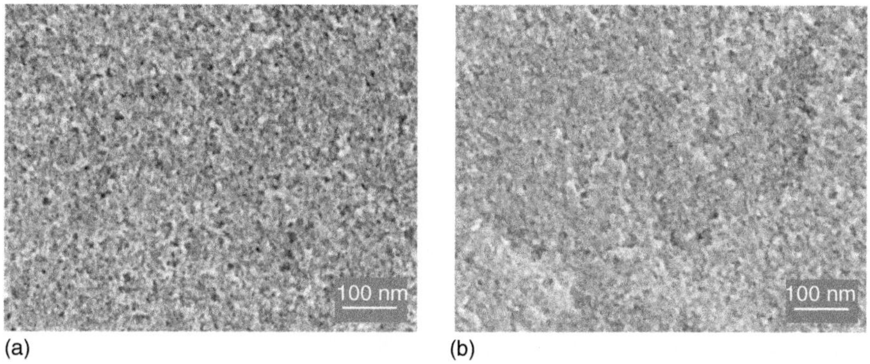

Figure 15.10 SEM images measured for ITO surfaces with (a) a pure ZnO EEL and (b) an Al (1.2 nm)-modified ZnO EEL. Source: Bo Wu.

enhancement in the PCE of the cells is originated primarily from the improved contact quality and the charge collection at the cathode/organic interface.

To better understand the origin of the improvement in the performance of the OSCs, the charge collection behavior at organic/cathode interface in the OSCs made with different cathode interlayers is analyzed. The double logarithmic plots of the J_{ph}–V_{eff} measured for two types of OSCs are shown in Figure 15.11. From Figure 15.11, it is clear that J_{ph} saturates at a large reverse bias voltage, e.g. V_{eff} = 1.8 V, for both devices, indicating that all photo-generated free carriers are collected at the electrodes with negligible loss due to the bimolecular recombination. Therefore, under a large reverse bias, J_{ph} is primarily determined by light absorption and the exciton dissociation rate in the photoactive layer, which is almost the same for both types of devices.

The inverted OSCs with an Al-modified ZnO interlayer possess a slightly higher J_{ph} (15.8 mA cm^{-2}) than that of the control device (15.5 mA cm^{-2}), caused by the favorable interfacial exciton dissociation at the Al/organic interface, therefore an improved charge collection efficiency in the inverted OSCs. As V_{eff} decreases, the decrease in the values of J_{ph} also reflects the influence of the charge collection efficiency because not all of the photo-generated carriers in the active layer are collected at the organic/electrode interfaces. For example, at V_{eff} = 0.8 V, J_{ph} measured for the

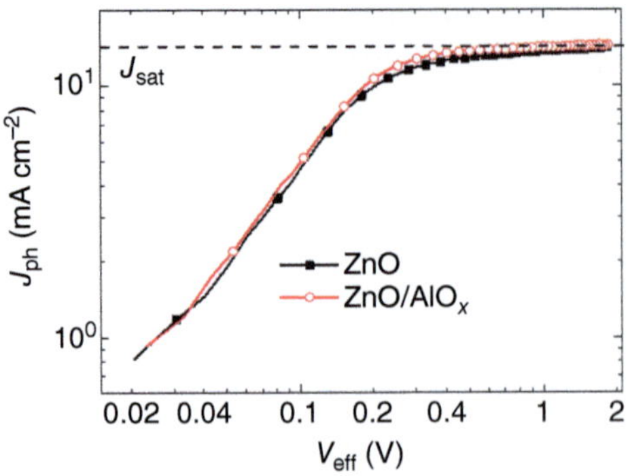

Figure 15.11 J_{ph}–V_{eff} characteristics of a set of structurally identical inverted OSCs made with the Al-modified ZnO interlayer (open circle) and pure ZnO (solid square), measured at the AM 1.5G condition (100 mW cm^{-2}). Source: Bo Wu.

OSCs with an Al-modified ZnO interlayer is 97.4% of the saturation current density obtained at the large reverse bias $V_{eff} = 1.8$ V, which is apparently higher than the J_{ph} (95.3%) obtained for the control cell. At a lower effective voltage, e.g. $V_{eff} = 0.2$ V, J_{ph} measured for the OSCs with Al-modified ZnO interlayer reaches 74% of the saturation current density, while that for the control cell is less than 70% of the saturation current density. As both OSCs have identical anode contacts, the higher J_{ph} observed at lower V_{eff} region reflects the superior charge collection efficiency in the OSCs with the Al-modified ZnO interlayer to that of the control cell.

15.4.2 Suppression of ZnO Sub-Gap States

Figure 15.12a,b show the double logarithmic plot of J_{ph} versus light intensity for both types of the devices as well as the corresponding linear fitting data, suggesting power law dependence between these two quantities. The ideal SCLC characteristics give rise to a power exponent α of 0.75 [83, 84]. However, the power exponent of both devices is about 1 at a high effective voltage $V_{eff} = 1.8$ V, indicating that almost all the photo-generated carriers are collected at the electrode without any space charge generation due to the balanced carrier transport. However, at lower V_{eff}, the less

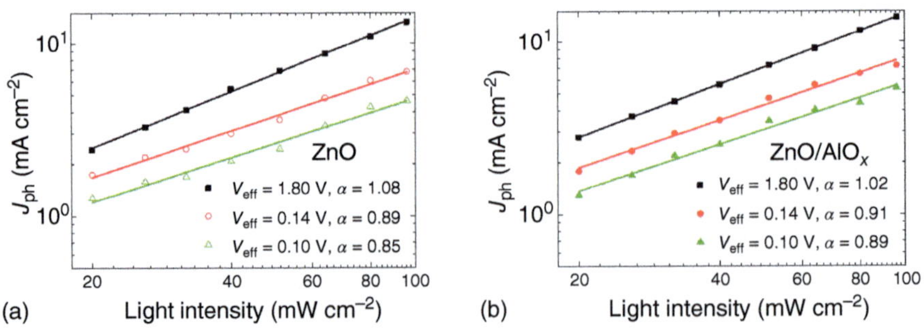

Figure 15.12 Double logarithmic plot of J_{ph} as a function of the intensity of light, illustrating the change in the power law exponent α for (a) control cell with a pure ZnO interlayer and (b) OSC with Al modified ZnO interlayer, measured under different V_{eff}. Solid lines are the linearly fitting results. Source: Bo Wu.

efficient charge transport and collection lead to the build-up of space charges that gives rise to higher bimolecular recombination and lower power exponent.

In Figure 15.12a, the exponent is 0.89 for the OSCs with Al-modified cathode interlayer, while for the control device, it decreases to 0.85 at the same V_{eff} of 0.1 V, as shown in Figure 15.12b, implying that more space charges are generated in the control cell. As the only difference between these two devices comes from the cathode interface, the results clearly reveal that the Al-modified ZnO interlayer helps to suppress the generation of space charges and the sequential bimolecular charge recombination, which otherwise would occur at the ZnO/PC_{71}BM interface, leading to an enhancement in FF, as shown in Table 15.3.

The sublevel electronic states in the bandgap of ZnO interlayer, induced by the defects in the ZnO interlayer, are crucial and play important role in determining both the charge collection at the organic/electrode interface and stability of OSCs. In this work, the interfacial properties at the organic/ZnO interfaces in the inverted OSCs were analyzed using photothermal deflection spectroscopy (PDS) and XPS measurements. In the PDS measurements, the Al-modified ZnO and pure ZnO samples were immersed in an inert liquid (Fluorinert FC-72) with a high refractive index. A monochromatic pump light is incident on the ZnO sample. The heat generated in the sample, caused by the absorption and subsequent non-radiative relaxation process, produces a thermal gradient in the liquid near its surface. The layers of liquid near the sample surface are hotter, less dense than the upper liquid and have lower refractive indices. A probe light, using a laser beam with the wavelength of 632 nm, parallel to the ZnO surface is refracted. By scanning the wavelength of the pump light, the correlation between the refraction angle of the probe light and the change in the refractive index of the liquid near the surface of the sample can then be obtained. The corresponding absorption coefficient of the samples at different photon energies, which provides the information of the excitation from the sub-band states in the ZnO, can then be measured. In this work, the PDS spectra of the specimens were obtained by measuring the refracted probe light using a position sensor and lock-in amplifier.

PDS is a high-sensitive absorption measurement technique, which can detect the change in the optical absorption due to the variation in sub-band gap states induced by the defects and energetic disorders [85, 86] in the specimens. Figure 15.13a plots the PDS spectra measured for the ZnO and Al-modified ZnO layers. The absorption coefficient of both samples saturates at the photon energy >3.37 eV, caused by the band-to-band absorption in the ZnO layer. The absorption of both samples drops rapidly when the photon energy is less than 3.4 eV. The Urbach Energy, E_u, can be extracted by the linear fitting of the cutoff edge in the PDS spectra, an important parameter that provides the information of sub-band gap states in the materials. When the photon energy, E, is less than bandgap E_g, the relationship between the absorption coefficient A and E_u can be expressed by [85]:

$$A = A_0 \exp((E - E_g)/E_u) \quad (15.9)$$

According to Eq. (15.9), the corresponding E_u values calculated for the pure ZnO and Al-modified ZnO films are 59 and 53 meV, respectively, implying that the

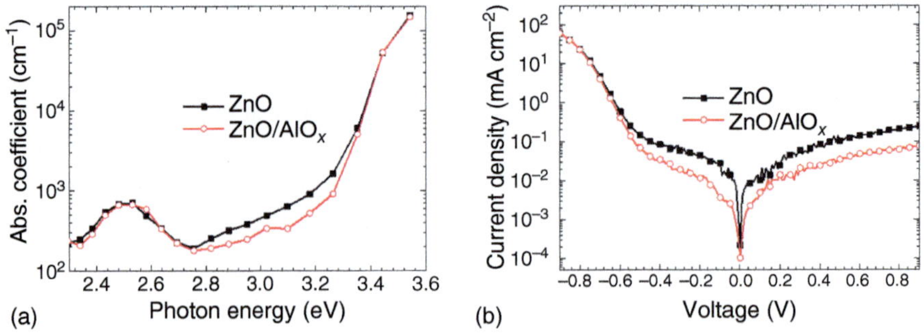

Figure 15.13 (a) PDS spectra measured for the pure ZnO and Al-modified ZnO layers. (b) Dark *J*–*V* characteristics measured for the inverted OSCs with the Al-modified ZnO and pure ZnO interlayers. Source: Bo Wu.

sub-band states in the solution-processed ZnO layer can be suppressed effectively by modification of its surface with an ultrathin Al layer.

As shown in Figure 15.13a, there is an obvious reduction in the absorption in the Al-modified ZnO layer as compared to the pure ZnO layer over the sub-gap energy region from 2.7 and 3.4 eV. The change in the PDS spectra suggests that the sub-gap states can be suppressed effectively in Al-modified ZnO interlayer. The suppression of the sub-gap states in the ZnO interlayer benefits the charge collection at the organic/electrode interface, which otherwise would act as the shallow electron traps to give rise to the bimolecular charge recombination and hamper the charge transport and charge extraction.

The relevance of PDS analyses can be seen by measuring the dark *J*–*V* characteristics of the inverted OSCs with the Al-modified ZnO and pure ZnO interlayers. As shown in Figure 15.13b, the decrease in the leakage current is clearly observed in the inverted OSCs with an Al-modified ZnO interlayer, demonstrating the effect of the suppression of the sub-gap states in the ZnO buffer layer on organic/electrode contact quality. The sub-band states in the ZnO interlayer provide additional sites to assist in charge injection, thereby an increase in the leakage current as seen in Figure 15.13b. The dark *J*–*V* characteristics clearly illustrate the change in the contact properties at the organic/ZnO and organic/Al-modified ZnO interfaces in the inverted OSCs. The leakage current can be reduced by suppressing the sub-gap states in ZnO interlayer in the OSCs, as shown in Figure 15.13a.

In a related work, the vertical phase separation in the polymer BHJ, based on the PTB7:$PC_{71}BM$ blend system, was studied by the XPS measurement, and its effect on the performance of the regular and inverted OSCs was examined [33]. The XPS results indicate that the segregation of donor/acceptor forming a $PC_{71}BM$-rich region at the bottom of the blend films can be moderated using solvent additive. The quantitative analyses show that the size of $PC_{71}BM$ domains in the BHJ decreases from 67 to 38 nm with the addition of the solvent additive. The use of the solvent additive helps to improve the homogeneous vertical phase separation in the PTB7:$PC_{71}BM$ BHJ, and thereby favoring an enhanced exciton dissociation process for attaining high-performance OSCs.

The hole collection and electron extraction in the cells are closely correlated to the interfacial electronic properties at the electron donor (PTB7)/anode and cathode/acceptor ($PC_{71}BM$) interfaces. As the OSCs with different cathodes of ZnO/ITO and Al-modified ZnO/ITO have the identical anode of MoO_3/Ag, the variation in charge collection of the cells can be associated with the modification in the contact property at the cathode/$PC_{71}BM$ interface. A better understanding on electron collection behavior with respect to the interfacial electronic properties at the cathode/$PC_{71}BM$ interface is important for achieving high-performance OSCs. Following our previous work of analyzing the interfacial exciton dissociation at the interface between the cathode and the surface of $PC_{71}BM$-rich PTB7:$PC_{71}BM$ blend layer in the OSCs [87], the interfacial electronic properties at the ZnO/$PC_{71}BM$ and Al-modified ZnO/$PC_{71}BM$ contacts were investigated by XPS measurements *in situ*. In this work, a thin layer of $PC_{71}BM$ was deposited using sublimation process stepwise onto the Al (1 nm)/ZnO (20 nm)/ITO sample.

For XPS characterizations, a ZnO (20 nm)/ITO substrate was loaded into an ultra-high vacuum (UHV) deposition chamber. A 1.2-nm Al film was deposited onto the substrate. The sample was then transferred to the UHV analysis chamber for XPS measurements. The bottom dashed line in Figure 15.14a shows an XPS Al 2p core level spectrum of the Al (1 nm)/ZnO (20 nm)/ITO sample. The sharp Al 2p peak at 72.8 eV can be attributed to metallic Al. Meanwhile, a broad peak emerges at 75.2 eV indicating the evaporated Al was partially oxidized forming AlO_X. Al 2p and C 1s XPS core level spectra of $PC_{71}BM$ (x nm)/Al (1 nm)/ZnO (20 nm)/ITO, with x of 0.2, 0.5, 1, 2, 5, and 10 nm, are shown as solid lines in Figure 15.14a,b, respectively. No obvious change in the Al 2p peak is observed as the $PC_{71}BM$ coverage increases. On the other hand, a strong C 1s peak can be found at ~284.8 eV corresponding to the C—C bond signal from $PC_{71}BM$. It is interesting to note that a small C 1s peak emerges at 287.1 eV during the initial coverage of $PC_{71}BM$. This C 1s peak later disappears when the coverage of $PC_{71}BM$ reaches 5.0 nm. The XPS results reveal an interfacial interaction between AlO_X and $PC_{71}BM$ during the initial deposition of $PC_{71}BM$ (0.5–2.0 nm), which provides a better contact and energy alignment

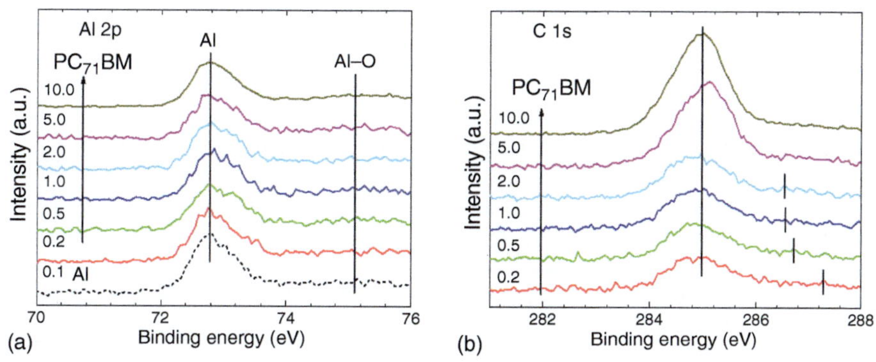

Figure 15.14 The high-resolution XPS core level spectra (a) Al 2p and (b) C 1s measured for the surface of Al (1 nm)/ZnO (20 nm)/ITO/glass substrate with different $PC_{71}BM$ thickness coverages from 0.2, 0.5, 1.0, 2.0, 5.0, and 10.0 nm. Source: Bo Wu.

of $PC_{71}BM$ and the interlayer, favoring the charge transport and collection at the cathode interface.

For device fabrication, the yield of high-performance OSCs is affected by fluctuation in morphology and surface electronic characteristics of the ZnO interlayer that vary due to many processing variables. There is thus a need to develop a robust and an effective interfacial modification approach that does not require any posttreatment for application in making high-performce OSCs via solution-processable route. The results of this work reveal clearly that the use of an Al-modified ZnO interlayer improves the performance reproducibility of PTB7:$PC_{71}BM$-based inverted OSCs fabricated in different batches as shown in Figure 15.15. Performance parameters of OSCs made with the pure ZnO interlayer and Al-modified ZnO interlayer are summarized in Table 15.4. OSCs with different cathode interlayers of Al (~1.2 nm)/ZnO (20 nm) and ZnO (20 nm), and different surface treatment conditions, e.g. with and without UV exposure prior to the cell fabrication, were made for comparison. It shows that pre-UV exposure of the ZnO (20 nm)/ITO cathode has a substantial influence on the performance of the cells, resulting in a much-reduced PCE of 5.9% for cells made with UV treated ZnO/ITO cathode, as compared to 6.8% for the structurally identical OSCs fabricated with non-UV-treated ZnO/ITO cathode. However, there is almost no change in the performance of the set of OSCs made with Al-modified ZnO/ITO cathode with and without UV exposure.

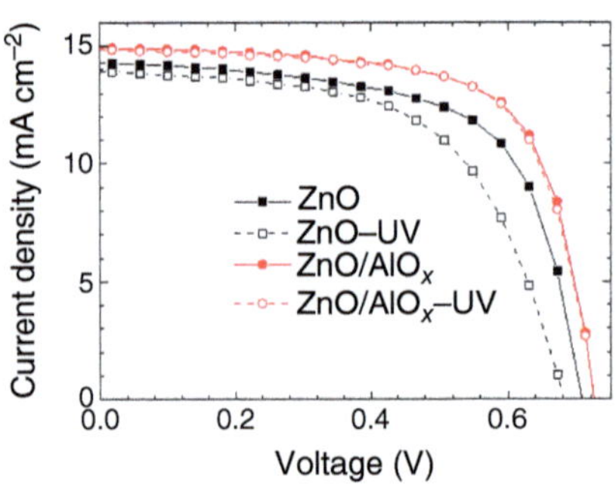

Figure 15.15 *J–V* characteristics of inverted OSCs with an Al-modified ZnO interlayer (circle) and a control cell with a pure ZnO interlayer (square), and the cells prepared with (solid symbol and line) and without (open symbol and dashed line) UV exposure. Source: Bo Wu.

Table 15.4 The performance of inverted OSCs made with the Al-modified ZnO and pure ZnO interlayers, and ZnO interlayers with and without UV exposure prior to the cell fabrication.

OSCs		V_{oc} (V)	J_{sc} (mA cm^{-2})	FF (%)	PCE (%)
Al-modified ZnO	Without UV	0.73 ± 0.01	15.7 ± 0.2	69 ± 1	7.8 ± 0.2
	With UV	0.73 ± 0.01	15.6 ± 0.3	68 ± 2	7.8 ± 0.2
Pure ZnO	Without UV	0.71 ± 0.01	15.1 ± 0.3	64 ± 2	6.8 ± 0.2
	With UV	0.68 ± 0.02	14.7 ± 0.3	59 ± 3	5.9 ± 0.3

The deviation in the performance of both types of OSCs is attributed to suppress of the sub-gap states in the ZnO interlayer, arising from the significant decrease in the interfacial defects at the organic/electrode interface. For example, the reaction between Al and loosely bound oxygen atoms in ZnO passivates the defects generated upon UV irradiation, which act as electron traps due to their deep energy level if they were not removed [88]. The density of such interfacial traps in the solution-processed ZnO interlayer may vary according to the different film preparation conditions and post-deposition treatments, e.g. with different UV exposures and post-annealing temperatures, causing an undesired increase in the charge recombination at the cathode/organic interface.

The surface condition of the ultrathin Al-modified ZnO is very stable, due to the formation of AlO_X even in the UHV condition, as illustrated by the Al 2p high-resolution XPS core-level spectra in Figure 15.14a. It is found that the cells made with a freshly prepared Al (~1.2 nm)-modified ZnO/ITO cathode and the ones that were prepared with the same condition but kept overnight in the glove box, with O_2 and H_2O levels below 0.1 ppm, remain essentially no observable change in the device performance. The Al-modified solution-processed ZnO interlayer assists in creation of a more stable cathode/organic interface for attaining high-performance OSCs. The above analyses, based on the PDS, XPS, and light-intensity-dependent J–V characteristic measurements, provide an insight on understanding the effect of suppression of ZnO sub-gap states on charge collection and performance reproducibility in inverted OSCs. This suggests that ultrathin Al (~1.2 nm)-modified ZnO is a simple and effective cathode modification approach, revealing the robustness and the process condition independent characteristics of the cathode that is suitable for high-performance OSC fabrication in a consistent and repeatable way, a process that can be easily adopted for mass production.

15.5 Effect of Built-in Potential on Stability of Organic Solar Cells

Apart from the steady progresses made in the increase of PCE, further improvement in the operational stability of OSCs is desired for practical applications. The device stability is closely associated with choice of the materials, material processing, morphology of the functional layers, the change in the stratification of the donor and acceptor in the BHJ [89, 90], interfacial chemical reaction at the BHJ/contact interfaces [25], balanced charge mobility in photoactive layer [24], and retaining a high built-in potential across the OSCs [91, 92]. Much research effort is needed to improve the stability of the OSCs through improved understanding of the degradation mechanisms.

PEDOT:PSS layer is one of the widely used hole contact materials for application in solution-processed OSCs [93]. However, a PEDOT:PSS HEL is coated using a highly acidic aqueous suspension, which results in corrosion of the ITO electrode. Its high hydroscopic property also leads to the encroachment of the moisture in the active layer, accelerating the cell degradation [82, 94]. To address this issue, a variety of chemical stable and low-cost hole extraction materials have been proposed to substitute PEDOT:PSS for more stable and efficient OSCs. Transition metal oxide, including NiO, WO_3, and MoO_3, represents an alternative class of materials that could potentially combine all-important attributes of an ideal HEL for OSC applications [95, 96]. In particular, solution-processable MoO_3 has been proved promising with excellent optical transparency and suitable high work function [97, 98]. However, the relatively high surface trap density and inferior charge transporting property of the pristine solution-processed MoO_3 HEL often cause poor contact at the organic/MoO_3 interface, resulting in high leakage current and undesired charge recombination loss [99]. The impact of a MoO_3/antimonene HEL on the stability of OSCs, e.g. prepared using a blend system of poly[(2,6-(4,8-bis(5-(2-ethylhexyl-3-fluoro)thiophen-2-yl)-benzo[1,2-*b*:4,5-*b*′] dithiophene))-*alt*-(5,5-(1′,3′-di-2-thienyl-5′,7′-bis(2-ethylhexyl)benzo[1′,2′-*c*:4′,5′-*c*′] dithiophene-4,8-dione)] (PM6) and 3,9-bis(2-methylene-((3-(1,1-dicyanomethylene) -6,7-difluoro)-indanone))-5,5,11,11-tetrakis(4-hexylphenyl)-dithieno[2,3-*d*:2′,3′-*d*′]-*s*-indaceno[1,2-*b*:5,6-*b*′]dithiophene (IT-4F), has been examined. It shows that removal of the interfacial chemical reactions between PEDOT:PSS HEL and the 2-(3-oxo-2,3-dihydroinden-1-ylidene)malononitrile (INCN) moieties in non-fullerene acceptor (NFA)-based OSCs is one of the critical factors for enhancing the stability of the OSCs. The use of bilayer MoO_3/antimonene HEL helps to eliminate the interfacial reaction between the PEDOT:PSS and INCN moieties in NFA, one of the recently found major degradation mechanisms in the NFA-based OSCs with a PEDOT:PSS HEL. The use of a bilayer MoO_3/antimonene HEL also favors the stable operation of the OSCs through retaining a high built-in potential.

15.5.1 Interlayer Modification

The J–V and EQE characteristics, measured for the OSCs with a MoO_3 HEL, a bilayer MoO_3/antimonene HEL and the control OSCs with a PEDOT:PSS HEL, are shown in Figure 15.16a,b. The OSCs with a MoO_3/antimonene HEL have an obvious higher PCE as compared to that of the cell with a pristine MoO_3 HEL, e.g. J_{sc} of 18.01 mA cm^{-2}, V_{oc} of 0.82 V, and FF of 66.92%. The combination of these improvements leads to a PCE of ~12% in OSCs with a MoO_3/antimonene HEL, achieving >17% increase in PCE as compared to that of the OSCs with a pristine MoO_3 HEL (10.02%). A device performance obtained for the PM6:IT-4F OSCs with a bilayer a MoO_3/antimonene HEL is comparable to that of a control cell with a PEDOT:PSS HEL. The work function of the PEDOT:PSS and MoO_3/antimonene layers was measured using the UPS. The corresponding UPS spectra measured for these films are presented in Figure 15.16c. The work function of −5.4 eV is obtained for the bilayer MoO_3/antimonene, which is slightly higher than that of a PEDOT:PSS layer (−5.2 eV). The corresponding schematic diagram illustrating the energy levels of the functional materials used in the OSCs is shown in Figure 15.16d. The transmission electron microscopy (TEM) images measured for a solution-processed antimonene layer are shown in Figure 15.17. The antimonene layer has a lattice fringe of 0.21 nm, corresponding to a (110) plane of the crystalline Sb as shown in Figure 15.17b.

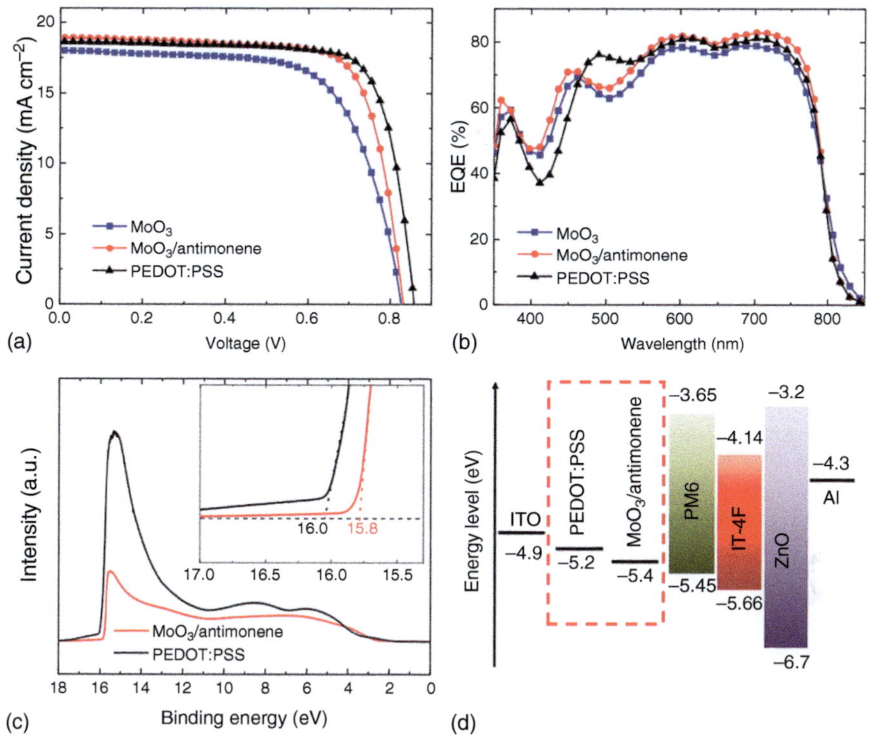

Figure 15.16 (a) *J*–*V* and (b) EQE characteristics measured for the PM6:IT-4F-based OSCs with different HELs, (c) USP spectra measured for the MoO_3/antimonene and PEDOT:PSS layers deposited on ITO substrates, and (d) schematic diagram illustrating the energy levels of the functional materials. Source: Weixia Lan.

It shows that the solution-processed bilayer MoO_3/antimonene HEL is energetically more favorable for hole extraction as compared to that of the PEDOT:PSS HEL for use in OSCs.

η_{CC} of the PM6:IT-4F OSCs with a bilayer MoO_3/antimonene HEL and a control cell with a PEDOT:PSS HEL has been analyzed, the results are presented in Figure 15.18. The efficient charge extraction can be realized in OSCs operated under a high effective voltage, facilitating an efficient sweeping-out of the charge carriers under a relatively high effective internal electric field. The charge collection efficiency η_{CC} of the OSCs reduces with the decrease of V_{eff}. This is because of the low efficiency in sweeping out of the photo-generated charges, thereby a high charge recombination to take place in the cells. η_{CC} at low V_{eff} also provides an insight revealing the charge collection capability of the cells. Apparently, a higher η_{CC} at a low V_{eff} reflects the superior charge collection capability and reduced charge recombination loss in the OSCs. As shown in Figure 15.18, OSCs with a bilayer MoO_3/antimonene HEL shows similar charge collection properties over the whole V_{eff} range, suggesting the favorable BHJ/HEL interfacial properties for assisting in charge extraction.

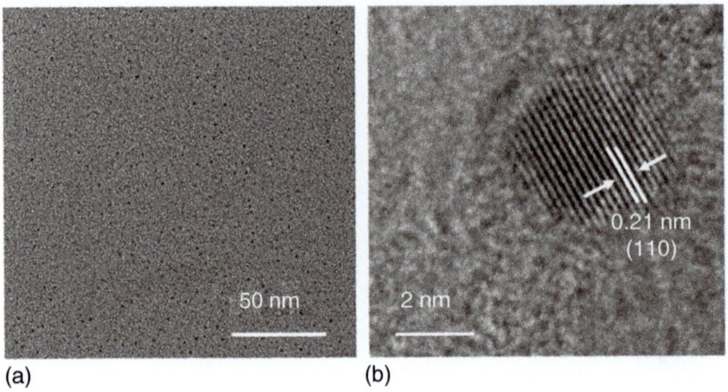

Figure 15.17 (a) TEM and (b) high-resolution TEM images measured for a solution-processed antimonene layer. Source: Weixia Lan.

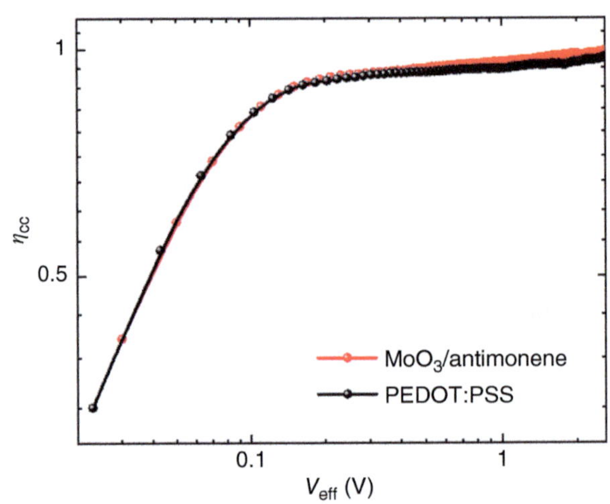

Figure 15.18 $\eta_{CC}-V_{eff}$ characteristics of the NFA-based OSCs with different HELs. Source: Weixia Lan.

15.5.2 Built-in Potential in Organic Solar Cells

E_{int} in the BHJ OSCs plays a crucial role in charge transfer, as the photo-generated charge carriers drift with the E_{int} in the OSCs. A larger E_{int} is desired for assisting the efficient charge transfer, thereby suppressing the bimolecular recombination. The device performance difference observed between the PM6:IT-4F OSCs with a bilayer MoO_3/antimonene HEL and ones with a PEDOT:PSS HEL is closely associated with the variation of the built-in potential V_{bi} in these OSCs during the aging test. The V_{bi} in the OSCs was analyzed using the TPC measurements. The TPC (I_{TPC}) is a function of E_{int} in a cell. I_{TPC} measured for the OSCs operated under different external biases are shown in Figure 15.19. A larger E_{int} in the OSCs helps for the efficient exciton dissociation and charge transfer, as the photo-generated charge carriers drift with the E_{int}. The relationship between the applied bias (V_a) and E_{int} has been described in Section 15.3 Eq. (15.8). V_{bi} can be determined by applying an

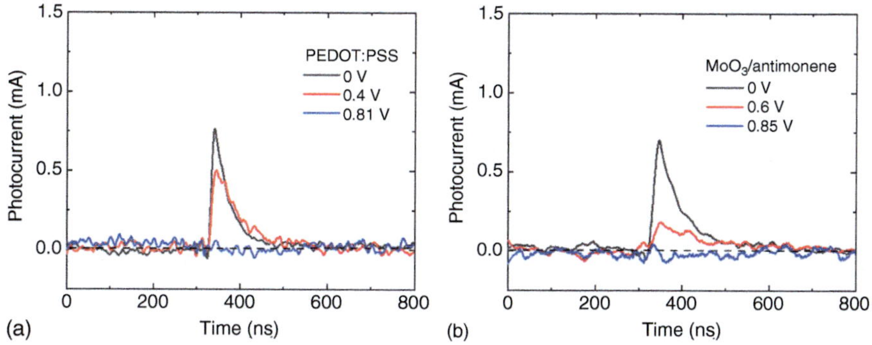

Figure 15.19 Transient photocurrents of the PM6:IT-4F-based OSCs with (a) a PEDOT:PSS HEL and (b) a MoO_3/antimonene HEL, measured under different reverse biases. Source: Weixia Lan.

opposite V_a to compensate V_{bi} for I_{TPC} approaching zero. Since d is the same for PM6:IT-4F OSCs with different HELs, E_{int} is proportional to V_{bi} in the OSCs. As shown in Figure 15.19, I_{TPC} decreases with increase in V_a and reaches zero when $V_a = 0.84\,V$ for the control OSCs with a PEDOT:PSS HEL. V_a of 0.81 V was required for compensating V_{bi} in the PM6:IT-4F OSCs with a MoO_3/antimonene HEL.

The effect of the HELs on the stability of the PM6:IT-4F OSCs has been investigated through examining the built-in potential in the OSCs using an aging test. V_{bi} of the PM6:IT-4F OSCs with different HELs as a function of the aging time was investigated. Normalized V_{bi} – aging time correlations measured for the PM6:IT-4F OSCs with a bilayer MoO_3/antimonene HEL and a PEDOT:PSS HEL are shown in Figure 15.20. A 10% drop in V_{bi} of the PM6:IT-4F OSCs with a MoO_3/antimonene HEL was observed during the aging test over a period of 30 days. However, a 22% drop in V_{bi} was observed for the control OSCs with a PEDOT:PSS HEL, dropping more than two times faster than the one observed in the PM6:IT-4F OSCs with a bilayer MoO_3/antimonene HEL. A steady V_{bi} or a constant large E_{int} is a prerequisite for the efficient and stable operation of OSCs. TPC results reveal clearly that a

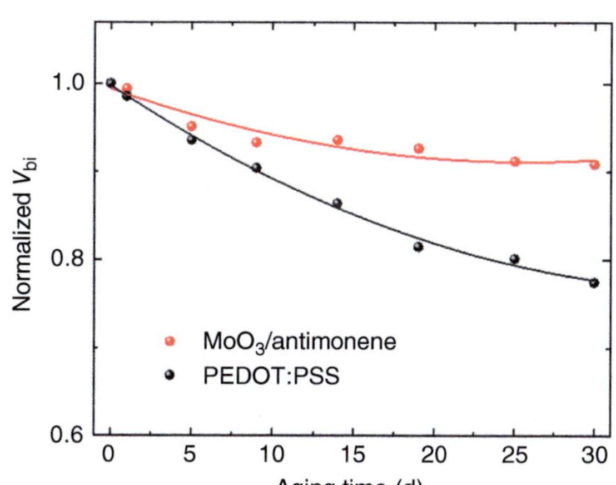

Figure 15.20 Normalized V_{bi} – aging time characteristics measured for the NFA-based OSCs with different HELs. Source: Weixia Lan.

gradual decrease in V_{bi} across the BHJ, as shown in Figure 15.20, is one main reason responsible for the deterioration in the performance of the PM6:IT-4F OSCs with a PEDOT:PSS HEL.

15.5.3 Stability of Organic Solar Cells

The stability of the encapsulated PM6:IT-4F OSCs made with different HELs has been analyzed in air. The V_{oc}, J_{sc}, FF, and PCE of different PM6:IT-4F OSCs as a function of the aging time are shown in Figure 15.21. The results show that the PM6:IT-4F OSCs with a bilayer MoO_3/antimonene HEL are more stable than the control cells with a PEDOT:PSS HEL. A 46% decrease in V_{oc} and a 25% drop in FF were observed for the control cells with a PEDOT:PSS HEL, leading to a 55% reduction in PCE after a 30-day aging test in air. In comparison, an 11% decrease in V_{oc} and a 13% drop in FF were observed in the PM6:IT-4F OSCs with a bilayer MoO_3/antimonene HEL, having a 25% drop in PCE for the OSCs aged over the same stability test.

The normalized η_{CC} obtained for the PM6:IT-4F OSCs with different HELs, operated at the maximum power point (P_{max}) and near V_{oc} conditions, as a function of the aging time are shown in Figure 15.22. It shows that the η_{CC}, monitored for the OSCs with a bilayer MoO_3/antimonene HEL operated at P_{max} (V_{eff} = 0.23 V), is more

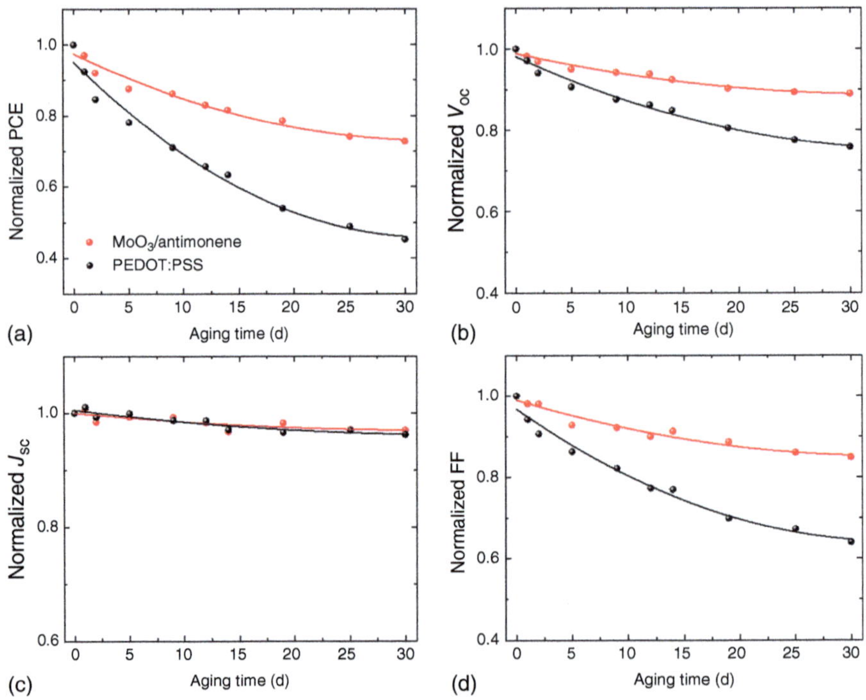

Figure 15.21 (a) PCE, (b) V_{oc}, (c) J_{sc}, and (d) FF measured for the NFA-based OSCs with a MoO_3/antimonene HEL and a PEDOT:PSS HEL as a function of the aging time. Source: Weixia Lan.

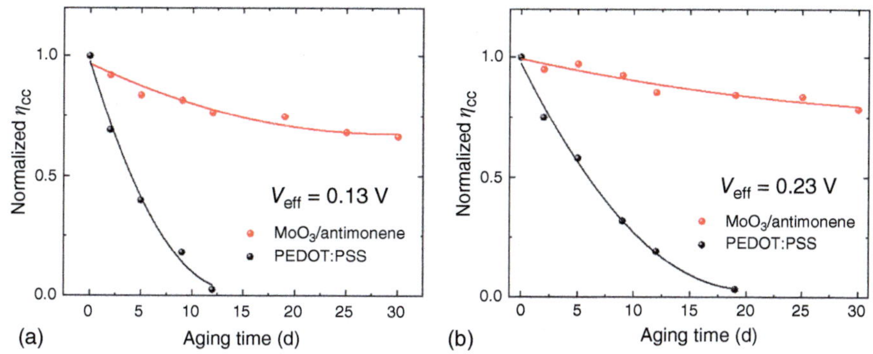

Figure 15.22 Normalized η_{CC} as a function of the aging time, measured for the different NFA-based OSCs operated at (a) P_{max} (V_{eff} = 0.23 V) and (b) near V_{oc} (V_{eff} = 0.13 V) conditions. Source: Weixia Lan.

stable as compared to the one observed for the OSCs with a PEDOT:PSS HEL aged over the same aging period. The advantage of the bilayer MoO_3/antimonene HEL for use in PM6:IT-4F OSCs is clearly manifested, as shown in Figure 15.22.

The possible interfacial reaction between HEL and BHJ photoactive layer was analyzed to reveal the degradation mechanism in the PM6:IT-4F OSCs. It is known that photostability of the polymer donors and NFAs used in OSCs are sensitive to the surrounding environment. The change in the absorption spectra of the functional organic semiconductors is associated with the deterioration in their functions, caused by the possible material degradation or imperfection. The absorption spectra measured for the PM6 layers, formed on the PEDOT:PSS and MoO_3/antimonene surfaces and exposed to the one sun illumination over an exposure period from zero to four hours, are shown in Figure 15.23a,b. There is no observable change found in the absorption spectra measured for the PM6 layer deposited on either a PEDOT:PSS HEL or a MoO_3/antimonene HEL, suggesting that there is no observable interfacial reaction between the PM6 donor and HELs. However, an obvious variation was observed in the absorption spectra measured for the IT-4F layers deposited on different HELs, as shown in Figure 15.23c,d. A much faster reduction was observed in the absorption spectra measured for the IT-4F layer formed on the PEDOT:PSS HEL as compared to the ones measured for the IT-4F layer deposited on the bilayer MoO_3/antimonene HEL over the wavelength range from 600 to 800 nm.

The significant change in the absorption behavior observed for the IT-4F layer formed on the PEDOT:PSS HEL is associated closely with the interfacial chemical reaction between the PEDOT structures and INCN moieties in IT-4F acceptor [91, 92], accompanied by free radical reaction during the photo-oxidation process. The concentration of free radicals in the NFA layer increases with the illumination time. IT-4F is an acceptor–donor–acceptor-type molecule, as shown in the inset in Figure 15.23c. The free radicals, generated by the INCN moieties under light illumination, would diffuse through to disturb conjugated bonds in the IT-4F molecules. The interaction between the free radicals on adjacent backbones and side chains induces a decrease in absorption due to the cross-linking of the polymer film.

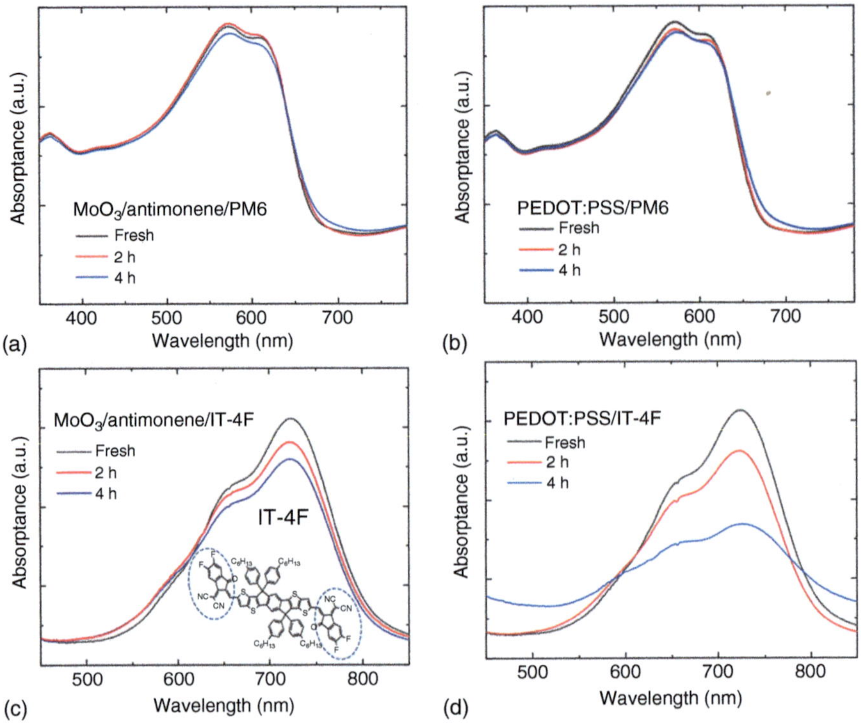

Figure 15.23 Absorption spectra measured for the (a) MoO_3/antimonene/PM6 layer, (b) PEDOT:PSS/PM6 layer, (c) MoO_3/antimonene/IT-4F layer, and (d) /PEDOT:PSS/IT-4F layer exposed to 100 mW cm^{-2} illumination over a light-soaking period from zero to four hours. Source: Weixia Lan.

The absorption peak at ~750 nm is related to the π–π* transition and intramolecular charge transfer between the donor and accepter moieties in the IT-4F molecules. The change in the absorption spectra measured for the IT-4F layer formed on the PEDOT:PSS HEL reveals a disruption of the intramolecular charge transfer between the donor and accepter moieties in the IT-4F molecules, caused by the interfacial reaction between the INCN moieties in the IT-4F molecules and the PEDOT quinoidal nanostructures in the PEDOT:PSS HEL. It shows the deterioration of the electron-withdrawing end moieties in NFA molecules, caused by the interfacial reaction between the INCN moieties in the IT-4F and the PEDOT quinoidal nanostructures in PEDOT:PSS [91, 92]. The fast decrease in the absorption measured for the IT-4F layer on PEDOT:PSS HEL, as shown in Figure 15.23d, is due to the interfacial interaction between the IT-4F acceptor and the PEDOT:PSS HEL, resulting in an undesired weakening in the π–π* transition and intramolecular charge transfer between the donor and accepter moieties in the IT-4F. The inevitable interfacial reaction between the IT-4F acceptor and the PEDOT:PSS HEL gives rise to a persistent decrease in V_{bi}, and hence a continuing drop in V_{oc} and FF in the PM6:IT-4F OSCs with a PEDOT:PSS HEL. The continuous deterioration in the IT-4F molecules, introduced by the inevitable interfacial reaction between IT-4F and PEDOT:PSS molecules, is apparently not favorable for the stable operation of PM6:IT-4F OSCs with a PEDOT:PSS HEL. This problem can be resolved by replacing a solution-processable NFA inert MoO_3/antimonene HEL for PEDOT:PSS HEL. A stable and high V_{bi} in the PM6:IT-4F OSCs with a MoO_3/antimonene HEL is clearly demonstrated, offering an encouraging option for making efficient and durable OSCs.

15.6 Summary

This chapter highlights the understanding of the organic/electrode interfacial exciton dissociation, charge recombination processes, and charge collection, which underpin the optimum device design and performance. The charge transport and recombination properties in the BHJ OSCs are investigated systematically using a combination of theoretical calculation, TPC measurements, morphology analyses, and device optimization. It is found that the dissociation of excitons at the organic/metal interface in regular geometry OSCs hampers the electron collection. The origin of unfavorable electron collection is mainly due to the trap-limited bimolecular recombination, resulting in the compensation of drifted photo-generated electrons at the organic/cathode interface. The undesired charge collection can be eliminated by inserting a thin metal oxide interlayer between the organic layer and electrode. It clearly reveals that the removal of the unfavorable interfacial exciton dissociation is a perquisite for a significant enhancement in overall performance of OSCs. Suppression of the metal oxide sub-gap states significantly improves the charge extraction and performance reproducibility. An enhanced stability of the OSCs can be realized through interfacial modification of the interlayer for retaining a high built-in potential. The use of a NFA inert HEL benefits the efficient operation of NFA-based OSCs through suppression of interfacial reaction between the HEL and the INCN moieties in NFA, which otherwise would occur in IT-4F-based NFA-based OSCs with a PEDOT:PSS HEL. The improved understanding on the organic/electrode interfacial modification and retaining a high built-in potential in OSCs provide a comprehensive insight for applications in efficient and stable operation of OSCs.

Acknowledgment

This work was supported by the National Natural Science Foundation of China (62005152), the Research Grants Council, University Grants Committee, Hong Kong Special Administrative Region, China, General Research Fund (12302419), Collaborative Research Fund (C5037-18GF), NSFC/RGC Joint Research Scheme (N_HKBU201/19), and Natural Science Foundation of Jiangsu Province (BK20180694).

References

1 Cao, L., Klimes, K., Ji, Y. et al. (2021). *Nat. Photonics* 15: 230–237.
2 Xu, C., Zhang, J., Xu, W., and Tian, H. (2021). *Mater. Chem. Front.* 5: 1060–1075.

3 Lan, Z., Lau, Y., Wang, Y. et al. (2020). *Adv. Opt. Mater.* 8: 202001388.
4 Lan, Z., Cai, L., Luo, D., and Zhu, F. (2021). *ACS Appl. Mater. Interfaces* 13: 981–988.
5 Wang, Z., Peng, Z., Xiao, Z. et al. (2020). *Adv. Mater.* 32: 2005386.
6 Li, D., Zhu, L., Liu, X. et al. (2020). *Adv. Mater.* 32: 2002344.
7 Clark, J. and Lanzani, G. (2010). *Nat. Photonics* 4: 438–446.
8 Li, D., Kim, K., Zhang, S. et al. (2015). *Sens. Actuators, A* 236: 364–368.
9 Sun, Q., Li, D., Dong, G. et al. (2013). *Sens. Actuators, B* 188: 879–885.
10 Amrani, A.E., Lucas, B., and Antony, R. (2015). *Superlattices Microstruct.* 85: 880–885.
11 Anichini, C., Czepa, W., Pakulski, D. et al. (2018). *Chem. Soc. Rev.* 47: 4860–4908.
12 Hu, Z., Deibert, B.J., and Li, J. (2014). *Chem. Soc. Rev.* 43: 5815–5840.
13 Choudhury, B., Shinar, R., and Shinar, J. (2004). *J. Appl. Phys.* 96: 2949–2954.
14 Ratcliff, E.L., Veneman, P.A., Simmonds, A. et al. (2010). *Anal. Chem.* 82: 2734–2742.
15 Kim, R.H., Kim, H.J., Bae, I. et al. (2014). *Nat. Commun.* 5: 3583.
16 Tam, H., Choi, W., and Zhu, F. (2015). *Electronics* 4: 623.
17 Ali, F., Periasamy, N., Patankar, M.P., and Narasimhan, K.L. (2011). *J. Phys. Chem. C* 115: 2462–2469.
18 Liu, Q., Jiang, Y., Jin, K. et al. (2020). *Sci. Bull.* 65: 272.
19 Jin, K., Xiao, Z., and Ding, L. (2021). *J. Semicond.* 42: 060502.
20 Wu, Z., Wu, B., Tam, H., and Zhu, F. (2016). *Org. Electron.* 31: 266–272.
21 Lan, W., Cui, Y., Yang, Q. et al. (2015). *AIP Adv.* 5: 057164.
22 Rafique, S., Abdullah, S.M., Shahid, M.M. et al. (2017). *Sci. Rep.* 7: 39555.
23 Qi, B., Zhang, Z., and Wang, J. (2015). *Sci. Rep.* 5: 7803.
24 Chellappan, V., Ng, G.M., Tan, M. et al. (2009). *Appl. Phys. Lett.* 95: 263305.
25 Wang, X., Zhao, C., Xu, G. et al. (2012). *Sol. Energy Mater. Sol. Cells* 104: 1–6.
26 Hau, S.K., Yip, H.L., Ma, H., and Jen, A.K.Y. (2008). *Appl. Phys. Lett.* 93: 233304.
27 Liao, H., Chen, L.M., Xu, Z. et al. (2008). *Appl. Phys. Lett.* 92: 173303.
28 Jiang, C., Sun, X., Zhao, D. et al. (2010). *Sol. Energy Mater. Sol. Cells* 94: 1618–1621.
29 Zhang, H. and Ouyang, J. (2010). *Appl. Phys. Lett.* 97: 063509.
30 Zhou, Y., Fuentes-Hernandez, C., Shim, J. et al. (2012). *Science* 336: 327–332.
31 Aydemir, M. (2021). *Polym. Bull.* 78: 2841–2847.
32 De, S., Pascher, T., Maiti, M. et al. (2007). *J. Am. Chem. Soc.* 129: 8466–8472.
33 Clarke, T., Ballantyne, A., Jamieson, F. et al. (2009). *Chem. Commun.* 1: 89–91.
34 Hwang, I.W., Moses, D., and Heeger, A.J. (2008). *J. Phys. Chem. C* 112: 4350–4354.
35 Peet, J., Kim, J.Y., Coates, N.E. et al. (2007). *Nat. Mater.* 6: 497–500.
36 Tang, Z., Tress, W., Bao, Q. et al. (2014). *Adv. Energy Mater.* 4: 1400643.
37 Yu, G., Gao, J., Hummelen, J.C. et al. (1995). *Science* 270: 1789–1791.
38 Hoppe, H. and Sariciftci, N.S. (2006). *J. Mater. Chem.* 16: 45–61.
39 Liu, A., Zhao, S., Rim, S.B. et al. (2008). *Adv. Mater.* 20: 1065–1070.
40 Shuttle, C.G., O'Regan, B., Ballantyne, A.M. et al. (2008). *Phys. Rev. B* 78: 113201.

41 Veldman, D., İpek, Ö., Meskers, S.C.J. et al. (2008). *J. Am. Chem. Soc.* 130: 7721–7735.
42 Street, R.A., Schoendorf, M., Roy, A., and Lee, J.H. (2010). *Phys. Rev. B* 81: 205307.
43 Park, S.H., Roy, A., Beaupre, S. et al. (2009). *Nat. Photonics* 3: 297–302.
44 Carsten, D. and Vladimir, D. (2010). *Rep. Prog. Phys.* 73: 096401.
45 Chen, S., Manders, J.R., Tsang, S.-W., and So, F. (2012). *J. Mater. Chem.* 22: 24202–24212.
46 Bin, H., Wang, J., Li, J. et al. (2021). *Adv. Mater.* 33: 2008429.
47 Hwang, J., Wan, A., and Kahn, A. (2009). *Mater. Sci. Eng., R* 64: 1–31.
48 Sorrentino, R., Kozma, E., Luzzati, S., and Po, R. (2021). *Energy Environ. Sci.* 14: 180–223.
49 Waldauf, C., Morana, M., Denk, P. et al. (2006). *Appl. Phys. Lett.* 89: 233517.
50 Sun, Y., Seo, J.H., Takacs, C.J. et al. (2011). *Adv. Mater.* 23: 1679–1683.
51 Hu, Z., Zhang, J., Liu, Y. et al. (2011). *Sol. Energy Mater. Sol. Cells* 95: 2126–2130.
52 Ibrahem, M.A., Wei, H.Y., Tsai, M.H. et al. (2013). *Sol. Energy Mater. Sol. Cells* 108: 156–163.
53 Ghosh, M., Ningthoujam, R.S., Vatsa, R.K. et al. (2011). *J. Appl. Phys.* 110: 054309.
54 Singh, S. and Chakrabarti, P. (2013). *Superlattices Microstruct.* 64: 283–293.
55 Kuwabara, T., Tamai, C., Omura, Y. et al. (2013). *Org. Electron.* 14: 649–656.
56 Mozer, A.J., Dennler, G., Sariciftci, N.S. et al. (2005). *Phys. Rev. B* 72: 035217.
57 Yang, Y., Chen, W., Dou, L. et al. (2015). *Nat. Photonics* 9: 190–198.
58 Albrecht, S., Schindler, W., Kurpiers, J. et al. (2012). *J. Phys. Chem. Lett.* 3: 640–645.
59 Mozer, A.J., Sariciftci, N.S., Lutsen, L. et al. (2005). *Appl. Phys. Lett.* 86: 112104.
60 Lan, W., Liu, Y., Wu, B. et al. (2019). *ACS Appl. Energy Mater.* 2: 7385–7392.
61 Eng, M.P., Barnes, P.R.F., and Durrant, J.R. (2010). *J. Phys. Chem. Lett.* 1: 3096–3100.
62 Clarke, T.M., Jamieson, F.C., and Durrant, J.R. (2009). *J. Phys. Chem. C* 113: 20934–20941.
63 Cowan, S.R., Roy, A., and Heeger, A.J. (2010). *Phys. Rev. B* 82: 245207.
64 Kyaw, A.K.K., Wang, D.H., Gupta, V. et al. (2013). *ACS Nano* 7: 4569–4577.
65 Bakulin, A.A., Rao, A., Pavelyev, V.G. et al. (2012). *Science* 335: 1340–1344.
66 Deibel, C., Strobel, T., and Dyakonov, V. (2010). *Adv. Mater.* 22: 4097–4111.
67 Qin, J., Zhang, L., Zuo, C. et al. (2021). *J. Semicond.* 42: 010501.
68 Liu, Q., Jin, K., Li, W. et al. (2020). *J. Mater. Chem. A* 8: 8857–8861.
69 Lo, M.F., Ng, T.W., Lai, S.L. et al. (2011). *Appl. Phys. Lett.* 99: 033302.
70 Lo, M.F., Ng, T.W., Mo, H.W., and Lee, C.S. (2013). *Adv. Funct. Mater.* 23: 1718–1723.
71 Mihailetchi, V.D., Koster, L.J.A., Hummelen, J.C., and Blom, P.W.M. (2004). *Phys. Rev. Lett.* 93: 216601.
72 Shuttle, C.G., O'Regan, B., Ballantyne, A.M. et al. (2008). *Appl. Phys. Lett.* 92: 093311.

73 Boix, P.P., Ajuria, J., Pacios, R., and Belmonte, G.G. (2011). *J. Appl. Phys.* 109: 074514.

74 Cheng, W., Wu, Z., Wen, S. et al. (2013). *Org. Electron.* 14: 2124–2131.

75 Wu, B., Wu, Z., Tam, H.L., and Zhu, F. (2014). *Appl. Phys. Lett.* 105: 103302.

76 Braun, S., Salaneck, W.R., and Fahlman, M. (2009). *Adv. Mater.* 21: 1450–1472.

77 Bisikalo, D.V., Zhilkin, A.G., and Kurbatov, E.P. (2019). *Phys. Usp.* 62: 1136.

78 Wang, W., Pröller, S., Niedermeier, M.A. et al. (2015). *ACS Appl. Mater. Interfaces* 7: 602–610.

79 Liang, Z., Zhang, Q., Jiang, L., and Cao, G. (2015). *Energy Environ. Sci.* 8: 3442–3476.

80 Yang, T., Wang, M., Duan, C. et al. (2012). *Energy Environ. Sci.* 5: 8208–8214.

81 Zhang, K., Zhong, C., Liu, S. et al. (2014). *ACS Appl. Mater. Interfaces* 6: 10429–10435.

82 Wu, B., Wu, Z., Yang, Q. et al. (2016). *ACS Appl. Mater. Interfaces* 8: 14717–14724.

83 He, Z., Zhong, C., Huang, X. et al. (2011). *Adv. Mater.* 23: 4636–4643.

84 Kazukauskas, V., Janonis, V., and Vertelis, V. (2021). *Opt. Mater.* 118: 111259.

85 Roy, A., Mandai, S., and Menon, R. (2021). *J. Appl. Phys.* 129: 195501.

86 Venkateshvaran, D., Nikolka, M., Sadhanala, A. et al. (2014). *Nature* 515: 384–388.

87 Xu, W., Wu, B., Zheng, F. et al. (2015). *Org. Electron.* 25: 266–274.

88 Bao, J., Shalish, I., Su, Z. et al. (2011). *Nanoscale Res. Lett.* 6: 1–7.

89 Yu, R., Yao, H., Chen, Z. et al. (2019). *Adv. Mater.* 31: 1900477.

90 Zhao, W., Zhang, S., and Hou, J. (2016). *Sci. China Chem.* 59: 1574–1582.

91 Wang, Y., Han, J., Cai, L. et al. (2020). *J. Mater. Chem. A* 8: 21255–21264.

92 Wang, Y., Lan, W., Li, N. et al. (2019). *Adv. Energy Mater.* 9: 1900157.

93 Lan, Z., Lei, Y., Chan, W.K.E. et al. (2020). *Sci. Adv.* 6: eaaw8065.

94 Lan, W., Gu, J., Wu, S. et al. (2021). *EcoMat.* 3: e12134.

95 Becker, T., Trost, S., Behrendt, A. et al. (2018). *Adv. Energy Mater.* 8: 1702533.

96 Chen, W., Liu, F., Feng, X. et al. (2017). *Adv. Energy Mater.* 7: 1700722.

97 Girotto, C., Voroshazi, E., Cheyns, D. et al. (2011). *ACS Appl. Mater. Interfaces* 3: 3244–3227.

98 Jasieniak, J.J., Seifter, J., Jo, J. et al. (2012). *Adv. Funct. Mater.* 22: 2594–2605.

99 Lou, Y., Wang, Z., Yuan, D. et al. (2014). *Appl. Phys. Lett.* 105: 113301.

16

Voltage Losses and Charge Transfer States in Donor–Acceptor Organic Solar Cells

Hongbo Wu, Mengyang Li, Zaifei Ma, and Zheng Tang

Donghua University, College of Materials Science and Engineering, Center for Advanced Low-dimension Materials, State Key Laboratory for Modification of Chemical Fibers and Polymer Materials, North Renmin road 2999, Songjiang District, Shanghai 201620, P.R. China

16.1 The Origin of V_{oc} of Solar Cells

16.1.1 Voltage Loss in an Ideal Solar Cell and the Upper Limit for V_{oc}

Performance of organic solar cells is improving rapidly, owing to the development of highly efficient photo-active material systems. Short-circuit current density (J_{sc}) and fill factor (FF) of state-of-the-art organic solar cells are already comparable to those of inorganic or organic/inorganic hybrid perovskite solar cells [1–15]. However, V_{oc} of organic solar cells is quite low, compared to that of the inorganic or perovskite solar cells, when active materials are with similar bandgap energy (E_g) [16, 17]. This is the main reason that performance of organic solar cells is still worse than that of inorganic or perovskite solar cells. In other words, a large V_{loss}, defined as the difference between V_{oc} and E_g/q, limits the power conversion efficiency (PCE) of organic solar cells [18–25].

To understand the reason for the high V_{loss} in organic solar cells, we first study the origin of V_{oc} and V_{loss} in an ideal solar cell. This was first derived by Shockley and Queisser, using the detailed balanced theory [26]. Assuming that for the ideal solar cell, the photo-absorbing active layer is a semiconductor with two energy levels for the ground state and the excited state. This gives rise to a clearly defined energy of bandgap (E_g). Thus, photons with energy higher than E_g can be absorbed by the active layer of the ideal solar cell, resulting in generation of electron–hole pairs. Meanwhile, the photo-generated electrons and holes can also recombine. Assuming that the photons are absorbed at a rate, N_{abs}, and the charge carriers recombine at a rate, N_{rec}, then, the rate for extracting the photo-generated charge carriers (N_{ext}) can be defined by the difference between N_{abs} and N_{rec}:

$$N_{ext} = N_{abs} - N_{rec} \tag{16.1}$$

According to the detailed balance theory [26], a lower limit for N_{rec} is when charge carriers recombine radiatively. Therefore, for the ideal solar cell, charge carriers can

Organic Solar Cells: Materials Design, Technology and Commercialization, First Edition. Edited by Liming Ding.

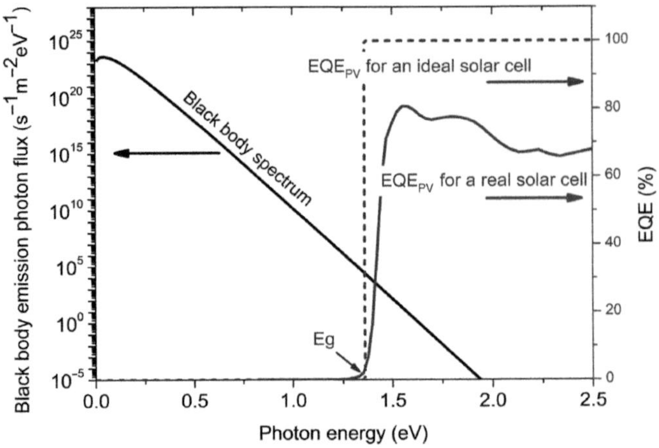

Figure 16.1 Blackbody radiation spectrum at $T = 300$ K, absorption spectrum of the ideal solar cell, represented by a step-like function, and absorption spectrum of a hypothetic solar cell. Source: Mengyang Li.

only recombine radiatively. Thus, N_{rec} is equal to the photon emission rate (N_{emi}) of the active material in the solar cell. According to Würfel's generalized Planck law [27], N_{emi} is related to the blackbody radiation spectrum ($BB(E)$, Figure 16.1) and the absorption spectrum ($A(E)$) of the photo-active layer, characterized by a step-like function, i.e. 100% absorbing for photons with energy larger than E_g and non-absorbing for photons with energy smaller than E_g (Figure 16.1); thus,

$$N_{rec} = N_{emi} = \int_{E_g}^{+\infty} BB(E)\mathrm{d}E \tag{16.2}$$

$BB(E)$ can be expressed as

$$BB(E) = \frac{2\pi E^2}{c^2 h^3 \left[\mathrm{e}^{\left(\frac{E-qV}{kT}\right)} - 1\right]} \tag{16.3}$$

Thus,

$$\begin{aligned} N_{rec} = N_{emi} &= \int_{E_g}^{+\infty} \frac{2\pi E^2}{c^2 h^3 \left[\mathrm{e}^{\left(\frac{E-qV}{kT}\right)} - 1\right]}\mathrm{d}E \\ &\approx \frac{2\pi E_g^2 kT}{c^2 h^3}\mathrm{e}^{\left(-\frac{E_g}{kT}\right)}\left[\mathrm{e}^{\left(\frac{qV}{kT}\right)} - 1\right] \end{aligned} \tag{16.4}$$

where q is the elementary charge, c is the light speed, h is the Planck's constant, k is the Boltzmann constant, E is the photon energy, and V is the potential voltage over the solar cell.

Assuming that the origin of dark current in the ideal solar cell is a recombination of charge carriers, thus, Eq. (16.4) can be rewritten as

$$J_{dark} = J_{0,sq}\left[\mathrm{e}^{\left(\frac{qV}{kT}\right)} - 1\right] \tag{16.5}$$

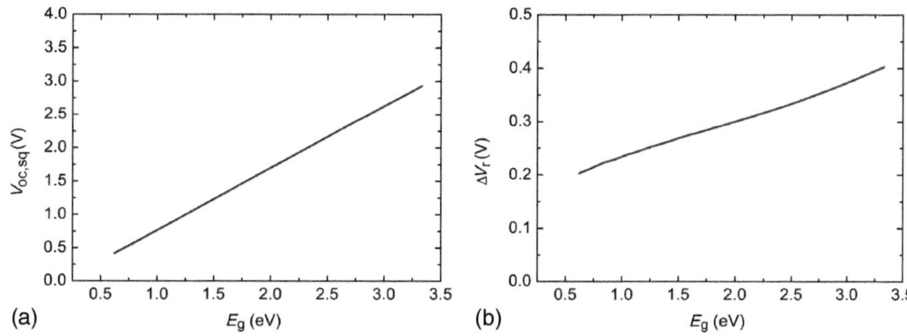

Figure 16.2 (a) $V_{oc,sq}$ and (b) ΔV_r as a function of E_g for an ideal solar cell, derived using the Shockley–Quessier theory. Source: Mengyang Li.

where J_{dark} is the dark current density, as a function of applied voltage, for the ideal solar cell, and $J_{0,sq}$ is often referred to as the dark saturation current density of the ideal Shockley–Quessier solar cell, related to a prefactor (J_{00}) and an exponential term dependent on E_g:

$$J_{0,sq} = J_{00} \exp\left(-\frac{E_g}{kT}\right) \tag{16.6}$$

$$J_{00} = q\frac{2\pi E_g^2 kT}{c^2 h^3} \tag{16.7}$$

Thus, the characteristic current density (J–V) curve of the ideal solar cell under illumination is formulated:

$$J = J_{ph} - J_{dark} = J_{ph} - J_{0,sq}\left[e^{\left(\frac{qV}{kT}\right)} - 1\right] \tag{16.8}$$

where the photocurrent density (J_{ph}) generated by the solar cell is related to N_{abs} via

$$J_{ph} = q\int_{E_g}^{+\infty} N_{abs}(E)\mathrm{d}E \tag{16.9}$$

Under an open-circuit condition, the total current density J is equal to zero,

$$J_{ph} = J_{0,sq}\left[e^{\left(\frac{qV_{oc}}{kT}\right)} - 1\right] \tag{16.10}$$

Thus,

$$V_{oc} \approx \frac{kT}{q}\ln\left(\frac{J_{ph}}{J_{0,sq}}\right) \tag{16.11}$$

$$\approx \frac{E_g}{q} + \frac{kT}{q}\ln\left(\frac{c^2h^3J_{ph}}{q2\pi E_g^2 kT}\right) \tag{16.12}$$

From the Eq. (16.12), it can be seen that V_{oc} of the solar cell increases with increasing E_g (Figure 16.2a). The second term on the right side of the Eq. (16.12) is a negative value. Thus, Eq. (16.12) shows that energy of the photons absorbed by the solar cell is not preserved, and V_{oc} of the ideal solar cell is always smaller than E_g/q.

Two different mechanisms can be deduced for the loss of photon energy in the ideal solar cell. The first type of energy loss is associated with rapid relaxations of the excited charge carriers, from higher energy states to the edge of the conduction band, releasing heat. The relaxation or thermalization loss of charge carriers can be reduced using multiply active materials with different E_g to construct multijunction solar cells [28–33]; however, for a single junction solar cell, the thermalization energy loss is high.

The second type of energy loss is related to the second term on the right-hand side of Eq. (16.12), which is mainly determined by the characteristics of the recombination current; thus, this loss is related to the rate of photon emission from the ideal solar cell, N_{emi}, or in other words, the rate of radiative recombination of charge carriers. The relationship between charge carrier recombination and V_{oc} or V_{loss} can be intuitively understood: under an open-circuit condition, charge carriers must recombine; therefore, concentration of charge carriers in the solar cell is limited under steady state. This reduces the degree of quasi-Fermi level splitting in the active layer, thus, limits V_{oc} [34, 35]. Because the second term on the right-hand side of Eq. (16.12) is related to radiative recombination of charge carriers in the ideal solar cell, it is often referred to as the radiative recombination loss (ΔV_r) term, and it is inevitable in all solar cells. According to Eq. (16.12), it can be deduced that ΔV_r increases with increasing E_g (Figure 16.2b).

16.1.2 V_{oc} and Voltage Loss in Non-ideal Solar Cells

In reality, the active layer of a solar cell is not a perfect absorber and absorption spectrum of the active layer is far from being a step-like function. Therefore, a useful equation, more general than the Würfel's generalized Planck law, was derived by Rau [36], which relates the emission rate and the dark current to the photovoltaic action of the solar cell:

$$J_{dark} = q \int BB(E)EQE_{PV}(E)dE \left[e^{\left(\frac{qV}{kT}\right)} - 1\right] \tag{16.13}$$

where $EQE_{PV}(E)$ represents the photovoltaic external quantum efficiency of the solar cell, as a function of photon energy E. Note that $EQE_{PV}(E)$ is equivalent to absorption in the active layer ($A(E)$), assuming that the ratio between the rate of extracted charge carriers and the rate of absorbed photons, i.e. the internal quantum efficiency (IQE), is 100% for the solar cell. Thus, $J_{0,sq}$ in Eq. (16.11) is replaced by $J_{0,rad}$ expressed as

$$J_{0,rad} = q \int BB(E)EQE_{PV}(E)dE \tag{16.14}$$

for the derivation of J–V characteristics of the non-ideal solar cell. We now have:

$$V_{oc} \approx \frac{kT}{q} \ln\left(\frac{J_{ph}}{J_{0,rad}}\right) \tag{16.15}$$

Because $BB(E)$ rapidly reduces with increasing photon energy, $J_{0,rad}$, related to the integral of the product of EQE_{PV} and $BB(E)$, is primarily determined by the low energy absorption tail of the solar cell (Eq. (16.14)). Therefore, both V_{oc} and ΔV_r

are expected to be strongly dependent on the spectral shape of the absorption onset of the active layer.

For a real solar cell with an absorption spectrum that differs from a step-like function, $J_{0,\text{rad}}$ is always higher compared to $J_{0,\text{sq}}$ for a given E_g (determination of E_g for a real-world organic semiconductor will be discussed later in this chapter). Thus, the radiative recombination voltage loss is always higher than that predicted by the Shockley–Quessier theory. Furthermore, charge carriers can also recombine nonradiatively in a real-world solar cell, leading to additional voltage losses. To analyze the additional voltage losses, the total recombination rate (N_{rec}) in Eq. (16.2) for deriving realistic J–V characteristics of the solar cell should be rewritten as a function of both radiative (N_{emi}) and non-radiative recombination rate (N_{nr}) of charge carriers:

$$N_{\text{rec}} = N_{\text{emi}} + N_{\text{nr}} \tag{16.16}$$

Because N_{emi} and N_{nr} are related to the electroluminescence external quantum efficiency (EQE_{EL}) of the solar cell,

$$EQE_{EL} = \frac{N_{\text{emi}}}{N_{\text{emi}} + N_{\text{nr}}} \tag{16.17}$$

Equation (16.16) is modified to

$$N_{\text{rec}} = \frac{N_{\text{emi}}}{EQE_{EL}} \tag{16.18}$$

Compared to Eq. (16.2), N_{rec} is not only determined by N_{emi} but also by EQE_{EL}.

The dark saturation current for a real-world solar cell (J_0) with both radiative and non-radiative recombination decay paths is then

$$J_0 = \frac{1}{EQE_{EL}} J_{0,\text{rad}} \tag{16.19}$$

Thus,

$$V_{\text{oc}} \approx \frac{kT}{q} \ln\left(\frac{J_{\text{ph}}}{J_0}\right) \tag{16.20}$$

$$\approx V_{\text{oc,rad}} + \frac{kT}{q} \ln(EQE_{EL}) \tag{16.21}$$

where

$$V_{\text{oc,rad}} \approx \frac{kT}{q} \ln\left(\frac{J_{\text{ph}}}{J_{0,\text{rad}}}\right) \tag{16.22}$$

Now we have a complete formula (Eq. (16.21)) for the V_{oc} and V_{loss} of a real-world solar cell, in which both radiative and non-radiative recombination of charge carriers can occur. In this formula, the non-radiative recombination voltage loss, represented by the second term, can be quantified by the measurement of EQE_{EL} of the solar cell. When EQE_{EL} is 100%, there is no non-radiative recombination voltage loss in the solar cell, thus Eq. (16.20) is reduced to Eq. (16.15). When the absorption spectrum of the active layer is engineered into a perfect step-like function described by Shockley and Quessier, this equation is further reduced to Eq. (16.11). Now we define V_{oc} of a solar cell with 100% EQE_{EL} as $V_{\text{oc,rad}}$, in addition, we define V_{oc} of a solar cell with

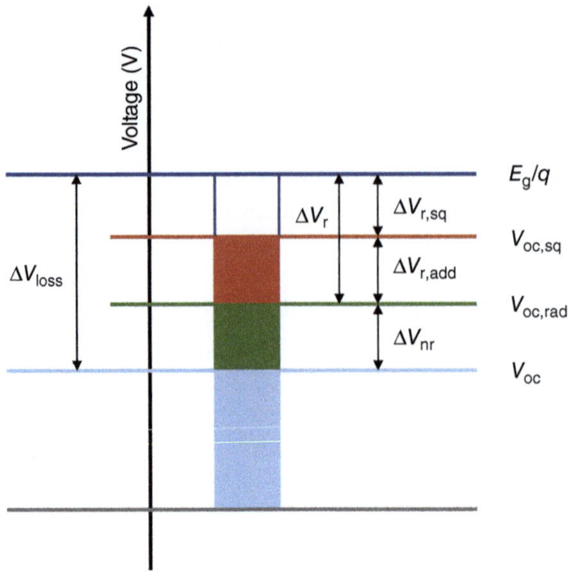

Figure 16.3 Definition of different voltage losses in a real-world solar cell. E_g is the optical bandgap energy of the active layer used in the solar cell; $V_{oc,sq}$ stands for the upper V_{oc} limit derived from the Shockley–Quessier theory; $V_{oc,rad}$ is the radiative recombination limit for V_{oc}; $\Delta V_{r,sq}$, $\Delta V_{r,add}$, and ΔV_{nr} are the inevitable radiative recombination loss, the radiative recombination loss due to broadening of the absorption tail of the active layer, and the non-radiative recombination loss, respectively. The total radiative recombination loss is indicated as ΔV_r, which is the sum of $\Delta V_{r,sq}$ and $\Delta V_{r,add}$. Source: Mengyang Li.

100% EQE_{EL} and a step-function absorption spectrum as $V_{oc,sq}$. Then, all voltage loss terms, and their relation to E_g and V_{oc}, as well as the total voltage loss (V_{loss}) due to charge carrier recombination, can be illustrated in one chart, as shown in Figure 16.3. From Figure 16.3, it is clear that in order to increase V_{oc} and reduce V_{loss}, we need to reduce both the non-radiative recombination loss (ΔV_{nr}) and the additional part of the radiative recombination loss ($\Delta V_{r,add}$).

16.2 V_{oc} of Organic Solar Cells

16.2.1 Charge Transfer States in Organic Solar Cells

In contrast to inorganic or perovskite solar cells in which free charge carriers are directly generated upon photon excitation, light absorption in organic solar cells results in generation of Frenkel excitons [37–43]. At room temperature, the excitons are tightly bound electron–hole pairs, thus additional energy is needed for the excitons to dissociate into free charge carriers.

In organic solar cells based on the bulk-heterojunction (BHJ) concept [44–47], the active layer is a mixture of an electron donor and an electron acceptor material (Figure 16.4a). At the interface between the donor and the acceptor material, a new

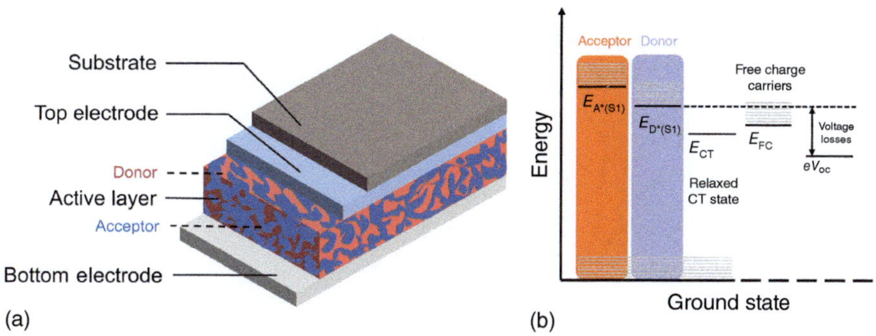

Figure 16.4 (a) A schematic picture for the device architecture of an organic solar cell based on the concept of BHJ. The photo-active layer is based on a blend of an electron donor and an electron acceptor material, sandwiched in between two electrodes. One of the electrodes is transparent to allow for light in-coupling. (b) Energetics of excited states and ground state in an organic BHJ solar cell; E_{A^*} and E_{D^*} represent the energies of the local excited states of the acceptor and the donor, respectively. E_{CT} represents the energy of charge transfer state formed at the interface between the donor and the acceptor. E_{FC} is the energy for free charge carriers in the solar cell. Source: Mengyang Li.

ground state often, referred to as the charge transfer (CT) complex, is formed [48–51]. Upon excitation of the new ground state, a new excited state, CT state, is created (Figure 16.4b).

Optical transitions between the ground state and the CT state are possible to detect using highly sensitive absorption or emission measurement techniques (Figure 16.5) [52–55], and the energy of CT state (E_{CT}) is related to the highest occupied molecular orbital (HOMO) and the lowest unoccupied molecular orbital (LUMO) levels of the donor and the acceptor materials in the blend. Because E_{CT} is generally lower than that of the local excited (LE) states in the pristine phases, an energetic offset defined as driving energy (ΔE_{CT}) or driving force exists, $\Delta E_{CT} = E_g - E_{CT}$ [56]. Excitons generated in the pristine phases and diffused to the donor–acceptor interface can dissociate into free charge carriers via the intermediate CT state, as illustrated by the schematic picture shown in Figure 16.5c. Thus, the CT states play a crucial role in determining the performance of the solar cell. In this chapter, we mainly focus on the impact of the property of CT state on V_{oc} and V_{loss} in organic solar cells.

16.2.2 Relation Between CT State and V_{oc} of Organic Solar Cells

CT state is the lowest energy state for a BHJ solar cell [49, 51, 57–61]. Therefore, CT state absorption determines the lower energy part, i.e. the tail of the absorption spectrum of the solar cell. Because $BB(E)$ spectrum decreases rapidly with increasing photon energy, the integral of the product of $BB(E)$ and $EQE_{PV}(E)$ spectra in Eq. (16.14), determining $J_{0,rad}$ for an organic solar cell, is primarily determined by the CT state absorption properties.

Vandewal et al. showed that CT state absorption is a Gaussian function by performing highly sensitivity EQE_{PV} measurements [62] and reported that the absorption feature and EQE_{PV} of CT state can be well explained in the framework

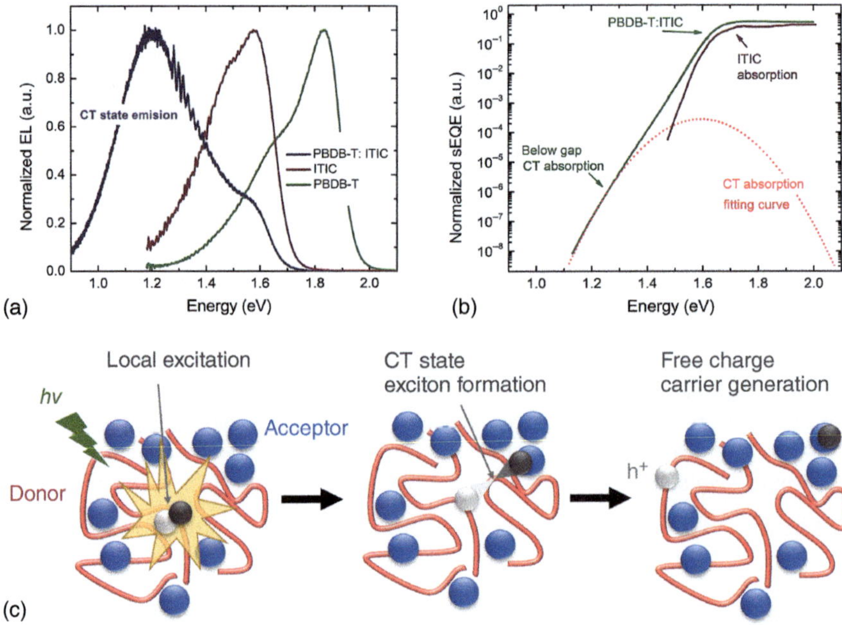

Figure 16.5 (a) Electroluminescence spectra of an organic solar cell based on PBDB-T as the donor material and ITIC as the acceptor material. The emission spectrum is red shifted compared to the emission of either PBDB-T or ITIC, suggesting that there is a new lower emissive energy state formed in the blend active layer of the solar cell, which is assigned to CT state. (b) Sensitive EQE_{PV} spectra of the solar cell based on PBDB-T and ITIC, an absorption band, red-shifted compared to that of the neat materials can be resolved in the lower energy part of the spectrum, corresponding to CT absorption. The CT absorption region can be fitted using Eq. (16.23), derived in the framework of Marcus theory. (c) A schematic picture illustrating the formation of local excited state, CT state, and generation of free charge carriers in an organic donor–acceptor blend. Source: Mengyang Li.

of Marcus theory [63],

$$EQE_{PV}(E) = \frac{fE}{\sqrt{4\pi\lambda kT}} \exp\left(\frac{-(E_{CT} + \lambda - E)^2}{4\lambda kT}\right) \tag{16.23}$$

where k is the Boltzmann constant, T is the temperature, λ is the reorganization energy, and f is a prefactor mainly determined by the density of the donor and acceptor interfaces (N_{CTC}) that allows for the formation of CT states and the oscillator strength (f_{osc}) for the transition from the ground state to the CT state.

Equation (16.18) can now be expanded for an analytical determination of J_0, using CT state-related molecular parameters.

$$J_0 \approx \frac{q}{EQE_{EL}} f \frac{2\pi}{h^3c^2}(E_{CT} - \lambda) \exp\left(-\frac{E_{CT}}{kT}\right) \tag{16.24}$$

By combining Eqs. (16.20) and (16.24), V_{oc} of an organic solar cell is expressed as a function of a number of parameters related CT state properties.

$$V_{\mathrm{oc}} = \frac{E_{CT}}{q} + \frac{kT}{q} \ln\left(\frac{J_{sc}h^3c^2}{fq2\pi(E_{CT} - \lambda)}\right) + \frac{kT}{q} \ln(EQE_{EL}) \tag{16.25}$$

Equation (16.25) shows that V_{oc} of a BHJ solar cell depends linearly on E_{CT}. Moreover, the radiative recombination voltage loss (ΔV_r), which is the second term on the right-hand side of the equation, is determined mainly by the prefactor f, related to N_{CTC} and f_{osc} of CT state transition. Another important implication from Eq. (16.25) is that for BHJ solar cells, it is the rate of CT state decay instead of free charge carrier recombination determining J_0 and thus V_{oc}. In addition, EQE_{EL}, determining non-radiative recombination voltage loss, represented by the third term on the right-hand side of the Eq. (16.25), is defined by the ratio between the radiative decay rate and the total decay rate of CT state. This was verified by Vandewal et al. by measuring V_{oc} as a function of temperature [62]: extrapolations of V_{oc} to 0 K were shown to match the value of E_{CT} determined by a fitting to the EQE_{PV} tail using Eq. (16.23).

The fact that V_{oc} of an organic solar cell is determined by the CT state properties is also observed in a simple three-state model [58, 64, 65]. As shown in Figure 16.6, in this model, free charge carriers can only be generated and recombined via CT state. Assuming that in steady state, free charge carriers are generated by dissociation of CT state with a density N^*_{CTC} and a rate constant K_{diss}. Under an open-circuit condition, this process needs to be balanced with the disappearing of free charge carriers via formation of CT states with a rate constant K_{fc}. Therefore, the following equation must be valid to balance the loss and the creation of free charge carriers:

$$K_{fc}np = K_{diss}N^*_{CTC} \tag{16.26}$$

where n and p represent the density of free electrons and the density of free holes, respectively. Furthermore, generation of CT states via dissociation of the LE states in the pure phases (at a rate G_{CT}) and via encounter of free charge carriers (at a rate $K_{fc}np$) must be balanced with the disappearing rate of the CT state via CT state dissociation (at a rate $K_{diss}N^*_{CTC}$) and via CT state recombination (at a rate $K_{rec}N^*_{CTC}$). Therefore, we have

$$G_{CT} + K_{fc}np = K_{diss}N^*_{CTC} + K_{rec}N^*_{CTC} \tag{16.27}$$

Combining Eqs. (16.26) and (16.27) gives

$$G_{CT} = K_{rec}N^*_{CTC} \tag{16.28}$$

Using Boltzmann statistics, generation of CT states (N^*_{CTC}) and free charge carriers (np) via thermal excitation depend mainly on E_{CT} and the energy for free charge carrier state (E_{fc}), respectively.

$$N^*_{CTC} = N_{CTC} \exp\left(-\frac{E_{CT}}{kT}\right) \tag{16.29}$$

$$np = N_A N_D \exp\left(-\frac{E_{fc}}{kT}\right) \tag{16.30}$$

where N_A and N_D are the effective density of state for the donor and the acceptor material, respectively. Combining Eqs. (16.26), (16.29), and (16.30) gives:

$$E_{fc} = E_{CT} - kT\ln\left(\frac{N_{CTC}}{N_A N_D}\frac{K_{diss}}{K_{fc}}\right) \tag{16.31}$$

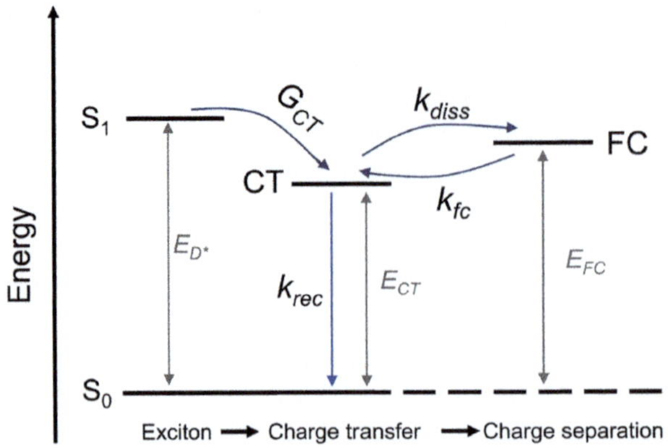

Figure 16.6 A three-state model illustrating the competition between CT state dissociation and recombination. LE (S_1) represents the local excited state in the donor or the acceptor phase. S_0 represents the ground state. G_{CT} is the rate for the transition from S_1 to CT state; K_{diss} and K_{fc} are the rates for the transitions between CT and free charge carrier (FC) states. K_{rec} is the decay rate of CT state. Source: Mengyang Li.

Because the expression for V_{oc}, derived in classic semiconductor physics [19, 66], should also be valid for the organic solar cells, we can derive:

$$V_{oc} = \frac{E_{fc}}{q} + \frac{kT}{q} \ln\left(\frac{np}{N_A N_D}\right) \tag{16.32}$$

Now, using Eqs. (16.26) and (16.31), the above Eq. (16.32) is reformulated as

$$V_{oc} = \frac{E_{CT}}{q} + \frac{kT}{q} \ln\left(\frac{N^*_{CTC}}{N_{CTC}}\right) \tag{16.33}$$

Using Eq. (16.28), we have

$$V_{oc} = \frac{E_{CT}}{q} + \frac{kT}{q} \ln\left(\frac{G_{CT}}{K_{rec} N_{CTC}}\right) \tag{16.34}$$

and Eq. (16.34) can be further modified as

$$V_{oc} = \frac{E_{CT}}{q} + \frac{kT}{q} \ln\left(\frac{G_{CT}}{K_r N_{CTC}}\right) + \frac{kT}{q} \ln(EQE_{EL}) \tag{16.35}$$

where K_r is the radiative decay rate constant of CT state, and EQE_{EL} is determined by the ratio between K_r and K_{rec}.

The expression for the V_{oc} of an organic solar cell derived by Vandewal et al. using the simple three-state model is similar to that derived using the detailed balance theory (Eq. (16.25)). They both predict that V_{oc} depends linearly on E_{CT} (not E_g) of the active materials. Thus, the energetic offset, ΔE_{CT}, is a limiting term for V_{oc}. In addition, V_{oc} is logarithmically dependent on EQE_{EL}, and there is a radiative recombination loss term related to the radiative decay rate of CT state and the density of CT complex.

16.2.3 Determining Factors of K_r and K_{nr} for Organic Solar Cells

From the above analyses, it is clear that the mechanism and the rate of CT state decay are critically important in determining V_{oc} and V_{loss} in organic solar cells. To reduce V_{loss}, K_r should be minimized and K_{nr} should be eliminated.

In the framework of Marcus theory, K_{nr} is expressed as [67]:

$$K_{nr} = \frac{2\pi}{h} V_c^2 \frac{1}{\sqrt{4\pi\lambda kT}} \exp\left(-\frac{(\lambda - E_{CT})^2}{4kT\lambda}\right) \tag{16.36}$$

where V_c is related to the electronic coupling between CT state and ground state, and reorganization energy (λ) is related to the electron–vibration coupling. Considering the quantum nature of electron vibration, λ should be divided into two parts, one for the classical low-frequency vibrations and the other for the quantum mechanical high-frequency vibrations [68]. Accordingly, using Born–Oppenheimer approximation, K_{nr} is related additionally to the Franck–Condon weighted density of states (FCWD) for a consideration of transitions from the lowest vibrational CT states to the ground states.

$$FCWD(g) = (4\pi\lambda_s kT)^{-0.5} \sum_{j=0}^{\infty} \frac{e^{-S} S^j}{j!} \exp\left[-\frac{(g - jh\bar{\nu}_v - \lambda_s)^2}{4\lambda_s kT}\right] \tag{16.37}$$

$$K_{nr} \propto V_c^2 FCWD(g = E_{CT}) \tag{16.38}$$

where $\bar{\nu}_v$ is the mean frequency of the high-frequency vibration modes, $S = \lambda_v / h\bar{\nu}_v$ is the Huang Rhys factor, and λ_s and λ_v are the high- and the low-frequency reorganization energies, respectively. Accordingly, the expression for K_r is also a function FCWD, similar to that of K_{nr},

$$\frac{K_r(\nu)}{\nu} \propto V_c^2 \Delta\mu^2 FCWD(g = E_{CT} - h\nu) \tag{16.39}$$

where $\Delta\mu$ is the difference between the static dipole moment of CT state and ground state.

Thus, both K_r and K_{nr} are proportional to ${V_c}^2$, and the non-radiative voltage loss (ΔV_{nr}) is strongly dependent on E_{CT} and the reorganization energies. Assuming λ_s and $\Delta\mu$ do not vary, Vandewal et al. showed that ΔV_{nr} defined by the ratio between K_r and K_{nr} can be expressed as

$$\Delta V_{nr} \propto A - B\frac{E_{CT} - \lambda_s}{h\bar{\nu}_v} \tag{16.40}$$

These results suggest that ΔV_{nr} is very high for systems with low E_{CT}. An intuitive understanding of the dependence of ΔV_{nr} on E_{CT} can be obtained from the diagram of potential energies of the CT and the ground state, as shown in Figure 16.7: the transition rate from the lowest vibrational state in the CT state to the higher vibrational ground state is expected to be significantly higher in the systems with a lower E_{CT} compared to that with a higher E_{CT}, due to a higher degree of overlapping between the vibrational wavefunctions in the CT and the ground states for the lower E_{CT} system.

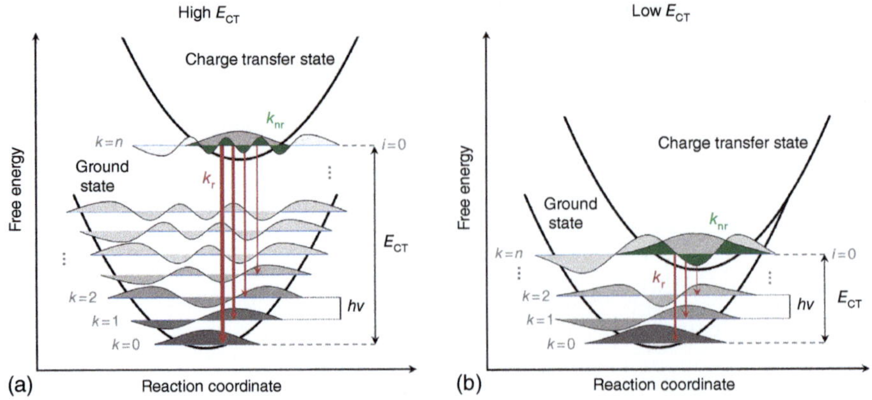

Figure 16.7 Potential energy diagram for the excited CT state and the ground state for (a) a high and (b) a low E_{CT} system. The sublevels indicate the vibrational states and their wavefunctions, and K_r and K_{nr} are indicated as k_r and k_{nr} in these graphs. The green areas represent the overlap of the wavefunctions between the lowest vibrational state in CT state and a higher vibrational state in the ground state. The degree of wavefunction overlapping determines the non-radiative transition rate from the CT state to the ground state. Source: Mengyang Li.

Considering the surrounding molecular environment in the organic BHJ active layer and its impact on CT state, Azzouzi et al. evaluated additional factors affecting ΔV_{nr} [69]. In their work, FCWD was expressed for all transitions possible between CT and ground states,

$$FCWD(hv) = \frac{1}{\sqrt{4\pi\lambda_s kT}} \sum_{w=0}^{\infty}\sum_{t=0}^{\infty} \frac{e^{-S}S^{w-t}}{w!}\left[L_t^{w-t}(S)\right]^2 e^{-\{[hv-\Delta G_0+\lambda_s+(w-t)*h\bar{v}_v]^2/4\lambda_s kT\}} e^{-th\bar{v}_v/kT} \tag{16.41}$$

where L is the generalized Laguerre polynomial, w and t are the quantum numbers of the vibrational levels in the ground and CT state, respectively, and ΔG_0 is the Gibbs-free energy between the initial and the final states, equivalent to E_{CT}, thus,

$$K_{nr} = \frac{4\pi^2}{h} V_c^2 FCWD(0) \tag{16.42}$$

For deriving the radiative recombination rate constant K_r, a dipole approximation was used. The recombination rate for a transition resulting in emission of photon energy hv, $k_r(hv)$ was expressed as

$$k_r(hv) = \frac{16\pi^3}{3\varepsilon_0 h^4}\left(\frac{hv}{c}\right)^3 M^2 \mathrm{FCWD}(hv) \tag{16.43}$$

where ε_0 is the permittivity of free space and M is the transition dipole moment, proportional to the square root of the oscillator strength f_{osc}. Then, the net radiative recombination rate K_r could be calculated by integrating $k_r(hv)$ over all the photon energies,

$$K_r = \int k_r(hv)dhv \tag{16.44}$$

The electronic coupling V_c and transition dipole moment M were related to each other using the generalized Mulliken–Hush (GMH) method.

$$V_c = \frac{ME_{CT}}{\sqrt{\Delta\mu^2 + 4M^2}} \tag{16.45}$$

In addition, the expression for EQE_{EL} was modified by the probability of the reabsorption of the emitted photons, expressed as [70]

$$p_e = \frac{\int A(E)BB(E)\mathrm{d}E}{4d \int n^2\alpha(E)BB(E)\mathrm{d}E} \tag{16.46}$$

where d is the thickness of the active layer, n is the refractive index, $A(E)$ is the absorption of photons in the active layer, and $\alpha(E)$ is the absorption coefficient, which can be reproduced theoretically based on calculation of Einstein coefficient [69] and photon-absorption rate.

As a result, Azzouzi et al. showed that $V_{oc,rad}$ for an organic solar cell could be modeled using Eq. (16.15), when CT state parameters needed by the FCWD were provided. In addition, ΔV_{nr} could be calculated as

$$\Delta V_{nr} = -kT\ln\left(p_e\frac{4\pi}{3\varepsilon_0 h^3}\frac{\left(\Delta\mu^2 + \frac{3h^2}{2\pi^2 h\nu_{avg}m_e}f_{osc}\right)\int FCWD(h\nu)(h\nu)^3\mathrm{d}h\nu}{FCWD(0)E_{CT}^2}\right) \tag{16.47}$$

where m_e is the electron mass and ν_{avg} is the average frequency of emitted photons. This equation predicts that ΔV_{nr} depends also on the additional molecular parameters, including reorganization energies, $\Delta\mu^2$, etc. Both the high and the low-frequency reorganization energies could play very important role in determining ΔV_{nr}, especially for organic BHJ systems with higher E_{CT}. For blend systems with smaller E_{CT}, ΔV_{nr} is more dominantly determined by the low-frequency reorganization energy. Furthermore, absorption oscillator strength is also important in determining ΔV_{nr} of the solar cells with lower E_{CT}.

The method proposed by Azzouzi et al. to analytically reproduce the absorption spectrum of the organic blends, allowing for the determination of $V_{oc,rad}$, also enabled computational studies on the entangled impact of these additional CT state parameters on both the ΔV_r (leading to reduced $V_{oc,rad}$) and the ΔV_{nr} of organic solar cells. An important result can be derived, i.e. a reduced ΔV_{nr} does not always lead to reduced V_{loss}, as ΔV_r could increase as a result of the reduced ΔV_{nr}. Furthermore, it can be seen that to reduce V_{loss} for a selected material system with a fixed E_g, the more effective methods are increasing E_{CT} and reducing the high-frequency reorganization energy, λ_ν. In addition, reducing N_{CTC} and reducing the low-frequency reorganization energy, λ_s, or increasing the difference between static dipole moment of the CT and the ground state can also lead to reduced V_{loss}. However, increasing f_{osc} is found to be least effective in reducing V_{loss}.

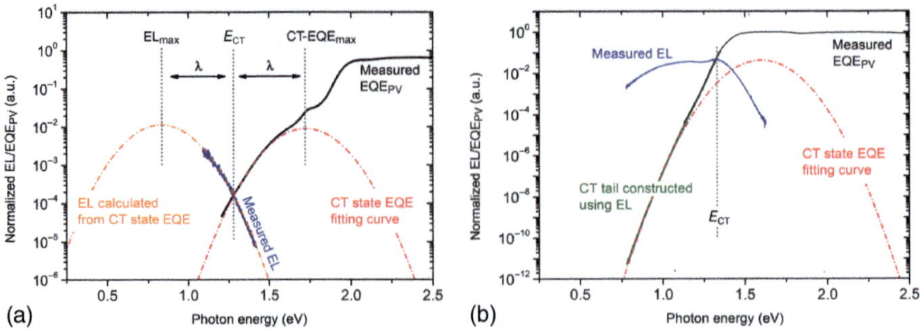

Figure 16.8 (a) Determination of CT state-related parameters for an organic solar cell from a fit based on the use of Eq. (16.23), derived in the framework of Marcus theory. The fitting gives the value for E_{CT}, f, and reorganization energy. Moreover, using the reciprocal relation between absorption and emission allows for a conversion of the CT state EQE_{PV} spectrum to the emission spectrum. (b) Reconstructing the tail of the EQE_{PV} spectrum from a measured EL spectrum, using the reciprocal relation, for a solar cell with a low ΔE_{CT}. The reconstructed EQE_{PV} values span for over 10 orders of magnitudes, enabling CT fitting for organic solar cells with small ΔE_{CT}. Source: Mengyang Li.

16.2.4 Experimental Determination of CT State-Related Parameters

Experimental evaluation on the relationship between V_{loss} and the CT state parameters in organic solar cells requires accurate determination of the different loss values, as well as the CT state-related parameters. A commonly used method is based on sensitive EQE_{PV} measurements for a determination of the characteristics of the CT state absorption. Equation (16.23) is used to fit the CT state-related lower energy part of EQE_{PV} spectra. The fitting allows for a quantitative determination of E_{CT} and total reorganization energy. In addition, the prefactor f related to the product of N_{CTC} and f_{osc} can also be determined, as illustrated in Figure 16.8a.

For a series of non-fullerene acceptor-based BHJ blends, CT state absorption is difficult to detect, due to a small energetic offset ΔE_{CT} leading to overlapping between the CT absorption and the LE state absorption [67, 71–73]. To solve this problem, electroluminescence spectroscopic measurements [74–76] are often employed, since the dominant contribution to the EL spectrum of a BHJ solar cell is from CT emission, as discussed before in Figure 16.8a. The measured EL spectrum can be used to reconstruct the lower energy part of the EQE_{PV} spectra (Figure 16.8b), using a reciprocal relation [36], extending the tail of the EQE_{PV} spectrum for a better determination of CT state absorption properties. However, a recent study pointed out that optical cavity effect in the solar cell device could lead to a distorted CT state EL spectrum, especially for solar cells with very thick active layers. Thus, the accuracy of the method of using EL to determine CT state parameters needs to be further evaluated [77].

The energetic offset between LE and CT state leads to reduced V_{oc}, thus it is a voltage loss term. Determination of the energetic offset ΔE_{CT} requires the determination of E_g. For organic materials, the physical definition of E_g is the energetic difference between the vibrationally relaxed excited state and ground states (Figure 16.9) [56]. Optical transitions from the relaxed ground state to high energy vibrational excited

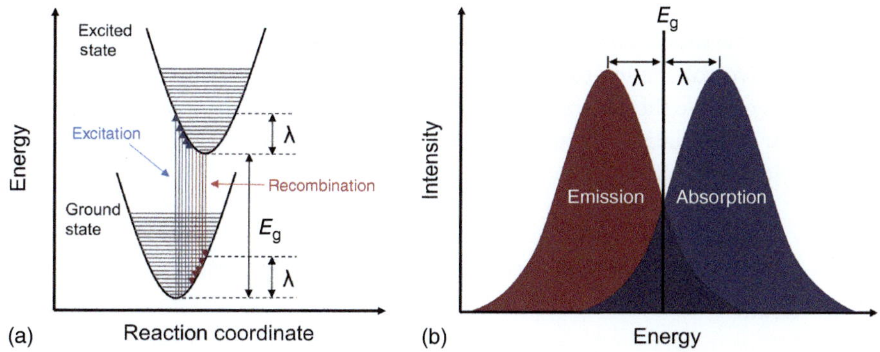

Figure 16.9 Definition of E_g for organic semiconductors. (a) A schematic picture showing the excited and the ground state of an organic material. Excitation and recombination occur between the vibrational states in the excited and the ground state. E_g is defined as the energy difference between the lowest vibrational states in the excited state and the ground state. (b) Reduced and normalized emission and absorption spectra of a hypothetic organic semiconductor material. The crossing point between the emission and the absorption spectrum gives the value for E_g. Source: Mengyang Li.

states lead to absorption of photon energy larger than E_g, and the transitions from the relaxed excited state to higher vibrational ground states lead to emission of photon energy smaller than E_g. Using harmonic oscillators to represent the vibrations in organic molecules, absorption ($A(E)$) and emission ($N(E)$) can be expressed as

$$A(E) \sim E \exp\left(-\frac{(E - E_g - \lambda)^2}{4\lambda_L kT}\right) \tag{16.48}$$

$$N(E) \sim E^3 \exp\left(-\frac{(E - E_g + \lambda)^2}{4\lambda kT}\right) \tag{16.49}$$

These equations indicate that when absorption and emission spectrum are reduced, i.e. divided by E for $A(E)$ and divided by E^3 for $N(E)$, the crossing point of the reduced spectra gives the exact value for E_g [56].

A quantitative determination of ΔE_{CT} can then be realized, which is the difference between E_{CT} and E_g. Meanwhile, the measurement used for characterizing CT state EQE_{PV} spectrum enables calculation of $J_{0,rad}$, using Eq. (16.14). Note that in Eq. (16.14), the blackbody spectral flux reduces rapidly with increasing photon energy E; thus, as discussed before, the integral of the product between $EQE_{PV}(E)$ and $BB(E)$ is primarily determined by the CT absorption related low energy part of the EQE_{PV} spectra. Consequently, for an accurate determination of $J_{0,rad}$, a sensitive EQE_{PV} measurement for a complete spectral coverage of CT absorption is critical. Accordingly, $V_{oc,rad}$ can be calculated using Eq. (16.15), and ΔV_r defined as the difference between E_{CT} and $V_{oc,rad}$ can be quantified. Using E_{CT} as the effective E_g in the Shockley-Queisser limit, we can additionally calculate the upper limit for $V_{oc,sq}(E_{CT})$ of BHJ solar cells, then the difference between $V_{oc,sq}(E_{CT})$ and $V_{oc,rad}$ represent the additional radiative recombination voltage losses ($\Delta V_{r,add}$) due to CT state absorption spectrum being a non-step-like function, as shown in Figure 16.3. This shows how much can V_{oc} be increased by reducing ΔV_r in organic solar cells.

Nonradiative voltage losses ΔV_{nr} can also be quantified, when $V_{oc,rad}$ is known, since ΔV_{nr} equals to the difference between $V_{oc,rad}$ and V_{oc} of the solar cell measured under solar illumination. In addition, ΔV_{nr} can be quantified by direct measurements of EQE_{EL} (Eq. (16.21)).

16.3 Strategies to Reduce V_{nr} and V_r in Organic Solar Cells

In reference [68], it was shown that ΔV_{nr} reduced with increasing E_{CT}, agreeing well with that predicted by Eq. (16.40). More importantly, via investigating a wide range of different organic materials systems, a common mechanism for the high-frequency vibrations in organic blends, being responsible for the large non-radiative recombination voltage losses in organic solar cells, was proposed. According the authors, this mechanism could be assigned to the vibration of carbon–carbon bonds.

To further demonstrate the possibility to achieve low non-radiative voltage loss in BHJ organic photovoltaic devices, Vandewal and coworkers employed organic materials with very high E_{CT} to construct organic solar cells [78]. Because of the high energy of CT state in these blends, solar cells based on these materials exhibited very high EQE_{EL}, thus very low ΔV_{nr}. This was a clear evidence proving that the vibronic coupling between the excited and the ground CT state in BHJ could be efficiently reduced using materials combinations with large E_{CT}. It was also observed that transport of charge carriers, thus the IQE and the fill factor (FF), could be problematic in the solar cells based on high E_{CT} materials systems, due to strong binding energies of CT state excitons. However, this problem was resolved using a better suited combination of materials with more desired microstructures. This implied that the electronic property of the highly emissive blend system with high E_{CT} was not an intrinsic limit for achieving good photovoltaic characteristics in organic solar cells. In fact, a major limit for the method of using high E_{CT} system is related to the poor optical property of the blend: to achieve high E_{CT}, active materials with high E_g must be used; thus, the spectral coverage of the photons to be absorbed in the active layer of the solar cell is limited, which leads to low J_{sc}.

Maximizing E_{CT} of a BHJ system with a given E_g, or in other words, reducing the energetic difference between E_g and E_{CT} (ΔE_{CT}) is a more appropriate strategy to achieve low ΔV_{nr} in organic solar cells with a desired spectral coverage for high photocurrent generation [65, 79–82]. For fullerene organic solar cells, this was observed in the isoindigo:PCBM systems [83]. Compared to an isoindigo-based polymer donor P3TI with three thiophene groups in the repeating unit, another isoindigo-based polymer PTI-1 with only one thiophene had deeper laying LUMO and HOMO levels. Therefore, E_{CT} of the blend based on PTI-1 and PCBM was higher compared to that of the P3TI:PCBM blend, despite that E_g of PTI-1 and P3TI was found very similar. As a result, the energetic offset between the donor and the acceptor in the PTI-1:PCBM blend was found extremely low, approaching 0 eV, which could be assigned as the primary reason for the considerably higher V_{oc} achieved in the PTI-1:PCBM-based solar cell, compared to that in P3TI:PCBM. More importantly, it was found that EQE_{EL} of the solar cell based on PTI-1:PCBM was much higher, over an order of magnitude higher than that in P3TI:PCBM, indicating a ΔE_{CT}-dependent ΔV_{nr}.

To evaluate that ΔV_{nr} is indeed dependent on ΔE_{CT}, two additional isoindigo-based polymer donors, P5TI and P6TI, were synthesized [84]. More thiophenes were added to the polymer backbones, and E_g of the isoindigo materials was found to be unaffected by the change of the number of thiophenes in the polymer backbone unit. However, it was noted that both the LUMO and the HOMO levels were shifted upward with the increasing number of the thiophenes , leading to reduced E_{CT} and larger ΔE_{CT}. A clear dependence of EQE_{EL} on ΔE_{CT} was found, confirming that low ΔE_{CT} could lead to lower ΔV_{nr}.

Despite the low ΔV_{nr} achieved using small ΔE_{CT} systems, IQE of the fullerene solar cells were found to be limited when ΔE_{CT} was very small. A possible reason for the low IQE was assigned to inefficient dissociation of CT state excitons into free charge carriers [85–87]. Thus, it was strongly debated whether the method of using the materials combinations with small ΔE_{CT} to achieve lower ΔV_{nr} was effective in improving the overall performance of organic solar cells, until the development of high efficiency non-fullerene organic solar cells.

For non-fullerene acceptor-based organic solar cells, ΔV_{nr} was also found to reduce with smaller ΔE_{CT}. This was first observed using P3TEA as the donor and SF-PDI2 as the acceptor: the electroluminescence spectrum of the solar cell based on the blend of P3TEA and SF-PDI2 was found to be identical to that of the device based on pure donor [79], suggesting that E_{CT} of the blend active layer was very close to E_g of the donor material, and ΔE_{CT} was negligible. The small ΔE_{CT} could also be seen from the fact that $V_{oc,rad}$ of the organic solar cell was similar to that predicted by the Shockley–Quessier theory. More importantly, EQE_{PV} achieved using this materials system was comparable to that of typical organic solar cells with large ΔE_{CT}, suggesting that dissociation of CT excitons was not necessarily prohibited by a low ΔE_{CT}. The benefit of using the low ΔE_{CT} system was clear, and the EQE_{EL} of the non-fullerene solar cell based on P3TEA and SF-PDI2 was over 10^{-5}, allowing to achieve ΔV_{nr} as low as 0.26 V, approaching that of perovskite solar cells operating without CT states. A comparative study was carried out, and it revealed that the use of a different donor material for a larger ΔE_{CT} resulted in reduced EQE_{EL} and increased ΔV_{nr}, suggesting that ΔV_{nr} reducing with ΔE_{CT} was a general trend for organic BHJ solar cells, regardless of the type of the acceptor, fullerene, or non-fullerene, used.

Similarly, high EQE_{PV} with low ΔV_{nr} was also achieved using several other non-fullerene materials combinations with small ΔE_{CT} by other groups [88]. To understand the mechanism for the reduced ΔV_{nr} in organic solar cells with small ΔE_{CT}, Qian et al. studied a low ΔE_{CT} system, PTB7-Th:IEICO with high EQE_{EL} and low ΔV_{nr} [65]. They employed a three-state model in which the electronic coupling for the transition rates between CT and LE state was considered for the determination of the CT state decay dynamics (Figure 16.10).

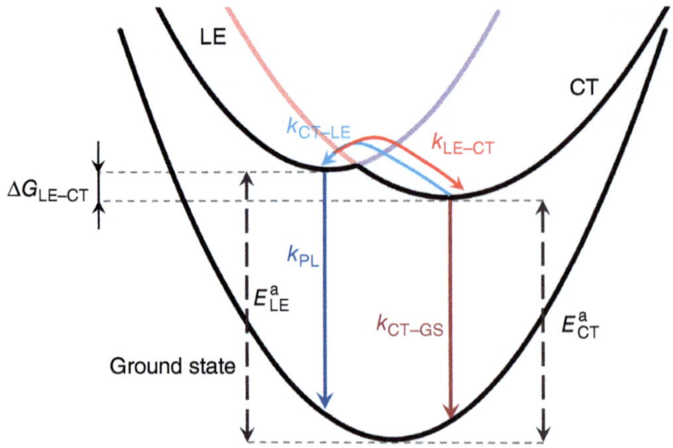

Figure 16.10 A schematic picture of a three-state model used to show the impact of LE excitation on CT state properties in organic solar cells with low ΔE_{CT}. Source: Mengyang Li.

Density functional theory (DFT) calculations were used to derive the parameters needed in Marcus theory for a determination of the transition rates of the possible transitions between the different states. As a result, a rapid back transfer from CT to LE state with a transition rate order of magnitudes higher than that of the CT to ground state was predicted for the BHJ systems with very small ΔE_{CT}. Thus, two mechanisms for the increased EQE_{EL} and reduced ΔV_{nr} in systems with small ΔE_{CT} were proposed: (i) a hybridization between the CT and the highly emissive LE state, allowing emission probability of CT state to increase through an intensity borrowing mechanism [89, 90]; (ii) the opening of an additional pathway for CT state to decay through the more emissive LE state. Similar conclusion was draw in reference [91].

The intensity borrowing mechanism for organic solar cells with small ΔE_{CT} was explained by Eisner et al. using a similar three-state model [72]. In their work, the interactions between the wave functions of LE state and CT state were allowed for deriving a mathematic expression for the transition dipole moment (M) of the CT to ground state transition. This gave rise to an effective CT to ground state oscillator strength ($f^{*}_{S_0,CT}$), related to the LE to ground state (f_{S_1,S_0}), the LE to CT state ($f_{S_1,CT}$), and the CT to ground state transition oscillator strength ($f_{S_0,CT}$) derived without considering the hybridization of the states,

$$f^{*}_{S_0,CT} = f_{S_0,CT} + \frac{f_{S_1,S_0} V^2_{S_1,CT}}{\Delta E^2_{CT}} + 2\frac{f^{1/2}_{S_1,S_0} f^{1/2}_{S_0,CT} V_{S_1,CT}}{\Delta E_{CT}} \quad (16.50)$$

where $V_{S_1,CT}$ represents the electronic coupling between the LE and the CT state.

An important observation from Eq. (16.50) was that the ratio between $f^{*}_{S_0,CT}$ and f_{CT,S_0} increased drastically with a reducing ΔE_{CT}, suggesting that the effective oscillator strength determining CT state decay rate was significantly higher in a low ΔE_{CT} system with hybridized states than that in the high ΔE_{CT} system. Furthermore, from Eq. (16.50), the dependence of ΔV_{nr} on ΔE_{CT} and on $V_{oc,rad}$ can be observed. For the systems in which LE state and CT state were hybridized, ΔV_{nr} was predicted to decrease with decreasing ΔE_{CT}, and it decreased much quicker when ΔE_{CT} was close to 0 eV. Nevertheless, it was noted that the reduced ΔV_{nr} did not always lead to reduced V_{loss} for the hybridized system because ΔV_r increased due to the increased effective oscillator strength $f^*_{S_0,CT}$. Further calculations revealed that K_r did increase with decreasing ΔE_{CT}.

The reduced ΔV_{nr}, associated with increased K_r leading to increased ΔV_r, was also observed experimentally in solar cells constructed using a solvent vapor annealing (SVA) method. Using different processing conditions [69], Azzouzi et al. showed that by varying the time used for the SVA treatment, ΔV_{nr} of the solar cells based on a series of BIT oligomer donors could be easily tuned: ΔV_r increased and ΔV_{nr} reduced with the reducing SVA time, leading to a rather constant V_{loss}. Because the energetics of the active layer and E_{CT} remained unaffected by the SVA treatment, the most likely reason for the increased ΔV_r and reduced ΔV_{nr} was thus assigned to increased f_{osc}, which was also confirmed by using the theoretical model proposed in that work [69]. Thus, increasing f_{osc} would not always be an effective method to reduce V_{loss} in organic solar cells.

The impact of the density of charge transfer complex (N_{CTC}) on V_{loss} was also found to be strong. Vandewal et al. studied a series of C_{60}-based vacuum deposited small molecule solar cells in which the concentration of the donor component (TAPC or MeO-TPD) in the active layer was varied from 1% to 10% [92]. Despite that the donor concentrations in these active layers were low and C_{60} was the main absorber, high EQE_{PV} were obtained from the solar cells with a low donor content. More importantly, the EQE_{PV} measurements revealed that N_{CTC} reduced with the decreasing concentration of the donor in the active layer, while E_{CT} remained constant. Thus, the impact of N_{CTC} on $V_{oc,rad}$ could be evaluated: $V_{oc,rad}$ was found to increase with the reducing N_{CTC}, a result of reduced K_r, and the increased $V_{oc,rad}$ could fully explain the increase in V_{oc} achieved in the solar cells with a lower donor concentration. Therefore, it was suggested that the reduced N_{CTC} leading to reduced ΔV_r could in turn lead to reduced V_{loss}. Moreover, the impact of N_{CTC} on ΔV_{nr} was expected to be weak because the change of N_{CTC} should equally influence both K_r and K_{nr}. Indeed, transient photovoltage decay measurement revealed that the charge carrier life time (inversely proportional to recombination rate) increased with reducing N_{CTC}.

16.4 Summary

Organic solar cells are highly promising as an alternative technology to inorganic solar cells. PCE of state-of-the-art organic solar cells are already over 18%, and further improvement in the solar cell performance will rely heavily on the elimination of unnecessary voltage losses. In this chapter, we discussed the origin of voltage losses in both ideal and nonideal solar cells using theoretical methods and demonstrated the importance of CT state properties in determining V_{oc} of organic solar cells. The dependences of both radiative and non-radiative recombination voltage losses on the radiative and non-radiative decay rates of CT state were discussed, and the determining factors for the decay rates of CT state were analyzed. Furthermore, experimental methods for evaluating the CT state-related properties and determining different types of voltage losses were introduced. Finally, the strategies used to reduce voltage losses in organic solar cells were reviewed.

Acknowledgments

Z. Tang acknowledges the National Natural Science Foundation of China (NSFC, project nos. 51973031, 51933001) and Shanghai Pujiang Talent Program (no. 19PJ1400500) for financial support. Z. Ma acknowledges financial support from NSFC (project no. 52073056) and the Natural Science Foundation of Shanghai (project no. 19ZR1401400).

References

1 Inganäs, O. (2018). *Adv. Mater.* 30: 1800388.
2 Meng, L., Zhang, Y., Wan, X. et al. (2018). *Science* 361: 1094.
3 Cheng, P., Li, G., Zhan, X., and Yang, Y. (2018). *Nat. Photonics* 12: 131.
4 Hou, J., Inganäs, O., Friend, R.H., and Gao, F. (2018). *Nat. Mater.* 17: 119.
5 Zheng, Z., Hu, Q., Zhang, S. et al. (2018). *Adv. Mater.* 30: 1801801.
6 Yuan, J., Zhang, Y., Zhou, L. et al. (2019). *Joule* 3: 1140.
7 Zheng, Z., Yao, H., Ye, L. et al. (2020). *Mater. Today* 35: 115.
8 Liu, Q., Jiang, Y., Jin, K. et al. (2020). *Sci. Bull.* 65: 272.
9 Green, M.A., Dunlop, E.D., Hohl-Ebinger, J. et al. (2020). *Prog. Photovoltaics Res. Appl.* 28: 629.
10 Chen, Y., Tan, S., Li, N. et al. (2020). *Joule* 4: 1961.
11 Xiao, K., Lin, R., Han, Q. et al. (2020). *Nat. Energy* 5: 870.
12 Kim, J.Y., Lee, J.-W., Jung, H.S. et al. (2020). *Chem. Rev.* 120: 7867.
13 Li, C., Zhou, J., Song, J. et al. (2021). *Nat. Energy* 6: 605–613.
14 Zhang, M., Zhu, L., Zhou, G. et al. (2021). *Nat. Commun.* 12: 309.
15 Zeng, Y., Li, D., Xiao, Z. et al. (2021). *Adv. Energy Mater.* 11: 2101338.
16 Chen, S., Wang, Y., Zhang, L. et al. (2018). *Adv. Mater.* 30: 1804215.
17 Liu, Z., Krückemeier, L., Krogmeier, B. et al. (2019). *ACS Energy Lett.* 4: 110.
18 Janssen, R.A.J. and Nelson, J. (2013). *Adv. Mater.* 25: 1847.
19 Elumalai, N.K. and Uddin, A. (2016). *Energy Environ. Sci.* 9: 391.
20 Liu, X., Du, X., Wang, J. et al. (2018). *Adv. Energy Mater.* 8: 1801699.
21 Azzouzi, M., Kirchartz, T., and Nelson, J. (2019). *Trends Chem.* 1: 49.
22 Menke, S.M., Ran, N.A., Bazan, G.C., and Friend, R.H. (2018). *Joule* 2: 25.
23 Ji, Y., Xu, L., Hao, X., and Gao, K. (2020). *Sol. RRL* 4: 2000130.
24 Wang, Y., Qian, D., Cui, Y. et al. (2018). *Adv. Energy Mater.* 8: 1801352.
25 Yuan, J., Zhang, H., Zhang, R. et al. (2020). *Chem* 6: 2147.
26 Shockley, W. and Queisser, H.J. (1961). *J. Appl. Phys.* 32: 510.
27 Würfel, P., Finkbeiner, S., and Daub, E. (1995). *Appl. Phys.* A 60: 67.

28 Ameri, T., Dennler, G., Lungenschmied, C., and Brabec, C.J. (2009). *Energy Environ. Sci.* 2: 347.
29 Salim, M.B., Nekovei, R., and Jeyakumar, R. (2020). *Sol. Energy* 198: 160.
30 Xiao, X., Lee, K., and Forrest, S.R. (2015). *Appl. Phys. Lett.* 106: 213301.
31 Hadipour, A., de Boer, B., and Blom, P.W.M. (2008). *Adv. Funct. Mater.* 18: 169.
32 Rau, U., Paetzold, U.W., and Kirchartz, T. (2014). *Phys. Rev.* B90: 035211.
33 Li, G., Chang, W.-H., and Yang, Y. (2017). *Nat. Rev. Mater.* 2: 17043.
34 Baruch, P., De Vos, A., Landsberg, P.T., and Parrott, J.E. (1995). *Sol. Energy Mater. Sol. Cells* 36: 201.
35 Luque, A., Martí, A., López, N. et al. (2005). *Appl. Phys. Lett.* 87: 083505.
36 Rau, U. (2007). *Phys. Rev.* B 76: 085303.
37 Jiang, Y., Wang, X., and Pan, A. (2019). *Adv. Mater.* 31: 1806671.
38 Caruso, D. and Troisi, A. (2012). *Proc. Natl. Acad. Sci. U.S.A.* 109: 13498.
39 Günes, S., Neugebauer, H., and Sariciftci, N.S. (2007). *Chem. Rev.* 107: 1324.
40 Zhu, X., Zhang, G., Zhang, J. et al. (2020). *Joule* 4: 2443.
41 Zhu, L., Tu, Z., Yi, Y., and Wei, Z. (2019). *J. Phys. Chem. Lett.* 10: 4888.
42 Tang, C.W. (1986). *Appl. Phys. Lett.* 48: 183.
43 Brédas, J.-L., Norton, J.E., Cornil, J., and Coropceanu, V. (2009). *Acc. Chem. Res.* 42: 1691.
44 Sariciftci, N.S., Smilowitz, L., Heeger, A.J., and Wudl, F. (1992). *Science* 258: 1474.
45 Dennler, G., Scharber, M.C., and Brabec, C.J. (2009). *Adv. Mater.* 21: 1323.
46 Yu, G., Gao, J., Hummelen, J.C. et al. (1995). *Science* 270: 1789.
47 Heeger, A.J. (2014). *Adv. Mater.* 26: 10.
48 Vandewal, K., Tvingstedt, K., Gadisa, A. et al. (2009). *Nat. Mater.* 8: 904.
49 Vandewal, K., Albrecht, S., Hoke, E.T. et al. (2014). *Nat. Mater.* 13: 63.
50 Vandewal, K. (2016). *Annu. Rev. Phys. Chem.* 67: 113.
51 Deibel, C., Strobel, T., and Dyakonov, V. (2010). *Adv. Mater.* 22: 4097.
52 Zeiske, S., Kaiser, C., Meredith, P., and Armin, A. (2020). *ACS Photonics* 7: 256.
53 Kaiser, C., Zeiske, S., Meredith, P., and Armin, A. (2020). *Adv. Opt. Mater.* 8: 1901542.
54 Vandewal, K., Goris, L., Haenen, K. et al. (2006). *Eur. Phys. J. Appl. Phys.* 36: 281.
55 Vandewal, K., Goris, L., Haeldermans, I. et al. (2008). *Thin Solid Films* 516: 7135.
56 Vandewal, K., Benduhn, J., and Nikolis, V.C. (2018). *Sustain. Energy Fuels* 2: 538.
57 Bakulin, A.A., Rao, A., Pavelyev, V.G. et al. (2012). *Science* 335: 1340.
58 Chen, X.-K., Coropceanu, V., and Brédas, J.-L. (2018). *Nat. Commun.* 9: 5295.
59 Zhao, Y. and Liang, W. (2012). *Chem. Soc. Rev.* 41: 1075.
60 Willems, R.E.M., Meskers, S.C.J., Wienk, M.M., and Janssen, R.A.J. (2019). *J. Phys. Chem.* C 123: 10253.
61 Scharber, M.C. and Sariciftci, N.S. (2013). *Prog. Polym. Sci.* 38: 1929.
62 Vandewal, K., Tvingstedt, K., Gadisa, A. et al. (2010). *Phys. Rev.* B81: 125204.
63 Marcus, R.A. (1989). *J. Phys. Chem.* 93: 3078.
64 Shoaee, S., Armin, A., Stolterfoht, M. et al. (2019). *Sol. RRL* 3: 1900184.
65 Qian, D., Zheng, Z., Yao, H. et al. (2018). *Nat. Mater.* 17: 703.

66 Nelson, J. (2003). *The Physics of Solar Cells*. River Edge, NJ: Imperial College Press; Distributed By World Scientific Pub. Co, London.

67 Coropceanu, V., Chen, X.-K., Wang, T. et al. (2019). *Nat. Rev. Mater.* 4: 689.

68 Benduhn, J., Tvingstedt, K., Piersimoni, F. et al. (2017). *Nat. Energy* 2: 17053.

69 Azzouzi, M., Yan, J., Kirchartz, T. et al. (2018). *Phys. Rev.* X 8: 031055.

70 Blank, B., Kirchartz, T., Lany, S., and Rau, U. (2017). *Phys. Rev. Appl.* 8: 024032.

71 Xu, Y., Yao, H., Ma, L. et al. (2020). *Angew. Chem. Int. Ed.* 59: 9004.

72 Eisner, F.D., Azzouzi, M., Fei, Z. et al. (2019). *J. Am. Chem. Soc.* 141: 6362.

73 Han, G. and Yi, Y. (2019). *J. Phys. Chem. Lett.* 10: 2911.

74 Tvingstedt, K., Vandewal, K., Gadisa, A. et al. (2009). *J. Am. Chem. Soc.* 131: 11819.

75 Kahle, F.-J., Rudnick, A., Bässler, H., and Köhler, A. (2018). *Mater. Horiz.* 5: 837.

76 Pasman, P., Rob, F., and Verhoeven, J.W. (1982). *J. Am. Chem. Soc.* 104: 5127.

77 Armin, A., Zarrabi, N., Sandberg, O.J. et al. (2020). *Adv. Energy Mater.* 10: 2001828.

78 Ullbrich, S., Benduhn, J., Jia, X. et al. (2019). *Nat. Mater.* 18: 459.

79 Liu, J., Chen, S., Qian, D. et al. (2016). *Nat. Energy* 1: 16089.

80 Xie, Y., Wang, W., Huang, W. et al. (2019). *Energy Environ. Sci.* 12: 3556.

81 Unger, T., Wedler, S., Kahle, F.-J. et al. (2017). *J. Phys. Chem.* C121: 22739.

82 Saito, T., Natsuda, S., Imakita, K. et al. (2020). *Sol. RRL* 4: 2000255.

83 Vandewal, K., Ma, Z., Bergqvist, J. et al. (2012). *Adv. Funct. Mater.* 22: 3480.

84 Ma, Z., Sun, W., Himmelberger, S. et al. (2014). *Energy Environ. Sci.* 7: 361.

85 An, N., Cai, Y., Wu, H. et al. (2020). *Adv. Mater.* 32: 2002122.

86 Hendriks, K.H., Wijpkema, A.S.G., van Franeker, J.J. et al. (2016). *J. Am. Chem. Soc.* 138: 10026.

87 Veldman, D., Meskers, S.C.J., and Janssen, R.A.J. (2009). *Adv. Funct. Mater.* 19: 1939.

88 Zhang, Y., Zhang, C., Huang, H. et al. (2021). *Dyes Pigm.* 184: 108789.

89 Bixon, M., Jortner, J., and Verhoeven, J.W. (1994). *J. Am. Chem. Soc.* 116: 7349.

90 Vandewal, K., Tvingstedt, K., and Inganäs, O. (2012). *Phys. Rev.* B86: 035212.

91 Zhou, Z., Xu, S., Song, J. et al. (2018). *Nat. Energy* 3: 952.

92 Vandewal, K., Widmer, J., Heumueller, T. et al. (2014). *Adv. Mater.* 26: 3839.

17

Stability of Organic Solar Cells: From Fullerene Derivatives to Non-fullerene Acceptors

Xiaoyan Du[1,3,4], Jing Guo[2], Jie Min[2], and Ning Li[3,4]

[1] *Shandong University, School of Physics, Shandananlu 27, Jinan, 250100, P.R. China*
[2] *Wuhan University, The Institute for Advanced Studies, Wuhan, 430072, P.R. China*
[3] *Friedrich-Alexander University Erlangen-Nürnberg (FAU), Institute of Materials for Electronics and Energy Technology (i-MEET), Department of Materials Science and Engineering, Martensstraße 7, 91058, Erlangen, Germany*
[4] *Helmholtz Institute Erlangen-Nürnberg (HI-ERN), Forchungszentrum Jülich GMBH (FZJ), Immerwahrstraße 2, 91058, Erlangen, Germany*

17.1 Introduction

Owing to the unique properties, such as low production cost, lightweight, and ease of continuous manufacturing on a large scale, organic photovoltaics (OPVs) technology is considered one of the most attractive renewable energy sources for off-grid power supply [1, 2]. Substantial efforts have been devoted to improving the performance of organic solar cells (OSCs) over the last two decades. Encouragingly, innovations in both material design and device engineering have recently driven the power conversion efficiency (PCE) of OSCs over 18% [3], approaching the benchmark efficiency of 20% [4, 5]. While the performance of OSCs has been continuously developed to reach the market readiness, the relatively poor stability (or short operational lifetime) of OSCs compared to other conventional photovoltaic technologies needs to be fully investigated. Today's OPV modules are targeting lifetimes of >5 years for flexible products and >10 years for glass-based products. Further increasing the lifetimes of OPV devices will directly contribute to the reduction of the levelized cost of energy (LCOE) and to broaden their applicability in various scenarios, for instance, the building- and mobile-integrated applications. Although OSCs with impressive device stability under various conditions have been reported in the literature, the stability of most state-of-the-art high-performing OPV systems is not very encouraging. As the stability of OSCs is one of the main obstacles toward their commercialization and industrial applications, it is imperative to investigate and understand the degradation mechanisms of OSCs under various conditions. In the following sections, we will discuss the major factors that are limiting the stability of OSCs, including extrinsic stresses and intrinsic ones. The evaluation protocols of device stability suggested by international commission and organizations are summarized and demonstrated. We will then review the recent progress on developing

Organic Solar Cells: Materials Design, Technology and Commercialization, First Edition. Edited by Liming Ding.

stable OSCs, specially focusing on the OPV material systems with stable microstructures and strategies to effectively enhance the morphological stability of OSCs.

17.2 Factors Limiting the Stability of Organic Solar Cells

In order to continuously develop the OPV technology with high efficiency and stability, it is very useful to ascertain the origin of degradation mechanisms. Is the degradation of OSCs fundamentally related to the organic materials or to the processing parameters? Is it possible to address or suppress the degradation of OSCs during operation? It has been found that many factors, including external stresses and intrinsic factors, can limit the stability of OSCs in a simulated environment or in the real operational conditions. For instance, a poor extrinsic stability could be caused by forming crack or delamination, whereas a poor intrinsic stability could be linked to an interface or phase separation inside a well-operating solar cell. In this section, we focus on the limiting factors behind the instability from the perspective of external stresses and intrinsic factors.

17.2.1 Extrinsic Stresses

OSCs with a multilayer structure suffer from external stresses under operation conditions, including irradiation, heating, oxidation, mechanical stress, etc. Here we will discuss the mechanisms of major external stresses as mentioned in the literature.

17.2.1.1 Light Effect

Light effect, also known as light-induced degradation or photo-degradation, is one of the well-known degradation mechanisms for OSCs, including photo-physical and photochemical degradations [6, 7]. The photo-physical degradation is a main pathway for photo-degradation of OSCs, which was found to be more sensitive to the ultraviolet (UV) light [7, 8]. On the one hand, UV light could induce decomposition of light-sensitive additives remained in the photoactive layer (e.g. 1,8-diiodooctane (DIO)), which creates more electron traps to reduce the device performance [9]. On the other hand, the solar irradiation could also increase the temperature of OSCs, resulting in the change of microscopic morphology of the photoactive layer as well as device performance [10].

Additionally, the photochemical degradation is another pathway for photo-degradation of OSCs, which is caused by the photo-oxidation of the active layer and interface layers in most of the cases. Both the donor and acceptor materials could face photo-oxidation process under illumination, which changes their molecular structures [11], and as a result, leads to reduced light absorption, charge generation, and mismatched energy level alignment [12–14]. In addition, the photo-oxidation process could generate sub-bandgap states, which induce traps and energetic disorders in the active layer and accelerate the non-radiative recombination, and thus attenuate device performance [12, 15, 16]. Besides, other photochemical processes such as the photo-oligomerization of fullerene acceptors are also responsible for the photochemical degradation in OSCs [17, 18]. Apart from the photo-physical degradation and photochemical degradation of active layer materials, stability of interfacial materials also plays a critical role in the photo-degradation process. Moreover, the interface property is also one of the most important factors that contribute to the photo-degradation behaviors of OSCs.

17.2.1.2 Thermal Effect

Most of high-performance OSCs suffer from the thermal instability issues in the real world [19–21], which is related to the physical degradation to a large extent [22, 23]. It is known that the optimal blend morphology is generally away from its thermodynamic equilibrium state and could slowly evolve to the equilibrium case without extrinsic factors [24], even at a low temperature [25]. Of particular note is that the working temperature of solar cells could be relatively high due to continuous illumination (>65 °C), which may significantly accelerate the evolution of blend microstructure morphology, for instance, the thermal-induced aggregation and crystallization of fullerene and non-fullerene acceptors (NFAs) in the active layer [23, 26–28]. The changes of blend morphology in terms of obvious phase separation in the active layer can reduce the interface area between the donor and acceptor materials, which can sharply decrease the performance of OSCs resulting from inefficient exciton separation and charge transfer by decreased D/A interfaces [23]. Moreover, heating-induced degradation can also occur at the interface between active layer and interface layers [29, 30] and between the interface layers and electrodes [7], normally leading to the increased energy barriers for hole and/or electron extraction [29–31].

17.2.1.3 Environmental Effect

Air instability is another important issue for OSCs, which is generally detrimental to the device performance [32]. Ambient water and oxygen can diffuse throughout the whole device and result in severe physical and chemical degradation in the active layers, interface layers as well as electrodes [33–35]. It has been reported that oxygen and water could take part in the photo-oxidation process of photovoltaic materials, which changes their molecular structures, optoelectronic properties, and charge transport mobilities [12, 14, 36, 37]. In addition, the permeation of water and oxygen could also cause the morphology changes of photoactive layers, leading to reduced open-circuit voltage (V_{oc}) and fill factor (FF). Besides, water and oxygen permeation can also cause oxidation in interface layers and at the interfaces [38], as well as low work function metal electrodes such as calcium and aluminum [39–41], which generally act as pinholes or electrically insulating layer to inhibit charge transport and extraction properties. Notably, water and oxygen always simultaneously lead to device degradation [42, 43].

17.2.1.4 Mechanical Stress Effect

Degradation induced by mechanical stress, which occurs in the fabrication, transportation, installation, and application, is generally unavoidable in OSCs [44–46]. Moreover, factors in real external conditions such as hail, wind, and rain can also cause mechanical stress to devices [47]. Thus, the mechanical robustness is one of the most crucial requirements for ensuring the long-term device stability. There are several types of degradation that can occur in OSCs under mechanical stress.

First, mechanical stress could cause delamination of the blend morphology resulting in the poor charge transport and extraction properties in devices [48]. Second, the linear mechanical tension can create strains of OPV materials which changes the blend morphology leading to the loss of device performance [49]. Third, the lateral mechanical strain can also cause degradation in the relative brittle layers (e.g. indium tin oxide [ITO] electrode and active layer), which further degrades the device performance [50, 51]. Fourth, mechanical stress can result in cracks and punctures in the functional layers of OSCs. These defects may facilitate the ingress of water and oxygen in the devices and thus accelerate the degradation of device performance [52]. Therefore, mechanical stress could induce degradation in photoactive layer, interface layers and electrodes. This is even critical when replacing the rigid glass substrates with flexible substrates for large-scale roll-to-roll production of OSCs.

17.2.2 Intrinsic Factors

OSCs with a multilayer stacked architecture show uncontrollable stability even without extrinsic degradation factors. It is known that the morphology changes in the active layer and the diffusion of interfacial and electrode materials into the photoactive layer are two main reasons for the inherent device instability [33, 34].

As aforementioned, the optimum morphology, which delivers the best performance, is usually away from its thermodynamic equilibrium. Even though the optimum bulk-heterojunction (BHJ) morphology is optimized for achieving high device performance, it will slowly evolve toward the equilibrium state and cause more or less microstructure changes in the active layer, or at the interfaces between the active layer and interfacial layers [24, 25]. Due to the different interfacial surface energy and high diffusion constant of donor and acceptor photovoltaic materials [24, 25], and sometimes the structural incompatibilities between donor and acceptor materials [22, 28, 53], the donor/acceptor phase separation is the main result of the inherent morphology evolution. Moreover, it has also been demonstrated that the high stability of BHJ photovoltaic systems can result from its ideal miscibility of donor and acceptor materials that the initially optimized morphology is located at the thermodynamic equilibrium [26, 54, 55].

Similar to the OPV materials, the interface materials and metal electrodes of OSCs also exhibit diffusivity. For instance, the commonly used hole transport layer PEDOT:PSS (poly(3,4-ethylenedioxythiophene):poly(styrenesulfonate)) could diffuse into the active layer [39]. Of note is that PEDOT:PSS layer possesses the hygroscopicity, acidity, and anisotropic charge injection properties, which could result in device instability. With respect to the commonly used electrodes, e.g., Ag and ITO, Ag and In atoms could diffuse into the functional semiconducting layers even without extrinsic factors [56–58]. The diffusion of electrodes can cause traps in the active layer and tune the relevant energy levels [11]. As a result, all of these intrinsic factors can accelerate the non-radiative charge recombination and lead to the inherent device instability.

It is worthwhile to mention that most of "degradation profiles" of OSCs can be divided into two regimes, which include an initial sharp drop in performance (also called as "burn-in" degradation) followed by a slow linear performance degradation trend [59]. Researchers speculated possible reasons for the origin of burn-in losses, including the increased trap mediated charge recombination, the broad polydispersity of polymer materials, the fullerene dimerization, the organic or inorganic impurities in the film, which have to be in-depth analyzed for given material systems [60–63]. In addition, many recent studies also revealed that the morphological change of photoactive layer is the reason for the initial sharp performance degradation of OSCs [23, 24, 64]. Notably, both intrinsic and extrinsic factors can cause morphology changes and accelerate the morphology evolution which present as a fast degradation at the initial stage. The rate of morphology evolution could be slowed down when the microstructure morphology is close to the equilibrium state, thus exhibiting the slow linear degradation behavior. As expected, the stability issues caused by the interface materials and electrodes could be effectively overcome by developing advanced functional materials. However, for efficient and stable OSCs, the control of the subtle nanometer-scale morphology is of the utmost importance, since efficient photo-to-electricity conversion processes depend crucially on the microstructure of the photoactive layer.

17.3 Stability Evaluation Protocols

Apart from cost and efficiency, the reliability assessment is regarded as a critical prerequisite for pre-commercial photovoltaic concepts. The lifetime for photovoltaic modules is expected to be around 20 years or longer for possible commercial applications. Thus, international committees have been developing standardized qualification test. In particular, a suite of standard documents were designed by International Electrotechnical Commission (IEC, Genf, Switzerland), including three most applied standards: terrestrial photovoltaic (PV) modules – design qualification and type approval (IEC 61215); thin-film terrestrial photovoltaic (PV) modules – design qualification and type approval (IEC 61646); and concentrator photovoltaic (CPV) modules and assemblies – design qualification and type approval (IEC 62108). These tests, however, are more applicable for the well-understood degradation modes generally related to the terrestrial thin film at the module level, which are not intended to interpret degradation pathways and identify the relative mechanism limiting the device lifetime. Thus, this must be considered when applying it directly to the emerging technologies such as OSCs, dye-sensitized solar cells (DSSCs), and perovskite solar cells (PSCs) because of their different material properties, various unapprehended failure modes, and immature product development for commercialization. In 2010, the International Summits on Organic PV Stability (ISOS) held in Roskilde, Denmark, discussed and established the standardized aging experiments (so-called ISOS protocols). These are summarized in a publication entitled "Consensus stability testing protocols for OPV materials and devices" in 2011 [65], offering a consensus in the OPV field on investigating and reporting degradation research in a controllable, comparable, and reproducible way, subsequently, allowing for a more comprehensive understanding of device degradation that is not fully elucidated or appreciated by IEC standards. Briefly, IEC standards (such as IEC technical specification [TS] 62876-2-2:2018) and ISOS stability protocols both describe the typical stress conditions including radiation, temperature, humidity, atmosphere, load, etc., but differing in some specific parameters. For example, IEC 61646 provides some extreme external stress conditions, like UV precondition, mechanical load, hail, or high external voltage. Further tests include bypass diode and hot spot endurance, and these tests intend to confirm device safety and qualify the device for commercial products.

Table 17.1 displays the newest existing ISOS protocols and suggests additional protocols that are discussed in the 11th International Summit on Organic and Hybrid Photovoltaics Stability. Generally, the ISOS protocols are grouped in terms of the applied stresses into seven categories, dark storage testing, outdoor testing, light soaking testing, thermal cycling testing, solar-humidity-thermal cycling testing, light–dark cycling and electrical bias in the dark testing, as well as three levels of execution with increasing complexity and requirement on equipment and accuracy, namely basic, intermediate, and advanced. In addition to these measurement conditions, many discussions regarding the light source used for artificial illumination, correct interpretation for stability data, and operational lifetimes are also discussed. Overall, these two frameworks outlined the hierarchy in stability analysis for module level (IEC) and cell level (ISOS), allowing the requirement of standardized testing conditions for industrial and scientific community [67].

17.4 Progress in Developing Stable Organic Solar Cells

17.4.1 Development of Organic Photovoltaic Materials with Stable Microstructure Morphology

The revolution and development of OPV materials with new molecule structures have greatly contributed to the rapid development of OPV technology [68, 69]. On the early stage of development of OSCs, some representative materials like polymer donor poly[2-methoxy-5-(2-ethylhexyloxy)-1,4-phenylenevinylene] (MEH-PPV), poly(3-hexylthiphene-2,5-diyl) (P3HT) and C_{60} derivative [6,6]-phenyl-C_{61}-butyricacid methyl ester ($PC_{61}BM$), and its corresponding C_{70} derivative [6,6]-phenyl-C_{71}-butyric acid methyl ester ($PC_{71}BM$) dominated the high-performance active layer materials, delivering a PCE of 4%–5%, thus attracting worldwide interests in OPV. Many more high-performance polymers have been developed subsequently. In 2009, Yu and coworkers developed a low bandgap polymer donor material PTB1 [70] and these serials PTBn (n = 1–7) boosted efficiency up to 7.4%. Following this work, PCEs of more than 10% were frequently reported for fullerene-based OSCs with new materials [71]. On

Table 17.1 Summary of testing guidelines for the different stress conditions and levels provided according to ISOS protocols.

Test ID	Light source	Temperature	Rel. humidity	Environment/ setup	Characterization light source	Load
Dark storage (ISOS-D)						
ISOS-D-1	None	Ambient (23 ± 4 °C)	Ambient	Ambient air	Solar simulator or sunlight	OC
ISOS-D-2	None	65, 85 °C	Ambient	Oven, ambient air	Solar simulator	OC
ISOS-D-3	None	65, 85 °C	85%	Env. chamber	Solar simulator	OC
Bias stability (ISOS-V)						
ISOS-V-1	None	Ambient (23 ± 4 °C)	Ambient	Ambient air	Solar simulator	Positive: V_{MPP}; V_{oc}; E_g/q;
ISOS-V-2	None	65, 85 °C	Ambient	Oven, ambient air	Solar simulator	J_{sc} Negative: $-V_{oc}$, J_{MPP}[a)]
ISOS-V-3	None	65, 85 °C	85%	Env. chamber	Solar simulator	
Light-soaking (ISOS-L)						
ISOS-L-1	Solar simulator	Ambient (23 ± 4 °C)	Ambient	Light only	Solar simulator	MPP or OC
ISOS-L-2	Solar simulator	65, 85 °C	Ambient	Light and temperature	Solar simulator	MPP or OC
ISOS-L-3	Solar simulator	65, 85 °C	~ 50%	Light, temperature, and RH	Solar simulator	MPP
Outdoor stability (ISOS-O)						
ISOS-O-1	Sunlight	Ambient	Ambient	Outdoor	Solar simulator	MPP or OC
ISOS-O-2	Sunlight	Ambient	Ambient	Outdoor	Sunlight	MPP or OC
ISOS-O-3	Sunlight	Ambient	Ambient	Outdoor	Sunlight and solar simulator	MPP

(*continued*)

Table 17.1 (Continued)

Test ID	Light source	Temperature	Rel. humidity	Environment/ setup	Characterization light source	Load
Thermal cycling (ISOS-T)						
ISOS-T-1	None	RT to 65, 85 °C	Ambient	Hot plate/oven	Solar simulator	OC
ISOS-T-2	None	RT to 65, 85 °C	Ambient	Oven/env. Chamber	Solar simulator	OC
ISOS-T-3	None	−40 to + 85 °C	< 55%[b)]	Env. chamber	Solar simulator	OC
Light cycling (ISOS-LC)						
ISOS-LC-1	Solar simulator/Dark cycle period: 2, 8, or 24 h Duty cycle: 1: 1 or 1: 2	Ambient (23 ± 4 °C)	Ambient	Light only	Solar simulator	MPP or OC
ISOS-LC-2		65, 85 °C	Ambient	Light and temperature	Solar simulator	MPP or OC
ISOS-LC-3		65, 85 °C	< 50%	Light, temperature, and RH	Solar simulator	MPP
Solar-thermal cycling (ISOS-LT)						
ISOS-LT-1	Solar simulator	Linear or step ramping between room temp. and 65 °C	Monitored, uncon-trolled	Weathering chamber	Solar simulator	MPP or OC
ISOS-LT-2	Solar simulator	Linear ramping between 5 °C and 65 °C	Monitored, controlled at 50% beyond 40 °C	Env. chamber with sun simulator	Solar simulator	MPP or OC
ISOS-LT-3	Solar simulator	Linear ramping between −25 °C and 65 °C	Monitored, controlled at 50% beyond 40 °C	Env. chamber with sun simulator and freezing	Solar simulator	MPP or OC

a) V_{oc}, V_{MPP}, and J_{MPP} are determined from light J–V curves measured under standard solar cell testing conditions on a fresh device. E_g and q are the bandgap of the active layer and elementary charge, respectively.

b) Relative humidity is controlled at temperatures above 40 °C and is not controlled for the remainder of the cycle. Env., environmental; OC, open-circuit condition; MPP, maximum power point; RT, room temperature; RH, relative humidity.

Source: Khenkin et al. [66]/Springer Nature/CC BY 4.0.

the other side, NFAs, including both polymeric and small molecular organic acceptors, attracted increasing attention in recent years. A breakthrough was made in 2015 by Zhan and coworkers, who reported an electron acceptor material (3,9-bis (2-methylene-(3-(1,1-dicyanomethylene) - indanone))-5,5,11,11- tetrakis(4-hexylphenyl)-dithieno[2,3-d:2′,3′-d′]-s-indaceno[1,2-b:5,6-b′]dithiophene [ITIC]) [72] showing a superior PCE to $PC_{61}BM$ in OSCs. This type of fused-ring electron acceptors (FREAs) remarkably boosted PCEs of OSCs to 15% [73]. Recently, high PCEs of 15%–18% attained for brand-new Y-series acceptors significantly prompted OSCs research [3, 74, 75] which in turn eagerly appeal for the improvement of stability and reducing synthetic costs.

17.4.1.1 Organic Solar Cells Based on Fullerene Acceptors

The previous investigation on improvement of PCE via subtle molecular structure design and modulation endows us with a new perceive into the interrelation among structure, morphology, and stability, which also attracted increasing researchers' attention and was typically used as a supplementary section in the article along with discussion about molecular design guidelines. In most cases of fullerenes-based devices, the initial loss up to 20%–50% of PCE (referred to as "burn-in loss") in the time period from a few hours to days imposes restrictions on industrialization. Tremendous studies revealed that such reported burn-in losses not barely stem from one single factor. The multiple degradation pathways occur and interact according to various environmental stresses, such as photo-induced dimerization [76], micro-phase separation caused by spinodal donor–acceptor demixing [24], fullerene diffusion induced formation of crystals via Ostwald ripening under thermal stress [77], and photochemical reaction with DIO [78, 79]. With these insights, the investigation of modification on molecular structure has been made to enhance the stability of fullerene-based OSCs. In addition, engineering of donor materials is one of the important research fields. McGehee and coworkers found that molecule size and crystallinity of active layer affected the initial exponential burn-in process. Heat- and light-induced burn-in was retarded by larger molecules and active layer morphology with higher crystallinity [80]. A similar result was demonstrated by Brabec and coworkers, as shown in Figure 17.1a [18]. With the enhancement of donor polymer crystallinity, leading to the formation of pure fullerene domains, fullerene dimers are more likely to be produced in the blend film during light soaking. By increasing fullerene ordering crystallinity, enhanced photo-stability against photo-induced dimerization was observed compared to the amorphous one. Briefly, these findings strongly suggested that phase morphology of OPV films, especially in terms of crystallization, has a strong impact on photo-stability. Thus, the knowledge of the effects of structural device stability is a prerequisite for superior stability by the rational structure design of donor materials. Also, it is now recognized that the stability of OSCs is influenced by the side chains of donor materials. Yang and coworkers synthesized three polymers, P1, P2, and P3 (as shown in Figure 17.1b) [81], containing different conjugated lengths of side chains on the same backbone and found that the longest side chain polymer, alkyl-thieno[3,2-b]thiophene (TT)-substituted P3, offered improved morphological stability relative to other

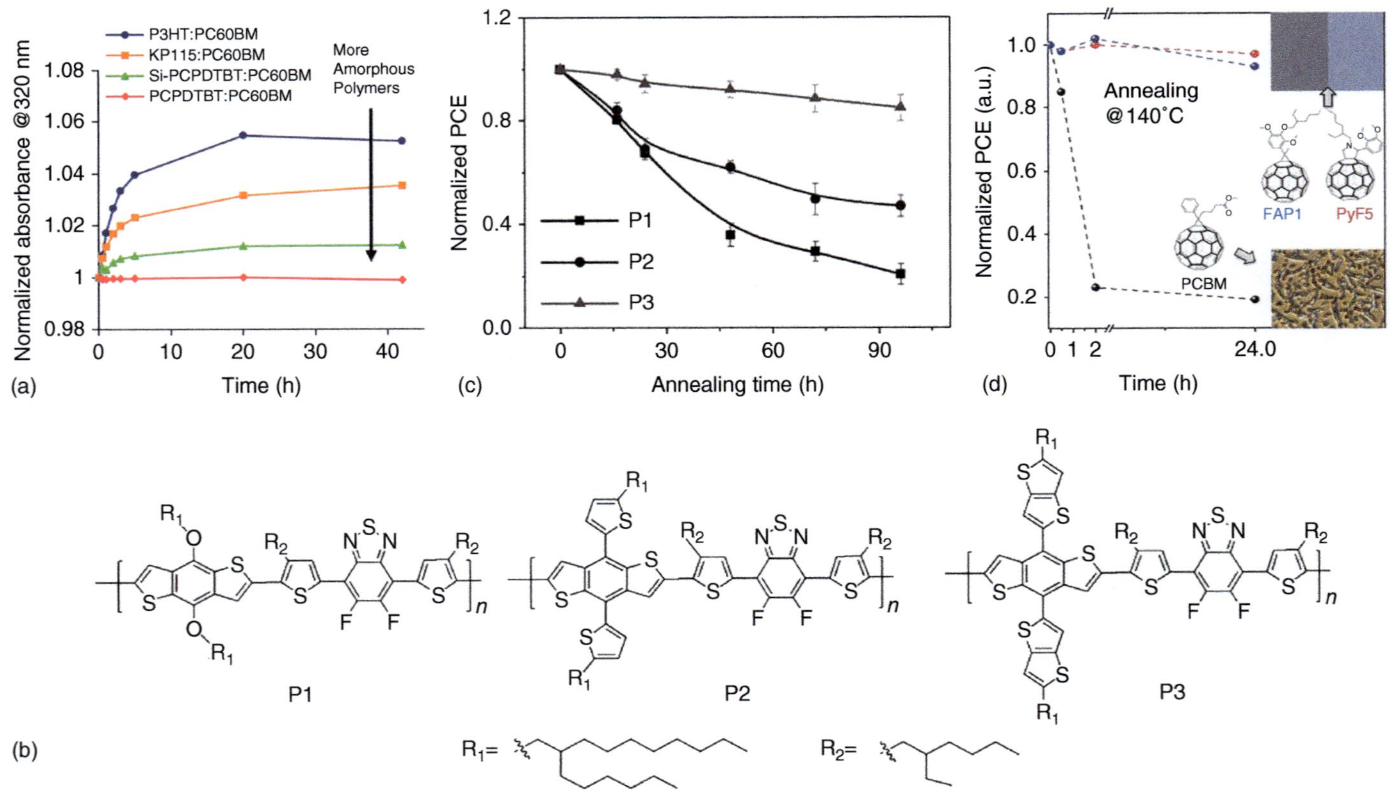

Figure 17.1 (a) Photo-induced dimerization measured with UV–vis absorption in polymer-fullerene (1: 2) blends with different degrees of polymer aggregation. Source: Heumueller et al. [18]/Royal Society of Chemistry/CC BY 3.0. (b and c) Chemical structures of P1, P2, and P3, and normalized PCE evolutions as a function of annealing time at 130 °C. Source: Li et al. [81]/Reproduced with permission of John Wiley & Sons. (d) Normalized photovoltaic performance of PTB7-Th: fullerene organic solar cells as a function of annealing time at 140 °C under inert atmosphere and the comparison of the corresponding morphology. Source: Zhang et al. [82]/with permission of John Wiley and Sons.

two counterparts. The outstanding morphological stability of P3 blend could be ascribed to continuous networks of polymer whiskers served as frames to "lock" the nanoscale PCBM domains, greatly reducing the nucleation rate of PCBM and impeding its further aggregations. As a result, P3-based devices exhibited the most stable performances and presented a slight PCE loss (Figure 17.1c). Furthermore, alkyl chain length and side chain placement could impact the thermal stability as well [83]. Besides, a number of researchers have reported that significant phase separation induced by PCBM dimerization/aggregation is responsible for performance deterioration due to the driven of thermodynamic non-equilibrium state. Thus some efforts have been devoted to the substitution of fullerene with modified derivatives [84]. Brabec and coworkers found that miscibility between donor and acceptor is also an essential parameter for the stability evaluation. Accordingly, two fullerene derivatives, pyrrolidinofullerene (PyF5) and methanofullerene (FAP1), showing improved miscibility with PTB7-Th than PCBM, exhibited excellent device thermal stability (Figure 17.1d) [82].

17.4.1.2 Organic Solar Cells Based on Non-fullerene Acceptors

Burn-in loss free has been reported for NFA-based OSCs, yielding excellent light- and thermal-stability [85, 86]. Therefore, the replacement of fullerene derivatives with NFAs in OPV has been recently proven to be useful to enhance stability. Furthermore, the diversity of the NFAs structure by rational design provides us with more options regarding prolonged intrinsic stability. Some valid strategies like altering the planarity and rigidity with the enhancement in crystallinity [87], incorporating the antioxidant additive [12], the substitution of unstable molecular side chains [88], and increasing the molecular weight [53] have been developed to significantly suppress the initial rapid attenuation. Recently, we investigated a group of ITIC derivatives and found that end-group and side-chain engineering of the NFAs strongly influenced the photo-stability of OSCs (Figure 17.2a) [88]. The energetic traps were raised with the breaking of conjugation during photo-aging. More specifically, fluorination of the end-group stabilized molecules against light soaking, while the addition of methyl groups showed an opposite trend, and side-chain modification can significantly influence the morphological stability, disclosing molecular orientation transformation from more face-on to edge-on after illumination in the case of ITIC (Figure 17.2b). Later, based on this observation, further investigations on the photo-stability of poly[(2,6-(4,8-bis(5-(2-ethylhexyl)thiophen-2-yl)benzo[1,2-b:4,5-b′]dithiophene)-co-(1,3-di(5-thiophene-2-yl)-5,7-bis(2-ethylhexyl)-benzo[1,2-c:4,5-c′]dithiophene-4,8-dione))] (PBDB-T) derivatives (PBDB-T-2F or PM6) incorporating IT-4F (also named as ITIC-2F) were discussed. We revealed that the initial morphology as to the ordering of both polymer and aggregated IT-4F domains significantly influenced degradation rate (Figure 17.2c).

Notably, a deeper insight demonstrated the formation of small-scale isolated NFA domains as a result of polymer reorganization and low diffusion coefficient of IT-4F (Figure 17.2d), which contributes to the morphological traps for electron transport and causes long-term FF degradation [89]. Moreover, the subtle changes in molecular structure, especially in the linkage difference, have

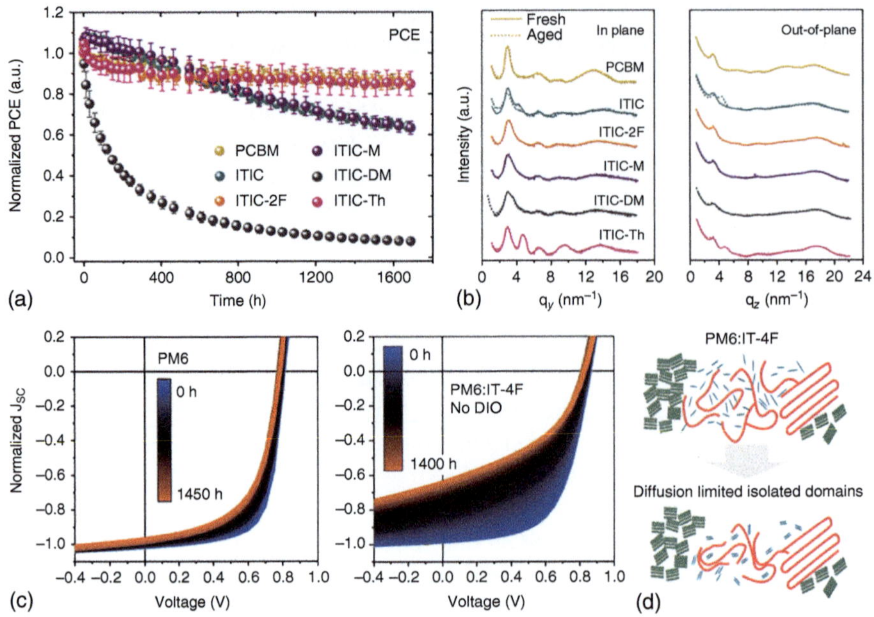

Figure 17.2 (a) Normalized photovoltaic performance based on different acceptors as a function of illumination time. (b) Line-cuts of grazing incidence wide-angle X-ray scattering (GIWAXS) profiles for various fresh and aged blend films. Source: Du et al. [88]/Reproduced with permission of Elsevier. (c) Periodically recorded current–voltage curves (with normalized initial short-circuit current density) for the device with DIO (left) and without DIO (right). (d) Schematic illustration of morphology changes of the mixed region for the polymer: IT-4F systems under illumination, showing the formation of small-scale isolated NFAs domains. Source: Du et al. [89]/John Wiley & Sons/CC BY 4.0.

been demonstrated to provide decent longevity under the external stress. Song and coworkers designed and synthesized a NFA i-IEICO-2F, an isomer of IEICO-2F [90]. The electron-withdrawing portion of i-IEICO-2F is connected at the 4-position of the alkoxythiophene unit, instead of the previously reported 5-position, leading to a more twisted main chain configuration and a significant blue shift of absorption peak by 132 nm. In return, this tiny structure variation with more twisted main chain has increased the intramolecular steric hindrance, maintained a consistent morphology and molecular orientation regardless of the continuous thermal stress, which decreased aggregation and phase separation. On the contrary, a higher-order packing and agglomeration of IEICO-4F and i-IEICO-4F generated excessive phase separation, resulting in poor charge transport, and eventually a rapid deterioration in J_{sc} and FF.

In summary, these continuous efforts provide insights into the effect of molecular structure, orientation, and morphology of the blended films on device stability; the more thorough understanding of molecular structure dependence of intrinsic material stability, materials compatibility between donor and acceptor, and morphology evolutions under environmental stresses should be emphasized in the following investigation.

17.4.1.3 Organic Solar Cells Based on Polymeric Acceptors

Another promising strategy for enhancing device stability was focused on the polymerization of small molecule acceptors or polymer acceptors. Many studies have shown that all-polymer solar cells possess unique characteristics, such as superior thermal stability, mechanical stability, and photo-stability in comparison to fullerene and small-molecule acceptor materials [91]. A direct comparison between all-polymers and polymer-fullerene-based OSCs was performed by Kim and coworkers [92]. The thermal-driven diffusion of PCBM and observable phase separation deteriorated the performance for fullerene-based OSCs. In contrast, an unchanged morphology under the same aging conditions was observed for all-polymer OSCs due to the low diffusion kinetics of the polymer acceptors (Figure 17.3a). Moreover, further exploration by Ma and coworkers disclosed that all-polymer photovoltaic devices displayed a stronger resistance against moisture over their polymer:PCBM counterparts [93]. They speculated that diffusion of water through the PCBM domains into the blend film changes the electrical properties, and further induces the aggregation of fullerene. Consequently, all-polymer devices almost remained the initial efficiency under ambient condition (≈25 °C, relative humidity: 20%), while polymer-fullerene device exhibited only 67% of its initial performance (Figure 17.3b). Other merits of all-polymer OSCs have also been proposed in recent years including excellent photo-stability and mechanical property [76, 94]. Under these scenarios, all-polymer systems are expected to be a promising candidate for further enhancing the long-term stability of OSCs under the real operational condition.

17.4.2 Strategies to Enhance the Morphological Stability of Organic Solar Cells

So far, most of the studies on stability of organic solar cells have been done with fullerene-based material systems. In recent years, dramatic progress was evidenced in NFA-based systems. Readers are referred to the recent review articles on this topic [95–97]. Here, we focus on the strategies specifically for improving morphological stability in both fullerene acceptor (FA) and NFA material systems.

For FA-based material systems, the main obstacles for stable morphology are phase separation between donor and FA acceptors and/or crystallization under thermal or illumination stress. For example, μm-sized PCBM crystals could form in P3HT:PCBM blends under thermal stress [98]. Strong demixing of donor and acceptor materials even under dark storage could lead to severe J_{sc} burn-in losses [24]. Over the past decades, significant efforts have been devoted to tackling these issues. Phase separation normally occurs in the amorphous donor/acceptor mixed region [24]. In general, increasing crystallinity, miscibility and/or compatibility between the components, and glass transition temperature of the polymer component could enhance the microstructure stability as well as the device stability [99]. Other strategies for stabilizing morphology are summarized in the following categories: (i) molecular level engineering of photo-active materials including side-chain

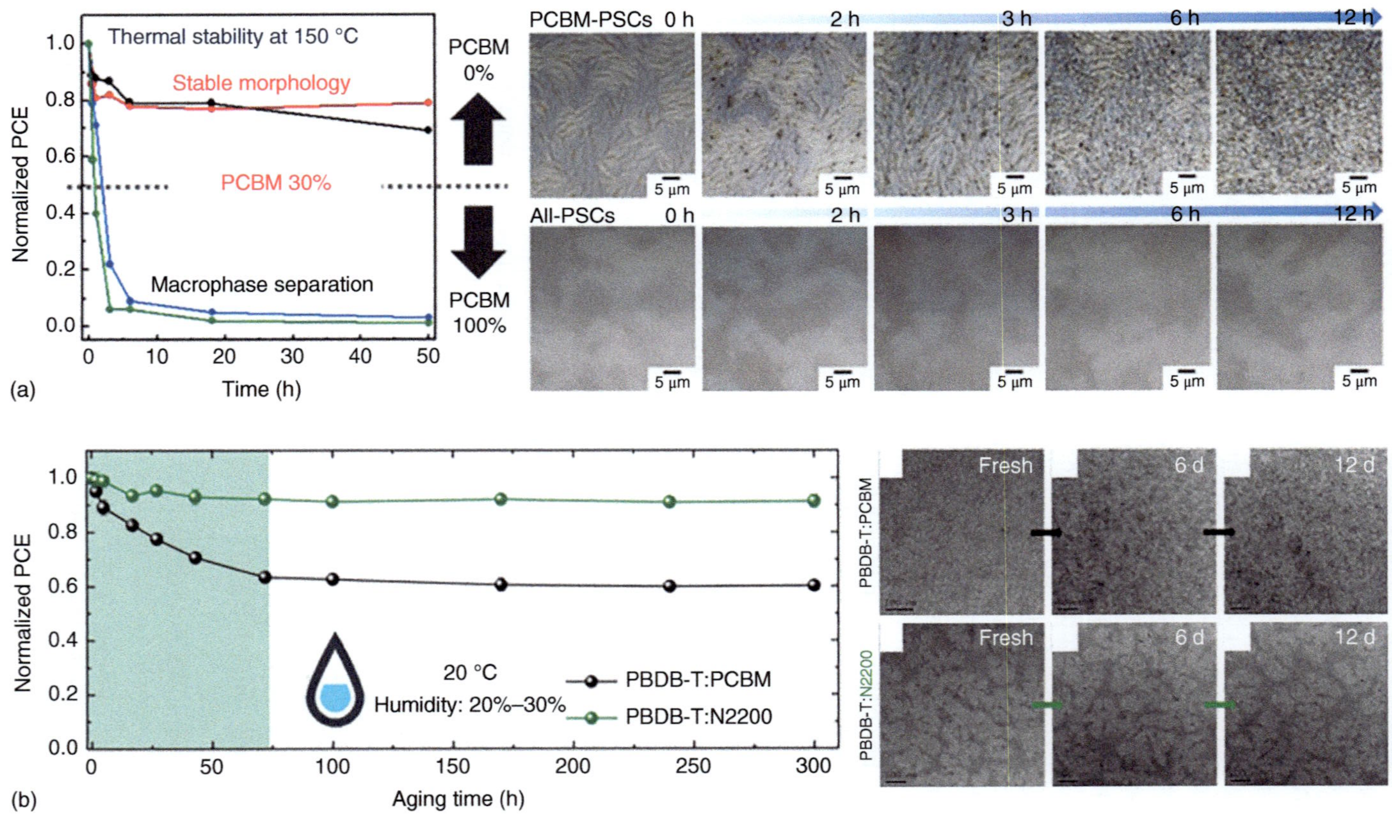

Figure 17.3 (a) Normalized photovoltaic performance in ternary devices with various ratios of polymer acceptors and PCBM annealed at 150 °C for different times, and morphological evolution from optical microscopy images in active layer films, polymer-PCBM and all-polymer blends annealed at 150 °C for 0, 2, 3, 6, and 12 hours. Source: Kim et al. [92]/with permission of American Chemical Society. (b) Normalized photovoltaic performance for the polymer-PCBM and all-polymer device stored in an ambient environment with 20%–30% relative humidity at 20 °C without encapsulation, and transmission electron microscopy (TEM) images for fresh polymer-PCBM and all-polymer blend films and aged films stored for 6 and 12 days. Source: Xu et al. [93]/with permission of John Wiley and Sons.

variations for introducing non-covalent interactions, or covalently linking donor and acceptor materials as a single-component active layer. (ii) Introducing cross-linking concept either by adding crosslinking groups in donor or acceptor side chains or adding extra crosslinkers. (iii) Adding solid additives to form extra bonding or as phase compatibilizers. (iv) Improving initial morphology by processing in terms of solvents, thermal annealing, solvent additives, etc. (v) Using ternary or multinary photo-active semiconductor components. (vi) Synthesizing organic nanoparticles to lock the nano-morphology. (vii) Novel processing methods for morphology control such as layer by layer sequential processing or bilayer heterojunction structures. We separately discuss the strategies with representative examples in the following.

17.4.2.1 Introducing Hydrogen Bonding in the Photo-Active Materials

In 2012, Watkins and coworkers reported the cooperative self-assembly of functionalized fullerenes and all-conjugated block copolymers (BCPs) containing polythiophene derivatives in both segments to yield solar cells with well-defined nanostructures and enhanced morphological stability [100]. The hydrogen bond interactions between bis-[6,6]-phenyl-C61-butyric acid (bis-PCBA) and P3TODT (Figure 17.4a) segments further suppressed crystallization and macro-phase separation of the fullerenes, even under harsh annealing conditions (150 °C for 12 hours). Similarly, in 2013, Qin and coworkers reported a polythiophene diblock copolymer selectively functionalized with 1-*n*-hexylisoorotic acid moieties (P4) and a 2,6-diaminopyridine tethered fullerene derivative (phenyl-C61-butyric acid 4-(4′-pyridinyl)-butyl ester [PCBP]) (Figure 17.4b) [101]. Self-assembly between P4 and PCBP through "three-point" complementary hydrogen bonding interactions was utilized to control and stabilize blend morphologies. Compared to P3HT:PCBM, the thermal stability of devices was improved when annealed at 110 °C for 120 hours in nitrogen atmosphere. In 2020, Liu et al. reported hydroxyl-functionalized NFAs, i.e. IT-OH with one hydroxyl and IT-DOH with two hydroxyls (Figure 17.4c), which could tune the molecular packing and crystallinity assisted by strong intermolecular hydrogen bonds [102]. The solar cells based on IT-DOH showed improved storage and light stabilities compared to those based on ITIC as acceptors, indicating that introducing hydroxyl groups is a promising strategy to tune the molecular packing and crystallinity to design NFAs in order to improve both device efficiency and stability.

17.4.2.2 Chemically Linked Donor and Acceptor as a Single-Component Photoactive Layer

An early strategy is to covalently link the donors and acceptors to prevent large-scale phase separation in fullerene-based systems. In 2012, Tajima and coworkers developed oligo(*p*-phenylenevinylene)-C_{70} dyad, which showed a thermally stable morphology compared to its BHJ counterpart (Figure 17.5a) [103]. In 2017, Woo and coworkers reported BDTRh–PCBM comprising three covalently linked blocks with a PCE of 2.4% showing no degradation at 80 °C for 100 hours (Figure 17.5b) [104]. In 2019, single-component solar cells based on a double-cable polymer, SCP3, were

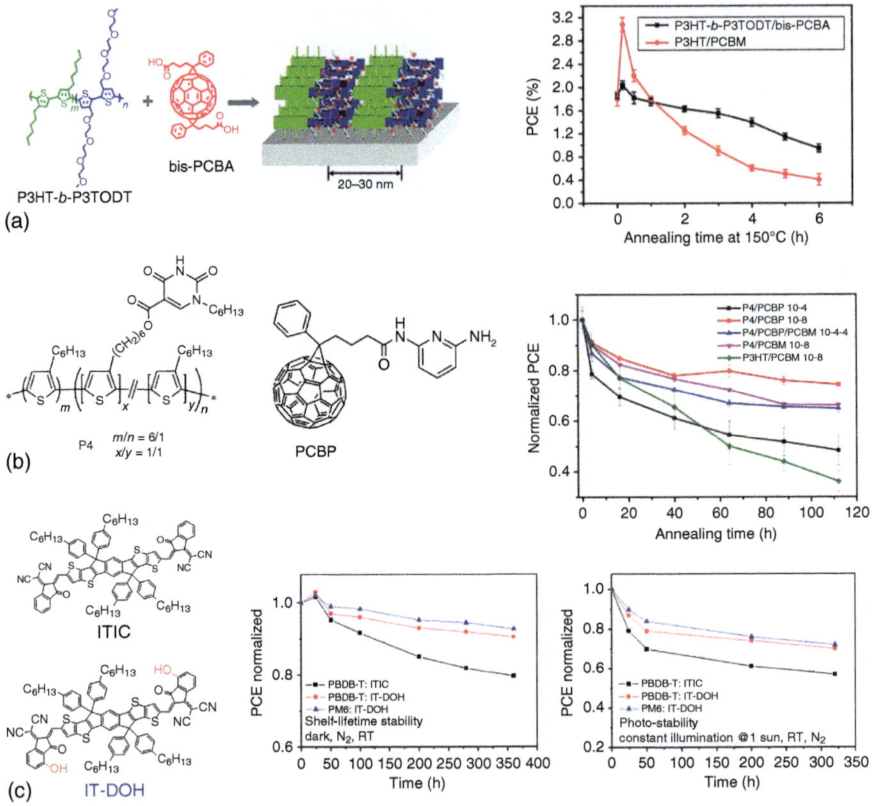

Figure 17.4 (a) Chemical structure of P3HT-b-P3TODT and bis-PCBA, which afford cooperative H-bonding between side chains, and PCE as a function of annealing time at 150 °C. Source: Lin et al. [100]/Reproduced with permission of American Chemical Society. (b) Chemical structure of P4 and PCBP with complementary hydrogen bonding and PCE as a function of annealing time at 110 °C. Source: Li et al. [101]/Reproduced with permission of American Chemical Society. (c) Chemical structure of ITIC and IT-DOH and the stability of devices stored in N_2 under dark or under constant one sun illumination in N_2. Source: Liu et al. [102]/Reproduced with permission of John Wiley & Sons.

reported to be super stable under thermal and illumination conditions (Figure 17.5c) [105]. The double-cable polymer was designed by covalently linking a diketopyrrolopyrrole (DPP)-based polymer backbone and pendant perylene bisimide (PBI) as electron acceptor with flexible alkyl side chains [107]. Solar cells based on this single-component active layer achieved a PCE of 2.7% and showed no performance losses under white light emitting diode (LED) illumination with 90 °C thermal stress after 300 hours. In a long time, the single-component solar cells have suffered from low performance despite of the promising morphological stability. Very recently, Li and coworkers demonstrated single-component solar cells based on PBDBPBI-S with a PCE of 6.32% after film annealed at 230 °C for 10 minutes, demonstrating high potential for further simultaneous improvement of performance and morphological stability using this concept (Figure 17.5d) [106].

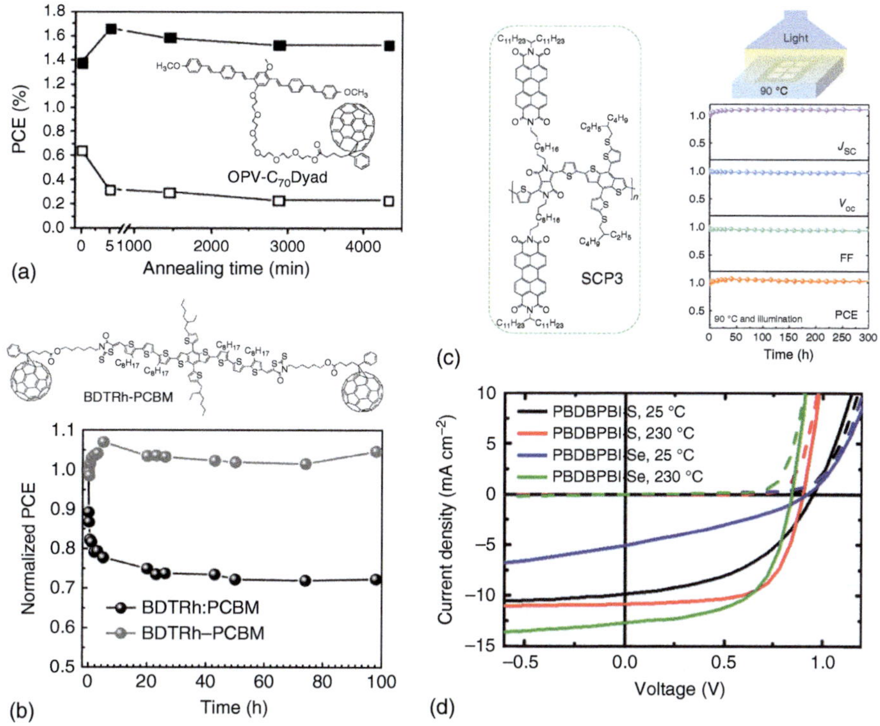

Figure 17.5 (a) PCE of solar cells with OPV-C_{70} dyad film annealed at 110 °C for different time. Source: Izawa et al. [103]/Reproduced with permission of Royal Society of Chemistry. (b) Thermal stability of devices based on BDTRh–PCBM at thermal annealing at 80 °C for 100 hours. Source: Nguyen et al. [104]/Reproduced with permission of John Wiley & Sons. (c) Stability test of SCP3-based organic solar cells in N_2 atmosphere for 300 hours under 90 °C and illumination condition. Source: He et al. [105]/Reproduced with permission of John Wiley & Sons. (d) *J*–*V* characteristics in the dark (dashed line) and under white light illumination (solid line) of PBDBPBI-S and PBDBPBI-Se based single-component OSCs with different active layer annealing conditions. Source: Yu et al. [106]/Reproduced with permission of Royal Society of Chemistry.

17.4.2.3 Cross-Linking

The idea behind crosslinking is to fix the nanoscale morphology of the active layer. Crosslinking has been explored in different ways including introducing cross linkable groups in the side-chains of donors and/or acceptors, and adding crosslinker in the BHJ blends [108–110]. The crosslinking can be initiated by external stresses like UV illumination and thermal annealing or by photo initiators [108]. Here we highlight some early and recent examples. In 2009, Fréchet and coworkers synthesized a series of bromine-functionalized poly(3-hexylthiophenes-2,5-diyl) copolymers (P3HT-Br) containing various amounts of UV crosslinkable brominated unit [111]. Solar cells based on the copolymer with 10% Br content showed comparable PCE with the reference cells without Br, but significantly improved thermal stability at 150 °C (Figure 17.6a). The crosslinkable polymer was also used

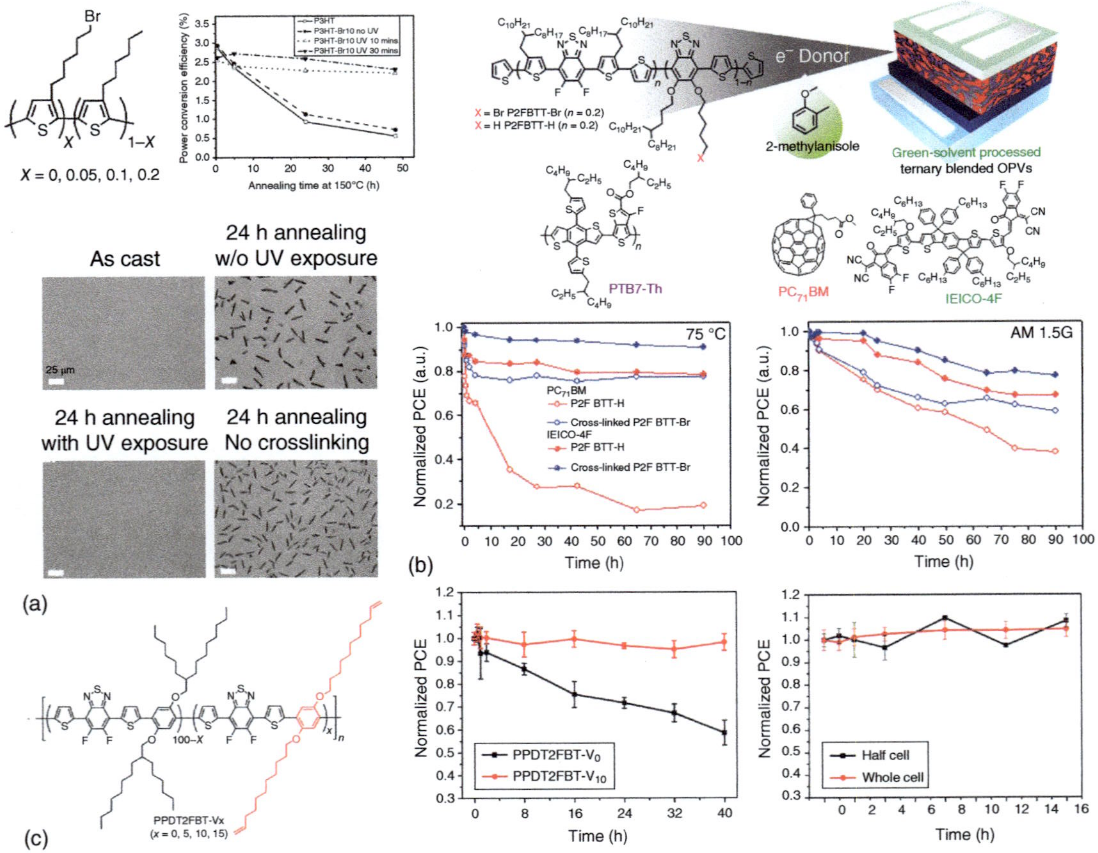

Figure 17.6 Stability enhancement using crosslinking. (a) Bromine-functionalized P3HT copolymers and its influence on PCE and optical microscopy images (scale bar = 25 μm) with annealing at 150 °C. Source: Kim et al. [111]/with permission of John Wiley and Sons. (b) Long-term stability of solar cells based on PTB7-Th:PCBM or PTB7-Th:IEICO-4F with P2FBTT-Br or P2FBTT-H as a third component. Source: Lee et al. [112]/Reproduced with permission of John Wiley & Sons. (c) Chemical structure of PPDT2FBT-Vx and the stability of solar cells based on PPDT2FBT-V10 after crosslinking by UV illumination. Middle: under thermal stress at 120 °C. Right: under air mass (AM) 1.5G spectrum and 100 mW cm^{-2} intensity. Source: Wu et al. [113]/Reproduced with permission of American Chemical Society.

in P3HT/PCBM bilayer devices with a PCE of 2.2% [111]. As a similar approach, in 2019, Park and coworkers synthesized P2FBTT-Br by random copolymerization using an asymmetric monomer with bromine-functionalized alkyl chains on one side and extended branching alkyl chains on the other (Figure 17.6b) [112]. A fast and highly efficient crosslinking can be triggered upon UV light irradiation through radical cleavage between bromine and carbon. This cross-linkable polymer was added into PTB7-Th:IEICO-4F or PTB7-Th:PCBM-based solar cells and effectively enhanced the device thermal and photo-stability (Figure 17.6b). It is noted that the UV-induced crosslinking might also cause photo-degradation of the photoactive materials. While the side-chains are used as crosslinking sites, the conjugated backbone has to be stable against UV irradiation [114]. For instance, several co-polymers (PPDT2FBT-Vx) based on PPDT2FBT and analogues with terminal vinyl-appended side chains for crosslinking showed excellent UV-stability (Figure 17.6c) [113]. The crosslinking concept has also been applied to double-cable single-component photoactive layers [115] and all-polymers solar cells [116] with significantly enhanced morphological stability.

17.4.2.4 Solid Additives

Many types of solid additives have been investigated to enhance the morphological stability of OPV materials. The mechanisms include formation of noncovalent chemical bonding (hydrogen and F—S bonding), compatibilizer with similar chemical structures [117], heat resistant insulating resins forming matrix to restrain the migration of molecular chain and freeze morphology [118] etc. Introducing solid additives which form hydrogen bonding between donor and PCBM can act as phase stabilizers. A classic example is 1,4-Piperazine (Figures 17.7a–d), which effectively suppresses demixing in P3HT:PCBM, PTB7-Th:PCBM, [119] and PCE11:PCBM systems [10]. Impressively, it is possible to rationally screen molecular phase stabilizers according to their interaction parameters/miscibility with the donor and acceptor materials [10]. In 2016, Zhan and coworkers demonstrated 4,4′-Biphenol (BPO), a common and cheap chemical, could be employed as a “molecular lock” in blends of fluorine-containing polymers or small molecule donors and fullerene acceptors to lock donors via hydrogen bond formed between the donor and BPO (Figure 17.7e) [120]. The molecular lock is a versatile strategy to enhance the efficiency and stability of OSCs simultaneously. PTB7-Th:ITIC systems with 10% coumarin 7 (C7) as a photoactive component also showed improved morphology stability due to the formed hydrogen bonding between C7 and ITIC [121].

17.4.2.5 Solvent Additives

Solvent additives are widely used for tuning nanoscale morphology of the BHJ blends, such as crystallinity, phase purity, and domain spacing, which are vital for both device performance and stability [122]. However, despite the positive effect on optimizing microstructure and the resulting morphological stability [89], the residual of high boiling point solvent additives in the active layer films was found to be detrimental for photochemical- [79, 123] and thermal-stability [124]. It is thus essential to completely remove the solvent additives after film formation. Several

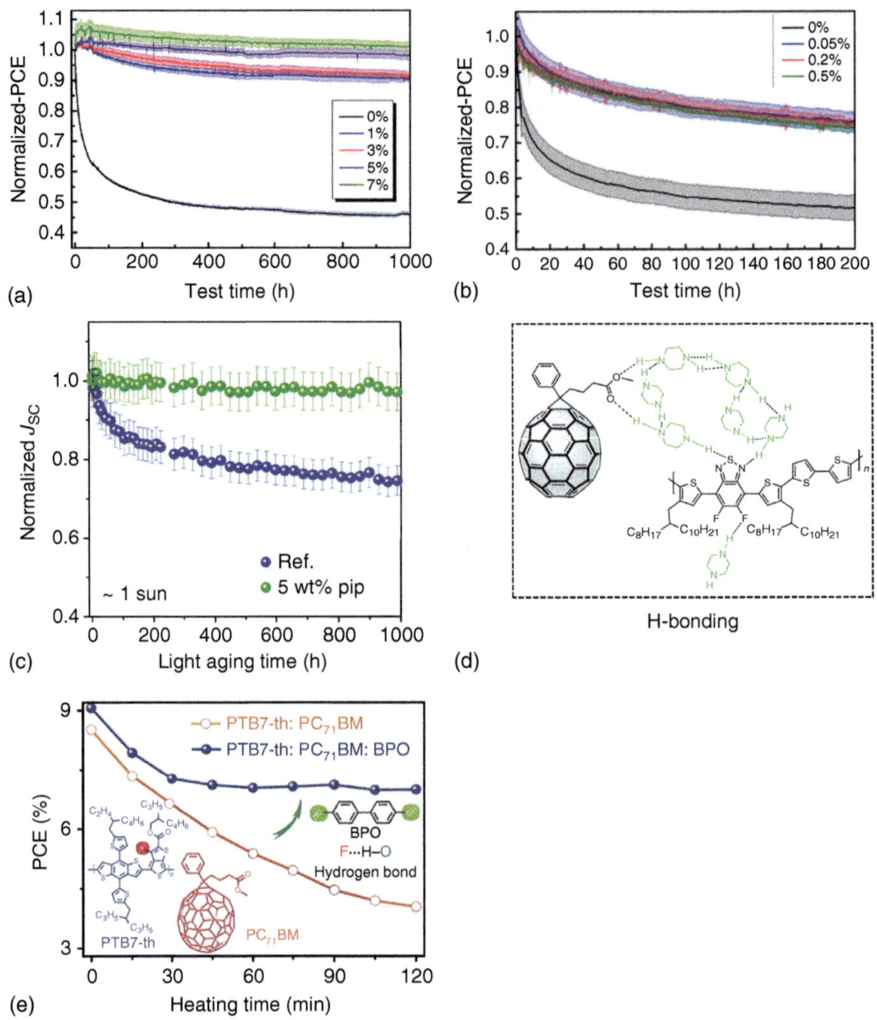

Figure 17.7 Stability of (a) P3HT:$PC_{61}BM$ and (b) PTB7-Th:$PC_{61}BM$ cells with different amount of 1,4-Piperazine in the active layer aged at maximum power point (MPP) with continuous light illumination. Source: Yan et al. [119]/Reproduced with permission of Royal Society of Chemistry. (c and d) J_{sc} losses as function of light aging time for PCE11:$PC_{71}BM$ systems with and without 1,4-Piperazine and the schematic diagram of hydrogen bonds between piperazine and PCE11/$PC_{71}BM$. Source: Zhang et al. [10]/Reproduced with permission of Royal Society of Chemistry. (e) BPO as a molecular lock in PTB7-Th:$PC_{71}BM$ based solar cells. PCE as a function of annealing time at 130 °C. Source: Cheng et al. [120]/Reproduced with permission of John Wiley & Sons.

methods such as vacuum treatment [125], solvent washing [126], and thermal annealing [79] have been explored. In this regard, solvent additives with low boiling point which can be easily removed from the active layers are favorable [125, 127]. Volatile solid additives which can tune the nanostructure morphology during film drying process and can be easily removed afterwards are also toward the same direction for improving both PCE and stability [128]. The elimination of using any solvent additives in many material systems also looks promising [129].

17.4.2.6 Ternary and Multiple Composites

The ternary strategy has been very successful in both FA and NFA systems to enhance the device performance [130–132]. The third and multicomponent composites in the host semiconductors might be able to increase the entropy and hence the operational stability [133]. Here we highlight some recent achievements by employing the ternary concept to simultaneously increase device performance and morphological stability in Figure 17.8. Kim and coworkers developed a new acceptor–donor–acceptor (A–D–A) triad-type small molecule, 5TRh-PCBM, consisting of an oligothiophene segment as the central core and fullerene derivatives as the end groups [117]. This small molecule as a third component in various material systems showed multiple beneficial effects and enhanced the thermal stability and mechanical properties of efficient OSCs while increasing their PCEs (Figure 17.8a,b). By introducing a small amount of fullerene ($PC_{71}BM$) into a narrow bandgap NFA-based BHJ nanocomposite (PTB7-Th:IEICO-4F), both efficiency and photo-stability were simultaneously enhanced (Figure 17.8c) [134]. Adding a small amount of polymer donor PZ1 significantly improved the thermal stability of PM6 and BTTT-2Cl, showing no degradation after annealing at 110 °C for 800 hours (Figure 17.8d) [135]. Using ((5Z,5′Z)-5,5′-((7,7′-(4,4,9,9-tetraoctyl-4,9-dihydro-s-indaceno[1,2-b:5,6-b′]dithiophene-2,7-diyl)bis(benzo[c][1,2,5]thiadiazole-7,4-diyl))bis(methanylylidene))bis(3-ethyl-2-thioxothiazolidin-4-one)) (O-IDTBR) as a third component, the stability of PM6:Y6-based solar cells could be enhanced under full solar spectrum illumination for 200 hours (Figure 17.8e) [136].

17.4.2.7 Organic Nanoparticles

The research on organic nanoparticles with embedded mixed donor and acceptor or separate donor/acceptor nanoparticles is largely driven by eco-friendly processing with water or alcohol-based solvents [137]. The synthesis routes include mini-emulsion technique [138, 139], surfactant-free precipitation method [140], or surfactant assisted nanoprecipitation [141]. It should be noted that the related approach also has the potential to freeze nanostructure and hence enhance morphological stability. Xie et al. demonstrated solar cells based on P3HT:IDTBR nanoparticles dispersed in water showing promising device efficiency and photo-stability [141]. It is noted that most of the research focused on synthesizing nanoparticles especially with stable dispersions as well as processing techniques to get the PCE on par with conventional aromatic solvent processed OSCs; investigation about microstructure stability is still limited and requires further research efforts.

17.4.2.8 Stratified Photoactive Layer Structure

From processing point of view, sequentially depositing donor and acceptor layers has the potential to better control the morphology than one-step coating of mixed donor and acceptor inks. In 2009, Schwartz and coworkers questioned the need

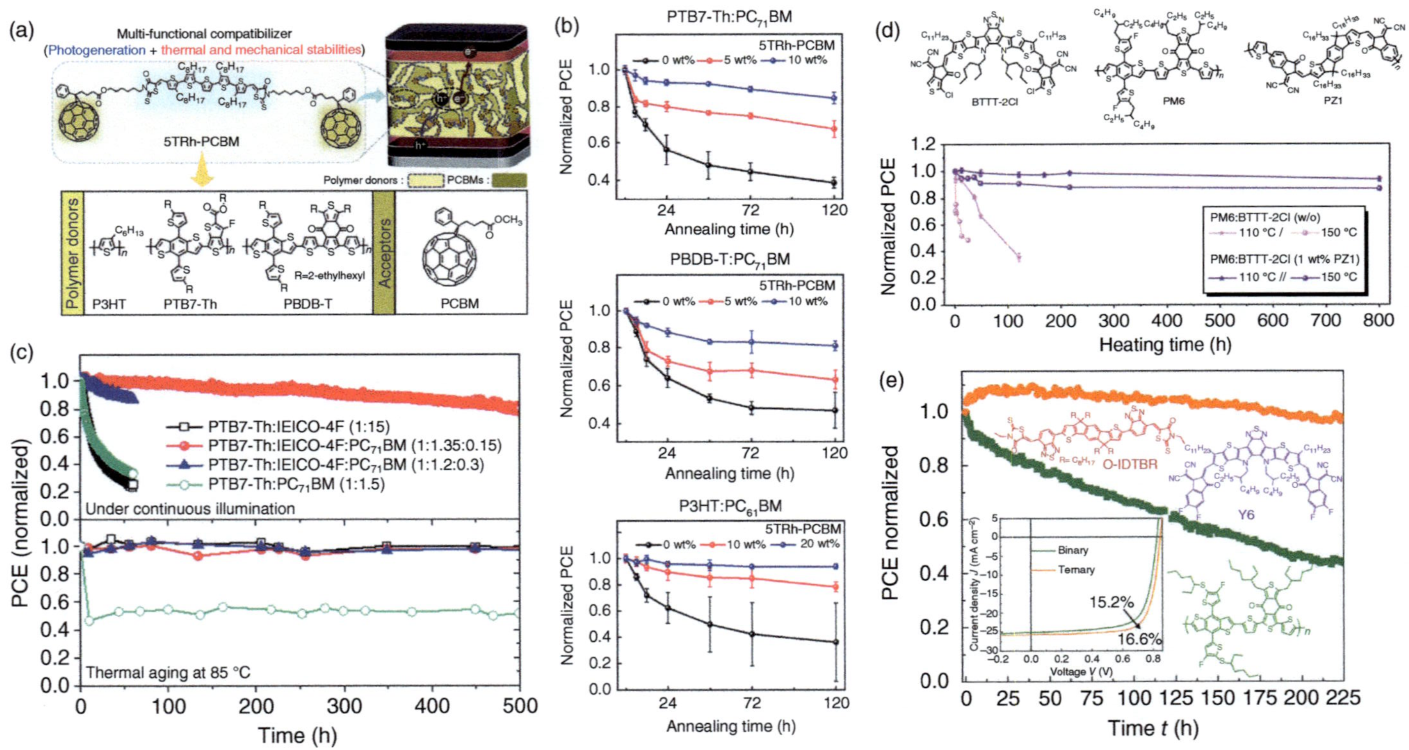

Figure 17.8 (a and b) The third component 5TRh-PCBM in PTB7-Th:$PC_{71}BM$ (aged at 80 °C), PBDB-T:$PC_{71}BM$ (aged at 80 °C) and P3HT:$PC_{61}BM$ (aged at 150 °C) with improved thermal stability. Source: Kim et al. [117]/Reproduced with permission of Royal Society of Chemistry. (c) Photo-stability (top) under continuous AM 1.5G illumination (100 mW cm^{-2}) in a nitrogen atmosphere and thermal stability (bottom) under 85 °C of PTB7-Th, IEICO-4F and PCBM based binary and ternary solar cells. Source: Lee et al. [134]/Reproduced with permission of Royal Society of Chemistry. (d) Normalized photovoltaic performance of PM6:BTTT-2Cl blends without and with 1 wt% PZ1 as a function of annealing time at 150 and 110 °C, respectively. Source: Yang et al. [135]/Springer Nature/CC BY 4.0. (e) PCE evolution of devices based on PM6:Y6 with or without O-IDTBR as a third component upon 1 sun illumination stress using a metal halide lamp after 225 hours. The inset depicts chemical structure of the materials and initial device J–V curves. Source: Gasparini et al. [136]/Reproduced with permission of American Chemical Society.

for BHJ in polymer-fullerene photovoltaics by showing that P3HT/PCBM bilayer sequentially processed from orthogonal solvents can have the similar performance [142]. Later in 2011, Burn and coworkers studied morphology of sequentially solution processed P3HT/PCBM layers and concluded that intermixed region formed initially and can be completed by thermal annealing. The origin for the initially intermixed domains is speculated to be the swelling of P3HT by PCBM solvent (dichloromethane) which allows PCBM interdiffusion without decreasing the P3HT crystalline domains [143]. In the past few years, several research groups have further investigated the sequential layer deposition to control morphology with successful results [144–152]. Among them, improved photo- and thermal stability compared to BHJ has been demonstrated. In 2019, Huang and coworkers demonstrated PffBT4T-2OD/PC_{71}BM-based bilayer devices with comparable PCE to BHJ devices and slightly improved photo-stability (Figure 17.9a) [145]. This bilayer structure outperformed the BHJ blend when processed with non-halogenated solvent. In 2019, Min and coworkers showed layer-by-layer (LbL) processed OSCs based on J71 and ITC6-IC outperformed BHJ control device in terms of light, thermal, and mechanical stability (Figure 17.9b) [146]. Of note is that this processing technique has also been applied successfully for fabrication of high-performance solar modules [153].

17.5 Recent Progress on Developing Organic Solar Cells with Excellent Stability

In 2011, McGehee and coworkers firstly reported polymer solar cells with seven years of T_{80} lifetime based on poly[[9-(1-octylnonyl)-9H-carbazole-2,7-diyl]-2,5-thiophenediyl-2,1,3-benzothiadiazole-4,7-diyl-2,5-thiophenediyl] (PCDTBT): PCBM with a PCE of ~5.5% (Figure 17.10a) [154]. Since 2017, the community evidenced several material systems with excellent device stability. Long-term stability under illumination is essential for solar cell operation. In many cases, UV part is filtered to mimic the employment of UV filters for real applications. Under UV-free simulated white light LED, P3HT:EH-IDTBR system with PCE of 6% survived 2000 hours illumination with only 5% losses (Figure 17.10b) [86]. For PBDB-T:ITIC-2F or ITIC-Th (PCE ~8%), devices were relatively stable within 1700 hours illumination after 10% burn-in losses in the first hundreds hours [88]. A linear extrapolation of stability data showed a T_{80} lifetime over 11 000 hours for PBDB-T:ITIC-2F (Figure 17.10c), which would lead to an impressive operational lifetime over 10 years considering an average of 3.33 sun hours per day for south Germany according to the National Renewable Energy Laboratory (NREL) database. With C_{60}-SAM modified ZnO as electron transport layer (ETL), the extrapolated T_{80} lifetime of inverted solar cells based on PTB7-Th:IEICO-4F operating under one sun illumination equivalent reached 34 000 hours (Figure 17.10d) [155]. Based on an estimation of solar irradiance of 1500 $kWh\,m^{-2}\,yr^{-1}$ for China, this would be equivalent to a potential of 22 years. For PffBT4T-2OD:EH-IDTBR with a PCE of 9.5%, no degradation was

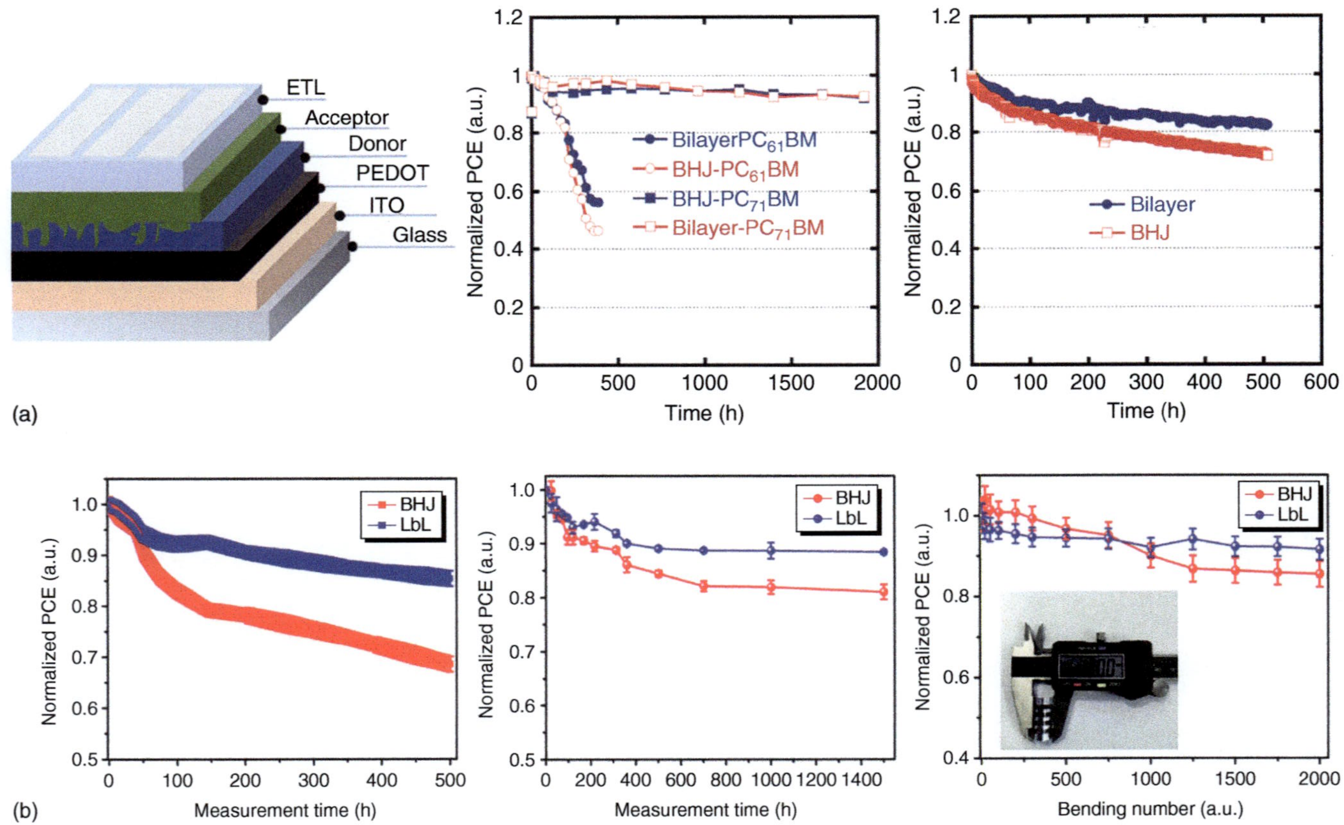

Figure 17.9 (a) Stability of bilayer and BHJ devices based on PffBT4T-2OD with fullerene derivatives. Left: device structure. Middle: continuously baked at 85 °C in dark. Right: continuously illuminated under sunlight simulator with intensity of 100 mW cm^{-2} at room temperature (25 °C) and relative humidity of 70%. Source: Dong et al. [145]/Reproduced with permission of John Wiley & Sons. (b) Stability of layer-by-layer (LbL) processed OSCs based on J71 and ITC6-IC compared to BHJ structures. Left: Variation in the normalized average PCE losses over illumination time over 500 hours measured in a dry nitrogen atmosphere. Middle: Variation of normalized PCE of the BHJ and LbL devices annealed at 120 °C over 1500 hours. Right: Bending test of the flexible BHJ- and LbL-based OSCs. The inset shows a photo of the bending instrument with a radius of 6 mm. Source: Sun et al. [146]/with permission of Royal Society of Chemistry.

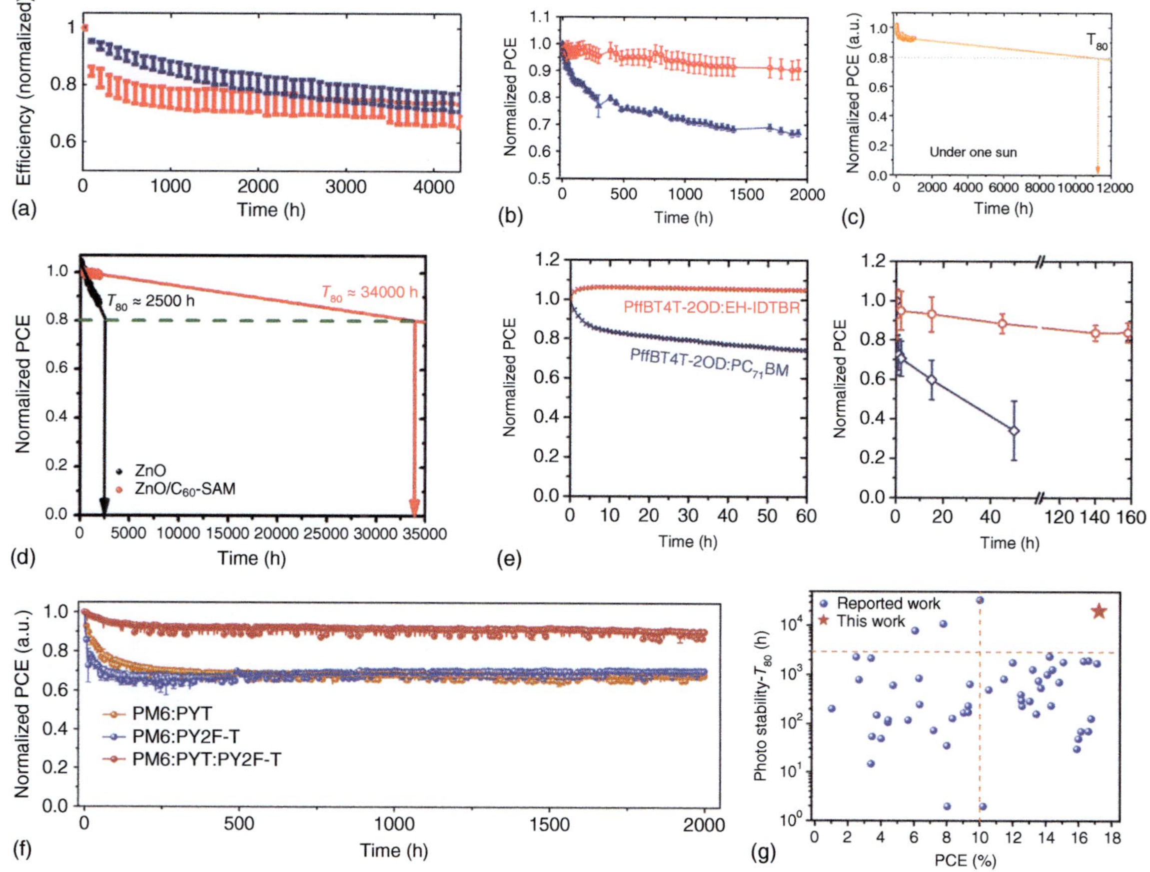

Figure 17.10 Reported organic solar cells with excellent stability. (a) Normalized PCE for PCDTBT (red) and P3HT (blue) solar cells over 4400 hours of continuous illumination at MPP tracking. Source: Peters et al. [154]/Reproduced with permission of John Wiley & Sons. (b) Stability of solar cells based on P3HT:PCBM (triangle symbol in blue) and P3HT:EH-IDTBR (sphere symbol in red) under illumination. Source: Gasparini et al. [86]/Reproduced with permission of John Wiley & Sons. (c) Linear extrapolation of lifetime for PBDB-T:ITIC-2F with T_{80} lifetime reaching 11 000 hours under equivalent one sun white LED illumination. Source: Du et al. [88]/Reproduced with permission of Elsevier. (d) Linear extrapolation of lifetime for PTB7-Th:IEICO-4F based inverted solar cells with or without C_{60}-SAM modification on ZnO with T_{80} lifetime reaching 34 000 hours under equivalent one sun illumination. Source: Xu et al. [155]/Reproduced with permission of Elsevier. (e) Normalized PCEs of the devices based on PffBT4T-2OD:EH-IDTBR or $PC_{71}BM$ under (left) light soaking without UV light maintained below 50 °C thermal stress or (right) annealing at 85 °C under nitrogen. Source: Cha et al. [85]/Reproduced with permission of John Wiley & Sons. (f) Light stability of solar cells based on PM6:PYT, PM6:PY2F-T and PM6:PYT:PY2F-T as the active layer and (g) the comparison of literature reported T_{80} and PCE. Source: Sun et al. [156]/Reproduced with permission of Elsevier.

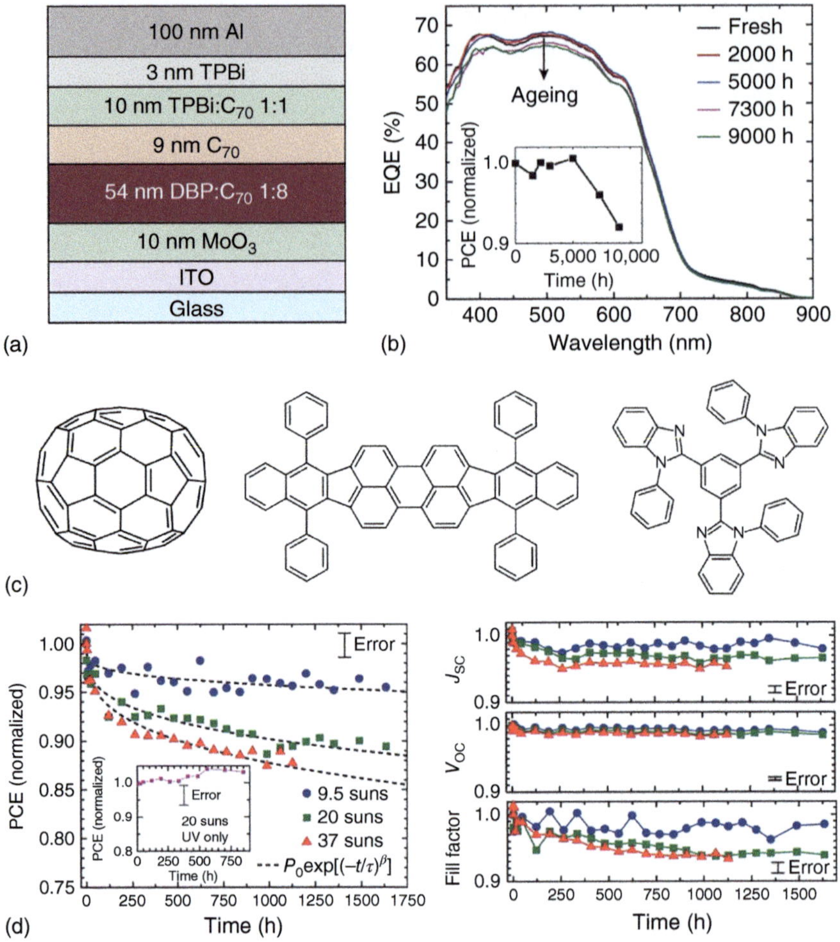

Figure 17.11 (a) Schematic of the device structure. (b) External quantum efficiency (EQE) spectra of an OPV before (fresh) and during aging under 1 kW m^{-2} of simulated AM1.5G illumination for more than one year (9000 hours). Inset, normalized PCE plotted against time. (c) Chemical structural of the molecules of C_{70}, DBP, and TPBi. (d) Normalized PCE, J_{sc}, V_{oc}, and fill factor plotted against aging time under illumination equivalent to 9.5 ± 1.4, 20 ± 3, and 37 ± 5.5 Suns. Stretched exponential fits to each PCE against time dataset are shown by dashed lines. The inset shows PCE plotted against time for an OPV aged under ultraviolet illumination equivalent to 20 suns. The error bars indicate the 1 s.d. uncertainty of each measurement. UV, ultraviolet. Source: Burlingame et al. [157]/Reproduced with permission of Springer Nature.

observed under 60 hours white LED illumination (Figure 17.10e) [85]. In 2021, Min and coworkers reported all polymer solar cells based on PM6:poly[(2,2′-((2Z,2′Z)-((12,13-bis(2-octyldodecyl)-3,9-diundecyl-12,13-dihydro[1,2,5]thiadiazolo [3,4e]thieno[2″,3″:4′,5′]thieno[2′,3′:4,5]pyrrolo[3,2-g]thieno[2′,3′:4,5]thieno[3,2-b]-indole-2,10-diyl)bis (methanylylidene))bis (3-oxo-2,3-dihydro-1H-indene-2,1-diylidene))dimalononitrile-alt-2,5-thiophene)] (PYT):PY2F-T achieved a PCE of

17.2% and an extrapolated T_{80} lifetime of 20 500 hours under white light soaking (Figure 17.10f–g) [156], indicating the bright future for OPV applications.

In 2019, Forrest group demonstrated extremely stable evaporated solar cells based on tetraphenyldibenzoperiflanthene (DBP):C_{70} with a PCE of 6% (Figure 17.11) [157]. Impressively, the devices measured under 20 suns of UV illumination showed no losses with dose equivalent to 9.3 years when accumulated outdoors, demonstrating that OPV materials could be super-photostable. The accelerated aging test under 37 suns (solar cells surface temperature went to 100 °C) even showed an extrapolated T_{80} operational lifetime extending to more than 27 000 years. Although being different with solution-processed OSCs in terms of material purity, crystallinity, and morphology, this work is an outstanding demonstration of intrinsic stability of the OSCs being reliable for long-term application [157].

17.6 Summary and Outlook

In this chapter, we reviewed the recent progress on developing stable OSCs based on fullerene derivatives and NFAs. The efficiency of OSCs has reached an impressively high level, which surpasses the benchmark value for commercialization and industrial applications. However, the operational lifetime of mature PV technologies based on Si and III-V semiconductors is generally more than 25 years, which is much longer than that of state-of-the-art OPV devices. Nevertheless, it is not required for OSCs to reach such a lifetime value, as they are anyway targeting at applications in different scenarios, such as building- and vehicle-integrated applications, portable flexible energy suppliers, etc. Therefore, it is generally accepted that OSCs should demonstrate at least 10 years lifetime in order to meet the requirements for the real-world applications. This is still very challenging for state-of-the-art high-performing OSCs, as the stability of OSCs is strongly influenced by many extrinsic and intrinsic factors, as discussed in this chapter. Although the extrinsic stresses could be minimized by performing excellent device encapsulation to improve the device operational stability, the intrinsic factors such as material stability and microstructure stability do play a decisive role in determining the operational lifetime of OSCs. It has to be mentioned that most OSCs that exhibit promising device stability are generally fabricated and evaluated in N_2 atmosphere, which is much different from the PV products operating in the real world. For large-scale roll-to-roll production, OSCs are processed using green solvents, in a fully printable device architecture (with printed interfacial layers and electrodes) and under ambient conditions. The achieved microstructure, interface properties as well as defect density are expected to be different from that of optimized small-scale devices reported in the literature. Therefore, intensive research efforts are required in order to continuously develop OPV products with promising device efficiency and stability. It is hoped that this chapter could trigger more research interests and efforts to successfully address the stability issues of OPV devices, not only for lab-scale small area devices but also for large-scale flexible organic solar modules.

References

1 Brabec, C.J., Distler, A., Du, X. et al. (2020). *Adv. Energy Mater.* 10: 2001864.
2 Li, N. and Brabec, C.J. (2017). *Nat. Energy* 2: 772.
3 Liu, Q., Jiang, Y., Jin, K. et al. (2020). *Sci. Bull.* 65: 272.
4 Li, S., Li, C.-Z., Shi, M., and Chen, H. (2020). *ACS Energy Lett.* 5: 1554.
5 Li, N., McCulloch, I., and Brabec, C.J. (2018). *Energy Environ. Sci.* 11: 1355.
6 Yamilova, O.R., Martynov, I.V., Brandvold, A.S. et al. (2020). *Adv. Energy Mater.* 10: 1903163.
7 Qin, M., Cheng, P., Mai, J. et al. (2017). *Solar RRL.* 1: 1700148.
8 Bartesaghi, D., Ye, G., Chiechi, R.C., and Koster, L.J.A. (2016). *Adv. Energy Mater.* 6: 1502338.
9 Lee, S., Kong, J., and Lee, K. (2016). *Adv. Energy Mater.* 6: 1600970.
10 Zhang, C., Heumueller, T., Leon, S. et al. (2019). *Energy Environ. Sci.* 12: 1078.
11 Deschler, F., De Sio, A., von Hauff, E. et al. (2012). *Adv. Funct. Mater.* 22: 1461.
12 Aygül, U., Hintz, H., Egelhaaf, H.-J. et al. (2013). *J. Phys. Chem. C* 117: 4992.
13 Hoke, E.T., Vandewal, K., Bartelt, J.A. et al. (2013). *Adv. Energy Mater.* 3: 220.
14 Speller, E.M., Clarke, A.J., Aristidou, N. et al. (2019). *ACS Energy Lett.* 4: 846.
15 Clarke, T.M., Lungenschmied, C., Peet, J. et al. (2013). *Adv. Energy Mater.* 3: 1473.
16 Min, J., Luponosov, Y.N., Cui, C. et al. (2017). *Adv. Energy Mater.* 7: 1700465.
17 Burlingame, Q., Tong, X., Hankett, J. et al. (2015). *Energy Environ. Sci.* 8: 1005.
18 Heumueller, T., Mateker, W.R., Distler, A. et al. (2016). *Energy Environ. Sci.* 9: 247.
19 Chander, N., Singh, S., and Iyer, S.S.K. (2017). *Sol. Energy Mater. Sol. Cells* 161: 407.
20 Chang, L., Lademann, H.W.A., Bonekamp, J.-B. et al. (2011). *Adv. Funct. Mater.* 21: 1779.
21 Wang, W., Sun, R., Guo, J., and Min, J. (2019). *Angew. Chem. Int. Ed.* 58: 14556.
22 Min, J., Cui, C., Heumueller, T. et al. (2016). *Adv. Energy Mater.* 6: 1600515.
23 Hu, H., Ghasemi, M., Peng, Z. et al. (2020). *Adv. Mater.* 32: 2005348.
24 Li, N., Perea, J.D., Kassar, T. et al. (2017). *Nat. Commun.* 8: 14541.
25 Sun, R., Deng, D., Guo, J. et al. (2019). *Energy Environ. Sci.* 12: 2518.
26 Schaffer, C.J., Palumbiny, C.M., Niedermeier, M.A. et al. (2013). *Adv. Mater.* 25: 6760.
27 Domínguez, I.F., Distler, A., and Lüer, L. (2016). *Adv. Energy Mater.* 7: 1601320.
28 Wang, W., Chen, B., Jiao, X. et al. (2019). *Org. Electron.* 70: 78.
29 Sachs-Quintana, I.T., Heumüller, T., Mateker, W.R. et al. (2014). *Adv. Funct. Mater.* 24: 3978.
30 Min, J., Luponosov, Y.N., Zhang, Z.-G. et al. (2014). *Adv. Energy Mater.* 4: 1400816.
31 Di Carlo Rasi, D. and Janssen, R.A.J. (2018). *Adv. Mater.* 31: 1806499.
32 Adams, J., Salvador, M., Lucera, L. et al. (2015). *Adv. Energy Mater.* 5: 1501065.
33 Cheng, P. and Zhan, X. (2016). *Chem. Soc. Rev.* 45: 2544.
34 Jorgensen, M., Norrman, K., Gevorgyan, S.A. et al. (2012). *Adv. Mater.* 24: 580.

35 Roesch, R., Faber, T., von Hauff, E. et al. (2015). *Adv. Energy Mater.* 5: 1501407.
36 Hoke, E.T., Sachs-Quintana, I.T., Lloyd, M.T. et al. (2012). *Adv. Energy Mater.* 2: 1351.
37 Guo, J., Wu, Y., Sun, R. et al. (2019). *J. Mater. Chem. A* 7: 25088.
38 Wang, F., Tan, Z.A., and Li, Y. (2015). *Energy Environ. Sci.* 8: 1059.
39 Jørgensen, M., Norrman, K., and Krebs, F.C. (2008). *Sol. Energy Mater. Sol. Cells* 92: 686.
40 Glatthaar, M., Riede, M., Keegan, N. et al. (2007). *Sol. Energy Mater. Sol. Cells* 91: 390.
41 Glen, T.S., Scarratt, N.W., Yi, H. et al. (2015). *Sol. Energy Mater. Sol. Cells* 140: 25.
42 Sapkota, S.B., Spies, A., Zimmermann, B. et al. (2014). *Sol. Energy Mater. Sol. Cells* 130: 144.
43 Balderrama, V.S., Estrada, M., Han, P.L. et al. (2014). *Sol. Energy Mater. Sol. Cells* 125: 155.
44 Krebs, F.C., Hösel, M., Corazza, M. et al. (2013). *Energy Technol.* 1: 378.
45 Guo, J. and Min, J. (2018). *Adv. Energy Mater.* 9: 1802521.
46 Berny, S., Blouin, N., Distler, A. et al. (2016). *Adv. Sci.* 3: 1500342.
47 Bruner, C., Novoa, F., Dupont, S., and Dauskardt, R. (2014). *ACS Appl. Mater. Interfaces* 6: 21474.
48 Tummala, N.R., Bruner, C., Risko, C. et al. (2015). *ACS Appl. Mater. Interfaces* 7: 9957.
49 Balcaen, V., Rolston, N., Dupont, S.R. et al. (2015). *Sol. Energy Mater. Sol. Cells* 143: 418.
50 Na, S.-I., Kim, S.-S., Jo, J., and Kim, D.-Y. (2008). *Adv. Mater.* 20: 4061.
51 Hösel, M., Søndergaard, R.R., Jørgensen, M., and Krebs, F.C. (2014). *Adv. Energy Mater.* 4: 1301625.
52 Juillard, S., Planes, E., Matheron, M. et al. (2018). *ACS Appl. Mater. Interfaces* 10: 29805.
53 Zhang, C., Heumueller, T., Gruber, W. et al. (2019). *ACS Appl. Mater. Interfaces* 11: 18555.
54 Liu, Y., Zhao, J., Li, Z. et al. (2014). *Nat. Commun.* 5: 5293.
55 Ye, L., Collins, B.A., Jiao, X. et al. (2018). *Adv. Energy Mater.* 8: 1703058.
56 Rösch, R., Tanenbaum, D.M., Jørgensen, M. et al. (2012). *Energy Environ. Sci.* 5: 6521.
57 Sharma, A., Andersson, G., and Lewis, D.A. (2011). *Phys. Chem. Chem. Phys.* 13: 4381.
58 Voroshazi, E., Uytterhoeven, G., Cnops, K. et al. (2014). *ACS Appl. Mater. Interfaces* 7: 618.
59 Kong, J., Song, S., Yoo, M. et al. (2014). *Nat. Commun.* 5: 5688.
60 Voroshazi, E., Cardinaletti, I., Conard, T., and Rand, B.P. (2014). *Adv. Energy Mater.* 4: 1400848.
61 Mateker, W.R., Douglas, J.D., Cabanetos, C. et al. (2013). *Energy Environ. Sci.* 6: 2529.

62 Fungura, F., Lindemann, W.R., Shinar, J., and Shinar, R. (2017). *Adv. Energy Mater.* 7: 1601420.
63 Wang, Y., Lee, J., Hou, X. et al. (2021). *Adv. Energy Mater.* 11: 2003002.
64 Ghasemi, M., Hu, H., Peng, Z. et al. (2019). *Joule* 3: 1328.
65 Reese, M.O., Gevorgyan, S.A., Jørgensen, M. et al. (2011). *Sol. Energy Mater. Sol. Cells* 95: 1253.
66 Khenkin, M.V., Katz, E.A., Abate, A. et al. (2020). *Nat. Energy* 5: 35.
67 Christians, J.A., Habisreutinger, S.N., Berry, J.J., and Luther, J.M. (2018). *ACS Energy Lett.* 3: 2136.
68 Yao, H., Ye, L., Zhang, H. et al. (2016). *Chem. Rev.* 116: 7397.
69 Zhang, G., Zhao, J., Chow, P.C.Y. et al. (2018). *Chem. Rev.* 118: 3447.
70 Liang, Y., Wu, Y., Feng, D. et al. (2009). *J. Am. Chem. Soc.* 131: 56.
71 He, Z., Xiao, B., Liu, F. et al. (2015). *Nat. Photonics* 9: 174.
72 Lin, Y., Wang, J., Zhang, Z.-G. et al. (2015). *Adv. Mater.* 27: 1170.
73 Cui, Y., Yao, H., Hong, L. et al. (2019). *Adv. Mater.* 31: 1808356.
74 Jiang, K., Wei, Q., Lai, J.Y.L. et al. (2019). *Joule* 3: 3020.
75 Yuan, J., Zhang, Y., Zhou, L. et al. (2019). *Joule* 3: 1140.
76 Kim, T., Younts, R., Lee, W. et al. (2017). *J. Mater. Chem. A* 5: 22170.
77 Dkhil, S.B., Pfannmöller, M., Saba, M.I. et al. (2016). *Adv. Energy Mater.* 6: 1601486.
78 Doumon, N.Y., Wang, G., Qiu, X. et al. (2019). *Sci. Rep.* 9: 4350.
79 Classen, A., Heumueller, T., Wabra, I. et al. (2019). *Adv. Energy Mater.* 9: 1902124.
80 Cheacharoen, R., Mateker, W.R., Zhang, Q. et al. (2017). *Sol. Energy Mater. Sol. Cells* 161: 368.
81 Li, Z., Wu, F., Lv, H. et al. (2015). *Adv. Mater.* 27: 6999.
82 Zhang, C., Mumyatov, A., Langner, S. et al. (2016). *Adv. Energy Mater.* 7: 1601204.
83 Saito, M. and Osaka, I. (2018). *J. Mater. Chem. C* 6: 3668.
84 Hsieh, Y.-J., Huang, Y.-C., Liu, W.-S. et al. (2017). *ACS Appl. Mater. Interfaces* 9: 14808.
85 Cha, H., Wu, J., Wadsworth, A. et al. (2017). *Adv. Mater.* 29: 1701156.
86 Gasparini, N., Salvador, M., Strohm, S. et al. (2017). *Adv. Energy Mater.* 7: 1700770.
87 Ye, P., Chen, Y., Wu, J. et al. (2019). *Mater. Chem. Front.* 3: 64.
88 Du, X., Heumueller, T., Gruber, W. et al. (2019). *Joule* 3: 215.
89 Du, X., Heumueller, T., Gruber, W. et al. (2020). *Adv. Mater.* 32: 1908305.
90 Cho, H.W., An, N.G., Park, S.Y. et al. (2020). *Adv. Energy Mater.* 10: 1903585.
91 Lee, C., Lee, S., Kim, G.-U. et al. (2019). *Chem. Rev.* 119: 8028.
92 Kim, T., Choi, J., Kim, H.J. et al. (2017). *Macromolecules* 50: 6861.
93 Xu, Y., Yuan, J., Zhou, S. et al. (2019). *Adv. Funct. Mater.* 29: 1806747.
94 Kim, T., Kim, J.-H., Kang, T.E. et al. (2015). *Nat. Commun.* 6: 8547.
95 Duan, L. and Uddin, A. (2020). *Adv. Sci.* 7: 1903259.
96 Wang, K., Li, Y., and Li, Y. (2020). *Macromol. Rapid Commun.* 41: 1900437.
97 Speller, E.M., Clarke, A.J., Luke, J. et al. (2019). *J. Mater. Chem. A* 7: 23361.

98 Collins, B.A., Tumbleston, J.R., and Ade, H. (2011). *J. Phys. Chem. Lett.* 2: 3135.

99 Heumueller, T., Mateker, W.R., Sachs-Quintana, I.T. et al. (2014). *Energy Environ. Sci.* 7: 2974.

100 Lin, Y., Lim, J.A., Wei, Q. et al. (2012). *Chem. Mater.* 24: 622.

101 Li, F., Yager, K.G., Dawson, N.M. et al. (2013). *Macromolecules* 46: 9021.

102 Liu, X., Wang, X., Xiao, Y. et al. (2020). *Adv. Energy Mater.* 10: 1903650.

103 Izawa, S., Hashimoto, K., and Tajima, K. (2012). *Phys. Chem. Chem. Phys.* 14: 16138.

104 Nguyen, T.L., Lee, T.H., Gautam, B. et al. (2017). *Adv. Funct. Mater.* 27: 1702474.

105 He, Y., Heumüller, T., Lai, W. et al. (2019). *Adv. Energy Mater.* 9: 1900409.

106 Yu, P., Feng, G., Li, J. et al. (2020). *J. Mater. Chem. C* 8: 2790.

107 Lai, W., Li, C., Zhang, J. et al. (2017). *Chem. Mater.* 29: 7073.

108 Wantz, G., Derue, L., Dautel, O. et al. (2014). *Polym. Int.* 63: 1346.

109 Rumer, J.W. and McCulloch, I. (2015). *Mater. Today* 18: 425.

110 Kahle, F.-J., Saller, C., Köhler, A., and Strohriegl, P. (2017). *Adv. Energy Mater.* 7: 1700306.

111 Kim, B.J., Miyamoto, Y., Ma, B., and Fréchet, J.M.J. (2009). *Adv. Funct. Mater.* 19: 2273.

112 Lee, J., Kim, J.W., Park, S.A. et al. (2019). *Adv. Energy Mater.* 9: 1901829.

113 Wu, S.-C., Strover, L.T., Yao, X. et al. (2018). *ACS Appl. Mater. Interfaces* 10: 35430.

114 Yau, C.P., Wang, S., Treat, N.D. et al. (2015). *Adv. Energy Mater.* 5: 1401228.

115 Lanzi, M., Salatelli, E., Marinelli, M., and Pierini, F. (2020). *Macromol. Chem. Phys.* 221: 1900433.

116 Lin, Z., Zhang, L., Tu, S. et al. (2020). *Sol. Energy* 201: 489.

117 Kim, G.-U., Lee, Y.W., Ma, B.S. et al. (2020). *J. Mater. Chem. A* 8: 13522.

118 Han, J., Bao, F., Huang, D. et al. (2020). *Adv. Funct. Mater.* 30: 2003654.

119 Yan, L., Wang, Y., Wei, J. et al. (2019). *J. Mater. Chem. A* 7: 7099.

120 Cheng, P., Yan, C., Lau, T.K. et al. (2016). *Adv. Mater.* 28: 5822.

121 Kong, X., Lin, H., Du, X. et al. (2018). *J. Mater. Chem. C* 6: 9691.

122 Liao, H.-C., Ho, C.-C., Chang, C.-Y. et al. (2013). *Mater. Today* 16: 326.

123 Tremolet de Villers, B.J., O'Hara, K.A., Ostrowski, D.P. et al. (2016). *Chem. Mater.* 28: 876.

124 Dkhil, S.B., Pfannmöller, M., Saba, M.I. et al. (2017). *Adv. Energy Mater.* 7: 1601486.

125 Holliday, S. and Luscombe, C.K. (2018). *Adv. Electron. Mater.* 4: 1700416.

126 Li, N. and Brabec, C.J. (2015). *Energy Environ. Sci.* 8: 2902.

127 Yin, J., Zhou, W., Zhang, L., et al. (2017). Macromol. Rapid Commun. 38: 1700428.

128 Yu, R., Yao, H., Hong, L. et al. (2018). *Nat. Commun.* 9: 4645.

129 Zhang, X., Tang, Y., Yang, K. et al. (2019). *ChemElectroChem* 6: 5547.

130 Gasparini, N., Jiao, X., Heumueller, T. et al. (2016). *Nat. Energy* 1: 16118.

131 Bi, P. and Hao, X. (2019). *Sol. RRL.* 3: 1800263.

132 Xu, W. and Gao, F. (2018). *Mater. Horiz.* 5: 206.

133 de Zerio, A.D. and Müller, C. (2018). *Adv. Energy Mater.* 8: 1702741.
134 Lee, J., Lee, J.-H., Yao, H. et al. (2020). *J. Mater. Chem. A* 8: 6682.
135 Yang, W., Luo, Z., Sun, R. et al. (2020). *Nat. Commun.* 11: 1218.
136 Gasparini, N., Paleti, S.H.K., Bertrandie, J. et al. (2020). *ACS Energy Lett.* 5: 1371.
137 Rammal, M., Lévêque, P., Schlatter, G. et al. (2020). *Mater. Chem. Front.* 4: 2904.
138 Gonsalves, K.E., Merhari, L., Wu, H., and Hu, Y. (2001). *Adv. Mater.* 13: 703.
139 D'Olieslaeger, L., Pirotte, G., Cardinaletti, I. et al. (2017). *Org. Electron.* 42: 42.
140 Gärtner, S., Christmann, M., Sankaran, S. et al. (2014). *Adv. Mater.* 26: 6653.
141 Xie, C., Heumuller, T., Gruber, W. et al. (2018). *Nat. Commun.* 9: 5335.
142 Ayzner, A.L., Tassone, C.J., Tolbert, S.H., and Schwartz, B.J. (2009). *J. Phys. Chem. C* 113: 20050.
143 Lee, K.H., Schwenn, P.E., Smith, A.R.G. et al. (2011). *Adv. Mater.* 23: 766.
144 Cui, Y., Zhang, S., Liang, N. et al. (2018). *Adv. Mater.* 30: 1802499.
145 Dong, S., Zhang, K., Xie, B. et al. (2018). *Adv. Energy Mater.* 9: 1802832.
146 Sun, R., Guo, J., Wu, Q. et al. (2019). *Energy Environ. Sci.* 12: 3118.
147 Arunagiri, L., Zhang, G., Hu, H. et al. (2019). *Adv. Funct. Mater.* 29: 1902478.
148 Zhang, J., Futscher, M.H., Lami, V. et al. (2019). *Adv. Energy Mater.* 9: 1902145.
149 Xu, Y., Yuan, J., Liang, S. et al. (2019). *ACS Energy Lett.* 4: 2277.
150 Zhang, X., Li, Y., Zhang, D. et al. (2021). *Sci. China Chem.* 64: 116.
151 Liu, S., Chen, D., Hu, X. et al. (2020). *Adv. Funct. Mater.* 30: 2003223.
152 Lee, T.H., Park, S.Y., Park, W.-W. et al. (2020). *ACS Energy Lett.* 5: 1628.
153 Sun, R., Wu, Q., Guo, J. et al. (2020). *Joule* 4: 407.
154 Peters, C.H., Sachs-Quintana, I.T., Kastrop, J.P. et al. (2011). *Adv. Energy Mater.* 1: 491.
155 Xu, X., Xiao, J., Zhang, G. et al. (2019). *Sci. Bull.* 65: 208.
156 Sun, R., Wang, W., Yu, H. et al. (2021). *Joule* 5: 1548.
157 Burlingame, Q., Huang, X., Liu, X. et al. (2019). *Nature* 573: 394.

18

Potential Applications of Organic Solar Cells

Chengyi Xiao and Weiwei Li

Beijing University of Chemical Technology, Beijing Advanced Innovation Center for Soft Matter Science and Engineering, State Key Laboratory of Organic-Inorganic Composites, North Third Ring East Road 15, Beijing, 100029, P.R. China

18.1 Introduction

Solar energy is regarded as promising renewable energy in the future and is expected to occupy an important place in future energy systems [1]. In its latest available World Energy Outlook, the *International Energy Agency* predicted that "*the increasing competitiveness of solar photovoltaic pushes its installed capacity beyond that of wind before 2025, past hydropower around 2030 and past coal before 2040*" [2]. Organic solar cells (OSCs), with the ability to directly converting sunlight to electricity, are regarded as a promising pathway to solve energy shocks and global warming [3]. In recent years, OSCs have made tremendous development with power conversion efficiencies (PCEs) beyond 17% by developing conjugated materials as electron donor/acceptor materials [4–10], controlling microphase separation in bulk-heterojunction (BHJ) photoactive layers [11, 12], and selecting suitable inorganic/organic materials as the interfacial layer to realize efficient charge transportation [13–15]. In particular, Lin et al. reported the fused-ring non-fullerene acceptor 3,9-bis(2-methylene-(3(1,1-dicyanomethylene) indanone))-5,5,11,11-tetrakis(4-hexylphenyl)-dithieno[2,3-*d*:2′,3′-*d*′]-s-indaceno [1,2-*b*:5,6-*b*′]dithiophene (ITIC) in 2015 that showed a PCE of 6.8% [16], and in 2019, Yuan et al. developed the non-fullerene electron acceptors Y6 that boosted PCEs up to 15.7% [17]. Since then, a record PCE of 18.22% by Liu et al. was rapidly realized in the state-of-the-art OSCs [4], making non-fullerene acceptors-based solar cells the most promising OSCs [4, 8, 9, 18–25].

Compared with the inorganic semiconductors, the versatility to design and synthesize of organic semiconductors enables a large-range tunability in chemical structures, absorption, and frontier orbital levels [26]. More importantly, organic semiconductors have the advantages of inherent flexibility and stretchability compared with the rigid inorganic counterparts [27–29]. Thus, flexible OSCs are lighter and have the compatibility of different surfaces, which make them have a

Organic Solar Cells: Materials Design, Technology and Commercialization, First Edition. Edited by Liming Ding.

much broader scope of applications [30–32]. Such as researchers of *Merck Chemicals* represented a >250 m^2 flexible power system reaching an average performance of ~4% PCE named "solar trees" [33]. In the viewpoint of manufacturing technologies, most of the organic semiconductors can be processed by solution-based roll-to-roll (R2R) coating or printing technologies, such as slot-die coating, spray coating, blade coating, inkjet printing, and screen printing, etc. which make the manufacturing more inexpensive, energy saving, and higher throughput [34, 35]. Most recently, researchers have demonstrated that OSC modules possessed operational stability for two years through monitoring of a pilot OSC (organic photovoltaics) solar park [36]. These advantages allow OSCs to be applied in many low-energy-consumption and maintenance-free scenarios. Considering their superiority, OSCs have been recognized as energy-harvesting systems for third-generation electronics.

Accompanied by the rapidly increased PCEs and established coating technique, the OSC modules are believed to be very close to commercialization. The application of OSCs will be a great complement to the existing photovoltaic market and expand the border of the photovoltaics industry. As we mentioned above, the performance and stability of the OSCs have been raised to a sufficient height through materials updates and processing optimizations. However, different targeted applications would result in distinct demands for performance, stability, and mechanical parameters, etc. For example, for a skin-wearable application, devices should possess a stretchability of at least 30% to be conformable with human skin [37]. Thus, one should consider a specific targeted application to make the relevant parameters straightforward before designing and constructing OSC devices, such as the demands in output power/voltage/current, encapsulation, and nontoxicity, etc.

Here, we discuss the numerous attempts on exploiting the possibility of OSCs for emerging unique industrialization. OSCs can be regarded as power solutions to several important fields, covering self-powered wearable sensors, electronic textiles, implantable/biological electronics, internet of things (IoTs), and potential indoor applications (Figure 18.1). Therefore, this chapter describes examples of where OSC technology can act as an energy source, potentially enabling larger deployment of low-power, low-maintenance, and environmentally friendly hardware solutions,

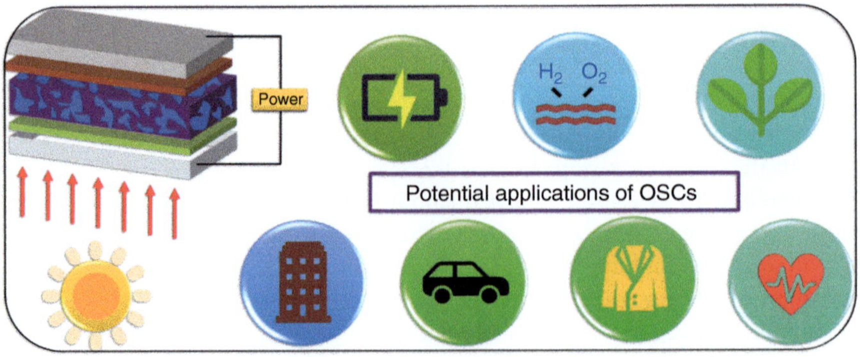

Figure 18.1 The potential application scenarios of OSCs. Source: Chengyi Xiao.

aiming to find the most appropriate ways to realize their commercialization. Notably, many companies such as *infinity PV*, *Merck Chemicals*, *BELECTRIC OSC*, and *ARMOR* have devoted numerous efforts to the actual products of OSCs.

18.2 Building-Integrated OSCs

The combination of global warming and energy shortage is expected to put significant stress on renewable energy systems. Among the clean powers, solar energy is supposed to take an important place in the energy systems [1]. Compared with perovskite solar cells and (poly- or mono-) crystalline silicon (Si)-based solar cells, the photoactive layers of OSCs have the superior advantages of low-temperature and solution-processing manufacturing. Owing to the large-range tunability of organic semiconductors in chemical structures, absorption, and frontier orbital levels, OSC devices can be tuned to create a wide range of choices in colors, transparency, and indoor efficiency [26, 38]. This allows OSCs to achieve semitransparency and meet many functional and esthetic demands in building-integrated OSCs (building-integrated organic solar cells [BIOSCs], Figure 18.2) [39, 40].

Uniform and large-scale printing technologies are another core issues for the commercialization of OSCs. Most of the photoactive materials exhibit impressive solubility in organic solvents, which enable cost-effective and eco-friendly R2R manufacturing compared to the typical evaporation or sputtering manufacturing process [3]. Therefore, OSCs are more suitable for large-scale, high-throughput, and uniform R2R process, and more promising than rigid cells in terms of cost and productivity [34, 41]. The thin, flexible and laminates features are other superiorities for OSCs. The state-of-the-art thin OSCs on plastic foil substrates have been demonstrated to be less than 2 μm thick, with equal PCEs to their glass-based counterparts [42, 43]. This feature enables ease of large-scale fabrication over arbitrary substrates and building materials, which are highly expected to offer functions of power supplying elements in the future BIOSCs products. Besides, utilizing OSCs in BIOSCs in greenhouses for agricultural applications have more obvious advantages over the traditional Si-based

(a) (b)

Figure 18.2 Building-integrated photovoltaics with semitransparent OSC modules (a) and with vivid colors (b). Source: (a) Debije [39]/with permission of John Wiley & Sons and (b) Brus et al. [40]/with permission of Springer Nature.

module due to the versatility of spectrum control [44]. Furthermore, OSC systems have been demonstrated their long-term stability in outdoor conditions [36] and even in space [45]. Here, we will elaborate on four promising applications of BIOSCs including solar panels as house energy suppliers, semitransparent windows, solar trees, and greenhouse integrated OSCs.

18.2.1 Solar Parks

Household installation of solar panels as energy suppliers is the common commercial product for Si-based solar cells. This application type is considered to reduce both global warming and urban heat island while generating electricity for human life [46, 47]. Organic semiconductors have the inherent advantages of flexibility and high mass-specific power in comparison with their inorganic counterparts. While OSCs showed both comparative PCE, long-term stability [48], and fast processing [34, 49] at the lab level, the OSC panels are supposed to be installed in building facades or roofs.

In order to compare different energy technologies, an analysis method named life cycle analysis (LCA), which defined in the *ISO14040* series [50], was introduced by Krebs et al. to evaluate how efficient the given technology is in converting the source energy into the form of energy that we would like to use [41]. The LCA is a complex method that can give parameters such as energy pay-back time (EPBT), embodied energy, and global warming potential (GWP). As shown in Table 18.1, it is clear that

Table 18.1 Lifecycle energy in MJ per kW h_{el}, energy payback time (EPBT) in years, and global warming potential (GWP) in kg CO_2 per kW h_{el} for several renewable energy technologies.

Energy source/ Technology	Location	EPBT (years)	Embedded energy (MJ per kW h_{el}[a])	GWP (kg CO_2 per kW h_{el}[a])	References
Wind	Global average	0.39	0.066	0.027	[51]
		PV technologies			
Mono- and poly-Si	US-Global average	1.65–4.14	11.9	0.032–0.36	[52, 53]
Amorphous Si	US-Global average	1.13	2.55	0.011–0.226	[54, 55]
CdTe	European average	0.73–1.61	9.22	0.025	[56, 57]
OSC[b]	Denmark/ Europe	0.2–4	44.50	0.01–0.06	[41, 58–60]
Hydropower	Germany–Canada	0.5	0.14	0.013–0.04	[61]
Geothermal	Germany	0.54	0.54	0.041	[61]
Biomass	China	5–10	—	0.045	[62]
Natural gas	Australia	—	12.91	0.751	[63]

a) The unit kW h_{el} in the table means the generated electrical energy of renewable energy technologies.
b) Note that these LCA values are given for the production of the modules and not for the whole installation.

Source: Carlé et al. [36]. Reproduced with permission of John Wiley & Sons.

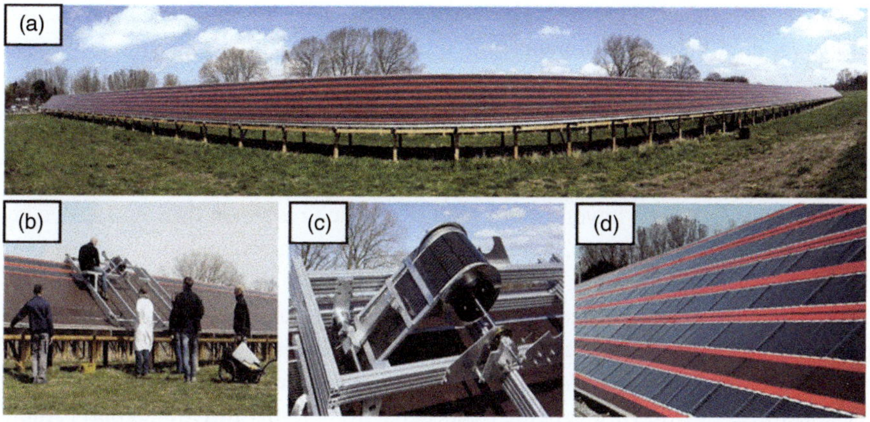

Figure 18.3 The solar panels were installed in a solar park. (a) The entire platform with the panels installed. (b) The roll mounting of the foil was shown (c) along with a close-up photograph of the application with a guide. (d) A view along the row showing dilation around connections in the platform. Source: Krebs et al. [41]/with permission of John Wiley & Sons.

the OSC technology exhibits faster EPBT, larger embodied energy, and producing less CO_2 in the whole manufacturing process than other competitors.

A solar park based on polymer solar panels was installed and operated by Krebs et al. in Denmark and southern Spain (Figure 18.3) [41]. The installation and de-installation rate were above 100 m min^{-1}, and considering the present performance and web width, the whole installation was of >200 W min^{-1}, which was much faster than any other photovoltaic technology. In terms of performance, the solar panels exhibited PCE of >2% for a large active area of over 14.7 m^2 (with six lanes of 100 m stretches of solar cell foil in a web width of 305 mm). And the solar panels stabilized relatively constant PCEs of 1.6%–1.8% in the long-time operation (277 days in Denmark and 180 days in southern Spain). When the six lanes connected in parallel (the active area >88.2 m^2), the PCEs of the solar panels dropped to 1.53% ascribed to the loss in fill factors (FFs) and open circuit voltages (V_{OC}s). Without any additional washing or cleaning procedure, they conjectured the initial drop in PCEs could be ascribed to shades by bird droppings. An LCA procedure was carried out by inputting the manufacturing and installation data, which revealed that the EPBT was 0.49 years in a southern European setting. For the GWP, the emissions of CO_2 in the installation of the solar panels include all the manufacturing process was 12 922 kg CO_{2eq}, which implied 420 g CO_2 per kW h in Denmark and the carbon footprint was 0.112 kg CO_2 per kW h. Similarly, the carbon footprint was 0.072 kg CO_2 per kW h in southern Spain.

From an OSC targeted costing perspective, Chatzisideris et al. demonstrated that OSC-battery systems (OSC integrated with battery storage) would offer the most cost-effective solution compared to OSC-only systems [64]. Eventually, they identified the OSC cost thresholds of 0.9 €/Wp for Denmark and 1.6 €/Wp for Greece. Furthermore, they built up an economic analysis model to demonstrate the environmental impacts of electricity self-consumption from OSC systems in

Denmark and Greece. The model proved that OSC systems tended to be associated with lower environmental impacts than other electricity sources.

18.2.2 Smart Windows

Compared with the traditional inorganic semiconductors, the organic semiconductor materials often have lower carrier mobility, which results in thin BHJ active layers with a thickness less than 200 nm [65] and brings to the optical semitransparency of the thin organic layer [66]. In another viewpoint, due to the large-range tunability of organic semiconductors in chemical structures, absorptions, and bandgaps, the active layers in OSCs can be designed effectively to weaken the absorption in the visible light range (370–740 nm) and result in a whole semitransparent device. These features of OSCs are generally used to achieve many functional and esthetic demands.

Smart windows (Figure 18.4) were demonstrated by Yao et al., which integrated both power-generation functions and color adjustment in a high-performance semitransparent organic solar cells (ST-OSCs) [67]. The smart windows were realized by introducing a gasochromic tungsten trioxide/platinum (WO_3/Pt) as a back-reflector layer. The WO_3/Pt film was switchable between colored and bleached states during hydrogen exposure [68, 69], thus dynamic and reversible adjustments in the colors of ST-OSCs could be achieved. Moreover, ultra-thin gasochromic WO_3/Pt film in the colored state could significantly increase the absorptions from the visible region to the near-infrared (NIR) region [70] and further improved the PCEs of ST-OSCs. As a result, the ST-OSCs integrated with the gasochromic WO_3/Pt film exhibited a PCE of 10.2% at the colored state and 9.1% at bleached states, as well as the corresponding

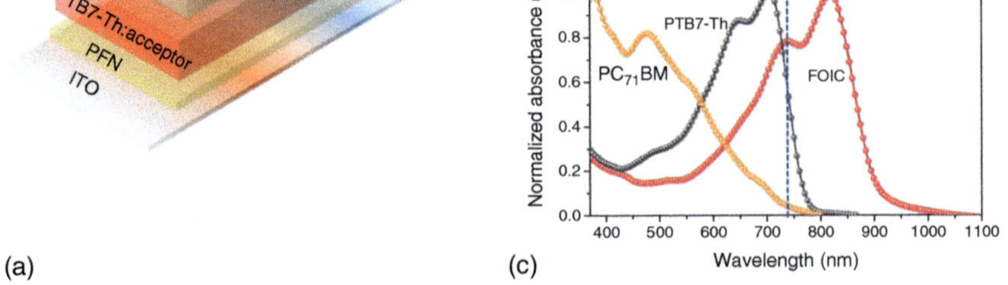

Figure 18.4 A color and transparency-switchable ST-OSC. (a) Illustration of the structure of an ST-OSC device. PFN, poly[(9,9-bis(3′-(*N*,*N*-dimethylamino)propyl)-2,7-fluorene)-*alt*-2,7-(9,9-dioctylfluorene)]. (b) Schematic diagram of the gasochromic WO_3/Pt layer. (c) Absorption spectra of $PC_{71}BM$, PTB7-Th, and acceptor as a thin film. Source: Yao et al. [67]/Reproduced with permission of Elsevier.

average visible-light transmittance (AVT) of 25.4% and 33.8% (Eq. (18.1)), respectively. The novel design concept of gasochromic materials integrated to the ST-OSCs has broadened the functions of OSCs as smart windows and sunroof of hydrogen fuel cell automobile.

In general, AVT is measured in the visible wavelength range (370–740 nm) and defined as follows [40]:

$$\mathrm{AVT} = \frac{\int T_{\mathrm{trans}}(\lambda)P_{\mathrm{r}}(\lambda)S_{\mathrm{AM1.5G}}(\lambda)\mathrm{d}(\lambda)}{\int P_{\mathrm{r}}(\lambda)S_{\mathrm{AM1.5G}}(\lambda)\mathrm{d}(\lambda)} \tag{18.1}$$

where T_{trans} is the transmittance, P_{r} is the photopic response (determined by the spectral sensitivity of the human eye), and $S_{\mathrm{AM1.5G}}$ is the solar photon flux under air mass (AM) 1.5G for window applications, λ is the wavelength.

18.2.3 Solar Trees

Researchers of *Merck Chemicals* and *BELECTRIC OSC* represented a paradigm shift for semitransparent (AVT ≈ 20%) OSCs modules by preparing large-area OSC products named "solar trees" and installing them at the German pavilion, *Universal Exhibition Milan 2015* (Figure 18.5) [33]. This attempt have addressed many of the current technical challenges and provided definite standards for OSCs using in building-integrated photovoltaics. Such as: the conjugated polymer active materials should have a good balance of PCE, solubility, acceptor miscibility, viscosity and coating uniformity, and better being entirely halogen-free. To approach the lab-to-manufacturing translation, all the fabrication processes, including ink formulations and device optimizations, should operate in a printing or R2R technique instead of the lab-level spin-coating.

In this case, all the layers of the devices were fabricated by doctor-blade coating technique under ambient conditions at lab-scale except the poly(3,4-ethylenedioxythiophene) polystyrene sulfonate (PEDOT:PSS) hole transporting layer (HTL), which was spin-coated onto the stacked devices. The manufacturing-scale OSCs were fabricated by a typical R2R process. Benefiting from the principles, the device transferred from lab-scale (0.04 cm^2) to manufacturing-scale (250 m^2) without additional optimization. Eventually, hexagonal "leaves" modules of the solar trees were obtained by connecting six groups in parallel, and each group had nine serially connected cells. With the areas growing larger, the PCEs of the devices decreased in sequence from 8.5% (0.04 cm^2) to 7.2% (1.00 cm^2), 4.3% (114.5 cm^2), and 4% (250 m^2).

They also tried to quantify the stability standards for OSCs integrating into the building-integrated photovoltaics. After heating two hours at 120 °C in the lamination process, the overall performance of a single cell showed no significant change. And after operating under 1 sun illumination for over 100 hours, there was not any significant decrease in PCEs. The results indicated that the whole cells had excellent thermal- and photo-stabilities. The solar trees appearing at the *Universal EXPO 2015* in Milan clearly demonstrated the possibility of large-scale solution-processable OSC systems.

18.2.4 Greenhouse and Photosynthesis

The greenhouse can provide a suitable environment for crops growing regardless of the outdoor environment, which greatly improves agricultural land-use efficiency. However, due to functions such as temperature regulation, they need to consume more energy than conventional farming. ST-OSCs with specific spectral absorption can be one of the approaches to relieve this problem.

ST-OSCs were employed on greenhouses to achieve net-zero energy greenhouses in warm and moderate climates by O'Connor and coworkers (Figure 18.6) through an elaborated dynamic energy fluxes balance model [71]. Two ST-OSCs with varying spectral absorption characteristics were employed. Also, both the generated solar power and the greenhouse energy load changes were considered in this model. Furthermore, distinct climates in different locations, including Phoenix (hot-dry), Raleigh (mixed-humid), and Antigo (cold), were selected in their work. The research revealed that the ST-OSCs-integrated greenhouse could afford sufficient energy in managing the thermal load of the system throughout the year in hot and

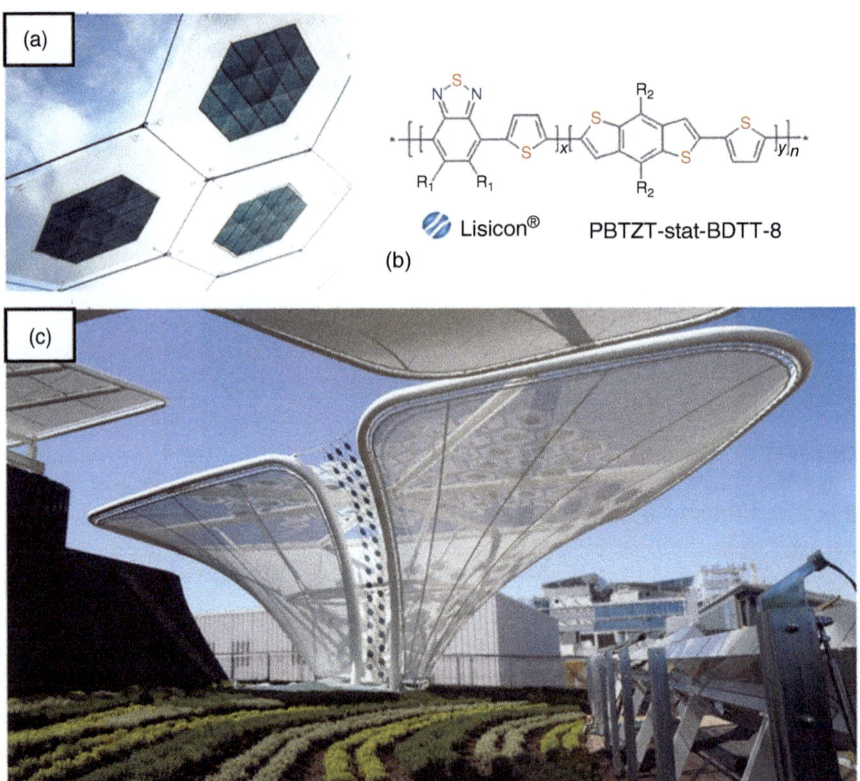

Figure 18.5 (a) Images of a commercial module fabricated by *BELECTRIC OSC*. (b) Chemical structure of the polymer PBTZT-stat-BDTT. (c) Large-scale deployment of *BELECTRIC OSC* modules at the *Universal Exhibition Milan 2015*. Source: Berny et al. [33]/John Wiley & Sons/CC BY 4.0.

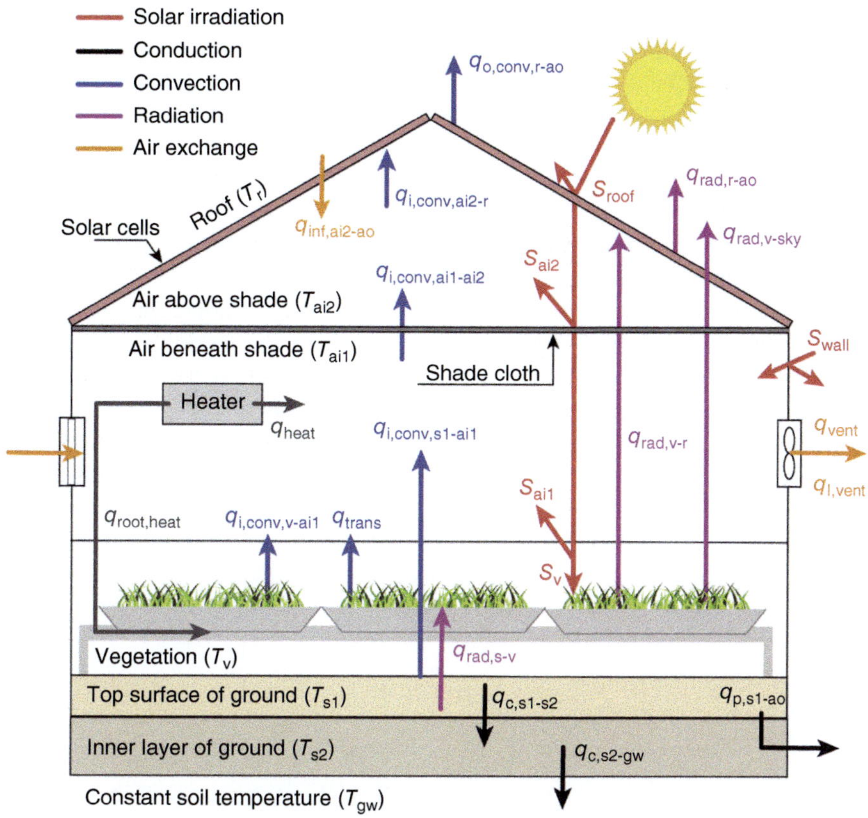

Figure 18.6 Schematic of the energy fluxes for OSC-greenhouse with shades deployed. The energy balance terms include the temperature of the various elements (T), the absorbed solar radiation (S), and sensible and latent heat transfer terms (q). Source: Ravishankar et al. [71]/Reproduced with permission of Elsevier.

mixed-humid climates. Despite the generated OSC power that cannot meet the energy load of greenhouse in a cold climate, the energy generated by the ST-OSCs would contribute to significant energy savings.

Another important issue of the research is related to the low-radiation nature of the ITO (indium tin oxide) electrodes and optical coating. This characteristic would avoid overheating at daylight by reducing IR (infrared ray) transmittance from sunlight while retaining the thermal energy at night through increasing IR reflectance from the thermal radiation of the greenhouse. It was found that the cooling demand of the conventional and OSC-greenhouses were almost the same, which was attributed to the transmittance of the ST-OSC. Overall, this modeling demonstrated that solar power can afford most of the energy needs of the greenhouse in warm and moderate climates by analyzing the realistic solar cell performance metrics and greenhouse energy demands. And there are significant opportunities for achieving net-zero energy OSC-greenhouse operation through reasonable energy fluxes for OSC-greenhouse.

Photosynthesis processes for plant growth generally absorb light irradiations in the visible light region (370–740 nm) because of the dominant status of chlorophyll a and b pigments [72]. And the photomorphogenesis process for plants also needs lights located at specific regions. For example, light irradiations with short-wavelength would facilitate vegetative leaf growth and with long-wavelength would accelerate for flowering. Thus, the semitransparent OSCs with tuned spectrum can be integrated into the greenhouse to provide additional electrical energy while transmitting enough sunlight for plant growth, or controlling the photomorphogenesis process.

Greenhouses for agricultural applications have been studied in many inorganic systems [73–76]. Recently, Song et al. exploited the feasibility of adopting foldable-flexible OSCs for photosynthesis on the lab-scale [77]. The ST-OSCs were fabricated with modified PEDOT:PSS as electrodes and PBDB-T-2F:Y6 as a photoactive layer to ensure enough transmittance in the visible region (Figure 18.7a). The modified PEDOT:PSS/PET (polyethylene terephthalate) exhibited low reflectance (<15%) across the wavelength region of 300–850 nm and overall AVT of ≈80% across 350–850 nm, while PBDB-T-2F:Y6 blend had the main absorptions located in the NIR region. The overall semitransparent foldable-flexible OSCs yielded

Figure 18.7 (a) Device structure of the flexible opaque OSCs. (b) An installation of FST-OSCs as roofs in the greenhouse. FTE, flexible transparent electrodes; PDINO, 2,9-bis[3-(dimethyloxidoamino)propyl]anthra[2,1,9-*def*:6,5,10-*d′e′f′*]diisoquinoline-1,3,8,10(2*H*,9*H*)-tetrone. (c) Optical photograph of mung bean leaves changing in nine days under direct sunlight (L), direct sunlight with ST-OSCs roof (S), and dark (D). Source: Song et al. [77]/Reproduced with permission of John Wiley & Sons.

PCEs of >10% and T_{trans} of 21% in the visible region. Excellent mechanical stability was proven through the bending and folding test. Furtherly, the semitransparent foldable-flexible OSCs were integrated as part of roofs into the simulated greenhouse (Figure 18.7b). The plants showed comparable growth rates under semitransparent foldable-flexible OSCs with those grown under sunlight after monitoring for nine days (Figure 18.7c). The researchers also prospected that additional power can be used for future multifunctional self-powered greenhouse, although they did not provide or store the output power.

18.3 Wearable-Integrated OSCs

The IoTs involve all aspects of the national economy and human life and are the third information technology revolution after computer and Internet [78, 79]. As the underlying hardware base of IoTs, billions of wireless sensors will be installed in human life within the next decade. In the IoTs landscape, wearable electronics are attracting a lot of attention for their potential applications in human health monitoring and smart healthcare. A recent report by *McKinsey Global Institute* summarized the potential commercial impact of IoTs and concluded that the IoTs market is expected to increase as much as 11 trillion dollars by 2025 [80].

Flexible IoT sensors generally consist of three parts: sensors, control circuits (CC), and energy supply modules [81]. Relevant researches at home and abroad mainly focus on independent sensor unit and free-standing power supply modules. Achieving wearable IoTs sensors requires independent and fewer power modules, which are generally rechargeable and replaceable lithium batteries. However, the range and frequency of data transmission can reduce battery life, thus limiting the use of conventional batteries. In addition, the battery replacement and charging process will incur higher operating and maintenance costs. As an emerging energy source, OSCs have the advantages of high quality-specific energy (over $10\,W\,g^{-1}$) [82], flexibility, and large-area preparation, etc., which are very suitable for being the energy module of the IoTs system and provide a reliable and continuous energy source for low-power wireless sensor network.

18.3.1 Portable Device Photovoltaics

In 2016, O'Connor et al. reported the fabrication of ultra-flexible and wearable OSCs [83]. In these ultra-flexible OSCs, the thin PI (polyimide) films (13 μm) adhesive substrates were spin-coated with PEDOT:PSS, polyethylenimine (PEI), and P3HpT (P3HT): $PC_{61}BM$ in sequence. Another PEDOT:PSS layer was then transferred on top of the overall device using thermal release tape, and silver paint was then applied to make the front electrode. An encapsulation layer (1-μm barrier of parylene C) was deposited on the whole devices by vacuum vapor evaporation before attaching them to the skin to increase the stability of the devices for outdoor applications.

The ultra-flexible and wearable OSCs have been adhered to human skin, exhibiting stable operational performance when compressively strained by about 75%.

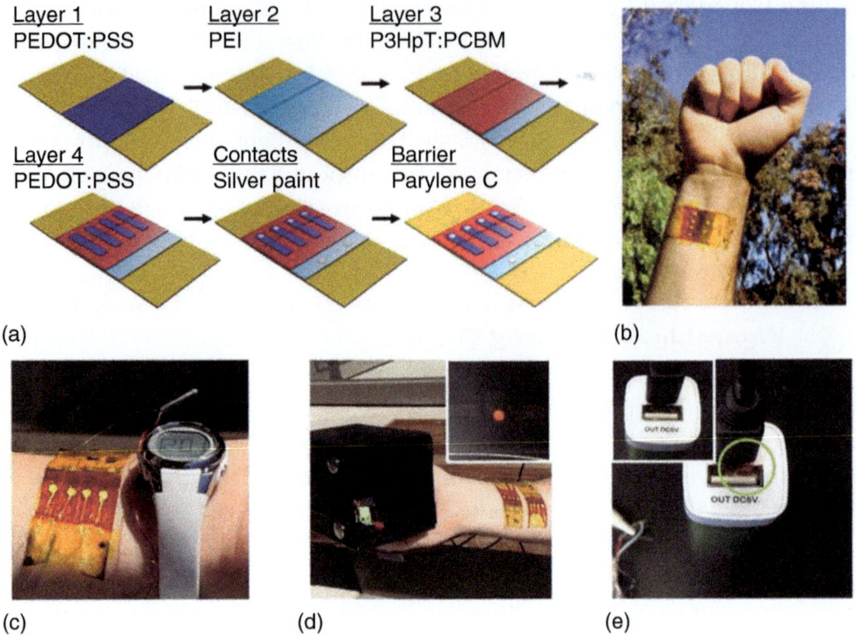

Figure 18.8 (a) Schematic summary of the process used to fabricate wearable OSCs. PCBM, phenyl-C_{61}-butyric acid methyl ester. (b) A photograph of the wearable solar cell on skin. (c) Wearable digital watch in outdoor sunlight and (d) LED in a minimalist Armstrong self-oscillating voltage booster circuit in outdoor sunlight (980 W m^{-2}). (e) Wearable OSCs on a flat substrate powering a portable battery under 1 sun illumination of a solar simulator (the green circle shows that the battery pack is in charging mode and the inset clarifies that the device is not charging when there is no incident light on the OSC). Source: O'Connor et al. [83]/Reproduced with permission of Elsevier.

These OSC devices based on P3HpT:PC_{61}BM exhibited less than 20% degradation in PCEs after bending for one thousand cycles, while devices based on P3HT:PC_{61}BM showed a huge drop after only five cycles. The usefulness of the wearable solar cells was to exploit their potential applications by directly powering an LED (light-emitting diode) and a digital watch under 980 W m^{-2} of natural sunlight after carrying the deliberate selection of materials based on mechanical properties rather than just efficiency for wearable electronics (Figure 18.8) [83]. The OSCs were wired in series to fulfill the required voltages (2.18 V) to power watch and 15 cells were wired in series to produce a V_{pp} (voltage peak to peak) of 4.66 V and an I_{max} (maximum discharge current) of 201 mA to charge the portable battery. Differently, two substrates and eight cells were connected in parallel to light up the LED (which needs 0.4 V and ~1 mA). These results demonstrated the possible use as the primary cells for OSCs devices.

18.3.2 Implantable and Wearable Self-Powered Sensors

Someya and coworkers at *The University of Tokyo* make a significant effort on the ultrathin and ultra-flexibility OSCs [42, 84, 85]. In 2012, they demonstrated ultrathin OSCs based on 1.4-μm-thickness PET substrate (which is commercially available Mylar 1.4 CW02) could be fabricated in a total <2 μm thickness device while exhibited comparative efficiency with ITO/glass-based counterparts [42]. The overall ultrathin OSCs yielded PCEs of 4.2% and an excellent specific weight value of $10\,W\,g^{-1}$. Also, high flexibility was measured by adhering the flexible device to a pre-stretched elastomer. The OSCs could survive quasi-linear compression to below 70% and cyclic compression and stretching to 50%, over more than 20 full cycles.

Further in 2018, they proved the possibility of OSCs integrating into organic electrochemical transistors (OECTs) and thus the self-powered cardiac sensors were achieved (Figure 18.9) [82]. The cardiac sensors were proven to have high transconductance (0.8 mS), high responsivity (10^5), high-speed responses (0.58–0.62 ms), and compatible with the frequency ranges of human biological signals (approximately 200 Hz) benefiting from the application of OECTs. The flexible OSC module in the system was demonstrated to have an enormous contribution to the signal-to-noise ratio (*S/N*) as it exhibited the lowest noise currents of 0.02 μA compared with that of the source meter (0.58 μA) and commercial battery (0.08 μA). The self-powered sensors were attached to the fingertip and the exposed surface of a rat heart with *S/N* of 25.9 and 40.02 dB, respectively.

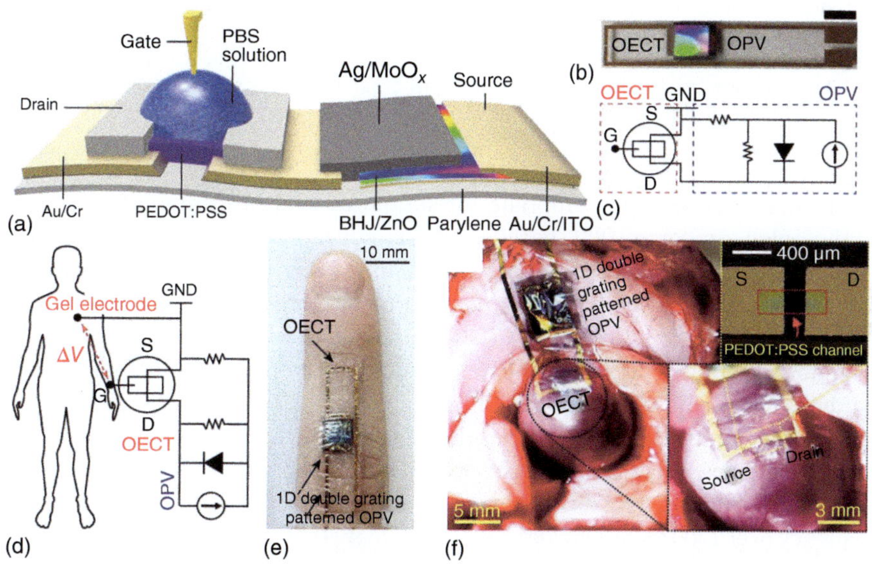

Figure 18.9 (a) Schematic of a double-grating-patterned OSC integrated with an OECT. See Methods for details. PBS, phosphate buffered saline; GND, ground. (b) Photograph of the integrated device. Scale bar, 5 mm. (c) Circuit diagram of the integrated device. (d) Wiring diagram for cardiac signal recording. The chest and source electrodes of the integrated device were grounded and the potential difference (ΔV) between the chest and active channel of the OECT functioned as the gate bias. (e) Photograph of the self-powered integrated electronic device attached to a finger. (f) Photograph of the self-powered integrated electronic device attached to the heart of a rat (left), and enlarged images of the channel area (right, top) and of the source–drain electrode (right, bottom). Source: Park et al. [82]/Reproduced with permission of Springer Nature.

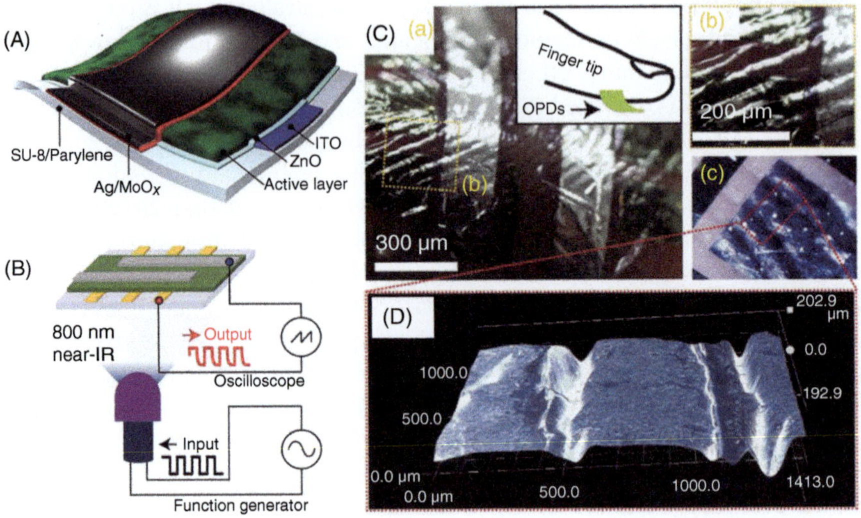

Figure 18.10 (A) The schematic device structure of near-IR photoresponsive ultra-flexible organic photodetector. OPD, organic photodetector. (B) Experimental setup to test near-IR light time response. (C) (a, left image) Photograph of fingerprint-conformal near-IR photodetectors and inset indicates the position of skin-conformal near-IR photodetector on the finger (b, right top image) enlarged image of (a), and (c, right bottom image) microscope image of the top silver electrode and skin surface, indicating high skin-conformability. (D) 3D microscope image of ultra-flexible near-IR organic detectors on the surface of the human index fingertip. Source: Park et al. [86]/Reproduced with permission of John Wiley & Sons.

NIR light is generally appropriate for skin-interfaced health monitoring, such as photoplethysmogram [87], because NIR is not harmful to humans and has a long propagation distance (>4 mm) within the tissue [88]. Thus, F-SOCs with narrow-bandgap active layer can be applied to NIR organic photodetectors. Park et al. fabricated an ultra-flexible OSCs-structured NIR skin-conformal photoplethysmogram sensor based on a narrow-bandgap regioregular polyindacenodithiophene-pyridyl[2,1,3]thiadiazole-cyclopentadithiophene (PIPCP) and $PC_{61}BM$ photovoltaic active layer (Figure 18.10) [86]. The ultrathin (3-μm thick) OSCs-structured NIR photodetector exhibited mechanical stability at bending radius of $r < 3$ μm for 1000 bending cycles. Also, the fabricated sensors exhibited outstanding cut-off response frequency over 1 kHz at −3 dB and better sensitivity than a normal rigid glass-based device. Therefore, OSCs would be capable as a human-compatible, wireless power source emerged in the next-generation electronics due to their high conformability and biocompatibility.

18.3.3 OSC Textile Toward Smart Clothing

Self-powered smart clothes integrated with multifunctional electronic devices have attracted much attention in the IoTs. For electronic wearable devices, it is essential to possess features of environmental stability, sufficient energy efficiency, and mechanical robustness. Textile-based wearable OSCs, which produce power from the surrounding environment and even the indoor circumstances, are promising to smart clothes. There are two strategies to realize smart clothes: the planar stacked structured OSCs based on textile substrate and the textile weaved by 1D (one-dimension) shaped OSCs.

In the planar stacked structured OSCs based on the textile substrate, there are many difficulties as the susceptibility of the traditional flexible OSC architecture and active layers to the surrounding moisture and oxygen. To address this problem, thin-film encapsulation barriers were utilized [89, 90]. Jinno et al. reported the single-/double-side-coated passivation layer, prepared by coating 500-μm-thick acrylic elastomers (3M VHB Y-4905J), was used to improve the operational stability of OSCs (Figure 18.11) [89]. The textile-based OSCs showed excellent washability with only a 5.4% decrease of PCEs after immersing in water for 120 minutes.

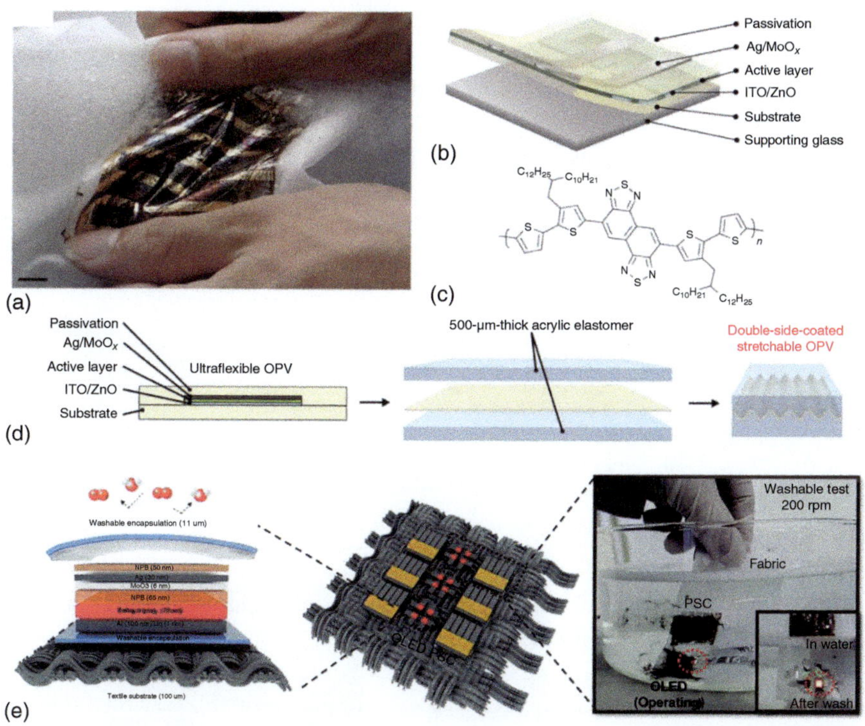

Figure 18.11 Textile-based washable OSCs. (a) Photograph of the washing process for the devices conforming to a dress shirt. Scale bar, 1 cm. (b) Structure of the free-standing OSCs. (c) Chemical structure of PNTz4T. (d) Schematic of the double-side-coated OSCs. (e) Schematic diagrams of the optoelectronic module and organic light-emitting diode (OLED) structure and the picture of the operating module during the washable test with the inset image indicating the module being dipped in water after washing. NPB, N,N′-Bis(naphthalen-1-yl)-N,N′-bis(phenyl)benzidine. Source: Jinno et al. [89]/with permission of Springer Nature, Jeong et al. [90]/Reproduced with permission of The Royal Society of Chemistry.

Furthermore, due to the elastomeric state of the passivation layer, PCEs of the OSCs remained at 80% of the initial values even after 52% mechanical compression for 20 cycles with 100 minutes of water exposure. Jeong et al. [90] incorporated the SiO_2 (silicon dioxide)-polymer composite capping layer to encapsulate the OSCs. The textile-based washable OSCs exhibited water resistance with only 2% degradation after washing 20 times for 10 minutes and mechanical stability when operating at a low curvature radius ($r = 3$ mm).

Compared with the stacked planar structure, the 1D structured OSCs possess many promising advantages, such as miniaturization, adaptability, and wearability, which make it more attractive for smart clothes [91–94]. The 1D structured OSCs consist of coaxial structure [95] and twisted structure [96]. Zhang et al. reported the coaxial structured flexible organic solar cell (F-OSC) textiles by dip-coated with the P3HT:PC_{61}BM photoactive layer, PEDOT:PSS HTL, and carbon nanotube-based transparent electrode on a metal textile electrode (Figure 18.12a) [95]. The flexible OSC device based on the coaxial structured textiles exhibited a PCE of 1.08% at every side and maintained >97% efficiencies after bending for >200 cycles. Also, the flexible textiles showed an excellent areal density of 5.9 mg cm^{-2}. The twisted structured textiles are regarded as lighter and more flexible than the coaxial structure textiles.

In 2009, Lee et al. demonstrated the twisted structured textiles by incorporating a second wire to the solar power wires [96]. The phase-separated photovoltaic layer was coated onto a thin stainless-steel wire. While the second stainless-steel wire, coated with a silver film and serving as the counter electrode, was wrapped around

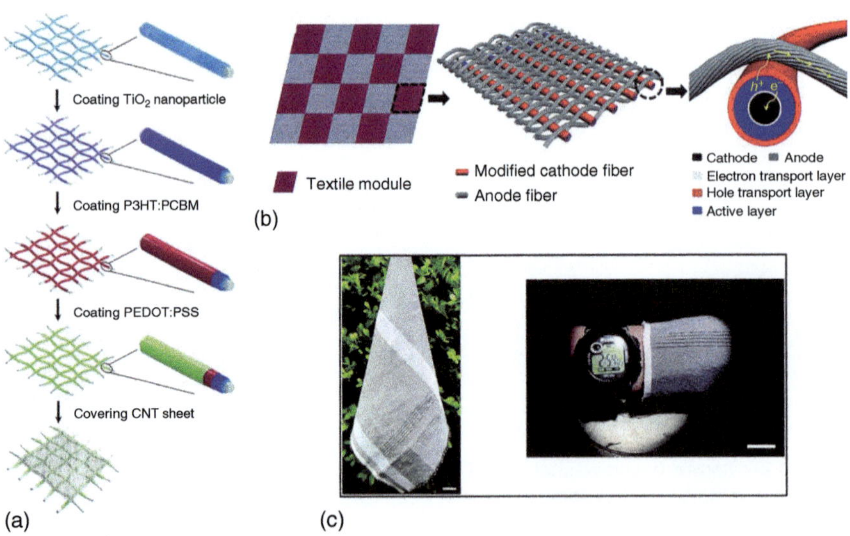

Figure 18.12 The 1D structured OSCs textiles. (a) Fabrication of the coaxial structured OSC textiles. CNT, carbon nanotube. (b) Schematic illustration of the twisted structured OSC textile. (c) Photograph of OSC textile, scale bar, 2 cm, left; Demonstration of the photovoltaic textiles used to power an electronic watch under solar irradiation. Scale bar, 2 cm. Source: (a) Zhang et al. [95]/Reproduced with permission of John Wiley & Sons, (b, c) Liu et al. [97]/Reproduced with permission of Royal Society of Chemistry.

the first wire. The two wires were added by a transparent polymer. The PCE values of the solar power wires ranged from 2.79% to 3.27%. The stretchable structure is usually achieved by intertwining two wires around an elastic substrate, which can enhance the stretchability of the textiles. In 2014, Zhang et al. have proven the twisted structured OSCs by incorporating a Ti (titanium) wire and an aligned multiwalled carbon nanotube (MWCNT) fiber as cathode and anode, respectively [98]. The as-prepared wire-shaped OSCs, with a maximum PCE of 1.78%, could be easily woven into various flexible structures without the necessity for sealing. Further in 2018, the ZnO (zinc oxide) electron transporting layer (ETL) was imported to the wire-shaped OSCs through immersing the Ti wire into a sol–gel precursor solution of zinc acetate dehydrates and the following thermal annealing process (Figure 18.12b) [97]. To demonstrate the practicality of the flexible textile as a power source for applications in fields such as wearable electronics, a piece of this photovoltaic textile was woven and used to drive an electronic device (Figure 18.12c).

The novel OSC textiles showed lightweight, thin, and flexible while high running stability, and revealed the possibility of OSCs attached or wove to daily clothes affording certain power. Considering these features, the OSC textiles will play an important part in many emerging new application fields, such as smart clothing, biomedical electronics, and facilitates the OSCs commercialization.

18.4 OSCs-Integrated Energy Storage System

The BIOSCs and wearable-integrated OSCs have partly demonstrated the possible applications for OSCs devices. Moreover, according to previous cost forecast [36, 64, 99], the costs of OSC modules could be cut down to 0.05 €/Wp due to the low-cost materials and the high-throughput R2R printing, while that of Si-based PV modules were 0.5–0.6 €/Wp in Europe. The colossal competitive advantage in price makes OSC-generated electricity more attractive than the Si-based PV modules. However, the performance of OSCs is sensitive to light intensity and incident angles, which means that the OSC modules are not a stable and constant energy source.

To address this issue, OSC-integrated energy storage systems (ESSs, Figure 18.13), such as photocharging (or named self-rechargeable) devices composed of OSCs and flexible supercapacitors [100, 101] or lithium battery [102] as energy storage units were reported. Also, Chatzisideristhe et al. [64] proved the economic costs and gains, and life cycle environmental impacts of the OSCs-integrated ESSs, which can offer a more cost-effective solution compared to OSC-only systems. Normally, there are two kinds of OSC structure integrated to combine ESS: a planar stacked OSCs-integrated ESS (Figures 18.13 and 18.14) and fiber-based OSCs-integrated ESS (Figure 18.15) [104].

To evaluate the performance of the whole OSCs-integrated ESS, there are various parameters in this system that need to be studied. The PCE of the OSC device represents the converted efficiency from solar energy into electric energy, which was calculated from the equation:

$$\mathrm{PCE} = \frac{P_{\mathrm{max}}}{P_{\mathrm{in}}} = \frac{J_{\mathrm{sc}} \times V_{\mathrm{oc}} \times \mathrm{FF}}{P_{\mathrm{in}}} \times 100\% \tag{18.2}$$

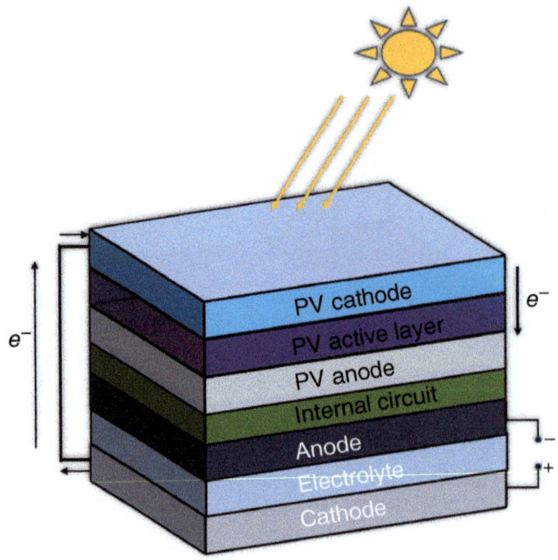

Figure 18.13 Schematic illustration of a photovoltaics-integrated energy storage system. Source: Chengyi Xiao.

where V_{oc}, J_{sc}, FF, P_{max}, and P_{in} represent the open-circuit voltage, short-circuit current densities, fill factor, maximum power output, and input light density (mW cm^{-2}), respectively.

The total efficiency of the OSCs-integrated ESS was calculated by dividing the total input energy from light by the stored energy in the battery or supercapacitor during the photocharging process; for total solar-to-PCE (η_{STP}) at a specific time with a discharging test, E_A (energy density) was calculated from the discharging curve:

$$\eta_{STP} = (E_A \times S_1)/(P_{in} \times t_2 \times S_2) \tag{18.3}$$

where P_{in} is the illuminated light density (mW cm^{-2}), t_2 is the photocharging time, and S_1 and S_2 are the effective area of the batteries/supercapacitors and OSC device, respectively.

18.4.1 Planar Stacked OSCs-Integrated ESS

The photocharging system can provide an independent power supply for next-generation wearable electronics or biomedical devices. A solid-state MWCNT-PEDOT:PSS-based supercapacitor with the PVA (polyvinyl alcohol)/H_2SO_4 (sulfuric acid) as gel electrolyte on the PET substrate was applied in the planar stacked OSCs-integrated ESS [100]. To balance the trade-off between performance and thickness, they demonstrated that the 50-μm-thick flexible photocharging capacitor showed a total efficiency (η_{STP}) of 5.92% (Eqs. (18.3), (18.4), (18.5)) by integrating 3-μm-thick OSCs on 40 μm-thick carbon nanotube/polymer-based supercapacitors with a PVA/H_2SO_4 gel electrolyte (Figure 18.14). The flexible photocharging capacitor was achieved by stacking the OSCs and the supercapacitor. The systems also exhibited superior operational stability (>96% efficiency retention after 100 charge/discharge cycles for one week) and mechanical robustness (94.66% efficiency

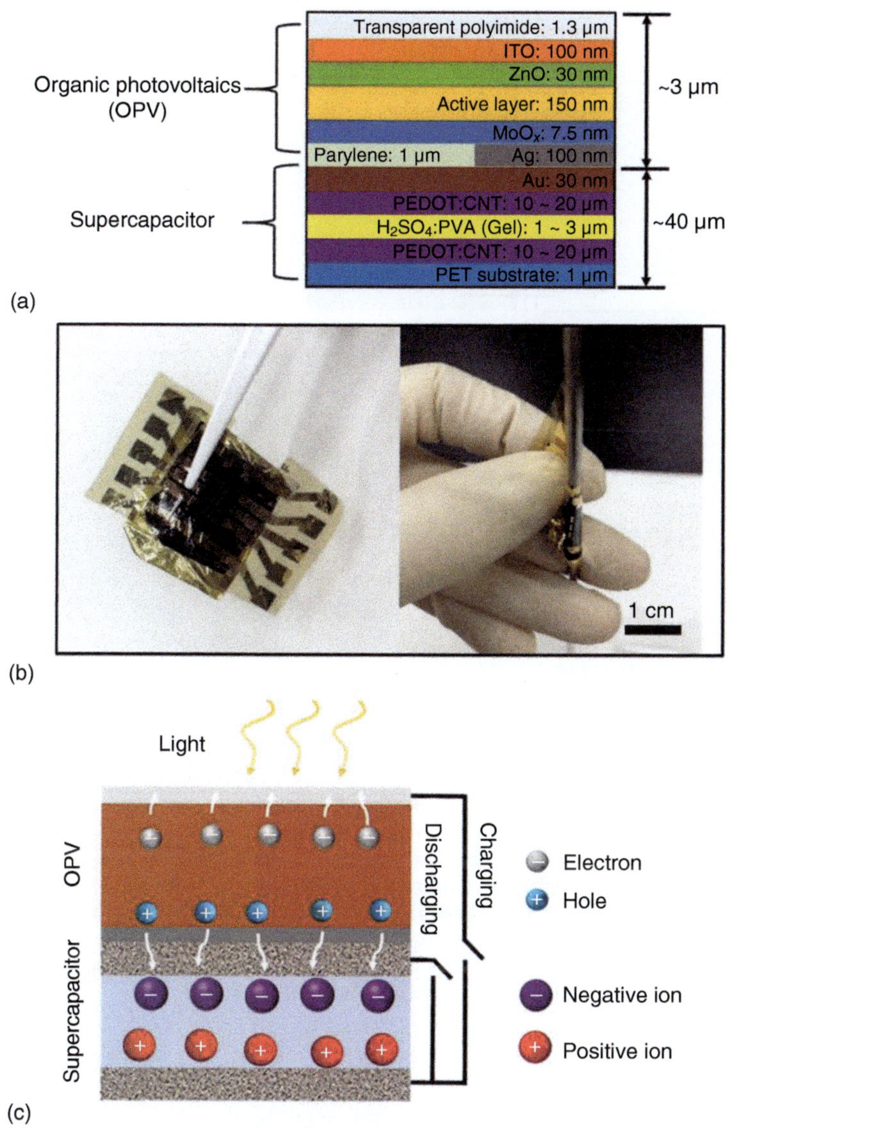

Figure 18.14 The schematic and working mechanism of the OSC-integrated photocharging capacitor. (a) Schematic illustration of the top-down integrated device. (b) Photographs of the as-fabricated ultra-thin device (left) and the device wrapped onto a rod (diameter: 2 mm). (c) The working mechanism of the photocharging process. The working electrode of the OSC and the anode of the supercapacitor were connected to start the photocharging process. Source: Liu et al. [100]/Reproduced with permission of John Wiley & Sons.

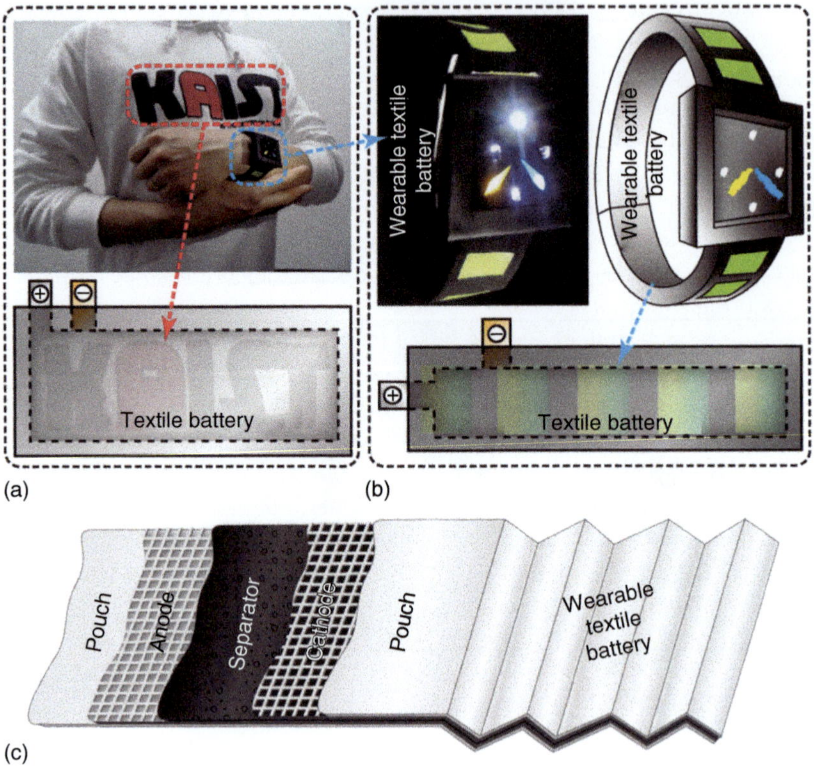

Figure 18.15 Fiber-based photocharging batteries. (a) A photograph of a wearable textile battery embedded in clothes together with its enlarged view of the inner cell structure. (b) Photograph and schematic representation of a watch with a wearable textile battery strap. (Left) six LEDs (1 yellow (= 0.042 W), 1 blue (= 0.062 W), and four whites (each = 0.062 W) were lighted up as a demonstration of a functioning watchstrap battery. (c) A schematic illustration of the cell configuration of the wearable textile battery. The key components are based on the materials suitable for flexibility and bendability. Source: Lee et al. [103]/Reproduced with permission of American Chemical Society.

retention after 5000 bending cycles at a $r \approx 2$ mm). However, the systems suffered the issues of undercharging or unbalanced charge storage of the supercapacitor arrays in practical application due to the instability of OSCs.

The energy density (E_A) was calculated using the equation:

$$E_A = 0.5 \times C_A \times V^2 \quad (18.4)$$

The areal capacitances (C_A) of the solid-state electrochemical supercapacitor with a PVA/H_2SO_4 gel electrolyte were calculated from the galvanostatic charge–discharge curves at different current densities using the following formula:

$$C_A = I \times t_1/S_1 \times V \quad (18.5)$$

where I is the constant current, t_1 is the discharge time, S_1 is the area of the supercapacitor, and V is the operating voltage window obtained from the discharge curves excluding the voltage drop for a full discharge.

Flexible and stretchable power sources were demonstrated by combing flexible silver–zinc batteries and rigid OSCs [105]. This approach adopted specialized device layouts, such as helical springs and serpentines, to achieve enhanced mechanical property for current collectors used as structural support and backbone for all battery components. The wire-shaped batteries exhibited robust deformability while retaining electrochemical performance over 17 000 flexure cycles at a 0.5 cm bending radius. An ITO/glass-based OSC was integrated with the wire-shaped batteries to construct an ESS. Also, varied light intensities (sunlight, LED, and compact fluorescent light) were employed to measure the performance of the ESS, and the resulting highest η is 2.7%.

Thostenson et al. [101] further incorporated a conversion and CC between OSCs and supercapacitors through the three dimension (3D)-printing technique, demonstrating its possiblity to solve the problems. The application of CC achieved more energy harvesting from the OSCs and stored in the supercapacitor bank. The photo-active layers in the OSCs absorbed light and generated electron–hole pairs. The electron–hole pairs were separated under a built-in electrical field and transferred to a corresponding electrode with electrons to the anode of the supercapacitor to attract the positive ions and holes to the cathode to attract the negative ions in the electrolyte. The charging of the supercapacitor via the OSC was realized by such a photo-charge generation and transfer process under continuous light illumination until the voltage on the supercapacitor approached the open-circuit voltage of the OSC.

18.4.2 Fiber-Based OSCs-Integrated ESS

Compared with the planar stacked OSCs-integrated ESS, the fiber-based OSCs-integrated ESS possesses many promising advantages, such as miniaturization, adaptability, and weavability [91–94].

In 2014, Zhang et al. combined fiber-based OSCs and fiber-based supercapacitors on the same treated Ti fiber substrate via electrochemical anodization to develop a novel, all-solid-state, flexible "energy fiber" [106]. For the OSCs part, perpendicularly aligned TiO_2 nanotubes array was grown on the Ti wire (back electrode) by electrochemical anodization, and the modified wire was dipped into a chlorobenzene solution of the active layer (P3HT:PC_{61}BM). The PEDOT:PSS layer was further coated on the active layer, following by wrapped with MWCNT to produce the OSC part as a front electrode. The PCEs of the fiber-like OSCs showed obvious lengths (ranging from 650 nm to 4 μm) dependence and the optimized PCE was 1.01% for the device length of 1.8 μm. For the supercapacitor part, the PVA/H_3PO_4 (phosphoric acid) gel electrolyte and MWCNT were coated on the TiO_2 nanotubes treated Ti fiber in sequence. The supercapacitor part exhibited a discharge rate from 0.1 to 5 μA and the scan rate from 100 to 5000 mV s^{-1}. Finally, the "energy fiber" achieved an overall η_{STP} of 0.82%.

In 2013, Lee et al. proposed mixed OSCs-integrated ESS, in which the textile battery and the planar flexible OSC were integrated (Figure 18.15) [103]. Polymer solar cells were fabricated on ITO-coated polyethylene naphthalate (PEN) substrates.

The PEDOT:PSS layer, PCDTBT/$PC_{71}BM$ active layer, and TiO_x (titanium oxide) were coated on the ITO–PEN substrate layer-by-layer and Al (aluminum) was deposited through a shadow mask by thermal evaporation on the devices. The overall flexible OSC showed a PCE of 5.49% on the PEN substrate. For the textile battery, nickel (Ni)-coated polyester yarn was adopted as a current collector to release stress; polyurethane (PU) binder was employed for strong adhesion of active materials. The wearable textile battery was embedded in clothes and watchstraps and exhibited comparable electrochemical performance to those of conventional metal foil-based cells even under severe bending conditions.

18.5 Other Applications

18.5.1 OSCs-Driven Water Splitting

Solution-processable OSCs have been well established as electric power for many integrating applications as we described above. Different from these applications, OSCs-driven water splitting, which converting solar energy to chemical energy (hydrogen, H_2, oxygen, O_2), has attracted very little attention recently. Owing to the diurnal and annual variations of irradiation, meteorological phenomena, and the nonuniformity through different regions, energy storage technology is needed to collect the OSC-generated power. Similar to the OSC-integrated battery or supercapacitor storage system, OSCs-driven water-splitting technology is another way to transfer and store the solar energy to renewable chemical energy [107–109]. In the OSCs-driven water splitting system, the solar-to-hydrogen conversion efficiency (η_{STH}) is defined as Eq. (18.6):

$$\eta_{STH} = (J_{OP} \times 1.23 \times \eta_F)/(P_{solar}) \tag{18.6}$$

where J_{OP} is the operating current density (mA/cm^2), the η_F is assumed to be 100%, and P_{solar} is the irradiance intensity of 100 mW cm^{-2}.

The mechanism of OSC-driven water splitting is shown in Figure 18.16 [107]. The system consists of an OSC device wired with an electrolysis cell filling with an aqueous electrolyte. The anode performs the oxygen evolution process (OER), while the cathode executes the hydrogen evolution reaction (HER) in the OSC-driven water splitting system. The photogenerated holes of the OSC are collected and transported to the anode, whereas the photogenerated electrons of the OSC are collected and transported to the cathode, and thus produce a photo potential. The OSC generated potentials are then acting as the potentials of electrolysis cell to perform the OER and HER reactions, and the electrochemical reactions are mediated by the OER and HER catalysts. Finally, a membrane separates the evolved H_2 and O_2 gasses.

Lin et al. reported a low-cost overall OSC-driven water splitting system combining a novel tandem OSC device and metal–organic framework (MOFs) derived bimetallic nanohybrids based flexible electrocatalysts [110]. The MOF-precursor Prussian blue analogs (PBAs) were used to synthesize the coupled electrocatalysts: nitrogen-doped carbon encased iron–cobalt bimetallic phosphides (CF@FeCoP/NC, an electrocatalyst for HER) and sulfides (CF@FeCoS/NC, an electrocatalyst for

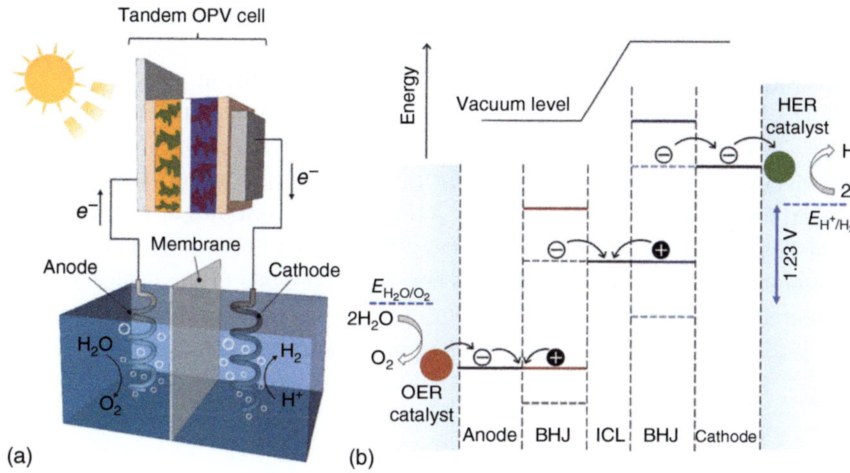

Figure 18.16 (a) Schematic representation for OSC-biased water splitting using a double BHJ junction tandem cell wired to an electrolysis cell. (b) The energy level diagram of a double BHJ junction tandem OSC device and the operation of water splitting reactions. Source: Yao et al. [107]/Reproduced with permission of John Wiley & Sons.

OER). The electrolyzer consisted of the electrode exhibited excellent catalysis property. A novel tandem OSC (ITO/ZnO/active layer/PEDOT/ZnO/PCE-10:IEIC-O-4F/V_2O_5/Al) was developed to power this electrolyzer, which showed the J_{sc}, V_{oc}, FF, and PCE of 10.84 mA cm^{-2}, 1.76 V, 67%, and 12.76%, respectively. The performance of the tandem OSCs illustrated that one device was sufficient to power the electrolyzer for overall water splitting. Eventually, the resulting OSC-driven water splitting system was proved to realize a current density of 10 mA cm^{-2} at a low cell voltage of 1.60 V, while maintaining rather stable in a long-time durability test in alkaline media. The overall OSC-driven water splitting system yielded a high water-splitting current density of 7.5 mA cm^{-2} with a η_{STH} of 9.2%.

In 2016, Esiner et al. investigated theoretical limits and experimental approach for η_{STH} in series-connected OSCs-driven electrochemical water splitting [111]. In the theoretical modeling, the electrochemical water splitting potential was assumed to be 1.50 V and the external quantum efficiency (EQE) and FF are both supposed as 0.65 of the polymer solar cell. Considering the realistic value of the polymer solar cells, the minimum photon energy loss (E_{loss}) which was determined by the energy difference between the optical band gap (E_g) and V_{oc} was set to 0.8 eV. With these assumptions in the modeling, the two series-connected single-junction cells with E_g of 1.73 eV and three series-connected cells with E_g of 1.44 eV were both expected to give a resulting η_{STH} of 6.9%, which slightly decreased to 6.2% for four series-connected cells at an optimal E_g = 1.33 eV. In the experimental approach, PTPTIBDT-OD (E_g = 1.89 eV), PTB7-Th (E_g = 1.56 eV), and PDPP5T-2 (E_g = 1.44 eV) were blended with $PC_{71}BM$ as BHJ active layer for two, three, and four series-connected OSC devices, respectively, giving the η_{STH} values of 4.1%, 6.1%, and 4.9%. The deviations between the theoretical limits and experimental approach were ascribed to the differences in E_g and E_{loss}.

The OSC devices fabricated by a large-scale slot-die printing method also have been employed in the OSCs-driven water splitting system. In 2019, Wu et al. demonstrated a high-performance non-fullerene OSC fabricating by a slot-die printing technology based on PBDB-T-SF:IT-4F photovoltaic blends [112]. The OSC showed a PCE of 12.27% for normal structure and 12.90% for inverted architecture after a detailed balancing of the flowing rate and coating speed combinations. Moreover, flexible OSCs based on a PET substrate were proved to a PCE of >12%, while a large-area OSC module with 0.4 cm^2 yielded a PCE of 12.12% for two series-connected flexible OSC devices. When the large-scale printed solar module was integrated into an existing water splitting system, η_{STH} of 6.15% was achieved. In this system, the commercial platinum (Pt) and Ni foam as OER and HER reaction electrodes were used.

18.5.2 OSCs-Integrated Glasses

In 2017, Landerer et al. integrated solution-processed ST-OSCs into smart sun glasses which were named solar glasses (Figure 18.17) [113]. The inverted ST-OSCs adopted the structure as glass/ITO/ZnO/active layer/PEDOT:PSS (transparent electrode), in which the ternary active layer was PBTZT-stat-benzo[1,2-b:4,5-b′] dithiophene (BDTT)-8:$PC_{61}BM$:$PC_{71}BM$ (seen in Figure 18.17). The larger substrates (17.8 cm^2) were shaped to fit the lens frames, thereby increasing the photoactive area to 15.5 cm^2, using about 90% of the substrate area for energy harvesting. The ST-OSCs were sandwiched between the shaped substrate and an epoxy encapsulation layer to achieve the solar glasses. In the solar glasses, the ST-OSC had been used to power a small signal processing sensor, providing constant information about the ambient temperature and the solar illumination intensity on two displays. All of the units, including power-supply ST-OSCs, signal processing sensors, and the displays were integrated into the model of the solar glasses. Moreover, the ST-OSCs unit was the only energy source for self-powered solar glasses. This system was designed to work under both low-light conditions (indoors) and high-light conditions (in the bright sun), and also under oblique light incidence and indirect or diffuse light.

18.6 Conclusion and Outlook

Owing to the speedy updates of the photoactive semiconductor materials and the innovation and optimization of the large-scale R2R techniques, the performance of the OSCs has been quickly prompted from 10% to over 17% in the past five years. Aiming to realize the actual commercial applications of OSCs, researchers have devoted many efforts to locate and extend their most appropriate application scenarios. As mentioned above, OSCs were integrated into building facades, smart clothes, batteries, water-splitting systems, sunglasses, etc., enabling solar-to-electricity or solar-to-hydrogen conversion. In all of the application fields, OSCs exhibited bright prospects due to its inherent features like flexibility, low-cost,

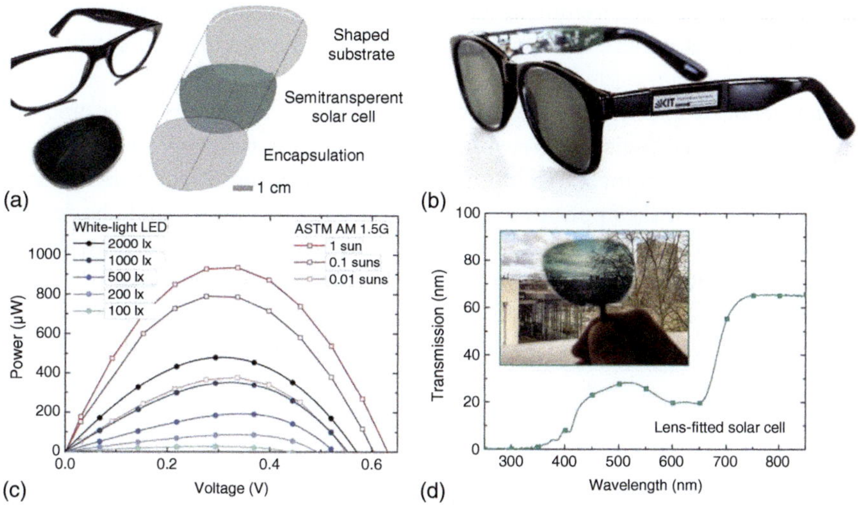

Figure 18.17 (a) Illustration of the solar glasses assembly with the organic solar cells sandwiched between glass substrates and encapsulation glass. (b) The solar glasses in operation. (c) The output power of a typical lens-fitted solar cell measured under different illumination spectra and intensities: an AM1.5G spectrum to represent outdoor illumination and a white-light LED spectrum to simulate artificial indoor lighting. (d) Representative transmission spectrum and photo of a lens-fitted solar cell. Source: Landerer et al. [113]/Reproduced with permission of John Wiley & Sons.

high mass-specific power, etc. However, different targeted applications would result in distinct demands for performance, stability, and mechanical parameters. For example, for a skin-wearable application, devices should possess a stretchability of at least 30% to be conformable with human skin [27]. Thus, one should consider a specific targeted application to make the relevant parameters straightforward before designing and constructing OSC devices and modules. Hence, we discuss the numerous attempts on exploiting the possibility of OSCs for emerging unique industrialization.

Wide attempts have been devoted to the practical use of OSCs, including building-integrated photovoltaics, portable electric chargers, smart clothing, and implantable/biological electronics. Among these applications, the solar parks and solar trees require a little comparable performance while strict operational and long-term stability were needed. The attended mode between devices in solar modules is also very important to reduce resistance or eliminate device failures. While for the OSCs used in smart windows and greenhouse, the photoactive layer of the device should be chosen based on their absorption, and frontier orbital levels to achieve the demand colors, transparency, and also the efficiency. The wearable OSCs demand ultra-flexibility, stretchability, and biocompatibility for implantable self-powered sensors while certain flexibility and excellent waterproofness. And as for the OSCs-integrated ESS, stable and efficient power charging is the chief issue. Besides, the innovative segment in OSCs varies in different applications. For example, a waterproof textile substrate and encapsulation are essential in washable electronics. Therefore, specific targeting applications should be fixed to make the relevant parameters straightforward before designing and constructing OSC devices.

However, most application attempts of OSCs are in a primary stage and just focus on a prototype device as shown in this section. Only very few researches have emphasized the importance of technical achievements, economic, and environmental performances for the application of OSCs. Thus, more efforts should be devoted to further extend the potential application of OSCs while taking technical craft, economic benefit, environmental issues into consideration, promoting the acceptance of OSCs by the end-consumer. The OSCs will be an important asset for other mobile and consumer electronics and occupy a major position in the future energy industry.

List of Symbols and Abbreviations

C_A	areal capacitances
E_A	energy density
E_g	optical band gap
E_{loss}	minimum photon energy loss
I_{max}	maximum discharge current
J_{OP}	operating current density
J_{sc}	short-circuit current density
P_{in}	input light density
P_{max}	maximum power output
P_r	photopic response
r	bending radius
S/N	signal-to-noise ratio
$S_{AM1.5G}$	solar photon flux under AM1.5G
t_2	photocharging time
T_{trans}	light transmittance
V_{oc}	open-circuit voltage
V_{pp}	voltage peak to peak
ΔV	potential difference
η_{STH}	total solar-to-hydrogen conversion efficiency
η_{STP}	total solar-to-power conversion efficiency
λ	light wavelength
1D	one-dimension
3D-printing	3-dimension printing
Al	aluminum
AVT	average visible-light transmittance
BHJ	bulk-heterojunction
BIOSCs	building-integrated organic solar cells
CC	control circuit

CF@FeCoP/NC	nitrogen-doped carbon encased iron-cobalt bimetallic phosphides
EPBT	energy pay-back time
EQE	external quantum efficiency
ESS	energy storage system
ETL	electron transporting layer
FF	fill factor
GWP	global warming potential
H_2	hydrogen
H_2SO_4	sulfuric acid
H_3PO_4	phosphoric acid
HER	hydrogen evolution reaction
HTL	hole transporting layer
IoTs	internet of things
IR	infrared ray
ITO	indium tin oxide
LCA	life cycle analysis
LED	light emitting diode
MOF	metal–organic frameworks
MWCNT	multiwalled carbon nanotube
near-IR	near-infrared light
Ni	nickel
O_2	oxygen
OECTs	organic electrochemical transistors
OER	oxygen evolution process
OSCs	organic solar cells
PBAs	Prussian blue analogs
PCE	power conversion efficiency
PEDOT:PSS	poly(3,4-ethylenedioxythiophene) polystyrene sulfonate
PEI	polyethylenimine
PEN	polyethylene naphthalate
PET	polyethylene terephthalate
PI	polyimide
Pt	platinum
PU	polyurethane
PVA	polyvinyl alcohol
R2R	roll-to-roll
Si	silicon
SiO_2	silicon dioxide
ST-OSCs	semitransparent organic solar cells
Ti	titanium
TiO_x	titanium oxide
V_2O_5	vanadium pentoxide
WO_3/Pt	tungsten trioxide/platinum
ZnO	zinc oxide

Molecular Formula

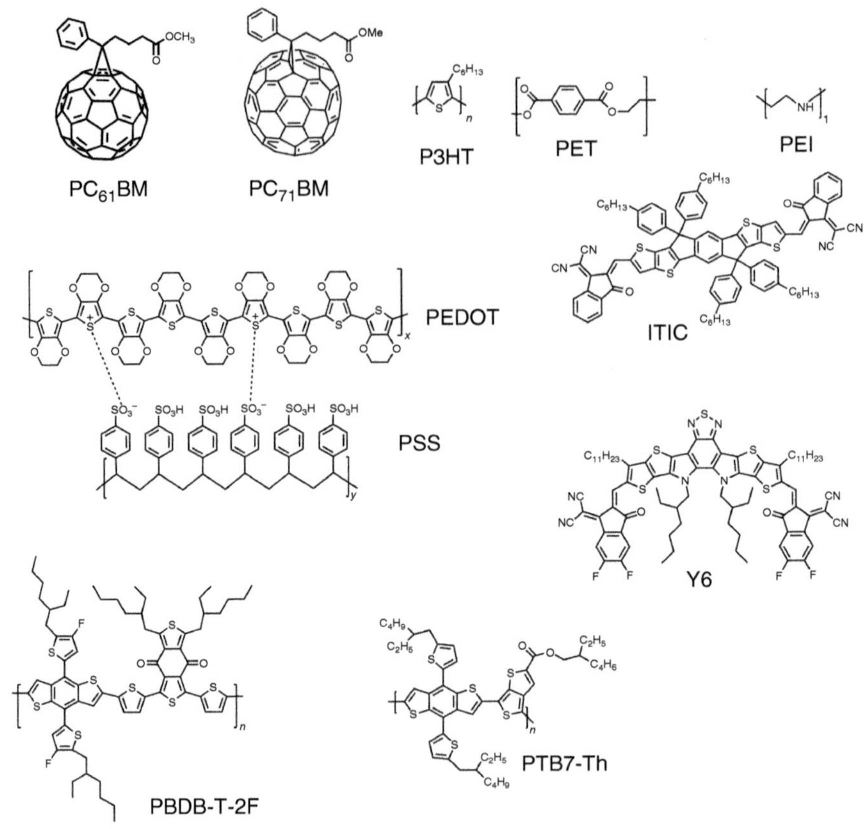

References

1 Breyer, C., Bogdanov, D., Gulagi, A. et al. (2017). *Prog. Photovoltaics Res. Appl.* 25: 727.

2 IEA (2018). *World Energy Outlook 2018*. International Energy Agency.

3 Inganäs, O. (2018). *Adv. Mater.* 30: 1800388.

4 Liu, Q.S., Jiang, Y.F., Jin, K. et al. (2020). *Sci. Bull.* 65: 272.

5 Wan, X., Li, C., Zhang, M., and Chen, Y. (2020). *Chem. Soc. Rev.* 49: 2828.

6 An, C., Zheng, Z., and Hou, J. (2020). *Chem. Commun.* 56: 4750.

7 Kini, G.P., Jeon, S.J., and Moon, D.K. (2020). *Adv. Mater.* 32: 1906175.

8 Cui, Y., Yao, H., Zhang, J. et al. (2020). *Adv. Mater.* 32: 1908205.

9 Yan, C., Barlow, S., Wang, Z. et al. (2018). *Nat. Rev. Mater.* 3: 18003.

10 Meng, L., Zhang, Y., Wan, X. et al. (2018). *Science* 361: 1094.

11 Huang, Y., Kramer, E.J., Heeger, A.J., and Bazan, G.C. (2014). *Chem. Rev.* 114: 7006.

12 Zhao, F.W., Wang, C.R., and Zhan, X.W. (2018). *Adv. Energy Mater.* 8, 1703147.
13 Xu, H.T., Yuan, F., Zhou, D. et al. (2020). *J. Mater. Chem. A* 8: 11478.
14 Liu, Y., Sheri, M., Cole, M.D. et al. (2019). *Angew. Chem. Int. Ed.* 58: 5677.
15 Yao, J., Qiu, B., Zhang, Z.G. et al. (2020). *Nat. Commun.* 11: 2726.
16 Lin, Y., Wang, J., Zhang, Z. et al. (2015). *Adv. Mater.* 27: 1170.
17 Yuan, J., Zhang, Y., Zhou, L. et al. (2019). *Joule* 3: 1140.
18 Jin, K., Xiao, Z., and Ding, L. (2021). *J. Semicond.* 42: 060502.
19 Lin, Y., Nugraha, M.I., Firdaus, Y. et al. (2020). *ACS Energy Lett.* 5: 3663.
20 Zhang, M., Zhu, L., Zhou, G. et al. (2021). *Nat. Commun.* 12: 309.
21 Lin, Y., Firdaus, Y., Isikgor, F.H. et al. (2020). *ACS Energy Lett.* 5: 2935.
22 Li, C., Zhou, J., Song, J. et al. (2021). *Nat. Energy* 6: 605.
23 Chen, S., Feng, L., Jia, T. et al. (2021). *Sci. China Chem.* 64: 1192.
24 Liu, L., Chen, S., Qu, Y. et al. (2021). *Adv. Mater.* 33: 2101279.
25 Liu, F., Zhou, L., Liu, W. et al. (2021). *Adv. Mater.* 33: 2100830.
26 Yeom, H.R., Song, S., Park, S.Y. et al. (2020). *Nano Energy* 77: 105146.
27 Root, S.E., Savagatrup, S., Printz, A.D. et al. (2017). *Chem. Rev.* 117: 6467.
28 Lin, Y.C., Chen, C.K., Chiang, Y.C. et al. (2020). *ACS Appl. Mater. Interfaces* 12: 33014.
29 Kim, T., Kim, J.H., Kang, T.E. et al. (2015). *Nat. Commun.* 6: 8547.
30 Pagliaro, M., Ciriminna, R., and Palmisano, G. (2008). *ChemSusChem* 1: 880.
31 Ramanujam, J., Bishop, D.M., Todorov, T.K. et al. (2020). *Prog. Mater Sci.* 110: 100619.
32 Li, Y., Xu, G., Cui, C., and Li, Y. (2018). *Adv. Energy Mater.* 8: 1701791.
33 Berny, S., Blouin, N., Distler, A. et al. (2016). *Adv. Sci.* 3: 1500342.
34 Wang, G., Adil, M.A., Zhang, J., and Wei, Z. (2019). *Adv. Mater.* 31: 1805089.
35 Alstrup, J., Jørgensen, M., Medford, A.J., and Krebs, F.C. (2010). *ACS Appl. Mater. Interfaces* 2: 2819.
36 Carlé, J.E., Helgesen, M., Hagemann, O. et al. (2014). *Joule* 1: 274.
37 Yang, W., Sherman, V.R., Gludovatz, B. et al. (2015). *Nat. Commun.* 6: 6649.
38 Ma, L.-K., Chen, Y., Chow, P.C.Y. et al. (2020). *Joule* 4: 1486.
39 Debije, M. (2015). *Nature* 519: 298.
40 Brus, V.V., Lee, J., Luginbuhl, B.R. et al. (2019). *Adv. Mater.* 31: 1900904.
41 Krebs, F.C., Espinosa, N., Hösel, M. et al. (2014). *Adv. Mater.* 26: 29.
42 Kaltenbrunner, M., White, M.S., Glowacki, E.D. et al. (2012). *Nat. Commun.* 3: 770.
43 Qin, F., Wang, W., Sun, L. et al. (2020). *Nat. Commun.* 11: 4508.
44 Chang, S.Y., Cheng, P., Li, G., and Yang, Y. (2018). *Joule* 2: 1039.
45 Reb, L.K., Bohmer, M., Predeschly, B. et al. (2020). *Joule* 4: 1880.
46 Palm, J. (2018). *Energy Policy* 113: 1.
47 Masson, V.r., Bonhomme, M., Salagnac, J.-L. et al. (2014). *Front. Environ. Sci.* 2: 14.
48 Burlingame, Q., Huang, X., Liu, X. et al. (2019). *Nature* 573: 394.
49 Hösel, M., Søndergaard, R.R., Jørgensen, M., and Krebs, F.C. (2013). *Energy Technol.* 1: 102.

50 EN ISO 14040:1997 (1997). *Environmental Management – Life Cycle Assessment – Principles and Framework*. Geneva, Switzerland, Geneva: International Organisation for Standardisation (ISO).

51 Schleisner, L. (2000). *Renewable Energy* 20: 279.

52 Wambach, K., Müller, A., and Alsema, E. (2005). Life cycle analysis of a solar module recycling process. Presented at *20th European Photovoltaic Solar Energy Conference*, Barcelona, Spain.

53 Knapp, K. and Jester, T. (2001). *Sol. Energy* 71: 165.

54 Keoleian, G.A. and Lewis, G.M. (2003). *Renewable Energy* 28: 271.

55 Crawford, R., Treloar, G.J., Fuller, R., and Bazilian, M. (2006). *Renewable Sustainable Energy Rev.* 10: 559.

56 Raugei, M., Bargigli, S., and Ulgiati, S. (2007). *Energy* 32: 1310.

57 Kato, K., Hibino, T., Komoto, K. et al. (2001). *Sol. Energy Mater. Sol. Cells* 67: 279.

58 Espinosa, N., Hösel, M., Angmo, D., and Krebs, F.C. (2012). *Energy Environ. Sci.* 5: 5117.

59 Roes, A.L., Alsema, E.A., Blok, K., and Patel, M.K. (2009). *Prog. Photovoltaics Res. Appl.* 17: 372.

60 García-Valverde, R., Cherni, J.A., and Urbina, A. (2010). *Prog. Photovoltaics Res. Appl.* 18: 535.

61 Pehnt, M. (2006). *Renewable Energy* 31: 55.

62 Leung, D.Y., Yin, X., and Wu, C. (2004). *Renewable Sustainable Energy Rev.* 8: 565.

63 Integrated Sustainability Analysis, The University of Sydney Consultancy Report (2006). Life-Cycle Energy Balance and Greenhouse Gas Emissions of Nuclear Energy in Australia.

64 Chatzisideris, M.D., Ohms, P.K., Espinosa, N. et al. (2019). *Appl. Energy* 256: 113977.

65 Zhang, G.C., Xia, R.X., Chen, Z. et al. (2018). *Adv. Energy Mater.* 8: 1801609.

66 Wang, Y.H., Chang, Y.L., Zhang, J.Q. et al. (2020). *Chem. Res. Chin. Univ.* 36: 343.

67 Yao, M.N., Li, T.F., Long, Y.B. et al. (2019). *Sci. Bull.* 65: 217.

68 Castillero, P., Rico-Gavira, V., López-Santos, C. et al. (2017). *J. Phys. Chem. C* 121: 15719.

69 Hu, C.-W., Yamada, Y., and Yoshimura, K. (2017). *Chem. Commun.* 53: 3242.

70 Luo, J.Y., Li, W., Chen, F. et al. (2014). *Sens. Actuators, B* 197: 81.

71 Ravishankar, E., Booth, R.E., Saravitz, C. et al. (2020). *Joule* 4: 490.

72 Singhal, G., Renger, G., Sopory, S., and Irrgang, K. (2012). *Concepts in Photobiology: Photosynthesis and Photomorphogenesis*. Springer Science & Business Media, Springer.

73 Corrado, C., Leow, S.W., Osborn, M. et al. (2016). *J. Renewable Sustainable Energy* 8: 043502.

74 Allardyce, C.S., Fankhauser, C., Zakeeruddin, S.M. et al. (2017). *Sol. Energy* 155: 517.

75 Yano, A., Onoe, M., and Nakata, J. (2014). *Biosystems Eng.* 122: 62.
76 Traverse, C.J., Pandey, R., Barr, M.C., and Lunt, R.R. (2017). *Nat. Energy* 2: 849.
77 Song, W., Fanady, B., Peng, R. et al. (2020). *Adv. Energy Mater.* 10: 2000136.
78 Mathews, I., Kantareddy, S.N., Buonassisi, T., and Peters, I.M. (2019). *Joule* 3: 1415.
79 Shi, W., Guo, Y., and Liu, Y. (2020). *Adv. Mater.* 32: 1901493.
80 Manyika, J. (2015). *The Internet of Things: Mapping the Value beyond the Hype.* Analysis & Policy Observatory.
81 Li, C., Cong, S., Tian, Z. et al. (2019). *Nano Energy* 60: 247.
82 Park, S., Heo, S.W., Lee, W. et al. (2018). *Nature* 561: 516.
83 O'Connor, T.F., Zaretski, A.V., Savagatrup, S. et al. (2016). *Sol. Energy Mater. Sol. Cells* 144: 438.
84 Kaltenbrunner, M., Sekitani, T., Reeder, J. et al. (2013). *Nature* 499: 458.
85 Kimura, H., Fukuda, K., Jinno, H. et al. (2019). *Adv. Mater.* 31: 1808033.
86 Park, S., Fukuda, K., Wang, M. et al. (2018). *Adv. Mater.* 30: 1802359.
87 Xu, H., Liu, J., Zhang, J. et al. (2017). *Adv. Mater.* 29: 1700975.
88 Bashkatov, A.N., Genina, E.A., Kochubey, V.I., and Tuchin, V.V. (2005). *J. Phys. D: Appl. Phys.* 38: 2543.
89 Jinno, H., Fukuda, K., Xu, X.M. et al. (2017). *Nat. Energy* 2: 780.
90 Jeong, E.G., Jeon, Y., Cho, S.H., and Choi, K.C. (2019). *Energy Environ. Sci.* 12: 1878.
91 Zhu, Y.H., Yang, X.Y., Liu, T., and Zhang, X.B. (2020). *Adv. Mater.* 32: 1901961.
92 Zeng, W., Shu, L., Li, Q. et al. (2014). *Adv. Mater.* 26: 5310.
93 Weng, W., Chen, P., He, S. et al. (2016). *Angew. Chem. Int. Ed.* 55: 6140.
94 Chen, G., Li, Y., Bick, M., and Chen, J. (2020). *Chem. Rev.* 120: 3668.
95 Zhang, Z., Li, X., Guan, G. et al. (2014). *Angew. Chem. Int. Ed.* 53: 11571.
96 Lee, M.R., Eckert, R.D., Forberich, K. et al. (2009). *Science* 324: 232.
97 Liu, P., Gao, Z., Xu, L.M. et al. (2018). *J. Mater. Chem. A* 6: 19947.
98 Zhang, Z.T., Yang, Z.B., Wu, Z.W. et al. (2014). *Adv. Energy Mater.* 4: 1301750.
99 Machui, F., Hösel, M., Li, N. et al. (2014). *Energy Environ. Sci.* 7: 2792.
100 Liu, R.Y., Takakuwa, M., Li, A.L. et al. (2020). *Adv. Energy Mater.* 10: 2000523.
101 Thostenson, J.O., Li, Z., Kim, C.H.J. et al. (2018). *J. Electrochem. Soc.* 165: B3122.
102 Dennler, G., Bereznev, S., Fichou, D. et al. (2007). *Sol. Energy* 81: 947.
103 Lee, Y.H., Kim, J.S., Noh, J. et al. (2013). *Nano Lett.* 13: 5753.
104 Hu, Y., Ding, S., Chen, P. et al. (2020). *Energy Storage Mater.* 32: 356.
105 Zamarayeva, A.M., Ostfeld, A.E., Wang, M. et al. (2017). *Sci. Adv.* 3: 1602051.
106 Zhang, Z., Chen, X., Chen, P. et al. (2014). *Adv. Mater.* 26: 466.
107 Yao, L., Rahmanudin, A., Guijarro, N., and Sivula, K. (2018). *Adv. Energy Mater.* 8: 1802585.
108 Steier, L. and Holliday, S. (2018). *J. Mater. Chem. A* 6: 21809.
109 Wang, D., Hu, J., Sherman, B.D. et al. (2020). *Proc. Natl. Acad. Sci. U.S.A.* 117: 13256.

110 Lin, R., Lei, H., Ruan, D. et al. (2019). *Nano Energy* 56: 82.
111 Esiner, S., van Eersel, H., van Pruissen, G.W. et al. (2016). *ACS Appl. Mater. Interfaces* 8: 26972.
112 Wu, Q., Guo, J., Sun, R. et al. (2019). *Nano Energy* 61: 559.
113 Landerer, D., Bahro, D., Röhm, H. et al. (2017). *Energy Technol.* 5: 1936.

Index

a

Organic Solar Cells: Materials Design, Technology and Commercialization, First Edition. Edited by Liming Ding.

b

f

g

o

q

p

r

u

v

w

x

y

z